Undergraduate Lecture Notes in Physics

Series Editors

Neil Ashby, University of Colorado, Boulder, CO, USA

William Brantley, Department of Physics, Furman University, Greenville, SC, USA

Matthew Deady, Physics Program, Bard College, Annandale-on-Hudson, NY, USA

Michael Fowler, Department of Physics, University of Virginia, Charlottesville, VA, USA

Morten Hjorth-Jensen, Department of Physics, University of Oslo, Oslo, Norway

Michael Inglis, Department of Physical Sciences, SUNY Suffolk County Community College, Selden, NY, USA

Barry Luokkala ⓘ, Department of Physics, Carnegie Mellon University, Pittsburgh, PA, USA

Lev Kantorovich

Mathematics for Natural Scientists II

Advanced Methods

Second Edition

 Springer

Lev Kantorovich
Department of Physics
School of Natural and Mathematical
Sciences
King's College London
London, UK

ISSN 2192-4791   ISSN 2192-4805   (electronic)
Undergraduate Lecture Notes in Physics
ISBN 978-3-031-46319-8   ISBN 978-3-031-46320-4   (eBook)
https://doi.org/10.1007/978-3-031-46320-4

This book is dedicated to my dear wife Tatiana who has never had anything to do with maths, never wanted to know, always considered all this unimportant in anybody's life in comparison with the family, writing poems and playing music, but without whom this book would have never been possible.

Preface to the Second Edition

In this Edition I have read the whole manuscript as carefully as I possibly could, checking the text and the problems. This enabled me to find various typos and occasional errors which were all corrected. Moreover, many improvements and further corrections and clarifications have been added into the text that hopefully made the text more transparent, logical and accessible to the reader. More importantly, additional material has been added, together with new problems. The length of the book grew by more than 33%, and the number of problems increased from 550 to 691. A number of physics problems illustrating the material have also been added.

In more detail, below the new material added to the text (ignoring relatively small additions and improvements) is listed:

- Chapter 1 (on linear algebra) has undergone significant changes and improvements. The material on vector spaces has been improved with several essential theorems added; relation of the determinant of a matrix to a linear antisymmetric function considered; we also considered in more detail linear systems of algebraic equations including the case of the zero vector in the right-hand side; Gaussian elimination is fully covered as a method of choice for solving linear systems of equations, finding the inverse of a matrix and its rank; we proved Abel's formula for the n–th order linear differential equation (DE) in the general case using the Wronskian and considered a curious method of using the latter in reverse engineering of linear DEs; the material related to eigenvectors/eigenvalues of a square matrix has been considerably generalised to considering not only Hermitian matrices, but all square matrices including normal and defective ones, algebraic and geometric multiplicities, left/right eigenvectors; correspondingly, the case of repeated eigenvalues in diagonalising of a square matrix has been also generalised, as well as the spectral theorem and a function of a matrix, including Jordan normal form and the related material; a generalised eigenproblem, frequently found in applications, is also discussed as well as a number of famous matrix identities frequently used in physics; systems of linear DEs are systematically considered as well including the ones that are related to matrices of lower rank and defective matrices; we also discuss perturbation methods in solving systems of linear DEs and thoroughly consider the pseudo-inverse of a general matrix. A section on 3D rotation is added. As new applications, we discuss (i) the least square method and its relation to pseudo-inverse and the determination of the force constant matrix from molecular dynamics simulations; (ii) Slater determinants and fermionic many-body wavefunctions in quantum mechanics; (iii) Dirac equation of relativistic quantum mechanics and (iv) exact solution of the classical 1D Ising model.

- Chapter 2 (on calculus of complex numbers and functions) has also undergone considerable modifications: (i) a proof of the fundamental theorem of algebra is presented; (ii) calculation of the Fresnel integral; (iii) an application of the theory of residues to calculating functional series added. In view of applications of the calculus of complex numbers and functions, the so-called Matsubara sums are considered in detail and the solution of the famous Landau–Zener problem. We have also discussed a beautiful counting problem based on complex numbers which I came across thanks to a tip from Dr. Filipe Louly Quinan Junqueira with whom I collaborate. I found this problem and its solution fascinating and thought that it may provide a fantastic illustration of the beauty of the mathematics and, at the same time, of usefulness of complex numbers in a problem that, at first sight, has nothing to do with them!

- In Chap. 3 (on Fourier series) (i) Gibbs–Wilbraham phenomenon is discussed, (ii) summation of a series of inverse powers of integers (iii) and, incidentally, Taylor expansions of tangent and cotangent functions are given; (iv) convergence of the Fourier series is carefully considered with some ideas on the convergence acceleration. As a new application, some properties of the free electron gas are calculated.

- In Chap. 4 (on special functions) a solution to a quantum-mechanical problem of a free particle confined in a cylinder is provided, as well as the full derivation of van Hove singularities in the electronic density of states of a crystal.

- Chapter 5 (on Fourier transform): (i) a more careful derivation of the Fourier integral is provided; (ii) we also consider the Fourier transform of an integral and (iii) the Poisson summation formula.

- In Chap. 6 (on Laplace transform, LT) the calculation of the transform of partial fraction is presented in detail, some improvements to the derivation of the inverse Laplace transform given, and applications of the LT to some DEs with variable coefficients considered.

- In Chap. 7 (on curvilinear coordinates) we (i) discuss more in detail the vector product of vectors expressed via curvilinear coordinates, (ii) calculation of a normal to a surface and (iii) of line and surface integrals; finally, (iii) multi-variable Gaussians are discussed including calculating general correlation functions of Gaussian processes.

- In Chap. 8 (on partial differential equations, PDEs) a Fourier transform method has been added for solving PDEs.

- In Chap. 9 (on variational calculus) an application of the method is added on calculating the width of a domain wall in a ferromagnet.

I'd also like to take this opportunity to thank my colleague at King's College London Alexander Pushnitski from the Maths Department who helped me several times during the work on this book.

London, UK   Lev Kantorovich

Preface to the First Edition

This is the second volume of the course of mathematics for natural scientists. It is loosely based on the mathematics course for second-year physics students at King's College London that I have been reading for more than 10 years. It follows the spirit of the first volume [1] by continuing a gradual build-up of the mathematical knowledge necessary for, but not exclusively, physics students.

This volume covers more advanced material, beginning with two essential components: linear algebra (Chapter 1) and theory of functions of complex variables (Chapter 2). These techniques are heavily used in the chapters that follow. Fourier series are considered in Chapter 3, special functions of mathematical physics (Dirac delta function, gamma and beta functions, detailed treatment of orthogonal polynomials, the hypergeometric differential equation, spherical and Bessel functions) in Chapter 4, and then Fourier (Chapter 5) and Laplace (Chapter 6) transforms. In Chapter 7, a detailed treatment of curvilinear coordinates is given, including the corresponding differential calculus. This is essential, as many physical problems possess symmetry and using appropriate curvilinear coordinates may significantly simplify the solution of the corresponding partial differential equations (studied in Chapter 8) if the symmetry of the problem at hand is taken into account. The book is concluded with variational calculus in Chapter 9.

As in the first volume, I have tried to introduce new concepts gradually and as clearly as possible, giving examples and problems to illustrate the material. Across the text, all the proofs necessary to understand and appreciate the mathematics involved are also given. In most cases, the proofs would satisfy the most demanding physicist or even a mathematician; only in a few cases have I had to sacrifice the "strict mathematical rigour" by presenting somewhat simplified derivations and/or proofs.

As in the first volume, many problems are given throughout the text. These are designed mainly to illustrate the theoretical material and require the reader to complete them in order to be in a position to move forward. In addition, other problems are offered for practice, although I have to accept, their number could have been larger. For more problems, the reader is advised to consult other texts, e.g. the books [2–6].

When working on this volume, I have mostly consulted a number of excellent classic Russian textbooks [7–12]. As far as I am aware, some of them are available in English, and I would advise a diligent student to continue his/her education by reading these. Concerning the others, there is of course an obvious language barrier. Unfortunately, as I cannot ask the reader to learn Russian purely for that purpose, these texts remain inaccessible for most readers. I hope that the reader would be able to find more specialised texts in English, which go beyond the scope of this (and the previous) book to further the development of their studies, e.g. books [13–26] represent a rather good selection which covers the topics of this volume, but this list of course by no means is complete.

The mathematics throughout the book is heavily illustrated by examples from condensed matter physics. In fact, probably over a quarter of the text of the whole volume is occupied by these. Every chapter, save Chapter 8, contains a large concluding section exploring physics topics that necessitate the mathematics presented in that chapter. Chapter 8, on partial differential equations, is somewhat special in this respect, as it is entirely devoted to solving equations of mathematical physics (wave, Laplace and heat transport equations). Consequently, it does not have a special section on applications. When selecting the examples from physics, I was mostly governed by my own experience and research interests as well as several texts, such as the books [27, 28]. In fact, examples from [27] have been used in both of these volumes.

As in the first volume, this book begins with a list of the names of all the scientists across the world, mathematicians, physicists and engineers, whose invaluable contribution formed the foundation of the beautiful sciences of mathematics and physics that have been enjoying a special bond throughout the centuries.

Should you find any errors or unnoticed misprints, please send your corrections either directly to myself (lev.kantorovitch@kcl.ac.uk) or to the publisher. Your general comments, suggestions and any criticism related to these two volumes would be greatly appreciated.

This book concludes the project I started about four years ago, mainly working in the evenings, at the weekends, as well as on the train going to and from work. Not everything I initially planned to include has appeared in the books, although the two volumes contain most of the essential ideas young physicists, engineers and computational chemists should become familiar with. Theory of operators, group theory, tensor calculus, stochastic theory (or theory of probabilities) and some other more specialised topics have not been included but can be found in multiple other texts. Still, I hope these volumes will serve as an enjoyable introduction to the beautiful world of mathematics for students, encouraging them to think more and ask for more. I am also confident that the books will serve as a rich source for lecturers.

I wish you happy reading!

London, UK   Lev Kantorovich

References

1. L. Kantorovich, "Mathematics for Natural Scientists: Fundamentals and Basics", Undergraduate Lecture Notes in Physics, Springer, 2015 (ASIN: B016CZWP36).
2. D. McQuarrie, "Mathematical methods for scientists and engineers", Univ. Sci. Books, 2003 (ISBN 1-891389-29-7).
3. K. F. Riley, M. P. Hobson, and S. J. Bence, "Mathematical Methods for Physics and Engineering", Cambridge Univ. Press, 2006 (ISBN 0521679710).
4. M. Boas, "Mathematical methods in the physical sciences", Wiley, 2nd Edition, 1983 (ISBN 0-471-04409-1).
5. K. Stroud, "Engineering mathematics", Palgrave, 5th Edition, 2001 (ISBN 0-333-919394).
6. G. B. Arfken, H. J. Weber and F. E. Harris, "Mathematical Methods for Physicists. A Comprehensive Guide", Academic Press, 7th Edition, 2013 (ISBN: 978-0-12-384654-9).
7. В. И. Смирнов, "Курс высшей матики" т. 1–5, Москва, Наука, 1974 (in Russian). Apparently, there is a rather old English translation: V. I. Smirnov, "A Course of Higher Mathematics: Adiwes International Series in Mathematics", Pergamon, 1964; Vols. 1 (ISBN 1483123944), 2 (ISBN 1483120171), 3-Part-1 (ISBN B0007ILX1K) and 3-Part-2 (ISBN B00GWQPPMO).
8. Г. М. Фихтенголц, "Курс цифференциального и интегрального исчисления", Москва, Физматлиит, 2001 (in Russian). [G. M. Fihtengolc, "A Course of Differentiation and Integration", Vols. 1–3., Fizmatgiz, 2001] I was not able to find an English translation of this book.
9. В. С. Шипачев, "Высшая математика", Москва, Высшая Школа, 1998 (in Russian). [V. S. Shipachev, "Higher Mathematics", Vishaja Shkola, 1998.] I was not able to find an English translation of this book.
10. А. Н. Тихонов и А. А. Самрский, "Уравнения математической физики", Москва, Наука, 1977 (in Russian). There is a resent reprint of a rather old translation: A. N. Tikhonov and A. A. Samarskii, "Equations of Mathematical Physics", Dover Books on Physics, 2011 (ISBN-10: 0486664228, ISBN-13: 978-0486664224).
11. М. А. Лаврентьв и Б. В. Шабат, "Методы теории функций комплексного переменного", УМН, 15:5(95), 1960 (in Russian). [M. A. Lavrent'ev and B. V. Shabat, Methods of theory of functions of complex variables", Uspekhi Mat. Nauk, 15:5(95) (1960).] I was not able to find an English translation of this book.
12. Ф. Никофоров и В. Б. Уваров, "Специальные функции математической физики", Москва, Haya, 1984 (in Russian). This excellent book is available in English: A. F. Nikiforov and V. B. Uvarov, "Special Functions of Mathematical Physics: A Unified Introduction With Applications", Springer, 2013. (ISBN-10: 1475715978, ISBN-13: 978-1475715972).
13. J. M. Howie, "Complex analysis", Springer, 2004 (ISBN-10: 1852337338, ISBN-13: 978-1852337339).
14. H. A. Priestley, "Introduction to Complex Analysis Paperback", OUP Oxford; 2 edition, 2003 (ISBN-10: 0198525621, ISBN-13: 978-0198525622)
15. Kevin W. Cassel, "Variational Methods with Applications in Science and Engineering Hardcover", Cambridge University Press, 2013 (ISBN-10: 1107022584, ISBN-13: 978-1107022584).
16. Stanley J. Farlow, "Partial Differential Equations for Scientists and Engineers", Dover Books on Mathematics, Dover, 2003 (ISBN-10: 048667620X, ISBN-13: 978-0486676203).
17. Geoffrey Stephenson, "Partial Differential Equations for Scientists and Engineers", Imperial College Press, 1996 (ISBN-10: 1860940242, ISBN-13: 978-1860940248).
18. Phil Dyke, "An Introduction to Laplace Transforms and Fourier Series", Springer Undergraduate Mathematics Series, Springer, 2014 (ISBN-10: 144716394X, ISBN-13: 978-1447163947).
19. Brian Davies, "Integral Transforms and Their Applications", Texts in Applied Mathematics, Springer, 2002 (ISBN-10: 0387953140, ISBN-13: 978-0387953144).

20. Allan Pinkus, "Fourier Series and Integral Transforms", Cambridge University Press, 1997 (ISBN-10: 0521597714, ISBN-13: 978-0521597715).

21. Serge Lang, "Linear Algebra", Undergraduate Texts in Mathematics, Springer, 2013 (ISBN-10: 1441930817, ISBN-13: 978-1441930811).

22. Richard Bellman, "Introduction to Matrix Analysis", Classics in Applied Mathematics, Society for Industrial and Applied Mathematics, 2 edition, 1987 (ISBN-10: 0898713994, ISBN-13: 978-0898713992).

23. George Andrews, Richard Askey and Ranjan Roy, "Special Functions", Encyclopedia of Mathematics and its Applications, Cambridge University Press, 2010 (ISBN-10: 0521789885, ISBN-13: 978-0521789882).

24. N. N. Lebedev, "Special Functions & Their Applications", Dover Books on Mathematics, Dover, 2003 (ISBN-10: 0486606244, ISBN-13: 978-0486606248).

25. Naismith Sneddon, "Special Functions of Mathematical Physics and Chemistry", University Mathematics Texts, Oliver & Boyd, 1966 (ISBN-10: 0050013343, ISBN-13: 978-0050013342).

26. W. W. Bell, "Special Functions for Scientists and Engineers", Dover Books on Mathematics, Dover, 2004 (ISBN-10: 0486435210, ISBN-13: 978-0486435213).

27. В. Г. Левич, "Курс теоретической физики", т. 1 ц 2, Наука, 1969 (in Russian). There is a translation: Benjamin G. Levich, "Theoretical physics: an advanced text", North-Holland Publishing in Amsterdam, London, 1970.

28. L. Kantorovich, "Quantum Theory of the Solid State: an Introduction", Springer, 2004 (ISBN 978-1-4020-1821-3 and 978-1-4020-2153-4).

Famous Scientists Mentioned in the Book

Throughout the book various people, both mathematicians and physicists, who are remembered for their outstanding contribution in developing science, will be mentioned. For reader's convenience their names (together with some information borrowed from their Wikipedia pages) are listed here in the alphabetical order:

Niels Henrik Abel (1802–1829) was a Norwegian mathematician.

Henry Frederick Baker (1866–1956) was a British mathematician.

Friedrich Wilhelm Bessel (1784–1846) was a German astronomer, mathematician, physicist and geodesist.

Gerd Binnig (born 1947) is a German physicist.

Arne Bjerhammar (1917–2011) was a Swedish geodesist.

Felix Bloch (1905–1983) was a Swiss physicist, who was awarded the 1952 Nobel Prize in physics.

Niels Henrik David Bohr (1885–1962) was a Danish physicist who made fundamental contributions to quantum theory. He received the Nobel Prize in Physics in 1922.

Ludwig Eduard Boltzmann (1844–1906) was an Austrian physicist and philosopher, one of the founders of statistical mechanics.

Max Born (1882–1970) was a German-British physicist and mathematician.

William Lawrence Bragg (1890–1971) was an Australian-born British physicist and X-ray crystallographer, discoverer (1912) of the Bragg law of X-ray diffraction, a joint winner (with his father, Sir William Bragg) of the Nobel Prize for Physics in 1915.

Léon Nicolas Brillouin (1889–1969) was a French physicist.

Named after Robert Brown (1773–1858) was a Scottish botanist and paleobotanist.

Felice Casorati (1835–1890), was an Italian mathematician.

Baron Augustin-Louis Cauchy (1789–1857) was a French mathematician widely reputed as a pioneer of analysis.

Pafnuty Lvovich Chebyshev (1821–1894) was a Russian mathematician.

Marc-Antoine Parseval des Chênes (1755–1836) was a French mathematician.

Gabriel Cramer (1704–1752) was a Swiss mathematician.

Jean-Baptiste le Rond d'Alembert (1717–1783) was a French mathematician, mechanician, physicist, philosopher, and music theorist.

Paul Adrien Maurice Dirac (1902–1984) was an English theoretical physicist who made fundamental contributions to the early development of both quantum mechanics and quantum electrodynamics. He shared the Nobel Prize in Physics for 1933 with Erwin Schrödinger "for the discovery of new productive forms of atomic theory".

Johann Peter Gustav Lejeune Dirichlet (1805–1859) was a German mathematician with deep contributions to number theory, the theory of Fourier series and other topics in mathematical analysis.

Freeman John Dyson (1923–2020) who was an English-American theoretical physicist and mathematician.

Albert Einstein (1879–1955) was a German-born theoretical physics who developed the general theory of relativity, and also contributed in many other areas of physics. He received the 1921 Nobel Prize in Physics for his "services to theoretical physics".

Leonhard Euler (1707–1783) was a pioneering Swiss mathematician and physicist.

Paul Peter Ewald (1888–1985) was a German-born U.S. crystallographer and physicist, a pioneer of X-ray diffraction methods.

Enrico Fermi (1901–1954) was an outstanding Italian physicist, the 1938 Nobel laureate in physics.

Richard Phillips Feynman (1918–1988) was an American theoretical physicist who made several fundamental contributions in physics. He received the Nobel Prize in Physics in 1965.

Vladimir Aleksandrovich Fock (1898–1974) was a Soviet physicist, who did foundational work on quantum mechanics and quantum electrodynamics.

Jean-Baptiste Joseph Fourier (1768–1830) was a French mathematician and physicist.

Joseph Fraunhofer (1787–1826), ennobled in 1824 as Ritter von Fraunhofer, was a German optician.

Augustin-Jean Fresnel (1788–1827) was a French civil engineer and physicist.

Ferdinand Georg Frobenius (1849–1917) was a German mathematician.

Johann Carl Friedrich Gauss (1777–1855) was a German mathematician and physicist.

Josiah Willard Gibbs (1839–1903) was an American scientist who made important theoretical contributions to physics, chemistry, and mathematics.

Jørgen Pedersen Gram (1850–1916) was a Danish actuary and mathematician.

George Green (1793–1841) was a British mathematical physics.

Sir William Rowan Hamilton (1805–1865) was an Irish physicist, astronomer, and mathematician.

Douglas Rayner Hartree (1897–1958) who was an English mathematician and physicist.

Felix Hausdorff (1868–1942) was a German mathematician.

Oliver Heaviside (1850–1925) was a self-taught English electrical engineer, mathematician, and physicist.

Hermann Ludwig Ferdinand von Helmholtz (1821–1894) was a German physician and physicist.

Charles Hermite (1822–1901) was a French mathematician.

Ludwig Otto Hesse (1811–1874) who was a German mathematician.

Józef Maria Hoene-Wroński (1776–1853) was a Polish Messianist philosopher, mathematician, physicist, inventor, lawyer, and economist.

Léon Charles Prudent Van Hove (1924–1990) was a Belgian physicist.

Ernst Ising (1900–1998) was a German physicist.

Carl Gustav Jacob Jacobi (1804–1851) was a German mathematician.

Marie Ennemond Camille Jordan (1838–1922) was a French mathematician.

Theodore von Kármán (1881–1963) was a Hungarian-American mathematician, aerospace engineer and physicist.

Aleksandr Yakovlevich Khinchin (1894–1959) was a Soviet mathematician.

Gustav Robert Kirchhoff (1824–1887) was a German physicist.

Walter Kohn (born in 1923) is an Austrian-born American theoretical physicist. He was awarded, with John Pople, the Nobel Prize in chemistry in 1998 for the development of the density functional theory.

Hendrik Anthony Kramers (1894–1952) was a Dutch physicist.

Leopold Kronecker (1823–1891) was a German mathematician.

Ralph Kronig (1904–1995) was a German-American physicist.

Aleksey Nikolaevich Krylov (1863–1945) was a Russian naval engineer, applied mathematician and memoirist.

Joseph-Louis Lagrange (1736–1813) was an Italian Enlightenment Era mathematician and astronomer who made significant contributions to the fields of analysis, number theory, and both classical and celestial mechanics.

Edmond Nicolas Laguerre (1834–1886) was a French mathematician.

Cornelius (Cornel) Lanczos (1893–1974) was a Hungarian mathematician and physicist.

Pierre-Simon, marquis de Laplace (1749–1827) was an influential French scholar whose work was important to the development of mathematics, statistics, physics, and astronomy.

Pierre Alphonse Laurent (1813–1854) was a French mathematician.

Adrien-Marie Legendre (1752–1833) was a French mathematician.

Gottfried Wilhelm von Leibniz (1646–1716) was a German polymath and philosopher, who to this day occupies a prominent place in the history of mathematics and the history of philosophy. Most scholars believe Leibniz developed calculus independently of Isaac Newton, and Leibniz's notation has been widely used ever since it was published.

Joseph Liouville (1809–1882) was a French mathematician and engineer.

Hendrik Antoon Lorentz (1853–1928) was a Dutch physicist.

Ettore Majorana (1906–possibly dying after 1959) was an Italian theoretical physicist.

Lorenzo Mascheroni (1750–1800) was an Italian mathematician.

Takeo Matsubara (1921–2014) was a Japanese theoretical physicist.

James Clerk Maxwell (1831–1879) was a Scottish scientist in the field of mathematical physics. His most notable achievement was to formulate the unified classical theory of electromagnetic radiation, bringing together for the first-time electricity and magnetism.

Hermann Minkowski (1864–1909) was a German mathematician.

Abraham de Moivre (1667–1754) was a French mathematician.

Eliakim Hastings Moore (1862–1932), usually cited as E. H. Moore or E. Hastings Moore, was an American mathematician.

Giacinto Morera (1856–1909), was an Italian engineer and mathematician.

Sir Isaac Newton (1642–1726/7) was a famous English physicist and mathematician who laid the foundations for classical mechanics and made seminal contributions to optics and (together with Gottfried Leibniz) the development of calculus.

Amalie Emmy Noether (1882–1935) was a German mathematician who made many important contributions to abstract algebra.

Mikhail Vasilyevich Ostrogradsky (1801–1862) was a Ukrainian mathematician and physicist.

Wolfgang Ernst Pauli (1900–1958) was an Austrian theoretical physicist, one of the pioneers of quantum physics.

William George Penney (1909–1991) was an English mathematician and mathematical physicist.

Roger Penrose (born 1931) is an English mathematician, mathematical physicist, philosopher of science and Nobel Laureate in Physics (2020).

Michel Plancherel (1885–1967) was a Swiss mathematician.

Max Karl Ernst Ludwig Planck (1858–1947) was a German theoretical physicist, one of the founders of the quantum theory. His discovery of energy quanta won him the Nobel Prize in Physics in 1918.

Josip Plemelj (1873–1967) was a Slovene mathematician.

Siméon Denis Poisson (1781–1840) was a French mathematician, geometer, and physicist.

Pythagoras of Samos (c. 570–c. 495 BC) was an Ionian Greek philosopher and mathematician.

Georg Friedrich Bernhard Riemann (1826–1866) was an influential German mathematician who made lasting and revolutionary contributions to analysis, number theory, and differential geometry.

Benjamin Olinde Rodrigues (1795–1851), more commonly known as Olinde Rodrigues, was a French banker, mathematician, and social reformer.

Heinrich Rohrer (born 1933) is a Swiss physicist.

Erhard Schmidt (1876–1959) was an Estonian-German mathematician.

Erwin Rudolf Josef Alexander Schrödinger (1887–1961) was a Nobel Prize-winning (1933) Austrian physicist who developed a number of fundamental results, which formed the basis of wave mechanics.

Lu Jeu Sham (born April 28, 1938) is a Chinese physicist.

John Clarke Slater (1900–1976) was a noted American physicist.

Julian Karol Sokhotski (1842–1927) was a Russian-Polish mathematician.

Arnold Johannes Wilhelm Sommerfeld (1868–1951) was a German theoretical physicist.

James Stirling (1692–1770) who was a Scottish mathematician.

Ernst Stueckelberg (1905–1984) was a Swiss mathematician and physicist.

Brook Taylor (1685–1731) was an English mathematician.

Llewellyn Hilleth Thomas (1903–1992) was a British physicist and applied mathematician.

Vito Volterra (1860–1940) was an Italian mathematician and physicist.

Karl Theodor Wilhelm Weierstrass (1815–1897) was a German mathematician often cited as the "father of modern analysis".

Norbert Wiener (1894–1964) was an American mathematician and philosopher.

Henry Wilbraham (1825–1883) was an English mathematician.

Max Woodbury (1917–2010) was an American mathematician.

Clarence Melvin Zener (1905–1993) was the American physicist.

Contents

1   **Elements of Linear Algebra** 1
    1.1   Vector Spaces 2
        1.1.1   Introduction to Multi-dimensional Complex
                Vector Spaces 2
        1.1.2   Analogy Between Functions and Vectors 10
        1.1.3   Orthogonalisation Procedure (Gram–Schmidt
                Method) 14
    1.2   Matrices: Definition and Properties 19
        1.2.1   Introduction of a Concept 19
        1.2.2   Operations with Matrices 20
        1.2.3   Inverse Matrix: An Introduction 27
        1.2.4   Linear Transformations and a Group 30
        1.2.5   Orthogonal and Unitary Transformations 32
    1.3   Determinant of a Square Matrix 42
        1.3.1   Formal Definition 42
        1.3.2   Properties of Determinants 47
        1.3.3   Relation to a Linear Antisymmetric Function 53
        1.3.4   Practical Method of Calculating Determinants.
                Minors 55
        1.3.5   Determinant of a Tridiagonal Matrix 60
    1.4   A Linear System of Algebraic Equations 61
        1.4.1   Cramer's Method 62
        1.4.2   Gaussian Elimination 70
    1.5   Rank of a Matrix 81
    1.6   Wronskian 84
        1.6.1   Linear Independence of Functions 84
        1.6.2   Linear ODE's: Abel's Formula 87
        1.6.3   Linear ODE Reverse Engineering 90
    1.7   Calculation of the Inverse Matrix 91

1.8   Eigenvectors and Eigenvalues of a Square Matrix 95
        1.8.1   General Formulation 95
        1.8.2   Algebraic and Geometric Multiplicities 107
        1.8.3   Left Eigenproblem 108
        1.8.4   Hermitian (Symmetric) Matrices 111
        1.8.5   Normal Matrices 114
1.9   Similarity Transformation 116
        1.9.1   Diagonalisation of Matrices 116
        1.9.2   General Case of Repeated Eigenvalues 123
        1.9.3   Simultaneous Diagonalisation of Two Matrices 127
1.10   Spectral Theorem and Function of a Matrix 130
        1.10.1   Normal Matrices 130
        1.10.2   General Matrices 136
1.11   Generalised Eigenproblem 138
1.12   Famous Identities with the Matrix Exponential 141
1.13   Quadratic Forms 145
1.14   Extremum of a Function of n Variables 148
1.15   Trace of a Matrix 149
1.16   Tridiagonalisation of a Matrix: The Lanczos Method 152
1.17   Dividing Matrices Into Blocks 156
1.18   Defective Matrices 161
        1.18.1   Jordan Normal Form 161
        1.18.2   Function of a General Matrix 167
1.19   Systems of Linear Differential Equations 171
        1.19.1   General Consideration 171
        1.19.2   Homogeneous Systems with Constant Coefficients 173
        1.19.3   Non-homogeneous Systems 179
        1.19.4   The Case of the Matrix Having a Lower Rank 182
        1.19.5   Higher Order Linear Differential Equations 187
        1.19.6   Homogeneous Systems with Defective Matrices 191
1.20   Pseudo-Inverse of a Matrix 198
1.21   General 3D Rotation 205
1.22   Examples in Physics 208
        1.22.1   Particle in a Magnetic Field 208
        1.22.2   Kinetics 211
        1.22.3   Vibrations in Molecules 213
        1.22.4   Least Square Method and MD Simulations 220
        1.22.5   Vibrations of Atoms in an Infinite Chain. A Point
                 Defect 224
        1.22.6   States of an Electron in a Solid 231
        1.22.7   Time Propagation of a Wavefunction 236
        1.22.8   Slater Determinants and Fermionic Many-Body
                 Wavefunctions 239
        1.22.9   Dirac Equation 246
        1.22.10 Classical 1D Ising Model 253

2 Complex Numbers and Functions 257
  2.1 Representation of Complex Numbers 257
  2.2 Functions on a Complex Plane 262
     2.2.1 Regions in the Complex Plane and Mapping 262
     2.2.2 Differentiation. Analytic Functions 266
  2.3 Main Elementary Functions 272
     2.3.1 Integer Power Function 273
     2.3.2 Integer Root Function 274
     2.3.3 Exponential and Hyperbolic Functions 282
     2.3.4 Logarithm 286
     2.3.5 Trigonometric Functions 289
     2.3.6 Inverse Trigonometric Functions 292
     2.3.7 General Power Function 294
  2.4 Integration in the Complex Plane 295
     2.4.1 Definition 295
     2.4.2 Integration of Analytic Functions 298
     2.4.3 Fundamental Theorem of Algebra 314
     2.4.4 Fresnel Integral 315
  2.5 Complex Functional Series 318
     2.5.1 Numerical Series 318
     2.5.2 General Functional Series 321
     2.5.3 Power Series 323
     2.5.4 The Laurent Series 328
     2.5.5 Zeros and Singularities of Functions 334
  2.6 Analytic Continuation 338
  2.7 Residues 342
     2.7.1 Definition 342
     2.7.2 Functional Series: An Example 347
     2.7.3 Applications of Residues in Calculating Real Axis
           Integrals 349
  2.8 Linear Differential Equations: Series Solutions 368
  2.9 Selected Applications in Physics 375
     2.9.1 Dispersion Relations 375
     2.9.2 Propagation of Electro-Magnetic Waves
           in a Material 378
     2.9.3 Electron Tunnelling in Quantum Mechanics 384
     2.9.4 Propagation of a Quantum State 392
     2.9.5 Matsubara Sums 394
     2.9.6 A Beautiful Counting Problem 399
     2.9.7 Landau-Zener Problem 403

3 Fourier Series 411
3.1   Trigonometric Series: An Intuitive Approach 412
3.2   Dirichlet Conditions 419
3.3   Gibbs–Wilbraham Phenomenon 422
3.4   Integration and Differentiation of the Fourier Series 425
3.5   Parseval's Theorem 429
3.6   Summing Inverse Powers of Integers 432
3.7   Complex (Exponential) form of the Fourier Series 435
3.8   Application to Differential Equations 440
3.9   A More Rigorous Approach to the Fourier Series 442
       3.9.1   Convergence 'on Average' 442
       3.9.2   A More Rigorous Approach to the Fourier Series:
                Dirichlet Theorem 447
       3.9.3   Expansion of a Function via Orthogonal Functions 450
3.10   Investigation of the Convergence of the FS 455
       3.10.1   The Second Mean-Value Integral Theorem 455
       3.10.2   First Estimates of the Fourier Coefficients 458
       3.10.3   More Detailed Investigation of FS Convergence 460
       3.10.4   Improvements of Convergence of the FS 464
3.11   Applications of Fourier Series in Physics 466
       3.11.1   Expansions of Functions Describing Crystal
                Properties 467
       3.11.2   Ewald's Formula 472
       3.11.3   Born and von Karman Boundary Conditions 475
       3.11.4   Atomic Force Microscopy 477
       3.11.5   Applications in Quantum Mechanics 480
       3.11.6   Free Electron Gas 482

4 Special Functions 487
4.1   Dirac Delta Function 487
4.2   The Gamma Function 501
       4.2.1   Definition and Main Properties 501
       4.2.2   Beta Function 508
4.3   Orthogonal Polynomials 509
       4.3.1   Legendre Polynomials 510
       4.3.2   General Theory of Orthogonal Polynomials 522
4.4   Differential Equation of Generalised Hypergeometric Type 537
       4.4.1   Transformation to a Standard Form 537
       4.4.2   Solutions of the Standard Equation 540
       4.4.3   Classical Orthogonal Polynomials 546
4.5   Associated Legendre Function 550
       4.5.1   Bound Solutions of the Associated Legendre
                Equation 551
       4.5.2   Orthonormality of Associated Legendre Functions 555
       4.5.3   Laplace Equation in Spherical Coordinates 557

4.6   Bessel Equation 564
        4.6.1   Bessel Differential Equation and its Solutions 564
        4.6.2   Half-Integer Bessel Functions 567
        4.6.3   Recurrence Relations for Bessel Functions 569
        4.6.4   Generating Function and Integral Representation
                 for Bessel Functions 571
        4.6.5   Orthogonality and Functional Series Expansion 573
4.7   Selected Applications in Physics 577
        4.7.1   Schrödinger Equation for a Harmonic Oscillator 577
        4.7.2   Schrödinger Equation for the Hydrogen Atom 579
        4.7.3   A Free Electron in a Cylindrical Ptential Well 582
        4.7.4   Stirling's Formula and Phase Transitions 585
        4.7.5   Band Structure of a Solid 590
        4.7.6   Oscillations of a Circular Membrane 594
        4.7.7   Multipole Expansion of the Electrostatic Potential 597
4.8   Van Hove Singularities 601

5   Fourier Transform 605
5.1   The Fourier Integral 605
        5.1.1   Intuitive Approach 605
        5.1.2   Alternative Forms of the Fourier Integral 607
        5.1.3   A More Rigorous Derivation of Fourier Integral 609
5.2   Fourier Transform 612
        5.2.1   General Idea 612
        5.2.2   Fourier Transform of Derivatives 618
        5.2.3   Fourier Transform of an Integral 620
        5.2.4   Convolution Theorem 620
        5.2.5   Parseval's Theorem 622
        5.2.6   Poisson Summation Formula 623
5.3   Applications of the Fourier Transform in Physics 628
        5.3.1   Various Notations and Multiple Fourier Transform 628
        5.3.2   Retarded Potentials 632
        5.3.3   Green's Function of a Differential Equation 637
        5.3.4   Time Correlation Functions 644
        5.3.5   Fraunhofer Diffraction 650

6   Laplace Transform 655
6.1   Definition 655
6.2   Method of Partial Fractions 660
6.3   Detailed Consideration of the LT 663
        6.3.1   Analyticity of the LT 663
        6.3.2   Relation to the Fourier Transform 665
        6.3.3   Inverse Laplace Transform 666
6.4   Properties of the Laplace Transform 679
        6.4.1   Derivatives of Originals and Images 679
        6.4.2   Shift in Images and Originals 682

	6.4.3	Integration of Images and Originals	685
	6.4.4	Convolution Theorem	688
6.5		Solution of Ordinary Differential Equations (ODEs)	689
6.6		Applications in Physics	697
	6.6.1	Application of the LT Method in Electronics	697
	6.6.2	Harmonic Particle with Memory	701
	6.6.3	Probabilities of Hops	708
	6.6.4	Inverse NC-AFM Problem	713

7 Curvilinear Coordinates   719
7.1		Definition of Curvilinear Coordinates	719
7.2		Unit Base Vectors	721
7.3		Line Elements and Line Integral	730
	7.3.1	Line Element	730
	7.3.2	Line Integrals	732
7.4		Surface Normal and Surface Integrals	735
7.5		Volume Element and Jacobian in 3D	738
7.6		Change of Variables in Multiple Integrals	741
7.7		Multi-variable Gaussian	746
7.8		N-Dimensional Sphere	755
7.9		Gradient of a Scalar Field	758
7.10		Divergence of a Vector Field	760
7.11		Laplacian	763
7.12		Curl of a Vector Field	765
7.13		Some Applications in Physics	769
	7.13.1	Partial Differential Equations of Mathematical Physics	769
	7.13.2	Classical Mechanics of a Particle	771
	7.13.3	Distribution Function of a Set of Particles	777

8 Partial Differential Equations of Mathematical Physics   781
8.1		General Consideration	782
	8.1.1	Characterisation of Second-Order PDEs	782
	8.1.2	Initial and Boundary Conditions	790
8.2		Wave Equation	791
	8.2.1	One-Dimensional String	791
	8.2.2	Propagation of Sound	793
	8.2.3	General Solution of PDE	795
	8.2.4	Uniqueness of Solution	800
	8.2.5	Fourier Method	802
	8.2.6	Forced Oscillations of the String	809
	8.2.7	General Boundary Problem	813
	8.2.8	Oscillations of a Rectangular Membrane	815
	8.2.9	General Remarks on the Applicability of the Fourier Method	819

	8.3	Heat-Conduction Equation	822
		8.3.1 Uniqueness of the Solution	823
		8.3.2 Fourier Method	825
		8.3.3 Stationary Boundary Conditions	827
		8.3.4 Heat Transport with Internal Sources	832
		8.3.5 Solution of the General Boundary Heat-Conduction Problem	833
	8.4	Problems Without Boundary Conditions	834
	8.5	Application of Fourier Method to Laplace Equation	837
	8.6	Method of Integral Transforms	840
		8.6.1 Fourier Transform	840
		8.6.2 Laplace Transform	843
9	**Calculus of Variations**		**849**
	9.1	Functions of a Single Variable	851
		9.1.1 Functionals Involving a Single Function	851
		9.1.2 Functionals Involving More Than One Function	868
		9.1.3 Functionals Containing Higher Derivatives	870
		9.1.4 Variation with Constraints Given by Zero Functions	872
		9.1.5 Variation with Constraints Given by Integrals	880
	9.2	Functions of Many Variables	884
	9.3	Applications in Physics	890
		9.3.1 Mechanics	890
		9.3.2 Functional Derivatives	899
		9.3.3 Many-Electron Theory	902
Index			**915**

Chapter 1
Elements of Linear Algebra

In practical problems it is often necessary to consider the behaviour of many functions; sometimes their number could be so large that one can easily drown in the calculations. For instance, in problems of chemical kinetics one is interested in calculating time evolution of concentrations $N_i(t)$ of many species i participating (and produced) in the reactions; including various intermediate species in complex reactions, the total number of all species (components) to consider can easily reach a hundred or even more. In all these cases it might be convenient to collect all such observables into a single object, $\mathbf{N}(t) = \{N_1(t), N_2(t), \ldots\}$, and then, instead of many equations governing evolution of each of its components, write a single equation for the time evolution of the whole object. This may enormously simplify calculations of many problems, provided, of course, that we develop special tools and a proper language for working with these collective objects.

One may say that this is nothing but a new convenient notation. However, it has made such an enormous impact on many branches of theory in mathematics, engineering, physics, chemistry, computing and others, that simply cannot be ignored.

Since this new language requires a detailed study of the objects it introduces, special attention should be paid to it. This is exactly what we are going to accomplish in this chapter. We shall start from spaces and vectors in them, then move on to matrices, their properties and multiple uses. We shall see how convenient and elegant vector and matrix notations are, and how powerful is algebra based on them. Then, at the end of the chapter, we will show how these objects may help in solving a number of physics problems.

© The Author(s), under exclusive license to Springer Nature Switzerland AG 2024
L. Kantorovich, *Mathematics for Natural Scientists II*, Undergraduate Lecture
Notes in Physics, https://doi.org/10.1007/978-3-031-46320-4_1

1.1 Vector Spaces

1.1.1 Introduction to Multi-dimensional Complex Vector Spaces

We gave the definition of vectors in *real* one, two, three and p dimensions in Sect. I.1.10.[1] Here we generalise vectors to complex p-dimensional spaces. It is straightforward to do so: we define a vector $\mathbf{x}$ in a p-dimensional space by specifying p (generally complex) numbers $x_1, x_2, \ldots, x_p$, called vector coordinates, which define uniquely the vector: $\mathbf{x} = (x_1, \ldots, x_p)$. Two such vectors can be added to each other, subtracted from each other or multiplied by a number c. In all these cases the operations are performed on the vectors coordinates:

$$\mathbf{x} + \mathbf{y} = \mathbf{g} \quad \text{means} \quad x_i + y_i = g_i \ , \quad i = 1, 2, \ldots, p \ ;$$

$$\mathbf{x} - \mathbf{y} = \mathbf{g} \quad \text{means} \quad x_i - y_i = g_i \ , \quad i = 1, 2, \ldots, p \ ;$$

$$c\mathbf{x} = \mathbf{g} \quad \text{means} \quad cx_i = g_i \ , \quad i = 1, 2, \ldots, p \ .$$

Consequently, a linear combination of two vectors $\mathbf{x}$ and $\mathbf{y}$ can also be defined as

$$\alpha\mathbf{x} + \beta\mathbf{y} = \mathbf{g} \ , \quad \text{where} \quad \alpha x_i + \beta y_i = g_i \ , \quad i = 1, 2, \ldots, p$$

for any (generally complex) numbers α and β. So in general coordinates of vectors could be complex and so it is said that the vectors are complex. A space in which a linear combination of vectors is defined as formulated above is called a vector (or linear) space. Since the algebraic operations between vectors basically correspond to p simultaneous operations on their components, that are usual algebraic operations with numbers, the summation of vectors is associative, $\mathbf{x} + (\mathbf{y} + \mathbf{z}) = (\mathbf{x} + \mathbf{y}) + \mathbf{z}$ and commutative, $\mathbf{x} + \mathbf{y} = \mathbf{y} + \mathbf{x}$, while a multiplication with a number is distributive, $a(\mathbf{x} + \mathbf{y}) = a\mathbf{x} + a\mathbf{y}$. Also, the zero vector $\mathbf{0}$ serves as an identity element in any vector space since for any $\mathbf{x}$ we have $\mathbf{x} + \mathbf{0} = \mathbf{x}$; also, for any $\mathbf{x}$ there is an inverse element (vector) $-\mathbf{x}$, since $\mathbf{x} + (-\mathbf{x}) = \mathbf{0}$.

A dot (scalar) product of two vectors $\mathbf{x}$ and $\mathbf{y}$ is defined as a (generally complex) number

$$(\mathbf{x}, \mathbf{y}) = \mathbf{x}^* \cdot \mathbf{y} = x_1^* y_1 + x_2^* y_2 + \cdots + x_p^* y_p = \sum_{i=1}^{p} x_i^* y_i \ , \tag{1.1}$$

---

[1] In the following, references to the first volume of this course (L. Kantorovich, Mathematics for natural scientists: fundamentals and basics, 2nd Edition, Springer, 2022) will be made by appending the Roman number I in front of the reference, e.g. Sect. I.1.8 or Eq. (I.5.18) refer to Sect. 1.8 and Eq. (5.18) of the first volume, respectively.

which generalises the corresponding definition for real vectors. Two vectors are called *orthogonal* if their dot product is zero. The length of a vector $\mathbf{x}$ is defined as $|\mathbf{x}| = \sqrt{(\mathbf{x}, \mathbf{x})}$, i.e. the dot product of the vector with itself is the square of its length. Obviously,

$$(\mathbf{x}, \mathbf{x}) = \sum_{i=1}^{p} x_i^* x_i = \sum_{i=1}^{p} |x_i|^2 > 0 \,,$$

i.e. the dot product of the vector with itself is real, non-negative and hence the vector length $|\mathbf{x}|$, as a square root of the dot product, is well defined.

The dot product defined above satisfies the following identity:

$$(\mathbf{x}, \mathbf{y})^* = \left(\sum_{i=1}^{p} x_i^* y_i \right)^* = \sum_{i=1}^{p} x_i y_i^* = (\mathbf{y}, \mathbf{x}) \,. \tag{1.2}$$

Problem 1.1 By using the definition (1.1), prove that the dot product is *distributive*:

$$(\mathbf{g}, \mathbf{u} + \mathbf{v}) = (\mathbf{g}, \mathbf{u}) + (\mathbf{g}, \mathbf{v}) \,.$$

Therefore, one can manipulate vectors algebraically.

It is convenient to introduce real unit base vectors $\mathbf{e}_1 = (1, 0, \ldots, 0)$, $\mathbf{e}_2 = (0, 1, 0, \ldots, 0)$, $\mathbf{e}_3 = (0, 0, 1, 0, \ldots, 0)$, etc. There are exactly p such vectors in a p-dimensional space; each of these vectors has only one component (coordinate) equal to one, all others are equal to zero. Specifically, $\mathbf{e}_i$ has all its coordinates equal to zero except for its ith coordinate which is equal to one. We may express this by writing $\mathbf{e}_i = (\delta_{ij})$, which means that the components $j = 1, \ldots, p$ of the base vector $\mathbf{e}_i$ for any $i = 1, \ldots, p$ are given by the familiar Kronecker symbol δ_{ij} ($\delta_{ij} = 1$ if $i = j$ and equal to 0 otherwise). Obviously, these unit base vectors are all of the unit length, $(\mathbf{e}_i, \mathbf{e}_i) = 1$, and are orthogonal to each other by construction, i.e. $(\mathbf{e}_i, \mathbf{e}_j) = 0$ for any $i \neq j$. In other words,

$$(\mathbf{e}_i, \mathbf{e}_j) = \delta_{ij} \,.$$

Then, any vector $\mathbf{x} = (x_1, \ldots, x_p)$ can be uniquely expanded in terms of the unit base vectors:

$$\mathbf{x} = (x_1, \ldots, x_p) = x_1 \mathbf{e}_1 + x_2 \mathbf{e}_2 + \cdots + x_p \mathbf{e}_p$$

$$= \sum_{i=1}^{p} x_i \mathbf{e}_i$$

with its (generally complex) coordinates $\{x_i\}$ serving as the expansion coefficients. Indeed, assuming that the expansion coefficients are different,

$$\mathbf{x} = \sum_{i=1}^{p} x_i' \mathbf{e}_i \; , \tag{1.3}$$

it is easy to see that solving for them gives simply $x_i' = x_i$ for any $i = 1, \ldots, p$.

Problem 1.2 Write Eq. (1.3) in components to prove this.

Next, we have to introduce the notion of a *linear independence* of vectors. Take two vectors $\mathbf{u}$ and $\mathbf{v}$ that are collinear, i.e. $\mathbf{u} = \lambda \mathbf{v}$ with λ being a generally complex number. It is said that $\mathbf{u}$ is linearly dependent of $\mathbf{v}$ (and *vice versa*). Now, consider two non-collinear vectors $\mathbf{u}$ and $\mathbf{v}$ (which are known to be not proportional to each other, i.e. $\mathbf{u} \neq \lambda \mathbf{v}$ for any complex λ). A third vector $\mathbf{g}$ is said to be *linearly dependent* of $\mathbf{u}$ and $\mathbf{v}$ if it can be written as their *linear combination*:

$$\mathbf{g} = \alpha \mathbf{u} + \beta \mathbf{v} \; ,$$

where at least one of the numerical coefficients α and β is non-zero. Otherwise, $\mathbf{g}$ is said to be *linearly independent* of the vectors $\mathbf{u}$ and $\mathbf{v}$.

The notion of linear dependence (independence) can be formulated more generally and in such a way that all three vectors enter on an equal footing: the three vectors $\mathbf{u}$, $\mathbf{v}$ and $\mathbf{g}$ are said to be linearly independent, if and only if their zero linear combination

$$\alpha \mathbf{x} + \beta \mathbf{y} + \gamma \mathbf{g} = \mathbf{0}$$

is only possible when $\alpha = \beta = \gamma = 0$. In other words, a linear combination of linearly independent vectors will always be a non-zero vector (unless, of course, the linear combination is constructed with zero coefficients). If it is possible to construct a zero linear combination of three non-zero vectors with at least two non-zero coefficients, that means that at least one vector can be expressed as a linear combination of the others. Obviously, the above definition is immediately generalised to any number of vectors: n vectors are said to be linearly independent if and only if their zero linear combination,

$$\alpha_1 \mathbf{x}_1 + \alpha_2 \mathbf{x}_2 + \cdots + \alpha_n \mathbf{x}_n = \mathbf{0} \tag{1.4}$$

is only possible if all the coefficients are equal to zero at the same time: $\alpha_1 = \alpha_2 = \cdots = \alpha_p = 0$. We shall see in the following that in a p-dimensional space the maximum number of linearly independent vectors cannot be larger than p.

It is easy to see that mutually orthogonal vectors are linearly independent. Indeed, by taking a dot product of both sides of Eq. (1.4) with $\mathbf{x}_1$ and using the fact that $\mathbf{x}_1$ is orthogonal to any other vector, we immediately get $\alpha_1 = 0$. Multiplying Eq. (1.4) with $\mathbf{x}_2$ we similarly obtain $\alpha_2 = 0$, and so on. In particular, the unit base vectors $\mathbf{e}_i = \left(\delta_{ij} \right)$ are linearly independent.

Example 1.1 ▶ Consider vectors $\mathbf{A} = \mathbf{i} + \mathbf{j} = (1, 1, 0)$, $\mathbf{B} = \mathbf{i} + \mathbf{k} = (1, 0, 1)$, where $\mathbf{i} = (1, 0, 0) \equiv \mathbf{e}_1$, $\mathbf{j} = (0, 1, 0) \equiv \mathbf{e}_2$ and $\mathbf{k} = (0, 0, 1) \equiv \mathbf{e}_3$ are the familiar unit base vectors of the 3D space. We shall show that the vector $\mathbf{C} = (2, 1, 1)$ is linearly dependent of $\mathbf{A}$ and $\mathbf{B}$. Indeed, trying a linear combination, $\mathbf{C} = \alpha\mathbf{A} + \beta\mathbf{B}$, we can write the following three equations in components:

$$\begin{cases} 2 = \alpha + \beta \\ 1 = \alpha \\ 1 = \beta \end{cases}$$

that have a unique solution $\alpha = \beta = 1$, i.e. $\mathbf{C} = \mathbf{A} + \mathbf{B}$, hence, $\mathbf{C}$ is linearly dependent of $\mathbf{A}$ and $\mathbf{B}$. Obviously, the same result is obtained if instead we solve the equation $\alpha\mathbf{A} + \beta\mathbf{B} + \gamma\mathbf{C} = \mathbf{0}$, in which case we have three equations:

$$\begin{cases} \alpha + \beta + 2\gamma = 0 \\ \alpha + \gamma = 0 \\ \beta + \gamma = 0 \end{cases}$$

that results in the solution $\beta = \alpha = -\gamma$ that satisfies all three equations. Hence, this system of equations has an infinite number of solutions. There is a zero solution $\alpha = \beta = \gamma = 0$. However, it is essential here that we also have (infinite number of) non-zero solutions indicating that the three vectors are linearly dependent. By taking specifically $\gamma = -1$, we shall obtain $\mathbf{C} = \mathbf{A} + \mathbf{B}$ as before. ◀

Example 1.2 ▶ Let us check the linear dependence of vectors $\mathbf{A} = (1, 0, 0)$, $\mathbf{B} = (1, 0, 1)$ and $\mathbf{C} = (0, 1, 2)$. Trying again $\mathbf{C} = \alpha\mathbf{A} + \beta\mathbf{B}$, we obtain in components:

$$\begin{cases} 0 = \alpha + \beta \\ 1 = 0 \\ 2 = \beta \end{cases} ,$$

which are obviously contradictive. If we, formally, take $\beta = 2$ and $\alpha = -2$ from the third and the first equations, respectively, then we can build a vector $\mathbf{C}_1 = -2\mathbf{A} + 2\mathbf{B}$ which is not equal to $\mathbf{C}$; in fact, $\mathbf{C} = \mathbf{C}_1 + \mathbf{j}$. One can see that $\mathbf{C}$ has an extra 'degree of freedom' along the $\mathbf{j}$ direction. Hence, $\mathbf{C}$ is linearly independent of $\mathbf{A}$ and $\mathbf{B}$.

The same result is obtained if we use the general definition of the linear independence and try to solve the equation $\alpha\mathbf{A} + \beta\mathbf{B} + \gamma\mathbf{C} = \mathbf{0}$ with respect to the coefficients α, β and γ. Writing this vector equation in components, we get

$$\begin{cases} \alpha + \beta = 0 \\ \gamma = 0 , \\ \beta + 2\gamma = 0 \end{cases}$$

which accepts only the zero solution $\alpha = \beta = \gamma = 0$, as required. Hence, the vectors **A**, **B** and **C** are indeed linearly independent. ◄

Example 1.3 ► Prove that vectors $\mathbf{u}_1 = (1, 0, 0, 0)$, $\mathbf{u}_2 = (0, 1, 1, 0)$, $\mathbf{u}_3 = (1, 0, 1, 0)$ and $\mathbf{u}_4 = (0, 0, 0, 1)$ are linearly independent.

Solution: Indeed, we have to solve the vector equation $\alpha_1 \mathbf{u}_1 + \alpha_2 \mathbf{u}_2 + \alpha_3 \mathbf{u}_3 + \alpha_4 \mathbf{u}_4 = \mathbf{0}$, which, if written in components, results in four algebraic equations with respect to the four coefficients α_1, α_2, α_3 and α_4:

$$\begin{cases} \alpha_1 \cdot 1 + \alpha_2 \cdot 0 + \alpha_3 \cdot 1 + \alpha_4 \cdot 0 = 0 \\ \alpha_1 \cdot 0 + \alpha_2 \cdot 1 + \alpha_3 \cdot 0 + \alpha_4 \cdot 0 = 0 \\ \alpha_1 \cdot 0 + \alpha_2 \cdot 1 + \alpha_3 \cdot 1 + \alpha_4 \cdot 0 = 0 \\ \alpha_1 \cdot 0 + \alpha_2 \cdot 0 + \alpha_3 \cdot 0 + \alpha_4 \cdot 1 = 0 \end{cases} \implies \begin{cases} \alpha_1 + \alpha_3 = 0 \\ \alpha_2 = 0 \\ \alpha_2 + \alpha_3 = 0 \\ \alpha_4 = 0 \end{cases},$$

which give a unique solution $\alpha_1 = \alpha_2 = \alpha_3 = \alpha_4 = 0$. ◄

Using linearly independent vectors called *basis*, one can construct multi-dimensional spaces. We shall illustrate this point assuming real vectors. Suppose, we are given as a basis p linearly independent *real* vectors $\mathbf{u}_1, \mathbf{u}_2, \ldots, \mathbf{u}_p$. These can be used to construct a p-dimensional (p-D) space using the following imaginative procedure:

- Start from $\mathbf{u}_1$. All vectors $c_1 \mathbf{u}_1$ with any numbers c_1 form a line (a 1D space) along $\mathbf{u}_1$.
- Take $\mathbf{u}_2$, which is not proportional to $\mathbf{u}_1$ (and hence is linearly independent of it), and form all linear combinations $c_1 \mathbf{u}_1 + c_2 \mathbf{u}_2$ with arbitrary numbers c_1 and c_2; these form a plane (2D space) with $\mathbf{u}_1$ and $\mathbf{u}_2$ within it.
- Similarly, linear combinations $c_1 \mathbf{u}_1 + c_2 \mathbf{u}_2 + c_3 \mathbf{u}_3$ would cover a 3D space (the volume), see Fig. 1.1, if the vector $\mathbf{u}_3$ is chosen out of the plane formed by $\mathbf{u}_1$ and $\mathbf{u}_2$, i.e. it is linearly independent of them.
- Continuing this procedure, the whole p-dimensional space p-D is built up by linear combinations

$$\mathbf{x} = c_1 \mathbf{u}_1 + \cdots + c_p \mathbf{u}_p = \sum_{i=1}^{p} c_i \mathbf{u}_i \tag{1.5}$$

of linearly independent vectors employing all possible numbers c_1, c_2, etc.

Fig. 1.1 A third vector goes beyond the plane formed by the first two vectors as it has an extra coordinate in the direction out of the plane (an extra 'degree of freedom'). The three vectors form a 3D space (volume)

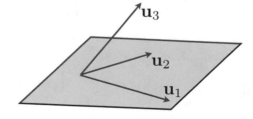

We shall now formulate a few rigorous statements that clarify the notion of the basis of a p-D linear space.

Theorem 1.1 *Let $\{\mathbf{u}_1, \ldots, \mathbf{u}_p\}$ be a set of linearly independent vectors. Then a vector $\mathbf{v}$ from the same space is given as a linear combination of the vectors $\{\mathbf{u}_i\}$ if and only if the extended system $\{\mathbf{u}_1, \ldots, \mathbf{u}_p, \mathbf{v}\}$ is linearly dependent.*

Proof If $\mathbf{v}$ is given as a linear combination of the vectors $\{\mathbf{u}_i\}$, then obviously, it is linearly dependent on them and the extended system of vectors is linearly dependent. Conversely, if the extended system is linearly dependent, then the vector equation

$$\sum_{i=1}^{p} c_i \mathbf{u}_i + \beta \mathbf{v} = \mathbf{0}$$

has at least one solution with non-zero coefficients. Clearly, $\beta \neq 0$ as otherwise we arrive at $\sum_{i=1}^{p} c_i \mathbf{u}_i = \mathbf{0}$ with at least some non-zero coefficients, which contradicts the linear independence of the vectors $\{\mathbf{u}_i\}$. Hence, one can solve the above equation for $\mathbf{v}$ to get it as a linear combination of $\{\mathbf{u}_i\}$. **Q.E.D.**

This theorem is a particular case of a more general statement.

Theorem 1.2 *Given a set of p linearly independent vectors $\{\mathbf{u}_i\}$, any $q > p$ set of vectors belonging to the same space are linearly dependent.*

Proof Indeed, if we construct q vectors

$$\mathbf{v}_j = \sum_{i=1}^{p} c_{ji} \mathbf{u}_i, \quad j = 1, \ldots, q > p,$$

then in order for them to be linearly independent, we need the vector equation

$$\sum_{j=1}^{q} \alpha_j \mathbf{v}_j = \sum_{i=1}^{p} \left(\sum_{j=1}^{q} \alpha_j c_{ji} \right) \mathbf{u}_i = \mathbf{0}$$

to have a unique zero solution

$$\sum_{j=1}^{q} \alpha_j c_{ji} = 0, \quad i = 1, \ldots, p$$

with respect to the coefficients α_j. The number q of the unknown coefficients α_j is however larger than the number p of the equations and, as will be shown in Sect. 1.4.2, this linear system of algebraic equations always has at least one non-zero solution. Hence, at least some of the coefficients α_j are non-zero and hence the vectors $\{\mathbf{v}_j\}$ must be linearly dependent, as required. **Q.E.D.**

Theorem 1.3 *An expansion*

$$\mathbf{v} = \sum_{i=1}^{p} c_i \mathbf{u}_i$$

of a vector $\mathbf{v}$ *via* $\{\mathbf{u}_i\}$ *is unique if and only if the vectors* $\{\mathbf{u}_1, \ldots, \mathbf{u}_p\}$ *are linearly independent.*

Proof Suppose, the expansion is not unique, and there is another vector with different expansion coefficients $\{c_i'\}$. Then, subtracting the two expansions we get $\sum_i (c_i - c_i') \mathbf{u}_i = \mathbf{0}$ with at least some of the coefficients $c_i - c_i' \neq 0$ leading to a linear dependence of the vectors $\{\mathbf{u}_i\}$. Conversely, assume that vectors $\{\mathbf{u}_i\}$ are linearly dependent, i.e. $\sum_{i=1}^{p} c_i \mathbf{u}_i = 0$ with at least some of the coefficients c_i being non-zero. If we construct a vector $\mathbf{v} = \sum_{i=1}^{p} \lambda_i \mathbf{u}_i$ using some coefficients $\{\lambda_i\}$, then one may obviously also construct another vector $\mathbf{v}' = \sum_{i=1}^{p} (c_i + \lambda_i) \mathbf{u}_i \neq \mathbf{v}$, i.e. the expansion with respect to the same set of vectors $\{\mathbf{u}_i\}$ is not unique. Hence, we have proved in both directions that linear dependence of the basis vectors $\{\mathbf{u}_i\}$ is equivalent to the expansion of $\mathbf{v}$ via them being non-unique. This proves the above-made statement. **Q.E.D.**

Hence, given the vector $\mathbf{x}$ and the basis $\{\mathbf{u}_i\}$, one can uniquely determine the coefficients $c_1, \ldots, c_p$ in its expansion in terms of p linearly independent vectors of the basis. It is instructive to illustrate this by considering a special case when the vectors $\mathbf{u}_1, \mathbf{u}_2, \ldots, \mathbf{u}_p$ are mutually orthogonal and normalised to unity (it is sometimes said that the vectors are *orthonormal*), i.e.

$$(\mathbf{u}_i, \mathbf{u}_j) = \begin{cases} 1, & \text{if } i = j \\ 0, & \text{if } i \neq j \end{cases} = \delta_{ij} \ . \tag{1.6}$$

As we shall see below in Sect. 1.1.3, it is always possible to form linear combinations of any n linearly independent vectors such that the constructed vectors are orthonormal.

In order to find the coefficients c_i for the given $\mathbf{x}$, let us calculate the dot product of both sides of Eq. (1.5) with $\mathbf{u}_j$:

$$(\mathbf{u}_j, \mathbf{x}) = \sum_{k=1}^{p} c_k (\mathbf{u}_j, \mathbf{u}_k) = \sum_{k=1}^{p} c_k \delta_{kj} = c_j \ ,$$

since only one term in the sum will survive, namely the one with the summation index $k = j$. This proves that for each $\mathbf{x}$ the coefficients c_i are uniquely determined, so that the appropriate (and unique) linear combination (1.5) is obtained.

Since any $\mathbf{x}$ from the p-D space can be uniquely specified by a linear combination of vectors $\mathbf{u}_1$, $\mathbf{u}_2$, ...,$\mathbf{u}_p$, the latter set is said to be *complete*. This means that it can be used to form a p-dimensional vector space by means of all possible linear combinations of these vectors. Note that the choice of the vectors of the basis $\mathbf{u}_1$, $\mathbf{u}_2$, ...,$\mathbf{u}_p$ is not unique; any set of p linearly independent vectors will do. Of course, in the new basis the expansion coefficients will be different. An important point is provided by the following

---

Theorem 1.4 *Any basis that builds the same vector space contains the same number of vectors.*

Proof Indeed, suppose we have two sets of basic vectors $\{\mathbf{u}_1, \ldots, \mathbf{u}_p\}$ and $\{\mathbf{v}_1, \ldots, \mathbf{v}_q\}$ with $q > p$. Since both sets belong to the same vector space, any vector $\mathbf{v}_j$ from the second set can be expanded into the vectors $\{\mathbf{u}_i\}$ from the first one. However, $q > p$ and hence, according to Theorem 1.2, the second set contains linearly dependent vectors. This, however, contradicts are assumption that the second set forms a basis, i.e. contains a set of linearly independent vectors. Hence, the other basis cannot have more vectors than p.

In the same vein, suppose the second set has a smaller number of vectors $q < p$. Then any vector from the first set (that contains more vectors) can be expanded into vectors from the second (that contains less vectors) and hence the vectors from the first set must be linearly dependent. Yet again we have arrived at a contradiction, hence the second basis set cannot contain less vectors than the first. Clearly, any basis can only contain the same number of vectors. **Q.E.D.**

Hence, any basis of the given vector space has the same number of vectors; this number defines the dimension of this space.

Example 1.4 ▶ Expand the vector $\mathbf{x} = (1, -1, 2, 3)$ of the 4D space in terms of the four vectors of Example 1.3.

Solution: Firstly, we note that vectors $\mathbf{u}_1, \ldots, \mathbf{u}_4$ are linearly independent as was proven in Example 1.3. Then, we write

$$\mathbf{x} = c_1\mathbf{u}_1 + c_2\mathbf{u}_2 + c_3\mathbf{u}_3 + c_4\mathbf{u}_4 . \tag{1.7}$$

Multiplying (in the sense of the dot product) both sides of this equations by $\mathbf{u}_1$, we obtain an algebraic equation; repeating this process with $\mathbf{u}_2$, $\mathbf{u}_3$ and $\mathbf{u}_4$, we obtain other three equations for the unknown coefficients, i.e. the required four equations:

$$\begin{cases} 1 = c_1 \cdot 1 + c_2 \cdot 0 + c_3 \cdot 1 + c_4 \cdot 0 \\ 1 = c_1 \cdot 0 + c_2 \cdot 2 + c_3 \cdot 1 + c_4 \cdot 0 \\ 3 = c_1 \cdot 1 + c_2 \cdot 1 + c_3 \cdot 2 + c_4 \cdot 0 \\ 3 = c_1 \cdot 0 + c_2 \cdot 0 + c_3 \cdot 0 + c_4 \cdot 1 \end{cases},$$

which are easily solved to give $c_1 = -2, c_2 = -1, c_3 = 3$ and $c_4 = 3$, i.e. the required expansion is

$$\mathbf{x} = -2\mathbf{u}_1 - \mathbf{u}_2 + 3\mathbf{u}_3 + 3\mathbf{u}_4 .$$

This can be verified directly by using the coordinates of the vectors.

Problem 1.3 Alternatively, solve this problem by writing Eq. (1.7) in components. This also gives a system of four linear algebraic equations for the unknown coefficients c_1, c_2, c_3 and c_4. This method is simpler than the one used above.

Problem 1.4 Check linear dependence of vectors $\mathbf{u}_1 = (1, 1, 0)$, $\mathbf{u}_2 = (1, 0, 1)$ and $\mathbf{u}_3 = (0, 1, -1)$. [Answer: *linearly dependent*]

Problem 1.5 Prove linear independence of vectors $\mathbf{u}_1 = (1, 1, 0)$, $\mathbf{u}_2 = (1, 0, 1)$ and $\mathbf{u}_3 = (0, 1, 1)$.

Problem 1.6 Expand the vector $\mathbf{u} = (-1, 1, 1)$ in terms of $\mathbf{u}_1 = (1, 0, 1)$, $\mathbf{u}_2 = (0, 1, 1)$ and $\mathbf{u}_3 = (1, 1, 0)$. [Answer: $\mathbf{u} = -\mathbf{u}_1/2 + 3/2\mathbf{u}_2 - \mathbf{u}_3/2$]

Problem 1.7 Find the expansion coordinates c_1, c_2 and c_3 of the vector $\mathbf{u} = (1, 2, 3)$ in terms of the basis vectors $\mathbf{u}_1 = (1, 1, 0)$, $\mathbf{u}_2 = (1, 0, 1)$ and $\mathbf{u}_3 = (1, 1, 1)$. [Answer: $c_1 = -2, c_2 = -1$ *and* $c_3 = 4$]

Problem 1.8 Check linear independence of vectors $\mathbf{u}_1 = (-1, 1, 1, 1)$, $\mathbf{u}_2 = (1, -1, 1, 1)$, $\mathbf{u}_3 = (1, 1, -1, 1)$ and $\mathbf{u}_4 = (1, 1, 1, 1)$. [Answer: *the vectors are linearly independent*]

1.1.2 Analogy Between Functions and Vectors

There is a close analogy between functions of a single variable and vectors which is frequently exploited. Indeed, consider a generally complex function $f(x)$ of a real variable defined on the interval $a \leq x \leq b$. We assume that the function is integrable in this interval. We divide the interval into N equidistant subintervals of the length $\Delta_N = (b - a)/N$ using division points $x_0 = a, x_1 = a + \Delta_N, x_2 = a + 2\Delta_N$, etc., and $x_N = a + N\Delta_N = b$, as shown in Fig. 1.2; generally, $x_i = a + i\Delta_N$. The values $(f(x_0), f(x_1), \ldots, f(x_N)) = (f_0, f_1, \ldots, f_N)$ of the function $f(x)$ at the $N + 1$ division points form a vector $\mathbf{f} = (f_0, f_1, \ldots, f_N)$ of dimension $N + 1$. Similarly

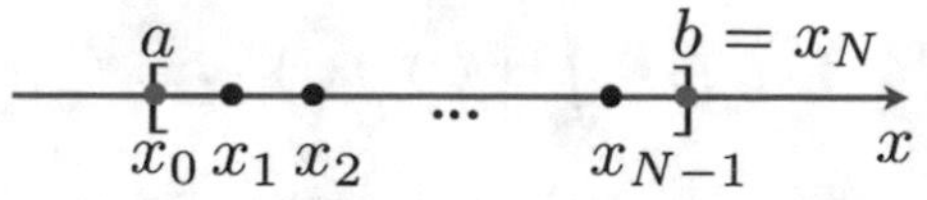

Fig. 1.2 The interval $a \leq x \leq b$ is divided into N equidistant subintervals by points $x_0 = a$, x_1, x_2, etc., $x_N = b$

we can form vectors **g**, **h**, etc. of the same dimension from other functions $g(x)$, $h(x)$, etc.

Then, similarly to vectors, we can sum and subtract vectors formed from functions as well as to multiply them by a number. All these operations will give values of thus obtained new functions at the division points. One can also consider the dot product of two vectors **f** and **g**, corresponding to the functions $f(x)$ and $g(x)$, respectively, in the usual way as

$$(\mathbf{f}, \mathbf{g})_N = \sum_{i=0}^{N} f(x_i)^* g(x_i) \ ,$$

where we indicated explicitly in our notations for the dot product that it is based on the division of the interval $a \leq x \leq b$ into N subintervals. The sum above diverges in the $N \to \infty$ limit. However, if we multiply the sum by the division interval $\Delta_N = \Delta x$, it would correspond to the Riemann integral sum (Sect. I.4.1) which in the limit becomes the definite integral of the product $f(x)^* g(x)$ between a and b, which may converge. Therefore, we define the dot product of two functions as the limit:

$$(f(x), g(x)) = \lim_{N \to \infty} (\mathbf{f}, \mathbf{g})_N \, \Delta_N = \lim_{N \to \infty} \left[\sum_{i=0}^{N} f(x_i)^* g(x_i) \, \Delta_N \right]$$

$$= \int_a^b f(x)^* g(x) dx \ . \tag{1.8}$$

In fact, it is convenient to generalise this definition a little by introducing the so-called *weight function* $w(x) > 0$:

$$(f(x), g(x)) = \int_a^b w(x) f(x)^* g(x) dx \ . \tag{1.9}$$

This integral, which is the closest to the dot product of vectors if $w(x) = 1$, is often called an *overlap integral* since its value depends crucially whether or not the two functions overlap within the interval: only if there is a subinterval where both functions are non-zero (i.e. they overlap there), the integral is non-zero. If the overlap integral is equal to zero, it is said that the two function are *orthogonal*. The overlap integral of the function with itself defines the 'length' of the function on the interval (also called *norm*):

$$(f, f) = \int_a^b w(x) \, |f(x)|^2 \, dx \ .$$

Assuming $f(x)$ is continuous, the norm is not equal to zero if and only if the function $f(x) \neq 0$ on at least one continuous subinterval of a finite length inside the original interval $a \leq x \leq b$. If $(f, f) = 0$, then $|f(x)| = 0$ at all points x within our interval. Therefore, the norm (f, f) characterises how strongly the function $f(x)$ is different from zero, while the dot product (f, g) demonstrates if the two functions $f(x)$ and $g(x)$ have appreciable overlap within the interval. If functions in the set $f_1(x)$, $f_2(x)$, etc., $f_n(x)$ satisfy $(f_i, f_j) = \delta_{ij}$, they are called orthonormal: orthogonal if $i \neq j$ (i.e. when functions are different), and each function is of the norm equal to one (when $i = j$).

Similarly to vectors, it is also possible to consider linear independence of functions. Functions $f_1(x)$, $f_2(x)$, ..., $f_k(x)$ are said to be linearly independent if any one of them cannot be expressed as a linear combination of the others for any value of x from a continuous interval. In other words, the equation

$$\alpha_1 f_1(x) + \alpha_2 f_2(x) + \cdots + \alpha_k f_k(x) = \sum_{i=1}^k \alpha_i f_i(x) = 0 \qquad (1.10)$$

is valid for any x from the specified interval if and only if there is only a unique trivial choice for the coefficients $\alpha_1 = \alpha_2 = \cdots = \alpha_k = 0$. This definition makes perfect sense: indeed, if a function *is* linearly dependent of some other functions, then one should be able to write this function as their linear combination with non-zero coefficients. For instance, $f(x) = 2x + 5x^2 - 1$ is linearly dependent on $f_0(x) = 1$, $f_1(x) = x$ and $f_2(x) = x^2$, since $f(x) = -f_0(x) + 2f_1(x) + 5f_2(x)$, and hence $f(x) + f_0(x) - 2f_1(x) - 5f_2(x) = 0$ for any x. It is seen that in this case the coefficients $\{\alpha_i\}$ are not zero, but equal to some real values. If one can only accommodate the linear combination (1.10) with all coefficients being equal to zero, the functions are indeed linearly independent. This definition of linear independence is exactly equivalent to that for vectors.

Example 1.5 ▶ As an example, let us prove, assuming the unit weight function, that the functions $f_1 = 1$, $f_2 = x$ and $f_3 = x^2$ are linearly independent on the interval $0 \leq x \leq 1$. Indeed, we need to solve the equation

$$\alpha_1 + \alpha_2 x + \alpha_3 x^2 = 0 \qquad (1.11)$$

with respect to the coefficients α_1, α_2 and α_3. To do that, we multiply both sides of the equation by $f_1 = 1$ and integrate both sides between 0 and 1:

$$\alpha_1 \int_0^1 dx + \alpha_2 \int_0^1 x\,dx + \alpha_3 \int_0^1 x^2 dx = 0 \quad \Longrightarrow \quad \alpha_1 + \frac{1}{2}\alpha_2 + \frac{1}{3}\alpha_3 = 0 \ .$$

Multiplying the original equation by $f_2 = x$ and performing a similar calculation, we obtain the second equation:

$$\frac{1}{2}\alpha_1 + \frac{1}{3}\alpha_2 + \frac{1}{4}\alpha_3 = 0 \, ,$$

while using $f_3 = x^2$ results in the third equation:

$$\frac{1}{3}\alpha_1 + \frac{1}{4}\alpha_2 + \frac{1}{5}\alpha_3 = 0 \, .$$

These three linear equations are solved e.g. by substitution and yield the unique trivial (zero) solution $\alpha_1 = \alpha_2 = \alpha_3 = 0$, proving that the three functions are indeed linearly independent. ◄

Note that the linear independence can also be verified by simply using several values of the x in the original Eq. (1.11):

$$x = 0 \implies \alpha_1 = 0 \, ;$$
$$x = 1 \implies \alpha_2 + \alpha_3 = 0 \, ;$$
$$x = \frac{1}{2} \implies \alpha_2 \frac{1}{2} + \alpha_3 \frac{1}{4} = 0 \, .$$

Solving the last two equations gives trivially $\alpha_2 = \alpha_3 = 0$; this, with $\alpha_1 = 0$ obtained from $x = 0$, gives immediately the desired result.

In the problems below the unit weight function is assumed.

Problem 1.9 Prove that the functions $f_0 = 1$, $f_1 = x^2 - 2$, $f_2 = -x^3 + x^2 - 1$ and $f_3 = 3x^3 - x^2 - 7$ are linearly dependent on the interval $-1 \leq x \leq 1$. [Answer: $f_3 = -3f_2 + 2f_1 - 6f_0$.]

Problem 1.10 Prove that the functions $f_1 = 1$, $f_2 = e^{-x}$, $f_3 = e^{-2x}$ and $f_4 = e^{-3x}$ are linearly independent on the interval $0 \leq x < \infty$.

Problem 1.11 Let us consider on the interval $a \leq x \leq b$ a set of orthonormal functions $f_1(x)$, $f_2(x)$, etc., $f_n(x)$. Prove, that they are linearly independent.

Later on in Sect. 1.6.1 we shall give a simpler method for checking linear independence of functions since it relies solely on differentiation, which can always be done analytically; neither orthogonality nor integration are required.

1.1.3   Orthogonalisation Procedure (Gram–Schmidt Method)

1.1.3.1   Orthogonalisation of Vectors

We mentioned above that if we are given a set of p linearly independent vectors $\mathbf{u}_1, \mathbf{u}_2, \ldots, \mathbf{u}_p$, new vectors, $\mathbf{v}_1, \mathbf{v}_2, \ldots, \mathbf{v}_p$, which are special linear combinations of the former set, can always be constructed such that they are orthonormal, i.e. $(\mathbf{v}_i, \mathbf{v}_j) = \delta_{ij}$. We shall prove here that this is possible by demonstrating how this can actually be done. This method is due to Gram and Schmidt. This kind of expansions is sometimes very useful.

First, we construct an intermediate orthogonal set of vectors which will not necessarily be normalised to one. This is done using the following *recurrence* procedure:

1. Chose the first vector $\mathbf{v}_1 = \mathbf{u}_1$.
2. Construct a vector $\mathbf{v}_2 = \mathbf{u}_2 + c_1^{(2)}\mathbf{v}_1$ and choose the coefficient $c_1^{(2)}$ in such a way that $(\mathbf{v}_2, \mathbf{v}_1) = 0$. Indeed, multiplying (using the dot product) both sides of the equation for $\mathbf{v}_2$ by $\mathbf{v}_1$, we obtain

$$(\mathbf{v}_2, \mathbf{v}_1) = (\mathbf{u}_2, \mathbf{v}_1) + c_1^{(2)}(\mathbf{v}_1, \mathbf{v}_1) = 0 \quad \Longrightarrow \quad c_1^{(2)} = -\frac{(\mathbf{u}_2, \mathbf{v}_1)}{(\mathbf{v}_1, \mathbf{v}_1)}.$$

   Thus, the vector $\mathbf{v}_2 = \mathbf{u}_2 + c_1^{(2)}\mathbf{v}_1$ with the special choice of the mixing coefficient $c_1^{(2)}$ as found above is made orthogonal to $\mathbf{v}_1$.
3. Construct a vector $\mathbf{v}_3 = \mathbf{u}_3 + c_1^{(3)}\mathbf{v}_1 + c_2^{(3)}\mathbf{v}_2$ that is orthogonal to both $\mathbf{v}_1$ and $\mathbf{v}_2$ by appropriately choosing the coefficients $c_1^{(3)}$ and $c_2^{(3)}$. To this end, we multiply both sides of this equation for $\mathbf{v}_3$ by $\mathbf{v}_1$ and use the fact that $\mathbf{v}_1$ and $\mathbf{v}_2$ have already been made orthogonal:

$$(\mathbf{v}_3, \mathbf{v}_1) = (\mathbf{u}_3, \mathbf{v}_1) + c_1^{(3)}(\mathbf{v}_1, \mathbf{v}_1) = 0 \quad \Longrightarrow \quad c_1^{(3)} = -\frac{(\mathbf{u}_3, \mathbf{v}_1)}{(\mathbf{v}_1, \mathbf{v}_1)};$$

   similarly, multiplying by $\mathbf{v}_2$, we get

$$(\mathbf{v}_3, \mathbf{v}_2) = (\mathbf{u}_3, \mathbf{v}_2) + c_2^{(3)}(\mathbf{v}_2, \mathbf{v}_2) = 0 \quad \Longrightarrow \quad c_2^{(3)} = -\frac{(\mathbf{u}_3, \mathbf{v}_2)}{(\mathbf{v}_2, \mathbf{v}_2)}.$$

4. This process can be continued until the last vector is obtained. All obtained new vectors $\mathbf{v}_1, \mathbf{v}_2, \ldots, \mathbf{v}_p$ will be mutually orthogonal by construction.

Finally, to make the new vectors orthonormal, it is just necessary to rescale each vector $\mathbf{v}_i$ by its own length, i.e. $\mathbf{v}_i \to \mathbf{v}_i/|\mathbf{v}_i|$. This step concludes the procedure.

Example 1.6 ▶ Given three linearly independent vectors $\mathbf{u}_1 = (1, 0, 1)$, $\mathbf{u}_2 = (1, 1, 0)$ and $\mathbf{u}_3 = (1, 1, 1)$, construct their orthonormal linear combination using the Gram–Schmidt method.

Solution: Using equations given above, we first take $\mathbf{v}_1 = \mathbf{u}_1 = (1, 0, 1)$. Then, $\mathbf{v}_2 = \mathbf{u}_2 + c_1^{(2)}\mathbf{v}_1$ with

$$c_1^{(2)} = -\frac{(\mathbf{u}_2, \mathbf{v}_1)}{(\mathbf{v}_1, \mathbf{v}_1)} = -\frac{1}{2} \implies \mathbf{v}_2 = \mathbf{u}_2 - \frac{1}{2}\mathbf{v}_1 = \left(\frac{1}{2}, 1, -\frac{1}{2}\right).$$

Next, consider the third vector $\mathbf{v}_3 = \mathbf{u}_3 + c_1^{(3)}\mathbf{v}_1 + c_2^{(3)}\mathbf{v}_2$ with

$$c_1^{(3)} = -\frac{(\mathbf{u}_3, \mathbf{v}_1)}{(\mathbf{v}_1, \mathbf{v}_1)} = -\frac{2}{2} = -1 \quad \text{and} \quad c_2^{(3)} = -\frac{(\mathbf{u}_3, \mathbf{v}_2)}{(\mathbf{v}_2, \mathbf{v}_2)} = -\frac{1}{3/2} = -\frac{2}{3},$$

so that

$$\mathbf{v}_3 = \mathbf{u}_3 - \mathbf{v}_1 - \frac{2}{3}\mathbf{v}_2 = \left(-\frac{1}{3}, \frac{1}{3}, \frac{1}{3}\right).$$

The constructed vectors $\mathbf{v}_1 = (1, 0, 1)$, $\mathbf{v}_2 = \left(\frac{1}{2}, 1, -\frac{1}{2}\right)$ and $\mathbf{v}_3 = \left(-\frac{1}{3}, \frac{1}{3}, \frac{1}{3}\right)$ are mutually orthogonal. To make them normalised, rescale them by their own length to finally obtain $\mathbf{v}_1 = \frac{1}{\sqrt{2}}(1, 0, 1)$, $\mathbf{v}_2 = \sqrt{\frac{2}{3}}\left(\frac{1}{2}, 1, -\frac{1}{2}\right)$ and $\mathbf{v}_3 = \frac{1}{\sqrt{3}}(-1, 1, 1)$. ◄

Eventually, the above procedure allows expanding new vectors via the old ones explicitly:

$$\mathbf{v}_1 = \mathbf{u}_1, \quad \mathbf{v}_2 = d_{21}\mathbf{u}_1 + \mathbf{u}_2, \quad \mathbf{v}_3 = d_{31}\mathbf{u}_1 + d_{32}\mathbf{u}_2 + \mathbf{u}_3, \quad \text{etc.,}$$

where d_{21}, d_{31}, etc. are some coefficients which are calculated during the course of the procedure. For instance, in the case of the previous example (before normalisation),

$$\mathbf{v}_1 = \mathbf{u}_1,$$
$$\mathbf{v}_2 = -\frac{1}{2}\mathbf{u}_1 + \mathbf{u}_2,$$
$$\mathbf{v}_3 = -\mathbf{u}_1 - \frac{2}{3}\mathbf{v}_2 + \mathbf{u}_3 = -\frac{2}{3}\mathbf{u}_1 - \frac{2}{3}\mathbf{u}_2 + \mathbf{u}_3.$$

We close this section by noting that the Gram–Schmidt method can also be used to check linear dependence of a set of vectors. If, during the course of this procedure, some of the new vectors come out to be zero, then there are linearly dependent vectors in the original set. The newly created set of vectors will then contain a smaller set of only linearly independent vectors.

To illustrate this point, consider, for instance, a situation in which we assume that the third vector is linearly dependent on the first two vectors: $\mathbf{u}_3 = \alpha\mathbf{u}_1 + \beta\mathbf{u}_2$. Let us run the first steps of the procedure to see how one of the vectors is going to be eliminated. Out of $\mathbf{u}_1$ and $\mathbf{u}_2$ we construct two orthonormal vectors $\mathbf{v}_1$ and $\mathbf{v}_2$ using the Gram–Schmidt method. The new vectors are some linear combinations of the old ones; reversely, the old ones are a linear combination of the new ones. Therefore, we can write that

$$\mathbf{u}_3 = \alpha \mathbf{u}_1 + \beta \mathbf{u}_2 = \gamma \mathbf{v}_1 + \delta \mathbf{v}_2 \, .$$

Then, if we are to construct the third vector, $\mathbf{v}_3 = \mathbf{u}_3 + c_1^{(3)} \mathbf{v}_1 + c_2^{(3)} \mathbf{v}_2$, that is to be orthogonal to both $\mathbf{v}_1$ and $\mathbf{v}_2$, then, following the Gram–Schmidt procedure, we shall obtain for the expansion coefficients:

$$c_1^{(3)} = -\frac{(\mathbf{u}_3, \mathbf{v}_1)}{(\mathbf{v}_1, \mathbf{v}_1)} = -\gamma \quad \text{and} \quad c_2^{(3)} = -\frac{(\mathbf{u}_3, \mathbf{v}_2)}{(\mathbf{v}_2, \mathbf{v}_2)} = -\delta \, ,$$

so that the new vector

$$\mathbf{v}_3 = \mathbf{u}_3 + c_1^{(3)} \mathbf{v}_1 + c_2^{(3)} \mathbf{v}_2 = (\gamma \mathbf{v}_1 + \delta \mathbf{v}_2) - \gamma \mathbf{v}_1 - \delta \mathbf{v}_2$$

is easily seen to be zero.

Problem 1.12 Prove generally that if a vector $\mathbf{u}_i$ is linearly dependent on some other vectors in the original set $\{\mathbf{u}_j\}$, then it will be eliminated during the construction of the new set, and hence the final set will be one vector less. [Hint: *order vectors in the set in such a way that all vectors which the vector $\mathbf{u}_i$ is linearly dependent upon to stand before it prior to applying the Gram–Schmidt procedure.*]

Problem 1.13 Orthogonalise vectors $\mathbf{u}_1 = (1, 1, 0)$, $\mathbf{u}_2 = (1, 0, 1)$ and $\mathbf{u}_3 = (0, 1, 1)$. [Answer: $\mathbf{v}_1 = (1, 1, 0)$, $\mathbf{v}_2 = (1/2, -1/2, 1)$ *and* $\mathbf{v}_3 = (-2/3, 2/3, 2/3)$.]

Problem 1.14 Construct orthogonal vectors from the set $\mathbf{u}_1 = (-1, 1, 1)$, $\mathbf{u}_2 = (1, -1, 1)$ and $\mathbf{u}_3 = (1, 1, -1)$. [Answer: the new vectors are $\mathbf{v}_1 = \mathbf{u}_1$, $\mathbf{v}_2 = (2/3, -2/3, 4/3)$ *and* $\mathbf{v}_3 = (1, 1, 0)$.]

1.1.3.2 Orthogonalisation of Functions

The procedure described above is especially useful for functions. Suppose, we have a set of functions $\{f_i(x)\}$ which are not mutually orthogonal. If we would like to construct a new set of functions,

$$g_i(x) = \sum_i a_{ij} f_j(x) \, ,$$

built as linear combinations of the original set with expansion coefficients α_{ij} (they bear two indices and form an object called *matrix*, see the next section), then the Gram–Schmidt procedure is the simplest method to achieve this goal. It goes exactly as described above for vectors; the only difference is in what we mean by the dot product in the case of the functions. The following example illustrates this point.

Example 1.7 ▶ Consider the first three powers of x on the interval $-1 \le x \le 1$ as our original set of functions, i.e. $f_1(x) = x^0 = 1$, $f_2(x) = x$ and $f_3(x) = x^2$. Assuming the unit weight function in the definition (1.9) of the dot product of functions, we immediately conclude by a direct calculation of the appropriate integrals that our functions are not orthogonal. Construct their linear combinations (i.e. polynomials of not higher than the second degree) which are orthogonal to each other.

Solution: Following the procedure outlined above, we choose the first function as $g_1(x) = f_1(x) = 1$. The second function is

$$g_2(x) = f_2(x) + c_1^{(2)} g_1(x) = x + c_1^{(2)} .$$

We would like it to be orthogonal to the first one, $g_1(x)$. Taking the dot product of both sides of the above equation for $g_2(x)$ with g_1 and setting it to zero, we obtain

$$(g_2, g_1) = (f_2, g_1) + c_1^{(2)} (g_1, g_1) = 0 .$$

Simple calculations show that $(f_2, g_1) = (x, 1) = \int_{-1}^{1} x \, dx = 0$ and similarly $(g_1, g_1) = (1, 1) = 2$. Hence, we get $c_1^{(2)} = -(f_2, g_1) / (g_1, g_1) = 0$, i.e. $g_2(x) = x$. Similarly, consider

$$g_3(x) = f_3(x) + c_1^{(3)} g_1(x) + c_2^{(3)} g_2(x) ,$$

which is to be made orthogonal to both $g_1(x)$ and $g_2(x)$ at the same time. Taking one dot product of both sides of the above equation for g_3 with g_1 and then another one with g_2 and setting both to zero, we obtain

$$(g_3, g_1) = (f_3, g_1) + c_1^{(3)} (g_1, g_1) + c_2^{(3)} (g_2, g_1) = (f_3, g_1) + c_1^{(3)} (g_1, g_1) = 0$$

$$\implies \quad c_1^{(3)} = -\frac{(f_3, g_1)}{(g_1, g_1)} = -\frac{2/3}{2} = -\frac{1}{3} ,$$

$$(g_3, g_2) = (f_3, g_2) + c_1^{(3)} (g_1, g_2) + c_2^{(3)} (g_2, g_2) = (f_3, g_2) + c_2^{(3)} (g_2, g_2) = 0$$

$$\implies \quad c_2^{(3)} = -\frac{(f_3, g_2)}{(g_2, g_2)} = -\frac{0}{2/3} = 0 ,$$

so that $g_3(x) = f_3(x) - g_1(x)/3 = x^2 - 1/3 = (3x^2 - 1)/3$. The obtained functions $g_1 = 1$, $g_2 = x$ and $g_3 = (3x^2 - 1)/3$ form mutually orthogonal functions; they are proportional to famous Legendre polynomials which we shall be studying in detail in Sect. 4.3. ◀

Problem 1.15 Continuing the above procedure, obtain the next two functions from the previous example related to Legendre polynomials. Use $f_4(x) = x^3$ and $f_5(x) = x^4$. [Answer: the new functions are $g_4 = (5x^3 - 3x)/5$ *and* $g_5 = (35x^4 - 30x^2 + 3)/35$.]

Problem 1.16 Assuming the weight function $w(x) = e^{-x}$ and the interval $0 \leq x < \infty$, show starting from the function $L_0(x) = 1$, that the next three orthogonal polynomials are

$$L_1(x) = x - 1 , \quad L_2(x) = x^2 - 4x + 2 , \quad L_3(x) = x^3 - 9x^2 + 18x - 6 .$$

The generated functions are directly related to Laguerre polynomials (Sect. 4.3).

Problem 1.17 Assuming the weight function $w(x) = e^{-x^2}$ and the interval $-\infty \leq x < \infty$, show starting from the function $H_0(x) = 1$, that the next three orthogonal polynomials are

$$H_1(x) = x , \quad H_2(x) = x^2 - \frac{1}{2} , \quad H_3(x) = x^3 - \frac{3}{2}x .$$

The generated functions are directly related to Hermite polynomials (Sect. 4.3). You need to use the fact (see Sect. 4.2.1 for more details) that

$$\gamma_{n+m} = (x^n, x^m) = (x^{n+m}, 1) = \int_{-\infty}^{\infty} x^{n+m} e^{-x^2} dx$$

is equal to zero if $n + m$ is odd, and $\gamma_{n+m} = \Gamma\left(p + \frac{1}{2}\right)$, if $n + m = 2p$ is even; here $\Gamma\left(p + \frac{1}{2}\right)$ is given by Eq. (4.36).

Problem 1.18 Assuming the weight function $w(x) = 1/\sqrt{1 - x^2}$ and the interval $-1 \leq x < 1$, show starting from the function $T_0(x) = 1$, that the next two orthogonal polynomials are

$$T_1(x) = x , \quad T_2(x) = x^2 - \frac{1}{2} , \quad T_3(x) = x^3 - \frac{3}{4}x .$$

The generated functions are directly related to Chebyshev polynomials (Sect. 4.3).

1.2 Matrices: Definition and Properties

1.2.1 Introduction of a Concept

Vectors can be considered, quite formally, as a *set of* (generally complex) *numbers* written vertically:

$$\mathbf{u} = \begin{pmatrix} u_1 \\ u_2 \\ \cdots \\ u_p \end{pmatrix} = \begin{Vmatrix} u_1 \\ u_2 \\ \cdots \\ u_p \end{Vmatrix}. \tag{1.12}$$

One can make a generalisation and stretch the set of numbers in the other dimension as well forming a *table of numbers*:

$$\mathbf{A} = \begin{pmatrix} a_{11} & a_{12} & a_{13} & \cdots & a_{1n} \\ a_{21} & a_{22} & a_{23} & \cdots & a_{2n} \\ \cdots & \cdots & \cdots & \cdots & \cdots \\ a_{m1} & a_{m2} & a_{m3} & \cdots & a_{mn} \end{pmatrix} = \begin{Vmatrix} a_{11} & a_{12} & a_{13} & \cdots & a_{1n} \\ a_{21} & a_{22} & a_{23} & \cdots & a_{2n} \\ \cdots & \cdots & \cdots & \cdots & \cdots \\ a_{m1} & a_{m2} & a_{m3} & \cdots & a_{mn} \end{Vmatrix}, \tag{1.13}$$

which is said to contain m rows and n columns, or is a $m \times n$ **matrix** (rows $\times$ columns). Many formulae and solutions of various problems can be simplified drastically by using vector and matrix notations. We shall use bold letters to indicate vectors and matrices.

Both notations, $(\cdots)$ and $\|\cdots\|$, are frequently used to represent vectors and matrices. In this book we shall adopt only the former one (with round brackets). We shall use capital letters for matrices and corresponding small letters for their matrix elements. An element of a matrix $\mathbf{A}$ is a_{ij}, where i refers to the row and j to the column in the table indicating the precise position of the element a_{ij}, see in Eq. (1.13) how these indices change if you go from one element to the other along a row or a column of a matrix. The matrix can also be written in a compact form via its elements as $\mathbf{A} = (a_{ij})$. Occasionally, to indicate a particular i, j element of a matrix $\mathbf{A}$, we may also use the following notation: $(\mathbf{A})_{ij}$.

A vector may be envisaged as a matrix with a single column ($n \times 1$ matrix), and hence sometimes is called *a vector-column*. One note is in order concerning adopted notations. Elements of any vector-column, e.g. $\mathbf{x} = (x_i)$, we usually show using a single index. In fact, since the vector is a matrix with a single column, hence as such its elements must be denoted as x_{i1} with 1 standing as its right index. However, since the right index has only a single value (equal to one) it is usually omitted and we arrive at the simplified notations used. But when performing operations on matrices and vectors it is essential to remember that the single value of the second (right) index in vector-columns is always implied.

A set of numbers written as a row can also be thought of as a vector and, at the same time, as a matrix with a single row, $\mathbf{x} = (x_{1i})$. The first (left) index equal to one is normally omitted. It is frequently called a *vector-row*.

Square matrices have equal number of rows and columns, $n = m$. *Diagonal square matrices* have only non-zero elements on their main diagonal:

$$\mathbf{A} = \begin{pmatrix} a_{11} & & & \\ & a_{22} & & \\ & & \ddots & \\ & & & a_{nn} \end{pmatrix} = (\delta_{ij}a_{ii}) \ . \tag{1.14}$$

All other, so-called *off-diagonal*, elements are equal to zero. In particular, if all elements of the diagonal matrix are equal to one, then this matrix is called a *unit* or *identity matrix*. We shall denote the unit matrix by symbol $\mathbf{E}$. Since non-diagonal elements of $\mathbf{E}$ are zeros, but all diagonal elements are equal to one, the element e_{ij} of the unit matrix is in fact the Kronecker symbol, $\mathbf{E} = (\delta_{ij})$.

For instance, the matrix

$$\mathbf{A} = \begin{pmatrix} 3 & 8 & 7 \\ -1 & 0 & -2 \end{pmatrix}$$

is the 2×3 matrix, i.e. it has 2 rows and 3 columns; its elements are $a_{11} = 3$, $a_{12} = 8$, $a_{13} = 7$ (the first row); $a_{21} = -1$, $a_{22} = 0$ and $a_{23} = -2$ (the second row).

1.2.2   Operations with Matrices

By summing up the corresponding elements of two matrices $\mathbf{A} = (a_{ij})$ and $\mathbf{B} = (b_{ij})$ having the same $m \times n$ structure, a new matrix $\mathbf{C} = (c_{ij}) = (a_{ij} + b_{ij})$ is obtained containing the same number of rows and columns. An example of this:

$$\begin{pmatrix} 1 & 3 & -2 \\ 4 & 7 & 1 \end{pmatrix} + \begin{pmatrix} 2 & -1 & 4 \\ 3 & -7 & -2 \end{pmatrix} = \begin{pmatrix} 3 & 2 & 2 \\ 7 & 0 & -1 \end{pmatrix} \ .$$

A matrix of any structure can be multiplied by a number. If c is a number and $\mathbf{A} = (a_{ij})$, then $\mathbf{B} = c\mathbf{A}$ is a new matrix with elements $b_{ij} = ca_{ij}$, e.g.

$$-3 \begin{pmatrix} 1 & 3 & -2 \\ 4 & 7 & 1 \end{pmatrix} = \begin{pmatrix} -3 & -9 & 6 \\ -12 & -21 & -3 \end{pmatrix} \ .$$

From two matrices $\mathbf{A} = (a_{ij})$ (which is $n \times m$) and $\mathbf{B} = (b_{ij})$ (which is $m \times p$) a new matrix

$$\mathbf{C} = \mathbf{AB} = (c_{ij}) \quad \text{with} \quad c_{ij} = \sum_{k=1}^{m} a_{ik}b_{kj} \tag{1.15}$$

can be constructed, which is a $n \times p$ matrix, see Fig. 1.3. This operation is a natural generalisation of the dot product of vectors. Thus, to obtain the i, j element c_{ij} of

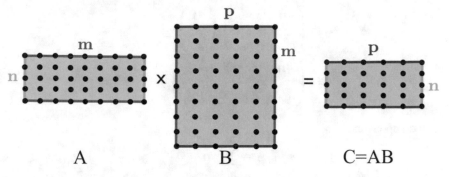

Fig. 1.3 Schematics of the matrix multiplication. Matrix **A** has $n = 5$ rows and $m = 9$ columns, while matrix **B** has $p = 6$ columns and the same number of rows $m = 9$ as **A** has columns. The resultant matrix $\mathbf{C} = \mathbf{AB}$ has $n = 5$ rows (as in **A**) and $p = 6$ columns (as in **B**)

the matrix **C**, it is necessary to calculate a dot product of the ith vector-row of A with the jth vector-column of **B**. This means, that not any matrices can be multiplied: the number of *columns* in **A** must be equal to the number of *rows* in **B**. Note how the indices in formula (1.15) for c_{ij} are written: i and j in the product $a_{ik}b_{kj}$ under the sum are written exactly in the same order as in c_{ij}, while the dump summation index k appears in between i and j twice, i.e. as the right index in a_{ik} and the left in b_{kj}. In physics literature the so-called Einstein summation convention is sometimes used whereby the sum sign is dropped, i.e. $\mathbf{C} = \mathbf{AB}$ is written in elements as $c_{ij} = a_{ik}b_{kj}$ and it is assumed that summation is performed over the repeated index (k in this case); we shall not be using this convention here however to avoid confusion.

As an example, consider the product of two matrices

$$\mathbf{A} = \begin{pmatrix} 1 & 3 & 1 \\ 2 & 3 & 4 \end{pmatrix} \quad \text{and} \quad \mathbf{B} = \begin{pmatrix} 1 & 4 \\ 10 & 3 \\ 1 & 2 \end{pmatrix}.$$

We obtain the matrix

$$\mathbf{C} = \mathbf{AB} = \begin{pmatrix} 1 \cdot 1 + 3 \cdot 10 + 1 \cdot 1 & 1 \cdot 4 + 3 \cdot 3 + 1 \cdot 2 \\ 2 \cdot 1 + 3 \cdot 10 + 4 \cdot 1 & 2 \cdot 4 + 3 \cdot 3 + 4 \cdot 2 \end{pmatrix} = \begin{pmatrix} 32 & 15 \\ 36 & 25 \end{pmatrix},$$

which is a 2×2 matrix.

In particular, one can multiply a matrix and a vector which results in a vector. Indeed, if $\mathbf{x} = (x_i)$ is a n-fold vector and $\mathbf{A} = (a_{ij})$ is a $m \times n$ matrix, then $\mathbf{y} = \mathbf{Ax}$ is an m-fold vector:

$$\mathbf{y} = \mathbf{Ax} = \begin{pmatrix} a_{11} & \cdots & a_{1n} \\ \cdots & \cdots & \cdots \\ a_{m1} & \cdots & a_{mn} \end{pmatrix} \begin{pmatrix} x_1 \\ \cdots \\ x_n \end{pmatrix}$$

$$= \begin{pmatrix} a_{11}x_1 + \cdots + a_{1n}x_n \\ \cdots \\ a_{m1}x_1 + \cdots + a_{mn}x_n \end{pmatrix} = \begin{pmatrix} y_1 \\ \cdots \\ y_m \end{pmatrix} = \mathbf{y} \, ,$$

(1.16)

or in components:

$$y_i = \sum_{j=1}^{n} a_{ij}x_j \; .$$

This is equivalent to the general rule (1.15) of matrix multiplication: since vectors are one-column matrices, then $\mathbf{x} = (x_i)$ can be written as $\mathbf{X} = (x_{i1})$ with $x_{i1} = x_i$ and hence $\mathbf{Y} = \mathbf{AX}$ in components means

$$y_{i1} = \sum_{j=1}^{n} a_{ij}x_{j1} = \sum_{j=1}^{n} a_{ij}x_j \; ,$$

which is the same as above because the elements y_{i1} form a single column matrix and therefore represent the vector $\mathbf{y}$ with components $y_i = y_{i1}$.

The matrix product is in general not commutative, i.e. $\mathbf{AB} \neq \mathbf{BA}$. However, some matrices may be commutative, e.g. diagonal matrices: if

$$\mathbf{A} = \begin{pmatrix} a_{11} & & \\ & a_{22} & \\ & & \ddots \\ & & & a_{nn} \end{pmatrix} \quad \text{and} \quad \mathbf{B} = \begin{pmatrix} b_{11} & & \\ & b_{22} & \\ & & \ddots \\ & & & b_{nn} \end{pmatrix} \, ,$$

then

$$\mathbf{AB} = \begin{pmatrix} a_{11}b_{11} & & \\ & a_{22}b_{22} & \\ & & \ddots \\ & & & a_{nn}b_{nn} \end{pmatrix} = \mathbf{BA} \; .$$

It is easy to see that a product of any matrix with a diagonal matrix is *not* necessarily commutative. Indeed, let $\mathbf{A} = (a_{ij})$ and $\mathbf{D} = (d_{ij}) = (d_{ii}\delta_{ij})$, then the i, j elements of the matrix $\mathbf{AD}$ is

$$(\mathbf{AD})_{ij} = \sum_{k} a_{ik}d_{kj} = \sum_{k} a_{ik}d_{kk}\delta_{kj} = a_{ij}d_{jj} \; ,$$

where the Kronecker delta symbol δ_{kj} cuts off all the terms in the sum except for the one with $k = j$. Similarly, the i, j element of the matrix $\mathbf{DA}$ works out to be

$$(\mathbf{DA})_{ij} = \sum_k d_{ik}a_{kj} = \sum_k d_{kk}\delta_{ik}a_{kj} = d_{ii}a_{ij} \ ,$$

which is different from $a_{ij}d_{jj}$ obtained when multiplying matrices in the reverse order, i.e. the two matrices are *not* generally commutative.

At the same time, the matrix product is *associative*: $(\mathbf{AB})\,\mathbf{C} = \mathbf{A}\,(\mathbf{BC})$. The simplest proof is again algebraic based on writing down the products of matrices by elements. Indeed, if $\mathbf{A} = \left(a_{ij}\right)$, $\mathbf{B} = \left(b_{ij}\right)$ and $\mathbf{C} = \left(c_{ij}\right)$, then $\mathbf{D} = \mathbf{AB}$ has elements

$$d_{ij} = \sum_k a_{ik}b_{kj} \ .$$

Then, the matrix $\mathbf{G} = (\mathbf{AB})\,\mathbf{C} = \mathbf{DC}$ has elements

$$g_{ij} = \sum_k d_{ik}c_{kj} = \sum_k \left(\sum_l a_{il}b_{lk}\right)c_{kj} \ .$$

Note that we introduced a new dump index l to indicate the product of elements $a_{il}b_{lk}$ in $\mathbf{D}$ since the index k has already been used in the product $d_{ik}c_{kj}$ of $\mathbf{DC}$. Also, note how indices have been used in the product: since d_{ik} has the left and right indices as i and k, these appear exactly in the same order in the product $a_{il}b_{lk}$ with the summation index l in between, as required. The final product of three matrices in elements becomes a double sum:

$$g_{ij} = \sum_{kl} a_{il}b_{lk}c_{kj} \ .$$

The indices i, j of the element g_{ij} of the final matrix $\mathbf{G}$ appear as the first and the last indices in the product $a_{il}b_{lk}c_{kj}$ of the elements under the sums; the summation indices follow one after the other and are both repeated twice.

On the other hand, the matrix $\mathbf{H} = \mathbf{A}\,(\mathbf{BC})$ has elements

$$h_{ij} = \sum_k a_{ik}\left(\sum_l b_{kl}c_{lj}\right) = \sum_{kl} a_{ik}b_{kl}c_{lj} \ .$$

This is the same as g_{ij} since we can always interchange the dumb indices $k \leftrightarrow l$ in the double sums as any symbols (letters) can be used for these indices. This proves the required statement.

When a square matrix $\mathbf{A}$ is multiplied from either side (from the left or from the right) by the unit matrix $\mathbf{E}$ of the same dimension, it does not change: $\mathbf{AE} = \mathbf{EA} = \mathbf{A}$. Indeed, in components:

$$(\mathbf{AE})_{ij} = \sum_k a_{ik}\delta_{kj} = a_{ij} \, ,$$

and similarly for the product $\mathbf{EA}$.

A new matrix $\mathbf{A}^T$, called the *transpose* of $\mathbf{A}$, can be obtained from the latter by writing rows as columns as shown in the example below:

$$\begin{pmatrix} 1 & 3 & -2 \\ 4 & 7 & 1 \end{pmatrix}^T = \begin{pmatrix} 1 & 4 \\ 3 & 7 \\ -2 & 1 \end{pmatrix} \, .$$

In other words, if $\mathbf{A} = (a_{ij})$, then

$$\mathbf{A}^T = (\tilde{a}_{ij}) \quad \text{with } \tilde{a}_{ij} = a_{ji} \, , \tag{1.17}$$

i.e. its indices are simply permuted. We shall be frequently using these notations here denoting elements of the transposed matrix with the same small letter as the original matrix, but putting a tilde (a wavy line) on top of it.

Theorem 1.5 *A transpose of a product of two matrices is equal to a product of transposed matrices taken in the reverse order:*

$$(\mathbf{AB})^T = \mathbf{B}^T\mathbf{A}^T \, . \tag{1.18}$$

Proof If $\mathbf{A} = (a_{ij})$ and $\mathbf{B} = (b_{ij})$, then $\mathbf{A}^T = (\tilde{a}_{ij})$ with $\tilde{a}_{ij} = a_{ji}$ and $\mathbf{B}^T = (\tilde{b}_{ij})$ with $\tilde{b}_{ij} = b_{ji}$. The product $\mathbf{AB} = (c_{ij})$ with elements $c_{ij} = \sum_k a_{ik}b_{kj}$ after transpose turns into the matrix $(\mathbf{AB})^T = (\tilde{c}_{ij})$ with elements

$$\tilde{c}_{ij} = c_{ji} = \sum_k a_{jk}b_{ki} = \sum_k \tilde{a}_{kj}\tilde{b}_{ik} = \sum_k \tilde{b}_{ik}\tilde{a}_{kj} = d_{ij} \, ,$$

where d_{ij} are exactly elements of $\mathbf{B}^T\mathbf{A}^T$. Note that here we were able to associate $d_{ij} = \sum_k \tilde{b}_{ik}\tilde{a}_{kj}$ with $\mathbf{B}^T\mathbf{A}^T$ only after making sure that indices in the product of the elements under the sum are properly ordered as required by the matrix multiplication with the dump summation index k appearing in the middle and the indices i and j as the first and the last ones, respectively, as in d_{ij}. **Q.E.D.**

The operation of transpose of a matrix appears in a dot product of two real vectors:

$$(\mathbf{Ax}, \mathbf{y}) = (\mathbf{x}, \mathbf{A}^T\mathbf{y}) \, . \tag{1.19}$$

Indeed, let $\mathbf{Ax} = \mathbf{c}$ be a vector with elements $c_i = \sum_k a_{ik}x_k$. Then, the dot product of $\mathbf{c}$ and another vector $\mathbf{y}$ is

$$(\mathbf{Ax}, \mathbf{y}) = \sum_l c_l y_l = \sum_l \sum_k a_{lk} x_k y_l = \sum_k x_k \left(\sum_l a_{lk} y_l \right)$$

$$= \sum_k x_k \left(\sum_l \tilde{a}_{kl} y_l \right) = (\mathbf{x}, \mathbf{A}^T \mathbf{y}) \;,$$

because the vector $\mathbf{d} = \mathbf{A}^T \mathbf{y}$ has elements $d_k = \sum_l \tilde{a}_{kl} y_l = \sum_l a_{lk} y_l$, as required.

If a square matrix is equal to its transpose, it is then symmetric with respect to its diagonal and hence is called a *symmetric matrix*:

$$\mathbf{A} = \mathbf{A}^T, \quad \text{i.e.} \quad a_{ij} = \tilde{a}_{ij} = a_{ji} \;. \tag{1.20}$$

For instance, consider the following two 3×3 matrices:

$$\begin{pmatrix} a & x^2+1 & 0 \\ x^2+1 & 1 & \sqrt{x+1} \\ 0 & \sqrt{x+1} & x \end{pmatrix} \quad \text{and} \quad \begin{pmatrix} a & x^2+1 & 0 \\ -(x^2+1) & 1 & \sqrt{x+1} \\ 0 & -\sqrt{x+1} & x \end{pmatrix} \;.$$

The first matrix is symmetric, while the second is not. In fact, the second matrix is *antisymmetric* as for it $a_{ij} = -a_{ji}$.

Problem 1.19 Let $\mathbf{A}$ be a symmetric matrix, $\mathbf{A}^T = \mathbf{A}$. Show that the matrix $\mathbf{C} = \mathbf{B}^T \mathbf{AB}$ with arbitrary $\mathbf{B}$ remains symmetric. The structures of the matrices here are such that this matrix multiplication makes sense.

Interestingly, a *dot product of two vectors* can also be written as a multiplication of two matrices:

$$(\mathbf{x}, \mathbf{y}) = \mathbf{X}^T \mathbf{Y} \;, \tag{1.21}$$

where $\mathbf{X} = (x_i)$ and $\mathbf{Y} = (y_i)$ are matrices consisting of a single column (i.e. vector-columns). Then, $\mathbf{X}^T$ will be a matrix containing a single row (a vector-row), so that $\mathbf{X}^T$ and $\mathbf{Y}$ can be multiplied using the usual matrix multiplication rules:

$$(\mathbf{x}, \mathbf{y}) = \mathbf{X}^T \mathbf{Y} = \begin{pmatrix} x_1 \\ \cdots \\ x_p \end{pmatrix}^T \begin{pmatrix} y_1 \\ \cdots \\ y_p \end{pmatrix} = (x_1 \cdots x_p) \begin{pmatrix} y_1 \\ \cdots \\ y_p \end{pmatrix}$$

$$= (x_1 y_1 + \cdots + x_p y_p) \;,$$

which is a 1×1 matrix (we enclosed its only element in the round brackets to stress this point); it is essentially a *scalar*. Note, however, that the operation in which the *second* vector is transposed, produces a *square matrix:*

$$C = XY^T = \begin{pmatrix} x_1 \\ \cdots \\ x_p \end{pmatrix} \begin{pmatrix} y_1 \\ \cdots \\ y_p \end{pmatrix}^T = \begin{pmatrix} x_1 \\ \cdots \\ x_p \end{pmatrix} (y_1 \cdots y_p)$$

$$= \begin{pmatrix} x_1 y_1 \cdots x_1 y_p \\ \cdots \cdots \cdots \\ x_p y_1 \cdots x_p y_p \end{pmatrix}$$

with elements $c_{ij} = x_i y_j$.

If a matrix $A = (a_{ij})$ contains complex numbers, the *complex conjugate matrix* $A^* = (\overline{a}_{ij})$ can be defined which contains elements $\overline{a}_{ij} = a_{ij}^*$, where $*$ indicates the operation of complex conjugation. For instance,

$$A = \begin{pmatrix} a+ib & 2 \\ 3i & a-ib \end{pmatrix} \implies A^* = \begin{pmatrix} a-ib & 2 \\ -3i & a+ib \end{pmatrix}.$$

Problem 1.20 Demonstrate by a direct calculation that the matrices

$$A = \begin{pmatrix} 1 & 3 & 0 \\ 0 & 4 & 0 \\ 1 & -1 & 0 \end{pmatrix} \text{ and } B = \begin{pmatrix} -1 & 1 & 0 \\ 0 & 2 & 1 \\ 1 & 0 & -1 \end{pmatrix}$$

are not commutative.

Problem 1.21 Check if matrices

$$\begin{pmatrix} 0 & 1 & 0 \\ 1 & 0 & 1 \\ 0 & 1 & 0 \end{pmatrix} \text{ and } \begin{pmatrix} -1 & 1 & 1 \\ 1 & -1 & 1 \\ 1 & 1 & -1 \end{pmatrix}$$

commute. [Answer: *they do not.*]

Problem 1.22 Consider the n-th power X^n of a square matrix X as a product of it with itself n times, i.e. $X^n = \underbrace{XX \cdots X}_{n}$. Show that matrices A^n and B^m commute if the matrices A and B do.

Problem 1.23 Show that the product of

$$A = \begin{pmatrix} 1 & -1 & 0 \\ 1 & 1 & 0 \\ 0 & 0 & 1 \end{pmatrix} \text{ and } X = \begin{pmatrix} 1 \\ 1 \\ 0 \end{pmatrix} \text{ is } Y = \begin{pmatrix} 0 \\ 2 \\ 0 \end{pmatrix}.$$

Problem 1.24 If $\mathbf{X} = \begin{pmatrix} 1 \\ 1 \\ 0 \end{pmatrix}$ and $\mathbf{Y} = \begin{pmatrix} 0 \\ 1 \\ 1 \end{pmatrix}$, then calculate, if allowed, the

following products: $\mathbf{X}^T\mathbf{Y}$, $\mathbf{X}\mathbf{Y}^T$, $\mathbf{X}\mathbf{Y}$ and $\mathbf{X}^T\mathbf{Y}^T$. Name the resultant matrices

where appropriate. [Answer: 1 *(a scalar)*, a 3×3 *matrix* $\begin{pmatrix} 0\ 1\ 1 \\ 0\ 1\ 1 \\ 0\ 0\ 0 \end{pmatrix}$; *the other*

two do not exist.]

Problem 1.25 The Pauli matrices are

$$\sigma_x = \begin{pmatrix} 0\ 1 \\ 1\ 0 \end{pmatrix}, \quad \sigma_y = \begin{pmatrix} 0\ -i \\ i\ 0 \end{pmatrix} \quad \text{and} \quad \sigma_z = \begin{pmatrix} 1\ 0 \\ 0\ -1 \end{pmatrix}. \tag{1.22}$$

Show that (i) $\sigma_x^2 = \sigma_y^2 = \sigma_z^2 = \mathbf{E}$, where $\mathbf{E}$ is the 2×2 identity matrix; (ii) any pair of matrices anti-commute, e.g. $\sigma_x\sigma_y = -\sigma_y\sigma_x$; (iii) a commutator of any two matrices, defined by $[\mathbf{A}, \mathbf{B}] = \mathbf{A}\mathbf{B} - \mathbf{B}\mathbf{A}$, is expressed via the third matrix, i.e. $[\mathbf{A}, \mathbf{B}] = 2i\mathbf{C}$.

Problem 1.26 Prove that any 2×2 matrix can be uniquely represented by a linear combination of the identity matrix $\mathbf{E}$ and the three Pauli matrices.

Problem 1.27 If $[\mathbf{A}, \mathbf{B}] = \mathbf{A}\mathbf{B} - \mathbf{B}\mathbf{A}$ is the commutator of two matrices, prove the following identity:

$$[\mathbf{A}, [\mathbf{B}, \mathbf{C}]] + [\mathbf{B}, [\mathbf{C}, \mathbf{A}]] + [\mathbf{C}, [\mathbf{A}, \mathbf{B}]] = 0 \,.$$

Problem 1.28 Consider two 3D vectors $\mathbf{a}$ and $\mathbf{b}$, and the Pauli matrix vector $\boldsymbol{\sigma} = \sigma_x\mathbf{i} + \sigma_y\mathbf{j} + \sigma_z\mathbf{k}$. Considering explicitly the Pauli matrices, prove the following useful identity:

$$(\mathbf{a} \cdot \boldsymbol{\sigma})(\mathbf{b} \cdot \boldsymbol{\sigma}) = (\mathbf{a} \cdot \mathbf{b})\mathbf{E} + i\boldsymbol{\sigma} \cdot [\mathbf{a} \times \mathbf{b}] \,. \tag{1.23}$$

Here $\mathbf{E}$ is the 2×2 identity matrix.

1.2.3 Inverse Matrix: An Introduction

We have already encountered *square, diagonal, unit* (or *identity*) and *symmetric* matrices. There are many other types of matrices that have special properties some of which to be considered later on. In the forthcoming subsections we shall encounter other types of matrices.

Some square matrices $\mathbf{A}$ have an *inverse* matrix, denoted $\mathbf{A}^{-1}$, that is defined in this way:

$$\mathbf{A}\mathbf{A}^{-1} = \mathbf{A}^{-1}\mathbf{A} = \mathbf{E} , \tag{1.24}$$

where $\mathbf{E}$ is the unit matrix. These matrix equations, when written via components of the matrices, $\mathbf{A} = (a_{ij})$ and $\mathbf{A}^{-1} = \mathbf{B} = (b_{ij})$, look like this:

$$\sum_{k=1}^{n} a_{ik} b_{kj} = \sum_{k=1}^{n} b_{ik} a_{kj} = \delta_{ij} , \tag{1.25}$$

where δ_{ij} is the Kronecker symbol. It is easy to see that $\mathbf{A}^{-1}$ standing on the left and right sides of $\mathbf{A}$ in the above equation is one and the same matrix. Indeed, assume these are different matrices called $\mathbf{B}$ and $\mathbf{C}$, respectively, with $\mathbf{B} \neq \mathbf{C}$, i.e.

$$\mathbf{B}\mathbf{A} = \mathbf{E} \tag{1.26}$$

and also, quite independently,

$$\mathbf{A}\mathbf{C} = \mathbf{E} . \tag{1.27}$$

Now multiply the second of these identities by $\mathbf{B}$ from the left: $\mathbf{B}\mathbf{A}\mathbf{C} = \mathbf{B}\mathbf{E}$. In the left-hand side: $\mathbf{B}\mathbf{A}\mathbf{C} = (\mathbf{B}\mathbf{A})\mathbf{C} = \mathbf{E}\mathbf{C} = \mathbf{C}$, where we used the first of the two identities above, Eq. (1.26), while in the right-hand side, $\mathbf{B}\mathbf{E} = \mathbf{B}$. Hence, we obtain $\mathbf{C} = \mathbf{B}$, which contradicts our initial assumption, i.e. $\mathbf{A}^{-1}$ on the left and right of $\mathbf{A}$ in Eq. (1.24) is indeed the *same* matrix.

Not all square matrices have an inverse. For instance, a matrix consisting of only zeros does not have one. Only *non-singular* matrices have an inverse and we shall explain what that means later on in Sect. 1.7.

Theorem 1.6 *Inverse of a product* $\mathbf{A}\mathbf{B}$ *of two square matrices is equal to a product of their inverse matrices taken in reverse:*

$$(\mathbf{A}\mathbf{B})^{-1} = \mathbf{B}^{-1}\mathbf{A}^{-1} . \tag{1.28}$$

Proof The matrix $\mathbf{C} = (\mathbf{A}\mathbf{B})^{-1}$ is defined as $\mathbf{C}\mathbf{A}\mathbf{B} = \mathbf{E}$. Multiply both sides of this matrix equation by $\mathbf{B}^{-1}$ *from the right*: $\mathbf{C}\mathbf{A}\mathbf{B}\mathbf{B}^{-1} = \mathbf{E}\mathbf{B}^{-1}$. Since $\mathbf{E}$ is the unit matrix, $\mathbf{E}\mathbf{B}^{-1} = \mathbf{B}^{-1}$. Also, $\mathbf{B}\mathbf{B}^{-1} = \mathbf{E}$ by definition of $\mathbf{B}^{-1}$. Thus, we obtain $\mathbf{C}\mathbf{A}\mathbf{E} = \mathbf{B}^{-1}$ or simply $\mathbf{C}\mathbf{A} = \mathbf{B}^{-1}$. Next multiply both sides of this matrix equation on $\mathbf{A}^{-1}$ *from the right* again; we obtain $\mathbf{C}\mathbf{A}\mathbf{A}^{-1} = \mathbf{B}^{-1}\mathbf{A}^{-1}$ or simply $\mathbf{C} = \mathbf{B}^{-1}\mathbf{A}^{-1}$. **Q.E.D.**

Problem 1.29 Prove that the inverse matrix, if exists, is unique. [Hint: *prove by contradiction.*]

Problem 1.30 Prove that the inverse of $\mathbf{A}^T$ is equal to the transpose of the inverse of $\mathbf{A}$, i.e.

$$\left(\mathbf{A}^T\right)^{-1} = \left(\mathbf{A}^{-1}\right)^T . \tag{1.29}$$

[Hint: transpose the equation $\mathbf{A}\mathbf{A}^{-1} = \mathbf{E}$.]

Problem 1.31 Prove that $\left(\mathbf{A}^{-1}\right)^{-1} = \mathbf{A}$.

Problem 1.32 Let a square matrix $\mathbf{G}$ be a solution of the equation $\mathbf{G} = \mathbf{G}_0 + \mathbf{G}_0 \mathbf{\Sigma} \mathbf{G}$, where $\mathbf{G}_0$ and $\mathbf{\Sigma}$ are known square matrices of the same dimension. Solve this equation iteratively to represent $\mathbf{G}$ as an infinite series

$$\mathbf{G} = \mathbf{G}_0 + \mathbf{G}_0 \mathbf{\Sigma} \mathbf{G} = \mathbf{G}_0 + \mathbf{G}_0 \mathbf{\Sigma} \mathbf{G}_0 + \mathbf{G}_0 \mathbf{\Sigma} \mathbf{G}_0 \mathbf{\Sigma} \mathbf{G}_0 + \dots .$$

Hence, show that $\mathbf{G}$ also satisfies the equation $\mathbf{G} = \mathbf{G}_0 + \mathbf{G} \mathbf{\Sigma} \mathbf{G}_0$.

Problem 1.33 Let a square matrix $\mathbf{G}$ be a solution of the equation $\mathbf{G} = \mathbf{G}_0 + \mathbf{G}_0 \left(\mathbf{\Sigma}_1 + \mathbf{\Sigma}_2\right) \mathbf{G}$, where $\mathbf{G}_0$, $\mathbf{\Sigma}_1$ and $\mathbf{\Sigma}_2$ are known square matrices of the same dimension. Show that $\mathbf{G}$ also satisfies the equation $\mathbf{G} = \mathbf{G}_1 + \mathbf{G}_1 \mathbf{\Sigma}_2 \mathbf{G}$, where $\mathbf{G}_1$ satisfies $\mathbf{G}_1 = \mathbf{G}_0 + \mathbf{G}_0 \mathbf{\Sigma}_1 \mathbf{G}_1$, and hence $\mathbf{G}$ can be found from $\mathbf{G} = \left(\mathbf{E} - \mathbf{G}_1 \mathbf{\Sigma}_2\right)^{-1} \mathbf{G}_1$.

Problem 1.34 Consider a left triangular square $(n \times n)$ matrix

$$\mathbf{A} = \begin{pmatrix} a_{11} & & & & \\ a_{21} & a_{22} & & & \\ a_{31} & a_{32} & a_{33} & & \\ \vdots & \vdots & \vdots & \ddots & \\ a_{n1} & a_{n2} & a_{n3} & \cdots & a_{nn} \end{pmatrix},$$

where all elements a_{ij} with $j > i$ are equal to zero while $a_{ij} \neq 0$ if $j \leq i$. Show by writing the identity $\mathbf{A}\mathbf{A}^{-1} = \mathbf{E}$ directly in components that the inverse matrix $\mathbf{A}^{-1}$ has exactly the same left triangular structure.

Example 1.8 ▶ Calculate the inverse

$$\mathbf{A}^{-1} = \begin{pmatrix} x & y \\ z & h \end{pmatrix} \quad \text{of the matrix} \quad \mathbf{A} = \begin{pmatrix} a & b \\ c & d \end{pmatrix} .$$

Solution: By definition,

$$\begin{pmatrix} a & b \\ c & d \end{pmatrix} \begin{pmatrix} x & y \\ z & h \end{pmatrix} = \begin{pmatrix} 1 & 0 \\ 0 & 1 \end{pmatrix} ,$$

which gives in components four equations:

$$\begin{cases} ax + bz = 1 \\ ay + bh = 0 \\ cx + dz = 0 \\ cy + dh = 1 \end{cases} ,$$

that can easily be solved with respect to x, y, z, h (the first and the third equations give x and z, while the other two give y and h) yielding:

$$x = \frac{d}{\Delta}, \quad y = -\frac{b}{\Delta}, \quad z = -\frac{c}{\Delta} \quad \text{and} \quad h = \frac{a}{\Delta}, \quad \text{where} \quad \Delta = da - bc .$$

Thus, the inverse of the square matrix A is

$$\mathbf{A}^{-1} = \frac{1}{\Delta} \begin{pmatrix} d & -b \\ -c & a \end{pmatrix} \quad \text{with} \quad \Delta = ad - bc . \tag{1.30}$$

Note that the inverse does not exist if $\Delta = 0$. ◄

1.2.4 Linear Transformations and a Group

We know that by applying a square matrix $\mathbf{A}$ to a vector $\mathbf{x}$, a new vector $\mathbf{z} = \mathbf{A}\mathbf{x}$ is obtained. By definition, the new vector belongs to the same p-D space as the old one. Thus, the operation $\mathbf{A}\mathbf{x}$ can be considered as a *transformation operation* $\mathbf{x} \to \mathbf{z}$. The transformation is a *linear* one since it transforms a linear combination of vectors into the linear combination of transformed vectors:

$$\mathbf{A} (\alpha \mathbf{x} + \beta \mathbf{y}) = \alpha (\mathbf{A}\mathbf{x}) + \beta (\mathbf{A}\mathbf{y}) , \tag{1.31}$$

where α, β are arbitrary constants.

> **Problem 1.35** Let $\mathbf{A} = (a_{ij})$, $\mathbf{x} = (x_i)$ and $\mathbf{y} = (y_i)$. Prove the above identity. [Hint: *write both sides of Eq. (1.31) in components.*]

Consider now linear matrix transformations within a more general context. Firstly, we note that for any non-singular transformation $\mathbf{y} = \mathbf{A}\mathbf{x}$ it must be possible to define an inverse transformation, $\mathbf{y} \to \mathbf{x}$, using the inverse matrix $\mathbf{A}^{-1}$, i.e. $\mathbf{x} = \mathbf{A}^{-1}\mathbf{y}$. Secondly, there is always a unitary transformation performed by the identity matrix $\mathbf{E}$ that does nothing: $\mathbf{Ex} = \mathbf{x}$ for any $\mathbf{x}$. Finally, let us now perform two consecutive transformations: first $\mathbf{A}$ and then $\mathbf{B}$:

$$\mathbf{y} = \mathbf{Ax}, \quad \text{then} \quad \mathbf{z} = \mathbf{By}, \quad \text{so that} \quad \mathbf{z} = (\mathbf{BA})\,\mathbf{x} \,. \tag{1.32}$$

We see that two transformations, performed one after another, act as some other transformation that is given by the product matrix $\mathbf{C} = \mathbf{BA}$, in which the order of matrices in the product follows the order of the transformations themselves from *right to left*. This operation, when one transformation is performed after another, is called a *multiplication operation*, and we see that in the case under consideration this operation corresponds exactly to the matrix multiplication.

A set of non-singular matrices may form an algebraic object called a *group*. A number of requirements exist which are necessary for a set of matrices to form such a group.[2] What are these requirements? There are three of them:

1. There is an identity element e in the set that does nothing; this is served by a unit matrix $\mathbf{E}$ in the case of matrices.
2. There exists a *multiplication operation* to be understood as an action of several operations one after another; such a combined operation must be equivalent to some other operation in the *same set*; in other words, if g_1 and g_2 belong to the set, then $g_1 g_2$ must also be in it (the *closure* condition); the multiplication is associative, i.e. $(g_1 g_2)\,g_3 = g_1\,(g_2 g_3)$. In the case of matrices, the multiplication operation corresponds to the usual matrix multiplication. It is clear that all matrices must be square matrices of the same dimension $n \times n$ since otherwise a product of matrices may only be allowed in a certain order (indeed, a product of two matrices $n \times m$ and $m \times p$ is perfectly legitimate yielding a matrix $n \times p$, however their product in the reverse order, i.e. $m \times p$ with $n \times m$, is only possible if $p = n$).
3. For each element g of the set, there exists its inverse g^{-1}, belonging to the same set, that is defined in such a way that a multiplication of g with g^{-1} (in any order) gives the unit element, i.e. $g g^{-1} = g^{-1} g = e$.

Example 1.9 ► As an example, consider a set of four matrices:

$$\mathbf{E} = \begin{pmatrix} 1 & 0 \\ 0 & 1 \end{pmatrix}, \quad \mathbf{A} = \begin{pmatrix} 0 & -1 \\ 1 & 0 \end{pmatrix},$$
$$\mathbf{B} = \begin{pmatrix} -1 & 0 \\ 0 & -1 \end{pmatrix} \quad \text{and} \quad \mathbf{C} = \begin{pmatrix} 0 & 1 \\ -1 & 0 \end{pmatrix}. \tag{1.33}$$

It is easily checked that a product of any two elements results in an element from the same set; all such products are shown in Table 1.1. Next, we see that each element has an inverse; indeed, by looking at the Table, we see that $\mathbf{A}^{-1} = \mathbf{C}$ (since

---

[2] Note, however, that groups can also be formed by other objects, not only matrices, although we shall limit ourselves only to matrices here.

Table 1.1 A group multiplication table showing the result of multiplication of each of the four elements of the set [**E, A, B, C**] in Eq. (1.33) with any other element

×	E	A	B	C
E	E	A	B	C
A	A	B	C	E
B	B	C	E	A
C	C	E	A	B

$\mathbf{AC} = \mathbf{CA} = \mathbf{E}$), $\mathbf{B}^{-1} = \mathbf{B}$ and $\mathbf{C}^{-1} = \mathbf{A}$. Also, it is obvious that $\mathbf{E}$ serves as the identity element. Hence, the four elements form a group with respect to matrix multiplication. Moreover, we can also notice that two elements $\mathbf{E}$ and $\mathbf{B}$ form a group of two elements on their own since $\mathbf{BB} = \mathbf{E}$ and $\mathbf{B}^{-1} = \mathbf{B}$. This smaller group consists of elements of a larger group and is called a *subgroup*. ◄

Groups possess a number of fascinating properties, however, we are not going to go into this theory here.

1.2.5 Orthogonal and Unitary Transformations

1.2.5.1 General Properties

Let us now consider a particular set of certain *real* transformation matrices $\mathbf{A}$. These are the matrices that conserve the dot product between any two *real* vectors: if $\mathbf{x}' = \mathbf{Ax}$ and $\mathbf{y}' = \mathbf{Ay}$, then

$$(\mathbf{x}', \mathbf{y}') = (\mathbf{Ax}, \mathbf{Ay}) = (\mathbf{x}, \mathbf{y}) \ . \tag{1.34}$$

In particular, by taking $\mathbf{y} = \mathbf{x}$, we obtain that $(\mathbf{Ax}, \mathbf{Ax}) = (\mathbf{x}, \mathbf{x})$, i.e. the length of a vector before and after this particular transformation does not change.

To uncover the appropriate condition the matrix $\mathbf{A}$ should satisfy, we use Eq. (1.19) and get $(\mathbf{Ax}, \mathbf{Ay}) = (\mathbf{x}, \mathbf{A}^T\mathbf{Ay})$, which will be equal to $(\mathbf{x}, \mathbf{y})$ only if

$$\mathbf{A}^T\mathbf{A} = \mathbf{E} \ . \tag{1.35}$$

One can also see that $\mathbf{AA}^T = \mathbf{E}$ as well (indeed, multiply both sides of Eq. (1.35) by $\mathbf{A}$ from the left and then by $\mathbf{A}^{-1}$ from the right). Matrices satisfying this condition are called *orthogonal*.[3] Transformations performed by orthogonal matrices are called

---

[3] This probably originates from the definition of two orthogonal vectors. Indeed, if we have two vectors-columns $\mathbf{X}$ and $\mathbf{Y}$ which are orthogonal, then one can write $\mathbf{X}^T\mathbf{Y} = 0$ which in some sense may be considered analogous to Eq. (1.35).

orthogonal transformations. One can see by comparing Eqs. (1.35) and (1.24), that for orthogonal matrices

$$\mathbf{A}^{-1} = \mathbf{A}^T , \qquad (1.36)$$

i.e. the transposed matrix $\mathbf{A}^T$ is at the same time the inverse of $\mathbf{A}$. Therefore, the inverse of the orthogonal matrix always exists and equal to $\mathbf{A}^T$, i.e. orthogonal matrices are non-singular.

The orthogonal matrices form a group: if $\mathbf{A}$ and $\mathbf{B}$ are orthogonal, then $\mathbf{C} = \mathbf{AB}$ is also orthogonal. Indeed, $\mathbf{A}^{-1} = \mathbf{A}^T$, $\mathbf{B}^{-1} = \mathbf{B}^T$, and

$$\mathbf{C}^{-1} = (\mathbf{AB})^{-1} = \mathbf{B}^{-1}\mathbf{A}^{-1} = \mathbf{B}^T\mathbf{A}^T = (\mathbf{AB})^T = \mathbf{C}^T .$$

Also, the set of orthogonal matrices contains the unit element $\mathbf{E}$ which is of course orthogonal; finally, each element has an inverse as $\mathbf{A}^{-1} = \mathbf{A}^T$ and $\mathbf{A}^T$ is also an orthogonal matrix because $\left(\mathbf{A}^T\right)^T = \mathbf{A} = \left(\mathbf{A}^{-1}\right)^{-1} = \left(\mathbf{A}^T\right)^{-1}$. This discussion proves the above-made statement.

As a simple example, consider the following matrix:

$$\mathbf{A} = \begin{pmatrix} \cos\theta & -\sin\theta \\ \sin\theta & \cos\theta \end{pmatrix} . \qquad (1.37)$$

Then, the matrix

$$\mathbf{A}^T = \begin{pmatrix} \cos\theta & \sin\theta \\ -\sin\theta & \cos\theta \end{pmatrix}$$

is its transpose. It is easily checked by direct multiplication that $\mathbf{A}$ is an orthogonal matrix since $\mathbf{A}^T\mathbf{A} = \mathbf{E}$. Indeed,

$$\begin{pmatrix} \cos\theta & \sin\theta \\ -\sin\theta & \cos\theta \end{pmatrix} \begin{pmatrix} \cos\theta & -\sin\theta \\ \sin\theta & \cos\theta \end{pmatrix} = \begin{pmatrix} 1 & 0 \\ 0 & 1 \end{pmatrix} ,$$

i.e. the transposed matrix is indeed the inverse of $\mathbf{A}$. Consider now a vector $\mathbf{x} = (x_1, x_2)$ of length square $|\mathbf{x}|^2 = (\mathbf{x}, \mathbf{x}) = x_1^2 + x_2^2$. It is easy to see that the length square $(\mathbf{y}, \mathbf{y})$ of the vector $\mathbf{y} = \mathbf{Ax}$ is also equal to $(\mathbf{x}, \mathbf{x})$ for any $\mathbf{x}$. Indeed,

$$\mathbf{y} = \begin{pmatrix} \cos\theta & -\sin\theta \\ \sin\theta & \cos\theta \end{pmatrix} \begin{pmatrix} x_1 \\ x_2 \end{pmatrix} = \begin{pmatrix} x_1\cos\theta - x_2\sin\theta \\ x_1\sin\theta + x_2\cos\theta \end{pmatrix} = \begin{pmatrix} y_1 \\ y_2 \end{pmatrix} ,$$

and its square length $y_1^2 + y_2^2$ is easily calculated (after simple manipulations) to be $x_1^2 + x_2^2$.

In the 3D space orthogonal matrices not only conserve the length of a vector after the transformation, they also conserve the angle between two vectors after transforming both of them. Indeed, according to Eq. (I.1.91), the dot product is equal to a product of vectors' lengths and the cosine of the angle between them. The lengths of the vectors are conserved after the transformation. Since the dot product is conserved as well, that means that the cosine of the angle is conserved, i.e. the angle between two transformed vectors remains the same.

What makes the matrix an orthogonal one? We shall now see that these matrices have a very special structure. Writing Eq. (1.35) element by element, we obtain (assuming that $\mathbf{A}$ is a square $n \times n$ matrix):

$$\sum_k \widetilde{a}_{ik} a_{kj} = \sum_k a_{ki} a_{kj} = \delta_{ij} \ .$$

Here elements $\{a_{ki}\} = (a_{1i}, a_{2i}, \ldots, a_{ni})$ with the fixed i form a vector which is composed of all elements of the ith column of $\mathbf{A}$; likewise, all elements $\{a_{kj}\}$ with the fixed j form a vector out of the jth column of $\mathbf{A}$. Therefore, the equation above tells us that the dot product of the ith and jth columns of $\mathbf{A}$ is equal to zero if $i \neq j$ and one if $i = j$. In other words, the columns of an orthogonal $n \times n$ matrix form a set of n orthonormal vectors.

Problem 1.36 Show that rows of an orthogonal matrix form a set of n orthonormal vectors. [Hint: *use the condition* $\mathbf{AA}^T = \mathbf{E}$.]

Similarly, we can consider a more general p-D space of *complex* vectors. But first, let us prove for that case an identity analogous to Eq. (1.19):

$$(\mathbf{Ax}, \mathbf{y}) = \sum_l \left(\sum_k a_{lk} x_k \right)^* y_l = \sum_k x_k^* \left(\sum_l a_{lk}^* y_l \right)$$

$$= \sum_k x_k^* \left(\sum_l \widetilde{a}_{kl}^* y_l \right) = \left(\mathbf{x}, \left(\mathbf{A}^T \right)^* \mathbf{y} \right) \ .$$

We see that when the matrix $\mathbf{A}$ goes from the left position in the dot product to the right position there, it changes in two ways: it is transposed (as in the previous case of the real space) *and* undergoes complex conjugation (obviously, in any order). The matrix $\left(\mathbf{A}^T \right)^* = \mathbf{A}^{T*}$ obtained by transposing $\mathbf{A}$ and then applying complex conjugation to all its elements is called *Hermitian conjugate* and is denoted with the dagger: $\left(\mathbf{A}^T \right)^* = \mathbf{A}^\dagger$. Hence, we can write

$$(\mathbf{Ax}, \mathbf{y}) = \left(\mathbf{x}, \mathbf{A}^\dagger \mathbf{y} \right) \ . \tag{1.38}$$

Matrices which satisfy

$$\mathbf{AA}^\dagger = \mathbf{E} \ , \tag{1.39}$$

are called *unitary matrices*. Transformations in complex vector spaces by unitary matrices are called *unitary transformations*. Physical quantities in quantum mechanics have a direct correspondence to unitary matrices. Comparing Eq. (1.39) with the definition of the inverse matrix, Eq. (1.24), one can see that for unitary matrices

$$\mathbf{A}^{-1} = \mathbf{A}^{\dagger} , \tag{1.40}$$

i.e. to calculate the inverse matrix, one has to simply transpose it and then take the complex conjugate of all its elements. Since for unitary matrices $\mathbf{A}^{\dagger}$ is the same as $\mathbf{A}^{-1}$, then $\mathbf{A}^{\dagger}\mathbf{A} = \mathbf{E}$ is true as well.

If we write down $\mathbf{A}\mathbf{A}^{\dagger} = \mathbf{A}^{\dagger}\mathbf{A} = \mathbf{E}$ in elements, we obtain the following equations:

$$\sum_k a_{ik} a_{jk}^* = \delta_{ij} \text{ and } \sum_k a_{ki} a_{kj}^* = \delta_{ij} . \tag{1.41}$$

These mean that rows and columns of a unitary matrix are orthonormal in the sense of the dot product of complex vectors.

It then follows from Eq. (1.38) that for unitary transformations,

$$(\mathbf{Ax}, \mathbf{Ay}) = \left(\mathbf{x}, \mathbf{A}^{\dagger}\left[\mathbf{Ay}\right]\right) = \left(\mathbf{x}, \mathbf{A}^{\dagger}\mathbf{Ay}\right) = (\mathbf{x}, \mathbf{Ey}) = (\mathbf{x}, \mathbf{y}) ,$$

i.e. the dot product $(\mathbf{x}, \mathbf{y})$ is conserved. We see that unitary transformations are generalisations of the orthogonal transformations for complex vector spaces.

Example 1.10 ▶ Show that

$$\mathbf{A} = \frac{1}{5} \begin{pmatrix} -1+2i & -4-2i \\ 2-4i & -2-i \end{pmatrix}$$

is unitary.

Solution: what we need to do is simply to check if $\mathbf{A}\mathbf{A}^{\dagger} = \mathbf{A}^{\dagger}\mathbf{A} = \mathbf{E}$:

$$\mathbf{A}^T = \frac{1}{5} \begin{pmatrix} -1+2i & 2-4i \\ -4-2i & -2-i \end{pmatrix}$$

and thus

$$\mathbf{A}^{T*} = \frac{1}{5} \begin{pmatrix} -1-2i & 2+4i \\ -4+2i & -2+i \end{pmatrix} = \mathbf{A}^{\dagger} ;$$

therefore,

$$\mathbf{A}^{\dagger}\mathbf{A} = \frac{1}{5} \begin{pmatrix} -1-2i & 2+4i \\ -4+2i & -2+i \end{pmatrix} \frac{1}{5} \begin{pmatrix} -1+2i & -4-2i \\ 2-4i & -2-i \end{pmatrix}$$

$$= \frac{1}{25} \begin{pmatrix} 25 & 0 \\ 0 & 25 \end{pmatrix} = \begin{pmatrix} 1 & 0 \\ 0 & 1 \end{pmatrix} = \mathbf{E} ,$$

$$\mathbf{AA}^\dagger = \frac{1}{5}\begin{pmatrix} -1+2i & -4-2i \\ 2-4i & -2-i \end{pmatrix} \frac{1}{5}\begin{pmatrix} -1-2i & 2+4i \\ -4+2i & -2+i \end{pmatrix}$$

$$= \frac{1}{25}\begin{pmatrix} 25 & 0 \\ 0 & 25 \end{pmatrix} = \begin{pmatrix} 1 & 0 \\ 0 & 1 \end{pmatrix} = \mathbf{E},$$

as it should be. ◀

Example 1.11 As another example, consider the following matrix:

$$\mathbf{U} = \begin{pmatrix} 0 & \sqrt{2/3} & 1/\sqrt{3} \\ 1/\sqrt{2} & i/\sqrt{6} & -i/\sqrt{3} \\ -1/\sqrt{2} & i/\sqrt{6} & -i/\sqrt{3} \end{pmatrix}.$$

Solution: this matrix is unitary. Indeed, the Hermitian conjugate of it is

$$\mathbf{U}^\dagger = \begin{pmatrix} 0 & 1/\sqrt{2} & -1/\sqrt{2} \\ \sqrt{2/3} & -i/\sqrt{6} & -i/\sqrt{6} \\ 1/\sqrt{3} & i/\sqrt{3} & i/\sqrt{3} \end{pmatrix},$$

and it is easily checked that $\mathbf{U}^\dagger\mathbf{U} = \mathbf{U}\mathbf{U}^\dagger = \mathbf{E}$, for instance:

$$\mathbf{U}^\dagger\mathbf{U} = \begin{pmatrix} 0 & 1/\sqrt{2} & -1/\sqrt{2} \\ \sqrt{2/3} & -i/\sqrt{6} & -i/\sqrt{6} \\ 1/\sqrt{3} & i/\sqrt{3} & i/\sqrt{3} \end{pmatrix}\begin{pmatrix} 0 & \sqrt{2/3} & 1/\sqrt{3} \\ 1/\sqrt{2} & i/\sqrt{6} & -i/\sqrt{3} \\ -1/\sqrt{2} & i/\sqrt{6} & -i/\sqrt{3} \end{pmatrix}$$

$$= \begin{pmatrix} 1 & 0 & 0 \\ 0 & 1 & 0 \\ 0 & 0 & 1 \end{pmatrix}.$$

It is also seen that row (columns) of $\mathbf{U}$ form a set of orthonormal vectors. For instance, consider the orthogonality of the rows:

$$\left(0\ \ \sqrt{2/3}\ \ 1/\sqrt{3}\right)^* \begin{pmatrix} 1/\sqrt{2} \\ i/\sqrt{6} \\ -i/\sqrt{3} \end{pmatrix} = 0,$$

$$\left(0\ \ \sqrt{2/3}\ \ 1/\sqrt{3}\right)^* \begin{pmatrix} -1/\sqrt{2} \\ i/\sqrt{6} \\ -i/\sqrt{3} \end{pmatrix} = 0,$$

$$\left(1/\sqrt{2}\ \ i/\sqrt{6}\ -i/\sqrt{3}\right)^* \begin{pmatrix} -1/\sqrt{2} \\ i/\sqrt{6} \\ -i/\sqrt{3} \end{pmatrix} = 0,$$

while they are also all normalised to unity:

$$\left(0 \ \sqrt{2/3} \ 1/\sqrt{3} \right)^* \begin{pmatrix} 0 \\ \sqrt{2/3} \\ 1/\sqrt{3} \end{pmatrix} = 1 \,,$$

$$\left(1/\sqrt{2} \ i/\sqrt{6} \ -i/\sqrt{3} \right)^* \begin{pmatrix} 1/\sqrt{2} \\ i/\sqrt{6} \\ -i/\sqrt{3} \end{pmatrix} = 1 \,,$$

$$\left(-1/\sqrt{2} \ i/\sqrt{6} \ -i/\sqrt{3} \right)^* \begin{pmatrix} -1/\sqrt{2} \\ i/\sqrt{6} \\ -i/\sqrt{3} \end{pmatrix} = 1 \,,$$

as required. ◄

If after applying the Hermitian conjugate to a matrix $\mathbf{A}$ it changes its sign, i.e. $\mathbf{A}^\dagger = -\mathbf{A}$, the matrix is called anti-Hermitian or skew-Hermitian.

Problem 1.37 In this problem we shall obtain one identity involving Pauli matrices (1.22) that is found useful in relativistic quantum theory. Define two row-vectors

$$\mathbf{A} = \begin{pmatrix} \mathbf{A}_1 \\ \mathbf{A}_2 \end{pmatrix},$$

components of which are ordinary vectors $\mathbf{A}_i = (A_{ix}, A_{iy}, A_{iz})$, where $i = 1, 2$. We then define dot and vector products of any two such vectors as

$$\mathbf{A}^\dagger \cdot \mathbf{B} = \sum_\alpha \left(A_{1\alpha}^* \ A_{2\alpha}^* \right) \begin{pmatrix} B_{1\alpha} \\ B_{2\alpha} \end{pmatrix} = \sum_{i=1}^2 \mathbf{A}_i^* \cdot \mathbf{B}_i \,,$$

$$\left[\mathbf{A}^\dagger \times \mathbf{B} \right] = \sum_{i=1}^2 \left[\mathbf{A}_i^* \times \mathbf{B}_i \right] \,.$$

The dot product is a scalar, while the defined vector product gives a usual 3D vector. The Pauli matrices would then operate on the whole vector components of the two row-vectors to produce new row-vectors, e.g.

$$\sigma_x \mathbf{A} = \begin{pmatrix} 0 & 1 \\ 1 & 0 \end{pmatrix} \begin{pmatrix} \mathbf{A}_1 \\ \mathbf{A}_2 \end{pmatrix} = \begin{pmatrix} \mathbf{A}_2 \\ \mathbf{A}_1 \end{pmatrix}.$$

Then, prove the following identity:

$$(\boldsymbol{\sigma} \cdot \mathbf{A})^{\dagger} (\boldsymbol{\sigma} \cdot \mathbf{B}) = \mathbf{A}^{\dagger} \cdot \mathbf{B} + i \sum_{\alpha} \left[(\boldsymbol{\sigma}_{\alpha} \cdot \mathbf{A})^{\dagger} \times \mathbf{B} \right]_{\alpha} ,$$

where we introduced the Pauli matrix-vector $\boldsymbol{\sigma} = \sigma_x \mathbf{i} + \sigma_y \mathbf{j} + \sigma_z \mathbf{k}$.

Problem 1.38 Let a two-row-vector function be defined via

$$\phi(\mathbf{r}) = \begin{pmatrix} \psi_1(\mathbf{r}) \\ \psi_2(\mathbf{r}) \end{pmatrix} ,$$

where $\psi_i(\mathbf{r})$ are two scalar functions ($i = 1, 2$), and an application of the curl to the construction like this one is defined simply via

$$\nabla \times \phi(\mathbf{r}) = \begin{pmatrix} \nabla \times \psi_1(\mathbf{r}) \\ \nabla \times \psi_2(\mathbf{r}) \end{pmatrix} ,$$

i.e. via its application to each of the two components. Then, prove the following identities also involving Pauli matrices:

$$\boldsymbol{\sigma} (\boldsymbol{\sigma} \cdot \nabla \phi) = \nabla \phi + i \nabla \times (\boldsymbol{\sigma} \phi) ,$$

$$(\nabla \phi^{\dagger} \cdot \boldsymbol{\sigma}) \boldsymbol{\sigma} = \nabla \phi^{\dagger} - i \nabla \times (\phi^{\dagger} \boldsymbol{\sigma}) .$$

Problem 1.39 (*Discrete Fourier transform*) Consider N numbers x_0, x_1, etc., x_{N-1}. One can generate another set of N numbers, y_j (with $j = 0, 1, \ldots, N - 1$), using the following recipe:

$$y_j = \frac{1}{\sqrt{N}} \sum_{k=0}^{N-1} e^{-i2\pi jk/N} x_k . \tag{1.42}$$

Form N-dimensional vectors $\mathbf{X}$ and $\mathbf{Y}$ from the quantities $\{x_j\}$ and $\{y_j\}$, respectively, and then write the above relationship in the matrix form as $\mathbf{Y} = \mathbf{U}\mathbf{X}$. Prove that rows (columns) of $\mathbf{U}$ form an orthonormal set of vectors, and hence that the matrix $\mathbf{U}$ is unitary. Finally, establish that the inverse transformation from $\mathbf{Y}$ to $\mathbf{X}$ reads

$$x_k = \frac{1}{\sqrt{N}} \sum_{j=0}^{N-1} e^{i2\pi jk/N} y_j . \tag{1.43}$$

Problem 1.40 Show that a product $U = U_1 U_2 \cdots U_p$ of unitary matrices U_1, U_2, etc., U_p is also a unitary matrix, i.e. $U^\dagger U = E$.

Problem 1.41 Consider an arbitrary square matrix A. It can always be split into two matrices as $A = B + C$, where

$$B = \frac{1}{2}\left(A + A^\dagger\right) \quad \text{and} \quad C = \frac{1}{2}\left(A - A^\dagger\right) .$$

Prove that while B is Hermitian, C is anti-Hermitian. This statement demonstrates that any square matrix can be represented as a sum of some Hermitian and anti-Hermitian matrices.

Problem 1.42 Let A be a Hermitian matrix, $A^\dagger = A$. Show that after a transformation, $C = B^\dagger A B$, with arbitrary B the matrix C remains Hermitian.

1.2.5.2 Rotations in 3D

An important particular set of orthogonal transformations is formed by 3D rotations.

Let us consider a clockwise rotation by an angle θ of the point A with coordinates (x, y, z) about the x axis, see Fig. 1.4. It goes into the point A' with coordinates (x', y', z'). Let us express the coordinates of the target point via the rotation angle and the coordinates of the initial point. Obviously, for any rotation about the x axis the x coordinate does not change: $x' = x$. Then, using the definitions given in the picture, we get

$$z = \rho \sin \phi, \quad y = \rho \cos \phi \quad \text{and} \quad z' = \rho \sin \phi', \quad y' = \rho \cos \phi' ,$$

where $\phi' = \phi - \theta$.

Fig. 1.4 For the derivation of the rotation matrix corresponding to the clockwise rotation about the x axis: $A \to A'$

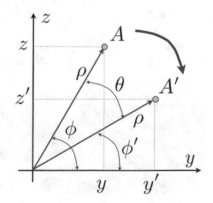

This simple result allows us to write down the transformation $A \rightarrow A'$ of the point A in the matrix form as follows:

$$\begin{pmatrix} x' \\ y' \\ z' \end{pmatrix} = \begin{pmatrix} 1 & 0 & 0 \\ 0 & \cos\theta & \sin\theta \\ 0 & -\sin\theta & \cos\theta \end{pmatrix} \begin{pmatrix} x \\ y \\ z \end{pmatrix} = \mathbf{R}_x(\theta) \begin{pmatrix} x \\ y \\ z \end{pmatrix} ,$$

i.e. the transformation is performed by the matrix

$$\mathbf{R}_x(\theta) = \begin{pmatrix} 1 & 0 & 0 \\ 0 & \cos\theta & \sin\theta \\ 0 & -\sin\theta & \cos\theta \end{pmatrix} . \tag{1.44}$$

Note that this matrix is orthogonal. Indeed, consider a rotation by the angle $-\theta$ about the same axis x:

$$\mathbf{R}_x(-\theta) = \begin{pmatrix} 1 & 0 & 0 \\ 0 & \cos\theta & -\sin\theta \\ 0 & \sin\theta & \cos\theta \end{pmatrix} = \mathbf{R}_x(\theta)^T ,$$

and this matrix can easily be checked is the inverse to that in Eq. (1.44):

$$\mathbf{R}_x(\theta)\mathbf{R}_x(\theta)^T = \begin{pmatrix} 1 & 0 & 0 \\ 0 & \cos\theta & \sin\theta \\ 0 & -\sin\theta & \cos\theta \end{pmatrix} \begin{pmatrix} 1 & 0 & 0 \\ 0 & \cos\theta & -\sin\theta \\ 0 & \sin\theta & \cos\theta \end{pmatrix}$$

$$= \begin{pmatrix} 1 & 0 & 0 \\ 0 & 1 & 0 \\ 0 & 0 & 1 \end{pmatrix} = \mathbf{E} ,$$

i.e. indeed

$$\mathbf{R}_x(-\theta) = \mathbf{R}_x(\theta)^T = \mathbf{R}_x(\theta)^{-1} . \tag{1.45}$$

$$\mathbf{R}_z(\theta) = \begin{pmatrix} \cos\theta & -\sin\theta & 0 \\ \sin\theta & \cos\theta & 0 \\ 0 & 0 & 1 \end{pmatrix}. \qquad (1.47)$$

Check that both matrices are orthogonal, i.e. the inverse of any of them coincides with their transpose and also corresponds to the rotation by the angle $-\theta$.

Problem 1.45 Consider now two consecutive rotations about the x axis: the first by θ_1 and the second by θ_2. The matrix $\mathbf{R}_x(\theta_2)\mathbf{R}_x(\theta_1)$ should correspond to a single rotation by the angle $\theta_1 + \theta_2$. Show by direct multiplication of the two matrices and using well-known trigonometric identities that this is indeed the case:

$$\mathbf{R}_x(\theta_2)\mathbf{R}_x(\theta_1) = \begin{pmatrix} 1 & 0 & 0 \\ 0 & \cos(\theta_1+\theta_2) & \sin(\theta_1+\theta_2) \\ 0 & -\sin(\theta_1+\theta_2) & \cos(\theta_1+\theta_2) \end{pmatrix} = \mathbf{R}_x(\theta_1+\theta_2).$$

It is seen now that 3D rotations about the x axis form a group. Indeed, any rotation has an inverse, the identity matrix is the identity element (rotation by $\theta = 0$ gives $\mathbf{E}$) and any two consecutive rotations by θ_1 and θ_2 correspond to a single rotation by $\theta_1 + \theta_2$. Obviously, all rotations about y or z axis also form groups. In fact, it is actually possible to establish that any arbitrary rotation in the 3D space (not necessarily about the same axis) can always be represented by no more than three elementary rotations described above, and all these rotations form a group of rotations of the 3D space.

A formula for the matrix of a general rotation will be derived in Sect. 1.21. You will be able to check that the formulae given above for the rotations around Cartesian axes represent particular cases of this general result.

1.2.5.3 Reflections in 3D

Consider another type of transformations in 3D space where a point A given by the vector $\mathbf{r}$ goes over to a point A' upon a reflection in a plane σ passing through the origin. The unit normal of the plane is given by the vector $\mathbf{n}$, see Fig. 1.5. The vector $\mathbf{r}$ forms an angle θ with the plane, and upon reflection goes over into the vector $\mathbf{r}'$ which has the same projection $\mathbf{a}$ on the plane, but the two vectors have opposite perpendicular components. The latter can be written as $(\mathbf{r} \cdot \mathbf{n})\,\mathbf{n}$ for $\mathbf{r}$ and as $-(\mathbf{r} \cdot \mathbf{n})\,\mathbf{n}$ for $\mathbf{r}'$. Therefore, one can write

$$\mathbf{r} = \mathbf{a} + (\mathbf{r} \cdot \mathbf{n})\,\mathbf{n} \quad \text{and} \quad \mathbf{r}' = \mathbf{a} - (\mathbf{r} \cdot \mathbf{n})\,\mathbf{n} \implies \mathbf{r}' - \mathbf{r} = -2\,(\mathbf{r} \cdot \mathbf{n})\,\mathbf{n}.$$

Correspondingly, the new vector $\mathbf{r}'$ can be written as a transformation $\mathbf{r}' = \mathbf{C}_\sigma \mathbf{r}$ with a reflection matrix $\mathbf{C}_\sigma$.

Fig. 1.5 For the derivation
of Eq. (1.48): the vector **r**
when reflected in the plane σ
with the unit normal **n** goes
into vector **r′**. The side view
is shown in which the plane
runs perpendicularly to the
surface of the page

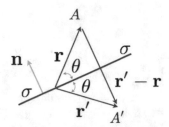

Problem 1.46 Show that

$$\mathbf{C}_{\sigma} = \begin{pmatrix} 1 - 2n_1^2 & -2n_1n_2 & -2n_1n_3 \\ -2n_2n_1 & 1 - 2n_2^2 & -2n_2n_3 \\ -2n_3n_1 & -2n_3n_2 & 1 - 2n_3^2 \end{pmatrix} . \tag{1.48}$$

Problem 1.47 Verify that $\mathbf{C}_{\sigma}\mathbf{C}_{\sigma} = \mathbf{E}$. This is to be expected as two consecutive reflections must return the point A back: $A \to A' \to A$.

In particular, the reflection in the $x - y$ plane with $\mathbf{n} = (0, 0, 1)$ is performed by the matrix

$$\mathbf{C}_{\sigma_{xy}} = \begin{pmatrix} 1 & 0 & 0 \\ 0 & 1 & 0 \\ 0 & 0 & -1 \end{pmatrix} ,$$

while the reflection in the plane with the unit normal $\mathbf{n} = \frac{1}{\sqrt{3}} (1, 1, 1)$ is given by the matrix

$$\mathbf{C}_{\sigma_{111}} = \frac{1}{3} \begin{pmatrix} 1 & -2 & -2 \\ -2 & 1 & -2 \\ -2 & -2 & 1 \end{pmatrix} ,$$

so that the vector along the normal, $\mathbf{r} = (1, 1, 1)$, transforms into $\mathbf{r}' = \mathbf{C}_{\sigma_{111}}\mathbf{r} = (-1, -1, -1)$, as expected.

1.3 Determinant of a Square Matrix

1.3.1 Formal Definition

One quantity which is of great importance in the theory of matrices is its *determinant*. We introduced the 2×2 and 3×3 determinants before in Sect. I.1.9. Here we shall generalise these definitions to determinants of arbitrary dimension and, most importantly, will relate the determinants to square matrices.

We start by recalling the two- and three-dimensional cases and then consider a general case of an arbitrary dimension. Consider a 2×2 and a 3×3 matrices

$$\mathbf{A} = \begin{pmatrix} a_{11} & a_{12} \\ a_{21} & a_{22} \end{pmatrix} \quad \text{and} \quad \mathbf{B} = \begin{pmatrix} b_{11} & b_{12} & b_{13} \\ b_{21} & b_{22} & b_{23} \\ b_{31} & b_{32} & b_{33} \end{pmatrix} .$$

The determinant of $\mathbf{A}$ is *defined* as

$$|\mathbf{A}| = \det \mathbf{A} = \begin{vmatrix} a_{11} & a_{12} \\ a_{21} & a_{22} \end{vmatrix} = a_{11}a_{22} - a_{12}a_{21} . \tag{1.49}$$

Note that the expression for $|\mathbf{A}|$ contains a sum of $2! = 2$ terms each containing products of two elements. Similarly, the determinant of $\mathbf{B}$ is *defined* as

$$|\mathbf{B}| = \det \mathbf{B} = \begin{vmatrix} b_{11} & b_{12} & b_{13} \\ b_{21} & b_{22} & b_{23} \\ b_{31} & b_{32} & b_{33} \end{vmatrix}$$

$$= b_{11}b_{22}b_{33} - b_{11}b_{23}b_{32} - b_{12}b_{21}b_{33} + b_{12}b_{23}b_{31} + b_{13}b_{21}b_{32} - b_{13}b_{22}b_{31} . \tag{1.50}$$

This expression contains $3! = 6$ terms.

Problem 1.48 Calculate determinants of matrices

$$\mathbf{A} = \begin{pmatrix} 1 & 0 & 2 \\ 3 & -1 & 0 \\ 0 & 5 & 1 \end{pmatrix}, \quad \mathbf{B} = \begin{pmatrix} 1 & 1 & 0 \\ 0 & 2 & 1 \\ 3 & -1 & 0 \end{pmatrix} \quad \text{and} \quad \mathbf{C} = \mathbf{AB} .$$

Check that $|\mathbf{C}| = |\mathbf{A}| \, |\mathbf{B}|$. [Answer: $|\mathbf{A}| = 29$, $|\mathbf{B}| = 4$ *and* $|\mathbf{AB}| = 116$.]

Let us have a look at the two expressions (1.49) and (1.50) more closely. Each expression contains a sum of products of elements of the corresponding matrix, such as $-b_{12}b_{21}b_{33}$ in Eq. (1.50). One can say that in every such a product each row is represented by a single element; at the same time, one may also independently say that each column is also represented by a single element of the matrix. This is because we have the left indices of all elements forming all possible integers between 1 and 3, and similarly for the right indices, without any repetitions or omissions.

This means that each elementary product in the cases of $|\mathbf{A}|$ and $|\mathbf{B}|$ can be written as $\pm a_{1j_1}a_{2j_2}$ and $\pm b_{1j_1}b_{2j_2}b_{3j_3}$, respectively, where indices j_1, j_2 or j_1, j_2, j_3 correspond to some *permutation* of numbers 1, 2 and 1, 2, 3 for the two determinants. There are $2! = 2$ and $3! = 6$ permutations of the second indices possible in the two cases clearly illustrating the number of terms in each of the two determinants.

Now, let us look at the sign attached to each of the products. It is defined by the *parity* of the permutation $1, 2 \rightarrow j_1, j_2$ or $1, 2, 3 \rightarrow j_1, j_2, j_3$. Every permutation of two indices contributes a factor of -1. By performing all necessary pair permutations one after another, starting from the first (perfectly ordered) term, such as $a_{11}a_{22}$ or $b_{11}b_{22}b_{33}$, the overall sign of each term in Eqs. (1.49) and (1.50) can be obtained. For instance, consider the second term in Eq. (1.50), $-b_{11}b_{23}b_{32}$. Only one permutation of the right indices in the third and the second elements is required in this case:

$$b_{11}b_{22}b_{33} \rightarrow b_{11}b_{23}b_{32} \, ,$$

where the permuted indices are underlined. Therefore, this term should appear with a single factor of -1. The fourth term, $+b_{12}b_{23}b_{31}$, on the other hand, requires two permutations:

$$b_{11}b_{22}b_{33} \rightarrow b_{12}b_{21}b_{33} \rightarrow b_{12}b_{23}b_{31} \, ,$$

and thus should acquire a factor of $(-1)^2 = 1$.

Once we have looked at the two cases that must be familiar to us, let us try to make a generalisation to an arbitrary $n \times n$ matrix. If

$$\mathbf{A} = \begin{pmatrix} a_{11} & \cdots & a_{1n} \\ \cdots & \cdots & \cdots \\ a_{n1} & \cdots & a_{nn} \end{pmatrix}$$

is such a matrix, then its determinant is *defined* in the following way:

$$\det \mathbf{A} = \begin{vmatrix} a_{11} & \cdots & a_{1n} \\ \cdots & \cdots & \cdots \\ a_{n1} & \cdots & a_{nn} \end{vmatrix} = \sum_P \epsilon_{j_1 j_2 \cdots j_n} a_{1j_1} a_{2j_2} \cdots a_{nj_n} \, , \qquad (1.51)$$

where the sum contains $n!$ terms corresponding to all possible arrangements (permutations) P of the right (column) indices of the elements of the matrix in the ordered product $a_{11}a_{22} \cdots a_{nn}$. The sign factor $\epsilon_{j_1 j_2 \cdots j_n} = \pm 1$ corresponds to the parity of the overall permutation: each pair permutation between two indices of the elements in the product brings a factor of -1, and one has to do as many pair permutations, starting from the ordered product, as necessary to get the required order of indices.

In the above consideration, we have initially ordered elements a_{ij} of the matrix $\mathbf{A}$ in the determinant in order of the left (row) index. In fact, a similar expression can also be written with the elements ordered with respect to the right (column) index:

$$\det \mathbf{A} = \begin{vmatrix} a_{11} & \cdots & a_{1n} \\ \cdots & \cdots & \cdots \\ a_{n1} & \cdots & a_{nn} \end{vmatrix} = \sum_P \epsilon_{j_1 j_2 \cdots j_n} a_{j_1 1} a_{j_2 2} \cdots a_{j_n n} \, . \qquad (1.52)$$

Exactly the same final expression for $\det A$ is obtained since in each term in the sum each index (either left or right) happens only once.

As the simplest application of our general definition of the determinant of a square matrix, let us calculate the determinant of a diagonal matrix:

$$\mathbf{D} = \left(d_{ii} \delta_{ij} \right) = \begin{pmatrix} d_{11} & \ddots & 0 & 0 \\ \ddots & d_{22} & \ddots & 0 \\ 0 & \ddots & \ddots & \ddots \\ 0 & 0 & \ddots & d_{nn} \end{pmatrix} .$$

We shall use Eq. (1.51). In each term in the product there at least one element is chosen from every row and column. However, all terms except those on the main diagonal are zeros in $\mathbf{D}$. Therefore, there is only one non-zero term in the sum, namely the ordered term $d_{11} d_{22} \cdots d_{nn}$; the parity of this term is obviously one, i.e. the determinant of a diagonal matrix is equal to a product of all its diagonal elements:

$$\begin{vmatrix} d_{11} & \ddots & 0 & 0 \\ \ddots & d_{22} & \ddots & 0 \\ 0 & \ddots & \ddots & \ddots \\ 0 & 0 & \ddots & d_{nn} \end{vmatrix} = d_{11} d_{22} \cdots d_{nn} = \prod_{k=1}^{n} d_{kk} . \tag{1.53}$$

In particular, the determinant of the unit matrix is equal to one: $\det \mathbf{E} = 1$.

Example 1.12 ▶ As a more elaborate example let us work out an expression for the determinant of a 4×4 matrix

$$\mathbf{A} = \begin{pmatrix} a_{11} & a_{12} & a_{13} & a_{14} \\ a_{21} & a_{22} & a_{23} & a_{24} \\ a_{31} & a_{32} & a_{33} & a_{34} \\ a_{41} & a_{42} & a_{43} & a_{44} \end{pmatrix} .$$

Solution: We have to construct, starting each time from the 'perfectly ordered' elementary product $a_{11} a_{22} a_{33} a_{44}$, all possible orderings of the right indices, keeping track of the parity of the permutation. There are $4! = 24$ terms to be expected. All 24 permutations of numbers 1, 2, 3, 4 and their parity together with the corresponding contribution to the $\det \mathbf{A}$ are shown in Table 1.2. Summing up all terms in the last column of the Table, we obtain the desired expression for the determinant. ◀

Problem 1.49 Let elements a_{ij} of a $n \times n$ matrix $\mathbf{A}$ are functions of a parameter λ, i.e. $a_{ij} = a_{ij}(\lambda)$. Prove the formula:

$$\frac{d}{d\lambda}|\mathbf{A}| = \sum_{k=1}^{n}|\mathbf{A}_k| \ ,$$

where the $n \times n$ matrix $\mathbf{A}_k$ is formed by all elements of the original matrix apart from those in the kth row (or column) which are derivatives of the corresponding elements of $\mathbf{A}$ with respect to λ.

Table 1.2 All contributions to the determinant of a 4×4 matrix

Permutation	No of pair permutations	Parity	Contributing term
$1, 2, 3, 4 \rightarrow 1, 2, 3, 4$	0	1	$a_{11}a_{22}a_{33}a_{44}$
$1, 2, 3, 4 \rightarrow 1, 2, 4, 3$	1	-1	$-a_{11}a_{22}a_{34}a_{43}$
$1, 2, 3, 4 \rightarrow 1, 3, 2, 4$	1	-1	$-a_{11}a_{23}a_{32}a_{44}$
$1, 2, 3, 4 \rightarrow 1, 3, 4, 2$	2	1	$a_{11}a_{23}a_{34}a_{42}$
$1, 2, 3, 4 \rightarrow 1, 4, 2, 3$	2	1	$a_{11}a_{24}a_{32}a_{43}$
$1, 2, 3, 4 \rightarrow 1, 4, 3, 2$	1	-1	$-a_{11}a_{24}a_{33}a_{42}$
$1, 2, 3, 4 \rightarrow 2, 1, 3, 4$	1	-1	$-a_{12}a_{21}a_{33}a_{44}$
$1, 2, 3, 4 \rightarrow 2, 1, 4, 3$	2	1	$a_{12}a_{21}a_{34}a_{43}$
$1, 2, 3, 4 \rightarrow 2, 3, 1, 4$	2	1	$a_{12}a_{23}a_{31}a_{44}$
$1, 2, 3, 4 \rightarrow 2, 3, 4, 1$	3	-1	$-a_{12}a_{23}a_{34}a_{41}$
$1, 2, 3, 4 \rightarrow 2, 4, 1, 3$	3	-1	$-a_{12}a_{24}a_{31}a_{43}$
$1, 2, 3, 4 \rightarrow 2, 4, 3, 1$	2	1	$a_{12}a_{24}a_{33}a_{41}$
$1, 2, 3, 4 \rightarrow 3, 1, 2, 4$	2	1	$a_{13}a_{21}a_{32}a_{44}$
$1, 2, 3, 4 \rightarrow 3, 1, 4, 2$	3	-1	$-a_{13}a_{21}a_{34}a_{42}$
$1, 2, 3, 4 \rightarrow 3, 2, 1, 4$	1	-1	$-a_{13}a_{22}a_{31}a_{44}$
$1, 2, 3, 4 \rightarrow 3, 2, 4, 1$	2	1	$a_{13}a_{22}a_{34}a_{41}$
$1, 2, 3, 4 \rightarrow 3, 4, 1, 2$	2	1	$a_{13}a_{24}a_{31}a_{42}$
$1, 2, 3, 4 \rightarrow 3, 4, 2, 1$	3	-1	$-a_{13}a_{24}a_{32}a_{41}$
$1, 2, 3, 4 \rightarrow 4, 1, 2, 3$	3	-1	$-a_{14}a_{21}a_{32}a_{43}$
$1, 2, 3, 4 \rightarrow 4, 1, 3, 2$	2	1	$a_{14}a_{21}a_{33}a_{42}$
$1, 2, 3, 4 \rightarrow 4, 2, 1, 3$	2	1	$a_{14}a_{22}a_{31}a_{43}$
$1, 2, 3, 4 \rightarrow 4, 2, 3, 1$	1	-1	$-a_{14}a_{22}a_{33}a_{41}$
$1, 2, 3, 4 \rightarrow 4, 3, 1, 2$	3	-1	$-a_{14}a_{23}a_{31}a_{42}$
$1, 2, 3, 4 \rightarrow 4, 3, 2, 1$	2	1	$a_{14}a_{23}a_{32}a_{41}$

1.3.2 Properties of Determinants

The formal definition of the determinant of a matrix is seen to be very cumbersome to use in practice; however, it is proven to be very handy in establishing various properties of the determinants, which is the subject of the current subsection. A convenient method of calculating determinants will be given at the end of this section.

Property 1.1 *Interchanging two rows in the determinant gives a factor of –1, e.g.*

$$\begin{vmatrix} 1 & 2 & -1 \\ 3 & 0 & 3 \\ 2 & 1 & 4 \end{vmatrix} = - \begin{vmatrix} 3 & 0 & 3 \\ 1 & 2 & -1 \\ 2 & 1 & 4 \end{vmatrix} .$$

Proof Consider a matrix $\mathbf{A}_{r \leftrightarrow t}$ obtained from $\mathbf{A}$ by interchanging the row r with the row t (assuming for certainty that $t > r$):

$$\mathbf{A} = \begin{pmatrix} a_{11} & \cdots & a_{1n} \\ \cdots & \cdots & \cdots \\ a_{r1} & \cdots & a_{rn} \\ \cdots & \cdots & \cdots \\ a_{t1} & \cdots & a_{tn} \\ \cdots & \cdots & \cdots \\ a_{n1} & \cdots & a_{nn} \end{pmatrix} \quad \text{and} \quad \mathbf{A}_{r \leftrightarrow t} = \begin{pmatrix} a_{11} & \cdots & a_{1n} \\ \cdots & \cdots & \cdots \\ a_{t1} & \cdots & a_{tn} \\ \cdots & \cdots & \cdots \\ a_{r1} & \cdots & a_{rn} \\ \cdots & \cdots & \cdots \\ a_{n1} & \cdots & a_{nn} \end{pmatrix} .$$

Then, we get

$$\det\mathbf{A} = \sum_P \epsilon_{j_1 j_2 \cdots j_n} a_{1 j_1} \cdots a_{r j_r} \cdots a_{t j_t} \cdots a_{n j_n} ,$$

$$\det\mathbf{A}_{r \leftrightarrow t} = \sum_P \epsilon_{j_1 j_2 \cdots j_n} a_{1 j_1} \cdots a_{t j_t} \cdots a_{r j_r} \cdots a_{n j_n} .$$

It is clear from here that if we make a single pair permutation $a_{t j_t} \leftrightarrow a_{r j_r}$ in every term of $\det \mathbf{A}_{r \leftrightarrow t}$, the expression will become exactly the same as for $\det \mathbf{A}$. However, a single permutation brings in an extra factor of -1 to each of the $n!$ terms of the sum. Thus, $\det \mathbf{A}_{r \leftrightarrow t} = -\det \mathbf{A}$, as required.

Property 1.2 *Interchanging two columns in the determinant gives a factor of -1, e.g.*

$$\begin{vmatrix} 1 & 2 & -1 \\ 3 & 0 & 3 \\ 2 & 1 & 4 \end{vmatrix} = - \begin{vmatrix} 1 & -1 & 2 \\ 3 & 3 & 0 \\ 2 & 4 & 1 \end{vmatrix} .$$

Proof Similar to the above; however, each term in the sum should be ordered with respect to the right rather than the left index of every element of the matrix in accordance with Eq. (1.52).

Property 1.3 *If a matrix has two identical rows (columns), its determinant is equal to zero, e.g.*

$$\begin{vmatrix} x & 1 & x \\ x^2 & -1 & x^2 \\ x^3 & 3 & x^3 \end{vmatrix} = 0 .$$

Proof Let the rows r and t be identical. If we interchange them, then, according to Property 1.1, $\det \mathbf{A}_{r \leftrightarrow t} = -\det \mathbf{A}$. However, because the two rows are identical, $\mathbf{A}_{r \leftrightarrow t}$ does not differ from $\mathbf{A}$, leading to $\det \mathbf{A} = -\det \mathbf{A}$, which means that $\det \mathbf{A} = 0$.

Property 1.4 *If every element of a single row (column) is multiplied by a factor c, the determinant is also multiplied by the same factor, e.g.*

$$\begin{vmatrix} a & 2 & -1 \\ 3a & 0 & 3 \\ 2a & 1 & 4 \end{vmatrix} = a \begin{vmatrix} 1 & 2 & -1 \\ 3 & 0 & 3 \\ 2 & 1 & 4 \end{vmatrix} .$$

Proof This property follows directly from Eq. (1.51) and the fact that in each product of elements there is only one element from that row (column).

Property 1.5 *If every element of a column is written as a sum (difference) of two terms, the determinant is equal to the sum (difference) of two determinants each containing one part of the column:*

$$\begin{vmatrix} \cdots & a_{1r} \pm b_{1r} & \cdots \\ \cdots & \cdots & \cdots \\ \cdots & a_{nr} \pm b_{nr} & \cdots \end{vmatrix} = \begin{vmatrix} \cdots & a_{1r} & \cdots \\ \cdots & \cdots & \cdots \\ \cdots & a_{nr} & \cdots \end{vmatrix} \pm \begin{vmatrix} \cdots & b_{1r} & \cdots \\ \cdots & \cdots & \cdots \\ \cdots & b_{nr} & \cdots \end{vmatrix} , \qquad (1.54)$$

e.g.

$$\begin{vmatrix} a & 2 & -1 \\ 3a & 0 & 3 \\ 2a & 1 & 4 \end{vmatrix} = \begin{vmatrix} a+x & 2 & -1 \\ 2a & 0 & 3 \\ 4a & 1 & 4 \end{vmatrix} + \begin{vmatrix} -x & 2 & -1 \\ a & 0 & 3 \\ -2a & 1 & 4 \end{vmatrix} .$$

Similarly if each element of a row is represented as a sum (difference) of two elements.

Proof Again, the proof follows directly from Eq. (1.51).

Property 1.6 *The value of the determinant is not changed if one of the rows is added (subtracted) to (from) another,*

$$\begin{vmatrix} \cdots & a_{1r} \pm a_{1t} & \cdots & a_{1t} & \cdots \\ \cdots & & \cdots & & \\ \cdots & a_{nr} \pm a_{nt} & \cdots & a_{nt} & \cdots \end{vmatrix} = \begin{vmatrix} \cdots & a_{1r} & \cdots & a_{1t} & \cdots \\ \cdots & & \cdots & & \\ \cdots & a_{nr} & \cdots & a_{nt} & \cdots \end{vmatrix} , \tag{1.55}$$

where, for certainty, we assumed that $t > r$. For instance,

$$\begin{vmatrix} 0 & x & x^2 \\ -1 & 0 & 1 \\ x^2 & x & 1 \end{vmatrix} = \begin{vmatrix} 1 & x & x^2 - 1 \\ -1 & 0 & 1 \\ x^2 & x & 1 \end{vmatrix} ,$$

where the first row in the determinant in the right-hand side is obtained by subtracting row 2 from row 1 in the determinant in the left-hand side. Similarly, for two columns.

Proof Follows from Properties 1.3 and 1.5.

Property 1.7 *The determinant is equal identically to zero if at least one row (column) is a linear combination of other rows (columns).*

Proof The idea of the proof can be clearly seen by considering a simple case when the first column is a linear combination of the 2nd and the 3rd columns:

$$\begin{vmatrix} \alpha a_{12} + \beta a_{13} & a_{12} & a_{13} & \cdots \\ \cdots & & \cdots & \\ \alpha a_{n2} + \beta a_{n3} & a_{n2} & a_{n3} & \cdots \end{vmatrix} = \begin{vmatrix} \alpha a_{12} & a_{12} & a_{13} & \cdots \\ \cdots & & \cdots & \\ \alpha a_{n2} & a_{n2} & a_{n3} & \cdots \end{vmatrix} + \begin{vmatrix} \beta a_{13} & a_{12} & a_{13} & \cdots \\ \cdots & & \cdots & \\ \beta a_{n3} & a_{n2} & a_{n3} & \cdots \end{vmatrix}$$

$$= \alpha \begin{vmatrix} a_{12} & a_{12} & a_{13} & \cdots \\ \cdots & & \cdots & \\ a_{n2} & a_{n2} & a_{n3} & \cdots \end{vmatrix} + \beta \begin{vmatrix} a_{13} & a_{12} & a_{13} & \cdots \\ \cdots & & \cdots & \\ a_{n3} & a_{n2} & a_{n3} & \cdots \end{vmatrix} ,$$

where we have used Properties 1.5 and 1.4. Finally, each of the determinants in the last line is equal to zero since it contains two identical columns (Property 1.3), as required. The proof in a general case is straightforward.

Property 1.8 *The determinants of the matrices* $\mathbf{A}$ *and* $\mathbf{A}^T$ *are equal.*

Proof The determinant of $\mathbf{A}^T = (\tilde{a}_{ij})$ with $\tilde{a}_{ij} = a_{ji}$ is given by

$$\det \left(\mathbf{A}^T \right) = \sum_P \epsilon_{j_1 j_2 \cdots j_n} \tilde{a}_{1j_1} \tilde{a}_{2j_2} \cdots \tilde{a}_{nj_n} = \sum_P \epsilon_{j_1 j_2 \cdots j_n} a_{j_1 1} a_{j_2 2} \cdots a_{j_n n} ,$$

which is an equivalent expression (1.52) for the determinant of $\mathbf{A}$.

As a simple illustration of the properties of determinants, we shall solve the system of two linear algebraic equations

$$\begin{cases} a_{11} x_1 + a_{12} x_2 = h_1 \\ a_{21} x_1 + a_{22} x_2 = h_2 \end{cases}$$

with respect to x_1 and x_2. To this end, consider the determinant of the coefficients in the left-hand side: $|\mathbf{A}| = \begin{vmatrix} a_{11} & a_{12} \\ a_{21} & a_{22} \end{vmatrix}$. Using Property 1.4, we write

$$x_1|\mathbf{A}| = \begin{vmatrix} a_{11}x_1 & a_{12} \\ a_{21}x_1 & a_{22} \end{vmatrix} .$$

Then, the determinant

$$\begin{vmatrix} a_{12}x_2 & a_{12} \\ a_{22}x_2 & a_{22} \end{vmatrix} = x_2 \begin{vmatrix} a_{12} & a_{12} \\ a_{22} & a_{22} \end{vmatrix} = 0$$

by virtue of Property 1.3. Thus, using Property 1.5, we can add two determinants together as follows (they both have the same second column):

$$x_1|\mathbf{A}| + 0 = \begin{vmatrix} a_{11}x_1 & a_{12} \\ a_{21}x_1 & a_{22} \end{vmatrix} + \begin{vmatrix} a_{12}x_2 & a_{12} \\ a_{22}x_2 & a_{22} \end{vmatrix} = \begin{vmatrix} a_{11}x_1 + a_{12}x_2 & a_{12} \\ a_{21}x_1 + a_{22}x_2 & a_{22} \end{vmatrix}$$

$$= \begin{vmatrix} h_1 & a_{12} \\ h_2 & a_{22} \end{vmatrix} ,$$

that gives the required solution for x_1 as the ratio of the determinant in the right-hand side and $|\mathbf{A}|$. Similarly, starting from

$$x_2|\mathbf{A}| = \begin{vmatrix} a_{11} & a_{12}x_2 \\ a_{21} & a_{22}x_2 \end{vmatrix}$$

and adding a zero determinant

$$\begin{vmatrix} a_{11} & a_{11}x_1 \\ a_{21} & a_{21}x_1 \end{vmatrix}$$

to it, we obtain

$$x_2|\mathbf{A}| = \begin{vmatrix} a_{11} & a_{11}x_1 + a_{12}x_2 \\ a_{21} & a_{21}x_1 + a_{22}x_2 \end{vmatrix} = \begin{vmatrix} a_{11} & h_1 \\ a_{21} & h_2 \end{vmatrix} ,$$

which gives x_2. The obtained solution is a particular case of Cramer's rule to be considered in more detail in Sect. 1.4.

Property 1.9 *The determinant of a product of two matrices is equal to the product of their determinants:*

$$\det (\mathbf{AB}) = \det \mathbf{A} \det \mathbf{B} . \tag{1.56}$$

Proof Let us first consider a two-dimensional case to see the main idea:

$$\det \mathbf{A} = \begin{vmatrix} a_{11} & a_{12} \\ a_{21} & a_{22} \end{vmatrix} = a_{11}a_{22} - a_{12}a_{21}$$

and

$$\det \mathbf{B} = \begin{vmatrix} b_{11} & b_{12} \\ b_{21} & b_{22} \end{vmatrix} = b_{11}b_{22} - b_{12}b_{21} \,,$$

and

$$\det (\mathbf{AB}) = \begin{vmatrix} a_{11}b_{11} + a_{12}b_{21} & a_{11}b_{12} + a_{12}b_{22} \\ a_{21}b_{11} + a_{22}b_{21} & a_{21}b_{12} + a_{22}b_{22} \end{vmatrix} .$$

We shall now use the properties of the determinants we discovered above. Using Properties 1.5 and 1.4 on the first column, we can split it:

$$\det (\mathbf{AB}) = \begin{vmatrix} a_{11}b_{11} & a_{11}b_{12} + a_{12}b_{22} \\ a_{21}b_{11} & a_{21}b_{12} + a_{22}b_{22} \end{vmatrix} + \begin{vmatrix} a_{12}b_{21} & a_{11}b_{12} + a_{12}b_{22} \\ a_{22}b_{21} & a_{21}b_{12} + a_{22}b_{22} \end{vmatrix}$$

$$= b_{11} \begin{vmatrix} a_{11} & a_{11}b_{12} + a_{12}b_{22} \\ a_{21} & a_{21}b_{12} + a_{22}b_{22} \end{vmatrix} + b_{21} \begin{vmatrix} a_{12} & a_{11}b_{12} + a_{12}b_{22} \\ a_{22} & a_{21}b_{12} + a_{22}b_{22} \end{vmatrix} .$$

Now we split the second column in the same way:

$$\det (\mathbf{AB}) = b_{11} \left(\begin{vmatrix} a_{11} & a_{11}b_{12} \\ a_{21} & a_{21}b_{12} \end{vmatrix} + \begin{vmatrix} a_{11} & a_{12}b_{22} \\ a_{21} & a_{22}b_{22} \end{vmatrix} \right)$$

$$+ b_{21} \left(\begin{vmatrix} a_{12} & a_{11}b_{12} \\ a_{22} & a_{21}b_{12} \end{vmatrix} + \begin{vmatrix} a_{12} & a_{12}b_{22} \\ a_{22} & a_{22}b_{22} \end{vmatrix} \right)$$

$$= b_{11} \left(b_{12} \underbrace{\begin{vmatrix} a_{11} & a_{11} \\ a_{21} & a_{21} \end{vmatrix}}_{=0} + b_{22} \begin{vmatrix} a_{11} & a_{12} \\ a_{21} & a_{22} \end{vmatrix} \right)$$

$$+ b_{21} \left(b_{12} \begin{vmatrix} a_{12} & a_{11} \\ a_{22} & a_{21} \end{vmatrix} + b_{22} \underbrace{\begin{vmatrix} a_{12} & a_{12} \\ a_{22} & a_{22} \end{vmatrix}}_{=0} \right) ,$$

where the selected two determinants are each equal to zero due to Property 1.3. Finally, collect the remaining terms:

$$\det (\mathbf{AB}) = b_{11}b_{22} \begin{vmatrix} a_{11} & a_{12} \\ a_{21} & a_{22} \end{vmatrix} + b_{21}b_{12} \begin{vmatrix} a_{12} & a_{11} \\ a_{22} & a_{21} \end{vmatrix}$$

$$= (b_{11}b_{22} - b_{21}b_{12}) \begin{vmatrix} a_{11} & a_{12} \\ a_{21} & a_{22} \end{vmatrix}$$

$$= \begin{vmatrix} b_{11} & b_{12} \\ b_{21} & b_{22} \end{vmatrix} \begin{vmatrix} a_{11} & a_{12} \\ a_{21} & a_{22} \end{vmatrix} ,$$

as required. Above we permuted the columns of the second determinant to have its elements in the correct order (as in the first one); this changed the sign before $b_{21}b_{12}$ to give exactly det $\mathbf{B}$.

Now we apply the same method to the general case of a $n \times n$ determinant. The general element of the matrix $\mathbf{C} = \mathbf{AB}$ is $c_{ij} = \sum_k a_{ik}b_{kj}$. Therefore, the determinant of $\mathbf{C}$ we would like to calculate is

$$
\det (\mathbf{AB}) = \begin{vmatrix} \sum_{k_1} a_{1k_1}b_{k_11} & \sum_{k_2} a_{1k_2}b_{k_22} & \cdots & \sum_{k_n} a_{1k_n}b_{k_nn} \\ \sum_{k_1} a_{2k_1}b_{k_11} & \sum_{k_2} a_{2k_2}b_{k_22} & \cdots & \sum_{k_n} a_{2k_n}b_{k_nn} \\ \cdots & \cdots & \cdots & \cdots \\ \sum_{k_1} a_{nk_1}b_{k_11} & \sum_{k_2} a_{nk_2}b_{k_22} & \cdots & \sum_{k_n} a_{nk_n}b_{k_nn} \end{vmatrix}.
$$

Notice that we used the same summation (dump) indices in each column as we shall immediately find it convenient. We start by splitting the first column term by term, then the second one, the third, and so on:

$$
\det (\mathbf{AB}) = \sum_{k_1} \begin{vmatrix} a_{1k_1}b_{k_11} & \sum_{k_2} a_{1k_2}b_{k_22} & \cdots & \sum_{k_n} a_{1k_n}b_{k_nn} \\ a_{2k_1}b_{k_11} & \sum_{k_2} a_{2k_2}b_{k_22} & \cdots & \sum_{k_n} a_{2k_n}b_{k_nn} \\ \cdots & \cdots & \cdots & \cdots \\ a_{nk_1}b_{k_11} & \sum_{k_2} a_{nk_2}b_{k_22} & \cdots & \sum_{k_n} a_{nk_n}b_{k_nn} \end{vmatrix}
$$

$$
= \sum_{k_1}\sum_{k_2} \begin{vmatrix} a_{1k_1}b_{k_11} & a_{1k_2}b_{k_22} & \cdots & \sum_{k_n} a_{1k_n}b_{k_nn} \\ a_{2k_1}b_{k_11} & a_{2k_2}b_{k_22} & \cdots & \sum_{k_n} a_{2k_n}b_{k_nn} \\ \cdots & \cdots & \cdots & \cdots \\ a_{nk_1}b_{k_11} & a_{nk_2}b_{k_22} & \cdots & \sum_{k_n} a_{nk_n}b_{k_nn} \end{vmatrix} = \cdots
$$

$$
= \sum_{k_1}\sum_{k_2}\cdots\sum_{k_n} \begin{vmatrix} a_{1k_1}b_{k_11} & a_{1k_2}b_{k_22} & \cdots & a_{1k_n}b_{k_nn} \\ a_{2k_1}b_{k_11} & a_{2k_2}b_{k_22} & \cdots & a_{2k_n}b_{k_nn} \\ \cdots & \cdots & \cdots & \cdots \\ a_{nk_1}b_{k_11} & a_{nk_2}b_{k_22} & \cdots & a_{nk_n}b_{k_nn} \end{vmatrix}
$$

$$
= \sum_{k_1}\sum_{k_2}\cdots\sum_{k_n} b_{k_11}b_{k_22}\cdots b_{k_nn} \begin{vmatrix} a_{1k_1} & a_{1k_2} & \cdots & a_{1k_n} \\ a_{2k_1} & a_{2k_2} & \cdots & a_{2k_n} \\ \cdots & \cdots & \cdots & \cdots \\ a_{nk_1} & a_{nk_2} & \cdots & a_{nk_n} \end{vmatrix}.
$$

Note that when splitting the columns, we took the summation signs out, keeping of course the same summation index in each term of the same column. The determinant in the last line of the expression above in the right-hand side contains only elements of the matrix $\mathbf{A}$; it would only be non-zero if all indices k_1, k_2, etc. are *different*. In other words, the n summations over the indices k_1, k_2, etc. in fact can be replaced with a single sum taken over all permutations P of the indices $(k_1, k_2, \ldots, k_n)$ running between 1 and n:

$$
\det (\mathbf{AB}) = \sum_P b_{k_11}b_{k_22}\cdots b_{k_nn} \begin{vmatrix} a_{1k_1} & a_{1k_2} & \cdots & a_{1k_n} \\ a_{2k_1} & a_{2k_2} & \cdots & a_{2k_n} \\ \cdots & \cdots & \cdots & \cdots \\ a_{nk_1} & a_{nk_2} & \cdots & a_{nk_n} \end{vmatrix}.
$$

Now, looking at the determinant: if the indices k_1, k_2, etc. were ordered correctly in the ascending order from 1 to n, then the determinant above would be equal exactly to $\det \mathbf{A}$. To put them in the correct order for an arbitrary arrangement of the indices k_1, k_2, etc., a permutation is required resulting in the sign $\epsilon_P = \epsilon_{k_1 k_2 \cdots k_n} = \pm 1$. Therefore,

$$\det(\mathbf{AB}) = \sum_P \epsilon_{k_1 k_2 \cdots k_n} b_{k_1 1} b_{k_2 2} \cdots b_{k_n n} \det \mathbf{A}$$

$$= \left(\sum_P \epsilon_{k_1 k_2 \cdots k_n} b_{k_1 1} b_{k_2 2} \cdots b_{k_n n} \right) \det \mathbf{A}$$

$$= \det \mathbf{B} \det \mathbf{A} \, ,$$

as required.

Problem 1.50 Prove that

$$\det \mathbf{A}^{-1} = \frac{1}{\det \mathbf{A}} \, . \tag{1.57}$$

[Hint: use $\mathbf{AA}^{-1} = \mathbf{E}$.]

Problem 1.51 Prove that the determinant of an orthogonal matrix $\mathbf{A}$ is equal to ± 1. [Hint: use $\mathbf{AA}^T = \mathbf{E}$.]

Problem 1.52 Prove by a direct calculation that $\det \mathbf{C}_\sigma = 1$, where $\mathbf{C}_\sigma$ is the reflection matrix (1.48). Is this result to be expected?

Problem 1.53 Prove that the absolute value of the determinant of a unitary matrix $\mathbf{A}$ is equal to 1. [Hint: use $\mathbf{AA}^\dagger = \mathbf{E}$.]

1.3.3 Relation to a Linear Antisymmetric Function

We shall show here that the determinant is the only linear antisymmetric function one can built, i.e. this specific construct is not accidental.

Let us define a linear antisymmetric function $f(\mathbf{a}_1, \ldots, \mathbf{a}_n)$ of n vectors $\mathbf{a}_i$ ($i = 1, \ldots, n$) as follows: (i) it is linear with respect to any of its arguments, i.e.

$$f(\ldots, \alpha \mathbf{a}_k + \beta \mathbf{a}'_k, \ldots) = \alpha \, f(\ldots, \mathbf{a}_k, \ldots) + \beta \, f(\ldots, \mathbf{a}'_k, \ldots) \, ,$$

where α and β are arbitrary (complex) constants, and (ii) it changes sign if any two of its arguments are permuted:

$$f(\ldots, \mathbf{a}_k, \ldots, \mathbf{a}_l, \ldots) = -f(\ldots, \mathbf{a}_l, \ldots, \mathbf{a}_k, \ldots).$$

Obviously, it follows from the last property that if any two arguments of the function are the same, the function is equal to zero:

$$f(\ldots, \mathbf{a}, \ldots, \mathbf{a}, \ldots) = -f(\ldots, \mathbf{a}, \ldots, \mathbf{a}, \ldots) \implies f(\ldots, \mathbf{a}, \ldots, \mathbf{a}, \ldots) = 0.$$

Theorem 1.7 *There exists only one linear antisymmetric function of* n *variables-vectors, satisfying* $f(\mathbf{e}_1, \ldots, \mathbf{e}_n) = 1$, *which is the determinant*

$$f(\mathbf{a}_1, \ldots, \mathbf{a}_n) = \sum_P \epsilon_{j_1 j_2 \ldots j_n} a_{1 j_1} a_{2 j_2} \cdots a_{n j_n}. \tag{1.58}$$

Here $\mathbf{e}_k$ *are unit base vectors defining the elements* a_{ij} *entering the function (and forming a matrix* $\mathbf{A}$) *in Eq. (1.58) via* $\mathbf{a}_i = \sum_{k=1}^{n} a_{ij} \mathbf{e}_j$.

Proof We first assume that there exists a function satisfying both conditions (linearity and antisymmetry). Let us obtain a formula for it. We have

$$f(\mathbf{a}_1, \ldots, \mathbf{a}_n) = f\left(\sum_{j_1=1}^{n} a_{1 j_1} \mathbf{e}_{j_1}, \sum_{j_2=1}^{n} a_{2 j_2} \mathbf{e}_{j_2}, \ldots, \sum_{j_n=1}^{n} a_{n j_n} \mathbf{e}_{j_n} \right)$$

$$= \sum_{j_1=1}^{n} \sum_{j_2=1}^{n} \cdots \sum_{j_n=1}^{n} a_{1 j_1} a_{2 j_2} \cdots a_{n j_n} f\left(\mathbf{e}_{j_1}, \mathbf{e}_{j_2}, \ldots, \mathbf{e}_{j_n} \right),$$

where we have made use of the linearity of this function. The expression $f\left(\mathbf{e}_{j_1}, \mathbf{e}_{j_2}, \ldots, \mathbf{e}_{j_n} \right)$ is zero if any of the vectors-arguments coincide, i.e. the indices $j_1, j_2, \ldots, j_n$ must all be different. Hence, the sums over these indices must correspond only to all possible $n!$ permutations of the numbers from 1 to n. It is easy to see that, given different indices $j_1, j_2, \ldots, j_n$ the function $f\left(\mathbf{e}_{j_1}, \mathbf{e}_{j_2}, \ldots, \mathbf{e}_{j_n} \right) = \pm 1$ depending on the particular permutation P (recall that for the ordered sequence $f(\mathbf{e}_1, \ldots, \mathbf{e}_n) = 1$). Hence, $f\left(\mathbf{e}_{j_1}, \mathbf{e}_{j_2}, \ldots, \mathbf{e}_{j_n} \right) = \epsilon_{j_1 j_2 \ldots j_n} = (-1)^p$, where p is the parity of the permutation, as required.

Conversely, assume that the function is given by Eq. (1.58), and let us check that it satisfies the two properties we require. The linearity follows immediately:

$$f(\mathbf{a}_1, \ldots, \alpha\mathbf{a}_k + \beta\mathbf{b}_k, \ldots, \mathbf{a}_n) = \sum_P \epsilon_{j_1 j_2 \ldots j_n} a_{1j_1} \ldots \left(\alpha a_{kj_k} + \beta b_{kj_k}\right) \ldots a_{nj_n}$$

$$= \alpha \sum_P \epsilon_{j_1 j_2 \ldots j_n} a_{1j_1} \ldots a_{kj_k} \ldots a_{nj_n}$$

$$+ \beta \sum_P \epsilon_{j_1 j_2 \ldots j_n} a_{1j_1} \ldots b_{kj_k} \ldots a_{nj_n}$$

$$= \alpha\, f(\mathbf{a}_1, \ldots, \mathbf{a}_k, \ldots, \mathbf{a}_n) + \beta\, f(\mathbf{a}_1, \ldots, \mathbf{b}_k, \ldots, \mathbf{a}_n)\,.$$

Let us now check that the function is antisymmetric. Swapping two its arguments, say, $\mathbf{a}_i$ and $\mathbf{a}_j$, would require one transposition to be performed to restore the order in each term, which would give a minus sign. **Q.E.D.**

1.3.4 *Practical Method of Calculating Determinants. Minors*

There is a very simple way of calculating determinants. The idea is to establish a recurrence relation in which a $n \times n$ determinant is expressed via determinants of a smaller dimension. Applying this formula a required number of times, the determinant of the given order can be finally written via determinants of low orders (e.g. 2 or 3), explicit expressions for which were given above.

To this end, let us re-examine expression (1.50) for the 3×3 determinant. By grouping terms by the first row elements of the matrix, we get

$$\begin{vmatrix} b_{11} & b_{12} & b_{13} \\ b_{21} & b_{22} & b_{23} \\ b_{31} & b_{32} & b_{33} \end{vmatrix} = b_{11}\,[b_{22}b_{33} - b_{23}b_{32}] - b_{12}\,[b_{21}b_{33} - b_{23}b_{31}]$$

$$+ b_{13}\,[b_{21}b_{32} - b_{22}b_{31}] \tag{1.59}$$

$$= b_{11} \begin{vmatrix} b_{22} & b_{23} \\ b_{32} & b_{33} \end{vmatrix} - b_{12} \begin{vmatrix} b_{21} & b_{23} \\ b_{31} & b_{33} \end{vmatrix} + b_{13} \begin{vmatrix} b_{21} & b_{22} \\ b_{31} & b_{32} \end{vmatrix}\,.$$

It can be seen that the 3×3 determinant is expressed via a sum of three 2×2 determinants with the pre-factors which are elements of the first row. Each of the 2×2 determinants is obtained by removing one row and one column corresponding to the first and the second indices of the pre-factor element. For instance, the determinant $\begin{vmatrix} b_{22} & b_{23} \\ b_{32} & b_{33} \end{vmatrix}$ is combined with the pre-factor b_{11} and can be obtained by removing the first row and the first column from the original 3×3 determinant, while the determinant $\begin{vmatrix} b_{21} & b_{23} \\ b_{31} & b_{33} \end{vmatrix}$ is obtained by removing the first row and the second column as these are the indices of its own pre-factor b_{12}. In other words, each element of the first row is multiplied by the 2×2 determinant which is obtained by removing the row and the column which cross at that particular element.

Table 1.3 Grouping permutations of numbers $(1, 2, 3)$ into three groups corresponding to a different fixed first number

Label	Sequence after permutation	Number of permutations	Sign/parity
a1	123	0	+1
a2	132	1	−1
b1	213	$0 + 1 = 1$	−1
b2	231	$1 + 1 = 2$	+1
c1	312	$1 + 1 = 2$	+1
c2	321	$2 + 1 = 3$	−1

Note that each 2×2 determinant in Eq. (1.59) is attached either a plus or minus sign. In fact, the signs alternate if the elements b_{1k} standing as pre-factors to the 2×2 determinants are ordered as shown with respect to $k = 1, 2, 3$. This fact can be illustrated by considering all six permutations of the integers $1, 2$ and 3. We can split these six permutations into three groups (the number of groups is equal to the number of elements in our number set) by selecting every one of the integers to stand at the first (left) position as shown in the second column of Table 1.3. All six permutations are now grouped, and in each group only the second and the third elements are allowed to be permuted, the first one is fixed. As we go from the first group (sequences a1, a2) to the second (b1, b2), only one more additional permutation is required giving an extra minus sign to the parity; going from the second to the third (c1, c2), a single additional permutation is added again bringing in another minus sign. Therefore, the element b_{11} in Eq. (1.59) corresponding to the first group (sequences a1, a2) has a plus sign, b_{12} acquired a minus, while b_{13} acquires another minus giving it plus in the end. In other words, the signs alternate starting from the plus for the very first term.

It appears that this method is very general and can be applied to a determinant of arbitrary order. To formulate the method, we introduce a new quantity. Consider a determinant $|\mathbf{A}|$ of order n. If we remove the ith row and the jth column from it (the row and the column cross at element a_{ij}), a determinant of order $n - 1$ is obtained. It is called a *minor* of a_{ij} and denoted M_{ij}. For example, the minor of the element $a_{34} = 8$ (bold underlined) of the determinant

$$
\begin{vmatrix} 1 & 2 & 3 & 4 \\ -4 & -3 & -2 & -1 \\ 0 & 2 & 4 & \mathbf{\underline{8}} \\ 8 & 4 & 2 & 0 \end{vmatrix} \quad \text{is} \quad M_{34} = \begin{vmatrix} 1 & 2 & 3 & | \\ -4 & -3 & -2 & | \\ - & - & - & - \\ 8 & 4 & 2 & | \end{vmatrix} = \begin{vmatrix} 1 & 2 & 3 \\ -4 & -3 & -2 \\ 8 & 4 & 2 \end{vmatrix}.
$$

A minor M_{ij} can be attached a sign $(-1)^{i+j}$ in which case it is called a *co-factor* of a_{ij} and denoted $A_{ij} = (-1)^{i+j} M_{ij}$. In the example above the co-factor of a_{34} is $(-1)^{3+4} M_{34} = -M_{34}$. The co-factor signs can be easily obtained by constructing a chess-board of plus and minus signs starting from the plus at the position 11:

$$
\begin{vmatrix}
+ & - & + & - & + \\
- & + & - & + & - \\
+ & - & + & - & + \\
- & + & - & + & - \\
+ & - & + & - & + \\
& & & & & \ddots \\
& & & & & & \ddots
\end{vmatrix}
$$

Thus, the correct sign $(-1)^{i+j}$ for the co-factor can be located on this chess-board at the same i, j position as the element a_{ij} itself in the original determinant.

Now we are ready to formulate the general result:

Theorem 1.8 *The determinant $\det \mathbf{A} = |\mathbf{A}|$ of a $n \times n$ matrix $\mathbf{A}$ can be expanded along its first row as follows:*

$$
|\mathbf{A}| = a_{11}A_{11} + a_{12}A_{12} + \cdots + a_{1n}A_{1n} = \sum_{k=1}^{n} a_{1k}A_{1k} , \qquad (1.60)
$$

where $A_{ij} = (-1)^{i+j} M_{ij}$ is the co-factor of the element a_{ij}.

Proof Consider all possible permutations P of the sequence $(1, 2, 3, \ldots, n)$ of integer numbers. There are $n!$ of such permutations. We shall split them into n groups by selecting the first element of the permuted set to be fixed, i.e. the first group has all permutations which start from 1, the second group starts from 2, and so on. Consider one such a group in which the first number is j_1. In this group we have all terms like $a_{1j_1}a_{2j_2}\cdots a_{nj_n}$ in the expansion of det $\mathbf{A}$, in which all possible values of the indices $j_2, \ldots, j_n$ are allowed (except for j_1). The parity $\epsilon_{j_1 j_2 \cdots j_n}$ of any member of this group of terms can be expressed as $(-1)^{j_1+1} \epsilon_{j_2 j_3 \cdots j_n}$ via the parity $\epsilon_{j_2 j_3 \cdots j_n}$ of all its elements but the first one. Indeed, if $j_1 = 1$, then the two parties are the same, for $j_1 = 2$ an extra minus sign is added required to bring this element with $j_1 = 2$ into the first position, for $j_1 = 3$ one more minus sign is added as two permutations are required, and so on. Therefore, we can present the sum over all permutations P of the original set via the sum over the groups and the permutations within each group of right indices (i.e. within the $(j_2, j_3, \ldots, j_n)$ set):

$$\det A = \sum_{P=\{j_1, j_2, \ldots, j_n\}} \epsilon_{j_1 j_2 \cdots j_n} a_{1 j_1} a_{2 j_2} \cdots a_{n j_n}$$

$$= \sum_{j_1=1}^{n} a_{1 j_1} (-1)^{j_1+1} \left[\sum_{P=\{j_2, \ldots, j_n\}} \epsilon_{j_2 \cdots j_n} a_{2 j_2} \cdots a_{n j_n} \right]$$

$$= \sum_{j_1=1}^{n} a_{1 j_1} (-1)^{j_1+1} M_{1 j_1} = \sum_{j_1=1}^{n} a_{1 j_1} A_{1 j_1} ,$$

as required since the expression in the square brackets is nothing but the determinant obtained by removing all elements of the first row and of the j_1-th column, i.e. it is the minor $M_{1 j_1}$. **Q.E.D.**

Problem 1.54 Prove that a similar formula can be written by expanding along any row or column.

Example 1.13 ▶ Solve the following equation with respect to x:

$$\begin{vmatrix} x & 1 & 1 & 1 \\ 1 & x & 0 & 0 \\ 1 & 0 & x & 0 \\ 1 & 0 & 0 & x \end{vmatrix} = 0 .$$

Solution: Expand the determinant with respect to its first row:

$$\begin{vmatrix} x & 1 & 1 & 1 \\ 1 & x & 0 & 0 \\ 1 & 0 & x & 0 \\ 1 & 0 & 0 & x \end{vmatrix} = x \begin{vmatrix} x & 0 & 0 \\ 0 & x & 0 \\ 0 & 0 & x \end{vmatrix} - 1 \cdot \begin{vmatrix} 1 & 0 & 0 \\ 1 & x & 0 \\ 1 & 0 & x \end{vmatrix} + 1 \cdot \begin{vmatrix} 1 & x & 0 \\ 1 & 0 & 0 \\ 1 & 0 & x \end{vmatrix}$$

$$- 1 \cdot \begin{vmatrix} 1 & x & 0 \\ 1 & 0 & x \\ 1 & 0 & 0 \end{vmatrix} .$$

It is convenient to open the 3×3 determinants along the rows or columns that contain more zeros:

$$\begin{vmatrix} x & 0 & 0 \\ 0 & x & 0 \\ 0 & 0 & x \end{vmatrix} = x^3 , \quad \begin{vmatrix} 1 & 0 & 0 \\ 1 & x & 0 \\ 1 & 0 & x \end{vmatrix} = x^2 , \quad \begin{vmatrix} 1 & x & 0 \\ 1 & 0 & 0 \\ 1 & 0 & x \end{vmatrix} = -x^2$$

and

$$\begin{vmatrix} 1 & x & 0 \\ 1 & 0 & x \\ 1 & 0 & 0 \end{vmatrix} = x^2 .$$

Therefore, we obtain

$$\begin{vmatrix} x & 1 & 1 & 1 \\ 1 & x & 0 & 0 \\ 1 & 0 & x & 0 \\ 1 & 0 & 0 & x \end{vmatrix} = x^4 - x^2 - x^2 - x^2 = x^2\left(x^2 - 3\right) = 0,$$

which has the following solutions: $x = 0, \pm\sqrt{3}.$ ◄

Problem 1.55 Show that the solutions of the following equation with respect to x,

$$\begin{vmatrix} x & 1 & 0 & 1 \\ 1 & x & 1 & 0 \\ 0 & 1 & x & 1 \\ 1 & 0 & 1 & x \end{vmatrix} = 0,$$

are $x = 0, \pm 2$.

Problem 1.56 Consider the following matrix:

$$\mathbf{A} = \begin{pmatrix} 2 & 0 & 1 \\ -1 & 1 & 3 \\ 0 & -2 & 1 \end{pmatrix}.$$

Show that all its minors M_{ij} and co-factors A_{ij}, if combined into 3×3 matrices, are

$$(M_{ij}) = \begin{pmatrix} 7 & -1 & 2 \\ 2 & 2 & -4 \\ -1 & 7 & 2 \end{pmatrix} \quad \text{and} \quad (A_{ij}) = \begin{pmatrix} 7 & 1 & 2 \\ -2 & 2 & 4 \\ -1 & -7 & 2 \end{pmatrix}.$$

Then verify that if all the co-factors are combined into a matrix $\mathbf{A}_{cof} = (A_{ij})$, then the matrix $\mathbf{A}_{cof}^T/|\mathbf{A}|$ gives the inverse $\mathbf{A}^{-1}$ of $\mathbf{A}$.

Problem 1.57 Show that the co-factors A_{ij} of the matrix

$$\mathbf{A} = \begin{pmatrix} 1 & 0 & 2 \\ 2 & 1 & 1 \\ -1 & 0 & 1 \end{pmatrix} \quad \text{are} \quad \mathbf{A}_{cof} = (A_{ij}) = \begin{pmatrix} 1 & -3 & 1 \\ 0 & 3 & 0 \\ -2 & 3 & 1 \end{pmatrix}.$$

Finally, verify that the inverse of $\mathbf{A}$ can be calculated via $\mathbf{A}_{cof}^T/|\mathbf{A}|$.

Problem 1.58 Show that the determinant of a right triangular matrix (its elements on the left of the diagonal are all zeros) is equal to the product of its diagonal elements:

$$\begin{vmatrix} a_{11} & a_{12} & \cdots & a_{1n} \\ 0 & a_{22} & \cdots & a_{2n} \\ \cdots & \cdots & \cdots & \cdots \\ 0 & 0 & \cdots & a_{nn} \end{vmatrix} = a_{11}a_{22}\cdots a_{nn} = \prod_{i=1}^{n} a_{ii}. \tag{1.61}$$

The same results is valid for the left triangular matrix as well.

1.3.5 Determinant of a Tridiagonal Matrix

As an interesting example of a determinant calculation, let us consider a matrix $\mathbf{A} = (a_{ij})$ of a special structure in which only elements along the diagonal ($i = j$) and next to it are non-zero, all other elements are equal to zero: $a_{ij} \neq 0$ with $j = i$ and $j = i \pm 1$. In other words, in the case of this so-called *tridiagonal matrix* $a_{ij} = 0$ as long as $|i - j| > 1$. The determinant we are about to calculate is shown in the left-hand side of the equation pictured in Fig. 1.6.

Let $\mathbf{A}_k$ be a matrix obtained from $\mathbf{A}$ by removing its first $(k-1)$ rows and columns, i.e. in $\mathbf{A}_k$ the diagonal elements are α_k, α_{k+1}, etc., α_n. In particular, $\mathbf{A} = \mathbf{A}_1$. Then, opening the determinant of $\mathbf{A}$ along the first row, we have $|\mathbf{A}_1| = \alpha_1 |\mathbf{A}_2| + \beta_1 \mathbf{R}_{12}$, where $\mathbf{R}_{12}$ is the corresponding co-factor to the $a_{12} = \beta_1$ element of $\mathbf{A}$, see the second term in the right-hand side in Fig. 1.6. Opening now $\mathbf{R}_{12}$ along its first column, we get $\mathbf{R}_{12} = -\beta_1 |\mathbf{A}_3|$, yielding:

$$|\mathbf{A}_1| = \alpha_1 |\mathbf{A}_2| - \beta_1^2 |\mathbf{A}_3| \implies \frac{|\mathbf{A}_1|}{|\mathbf{A}_2|} = \alpha_1 - \frac{\beta_1^2}{|\mathbf{A}_2|/|\mathbf{A}_3|}.$$

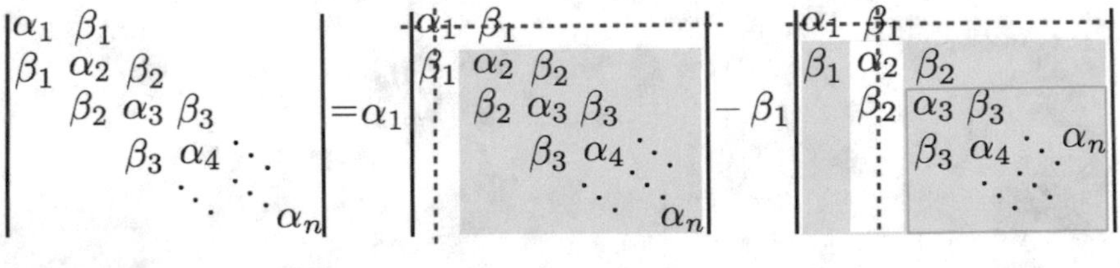

Fig. 1.6 For the calculation of the determinant of a tridiagonal matrix: opening along the first (upper) row. At the next step, the second determinant in the right-hand side is opened along its first column leading to a product of β_1 with the minor enclosed in the green box

Let $\lambda_k = |\mathbf{A}_k| / |\mathbf{A}_{k+1}|$ be the ratio of two consecutive determinants containing α_k and α_{k+1} as their first diagonal elements, respectively. Then the above expression can be written simply as $\lambda_1 = \alpha_1 - \beta_1^2/\lambda_2$. If we started from the matrix $\mathbf{A}_2$, a similar calculation would result in $\lambda_2 = \alpha_2 - \beta_2^2/\lambda_3$. It is obvious now that generally we would have

$$\lambda_k = \alpha_k - \beta_k^2/\lambda_{k+1} \tag{1.62}$$

for any $k = 1, 2, \ldots, n-2$. In particular, at the last step we get $\lambda_{n-2} = \alpha_{n-2} - \beta_{n-2}^2/\lambda_{n-1}$. However, λ_{n-1} is calculated explicitly:

$$\lambda_{n-1} = \frac{|\mathbf{A}_{n-1}|}{|\mathbf{A}_n|} = \frac{\begin{vmatrix} \alpha_{n-1} & \beta_{n-1} \\ \beta_{n-1} & \alpha_n \end{vmatrix}}{|\alpha_n|} = \frac{\alpha_{n-1}\alpha_n - \beta_{n-1}^2}{\alpha_n},$$

so that, going backwards in Eq. (1.62) for $k = n-2$, it is possible to calculate λ_{n-2}; once λ_{n-2} is known, then λ_{n-3} is within reach, and so on, until at the final step λ_1 is calculated. Formally λ_1 can be written as a so-called *continued fraction*:

$$\lambda_1 = \alpha_1 - \frac{\beta_1^2}{\lambda_2} = \alpha_1 - \frac{\beta_1^2}{\alpha_2 - \frac{\beta_2^2}{\lambda_3}} = \alpha_1 - \frac{\beta_1^2}{\alpha_2 - \frac{\beta_2^2}{\alpha_3 - \frac{\beta_3^2}{\alpha_4 - \cdots}}}.$$

This fraction has a finite number of terms (as the matrix $\mathbf{A}$ is of a finite dimension) and can be denoted in several different ways, e.g.

$$\lambda_1 = \alpha_1 - \frac{\beta_1^2|}{|\alpha_2} - \frac{\beta_2^2|}{|\alpha_3} - \frac{\beta_3^2|}{|\alpha_4} - \cdots - \frac{\beta_{n-1}^2|}{\alpha_n}$$

or

$$\lambda_1 = \alpha_1 - \frac{\beta_1^2}{\alpha_2-} \frac{\beta_2^2}{\alpha_3-} \frac{\beta_3^2}{\alpha_4-} \cdots \frac{\beta_{n-1}^2}{\alpha_n}.$$

1.4 A Linear System of Algebraic Equations

Consider a system of n linear algebraic equations with respect to the *same number* of unknown numbers x_1, x_2, etc., x_n:

$$\begin{cases} a_{11}x_1 + a_{12}x_2 + \cdots + a_{1n}x_n = b_1 \\ a_{21}x_1 + a_{22}x_2 + \cdots + a_{2n}x_n = b_2 \\ \quad \cdots \quad \cdots \\ a_{n1}x_1 + a_{n2}x_2 + \cdots + a_{nn}x_n = b_n \end{cases} \tag{1.63}$$

It can be rewritten in a compact form using the matrix notations. Let us collect all the unknown quantities $x_1, \ldots, x_n$ into a vector-column $\mathbf{X} = (x_i)$, the coefficients a_{ij} into a square matrix $\mathbf{A} = (a_{ij})$ and, finally, the quantities in the right-hand side $b_1, \ldots, b_n$ into a vector-column $\mathbf{B} = (b_i)$. Then, instead of Eq. (1.63) we write simply

$$\mathbf{AX} = \mathbf{B} . \tag{1.64}$$

If we multiply both sides of this equation *from the left* by $\mathbf{A}^{-1}$, we obtain in the left-hand side $\mathbf{A}^{-1}\mathbf{AX} = \mathbf{EX} = \mathbf{X}$, and hence we get a formal solution:

$$\mathbf{X} = \mathbf{A}^{-1}\mathbf{B} . \tag{1.65}$$

Thus, the solution of the system of linear equations (1.64) is expressed via the inverse of the matrix $\mathbf{A}$ of the coefficients. Although this solution is in many cases useful, especially for analytical work, it does not give a simple practical way of calculating $\mathbf{X}$ since it is not always convenient to find the inverse of a matrix, especially if the dimension n of the problem is large.

1.4.1 Cramer's Method

Instead, we shall employ a different method due to Cramer. Consider

$$x_1|\mathbf{A}| = \begin{vmatrix} a_{11}x_1 & \cdots & a_{1n} \\ \cdots & \cdots & \cdots \\ a_{n1}x_1 & \cdots & a_{nn} \end{vmatrix} .$$

Now we add to this determinant a zero determinant,

$$0 = x_2 \begin{vmatrix} a_{12} & a_{12} & \cdots & a_{1n} \\ \cdots & \cdots & \cdots & \cdots \\ a_{n2} & a_{n2} & \cdots & a_{nn} \end{vmatrix} = \begin{vmatrix} a_{12}x_2 & a_{12} & \cdots & a_{1n} \\ \cdots & \cdots & \cdots & \cdots \\ a_{n2}x_2 & a_{n2} & \cdots & a_{nn} \end{vmatrix} ,$$

in which the first column is proportional to the second one. Now sum up both expressions above, which yields

$$x_1|\mathbf{A}| = \begin{vmatrix} a_{11}x_1 + a_{12}x_2 & \cdots & a_{1n} \\ \cdots & \cdots & \cdots \\ a_{n1}x_1 + a_{n2}x_2 & \cdots & a_{nn} \end{vmatrix} .$$

Next, we consider the zero determinant with the first column equal to the third one times x_3, and then add this to the determinant above; we obtain in this way:

$$x_1|\mathbf{A}| = \begin{vmatrix} a_{11}x_1 + a_{12}x_2 + a_{13}x_3 & \cdots & a_{1n} \\ \cdots & & \cdots \cdots \\ a_{n1}x_1 + a_{n2}x_2 + a_{n3}x_3 & \cdots & a_{nn} \end{vmatrix}.$$

This process is repeated until the last column is added:

$$x_1|\mathbf{A}| = \begin{vmatrix} a_{11}x_1 + a_{12}x_2 + a_{13}x_3 + \cdots + a_{1n}x_n & \cdots & a_{1n} \\ \cdots & & \cdots \cdots \\ a_{n1}x_1 + a_{n2}x_2 + a_{n3}x_3 + \cdots + a_{nn}x_n & \cdots & a_{nn} \end{vmatrix}.$$

The elements along the 1st column can now be recognised to be elements of the vector $\mathbf{B}$ in Eq. (1.63), i.e. we can write

$$\begin{aligned} x_1|\mathbf{A}| &= \begin{vmatrix} a_{11}x_1 + a_{12}x_2 + a_{13}x_3 + \cdots + a_{1n}x_n & \cdots & a_{1n} \\ \cdots & & \cdots \cdots \\ a_{n1}x_1 + a_{n2}x_2 + a_{n3}x_3 + \cdots + a_{nn}x_n & \cdots & a_{nn} \end{vmatrix} \\ &= \begin{vmatrix} b_1 & a_{12} & \cdots & a_{1n} \\ \cdots & \cdots & \cdots & \cdots \\ b_n & a_{n2} & \cdots & a_{nn} \end{vmatrix}, \end{aligned} \qquad (1.66)$$

that gives the required closed solution for x_1 as a ratio of two determinants.

Multiplying $|\mathbf{A}|$ by x_2 and inserting it into the 2nd column and repeating the above procedure, we obtain

$$x_2|\mathbf{A}| = \begin{vmatrix} a_{11} & b_1 & \cdots & a_{1n} \\ \cdots & \cdots & \cdots & \cdots \\ a_{n1} & b_n & \cdots & a_{nn} \end{vmatrix}, \qquad (1.67)$$

which enables one to express x_2 as a ratio of two determinants. Generally, in order to find x_i, one has to replace the ith column in the determinant of the matrix $\mathbf{A}$ by vector $\mathbf{B}$ and divide it by $|\mathbf{A}|$. This method is called *Cramer's rule:*

$$x_i = \frac{\begin{vmatrix} a_{11} & \cdots & a_{1,i-1} & b_1 & a_{1,i+1} & \cdots & a_{1n} \\ \cdots & \cdots & \cdots & \cdots & \cdots & \cdots & \cdots \\ a_{n1} & \cdots & a_{n,i-1} & b_n & a_{n,i+1} & \cdots & a_{nn} \end{vmatrix}}{\begin{vmatrix} a_{11} & \cdots & a_{n1} \\ \cdots & \cdots & \cdots \\ a_{n1} & \cdots & a_{nn} \end{vmatrix}}. \qquad (1.68)$$

Example 1.14 ▶ Use Cramer's rules to solve the system of equations

$$\begin{cases} x + y + z = 2 \\ 2x - y - z = 1 \\ x + 2y - z = -3 \end{cases}.$$

Solution: The matrix **A** and vector **B** here are given by

$$\mathbf{A} = \begin{pmatrix} 1 & 1 & 1 \\ 2 & -1 & -1 \\ 1 & 2 & -1 \end{pmatrix} \quad \text{and} \quad \mathbf{B} = \begin{pmatrix} 2 \\ 1 \\ -3 \end{pmatrix}.$$

Calculating the determinant of **A**, we get $|\mathbf{A}| = 9$. Therefore, using Cramer's rule, we have

$$x = \frac{\begin{vmatrix} 2 & 1 & 1 \\ 1 & -1 & -1 \\ -3 & 2 & -1 \end{vmatrix}}{9} = \frac{9}{9} = 1, \quad y = \frac{\begin{vmatrix} 1 & 2 & 1 \\ 2 & 1 & -1 \\ 1 & -3 & -1 \end{vmatrix}}{9} = \frac{-9}{9} = -1$$

$$\text{and} \quad z = \frac{\begin{vmatrix} 1 & 1 & 2 \\ 2 & -1 & 1 \\ 1 & 2 & -3 \end{vmatrix}}{9} = \frac{18}{9} = 2. \quad \blacktriangleleft$$

Problem 1.59 Solve the system of equations:

$$(a) \begin{cases} x + y - z = 2 \\ 2x - y + 3z = 5 \\ 3x + 2y - 2z = 5 \end{cases} ; \quad (b) \begin{cases} x + 2y + 3z = -5 \\ -x - 3y + z = -14 \\ 2x + y + z = 1 \end{cases},$$

using Cramer's rules. [Answer: (a) $x = 1$, $y = 3$, $z = 2$; (b) $x = 1$, $y = 3$, $z = -4$.]

Problem 1.60 In special relativity, coordinates of particles and time associated with physical events depend on the particular coordinate system (frame) chosen. Consider a (x', t') frame moving with velocity v along the positive x direction of a fixed (laboratory) frame (x, t). By virtue of the *Lorenz transformation*,

$$x' = \gamma(x - vt) \quad \text{and} \quad t' = \gamma(t - xv/c^2),$$

where $\gamma = (1 - v^2/c^2)^{-1/2}$ and c is the speed of light, one can calculate (x', t') in the moving frame from (x, t) in the fixed one. Using Cramer's rules, show that the inverse transformation $(x', t') \rightarrow (x, t)$ is

$$x = \gamma(x' + vt') \quad \text{and} \quad t = \gamma(t' + x'v/c^2).$$

Formula (1.68) allows us to draw one important conclusion that is proven to be extremely useful in practice. Consider an algebraic system of Eqs. (1.63) or (1.64) with the *zero right-hand side*:

$$\mathbf{AX} = \mathbf{0} \,, \qquad\qquad (1.69)$$

where $\mathbf{0}$ is the vector-column consisting of zeros. According to the general result (1.68), if $|\mathbf{A}| \neq 0$ and all $b_k = 0$, then all $x_i = 0$ as well. Note that the determinant in the numerator of Eq. (1.68) is zero since the whole ith column contains zeros. This follows from Property 1.4 of the determinants (Sect. 1.3.2): if we choose a common factor $c = 0$ to all elements of the ith column, then the whole determinant is multiplied by it and thus in our case the determinant is indeed zero.

Thus, if $|\mathbf{A}| \neq 0$ the system of Eq. (1.69) has only a single (trivial or zero) solution, i.e. the solution is unique. The situation is different, however, if $|\mathbf{A}| = 0$ (the matrix $\mathbf{A}$ is *singular*). As we know from Property 1.7 of determinants (Sect. 1.3.2), this means that some of the rows or columns are linearly dependent of each other. Thus, if $|\mathbf{A}| = 0$, our previous reasoning becomes false (as we cannot divide by zero in Eq. (1.68)) and hence a non-trivial solution of Eq. (1.69) may exist.

In fact, the case of $|\mathbf{A}| = 0$ deserves special consideration. We shall show in what follows that the number of solutions of Eq. (1.64) in the case of $|\mathbf{A}| = 0$ is either zero (no solutions, i.e. the equations in the system (1.64) are inconsistent) or infinite (an infinite number of vectors $\mathbf{X}$ satisfying the equation $\mathbf{AX} = \mathbf{B}$). We shall prove this by means of the following two theorems.

Theorem 1.9 *The linear system of algebraic equations* $\mathbf{AX} = \mathbf{0}$ *has an infinite number of solutions if* $|\mathbf{A}| = 0$.

Proof Let us first consider the case when there is at least one minor of the matrix A that is not equal to zero. Without loss of generality,[4] we can assume that this minor is $\mathbf{A}_{nn}$, it is obtained by crossing out along the element a_{nn}. Consider then our system of equations without the last one in and moving the terms containing x_n to the right-hand side:

$$\begin{cases} a_{11}x_1 + a_{12}x_2 + \cdots + a_{1,n-1}x_{n-1} = -a_{1n}x_n \\ a_{21}x_1 + a_{22}x_2 + \cdots + a_{2,n-1}x_{n-1} = -a_{2n}x_n \\ \qquad\qquad \cdots \qquad = \cdots \\ a_{n-1,1}x_1 + a_{n-1,2}x_2 + \cdots + a_{n-1,n-1}x_{n-1} = -a_{n-1,n}x_n \end{cases} \qquad (1.70)$$

It is a system of linear algebraic equations with respect to the variables $x_1, \ldots, x_{n-1}$ whose determinant is $|\mathbf{A}_{nn}| \neq 0$. Hence, it has a unique solution

---

[4] After an appropriate reordering of the equations and/or renumbering of the variables.

$$x_{j=}\frac{\left|\widetilde{\mathbf{A}}_j\right|}{\left|\mathbf{A}_{nn}\right|}\,,\quad j=1,\ldots,n-1\,,$$

where $\widetilde{\mathbf{A}}_j$ is the $(n-1)\times(n-1)$ determinant obtained by replacing the jth column of the matrix $\mathbf{A}_{nn}$ with the right-hand side column of Eq. (1.70):

$$\left|\widetilde{\mathbf{A}}_j\right| = \begin{vmatrix} a_{11} & \cdots & -a_{1n}x_n & \cdots & a_{1,n-1} \\ \cdots & \cdots & \cdots & \cdots & \cdots \\ a_{n-1,1} & \cdots & \underbrace{-a_{n-1,n}x_n}_{j-\text{th column}} & \cdots & a_{n-1,n-1} \end{vmatrix}$$

$$= -x_n \begin{vmatrix} a_{11} & \cdots & a_{1n} & \cdots & a_{1,n-1} \\ \cdots & \cdots & \cdots & \cdots & \cdots \\ a_{n-1,1} & \cdots & a_{n-1,n} & \cdots & a_{n-1,n-1} \end{vmatrix}\,.$$

Then we shall move the jth column to the right placing it at the end after the $(n-1)$th column; this requires $n-j-1$ permutations, and hence a factor of $(-1)^{n-j-1}$ will appear

$$\left|\widetilde{\mathbf{A}}_j\right| = x_n(-1)^{n-j} \begin{vmatrix} a_{11} & \cdots & & \cdots & a_{1,n-1} & a_{1n} \\ \cdots & \cdots & & \cdots & \cdots & \cdots \\ a_{n-1,1} & \cdots & & \cdots & a_{n-1,n-1} & a_{n-1,n} \end{vmatrix}\,.$$

$$\underset{j-\text{th column}}{}$$

The obtained determinant is in fact the minor $\left|\mathbf{A}_{nj}\right|$ because it is obtained by crossing at the a_{nj} element in $\mathbf{A}$. Hence, $\left|\widetilde{\mathbf{A}}_j\right| = x_n(-1)^{n-j}\left|\mathbf{A}_{nj}\right|$. This enables us to obtain for the solution:

$$x_j = \frac{x_n}{\left|\mathbf{A}_{nn}\right|}(-1)^{n-j}\left|\mathbf{A}_{nj}\right|\,,\quad j=1,\ldots,n-1\,. \tag{1.71}$$

Note that formally the above formula is valid for $j=n$ as well. It is easy to see then that these values also satisfy the last (the nth) equation in the original system (1.70):

$$\sum_{j=1}^{n} a_{nj}x_j = \sum_{j=1}^{n} a_{nj}\frac{x_n}{\left|\mathbf{A}_{nn}\right|}(-1)^{n-j}\left|\mathbf{A}_{nj}\right|$$

$$= \frac{x_n}{\left|\mathbf{A}_{nn}\right|}\left[\sum_{j=1}^{n} a_{nj}(-1)^{n-j}\left|\mathbf{A}_{nj}\right|\right]\,.$$

The sum in the square brackets represents an expansion of the determinant $|\mathbf{A}|$ over the last nth row; hence it is equal to $|\mathbf{A}|$ that is zero via our initial assumption. This means that, indeed, the n-the equation is satisfied, and, therefore, the values of x_j for all $j=1,\ldots,n$ given by Eq. (1.71) are solutions of Eq. (1.70). Since x_n cannot be determined, it is arbitrary, leading to an infinite number of solutions. Hence, if at least one minor of the matrix $\mathbf{A}$ is not equal to zero, there will be an infinite number of solutions of the system $\mathbf{AX} = \mathbf{0}$. Note that an obvious zero solution $\mathbf{X} = \mathbf{0}$ corresponds to selecting $x_n = 0$.

Now, we should consider the case when all minors of $\mathbf{A}$ are equal to zero. It is easy to see that this is only possible if all equations are linearly dependent, i.e. there is only one independent equation to exist, others are obtained by multiplying it by different numbers. Obviously, if just one equation is left, we have an infinite number of solutions again. **Q.E.D.**

A by-product of this theorem is a statement that all solutions of the system $\mathbf{AX} = \mathbf{0}$ in the case of $|\mathbf{A}| = 0$ can be written as $x_j = c_j t$, where c_j are constants and t is an arbitrary parameter.

Theorem 1.10 *The linear system of equations* $\mathbf{AX} = \mathbf{B}$ *given by Eq. (1.63) in the case of* $|\mathbf{A}| = 0$ *has either none (the equations are inconsistent) or an infinite number of solutions.*

Proof Let $|\mathbf{A}_j|$ be the $n \times n$ determinant obtained by replacing the j-the column in $|\mathbf{A}|$ with the vector $\mathbf{B}$ of the right-hand side; it is the numerator in Cramer's solution in Eq. (1.68). Consider first the case in which at least one such determinant $|\mathbf{A}_j|$ is non-zero. The derivation performed when deriving Eq. (1.68) led us to the equation $x_j |\mathbf{A}| = |\mathbf{A}_j|$. Since $|\mathbf{A}| = 0$, then this equation is inconsistent with the assumption that $|\mathbf{A}_j| \neq 0$. Hence, in this case the system of Eq. (1.63) does not have solutions, it is inconsistent.

Consider now the case of all such determinants being zero, i.e. $|\mathbf{A}_j| = 0$ for all $j = 1, \ldots, n$. Let us assume that there is a solution $\mathbf{X}_0$, i.e. $\mathbf{AX}_0 = \mathbf{B}$. Subtracting this equation from $\mathbf{AX} = \mathbf{B}$, we arrive at $\mathbf{A}(\mathbf{X} - \mathbf{X}_0) = \mathbf{0}$, for which we know that in the case of $|\mathbf{A}| = 0$ we have an infinite number of solutions. Hence, the vector $\mathbf{X} - \mathbf{X}_0$ may take an infinite number of values, and so is the vector $\mathbf{X}$. **Q.E.D.**

To illustrate the proven statements, let us consider the following three examples:

$$(a) \begin{cases} 2x_1 + x_2 + x_3 = 2 \\ 3x_1 + 3x_2 + 3x_3 = 1 \, , \\ 4x_1 + 6x_2 + 6x_3 = 4 \end{cases}$$

$$(b) \begin{cases} x_1 + 5x_2 + 5x_3 = 1 \\ 2x_1 + 5x_2 + 5x_3 = 2 \, , \\ 3x_1 + 6x_2 + 6x_3 = 3 \end{cases}$$

$$(c) \quad \begin{cases} x_1 + x_2 + x_3 = 1 \\ 3x_1 + 3x_2 + 3x_3 = 3 \\ 5x_1 + 5x_2 + 5x_3 = 4 \end{cases}.$$

In the first case (a) the determinant $|\mathbf{A}|$ of the 3×3 matrix $\mathbf{A}$ of the coefficients is zero (the second and third columns of $\mathbf{A}$ are identical); at the same time, $|\mathbf{A}_2| = -16$ and $|\mathbf{A}_3| = 16$ are non-zero. This means that the system (a) does not have solutions. In the case (b) $|\mathbf{A}| = 0$, but, as can easily be seen, $\left|\mathbf{A}_j\right| = 0$ as well for all $j = 1, 2, 3$. Hence, this system of equations has an infinite number of solutions. In the third case (c) we have $|\mathbf{A}| = |\mathbf{A}_1| = |\mathbf{A}_2| = |\mathbf{A}_3| = 0$; however, the equations are incompatible and hence there is no solution (if the first two equations are dependent, the third one contradicts them because we have 4, not 5, in the right-hand side).

Problem 1.61 Solve the system (b) to show that its solutions are given by $x_1 = 1$, $x_2 = -t$ and $x_3 = t$, where t is an arbitrary parameter.

We shall come across the situation of the zero determinant when solving the system of linear equations in Sect. 1.8.

In the meantime, we illustrate the use of this statement by solving a problem.

Problem 1.62 Determine the values of x for which the system of equations with respect to c_1, c_2, c_3, c_4 has a non-trivial solution:

$$\begin{cases} xc_1 + c_2 + c_3 + c_4 = 0 \\ c_1 + xc_2 + c_4 = 0 \\ c_1 + xc_3 + c_4 = 0 \\ c_1 + c_2 + c_3 + xc_4 = 0 \end{cases}.$$

[Answer: $x = 0, 1, (-1 \pm \sqrt{17})/2$.]

Finally, we shall consider two theorems that are related to the vector spaces introduced in Sect. 1.1.

Theorem 1.11 *The matrix* $\mathbf{A} = \left(a_{ij}\right)$ *relating two different sets of basic vectors* $\{\mathbf{e}_i\}$ *and* $\{\mathbf{e}_i'\}$ *of the same n-dimensional vector space is always non-singular.*

Proof Indeed, the two basis sets are related via $\mathbf{e}_j' = \sum_i a_{ji}\mathbf{e}_i$. Since each set forms a basis of the space, its vectors are linearly independent, i.e. the vector equation

$$\sum_{j=1}^{n} \lambda_j \mathbf{e}'_j = 0$$

must have a unique zero solution for the vector $\boldsymbol{\lambda} = (\lambda_j)$. On the other hand, replacing the vectors $\mathbf{e}'_i$ in this equation and rearranging, we arrive at the condition

$$\sum_{i=1}^{n} \underbrace{\left(\sum_{j=1}^{n} a_{ji} \lambda_j \right)}_{\lambda'_i} \mathbf{e}_i = \sum_{i=1}^{n} \lambda'_j \mathbf{e}_i = 0 .$$

Since vectors $\mathbf{e}_i$ are also linearly independent, the expansion coefficients λ'_j (that form a vector column $\boldsymbol{\lambda}' = \mathbf{A}^T \boldsymbol{\lambda}$) must also be all equal to zero. In other words, the equation $\mathbf{A}^T \boldsymbol{\lambda} = 0$ must have a unique zero solution $\boldsymbol{\lambda} = 0$. We know, however, that even though the zero solution always exists for the linear homogenous problem $\mathbf{A}^T \boldsymbol{\lambda} = 0$, it is unique if and only if $|\mathbf{A}^T| = |\mathbf{A}| \neq 0$, i.e. the matrix $\mathbf{A}$ is not singular. **Q.E.D.**

Theorem 1.12 *If vectors $\mathbf{d}_1, \mathbf{d}_2, \ldots, \mathbf{d}_p$ are orthogonal, then the determinant of the matrix $\mathbf{D} = \begin{pmatrix} \mathbf{d}_1 & \mathbf{d}_2 & \ldots & \mathbf{d}_p \end{pmatrix}$ composed of them as columns are not equal to zero.*

Proof Indeed, consider the determinant of the matrix $\mathbf{D}$ formed by the coordinates of vectors $\mathbf{d}_1, \mathbf{d}_2, \ldots, \mathbf{d}_p$;

$$|\mathbf{D}| = \begin{vmatrix} d_{11} & \cdots & d_{1p} \\ \cdots & \cdots & \cdots \\ d_{p1} & \cdots & d_{pp} \end{vmatrix} . \tag{1.72}$$

Note that the right index in any element d_{ij} of the determinant indicates the vector number, while the left—its component. Let us first assume that the vectors $\mathbf{d}_1, \mathbf{d}_2, \ldots, \mathbf{d}_p$ are orthonormal, i.e. not only orthogonal, but have the unit length. Then, we can write

$$(\mathbf{d}_i, \mathbf{d}_j) = \delta_{ij} \quad \text{or} \quad \sum_{k=1}^{p} d^*_{ki} d_{kj} = \delta_{ij} . \tag{1.73}$$

If we introduce the Hermitian conjugate matrix $\mathbf{D}^\dagger = (\overline{d}_{ij})$ with $\overline{d}_{ij} = d^*_{ji}$, then it is seen that Eq. (1.73) can be rewritten simply as

$$\sum_{k=1}^{p} \overline{d}_{ik} d_{kj} = \delta_{ij} \quad \text{or} \quad \mathbf{D}^\dagger \mathbf{D} = \mathbf{E} ,$$

i.e. $\mathbf{D}$ is a unitary matrix whose determinant (see Problem 1.53) $|\mathbf{D}| = \pm 1$.

Returning now to our original vectors whose lengths are not necessarily equal to one, we can always write them as $\mathbf{d}_i = \lambda_i \tilde{\mathbf{d}}_i$, where $\lambda_i \neq 0$ is a scaling factor and the vector $\tilde{\mathbf{d}}_i$ is of unit length. Since the scaling factors multiply equally all components of the vector in the determinant $|\mathbf{D}|$, we can write $|\mathbf{D}| = \lambda_1 \lambda_2 \ldots \lambda_p |\tilde{\mathbf{D}}|$, where $\tilde{\mathbf{D}}$ is made of normalised vectors as its columns. The determinant of the latter matrix is ± 1, which makes the determinant of $\mathbf{D}$ *not* equal to zero. **Q.E.D.**

Theorem 1.13 *If vectors* $\mathbf{d}_1, \mathbf{d}_2, \ldots, \mathbf{d}_p$ *are orthonormal, then they are linearly independent.*

Proof To prove that the vectors are linearly independent, we have to demonstrate that the system of equations

$$c_1 \mathbf{d}_1 + c_2 \mathbf{d}_2 + \cdots + c_p \mathbf{d}_p = \mathbf{0}$$

with respect to the coefficients $c_1, c_2, \ldots, c_p$ has only the trivial (zero) solution. Write these equations in components of the vectors $\mathbf{d}_k = (d_{ik})$ using the right index in d_{ik} to indicate the vector number k and the left one for the component:

$$\begin{cases} d_{11}c_1 + d_{12}c_2 + \cdots + d_{1p}c_p = 0 \\ \qquad\qquad \cdots \\ d_{p1}c_1 + d_{p2}c_2 + \cdots + d_{pp}c_p = 0 \end{cases} . \tag{1.74}$$

This is a set of p linear algebraic equations with respect to the unknown coefficients $c_1, c_2, \ldots, c_p$ with the zero right-hand side. It has a trivial solution if the determinant of a matrix formed by the coordinates of vectors $\mathbf{d}_1, \mathbf{d}_2, \ldots, \mathbf{d}_p$ is not equal to zero. According to Theorem 1.12 this is indeed the case. This means that the only solution of Eq. (1.74) is the trivial solution which, in turn, means that the set of vectors $\mathbf{d}_1, \mathbf{d}_2, \ldots, \mathbf{d}_p$ is indeed linearly independent. **Q.E.D.**

Note that requirement of the theorem that the vectors are normalised to unity is not essential and was only assumed for convenience. It is sufficient to have vectors orthogonal to guarantee their linear independence.

The theorem just proven is important as it says that if we have a set of p orthogonal vectors, they can form a basis of a p-D vector space, and any vector from this space can be expanded in terms of them. Correspondingly, that means that no more than p linearly independent vectors can be constructed for the $p - D$ space: any additional vector within the same space will necessarily be linearly dependent of them (Theorem 1.2).

1.4.2 Gaussian Elimination

Above, we have considered solving a system of n linear algebraic equation with respect to the same number of unknown variables x_i. Here we shall discuss a general problem of solving a system of linear algebraic equation when the number of equation may be either equal, smaller or bigger than the number of the unknowns.

The method we shall consider is due to Gauss. Even though this method is not the most efficient from the practical point of view, it has a number of important features that deserve its careful consideration: (i) in this method any system of equations can be solved by means of a set of very simple transformations on the vectors of the coefficients of the equations and their right-hand sides; (ii) one can deal with any number of equations in a rather formal and unified way and conclude if the equations either do not have, have a unique or many solutions; (iii) the inverse matrix can be calculated using an appropriate extension of the method (Sect. 1.7), (iv) it is possible to analyse rectangular matrices and (v) a set of linear differential equations can also be considered with ease (Sect. 1.19). Importantly, the very general cases of the number of equations (either algebraic or differential) being different to the number of unknowns (correspondingly, variables or functions) can be studied.

We shall start our discussion from an example of the following system of linear equations:

$$\begin{cases} 2x_1 + 2x_2 - x_3 + 2x_4 = -6 \\ x_1 - 2x_4 = 5 \\ x_2 + 2x_3 - x_4 = 5 \\ x_1 + 4x_3 + x_4 = 7 \end{cases}$$

or

$$\underbrace{\begin{pmatrix} 2 & 2 & -1 & 2 \\ 1 & 0 & 0 & -2 \\ 0 & 1 & 2 & -1 \\ 1 & 0 & 4 & 1 \end{pmatrix}}_{A} \underbrace{\begin{pmatrix} x_1 \\ x_2 \\ x_3 \\ x_4 \end{pmatrix}}_{X} = \underbrace{\begin{pmatrix} -6 \\ 5 \\ 5 \\ 7 \end{pmatrix}}_{B},$$

which we have also rewritten in the matrix form. Let us now make an observation. We can always permute equations i and j, an equivalent system of equations is obtained with the same solutions; we shall denote this operation $R_i \longleftrightarrow R_j$, where R_i stands for the 'row i'; in will be clear in a moment why we use these notations. We can also multiply equation i by a non-zero number c, to be denoted $cR_i \longleftrightarrow R_i$; again, this will give a system with an identical solution. Finally, it is also possible to make a linear combination of two equations and then replace one of them with the resulting one. Let us illustrate this operation: let us multiply the second equation by -2 to obtain $-2x_1 + 4x_4 = -10$, and then add up to the first equation to get $2x_2 - x_3 + 6x_4 = -16$, and finally replace the second equation with this one; we then obtain an equivalent system of equations

$$\begin{cases} 2x_1 + 2x_2 - x_3 + 2x_4 = -6 \\ 2x_2 - x_3 + 6x_4 = -16 \\ x_2 + 2x_3 - x_4 = 5 \\ x_1 + 4x_3 + x_4 = 7 \end{cases}$$

or

$$\begin{pmatrix} 2 & 2 & -1 & 2 \\ 0 & 2 & -1 & 6 \\ 0 & 1 & 2 & -1 \\ 1 & 0 & 4 & 1 \end{pmatrix} \begin{pmatrix} x_1 \\ x_2 \\ x_3 \\ x_4 \end{pmatrix} = \begin{pmatrix} -6 \\ -16 \\ 5 \\ 7 \end{pmatrix}.$$

If equation j is multiplied by a number c, the result is added to equation i and then equation is j is replaced with the result, this operation will be denoted $R_i + cR_j \rightarrow R_j$.

Clearly, every time we perform the elementary operations mentioned above, we only change the elements of the matrix $\mathbf{A}$ and components of the vector-column $\mathbf{B}$ in the right-hand side; therefore, there is no need to write all the time the full set of equations; it is sufficient to write only the elements of $\mathbf{A}$ and $\mathbf{B}$. It is customary to combine the two into the so-called augmented matrix that is obtained by attaching the vector-column $\mathbf{B}$ on the right of the matrix $\mathbf{A}$. In our case the augmented matrix reads

$$\begin{pmatrix} 2 & 2 & -1 & 2 & -6 \\ 1 & 0 & 0 & -2 & 5 \\ 0 & 1 & 2 & -1 & 5 \\ 1 & 0 & 4 & 1 & 7 \end{pmatrix}. \tag{1.75}$$

It is a rectangular matrix with 4 rows and 5 columns. Hence, the three elementary operations introduced above are actually executed on the rows of the augmented matrix, hence the letter R when denoting a particular operation.

The idea of the Gauss elimination is to apply a set of linear combinations of the rows of the augmented matrix (or a permutation) such that the system of equations is brought into the so-called *row echelon form* that enables one to easily solve it. Let us show in this example how this works. The first step is to eliminate the elements in the first column for the rows 2 to 4. To do this, we perform the following operations:

$$\begin{pmatrix} 2 & 2 & -1 & 2 & -6 \\ 1 & 0 & 0 & -2 & 5 \\ 0 & 1 & 2 & -1 & 5 \\ 1 & 0 & 4 & 1 & 7 \end{pmatrix} \rightarrow \boxed{R_1 - 2R_2 \rightarrow R_2} \rightarrow \begin{pmatrix} 2 & 2 & -1 & 2 & -6 \\ 0 & 2 & -1 & 6 & -16 \\ 0 & 1 & 2 & -1 & 5 \\ 1 & 0 & 4 & 1 & 7 \end{pmatrix}$$

$$\boxed{R_1 - 2R_4 \rightarrow R_4} \rightarrow \begin{pmatrix} 2 & 2 & -1 & 2 & -6 \\ 0 & 2 & -1 & 6 & -16 \\ 0 & 1 & 2 & -1 & 5 \\ 0 & 2 & -9 & 0 & -20 \end{pmatrix}$$

$$\boxed{R_2 - 2R_3 \rightarrow R_3} \rightarrow \begin{pmatrix} 2 & 2 & -1 & 2 & -6 \\ 0 & 2 & -1 & 6 & -16 \\ 0 & 0 & -5 & 8 & -26 \\ 0 & 2 & -9 & 0 & -20 \end{pmatrix}$$

$$\boxed{R_2 - R_4 \to R_4} \to \begin{pmatrix} 2 & 2 & -1 & 2 & -6 \\ 0 & 2 & -1 & 6 & -16 \\ 0 & 0 & -5 & 8 & -26 \\ 0 & 0 & 8 & 6 & 4 \end{pmatrix}$$

$$\boxed{R_3 + \frac{5}{8}R_4 \to R_4} \to \begin{pmatrix} 2 & 2 & -1 & 2 & -6 \\ 0 & 2 & -1 & 6 & -16 \\ 0 & 0 & -5 & 8 & -26 \\ 0 & 0 & 0 & 47/4 & -47/2 \end{pmatrix}.$$

The obtained triangular form of the augmented matrix, in which the lower part contains only zeros, is the required row echelon form. It allows for a very easy calculation of the unknowns. Indeed, in each successive row we have one variable less, so that the last equation is left with only one (last) variable. Translating this result into the actual form of the equations, we get

$$\begin{cases} 2x_1 + 2x_2 - x_3 + 2x_4 = -6 \\ 2x_2 - x_3 + 6x_4 = -16 \\ -5x_3 + 8x_4 = -26 \\ \dfrac{47}{4}x_4 = -\dfrac{47}{2} \end{cases} .$$

We see that, because of the triangular form of the matrix of the coefficients we arrived at, from the last equation $x_4 = -2$ is immediately obtained. Substituting this result into the third equation, we are able to determine $x_3 = 2$. Repeating this process, we successively obtain $x_2 = -1$ and then, at the last step, $x_1 = 1$. The process is called *back substitution*.

It is possible, however, to continue this process to reveal the unknowns. For this to happen we need to make sure that the matrix of the coefficients becomes the identity matrix. First of all, we multiply the last equation by $4/47$ to obtain 1 at the fourth position:

$$\boxed{\frac{4}{47}R_4 \to R_4} \to \begin{pmatrix} 2 & 2 & -1 & 2 & -6 \\ 0 & 2 & -1 & 6 & -16 \\ 0 & 0 & -5 & 8 & -26 \\ 0 & 0 & 0 & 1 & -2 \end{pmatrix}.$$

Then, we perform the appropriate linear combinations of rows 1 to 3 so that these rows have zeros in their fourth position:

$$\boxed{R_4 - \frac{1}{8}R_3 \to R_3} \to \begin{pmatrix} 2 & 2 & -1 & 2 & -6 \\ 0 & 2 & -1 & 6 & -16 \\ 0 & 0 & 5/8 & 0 & 5/4 \\ 0 & 0 & 0 & 1 & -2 \end{pmatrix},$$

$$
\boxed{R_4 - \frac{1}{6}R_2 \to R_2} \to
\begin{pmatrix}
2 & 2 & -1 & 2 & -6 \\
0 & -1/3 & 1/6 & 0 & 2/3 \\
0 & 0 & 5/8 & 0 & 5/4 \\
0 & 0 & 0 & 1 & -2
\end{pmatrix},
$$

$$
\boxed{R_4 - \frac{1}{2}R_1 \to R_1} \to
\begin{pmatrix}
-1 & -1 & 1/2 & 0 & 1 \\
0 & -1/3 & 1/6 & 0 & 2/3 \\
0 & 0 & 5/8 & 0 & 5/4 \\
0 & 0 & 0 & 1 & -2
\end{pmatrix}.
$$

Finally, we shall rescale the third row to have 1 in the third position:

$$
\boxed{\frac{8}{5}R_3 \to R_3} \to
\begin{pmatrix}
-1 & -1 & 1/2 & 0 & 1 \\
0 & -1/3 & 1/6 & 0 & 2/3 \\
0 & 0 & 1 & 0 & 2 \\
0 & 0 & 0 & 1 & -2
\end{pmatrix}.
$$

Next, we shall make sure that at the third position we have zero in the first and second rows by combining them with the third row:

$$
\boxed{R_3 - 2R_1 \to R_1} \to
\begin{pmatrix}
2 & 2 & 0 & 0 & 0 \\
0 & -1/3 & 1/6 & 0 & 2/3 \\
0 & 0 & 1 & 0 & 2 \\
0 & 0 & 0 & 1 & -2
\end{pmatrix},
$$

$$
\boxed{R_3 - 6R_2 \to R_2} \to
\begin{pmatrix}
2 & 2 & 0 & 0 & 0 \\
0 & 2 & 0 & 0 & -2 \\
0 & 0 & 1 & 0 & 2 \\
0 & 0 & 0 & 1 & -2
\end{pmatrix},
$$

and, finally, we remove the 2 at the second position in the first row by combining it with the second row:

$$
\boxed{R_2 - R_1 \to R_1} \to
\begin{pmatrix}
-2 & 0 & 0 & 0 & -2 \\
0 & 2 & 0 & 0 & -2 \\
0 & 0 & 1 & 0 & 2 \\
0 & 0 & 0 & 1 & -2
\end{pmatrix}.
$$

What is left to do is to rescale the first two rows,

$$
\boxed{\begin{array}{c} -\dfrac{1}{2}R_1 \to R_1 \\[2mm] \dfrac{1}{2}R_2 \to R_2 \end{array}} \to
\begin{pmatrix}
1 & 0 & 0 & 0 & 1 \\
0 & 1 & 0 & 0 & -1 \\
0 & 0 & 1 & 0 & 2 \\
0 & 0 & 0 & 1 & -2
\end{pmatrix},
$$

leading to the values of the unknowns in the last column, the same as before. The obtained form in which there is an identity matrix to the left of the last column is called the *reduced echelon form*.

So far, we have considered an example of a system of equations that have a unique solution. Let us now consider other types of systems of equations that may have more than one solution:

$$\begin{cases} 2x_1 + 2x_2 - x_3 + 2x_4 = -6 \\ x_1 - 2x_4 = 5 \\ x_2 + 2x_3 - x_4 = 5 \end{cases}$$

or

$$\underbrace{\begin{pmatrix} 2 & 2 & -1 & 2 \\ 1 & 0 & 0 & -2 \\ 0 & 1 & 2 & -1 \end{pmatrix}}_{A} \underbrace{\begin{pmatrix} x_1 \\ x_2 \\ x_3 \\ x_4 \end{pmatrix}}_{X} = \underbrace{\begin{pmatrix} -6 \\ 5 \\ 5 \end{pmatrix}}_{B}.$$

This is the same system as before from which the last equation has been removed. Hence, we have three equations and four unknowns to determine.

As we saw above, the elementary transformations $R_1 - 2R_2 \rightarrow R_2$ and $R_2 - 2R_3 \rightarrow R_3$ bring the augmented matrix into the following form:

$$\begin{pmatrix} 2 & 2 & -1 & 2 & -6 \\ 0 & 2 & -1 & 6 & -16 \\ 0 & 0 & -5 & 8 & -26 \end{pmatrix}.$$

This is the required echelon (triangular) form; any additional transformations would not produce any additional zeros in the lower triangle. Hence, we have arrived at the system of equations

$$\begin{cases} 2x_1 + 2x_2 - x_3 + 2x_4 = -6 \\ 2x_2 - x_3 + 6x_4 = -16 \\ -5x_3 + 8x_4 = -26 \end{cases}.$$

It still contains four variables. What we can do is to take the terms with, say, x_4, into the right-hand side and then consider it as a new vector **B**:

$$\begin{cases} 2x_1 + 2x_2 - x_3 = -6 - 2x_4 \\ 2x_2 - x_3 = -16 - 6x_4 \\ -5x_3 = -26 - 8x_4 \end{cases}.$$

Now we have three equations for three variables x_1, x_2, x_3, which is already in the required echelon form. It is clear, however, that the variable x_4 cannot be determined at all, it is left as an arbitrary parameter. Replacing $x_4 \rightarrow t$ and solving the obtained system of three equations successively for x_3, then x_2 and finally for x_1, we shall obtain its solutions.

Problem 1.63 Show that the solution of this system of equations is

$$x_3 = \frac{2}{5}(13 + 4t), \quad x_2 = -\frac{1}{5}(27 + 11t), \quad x_1 = 5 + 2t.$$

So, as we can see, in this case of the system being incomplete, one variable (any) has to be considered as arbitrary and treated as a parameter, and other variables are obtained as a linear function of this parameter; we have an infinite number of solutions.

Next, we shall consider our original example again in which we only modify the 2nd equation:

$$\begin{cases} 2x_1 + 2x_2 - x_3 + 2x_4 = -6 \\ x_1 - 2x_2 + 3x_4 = -3 \\ x_2 + 2x_3 - x_4 = 5 \\ x_1 + 4x_3 + x_4 = 7 \end{cases}$$

or

$$\underbrace{\begin{pmatrix} 2 & 2 & -1 & 2 \\ 1 & -2 & 0 & 3 \\ 0 & 1 & 2 & -1 \\ 1 & 0 & 4 & 1 \end{pmatrix}}_{A} \underbrace{\begin{pmatrix} x_1 \\ x_2 \\ x_3 \\ x_4 \end{pmatrix}}_{X} = \underbrace{\begin{pmatrix} -6 \\ -3 \\ 5 \\ 7 \end{pmatrix}}_{B}. \tag{1.76}$$

Problem 1.64 Show that a set of elementary operations (a) $R_3 \longleftrightarrow R_4$; (b) $R_1 - 2R_2 \to R_2$; (c) $R_1 - 2R_3 \to R_3$; (d) $R_2 - 3R_3 \to R_3$; (e) $R_2 - 6R_4 \to R_4$ and, finally, (f) $R_3 - 2R_4 \to R_4$ lead to the augmented matrix

$$\begin{pmatrix} 2 & 2 & -1 & 2 & -6 \\ 0 & 6 & -1 & -4 & 0 \\ 0 & 0 & 26 & -4 & 60 \\ 0 & 0 & 0 & 0 & 0 \end{pmatrix}.$$

This augmented matrix has been brought into the echelon form. But it has a specific feature: the last row is the zero row. What does it mean? To understand this it is sufficient to realise that what we did are a set of linear transformations; if after these transformation we got a zero row, this means that the vectors involved in these transformations must be linearly dependent. Indeed, let us write again our transformations explicitly as a set of equations (we shall ignore the initial swap for the moment):

$$\begin{cases} R_1 - 2R_2 = R_2' \\ R_1 - 2R_3 = R_3' \\ R_2' - 3R_3' = R_3'' \\ R_2' - 6R_4 = R_4' \\ R_3'' + 2R_4' = 0 \end{cases}.$$

In the last equation we have put zero in the right-hand side as the last row (which could have been denoted as R_4'') consists of zeros. Substituting recursively all the quantities into the last equation,

$$[(R_1 - 2R_2) - 3(R_1 - 2R_3)] + 2[(R_1 - 2R_2) - 6R_4] = 0,$$

we immediately obtain that $-R_2 + R_3 - 2R_4 = 0$. Recalling now that the rows 3 and 4 have been swapped, we obtain that the second, third and fourth rows of the original augmented matrix are related via $-R_2 + R_4 - 2R_3 = 0$, i.e. they are indeed linearly dependent, and this can be directly verified from the matrix itself:

$$-\underbrace{\begin{pmatrix} 1 \\ -2 \\ 0 \\ 3 \\ -3 \end{pmatrix}}_{R_2} + \underbrace{\begin{pmatrix} 1 \\ 0 \\ 4 \\ 1 \\ 7 \end{pmatrix}}_{R_4} - 2\underbrace{\begin{pmatrix} 0 \\ 1 \\ 2 \\ -1 \\ 5 \end{pmatrix}}_{R_4} = \begin{pmatrix} 0 \\ 0 \\ 0 \\ 0 \\ 0 \end{pmatrix}.$$

So, the Gauss method resulted in a zero row of the echelon matrix indicating immediately that one equation in the set is linearly dependent. Hence, effectively we have only three independent equations, and hence one variable must be treated as a parameter; in this case we have an infinite number of solutions as in the example that preceded this one.

In our last example we return to our previous example in which we shall change the number in the right-hand side of the last equation from 7 to 8:

$$\begin{cases} 2x_1 + 2x_2 - x_3 + 2x_4 = -6 \\ x_1 - 2x_2 + 3x_4 = -3 \\ x_2 + 2x_3 - x_4 = 5 \\ x_1 + 4x_3 + x_4 = 8 \end{cases}$$

or

$$\underbrace{\begin{pmatrix} 2 & 2 & -1 & 2 \\ 1 & -2 & 0 & 3 \\ 0 & 1 & 2 & -1 \\ 1 & 0 & 4 & 1 \end{pmatrix}}_{A} \underbrace{\begin{pmatrix} x_1 \\ x_2 \\ x_3 \\ x_4 \end{pmatrix}}_{X} = \underbrace{\begin{pmatrix} -6 \\ -3 \\ 5 \\ 8 \end{pmatrix}}_{B}. \tag{1.77}$$

Problem 1.65 Show that a set of elementary operations (the same as in the previous example) (a) $R_3 \longleftrightarrow R_4$; (b) $R_1 - 2R_2 \to R_2$; (c) $R_1 - 2R_3 \to R_3$; (d) $R_2 - 3R_3 \to R_3$; (e) $R_2 - 6R_4 \to R_4$ and, finally, (f) $R_3 - 2R_4 \to R_4$ lead to the augmented matrix

$$\begin{pmatrix} 2 & 2 & -1 & 2 & -6 \\ 0 & 6 & -1 & -4 & 0 \\ 0 & 0 & 26 & -4 & 66 \\ 0 & 0 & 0 & 0 & 6 \end{pmatrix}.$$

Clearly, the system of equations is not consistent as the last row corresponds to the equation $0 = 6$ that is impossible. Hence, the system of Eq. (1.77) does not have solutions.

If the number of equations n is larger than the number of unknown variables m, then after the Gaussian procedure we must first eliminate linear dependencies by removing all zero rows. After that, we either have a unique solution (if there are exactly $n - m$ zero rows), no solutions (some of the rows are contradictive) and an infinite number of solutions (if there are more than $n - m$ zero rows).

At this point we are ready to make our discussion more formal. We shall start by stating the three elementary row operations on the augmented matrix:

- swap rows i and j ($R_i \longleftrightarrow R_j$);
- add row i with the row j multiplied by a number c; the resulting vector replaces the row j ($R_i + cR_j \to R_j$);
- multiply row i by a number c ($cR_i \to R_i$).

It then follows from the above procedure that any augmented matrix can be brought into a triangular (the row echelon) form by means of the first two elementary transformations:

$$\begin{pmatrix} a_{11} & \dots & a_{1n} & b_1 \\ \dots & \dots & \dots & \dots \\ a_{m1} & \dots & a_{mn} & b_m \end{pmatrix} \implies \begin{pmatrix} a_{1k_1} & \dots & \dots & \dots & \dots & \dots \\ & a_{2k_2} & \dots & \dots & \dots & \dots \\ & & \dots & \dots & \dots & \dots \\ & & & a_{rk_r} & \dots & \\ 0 & 0 & \dots & \dots & \dots & 0 \\ 0 & 0 & \dots & \dots & \dots & 0 \end{pmatrix}.$$

The matrix on the right is said to be in the row echelon form if the first (reading from left to right) non-zero element a_{rk_k} of any row $r \geqslant 1$ (that is placed in the k_r column) is positioned *to the right* of such element $a_{r-1,k_{r-1}}$ of the previous row, i.e. if $k_{r-1} > k_r$. The first (left-most) non-zero elements in each row is called the leading element or *pivot*. Hence, the columns in which pivots are placed should form a sequence $k_1 > k_2 > k_3 > \cdots > k_m$. There might be last rows with all zeroes; these correspond to linearly dependent rows of the matrix and indicate on linear dependencies of the equations. If, however, at the last (right-most) position of the otherwise zero row we have a non-zero number, the set of equations is not consistent and does not have solutions.

The matrix is said to be in the *reduced row echelon form* if all pivoting elements are equal to one, while any other element in the matrix (apart possibly of its last column) are equal to zero, i.e. in each column, apart possibly from the last one, only leading element is non-zero.

Formally, the three elementary transformations introduced above can be described by an action with a specific square matrix $\mathbf{U}$. Let n be the number of equations and m the number of unknowns, as before. If $n = m$, then the dimension of $\mathbf{U}$ is also n. If $\mathbf{A}$ is a rectangular matrix, then one can always augment it appropriately by either introducing new variables with zero coefficients if $n > m$ or adding zero rows (and the corresponding additional zero components in the right-hand side vector $\mathbf{B}$) if $n < m$, so that the modified set of n' equation would contain $m' = n'$ variables, and the modified matrix $\mathbf{A}$ would be a square matrix. Obviously, the new set of equations is equivalent to the original one. By multiplying the equation $\mathbf{AX} = \mathbf{B}$ with $\mathbf{U}$ from the left, a new set of equations $(\mathbf{UA})\,\mathbf{X} = (\mathbf{UB})$ is obtained. Obviously, all solutions of the original set satisfy this equation as well. If the matrix $\mathbf{U}$ has an inverse (is not singular, Sect. 1.7), then multiplying from the left with $\mathbf{U}^{-1}$ we return back to the original equations and hence the equations $(\mathbf{UA})\,\mathbf{X} = (\mathbf{UB})$ are indeed fully equivalent to the former ones, i.e. do not have extra solutions. We can conclude that one may use any non-singular $\mathbf{U}$ to transform the original set of equations $\mathbf{AX} = \mathbf{B}$ into the form $(\mathbf{UA})\,\mathbf{X} = (\mathbf{UB})$ and solve it instead; we shall not acquire new or loose any solutions.

Then, indeed each of the elementary row operations we have introduced above can be described by a non-singular matrix:

- the swap $R_i \longleftrightarrow R_j$ can be described by

$$
\mathbf{U} \to \mathbf{P}_{ij} =
\begin{pmatrix}
\ddots & & & & \\
 & 0 & 1 & & \\
 & & \ddots & & \\
 & 1 & 0 & & \\
 & & & \ddots &
\end{pmatrix}
\begin{matrix}
\\ \leftarrow i \\ \\ \leftarrow j \\ \\
\end{matrix}
$$
$$
\begin{matrix}
\quad\uparrow\ \ \uparrow \\
\quad i\ \ \ j
\end{matrix}
$$

where we specifically indicated elements on the ith and jth rows and columns
$(i < j)$;

- $R_i + cR_j \rightarrow R_j$ can be described by

$$
\mathbf{U} \rightarrow \mathbf{G}_{ij}(c) = \begin{pmatrix} \ddots & & & \\ & 1 & & 0 \\ & & \ddots & \\ & 1 & & c \\ & & & & \ddots \end{pmatrix} \begin{matrix} \\ \leftarrow i \\ \\ \leftarrow j \\ \\ \end{matrix}
$$

$$
\begin{matrix} \uparrow & \uparrow \\ i & j \end{matrix}
$$

- $cR_i \rightarrow R_i$ is described by

$$
\mathbf{U} \rightarrow \mathbf{M}_i(c) = \begin{pmatrix} \ddots & & \\ & c & \\ & & \ddots \end{pmatrix} \leftarrow i
$$

$$
\begin{matrix} \uparrow \\ i \end{matrix}
$$

In all three cases elements that are not shown are the same as in the identity matrix;
in particular, not shown elements on the main diagonal are equal to one.

Problem 1.66 Verify that the matrices given above do indeed perform the
appropriate operations (you may simply use 2×2 matrices for the first two
cases, the third case is trivial).

The introduced elementary matrices are actually non-singular.

Problem 1.67 Verify by direct multiplication that $\mathbf{P}_{ij}^{-1} = \mathbf{P}_{ij}$,

$$
\mathbf{G}_{ij}^{-1}(c) = \begin{pmatrix} \ddots & & & \\ & 1 & & 0 \\ & & \ddots & \\ & 1/c & & 1/c \\ & & & & \ddots \end{pmatrix} \begin{matrix} \\ \leftarrow i \\ \\ \leftarrow j \\ \\ \end{matrix}
$$

$$
\begin{matrix} \uparrow & \uparrow \\ i & j \end{matrix}
$$

and

$$\mathbf{M}_i^{-1}(c) = \begin{pmatrix} \ddots & & \\ & 1/c & \\ & & \ddots \end{pmatrix} \leftarrow i$$

$$\uparrow$$
$$i$$

Therefore, as anticipated, the elementary operations we have introduced cannot modify the set of solutions of the original set of equations.

In the problems below use the Gaussian elimination method to find all solutions.

Problem 1.68 Show that the system of equations

$$\begin{cases} x_1 + x_2 - x_3 + x_4 = 1 \\ 2x_2 - 2x_3 - x_4 = 0 \\ 3x_1 - 3x_2 - 5x_3 = 10 \end{cases}$$

has an infinite number of solutions

$$x_1 = \frac{1}{2}(2 - 3t), \quad x_2 = \frac{1}{8}(-7 - 2t), \quad x_3 = \frac{1}{8}(-7 - 6t), \quad \text{and } x_4 = t.$$

Problem 1.69 Show that the system of equations

$$\begin{cases} x_1 + x_2 - x_3 + x_4 = 1 \\ 2x_2 - 2x_3 - x_4 = 0 \\ 3x_1 - 3x_2 - 5x_3 = 10 \\ -x_1 + x_2 - 3x_3 + 4x_4 = -1 \end{cases}$$

has the unique solution $x_1 = 27/20$, $x_2 = -49/60$, $x_3 = -7/10$ and $x_4 = -7/30$.

1.5 Rank of a Matrix

The rank (the row-rank) of a general rectangular $n \times m$ matrix $\mathbf{A} = (a_{ij})$ is the dimension of a space spanned by its row-vectors. In other words, it is the number of linearly independent rows of the matrix. It is denoted as $\mathrm{rk}(\mathbf{A})$ or $\mathrm{rank}(\mathbf{A})$. One can also introduce the rank via the columns of the matrix, the column-rank. As can be shown, these are actually the same. This can be intuitively understood as follows (although this is not a proof). Consider a set of n linear algebraic equations with respect to m variables forming a vector $\mathbf{X}$, i.e. $\mathbf{AX} = \mathbf{B}$. The rows of $\mathbf{A}$ correspond to the number of equations, and the row-rank gives the number of linearly independent equations. On the other hand, the columns of $\mathbf{A}$ can be associated with the number of different variables. We can only determine as many variables as the number of independent equations, i.e. the two must be the same.

The matrix has full rank if it is equal to the lowest of n and m; otherwise the matrix is called rank-deficient.

The rank of a matrix can be determined using the Gaussian elimination. Indeed, if at the end of the procedure of constructing the row echelon matrix we find r non-zero rows, then the remaining $n - r$ rows form a set of linearly dependent vectors.

Indeed, let $a_{1k_1}, a_{2k_2}, \ldots, a_{rk_r}$ be pivots of the r non-zero rows of the row echelon matrix. Here $k_1 > k_2 > \cdots > k_r$. Consider the vectors $\mathbf{a}_1 = (a_{1k})$, $\mathbf{a}_2 = (a_{2k})$, etc. formed by the rows, and let us form their linear combination

$$\sum_{i=1}^{r} \lambda_i \mathbf{a}_i = 0 \,.$$

If the vector-rows are linearly independent, all coefficients λ_i in this expansion must be zero. To solve for the coefficients, we shall write this equation in components:

$$\sum_{i=1}^{r} \lambda_i a_{ik} = 0 \,, \quad k = 1, \ldots, m \,.$$

First, consider this equation for $k = k_1$. This value of k specifies the k_1-th column of $\mathbf{A}$. Since in this column only one non-zero pivot $a_{1k_1} \neq 0$, there is only a single term $i = 1$ in this equation leading to $\lambda_1 a_{1k_1} = 0$, which gives $\lambda_1 = 0$. Next, let us consider $k = k_2$. Two terms will contribute in this case, $i = 1$ and $i = 2$:

$$\lambda_1 a_{1k_2} + \lambda_2 a_{2k_2} = 0 \implies \lambda_2 = 0 \,,$$

since $\lambda_1 = 0$ and the pivot of the second row, $a_{2k_2} \neq 0$. Continuing this process, we find that $\lambda_i = 0$ for all values of i, which proves the made above statement. Hence, one can calculate the rank of a matrix by running the Gaussian elimination and then counting the number of non-zero rows.

For instance, the rank of the matrix Eq. (1.75) is four since its row echelon form does not contain zero rows. On the other hand, the rank of the matrix

$$\begin{pmatrix} 2 & 2 & -1 & 2 \\ 1 & -2 & 0 & 3 \\ 0 & 1 & 2 & -1 \\ 1 & 0 & 4 & 1 \end{pmatrix}$$

from Eq. (1.76) is only 3 since there is one zero row in its row echelon form, see Problem 1.64.

Problem 1.70 Show that the rank of the matrix

$$\begin{pmatrix} 1 & -1 & 2 & 5 \\ 2 & 2 & 0 & -1 \\ 3 & 0 & -3 & 6 \end{pmatrix}$$

is three.

The above discussion enables one to conclude that the following statements concerning a square $n \times n$ matrix $\mathbf{A}$ are equivalent:

- $\mathbf{A}$ is singular;
- $\mathbf{A}^{-1}$ does not exist;
- $\det \mathbf{A} = 0$;
- $rk(\mathbf{A}) < n$, and hence
- the rows (columns) of $\mathbf{A}$ are linearly dependent;
- the system of equations $\mathbf{AX} = \mathbf{0}$ has an infinite number of non-trivial solutions.

It is easy to establish the following

Theorem 1.14 *A square matrix $\mathbf{A}$ is not singular if and only if its rows (columns) form linearly independent vectors, i.e. the rank of $\mathbf{A}$ equals its dimension (i.e. $\mathbf{A}$ is a full-rank matrix).*

Proof Assume first that the square $n \times n$ matrix $\mathbf{A}$ is non-singular, i.e. $\det \mathbf{A} \neq 0$. Consider all its row-vectors $\mathbf{a}_i = (a_{ij})$ and let us set their linear combination to zero:

$$\mathbf{a}_1 x_1 + \mathbf{a}_2 x_2 + \cdots + x_n \mathbf{a}_n = \mathbf{0},$$

where the numerical coefficients x_j form a vector $\mathbf{X} = (x_j)$. The equation above is equivalent to n scalar equations

$$a_{1i} x_1 + a_{2i} x_2 + \cdots + a_{ni} x_n = 0, \quad i = 1, \ldots, n.,$$

which can be written simply as $\mathbf{A}^T \mathbf{X} = \mathbf{0}$. Since the matrix $\mathbf{A}$ is non-singular, this system of equations has a single solution $\mathbf{X} = \mathbf{0}$, as required, i.e. the vector-rows of $\mathbf{A}$ are indeed linearly independent.

Conversely, assume that the rows of $\mathbf{A}$ are linearly independent. Then the solution of the equation $\mathbf{A}^T \mathbf{X} = \mathbf{0}$ must only give the zero solution $\mathbf{X} = \mathbf{0}$, which requires $\det \mathbf{A} \neq 0$, i.e. the matrix $\mathbf{A}$ is non-singular. **Q.E.D.**

Next, let us consider a $m \times r$ matrix $\mathbf{B}$ of rank r (it is a full-rank matrix). Its Hermitian transpose $\mathbf{B}^\dagger$ is a $r \times m$ matrix. Hence, the matrix $\mathbf{B}^\dagger\mathbf{B}$ is a square $r \times r$ matrix, and one might infer that it is not singular. This is indeed the case.

Theorem 1.15 *Consider a $m \times r$ matrix $\mathbf{B}$ of rank r. Then the square $r \times r$ matrix $\mathbf{B}^\dagger\mathbf{B}$ is not singular and hence its inverse $(\mathbf{B}^\dagger\mathbf{B})^{-1}$ exists.*

Proof Consider an r-row vector $\mathbf{X}$ that is a solution of the equation $\mathbf{B}^\dagger\mathbf{B}\mathbf{X} = \mathbf{0}$. Multiply both sides by $\mathbf{X}^\dagger$ from the left:

$$\mathbf{X}^\dagger\mathbf{B}^\dagger\mathbf{B}\mathbf{X} = 0 \implies (\mathbf{B}\mathbf{X})^\dagger (\mathbf{B}\mathbf{X}) = 0 \,.$$

The left-hand side here gives the lengths of the vector $\mathbf{B}\mathbf{X}$ squared. Since it is zero, it must be that it is the null vector, $\mathbf{B}\mathbf{X} = \mathbf{0}$. At the same time,

$$\mathbf{B}\mathbf{X} = \begin{pmatrix} b_{11} & b_{12} & \ldots & b_{1r} \\ b_{21} & b_{22} & \ldots & b_{2r} \\ \ldots & \ldots & \ddots & \\ b_{m1} & b_{m2} & \ldots & b_{mr} \end{pmatrix} \begin{pmatrix} x_1 \\ x_2 \\ \ldots \\ x_r \end{pmatrix} = \begin{pmatrix} \mathbf{b}_1 & \mathbf{b}_2 & \ldots & \mathbf{b}_r \end{pmatrix} \begin{pmatrix} x_1 \\ x_2 \\ \ldots \\ x_r \end{pmatrix}$$

$$= x_1\mathbf{b}_1 + x_2\mathbf{b}_2 + \cdots + x_r\mathbf{b}_r = \mathbf{0} \,,$$

where we presented $\mathbf{B}$ as a collection of linearly independent vector-columns $\mathbf{b}_1, \mathbf{b}_2, \ldots, \mathbf{b}_r$. It is seen that the zero product $\mathbf{B}\mathbf{X}$ can be written as a linear combination of r linearly independent vectors, and hence the expansion coefficients $x_1, \ldots, x_r$ must all be equal to zero. Hence, $\mathbf{X} = 0$. To ensure the zero solution in the equation $(\mathbf{B}^\dagger\mathbf{B})\mathbf{X} = \mathbf{0}$ that we have started from, the matrix $\mathbf{B}^\dagger\mathbf{B}$ must be non-singular with a well-defined inverse. **Q.E.D.**

Problem 1.71 Consider a $r \times m$ matrix $\mathbf{C}$ of rank r (this is also a full-rank matrix). Prove that the $r \times r$ square matrix $\mathbf{C}\mathbf{C}^\dagger$ is non-singular and hence has an inverse.

1.6   Wronskian

1.6.1   Linear Independence of Functions

The notion of the determinant appears to be very useful in verifying whether a set of functions $f_1(x), f_2(x), \ldots, f_n(x)$ is linearly independent. Let us establish a simple criterion for that. We know from Section 1.1.2 that the functions are linearly independent if the equation

$$\alpha_1 f_1(x) + \alpha_2 f_2(x) + \cdots + \alpha_n f_n(x) = 0 , \tag{1.78}$$

valid for any continuous interval of the values of x, has only the trivial solution with respect to the coefficients α_1, α_2, etc.:

$$\alpha_1 = \alpha_2 = \cdots = \alpha_n = 0 .$$

It is possible to work out a simple method for verifying this. To this end, let us generate $(n-1)$ more equations by differentiating both sides of Eq. (1.78) once, twice, etc., $(n-1)$ times. We obtain $n-1$ additional equations:

$$\alpha_1 f_1^{(1)}(x) + \alpha_2 f_2^{(1)}(x) + \cdots + \alpha_n f_n^{(1)}(x) = 0, ,$$
$$\alpha_1 f_1^{(2)}(x) + \alpha_2 f_2^{(2)}(x) + \cdots + \alpha_n f_n^{(2)}(x) = 0,$$
$$\cdots \quad \cdots \quad \cdots$$
$$\alpha_1 f_1^{(n-1)}(x) + \alpha_2 f_2^{(n-1)}(x) + \cdots + \alpha_n f_n^{(n-1)}(x) = 0,$$

where $f_i^{(k)}(x) = d^k f_i / dx^k$. These equations, together with Eq. (1.78), form a system of n linear algebraic equations with respect to the coefficients α_1, α_2, etc.:

$$\widetilde{\mathbf{W}} \boldsymbol{\alpha} = \mathbf{0} , \tag{1.79}$$

where $\boldsymbol{\alpha}$ is the vector of the coefficients and

$$\widetilde{\mathbf{W}} = \begin{pmatrix} f_1 & f_2 & \cdots & f_n \\ f_1^{(1)} & f_2^{(1)} & \cdots & f_n^{(1)} \\ \cdots & \cdots & \cdots & \cdots \\ f_1^{(n-1)} & f_2^{(n-1)} & \cdots & f_n^{(n-1)} \end{pmatrix} \tag{1.80}$$

is a matrix the determinant of which is called Wronskian, $\mathbf{W} = \det \widetilde{\mathbf{W}}$. We have already come across it before, e.g. in Sect. I.8.2. The point is that Eq. (1.79) has the form of a set of linear algebraic equations with the zero right-hand side which we have already discussed above. Therefore, it has a non-zero solution if and only if the determinant of the Wronskian is equal to zero for any x within the given interval. Conversely, if $\mathbf{W} \neq 0$ for all x from a given interval, then there is only a zero solution and hence the functions are linearly independent in that interval.

As an example, let us prove that power functions x^i with different powers are linearly independent. We shall specifically consider a set of $(n+1)$ functions with $i = 0, 1, \ldots, n$. The Wronskian in this case is

$$\mathbf{W} = \begin{vmatrix} 1 & x & x^2 & x^3 & \cdots & & x^n \\ 0 & 1 & 2x & 3x^2 & \cdots & & nx^{n-1} \\ 0 & 0 & 2 & 3 \cdot 2x & \cdots & n(n-1)x^{n-2} \\ 0 & 0 & 0 & 3! & \cdots & n!/(n-3)!x^{n-1} \\ \cdots\cdots\cdots & \cdots & \cdots & & \cdots & \\ 0 & 0 & 0 & 0 & \cdots & & n! \end{vmatrix}.$$

It is easy to see that $\mathbf{W}$ has a triangular form with the elements along its diagonal being $1, 1, 2!, 3!, 4!$, etc., $n!$. Calculating the determinant $|\mathbf{W}|$ along the first column followed by the calculation of all the consecutive minors also along the first column (cf. Problem 1.58), results in that

$$|\mathbf{W}| = 1 \cdot 2! \cdot 3! \cdot \cdots \cdot n! = \prod_{i=1}^{n} i! \neq 0 \,.$$

Therefore, the power functions are linearly independent for any real x.

Problem 1.72 Demonstrate, using the method based on the calculation of the Wronskian, that the functions $\sin x$, $\cos x$ and e^{ix} are linearly dependent.

Problem 1.73 The same for functions e^x, e^{-x} and $\sinh(x)$.

In practice, however, when considering linear independence (dependence) of many functions, calculation of determinants of large sizes could be impractical. A simple method exists that enables one to avoid this. Consider n functions $f_1(x), f_2(x), \ldots, f_n(x)$, and let us assume that for any pair of the functions one can prove their linear independence. Then the linear independence of the whole set of n functions follows. Indeed, let us start from f_1 and f_2, and let us suppose that they are linearly independent, i.e. the Wronskian built upon them is non-zero for x within some interval. Let us also suppose that, separately, f_1 and f_2 are proven to be linearly independent from f_3. Then it follows that these three functions are mutually linearly independent. Indeed, suppose there exists a non-zero solution of the equation

$$\alpha_1 f_1(x) + \alpha_2 f_2(x) + \alpha_3 f_3(x) = 0$$

with respect to the coefficients. Obviously, α_3 cannot be equal to zero as this would mean that there exists a non-zero solution of the equation without $f_3(x)$ which contradicts our assumption that f_1 and f_2 are known to be linearly independent. Hence, we can solve this equation for $f_3(x)$ expressing it via the first two functions. But this is also impossible since it contradicts the fact that f_3 is separately independent from each of them. Hence, there is only a zero solution possible. Similarly, we prove that if f_4 is separately known to be linearly independent from the first three functions, then all four are linearly independent.

Problem 1.74 Show that the exponential functions e^x, e^{2x}, etc., e^{nx} are linearly independent.

1.6.2 Linear ODE's: Abel's Formula

There exists an important formula due to Abel related to an n-th order linear ordinary differential equation (ODE)

$$a_n(x)y^{(n)}(x) + a_{n-1}(x)y^{(n-1)}(x) + \cdots + a_1(x)y'(x) + a_0(x)y(x) = 0, \quad (1.81)$$

in which the Wronskian plays a pivotal role.

In fact, we have already considered the case of $n = 2$ in Sect. I.8.2. To derive Abel's formula in the general case, it is instructive to consider first a particular case of $n = 3$. Let functions $y_1(x)$, $y_2(x)$ and $y_3(x)$ form a fundamental set of solutions of the ODE

$$a_3(x)y''' + a_2(x)y'' + a_1(x)y' + a_0(x)y = 0 \quad (1.82)$$

on the interval $a < x < b$. These functions are linearly independent and hence their Wronskian

$$W(x) = \begin{vmatrix} y_1 & y_2 & y_3 \\ y_1' & y_2' & y_3' \\ y_1'' & y_2'' & y_3'' \end{vmatrix} \neq 0$$

on that interval.

Let us calculate the derivative of $W(x)$. To this end, it is convenient to rewrite the determinant as a sum over permutations from its definition, Eq. (1.52),

$$W(x) = \sum_P \epsilon_P y_{p_1} y_{p_2}' y_{p_3}'' ,$$

where we sum over all 3! permutations of integer numbers 1, 2 and 3, each such combination forming the indices p_1, p_2 and p_3 of the solutions of the ODE, i.e.

$$W = y_1 y_2' y_3'' - y_2 y_1' y_3'' - y_3 y_2' y_1'' + \ldots .$$

Let us now differentiate the Wronskian:

$$W'(x) = \frac{d}{dx} \sum_P \epsilon_P y_{p_1} y'_{p_2} y''_{p_3}$$

$$= \sum_P \epsilon_P \left[y'_{p_1} y'_{p_2} y''_{p_3} + y_{p_1} y''_{p_2} y''_{p_3} + y_{p_1} y'_{p_2} y'''_{p_3} \right]$$

$$= \begin{vmatrix} y'_1 & y'_2 & y'_3 \\ y'_1 & y'_2 & y'_3 \\ y''_1 & y''_2 & y''_3 \end{vmatrix} + \begin{vmatrix} y_1 & y_2 & y_3 \\ y''_1 & y''_2 & y''_3 \\ y''_1 & y''_2 & y''_3 \end{vmatrix} + \begin{vmatrix} y_1 & y_2 & y_3 \\ y'_1 & y'_2 & y'_3 \\ y'''_1 & y'''_2 & y'''_3 \end{vmatrix}.$$

It is easy to see that the first two determinants are equal to zero since they contain identical rows. Hence, only the last determinant remains

$$W'(x) = \begin{vmatrix} y_1 & y_2 & y_3 \\ y'_1 & y'_2 & y'_3 \\ y'''_1 & y'''_2 & y'''_3 \end{vmatrix} = \sum_P \epsilon_P y_{p_1} y'_{p_2} y'''_{p_3}.$$

The functions differentiated three times, y'''_{p_3}, can be expressed via their lower derivatives from the ODE (1.82) that these functions satisfy. Hence,

$$W'(x) = \sum_P \epsilon_P y_{p_1} y'_{p_2} \left[-p_2(x) y''_{p_3} + p_1(x) y'_{p_3} + p_0(x) y_{p_3} \right]$$

$$= -p_2 \left[\sum_P \epsilon_P y_{p_1} y'_{p_2} y''_{p_3} \right] - p_1 \left[\sum_P \epsilon_P y_{p_1} y'_{p_2} y'_{p_3} \right] - p_0 \left[\sum_P \epsilon_P y_{p_1} y'_{p_2} y_{p_3} \right]$$

$$= -p_2 \begin{vmatrix} y_1 & y_2 & y_3 \\ y'_1 & y'_2 & y'_3 \\ y''_1 & y''_2 & y''_3 \end{vmatrix} - p_1 \begin{vmatrix} y_1 & y_2 & y_3 \\ y'_1 & y'_2 & y'_3 \\ y'_1 & y'_2 & y'_3 \end{vmatrix} - p_0 \begin{vmatrix} y_1 & y_2 & y_3 \\ y'_1 & y'_2 & y'_3 \\ y_1 & y_2 & y_3 \end{vmatrix},$$

where we introduced functions $p_i(x) = a_i(x)/a_3(x)$ (for $i = 0, 1, 2$). As before, the last two determinants are identically zero as containing identical rows. Hence, only the first determinant remains which is in fact exactly $W(x)$. Hence, we have obtained an ODE for the Wronskian, $W' + p_2 W = 0$, whose solution, satisfying the initial condition $W(x) = W(x_0)$ at $x = x_0$ is

$$W(x) = W(x_0) \exp \left[-\int_{x_0}^x p_2(x') dx' \right]$$

$$= W(x_0) \exp \left[-\int_{x_0}^x \frac{a_2(x')}{a_3(x')} dx' \right].$$

Equipped by this example, it is not difficult now to consider the general n-th order case. Indeed, in this case

$$W(x) = \sum_P \epsilon_P y_{p_1} y'_{p_2} y''_{p_3} \cdots y_{p_n}^{(n-1)},$$

so that

$$W'(x) = \sum_P \epsilon_P y'_{p_1} y'_{p_2} y''_{p_3} \cdots y^{(n-2)}_{p_{n-1}} y^{(n-1)}_{p_n} + \sum_P \epsilon_P y_{p_1} y''_{p_2} y''_{p_3} \cdots y^{(n-2)}_{p_{n-1}} y^{(n-1)}_{p_n} + \cdots$$

$$+ \sum_P \epsilon_P y_{p_1} y'_{p_2} y''_{p_3} \cdots y^{(n-1)}_{p_{n-1}} y^{(n-1)}_{p_n} + \sum_P \epsilon_P y_{p_1} y'_{p_2} y''_{p_3} \cdots y^{(n-2)}_{p_{n-1}} y^{(n)}_{p_n}.$$

Only the last term remains as the others correspond to determinants with two rows being identical,

$$W' = \sum_P \epsilon_P y_{p_1} y'_{p_2} y''_{p_3} \cdots y^{(n-2)}_{p_{n-1}} y^{(n)}_{p_n}.$$

The function $y^{(n)}_{p_n}$ is replaced from the ODE (1.81) leading to

$$W' = \sum_P \epsilon_P y_{p_1} y'_{p_2} y''_{p_3} \cdots y^{(n-2)}_{p_{n-1}} \left[-\sum_{k=1}^n p_{n-k}(x) y^{(n-k)}_{p_n} \right]$$

$$= -\sum_{k=1}^n p_{n-k}(x) \left[\sum_P \epsilon_P y_{p_1} y'_{p_2} y''_{p_3} \cdots y^{(n-2)}_{p_{n-1}} y^{(n-k)}_{p_n} \right],$$

where $p_i(x) = a_i(x)/a_n(x)$ for any $i = 0, \ldots, n-1$. Each expression in the square brackets corresponds to a n-th order determinant. However, for all values of $k = 2, 3, \ldots, n$ the corresponding determinants contain two identical rows and can be dropped. Hence, only the determinant for $k = 1$ needs to be kept, which is equal to the original Wronskian. We therefore obtain $W' = -p_{n-1} W = -(a_{n-1}/a_n) W$ whose solution is

$$W(x) = W(x_0) \exp\left[-\int_{x_0}^x \frac{a_{n-1}(x')}{a_n(x')} dx' \right], \qquad (1.83)$$

which is Abel's formula.

It shows, in particular, that if $W(x_0) \neq 0$ for some x_0, then $W(x) \neq 0$ for any x; conversely, if $W(x_0) = 0$, then $W(x) = 0$ for any x.

Another consequence of Abel's formula is the following: suppose we only know $n-1$ solutions $y_1(x), y_2(x), \ldots, y_{n-1}(x)$ of the ODE (1.81) and would like to determine the last solution $y_n(x)$. Then, the explicit expression for the Wronskian (1.83) allows obtaining an ODE of order $n-1$ for the last fundamental solution $y_n(x)$. We saw in Sect. I.8.2.1 that in the case of the second-order ODE ($n = 2$), Abel's formula gives a first-order ODE for the second solution $y_2(x)$ (for $n = 2$ we have $n-1 = 1$) and hence it can be easily found. In the cases of higher values of $n \geq 3$ a higher order ODE is obtained; however, it is one order less than the original ODE which could be of some help.

1.6.3   Linear ODE Reverse Engineering

Finally, we shall mention yet another point related to high-order linear ODEs that is based on an 'extended' Wronskian. Consider a $(n+1) \times (n+1)$ determinant

$$W_{ext}(x) = \begin{vmatrix} y_1(x) & y_2(x) & \cdots & y_n(x) & y(x) \\ y_1'(x) & y_2'(x) & \cdots & y_n'(x) & y'(x) \\ y_1''(x) & y_2''(x) & \cdots & y_n''(x) & y''(x) \\ \cdots & \cdots & \cdots & \cdots \\ y_1^{(n)}(x) & y_2^{(n)}(x) & \cdots & y_n^{(n)}(x) & y^{(n)}(x) \end{vmatrix},$$

in which we assumed that the n functions $y_1(x), \ldots, y_n(x)$ are linearly independent. This determinant can be considered as a functional[5] of the function $y(x)$ as using a different such function would produce a different function $W_{ext}(x)$. An interesting property of this determinant is that it is equal to zero for any function $y(x) = y_i(x)$ $(i = 1, \ldots, n)$ since replacing $y(x)$ with $y_i(x)$ produces a determinant with two identical columns. Moreover, it is easy to see that replacing $y(x)$ with any linear combination of the functions $\{y_i(x)\}$ would also nullify the determinant. On the other hand, if we open this determinant using usual rules, it will become a linear n-th order ODE. Hence, building such an extended Wronskian enables one to construct a linear ODE with pre-defined solutions, which might be useful in applications.

Problem 1.75 Using this method, show that the ODE with solutions $y_1(x) = \sin x$ and $y_2(x) = \cos x$ is $y'' + y = 0$, as expected.

Problem 1.76 Show that the ODE with solutions $y_1(x) = e^x$ and $y_2(x) = xe^x$ is $y'' - 2y' + y = 0$. This result is also to be expected.

Problem 1.77 Show that the ODE with solutions $y_1(x) = e^x$ and $y_2(x) = \cos x$ must be

$$(\cos x + \sin x)y'' - (2\cos x)\, y' - (\sin x - \cos x)\, y = 0.$$

The ODE in the last problem contains only variable coefficients. Remarkably, it is not possible to obtain these two solutions from a single ODE with constant coefficients even though each of them, separately, can be obtained from an appropriately chosen ODE with constant coefficients; when combined together as solutions of the same ODE, it appears to be rather impossible.

---

[5] We shall discuss functionals in more detail in Chap. 9.

1.7   Calculation of the Inverse Matrix

The formulae obtained in Sect. 1.4.1 allow us to derive a general formula for the inverse matrix. We should also be able to establish a necessary condition for the inverse of a matrix to exist.

To accomplish this programme, we have to compare Cramer's solution of Eq. (1.68) with that given by Eq. (1.65). To this end, let us first rewrite the solution (1.68) in a slightly different form. We expand the determinant in the numerator of the expression for x_i along its ith column (the one which contains elements $\{b_j\}$):

$$\begin{vmatrix} a_{11} & \cdots & a_{1,i-1} & b_1 & a_{1,i+1} & \cdots & a_{1n} \\ \cdots & \cdots & \cdots & \cdots & \cdots & & \cdots \\ a_{n1} & \cdots & a_{n,i-1} & b_n & a_{n,i+1} & \cdots & a_{nn} \end{vmatrix} = b_1 A_{1i} + b_2 A_{2i} + \cdots + b_n A_{ni}$$

$$= \sum_{k=1}^{n} b_k A_{ki} \; ,$$

where A_{ki} are the corresponding co-factors of the matrix $\mathbf{A}$ of the coefficients of the system of equations: indeed, by removing the ith column and the kth row we arrive at the k, i co-factor A_{ki} of $\mathbf{A}$. Therefore, combining the last equation with Eq. (1.68), we have

$$x_i = \frac{1}{|A|} \sum_{k=1}^{n} b_k A_{ki} \; . \tag{1.84}$$

On the other hand, we also formally have the solution in the form of Eq. (1.65) which contains the inverse matrix; it can be written in components as

$$x_i = \sum_{k=1}^{n} \left(\mathbf{A}^{-1} \right)_{ik} b_k \; .$$

The last two expressions should give the same answer for *any* numbers b_k. Therefore, the following expression must be generally valid:

$$\left(\mathbf{A}^{-1} \right)_{ik} = \frac{A_{ki}}{|A|} \; , \tag{1.85}$$

which gives a general expression for the elements of the inverse matrix sought for. It can be used to calculate it in the case of arbitrary dimension of the matrix $\mathbf{A}$. Note the reverse order of indices in Eq. (1.85): if the denote by $\mathbf{A}_{cof} = \left(A_{ij} \right)$ the matrix of co-factors, then

$$\mathbf{A}^{-1} = \frac{\mathbf{A}_{cof}^T}{|A|} \; . \tag{1.86}$$

This is the general result we have been looking for.

It also follows from the above formula that $\mathbf{A}^{-1}$ exists if and only if $|\mathbf{A}| \neq 0$. Matrices that have a non-zero determinant are called *non-singular* as opposite to *singular* matrices that have a zero determinant.

Example 1.15 ▶ Find the inverse of the matrix

$$\mathbf{A} = \begin{pmatrix} 2 & 1 & 1 \\ -1 & 3 & 2 \\ 2 & 0 & 1 \end{pmatrix} . \tag{1.87}$$

Solution: First of all, we calculate all necessary co-factors:

$$A_{11} = + \begin{vmatrix} 3 & 2 \\ 0 & 1 \end{vmatrix} = 3 , \quad A_{12} = - \begin{vmatrix} -1 & 2 \\ 2 & 1 \end{vmatrix} = 5 , \quad A_{13} = + \begin{vmatrix} -1 & 3 \\ 2 & 0 \end{vmatrix} = -6 ,$$

$$A_{21} = - \begin{vmatrix} 1 & 1 \\ 0 & 1 \end{vmatrix} = -1 , \quad A_{22} = + \begin{vmatrix} 2 & 1 \\ 2 & 1 \end{vmatrix} = 0 , \quad A_{23} = - \begin{vmatrix} 2 & 1 \\ 2 & 0 \end{vmatrix} = 2 ,$$

$$A_{31} = + \begin{vmatrix} 1 & 1 \\ 3 & 2 \end{vmatrix} = -1 , \quad A_{32} = - \begin{vmatrix} 2 & 1 \\ -1 & 2 \end{vmatrix} = -5 , \quad A_{33} = + \begin{vmatrix} 2 & 1 \\ -1 & 3 \end{vmatrix} = 7 .$$

Using already available co-factors, the calculation of the determinant is straightforward as one can, e.g. expand alone the first row:

$$|\mathbf{A}| = 2 \cdot A_{11} + 1 \cdot A_{12} + 1 \cdot A_{13} = 2 \cdot 3 + 1 \cdot 5 + 1 \cdot (-6) = 5 .$$

Therefore, noting the reverse order of indices between the elements of the inverse matrix and the corresponding co-factors (i.e. using the transpose of the co-factor matrix), we get

$$\mathbf{A}^{-1} = \frac{1}{5} \begin{pmatrix} A_{11} & A_{21} & A_{31} \\ A_{12} & A_{22} & A_{32} \\ A_{13} & A_{23} & A_{33} \end{pmatrix} = \begin{pmatrix} 3/5 & -1/5 & -1/5 \\ 1 & 0 & -1 \\ -6/5 & 2/5 & 7/5 \end{pmatrix} .$$

It is easy to check that the matrix $\mathbf{A}^{-1}$ we found is indeed the inverse of the matrix $\mathbf{A}$. Using the matrix multiplication, we get

$$\mathbf{A}^{-1}\mathbf{A} = \begin{pmatrix} 3/5 & -1/5 & -1/5 \\ 1 & 0 & -1 \\ -6/5 & 2/5 & 7/5 \end{pmatrix} \begin{pmatrix} 2 & 1 & 1 \\ -1 & 3 & 2 \\ 2 & 0 & 1 \end{pmatrix} = \begin{pmatrix} 1 & 0 & 0 \\ 0 & 1 & 0 \\ 0 & 0 & 1 \end{pmatrix} = \mathbf{E} ,$$

$$\mathbf{A}\mathbf{A}^{-1} = \begin{pmatrix} 2 & 1 & 1 \\ -1 & 3 & 2 \\ 2 & 0 & 1 \end{pmatrix} \begin{pmatrix} 3/5 & -1/5 & -1/5 \\ 1 & 0 & -1 \\ -6/5 & 2/5 & 7/5 \end{pmatrix} = \begin{pmatrix} 1 & 0 & 0 \\ 0 & 1 & 0 \\ 0 & 0 & 1 \end{pmatrix} = \mathbf{E} . ◀$$

Another method for calculating the inverse is based on the Gaussian elimination procedure considered in Sect. 1.4.2. Suppose we would like to calculate the inverse $\mathbf{A}^{-1}$ of a square matrix $\mathbf{A}$ of dimension n. Let us construct an augmented matrix by placing the identity matrix $\mathbf{E}$ on the right of $\mathbf{A}$ as

$$(\mathbf{A} \vert \mathbf{E}) = \begin{pmatrix} a_{11} \ldots a_{1k} \ldots a_{1n} & 1 \ldots 0 \ldots 0 \\ \ldots \ldots \ldots \ldots \ldots \ldots & \ldots \ldots \ldots \ldots \ldots \\ a_{k1} \ldots a_{kk} \ldots a_{kn} & 0 \ldots 1 \ldots 0 \\ \ldots \ldots \ldots \ldots \ldots \ldots & \ldots \ldots \ldots \ldots \ldots \\ a_{n1} \ldots a_{nk} \ldots a_{nn} & 0 \ldots 0 \ldots 1 \end{pmatrix} .$$

Next we shall perform the necessary elementary operations of Sect. 1.4.2 on full rows of the augmented matrix in order to bring the non-singular matrix $\mathbf{A}$ in the left half into the reduced row echelon form, i.e. so that it becomes an identity matrix $\mathbf{E}$. Each such operation would affect exactly in the same way the right half of the augmented matrix as well. We know that each elementary operation is described by a certain square matrix $\mathbf{U}$. When we apply the first elementary operation $\mathbf{U}_1$, the augmented matrix changes to $(\mathbf{U}_1 \mathbf{A} \vert \mathbf{U}_1 \mathbf{E}) = (\mathbf{U}_1 \mathbf{A} \vert \mathbf{U}_1)$. Applying the second operation $\mathbf{U}_2$ we arrive at $(\mathbf{U}_2 \mathbf{U}_1 \mathbf{A} \vert \mathbf{U}_2 \mathbf{U}_1)$. Repeating this process p times necessary to transform the left half into the identity matrix $\mathbf{E}$, we obtain $(\mathbf{U} \mathbf{A} \vert \mathbf{U}) = (\mathbf{E} \vert \mathbf{U})$, where $\mathbf{U} = \mathbf{U}_p \mathbf{U}_{p-1} \ldots \mathbf{U}_2 \mathbf{U}_1$ is the matrix of the whole transformation. Since $\mathbf{U} \mathbf{A} = \mathbf{E}$, then the matrix $\mathbf{U}$ of the transformation is the required inverse matrix $\mathbf{A}^{-1}$, and it appears in the right half of the augmented matrix.

As an illustration of this method, let us calculate again the inverse of the matrix Eq. (1.87):

$$\begin{pmatrix} 2 & 1 & 1 & 1 & 0 & 0 \\ -1 & 3 & 2 & 0 & 1 & 0 \\ 2 & 0 & 1 & 0 & 0 & 1 \end{pmatrix} \rightarrow \boxed{R_1 + 2R_2 \rightarrow R_2} \rightarrow \begin{pmatrix} 2 & 1 & 1 & 1 & 0 & 0 \\ 0 & 7 & 5 & 1 & 2 & 0 \\ 2 & 0 & 1 & 0 & 0 & 1 \end{pmatrix},$$

$$\boxed{R_1 - R_3 \rightarrow R_3} \rightarrow \begin{pmatrix} 2 & 1 & 1 & 1 & 0 & 0 \\ 0 & 7 & 5 & 1 & 2 & 0 \\ 0 & 1 & 0 & 1 & 0 & -1 \end{pmatrix},$$

$$\boxed{R_2 - 7R_3 \rightarrow R_3} \rightarrow \begin{pmatrix} 2 & 1 & 1 & 1 & 0 & 0 \\ 0 & 7 & 5 & 1 & 2 & 0 \\ 0 & 0 & 5 & -6 & 2 & 7 \end{pmatrix},$$

$$\boxed{\frac{1}{5} R_3 \rightarrow R_3} \rightarrow \begin{pmatrix} 2 & 1 & 1 & 1 & 0 & 0 \\ 0 & 7 & 5 & 1 & 2 & 0 \\ 0 & 0 & 1 & -1.2 & 0.4 & 1.4 \end{pmatrix},$$

$$\boxed{R_3 - \frac{1}{5} R_2 \rightarrow R_2} \rightarrow \begin{pmatrix} 2 & 1 & 1 & 1 & 0 & 0 \\ 0 & -1.4 & 0 & -1.4 & 0 & 1.4 \\ 0 & 0 & 1 & -1.2 & 0.4 & 1.4 \end{pmatrix},$$

$$\boxed{R_3 - R_1 \rightarrow R_1} \rightarrow \begin{pmatrix} -2 & -1 & 0 & | & -2.2 & 0.4 & 1.4 \\ 0 & -1.4 & 0 & | & -1.4 & 0 & 1.4 \\ 0 & 0 & 1 & | & -1.2 & 0.4 & 1.4 \end{pmatrix},$$

$$\boxed{-\frac{5}{7}R_2 \rightarrow R_2} \rightarrow \begin{pmatrix} -2 & -1 & 0 & | & -2.2 & 0.4 & 1.4 \\ 0 & 1 & 0 & | & 1 & 0 & -1 \\ 0 & 0 & 1 & | & -1.2 & 0.4 & 1.4 \end{pmatrix},$$

$$\boxed{R_2 + R_1 \rightarrow R_1} \rightarrow \begin{pmatrix} -2 & 0 & 0 & | & -1.2 & 0.4 & 0.4 \\ 0 & 1 & 0 & | & 1 & 0 & -1 \\ 0 & 0 & 1 & | & -1.2 & 0.4 & 1.4 \end{pmatrix},$$

$$\boxed{-\frac{1}{2}R_1 \rightarrow R_1} \rightarrow \begin{pmatrix} 1 & 0 & 0 & | & 0.6 & -0.2 & -0.2 \\ 0 & 1 & 0 & | & 1 & 0 & -1 \\ 0 & 0 & 1 & | & -1.2 & 0.4 & 1.4 \end{pmatrix},$$

and we obtain exactly the same result in the right half of the augmented matrix for the inverse $\mathbf{A}^{-1}$ as before, when we used the method of co-factors.

Problem 1.78 The matrix of rotations by the angle ϕ around the y axis is

$$\mathbf{R}_y(\phi) = \begin{pmatrix} \cos\phi & 0 & \sin\phi \\ 0 & 1 & 0 \\ -\sin\phi & 0 & \cos\phi \end{pmatrix}.$$

Calculate the inverse matrix $\mathbf{R}_y(\phi)^{-1}$ via building the corresponding co-factors. Interpret your result. Repeat the calculation using the method of Gaussian elimination.

Problem 1.79 Show using the method of co-factors that the inverse of the matrix

$$\mathbf{A} = \begin{pmatrix} a & 0 & f \\ 0 & b & d \\ c & 0 & 1 \end{pmatrix} \quad \text{is} \quad \mathbf{A}^{-1} = \frac{1}{b(a-cf)} \begin{pmatrix} b & 0 & -bf \\ cd & a-cf & -ad \\ -cb & 0 & ab \end{pmatrix}.$$

Repeat the calculation using the method of Gaussian elimination.

Problem 1.80 Consider a tridiagonal $n \times n$ matrix $\mathbf{A}$ shown in the left-hand side of the equation depicted in Fig. 1.6. Show that the element $\left(\mathbf{A}^{-1}\right)_{11}$ of its inverse can be represented as a continued fraction:

$$\left(\mathbf{A}^{-1}\right)_{11} = \frac{1|}{|\alpha_1} - \frac{\beta_1^2|}{|\alpha_2} - \frac{\beta_2^2|}{|\alpha_3} - \frac{\beta_3^2|}{|\alpha_4} - \cdots - \frac{\beta_{n-1}^2|}{\alpha_n}.$$

Problem 1.81 For the same tridiagonal matrix $\mathbf{A}$, show that the element $\left(\mathbf{A}^{-1}\right)_{12}$ of its inverse can be represented as

$$\left(\mathbf{A}^{-1}\right)_{12} = \frac{1}{\beta_1} - \frac{\alpha_1}{\beta_1}\left[\alpha_1 - \frac{\beta_1^2|}{|\alpha_2} - \frac{\beta_2^2|}{|\alpha_3} - \frac{\beta_3^2|}{|\alpha_4} - \cdots - \frac{\beta_{n-1}^2|}{\alpha_n}\right]^{-1}.$$

1.8 Eigenvectors and Eigenvalues of a Square Matrix

1.8.1 General Formulation

When a pth order square matrix $\mathbf{A}$ is multiplied with a vector $\mathbf{x}$, this gives another vector $\mathbf{y} = \mathbf{A}\mathbf{x}$. However, in many applications, most notably in quantum mechanics, it is important to know particular vectors $\mathbf{x}$ for which the transformation $\mathbf{A}\mathbf{x}$ gives a vector in *the same* direction as $\mathbf{x}$, i.e. different from $\mathbf{x}$ only by some (generally complex) constant factor λ:

$$\mathbf{A}\mathbf{x} = \lambda\mathbf{x} . \tag{1.88}$$

Here *both* $\mathbf{x}$ and λ are to be found; this problem is usually called an *eigenproblem*. The number λ is called an *eigenvalue*, and the corresponding to it vector $\mathbf{x}$ *eigenvector*. More than one pair of eigenvectors and eigenvalues may exist for the given square matrix $\mathbf{A}$. Note that the vectors $\mathbf{x}$ and numbers λ are necessarily related to each other by the nature of Eq. (1.88) defining them. Also, note that if $\mathbf{x}$ is a solution of Eq. (1.88) with some λ or, as it is usually said, it is an eigenvector corresponding to the eigenvalue λ, then any vector $c\mathbf{x}$ with an arbitrary complex factor c is also an eigenvector with the same eigenvalue. In other words, eigenvectors are always defined up to an arbitrary pre-factor.

Obviously, the vector $\mathbf{x} = \mathbf{0}$ is a solution of Eq. (1.88) with $\lambda = 0$. However, in applications this trivial solution is never of a value, and so we shall only be interested in non-trivial solutions of this problem.

To solve the problem, we shall rewrite it in the following way: let us take the vector $\lambda\mathbf{x}$ to the left-hand side and rewrite it in a matrix form as $-\lambda\mathbf{E}\mathbf{x}$ using the identity matrix $\mathbf{E}$. We get

$$(\mathbf{A} - \lambda\mathbf{E})\mathbf{x} = \mathbf{0} . \tag{1.89}$$

This equation should ring a bell as it is simply a set of linear algebraic equations with respect to $\mathbf{x}$ with the zero right-hand side. We know from Sect. 1.4 that this system of equations has a non-trivial solution if

$$\det(\mathbf{A} - \lambda\mathbf{E}) = |\mathbf{A} - \lambda\mathbf{E}| = 0 . \tag{1.90}$$

This gives an equation for the eigenvalues λ. By expanding the determinant $|\mathbf{A} - \lambda\mathbf{E}|$, we shall obtain a pth order polynomial in λ where p is the order of the determinant. Therefore, allowed values of λ are obtained from the polynomial equation (1.90) which is called the *secular* or *characteristic* equation. It is known from elementary algebra (Sect. I.1.4.1) that a polynomial equation of order p has always p (generally complex) solutions (roots of the polynomial) of which some may coincide, so that the number of *different* values of λ may in reality be smaller than p, see below.

Theorem 1.16 *The full-rank square matrix always has an inverse.*

Proof Consider a full-rank square matrix $\mathbf{A}$. It is clear that all its eigenvalues a_λ must be non-zero. Indeed, if there was a zero eigenvalue $a_{\lambda_0} = 0$, then the corresponding eigenvector $\mathbf{x}_{\lambda_0} \neq \mathbf{0}$ would satisfy the equation $\mathbf{A}\mathbf{x}_{\lambda_0} = \mathbf{0}$. However, this homogeneous linear system of equations has a non-trivial solution if and only if the matrix $\mathbf{A}$ has zero determinant (Sect. 1.4.1), i.e. it contains linearly dependent rows and hence is of a lower rank than its dimension contradicting our assumption. Hence $|\mathbf{A}| \neq 0$. As $\mathbf{A}$ is non-singular, it always has an inverse. **Q.E.D.**

Problem 1.82 Prove that the square matrices $\mathbf{A}$ and $\mathbf{A}^T$ share the same eigenvalues.

Problem 1.83 Prove that the square matrices $\mathbf{A}$ and $\mathbf{U}^{-1}\mathbf{A}\mathbf{U}$ share the same eigenvalues.

Problem 1.84 Let the $n \times n$ matrix $\mathbf{A}$ has eigenvalues $\lambda_1, \lambda_2, \ldots, \lambda_n$. Show that the eigenvalues of the matrix $\mathbf{A}^m = \underbrace{\mathbf{A}\mathbf{A}\cdots\mathbf{A}}_{m}$ are $\lambda_1^m, \lambda_2^m, \ldots, \lambda_n^m$.

The determinant of a square matrix $\mathbf{A}$ can be directly expressed via its eigenvalues as demonstrated by the following theorem.

Theorem 1.17 *The determinant of a matrix is equal to the product of all its eigenvalues:*

$$det\,\mathbf{A} = \prod_{i=1}^{p} \lambda^{(i)} . \tag{1.91}$$

Proof We know that, after opening up the determinant of $\mathbf{A} - \lambda\mathbf{E}$, we get a polynomial algebraic equation in λ which has roots $\lambda^{(1)}, \ldots, \lambda^{(p)}$, i.e. one can write

$$\det(\mathbf{A} - \lambda\mathbf{E}) = (\lambda^{(1)} - \lambda) \cdots (\lambda^{(p)} - \lambda) .$$

By putting $\lambda = 0$ in the above equation, we obtain the desired result. **Q.E.D.**

An important criterion for checking if a matrix $\mathbf{A}$ is singular, i.e. has a zero determinant, follows immediately from this: if at least one of the eigenvalues of $\mathbf{A}$ is zero, then $|\mathbf{A}| = 0$ and the matrix $\mathbf{A}$ is indeed a singular matrix. In particular, this means that such a matrix does *not* have an inverse.

> **Problem 1.85** Consider a triangular matrix $\mathbf{A}$. Show that its eigenvalues are given by the elements on its diagonal. [Hint: *the result of Problem 1.58 you may find useful.*]

The expression

$$\Delta(\lambda) = \det(\mathbf{A} - \lambda\mathbf{E}) = (\lambda^{(1)} - \lambda) \cdots (\lambda^{(p)} - \lambda) \tag{1.92}$$

is a polynomial of order p, which is the same as the dimension of the square matrix $\mathbf{A}$; it is called the *characteristic polynomial* of matrix $\mathbf{A}$ and is an important object in theory of matrices.

Generally complex matrices would have complex eigenvalues and eigenvectors. Real matrices may have real and complex eigenvalues as solutions of the polynomial equation $\Delta(\lambda) = 0$; if there are complex eigenvalues, they always come in complex conjugate pairs (Sect. I.1.11.5). Correspondingly, eigenvectors of a pair of complex conjugate eigenvalues can also be chosen as complex conjugate of each other. Hence, if the order p of a real matrix $\mathbf{A}$ is even, there could be both real and complex eigenvalues; however, if p is odd, at least one eigenvalue is guaranteed to be real.

Once the eigenvalues are determined, the corresponding eigenvectors $\mathbf{x}$ are found: by taking a particular eigenvalue λ and substituting it into Eq. (1.89), the corresponding eigenvector $\mathbf{x}$, corresponding to this eigenvalue, is obtained by solving this system of linear algebraic equations. Each value of λ gives one vector $\mathbf{x}$ associated with it.

Example 1.16 ▶ Find all eigenvalues and eigenvectors of the matrix

$$\mathbf{A} = \begin{pmatrix} -1 & 1 \\ 4 & 2 \end{pmatrix} .$$

Solution: First of all, we need to solve the secular equation to find the eigenvalues:

$$\left| \begin{pmatrix} -1 & 1 \\ 4 & 2 \end{pmatrix} - \lambda \begin{pmatrix} 1 & 0 \\ 0 & 1 \end{pmatrix} \right| = \left| \begin{pmatrix} -1 & 1 \\ 4 & 2 \end{pmatrix} - \begin{pmatrix} \lambda & 0 \\ 0 & \lambda \end{pmatrix} \right|$$

$$= \begin{vmatrix} -1 - \lambda & 1 \\ 4 & 2 - \lambda \end{vmatrix} = \lambda^2 - \lambda - 6 = 0$$

that has two solutions: $\lambda^{(1)} = 3$ and $\lambda^{(2)} = -2$. Substituting the value of $\lambda^{(1)}$ into the matrix equation $\mathbf{A}\mathbf{x} = \lambda^{(1)}\mathbf{x}$, we obtain two equations for the two components x_1, x_2 of the eigenvector $\mathbf{x}$:

$$\mathbf{A}\mathbf{x} = 3\mathbf{x} \implies \begin{cases} -x_1 + x_2 = 3x_1 \\ 4x_1 + 2x_2 = 3x_2 \end{cases} \text{ or } \begin{cases} -4x_1 + x_2 = 0 \\ 4x_1 - x_2 = 0 \end{cases}.$$

The two equations are equivalent since, by construction, the corresponding rows of the matrix $\mathbf{A} - \lambda^{(1)}\mathbf{E}$ are linearly dependent because its determinant is zero (recall Property 1.7 of the determinant in Sect. 1.3.2). Solving either of the two yields $x_2 = 4x_1$, i.e. the eigenvector corresponding to the eigenvalue $\lambda^{(1)} = 3$ is $\mathbf{x}^{(1)} = \begin{pmatrix} a \\ 4a \end{pmatrix} = a \begin{pmatrix} 1 \\ 4 \end{pmatrix}$, where a is an arbitrary constant.[6] Similarly, using $\lambda^{(2)} = -2$ in the matrix equation $\mathbf{A}\mathbf{x} = \lambda^{(2)}\mathbf{x}$ yields two equations

$$\begin{cases} -x_1 + x_2 = -2x_1 \\ 4x_1 + 2x_2 = -2x_2 \end{cases} \text{ or } \begin{cases} x_1 + x_2 = 0 \\ 4x_1 + 4x_2 = 0 \end{cases}.$$

Again, both equations are equivalent, and thus only $x_2 = -x_1$ follows from either of them, giving the eigenvector $\mathbf{x}^{(2)} = \begin{pmatrix} b \\ -b \end{pmatrix} = b \begin{pmatrix} 1 \\ -1 \end{pmatrix}$, where b is another arbitrary constant. The two arbitrary constants a and b can be given a certain value (for certainty) by imposing an additional condition on the eigenvectors. A usual choice in physics is to construct them of unit length. This gives $a = 1/\sqrt{17}$ and $b = 1/\sqrt{2}$. Notice that the two eigenvectors $\mathbf{x}^{(1)} = \frac{1}{\sqrt{17}} \begin{pmatrix} 1 \\ 4 \end{pmatrix}$ and $\mathbf{x}^{(2)} = \frac{1}{\sqrt{2}} \begin{pmatrix} 1 \\ -1 \end{pmatrix}$ are linearly independent.

To check that the obtained eigenvectors and eigenvalues are correct, it is advisable to use them directly in the equation $\mathbf{A}\mathbf{x} = \lambda\mathbf{x}$, e.g.

$$\begin{pmatrix} -1 & 1 \\ 4 & 2 \end{pmatrix} \begin{pmatrix} a \\ 4a \end{pmatrix} = \begin{pmatrix} -a + 4a \\ 4a + 8a \end{pmatrix} = \begin{pmatrix} 3a \\ 12a \end{pmatrix} = 3 \begin{pmatrix} a \\ 4a \end{pmatrix},$$

as it should be! ◄

Example 1.17 ► Find eigenvectors and eigenvalues of the matrix

$$\mathbf{A} = \begin{pmatrix} 1 & 1 & 1 \\ 1 & 0 & -2 \\ 1 & -1 & 1 \end{pmatrix}.$$

Solution: The secular equation in this case

---

[6] Recall that each eigenvector is defined up to an arbitrary constant pre-factor anyway.

$$\begin{vmatrix} 1-\lambda & 1 & 1 \\ 1 & 0-\lambda & -2 \\ 1 & -1 & 1-\lambda \end{vmatrix} = (1-\lambda)\,[-\lambda(1-\lambda)-2] - [(1-\lambda)+2]$$

$$+\,[-1+\lambda] = -(\lambda^2-3)(\lambda-2) = 0$$

(we have expanded the determinant with respect to its first row), yielding $\lambda^{(1,2)} = \pm\sqrt{3}$ and $\lambda^{(3)} = 2$. The system of equations for the first eigenvector $\mathbf{x}^{(1)}$, which corresponds to $\lambda^{(1)} = \sqrt{3}$, reads

$$\begin{cases} (1-\sqrt{3})x_1 + x_2 + x_3 = 0 \\ x_1 - \sqrt{3}x_2 - 2x_3 = 0 \\ x_1 - x_2 + (1-\sqrt{3})x_3 = 0 \end{cases},$$

which yields the vector $\mathbf{x}^{(1)} = a\begin{pmatrix} -1 \\ -\sqrt{3} \\ 1 \end{pmatrix} = a\left(-1, -\sqrt{3}, 1\right)^T$ with arbitrary a.

Again, one of the equations must be equivalent to two others, so we set $x_3 = a$ and expressed x_1 and x_2 from the first two equations via x_3. Similarly, for the second eigenvalue $\lambda^{(2)} = -\sqrt{3}$ we get the system of equations

$$\begin{cases} (1+\sqrt{3})x_1 + x_2 + x_3 = 0 \\ x_1 + \sqrt{3}x_2 - 2x_3 = 0 \\ x_1 - x_2 + (1+\sqrt{3})x_3 = 0 \end{cases},$$

that yields the eigenvector $\mathbf{x}^{(2)} = b\begin{pmatrix} -1 \\ \sqrt{3} \\ 1 \end{pmatrix} = b\left(-1, \sqrt{3}, 1\right)^T$. Finally, performing similar calculations for the third eigenvalue $\lambda^{(3)} = 2$, we get the equations

$$\begin{cases} -x_1 + x_2 + x_3 = 0 \\ x_1 - 2x_2 - 2x_3 = 0 \\ x_1 - x_2 - x_3 = 0 \end{cases},$$

which results in $\mathbf{x}^{(3)} = c\begin{pmatrix} 0 \\ -1 \\ 1 \end{pmatrix} = c\,(0, -1, 1)^T$. In this case the first and the third equations are obviously equivalent and hence one of them should be dropped. We fixed $x_3 = c$ and expressed x_1 and x_2 via x_3 from the second and third equations.

Hence, in this case we obtained three solutions of the eigenproblem, i.e. three pairs of eigenvectors and corresponding to them eigenvalues. ◄

In some cases care is needed in solving equations for the eigenvectors as illustrated by the following example.

Example 1.18 ▶ Find eigenvectors and eigenvalues of a matrix

$$
\mathbf{A} = \begin{pmatrix} 1 & 0 & 0 \\ 0 & 4 & 2 \\ 0 & 2 & 4 \end{pmatrix} .
$$

Solution: The secular equation in this case:

$$
\begin{vmatrix} 1-\lambda & 0 & 0 \\ 0 & 4-\lambda & 2 \\ 0 & 2 & 4-\lambda \end{vmatrix} = (1-\lambda)\left[(4-\lambda)^2 - 4\right] = 0 ,
$$

giving $\lambda_1 = 1$ and $4 - \lambda = \pm 2$ for the other two solutions, i.e. $\lambda_2 = 2$ and $\lambda_3 = 6$. Consider now the first eigenvector:

$$
\begin{pmatrix} 1 & 0 & 0 \\ 0 & 4 & 2 \\ 0 & 2 & 4 \end{pmatrix} \begin{pmatrix} x_1 \\ x_2 \\ x_3 \end{pmatrix} = \begin{pmatrix} x_1 \\ x_2 \\ x_3 \end{pmatrix} \implies \begin{cases} x_1 = x_1 \\ 3x_2 + 2x_3 = 0 \\ 2x_2 + 3x_3 = 0 \end{cases}
$$

$$
\implies \begin{cases} x_1 \text{ is arbitrary} \\ x_2 = 0 \\ x_3 = 0 \end{cases} .
$$

Note that the first equation has as a solution any x_1; the second and the third equations should be solved simultaneously (e.g. solve for x_2 from the second equation and substitute into the third), resulting in the zero solution for x_2 and x_3.[7] So, the first eigenvector corresponding to $\lambda_1 = 1$ is $\mathbf{x}^{(1)} = (a, 0, 0)^T$ with an arbitrary constant a. Finding eigenvectors for the other two eigenvalues is more straightforward. Consider for instance the second one, $\lambda_2 = 2$:

$$
\begin{pmatrix} 1 & 0 & 0 \\ 0 & 4 & 2 \\ 0 & 2 & 4 \end{pmatrix} \begin{pmatrix} x_1 \\ x_2 \\ x_3 \end{pmatrix} = 2 \begin{pmatrix} x_1 \\ x_2 \\ x_3 \end{pmatrix} \implies \begin{cases} x_1 = 0 \\ 2x_2 + 2x_3 = 0 \\ 2x_2 + 2x_3 = 0 \end{cases}
$$

$$
\implies \begin{cases} x_1 = 0 \\ x_2 = -x_3 \\ x_3 = -x_2 \end{cases} .
$$

---

[7] This situation can also be considered as a solution of two algebraic linear equations with the zero right-hand side:

$$
\begin{cases} 3x_2 + 2x_3 = 0 \\ 2x_2 + 3x_3 = 0 \end{cases} .
$$

As we know, this system of equations has a non-trivial solution only if the determinant of its matrix $\begin{pmatrix} 3 & 2 \\ 2 & 3 \end{pmatrix}$ is equal to zero. Obviously, this is not the cases and hence only the zero solution exists.

We see that in this case the second and the third equations give identical information about x_2 and x_3, which is that $x_2 = -x_3$; nothing can be said about the absolute values of them. Therefore, we can take $x_2 = b$ with an arbitrary b and then write the eigenvector as $\mathbf{x}^{(2)} = (0, b, -b)^T = b\,(0, 1, -1)^T$. Similarly the third eigenvector is found to be $\mathbf{x}^{(3)} = (0, c, c)^T = c\,(0, 1, 1)^T$ with an arbitrary c. ◄

The method we considered so far relies on some intuition. Indeed, it is not always clear which of the equations that need to be solved to find eigenvectors are linearly dependent and hence which variables (and how many of them) need to be considered as arbitrary. At this junction it is useful to recall that the Gaussian elimination method considered in Sect. 1.4.2 is extremely convenient in solving linear equations in general, and of course it is indispensable in finding eigenvectors as it allows for an automatic determination of linear dependencies. For instance, it can tell you straight away if, e.g. in a three-dimensional problem we deal with one or two linearly dependent equations avoiding any uncertainty there.

Consider a problem of finding the eigenvector $\mathbf{x}$ of the eigenproblem $\mathbf{A}\mathbf{x} = \lambda\mathbf{x}$ for a certain eigenvalue λ. This system of equations is equivalent to $\mathbf{D}\mathbf{x} = \mathbf{0}$ with $\mathbf{D} = \mathbf{A} - \lambda\mathbf{E}$, to which we apply the Gauss elimination. Solving it, we should get a general solution via arbitrary parameters, the number of which in each particular case (equal to the number of linearly dependent equations) would follow automatically from the procedure itself; the intuition that we exercised above in our examples to do this would not be required.

Consider as an illustration our Example 1.17 with the eigenvalue $\lambda^{(3)} = 2$. The required equations and the corresponding matrix $\mathbf{D}$ are

$$\begin{cases} -x_1 + x_2 + x_3 = 0 \\ x_1 - 2x_2 - 2x_3 = 0 \\ x_1 - x_2 - x_3 = 0 \end{cases} \quad \text{and } \mathbf{D} = \begin{pmatrix} -1 & 1 & 1 \\ 1 & -2 & -2 \\ 1 & -1 & -1 \end{pmatrix}.$$

After elementary row operations $R_1 + R_2 \to R_2$ and $R_1 + R_3 \to R_3$ we obtain the matrix

$$\tilde{\mathbf{D}} = \begin{pmatrix} -1 & 1 & 1 \\ 0 & -1 & -1 \\ 0 & 0 & 0 \end{pmatrix}$$

in the row echelon form associated with the equations

$$\begin{cases} -x_1 + x_2 + x_3 = 0 \\ -x_2 - x_3 = 0 \\ 0 = 0 \end{cases}.$$

Choosing $x_3 \equiv c$ as arbitrary, we obtain instead

$$\begin{cases} -x_1 + x_2 = -c \\ -x_2 = c \end{cases} \implies \begin{cases} x_1 = 0 \\ x_2 = -c \end{cases} \quad \text{and } x_3 = c.$$

Hence, the required solution $\mathbf{x}^{(3)} = c\,(0, -1, 1)^T$, the same as before. The constant c can be chosen $c = 1/\sqrt{2}$ to normalise the eigenvector to one if required; otherwise, any non-zero value of c will do, e.g. $c = 1$.

If eigenvalues of a matrix $\mathbf{A}$ are all distinct, they are said to be *non-degenerate*. If an eigenvalue repeats itself, it is said to be *degenerate*. We have seen in the examples above that if a matrix $\mathbf{A}$ of dimension p has all different eigenvalues, it has exactly p linearly independent eigenvectors. In fact, this is a very general result that is proven by the following theorem.

Theorem 1.18 *If all eigenvalues of a square matrix $\mathbf{A}$ are different, then all its eigenvectors are linearly independent.*

Proof Let eigenvalues and eigenvectors of the matrix $\mathbf{A}$ be $\lambda^{(i)}$ and $\mathbf{x}^{(i)}$ ($i = 1, \ldots, p$), respectively. If the vectors are all linearly independent, then the equation

$$c_1 \mathbf{x}^{(1)} + \cdots + c_p \mathbf{x}^{(p)} = \sum_{i=1}^{p} c_i \mathbf{x}^{(i)} = 0 \tag{1.93}$$

has only the trivial solution $c_1 = c_2 = \cdots = c_p = 0$. This is what we have to prove. Let us first show that $c_1 = 0$. To this end, multiply the equation above with the matrix $\mathbf{A} - \lambda^{(2)}\mathbf{E}$ from the left:

$$\left(\mathbf{A} - \lambda^{(2)}\mathbf{E}\right) \sum_{i=1}^{p} c_i \mathbf{x}^{(i)} = c_2 \left(\mathbf{A} - \lambda^{(2)}\mathbf{E}\right) \mathbf{x}^{(2)} + \left(\mathbf{A} - \lambda^{(2)}\mathbf{E}\right) \sum_{i \neq 2} c_i \mathbf{x}^{(i)} = 0 \,,$$

where in the right-hand side we sum over all values of i between 1 and p except $i = 2$. Because $\mathbf{x}^{(2)}$ is an eigenvector of $\mathbf{A}$ corresponding to the eigenvalue $\lambda^{(2)}$, the second term in the right-hand side is equal to zero, $\left(\mathbf{A} - \lambda^{(2)}\mathbf{E}\right) \mathbf{x}^{(2)} = 0$, so that we can write

$$\left(\mathbf{A} - \lambda^{(2)}\mathbf{E}\right) \sum_{i \neq 2} c_i \mathbf{x}^{(i)} = 0 \,.$$

Now act from the left with the matrix $\left(\mathbf{A} - \lambda^{(3)}\mathbf{E}\right)$. First, we notice that the matrices $\left(\mathbf{A} - \lambda^{(2)}\mathbf{E}\right)$ and $\left(\mathbf{A} - \lambda^{(3)}\mathbf{E}\right)$ commute since $\mathbf{A}$ and $\mathbf{E}$ do. Thus, we have

$$\left(\mathbf{A} - \lambda^{(3)}\mathbf{E}\right) \left(\mathbf{A} - \lambda^{(2)}\mathbf{E}\right) \sum_{i \neq 2} c_i \mathbf{x}^{(i)} = \left(\mathbf{A} - \lambda^{(2)}\mathbf{E}\right) \left(\mathbf{A} - \lambda^{(3)}\mathbf{E}\right) \sum_{i \neq 2} c_i \mathbf{x}^{(i)} = 0 \,.$$

Similarly to the above, the term $i = 3$ in the sum will disappear since $\mathbf{x}^{(3)}$ is the eigenvector corresponding to the eigenvalue $\lambda^{(3)}$. Thus we get

$$\left(\mathbf{A} - \lambda^{(3)} E\right)\left(\mathbf{A} - \lambda^{(2)} E\right) \sum_{i \neq 2,3} c_i \mathbf{x}^{(i)} = 0 \, .$$

Repeating this procedure, we finally remove all the terms in the sum except for the very first one:

$$\left[\prod_{i=2}^{n} \left(\mathbf{A} - \lambda^{(i)} E\right)\right] c_1 \mathbf{x}^{(1)} = \left(\mathbf{A} - \lambda^{(p)} E\right) \cdots \left(\mathbf{A} - \lambda^{(3)} E\right)\left(\mathbf{A} - \lambda^{(2)} E\right) c_1 \mathbf{x}^{(1)} \tag{1.94}$$
$$= 0 \, .$$

However, for any $i \neq 1$, we get

$$\left(\mathbf{A} - \lambda^{(i)} E\right) \mathbf{x}^{(1)} = \mathbf{A} \mathbf{x}^{(1)} - \lambda^{(i)} \mathbf{x}^{(1)} = \lambda^{(1)} \mathbf{x}^{(1)} - \lambda^{(i)} \mathbf{x}^{(1)} = \left(\lambda^{(1)} - \lambda^{(i)}\right) \mathbf{x}^{(1)} \, ,$$

Therefore, after repeatedly using the above identity, Eq. (1.94) turns into

$$c_1 \left(\lambda^{(1)} - \lambda^{(p)}\right) \cdots \left(\lambda^{(1)} - \lambda^{(3)}\right)\left(\lambda^{(1)} - \lambda^{(2)}\right) \mathbf{x}^{(1)} = 0 \, .$$

If all the eigenvalues are different, then this equation can only be satisfied if $c_1 = 0$.

Similarly, by operating with the matrix $\prod_{i \neq 2} \left(\mathbf{A} - \lambda^{(i)} E\right)$ on the left-hand side of Eq. (1.93), we obtain $c_2 = 0$. All other coefficients c_i are found to be zero in the same way. **Q.E.D.**

Note that if there are some repeated (degenerate) eigenvalues, the number of distinct eigenvectors may be smaller than the dimension of the matrix p. If the number of linearly independent eigenvectors is smaller than the dimension of the matrix, the latter is called *defective*. The two examples below illustrate this point.

Example 1.19 ▶ Find eigenvalues and eigenvectors of a matrix

$$\mathbf{A} = \begin{pmatrix} 1 & -1 \\ 1 & 3 \end{pmatrix} \, .$$

Solution: The characteristic equation gives

$$\begin{vmatrix} 1 - \lambda & -1 \\ 1 & 3 - \lambda \end{vmatrix} = (\lambda - 2)^2 = 0 \, ,$$

leading to a single (degenerate) eigenvalue $\lambda = 2$. The corresponding eigenvectors are obtained from the equation $(\mathbf{A} - 2E)\mathbf{x} = 0$, which has the form:

$$\begin{cases} -x_1 - x_2 = 0 \\ x_1 + x_2 = 0 \end{cases} ,$$

from where we get $x_1 = -x_2$. Hence only a single eigenvector $\mathbf{x}^{(1)} = a(1, -1)^T$ is found. Of course, formally, one can always construct the second eigenvector, $\mathbf{x}^{(2)} = b(1, -1)^T$, but this is linearly dependent of the first one. Thus, there is only a single linearly independent eigenvector. Hence, this matrix is defective.◄

In the next example we have repeated eigenvalues, however, it is possible to construct as many linearly independent eigenvectors as the dimensionality of the matrix.

Example 1.20 ► Find eigenvectors and eigenvalues of the matrix

$$\mathbf{A} = \begin{pmatrix} 3 & -2 & -1 \\ 3 & -4 & -3 \\ 2 & -4 & 0 \end{pmatrix}$$

Solution: The eigenvalues of $\mathbf{A}$ are obtained via

$$\begin{vmatrix} 3 - \lambda & -2 & -1 \\ 3 & -4 - \lambda & -3 \\ 2 & -4 & -\lambda \end{vmatrix} = (-5 - \lambda)(2 - \lambda)^2 = 0 ,$$

yielding $\lambda^{(1)} = -5$, $\lambda^{(2,3)} = 2$. For the first eigenvalue, the eigenvector is obtained in the usual way from

$$(\mathbf{A} + 5\mathbf{E})\mathbf{x} = \begin{cases} 8x_1 - 2x_2 - x_3 = 0 \\ 3x_1 + x_2 - 3x_3 = 0 \\ 2x_1 - 4x_2 + 5x_3 = 0 \end{cases} ,$$

yielding $\mathbf{x}^{(1)} = a(1, 3, 2)^T$. The situation with the other two eigenvectors is a bit peculiar: since both eigenvalues are degenerate, the eigenvector equations are the same:

$$(\mathbf{A} - 2\mathbf{E})\mathbf{x} = \begin{cases} x_1 - 2x_2 - x_3 = 0 \\ 3x_1 - 6x_2 - 3x_3 = 0 \\ 2x_1 - 4x_2 - 2x_3 = 0 \end{cases}$$

or, after simplification,

$$\begin{cases} x_1 - 2x_2 - x_3 = 0 \\ x_1 - 2x_2 - x_3 = 0 \\ x_1 - 2x_2 - x_3 = 0 \end{cases} ,$$

i.e. all *three* equations are identical! Thus, the only relationship between components of either of the eigenvectors $\mathbf{x}^{(2)}$ and $\mathbf{x}^{(3)}$ is that $x_1 = 2x_2 + x_3$, i.e. there are *two* arbitrary constants possible. This means that *two* linearly independent vectors *can* be constructed. There is an infinite number of possibilities. For instance, if we take $x_2 = 0$, then we obtain $x_1 = x_3$ and hence $\mathbf{x}^{(2)} = a(1, 0, 1)^T$, and by setting instead

$x_3 = 0$, we obtain another vector $\mathbf{x}^{(3)} = b(2, 1, 0)^T$. It is easily seen that all three vectors are linearly independent. Hence, in this case the matrix $\mathbf{A}$ is not defective.

We see that the eigenvectors $\mathbf{x}^{(2)}$ and $\mathbf{x}^{(3)}$ correspond to the same eigenvalue $\lambda^{(2,3)} = 2$. As shown in the following problem, any linear combination of vectors $\mathbf{x}^{(2)}$ and $\mathbf{x}^{(3)}$,

$$\mathbf{x}^{(2)\prime} = \alpha_2 \mathbf{x}^{(2)} + \beta_2 \mathbf{x}^{(3)} \quad \text{and} \quad \mathbf{x}^{(3)\prime} = \alpha_3 \mathbf{x}^{(2)} + \beta_3 \mathbf{x}^{(3)} \; ,$$

will also serve as the second and third eigenvectors provided that the new vectors are linearly independent.

Problem 1.86 Let an eigenvalue λ of a matrix $\mathbf{A}$ be p-fold degenerate, and there are $k \le p$ linearly independent eigenvectors $\mathbf{x}^{(i)}$, $i = 1, \ldots, k$. Demonstrate that any linear combination

$$\mathbf{y}^{(i)} = \sum_{j=1}^{k} \alpha_{ij} \mathbf{x}^{(j)}$$

of the eigenvectors, corresponding to the same eigenvalue λ, are also eigenvectors of $\mathbf{A}$ with the same λ.

Let us apply now the Gauss elimination procedure to the last (degenerate) case as well for comparison. The matrix

$$\mathbf{D} = \mathbf{A} - 2\mathbf{E} = \begin{pmatrix} 1 & -2 & -1 \\ 3 & -6 & -3 \\ 2 & -4 & -2 \end{pmatrix}$$

is brought into the row echelon form

$$\begin{pmatrix} 1 & -2 & -1 \\ 0 & 0 & 0 \\ 0 & 0 & 0 \end{pmatrix}$$

by the elementary row operations $R_1 - \frac{1}{3} R_2 \to R_2$ and $R_1 - \frac{1}{2} R_3 \to R_3$. Taking $x_2 \equiv t_1$ and $x_3 \equiv t_2$ with t_1 and t_2 being arbitrary parameters, we obtain $x_1 = 2t_1 + t_2$. Hence, we obtain an infinite family of vectors

$$\mathbf{x} = \begin{pmatrix} 2t_1 + t_2 \\ t_1 \\ t_2 \end{pmatrix}$$

that span a 2D space (a plane). From these we need to pick up two linearly independent ones. The first vector can be chosen by setting, e.g. $t_1 = 0$ and $t_2 = 1$ resulting in $\mathbf{x}_2 = (1, 0, 1)^T$. The second vector can be chosen using $t_1 = 1$ and $t_2 = 0$, yielding $\mathbf{x}_3 = (2, 1, 0)^T$. Both vectors are the sane as before. The two vectors can be made orthogonal to each other using the Gram–Schmidt procedure of Sect. 1.1.3 if required.   ◀

Problem 1.87 Show that the normalised to unity eigenvectors of the matrix

$$\mathbf{A} = \begin{pmatrix} 1 & 2 \\ 2 & 1 \end{pmatrix} \quad \text{are} \quad \mathbf{x}^{(1)} = \frac{1}{\sqrt{2}} \begin{pmatrix} 1 \\ 1 \end{pmatrix} \quad \text{and} \quad \mathbf{x}^{(2)} = \frac{1}{\sqrt{2}} \begin{pmatrix} 1 \\ -1 \end{pmatrix},$$

they correspond to the eigenvalues 3 and -1, respectively.

Problem 1.88 Show that eigenvectors of the matrix

$$\mathbf{A} = \begin{pmatrix} 3 & 1 & 0 \\ 1 & 3 & 0 \\ 0 & 0 & 2 \end{pmatrix}$$

are

$$\mathbf{x}^{(1)} = a \begin{pmatrix} 1 \\ -1 \\ 0 \end{pmatrix}, \quad \mathbf{x}^{(2)} = b \begin{pmatrix} 1 \\ 1 \\ 0 \end{pmatrix} \quad \text{and} \quad \mathbf{x}^{(3)} = c \begin{pmatrix} 0 \\ 0 \\ 1 \end{pmatrix},$$

where a, b and c are arbitrary constants. The corresponding eigenvalues are 2, 4 and 2.

Problem 1.89 Find the eigenvectors and eigenvalues of the matrix

$$\mathbf{A} = \begin{pmatrix} 2 & 2 & 1 \\ 1 & 3 & 1 \\ 1 & 2 & 2 \end{pmatrix}.$$

[Answer: *the eigenvalues are* 1, 1, 5, *while the corresponding eigenvectors are, for instance,* $(-1, 0, 1)$, $(-2, 1, 0)$ *and* $(1, 1, 1)$, *respectively.*]

Problem 1.90 Let $\mathbf{x}$ and λ be the eigenvector and eigenvalue of a unitary matrix $\mathbf{A}$, i.e. $\mathbf{A}\mathbf{x} = \lambda\mathbf{x}$ and $\mathbf{A}^\dagger = \mathbf{A}^{-1}$. Show that $\mathbf{A}^\dagger$ has the same eigenvector with the eigenvalue $1/\lambda$.

Problem 1.91 Use the result of the previous problem to show that $|\lambda|^2 = 1$, i.e. any eigenvalue of a unitary matrix has the absolute value of one, and correspondingly eigenvalues of an orthogonal matrix can only be ± 1. [Hint: consider the dot product $\left(\mathbf{A}^\dagger\mathbf{x}, \mathbf{x}\right)$.]

Problem 1.92 Prove that eigenvalues of an anti-Hermitian (or skew-Hermitian) matrix $\mathbf{A}$ defined via $\mathbf{A}^{\dagger} = -\mathbf{A}$ are purely imaginary (including zero).

Problem 1.93 Show that determinant of the matrix

$$
\mathbf{A} = \begin{pmatrix}
\alpha_0 & \beta_1 & \beta_2 & \cdots & \beta_N \\
\beta_1 & \alpha_1 & 0 & \cdots & 0 \\
\beta_2 & 0 & \alpha_2 & \cdots & 0 \\
\cdots\cdots\cdots & & & \ddots & \cdots \\
\beta_N & 0 & 0 & \cdots & \alpha_N
\end{pmatrix}
$$

can be written as

$$
\det \mathbf{A} = \left(\alpha_0 - \sum_{j=1}^{N} \frac{\beta_j^2}{\alpha_j} \right) \prod_{i=1}^{N} \alpha_i \,.
$$

Show that all eigenvalues λ of the matrix $\mathbf{A}$ are obtained by solving the transcendental equation

$$
\alpha_0 - \lambda = \sum_{j=1}^{N} \frac{\beta_j^2}{\alpha_j - \lambda} \,.
$$

Prove specifically that the eigenvalues are not given by the diagonal elements α_j. Then, show that the normalised to unity eigenvector of $\mathbf{A}$ corresponding to the eigenvalue λ is given by

$$
\mathbf{e}_{\lambda} = \frac{1}{\sqrt{1 + \sum_{j=1}^{N} \gamma_j^2}} \begin{pmatrix}
1 \\
-\gamma_1 \\
-\gamma_2 \\
\cdots \\
-\gamma_N
\end{pmatrix} ,
$$

where $\gamma_i = \beta_i / (\alpha_i - \lambda)$. Finally, show that any two eigenvectors, $\mathbf{e}_{\lambda}$ and $\mathbf{e}_{\lambda'}$, corresponding to different eigenvalues $\lambda \neq \lambda'$, are orthogonal.

1.8.2 Algebraic and Geometric Multiplicities

Consider a square matrix $\mathbf{A}$ of dimension p. Its eigenvalues λ_k are determined by finding the p roots of the characteristic polynomial, $\Delta(\lambda) = 0$. We know from algebra (see Sects. I.1.4.1 and I.2.3.1) that roots of a polynomial may be repeated. In general, there will be $k \leq p$ distinct roots $\{\lambda_i; i = 1, \ldots, k\}$, and we can write

$$\Delta(\lambda) = \prod_{i=1}^{k} (\lambda_i - \lambda)^{n_i} \, ,$$

where n_i is the degeneracy of the root λ_i. It is also called the *algebraic multiplicity* of the eigenvalue λ_i. Of course, the total number of all eigenvalues, including their repetitions, must be equal to the size of the matrix:

$$n_1 + n_2 + \cdots + n_k = p \, .$$

If all eigenvectors are distinct, it is always possible to construct exactly p linearly independent eigenvectors $\{\mathbf{x}_i\}$ (Theorem 1.18). These vectors can be used as a basis of a p-dimensional vector space. The same is true for many other matrices that may even contain repeated eigenvalues, one can still construct p linearly independent eigenvectors. For instance, we shall see below that this is always possible for a wide class of the so-called normal matrices, and their particular cases such as symmetric and Hermitian matrices.

However, we have seen above by considering different examples that in the cases of defective matrices it is not possible to construct exactly p linearly independent eigenvectors. This may only happen if the eigenvalues are repeated, i.e. their algebraic multiplicity is larger than one. The number of linearly independent eigenvectors that one can construct for the repeated eigenvalue λ_i is called its *geometric multiplicity*. If the given eigenvalue has algebraic multiplicity n_i, its geometric multiplicity g_i lies between 1 and n_i, i.e. $1 \le g_i \le n_i$. If $g_i < n_i$ at least for one of the eigenvalues λ_i, then the vector space constructed from all linearly independent eigenvectors will have the dimension $g = g_1 + g_2 + \cdots + g_k$ strictly smaller than the dimensionality p of the matrix, $1 \le g < p$. Still, as will be discussed below (see Sect. 1.18) it is still possible to generate missing $p - g$ linearly independent vectors to supplement the vector space; these extra vectors will not be, however, eigenvectors of the matrix $\mathbf{A}$, although they will be linearly independent to all its eigenvectors. Hence, for any square $p \times p$ matrix one can always construct a p-dimensional space associated with it.

1.8.3   Left Eigenproblem

So far we have been discussing the so-called right eigenproblem in which the eigenvector appears on the right-hand side of a $p \times p$ matrix, $\mathbf{A}\mathbf{x} = \lambda \mathbf{x}$. For general square matrices one may also consider left eigenproblem as well,

$$\mathbf{y}^{\dagger} \mathbf{A} = \mu \mathbf{y}^{\dagger} \, . \tag{1.95}$$

Here $\mathbf{y}$ is the left eigenvector and μ is the corresponding eigenvalue.

The left eigenproblem is solved exactly in the same way as the right one: first, the eigenvalues μ are determined by solving the same polynomial equation $|\mathbf{A} - \mu\mathbf{E}| = 0$ as for the right eigenvalues ensuring that the left and right eigenvalues are exactly the same, $\mu_i = \lambda_i$ $(i = 1, \ldots, p)$, and then, for each eigenvalue μ_i, the corresponding eigenvector is found by solving the corresponding linear algebraic equations $\mathbf{y}_i^\dagger (\mathbf{A} - \mu\mathbf{E}) = \mathbf{0}$.

Alternatively, we note that the left eigenproblem can be easily converted into a more conventional right one. Indeed, by taking the Hermitian conjugate of Eq. (1.95) one obtains a right eigenproblem, $\mathbf{A}^\dagger \mathbf{y} = \mu^* \mathbf{y}$, for the matrix $\mathbf{A}^\dagger$ which can be solved using the methods discussed above. Hence, the left eigenvectors of $\mathbf{A}$ coincide with the right eigenvectors of its Hermitian conjugate matrix.

Problem 1.94 Consider a square matrix

$$\mathbf{A} = \begin{pmatrix} 1 & -1 & 2 \\ 2 & 1 & -1 \\ -1 & 2 & 1 \end{pmatrix}.$$

Show that the right eigenvectors, corresponding to eigenvalues $\lambda_1 = \frac{1}{2}\left(1 + i3\sqrt{3}\right)$, $\lambda_2 = \frac{1}{2}\left(1 - i3\sqrt{3}\right)$ and $\lambda_3 = 2$, are

$$\mathbf{x}_1 = \begin{pmatrix} -1 - i\sqrt{3} \\ -1 + i\sqrt{3} \\ 2 \end{pmatrix}, \quad \mathbf{x}_2 = \begin{pmatrix} -1 + i\sqrt{3} \\ -1 - i\sqrt{3} \\ 2 \end{pmatrix}, \quad \mathbf{x}_3 = \begin{pmatrix} 1 \\ 1 \\ 1 \end{pmatrix},$$

while the left eigenvectors can be chosen as follows: $\mathbf{y}_1 = \mathbf{x}_2$, $\mathbf{y}_2 = \mathbf{x}_1$ and $\mathbf{y}_3 = \mathbf{x}_3$.

So, in this example the left eigenvectors are composed out of the right ones. In the following example the eigenvectors are different.

Problem 1.95 Consider a square matrix

$$\mathbf{A} = \begin{pmatrix} 2 & 0 & 0 \\ 0 & 1 & 3 \\ 0 & 2 & 3 \end{pmatrix}.$$

Show that the right eigenvalues $\lambda_{1,2} = 2 \pm \sqrt{7}$ and $\lambda_3 = 2$ correspond to the eigenvectors

$$\mathbf{x}_{1,2} = \begin{pmatrix} 0 \\ \left(-1\pm\sqrt{7}\right)/2 \\ 1 \end{pmatrix} \text{ and } \mathbf{x}_3 = \begin{pmatrix} 1 \\ 0 \\ 0 \end{pmatrix},$$

while the left eigenvectors are different,

$$\mathbf{y}_{1,2} = \begin{pmatrix} 0 \\ \left(-1\pm\sqrt{7}\right)/3 \\ 1 \end{pmatrix} \text{ and } \mathbf{y}_3 = \begin{pmatrix} 1 \\ 0 \\ 0 \end{pmatrix},$$

while corresponding to the same eigenvalues.

It is easy to show that the left and right eigenvectors corresponding to different eigenvalues are orthogonal. Indeed, consider two eigensolutions with two different eigenvalues:

$$\mathbf{A}\mathbf{x}_i = \lambda_i \mathbf{x}_i \text{ and } \mathbf{y}_j^\dagger \mathbf{A} = \lambda_j \mathbf{y}_j^\dagger.$$

Multiply from the left the first equation by $\mathbf{y}_j^\dagger$ and make use of the second equation:

$$\left(\mathbf{y}_j^\dagger \mathbf{A}\right)\mathbf{x}_i = \lambda_i \mathbf{y}_j^\dagger \mathbf{x}_i \Rightarrow \lambda_j \mathbf{y}_j^\dagger \mathbf{x}_i = \lambda_i \mathbf{y}_j^\dagger \mathbf{x}_i \Rightarrow (\lambda_j - \lambda_i)\mathbf{y}_j^\dagger \mathbf{x}_i = 0.$$

Since the two eigenvalues are different, the eigenvectors must be orthogonal, $\mathbf{y}_j^\dagger \mathbf{x}_i = 0$, as required. The reader is asked to check this statement using the two examples given above.

Consider now eigenvectors corresponding to a degenerate eigenvalue. A linear combination of either right or left eigenvectors associated with this single eigenvalue would also be eigenvectors of the same matrix with this eigenvalue. If we have n degenerate eigenvectors, one may construct n new right and left eigenvectors,

$$\widetilde{\mathbf{x}}_k = \sum_{l=1}^n \alpha_{kl}\mathbf{x}_l \text{ and } \widetilde{\mathbf{y}}_k = \sum_{l=1}^n \beta_{kl}\mathbf{y}_l, \quad k = 1,\ldots,n.$$

These linear combinations are built using two $n \times n$ matrices $\boldsymbol{\alpha} = (\alpha_{kl})$ and $\boldsymbol{\beta} = (\beta_{kl})$, which contain altogether $2n^2$ parameters. Therefore, it is clear that one can choose these parameters in such a way that new left and right eigenvectors would satisfy n^2 orthonormality conditions $\widetilde{\mathbf{y}}_k^\dagger \widetilde{\mathbf{x}}_{k'} = \delta_{kk'}$; since the number of conditions is twice as small as the number of parameters one can choose, this orthogonalisation seems to be possible. This consideration shows that one can always choose left and right eigenvectors of a degenerate eigenvalue to be mutually orthonormal. We also know that the left and right eigenvectors corresponding to *different* eigenvalues are orthogonal automatically. Finally, if the $p \times p$ matrix $\mathbf{A}$ is non-defective, there will be exactly p left and right eigenvectors. Hence, we can conclude that for any non-defective matrix one can always choose its right and left eigenvectors to be orthonormal to each other in the following sense:

$$\mathbf{y}_k^\dagger \mathbf{x}_{k'} = \delta_{kk'} . \tag{1.96}$$

Problem 1.96 Let $\mathbf{x}$ and $\mathbf{y}$ be the right and left eigenvectors of a matrix $\mathbf{A}$ corresponding to the common eigenvalue λ. Show that the right and left eigenvectors of $\mathbf{A}^\dagger$ are $\mathbf{y}$ and $\mathbf{x}$, respectively, both corresponding to the eigenvalue λ^*.

1.8.4 Hermitian (Symmetric) Matrices

We have seen above that in the case of degenerate eigenvalues of a $p \times p$ matrix $\mathbf{A}$ it is *not* always possible to find all its p linearly independent eigenvectors. We shall show in this section that for Hermitian, $\mathbf{A}^{T*} = \mathbf{A}^\dagger = \mathbf{A}$ matrices[8] all their eigenvectors can always be chosen linearly independent even if there are some degenerate eigenvalues. The case of symmetric matrices, $\mathbf{A}^T = \mathbf{A}$, can be considered as a particular case of the Hermitian ones when all elements of the matrices are real, so that there is no need to consider the case of the symmetric matrices separately. Also, quantum mechanics is based upon Hermitian matrices, so that their consideration is of special importance.

Theorem 1.19 *All eigenvalues of a Hermitian matrix* $\mathbf{A}^\dagger = \mathbf{A}$ *are real.*

Proof Let $\mathbf{x}$ be an eigenvector of $\mathbf{A}$ with the eigenvalue λ, i.e. $\mathbf{A}\mathbf{x} = \lambda\mathbf{x}$. Next, we multiply the first equation by $\mathbf{x}^\dagger$ from the left:

$$\mathbf{x}^\dagger \mathbf{A}\mathbf{x} = \lambda \mathbf{x}^\dagger \mathbf{x} . \tag{1.97}$$

Alternatively, take the Hermitian conjugate of both sides of $\mathbf{A}\mathbf{x} = \lambda\mathbf{x}$ to get $\mathbf{x}^\dagger \mathbf{A}^\dagger = \lambda^* \mathbf{x}^\dagger$. Next, multiply both sides from the right by $\mathbf{x}$:

$$\mathbf{x}^\dagger \mathbf{A}^\dagger \mathbf{x} = \lambda^* \mathbf{x}^\dagger \mathbf{x} . \tag{1.98}$$

---

[8] They often also called *self-adjoint*.

As the matrix $\mathbf{A}$ is Hermitian, the left-hand sides of the two Eqs. (1.97) and (1.98) are the same. Therefore,

$$\left(\lambda - \lambda^*\right) \mathbf{x}^\dagger \mathbf{x} = 0 \ .$$

Since $\mathbf{x} \neq 0$, $\mathbf{x}^\dagger \mathbf{x} > 0$, so that the only way to satisfy the above equation is to admit that λ is real: $\lambda = \lambda^*$. **Q.E.D.**

Note that in quantum mechanics measurable quantities correspond to eigenvalues of Hermitian matrices associated with them. This theorem guarantees that all measurable quantities are real.

Another important theorem deals with eigenvectors of a Hermitian matrix:

Theorem 1.20 *Eigenvectors of a Hermitian matrix corresponding to* different eigenvalues *can always be chosen orthogonal.*

Proof Consider two distinct eigenvalues α and β of a Hermitian matrix $\mathbf{A}$. Corresponding to them eigenvectors are $\mathbf{x}$ and $\mathbf{y}$, i.e. $\mathbf{A}\mathbf{x} = \alpha\mathbf{x}$ and $\mathbf{A}\mathbf{y} = \beta\mathbf{y}$. Similarly to the steps taken when proving the previous theorem, multiply the first equation from the left by $\mathbf{y}^\dagger$; then, take the Hermitian conjugate of the second equation and then multiply it from the right by $\mathbf{x}$. You will get two equations:

$$\mathbf{y}^\dagger \mathbf{A}\mathbf{x} = \alpha \mathbf{y}^\dagger \mathbf{x} \quad \text{and} \quad \mathbf{y}^\dagger \mathbf{A}^\dagger \mathbf{x} = \beta \mathbf{y}^\dagger \mathbf{x} \ .$$

Since $\mathbf{A}^\dagger = \mathbf{A}$, the left-hand sides of these equations are identical and so $(\alpha - \beta)\, \mathbf{y}^\dagger \mathbf{x} = 0$. Since the two eigenvalues are different, $\alpha \neq \beta$, then $\mathbf{y}^\dagger \mathbf{x} = (\mathbf{y}, \mathbf{x}) = 0$, i.e. the two eigenvectors are orthogonal. **Q.E.D.**

In fact, even if a matrix has repeated eigenvalues, one can always choose linearly independent set of eigenvectors for each of them which (by virtue of the Gram–Schmidt procedure of Sect. 1.1.3) can always be made into linear combinations which are mutually orthogonal. We also note that the vectors corresponding to generate eigenvalue are orthogonal to other eigenvectors since they correspond to different eigenvalues. Therefore, if all eigenvectors corresponding to the same eigenvalue can be made orthogonal, then we can conclude that all eigenvectors of a Hermitian matrix can always be made orthogonal to each other.

How many eigenvectors do exist for a Hermitian (or symmetric) $p \times p$ matrix? It is possible to show that a Hermitian $p \times p$ matrix has always exactly p eigenvectors, no matter if its eigenvalues are all distinct or there are repeated ones. The proof of this more general result corresponding to a Hermitian matrix is more involved and will be given later.

Thus, a Hermitian (or symmetric) $p \times p$ matrix has exactly p linearly independent orthogonal eigenvectors corresponding to real eigenvalues. In fact, by properly choosing an arbitrary multiplier associated with each eigenvector, one can always construct an *orthonormal* set of eigenvectors, which are obtained by normalising each eigenvector $\mathbf{x}$ with the pre-factor $1/\sqrt{(\mathbf{x}, \mathbf{x})}$.

Example 1.21 ► Find eigenvalues and eigenvectors of a Hermitian matrix

$$\mathbf{A} = \begin{pmatrix} 0 & i & i \\ -i & 1 & 0 \\ -i & 0 & 1 \end{pmatrix} .$$

Solution: Eigenvalues are found from

$$\begin{vmatrix} -\lambda & i & i \\ -i & 1-\lambda & 0 \\ -i & 0 & 1-\lambda \end{vmatrix} = -\left(\lambda^2 - 1\right)\left(\lambda - 2\right) = 0$$

are $\lambda^{(1)} = 1$, $\lambda^{(2)} = -1$ and $\lambda^{(3)} = 2$. Note that all are indeed real. The eigenvector $\mathbf{x}^{(1)}$ is obtained from

$$\begin{pmatrix} -1 & i & i \\ -i & 0 & 0 \\ -i & 0 & 0 \end{pmatrix} \begin{pmatrix} x_1 \\ x_2 \\ x_3 \end{pmatrix} = 0 \implies \begin{cases} -x_1 + ix_2 + ix_3 = 0 \\ -ix_1 = 0 \\ -ix_1 = 0 \end{cases}$$

that gives $\mathbf{x}^{(1)} = a\,(0, 1, -1)^T$, where a is an arbitrary complex number. Similarly, $\mathbf{x}^{(2)}$ is found via solving

$$\begin{pmatrix} 1 & i & i \\ -i & 2 & 0 \\ -i & 0 & 2 \end{pmatrix} \begin{pmatrix} x_1 \\ x_2 \\ x_3 \end{pmatrix} = 0 \implies \begin{cases} x_1 + ix_2 + ix_3 = 0 \\ -ix_1 + 2x_2 = 0 \\ -ix_1 + 2x_3 = 0 \end{cases} ,$$

yielding $\mathbf{x}^{(2)} = b\,(1, i/2, i/2)^T$. Finally, $\mathbf{x}^{(3)}$ is determined from equations

$$\begin{pmatrix} -2 & i & i \\ -i & -1 & 0 \\ -i & 0 & -1 \end{pmatrix} \begin{pmatrix} x_1 \\ x_2 \\ x_3 \end{pmatrix} = 0 \implies \begin{cases} -2x_1 + ix_2 + ix_3 = 0 \\ -ix_1 - x_2 = 0 \\ -ix_1 - x_3 = 0 \end{cases} ,$$

and results in $\mathbf{x}^{(3)} = c\,(1, -i, -i)^T$. The obtained eigenvectors are all orthogonal:

$$\mathbf{x}^{(1)\dagger}\mathbf{x}^{(2)} = a^*b\,(0\ 1\ -1) \begin{pmatrix} 1 \\ i/2 \\ i/2 \end{pmatrix} = a^*b \left[\frac{i}{2} - \frac{i}{2} \right] = 0 ,$$

$$\mathbf{x}^{(2)\dagger}\mathbf{x}^{(3)} = b^*c\,(1\ -\tfrac{i}{2}\ -\tfrac{i}{2}) \begin{pmatrix} 1 \\ -i \\ -i \end{pmatrix} = b^*c \left[1 + \frac{i^2}{2} + \frac{i^2}{2} \right] = 0,$$

$$\mathbf{x}^{(1)\dagger}\mathbf{x}^{(3)} = a^*c\,(0\ 1\ -1) \begin{pmatrix} 1 \\ -i \\ -i \end{pmatrix} = a^*c\,[-i + i] = 0 . ◄$$

Problem 1.97 Obtain all three eigenvectors of the previous example using Gauss elimination method instead.

1.8.5 Normal Matrices

There exist a wider family of p-dimensional square matrices for which one can always construct exactly p linearly independent eigenvectors that can be orthogonalised. These are called normal matrices. These are the matrices that commute with their Hermitian conjugate counterpart[9]:

$$\mathbf{A}\mathbf{A}^\dagger = \mathbf{A}^\dagger\mathbf{A}. \tag{1.99}$$

Of course, Hermitian matrices $\mathbf{A} = \mathbf{A}^\dagger$ are normal; unitary matrices $\mathbf{A}^{-1} = \mathbf{A}^\dagger$ as well.

Problem 1.98 Prove that Hermitian and unitary matrices are normal.

However, the family of normal matrices is much wider. For instance, consider the following matrix:

$$\mathbf{A} = \begin{pmatrix} 1 & 1 & 0 \\ 0 & 1 & 1 \\ 1 & 0 & 1 \end{pmatrix}.$$

This matrix is neither Hermitian nor unitary as

$$\mathbf{A}^{-1} = \frac{1}{2}\begin{pmatrix} 1 & -1 & 1 \\ 1 & 1 & -1 \\ -1 & 1 & 1 \end{pmatrix} \neq \mathbf{A}^\dagger,$$

as can easily be checked. At the same time, it is normal since

$$\mathbf{A}\mathbf{A}^\dagger = \begin{pmatrix} 1 & 1 & 0 \\ 0 & 1 & 1 \\ 1 & 0 & 1 \end{pmatrix}\begin{pmatrix} 1 & 0 & 1 \\ 1 & 1 & 0 \\ 0 & 1 & 1 \end{pmatrix} = \begin{pmatrix} 2 & 1 & 1 \\ 1 & 2 & 1 \\ 1 & 1 & 2 \end{pmatrix}$$

and the same result is obtained after calculating $\mathbf{A}^\dagger\mathbf{A}$.

---

[9] Of course, if the matrix is real, then the Hermitian conjugate operation can be replaced with the transposition.

Problem 1.99 Show that the matrix

$$\mathbf{A} = \begin{pmatrix} 1 & 2 & -1 \\ -1 & 1 & 2 \\ 2 & -1 & 1 \end{pmatrix}$$

is normal.

Problem 1.100 Prove that if $\mathbf{A}$ is normal, then its inverse $\mathbf{A}^{-1}$ is also normal.

As was stated above, the normal matrices share many properties of the Hermitian matrices. Here we shall consider corresponding to them eigenvectors and eigenvalues.

Let $\mathbf{x}$ be an eigenvector of a normal matrix $\mathbf{A}$ with the eigenvalue λ, i.e. $\mathbf{A}\mathbf{x} = \lambda\mathbf{x}$. We shall now show that the matrix $\mathbf{A}^\dagger$ also has this vector as its eigenvector, but with the eigenvalue λ^*. Indeed, we have that $(\mathbf{A} - \lambda\mathbf{E})\mathbf{x} = \mathbf{0}$. On both sides here we have a vector-column. One can also construct a vector-row on both sides by taking the Hermitian conjugate of it: $\mathbf{x}^\dagger \left(\mathbf{A}^\dagger - \lambda^*\mathbf{E}\right) = \mathbf{0}$. One can multiply a vector-row on a vector-column; this is a dot product, Hence, we can also write

$$\mathbf{x}^\dagger \left(\mathbf{A}^\dagger - \lambda^*\mathbf{E}\right) (\mathbf{A} - \lambda\mathbf{E})\mathbf{x} = 0 \,,$$

which is a scalar equal to zero. Let us multiply the two matrices in the round brackets with each other and use the fact that $\mathbf{A}\mathbf{A}^\dagger = \mathbf{A}^\dagger\mathbf{A}$:

$$\begin{aligned}
\left(\mathbf{A}^\dagger - \lambda^*\mathbf{E}\right)(\mathbf{A} - \lambda\mathbf{E}) &= \mathbf{A}^\dagger\mathbf{A} - \lambda\mathbf{A}^\dagger - \lambda^*\mathbf{A} + \lambda\lambda^* \\
&= \mathbf{A}\mathbf{A}^\dagger - \lambda\mathbf{A}^\dagger - \lambda^*\mathbf{A} + \lambda\lambda^* \\
&= (\mathbf{A} - \lambda\mathbf{E})\left(\mathbf{A}^\dagger - \lambda^*\mathbf{E}\right) \,,
\end{aligned}$$

so that we obtain that

$$\mathbf{x}^\dagger (\mathbf{A} - \lambda\mathbf{E})\left(\mathbf{A}^\dagger - \lambda^*\mathbf{E}\right)\mathbf{x} = \left[\left(\mathbf{A}^\dagger - \lambda^*\mathbf{E}\right)\mathbf{x}\right]^\dagger \underbrace{\left(\mathbf{A}^\dagger - \lambda^*\mathbf{E}\right)\mathbf{x}}_{\mathbf{y}}$$

$$= \mathbf{y}^\dagger\mathbf{y} = 0 \,,$$

which can only mean that the vector $\mathbf{y} = \left(\mathbf{A}^\dagger - \lambda^*\mathbf{E}\right)\mathbf{x}$ is the null vector, i.e. $\mathbf{A}^\dagger\mathbf{x} = \lambda^*\mathbf{x}$, as required.

Next, consider two eigenvectors $\mathbf{x}_i$ and $\mathbf{x}_j$ corresponding to two distinct eigenvalues $\lambda_i \neq \lambda_j$, i.e.

$$\mathbf{A}\mathbf{x}_i = \lambda_i\mathbf{x}_i \quad \text{and} \quad \mathbf{A}\mathbf{x}_j = \lambda_j\mathbf{x}_j \,.$$

Multiply both sides of the first equation from the left by $\mathbf{x}_j^\dagger$:

$$\mathbf{x}_j^\dagger \mathbf{A} \mathbf{x}_i = \lambda_i \mathbf{x}_j^\dagger \mathbf{x}_i \quad \Rightarrow \quad \left(\mathbf{A}^\dagger \mathbf{x}_j\right)^\dagger \mathbf{x}_i = \lambda_i \mathbf{x}_j^\dagger \mathbf{x}_i \, .$$

However, $\mathbf{A}^\dagger \mathbf{x}_j = \lambda^* \mathbf{x}_j$, as was shown above; hence, we obtain

$$\left(\lambda_j^* \mathbf{x}_j\right)^\dagger \mathbf{x}_i = \lambda_i \mathbf{x}_j^\dagger \mathbf{x}_i \quad \Rightarrow \quad \lambda_j \mathbf{x}_j^\dagger \mathbf{x}_i = \lambda_i \mathbf{x}_j^\dagger \mathbf{x}_i \quad \Rightarrow \quad \left(\lambda_j - \lambda_i\right) \mathbf{x}_j^\dagger \mathbf{x}_i = 0 \, .$$

Since the two eigenvalues are distinct, the two eigenvectors are orthogonal, $\mathbf{x}_j^\dagger \mathbf{x}_i = 0$, as required. Hence, we conclude that eigenvectors corresponding to different eigenvalues are always orthogonal.

If there are eigenvectors corresponding to the same eigenvalue, we can always construct their linear combinations (using the Gram–Schmidt procedure, Sect. 1.1.3) that will be mutually orthogonal. At the same time, as was proven above in Problem 1.86, they will remain eigenvectors corresponding to the same eigenvalue. Hence, all eigenvectors of a normal matrix can be made orthogonal to each other (orthonormal, if required).

We shall prove later on in Sect. 1.9.2 that a normal $p \times p$ matrix has exactly p eigenvectors even though some of the eigenvalues are repeated (degenerate). We have already mentioned that this is true for Hermitian matrices; in fact, it appears to be true for a wider class of matrices, the normal matrices.

Finally, we note that eigenvalues of a normal matrix may in general be complex. This is in contrast to Hermitian matrices whose eigenvalues are always real.

1.9 Similarity Transformation

1.9.1 Diagonalisation of Matrices

Consider two matrices: $\mathbf{A}$ and non-singular $\mathbf{B}$. Then, the special product $\mathbf{B}^{-1}\mathbf{A}\mathbf{B}$ is called a *similarity transformation* of $\mathbf{A}$ by $\mathbf{B}$; the matrices $\mathbf{A}$ and $\mathbf{B}^{-1}\mathbf{A}\mathbf{B}$ are called *similar*.

Next, consider an eigenvalue/eigenvector problem $\mathbf{A}\mathbf{x} = \lambda \mathbf{x}$, where $\mathbf{A}$ is a square $p \times p$ matrix. It is convenient in this section to denote eigenvectors of $\mathbf{A}$ as $\mathbf{x}_j = \left(x_{ij}\right)$, where in the components x_{ij} the second index corresponds to the vector number and the first to its components. We assume that the matrix $\mathbf{A}$ has all its eigenvectors linearly independent. This can be guaranteed for normal matrices which include Hermitian ones as a special case because their eigenvectors can always be chosen linearly independent. Moreover, normal matrices have as many eigenvectors are their dimension, as this is also an important property that we shall need here. However, the consideration given below can be applied to all non-defective matrices.

Let us have all eigenvectors of $\mathbf{A}$ arranged as columns,

$$\mathbf{U} = \begin{pmatrix} | & | & \cdots & | \\ \mathbf{x}_1 & \mathbf{x}_2 & \cdots & \mathbf{x}_p \\ | & | & \cdots & | \end{pmatrix} = \begin{pmatrix} x_{11} & x_{12} & \cdots & x_{1p} \\ \cdots & \cdots & \cdots & \cdots \\ x_{p1} & x_{p2} & \cdots & x_{pp} \end{pmatrix} . \tag{1.100}$$

The matrix thus defined is called the *modal matrix* of $\mathbf{A}$.

It appears that the modal matrix has a very useful property which is stated by the following theorem.

Theorem 1.21 *If a $p \times p$ matrix $\mathbf{A}$ has p (linearly independent) eigenvectors $\mathbf{x}_j$ ($j = 1, \ldots, p$) which form the modal matrix $\mathbf{U}$ given by Eq. (1.100) (recall that due to Theorem 1.16 the matrix $\mathbf{U}$ is non-singular and hence has an inverse), then the matrix $\mathbf{A}$ can be diagonalised by the similarity transformation with $\mathbf{U}$:*

$$\mathbf{D} = \mathbf{U}^{-1}\mathbf{A}\mathbf{U} , \tag{1.101}$$

where

$$\mathbf{D} = \begin{pmatrix} \lambda_1 & \cdots & 0 \\ \cdots & \cdots & \cdots \\ 0 & \cdots & \lambda_p \end{pmatrix} \tag{1.102}$$

is a diagonal matrix containing all eigenvalues of $\mathbf{A}$ on its main diagonal. The order in which the eigenvalues appear on the main diagonal of $\mathbf{D}$ corresponds to the order in which the eigenvectors are placed in the modal matrix $\mathbf{U}$.

Proof Consider the product of two matrices, $\mathbf{A}$ and $\mathbf{U}$:

$$\begin{aligned}
\mathbf{A}\mathbf{U} &= \begin{pmatrix} a_{11} & \cdots & a_{1p} \\ \cdots & a_{ik} & \cdots \\ a_{p1} & \cdots & a_{pp} \end{pmatrix} \begin{pmatrix} x_{11} & \cdots & x_{1p} \\ \cdots & x_{kj} & \cdots \\ x_{p1} & \cdots & x_{pp} \end{pmatrix} \\
&= \begin{pmatrix} \sum_k a_{1k}x_{k1} & \cdots & \sum_k a_{1k}x_{kp} \\ \cdots & \sum_k a_{ik}x_{kj} & \cdots \\ \sum_k a_{pk}x_{k1} & \cdots & \sum_k a_{pk}x_{kp} \end{pmatrix} \\
&= \begin{pmatrix} \lambda_1 x_{11} & \cdots & \lambda_p x_{1p} \\ \cdots & \lambda_j x_{ij} & \cdots \\ \lambda_1 x_{p1} & \cdots & \lambda_p x_{pp} \end{pmatrix} ,
\end{aligned}$$

where use has been made of the fact that $\mathbf{x}_j$ are eigenvectors of $\mathbf{A}$. This result can also be written as a product $\mathbf{U}\mathbf{D}$ of the two matrices, this can be checked by direct multiplication:

$$\mathbf{UD} = \begin{pmatrix} x_{11} & \cdots & x_{1p} \\ \cdots & x_{ij} & \cdots \\ x_{p1} & \cdots & x_{pp} \end{pmatrix} \begin{pmatrix} \lambda_1 & \cdots & 0 \\ \cdots & \lambda_j & \cdots \\ 0 & \cdots & \lambda_p \end{pmatrix}$$

$$= \begin{pmatrix} \lambda_1 x_{11} & \cdots & \lambda_p x_{1p} \\ \cdots & \lambda_j x_{ij} & \cdots \\ \lambda_1 x_{p1} & \cdots & \lambda_p x_{pp} \end{pmatrix} .$$

Hence, we obtained the matrix identity $\mathbf{AU} = \mathbf{UD}$. Multiplying it by $\mathbf{U}^{-1}$ from the left, we obtain the required result, Eq. (1.101).

It is instructive to illustrate the above derivation algebraically as well. Consider the i, j element of the matrix $\mathbf{AU}$:

$$(\mathbf{AU})_{ij} = \sum_k a_{ik} u_{kj} = \sum_k a_{ik} x_{kj} = \sum_k a_{ik} \left(\mathbf{x}_j\right)_k = \left(\mathbf{A}\mathbf{x}_j\right)_i$$

$$= \left(\lambda_j \mathbf{x}_j\right)_i = \lambda_j x_{ij} ,$$

which is the same as the i, j element of the matrix $\mathbf{UD}$:

$$(\mathbf{UD})_{ij} = \sum_k u_{ik} d_{kj} = \sum_k x_{ik} \lambda_k \delta_{kj} = \lambda_j x_{ij} .$$

Note that it was not required for this prove to assume that the eigenvectors $\mathbf{x}_j$ are orthogonal. **Q.E.D.**

We see from Eq. (1.101) that the similarity transformation of $\mathbf{A}$ by the matrix $\mathbf{U}$, that contains as its columns the eigenvectors of $\mathbf{A}$, transforms $\mathbf{A}$ into the diagonal form.

Inversely, let us multiply the matrix identity $\mathbf{AU} = \mathbf{UD}$ *from the right* by $\mathbf{U}^{-1}$. We get

$$\mathbf{A} = \mathbf{UDU}^{-1} . \tag{1.103}$$

Since $\mathbf{D}$ consists of eigenvalues of $\mathbf{A}$ and $\mathbf{U}$ of its eigenvectors, the last formula shows that any square $p \times p$ matrix $\mathbf{A}$ that has p linearly independent eigenvectors can actually be written via its eigenvectors and eigenvalues (see also Sect. 1.10).

It is essential that the matrix $\mathbf{A}$ has p linearly independent eigenvectors, the same number as its dimension, so that a square non-singular modal matrix could be constructed. In that case the matrix $\mathbf{A}$ is said to be diagonalisable. Otherwise, the matrix $\mathbf{A}$ is said to be not diagonalisable. That is why the above consideration is only applicable to non-defective matrices that are diagonalisable.

Consider now an important special case of a normal matrix $\mathbf{A}$. In that case we are guaranteed that its eigenvectors $\mathbf{x}_j$ are orthogonal, $\mathbf{x}_i^\dagger \mathbf{x}_j = \delta_{ij}$. Moreover, we can always normalise each of them to unity. This means that the modal matrix $\mathbf{U}$ is unitary (orthogonal if eigenvectors are real), i.e. $\mathbf{U}^{-1} = \mathbf{U}^\dagger$ (or $\mathbf{U}^{-1} = \mathbf{U}^T$). This follows immediately from the fact the columns of $\mathbf{U}$ are made of orthogonal vectors $\mathbf{x}_j$. There are a few consequences of this: (i) the determinant of $\mathbf{U}$ is equal to ± 1; (ii) since $\mathbf{U}$ is unitary (orthogonal), its rows also form a set of orthonormal vectors, i.e.

$$\sum_k x_{ik}x_{jk}^* = \delta_{ij} \, ,$$

which can be written in the matrix form as

$$\sum_k \mathbf{x}_k \mathbf{x}_k^\dagger = \mathbf{E} \, . \tag{1.104}$$

(iii) Finally, in this particular case Eqs. (1.101) and (1.103) are simplified: $\mathbf{D} = \mathbf{U}^\dagger \mathbf{A}\mathbf{U}$ and $\mathbf{A} = \mathbf{U}\mathbf{D}\mathbf{U}^\dagger$.

The Property (1.104) is sometimes referred to as the completeness of the eigenvectors.

We have shown in Theorem 1.17 that the determinant of any square matrix $\mathbf{A}$ is equal to the product of all its eigenvalues. For non-defective matrices this statement is simply a trivial consequence of Eq. (1.103).

Theorem 1.22 *The determinant of any non-defective matrix* $\mathbf{A}$ *is equal to the product of all its eigenvalues:*

$$\det \mathbf{A} = \prod_i \lambda_i \, . \tag{1.105}$$

Proof We write first $\mathbf{A} = \mathbf{U}\mathbf{D}\mathbf{U}^{-1}$ and then calculate the determinant on both sides: in the left-hand side we simply have $\det \mathbf{A}$, while in the right we get

$$\det\left(\mathbf{U}\mathbf{D}\mathbf{U}^{-1}\right) = \det \mathbf{U} \det \mathbf{D} \det \mathbf{U}^{-1} = \det \mathbf{U} \det \mathbf{U}^{-1} \det \mathbf{D}$$

$$= \underbrace{\det\left(\mathbf{U}\mathbf{U}^{-1}\right)}_{=1} \det \mathbf{D} = \det \mathbf{D} = \prod_i \lambda_i \, ,$$

which is the result we seek. **Q.E.D.**◄

Example 1.22 ►Diagonalise the matrix

$$\mathbf{A} = \begin{pmatrix} 0 & i & i \\ -i & 1 & 0 \\ -i & 0 & 1 \end{pmatrix} \, .$$

Solution: We found in Example 1.21 that $\mathbf{A}$ has three eigenvectors

$$\mathbf{x}_1 = a \begin{pmatrix} 0 \\ 1 \\ -1 \end{pmatrix} \, , \quad \mathbf{x}_2 = b \begin{pmatrix} 1 \\ i/2 \\ i/2 \end{pmatrix} \, , \quad \mathbf{x}_3 = c \begin{pmatrix} 1 \\ -i \\ -i \end{pmatrix}$$

that correspond to three eigenvalues $\lambda_1 = 1$, $\lambda_2 = -1$ and $\lambda_3 = 2$. To ensure that the eigenvectors are normalised to unity, we shall choose appropriately the values of the arbitrary constants a, b and c. This would enable us to calculate easily $\mathbf{U}^{-1}$ as simply $\mathbf{U}^\dagger$. Then,

$$
\mathbf{x}_1 = \frac{1}{\sqrt{2}} \begin{pmatrix} 0 \\ 1 \\ -1 \end{pmatrix} , \quad \mathbf{x}_2 = \sqrt{\frac{2}{3}} \begin{pmatrix} 1 \\ i/2 \\ i/2 \end{pmatrix} , \quad \mathbf{x}_3 = \frac{1}{\sqrt{3}} \begin{pmatrix} 1 \\ -i \\ -i \end{pmatrix} ,
$$

and hence the modal matrix is

$$
\mathbf{U} = \begin{pmatrix} 0 & \sqrt{2/3} & 1/\sqrt{3} \\ 1/\sqrt{2} & i/\sqrt{6} & -i/\sqrt{3} \\ -1/\sqrt{2} & i/\sqrt{6} & -i/\sqrt{3} \end{pmatrix} .
$$

Now the modal matrix consists of orthonormal rows (columns) and is hence unitary. The inverse of this matrix can be now calculated simply as

$$
\mathbf{U}^{-1} = \mathbf{U}^\dagger = \begin{pmatrix} 0 & 1/\sqrt{2} & -1/\sqrt{2} \\ \sqrt{2/3} & -i/\sqrt{6} & -i/\sqrt{6} \\ 1/\sqrt{3} & i/\sqrt{3} & i/\sqrt{3} \end{pmatrix} .
$$

Therefore, the similarity transformation $\mathbf{U}^{-1}\mathbf{A}\mathbf{U} = \mathbf{U}^\dagger\mathbf{A}\mathbf{U}$ should diagonalise $\mathbf{A}$. We can check that by a direct calculation:

$$
\begin{pmatrix} 0 & 1/\sqrt{2} & -1/\sqrt{2} \\ \sqrt{2/3} & -i/\sqrt{6} & -i/\sqrt{6} \\ 1/\sqrt{3} & i/\sqrt{3} & i/\sqrt{3} \end{pmatrix} \begin{pmatrix} 0 & i & i \\ -i & 1 & 0 \\ -i & 0 & 1 \end{pmatrix} \begin{pmatrix} 0 & \sqrt{2/3} & 1/\sqrt{3} \\ 1/\sqrt{2} & i/\sqrt{6} & -i/\sqrt{3} \\ -1/\sqrt{2} & i/\sqrt{6} & -i/\sqrt{3} \end{pmatrix}
$$

$$
= \begin{pmatrix} 0 & 1/\sqrt{2} & -1/\sqrt{2} \\ \sqrt{2/3} & -i/\sqrt{6} & -i/\sqrt{6} \\ 1/\sqrt{3} & i/\sqrt{3} & i/\sqrt{3} \end{pmatrix} \begin{pmatrix} 0 & -\sqrt{2/3} & 2/\sqrt{3} \\ 1/\sqrt{2} & -i/\sqrt{6} & -2i/\sqrt{3} \\ -1/\sqrt{2} & -i/\sqrt{6} & -2i/\sqrt{3} \end{pmatrix}
$$

$$
= \begin{pmatrix} 1 & 0 & 0 \\ 0 & -1 & 0 \\ 0 & 0 & 2 \end{pmatrix}
$$

with the eigenvalues on the main diagonal, as required. Note that the eigenvalues run along the diagonal exactly in the same order as eigenvectors chosen in the modal matrix $\mathbf{U}$. ◄

Note that in the above example the product of eigenvalues is -2; a direct calculation of the determinant det $\mathbf{A}$ yields exactly the same result.

The following theorem establishes an important property of the similar matrices.

Theorem 1.23 *Similar matrices* $\mathbf{A}$ *and* $\mathbf{B} = \mathbf{U}^{-1}\mathbf{A}\mathbf{U}$ *have the same eigenvalues, and their eigenvectors are related via the transformation* $\mathbf{U}$.

Proof Let two matrices $\mathbf{A}$ and $\mathbf{B} = \mathbf{U}^{-1}\mathbf{A}\mathbf{U}$ be similar, $\mathbf{U}$ being some transformation (i.e. some non-singular matrix). The eigenvalues of $\mathbf{A}$ and $\mathbf{B}$ are determined from the following secular equations:

$$|\mathbf{A} - \lambda\mathbf{E}| = 0 \ \text{ and } \ |\mathbf{B} - \mu\mathbf{E}| = \left|\mathbf{U}^{-1}\mathbf{A}\mathbf{U} - \mu\mathbf{E}\right| = 0 \ .$$

The matrix inside the determinant in the second equation can be rearranged by using $\mathbf{E} = \mathbf{U}^{-1}\mathbf{U}$, i.e.

$$\mathbf{U}^{-1}\mathbf{A}\mathbf{U} - \mu\mathbf{E} = \mathbf{U}^{-1}\mathbf{A}\mathbf{U} - \mu\mathbf{U}^{-1}\mathbf{U} = \mathbf{U}^{-1}\left(\mathbf{A} - \mu\mathbf{E}\right)\mathbf{U} \ ,$$

and hence its determinant,

$$\left|\mathbf{U}^{-1}\mathbf{A}\mathbf{U} - \mu\mathbf{E}\right| = \left|\mathbf{U}^{-1}\left(\mathbf{A} - \mu\mathbf{E}\right)\mathbf{U}\right| = \left|\mathbf{U}^{-1}\right||\mathbf{A} - \mu\mathbf{E}|\,|\mathbf{U}| = |\mathbf{A} - \mu\mathbf{E}| \ .$$

since $\left|\mathbf{U}^{-1}\right|\,|\mathbf{U}| = 1$. Thus, both $\mathbf{A}$ and $\mathbf{B}$ have the same characteristic (secular) equation and, therefore, share the same eigenvalues (cf. Problem 1.83).

Next, let us establish a connection between the eigenvectors of the two eigenproblems. Consider the eigenproblems for the two matrices corresponding to the same eigenvalue (assuming for simplicity that all eigenvalues are different, i.e. there is no degeneracy):

$$\mathbf{A}\mathbf{u} = \lambda\mathbf{u} \ \text{ and } \ \mathbf{B}\mathbf{v} = \mathbf{U}^{-1}\mathbf{A}\mathbf{U}\mathbf{v} = \lambda\mathbf{v} \ .$$

Multiply the second equation by $\mathbf{U}$ from the left to get $\mathbf{A}\left(\mathbf{U}\mathbf{v}\right) = \lambda\left(\mathbf{U}\mathbf{v}\right)$. It is seen that the vector $\mathbf{U}\mathbf{v}$ is an eigenvector of $\mathbf{A}$ with the eigenvalue λ. However, we know that $\mathbf{u}$ is also an eigenvector of $\mathbf{A}$ with the same λ, and there is only one such eigenvalue. Therefore, $\mathbf{u} = \mathbf{U}\mathbf{v}$ (up to an arbitrary multiplier).

In the general case of repeated eigenvalues (degeneracy) one can always construct linear combinations of eigenvectors $\mathbf{u}_i$ (of $\mathbf{A}$) or $\mathbf{v}_i$ (of $\mathbf{B}$) corresponding to the same eigenvalue λ (see Problem 1.86); these will still serve as accepted eigenvectors for that eigenvalue. Therefore, in this case one can write that

$$\mathbf{U}\mathbf{v}_i = \sum_k c_{ik}\mathbf{u}_k$$

with some coefficients c_{ik} forming a matrix $\mathbf{C}$; here we sum up over all eigenvectors corresponding to the same λ. Inversely,

$$\mathbf{u}_k = \sum_i c_{ki}^{-1} \mathbf{U} \mathbf{v}_i = \mathbf{U} \sum_i c_{ki}^{-1} \mathbf{v}_i = \mathbf{U} \mathbf{v}'_k ,$$

where c_{ki}^{-1} are elements of the inverse matrix $\mathbf{C}^{-1}$, and $\mathbf{v}'_k = \sum_i c_{ki}^{-1} \mathbf{v}_i$ are linear combinations of the eigenvectors of $\mathbf{B}$. **Q.E.D.**

Example 1.23 As a simple example, let us find all eigenvalues and eigenvectors of the matrix

$$\mathbf{A} = \begin{pmatrix} -1 & 1 \\ 4 & 2 \end{pmatrix}$$

(see Example 1.16) after a similarity transformation with the matrix

$$\mathbf{U} = \begin{pmatrix} 1 & -1 \\ 1 & 1 \end{pmatrix}, \quad \mathbf{U}^{-1} = \begin{pmatrix} 1/2 & 1/2 \\ -1/2 & 1/2 \end{pmatrix}.$$

Solution: As we found in Example 1.16, the matrix $\mathbf{A}$ has two eigenvalues $\lambda^{(1)} = 3$ and $\lambda^{(2)} = -2$, the corresponding normalised eigenvectors being $\mathbf{x}^{(1)} = \frac{1}{\sqrt{17}} \begin{pmatrix} 1 \\ 4 \end{pmatrix}$ and $\mathbf{x}^{(2)} = \frac{1}{\sqrt{2}} \begin{pmatrix} 1 \\ -1 \end{pmatrix}$. Note that these are not orthogonal (they do not need be as $\mathbf{A}$ is neither normal, nor symmetric or Hermitian). Then, the matrix $\mathbf{A}$ after the similarity transformation becomes

$$\mathbf{B} = \mathbf{U}^{-1} \mathbf{A} \mathbf{U} = \begin{pmatrix} 1/2 & 1/2 \\ -1/2 & 1/2 \end{pmatrix} \begin{pmatrix} -1 & 1 \\ 4 & 2 \end{pmatrix} \begin{pmatrix} 1 & -1 \\ 1 & 1 \end{pmatrix} = \begin{pmatrix} 3 & 0 \\ 3 & -2 \end{pmatrix}.$$

The secular problem for $\mathbf{B}$ yields

$$\begin{vmatrix} 3 - \lambda & 0 \\ 3 & -2 - \lambda \end{vmatrix} = (3 - \lambda)(-2 - \lambda) = 0 ,$$

leading to the same eigenvalues, 3 and -2. The first eigenvector $\mathbf{y}^{(1)}$ is obtained from

$$\begin{cases} 0y_1 + 0y_2 = 0 \\ 3y_1 - 5y_2 = 0 \end{cases},$$

so that $y_1 = \frac{5}{3} y_2$ and the normalised vector $\mathbf{y}^{(1)} = \frac{3}{\sqrt{34}} \begin{pmatrix} 5/3 \\ 1 \end{pmatrix}$. It is indeed directly proportional to the vector $\mathbf{x}^{(1)}$ via $\mathbf{U}$ (we can omit the unnecessary multipliers):

$$\mathbf{U} \mathbf{y}^{(1)} = \begin{pmatrix} 1 & -1 \\ 1 & 1 \end{pmatrix} \begin{pmatrix} 5/3 \\ 1 \end{pmatrix} = \begin{pmatrix} 2/3 \\ 8/3 \end{pmatrix} = \frac{2}{3} \begin{pmatrix} 1 \\ 4 \end{pmatrix} \sim \mathbf{x}^{(1)} .$$

The second eigenvector is found by solving

$$\begin{cases} 5y_1 + 0y_2 = 0 \\ 3y_1 + 0y_2 = 0 \end{cases},$$

yielding $y_1 = 0$ and the normalised eigenvector $\mathbf{y}^{(2)} = \begin{pmatrix} 0 \\ 1 \end{pmatrix}$. It is also related to $\mathbf{x}^{(2)}$

via $\mathbf{U}$:

$$\mathbf{U}\mathbf{y}^{(2)} = \begin{pmatrix} 1 & -1 \\ 1 & 1 \end{pmatrix} \begin{pmatrix} 0 \\ 1 \end{pmatrix} = \begin{pmatrix} -1 \\ 1 \end{pmatrix} = -\begin{pmatrix} 1 \\ -1 \end{pmatrix} \sim \mathbf{x}^{(2)} \; . \; \blacktriangleleft$$

1.9.2 General Case of Repeated Eigenvalues

Now we are ready to prove the statement made at the end of Sect. 1.8.5 that any normal $p \times p$ matrix $\mathbf{A}$ has exactly p linearly independent eigenvectors no matter whether some of the eigenvalues are repeated or not. Moreover, we shall show that these eigenvectors can always be chosen mutually orthogonal. This will also conclude the proof of Theorem 1.20 for Hermitian matrices that are a particular class of the normal ones.

Theorem 1.24 *For any normal $p \times p$ matrix $\mathbf{A}$ (even with repeated eigenvalues) there exists a unitary matrix $\mathbf{U}$ such that the similarity transformation $\mathbf{U}^\dagger \mathbf{A} \mathbf{U} = \mathbf{D}$ results in a diagonal matrix $\mathbf{D}$ containing all eigenvalues of $\mathbf{A}$.*

Proof Let us pick up one eigenvalue λ_1 of $\mathbf{A}$ with the corresponding eigenvector $\mathbf{x}_1$. Next, we choose any $p - 1$ orthonormal (and hence linearly independent) vectors $\mathbf{x}_2$, $\mathbf{x}_3$, etc., $\mathbf{x}_n$ and form the matrix

$$\mathbf{U}_1 = \begin{pmatrix} | & | & \cdots & | \\ \mathbf{x}_1 & \mathbf{x}_2 & \cdots & \mathbf{x}_p \\ | & | & \cdots & | \end{pmatrix} = \begin{pmatrix} x_{11} & x_{12} & \cdots & x_{1p} \\ \cdots & \cdots & \cdots & \cdots \\ x_{p1} & x_{p2} & \cdots & x_{pp} \end{pmatrix},$$

which columns are composed of the chosen vectors $\mathbf{x}_j = (x_{ij})$. Note that, according to the notations above, the first (left) index in x_{ij} corresponds to the component i of the vector $\mathbf{x}_j$ which number j is given by the second (right) index. Also note that $\mathbf{U}_1$ is *not* the modal matrix of $\mathbf{A}$ as only the first vector (the first column) is the actual eigenvector of $\mathbf{A}$.

Since $\mathbf{x}_1$ is an eigenvector of $\mathbf{A}$, we can write

$$\sum_l a_{kl} (\mathbf{x}_1)_l = \sum_l a_{kl} x_{l1} = \lambda_1 (\mathbf{x}_1)_k = \lambda_1 x_{k1} \; . \tag{1.106}$$

Then, consider the similarity transformation $\mathbf{U}_1^\dagger \mathbf{A} \mathbf{U}_1$. The i, j components of the matrix formed in this way are

$$\left(\mathbf{U}_1^\dagger \mathbf{A} \mathbf{U}_1\right)_{ij} = \sum_{kl} x_{ki}^* a_{kl} x_{lj} = \sum_k x_{ki}^* \left(\sum_l a_{kl} x_{lj}\right) .$$

Now, consider specifically $j = 1$ which corresponds to the first column of the matrix $\mathbf{U}_1^\dagger \mathbf{A} \mathbf{U}_1$. Using Eq. (1.106), we have

$$\left(\mathbf{U}_1^\dagger \mathbf{A} \mathbf{U}_1\right)_{i1} = \sum_k x_{ki}^* \left(\sum_l a_{kl} x_{l1}\right) = \sum_k x_{ki}^* \left(\lambda_1 x_{k1}\right)$$
$$= \lambda_1 \sum_k x_{ki}^* x_{k1} = \lambda_1 \delta_{i1} ,$$

where in the last step we used the fact that, by construction, the vectors $\mathbf{x}_i = (x_{ki})$ and $\mathbf{x}_1 = (x_{k1})$ are orthonormal. We see from the above, that all elements of the first column of $\mathbf{U}_1^\dagger \mathbf{A} \mathbf{U}_1$ are zeros, apart from the 1, 1 element which is equal to λ_1, the first eigenvalue of $\mathbf{A}$.

Similarly, consider now specifically $i = 1$ which corresponds to the first row of the matrix $\mathbf{U}_1^\dagger \mathbf{A} \mathbf{U}_1$. Using the fact that $\mathbf{x}_1$ is the eigenvector of the normal matrix and hence $\mathbf{A}^\dagger \mathbf{x}_1 = \lambda_1^* \mathbf{x}_1$, we can write

$$\left(\mathbf{U}_1^\dagger \mathbf{A} \mathbf{U}_1\right)_{1j} = \sum_l \left(\sum_k x_{k1}^* a_{kl}\right) x_{lj} = \sum_l \left(\sum_k a_{kl}^* x_{k1}\right)^* x_{lj}$$
$$= \sum_l \left(\sum_k \left(\mathbf{A}^\dagger\right)_{lk} x_{k1}\right)^* x_{lj} = \sum_l \left(\mathbf{A}^\dagger \mathbf{x}_1\right)_{l1}^* x_{lj}$$
$$= \sum_l \left(\lambda_1^* x_{l1}\right)^* x_{lj} = \lambda_1 \sum_l x_{l1}^* x_{lj} = \lambda_1 \delta_{1j} .$$

Therefore, all elements on the first row, apart from the first one (which is again λ_1), should also be zeros. In other words, the structure of the matrix $\mathbf{U}_1^\dagger \mathbf{A} \mathbf{U}_1$ must be this:

$$\mathbf{U}_1^\dagger \mathbf{A} \mathbf{U}_1 = \begin{pmatrix} \lambda_1 & 0 & \cdots & 0 \\ 0 & & & \\ \cdots & & \mathbf{A}_{p-1} & \\ 0 & & & \end{pmatrix}$$

with its elements, apart from the ones in the first row and column, forming a $(p - 1) \times (p - 1)$ matrix $\mathbf{A}_{p-1}$ (this will be the 1, 1-minor of the matrix $\mathbf{U}_1^\dagger \mathbf{A} \mathbf{U}_1$). Note that $\left|\mathbf{U}_1^\dagger \mathbf{A} \mathbf{U}_1\right| = \left|\mathbf{U}_1^\dagger\right| |\mathbf{A}| |\mathbf{U}_1| = |\mathbf{A}|$, since $\mathbf{U}_1$ is unitary by construction. On the other hand, because of the special structure of the matrix $\mathbf{U}_1^\dagger \mathbf{A} \mathbf{U}_1$, we can also write $\left|\mathbf{U}_1^{\dagger}\mathbf{A}\mathbf{U}_1\right| = \lambda_1 \left|\mathbf{A}_{p-1}\right|$, and hence (see Eq. (1.91) in Theorem 1.22) the matrix $\mathbf{A}_{p-1}$ must have as its eigenvalues all other eigenvalues of $\mathbf{A}$. In other words, if λ_1 is not repeated, $\mathbf{A}_{p-1}$ does not have this one but has all others; if λ_1 is repeated, then $\mathbf{A}_{p-1}$ has this one as well, but repeated one time less.

At the next step, we consider one eigenvector of $\mathbf{A}_{p-1}$, let us call it $\mathbf{y}_2$, corresponding to the eigenvalue λ_2 of $\mathbf{A}_{p-1}$ (and hence of $\mathbf{A}$). Note that at least one eigenvector can always be found for any square matrix. Repeating the above procedure, we can construct a unitary $(p-1) \times (p-1)$ matrix

$$\mathbf{S}_2 = \begin{pmatrix} | & \cdots & | \\ \mathbf{y}_2 & \cdots & \mathbf{y}_p \\ | & \cdots & | \end{pmatrix},$$

such that it brings the matrix $\mathbf{A}_{p-1}$ into the form:

$$\mathbf{S}_2^{\dagger}\mathbf{A}_{p-1}\mathbf{S}_2 = \begin{pmatrix} \lambda_2 & 0 & \cdots & 0 \\ 0 & & & \\ \cdots & & \mathbf{A}_{p-2} & \\ 0 & & & \end{pmatrix},$$

where $\mathbf{A}_{p-2}$ is a square matrix of dimension $p-2$. Therefore, if one constructs a $p \times p$ matrix

$$\mathbf{U}_2 = \begin{pmatrix} 1 & 0 \cdots 0 \\ 0 & \\ \cdots & \mathbf{S}_2 \\ 0 & \end{pmatrix},$$

then it will bring the matrix $\mathbf{U}_1^{\dagger}\mathbf{A}\mathbf{U}_1$ into the following form:

$$\mathbf{U}_2^{\dagger}\left(\mathbf{U}_1^{\dagger}\mathbf{A}\mathbf{U}_1\right)\mathbf{U}_2 = \begin{pmatrix} 1 & 0 \cdots 0 \\ 0 & \\ \cdots & \mathbf{S}_2^{\dagger} \\ 0 & \end{pmatrix} \begin{pmatrix} \lambda_1 & 0 & \cdots & 0 \\ 0 & & & \\ \cdots & & \mathbf{A}_{p-1} & \\ 0 & & & \end{pmatrix} \begin{pmatrix} 1 & 0 \cdots 0 \\ 0 & \\ \cdots & \mathbf{S}_2 \\ 0 & \end{pmatrix}$$

$$= \begin{pmatrix} \lambda_1 & 0 & \cdots & & 0 \\ 0 & & & & \\ \cdots & & \mathbf{S}_2^{\dagger}\mathbf{A}_{p-1}\mathbf{S}_2 & \\ 0 & & & \end{pmatrix} = \begin{pmatrix} \lambda_1 & 0 & 0 & \cdots & 0 \\ 0 & \lambda_2 & 0 & \cdots & 0 \\ 0 & 0 & & & \\ \cdots & \cdots & & \mathbf{A}_{p-2} & \\ 0 & 0 & & & \end{pmatrix}$$

$$= (\mathbf{U}_1\mathbf{U}_2)^{\dagger}\,\mathbf{A}\,(\mathbf{U}_1\mathbf{U}_2)\ .$$

As before, the determinant of $\left|(\mathbf{U}_1\mathbf{U}_2)^{\dagger}\,\mathbf{A}\,(\mathbf{U}_1\mathbf{U}_2)\right| = |\mathbf{A}|$ is also equal to $\lambda_1\lambda_2\left|\mathbf{A}_{p-2}\right|$ because of the special structure of the matrix above obtained after the similarity transformation. Therefore, $\mathbf{A}_{p-2}$ has all eigenvalues of $\mathbf{A}$ apart from λ_1 and λ_2.

This process can be repeated another $p - 3$ times by constructing a sequence of unitary matrices $\mathbf{U}_3$, $\mathbf{U}_4$, etc., $\mathbf{U}_{p-1}$, until only the last element at the p, p position is left which is a scalar $\mathbf{A}_{p-(p-1)} = A_1$ that ought to be equal to the last remaining eigenvalue λ_p. This way the matrix $\mathbf{A}$ would finally be brought into the diagonal form:

$$
\left(\mathbf{U}_1 \mathbf{U}_2 \cdots \mathbf{U}_{p-1}\right)^\dagger \mathbf{A} \left(\mathbf{U}_1 \mathbf{U}_2 \cdots \mathbf{U}_{p-1}\right) =
\begin{pmatrix}
\lambda_1 & 0 & 0 & \cdots & 0 \\
0 & \lambda_2 & 0 & \cdots & 0 \\
0 & 0 & \lambda_3 & \cdots & 0 \\
\multicolumn{5}{c}{\dotfill} \\
0 & 0 & 0 & \cdots & \lambda_p
\end{pmatrix} = \mathbf{D}.
$$

Thus, we managed to construct a unitary matrix $\mathbf{U} = \mathbf{U}_1 \mathbf{U}_2 \cdots \mathbf{U}_{p-1}$ which upon performing the similarity transformation, $\mathbf{U}^\dagger \mathbf{A} \mathbf{U}$, brings the original matrix $\mathbf{A}$ into the diagonal form $\mathbf{D}$ with its eigenvalues on the principal diagonal, as required. Our proof is general as it was not based on assuming that all eigenvalues are different. **Q.E.D.**

Now we should be able to finally prove that even in the case of a normal $p \times p$ matrix $\mathbf{A}$ having degenerate (repeated) eigenvalues it is always possible to choose exactly p linearly independent eigenvectors $\{\mathbf{x}_i\}$ $(i = 1, \ldots, p)$ corresponding to its eigenvalues $\{\lambda_i\}$. This is demonstrated by the following theorem.

Theorem 1.25 *Consider a normal $p \times p$ matrix $\mathbf{A}$ which has p eigenvalues $\{\lambda_i\}$ amongst which there could be repeated ones. Then, it is always possible to choose exactly p orthogonal eigenvectors $\{\mathbf{x}_i\}$ $(i = 1, \ldots, p)$ of $\mathbf{A}$.*

Proof We know from the previous theorem that there exists a unitary matrix

$$
\mathbf{U} = \begin{pmatrix} | & | & \cdots & | \\ \mathbf{x}_1 & \mathbf{x}_2 & \cdots & \mathbf{x}_p \\ | & | & \cdots & | \end{pmatrix} = \left(x_{ij}\right) ,
$$

such that $\mathbf{U}^\dagger \mathbf{A} \mathbf{U} = \mathbf{D}$ with $\mathbf{D} = \left(d_{ij}\right) = \left(\lambda_i \delta_{ij}\right)$ only containing all eigenvalues of $\mathbf{A}$ on its diagonal. Recall that the second (column) index of x_{ij} in $\mathbf{U}$ indicates the vector number, while the first index corresponds to its components. But, multiplying $\mathbf{U}^\dagger \mathbf{A} \mathbf{U} = \mathbf{D}$ from the left with $\mathbf{U}$, we obtain $\mathbf{A} \mathbf{U} = \mathbf{U} \mathbf{D}$. Let us write the last equation in components. In the left-hand side we have $\sum_k a_{ik} x_{kj} = \sum_k a_{ik} \left(\mathbf{x}_j\right)_k$, while the right-hand side reads

$$
\sum_k x_{ik} d_{kj} = \sum_k x_{ik} \left(\lambda_j \delta_{kj}\right) = \lambda_j x_{ij} = \lambda_j \left(\mathbf{x}_j\right)_i
$$

$$
\implies \quad \sum_k a_{ik} \left(\mathbf{x}_j\right)_k = \lambda_j \left(\mathbf{x}_j\right)_i ,
$$

i.e. the jth column $\mathbf{x}_j = (x_{ij})$ of the matrix $\mathbf{U}$ is an eigenvector of $\mathbf{A}$ corresponding to the eigenvalue λ_j. Since the matrix $\mathbf{U}$ is unitary, all the eigenvectors are orthogonal (and hence linearly independent). **Q.E.D.**

Problem 1.101 Consider a Hermitian matrix $\mathbf{A}$, and let the eigenvalue λ be repeated d times (d-fold degenerate). Also, let $\mathbf{x}_1$, etc., $\mathbf{x}_d$ be the corresponding d eigenvectors associated with this eigenvalue. It is known (Problem 1.86) that any linear combinations

$$\mathbf{y}_i = \sum_{k=1}^{d} c_{ki}\mathbf{x}_k , \quad i = 1, \ldots, p$$

of the eigenvectors can also serve as eigenvectors of $\mathbf{A}$ with the same eigenvalue λ (note the 'reverse' order of indices in the coefficients above!). Show that the vectors $\{\mathbf{y}_i\}$ will be orthonormal if the matrix $\mathbf{C} = (c_{ik})$ of the coefficients of the linear transformation satisfy the matrix equation: $\mathbf{C}^\dagger \mathbf{S} \mathbf{C} = \mathbf{E}$, where $\mathbf{S} = (s_{ij})$ is the matrix of dot products of the original vectors, $s_{ij} = (\mathbf{x}_i, \mathbf{x}_j)$. In particular, it is seen that if $\mathbf{S} = \mathbf{E}$ (the original set is orthonormal), then $\mathbf{C}$ must be unitary to preserve this property for the new eigenvectors.

1.9.3 Simultaneous Diagonalisation of Two Matrices

A question one frequently asks in quantum mechanics is what are the conditions at which two (or more) physical quantities can be observed (measured) at the same time. The answer to this question, from the mathematical point of view, essentially boils down to the following question: is it possible to diagonalise two (or more) matrices associated with the chosen physical quantities using the same similarity transformation $\mathbf{U}$? This problem is solved with the following theorem.

Theorem 1.26 *Two Hermitian matrices can be diagonalised with the same similarity transformation $\mathbf{U}$ if and only if they commute.*

Proof The necessary condition of the proof we shall do first. Consider two $p \times p$ matrices $\mathbf{A}$ and $\mathbf{B}$. Let us assume that there exists the same similarity transformation that diagonalises both of them:

$$\mathbf{D}_a = \mathbf{U}^{-1}\mathbf{A}\mathbf{U} \text{ and } \mathbf{D}_b = \mathbf{U}^{-1}\mathbf{B}\mathbf{U} .$$

The two diagonal matrices obviously commute:

$$\mathbf{D}_a\mathbf{D}_b = \mathbf{D}_b\mathbf{D}_a \quad \Longrightarrow \quad \mathbf{U}^{-1}\mathbf{A}\mathbf{U}\mathbf{U}^{-1}\mathbf{B}\mathbf{U} = \mathbf{U}^{-1}\mathbf{B}\mathbf{U}\mathbf{U}^{-1}\mathbf{A}\mathbf{U}$$

$$\Longrightarrow \quad \mathbf{U}^{-1}\mathbf{A}\mathbf{B}\mathbf{U} = \mathbf{U}^{-1}\mathbf{B}\mathbf{A}\mathbf{U} \quad \Longrightarrow \quad \mathbf{A}\mathbf{B} = \mathbf{B}\mathbf{A} \ ,$$

i.e. the matrices $\mathbf{A}$ and $\mathbf{B}$ must commute.

To proof the sufficient condition we shall first make an assumption that the eigenvalues of $\mathbf{A}$ are all distinct. Then, assuming this time that the matrices commute, $\mathbf{A}\mathbf{B} = \mathbf{B}\mathbf{A}$, we can write for an eigenvector $\mathbf{x}_j$ of $\mathbf{A}$ corresponding to the eigenvalue λ_j:

$$\mathbf{A}\left(\mathbf{B}\mathbf{x}_j\right) = \mathbf{B}\mathbf{A}\mathbf{x}_j = \mathbf{B}\lambda_j\mathbf{x}_j = \lambda_j\left(\mathbf{B}\mathbf{x}_j\right) \ ,$$

i.e. the vector $\mathbf{y} = \mathbf{B}\mathbf{x}_j$ must also be an eigenvector of $\mathbf{A}$ with the same eigenvalue λ_j as $\mathbf{x}_j$. Since this is a unique eigenvalue, this is only possible if $\mathbf{y}$ is proportional to the $\mathbf{x}_j$, i.e. $\mathbf{y} = \beta\mathbf{x}_j$. This means that $\mathbf{B}\mathbf{x}_j = \beta\mathbf{x}_j$, i.e. $\mathbf{x}_j$ is also an eigenvector of $\mathbf{B}$. Hence, both matrices $\mathbf{A}$ and $\mathbf{B}$ share the same eigenvalues and eigenvectors, and hence the same modal matrix $\mathbf{U} = \left(\mathbf{x}_1 \ \cdots \ \mathbf{x}_p\right)$ will diagonalise both of them.

If $\mathbf{A}$ has a d-fold ($d > 1$) repeated eigenvalue λ, then $\mathbf{B}\mathbf{x}_j$ must be a linear combination of all eigenvectors associated with this particular eigenvalue. For simplicity let us assume that the *first* d eigenvectors of $\mathbf{A}$ correspond to the same eigenvalue λ. Then, for any j between 1 and d,

$$\mathbf{B}\mathbf{x}_j = \sum_{k=1}^{d} c_{jk}\mathbf{x}_k \ , \quad j = 1, \ldots, d$$

with some coefficients c_{jk} forming a square $d \times d$ matrix $\mathbf{C}$. This matrix can easily be seen to be Hermitian. Indeed,

$$\left(\mathbf{x}_i, \mathbf{B}\mathbf{x}_j\right) = \sum_{k=1}^{d} c_{jk}\left(\mathbf{x}_i, \mathbf{x}_k\right) = \sum_{k=1}^{d} c_{jk}\delta_{ik} = c_{ji}$$

and also

$$\left(\mathbf{B}\mathbf{x}_j, \mathbf{x}_i\right) = \left(\mathbf{x}_j, \mathbf{B}^\dagger\mathbf{x}_i\right) = \left(\mathbf{x}_j, \mathbf{B}\mathbf{x}_i\right) = \sum_{k=1}^{d} c_{ik}\left(\mathbf{x}_j, \mathbf{x}_k\right) = \sum_{k=1}^{d} c_{ik}\delta_{jk} = c_{ij} \ ,$$

since $\mathbf{B} = \mathbf{B}^\dagger$. However, $\left(\mathbf{B}\mathbf{x}_j, \mathbf{x}_i\right)^* = \left(\mathbf{x}_i, \mathbf{B}\mathbf{x}_j\right)$, from which it immediately follows that $c_{ji} = c_{ij}^*$, i.e. $\mathbf{C} = \mathbf{C}^\dagger$, i.e. it is indeed a Hermitian matrix. Once $\mathbf{C}$ is Hermitian, it can be diagonalised, i.e. there exists a $d \times d$ unitary matrix $\mathbf{S} = \left(s_{ij}\right)$ such that $\mathbf{C} = \mathbf{S}^\dagger\mathbf{D}\mathbf{S}$ with $\mathbf{D}$ being a diagonal matrix of eigenvalues $\{\zeta_k\}$ of $\mathbf{C}$.

Now it is easy to see that the vectors $\mathbf{y}_i = \sum_{k=1}^{d} s_{ik}\mathbf{x}_k$ (with the coefficients s_{ik} from $\mathbf{S}$) are eigenvectors of both $\mathbf{A}$ and $\mathbf{B}$. Indeed,

$$\mathbf{A}\mathbf{y}_i = \sum_k s_{ik}\mathbf{A}\mathbf{x}_k = \sum_k s_{ik}\lambda\mathbf{x}_k = \lambda\sum_k s_{ik}\mathbf{x}_k = \lambda\mathbf{y}_i$$

and

$$\mathbf{B}\mathbf{y}_i = \sum_k s_{ik}\mathbf{B}\mathbf{x}_k = \sum_k s_{ik}\sum_l c_{kl}\mathbf{x}_l = \sum_l (\mathbf{S}\mathbf{C})_{il}\,\mathbf{x}_l$$

$$= \sum_l \left(\mathbf{S}\mathbf{S}^\dagger\mathbf{D}\mathbf{S}\right)_{il}\,\mathbf{x}_l = \sum_l (\mathbf{D}\mathbf{S})_{il}\,\mathbf{x}_l = \zeta_i\sum_l s_{il}\mathbf{x}_l = \zeta_i\mathbf{y}_i\ .$$

Therefore, even in the case of repeated eigenvalues there exists a *common set of eigenvectors* for both matrices. Repeating this procedure for all repeated eigenvectors, we can collect all eigenvectors which are common to both matrices. Therefore, collecting all eigenvectors (corresponding both to repeated and distinct eigenvalues) into a matrix $\mathbf{U} = \left(\mathbf{y}_1 \cdots \mathbf{y}_p\right)$ (we should use $\mathbf{y}_j = \mathbf{x}_j$ in the case of a distinct jth eigenvalue) and following the statement of Theorem 1.21, we conclude that one can diagonalise both $\mathbf{A}$ and $\mathbf{B}$ at the same time via the same similarity transformation: $\mathbf{D}_a = \mathbf{U}^{-1}\mathbf{A}\mathbf{U}$ and $\mathbf{D}_b = \mathbf{U}^{-1}\mathbf{B}\mathbf{U}$. This proves the theorem. **Q.E.D.**

Example 1.24 ▶ Show that the matrices

$$\mathbf{A} = \begin{pmatrix} 3 & 1 \\ 1 & 3 \end{pmatrix} \ \text{and} \ \mathbf{B} = \begin{pmatrix} 1 & 2 \\ 2 & 1 \end{pmatrix}$$

commute and then diagonalise them by the same similarity transformation.
Solution: First of all, we make sure that they indeed commute:

$$\begin{pmatrix} 3 & 1 \\ 1 & 3 \end{pmatrix}\begin{pmatrix} 1 & 2 \\ 2 & 1 \end{pmatrix} = \begin{pmatrix} 5 & 7 \\ 7 & 5 \end{pmatrix} \ \text{and} \ \begin{pmatrix} 1 & 2 \\ 2 & 1 \end{pmatrix}\begin{pmatrix} 3 & 1 \\ 1 & 3 \end{pmatrix} = \begin{pmatrix} 5 & 7 \\ 7 & 5 \end{pmatrix}\ .$$

Then, in order to find the required similarity transformation, we should find the modal matrix of one of the matrices. For instance, consider $\mathbf{A}$. The secular equation,

$$\begin{vmatrix} 3-\lambda & 1 \\ 1 & 3-\lambda \end{vmatrix} = (3-\lambda)^2 - 1 = \lambda^2 - 6\lambda + 8 = 0\ ,$$

has solutions $\lambda_1 = 2$ and $\lambda_2 = 4$. The normalised eigenvectors, corresponding to these eigenvalues are easily found to be $\mathbf{x}_1 = \frac{1}{\sqrt{2}}\begin{pmatrix} 1 \\ -1 \end{pmatrix}$ and $\mathbf{x}_2 = \frac{1}{\sqrt{2}}\begin{pmatrix} 1 \\ 1 \end{pmatrix}$. The eigenvectors are orthonormal (as $\mathbf{A}$ is symmetric and eigenvalues are distinct):

$$(\mathbf{x}_1, \mathbf{x}_2) = \mathbf{x}_1^T\mathbf{x}_2 = \frac{1}{\sqrt{2}}\left(1 \ {-1}\right)\frac{1}{\sqrt{2}}\begin{pmatrix} 1 \\ 1 \end{pmatrix} = \frac{1}{2}(1-1) = 0\ .$$

The corresponding modal matrix is then orthogonal:

$$\mathbf{U} = \begin{pmatrix} 1/\sqrt{2} & 1/\sqrt{2} \\ -1/\sqrt{2} & 1/\sqrt{2} \end{pmatrix} = \frac{1}{\sqrt{2}} \begin{pmatrix} 1 & 1 \\ -1 & 1 \end{pmatrix} \quad \text{and} \quad \mathbf{U}^{-1} = \mathbf{U}^T = \frac{1}{\sqrt{2}} \begin{pmatrix} 1 & -1 \\ 1 & 1 \end{pmatrix}.$$

Therefore, both matrices $\mathbf{A}$ and $\mathbf{B}$ should be diagonalisable by the same $\mathbf{U}$. And, indeed:

$$\mathbf{U}^T \mathbf{A} \mathbf{U} = \frac{1}{\sqrt{2}} \begin{pmatrix} 1 & -1 \\ 1 & 1 \end{pmatrix} \begin{pmatrix} 3 & 1 \\ 1 & 3 \end{pmatrix} \frac{1}{\sqrt{2}} \begin{pmatrix} 1 & 1 \\ -1 & 1 \end{pmatrix}$$

$$= \frac{1}{2} \begin{pmatrix} 1 & -1 \\ 1 & 1 \end{pmatrix} \begin{pmatrix} 2 & 4 \\ -2 & 4 \end{pmatrix} = \frac{1}{2} \begin{pmatrix} 4 & 0 \\ 0 & 8 \end{pmatrix} = \begin{pmatrix} 2 & 0 \\ 0 & 4 \end{pmatrix},$$

$$\mathbf{U}^T \mathbf{B} \mathbf{U} = \frac{1}{\sqrt{2}} \begin{pmatrix} 1 & -1 \\ 1 & 1 \end{pmatrix} \begin{pmatrix} 1 & 2 \\ 2 & 1 \end{pmatrix} \frac{1}{\sqrt{2}} \begin{pmatrix} 1 & 1 \\ -1 & 1 \end{pmatrix}$$

$$= \frac{1}{2} \begin{pmatrix} 1 & -1 \\ 1 & 1 \end{pmatrix} \begin{pmatrix} -1 & 3 \\ 1 & 3 \end{pmatrix} = \frac{1}{2} \begin{pmatrix} -2 & 0 \\ 0 & 6 \end{pmatrix} = \begin{pmatrix} -1 & 0 \\ 0 & 3 \end{pmatrix}.$$

As one can see, the eigenvalues of $\mathbf{B}$ must be $\mu_1 = -1$ and $\mu_2 = 3$, that can be confirmed by an independent calculation:

$$\begin{vmatrix} 1 - \lambda & 2 \\ 2 & 1 - \lambda \end{vmatrix} = (1 - \lambda)^2 - 4 = \lambda^2 - 2\lambda - 3 = 0,$$

as this equation obviously has the same roots. ◄

1.10   Spectral Theorem and Function of a Matrix

1.10.1   Normal Matrices

We start by rewriting in components Eq. (1.103) obtained previously.

Problem 1.102 Show that any normal matrix $\mathbf{A} = (a_{ij})$ can be written via its eigenvalues λ_i and eigenvectors $\mathbf{x}_i = (x_{ki})$ as a sum

$$\mathbf{A} = \sum_k \lambda_k \mathbf{x}_k \mathbf{x}_k^\dagger \quad \text{or} \quad a_{ij} = \sum_k \lambda_k x_{ik} x_{jk}^* . \tag{1.107}$$

Here in x_{ik} and x_{jk} the right index corresponds to the eigenvector number, while the left to its component, i.e. $\mathbf{x}_k = (x_{ik})$.

Problem 1.103 By acting with $\mathbf{A}$ on the matrix $\sum_i \mathbf{x}_i \mathbf{x}_i^\dagger$ and using the above formula, give another prove of the completeness condition (1.104).

One can see that any normal matrix (including Hermitian and hence symmetric matrices) can actually be written as a sum over its all eigenvalues and eigenvectors (the so-called *spectral theorem*). This theorem opens a way to define *functions of matrices*. Indeed, consider a square of a symmetric matrix $\mathbf{A}$ as the matrix product $\mathbf{AA}$:

$$\mathbf{A}^2 = \mathbf{AA} = \mathbf{UDU}^{-1}\mathbf{UDU}^{-1} = \mathbf{UDDU}^{-1} \, .$$

Since the matrix $\mathbf{D}$ is diagonal, the product $\mathbf{DD} = \mathbf{D}^2$ is also a diagonal matrix with the squares of the eigenvalues of $\mathbf{A}$ on its main diagonal, i.e.

$$\mathbf{D}^2 = \mathbf{DD} = \begin{pmatrix} \lambda_1^2 & \cdots & 0 \\ \cdots & \cdots & \cdots \\ 0 & \cdots & \lambda_p^2 \end{pmatrix} \, .$$

One can see that $\mathbf{A}^2$ has the same form as $\mathbf{A}$ of Eq. (1.103), but with squares of its eigenvalues (instead of the eigenvalues themselves) in the sum. Therefore, we can immediately write

$$\mathbf{A}^2 = \sum_i \lambda_i^2 \mathbf{x}_i \mathbf{x}_i^\dagger \, .$$

Similarly, one can show (e.g. by induction) that for any power n,

$$\mathbf{A}^n = \sum_i \lambda_i^n \mathbf{x}_i \mathbf{x}_i^\dagger \, .$$

Therefore, it is possible to define a (scalar) function of a matrix $f(\mathbf{A})$ using a Taylor expansion of the function $f(x)$ around $x = 0$ (if exists):

$$f(\mathbf{A}) = f(0)\mathbf{E} + f'(0)\mathbf{A} + \frac{1}{2}f''(0)\mathbf{A}^2 + \frac{1}{3!}f'''(0)\mathbf{A}^3 + \cdots \, . \tag{1.108}$$

This infinite expansion defines a matrix which can be written using the spectral theorem as

$$f(\mathbf{A}) = \sum_i \left[f(0) + f'(0)\lambda_i + \frac{1}{2}f''(0)\lambda_i^2 + \frac{1}{3!}f'''(0)\lambda_i^3 + \cdots \right] \mathbf{x}_i \mathbf{x}_i^\dagger$$
$$= \sum_i f(\lambda_i)\mathbf{x}_i \mathbf{x}_i^\dagger \, . \tag{1.109}$$

It is seen that a function of the matrix shares the same eigenvectors, but has the eigenvalues $f(\lambda_i)$.

We have come across natural powers of a matrix $\mathbf{A}$; however, once a function of a matrix was defined, one can define any power α of the matrix, e.g.

$$\sqrt{\mathbf{A}} = \mathbf{A}^{1/2} = \sum_i \sqrt{\lambda_i}\mathbf{x}_i\mathbf{x}_i^\dagger \quad \text{or} \quad \frac{1}{\sqrt{\mathbf{A}}} = \mathbf{A}^{-1/2} = \sum_i \frac{1}{\sqrt{\lambda_i}}\mathbf{x}_i\mathbf{x}_i^\dagger .$$

The latter formula is valid as long as the matrix $\mathbf{A}$ is not singular (and hence does not have zero eigenvalues).

Problem 1.104 If λ_i and $\mathbf{x}_i$ are the eigenvalues and eigenvectors of a matrix $\mathbf{A}$, prove that the eigenvalues and eigenvectors of the inverse matrix $\mathbf{A}^{-1}$ are $1/\lambda_i$ and $\mathbf{x}_i$, respectively. Hence write the inverse matrix $\mathbf{A}^{-1}$ via the eigenvectors and eigenvalues of $\mathbf{A}$.

Problem 1.105 Show that $\mathbf{A}^{1/2}\mathbf{A}^{1/2} = \mathbf{A}$ and $\mathbf{A}^{1/2}\mathbf{A}^{-1/2} = \mathbf{E}$.

As an example, consider the matrix

$$\mathbf{A} = \begin{pmatrix} 1 & -1 \\ -1 & 1 \end{pmatrix},$$

whose normalised eigenvectors are $\mathbf{x}_1 = \frac{1}{\sqrt{2}}\begin{pmatrix} -1 \\ 1 \end{pmatrix}$ and $\mathbf{x}_2 = \frac{1}{\sqrt{2}}\begin{pmatrix} 1 \\ 1 \end{pmatrix}$, corresponding to the eigenvalues $a_1 = 2$ and $a_2 = 0$. Correspondingly, the matrix exponential function,

$$e^{\mathbf{A}t} = \sum_{j=1}^2 e^{a_j t}\mathbf{x}_j\mathbf{x}_j^T$$

$$= e^{2t}\frac{1}{\sqrt{2}}\begin{pmatrix} -1 \\ 1 \end{pmatrix}\frac{1}{\sqrt{2}}(-1\ 1) + e^{0t}\frac{1}{\sqrt{2}}\begin{pmatrix} 1 \\ 1 \end{pmatrix}\frac{1}{\sqrt{2}}(1\ 1)$$

$$= \frac{1}{2}e^{2t}\begin{pmatrix} 1 & -1 \\ -1 & 1 \end{pmatrix} + \frac{1}{2}\begin{pmatrix} 1 & 1 \\ 1 & 1 \end{pmatrix} = \frac{1}{2}\begin{pmatrix} 1+e^{2t} & 1-e^{2t} \\ 1-e^{2t} & 1+e^{2t} \end{pmatrix}.$$

Similarly,

$$\sqrt{\mathbf{A}} = \sqrt{2}\frac{1}{2}\begin{pmatrix} -1 \\ 1 \end{pmatrix}(-1\ 1) + \sqrt{0}\frac{1}{2}\begin{pmatrix} 1 \\ 1 \end{pmatrix}(1\ 1)$$

$$= \frac{1}{2}\begin{pmatrix} \sqrt{2} & -\sqrt{2} \\ -\sqrt{2} & \sqrt{2} \end{pmatrix} = \begin{pmatrix} 1/\sqrt{2} & -1/\sqrt{2} \\ -1/\sqrt{2} & 1/\sqrt{2} \end{pmatrix}.$$

It is easy to see that $\sqrt{\mathbf{A}}\sqrt{\mathbf{A}} = \mathbf{A}$ and $e^{\mathbf{A}t}e^{-\mathbf{A}t} = \mathbf{E}$, as required. The latter identity is in fact quite general as is shown in the following problem.

Problem 1.106 Using the expansion of a normal matrix **A** via its eigenvectors and eigenvalues, Eq. (1.107), prove that

$$\left(e^{\mathbf{A}t}\right)^{-1} = e^{-\mathbf{A}t}. \tag{1.110}$$

Problem 1.107 Consider two similar matrices, **A** and $\mathbf{B} = \mathbf{UAU}^{-1}$, related by a non-singular transformation matrix **U**. Show that

$$f(\mathbf{B}) = f\left(\mathbf{UAU}^{-1}\right) = \mathbf{U}f(\mathbf{A})\mathbf{U}^{-1} \tag{1.111}$$

is valid for any square matrix **A**.

Problem 1.108 Prove the following matrix identities:

$$e^{\mathbf{A}t} = e^{3t}\left(\begin{array}{cc} \cosh(2t) & i\sinh(2t) \\ -i\sinh(2t) & \cosh(2t) \end{array}\right);$$

$$\sqrt{\mathbf{A}} = \left(\begin{array}{cc} \left(1+\sqrt{5}\right)/2 & i\left(-1+\sqrt{5}\right)/2 \\ -i\left(-1+\sqrt{5}\right)/2 & \left(1+\sqrt{5}\right)/2 \end{array}\right);$$

$$\ln\mathbf{A} = \frac{\ln 5}{2}\left(\begin{array}{cc} 1 & i \\ -i & 1 \end{array}\right),$$

where

$$\mathbf{A} = \left(\begin{array}{cc} 3 & 2i \\ -2i & 3 \end{array}\right).$$

Differentiation of a matrix $\mathbf{A}(t) = \left(a_{ij}(t)\right)$ with respect to the parameter t is understood as the differentiation of every its element, i.e.

$$\frac{d}{dt}\mathbf{A}(t) = \frac{d}{dt}\left(a_{ij}(t)\right) = \left(\frac{da_{ij}(t)}{dt}\right).$$

Similarly, one can also integrate a matrix with respect to the parameter the matrix depends upon:

$$\int_a^b \mathbf{A}(t)dt = \left(\int_a^b a_{ij}(t)dt\right),$$

i.e. the integration is performed of each element $a_{ij}(t)$ of the matrix **A**.

Problem 1.109 Using the definition Eq. (1.109) of the function of a matrix, prove that

$$\frac{d}{dt}e^{\mathbf{A}t} = \mathbf{A}e^{\mathbf{A}t} = e^{\mathbf{A}t}\mathbf{A}.\qquad(1.112)$$

Problem 1.110 In quantum statistical mechanics a system is described by a time-dependent matrix $\rho(t)$ that depends on time t according to the Liouville equation:

$$i\frac{d\rho}{dt} = \mathbf{H}\rho - \rho\mathbf{H},$$

where $\mathbf{H}$ is the time-independent (constant) Hamiltonian matrix of the system under consideration. By defining an auxiliary density matrix

$$\widetilde{\rho}(t) = e^{i\mathbf{H}t}\rho(t)e^{-i\mathbf{H}t}$$

show that it is constant in time and hence the Liouville equation accepts the general solution

$$\rho(t) = e^{-i\mathbf{H}t}\rho(0)e^{i\mathbf{H}t}.$$

Note that this solution is only valid for time-independent $\mathbf{H}$. If $\mathbf{H}$ depends on time, a more general concept of the time-ordered exponential matrix needs to be introduced instead of the exponential matrices $e^{\pm i\mathbf{H}t}$. We shall not do it here, the interested reader should consult any book on many-body quantum theory.

Problem 1.111 Similarly prove that

$$\int e^{\mathbf{A}t}\,dt = \mathbf{A}^{-1}e^{\mathbf{A}t} + \mathbf{C},$$

where $\mathbf{C}$ is an arbitrary matrix of the same size as $\mathbf{A}$.

Problem 1.112 Using any of the above examples of $\mathbf{A}$, check that both sides of Eq. (1.112) give identical results.

Problem 1.113 The Pauli matrices (1.22) can form the Pauli matrix-vector $\boldsymbol{\sigma} = \sigma_x\mathbf{i} + \sigma_y\mathbf{j} + \sigma_z\mathbf{k}$. Let $\mathbf{n}$ be a unit vector. Next, using Eq. (1.23) show that $(\mathbf{n}\cdot\boldsymbol{\sigma})^{2k} = \mathbf{E}$ is the 2×2 identity matrix, while, correspondingly, $(\mathbf{n}\cdot\boldsymbol{\sigma})^{2k+1} = \mathbf{n}\cdot\boldsymbol{\sigma}$, for any integer $k = 0, 1, 2, \ldots$. Hence, prove the following identity:

$$e^{i\alpha(\mathbf{n}\cdot\boldsymbol{\sigma})} = \mathbf{E}\cos\alpha + i(\mathbf{n}\cdot\boldsymbol{\sigma})\sin\alpha,$$

where α is a parameter.

Problem 1.114 Consider a two-level quantum system whose Hamiltonian matrix

$$\mathbf{H} = \begin{pmatrix} \epsilon_1 & T \\ T^* & \epsilon_2 \end{pmatrix}.$$

Show that eigenvalues $\lambda_{1,2}$ and the corresponding normalised eigenvectors $\mathbf{x}_{1,2}$ of $\mathbf{H}$ are

$$\lambda_1 = \frac{1}{2}\left(\epsilon_1 + \epsilon_2 + \sqrt{(\epsilon_1 - \epsilon_2)^2 + 4|T|^2}\right) \quad \text{and} \quad \mathbf{x}_1 = \frac{1}{\sqrt{a^2 + |T|^2}}\begin{pmatrix} T \\ -a \end{pmatrix},$$

$$\lambda_2 = \frac{1}{2}\left(\epsilon_1 + \epsilon_2 - \sqrt{(\epsilon_1 - \epsilon_2)^2 + 4|T|^2}\right) \quad \text{and} \quad \mathbf{x}_2 = \frac{1}{\sqrt{b^2 + |T|^2}}\begin{pmatrix} T \\ b \end{pmatrix},$$

where

$$a = \frac{1}{2}\left(\epsilon_1 - \epsilon_2 - \sqrt{(\epsilon_1 - \epsilon_2)^2 + 4|T|^2}\right) \quad \text{and} \quad b = -\frac{1}{2}\left(\epsilon_1 - \epsilon_2 + \sqrt{(\epsilon_1 - \epsilon_2)^2 + 4|T|^2}\right).$$

Then verify explicitly, that the two eigenvectors are orthogonal and that $\mathbf{H}$ can be indeed written in its spectral form as $\mathbf{H} = \lambda_1 \mathbf{x}_1^{\dagger} \mathbf{x}_1 + \lambda_2 \mathbf{x}_2^{\dagger} \mathbf{x}_2$.

Special significance in physics has the matrix $\mathbf{G}(z) = (z\mathbf{E} - \mathbf{\Lambda})^{-1}$, called *resolvent* of a Hermitian matrix $\mathbf{A}$; it is a *function* of (generally complex) number z. It is defined via the scalar function $f(x) = (z - x)^{-1} = 1/(z - x)$. Using the spectral theorem, we can write down $\mathbf{G}(z)$ explicitly as

$$\mathbf{G}(z) = \sum_i \frac{1}{z - \lambda_i}\mathbf{x}_i \mathbf{x}_i^{\dagger} = \sum_i \frac{\mathbf{x}_i \mathbf{x}_i^{\dagger}}{z - \lambda_i}, \tag{1.113}$$

where $\mathbf{x}_i$ and λ_i are eigenvectors and eigenvalues of $\mathbf{A}$. It is also often called Green's function.

Problem 1.115 Consider a matrix $\mathbf{A}_0$ for which one can define Green's function (resolvent matrix) $\mathbf{G}_0(z) = (z\mathbf{E} - \mathbf{A}_0)^{-1}$. Show that Green's function $\mathbf{G}(z) = (z\mathbf{E} - \mathbf{A})^{-1}$ of the perturbed matrix $\mathbf{A} = \mathbf{A}_0 + \mathbf{W}$ satisfies the equation

$$\mathbf{G}(z) = \mathbf{G}_0(z) + \mathbf{G}_0(z)\mathbf{W}\mathbf{G}(z) \quad \text{or} \quad \mathbf{G}(z) = \mathbf{G}_0(z) + \mathbf{G}(z)\mathbf{W}\mathbf{G}_0(z) \tag{1.114}$$

that is normally referred to in the physics literature as the Dyson equation.

A simple application of the Dayson equation is given in Sect. 1.22.5.

The Dyson equation can be used for obtaining successful approximations for Green's function by expanding it in terms of the perturbation $\mathbf{W}$. This is done by iterating the right-hand side:

$$
\begin{aligned}
\mathbf{G}(z) &= \mathbf{G}_0(z) + \mathbf{G}_0(z)\mathbf{W}\mathbf{G}(z) \\
&= \mathbf{G}_0(z) + \mathbf{G}_0(z)\mathbf{W}\{\mathbf{G}_0(z) + \mathbf{G}_0(z)\mathbf{W}\mathbf{G}(z)\} \\
&= \mathbf{G}_0(z) + \mathbf{G}_0(z)\mathbf{W}\{\mathbf{G}_0(z) + \mathbf{G}_0(z)\mathbf{W}[\mathbf{G}_0(z) + \mathbf{G}_0(z)\mathbf{W}\mathbf{G}(z)]\} \quad (1.115) \\
&= \ldots \\
&= \mathbf{G}_0(z) + \mathbf{G}_0(z)\mathbf{W}\mathbf{G}_0(z) + \mathbf{G}_0(z)\mathbf{W}\mathbf{G}_0(z)\mathbf{W}\mathbf{G}_0(z) + \ldots
\end{aligned}
$$

that is usually called in physics the Born series.

Problem 1.116 Using the spectral representation of a normal matrix $\mathbf{A}$, show that

$$
\lim_{n \to \infty} \left(\mathbf{E} + \frac{x}{n}\mathbf{A}\right)^n = e^{\mathbf{A}x},
$$

where $\mathbf{E}$ is, as usual, the identity matrix.

1.10.2 General Matrices

So far we have been interested in normal matrices and have been able to write them via their eigenvectors and eigenvalues; moreover, definitions of functions of such matrices then naturally followed. In applications one has to deal also with general non-defective matrices, and it is important to understand whether the spectral theorem can also be written for them and whether one can define in a similar manner functions of such matrices.

We start by stating that the formula $\mathbf{A} = \mathbf{U}\mathbf{D}\mathbf{U}^{-1}$ we derived in Sect. 1.9.1 is valid for any matrix $\mathbf{A}$ that has as many linearly independent eigenvectors as its dimension. Therefore, one can calculate its powers, $\mathbf{A}^n = \mathbf{U}\mathbf{D}^n\mathbf{U}^{-1}$ in the same way as for normal matrices, and hence define the function of $\mathbf{A}$ as $f(\mathbf{A}) = \mathbf{U}f(\mathbf{D})\mathbf{U}^{-1}$, where $f(\mathbf{D})$ is a diagonal matrix with $f(\lambda_i)$ standing on its main diagonal, λ_i being eigenvalues of $\mathbf{A}$. However, writing an expansion like in Eq. (1.109) for the function of the matrix using exclusively the right eigenvectors of $\mathbf{A}$ in general is not possible since if the eigenvectors of $\mathbf{A}$ are not orthogonal (and hence cannot be made orthonormal), it is not possible to write $\mathbf{U}^{-1} = \mathbf{U}^{\dagger}$, and hence equations like (1.107) and (1.109) will not follow. However, this type of equation can actually be derived in general using both right and left eigenvectors.

Consider a general $p \times p$ non-defective matrix $\mathbf{A}$. We can define its right and left eigenvectors $\mathbf{x}_i$ and $\mathbf{y}_j$ such that $\mathbf{A}\mathbf{x}_i = \lambda_i \mathbf{x}_i$ and $\mathbf{y}_j^\dagger \mathbf{A} = \lambda_j \mathbf{y}_j^\dagger$. Recall that both eigenproblems share the same eigenvalues. Let us order the eigenvalues in some way and correspondingly order both the right and left eigenvectors in the same way. Next we define two square matrices:

$$
\mathbf{U} = \begin{pmatrix} | & | & \cdots & | \\ \mathbf{x}_1 & \mathbf{x}_2 & \cdots & \mathbf{x}_p \\ | & | & \cdots & | \end{pmatrix} = (x_{ij}) ,
$$

containing the right eigenvectors along its columns (with the right index in x_{ij} indicating the vector number), and another one,

$$
\mathbf{V} = \begin{pmatrix} - & \mathbf{y}_1^\dagger & - \\ - & \mathbf{y}_2^\dagger & - \\ \cdots & \cdots & \cdots \\ - & \mathbf{y}_p^\dagger & - \end{pmatrix} = (y_{ji}^*) ,
$$

in which the left eigenvectors are placed as its rows; each element y_{ji} corresponds to the ith component of the eigenvector $\mathbf{y}_j$. Note that the vectors $\mathbf{y}_i^\dagger$ appear as rows ($\mathbf{y}_i$ would correspond to the vector-column), that is why we have written the matrix $\mathbf{V}$ in this way. Since, according to Sect. 1.8.3, the left and right eigenvectors can always be chosen mutually orthonormal, i.e. $\mathbf{y}_k^\dagger \mathbf{x}_{k'} = \delta_{kk'}$, it is easy to see that $\mathbf{VU} = \mathbf{E}$, i.e. their product is the identity matrix. In other words, $\mathbf{V} = \mathbf{U}^{-1}$. Note that generally the matrices $\mathbf{U}$ and $\mathbf{V}$ are not unitary since the right (left) eigenvectors may not be orthogonal between themselves.

Next, consider the product $\mathbf{VAU}$. This is easier to calculate in components:

$$
(\mathbf{VAU})_{ij} = \sum_{kk'} y_{ik}^* a_{kk'} x_{k'j} = \sum_{k'} \left(\mathbf{y}_i^\dagger \mathbf{A} \right)_{k'} (\mathbf{x}_j)_{k'}
$$
$$
= \sum_{k'} \left(\lambda_i \mathbf{y}_i^\dagger \right)_{k'} (\mathbf{x}_j)_{k'} = \lambda_i \mathbf{y}_i^\dagger \mathbf{x}_j = \lambda_i \delta_{ij} = (\mathbf{D})_{ij} ,
$$

where $\mathbf{D} = (\delta_{ij} \lambda_i)$ is the diagonal matrix of all p eigenvalues of $\mathbf{A}$. Hence, we find that $\mathbf{VAU} = \mathbf{D}$. Multiplying from both sides by the corresponding inverse matrices (these must exist as we assumed that $\mathbf{A}$ has exactly p right and left eigenvectors), we shall finally get a generalisation of Eq. (1.103) for general non-defective matrices:

$$
\mathbf{A} = \mathbf{V}^{-1} \mathbf{D} \mathbf{U}^{-1} = \mathbf{UDV} . \tag{1.116}
$$

If rewritten explicitly via the right and left eigenvectors, we obtain a generalisation of Eq. (1.107):

$$
\mathbf{A} = \sum_j \lambda_j \mathbf{x}_j \mathbf{y}_j^\dagger . \tag{1.117}
$$

Indeed, we can easily check that

$$\mathbf{A}\mathbf{x}_k = \sum_j \lambda_j \mathbf{x}_j \underbrace{\mathbf{y}_j^\dagger \mathbf{x}_k}_{\delta_{jk}} = \lambda_k \mathbf{x}_k$$

and

$$\mathbf{y}_k^\dagger \mathbf{A} = \sum_j \lambda_j \underbrace{\mathbf{y}_k^\dagger \mathbf{x}_j}_{\delta_{jk}} \mathbf{y}_j^\dagger = \lambda_k \mathbf{y}_k^\dagger ,$$

as expected.

The obtained formulation of the spectral theorem for general matrices can now be naturally employed to define functions of such matrices along the same line of logic as used above for normal matrices.

Problem 1.117 Show explicitly that the integer power of a general non-defective matrix $\mathbf{A}$ can be written as

$$\mathbf{A}^n = \sum_j \lambda_j^n \mathbf{x}_j \mathbf{y}_j^\dagger \qquad (1.118)$$

and hence the function of the matrix can be defined as

$$f(\mathbf{A}) = \sum_j f(\lambda_j) \mathbf{x}_j \mathbf{y}_j^\dagger . \qquad (1.119)$$

Problem 1.118 Let $\mathbf{x}_j$ and $\mathbf{y}_j$ be the right and left eigenvectors of a matrix $\mathbf{A}$. Show that for any Taylor-expandable function $f(x)$,

$$f(\mathbf{A}^\dagger) = \sum_j f(\lambda_j^*) \mathbf{y}_j \mathbf{x}_i^\dagger .$$

Problem 1.119 Generalise the result of Problem 1.116 to general non-defective matrices:

$$\lim_{n \to \infty} \left(\mathbf{E} + \frac{x}{n}\mathbf{A} \right)^n = e^{\mathbf{A}x} .$$

1.11   Generalised Eigenproblem

In various physics applications, e.g. in problems on vibrations or on electronic structure, one needs to solve the so-called generalised eigenproblem in which not one, but two square matrices $\mathbf{A}$ and $\mathbf{S}$, participate:

$$\mathbf{A}\mathbf{x}_k = \lambda_k \mathbf{S}\mathbf{x}_k, \tag{1.120}$$

where λ_k is an eigenvalue and $\mathbf{x}_k$ corresponding to it eigenvector; the index k numbers different solutions. We shall assume that $\mathbf{S}$ is not singular. In the ordinary eigenproblem the matrix $\mathbf{S}$ is the identity matrix $\mathbf{E}$. But how one could solve this problem if $\mathbf{S}$ is not the identity matrix?

We shall assume $\mathbf{S}$ being a positive-definite[10] matrix.

The simplest formal solution relies on the fact that this problem can be mapped onto an ordinary eigenproblem. Indeed, if $\mathbf{S}$ is non-singular, we can define two matrices $\mathbf{S}^{1/2}$ and its inverse, $\mathbf{S}^{-1/2}$, such that $\mathbf{S}^{1/2}\mathbf{S}^{1/2} = \mathbf{S}$, $\mathbf{S}^{-1/2}\mathbf{S}^{-1/2} = \mathbf{S}^{-1}$ and $\mathbf{S}^{1/2}\mathbf{S}^{-1/2} = \mathbf{E}$. The fact that $\mathbf{S}$ is positive-definite guarantees that $\mathbf{S}^{-1/2}$ exists (see Sect. 1.10). We can then write $\mathbf{S}$ as the product of two $\mathbf{S}^{1/2}$ matrices, insert the identity $\mathbf{S}^{1/2}\mathbf{S}^{-1/2}$ between $\mathbf{A}$ and $\mathbf{x}_k$ in the left-hand side and then multiply the obtained equation from the left by $\mathbf{S}^{-1/2}$:

$$\mathbf{A}\left(\mathbf{S}^{-1/2}\mathbf{S}^{1/2}\right)\mathbf{x}_k = \lambda_k\left(\mathbf{S}^{1/2}\mathbf{S}^{1/2}\right)\mathbf{x}_k,$$
$$\mathbf{S}^{-1/2}\mathbf{A}\mathbf{S}^{-1/2}\mathbf{S}^{1/2}\mathbf{x}_k = \lambda_k\underbrace{\mathbf{S}^{-1/2}\mathbf{S}^{1/2}}_{\mathbf{E}}\mathbf{S}^{1/2}\mathbf{x}_k,$$
$$\left(\mathbf{S}^{-1/2}\mathbf{A}\mathbf{S}^{-1/2}\right)\left(\mathbf{S}^{1/2}\mathbf{x}_k\right) = \lambda_k\left(\mathbf{S}^{1/2}\mathbf{x}_k\right),$$

which is seen to be an ordinary eigenproblem $\widetilde{\mathbf{A}}\widetilde{\mathbf{x}}_k = \lambda_k\widetilde{\mathbf{x}}_k$ for the matrix $\widetilde{\mathbf{A}} = \mathbf{S}^{-1/2}\mathbf{A}\mathbf{S}^{-1/2}$. The eigenvectors $\widetilde{\mathbf{x}}_k$ of the new problem are linearly related to the eigenvectors $\mathbf{x}_k$ of the original problem via $\widetilde{\mathbf{x}}_k = \mathbf{S}^{1/2}\mathbf{x}_k$, so that $\mathbf{x}_k = \mathbf{S}^{-1/2}\widetilde{\mathbf{x}}_k$. Notably, the eigenvalues are the same. Moreover, the number of solutions of the two problems is the same.

Problem 1.120 Prove that if the matrices $\mathbf{A}$ and $\mathbf{S}$ are both Hermitian, so is the new matrix $\widetilde{\mathbf{A}}$.

We shall now limit ourselves to the case of Hermitian $\mathbf{A}$ and $\mathbf{S}$ as this case is met very often in applications. In this case we know that one can always find exactly n linearly independent eigenvectors $\widetilde{\mathbf{x}}_k$ of the reduced problem, where n is the dimension of the square matrix $\widetilde{\mathbf{A}}$; hence there will also be exactly n linearly independent eigenvectors of the generalised problem.

Consider now, using the proved correspondence between the two eigenproblems, properties that the eigenvectors $\mathbf{x}_k = \mathbf{S}^{-1/2}\widetilde{\mathbf{x}}_k$ of the generalised problem must satisfy. Firstly, we know that the eigenvectors $\widetilde{\mathbf{x}}_k$ and $\widetilde{\mathbf{x}}_{k'}$ of an ordinary eigenproblem with a Hermitian matrix can always be chosen orthonormal:

$$\widetilde{\mathbf{x}}_k^\dagger\widetilde{\mathbf{x}}_{k'} = \delta_{kk'} \quad\Rightarrow\quad \left(\mathbf{S}^{1/2}\mathbf{x}_k\right)^\dagger\left(\mathbf{S}^{1/2}\mathbf{x}_{k'}\right) = \delta_{kk'},$$

---

[10] All its eigenvalues are strictly positive.

which results in the corresponding modification of the orthonormality relation between the eigenvectors:

$$\mathbf{x}^\dagger{}_k \underbrace{\left(\mathbf{S}^{1/2}\right)^\dagger \mathbf{S}^{1/2}}_{\mathbf{S}^{1/2}} \mathbf{x}_{k'} = \delta_{kk'} \implies \mathbf{x}_k^\dagger \mathbf{S} \mathbf{x}_{k'} = \delta_{kk'} . \tag{1.121}$$

Secondly, the eigenvectors $\widetilde{\mathbf{x}}_k$ must satisfy the completeness relation (1.104); hence,

$$\sum_k \widetilde{\mathbf{x}}_k \widetilde{\mathbf{x}}_k^\dagger = \mathbf{E} \;\Rightarrow\; \sum_k \left(\mathbf{S}^{1/2}\mathbf{x}_k\right)\left(\mathbf{S}^{1/2}\mathbf{x}_k\right)^\dagger = \mathbf{E}$$

$$\Rightarrow\; \sum_k \mathbf{S}^{1/2}\mathbf{x}_k\mathbf{x}_k^\dagger\mathbf{S}^{1/2} = \mathbf{E},$$

and we obtain after multiplying from the left and from the right by $\mathbf{S}^{-1/2}$:

$$\sum_k \mathbf{x}_k\mathbf{x}_k^\dagger = \mathbf{S}^{-1} . \tag{1.122}$$

Hence, the matrix $\mathbf{S}$ modifies the completeness relation as well.

Problem 1.121 Consider the generalised eigenproblem with

$$\mathbf{A} = \begin{pmatrix} 2 & 1 \\ 1 & 2 \end{pmatrix} \quad \text{and} \quad \mathbf{S} = \begin{pmatrix} 5/2 & 3/2 \\ 3/2 & 5/2 \end{pmatrix} .$$

Show first that the normalised eigenvectors of $\mathbf{S}$ corresponding to the eigenvalues $s_1 = 4$ and $s_2 = 1$ are $\mathbf{y}_1 = \frac{1}{\sqrt{2}}(1\ 1)^T$ and $\mathbf{y}_2 = \frac{1}{\sqrt{2}}(1\ -1)^T$. Next, show that

$$\mathbf{S}^{1/2} = \begin{pmatrix} 3/2 & 1/2 \\ 1/2 & 3/2 \end{pmatrix} \quad \text{and} \quad \mathbf{S}^{-1/2} = \begin{pmatrix} 3/4 & -1/4 \\ -1/4 & 3/4 \end{pmatrix} .$$

Then, demonstrate that

$$\widetilde{\mathbf{A}} = \mathbf{S}^{-1/2}\mathbf{A}\mathbf{S}^{-1/2} = \begin{pmatrix} 7/8 & -1/8 \\ -1/8 & 7/8 \end{pmatrix}$$

and its eigenvalues and eigenvectors are $\lambda_1 = 1$, $\widetilde{\mathbf{x}}_1 = \frac{1}{\sqrt{2}}(1\ -1)^T$ and $\lambda_2 = 3/4$, $\widetilde{\mathbf{x}}_2 = \frac{1}{\sqrt{2}}(1\ 1)^T$. Finally, show that the normalised eigenvectors of the generalised problem are $\mathbf{x}_1 = \frac{1}{\sqrt{2}}(1\ -1)^T$ and $\mathbf{x}_2 = \frac{1}{\sqrt{2}}(1\ 1)^T$.

The proved mapping serves mostly to make the direct correspondence of the generalised eigenproblem to the ordinary one, but it is not very efficient in practical calculations. There exist other methods, not based on the direct calculation of $\mathbf{S}^{\pm 1/2}$, that enable one to perform the determination of the eigenvalues and eigenvectors of the generalised problem cheaper computationally. The interested reader should refer to specialised literature on numerical methods.

1.12 Famous Identities with the Matrix Exponential

The exponential matrix function $e^{\mathbf{A}}$ is frequently encountered in physics. There are a number of identities related to this matrix which we shall now consider. We start from the Baker–Hausdorff identity:

$$e^{\mathbf{A}}\mathbf{B}e^{-\mathbf{A}} = \mathbf{B} + [\mathbf{A}, \mathbf{B}] + \frac{1}{2!}[\mathbf{A}, [\mathbf{A}, \mathbf{B}]] + \frac{1}{3!}[\mathbf{A}, [\mathbf{A}, [\mathbf{A}, \mathbf{B}]]] + \ldots , \quad (1.123)$$

where $[\mathbf{X}, \mathbf{Y}] = \mathbf{X}\mathbf{Y} - \mathbf{Y}\mathbf{X}$ is the commutator of the two matrices $\mathbf{X}$ and $\mathbf{Y}$.

We shall start from the definition of the matrix exponentials as a Taylor expansion:

$$e^{\mathbf{A}}\mathbf{B}e^{-\mathbf{A}} = \left(\sum_{n=0}^{\infty}\frac{\mathbf{A}^n}{n!}\right)\mathbf{B}\left(\sum_{m=0}^{\infty}\frac{(-\mathbf{A})^m}{m!}\right) = \sum_{n=0}^{\infty}\sum_{m=0}^{\infty}\frac{(-1)^m}{n!m!}\mathbf{A}^n\mathbf{B}\mathbf{A}^m . \quad (1.124)$$

Using an obvious identity (that can easily be proven term-by-term by expanding both sides),

$$\sum_{n=0}^{\infty}\sum_{m=0}^{\infty} f_n g_m = \sum_{n=0}^{\infty}\sum_{m=0}^{n} f_{n-m} g_m ,$$

we can manipulate expression (1.124) as follows:

$$\begin{aligned} e^{\mathbf{A}}\mathbf{B}e^{-\mathbf{A}} &= \sum_{n=0}^{\infty}\sum_{m=0}^{n}\frac{(-1)^m}{(n-m)!m!}\mathbf{A}^{n-m}\mathbf{B}\mathbf{A}^m \\ &= \sum_{n=0}^{\infty}\frac{1}{n!}\left[\sum_{m=0}^{n}(-1)^m\binom{n}{m}\mathbf{A}^{n-m}\mathbf{B}\mathbf{A}^m\right] = \sum_{n=0}^{\infty}\frac{1}{n!}\mathbf{C}_n , \end{aligned} \quad (1.125)$$

where the matrix $\mathbf{C}_n$ is the expression in the square brackets on the second line. We expect that it should contain the required nested commutators. Indeed, as can be checked by direct calculation,

$$\mathbf{C}_0 = \mathbf{B}, \quad \mathbf{C}_1 = \mathbf{A}\mathbf{B} - \mathbf{B}\mathbf{A} = [\mathbf{A}, \mathbf{B}] ,$$

$$\mathbf{C}_2 = [\mathbf{A}, [\mathbf{A}, \mathbf{B}]] = \mathbf{A}^2\mathbf{B} - 2\mathbf{A}\mathbf{B}\mathbf{A} + \mathbf{B}\mathbf{A}^2 ,$$

and so on. In order to prove that $\mathbf{C}_n$ for any integer n is given by the n-nested commutators,

$$\mathbf{C}_n = \underbrace{[\mathbf{A}, [\mathbf{A}, [\mathbf{A}, \ldots [\mathbf{A}, \mathbf{B}]]] \ldots]}_{n \text{ times}},$$

it is convenient to use mathematical induction. Our assumption is correct at $n = 1$ and 2; we shall assume that it is valid for a general n and consider the case of $n + 1$. We have to prove that

$$\mathbf{C}_{n+1} = [\mathbf{A}, \mathbf{C}_n] = \mathbf{A}\mathbf{C}_n - \mathbf{C}_n\mathbf{A}$$

is given by

$$\mathbf{C}_{n+1} = \sum_{m=0}^{n+1} (-1)^m \binom{n+1}{m} \mathbf{A}^{n+1-m}\mathbf{B}\mathbf{A}^m \tag{1.126}$$

that is obtained from the expression in the square brackets in Eq. (1.125) by replacing $n \mapsto n + 1$. To this end, consider the difference $\mathbf{A}\mathbf{C}_n - \mathbf{C}_n\mathbf{A}$ in which we replace $\mathbf{C}_n$ by its corresponding expression since we assumed that it is valid for the given value of n:

$$\mathbf{A}\mathbf{C}_n - \mathbf{C}_n\mathbf{A} = \sum_{m=0}^{n} (-1)^m \binom{n}{m} \mathbf{A}^{n+1-m}\mathbf{B}\mathbf{A}^m$$

$$+ \sum_{m=0}^{n} (-1)^{m+1} \binom{n}{m} \mathbf{A}^{n-m}\mathbf{B}\mathbf{A}^{m+1}.$$

In the second sum we shall replace the summation variable $m + 1 \mapsto m$ and then separate out the $m = n + 1$ term; in the first sum we shall separate out the $m = 0$ term:

$$\mathbf{A}\mathbf{C}_n - \mathbf{C}_n\mathbf{A} = \mathbf{A}^{n+1}\mathbf{B} + \sum_{m=1}^{n} (-1)^m \binom{n}{m} \mathbf{A}^{n+1-m}\mathbf{B}\mathbf{A}^m$$

$$+ \sum_{m=1}^{n} (-1)^m \binom{n}{m-1} \mathbf{A}^{n+1-m}\mathbf{B}\mathbf{A}^m + (-1)^{n+1}\mathbf{B}\mathbf{A}^{n+1}$$

$$= \mathbf{A}^{n+1}\mathbf{B} + \sum_{m=1}^{n} (-1)^m \left[\binom{n}{m} + \binom{n}{m-1} \right] \mathbf{A}^{n+1-m}\mathbf{B}\mathbf{A}^m$$

$$+ (-1)^{n+1}\mathbf{B}\mathbf{A}^{n+1}.$$

Since the sum of the two binomial coefficients within the square brackets is simply $\binom{n+1}{m}$, see Eq. (I.1.157), the obtained expression is easily seen to be identical to the required form of Eq. (1.126), as required.

Note that by setting $\mathbf{B} = \mathbf{E}$ as an identity matrix, the identity $e^{\mathbf{A}} e^{-\mathbf{A}} = \mathbf{E}$ immediately follows. Another useful consequence of the obtained identity is the case in which the commutator $\mathbf{D} = [\mathbf{A}, \mathbf{B}]$ of $\mathbf{A}$ and $\mathbf{B}$ commutes with $\mathbf{A}$; then

$$e^{\mathbf{A}} \mathbf{B} e^{-\mathbf{A}} = \mathbf{B} + [\mathbf{A}, \mathbf{B}] = \mathbf{B} + \mathbf{D}. \tag{1.127}$$

Problem 1.122 The Pauli matrices (1.22) satisfy the following commutator identities (see Problem 1.25):

$$[\sigma_x, \sigma_y] = 2i\sigma_z, \quad [\sigma_y, \sigma_z] = 2i\sigma_x, \quad [\sigma_z, \sigma_x] = 2i\sigma_y.$$

Use the Baker–Hausdorff identity (1.123) to prove that

$$e^{i\sigma_x t} \sigma_z e^{-i\sigma_x t} = \sigma_z \cos(2t) + \sigma_y \sin(2t),$$

$$e^{i\sigma_y t} \sigma_z e^{-i\sigma_y t} = \sigma_z \cos(2t) - \sigma_x \sin(2t).$$

Problem 1.123 Derive the identities of the previous problem using a different method. Define the left-hand side as a function $f(t)$ and obtain the second-order differential equation $f'' + 4f = 0$ for it. Then, use the initial conditions at $t = 0$ to find the function $f(t)$ in each case.

If the commutator $\mathbf{D} = [\mathbf{A}, \mathbf{B}]$ of $\mathbf{A}$ and $\mathbf{B}$ commutes with $\mathbf{A}$, a more powerful identity exists:

$$e^{\mathbf{A}} f(\mathbf{B}) = f(\mathbf{B} + \mathbf{D}) e^{\mathbf{A}}, \tag{1.128}$$

where $f(x)$ is an arbitrary function that accepts the Taylor expansion. To prove it, we shall manipulate Eq. (1.127) by multiplying from the right with $e^{\mathbf{A}}$:

$$e^{\mathbf{A}} \mathbf{B} = (\mathbf{B} + \mathbf{D}) e^{\mathbf{A}}.$$

Next, consider

$$e^{\mathbf{A}} \mathbf{B}^2 = \left(e^{\mathbf{A}} \mathbf{B}\right) \mathbf{B} = (\mathbf{B} + \mathbf{D}) \left(e^{\mathbf{A}} \mathbf{B}\right) = (\mathbf{B} + \mathbf{D})^2 e^{\mathbf{A}},$$

$$e^{\mathbf{A}} \mathbf{B}^3 = \left(e^{\mathbf{A}} \mathbf{B}^2\right) \mathbf{B} = (\mathbf{B} + \mathbf{D})^2 \left(e^{\mathbf{A}} \mathbf{B}\right) = (\mathbf{B} + \mathbf{D})^3 e^{\mathbf{A}},$$

and so on. It is straightforward to see (e.g. by induction) that

$$e^{\mathbf{A}} \mathbf{B}^n = (\mathbf{B} + \mathbf{D})^n e^{\mathbf{A}}$$

for any integer n and hence

$$e^{\mathbf{A}} f(\mathbf{B}) = e^{\mathbf{A}} \sum_{n=0}^{\infty} \frac{f^{(n)}(0)}{n!} \mathbf{B}^n = \sum_{n=0}^{\infty} \frac{f^{(n)}(0)}{n!} e^{\mathbf{A}} \mathbf{B}^n$$

$$= \underbrace{\left[\sum_{n=0}^{\infty} \frac{f^{(n)}(0)}{n!} (\mathbf{B} + \mathbf{D})^n \right]}_{f(\mathbf{B}+\mathbf{D})} e^{\mathbf{A}},$$

which proves identity (1.128).

Next, we shall prove another important identity

$$e^{\mathbf{A}+\mathbf{B}} = e^{\mathbf{A}} e^{\mathbf{B}} e^{-[\mathbf{A},\mathbf{B}]/2}, \tag{1.129}$$

which is valid if both matrices $\mathbf{A}$ and $\mathbf{B}$ commute with their commutator $\mathbf{D} = [\mathbf{A}, \mathbf{B}]$.
Consider a matrix function of a scalar variable x:

$$\mathbf{F}(x) = e^{x\mathbf{A}} e^{x\mathbf{B}} e^{-x(\mathbf{A}+\mathbf{B})}.$$

Differentiating both sides with respect to x, we obtain three terms:

$$\frac{d\mathbf{F}(x)}{dx} = \mathbf{A} e^{x\mathbf{A}} e^{x\mathbf{B}} e^{-x(\mathbf{A}+\mathbf{B})} + e^{x\mathbf{A}} e^{x\mathbf{B}} \mathbf{B} e^{-x(\mathbf{A}+\mathbf{B})}$$

$$- e^{x\mathbf{A}} e^{x\mathbf{B}} (\mathbf{A} + \mathbf{B}) e^{-x(\mathbf{A}+\mathbf{B})}$$

$$= \mathbf{A} \mathbf{F}(x) - e^{x\mathbf{A}} e^{x\mathbf{B}} \mathbf{A} e^{-x(\mathbf{A}+\mathbf{B})}$$

$$= \mathbf{A} \mathbf{F}(x) - e^{x\mathbf{A}} \left(e^{x\mathbf{B}} \mathbf{A} e^{-x\mathbf{B}} \right) e^{x\mathbf{B}} e^{-x(\mathbf{A}+\mathbf{B})}.$$

Note that when differentiating a matrix exponential, e.g. $e^{x\mathbf{A}}$, with respect to x, the
matrix $\mathbf{A}$ can be placed on either side of the exponential as $\mathbf{A}$ commutes with the
exponential of itself: $\frac{d}{dx} e^{x\mathbf{A}} = \mathbf{A} e^{x\mathbf{A}} = e^{x\mathbf{A}} \mathbf{A}$.

The expression in the round brackets can be simplified using identity (1.127),
leading to

$$\frac{d\mathbf{F}(x)}{dx} = \mathbf{A} \mathbf{F}(x) - e^{x\mathbf{A}} (\mathbf{A} + x [\mathbf{B}, \mathbf{A}]) e^{x\mathbf{B}} e^{-x(\mathbf{A}+\mathbf{B})}$$

$$= \mathbf{A} \mathbf{F}(x) - \mathbf{A} \underbrace{e^{x\mathbf{A}} e^{x\mathbf{B}} e^{-x(\mathbf{A}+\mathbf{B})}}_{\mathbf{F}(x)}$$

$$- x e^{x\mathbf{A}} [\mathbf{B}, \mathbf{A}] e^{x\mathbf{B}} e^{-x(\mathbf{A}+\mathbf{B})}$$

$$= -x e^{x\mathbf{A}} [\mathbf{B}, \mathbf{A}] e^{x\mathbf{B}} e^{-x(\mathbf{A}+\mathbf{B})}.$$

However, the commutator $[\mathbf{B}, \mathbf{A}]$ commutes with $\mathbf{A}$ and hence can be swapped with
the matrix $e^{x\mathbf{A}}$ on the left of it leading to the differential equation

$$\frac{d\mathbf{F}(x)}{dx} = -x [\mathbf{B}, \mathbf{A}] \mathbf{F}(x) = x [\mathbf{A}, \mathbf{B}] \mathbf{F}(x).$$

The matrix $\mathbf{F}(x)$ contains matrices $\mathbf{A}$ and $\mathbf{B}$ with which the commutator $[\mathbf{B}, \mathbf{A}]$ commutes; hence, this differential equation can be integrated as in the scalar case subject to the obvious initial condition $\mathbf{F}(0) = \mathbf{E}$, yielding

$$\mathbf{F}(x) = \exp\left(\frac{x^2}{2}[\mathbf{A}, \mathbf{B}]\right) .$$

The required identity (1.129) immediately follows by setting $x = 1$,

$$\mathbf{F}(1) \equiv e^{[\mathbf{A},\mathbf{B}]/2} = e^{\mathbf{A}}e^{\mathbf{B}}e^{-(\mathbf{A}+\mathbf{B})} ,$$

multiplying from the right by $e^{\mathbf{A}+\mathbf{B}}$, swapping (commuting) it with $e^{[\mathbf{A},\mathbf{B}]/2}$ in the left-hand side (we can do it since the commutator $\mathbf{D}$ commutes with both $\mathbf{A}$ and $\mathbf{B}$) and finally multiplying from the right by $e^{-[\mathbf{A},\mathbf{B}]/2}$:

$$e^{[\mathbf{A},\mathbf{B}]/2}e^{\mathbf{A}+\mathbf{B}} = e^{\mathbf{A}}e^{\mathbf{B}} \implies e^{\mathbf{A}+\mathbf{B}}e^{[\mathbf{A},\mathbf{B}]/2} = e^{\mathbf{A}}e^{\mathbf{B}}$$

$$\implies e^{\mathbf{A}+\mathbf{B}} = e^{\mathbf{A}}e^{\mathbf{B}}e^{-[\mathbf{A},\mathbf{B}]/2} .$$

It follows from identity (1.129) that generally $e^{\mathbf{A}+\mathbf{B}} \neq e^{\mathbf{A}}e^{\mathbf{B}}$; only if the two matrices commute, then $e^{\mathbf{A}+\mathbf{B}} = e^{\mathbf{A}}e^{\mathbf{B}}$, i.e. the exponential function of a sum of two matrices can be factorised, exactly as with numbers.

1.13 Quadratic Forms

A real scalar function

$$Q = \sum_{ij} a_{ij}x_i^* x_j = \sum_{ij} x_i^* a_{ij}x_j = \mathbf{X}^\dagger \mathbf{A}\mathbf{X} \tag{1.130}$$

of p variables $x_1, \cdots, x_p$ is called a *quadratic form*. The (generally complex) coefficients a_{ij} form a square $p \times p$ matrix $\mathbf{A} = (a_{ij})$, while all the variables form a vector-column $\mathbf{X} = (x_i)$. An example of such an expression in physics is, e.g. the kinetic energy of a set of particles,

$$K = \sum_i \frac{1}{2}m\dot{x}_i^2 ,$$

which is a quadratic form with respect to particles velocities $\dot{x}_i$, or the potential energy of oscillating atoms in a molecule as in Eq. (1.217) in Sect. 1.22.3.

In most applications the quadratic form Q is real.

Problem 1.124 Show that in order for Q to be real, the matrix $\mathbf{A}$ must be Hermitian, $\mathbf{A}^\dagger = \mathbf{A}$.

Therefore, we shall assume in this section that $\mathbf{A}$ is Hermitian. Note that if the matrix $\mathbf{A}$ and the variables x_i are real, then $\mathbf{A}$ in Eq. (1.130) can always be made symmetric, i.e. satisfying $a_{ij} = a_{ji}$. Indeed, if the form Q contains an expression $a x_i x_j + b x_j x_i$ with different coefficients, $a \neq b$, it is always possible to write this sum as $c x_i x_j + c x_j x_i$ with $c = (a + b)/2$.

It is sometimes necessary to find a special linear combination $\mathbf{Y} = \mathbf{UX}$ of the original variables $\mathbf{X}$, such that the quadratic form be diagonal in the new variables $\mathbf{Y}$, i.e. it would not contain off-diagonal elements at all (the so-called *canonical form*):

$$Q = \sum_i c_i \, |y_i|^2 \; . \tag{1.131}$$

Transformation of a quadratic form to the canonical form is sometimes also called diagonalisation of the quadratic form. What is necessary for us to find is the right transformation $\mathbf{U}$ that would do the trick. The tools we have developed so far allow us to solve this problem with ease.

Theorem 1.27 *A quadratic form (1.130) can be diagonalised by a modal matrix* $\mathbf{U}$ *of* $\mathbf{A}$.

Proof Let $\mathbf{U}$ be the modal matrix of $\mathbf{A}$, i.e. $\mathbf{A} = \mathbf{UDU}^{-1} = \mathbf{UDU}^\dagger$, where $\mathbf{D} = (\delta_{ij}\lambda_i)$ is the diagonal matrix with all (real) eigenvalues λ_i of $\mathbf{A}$. Note that, because $\mathbf{A}$ is Hermitian, $\mathbf{U}$ is unitary, i.e. $\mathbf{U}^{-1} = \mathbf{U}^\dagger$. Then,

$$Q = \mathbf{X}^\dagger \mathbf{A} \mathbf{X} = \mathbf{X}^\dagger \mathbf{U D U}^\dagger \mathbf{X} = \left(\mathbf{U}^\dagger \mathbf{X}\right)^\dagger \mathbf{D} \left(\mathbf{U}^\dagger \mathbf{X}\right) = \mathbf{Y}^\dagger \mathbf{D} \mathbf{Y}$$
$$= \sum_{ij} y_i^* \left(\delta_{ij}\lambda_i\right) y_j = \sum_i \lambda_i \, |y_i|^2 \; ,$$

as required, where $\mathbf{Y} = \mathbf{U}^\dagger \mathbf{X}$ is a new set of variables. **Q.E.D.**

Example 1.25 ▶ Diagonalise the real quadratic form

$$Q = 3x^2 + 8xy + 3y^2 \; .$$

Solution: First, write Q using matrix notations paying specific attention to the off-diagonal elements:

$$Q = 3x^2 + 4xy + 4yx + 3y^2$$

$$= \begin{pmatrix} x \\ y \end{pmatrix}^T \begin{pmatrix} 3 & 4 \\ 4 & 3 \end{pmatrix} \begin{pmatrix} x \\ y \end{pmatrix} = \begin{pmatrix} x \\ y \end{pmatrix}^T \mathbf{A} \begin{pmatrix} x \\ y \end{pmatrix} .$$

The symmetric matrix $\mathbf{A} = \begin{pmatrix} 3 & 4 \\ 4 & 3 \end{pmatrix}$ has two eigenvalues $\lambda_1 = -1$ and $\lambda_2 = 7$ and

the corresponding normalised eigenvectors are $\mathbf{u}_1 = \frac{1}{\sqrt{2}} \begin{pmatrix} -1 \\ 1 \end{pmatrix}$ and $\mathbf{u}_2 = \frac{1}{\sqrt{2}} \begin{pmatrix} 1 \\ 1 \end{pmatrix}$,

so that the modal matrix is orthogonal,

$$\mathbf{U} = \begin{pmatrix} -1/\sqrt{2} & 1/\sqrt{2} \\ 1/\sqrt{2} & 1/\sqrt{2} \end{pmatrix} , \quad \text{and hence} \quad \mathbf{U}^T \mathbf{A} \mathbf{U} = \begin{pmatrix} -1 & 0 \\ 0 & 7 \end{pmatrix} .$$

Thus, the new variables are

$$\mathbf{Y} = \begin{pmatrix} y_1 \\ y_2 \end{pmatrix} = \mathbf{U}^T \mathbf{X} = \begin{pmatrix} -1/\sqrt{2} & 1/\sqrt{2} \\ 1/\sqrt{2} & 1/\sqrt{2} \end{pmatrix} \begin{pmatrix} x \\ y \end{pmatrix}$$

$$\longrightarrow \quad \begin{cases} y_1 = (-x + y)/\sqrt{2} \\ y_2 = (x + y)/\sqrt{2} \end{cases} ,$$

and the quadratic form Q in the new variables reads $Q = -y_1^2 + 7y_2^2$, i.e. it has the canonical form. Note that by a direct substitution of the new variables y_1 and y_2 into the Q it is returned back into its original form via x and y. ◄

It can easily be seen that any unitary transformation of the modal matrix of $\mathbf{A}$ also diagonalises a quadratic form based on it. Indeed, consider $Q = \mathbf{X}^\dagger \mathbf{A} \mathbf{X}$ and $\mathbf{A} = \mathbf{U} \mathbf{D} \mathbf{U}^\dagger$ with $\mathbf{U}$ being the modal matrix of $\mathbf{A}$. Since $\mathbf{A}$ is Hermitian, $\mathbf{U}$ can always be chosen unitary. If $\mathbf{V}$ is another unitary matrix, then the transformation $\mathbf{U}\mathbf{V}$ also diagonalises Q. Indeed, define auxiliary variables $\mathbf{Y} = (\mathbf{U}\mathbf{V})^\dagger \mathbf{X} = \mathbf{V}^\dagger \mathbf{U}^\dagger \mathbf{X}$, then $\mathbf{X} = \mathbf{U}\mathbf{V}\mathbf{Y}$, so that

$$Q = \mathbf{X}^\dagger \mathbf{A} \mathbf{X} = (\mathbf{U}\mathbf{V}\mathbf{Y})^\dagger \mathbf{A} (\mathbf{U}\mathbf{V}\mathbf{Y}) = \left((\mathbf{V}\mathbf{Y})^\dagger \mathbf{U}^\dagger\right) \mathbf{A} \mathbf{U} (\mathbf{V}\mathbf{Y})$$

$$= (\mathbf{V}\mathbf{Y})^\dagger \left(\mathbf{U}^\dagger \mathbf{A} \mathbf{U}\right) (\mathbf{V}\mathbf{Y}) = (\mathbf{V}\mathbf{Y})^\dagger \mathbf{D} (\mathbf{V}\mathbf{Y}) = \mathbf{Z}^\dagger \mathbf{D} \mathbf{Z} ,$$

which appears to be in the canonical form with respect to the new variables

$$\mathbf{Z} = \mathbf{V}\mathbf{Y} = \left(\mathbf{V}\mathbf{V}^\dagger\right) \mathbf{U}^\dagger \mathbf{X} = \mathbf{U}^\dagger \mathbf{X} ,$$

since $\mathbf{V}$ is a unitary matrix. Thus, $\mathbf{V}$ disappears completely from our final result, and the transformed form Q is the same as given by the modal matrix alone.

Therefore, any Hermitian matrix can be diagonalised with a similarity transformation. The latter is defined up to a unitary transformation.

Problem 1.125 Show that the quadratic form $Q = 2x_1^2 - 2x_1x_2 + 2x_2^2$ is $Q = y_1^2 + 3y_2^2$ in the canonical form, and find the new variables y_i via the old ones. [Answer: $y_1 = \frac{1}{\sqrt{2}}(x_1 + x_2)$ *and* $y_2 = \frac{1}{\sqrt{2}}(x_1 - x_2)$.]

Problem 1.126 Show that the quadratic form $Q = x_1^2 + 2x_1x_2 + x_2^2$ can be brought into the diagonal form $Q = 2y_2^2$ by means of an orthogonal transformation. Find y_2 via x_1 and x_2.

Problem 1.127 Prove that if a matrix $\mathbf{A}$ has non-negative eigenvalues, the quadratic form $\mathbf{X}^T \mathbf{A} \mathbf{X}$ for any real vectors $\mathbf{X}$ is non-negative as well.

1.14   Extremum of a Function of n Variables

We have formulated the sufficient condition for a function of two variables to have a minimum or a maximum at a point $\mathbf{r}_0 = (x_0, y_0)$ in Sect. I.5.10. Here we shall generalise this result to the case of a function of n variables.

Consider a function $y = y(\mathbf{x}) = y(x_1, x_2, \ldots, x_n)$ around a point specified by the vector $\mathbf{x}^0 = \left(x_1^0, x_2^0, \ldots, x_n^0\right)$. We shall expand y around that point into Taylor's series:

$$
\begin{aligned}
y(\mathbf{x}) &= y\left(\mathbf{x}^0 + \Delta\mathbf{x}\right) \\
&= y\left(\mathbf{x}^0\right) + \sum_{i=1}^{n}\left(\frac{\partial y}{\partial x_i}\right)_{\mathbf{x}^0} \Delta x_i + \frac{1}{2}\sum_{i,j=1}^{n}\left(\frac{\partial^2 y}{\partial x_i \partial x_j}\right)_{\mathbf{x}^0} \Delta x_i \Delta x_j + \cdots,
\end{aligned}
$$

where $\Delta\mathbf{x} = \mathbf{x} - \mathbf{x}^0$ (i.e. $\Delta x_i = x_i - x_i^0$). If the point $\mathbf{x}^0$ is a stationary point, the first-order partial derivatives are all equal to zero, so that the first non-vanishing term in the above expansion is the one containing second derivatives. We shall now write $y(\mathbf{x})$ above using matrix notations:

$$
\Delta y = y\left(\mathbf{x}^0 + \Delta\mathbf{x}\right) - y\left(\mathbf{x}^0\right) = \frac{1}{2}\Delta\mathbf{x}^T \mathbf{H} \Delta\mathbf{x} + \cdots, \qquad (1.132)
$$

where $\mathbf{H} = \left(h_{ij}\right)$ is the Hessian[11] matrix of second derivatives, which is symmetric due to the well-known property of mixed derivatives (Sect. I.5.3):

$$
h_{ij} = \left(\frac{\partial^2 y}{\partial x_i \partial x_j}\right)_{\mathbf{x}^0} = \left(\frac{\partial^2 y}{\partial x_j \partial x_i}\right)_{\mathbf{x}^0} = h_{ji}.
$$

---

[11] Named after Ludwig Otto Hesse.

The change Δy of the function $y(\mathbf{x})$ is a quadratic form with respect to the changes of the variables, Δx_i. By choosing a proper transformation of the vector $\Delta \mathbf{x}$ into new variables $\Delta \mathbf{z}$ which diagonalises the matrix $\mathbf{H}$, we obtain

$$\Delta y = \frac{1}{2} \sum_j \lambda_j \Delta z_j^2 + \cdots , \qquad (1.133)$$

where λ_j are eigenvalues of the Hessian matrix. This is our final result: it shows that if all eigenvalues of the Hessian matrix at the stationary point $\mathbf{x}^0$ are positive, then a small deviation from the stationary point $\mathbf{x}^0 \to \mathbf{x} = \mathbf{x}^0 + \Delta \mathbf{x}$ can only increase the function $y(\mathbf{x})$, i.e. the stationary point is a minimum. If, however, all the eigenvalues λ_j are negative, then $\mathbf{x}^0$ corresponds to a maximum instead. If at least one of the eigenvalues has a different sign to others, this is neither minimum nor maximum.

> **Problem 1.128**   Consider a function $z = z(x, y)$ of two variables and consider conditions for the Hessian matrix $\mathbf{H} = \begin{pmatrix} a & c \\ c & b \end{pmatrix}$ to have both its eigenvalues positive. Hence demonstrate that the sufficient conditions for the function to have a minimum at point $\left(x^0, y^0\right)$ are the same as those derived in Sect. I.5.10.

1.15   Trace of a Matrix

A sum of diagonal elements of a matrix is called its *trace*:

$$\mathrm{Tr}\,(\mathbf{A}) = \sum_{k=1}^{p} a_{kk} . \qquad (1.134)$$

Traces of matrices are frequently used in physics, e.g. quantum statistical mechanics is based on them. They possess a number of useful and important properties which we shall consider here.

Firstly, the trace of a product of matrices is *invariant* (i.e. does not change) under any *cyclic permutation* of them, i.e.

$$\mathrm{Tr}\,(\mathbf{AB}) = \mathrm{Tr}\,(\mathbf{BA}) \quad \text{or} \quad \mathrm{Tr}\,(\mathbf{ABC}) = \mathrm{Tr}\,(\mathbf{BCA}) ,$$

but

$$\mathrm{Tr}\,(\mathbf{ABC}) \neq \mathrm{Tr}\,(\mathbf{ACB})$$

in general. Indeed, consider first just two matrices:

$$\text{Tr}\,(\mathbf{AB}) = \sum_k (\mathbf{AB})_{kk} = \sum_{ki} a_{ki} b_{ik} = \sum_{ik} b_{ik} a_{ki} = \sum_i (\mathbf{BA})_{ii} = \text{Tr}\,(\mathbf{BA}) \ ,$$

as required. This result can be used to prove the statement for three (and more) matrices. We shall consider the case of three for simplicity, a more general case can be considered similarly:

$$\text{Tr}\,(\mathbf{ABC}) = \text{Tr}\,(\mathbf{A}\,(\mathbf{BC})) = \text{Tr}\,(\mathbf{AD}) = \text{Tr}\,(\mathbf{DA}) = \text{Tr}\,(\mathbf{BCA}) \ ,$$

where $\mathbf{D} = \mathbf{BC}$ and we have used just proven statement for two matrices. Similarly,

$$\text{Tr}\,(\mathbf{ABC}) = \text{Tr}\,(\mathbf{GC}) = \text{Tr}\,(\mathbf{CG}) = \text{Tr}\,(\mathbf{CAB}) \ .$$

Therefore,

$$\text{Tr}\,(\mathbf{CAB}) = \text{Tr}\,(\mathbf{ABC}) = \text{Tr}\,(\mathbf{BCA}) \ .$$

Note the cyclic order of matrices, $\mathbf{A} \to \mathbf{B} \to \mathbf{C} \to \mathbf{A} \to \cdots$, in each term is the same!

Next, it is also easy to see that the trace of a $p \times p$ matrix, which has p linearly independent eigenvectors, is equal to the sum of its eigenvalues:

$$\text{Tr}\,(\mathbf{A}) = \sum_i \lambda_i \ . \tag{1.135}$$

Indeed, let the square matrix $\mathbf{A}$ has $\{\lambda_i\}$ as its eigenvalues. Then, there exists a modal matrix $\mathbf{U}$ that diagonalises $\mathbf{A}$, i.e. $\mathbf{A} = \mathbf{UDU}^{-1}$ with the diagonal matrix $\mathbf{D}$ containing all eigenvalues λ_i on its main diagonal. Therefore,

$$\text{Tr}\,(\mathbf{A}) = \text{Tr}\left(\mathbf{UDU}^{-1}\right) = \text{Tr}\left(\mathbf{U}^{-1}\mathbf{UD}\right) = \text{Tr}\,(\mathbf{D}) = \sum_i \lambda_i \ ,$$

as required. More generally, the trace of a matrix $\mathbf{A}$ does not change after a similarity transformation, i.e.

$$\text{Tr}\,\mathbf{A} = \text{Tr}\left(\mathbf{UAU}^{-1}\right) \ .$$

Problem 1.129 Check by direct calculation of the matrices $\mathbf{A}$ and $\mathbf{U}^{-1}\mathbf{AU}$, where

$$\mathbf{A} = \begin{pmatrix} 1 & 1 \\ 1 & -1 \end{pmatrix} \quad \text{and} \quad \mathbf{U} = \begin{pmatrix} 2 & -3 \\ 3 & 2 \end{pmatrix} \ ,$$

that their traces are both equal to zero.

Problem 1.130 Consider the three matrices:

$$\mathbf{A} = \begin{pmatrix} 1 & -1 \\ 1 & 1 \end{pmatrix}, \quad \mathbf{B} = \begin{pmatrix} 0 & 2 \\ 2 & 0 \end{pmatrix} \quad \text{and} \quad \mathbf{C} = \begin{pmatrix} 1 & 0 \\ 3 & 2 \end{pmatrix}.$$

Calculate the products of matrices **ABC**, **CAB** and **BAC** and then the traces of them. Compare your results and explain your findings.

The following identity is often found useful:

$$\mathrm{Tr}\,(\ln \mathbf{A}) = \ln\,(\det \mathbf{A}) \quad \text{or} \quad \det \mathbf{A} = \exp\,(\mathrm{Tr}\,(\ln \mathbf{A})). \tag{1.136}$$

Problem 1.131 Using the spectral representation of a non-singular Hermitian matrix **A**, prove the identity (1.136).

Problem 1.132 A two-level quantum system can be represented by the density matrix

$$\rho = \begin{pmatrix} \rho_{11} & \rho_{12} \\ \rho_{21} & \rho_{22} \end{pmatrix}.$$

The density matrix must be normalised to unity, $\mathrm{Tr}\,(\rho) = 1$. Defining three observables $r_\alpha = \mathrm{Tr}\,(\sigma_\alpha \rho)$, where the index $\alpha = x, y, z$ indicates Cartesian components, and σ_α are three Pauli matrices given in Eq. (1.22), determine the four matrix elements of ρ to show that the latter can be compactly written as

$$\rho = \frac{1}{2}\,(\mathbf{E} + \mathbf{r} \cdot \boldsymbol{\sigma}), \tag{1.137}$$

where $\mathbf{r}$ and $\boldsymbol{\sigma}$ are the vectors defined by their components given above and **E** is the identity matrix. Show next using the derived representation of the density matrix that from the condition $\mathbf{r}^2 = 1$ follows $\mathrm{Tr}\,(\rho^2) = 1$. Also show the same directly without using Eq. (1.137) by simply employing the definition of the vector $\mathbf{r}$ and the normalisation condition for the density matrix.

Problem 1.133 The density matrix written via its eigenvalues ρ_λ and eigenvectors $\mathbf{X}_\lambda$ is

$$\rho = \sum_\lambda \rho_\lambda \mathbf{X}_\lambda \mathbf{X}_\lambda^\dagger.$$

Show that if the density matrix is idempotent, i.e. $\rho^2 = \rho$, then only exactly one eigenvalue is equal to one and all others to zero, i.e. in this case $\rho = \mathbf{X}_{\lambda_0} \mathbf{X}_{\lambda_0}^\dagger$, i.e. ρ_{λ_0} is the only non-zero eigenvalue. Note that $\mathrm{Tr}(\rho)=1$. It then follows that the condition $\rho^2 = \rho$ corresponds to that of the *purity* of the quantum state, when it is described by a single wavefunction instead of a mixture of those.

1.16   Tridiagonalisation of a Matrix: The Lanczos Method

Here we shall consider a special procedure which allows constructing recurrently a similarity transformation which transforms a Hermitian matrix into a tridiagonal form. This procedure developed by Lanczos has found numerous applications in physics.

Consider a Hermitian matrix $\mathbf{A}$ of dimension n and some vector $\mathbf{x}_1$ of the same dimensionality and of unit length, i.e. $\mathbf{x}_1^\dagger \mathbf{x}_1 = (\mathbf{x}_1, \mathbf{x}_1) = 1$ (the vector x_1 can be complex). Now, we construct a vector $\widetilde{\mathbf{x}}_2$ (also generally complex) using the recipe:

$$\widetilde{\mathbf{x}}_2 = \mathbf{A}\mathbf{x}_1 - \alpha_1 \mathbf{x}_1 . \tag{1.138}$$

Let us choose the parameter α_1 in such a way that the vector $\widetilde{\mathbf{x}}_2$ be orthogonal to $\mathbf{x}_1$. Calculating the dot product of both sides of the above equation with $\mathbf{x}_1$, i.e. multiplying the equation from the left by $\mathbf{x}_1^\dagger$, we have

$$\mathbf{x}_1^\dagger \widetilde{\mathbf{x}}_2 = \mathbf{x}_1^\dagger \mathbf{A}\mathbf{x}_1 - \alpha_1 \mathbf{x}_1^\dagger \mathbf{x}_1 \quad \Longrightarrow \quad \mathbf{x}_1^\dagger \widetilde{\mathbf{x}}_2 = \mathbf{x}_1^\dagger \mathbf{A}\mathbf{x}_1 - \alpha_1 .$$

It is seen that $\mathbf{x}_1^\dagger \widetilde{\mathbf{x}}_2 = 0$ if

$$\alpha_1 = \mathbf{x}_1^\dagger \mathbf{A}\mathbf{x}_1. \tag{1.139}$$

Since the matrix $\mathbf{A}$ is Hermitian, $\mathbf{A}^\dagger = \mathbf{A}$, the number α_1 is real:

$$\alpha_1^* = \left(\mathbf{x}_1^\dagger \mathbf{A}\mathbf{x}_1\right)^* = \left(\mathbf{x}_1^\dagger \mathbf{A}\mathbf{x}_1\right)^\dagger = \mathbf{x}_1^\dagger \mathbf{A}^\dagger \left(\mathbf{x}_1^\dagger\right)^\dagger = \mathbf{x}_1^\dagger \mathbf{A}\mathbf{x}_1 = \alpha_1 .$$

The vector $\widetilde{\mathbf{x}}_2$ may not be of unit length; therefore, we normalise it to unity, i.e. we introduce a (real) scaling factor β_1 such that the vector $\mathbf{x}_2 = \widetilde{\mathbf{x}}_2/\beta_1$ be of unit length: $\mathbf{x}_2^\dagger \mathbf{x}_2 = 1$. Obviously, $\beta_1^2 = \widetilde{\mathbf{x}}_2^\dagger \widetilde{\mathbf{x}}_2$. However, this expression for β_1 is not really useful. Another expression for the parameter β_1 can formally be derived which is directly related to the vectors $\mathbf{x}_2$ and $\mathbf{x}_1$. Indeed, Eq. (1.138) can be rewritten as

$$\beta_1 \mathbf{x}_2 = \mathbf{A}\mathbf{x}_1 - \alpha_1 \mathbf{x}_1 . \tag{1.140}$$

Then, multiplying both sides of (1.140) from the left by $\mathbf{x}_2^\dagger$, we obtain

$$\beta_1 \mathbf{x}_2^\dagger \mathbf{x}_2 = \mathbf{x}_2^\dagger \mathbf{A}\mathbf{x}_1 - \alpha_1 \mathbf{x}_2^\dagger \mathbf{x}_1 .$$

Since the two vectors are orthogonal and $\mathbf{x}_2$ is normalised to unity, we have $\beta_1 = \mathbf{x}_2^\dagger \mathbf{A}\mathbf{x}_1$. Since β_1 is real and $\mathbf{A}$ Hermitian, we can also write

$$\beta_1 = \mathbf{x}_2^\dagger \mathbf{A}\mathbf{x}_1 = \mathbf{x}_1^\dagger \mathbf{A}\mathbf{x}_2 . \tag{1.141}$$

Next, we construct the third vector $\mathbf{x}_3$ using a linear combination of the vector $\mathbf{A}\mathbf{x}_2$ and the vectors $\mathbf{x}_1$ and $\mathbf{x}_2$:

$$\beta_2 \mathbf{x}_3 = \mathbf{A}\mathbf{x}_2 - \alpha_2 \mathbf{x}_2 - \beta_1 \mathbf{x}_1 \ . \tag{1.142}$$

Above we immediately introduced the real scaling factor β_2 to ensure normalisation of $\mathbf{x}_3$. Let us show that one can choose the real constant α_2 such that $\mathbf{x}_3$ be orthogonal to both $\mathbf{x}_1$ and $\mathbf{x}_2$. Indeed, multiplying both sides of (1.142) from the left by $\mathbf{x}_2^{\dagger}$ and using the fact that $\mathbf{x}_2$ is of unit length and has already been made orthogonal to $\mathbf{x}_1$, we obtain

$$\beta_2 \mathbf{x}_2^{\dagger}\mathbf{x}_3 = \mathbf{x}_2^{\dagger}A\mathbf{x}_2 - \alpha_2 \mathbf{x}_2^{\dagger}\mathbf{x}_2 - \beta_1 \mathbf{x}_2^{\dagger}\mathbf{x}_1 \implies \beta_2 \mathbf{x}_2^{\dagger}\mathbf{x}_3 = \mathbf{x}_2^{\dagger}A\mathbf{x}_2 - \alpha_2 \ .$$

It is seen that $\mathbf{x}_2^{\dagger}\mathbf{x}_3 = 0$ if $\alpha_2 = \mathbf{x}_2^{\dagger}\mathbf{A}\mathbf{x}_2$ is chosen. Similarly, we can find that $\mathbf{x}_3$ and $\mathbf{x}_1$ are orthogonal by this construction automatically:

$$\beta_2 \mathbf{x}_1^{\dagger}\mathbf{x}_3 = \mathbf{x}_1^{\dagger}\mathbf{A}\mathbf{x}_2 - \alpha_2 \mathbf{x}_1^{\dagger}\mathbf{x}_2 - \beta_1 \mathbf{x}_1^{\dagger}\mathbf{x}_1$$
$$\implies \beta_2 \mathbf{x}_1^{\dagger}\mathbf{x}_3 = \mathbf{x}_1^{\dagger}\mathbf{A}\mathbf{x}_2 - \beta_1$$
$$\implies \mathbf{x}_2^{\dagger}\mathbf{x}_3 = 0 \ ,$$

because of the expression (1.141) for β_1. Finally, β_2 is chosen in such a way that the vector $\mathbf{x}_3$ be of unit length. This constant formally satisfies the equation $\beta_2 = \mathbf{x}_3^{\dagger}\mathbf{A}\mathbf{x}_2$ which can be obtained by multiplying Eq. (1.142) from the left by $\mathbf{x}_3^{\dagger}$:

$$\beta_2 \mathbf{x}_3^{\dagger}\mathbf{x}_3 = \mathbf{x}_3^{\dagger}\mathbf{A}\mathbf{x}_2 - \alpha_2 \mathbf{x}_3^{\dagger}\mathbf{x}_2 - \beta_1 \mathbf{x}_3^{\dagger}\mathbf{x}_1 \implies \beta_2 \mathbf{x}_3^{\dagger}\mathbf{x}_3 = \mathbf{x}_3^{\dagger}\mathbf{A}\mathbf{x}_2 \ ,$$

since $\mathbf{x}_3$ has already been made orthogonal to $\mathbf{x}_1$ and $\mathbf{x}_2$.

The next vector, $\mathbf{x}_4$, is obtained from $\mathbf{x}_2$ and $\mathbf{x}_3$ via

$$\beta_3 \mathbf{x}_4 = \mathbf{A}\mathbf{x}_3 - \alpha_3 \mathbf{x}_3 - \beta_2 \mathbf{x}_2 \ . \tag{1.143}$$

Problem 1.134 Show that by choosing $\alpha_3 = \mathbf{x}_3^{\dagger}\mathbf{A}\mathbf{x}_3$, the vector $\mathbf{x}_4$ is made orthogonal to both $\mathbf{x}_2$ and $\mathbf{x}_3$. Next, demonstrate that $\beta_3 = \mathbf{x}_4^{\dagger}\mathbf{A}\mathbf{x}_3 = \mathbf{x}_3^{\dagger}\mathbf{A}\mathbf{x}_4$ if $\mathbf{x}_4$ is to be made of unit length.

We see that, by construction, $\mathbf{x}_4$ is orthogonal to two previous vectors, $\mathbf{x}_2$ and $\mathbf{x}_3$. Remarkably, it can also be shown that it is automatically orthogonal to $\mathbf{x}_1$ as well. Indeed, let us take the Hermitian conjugate of Eq. (1.143) and then multiply its both sides from the right by $\mathbf{x}_1$:

$$\beta_3 \mathbf{x}_4^{\dagger}\mathbf{x}_1 = \mathbf{x}_3^{\dagger}\mathbf{A}\mathbf{x}_1 - \alpha_3 \mathbf{x}_3^{\dagger}\mathbf{x}_1 - \beta_2 \mathbf{x}_2^{\dagger}\mathbf{x}_1 \implies \beta_3 \mathbf{x}_4^{\dagger}\mathbf{x}_1 = \mathbf{x}_3^{\dagger}\mathbf{A}\mathbf{x}_1 \ ,$$

the last step was legitimate because of the orthogonality of $\mathbf{x}_3$ to $\mathbf{x}_2$ and of $\mathbf{x}_2$ to $\mathbf{x}_1$, ensured by the previous steps of the procedure. Next, from Eq. (1.140) it follows that $\mathbf{Ax}_1 = \alpha_1 \mathbf{x}_1 + \beta_1 \mathbf{x}_2$, i.e. $\mathbf{Ax}_1$ it is a linear combination of $\mathbf{x}_1$ and $\mathbf{x}_2$. However, $\mathbf{x}_3$ is orthogonal to both $\mathbf{x}_1$ and $\mathbf{x}_2$ which ensures that $\mathbf{x}_4$ is orthogonal to $\mathbf{x}_1$:

$$\beta_3 \mathbf{x}_4^\dagger \mathbf{x}_1 = \mathbf{x}_3^\dagger \mathbf{Ax}_1 \implies \beta_3 \mathbf{x}_4^\dagger \mathbf{x}_1 = \mathbf{x}_3^\dagger (\alpha_1 \mathbf{x}_1 + \beta_1 \mathbf{x}_2)$$
$$\implies \mathbf{x}_4^\dagger \mathbf{x}_1 = 0 \,,$$

as required. Since $\mathbf{x}_3^\dagger \mathbf{Ax}_1$ was found to be proportional to $\mathbf{x}_4^\dagger \mathbf{x}_1$, we also conclude that $\mathbf{x}_3^\dagger \mathbf{Ax}_1 = 0$.

The subsequent vectors $\mathbf{x}_5, \ldots, \mathbf{x}_n$ are obtained in a similar way by using at each step two previously constructed vectors, and all vectors constructed in this way form an orthonormal set, i.e. they are mutually orthogonal and are all of unit length.

We are now ready to formulate the general procedure. Starting from a unit vector $\mathbf{x}_1$, a vector $\mathbf{x}_2$ is constructed using (1.140) with the constants α_1 and β_1 satisfying Eqs. (1.139) and (1.141). Then, each subsequent vector $\mathbf{x}_{i+1}$ for $i = 2, 3, \ldots, n - 1$ is built using the following rule:

$$\beta_i \mathbf{x}_{i+1} = \mathbf{Ax}_i - \alpha_i \mathbf{x}_i - \beta_{i-1} \mathbf{x}_{i-1} \,. \tag{1.144}$$

It is seen that each consecutive vector is obtained from two preceding ones.

Problem 1.135 Show that by choosing

$$\alpha_i = \mathbf{x}_i^\dagger \mathbf{Ax}_i \tag{1.145}$$

the vector $\mathbf{x}_{i+1}$ is made orthogonal to $\mathbf{x}_{i-1}$ and $\mathbf{x}_i$. Then, demonstrate that the scaling factor

$$\beta_{i-1} = \mathbf{x}_{i-1}^\dagger \mathbf{Ax}_i = \mathbf{x}_i^\dagger \mathbf{Ax}_{i-1} \,. \tag{1.146}$$

Problem 1.136 Assume that vectors $\mathbf{x}_1, \mathbf{x}_2, \ldots, \mathbf{x}_i$ ($i \geq 3$) constructed by the Lanczos procedure are mutually orthogonal and normalised to one. Then, prove by mathematical induction, that $\mathbf{x}_{i+1}$ is orthogonal to $\mathbf{x}_1, \mathbf{x}_2, \ldots, \mathbf{x}_{i-2}$. Therefore, establish that

$$\mathbf{x}_j^\dagger \mathbf{Ax}_i = \mathbf{x}_i^\dagger \mathbf{Ax}_j = 0 \quad \text{if} \quad |i - j| \geq 2 \,. \tag{1.147}$$

Thus, the matrix $\mathbf{A}$ and the unit vector $\mathbf{x}_1$ generate a set of n mutually orthogonal unit vectors $\mathbf{x}_1, \ldots, \mathbf{x}_n$. By taking a different first vector, another set $\{\mathbf{x}_i'\}$ of n vectors is generated. These new vectors belong to the same n-dimensional vector space, are orthogonal to each other (and hence are linearly independent), and therefore are obtained as a linear combination of the vectors from the first set:

$$\mathbf{x}_i' = \sum_{j=1}^{n} w_{ij}\mathbf{x}_j \,,$$

where expansion coefficients w_{ij} form a square matrix $\mathbf{W}$. Since both sets are orthonormal, the matrix $\mathbf{U}$ must be unitary (Sect. 1.2.5).

The Lanczos procedure described above has an interesting implication. Let us construct a square matrix $\mathbf{U} = (\mathbf{x}_1\,\mathbf{x}_2 \cdots \mathbf{x}_n) = \left(u_{ij}\right)$ by placing the vectors $\mathbf{x}_1, \mathbf{x}_2, \ldots, \mathbf{x}_n$ generated using the Lanczos algorithm as its columns. Obviously, the u_{ij} element of $\mathbf{U}$ is then equal to the ith component of the vector $\mathbf{x}_j$, i.e. $u_{ij} = \left(\mathbf{x}_j\right)_i$. Recall that in Sect. 1.9.1 we built the modal matrix in exactly the same way. Since the vectors $\mathbf{x}_1, \mathbf{x}_2, \ldots, \mathbf{x}_n$ are orthonormal, the matrix $\mathbf{U}$ is unitary. We shall now show that the similarity transformation with the matrix $\mathbf{U}$ results in a matrix

$$\mathbf{U}^\dagger\mathbf{A}\mathbf{U} = \mathbf{T} = \begin{pmatrix} \alpha_1 & \beta_1 & & & & & & & \\ \beta_1 & \alpha_2 & \beta_2 & & & & & & \\ & \beta_2 & \alpha_3 & \beta_3 & & & & & \\ & & \ddots & \ddots & \ddots & & & & \\ & & & \beta_{i-1} & \alpha_i & \beta_i & & & \\ & & & & \beta_i & \alpha_{i+1} & \beta_{i+1} & & \\ & & & & & \ddots & \ddots & \ddots & \\ & & & & & & \beta_{n-3} & \alpha_{n-2} & \beta_{n-2} \\ & & & & & & & \beta_{n-2} & \alpha_{n-1} & \beta_{n-1} \\ & & & & & & & & \beta_{n-1} & \alpha_n \end{pmatrix}, \qquad (1.148)$$

which is tridiagonal. Note that $\mathbf{T}$ contains as diagonal and off-diagonal elements numbers α_i and β_i which are generated during the Lanczos procedure.

To prove this, let us write $\mathbf{U}^\dagger\mathbf{A}\mathbf{U} = \mathbf{T}$ explicitly in components:

$$\sum_{kl} u_{ki}^* a_{kl} u_{lj} = t_{ij} \quad\implies\quad \sum_{kl} \left(\mathbf{x}_i^\dagger\right)_k a_{kl} \left(\mathbf{x}_j\right)_l = t_{ij} \quad\implies\quad \mathbf{x}_i^\dagger\mathbf{A}\mathbf{x}_j = t_{ij}\,,$$

where at the last step we returned back to the vector and matrix notations. The numbers $\mathbf{x}_i^\dagger\mathbf{A}\mathbf{x}_j$ are however all equal to zero if the indices i and j differ by more than one according to Eq. (1.147). The diagonal elements $t_{ii} = \mathbf{x}_i^\dagger\mathbf{A}\mathbf{x}_i$ coincide with α_i, see Eqs. (1.139) and (1.145), and the nearest off-diagonal elements $t_{i,i+1} = \mathbf{x}_i^\dagger\mathbf{A}\mathbf{x}_{i+1}$ and $t_{i+1,i} = \mathbf{x}_{i+1}^\dagger\mathbf{A}\mathbf{x}_i$ both coincide with β_i, see Eqs. (1.141) and (1.146). This finally proves formula (1.148).

1.17 Dividing Matrices Into Blocks

In applications it is often needed to partition a matrix into blocks which are themselves matrices of smaller dimensions. In Fig. 1.7a a $p \times p$ matrix is split into four blocks by picking up the equal number n_1 of first rows and columns. Then, the block $A_{11} = (a_{ij})$ with $1 \leq i \leq n_1$ and $1 \leq j \leq n_1$ becomes a square $n_1 \times n_1$ matrix; other blocks are defined similarly. Note that with this partition the 22 block A_{22} is also a square $n_2 \times n_2$ matrix, while the blocks A_{12} and A_{21} are rectangular matrices if $n_1 \neq n_2$. Similarly, more partitions can be made; an example of nine blocks (three partitions along each side) is shown in Fig. 1.7b.

It is useful to be aware of the fact that one can operate with blocks as with matrix elements. Consider a $p \times p$ matrix $A = (a_{ij})$ and another matrix $B = (b_{ij})$ of the same dimension, and let us introduce for each of them the same block structure as shown in Fig. 1.7a. Then, the product of the two matrices $C = AB$ will also be a $p \times p$ matrix for which the same partition can be made. Then, one can write

$$C_{11} = A_{11}B_{11} + A_{12}B_{21} , \quad C_{12} = A_{11}B_{12} + A_{12}B_{22} ,$$

$$C_{21} = A_{21}B_{11} + A_{22}B_{21} , \quad C_{22} = A_{21}B_{12} + A_{22}B_{22} ,$$

which all can formally be written as

$$C_{IJ} = A_{I1}B_{1J} + A_{I2}B_{2J} = \sum_{K=1}^{2} A_{IK}B_{KJ} , \tag{1.149}$$

which is identical to the common rule of matrix multiplication.

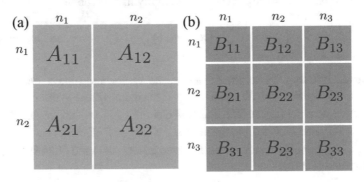

Fig. 1.7 Examples of partitioning of two square $p \times p$ matrices into blocks: **a** the matrix A is divided into four blocks A_{11}, A_{12}, A_{21} and A_{22} with their dimensions indicated, where $n_1 + n_2 = p$, and **b** B is split into nine blocks with $n_1 + n_2 + n_3 = p$

Problem 1.137 Prove the above identities by writing down explicitly all matrix multiplications using elements a_{ij}, b_{ij} and c_{ij} of the matrices $\mathbf{A}$, $\mathbf{B}$ and $\mathbf{C}$.

Problem 1.138 Consider the inverse $\mathbf{G} = \mathbf{A}^{-1}$ of the matrix $\mathbf{A}$ from the previous problem. By writing explicitly in blocks the identity $\mathbf{AG} = \mathbf{E}$, show that the blocks of $\mathbf{G}$ can be written as follows

$$
\begin{pmatrix} \mathbf{A}_{11} & \mathbf{A}_{12} \\ \mathbf{A}_{21} & \mathbf{A}_{22} \end{pmatrix}^{-1} = \begin{pmatrix} \mathbf{G}_{11} & -\mathbf{A}_{11}^{-1}\mathbf{A}_{12}\mathbf{G}_{22} \\ -\mathbf{A}_{22}^{-1}\mathbf{A}_{21}\mathbf{G}_{11} & \mathbf{G}_{22} \end{pmatrix}, \tag{1.150}
$$

where

$$
\mathbf{G}_{11} = \left(\mathbf{A}_{11} - \mathbf{A}_{12}\mathbf{A}_{22}^{-1}\mathbf{A}_{21}\right)^{-1}, \quad \mathbf{G}_{22} = \left(\mathbf{A}_{22} - \mathbf{A}_{21}\mathbf{A}_{11}^{-1}\mathbf{A}_{12}\right)^{-1}.
$$

Problem 1.139 Repeat the previous problem by starting from $\mathbf{GA} = \mathbf{E}$ instead. The same result should be obtained.

Problem 1.140 The inverse of a 2×2 block square matrix can also be written slightly differently (due to Frobenius):

$$
\mathbf{A}^{-1} = \begin{pmatrix} \mathbf{A}_{11} & \mathbf{A}_{12} \\ \mathbf{A}_{21} & \mathbf{A}_{22} \end{pmatrix}^{-1} = \begin{pmatrix} \mathbf{A}_{11}^{-1} + \mathbf{A}_{11}^{-1}\mathbf{A}_{12}\mathbf{G}_{22}\mathbf{A}_{21}\mathbf{A}_{11}^{-1} & -\mathbf{A}_{11}^{-1}\mathbf{A}_{12}\mathbf{G}_{22} \\ -\mathbf{G}_{22}\mathbf{A}_{21}\mathbf{A}_{11}^{-1} & \mathbf{G}_{22} \end{pmatrix}.
$$
$$\tag{1.151}$$

Verify this result by checking that the multiplication of the direct matrix $\mathbf{A}$ with the matrix in the right-hand side above results in the identity matrix.

The matrix identities given above might be useful in finding inverse of a large matrix as they enable one to replace this operation with simpler problems of finding inverse of matrices of smaller size.

Problem 1.141 Check by direct multiplication that any square matrix $\mathbf{X} = \begin{pmatrix} \mathbf{X}_{11} & \mathbf{X}_{12} \\ \mathbf{X}_{21} & \mathbf{X}_{22} \end{pmatrix}$, which is split into four blocks, can be represented in the following form:

$$
\begin{pmatrix} \mathbf{X}_{11} & \mathbf{X}_{12} \\ \mathbf{X}_{21} & \mathbf{X}_{22} \end{pmatrix} = \begin{pmatrix} \mathbf{X}_{11} & \mathbf{0}_{12} \\ \mathbf{0}_{21} & \mathbf{E}_{22} \end{pmatrix} \begin{pmatrix} \mathbf{E}_{11} & \mathbf{0}_{12} \\ \mathbf{X}_{21} & \mathbf{E}_{22} \end{pmatrix} \begin{pmatrix} \mathbf{E}_{11} & \mathbf{X}_{11}^{-1}\mathbf{X}_{12} \\ \mathbf{0}_{21} & \mathbf{X}_{22} - \mathbf{X}_{21}\mathbf{X}_{11}^{-1}\mathbf{X}_{12} \end{pmatrix},
$$
$$\tag{1.152}$$

where zeros means zero elements in the whole block, and $\mathbf{E}_{11}$, etc. mean the corresponding identity matrices. This form is sometimes called triangular decomposition since the second and the third matrices in the right-hand side are left and right triangular, respectively.

Problem 1.142 Similarly, prove another, the so-called LDE decomposition, of a matrix:

$$\begin{pmatrix} \mathbf{X}_{11} & \mathbf{X}_{12} \\ \mathbf{X}_{21} & \mathbf{X}_{22} \end{pmatrix} = \begin{pmatrix} \mathbf{E}_{11} & \mathbf{X}_{12}\mathbf{X}_{22}^{-1} \\ \mathbf{0}_{21} & \mathbf{E}_{22} \end{pmatrix} \begin{pmatrix} \mathbf{X}_{11} - \mathbf{X}_{12}\mathbf{X}_{22}^{-1}\mathbf{X}_{21} & \mathbf{0}_{12} \\ \mathbf{0}_{21} & \mathbf{X}_{22} \end{pmatrix} \begin{pmatrix} \mathbf{E}_{11} & \mathbf{0}_{12} \\ \mathbf{X}_{22}^{-1}\mathbf{X}_{21} & \mathbf{E}_{22} \end{pmatrix}.$$
(1.153)

Here the matrix is presented as a product of two triangular and one quasi-diagonal matrices.

Problem 1.143 Inverting the decomposition (1.153), re-derive formula (1.150).

Problem 1.144 Consider a 2×2 block matrix $\mathbf{X} = \begin{pmatrix} \mathbf{E}_{11} & \mathbf{0}_{12} \\ \mathbf{X}_{21} & \mathbf{X}_{22} \end{pmatrix}$. Prove that $|\mathbf{X}| = |\mathbf{X}_{22}|$.

Problem 1.145 Using the matrix decomposition Eq. (1.152) and the result of the previous problem, show that the determinant of $\mathbf{X}$ can be written as a product of determinants of two matrices of smaller size:

$$\begin{vmatrix} \mathbf{X}_{11} & \mathbf{X}_{12} \\ \mathbf{X}_{21} & \mathbf{X}_{22} \end{vmatrix} = |\mathbf{X}_{11}| \, |\mathbf{X}_{22} - \mathbf{X}_{21}\mathbf{X}_{11}^{-1}\mathbf{X}_{12}| \; .$$
(1.154)

Prove the same result from the LDE decomposition (1.153).

Next, let us now prove a simple fact that is sometimes useful. Consider a square matrix with 2×2 block structure containing a zero matrix in an off-diagonal block position, e.g.

$$\mathbf{A} = \begin{pmatrix} \mathbf{B} & \mathbf{D} \\ \mathbf{0} & \mathbf{C} \end{pmatrix}.$$

Let us prove that $|\mathbf{A}| = |\mathbf{B}| \cdot |\mathbf{C}|$. This result generalises that of Problem 1.144. Indeed, let us perform various elementary operations of Sect. 1.4.2 on the bottom rows of the matrix corresponding to the rows of its block $\mathbf{C}$. These transformations can bring the matrix $\mathbf{C}$ into the upper triangular form. Let

$$\mathbf{U} = \begin{pmatrix} \mathbf{E} & \mathbf{0} \\ \mathbf{0} & \mathbf{U}_1 \end{pmatrix}, \quad |\mathbf{U}| = |\mathbf{U}_1| \; ,$$

corresponds to that transformation, written in the same block form, with $\mathbf{E}$ being an appropriate identity matrix. Then

$$\mathbf{U}\mathbf{A} = \begin{pmatrix} \mathbf{B} & \mathbf{D} \\ \mathbf{0} & \mathbf{U}_1\mathbf{C} \end{pmatrix}.$$

Note that these operations do not change the upper rows of $\mathbf{A}$. Next, consider various elementary operations that work on the upper rows of $\mathbf{A}$ to bring $\mathbf{B}$ also into the upper triangular form. These are accomplished by the matrix (of the same block structure)

$$\mathbf{U}' = \begin{pmatrix} \mathbf{U}_2 & \mathbf{0} \\ \mathbf{0} & \mathbf{E}' \end{pmatrix}, \quad |\mathbf{U}'| = |\mathbf{U}_2| ,$$

(with $\mathbf{E}'$ being an appropriate identity matrix) and the result is

$$\mathbf{U}'\mathbf{U}\mathbf{A} = \begin{pmatrix} \mathbf{U}_2\mathbf{B} & \mathbf{D}' \\ \mathbf{0} & \mathbf{U}_1\mathbf{C} \end{pmatrix} .$$

The obtained matrix is now of a triangular form, and its determinant is equal to the product of diagonal elements of the matrices $\mathbf{U}_2\mathbf{B}$ and $\mathbf{U}_1\mathbf{C}$. However, these matrices are themselves triangular with their determinants being equal to the product of their diagonal elements. Hence, the determinant of $\mathbf{U}'\mathbf{U}\mathbf{A}$ is equal to the product of the determinants of these two matrices, and we can write

$$|\mathbf{U}'\mathbf{U}\mathbf{A}| = |\mathbf{U}_2\mathbf{B}| \cdot |\mathbf{U}_1\mathbf{C}| ,$$
$$|\mathbf{U}'| \cdot |\mathbf{U}| \cdot |\mathbf{A}| = |\mathbf{U}_2| \cdot |\mathbf{B}| \cdot |\mathbf{U}_1| \cdot |\mathbf{C}| .$$

Cancelling out on the determinants of the transformation matrices on both sides, we get the desired result.

Finally, we shall prove a matrix identity due to Woodbury:

$$(\mathbf{A} + \mathbf{U}\mathbf{C}\mathbf{V})^{-1} = \mathbf{A}^{-1} - \mathbf{A}^{-1}\mathbf{U}\left(\mathbf{C}^{-1} + \mathbf{V}\mathbf{A}^{-1}\mathbf{U}\right)^{-1}\mathbf{V}\mathbf{A}^{-1} . \tag{1.155}$$

Here $\mathbf{A}$ is a $n \times n$ matrix, $\mathbf{C}$ is a $k \times k$ matrix with $k \leq n$, while $\mathbf{U}$ and $\mathbf{V}$ are generally rectangular $n \times k$ and $k \times n$ matrices. The matrix $\mathbf{U}\mathbf{C}\mathbf{V}$ can be considered as a 'perturbation' to the matrix $\mathbf{A}$; then the formula explicitly stipulates that the inverse of the perturbed matrix $\mathbf{A} + \mathbf{U}\mathbf{C}\mathbf{V}$ can be written as the inverse of $\mathbf{A}$ plus a correction (the second term in the right-hand side).

The simplest method to prove this result is based on the following auxiliary linear system of equations for vectors $\mathbf{X}$ (of length n) and $\mathbf{Y}$ (of length k):

$$\begin{pmatrix} \mathbf{A} & \mathbf{U} \\ \mathbf{V} & -\mathbf{C}^{-1} \end{pmatrix} \begin{pmatrix} \mathbf{X} \\ \mathbf{Y} \end{pmatrix} = \begin{pmatrix} \mathbf{E} \\ \mathbf{0} \end{pmatrix} .$$

We shall start by working out the multiplication in the left-hand side and setting the result to the right-hand side, the procedure leading to two equations:

$$\mathbf{A}\mathbf{X} + \mathbf{U}\mathbf{Y} = \mathbf{E} \quad \text{and} \quad \mathbf{V}\mathbf{X} - \mathbf{C}^{-1}\mathbf{Y} = \mathbf{0} .$$

Since $\mathbf{C}^{-1}$ is a square matrix, we can solve the second equation for $\mathbf{Y} = \mathbf{CVX}$ and substitute the result into the first equation which gives simply $\mathbf{X} = (\mathbf{A} + \mathbf{UCV})^{-1}$, i.e. the required matrix in the left-hand side of (1.155) is actually $\mathbf{X}$. On the other hand, we can also express $\mathbf{X}$ from the first equation,

$$\mathbf{X} = \mathbf{A}^{-1} - \mathbf{A}^{-1}\mathbf{UY}, \tag{1.156}$$

and substitute this into the second, resulting in the equation $\mathbf{VA}^{-1} - \mathbf{VA}^{-1}\mathbf{UY} = \mathbf{C}^{-1}\mathbf{Y}$ for $\mathbf{Y}$, that can easily be solved:

$$\mathbf{Y} = \left(\mathbf{C}^{-1} + \mathbf{VA}^{-1}\mathbf{U}\right)^{-1}\mathbf{VA}^{-1}.$$

Substituting this back into Eq. (1.156) leads to the expression for $\mathbf{X}$ that is identical to the right-hand side of the Woodbury formula (1.155).

We discussed above in Sect. 1.10.1 how Green's function (the resolvent matrix) of a matrix $\mathbf{A}(z) = \mathbf{A}^0(z) + \mathbf{W}$ can be expanded into the Born series in powers of the 'perturbation' $\mathbf{W}$. This expansion can be useful if the perturbation is 'small'. If this is not the case, one may attempt expanding Green's function in powers of the inverse perturbation $\mathbf{W}^{-1}$. We shall now illustrate how the Woodbury identity can be used to do this. Let the perturbation $\mathbf{W}$ be of a smaller size (the block 11) than the full matrix $\mathbf{A}^0(z) = z\mathbf{E} - \mathbf{H}^0$, i.e. using the block structure,

$$\begin{aligned}
\mathbf{A}(z) &= \begin{pmatrix} \mathbf{A}_{11}^0(z) + \mathbf{W}_{11} & \mathbf{A}_{12}^0(z) \\ \mathbf{A}_{21}^0(z) & \mathbf{A}_{22}^0(z) \end{pmatrix} \\
&= \underbrace{\begin{pmatrix} \mathbf{A}_{11}^0(z) & \mathbf{A}_{12}^0(z) \\ \mathbf{A}_{21}^0(z) & \mathbf{A}_{22}^0(z) \end{pmatrix}}_{\mathbf{A}^0(z)} + \begin{pmatrix} \mathbf{W}_{11} & \mathbf{0}_{12} \\ \mathbf{0}_{21} & \mathbf{0}_{22} \end{pmatrix}.
\end{aligned}$$

The inverse of $\mathbf{A}^0(z)$ represents an unperturbed Green's function (matrix) $\mathbf{G}^0(z) = \left(z\mathbf{E} - \mathbf{H}^0\right)^{-1}$. We would like to generate a series in terms of $\mathbf{W}^{-1}$ for the full Green's function $\mathbf{G}(z) = (z\mathbf{E} - \mathbf{A})^{-1}$. Let the rectangular matrices $\mathbf{U}$ and $\mathbf{V}$ are chosen as follows:

$$\mathbf{U} = \begin{pmatrix} \mathbf{E}_{11} \\ \mathbf{0}_{21} \end{pmatrix} \quad \text{and} \quad \mathbf{V} = \begin{pmatrix} \mathbf{E}_{11} & \mathbf{0}_{12} \end{pmatrix}.$$

It is easy to see now that $\mathbf{A}^0 + \mathbf{UW}_{11}\mathbf{V}$ is exactly equal to $\mathbf{A}$.

Problem 1.146 Using Woodbury formula (1.155), prove the following result:

$$\mathbf{G}(z) = \mathbf{G}^0(z) - \begin{pmatrix} \mathbf{G}_{11}^0 \\ \mathbf{G}_{21}^0 \end{pmatrix} \left(\mathbf{G}_{11}^0 + \mathbf{W}_{11}^{-1}\right)^{-1} \begin{pmatrix} \mathbf{G}_{11}^0 & \mathbf{G}_{12}^0 \end{pmatrix}.$$

Because of an obvious matrix identity (that can be checked, e.g. by multiplying both sides by $\mathbf{A} + \mathbf{B}$)

$$(\mathbf{A} + \mathbf{B})^{-1} = \mathbf{A}^{-1} - \mathbf{A}^{-1}\mathbf{B}(\mathbf{A} + \mathbf{B})^{-1}$$
$$= \mathbf{A}^{-1} - \mathbf{A}^{-1}\mathbf{B}\mathbf{A}^{-1} + \mathbf{A}^{-1}\mathbf{B}\mathbf{A}^{-1}\mathbf{B}\mathbf{A}^{-1} - \dots$$

(*cf.* Problem 1.115 and Eq. (1.115)), the inverse matrix $\left(\mathbf{G}_{11}^{0} + \mathbf{W}_{11}^{-1}\right)^{-1}$ can be expanded into a Born series in terms of $\mathbf{W}_{11}^{-1}$. This proves the above-made statement.

1.18 Defective Matrices

We have seen on multiple occasions above that all square matrices can be divided into two broad classes: diagonalisable that have as many linearly independent eigenvectors as their dimension, and defective ones, that have a smaller number of linearly independent eigenvectors one can built, and hence they are not diagonalisable. In other words, their geometric multiplicity is smaller than the algebraic one (Sect. 1.8.2). We also know that eigenvectors with different eigenvalues are always linearly independent (Theorem 1.18). Hence, matrices may become defective if they have repeated eigenvalues, and when at least for some of these eigenvalues a smaller number of linearly independent eigenvectors one can built than their algebraic multiplicity.

In this subsection we shall consider the case of defective matrices in much more detail by discussing the so-called Jordan normal form and the notion of generalised eigenvectors. The latter enable one to construct a linear vector space, of the same dimension as the dimensionality of the matrix in question, by appending to the existing eigenvectors additional vectors built from them in a specific way (via the so-called Jordan chains).

1.18.1 Jordan Normal Form

Let $\mathbf{A}$ be a defective square $p \times p$ matrix with distinct eigenvalues λ_i, where $i = 1, \dots, q$, and $q \leq p$. If $q < p$, there are repeated eigenvalues. Let us next construct a square $p \times p$ matrix $\mathbf{J}$ of a very specific structure called the Jordan normal form. This matrix has a block-diagonal structure; each its block, also called the Jordan block, corresponds to one eigenvalue. If an eigenvalue λ_i is unique, the corresponding to it block $\mathbf{J}_i$ contains a single element with λ_i in it. If λ_i is repeated q_i times, but only a single eigenvector can be constructed, the corresponding Jordan block $\mathbf{J}_i$ is a $q_i \times q_i$ matrix that has the following structure:

Fig. 1.8 Jordan normal form. Note that eigenvalues in different blocks may be identical. The elements that are not shown are zeros

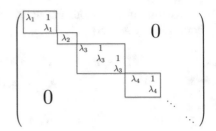

$$\mathbf{J}_i = \begin{pmatrix} \lambda_i & 1 & & \\ & \lambda_i & 1 & \\ & & \ddots & 1 \\ & & & \lambda_i \end{pmatrix}. \tag{1.157}$$

It contains the repeated eigenvalue on its main diagonal and the unity on all elements of its superdiagonal (these are the elements immediately to the right of the diagonal elements[12]); all other elements are filled up by zeros. If for the given repeated eigenvalue λ_i one can construct $r > 1$ linearly independent eigenvectors, then r Jordan blocks can be constructed with the overall dimension of the algebraic multiplicity of that eigenvalue.

The general structure of the Jordan normal form $\mathbf{J}$ is shown in Fig. 1.8. Each Jordan block is shown as a square block.[13] The first block is due to the eigenvalue λ_1 that is repeated twice; the second block is constructed from the eigenvalue λ_2, the third 3×3 block is generated by the eigenvalue λ_3, and so on. Note that we have elements 1 on the superdiagonal only within the Jordan blocks; such elements are zeros on the superdiagonal between Jordan blocks. Hence, $\mathbf{J}$ for a matrix that contains as many linearly independent eigenvectors as its dimension has truly diagonal form, i.e. its whole superdiagonal contains only zeros. One may say that in general $\mathbf{J}$ has nearly diagonal form. It is easy to see that $\mathbf{J}$ has exactly the same characteristic polynomial (the same eigenvalues) as the original matrix $\mathbf{A}$ since it is an upper triangular matrix.

Now we shall show that any square matrix $\mathbf{A}$ (defective or not) can be brought into the Jordan normal form via a similarity transformation using a non-singular matrix $\mathbf{U}$, i.e.

$$\mathbf{A} = \mathbf{U}\mathbf{J}\mathbf{U}^{-1} \quad \text{or} \quad \mathbf{J} = \mathbf{U}^{-1}\mathbf{A}\mathbf{U} \quad \text{or} \quad \mathbf{A}\mathbf{U} = \mathbf{U}\mathbf{J}. \tag{1.158}$$

The transformation is not unique as the Jordan blocks may appear along the diagonal of $\mathbf{J}$ in a different order. The existence of the similarity transformation is to be expected as similar matrices share the same eigenvalues.

In the following, we shall determine the form of the matrix $\mathbf{U}$ that 'nearly diagonalises' $\mathbf{A}$. Similarly to what we did in the case of the modal matrix (Sect. 1.9.1), let

---

[12] The Jordan block with subdiagonal elements that stay immediately to the left of the main diagonal may also be considered.

[13] It may look rectangular on the picture, but it is actually a square.

us represent $\mathbf{U}$ as a collection of vectors $\mathbf{v}_i = (v_{ji})$ along its columns:

$$\mathbf{U} = (\mathbf{v}_1 \ \mathbf{v}_2 \ \dots \ \mathbf{v}_p) = \begin{pmatrix} v_{11} & v_{12} & \dots & v_{1p} \\ v_{21} & v_{22} & \dots & v_{2p} \\ \dots & \dots & \ddots & \dots \\ v_{p1} & v_{p2} & \dots & v_{pp} \end{pmatrix}.$$

Note that the components v_{ji} of each vector $\mathbf{v}_i$ have the right index indicating the vector number and the left—its component. Let us order the eigenvalues in a particular way as λ_1, λ_2, and so on, repeating them as many times as their algebraic multiplicity. Consider the equation $\mathbf{AU} = \mathbf{UJ}$, or

$$\mathbf{A} (\mathbf{v}_1 \ \mathbf{v}_2 \ \dots \ \mathbf{v}_p) = (\mathbf{v}_1 \ \mathbf{v}_2 \ \dots \ \mathbf{v}_p) \mathbf{J}.$$

Consider the very first eigenvalue λ_1 assuming that there is only one linearly independent eigenvector possible for that eigenvalue; its algebraic multiplicity is $q_1 > 1$. That eigenvector we shall place as the first column in $\mathbf{U}$, i.e. it is $\mathbf{v}_1$. Let us start from the right-hand side of the equation $\mathbf{AU} = \mathbf{UJ}$ and determine the first q_1 columns of the matrix $\mathbf{UJ}$ by considering the multiplication of the first q_1 columns of $\mathbf{U}$ (i.e. of the vectors $\mathbf{v}_1, \dots, \mathbf{v}_{q_1}$) on $\mathbf{J}$. In that case only the first Jordan block $\mathbf{J}_1$ will participate. Note that the elements $(\mathbf{J}_1)_{kl}$ of the first Jordan block $\mathbf{J}_1$ (with $1 \le k \le p$ and $1 \le l \le q_1$) can generally be written as

$$(\mathbf{J}_1)_{kl} = \begin{cases} \lambda_1 \delta_{kl} + \delta_{k+1,l} & , \ l > 1 \\ \lambda_1 \delta_{kl} & , \ l = 1 \end{cases}.$$

Therefore,

$$(\mathbf{UJ})_{jl} = \sum_{k=1}^{p} v_{jk} (\mathbf{J}_1)_{kl} = \sum_{k=1}^{p} v_{jk} [\delta_{kl} \lambda_1 + \delta_{k+1,l}] = \lambda_1 v_{jl} + v_{j,l-1}$$
$$= \lambda_1 (\mathbf{v}_l)_j + (\mathbf{v}_{l-1})_j$$

for $l > 1$ and $(\mathbf{UJ})_{jl} = \lambda_1 (\mathbf{v}_l)_j$ for $l = 1$. In other words, the first column of the matrix $\mathbf{UJ}$ contains the vector $\lambda_1 \mathbf{v}_1$ and any other column $l > 1$ is the linear combination $\lambda_1 \mathbf{v}_l + \mathbf{v}_{l-1}$.

On the other hand, the first q_1 columns of the product $\mathbf{AU}$ are simply built by vectors $\mathbf{Av}_l$, where $l = 1, \dots, q_1$. Hence, the vectors $\mathbf{v}_1, \dots, \mathbf{v}_{q_1}$ satisfy the following equations: the first vector $\mathbf{v}_1$ is the eigenvector satisfying $\mathbf{Av}_1 = \lambda_1 \mathbf{v}_1$ or $(\mathbf{A} - \lambda_1 \mathbf{E}) \mathbf{v}_1 = 0$; the vector $\mathbf{v}_2$ satisfies

$$\mathbf{Av}_2 = \lambda_1 \mathbf{v}_2 + \mathbf{v}_1 \implies (\mathbf{A} - \lambda_1 \mathbf{E}) \mathbf{v}_2 = \mathbf{v}_1,$$

the next vector satisfies

$$\mathbf{A}\mathbf{v}_3 = \lambda_1 \mathbf{v}_3 + \mathbf{v}_2 \implies (\mathbf{A} - \lambda_1 \mathbf{E})\,\mathbf{v}_3 = \mathbf{v}_2 \, ,$$

and so on; for the last vector we have $(\mathbf{A} - \lambda_1 \mathbf{E})\,\mathbf{v}_{q_1} = \mathbf{v}_{q_1 - 1}$. Generally, for any $k > 1$ we have

$$(\mathbf{A} - \lambda_1 \mathbf{E})^k\,\mathbf{v}_k = (\mathbf{A} - \lambda_1 \mathbf{E})^{k-1}\,\mathbf{v}_{k-1} = \cdots = (\mathbf{A} - \lambda_1 \mathbf{E})\,\mathbf{v}_1 = \mathbf{0} \, .$$

The vectors $\mathbf{v}_1, \ldots, \mathbf{v}_{q_1}$ form the so-called Jordan chain. The last vector in the chain, $\mathbf{v}_{q_1}$, can be considered as its generator:

$$\mathbf{v}_{q_1 - 1} = (\mathbf{A} - \lambda_1 \mathbf{E})\,\mathbf{v}_{q_1} \, ,$$

$$\mathbf{v}_{q_1 - 2} = (\mathbf{A} - \lambda_1 \mathbf{E})\,\mathbf{v}_{q_1 - 1} = (\mathbf{A} - \lambda_1 \mathbf{E})^2\,\mathbf{v}_{q_1} \, ,$$

and so on;

$$\mathbf{v}_k = (\mathbf{A} - \lambda_1 \mathbf{E})^{q_1 - k}\,\mathbf{v}_{q_1} \tag{1.159}$$

for any $k = 1, \ldots, q_1.$,

The constructed vectors of the chain are linearly independent. Indeed, let us introduce the shorthand $\mathbf{G}_{\lambda_1} = \mathbf{A} - \lambda_1 \mathbf{E}$ and consider the following equation for the coefficients $\alpha_1, \ldots, \alpha_{q_1}$:

$$\alpha_1 \mathbf{v}_1 + \alpha_2 \mathbf{v}_2 + \cdots + \alpha_{q_1} \mathbf{v}_{q_1} = \mathbf{0} \, . \tag{1.160}$$

Note that, since

$$(\mathbf{A} - \lambda_1 \mathbf{E})^k\,\mathbf{v}_k = \mathbf{G}_{\lambda_1}^k\,\mathbf{v}_k = \mathbf{0} \, ,$$

then $\mathbf{G}_{\lambda_1}^n\,\mathbf{v}_k = \mathbf{G}_{\lambda_1}^{n-k}\left(\mathbf{G}_{\lambda_1}^k\,\mathbf{v}_k\right) = \mathbf{0}$ for any $n \geq k$. Therefore, acting with the matrix $\mathbf{G}_{\lambda_1}^{q_1 - 1}$ on the vector in the left-hand side, we obtain

$$\alpha_1 \underbrace{\mathbf{G}_{\lambda_1}^{q_1 - 1}\mathbf{v}_1}_{0} + \alpha_2 \underbrace{\mathbf{G}_{\lambda_1}^{q_1 - 1}\mathbf{v}_2}_{0} + \cdots + \alpha_{q_1} \underbrace{\mathbf{G}_{\lambda_1}^{q_1 - 1}\mathbf{v}_{q_1 - 1}}_{0} + \alpha_{q_1} \underbrace{\mathbf{G}_{\lambda_1}^{q_1 - 1}\mathbf{v}_{q_1}}_{\mathbf{v}_1} = \mathbf{0} \implies \alpha_{q_1}\mathbf{v}_1 = \mathbf{0} \, .$$

Since the vector $\mathbf{v}_1$ is not the null vector, we should have the coefficient $\alpha_{q_1} = 0$. Multiplying both sides of Eq. (1.160) by $\mathbf{G}_{\lambda_1}^{q_1 - 2}$, we similarly obtain that $\alpha_{q_1 - 1} = 0$. Repeating this procedure, we shall find that all coefficients $\alpha_k = 0$ for any $k = 1, \ldots, q_1$. This proves the linear independence of the vectors in the chain.

In a similar way one can consider the next bunch of columns of the matrix $\mathbf{U}$ corresponding to the second eigenvalue λ_2. If that eigenvalue is repeated q_2 times and again only one eigenvector exists, then the next q_2 columns of $\mathbf{U}$ are formed by vectors $\mathbf{w}_1 = \mathbf{v}_{q_1 + 1}$, $\mathbf{w}_2 = \mathbf{v}_{q_1 + 2}$, and so on, $\mathbf{w}_{q_2} = \mathbf{v}_{q_1 + q_2}$. These vectors satisfy the same type of equations as above (as in the previous case, the first vector $\mathbf{w}_1$ is chosen as the only eigenvector of $\mathbf{A}$ with the eigenvalue λ_2): $\mathbf{G}_{\lambda_2}\mathbf{w}_1 = \mathbf{0}$ and $\mathbf{G}_{\lambda_2}\mathbf{w}_k = \mathbf{w}_{k-1}$ for any $k = 2, \ldots, q_2$. Only the second Jordan block $\mathbf{J}_2$ would be active in this case.

It is easy to show that both sets of vectors $\mathbf{v}_1, \ldots, \mathbf{v}_{q_1}$ and $\mathbf{w}_1, \ldots, \mathbf{w}_{q_2}$ are linearly independent. Indeed, consider the solution of the equation

$$\alpha_1 \mathbf{v}_1 + \cdots + \alpha_{q_1} \mathbf{v}_{q_1} + \beta_1 \mathbf{w}_1 + \cdots + \beta_{q_2} \mathbf{w}_{q_2} = \mathbf{0}. \tag{1.161}$$

Without loss of generality, we can assume that $q_2 \leq q_1$. Then, multiply this equation from the left by $\mathbf{G}_{\lambda_2}^{q_2}$. This operation will eliminate all $\mathbf{w}$ vectors from Eq. (1.161), leading to the equation

$$\mathbf{G}_{\lambda_2}^{q_2} \left(\alpha_1 \mathbf{v}_1 + \cdots + \alpha_{q_1} \mathbf{v}_{q_1} \right) = \mathbf{0}. \tag{1.162}$$

Since the matrix $\mathbf{G}_{\lambda_2} = \mathbf{A} - \lambda_2 \mathbf{E} = \mathbf{G}_{\lambda_1} + (\lambda_1 - \lambda_2) \mathbf{E}$, we obtain instead

$$\left(\mathbf{G}_{\lambda_1} + \mathbf{E}\Delta \right)^{q_2} \left(\alpha_1 \mathbf{v}_1 + \cdots + \alpha_{q_1} \mathbf{v}_{q_1} \right) = \mathbf{0}, \tag{1.163}$$

where $\Delta = \lambda_1 - \lambda_2$. Because $\mathbf{G}_{\lambda_1}$ commutes with itself and with the identity matrix $\mathbf{E}$, we can use the binomial expansion for the matrix in the left-hand side:

$$\left(\mathbf{G}_{\lambda_1} + \mathbf{E}\Delta \right)^{q_2} = \mathbf{G}_{\lambda_1}^{q_2} + q_2 \mathbf{G}_{\lambda_1}^{q_2-1} \Delta + \cdots + q_2 \mathbf{G}_{\lambda_1} \Delta^{q_2-1} + \Delta^{q_2}.$$

For any $q \geq k$, we have $\mathbf{G}_{\lambda_1}^q \mathbf{v}_k = \mathbf{0}$, while

$$\mathbf{G}_{\lambda_1}^q \mathbf{v}_k = \mathbf{G}_{\lambda_1}^{q-1} \mathbf{v}_{k-1} = \mathbf{G}_{\lambda_1}^{q-2} \mathbf{v}_{k-2} = \cdots = \mathbf{v}_{k-q}$$

for $q < k$. It is seen that the left-hand side of Eq. (1.163) is simply a linear combination of the vectors $\mathbf{v}_1, \ldots, \mathbf{v}_{q_1}$, that can only be equal to zero when all coefficients α_k are zero. Returning back to the original Eq. (1.161), we are left with the linear combination of the $\mathbf{w}$ vectors in the left-hand side which we know can only be equal to zero if all the β_k coefficients are zero. This finalises the proof that both sets of vectors $\{\mathbf{v}_k\}$ and $\{\mathbf{w}_k\}$ are indeed linearly independent.

Repeating this process for all eigenvalues, the whole matrix $\mathbf{U}$ is constructed. It consists of vector-columns of all Jordan chains associated with every Jordan block. The vector-columns of $\mathbf{U}$ form generalised eigenvectors of $\mathbf{A}$. All these vectors are linearly independent. Indeed, we know that vectors within any of the sets are such, and also of any pair of sets; clearly, by considering all possible pairs we arrive at the conclusion that all vectors are linearly independent. Hence, the matrix $\mathbf{U}$ is non-singular and has an inverse (see Theorem 1.14).

Concluding the formal part of our consideration, we can state that for any defective square matrix $\mathbf{A}$ for which the number of eigenvectors is smaller than its dimension, one can always construct additional vectors that, together with the existing eigenvectors, form a linearly independent set of vectors of the same dimension as the matrix itself. The new vectors are grouped by repeated eigenvalues into sets,

$$\underbrace{\mathbf{v}_1, \ldots, \mathbf{v}_{q_1}}_{\lambda_1}; \ \underbrace{\mathbf{w}_1, \ldots, \mathbf{w}_{q_2}}_{\lambda_2}; \ \underbrace{\mathbf{s}_1, \ldots, \mathbf{s}_{q_3}}_{\lambda_3}; \ \ldots$$

and are generated considering each existing eigenvector $\mathbf{v}_1$, $\mathbf{w}_1$, $\mathbf{s}_1$, and so on.

In practice, to calculate the Jordan chain in each case one has to solve a chain of linear algebraic equations. Consider as an example the following matrix

$$\mathbf{A} = \begin{pmatrix} 2 & 2 & 0 \\ -1 & 5 & 1 \\ -1 & 1 & 3 \end{pmatrix}.$$

Problem 1.147 Show that the characteristic polynomial of this matrix is $\Delta(\lambda) = (4 - \lambda)^2 (2 - \lambda)$, i.e. it has two eigenvalues: $\lambda_1 = 4$ that has algebraic multiplicity 2 and a single eigenvector $\mathbf{v}_1 = \begin{pmatrix} 1 & 1 & 0 \end{pmatrix}^T$, and $\lambda_2 = 2$ with the algebraic multiplicity 1 and a single eigenvector $\mathbf{w}_1 = \begin{pmatrix} 1 & 0 & 1 \end{pmatrix}^T$.

As we only have one eigenvector for the doubly degenerate eigenvalue $\lambda_1 = 4$, we should construct the second vector $\mathbf{v}_2$ by solving the equation $(\mathbf{A} - 4\mathbf{E}) \mathbf{v}_2 = \mathbf{v}_1$.

Problem 1.148 Solve this equation (e.g. by using Gaussian elimination) to show that the vector $\mathbf{v}_2$ can be chosen, e.g. as $\begin{pmatrix} 0 & 1/2 & 1/2 \end{pmatrix}^T$.

All three vectors are obviously linearly independent. Using them, we can construct the matrix

$$\mathbf{U} = \begin{pmatrix} \mathbf{v}_1 & \mathbf{v}_2 & \mathbf{w}_1 \end{pmatrix} = \begin{pmatrix} 1 & 0 & 1 \\ 1 & 1/2 & 0 \\ 0 & 1/2 & 1 \end{pmatrix}$$

that relates the original matrix $\mathbf{A}$, its Jordan normal form

$$\mathbf{J} = \begin{pmatrix} 4 & 1 & 0 \\ 0 & 4 & 0 \\ 0 & 0 & 2 \end{pmatrix},$$

and the inverse,

$$\mathbf{U}^{-1} = \begin{pmatrix} 1/2 & 1/2 & -1/2 \\ -1 & 1 & 1 \\ 1/2 & -1/2 & 1/2 \end{pmatrix}.$$

Indeed,

$$\begin{pmatrix} 1 & 0 & 1 \\ 1 & 1/2 & 0 \\ 0 & 1/2 & 1 \end{pmatrix} \begin{pmatrix} 4 & 1 & 0 \\ 0 & 4 & 0 \\ 0 & 0 & 2 \end{pmatrix} \begin{pmatrix} 1/2 & 1/2 & -1/2 \\ -1 & 1 & 1 \\ 1/2 & -1/2 & 1/2 \end{pmatrix} = \begin{pmatrix} 2 & 2 & 0 \\ -1 & 5 & 1 \\ -1 & 1 & 3 \end{pmatrix}.$$

$$\underbrace{}_{\mathbf{U}} \quad \underbrace{}_{\mathbf{J}} \quad \underbrace{}_{\mathbf{U}^{-1}} \quad \underbrace{}_{\mathbf{A}}$$

Problem 1.149 Consider matrix

$$\mathbf{A} = \begin{pmatrix} 4 & 1 & -3 \\ -2 & 3 & 1 \\ 2 & 1 & -1 \end{pmatrix}. \tag{1.164}$$

Show that its characteristic polynomial $\Delta(\lambda) = (2 - \lambda)^3$, and the Jordan chain contains three vectors forming three columns of the transformation matrix

$$\mathbf{U} = \begin{pmatrix} 1 & 0 & -1/4 \\ 1 & 1 & 1/2 \\ 1 & 0 & 0 \end{pmatrix},$$

while the Jordan normal form of $\mathbf{A}$ is

$$\mathbf{J} = \begin{pmatrix} 2 & 1 & 0 \\ 0 & 2 & 1 \\ 0 & 0 & 2 \end{pmatrix}.$$

1.18.2 Function of a General Matrix

We have introduced a matrix function in Sect. 1.10 only for general non-defective matrices. In fact, a matrix function can also be defined for defective matrices as well using the similarity transformation (1.158) that relates the matrix in question with its Jordan normal form.

We shall start by considering the natural powers of the Jordan matrix. Since the latter has the block-diagonal form, it is sufficient to consider only the powers of its single block corresponding to some eigenvalue λ (that is q times degenerate), as the n-th power of the Jordan normal form will also have the block-diagonal form with Jordan blocks in the same power. The Jordan block is the $q \times q$ matrix

$$\mathbf{J}_b = \begin{pmatrix} \lambda & 1 & & & \\ & \lambda & 1 & & \\ & & \ddots & 1 & \\ & & & \lambda \end{pmatrix}.$$

A direct calculation shows that its square and the cube are

$$\mathbf{J}_b^2 = \begin{pmatrix} \lambda^2 & 2\lambda & 1 & & & \\ & \lambda^2 & 2\lambda & 1 & & \\ & & \lambda^2 & 2\lambda & 1 & \\ & & & \ddots & & \\ & & & & \lambda^2 & 2\lambda \\ & & & & & \lambda^2 \end{pmatrix}, \quad \mathbf{J}_b^3 = \begin{pmatrix} \lambda^3 & 3\lambda^2 & 3\lambda & 1 & & \\ & \lambda^3 & 3\lambda^2 & 3\lambda & 1 & \\ & & \ddots & & & \\ & & & \lambda^3 & 3\lambda^2 & 3\lambda \\ & & & & \lambda^3 & 3\lambda^2 \\ & & & & & \lambda^3 \end{pmatrix}.$$

One may recognise, respectively, terms in the binomial formulae for $(\lambda + 1)^2$ and $(\lambda + 1)^3$ in the non-zero elements of the square and cube of the Jordan blocks along each row of them. Correspondingly, we may propose that a general integer power n of the Jordan block $\mathbf{J}_b$ would be

$$\mathbf{J}_b^n = \begin{pmatrix} \binom{n}{0}\lambda^n & \binom{n}{1}\lambda^{n-1} & \binom{n}{2}\lambda^{n-2} & \binom{n}{3}\lambda^{n-3} & \cdots & & \\ & \binom{n}{0}\lambda^n & \binom{n}{1}\lambda^{n-1} & \binom{n}{2}\lambda^{n-2} & \cdots & & \\ & & \binom{n}{0}\lambda^n & \binom{n}{1}\lambda^{n-1} & \cdots & & \\ & & & \binom{n}{0}\lambda^n & \cdots & & \\ & & & & \ddots & & \\ & & & & & \binom{n}{0}\lambda^n & \binom{n}{1}\lambda^{n-1} \\ & & & & & & \binom{n}{0}\lambda^n \end{pmatrix}.$$

This matrix has the right triangular form. Note, that if we formally continue writing elements in each row in the right triangular part of the matrix, the last non-diagonal element will contain λ^0 in the zero power. Formally, we can continue writing elements along the row after this last element (if there are still elements left to be written), then these will contain λ in a negative power; we shall set their corresponding pre-factors to zero. Generally, we may infer that a general element of this matrix is

$$(\mathbf{J}_b^n)_{i,i+l} = \binom{n}{l}\lambda^{n-l}, \quad l = 0, \ldots, q - i; \tag{1.165}$$

all other elements are equal to zero. We shall prove this by induction. Indeed, consider $(0 \leq l \leq q - i)$

$$
\begin{aligned}
\left(\mathbf{J}_b^{n+1}\right)_{i,i+l} &= \sum_{k=i}^{q} \left(\mathbf{J}_b^n\right)_{i,k} \left(\mathbf{J}_b\right)_{k,i+l} = \sum_{m=0}^{q-i} \left(\mathbf{J}_b^n\right)_{i,i+m} \left(\mathbf{J}_b\right)_{i+m,i+l} \\
&= \sum_{m=0}^{\min(q-i,n)} \binom{n}{m} \lambda^{n-m} \left(\mathbf{J}_b\right)_{i+m,i+l} ,
\end{aligned}
\tag{1.166}
$$

where, we recall, q is the size of the block. In the first equality we have taken into account that we are dealing with the right triangular matrix and hence k starts from i, while in the second equality we changed to a new summation index $m = k - i$. For the given i and l, the Jordan block elements $(\mathbf{J}_b)_{i+m,i+l}$ are non-zero only in two cases: (i) it is equal to λ when $i + m = i + l$ or $m = l$ and (ii) it is equal to 1 when $i + m + 1 = i + l$ or $m = l - 1$. The case (ii) is to be dropped for the last diagonal element when $i = q$ and $m = l = 0$.

Consider first the case of $i + l < q$. Then we have the two values of m that contribute in the sum in Eq. (1.166):

$$
\begin{aligned}
\left(\mathbf{J}_b^{n+1}\right)_{i,i+l} &= \binom{n}{l} \lambda^{n-l} \lambda + \binom{n}{l-1} \lambda^{n-(l-1)} \\
&= \left[\binom{n}{l} + \binom{n}{l-1} \right] \lambda^{(n+1)-l} = \binom{n+1}{l} \lambda^{(n+1)-l} ,
\end{aligned}
$$

which is the required result, cf. Eq. (1.165). We have used here a well-known identity valid for binomial coefficients (see Eq. (I. 1.157)).

It is left to consider the case of $i = q$ and correspondingly $l = 0$. Only one term $m = l = 0$ contributes to the sum in Eq. (1.166), and we obtain

$$
\left(\mathbf{J}_b^{n+1}\right)_{q,q} = \binom{n}{0} \lambda^n \lambda = \binom{n+1}{0} \lambda^{n+1} ,
$$

as required.

Given a function $f(x)$, we can define the matrix function $f(\mathbf{J}_b)$ as an infinite series

$$
f(\mathbf{J}_b) = \sum_{n=0}^{\infty} \frac{f^{(n)}(0)}{n} \mathbf{J}^n .
$$

Since each power of $\mathbf{J}$ is the right triangular matrix, so will be $f(\mathbf{J})$. Consider then its $i, i + l$ matrix element and use Eq. (1.165):

$$[f(\mathbf{J}_b)]_{i,i+l} = \sum_{n=0}^{\infty} \frac{f^{(n)}(0)}{n!} (\mathbf{J}_b^n)_{i,i+l} = \sum_{n=l}^{\infty} \frac{f^{(n)}(0)}{n!} \binom{n}{l} \lambda^{n-l}$$

$$= \sum_{n=l}^{\infty} \frac{f^{(n)}(0)}{l!(n-l)!} \lambda^{n-l} = \frac{1}{l!} \left[\sum_{n=l}^{\infty} \frac{f^{(n)}(0)}{(n-l)!} \lambda^{n-l} \right] = \frac{1}{l!} f^{(l)}(\lambda) . \tag{1.167}$$

Note that in the second equality we started the summation at $n = l$ to ensure the non-negative power of λ. Let us write the form of this matrix block in a more explicit form:

$$f(\mathbf{J}_b) = \begin{pmatrix} f(\lambda) & f'(\lambda) & f^{(2)}(\lambda)/2! & f^{(3)}(\lambda)/3! & f^{(4)}(\lambda)/4! & \cdots \\ & f(\lambda) & f'(\lambda) & f^{(2)}(\lambda)/2! & f^{(3)}(\lambda)/3! & \cdots \\ & & f(\lambda) & f'(\lambda) & f^{(2)}(\lambda)/2! & \cdots \\ & & & f(\lambda) & f'(\lambda) & \cdots \\ & & & & \ddots & \\ & & & & & f(\lambda) \ f'(\lambda) \\ & & & & & f(\lambda) \end{pmatrix} . \tag{1.168}$$

The matrix $f(\mathbf{J})$ of the whole Jordan normal form $\mathbf{J}$ of the matrix $\mathbf{A}$ will have the block-diagonal form consisting of blocks like the one above.

Now the matrix function $f(\mathbf{A})$ can be defined following Eq. (1.111):

$$f(\mathbf{A}) = f(\mathbf{U}\mathbf{J}\mathbf{U}^{-1}) = \mathbf{U}f(\mathbf{J})\mathbf{U}^{-1} .$$

Note that we have defined here a matrix function without using the spectral decomposition of the matrix, and hence this definition is valid in a general case of *any* matrix including defective ones.

As an example, let us calculate the matrix function $e^{\mathbf{A}t}$ for the matrix $\mathbf{A}$ from Problem 1.149. There is just one Jordan block corresponding to the eigenvalue $\lambda = 2$, the scalar function is $f(\lambda) = e^{\lambda t}$, hence

$$e^{\mathbf{J}t} = \exp \left[\begin{pmatrix} 2 & 1 & 0 \\ 0 & 2 & 1 \\ 0 & 0 & 2 \end{pmatrix} t \right] = \begin{pmatrix} e^{2t} & te^{2t} & t^2 e^{2t}/2 \\ 0 & e^{2t} & te^{2t} \\ 0 & 0 & e^{2t} \end{pmatrix} .$$

Hence,

$$f(\mathbf{A}) = \underbrace{\begin{pmatrix} 1 & 0 & -1/4 \\ 1 & 1 & 1/2 \\ 1 & 0 & 0 \end{pmatrix}}_{\mathbf{U}} \underbrace{\begin{pmatrix} e^{2t} & te^{2t} & t^2 e^{2t}/2 \\ 0 & e^{2t} & te^{2t} \\ 0 & 0 & e^{2t} \end{pmatrix}}_{e^{\mathbf{J}t}} \underbrace{\begin{pmatrix} 0 & 0 & 1 \\ 2 & 1 & -3 \\ -4 & 0 & 4 \end{pmatrix}}_{\mathbf{U}^{-1}}$$

$$= e^{2t} \begin{pmatrix} -t^2 + 2t + 1 & t & t(2t - 3) \\ -2t(t+1) & 1+t & t(2t+1) \\ -2t(t-1) & t & 2t^2 - 3t + 1 \end{pmatrix}$$

> **Problem 1.150** Show that, for the same matrix $\mathbf{A}$ as in the example above,
>
> $$\cos(\mathbf{A}t) = \begin{pmatrix} (2t^2+1)\cos(2t) - 2t\sin(2t) & -t\sin(2t) & -2t^2\cos(2t) + 3t\sin(2t) \\ 2t^2\cos(2t) + 2t\sin(2t) & \cos(2t) - t\sin(2t) & -2t^2\cos(2t) - t\sin(2t) \\ 2t^2\cos(2t) - 2t\sin(2t) & -t\sin(2t) & (1-2t^2)\cos(2t) + 3t\sin(2t) \end{pmatrix}.$$

1.19 Systems of Linear Differential Equations

1.19.1 General Consideration

In this section we shall discuss methods of solving the first-order linear systems of m differential equations (DEs)

$$\begin{cases} \widetilde{a}_{11}\dot{x}_1 + \widetilde{a}_{12}\dot{x}_2 + \cdots + \widetilde{a}_{1n}\dot{x}_n + \widetilde{b}_{11}x_1 + \widetilde{b}_{12}x_2 + \cdots + \widetilde{b}_{1n}x_n = \widetilde{f}_1 \\ \widetilde{a}_{21}\dot{x}_1 + \widetilde{a}_{22}\dot{x}_2 + \cdots + \widetilde{a}_{2n}\dot{x}_n + \widetilde{b}_{21}x_1 + \widetilde{b}_{22}x_2 + \cdots + \widetilde{b}_{2n}x_n = \widetilde{f}_2 \\ \qquad\qquad\qquad\qquad \cdots\cdots \\ \widetilde{a}_{m1}\dot{x}_1 + \widetilde{a}_{m2}\dot{x}_2 + \cdots + \widetilde{a}_{mn}\dot{x}_n + \widetilde{b}_{m1}x_1 + \widetilde{b}_{m2}x_2 + \cdots + \widetilde{b}_{mn}x_n = \widetilde{f}_m \end{cases} \quad (1.169)$$

with respect to n functions $x_i(t)$ ($i = 1, \ldots, n$) of a single variable t. The numbers n and m may be different in general. In the matrix form these equations can be written as

$$\widetilde{\mathbf{A}}\dot{\mathbf{X}} + \widetilde{\mathbf{B}}\mathbf{X} = \widetilde{\mathbf{F}}(t), \qquad (1.170)$$

where $\widetilde{\mathbf{A}} = (\widetilde{a}_{ij})$ and $\widetilde{\mathbf{B}} = (\widetilde{b}_{ij}(t))$ are rectangular matrices of coefficients. We shall assume that the matrix $\widetilde{\mathbf{A}}$ is a constant matrix, whereas the elements of the matrix $\widetilde{\mathbf{B}}$ may depend on t. $\mathbf{X} = (x_i(t))$ is the vector-column of unknown functions, $\dot{\mathbf{X}} = (\dot{x}_i(t)) = (dx_i/dt)$ is the vector-column of their derivatives, and, finally, $\widetilde{\mathbf{F}}(t) = (\widetilde{f}_i(t))$ is the vector-column of given functions of t that appear in the right-hand side. For convenience, we shall call the variable t 'time'.

We shall start from a general discussion. The first observation that one can make here is that the system (1.170) can be somewhat simplified. Indeed, rewriting our equations as $\widetilde{\mathbf{A}}\dot{\mathbf{X}} = \widetilde{\mathbf{F}}(t) - \widetilde{\mathbf{B}}\mathbf{X}$ and applying the Gauss elimination procedure, one can bring the left-hand side matrix $\widetilde{\mathbf{A}}$ into a triangular (row echelon) form, which would enable one to write $r = \operatorname{rank}(\widetilde{\mathbf{A}}) \leq m$ derivatives $\dot{x}_1, \ldots, \dot{x}_r$ via a linear combination of the functions $x_1, \ldots, x_r$ and $\widetilde{f}_1, \ldots, \widetilde{f}_m$, as well as the rest of the unknown functions $x_{r+1}, \ldots, x_n$ and their derivatives $\dot{x}_{r+1}, \ldots, \dot{x}_n$. Clearly, these latter functions x_i ($i = r+1, \ldots, n$) cannot be determined and remain arbitrary. Hence, we arrive at a simpler problem

$$\dot{\mathbf{X}} = \mathbf{A}(t)\mathbf{X} + \mathbf{F}(t), \qquad (1.171)$$

in which $\mathbf{A} = (a_{ij})$ is a new quadratic $r \times r$ matrix of coefficients and $\mathbf{F}(t) = (f_i)$ is a new vector of r functions that are to be considered as known. If the latter functions are present in the system of Eq. (1.171), the latter is called non-homogeneous; if $\mathbf{F} = \mathbf{0}$, the system is homogeneous.

Higher order linear systems of DEs can be introduced in a similar way. They can be resolved with respect to their highest derivatives; however, no more simplification is possible in the general case.

If $\mathbf{X}_1(t), \mathbf{X}_2(t), \ldots, \mathbf{X}_r(t)$ are linearly independent solutions of the homogeneous system $\dot{\mathbf{X}}_k = \mathbf{A}\mathbf{X}_k$ ($k = 1, \ldots, r$), then their arbitrary linear combination

$$\mathbf{X}(t) = C_1 \mathbf{X}_1(t) + \cdots + C_r \mathbf{X}_r(t)$$

is also its solution, where $C_1, \ldots, C_r$ are arbitrary (in general, complex) coefficients. This is because the homogeneous system is *linear*. Indeed,

$$\dot{\mathbf{X}} = \sum_{k=1}^{r} C_k \dot{\mathbf{X}}_k = \sum_{k=1}^{r} C_k \mathbf{A}\mathbf{X}_k = \mathbf{A} \sum_{k=1}^{r} C_k \mathbf{X}_k = \mathbf{A}\mathbf{X}.$$

The elementary solutions $\mathbf{X}_1(t), \mathbf{X}_2(t), \ldots, \mathbf{X}_r(t)$ are said to form the *fundamental set of solutions* of the homogeneous system.

Let us establish a condition for these elementary solutions to be linearly independent. Vector-columns $\mathbf{X}_k = (x_{ik}(t))$ each consist of r functions $x_{ik}(t)$ of t, where the left subscript (i) indicates the component of the vector $\mathbf{X}_k$ and the right one—the vector number (k). Then the zero solution of the vector equation

$$\alpha_1 \mathbf{X}_1(t) + \cdots + \alpha_r \mathbf{X}_r(t) = \begin{pmatrix} x_{11} & x_{12} & \ldots & x_{1r} \\ x_{21} & x_{22} & \ldots & x_{2r} \\ \ldots & \ldots & \ddots & \ldots \\ x_{r1} & x_{r2} & \ldots & x_{rr} \end{pmatrix} \begin{pmatrix} \alpha_1 \\ \alpha_2 \\ \ldots \\ \alpha_r \end{pmatrix} = \mathbf{0}$$

with respect to the coefficients $\alpha_1, \ldots, \alpha_r$ is only possible if the determinant, composed of the vectors $\mathbf{X}_1(t), \mathbf{X}_2(t), \ldots, \mathbf{X}_r(t)$ as its columns, is not equal to zero,

$$\left| \mathbf{X}_1 \, \mathbf{X}_2 \, \ldots \, \mathbf{X}_r \right| = \begin{vmatrix} x_{11} & x_{12} & \ldots & x_{1r} \\ x_{21} & x_{22} & \ldots & x_{2r} \\ \ldots & \ldots & \ddots & \ldots \\ x_{r1} & x_{r2} & \ldots & x_{rr} \end{vmatrix} \neq 0$$

for all values of t.

If one can find a particular integral $\mathbf{X}_p(t)$ of the non-homogeneous system (1.171), then the full its solution is provided by the vector-function

$$\mathbf{X}(t) = \underbrace{C_1\mathbf{X}_1(t) + \cdots + C_r\mathbf{X}_r(t)}_{\mathbf{X}_c(t)} + \mathbf{X}_p(t).$$

The linear combination of the elementary solutions of the homogeneous system serves as the complementary solution $\mathbf{X}_c(t)$ of the problem, and the full solution is given by the sum of the complementary and particular solutions, exactly along the same lines as in the case of the ODEs (see Sect. I.8.2.3). The r constants $C_1, \ldots, C_r$ are obtained from the initial conditions, $\mathbf{X}(t = 0) = \mathbf{X}_0 = \left(x_i^{(0)}\right)$, that provide exactly r scalar equations $x_1(0) = x_1^{(0)}$, $x_2(0) = x_2^{(0)}$, etc., $x_r(0) = x_r^{(0)}$ for the initial components $x_i(0)$ of the vector $\mathbf{X}$.

1.19.2 Homogeneous Systems with Constant Coefficients

Consider first the system of homogeneous DEs of the first order:

$$\dot{\mathbf{X}} = \mathbf{A}\mathbf{X}, \tag{1.172}$$

where $\mathbf{X}(t) = (x_i(t))$ is an n-fold vector column of n scalar functions $x_i(t)$ to be determined, and $\mathbf{A} = \left(a_{ij}\right)$ is a constant square $n \times n$ matrix. We first assume that the matrix $\mathbf{A}$ has exactly n linearly independent eigenvectors (i.e. its rank is n). This is guaranteed, for instance, if $\mathbf{A}$ is normal, symmetric or Hermitian. Hence, we shall limit ourselves here with any of these cases. Guided by a one-dimensional ($n = 1$) case of the differential equation $\dot{x}_1 = a_{11}x_1$ that has an exponential solution $x_1(t) = Ce^{a_{11}t}$ (here C is an arbitrary constant), let us seek the solution of Eq. (1.172) in the following form: $\mathbf{X}(t) = \mathbf{Y}e^{at}$, where a is a constant and $\mathbf{Y}$ is a constant vector-column, both to be determined. Here the whole time dependence is contained in the scalar exponential function. Substituting this trial solution into Eq. (1.172), we obtain

$$a\mathbf{Y}e^{at} = \mathbf{A}\mathbf{Y}e^{at}.$$

One can see that the only time-dependent term (the exponential function) can be cancelled out, and we arrive at the eigenproblem for the vector $\mathbf{Y}$ and the number a:

$$\mathbf{A}\mathbf{Y} = a\mathbf{Y}. \tag{1.173}$$

These steps demonstrate that the exponential solution $\mathbf{X}(t) = \mathbf{Y}e^{at}$ just attempted actually works as we have been able to get rid of the time dependence completely when substituting this solution into our equation (1.172); in other words, our exponential guess was intelligent enough.

Leaving for the later to justify this choice, we shall now proceed to establish a general solution of our matrix equation. To this end, we note that the eigenproblem (1.173) has n eigenvalues a_λ (these may not be all distinct, we accept repeated ones as well) and correspondingly n linearly independent eigenvectors $\mathbf{Y}_\lambda = (y_{i\lambda})$, where $\lambda = 1, \ldots, n$ labels different eigenvalue–eigenvector pairs. Hence, any of the n vectors $\mathbf{X}_1(t) = \mathbf{Y}_1 e^{a_1 t}$, $\mathbf{X}_2(t) = \mathbf{Y}_2 e^{a_2 t}$, etc., $\mathbf{X}_n(t) = \mathbf{Y}_n e^{a_n t}$ satisfies the original differential equation (1.172) and forms its elementary solution. The arbitrary linear combination of such elementary solutions forms a general solution of our matrix differential equation,

$$
\begin{aligned}
\mathbf{X}(t) &= C_1 \mathbf{X}_1(t) + C_2 \mathbf{X}_2(t) + \cdots + C_n \mathbf{X}_n(t) \\
&= \sum_{\lambda=1}^{n} C_\lambda \mathbf{X}_\lambda(t) = \sum_{\lambda=1}^{n} C_\lambda \mathbf{Y}_\lambda e^{a_\lambda t}.
\end{aligned}
\tag{1.174}
$$

One can see that the general solution contains n arbitrary constants.

It is easy to see that the elementary solutions comprising the fundamental set are indeed linearly independent, since the determinant composed of them,

$$
\left| \mathbf{X}_1(t) \ldots \mathbf{X}_n(t) \right| =
\begin{vmatrix}
x_{11}(t) & \ldots & x_{1n}(t) \\
\ldots & \ddots & \ldots \\
x_{n1}(t) & \ldots & x_{nn}(t)
\end{vmatrix}
=
\begin{vmatrix}
y_{11} e^{a_1 t} & \ldots & y_{1n} e^{a_n t} \\
\ldots & \ddots & \ldots \\
y_{n1} e^{a_1 t} & \ldots & y_{nn} e^{a_n t}
\end{vmatrix}
$$

$$
= e^{a_1 t} \ldots e^{a_n t}
\begin{vmatrix}
y_{11} & \ldots & y_{1n} \\
\ldots & \ddots & \ldots \\
y_{n1} & \ldots & y_{nn}
\end{vmatrix}
$$

$$
= e^{a_1 t} \ldots e^{a_n t} \left| \mathbf{Y}_1 \ldots \mathbf{Y}_n \right| \neq 0
$$

for all values of t. This is because the determinant formed by the vectors $\mathbf{Y}_1, \ldots, \mathbf{Y}_n$ cannot be equal to zero as the vectors are linearly independent (Theorem 1.14).

Note that the solution can also be written in a very simple matrix exponential form. To show this (and also prove the exponential trial solution we introduced above), we benefit from the *spectral theorem* (Sect. 1.10) that enables one to write a square matrix $\mathbf{A}$ via its all n eigenvectors:

$$
\mathbf{A} = \sum_{\lambda=1}^{n} a_\lambda \mathbf{Y}_\lambda \mathbf{Y}_\lambda^\dagger.
\tag{1.175}
$$

The eigenvectors form a unitary square matrix (the *modal matrix* of $\mathbf{A}$, Sect. 1.9.1) with rows forming an orthonormal set of vectors[14]; its columns form another orthonormal set. Hence, the eigenvectors satisfy the orthogonality,

$$
\mathbf{Y}_\lambda^\dagger \mathbf{Y}_{\lambda'} = \delta_{\lambda \lambda'}
$$

(for columns of the modal matrix), and completeness,

---

[14] The eigenvectors staying as columns in the modal matrix do not need to be normalised to one; however, here for convenience we shall assume that they are.

$$\sum_{\lambda} \mathbf{Y}_{\lambda} \mathbf{Y}_{\lambda}^{\dagger} = \mathbf{E}$$

(for rows of the modal matrix), conditions. Substitute Eq. (1.175) into our Eq. (1.172):

$$\dot{\mathbf{X}} = \sum_{\lambda=1}^{n} a_{\lambda} \mathbf{Y}_{\lambda} \left(\mathbf{Y}_{\lambda}^{\dagger} \mathbf{X} \right),$$

then multiply it from the left by $\mathbf{Y}_{\lambda'}^{\dagger}$, and define scalar functions (dot products) $z_{\lambda}(t) = \mathbf{Y}_{\lambda}^{\dagger} \mathbf{X}(t)$. We find, using the mentioned orthogonality of the eigenvectors, that all equations decouple:

$$\frac{d}{dt} \underbrace{\left(\mathbf{Y}_{\lambda'}^{\dagger} \mathbf{X} \right)}_{z_{\lambda'}} = \sum_{\lambda=1}^{n} a_{\lambda} \underbrace{\left(\mathbf{Y}_{\lambda'}^{\dagger} \mathbf{Y}_{\lambda} \right)}_{\delta_{\lambda\lambda'}} \underbrace{\left(\mathbf{Y}_{\lambda}^{\dagger} \mathbf{X} \right)}_{z_{\lambda}} \implies \dot{z}_{\lambda'} = \sum_{\lambda=1}^{n} a_{\lambda} \delta_{\lambda\lambda'} z_{\lambda} = a_{\lambda'} z_{\lambda'}.$$

Each of these scalar equations is solved trivially yielding $z_{\lambda}(t) = C_{\lambda} e^{a_{\lambda} t}$, where C_{λ} is a constant. Hence, we obtain that $z_{\lambda}(t) = \mathbf{Y}_{\lambda}^{\dagger} \mathbf{X} = C_{\lambda} e^{a_{\lambda} t}$.

We have exactly n such equations, one for each value of λ. To find the vector $\mathbf{X}$, we need to solve all these equations and then combine the solutions in some way to get $\mathbf{X}$. This plan can be accomplished by noticing that the sum $\sum_{\lambda} \mathbf{Y}_{\lambda} z_{\lambda}$ is exactly equal to $\mathbf{X}$:

$$\sum_{\lambda} \mathbf{Y}_{\lambda} z_{\lambda} = \sum_{\lambda} \mathbf{Y}_{\lambda} \left(\mathbf{Y}_{\lambda}^{\dagger} \mathbf{X} \right) = \underbrace{\left(\sum_{\lambda} \mathbf{Y}_{\lambda} \mathbf{Y}_{\lambda}^{\dagger} \right)}_{\mathbf{E}} \mathbf{X} = \mathbf{X},$$

where this time we have used the completeness of the eigenvectors $\mathbf{Y}_{\lambda}$. Hence,

$$\mathbf{X} = \sum_{\lambda} \mathbf{Y}_{\lambda} z_{\lambda} = \sum_{\lambda} \mathbf{Y}_{\lambda} C_{\lambda} e^{a_{\lambda} t},$$

which coincides with our previous result (1.174).

So far we have only proved that the trial solution must be sought in the exponential form. The final step is related to noticing that the sum $\sum_{\lambda} \mathbf{Y}_{\lambda} \mathbf{Y}_{\lambda}^{\dagger} e^{a_{\lambda} t}$ is actually a spectral representation of the exponential function of the matrix $\mathbf{A}$ (Sect. 1.10):

$$e^{\mathbf{A}t} = \sum_{\lambda} e^{a_{\lambda} t} \mathbf{Y}_{\lambda} \mathbf{Y}_{\lambda}^{\dagger},$$

since any function of a matrix $\mathbf{B} = f(\mathbf{A})$ shares its eigenvectors, and the eigenvalues of the matrix $\mathbf{B}$ are given by $f(a_{\lambda})$. Next, multiply this exponential matrix function with an arbitrary vector-column $\mathbf{C}$:

$$e^{\mathbf{A}t}\mathbf{C} = \sum_{\lambda} e^{a_\lambda t} \mathbf{Y}_\lambda \left(\mathbf{Y}_\lambda^\dagger \mathbf{C} \right).$$

Here $\mathbf{Y}_\lambda^\dagger \mathbf{C}$ is the dot product of two vectors. As the eigenvector $\mathbf{Y}_\lambda$ of $\mathbf{A}$ is a constant vector and the vector-column $\mathbf{C}$ is arbitrary, the dot product $\mathbf{Y}_\lambda^\dagger \mathbf{C}$ is a scalar which we can denote as C_λ as it is an arbitrary number. Hence,

$$e^{\mathbf{A}t}\mathbf{C} = \sum_{\lambda} e^{a_\lambda t} \mathbf{Y}_\lambda C_\lambda,$$

which is exactly the formula for $\mathbf{X}$ we obtained previously. Hence, quite formally, the solution of Eq. (1.172) can be elegantly (and quite formally) written as

$$\mathbf{X} = e^{\mathbf{A}t}\mathbf{C}. \tag{1.176}$$

The constant vector $\mathbf{C}$ trivially follows from the initial ($t = 0$) conditions as $\mathbf{X}(0) = \mathbf{C}$.

Example 1.26 ► Solve the following system of two differential equations

$$\begin{cases} \dot{x}_1 = x_1 + 2x_2 \\ \dot{x}_2 = 2x_1 + x_2 \end{cases}$$

subject to the initial conditions $x_1(0) = 1$ and $x_2(0) = 2$.

Solution: Following the prescription given above, we first recast the equations in the matrix form:

$$\frac{d}{dt}\begin{pmatrix} x_1 \\ x_2 \end{pmatrix} = \begin{pmatrix} 1 & 2 \\ 2 & 1 \end{pmatrix}\begin{pmatrix} x_1 \\ x_2 \end{pmatrix}, \quad \text{or simply } \dot{\mathbf{X}} = \mathbf{A}\mathbf{X}.$$

Next, we attempt a trial solution in the form suggested by the appropriate scalar problem $\dot{x} = ax$, i.e. $\mathbf{X}(t) = \mathbf{Y}e^{\lambda t}$. Substituting this into the matrix equation above, we obtain, upon cancelling on the $e^{\lambda t}$, the appropriate eigenproblem $\mathbf{A}\mathbf{Y} = \lambda \mathbf{Y}$. A simple calculation yields the following two pairs of eigenvectors and eigenvalues:

$$\mathbf{Y}_1 = \begin{pmatrix} 1 \\ 1 \end{pmatrix}, \ \lambda_1 = 3 \quad \text{and} \quad \mathbf{Y}_2 = \begin{pmatrix} 1 \\ -1 \end{pmatrix}, \ \lambda_2 = -1,$$

so that we have two elementary solutions:

$$\mathbf{X}_1(t) = \mathbf{Y}_1 e^{\lambda_1 t} = \begin{pmatrix} 1 \\ 1 \end{pmatrix} e^{3t} \quad \text{and} \quad \mathbf{X}_2(t) = \mathbf{Y}_2 e^{\lambda_2 t} = \begin{pmatrix} 1 \\ -1 \end{pmatrix} e^{-t}.$$

They form the fundamental set. Hence, the general solution of our system of two DEs is then the vector function

$$\mathbf{X}(t) = C_1\mathbf{X}_1(t) + C_2\mathbf{X}_2(t) = C_1 \begin{pmatrix} 1 \\ 1 \end{pmatrix} e^{3t} + C_2 \begin{pmatrix} 1 \\ -1 \end{pmatrix} e^{-t}$$

$$= \begin{pmatrix} C_1 e^{3t} + C_2 e^{-t} \\ C_1 e^{3t} - C_2 e^{-t} \end{pmatrix} \equiv \begin{pmatrix} x_1(t) \\ x_2(t) \end{pmatrix}.$$

Hence, the two unknown functions that we seek are

$$x_1(t) = C_1 e^{3t} + C_2 e^{-t} \quad \text{and} \quad x_2(t) = C_1 e^{3t} - C_2 e^{-t},$$

as required. Finally, applying the initial conditions, we get

$$\begin{pmatrix} 1 \\ 2 \end{pmatrix} = \begin{pmatrix} C_1 e^{3t} + C_2 e^{-t} \\ C_1 e^{3t} - C_2 e^{-t} \end{pmatrix}_{t=0} = \begin{pmatrix} C_1 + C_2 \\ C_1 - C_2 \end{pmatrix},$$

leading to $C_1 = \frac{3}{2}$ and $C_2 = -\frac{1}{2}$. Hence, we obtain the required solution satisfying the initial conditions as

$$x_1(t) = \frac{3}{2} e^{3t} - \frac{1}{2} e^{-t} \quad \text{and} \quad x_2(t) = \frac{3}{2} e^{3t} + \frac{1}{2} e^{-t} \quad \blacktriangleleft$$

It is easy to see that the system of equations we considered in the above example is actually equivalent to a single second-order DE. Indeed, solve the second DE with respect to x_1 and substitute into the first:

$$x_1 = \frac{1}{2}(\dot{x}_2 - x_2) \implies \frac{1}{2}(\ddot{x}_2 - \dot{x}_2) = \frac{1}{2}(\dot{x}_2 - x_2) + 2x_2$$

$$\implies \ddot{x}_2 - 2\dot{x}_2 - 3x_2 = 0.$$

Solving this second-order DE gives exactly the same solution for $x_2(t)$ as the above method based on the eigenproblem for the matrix $\mathbf{A}$. Indeed, using an exponential trial solution $x_2 \sim e^{pt}$, we get $p^2 - 2p - 3 = 0$ with the roots $p_1 = -1$ and $p_2 = 3$ and the general solution $x_2 = C_1 e^{3t} + C_2 e^{-t}$. It is identical to the one obtained earlier (apart from the sign to an arbitrary constant which is irrelevant). Substituting this into the first equation $\dot{x}_1 = x_1 + 2x_2$ and solving for $x_1(t)$ we again obtain the same result as before.

It can be concluded without difficulty that a system of n coupled DEs is equivalent to a single nth order DE for any of the functions.

Problem 1.151 Show that the general solution of the system of equations

$$\begin{cases} \dot{x} = 2x + 5y \\ \dot{y} = 5x + 2y \end{cases}$$

is given by $x(t) = C_1 e^{-3t} + C_2 e^{7t}$ and $y(t) = -C_1 e^{-3t} + C_2 e^{7t}$.

Problem 1.152 Show that $x(t) = C_1 e^{-4t} + C_2 e^{6t}$ and $y(t) = -C_1 e^{-4t} + C_2 e^{6t}$ give the general solution of the system of equations

$$\begin{cases} \dot{x} = x + 5y \\ \dot{y} = 5x + y \end{cases}.$$

Problem 1.153 Show that $x(t) = C_1 e^{\lambda t} - 3C_2$ and $y(t) = 4C_1 e^{\lambda t} + C_2$, where $\lambda = 13/2$, give the general solution of the system

$$\begin{cases} 2\dot{x} = x + 3y \\ \dot{y} = 2x + 6y \end{cases}.$$

Problem 1.154 Solve the following system of linear differential equations:

$$\begin{cases} \dot{x}_1 = x_1 + 2x_2 \\ \dot{x}_2 = 2x_1 + x_2 \end{cases}$$

subject to the following initial conditions: $x_1(0) = 0$ and $x_2(0) = 1$. [Answer: $x_1(t) = \left(-e^{-t} + e^{3t} \right)/2$, $x_2(t) = \left(e^{-t} + e^{3t} \right)/2$.]

Problem 1.155 Similarly, solve the system of equations

$$\begin{cases} \dot{x}_1 = -3x_1 - 2x_2 \\ \dot{x}_2 = 2x_1 - x_2 \end{cases}$$

subject to the initial conditions $x_1(0) = 1$, $x_2(0) = 0$. [Answer: $x_1(t) = e^{-2t} \left[-\sin\left(\sqrt{3}t\right) + \sqrt{3}\cos\left(\sqrt{3}t\right) \right]/\sqrt{3}$ and $x_2(t) = \left(2/\sqrt{3} \right) e^{-2t} \sin\left(\sqrt{3}t\right)$.]

Note that in the last problem the eigenvalues of the matrix $\mathbf{A}$ are complex, $\lambda_{1,2} = -2 \pm i\sqrt{3}$.

Similarly one can consider a system of linear equations

$$\ddot{\mathbf{X}} = -\mathbf{A}\mathbf{X} \tag{1.177}$$

that appear in such physical applications as atomic vibrations, Sect.1.22.3 (the minus sign is introduced for convenience as the square matrix $\mathbf{A}$ is positively defined, i.e. has positive eigenvalues ω_λ^2).

Problem 1.156 Using the spectral decomposition of this matrix,

$$\mathbf{A} = \sum_{\lambda=1}^{n} \omega_\lambda^2 \mathbf{Y}_\lambda \mathbf{Y}_\lambda^\dagger$$

with the eigenvectors $\mathbf{Y}_\lambda$, show that this time the equations for the scalars $z_\lambda(t) = \mathbf{Y}_\lambda^\dagger \mathbf{X}$ are the equations for a single harmonic oscillator, $\ddot{z}_\lambda + \omega_\lambda^2 z_\lambda = 0$. Hence, demonstrate that the general real solution of Eq. (1.177) can formally be written as

$$\mathbf{X} = \sum_\lambda \mathbf{Y}_\lambda \left(B_\lambda e^{i\omega_\lambda t} + B_\lambda^* e^{-i\omega_\lambda t} \right) = e^{i\sqrt{\mathbf{A}}t} \mathbf{B} + e^{-i\sqrt{\mathbf{A}}t} \mathbf{B}^*,$$

where B_λ are complex arbitrary constants and $\mathbf{B}$ is a vector composed out of them.

Note that, to ensure the real solution $\mathbf{X}$, the vectors of arbitrary constants are complex here. The important observation that we can make here is that solutions of Eq. (1.177) can be sought as $\mathbf{X} = \mathbf{F}e^{iat}$, where $\mathbf{F}$ is an unknown vector and a unknown real constant.

Problem 1.157 Let us assume that the initial conditions for $\mathbf{X}(t)$ in Eq. (1.177) are vectors $\mathbf{X}_0 = \mathbf{X}(0)$ and $\dot{\mathbf{X}}_0 = \dot{\mathbf{X}}(0)$. Writing the vector $\mathbf{B}$ of the constants as the sum of two real vectors $\mathbf{C}_1 + i\mathbf{C}_2$, solve for the constants and show that the solution satisfying the initial conditions is

$$\mathbf{X}(t) = \cos\left(\sqrt{\mathbf{A}}t\right)\mathbf{X}_0 + \left(\sqrt{\mathbf{A}}\right)^{-1}\sin\left(\sqrt{\mathbf{A}}t\right)\dot{\mathbf{X}}_0.$$

Here the sine and cosine functions of the matrix $\sqrt{\mathbf{A}}t$ are defined in the usual way via the corresponding Taylor expansions.

1.19.3 Non-homogeneous Systems

It is also easy to extend our method to solving the corresponding non-homogeneous matrix equation that contains a constant vector $\mathbf{F}$ in the right-hand side:

$$\dot{\mathbf{X}} = \mathbf{A}\mathbf{X} + \mathbf{F}, \tag{1.178}$$

where $\mathbf{A}$ is a $n \times n$ matrix of rank n. Similarly to the case of linear homogeneous DEs (see Sect. I.8.2.3), the solution of this matrix equation is composed of two components: the complementary solution $\mathbf{X}_c$ of the homogeneous equation $\dot{\mathbf{X}} = \mathbf{A}\mathbf{X}$, and a particular integral $\mathbf{X}_p$ that solves the whole Eq. (1.178). The particular integral can be found straightforwardly by noticing that the *stationary solution* of Eq. (1.178), i.e. the one for which the left-hand side is zero ($\dot{\mathbf{X}} = 0$) is easily obtained as (assuming that $\mathbf{A}$ is the full-rank matrix, it is always non-singular and hence its inverse $\mathbf{A}^{-1}$ exists, see Theorem 1.16):

$$\mathbf{A}\mathbf{X} + \mathbf{F} = 0 \implies \mathbf{X} = -\mathbf{A}^{-1}\mathbf{F}.$$

This constant solution also satisfies the whole Eq. (1.178) and hence can be chosen as its particular integral. Therefore, the solution of the whole inhomogeneous problem is

$$\mathbf{X} = e^{\mathbf{A}t}\mathbf{C} - \mathbf{A}^{-1}\mathbf{F}.$$

Applying initial conditions, one obtains

$$\mathbf{X}(t) = e^{\mathbf{A}t}\left(\mathbf{X}(0) + \mathbf{A}^{-1}\mathbf{F}\right) - \mathbf{A}^{-1}\mathbf{F}.$$

One can obtain the particular integral also in a more general case when the right-hand side vector $\mathbf{F}(t)$ is time dependent. We shall use here the same idea as in the method of variation of parameters that was discussed in Sect. I.8.2.3.2 for ODEs. The complementary solution of the homogeneous system $\dot{\mathbf{X}} = \mathbf{A}\mathbf{X}$ contains a linear combination of the elementary solutions $\mathbf{X}_1(t) = \mathbf{Y}_1 e^{\lambda_1 t}, \ldots, \mathbf{X}_n(t) = \mathbf{Y}_n e^{\lambda_n t}$ with arbitrary coefficients $C_1, \ldots, C_n$. Here $\mathbf{Y}_i$ and λ_i are eigenvectors and eigenvalues of $\mathbf{A}$. Replacing the coefficients with the unknown scalar functions $C_k \mapsto u_k(t)$, we seek the solution of the non-homogeneous equation as

$$\mathbf{X}_p(t) = \sum_{k=1}^{n} u_k(t)\mathbf{X}_k(t).$$

Substituting this solution into Eq. (1.178) , we obtain

$$\sum_k \left(u_k\dot{\mathbf{X}}_k + \dot{u}_k\mathbf{X}_k\right) = \mathbf{A}\sum_k u_k\mathbf{X}_k + \mathbf{F}(t).$$

The first terms on both sides cancel out since $\mathbf{X}_k$ satisfies the homogeneous equation. We obtain

$$\sum_k \dot{u}_k\mathbf{X}_k = \mathbf{F}(t) \implies \sum_k \dot{u}_k\mathbf{Y}_k e^{\lambda_k t} = \mathbf{F}(t).$$

Multiplying both sides of this equation by the vector-row $\mathbf{Y}_{k'}^{\dagger}$ and employing the orthogonality of the eigenvectors, we get

$$\sum_k \dot{u}_k \underbrace{\mathbf{Y}_{k'}^\dagger \mathbf{Y}_k}_{\delta_{kk'}} e^{\lambda_k t} = \mathbf{Y}_{k'}^\dagger \mathbf{F}(t) \implies \dot{u}_k = e^{-\lambda_k t} \mathbf{Y}_k^\dagger \mathbf{F}(t)$$

$$\implies u_k(t) = \int_{t_0}^t e^{-\lambda_k t'} \mathbf{Y}_k^\dagger \mathbf{F}(t') dt' ,$$

where t_0 is the time used for specifying initial conditions. Correspondingly, the particular solutions we seek is

$$\mathbf{X}_p(t) = \sum_k u_k(t) \mathbf{Y}_k e^{\lambda_k t} = \sum_k \int_{t_0}^t e^{\lambda_k (t-t')} \left(\mathbf{Y}_k^\dagger \mathbf{F}(t') \right) \mathbf{Y}_k \, dt' .$$

This formula formally solves the problem. Note, however, that it can also be written in a simpler, more transparent form. To this end, let us consider the ith component $x_i(t)$ of the vector $\mathbf{X}_p$:

$$x_i(t) = \sum_{k,j} \int_{t_0}^t e^{\lambda_k (t-t')} y_{jk}^* f_j(t') y_{ik} \, dt' = \sum_j \int_{t_0}^t \left[\sum_k e^{\lambda_k (t-t')} y_{ik} y_{jk}^* \right] f_j(t') \, dt' .$$

It is easy to recognise in the expression in the square brackets the spectral representation of the $i - j$ element of the matrix $e^{\mathbf{A}(t-t')}$, which enables one to finally write the particular integral as

$$\mathbf{X}_p(t) = \int_{t_0}^t dt' \, e^{\mathbf{A}(t-t')} \mathbf{F}(t') .$$

Problem 1.158   Consider the non-homogeneous problem

$$\ddot{\mathbf{X}} + \mathbf{A}\mathbf{X} = \mathbf{F}(t) .$$

In this problem we shall find the particular integral of this system using the method of variation of parameters.
  (i) Using the trial function

$$\mathbf{X}_p(t) = \sum_k \left[u_k(t) e^{i\omega_k t} + v_k(t) e^{-i\omega_k t} \right] \mathbf{Y}_k , \qquad (1.179)$$

where $\omega_k = \sqrt{\lambda_k}$ and λ_k and $\mathbf{Y}_k$ are eigenvalues and eigenvectors of $\mathbf{A}$, establish differential equations for the auxiliary functions $u_k(t)$ and $v_k(t)$:

$$\begin{cases} \dot{u}_k e^{i\omega_k t} + \dot{v}_k e^{-i\omega_k t} = 0 \\ \dot{u}_k e^{i\omega_k t} - \dot{v}_k e^{-i\omega_k t} = \frac{1}{i\omega_k} \mathbf{Y}_k^\dagger \mathbf{F}(t) \end{cases};$$

(ii) Solve these equations to find the auxiliary functions

$$u_k(t) = \frac{1}{2i\omega_k} \int_{t_0}^t dt' \, e^{-i\omega_k t'} \mathbf{Y}_k^\dagger \mathbf{F}(t'),$$

$$v_k(t) = -\frac{1}{2i\omega_k} \int_{t_0}^t dt' \, e^{i\omega_k t'} \mathbf{Y}_k^\dagger \mathbf{F}(t');$$

(iii) Finally, substitute these solutions into Eq. (1.179) to show that the particular integral can be written as

$$\mathbf{X}_p(t) = \left(\sqrt{\mathbf{A}}\right)^{-1} \int_{t_0}^t dt' \, \sin\left(\sqrt{\mathbf{A}} \left(t - t'\right)\right) \mathbf{F}(t'). \qquad (1.180)$$

1.19.4 The Case of the Matrix Having a Lower Rank

So far we have assumed that the rank of the matrix $\mathbf{A}$ in Eq. (1.172) is equal to its dimension n. What if this is *not* the case and the rank r of $\mathbf{A}$ is smaller ($r < n$)? In this case some of the unknown functions (components of the vector $\mathbf{X}$) will be linearly related to others, i.e. they will be linearly dependent. We shall first illustrate the main ideas here by means of an example.

Example 1.27 ▶ Solve the following set of two coupled differential equations:

$$\begin{cases} \dot{x}_1 = 2x_1 + 2x_2 \\ \dot{x}_2 = x_1 + x_2 \end{cases} \rightarrow \frac{d}{dt}\begin{pmatrix} x_1 \\ x_2 \end{pmatrix} = \begin{pmatrix} 2 & 2 \\ 1 & 1 \end{pmatrix}\begin{pmatrix} x_1 \\ x_2 \end{pmatrix}. \qquad (1.181)$$

Solution: The equations in the matrix form read $\dot{\mathbf{X}} = \mathbf{A}\mathbf{X}$ with the matrix $\mathbf{A} = \begin{pmatrix} 2 & 2 \\ 1 & 1 \end{pmatrix}$ being of rank 1 (the first row is linearly dependent of the second): indeed an elementary row operation $R_1 - 2R_2 \rightarrow R_2$ (Sect. 1.4.2) will bring the matrix into the form containing a zero row: $\begin{pmatrix} 2 & 2 \\ 1 & 1 \end{pmatrix} \rightarrow \begin{pmatrix} 2 & 2 \\ 0 & 0 \end{pmatrix}$. Therefore, multiplying the second equation in the coupled system (1.181) by 2 and subtracting from the first, we obtain an equivalent set of equations:

$$\begin{cases} \dot{x}_1 = 2x_1 + 2x_2 \\ \dot{x}_1 - 2\dot{x}_2 = 0 \end{cases} \implies \begin{pmatrix} 1 & 0 \\ 1 & -2 \end{pmatrix}\begin{pmatrix} \dot{x}_1 \\ \dot{x}_2 \end{pmatrix} = \begin{pmatrix} 2 & 2 \\ 0 & 0 \end{pmatrix}\begin{pmatrix} x_1 \\ x_2 \end{pmatrix}.$$

The first equation here (corresponding to the upper row) is our original equation; the second equation gives

$$\dot{x}_1 - 2\dot{x}_2 = 0 \implies \frac{d}{dt}(x_1 - 2x_2) = 0 \implies x_1 - 2x_2 = C_2, \qquad (1.182)$$

where C_2 is a constant. We see that the two solutions, $x_1(t)$ and $x_2(t)$ are linearly dependent. This is a consequence of the matrix $\mathbf{A}$ having linearly dependent rows (columns). Eliminating $x_2 = \frac{1}{2}(x_1 - C_2)$ from Eq. (1.182) and substituting into the first equation, we obtain

$$\begin{aligned} \dot{x}_1 &= 2x_1 + 2x_2 \\ &= 2x_1 + (x_1 - C_2) = 3x_1 - C_2, \end{aligned}$$

which is the first-order linear DE with constant coefficients and a constant right-hand side (non-homogeneous). Its solution can be found, e.g. as a sum of a constant partial integral $x_{1p} = \frac{1}{3}C_2$, and a complementary solution $x_{1c} = C_1 e^{3t}$ of the homogeneous equation $\dot{x}_1 = 3x_1$, and we have introduced the second arbitrary constant C_1. Hence,

$$x_1(t) = x_{1c} + x_{1p} = C_1 e^{3t} + \frac{1}{3}C_2 \text{ and } x_2(t) = \frac{1}{2}C_1 e^{3t} - \frac{1}{3}C_2.$$

It is easy to check that these two functions do indeed satisfy the original system of Eq. (1.181). ◄

It is possible to generalise this consideration for the case of an arbitrary square[15] matrix $\mathbf{A}$ in the equation $\dot{\mathbf{X}} = \mathbf{A}\mathbf{X}$. What we are going to do next is to perform the necessary number of elementary row operations described in Sect. 1.4.2 to reduce the matrix $\mathbf{A}$ to the row echelon form. As any of the row operations is equivalent to acting with a certain elementary matrix $\mathbf{U}_i$ from the left, eventually, we shall arrive at the equation

$$\mathbf{U}\dot{\mathbf{X}} = \mathbf{A}_{ech}\mathbf{X}, \qquad (1.183)$$

where $\mathbf{U} = \mathbf{U}_l \mathbf{U}_{l-1} \cdots \mathbf{U}_1 = (u_{ij})$ is the final matrix appearing in the left-hand side after l row operations, and $\mathbf{A}_{ech} = \mathbf{U}\mathbf{A}$ is the matrix $\mathbf{A}$ in the row echelon form. If that form is triangular without zero rows, then the matrix $\mathbf{A}$ is of rank n (equals to its dimension) and hence may possess exactly n linearly independent eigenvectors.[16] The method developed above (based on solving the eigenproblem for $\mathbf{A}$) can be applied and the solution has the form of Eqs. (1.174) or (1.176).

---

[15] The case of a rectangular matrix will not be considered for simplicity.

[16] The case of repeated eigenvalues and defective matrices will be considered separately in Sect. 1.19.6.

If, however, we find one or more zero rows after this procedure, then the rank r of $\mathbf{A}$ is less than n and hence there will only be r linearly independent solutions of the system of equations. Other $k = n - r$ solutions can be written as a linear combination of the linearly independent ones. Indeed, let the ith row of $\mathbf{A}_{ech}$ consists of zeros. Writing explicitly that ith row in the system (1.183), we obtain

$$\sum_{j=1}^{n} u_{ij}\dot{x}_j = 0 \Rightarrow \frac{d}{dt}\left(\sum_{j=1}^{n} u_{ij}x_j\right) = 0$$

$$\Rightarrow \sum_{j=1}^{n} u_{ij}x_j = C_i$$

$$\Rightarrow \sum_{j=r+1}^{n} u_{ij}x_j = C_i - \sum_{j=1}^{r} u_{ij}x_j,$$

where C_i is an arbitrary constant. If we have k zero rows in the matrix $\mathbf{A}_{ech}$, there will then be k such relationships linearly relating different functions to each other. Hence, k functions $x_j(t)$ for $j = r + 1, \ldots, n$ can be linearly expressed via the first r functions; these formulae would include k arbitrary constants C_i. In the example above $n = 2$ and $r = 1$, so that $k = 1$ solution is linearly dependent on the other one. Once the linear dependence has been established and the r linearly independent equations identified (these correspond to the first r rows of $\mathbf{A}_{ech}$ that contain non-zero pivots), the k functions selected as being linearly dependent ($x_j(t)$ for $j = r + 1, \ldots, n$) are substituted into the selected r linearly independent DEs. A new (smaller) set of r equations of the form $\dot{\widetilde{\mathbf{X}}} = \widetilde{\mathbf{A}}\widetilde{\mathbf{X}} + \widetilde{\mathbf{F}}$ is obtained with $\widetilde{\mathbf{A}}$ being a full-rank $r \times r$ matrix, $\widetilde{\mathbf{X}}$ an r-fold vector of unknown functions ($x_i(t)$) and $\widetilde{\mathbf{F}}$ an r-fold vector containing (in specific combinations) k arbitrary constants. The obtained equation is solved by determining first the complementary solution $\widetilde{\mathbf{X}}_c$ of the homogeneous equation $\dot{\widetilde{\mathbf{X}}} = \widetilde{\mathbf{A}}\widetilde{\mathbf{X}}$ (that we know how to solve using, e.g. the method of eigenvectors) and then adding to it the particular integral $\widetilde{\mathbf{X}}_p = -\widetilde{\mathbf{A}}^{-1}\widetilde{\mathbf{F}}$.

Example 1.28 ► Solve the following system of four DEs:

$$\begin{cases} \dot{x}_1 = x_1 - 3x_2 + 4x_3 \\ \dot{x}_2 = -2x_1 + x_2 + 2x_4 \\ \dot{x}_3 = -x_1 - 2x_2 + 4x_3 + 2x_4 \\ \dot{x}_4 = 4x_1 + 3x_2 - 8x_3 - 6x_4 \end{cases}. \tag{1.184}$$

Solution: To solve it, we first write it down in the matrix form:

$$\begin{pmatrix} 1\,0\,0\,0 \\ 0\,1\,0\,0 \\ 0\,0\,1\,0 \\ 0\,0\,0\,1 \end{pmatrix}\begin{pmatrix} \dot{x}_1 \\ \dot{x}_2 \\ \dot{x}_3 \\ \dot{x}_4 \end{pmatrix} = \begin{pmatrix} 1 & -3 & 4 & 0 \\ -2 & 1 & 0 & 2 \\ -1 & -2 & 4 & 2 \\ 4 & 3 & -8 & -6 \end{pmatrix}\begin{pmatrix} x_1 \\ x_2 \\ x_3 \\ x_4 \end{pmatrix}.$$

Next we perform several row operations in both sides of the equation; obviously, we only need to change the matrices themselves, in the same fashion and both at the same time, so it is convenient to consider the augmented matrix

$$\begin{pmatrix} 1\ 0\ 0\ 0 & 1 & -3 & 4 & 0 \\ 0\ 1\ 0\ 0 & -2 & 1 & 0 & 2 \\ 0\ 0\ 1\ 0 & -1 & -2 & 4 & 2 \\ 0\ 0\ 0\ 1 & 4 & 3 & -8 & -6 \end{pmatrix}.$$

Next we shall perform a number of row operations aiming at bringing the right half of the augmented matrix into the row echelon form:

$$\boxed{2R_1 + R_2 \to R_2} \longrightarrow \begin{pmatrix} 1\ 0\ 0\ 0 & 1 & -3 & 4 & 0 \\ 2\ 1\ 0\ 0 & 0 & -5 & 8 & 2 \\ 0\ 0\ 1\ 0 & -1 & -2 & 4 & 2 \\ 0\ 0\ 0\ 1 & 4 & 3 & -8 & -6 \end{pmatrix},$$

$$\boxed{R_1 + R_3 \to R_3} \longrightarrow \begin{pmatrix} 1\ 0\ 0\ 0 & 1 & -3 & 4 & 0 \\ 2\ 1\ 0\ 0 & 0 & -5 & 8 & 2 \\ 1\ 0\ 1\ 0 & 0 & -5 & 8 & 2 \\ 0\ 0\ 0\ 1 & 4 & 3 & -8 & -6 \end{pmatrix},$$

$$\boxed{R_2 - R_3 \to R_3} \longrightarrow \begin{pmatrix} 1\ 0\ 0\ 0 & 1 & -3 & 4 & 0 \\ 2\ 1\ 0\ 0 & 0 & -5 & 8 & 2 \\ 1\ 1\ -1\ 0 & 0 & 0 & 0 & 0 \\ 0\ 0\ 0\ 1 & 4 & 3 & -8 & -6 \end{pmatrix},$$

$$\boxed{4R_1 - R_4 \to R_4} \longrightarrow \begin{pmatrix} 1\ 0\ 0\ 0 & 1 & -3 & 4 & 0 \\ 2\ 1\ 0\ 0 & 0 & -5 & 8 & 2 \\ 1\ 1\ -1\ 0 & 0 & 0 & 0 & 0 \\ 4\ 0\ 0\ -1 & 0 & -15 & 24 & 6 \end{pmatrix},$$

$$\boxed{3R_2 - R_4 \to R_4} \longrightarrow \begin{pmatrix} 1\ 0\ 0\ 0 & 1 & -3 & 4 & 0 \\ 2\ 1\ 0\ 0 & 0 & -5 & 8 & 2 \\ 1\ 1\ -1\ 0 & 0 & 0 & 0 & 0 \\ 2\ 3\ 0\ 1 & 0 & 0 & 0 & 0 \end{pmatrix}. \qquad (1.185)$$

The matrix in the right-hand side has been finally converted into the row echelon form, from which it is seen that only the first two equations are linearly independent, i.e. $r = 2$ and $k = 2$. Writing explicitly the last two equations in the system (1.185), we obtain

$$x_1 + x_2 - x_3 = C_3 \quad \text{and} \quad 2x_1 + 3x_2 + x_4 = C_4.$$

We can choose x_3 and x_4 as linearly dependent functions, for which we obtain their relationship with x_1 and x_2 as follows:

$$x_3 = x_1 + x_2 - C_3 \quad \text{and} \quad x_4 = -2x_1 - 3x_2 + C_4. \qquad (1.186)$$

Substituting these into the first two equations of the original set (1.184) and simplifying,[17] we obtain

$$\begin{cases} \dot{x}_1 = 5x_1 + x_2 - 4C_3 \\ \dot{x}_2 = -6x_1 - 5x_2 + 2C_4 \end{cases} \implies \begin{pmatrix} \dot{x}_1 \\ \dot{x}_2 \end{pmatrix} = \begin{pmatrix} 5 & 1 \\ -6 & -5 \end{pmatrix} \begin{pmatrix} x_1 \\ x_2 \end{pmatrix} + \begin{pmatrix} -4C_3 \\ 2C_4 \end{pmatrix}.$$

First we have to solve the homogeneous problem

$$\begin{pmatrix} \dot{x}_1 \\ \dot{x}_2 \end{pmatrix} = \begin{pmatrix} 5 & 1 \\ -6 & -5 \end{pmatrix} \begin{pmatrix} x_1 \\ x_2 \end{pmatrix}.$$

The matrix $\tilde{\mathbf{A}} = \begin{pmatrix} 5 & 1 \\ -6 & -5 \end{pmatrix}$ has the following two pairs of eigenvalues–eigenvectors:

$a_1 = \sqrt{19}$, $a_2 = -\sqrt{19}$ and $\tilde{\mathbf{Y}}_{1,2} = \begin{pmatrix} 1 \\ a_{1,2} - 5 \end{pmatrix}$. Therefore, the complementary solution reads

$$\begin{pmatrix} x_{1c} \\ x_{2c} \end{pmatrix} = C_1 \begin{pmatrix} 1 \\ a_1 - 5 \end{pmatrix} e^{a_1 t} + C_2 \begin{pmatrix} 1 \\ a_2 - 5 \end{pmatrix} e^{a_2 t}.$$

The necessary particular integral

$$\begin{pmatrix} x_{1p} \\ x_{2p} \end{pmatrix} = -\begin{pmatrix} 5 & 1 \\ -6 & -5 \end{pmatrix}^{-1} \begin{pmatrix} -4C_3 \\ 2C_4 \end{pmatrix}$$

$$= -\frac{1}{19} \begin{pmatrix} 5 & 1 \\ -6 & -5 \end{pmatrix} \begin{pmatrix} -4C_3 \\ 2C_4 \end{pmatrix} = \frac{1}{19} \begin{pmatrix} 20C_3 - 2C_4 \\ -24C_3 + 10C_4 \end{pmatrix},$$

so that the sum of the two yields the general solution for the two linearly independent unknown functions $x_1(t)$ and $x_2(t)$:

$$\begin{pmatrix} x_1 \\ x_2 \end{pmatrix} = C_1 \begin{pmatrix} 1 \\ a_1 - 5 \end{pmatrix} e^{a_1 t} + C_2 \begin{pmatrix} 1 \\ a_2 - 5 \end{pmatrix} e^{a_2 t} + \frac{1}{19} \begin{pmatrix} 20C_3 - 2C_4 \\ -24C_3 + 10C_4 \end{pmatrix}$$

or simply

$$x_1(t) = C_1 e^{a_1 t} + C_2 e^{a_2 t} + \frac{1}{19} (20C_3 - 2C_4),$$

$$x_2(t) = C_1 (a_1 - 5) e^{a_1 t} + C_2 (a_2 - 5) e^{a_2 t} + \frac{1}{19} (-24C_3 + 10C_4).$$

---

[17] Alternatively, we can use the transformed equations corresponding to the augmented matrix in Eq. (1.185).

The linearly dependent functions are obtained from Eq. (1.186):

$$x_3(t) = C_1 (a_1 - 4) e^{a_1 t} + C_2 (a_2 - 4) e^{a_2 t} + \frac{1}{19} (-23C_3 + 8C_4) \,,$$

$$x_4(t) = C_1 (-3a_1 + 13) e^{a_1 t} + C_2 (-3a_2 + 13) e^{a_2 t} + \frac{1}{19} (32C_3 - 7C_4) \,.$$

The obtained equations fully solve the problem. It is important to note the following: although it may be tempting to replace the combinations of the constants C_3 and C_4 we obtained in the four functions by some other constants, this is not possible as the four combinations are linearly dependent and have to be used precisely as they appear above. Therefore, we have four arbitrary constants to be obtained from the initial conditions.◄

Problem 1.159 Show that the functions

$$x_1(t) = 3C_1 e^t - C_2 e^{-7t} - \frac{1}{7} C_3 \,,$$

$$x_2(t) = 2C_1 e^t + 2C_2 e^{-7t} + \frac{2}{7} C_3 \,,$$

$$x_3(t) = 4C_1 e^t - 4C_2 e^{-7t} + \frac{3}{7} C_3$$

are solutions of the system

$$\begin{cases} \dot{x}_1 = x_1 + 2x_2 - x_3 \\ \dot{x}_2 = -3x_2 + 2x_3 \\ \dot{x}_3 = 2x_1 + 7x_2 - 4x_3 \end{cases} \,.$$

1.19.5 Higher Order Linear Differential Equations

It can be easily shown that linear differential equations and systems of such equations with constant coefficients (but arbitrary right hand sides) can always be brought into the form (1.178) after an appropriate choice of the vector $\mathbf{X}(t)$, matrix $\mathbf{A}$ and vector-column $\mathbf{F}(t)$ of the right-hand side has been made.

It is instructive to start from a single DE of a high order, e.g.

$$y''' - 2y'' - y' + 2y = 0 \,.$$

This equation can be solved using the trial solution $y(t) \sim e^{pt}$, leading to the algebraic equation for the constant p,

$$p^3 - 2p^2 - p + 2 = 0$$

that has three distinct roots $p_{1,2} = \pm 1$ and $p_3 = 2$, and hence its general solution is

$$y(t) = C_1 e^t + C_2 e^{-t} + C_3 e^{2t}. \tag{1.187}$$

Alternatively, this single DE of the third order can also be rewritten as a first-order system of three DEs as follows. Define three functions:

$$x_1(t) = y(t), \quad x_2(t) = y'(t) = x_1'(t), \quad \text{and} \quad x_3(t) = y''(t) = x_2'(t).$$

This leads to

$$\begin{cases} x_1' = x_2 \\ x_2' = x_3 \\ x_3' = 2x_3 + x_2 - 2x_1 \end{cases} \implies \underbrace{\frac{d}{dt} \begin{pmatrix} x_1 \\ x_2 \\ x_3 \end{pmatrix}}_{\mathbf{X}} = \underbrace{\begin{pmatrix} 0 & 1 & 0 \\ 0 & 0 & 1 \\ -2 & 1 & 2 \end{pmatrix}}_{\mathbf{A}} \underbrace{\begin{pmatrix} x_1 \\ x_2 \\ x_3 \end{pmatrix}}_{\mathbf{X}}.$$

The eigenvalues of $\mathbf{A}$ are $\lambda_1 = 1$, $\lambda_2 = -1$ and $\lambda_3 = 2$, while the corresponding eigenvectors are

$$\mathbf{e}_1 = \begin{pmatrix} 1 \\ 1 \\ 1 \end{pmatrix}, \quad \mathbf{e}_2 = \begin{pmatrix} 1 \\ -1 \\ 1 \end{pmatrix}, \quad \mathbf{e}_3 = \begin{pmatrix} 1 \\ 2 \\ 4 \end{pmatrix},$$

so that the final solution is provided by

$$\mathbf{X}(t) = \begin{pmatrix} x_1 \\ x_2 \\ x_3 \end{pmatrix} = C_1 \begin{pmatrix} 1 \\ 1 \\ 1 \end{pmatrix} e^t + C_2 \begin{pmatrix} 1 \\ -1 \\ 1 \end{pmatrix} e^{-t} + C_3 \begin{pmatrix} 1 \\ 2 \\ 4 \end{pmatrix} e^{2t}$$

$$\equiv \begin{pmatrix} y \\ y' \\ y'' \end{pmatrix},$$

where the first row gives the required function $y(t)$ identical to the one we obtained using the exponential trial function, see Eq. (1.187), while the second and third rows give its first and second derivatives. It can be easily seen they are indeed correct.

Problem 1.160 Use both methods to show that the solution of the DE

$$y''' + 10y'' + 19y' - 30y = 0$$

is

$$y(t) = C_1 e^{-5t} + C_2 e^{-6t} + C_3 e^t .$$

Problem 1.161 Show that the solution of

$$4y'''' - 5y'' + y = 0$$

is

$$y(t) = C_1 e^{t/2} + C_2 e^{-t/2} + C_3 e^t + C_4 e^{-t} .$$

In a similar vein, linear systems of differential equations of any order can also be mapped onto the first-order differential equation of the type (1.178) of an appropriate dimension. As an example, let us consider a system of two DEs of the second order:

$$\begin{cases} 2y'' - z'' + 2y' - 2z' + 3z - y = 1 \\ -3z'' + y'' - z' - y' + z + y = -1 \end{cases} .$$

We define new functions $x_1(t) = y(t)$, $x_2(t) = y'(t)$, $x_3(t) = z(t)$ and $x_4(t) = z'(t)$, so that the two equations are equivalently rewritten as

$$\begin{cases} 2x_2' - x_4' + 2x_2 - 2x_4 + 3x_3 - x_1 = 1 \\ -3x_4' + x_2' - x_4 - x_2 + x_3 + x_1 = -1 \\ x_3' - x_4 = 0 \\ x_1' - x_2 = 0 \end{cases}$$

that in the matrix form is

$$\underbrace{\begin{pmatrix} 0 & 2 & 0 & -1 \\ 0 & 1 & 0 & -3 \\ 0 & 0 & 1 & 0 \\ 1 & 0 & 0 & 0 \end{pmatrix}}_{\mathbf{D}} \frac{d}{dt} \underbrace{\begin{pmatrix} x_1 \\ x_2 \\ x_3 \\ x_4 \end{pmatrix}}_{\mathbf{X}} = \underbrace{\begin{pmatrix} 1 & -2 & -3 & 2 \\ -1 & 1 & -1 & 1 \\ 0 & 0 & 0 & 1 \\ 0 & 1 & 0 & 0 \end{pmatrix}}_{\mathbf{G}} \underbrace{\begin{pmatrix} x_1 \\ x_2 \\ x_3 \\ x_4 \end{pmatrix}}_{\mathbf{X}} + \underbrace{\begin{pmatrix} 1 \\ -1 \\ 0 \\ 0 \end{pmatrix}}_{\mathbf{R}} ,$$

or

$$\mathbf{D} \frac{d}{dt} \mathbf{X} + \mathbf{G} \mathbf{X} = \mathbf{R} \implies \frac{d}{dt} \mathbf{X} + \mathbf{A} \mathbf{X} = \mathbf{F}$$

with

$$\mathbf{A} = \mathbf{D}^{-1}\mathbf{G} = \begin{pmatrix} 0 & 1 & 0 & 0 \\ 4/5 & -7/5 & -8/5 & 1 \\ 0 & 0 & 0 & 1 \\ 3/5 & -4/5 & -1/5 & 0 \end{pmatrix} \quad \text{and} \quad \mathbf{F} = \mathbf{D}^{-1}\mathbf{R} = \begin{pmatrix} 0 \\ 4/5 \\ 0 \\ 3/5 \end{pmatrix}.$$

We can see that a rather complicated system of two-second-order differential equations can be mapped equivalently into a system of the first-order four-dimensional matrix equation for which we have developed simple solution methods in the above sections. The eigenvalues and eigenvectors of $\mathbf{A}$ appear to be complex and can only be found numerically; they appear as two complex conjugate pairs[18]: $\lambda_1 \approx -1.21 + 0.948i$, $\lambda_2 = \lambda_1^*$, $\lambda_3 \approx 0.514 + 0.274i$, and $\lambda_4 = \lambda_3^*$, corresponding to the eigenvectors

$$\mathbf{X}_1 = \begin{pmatrix} a + ib \\ c + id \\ g + ip \\ 1 \end{pmatrix}, \quad \mathbf{X}_2 = \mathbf{X}_1^*, \quad \mathbf{X}_3 = \begin{pmatrix} a' + ib' \\ c' + id' \\ g' + ip' \\ 1 \end{pmatrix}, \quad \mathbf{X}_4 = \mathbf{X}_3^*$$

with $a \approx -0.895$, $b \approx 0.119$, $c \approx 0.975$, $d \approx -0.986$, $g \approx -0.515$, $p \approx -0.399$ and $a' \approx 1.55$, $b' \approx 2.39$, $c' \approx 0.140$, $d' \approx 1.66$, $g' \approx 1.52$, $p' \approx -0.807$. Correspondingly, the solution of the homogeneous equation is

$$\mathbf{X}_c(t) = C_1\mathbf{X}_1 e^{\lambda_1 t} + C_2\mathbf{X}_2 e^{\lambda_2 t} + C_3\mathbf{X}_3 e^{\lambda_3 t} + C_4\mathbf{X}_4 e^{\lambda_4 t}.$$

Generally, the constants C_1, C_2, etc., are complex. Since the first two and the last two elementary solutions form complex conjugate pairs, real constants can be introduced instead if we set $C_2 = C_1^*$ and $C_4 = C_3^*$, and combining the two terms in each pair:

$$\mathbf{X}_c(t) = 2\mathrm{Re}\left(C_1\mathbf{X}_1 e^{\lambda_1 t}\right) + 2\mathrm{Re}\left(C_3\mathbf{X}_3 e^{\lambda_3 t}\right),$$

leading to the real solution containing a sum of exponential functions multiplied by sines and cosines, and four real arbitrary coefficients B_i ($i = 1, \ldots, 4$) originating from $C_1 = B_1 + iB_2$ and $C_3 = B_3 + iB_4$. The final solution is obtained by adding to this complementary solution $\mathbf{X}_c(t)$ the particular solution

$$\mathbf{X}_p = \mathbf{A}^{-1}\mathbf{F} = \begin{pmatrix} 1 \\ 0 \\ 0 \\ 0 \end{pmatrix}.$$

This way all four functions $x_i(t)$ are obtained as components of $\mathbf{X}(t) = \mathbf{X}_c(t) + \mathbf{X}_p(t)$, and the required solutions $y(t)$ and $z(t)$ are provided by $x_1(t)$ and $x_3(t)$, respectively.

---

[18] Which could be found, e.g. using Mathematica software.

Problem 1.162 Show that the system of DEs

$$\begin{cases} y'' - 2x' - y = 2 \\ y' - x' + 2x = -1 \end{cases}$$

can equivalently be written in the matrix form as

$$\frac{d}{dt}\begin{pmatrix} x_1 \\ x_2 \\ x_3 \end{pmatrix} = \begin{pmatrix} 2 & 0 & 1 \\ 0 & 0 & 1 \\ 4 & 1 & 2 \end{pmatrix}\begin{pmatrix} x_1 \\ x_2 \\ x_3 \end{pmatrix} + \begin{pmatrix} 1 \\ 0 \\ 0 \end{pmatrix},$$

where $x_1(t) = x(t), x_2(t) = y(t)$ and $x_3(t) = y'(t)$. Hence, demonstrate, using a numerical calculation, e.g. Mathematica, that its solution is

$$\begin{pmatrix} x_1 \\ x_2 \\ x_3 \end{pmatrix} = C_1\begin{pmatrix} a_1 \\ b_1 \\ 1 \end{pmatrix}e^{\lambda_1 t} + C_2\begin{pmatrix} a_2 \\ b_2 \\ 1 \end{pmatrix}e^{\lambda_2 t} + C_3\begin{pmatrix} a_3 \\ b_3 \\ 1 \end{pmatrix}e^{\lambda_3 t} + \begin{pmatrix} -1/2 \\ 2 \\ 0 \end{pmatrix},$$

where $\lambda_1 \approx 4.12$, $\lambda_2 \approx -0.762$, $\lambda_3 \approx 0.637$ and $a_1 \approx 0.471$, $b_1 \approx 0.242$, $a_2 \approx -0.362$, $b_2 \approx -1.31$, $a_3 \approx -0.733$ and $b_3 \approx 1, 57$.

1.19.6 Homogeneous Systems with Defective Matrices

Special consideration is deserved for the case of the matrix $\mathbf{A}$ in the system $\dot{\mathbf{X}} = \mathbf{A}\mathbf{X}$ being defective. Indeed, if a certain eigenvalue λ of $\mathbf{A}$ is repeated n times, but the number of eigenvectors $\mathbf{Y}_{\lambda i}$ one can construct ($i = 1, \ldots, m$) is smaller than its algebraic multiplicity n (i.e. $m < n$), then the method discussed earlier would not work: the number m of linearly independent solutions based on elementary solutions $\mathbf{Y}_{\lambda i}e^{\lambda t}$ would simply be smaller than n. We must expect exactly n linearly independent solutions.

Our method can be however modified to construct all the necessary solutions. We know that each linearly independent eigenvector of $\mathbf{A}$ corresponding to the given eigenvalue λ forms its own Jordan chain (see Sect. 1.18). Consider one such eigenvector for which we shall use a simplifying notation $\mathbf{v}_1$ (as in Sect. 1.18). This vector can be used to form $p - 1$ more linearly independent vectors $\mathbf{v}_2, \ldots, \mathbf{v}_p$ that satisfy the following equations:

$$\mathbf{A}\mathbf{v}_i = \lambda\mathbf{v}_i + \mathbf{v}_{i-1}, \quad i = 2, \ldots, p. \tag{1.188}$$

Our job is to find, apart from the elementary solution $\mathbf{X}_1(t) = \mathbf{v}_1 e^{\lambda t}$, the other $p - 1$ linearly independent solutions of our system of DEs. It seems natural to build these extra solutions using a linear combination of the vectors $\mathbf{v}_i$ of the Jordan chain with coefficients $u_i(t)$ that may possibly be also functions of t. Hence, consider the following p trial solutions:

$$\widetilde{\mathbf{X}}_k(t) = [u_1(t)\mathbf{v}_1 + \cdots + u_k(t)\mathbf{v}_k] e^{\lambda t} = \sum_{i=1}^{k} u_i(t) e^{\lambda t} \mathbf{v}_i, \quad k = 1, \ldots, p. \quad (1.189)$$

Note that in the kth trial solution we sum only over first k terms. Also note that the coefficients $u_i(t)$ may be different for different trial solutions k. Substituting one such trial solution into the equation $\dot{\widetilde{\mathbf{X}}} = \mathbf{A}\widetilde{\mathbf{X}}$, we get for any $k > 1$ (in the case of $k = 1$ we simply take $u_1 = 1$ as we know that $\mathbf{v}_1 e^{\lambda t}$ is already an elementary solution; we only need to find additional solutions):

$$\lambda \sum_{i=1}^{k} u_i(t) e^{\lambda t} \mathbf{v}_i + \sum_{i=1}^{k} \dot{u}_i(t) e^{\lambda t} \mathbf{v}_i = \mathbf{A} \sum_{i=1}^{k} u_i(t) e^{\lambda t} \mathbf{v}_i$$

$$= u_1(t) e^{\lambda t} \underbrace{\mathbf{A}\mathbf{v}_1}_{\lambda \mathbf{v}_1} + \sum_{i=2}^{k} u_i(t) e^{\lambda t} \underbrace{\mathbf{A}\mathbf{v}_i}_{\lambda \mathbf{v}_i + \mathbf{v}_{i-1}}$$

$$= \lambda \sum_{i=1}^{k} u_i(t) e^{\lambda t} \mathbf{v}_i + \sum_{i=2}^{k} u_i(t) e^{\lambda t} \mathbf{v}_{i-1}.$$

The exponential function and the first terms on both sides can be cancelled out, and we obtain

$$\sum_{i=1}^{k} \dot{u}_i(t) \mathbf{v}_i = \sum_{i=2}^{k} u_i(t) \mathbf{v}_{i-1},$$

that can also be rewritten as follows:

$$(\dot{u}_1 - u_2)\,\mathbf{v}_1 + (\dot{u}_2 - u_3)\,\mathbf{v}_2 + \cdots + (\dot{u}_{k-1} - u_k)\,\mathbf{v}_{k-1} + \dot{u}_k \mathbf{v}_k = 0.$$

Now, the vectors $\mathbf{v}_1, \ldots, \mathbf{v}_k$ of the Jordan chain are linearly independent. Hence, for this equation to be satisfied for all values of t, the auxiliary functions $u_i(t)$ must satisfy the following set of coupled differential equations:

$$\dot{u}_k = 0 \implies u_k(t) = C_k,$$

$$\dot{u}_{k-1} = u_k \implies u_{k-1}(t) = C_k t + C_{k-1},$$

$$\dot{u}_{k-2} = u_{k-1} \implies u_{k-2}(t) = C_k \frac{t^2}{2} + C_{k-1}t + C_{k-2},$$

$$\cdots \qquad \cdots$$

$$\dot{u}_1 = u_2 \implies u_1(t) = C_k \frac{t^{k-1}}{(k-1)!} + C_{k-1}\frac{t^{k-2}}{(k-2)!} + \cdots + C_2 t + C_1.$$

Here C_i are arbitrary constants. Substituting these expressions into Eq. (1.189) and grouping the terms by the arbitrary coefficients, we obtain

$$\widetilde{\mathbf{X}}_k(t) = C_1 \mathbf{v}_1 e^{\lambda t} + C_2 \left(\mathbf{v}_1 t + \mathbf{v}_2\right) e^{\lambda t} + C_3 \left(\mathbf{v}_1 \frac{t^2}{2} + \mathbf{v}_2 t + \mathbf{v}_3\right) e^{\lambda t}$$

$$+ \cdots + C_i \sum_{j=1}^{i} \frac{t^{i-j}}{(i-j)!} \mathbf{v}_j e^{\lambda t} + \cdots + C_k \sum_{j=1}^{k} \frac{t^{k-j}}{(k-j)!} \mathbf{v}_j e^{\lambda t}.$$

This kind of expressions is obtained for any $k = 2, \ldots, p$. It follows from this consideration that one may choose, as the required linearly independent solutions corresponding to the repeating eigenvalue λ, the following p functions:

$$\mathbf{X}_1(t) = \mathbf{v}_1 e^{\lambda t}, \quad \mathbf{X}_2(t) = (\mathbf{v}_1 t + \mathbf{v}_2) e^{\lambda t}, \quad \mathbf{X}_3(t) = \left(\mathbf{v}_1 \frac{t^2}{2} + \mathbf{v}_2 t + \mathbf{v}_3\right) e^{\lambda t},$$

$$(1.190)$$

and so on, or, more formally,

$$\mathbf{X}_k(t) = \sum_{j=1}^{k} \frac{t^{k-j}}{(k-j)!} \mathbf{v}_j e^{\lambda t}, \quad k = 1, \ldots, p. \tag{1.191}$$

These results fully solve the problem, contribution to the general solution form that eigenvalue λ will be

$$\mathbf{X}(t) = \sum_{k=1}^{p} C_k \mathbf{X}_k(t). \tag{1.192}$$

As an example, consider the system

$$\begin{cases} \dot{x}_1 = 2x_1 + 2x_2 \\ \dot{x}_2 = -x_1 + 5x_2 + x_3 \\ \dot{x}_3 = -x_1 + x_2 + 3x_3 \end{cases}$$

with the matrix

$$\mathbf{A} = \begin{pmatrix} 2 & 2 & 0 \\ -1 & 5 & 1 \\ -1 & 1 & 3 \end{pmatrix}$$

from Sect. 1.18.1. We found there that there are two eigenvalues: $\lambda_1 = 4$, which has algebraic multiplicity 2 and the geometric multiplicity 1, and $\lambda_2 = 2$ of both the algebraic and geometric multiplicities being 1. For the second eigenvalue the eigenvector is $\mathbf{w}_1 = \begin{pmatrix} 1 & 0 & 1 \end{pmatrix}^T$ bringing an obvious solution

$$\mathbf{X}_3(t) = \mathbf{w}_1 e^{2t} = \begin{pmatrix} 1 \\ 0 \\ 1 \end{pmatrix} e^{2t},$$

while in the case of the first eigenvalue we have two vectors of the Jordan chain, $\mathbf{v}_1 = \begin{pmatrix} 1 & 1 & 0 \end{pmatrix}^T$ and $\mathbf{v}_2 = \begin{pmatrix} 0 & 1/2 & 1/2 \end{pmatrix}^T$. Hence, we can build two linearly independent solutions

$$\mathbf{X}_1(t) = \mathbf{v}_1 e^{4t} = \begin{pmatrix} 1 \\ 1 \\ 0 \end{pmatrix} e^{4t} \text{ and } \mathbf{X}_2(t) = (\mathbf{v}_1 t + \mathbf{v}_2) e^{4t} = \begin{pmatrix} t \\ t + 1/2 \\ 1/2 \end{pmatrix} e^{4t}.$$

Correspondingly, the general solution is

$$\mathbf{X}(t) = C_1 \begin{pmatrix} 1 \\ 1 \\ 0 \end{pmatrix} e^{4t} + C_2 \begin{pmatrix} t \\ t + 1/2 \\ 1/2 \end{pmatrix} e^{4t} + C_3 \begin{pmatrix} 1 \\ 0 \\ 1 \end{pmatrix} e^{2t},$$

so that the three functions that solve the above system are

$$\begin{cases} x_1(t) & = C_1 e^{4t} + C_2 t e^{4t} + C_3 e^{2t} \\ x_2(t) & = C_1 e^{4t} + C_2 \left(t + \frac{1}{2}\right) e^{4t} \\ x_3(t) & = C_2 \frac{1}{2} e^{4t} + C_3 e^{2t} \end{cases}.$$

Problem 1.163 Show that the functions

$$\begin{cases} x_1(t) & = \left(C_1 + C_2 t + C_3 \left(\frac{t^2}{2} - \frac{1}{4}\right)\right) e^{2t} \\ x_2(t) & = \left(C_1 + C_2(t + 1) + C_3 \left(\frac{t^2}{2} + t + \frac{1}{2}\right)\right) e^{2t} \\ x_3(t) & = \left(C_1 + C_2 t + C_3 \frac{t^2}{2}\right) e^{2t} \end{cases}$$

form the solution of the system of equations (see specifically Problem 1.149):

$$\begin{cases} \dot{x}_1 = 4x_1 + x_2 - 3x_3 \\ \dot{x}_2 = -2x_1 + 3x_2 + x_3 \ . \\ \dot{x}_3 = 2x_1 + x_2 - x_3 \end{cases}$$

To conclude this subsection, two comments are in order. Firstly, the appearance of the t-polynomial in front of the exponential function $e^{\lambda t}$ in the case of the repeated eigenvalues should not sound alarming. Indeed, we know that the linear system $\dot{\mathbf{X}} = \mathbf{AX}$ of n scalar equations is equivalent to a single nth order linear ODE with constant coefficients. Substitution of the exponential trial solution $e^{\lambda t}$ into it results in an algebraic equation for λ that is identical to the zero of the characteristic polynomial of $\mathbf{A}$, i.e. $\Delta(\lambda) = 0$, and hence the values of λ are identical to the eigenvalues of $\mathbf{A}$. On the other hand, it is known (see Sect. I.8.2.2) that in the case of the second-order linear ODE with constant coefficients and a single root λ one has to multiply the solution $e^{\lambda t}$ with t to construct the second linearly independent solution $te^{\lambda t}$. It is easy to show that in the case of the nth order linear ODE with constant coefficients the additional solutions form the sequence $e^{\lambda t}, te^{\lambda t}, t^2 e^{\lambda t}$, and so on. In the end, both methods result in identical solutions.

Secondly, we learned above, quite formally, that the general solution of the system $\dot{\mathbf{X}} = \mathbf{AX}$ for the matrix $\mathbf{A}$ that has precisely as many eigenvectors as its dimension can be written in the exponential form $\mathbf{X} = e^{\mathbf{A}t}\mathbf{C}$, where $\mathbf{C}$ is the vector of arbitrary constants. It is remarkable that even in the case of a defective $\mathbf{A}$ the same formal solution is valid. This is easy to understand as the solution $\mathbf{X}(t) = e^{\mathbf{A}t}\mathbf{C}$ formally satisfies the system of equations for any $\mathbf{A}$. Indeed, let us differentiate this solution,

$$\mathbf{X}(t) = \left[\mathbf{E} + t\mathbf{A} + \frac{t^2}{2}\mathbf{A}^2 + \frac{t^3}{3!}\mathbf{A}^3 + \dots \right] \mathbf{C}$$

with respect to time:

$$\frac{d}{dt}\mathbf{X}(t) = \left[\mathbf{A} + t\mathbf{A}^2 + \frac{t^2}{2!}\mathbf{A}^3 + \frac{t^3}{3!}\mathbf{A}^4 \dots \right] \mathbf{C}$$

$$= \mathbf{A} \left[\mathbf{E} + t\mathbf{A} + \frac{t^2}{2!}\mathbf{A}^2 + \frac{t^3}{3!}\mathbf{A}^3 \dots \right] \mathbf{C}$$

$$= \mathbf{A}e^{\mathbf{A}t}\mathbf{C} \equiv \mathbf{AX} \ .$$

It is, however, not difficult to check explicitly that the exponential form coincides with the solution we have obtained previously. To illustrate this, consider the case of the matrix $\mathbf{A}$ having a p times repeated single eigenvalue λ with a single eigenvector, i.e. the Jordan chain contains p members. The solution becomes

$$\mathbf{X} = e^{\mathbf{A}t}\mathbf{C} = \mathbf{U}e^{\mathbf{J}t}\mathbf{U}^{-1}\mathbf{C} = \mathbf{U}e^{\mathbf{J}t}\widetilde{\mathbf{C}},$$

where $\widetilde{\mathbf{C}}$ is the vector of new arbitrary constants, and $\mathbf{U} = \begin{pmatrix} \mathbf{v}_1 & \mathbf{v}_2 & \dots & \mathbf{v}_p \end{pmatrix}$ is the transformation matrix containing vectors $\mathbf{v}_i$ of the Jordan chain as its columns. The exponential function of the Jordan matrix $\mathbf{J}$ can be calculated explicitly using Eq. (1.168) as

$$e^{\mathbf{J}t} = \begin{pmatrix} e^{\lambda t} & te^{\lambda t} & (t^2/2)e^{\lambda t} & (t^3/3!)e^{\lambda t} & (t^4/4!)e^{\lambda t} & \dots \\ & e^{\lambda t} & te^{\lambda t} & (t^2/2)e^{\lambda t} & (t^3/3!)e^{\lambda t} & \dots \\ & & e^{\lambda t} & te^{\lambda t} & (t^2/2)e^{\lambda t} & \dots \\ & & & e^{\lambda t} & te^{\lambda t} & \dots \\ & & & & \ddots \\ & & & & & e^{\lambda t} & te^{\lambda t} \\ & & & & & & e^{\lambda t} \end{pmatrix},$$

yielding

$$\mathbf{X} = \begin{pmatrix} \mathbf{v}_1 & \dots & \mathbf{v}_p \end{pmatrix} e^{\mathbf{J}t} \begin{pmatrix} \widetilde{C}_1 \\ \dots \\ \widetilde{C}_p \end{pmatrix}$$

$$= e^{\lambda t} \left(\mathbf{v}_1 \; t\mathbf{v}_1 + \mathbf{v}_2 \; \frac{t^2}{2}\mathbf{v}_1 + t\mathbf{v}_2 + \mathbf{v}_3 \; \dots \; \sum_{j=1}^{p} \frac{t^{p-j}}{(p-j)!}\mathbf{v}_j \right) \begin{pmatrix} \widetilde{C}_1 \\ \widetilde{C}_2 \\ \widetilde{C}_3 \\ \dots \\ \widetilde{C}_p \end{pmatrix}$$

$$= e^{\lambda t} \left[\widetilde{C}_1\mathbf{v}_1 + \widetilde{C}_2 \left(t\mathbf{v}_1 + \mathbf{v}_2 \right) + \widetilde{C}_1 \left(\frac{t^2}{2}\mathbf{v}_1 + t\mathbf{v}_2 + \mathbf{v}_3 \right) + \dots + \widetilde{C}_p \sum_{j=1}^{p} \frac{t^{p-j}}{(p-j)!}\mathbf{v}_j \right],$$

which is the same as before, cf. Eqs. (1.190)–(1.192). Perturbation method

We shall finish this section by a short note as to how one may approach a more complicated set of linear DEs $\dot{\mathbf{X}} = \mathbf{A}\mathbf{X}$ in which the matrix $\mathbf{A}$ is not constant anymore and depends on variable t. Specifically, we shall consider a case,[19] often met in applications, in which $\mathbf{A} = \mathbf{A}_0 + \mathbf{V}(t)$, i.e. it consists of two parts: the first part $\mathbf{A}_0$ is a constant matrix and $\mathbf{V}(t)$ is a relatively small 'perturbation' that explicitly depends on time.

We know that the solution of the 'unperturbed' equation $\dot{\mathbf{X}} = \mathbf{A}_0\mathbf{X}$ is given by the matrix exponential $\mathbf{X}(t) = e^{\mathbf{A}_0 t}\mathbf{X}(0)$. Inspired by this result, let us try to seek the solution of the full problem as $\mathbf{X}(t) = e^{\mathbf{A}_0 t}\mathbf{Y}(t)$, where $\mathbf{Y}(t)$ is an unknown auxiliary vector. We have

---

[19] Our consideration is rooted in a perturbation method of quantum many-body theory when the original equation for a propagator is transformed, using the so-called interaction picture, into an expression for a redefined propagator that contains exclusively the perturbation. This way one can 'localise' the perturbation and develop the appropriate perturbation theory leading in the end to Feynman diagrams.

$$\dot{\mathbf{X}} = \frac{d}{dt} e^{\mathbf{A}_0 t} \mathbf{Y}(t) = \mathbf{A}_0 \underbrace{e^{\mathbf{A}_0 t} \mathbf{Y}(t)}_{\mathbf{X}} + e^{\mathbf{A}_0 t} \dot{\mathbf{Y}}(t).$$

On the other hand, this is to be equal to $\mathbf{AX} = (\mathbf{A}_0 + \mathbf{V})\mathbf{X}$. Then, an obvious cancellation yields

$$e^{\mathbf{A}_0 t} \dot{\mathbf{Y}}(t) = \mathbf{VX} = \mathbf{V} e^{\mathbf{A}_0 t} \mathbf{Y} \quad \Rightarrow \quad \dot{\mathbf{Y}} = \tilde{\mathbf{V}}(t)\mathbf{Y} \tag{1.193}$$

with the matrix $\tilde{\mathbf{V}}$ being defined as[20]

$$\tilde{\mathbf{V}}(t) = e^{-\mathbf{A}_0 t} \mathbf{V}(t) e^{\mathbf{A}_0 t}. \tag{1.194}$$

Since the matrix $\mathbf{A}_0$ is a constant matrix (which even may defective), the matrix exponentials can be calculated and presented in the standard matrix form, which enables, at least in principle, the calculation of the matrix $\tilde{\mathbf{V}}(t)$ (if required, see below).

Next, we shall develop a perturbation theory for solving the matrix equation (1.193). Let us first integrate both sides of the equation between $t = 0$ and t:

$$\mathbf{Y}(t) = \mathbf{Y}(0) + \int_0^t dt_1 \, \tilde{\mathbf{V}}(t_1)\mathbf{Y}(t_1). \tag{1.195}$$

We shall now iterate this equation by replacing $\mathbf{Y}(t_1)$ in the right-hand side with the right-hand side of the same equation:

$$\mathbf{Y}(t) = \mathbf{Y}(0) + \int_0^t dt_1 \, \tilde{\mathbf{V}}(t_1) \left[\mathbf{Y}(0) + \int_0^{t_1} dt_2 \, \tilde{\mathbf{V}}(t_2)\mathbf{Y}(t_2) \right]$$

$$= \left[\mathbf{E} + \int_0^t dt_1 \, \tilde{\mathbf{V}}(t_1) \right] \mathbf{Y}(0) + \int_0^t dt_1 \int_0^{t_1} dt_2 \, \tilde{\mathbf{V}}(t_1)\tilde{\mathbf{V}}(t_2)\mathbf{Y}(t_2).$$

Repeating this process again by replacing $\mathbf{Y}(t_2)$ in the integral in the right-hand side by the right-hand side of Eq. (1.195), we obtain

$$\mathbf{Y}(t) = \left[\mathbf{E} + \int_0^t dt_1 \, \tilde{\mathbf{V}}(t_1) + \int_0^t dt_1 \int_0^{t_1} dt_2 \, \tilde{\mathbf{V}}(t_1)\tilde{\mathbf{V}}(t_2) \right] \mathbf{Y}(0)$$

$$+ \int_0^t dt_1 \int_0^{t_1} dt_2 \int_0^{t_2} dt_3 \, \tilde{\mathbf{V}}(t_1)\tilde{\mathbf{V}}(t_2)\tilde{\mathbf{V}}(t_3)\mathbf{Y}(t_3).$$

It can be seen that repeating this procedure indefinitely, we shall arrive at an infinite series with respect to the 'small' perturbation matrix $\tilde{\mathbf{V}}$:

---

[20] This transformation of the original matrix is analogous to the 'interaction picture' of the many-body theory.

$$\mathbf{Y}(t) = \left[\mathbf{E} + \sum_{n=1}^{\infty} \int_0^t dt_1 \int_0^{t_1} dt_2 \ldots \int_0^{t_{n-1}} dt_n \, \widetilde{\mathbf{V}}(t_1)\widetilde{\mathbf{V}}(t_2) \ldots \widetilde{\mathbf{V}}(t_n) \right] \mathbf{Y}(0) \, .$$

The order of the perturbation matrices in the integrand is important as we do not assume that at different times they commute with each other. Finally, noticing that $\mathbf{Y}(0) = \mathbf{X}(0)$ and employing the definition of the $\mathbf{X}$ via $\mathbf{Y}$, we can write the final solution as an infinite series:

$$\mathbf{X}(t) = e^{\mathbf{A}_0 t} \left[\mathbf{E} + \sum_{n=1}^{\infty} \int_0^t dt_1 \ldots \int_0^{t_{n-1}} dt_n \, \widetilde{\mathbf{V}}(t_1)\widetilde{\mathbf{V}}(t_2) \ldots \widetilde{\mathbf{V}}(t_n) \right] \mathbf{X}(0) \, .$$

In practical calculations one may keep a certain number of terms in the expansion and this way obtain an approximate solution of the original matrix DE.

1.20   Pseudo-Inverse of a Matrix

We know that if a square matrix $\mathbf{A}$ is singular or is rectangular $m \times n$ matrix ($m \neq n$), then it is not possible to define its inverse $\mathbf{A}^{-1}$. One can introduce, however, a matrix called pseudo-inverse that has some properties of the ordinary inverse of a non-singular square matrix. This matrix has a number of useful applications, e.g. in solving an over-complete system of linear algebraic equations.[21]

Consider a generally rectangular $m \times n$ matrix $\mathbf{A} = (a_{ij})$ of rank r; of course, $r \leq \min(n, m)$. This matrix can always be represented as a product of $m \times r$ and $r \times n$ matrices, what we shall call a *rank decomposition* of $\mathbf{A}$ in the following:

$$\underbrace{\begin{pmatrix} a_{11} & a_{12} & \ldots & a_{1n} \\ a_{21} & a_{22} & \ldots & a_{2n} \\ \ldots & \ldots & \ddots & \ldots \\ a_{m1} & a_{m2} & \ldots & a_{mn} \end{pmatrix}}_{\mathbf{A}} = \underbrace{\begin{pmatrix} b_{11} & b_{12} & \ldots & b_{1r} \\ b_{21} & b_{22} & \ldots & b_{2r} \\ \ldots & \ldots & \ddots & \ldots \\ b_{m1} & b_{m2} & \ldots & b_{mr} \end{pmatrix}}_{\mathbf{B}} \underbrace{\begin{pmatrix} c_{11} & c_{12} & \ldots & c_{1n} \\ c_{12} & c_{22} & \ldots & c_{2n} \\ \ldots & \ldots & \ddots & \ldots \\ c_{r1} & c_{r2} & \ldots & c_{rn} \end{pmatrix}}_{\mathbf{C}} \, , \quad (1.196)$$

or simply $\mathbf{A} = \mathbf{BC}$, where the matrix

$$\mathbf{B} = \begin{pmatrix} \mathbf{b}_1 & \ldots & \mathbf{b}_r \end{pmatrix}$$

is composed of r linearly independent vectors $\mathbf{b}_j = (b_{kj})$ ($j = 1, \ldots, r$ is the right index of b_{kj}) forming its r columns. These vectors can be constructed from r linearly independent vectors formed by elements of $\mathbf{A}$. This can always be done as r is the rank of $\mathbf{A}$. For instance, one can take r linearly independent columns of $\mathbf{A}$ (or their

---

[21] It was independently described by E. H. Moore in 1920, Arne Bjerhammar in 1951, and Roger Penrose in 1955.

linear combinations). Then the elements of $\mathbf{C} = (c_{ji})$ can then be easily obtained. Indeed, writing explicitly $\mathbf{A} = \mathbf{BC}$ in components,

$$\sum_{k=1}^{r} b_{ik} c_{kj} = a_{ij}, \quad (i = 1, \ldots, m \text{ and } j = 1, \ldots, n),$$

we obtain, for each fixed value of the index j, the vector-column $\mathbf{a}_i = (a_{ij})$ of $\mathbf{A}$ as a linear combination $\sum_k c_{kj} \mathbf{b}_k$ of vector-columns $\mathbf{b}_k = (b_{ik})$ of $\mathbf{B}$, with the coefficients $\mathbf{c}_j = (c_{kj})$ forming the jth vector-column of $\mathbf{C}$. Given the particular choice of the matrix $\mathbf{B}$, this uniquely determines the coefficients $\mathbf{c}_j$ in this decomposition since the ranks of all three matrices must be the same and, as can be shown,[22] are all equal to r. Since all r vector-columns $\mathbf{b}_k$ of $\mathbf{B}$ are linearly independent, they form a basis of a space of r dimensions, and any vector-column of $\mathbf{A}$ can be expanded into them uniquely since, out of n columns of $\mathbf{A}$, also r are linearly independent. Running this procedure for all values of j, the whole matrix $\mathbf{C}$ is determined uniquely. This shows, that the decomposition (1.196) is entirely determined by the choice of r linearly independent vectors forming columns of $\mathbf{B}$.

The simplest choice of the matrix $\mathbf{B}$ would be to take as its r columns, r linearly independent columns of $\mathbf{A}$; this would determine immediately r columns of $\mathbf{C}$. Consider, as an example, the decomposition of the matrix

$$\mathbf{A} = \begin{pmatrix} 1 & -1 & 2 & 5 \\ 2 & 2 & 0 & -1 \\ 3 & 0 & -3 & 6 \end{pmatrix}. \tag{1.197}$$

It has the rank $r = 3$ (see Problem 1.70). For instance, its first three columns can be chosen as linearly independent. Hence, it can be decomposed, e.g. as follows:

$$\begin{pmatrix} 1 & -1 & 2 & 5 \\ 2 & 2 & 0 & -1 \\ 3 & 0 & -3 & 6 \end{pmatrix} = \begin{pmatrix} 1 & -1 & 2 \\ 2 & 2 & 0 \\ 3 & 0 & -3 \end{pmatrix} \begin{pmatrix} 1 & 0 & 0 & c_{14} \\ 0 & 1 & 0 & c_{24} \\ 0 & 0 & 1 & c_{34} \end{pmatrix},$$

where the first three columns of $\mathbf{C}$ represent the identity sub-matrix, and now we only need to determine the three elements of $\mathbf{C}$ in its last column. Performing the multiplication, we obtain a system of three linear equations,

$$\begin{cases} c_{14} - c_{24} + 2c_{34} = 5 \\ 2c_{14} + 2c_{24} = -1 \\ 3c_{14} - 3c_{34} = 6 \end{cases} \implies \begin{cases} c_{14} = 17/8 \\ c_{24} = -21/8 \,, \\ c_{34} = 1/8 \end{cases}$$

---

[22] See, for instance, the book by F. R. Gantmacher 'Theory of matrices', Vol. 1, Chelsea Pub Co (1984), that we closely follow here.

so that the final decomposition is

$$\underbrace{\begin{pmatrix} 1 & -1 & 2 & 5 \\ 2 & 2 & 0 & -1 \\ 3 & 0 & -3 & 6 \end{pmatrix}}_{\mathbf{A}} = \underbrace{\begin{pmatrix} 1 & -1 & 2 \\ 2 & 2 & 0 \\ 3 & 0 & -3 \end{pmatrix}}_{\mathbf{B}} \underbrace{\begin{pmatrix} 1 & 0 & 0 & 17/8 \\ 0 & 1 & 0 & -21/8 \\ 0 & 0 & 1 & 1/8 \end{pmatrix}}_{\mathbf{C}} . \tag{1.198}$$

Note that the decomposition of a matrix $\mathbf{A}$ of rank r into a product of two full-rank matrices $\mathbf{B}$ and $\mathbf{C}$ is always possible, but it is not unique. But having obtained this decomposition, we can define and calculate the Moore–Penrose pseudo-inverse. A $n \times m$ matrix $\mathbf{A}^+$ (with a plus sign in its superscript) is called pseudo-inverse of the given $m \times n$ matrix $\mathbf{A}$ if it satisfies the following three conditions:

• $\mathbf{A}\mathbf{A}^+\mathbf{A} = \mathbf{A}$;
• $\mathbf{A}^+ = \mathbf{U}\mathbf{A}^\dagger$, where $\mathbf{U}$ is a square $n \times n$ matrix, and
• $\mathbf{A}^+ = \mathbf{A}^\dagger\mathbf{V}$, where $\mathbf{V}$ is a square $m \times m$ matrix.

Obviously, if a matrix $\mathbf{A}$ is a $n \times n$ matrix of rank n, $\mathbf{A}^{-1}$ as $\mathbf{A}^+$ satisfies all three conditions; in fact in that case $\mathbf{A}^+$ coincides with $\mathbf{A}^{-1}$. Indeed, let us check first the conditions. The second condition gives $\mathbf{A}^{-1} = \mathbf{U}\mathbf{A}^\dagger$, yielding $\mathbf{U} = \mathbf{A}^{-1}\left(\mathbf{A}^\dagger\right)^{-1} = \left(\mathbf{A}^\dagger\mathbf{A}\right)^{-1}$ (and so it exists) and similarly the third condition is also satisfied with $\mathbf{V} = \left(\mathbf{A}^\dagger\right)^{-1}\mathbf{A}^{-1} = \left(\mathbf{A}\mathbf{A}^\dagger\right)^{-1}$. Finally, since $\mathbf{A}^{-1}$ exists in this case, multiplying the first condition with $\mathbf{A}^{-1}$ from the left or from the right, we obtain $\mathbf{A}^+\mathbf{A} = \mathbf{A}\mathbf{A}^+ = \mathbf{E}$, which is indeed only satisfied by the inverse matrix $\mathbf{A}^+ = \mathbf{A}^{-1}$.

Using this definition, we can derive an explicit formula for the pseudo-inverse of a general rectangular matrix $\mathbf{A}$. We shall start from writing the matrix $\mathbf{A}$ of rank r as a product $\mathbf{B}\mathbf{C}$ of two full-rank matrices, as discussed above. Let us calculate the pseudo-inverse of matrices $\mathbf{B}$ and $\mathbf{C}$ first. According to the definition,

$$\mathbf{B}\mathbf{B}^+\mathbf{B} = \mathbf{B} \text{ and } \mathbf{B}^+ = \mathbf{U}\mathbf{B}^\dagger \implies \mathbf{B}\mathbf{U}\mathbf{B}^\dagger\mathbf{B} = \mathbf{B}$$
$$\implies \mathbf{B}^\dagger\mathbf{B}\mathbf{U}\mathbf{B}^\dagger\mathbf{B} = \mathbf{B}^\dagger\mathbf{B},$$

where on the second line we multiplied both sides from the left by $\mathbf{B}^\dagger$. The square matrix $\mathbf{B}^\dagger\mathbf{B}$ is non-singular and has an inverse (see Sect. 1.5). Hence, multiplying from the left and from the right with $\left(\mathbf{B}^\dagger\mathbf{B}\right)^{-1}$, we obtain $\mathbf{U} = \left(\mathbf{B}^\dagger\mathbf{B}\right)^{-1}$ and hence

$$\mathbf{B}^+ = \left(\mathbf{B}^\dagger\mathbf{B}\right)^{-1}\mathbf{B}^\dagger . \tag{1.199}$$

Problem 1.164 Use the similar argument to show that

$$\mathbf{C}^+ = \mathbf{C}^\dagger\left(\mathbf{C}\mathbf{C}^\dagger\right)^{-1} . \tag{1.200}$$

Then, it is verified that

$$\mathbf{A}^+ = (\mathbf{BC})^+ = \mathbf{C}^+\mathbf{B}^+ = \mathbf{C}^\dagger \left(\mathbf{CC}^\dagger\right)^{-1} \left(\mathbf{B}^\dagger\mathbf{B}\right)^{-1} \mathbf{B}^\dagger. \tag{1.201}$$

To prove this, we have to check that the matrix $\mathbf{A}^+$ defined in this way satisfies the conditions for the pseudo-inverse stated above. First of all, we check that it satisfies the first condition, $\mathbf{AA}^+\mathbf{A} = \mathbf{A}$. We have

$$\mathbf{AA}^+\mathbf{A} = \underbrace{\mathbf{BC}}_{\mathbf{A}} \underbrace{\mathbf{C}^\dagger \left(\mathbf{CC}^\dagger\right)^{-1} \left(\mathbf{B}^\dagger\mathbf{B}\right)^{-1} \mathbf{B}^\dagger}_{\mathbf{A}^+} \underbrace{\mathbf{BC}}_{\mathbf{A}}$$

$$= \mathbf{B} \underbrace{\left(\mathbf{CC}^\dagger\right) \left(\mathbf{CC}^\dagger\right)^{-1}}_{\mathbf{E}} \underbrace{\left(\mathbf{B}^\dagger\mathbf{B}\right)^{-1} \left(\mathbf{B}^\dagger\mathbf{B}\right)}_{\mathbf{E}} \mathbf{C}$$

$$= \mathbf{BC} = \mathbf{A},$$

as required. To check if one can write $\mathbf{A}^+$ via square matrices $\mathbf{U}$ or $\mathbf{V}$, we can rewrite Eq. (1.201) either as

$$\mathbf{A}^+ = \mathbf{C}^\dagger \underbrace{\left(\mathbf{B}^\dagger\mathbf{B}\right) \left(\mathbf{B}^\dagger\mathbf{B}\right)^{-1}}_{\mathbf{E}\ \text{inserted}} \left(\mathbf{CC}^\dagger\right)^{-1} \left(\mathbf{B}^\dagger\mathbf{B}\right)^{-1} \mathbf{B}^\dagger$$

$$= \underbrace{\mathbf{C}^\dagger\mathbf{B}^\dagger}_{\mathbf{A}^\dagger} \underbrace{\mathbf{B} \left(\mathbf{B}^\dagger\mathbf{B}\right)^{-1} \left(\mathbf{CC}^\dagger\right)^{-1} \left(\mathbf{B}^\dagger\mathbf{B}\right)^{-1} \mathbf{B}^\dagger}_{\mathbf{V}}$$

or

$$\mathbf{A}^+ = \underbrace{\mathbf{C}^\dagger \left(\mathbf{CC}^\dagger\right)^{-1} \left(\mathbf{B}^\dagger\mathbf{B}\right)^{-1} \left(\mathbf{CC}^\dagger\right)^{-1} \mathbf{C}}_{\mathbf{U}} \underbrace{\mathbf{C}^\dagger\mathbf{B}^\dagger}_{\mathbf{A}^\dagger},$$

from which both square matrices $\mathbf{V}$ and $\mathbf{U}$ follow.

It may seem that formula (1.201) proposed above is heavily based on the particular rank decomposition of $\mathbf{A}$ that we know is not unique; hence, that particular representation of $\mathbf{A}^+$ may not be unique as well. Actually, it appears that the pseudo-inverse matrix is in fact unique, and hence one may use Eq. (1.201) for its calculation starting from a particular rank decomposition $\mathbf{BC}$; starting from another rank decomposition (different $\mathbf{B}$ and $\mathbf{C}$) must lead to exactly the same result.

Theorem 1.28 *The pseudo-inverse matrix is unique.*

Proof: we shall prove this by contradiction. Let us assume that there exist two different matrices $\mathbf{A}_1^+ \neq \mathbf{A}_2^+$, satisfying each of the respective conditions:

$$\mathbf{AA}_1^+\mathbf{A} = \mathbf{A}, \quad \mathbf{A}_1^+ = \mathbf{U}_1\mathbf{A}^\dagger \ \text{and} \ \mathbf{A}_1^+ = \mathbf{A}^\dagger\mathbf{V}_1$$

and

$$\mathbf{A}\mathbf{A}_2^+\mathbf{A} = \mathbf{A}, \quad \mathbf{A}_2^+ = \mathbf{U}_2\mathbf{A}^\dagger \quad \text{and} \quad \mathbf{A}_2^+ = \mathbf{A}^\dagger\mathbf{V}_2.$$

Subtracting these expressions from each other and introducing matrices $\mathbf{X} = \mathbf{A}_1^+ - \mathbf{A}_2^+$, $\mathbf{U} = \mathbf{U}_1 - \mathbf{U}_2$ and $\mathbf{V} = \mathbf{V}_1 - \mathbf{V}_2$, we obtain

$$\mathbf{A}\mathbf{X}\mathbf{A} = \mathbf{0} \quad \text{and} \quad \mathbf{X} = \mathbf{A}^\dagger\mathbf{V} = \mathbf{U}\mathbf{A}^\dagger.$$

Let us calculate

$$(\mathbf{X}\mathbf{A})^\dagger (\mathbf{X}\mathbf{A}) = \mathbf{A}^\dagger\mathbf{X}^\dagger\mathbf{X}\mathbf{A} = \mathbf{A}^\dagger \left(\mathbf{A}^\dagger\mathbf{V}\right)^\dagger \mathbf{X}\mathbf{A} = \mathbf{A}^\dagger\mathbf{V}^\dagger \underbrace{\mathbf{A}\mathbf{X}\mathbf{A}}_{=0} = \mathbf{0}.$$

From this it immediately follows that the vector $\mathbf{X}\mathbf{A} = \mathbf{0}$. Moreover,

$$\mathbf{X}\mathbf{X}^\dagger = \mathbf{X}\left(\mathbf{U}\mathbf{A}^\dagger\right)^\dagger = \underbrace{\mathbf{X}\mathbf{A}}_{=0}\,\mathbf{U}^\dagger = \mathbf{0}$$

is zero as well, leading immediately to $\mathbf{X} = \mathbf{0}$, i.e. both matrices being the same and hence contradicting our assumption. **Q.E.D.**

The significance of these results is difficult to underestimate: starting from an arbitrary rank decomposition of $\mathbf{A}$, we can always calculate its pseudo-inverse $\mathbf{A}^+$.

The pseudo-inverse operation on a matrix possesses some important properties which we shall now consider,[23]

The first identity is

$$\left(\mathbf{A}^\dagger\right)^+ = \left(\mathbf{A}^+\right)^\dagger. \tag{1.202}$$

Indeed, $\mathbf{A}^\dagger = (\mathbf{B}\mathbf{C})^\dagger = \mathbf{C}^\dagger\mathbf{B}^\dagger$ can be considered as its rank decomposition, i.e. when calculating the pseudo-inverse of this matrix, $\mathbf{C}^\dagger$ serves as $\mathbf{B}$, while $\mathbf{B}^\dagger$ as $\mathbf{C}$ in its rank decomposition. Therefore, using the explicit formula (1.201), we get

$$\left(\mathbf{A}^\dagger\right)^+ = \mathbf{B}\left(\mathbf{B}^\dagger\mathbf{B}\right)^{-1}\left(\mathbf{C}\mathbf{C}^\dagger\right)^{-1}\mathbf{C},$$

which is exactly the same as

$$\left(\mathbf{A}^+\right)^\dagger = \left[\mathbf{C}^\dagger\left(\mathbf{C}\mathbf{C}^\dagger\right)^{-1}\left(\mathbf{B}^\dagger\mathbf{B}\right)^{-1}\mathbf{B}^\dagger\right]^\dagger = \mathbf{B}\left(\mathbf{B}^\dagger\mathbf{B}\right)^{-1}\left(\mathbf{C}\mathbf{C}^\dagger\right)^{-1}\mathbf{C},$$

The second identity:

$$\left(\mathbf{A}^+\right)^+ = \mathbf{A}. \tag{1.203}$$

Indeed, according to Eq. (1.201), if $\mathbf{A} = \mathbf{B}\mathbf{C}$, then $\mathbf{A}^+ = \mathbf{C}^+\mathbf{B}^+$. If we would like to calculate the pseudo-inverse of $\mathbf{A}^+$, we can consider $\mathbf{C}^+$ and $\mathbf{B}^+$ in the rank

---

[23] Some of them are known as Moore–Penrose conditions.

decomposition of $\mathbf{A}^+$ as $\mathbf{B}$ and $\mathbf{C}$, respectively, in the rank decomposition of $\mathbf{A}$. Hence,

$$\left(\mathbf{A}^+\right)^+ = \left(\mathbf{C}^+\mathbf{B}^+\right)^+ = \left(\mathbf{B}^+\right)^+ \left(\mathbf{C}^+\right)^+ .$$

To calculate $\left(\mathbf{B}^+\right)^+$, let us recall that $\mathbf{B}^+ = \left(\mathbf{B}^\dagger\mathbf{B}\right)^{-1}\mathbf{B}^\dagger$, see Eq. (1.199), hence, $\left(\mathbf{B}^\dagger\mathbf{B}\right)^{-1}$ must be treated as $\mathbf{B}$, and $\mathbf{B}^\dagger$ as $\mathbf{C}$ in the rank decomposition of $\mathbf{B}^+$, and hence

$$\left(\mathbf{B}^+\right)^+ = \left(\mathbf{B}^\dagger\right)^+ \left[\left(\mathbf{B}^\dagger\mathbf{B}\right)^{-1}\right]^+ = \left(\mathbf{B}^\dagger\right)^+ \left(\mathbf{B}^\dagger\mathbf{B}\right) = \left(\mathbf{B}^+\right)^\dagger \mathbf{B}^\dagger\mathbf{B} ,$$

where in the second passage we have made use of the fact that $\mathbf{B}^\dagger\mathbf{B}$ is a full-rank square matrix (see Sect. 1.5), whose inverse therefore coincides with the pseudo-inverse (both are unique). Similarly,

$$\left(\mathbf{C}^+\right)^+ = \left[\mathbf{C}^\dagger \left(\mathbf{C}\mathbf{C}^\dagger\right)^{-1}\right]^+ = \left[\left(\mathbf{C}\mathbf{C}^\dagger\right)^{-1}\right]^+ \left(\mathbf{C}^\dagger\right)^+ = \mathbf{C}\mathbf{C}^\dagger \left(\mathbf{C}^+\right)^\dagger .$$

Therefore,

$$\begin{aligned}
\left(\mathbf{A}^+\right)^+ &= \left(\mathbf{B}^+\right)^+ \left(\mathbf{C}^+\right)^+ = \left[\left(\mathbf{B}^+\right)^\dagger \left(\mathbf{B}^\dagger\mathbf{B}\right)\right]\left[\left(\mathbf{C}\mathbf{C}^\dagger\right)\left(\mathbf{C}^+\right)^\dagger\right] \\
&= \left[\left(\mathbf{B}^\dagger\mathbf{B}\right)^{-1}\mathbf{B}^\dagger\right]^\dagger \left(\mathbf{B}^\dagger\mathbf{B}\right)\left(\mathbf{C}\mathbf{C}^\dagger\right)\left[\mathbf{C}^\dagger\left(\mathbf{C}\mathbf{C}^\dagger\right)^{-1}\right]^\dagger \\
&= \left[\mathbf{B}\left(\mathbf{B}^\dagger\mathbf{B}\right)^{-1}\right]\left(\mathbf{B}^\dagger\mathbf{B}\right)\left(\mathbf{C}\mathbf{C}^\dagger\right)\left[\left(\mathbf{C}\mathbf{C}^\dagger\right)^{-1}\mathbf{C}\right] = \mathbf{B}\mathbf{C} = \mathbf{A} ,
\end{aligned}$$

as required. Above, in the second line we replaced $\mathbf{B}^+$ and $\mathbf{C}^+$ with their explicit expressions (1.199) and (1.200).

The other four identities we shall formulate as a problem for the reader to solve.

Problem 1.165 Using the explicit expression (1.201) for the pseudo-inverse, show that

$$\left(\mathbf{A}\mathbf{A}^+\right)^\dagger = \mathbf{A}\mathbf{A}^+ ,$$

$$\left(\mathbf{A}\mathbf{A}^+\right)^2 = \mathbf{A}\mathbf{A}^+ ,$$

$$\left(\mathbf{A}^+\mathbf{A}\right)^\dagger = \mathbf{A}^+\mathbf{A} ,$$

$$\left(\mathbf{A}^+\mathbf{A}\right)^2 = \mathbf{A}^+\mathbf{A} .$$

These identities show that the matrices $\mathbf{A}^+\mathbf{A}$ and $\mathbf{A}\mathbf{A}^+$ are both Hermitian and idempotent. A matrix $\mathbf{P}$ is called idempotent, if $\mathbf{P}^2 = \mathbf{P}$.

Problem 1.166 Prove that

$$\mathbf{A}^{\dagger}\mathbf{A}\mathbf{A}^{+} = \mathbf{A}^{\dagger}, \tag{1.204}$$

$$\left(\mathbf{A}^{+}\right)^{\dagger}\mathbf{A}^{+}\mathbf{A} = \left(\mathbf{A}^{+}\right)^{\dagger}. \tag{1.205}$$

Let us work out the pseudo-inverse $\mathbf{A}^{+}$ of the matrix (1.197) whose rank decomposition is given in Eq. (1.198). We have

$$\mathbf{B}^{\dagger}\mathbf{B} = \begin{pmatrix} 1 & 2 & 3 \\ -1 & 2 & 0 \\ 2 & 0 & -3 \end{pmatrix} \begin{pmatrix} 1 & -1 & 2 \\ 2 & 2 & 0 \\ 3 & 0 & -3 \end{pmatrix} = \begin{pmatrix} 14 & 3 & -7 \\ 3 & 5 & -2 \\ -7 & -2 & 13 \end{pmatrix},$$

$$\left(\mathbf{B}^{\dagger}\mathbf{B}\right)^{-1} = \frac{1}{576} \begin{pmatrix} 61 & -25 & 29 \\ -25 & 133 & 7 \\ 29 & 7 & 61 \end{pmatrix},$$

$$\mathbf{C}\mathbf{C}^{\dagger} = \begin{pmatrix} 1 & 0 & 0 & 17/8 \\ 0 & 1 & 0 & -21/8 \\ 0 & 0 & 1 & 1/8 \end{pmatrix} \begin{pmatrix} 1 & 0 & 0 \\ 0 & 1 & 0 \\ 0 & 0 & 1 \\ 17/8 & -21/8 & 1/8 \end{pmatrix} = \frac{1}{64} \begin{pmatrix} 353 & -357 & 17 \\ -357 & 505 & -21 \\ 17 & -21 & 65 \end{pmatrix},$$

$$\left(\mathbf{C}\mathbf{C}^{\dagger}\right)^{-1} = \frac{1}{795} \begin{pmatrix} 506 & 357 & -17 \\ 357 & 354 & 21 \\ -17 & 21 & 794 \end{pmatrix},$$

so that

$$\mathbf{A}^{+} = \mathbf{C}^{\dagger}\left(\mathbf{C}\mathbf{C}^{\dagger}\right)^{-1}\left(\mathbf{B}^{\dagger}\mathbf{B}\right)^{-1}\mathbf{B}^{\dagger} = \frac{1}{2385} \begin{pmatrix} 99 & 585 & 83 \\ 18 & 540 & -9 \\ 567 & 315 & -416 \\ 234 & -135 & 148 \end{pmatrix}.$$

It is a 4×3 matrix, as expected.

Problem 1.167 Show that

$$\begin{pmatrix} 1 & 1 & -1 & 1 \\ -2 & 2 & -2 & 2 \\ -1 & 2 & -3 & 0 \end{pmatrix}^{+} = \frac{1}{28} \begin{pmatrix} 14 & -7 & 0 \\ 4 & 1 & 2 \\ -2 & 3 & -8 \\ 8 & 9 & -10 \end{pmatrix}.$$

1.21 General 3D Rotation

Here we shall discuss a derivation, using several closely related methods, of an explicit expression for the 3×3 matrix of a general 3D rotation by angle φ around an arbitrarily oriented axis given by the unit vector $\mathbf{w} = (w_1, w_2, w_3)$. The first method is based on solving a system of linear differential equations which we shall do by first rewriting them as a matrix equation. The second method will require, first, presenting the solution in an exponential form, and then performing a spectral decomposition of a general matrix in order to present the matrix exponent as an ordinary matrix.

Consider a rotation by an infinitesimal angle $\delta\varphi$ around $\mathbf{w}$ as shown in Fig. 1.9. The vector $\mathbf{r}$ goes over into a vector $\mathbf{r}'$ so that the direction of the difference vector $\delta\mathbf{r} = \mathbf{r}' - \mathbf{r}$ coincides with the direction of the cross product $\mathbf{w} \times \mathbf{r}$ as $\delta\mathbf{r}$ is perpendicular to both $\mathbf{w}$ and $\mathbf{r}$. Consider now the length of the difference vector $\overrightarrow{AA'}$. Looking at the figure, it is easy to see that $AA' = (r \sin\theta)\, \delta\varphi$. Hence, we can write

$$\delta\mathbf{r} = [\mathbf{w} \times \mathbf{r}]\, \delta\varphi \ \text{ or } \ \mathbf{r}' = \mathbf{r} + [\mathbf{w} \times \mathbf{r}]\, \delta\varphi. \tag{1.206}$$

This equation can be rewritten in the matrix form as

$$\frac{d\mathbf{r}}{d\varphi} = \mathbf{A}\mathbf{r}, \ \text{ where } \ \mathbf{A} = \begin{pmatrix} 0 & -w_3 & w_2 \\ w_3 & 0 & -w_1 \\ -w_2 & w_1 & 0 \end{pmatrix} \tag{1.207}$$

and $\mathbf{r}$ is the vector-column of the point A in Fig. 1.9 obtained after the full rotation. Note that the non-defective (see below) matrix $\mathbf{A}$ depends only on the components of the unit vector $\mathbf{w}$.

Fig. 1.9 For the derivation of Eq. (1.206): vector $\mathbf{r}$ goes into vector $\mathbf{r}'$ upon rotation by an infinitesimal angle $\delta\varphi$ about the axis given by the unit vector $\mathbf{w}$. The angle between vectors $\mathbf{w}$ and $\mathbf{r}$ (and $\mathbf{r}'$) is θ

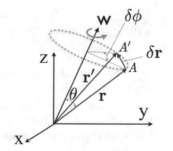

Problem 1.168 Using a substitution $\mathbf{r} = \mathbf{v}e^{\lambda\varphi}$, show that the constant vector $\mathbf{v}$ and the scalar λ are obtained as eigenvectors and eigenvalues of the matrix $\mathbf{A}$. Demonstrate by a direct calculation that the three eigenvalues are $\lambda_0 = 0$ and $\lambda_{\pm} = \pm i$. Show that the corresponding normalised right eigenvectors are

$$\mathbf{v}_0 = \begin{pmatrix} w_1 \\ w_2 \\ w_3 \end{pmatrix}, \quad \mathbf{v}_{\pm} = \frac{1}{\sqrt{2\left(1 - w_3^2\right)}} \begin{pmatrix} -w_2 \pm i w_1 w_3 \\ w_1 \pm i w_2 w_3 \\ \pm i \left(w_3^2 - 1\right) \end{pmatrix}. \tag{1.208}$$

Note that $\mathbf{v}_- = \mathbf{v}_+^*$ and $\lambda_- = \lambda_+^*$. Next, construct the general solution

$$\mathbf{r}(\varphi) = D_0 \mathbf{v}_0 + D_+ \mathbf{v}_+ e^{i\varphi} + D_- \mathbf{v}_- e^{-i\varphi}$$

as a linear combination of the three elementary ones using three arbitrary complex constants, D_0 and $D_{\pm}$. Noticing that the two elementary solutions corresponding to $\lambda_{\pm} = \pm i$ are complex conjugate of each other, argue that in order for the vector $\mathbf{r}(\varphi)$ to be real, D_- must be equal to D_+^*. Hence, altogether there are only three real arbitrary constants, e.g. D_0, $\operatorname{Re} D_+$ and $\operatorname{Im} D_+$. Finally, using the initial condition $\mathbf{r}(0) = \mathbf{r}_0$, find the constants. These manipulations yield $\mathbf{r}(\varphi) = \mathbf{C}(\varphi)\mathbf{r}_0$, where

$$\mathbf{C}(\varphi) = \begin{pmatrix} w_1^2 + \left(1 - w_1^2\right)c & w_1 w_2 (1 - c) - w_3 s & w_1 w_3 (1 - c) + w_2 s \\ w_1 w_2 (1 - c) + w_3 s & w_2^2 + \left(1 - w_2^2\right)c & w_2 w_3 (1 - c) - w_1 s \\ w_1 w_3 (1 - c) - w_2 s & w_2 w_3 (1 - c) + w_1 s & w_3^2 + \left(1 - w_3^2\right)c \end{pmatrix},$$

$$\tag{1.209}$$

where $c = \cos\varphi$ and $s = \sin\varphi$. Show next that $\mathbf{C}(0) = \mathbf{E}$, $\mathbf{C}(\varphi_2)\mathbf{C}(\varphi_1) = \mathbf{C}(\varphi_1 + \varphi_2)$ and $\mathbf{C}(\varphi_1)^{-1} = \mathbf{C}(-\varphi_1) = \mathbf{C}(\varphi_1)^T$, so that the matrix $\mathbf{C}(\varphi)$ is orthogonal. Finally, demonstrate that $|\mathbf{C}(\varphi)| = 1$.

It is easy to see that $\mathbf{C}(\varphi)$ is a generalisation of the rotational matrices around the Cartesian axes derived in Sect. 1.2.5.2.

Another approach is based on presenting the rotated vector $\mathbf{r}(\varphi)$ in the exponential form. To this end, let us return to the system of DEs (1.207). This is an equation of the form of Eq. (1.172) whose solution can be written as a matrix exponential, see Eq. (1.176):

$$\mathbf{r}(\varphi) = e^{\mathbf{A}\varphi}\mathbf{r}(0). \tag{1.210}$$

It is instructive to obtain the same result using a different method in which we shall build the complete rotation on φ by performing a large number n of small rotations, each by the same angle $\Delta\varphi = \varphi/n$; we intend to consider in the end the limit of $n \to \infty$. Let us use Eq. (1.207):

$$\mathbf{r}(\phi + \Delta\varphi) \approx \mathbf{r}(\phi) + \mathbf{A}\mathbf{r}(\phi)\Delta\varphi = (\mathbf{E} + \mathbf{A}\Delta\varphi)\,\mathbf{r}(\phi) = \left(\mathbf{E} + \mathbf{A}\frac{\varphi}{n}\right)\mathbf{r}(\phi).$$

This expression is approximate since $\Delta\varphi$ is a finite (albeit small) angle, that is why we have used $\approx$ instead of $=$ here. However, as the number of divisions n becomes larger and larger, this equation becomes more and more accurate, becoming exact in the $n \to \infty$ limit. Having this in mind, we shall use $=$ from now on.

By applying this small rotations one after another n times, we will reach the required full rotation by angle $\varphi = n\Delta\varphi$, i.e.

$$\mathbf{r}(\Delta\varphi) = \left(\mathbf{E} + \mathbf{A}\frac{\varphi}{n}\right)\mathbf{r}(0)$$

$$\mathbf{r}(2\Delta\varphi) = \left(\mathbf{E} + \mathbf{A}\frac{\varphi}{n}\right)\mathbf{r}(\Delta\varphi) = \left(\mathbf{E} + \mathbf{A}\frac{\varphi}{n}\right)^2 \mathbf{r}(0)$$

$$\mathbf{r}(3\Delta\varphi) = \left(\mathbf{E} + \mathbf{A}\frac{\varphi}{n}\right)\mathbf{r}(2\Delta\varphi) = \left(\mathbf{E} + \mathbf{A}\frac{\varphi}{n}\right)^3 \mathbf{r}(0)$$

$$\cdots \quad \cdots$$

$$\mathbf{r}(n\Delta\phi) = \left(\mathbf{E} + \mathbf{A}\frac{\varphi}{n}\right)^n \mathbf{r}(0) \equiv \mathbf{r}(\phi),$$

Now we perform the limit using the result of Problem 1.119 to obtain the same exponential solution as before, Eq. (1.210).

In order to obtain the expression (1.209) for the rotational matrix, we now have to transform the exponential matrix expression $e^{\mathbf{A}\varphi}$ into an standard form of a 3×3 matrix. This can be done, e.g. by expanding the matrix $\mathbf{A}$ into its spectral decomposition via the right eigenvectors. Consider first a simple case of $\mathbf{w}^T = (0, 0, 1)$ when the rotational axis is directed along the z axis. In this case

$$\mathbf{A} = \begin{pmatrix} 0 & -1 & 0 \\ 1 & 0 & 0 \\ 0 & 0 & 0 \end{pmatrix}$$

and its normalised right eigenvectors are

$$\mathbf{x}_1 = \begin{pmatrix} 0 \\ 0 \\ 1 \end{pmatrix}, \quad \mathbf{x}_2 = \frac{1}{\sqrt{2}}\begin{pmatrix} 1 \\ -i \\ 0 \end{pmatrix}, \quad \mathbf{x}_3 = \frac{1}{\sqrt{2}}\begin{pmatrix} 1 \\ i \\ 0 \end{pmatrix}$$

corresponding to eigenvalues $\lambda_1 = 0$, $\lambda_2 = i$ and $\lambda_3 = -i$. Hence,

$$e^{\mathbf{A}\varphi} = \sum_{k=1}^{3} e^{\lambda_k \varphi} \mathbf{x}_k \mathbf{x}_k^\dagger = \mathbf{x}_1 \mathbf{x}_1^\dagger + e^{i\varphi} \mathbf{x}_2 \mathbf{x}_2^\dagger + e^{-i\varphi} \mathbf{x}_3 \mathbf{x}_3^\dagger$$

$$= \begin{pmatrix} 0 & 0 & 0 \\ 0 & 0 & 0 \\ 0 & 0 & 1 \end{pmatrix} + \frac{e^{i\varphi}}{2} \begin{pmatrix} 1 & i & 0 \\ -i & 1 & 0 \\ 0 & 0 & 0 \end{pmatrix} + \frac{e^{-i\varphi}}{2} \begin{pmatrix} 1 & -i & 0 \\ i & 1 & 0 \\ 0 & 0 & 0 \end{pmatrix}$$

$$= \begin{pmatrix} \cos\varphi & -\sin\varphi & 0 \\ \sin\varphi & \cos\varphi & 0 \\ 0 & 0 & 1 \end{pmatrix},$$

which is the correct rotational matrix, Eq. (1.47).

The calculation in the general case requires a rather lengthy manipulations, although it is rather straightforward, and we shall leave it as a problem for the reader.

Problem 1.169 Consider a general case of the matrix $\mathbf{A}$ of Eq. (1.207). The normalised right eigenvectors corresponding to the eigenvalues $\lambda_1 = 0$, $\lambda_2 = i$ and $\lambda_3 = -i$, are given by Eq. (1.208). Demonstrate then that the expression

$$e^{\mathbf{A}\varphi} = \sum_{k=1}^{3} e^{\lambda_k \varphi} \mathbf{x}_k \mathbf{x}_k^\dagger = \mathbf{x}_1 \mathbf{x}_1^\dagger + e^{i\varphi} \mathbf{x}_2 \mathbf{x}_2^\dagger + e^{-i\varphi} \mathbf{x}_3 \mathbf{x}_3^\dagger$$

results in the matrix (1.209).

1.22   Examples in Physics

1.22.1   Particle in a Magnetic Field

Consider a particle of charge q (e.g. an electron) and mass m in a constant magnetic field $\mathbf{B}$. The equation of motion reads

$$m\frac{d\mathbf{v}}{dt} = q\,(\mathbf{v} \times \mathbf{B}) .$$

We would like to know the trajectory of the particle $\mathbf{r}(t)$ and its velocity $\mathbf{v}(t)$ as a function of time subject to the known initial conditions, $\mathbf{r}(0)$ and $\mathbf{v}(0)$.

In fact, there are three equations for each of the Cartesian components of the velocity, i.e. we have a system of three linear differential equations:

$$\begin{cases} m\dot{v}_1 = q\,(v_2 B_3 - v_3 B_2) \\ m\dot{v}_2 = q\,(v_3 B_1 - v_1 B_3) \\ m\dot{v}_3 = q\,(v_1 B_2 - v_2 B_1) \end{cases}.$$

To solve the problem, we shall rewrite the equations as a single matrix equation:

$$\frac{d}{dt}\begin{pmatrix} v_1 \\ v_2 \\ v_3 \end{pmatrix} = \frac{q}{m}\begin{pmatrix} 0 & B_3 & -B_2 \\ -B_3 & 0 & B_1 \\ B_2 & -B_1 & 0 \end{pmatrix}\begin{pmatrix} v_1 \\ v_2 \\ v_3 \end{pmatrix}, \qquad (1.211)$$

i.e.

$$\frac{d\mathbf{v}}{dt} = \mathbf{G}\mathbf{v}, \quad \text{where} \quad \mathbf{G} = \frac{q}{m}\begin{pmatrix} 0 & B_3 & -B_2 \\ -B_3 & 0 & B_1 \\ B_2 & -B_1 & 0 \end{pmatrix}. \qquad (1.212)$$

We shall attempt to solve the corresponding matrix (three dimensional) equation using the following trial solution: $\mathbf{v}(t) = \mathbf{u}e^{\zeta t}$ with $\mathbf{u}$ and ζ being an unknown vector and scalar. Using the trial solution in the equation of motion, we first calculate its time derivative, $d\mathbf{v}/dt - \zeta\mathbf{u}e^{\zeta t}$, and then, substituting it into Eq. (1.212), we obtain

$$\zeta\mathbf{u}e^{\zeta t} = \mathbf{G}\mathbf{u}e^{\zeta t} \quad \Longrightarrow \quad \zeta\mathbf{u} = \mathbf{G}\mathbf{u}, \qquad (1.213)$$

after cancelling on the exponent in both sides. We have obtained a familiar eigenvalue–eigenvector problem which should allow us to find both ζ and $\mathbf{u}$. Since the matrix $\mathbf{G}$ is antisymmetric, nothing can be said about the eigenvectors in advance apart from the fact that they could be either zero or purely imaginary (see Problem 1.92).

To simplify the problem, let us assume that the magnetic field $\mathbf{B}$ acts along the z axis only:

$$\mathbf{G} = \frac{q}{m}\begin{pmatrix} 0 & B & 0 \\ -B & 0 & 0 \\ 0 & 0 & 0 \end{pmatrix} = \begin{pmatrix} 0 & \omega & 0 \\ -\omega & 0 & 0 \\ 0 & 0 & 0 \end{pmatrix}, \quad \text{where} \quad \omega = \frac{qB}{m}.$$

The eigenvalues are found by solving the corresponding characteristic equation:

$$|\mathbf{G} - \lambda\mathbf{E}| = \begin{vmatrix} -\lambda & \omega & 0 \\ -\omega & -\lambda & 0 \\ 0 & 0 & -\lambda \end{vmatrix} = -\lambda\left(\lambda^2 + \omega^2\right) = 0,$$

which gives three eigenvalues: $\lambda = 0, i\omega, -i\omega$. As expected, they are purely imaginary and zero. The normalised eigenvectors (generally complex) are easily obtained to be

$$\mathbf{u}_1 = \begin{pmatrix} 0 \\ 0 \\ 1 \end{pmatrix}, \quad \mathbf{u}_2 = \frac{1}{\sqrt{2}}\begin{pmatrix} 1 \\ i \\ 0 \end{pmatrix}, \quad \mathbf{u}_3 = \frac{1}{\sqrt{2}}\begin{pmatrix} 1 \\ -i \\ 0 \end{pmatrix}.$$

Therefore, a general solution of the system of three linear differential equation can be written as a linear combination of the three functions with three arbitrary constants c_1, c_2 and c_3:

$$\mathbf{v}(t) = c_1\mathbf{u}_1 + c_2\mathbf{u}_2 e^{i\omega t} + c_3\mathbf{u}_3 e^{-i\omega t} . \tag{1.214}$$

Note that the first term is not time dependent as it corresponds to the zero eigenvalue. Since $\mathbf{u}_2$ and $\mathbf{u}_3$ are within the x, y plane and $\mathbf{u}_3$ is directed along z, we may anticipate that if the magnetic field is along the z direction the particle moves with a constant speed along the z axis and performs an oscillatory motion in the x, y plane, i.e. perpendicular to the magnetic field.

To see this explicitly, we need to apply the initial conditions which determine the undefined constants. Let us assume that $\mathbf{v}(0) = (0, v_\perp, v_\parallel)$, i.e. the particle enters the field with a velocity $v_\parallel$ parallel to the field $\mathbf{B} = (0, 0, B)$ and a velocity $v_\perp$ perpendicular to it. Then at $t = 0$ we obtain

$$\begin{pmatrix} 0 \\ v_\perp \\ v_\parallel \end{pmatrix} = c_1 \begin{pmatrix} 0 \\ 0 \\ 1 \end{pmatrix} + \frac{c_2}{\sqrt{2}} \begin{pmatrix} 1 \\ i \\ 0 \end{pmatrix} + \frac{c_3}{\sqrt{2}} \begin{pmatrix} 1 \\ -i \\ 0 \end{pmatrix} ,$$

which can be solved to give: $c_1 = v_\parallel$, $c_2 = -c_3 = -iv_\perp/\sqrt{2}$. Substituting these constants into solution (1.214), we obtain

$$\begin{aligned}
\mathbf{v}(t) &= \begin{pmatrix} 0 \\ 0 \\ v_\parallel \end{pmatrix} - \frac{iv_\perp}{2}\begin{pmatrix} 1 \\ i \\ 0 \end{pmatrix} e^{i\omega t} + \frac{iv_\perp}{2}\begin{pmatrix} 1 \\ -i \\ 0 \end{pmatrix} e^{-i\omega t} \\
&= \begin{pmatrix} 0 \\ 0 \\ v_\parallel \end{pmatrix} + \frac{v_\perp}{2}\begin{pmatrix} -ie^{i\omega t} + ie^{-i\omega t} \\ e^{-i\omega t} + e^{-i\omega t} \\ 0 \end{pmatrix} \\
&= \begin{pmatrix} 0 \\ 0 \\ v_\parallel \end{pmatrix} + \frac{v_\perp}{2}\begin{pmatrix} 2\sin(\omega t) \\ 2\cos(\omega t) \\ 0 \end{pmatrix} = \begin{pmatrix} v_\perp \sin(\omega t) \\ v_\perp \cos(\omega t) \\ v_\parallel \end{pmatrix} .
\end{aligned} \tag{1.215}$$

To obtain the position vector $\mathbf{r}(t)$ of the particle, we should integrate the velocity vector:

$$\mathbf{r}(t) = \int_0^t \mathbf{v}(t_1)dt_1 = \begin{pmatrix} (v_\perp/\omega)[1 - \cos(\omega t)] \\ (v_\perp/\omega)\sin(\omega t) \\ v_\parallel t \end{pmatrix} , \tag{1.216}$$

where we have assumed that initially the particle was in the centre of the coordinate system, $\mathbf{r}(0) = \mathbf{0}$. Thus, indeed, the particle performs a circular motion in the plane perpendicular to the magnetic field and, at the same time, it moves with the constant speed along the field direction. The circle radius $R = v_\perp/\omega = mv_\perp/qB$ and the rotation frequency $\omega = qB/m$.

The kinetic energy of the particle,

$$K(t) = \frac{m}{2}\left(v_1^2 + v_2^2 + v_3^2\right) = \frac{m}{2}\left(v_\perp^2 + v_\parallel^2\right) = K(0)$$

is conserved, as the magnetic field, as it is well-known, does not do any work on the particle.

Problem 1.170 Assume a general direction of the magnetic field. Show that the general solution for the velocity in this case is

$$\mathbf{v}(t) = c_1 \begin{pmatrix} \omega_1 \\ \omega_2 \\ \omega_3 \end{pmatrix} + 2\,\mathrm{Re}\left[c_1 \begin{pmatrix} \omega_1\omega_2 - i\omega_3 \\ -1 + \omega_2^2 \\ \omega_2\omega_3 + i\omega_1 \end{pmatrix} e^{-i\omega t} \right],$$

where $\omega_i = eB_i/m$.

1.22.2 Kinetics

In chemical, biological and physical kinetics one often comes across systems of linear (and no only) ODEs. In this section we shall consider a few examples to illustrate this.

Problem 1.171 Upon an electronic bombardment, molecules A turn into an unstable species B that at room temperature breaks down into two stable C molecules: $A \to B \to C + C$. The concentrations $n_A(t)$, $n_B(t)$ and $n_C(t)$ of the molecules A, B and C satisfy the following kinetics equations:

$$\dot{n}_A = -3n_A, \quad \dot{n}_B = 3n_A - 2n_B \quad \text{and} \quad \dot{n}_C = 4n_B \ .$$

Initially $n_A(0) = n_0$ and $n_B(0) = n_C(0) = 0$. Determine the time evolution of the three concentrations. Plot the obtained concentrations and comment on your findings. [Answer: $n_A(t) = n_0 e^{-3t}$, $n_B(t) = 3n_0\left(e^{-2t} - e^{-3t}\right)$ and $n_C(t) = 2n_0\left(1 - 3e^{-2t} + 2e^{-3t}\right)$]

Problem 1.172 The populations of three species in a forest are denoted by the time-dependent functions $N_1(t)$, $N_2(t)$ and $N_3(t)$, respectively. The time dependence of the populations is determined by the rates at which the species eat each other as well as by the availability of food. Suppose that this ecosystem can approximately be described by the following system of coupled differential equations:

$$\begin{cases} \dot{N}_1 = -N_1 + 2N_2 + N_3 \\ \dot{N}_2 = N_1 - N_2 \\ \dot{N}_3 = 2N_1 - N_3 \end{cases}.$$

Assuming that initially all species had identical populations, $N_i(0) = N_0$, where $i = 1, 2, 3$, show that the corresponding particular solution is $N_1(t) = N_0 \left(5e^t - e^{-3t}\right)/4$, $N_2(t) = N_0 \left(5e^t + 2e^{-t} + e^{-3t}\right)/8$ and $N_3(t) = N_0 \left(5e^t - 2e^{-t} + e^{-3t}\right)/4$. Show that after a long period of time all three populations grow exponentially and become related to each other as $N_1 \simeq N_3 \simeq 2N_2$.

Problem 1.173 Consider kinetics of formation of two different 2D molecular phases on a surface of a crystal during the course of molecular self-assembly, see Fig. 1.10. Both phases are formed from freely diffusing molecules which may attach and detach from islands of either of the phases. We assume that the concentration of islands is small and hence exchange of molecules between them can only happen via the free phase between islands, i.e. when islands loose molecules, they freely diffuse along the surface to attach to another island(s). This rather general formulation of the problem of growth we shall simplify, however. We shall assume that islands of the first phase experience a *reversible growth*, i.e. molecules can attach (with the rate k_0) and also detach from them (with the rate k_1), while the islands of the second phase experience an *irreversible growth* with the same attachment rate. Their growth is irreversible since the detachment barrier for the molecules from the phase 2 islands is relatively large, rendering the corresponding detachment rate k_2 being very small and hence this process negligible. In summary, phase 2 islands can only grow, while phase 1 islands can both grow and diminish in size. The corresponding rate equations for the number of free molecules N_f, and the number of molecules N_1 and N_2 in the two phases (the totals in all islands) are

$$\begin{cases} \dot{N}_1 = -k_1 N_1 + k_0 N_f \\ \dot{N}_f = k_1 N_1 - 2k_0 N_f \\ \dot{N}_2 = k_0 N_f \end{cases}.$$

Show that the solution of these equations subject to the initial conditions that initially ($t = 0$) only N_0 free molecules existed, is

$$N_1(t) = \frac{N_0 k_0}{\kappa} \left(e^{\lambda_- t} - e^{\lambda_+ t}\right),$$

$$N_f(t) = \frac{N_0}{\kappa} \left[(k_1 + \lambda_-) e^{\lambda_- t} - (k_1 + \lambda_+) e^{\lambda_+ t}\right],$$

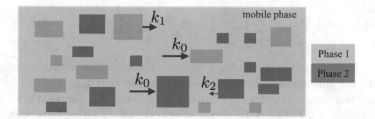

Fig. 1.10 Rectangular islands of two phases growing on a surface of a crystal (Problem 1.173). Between islands is a mobile phase consisting of mobile molecules which can attach (with the rate k_0) to any of the islands of either of the two phases, and/or detach (with the rates k_1 and k_2) from the islands of the phases 1 and 2, respectively. However, it is assumed in the problem that the rate $k_2 \ll k_1$ and hence the process of detachment from the islands of phase 2 is neglected

$$N_2(t) = N_0 \left\{ 1 + \frac{k_0}{\kappa} \left[\left(1 + \frac{k_1}{\lambda_-} \right) e^{\lambda_- t} - \left(1 + \frac{k_1}{\lambda_+} \right) e^{\lambda_+ t} \right] \right\} ,$$

where

$$\lambda_\pm = -\frac{1}{2} (k_1 + 2k_0 \pm \kappa) , \quad \kappa = \sqrt{k_1^2 + 4k_0^2} .$$

Prove that both eigenvalues $\lambda_\pm < 0$ and that the first phase and the free molecules completely disappear with time, while the second phase will consume all the molecules, i.e. $N_1(\infty) = N_f(\infty) = 0$, while $N_2(\infty) = N_0$. Explain this result.

1.22.3 Vibrations in Molecules

The notion of eigenvectors and eigenvalues is of huge importance in many areas of physics; quantum mechanics is almost entirely based on it. We shall illustrate this concept here using a classical example of atomic vibrations in a molecule.

From the point of view of classical mechanics atomic displacements $x_i(t)$ in a molecule (here i counts all its degrees of freedom) perform a complicated oscillatory motion in time that is governed by the equations of motion. These are derived from the potential energy $V(x_1, \ldots, x_n)$ which is obtained by assuming that atomic displacements x_i are small. Adopting that equilibrium geometry of the system is given by $\mathbf{x} = (x_i) = 0$, we expand the potential energy $V(x_1, \ldots, x_n)$ in a Taylor series around the point $\mathbf{x} = 0$:

$$V(x_1, \ldots, x_n) = V_0 + \sum_{i=1}^{n} \left(\frac{\partial V}{\partial x_i} \right)_{\mathbf{x}=0} x_i + \frac{1}{2} \sum_{i,j=1}^{n} \left(\frac{\partial^2 V}{\partial x_i \partial x_j} \right)_{\mathbf{x}=0} x_i x_j + \cdots .$$

Here n is the total number of degrees of freedom. Since the point $\mathbf{x} = 0$ corresponds to mechanical equilibrium, the first derivatives of the energy at this point, $(\partial V/\partial x_i)_{\mathbf{x}=0}$, must be equal to zero for every degree of freedom. Therefore, assuming small atomic displacements, we can keep the first non-zero (quadratic) term (this is called *the harmonic approximation*) and write

$$\begin{aligned} V(x_1, \ldots, x_n) &\simeq V_0 + \frac{1}{2} \sum_{i,j=1}^{n} \left(\frac{\partial^2 V}{\partial x_i \partial x_j} \right)_{\mathbf{x}=0} x_i x_j \\ &= V_0 + \frac{1}{2} \sum_{i,j=1}^{n} \Phi_{ij} x_i x_j . \end{aligned} \tag{1.217}$$

This is nothing but a quadratic form with respect to atomic displacements. The constants $\Phi_{ij} = \Phi_{ji}$ form a symmetric matrix, which is called the force constant matrix $\mathbf{\Phi} = (\Phi_{ij})$. The force acting on the degree of freedom i is given by

$$\begin{aligned} F_k &= -\frac{\partial V}{\partial x_k} = -\frac{\partial}{\partial x_k} \left(\frac{1}{2} \sum_{i,j=1}^{n} \Phi_{ij} x_i x_j \right) = -\frac{1}{2} \sum_{i,j=1}^{n} \Phi_{ij} \frac{\partial}{\partial x_k} (x_i x_j) \\ &= -\frac{1}{2} \sum_{i,j=1}^{n} \Phi_{ij} \left(\delta_{ik} x_j + x_i \delta_{jk} \right) = -\frac{1}{2} \left(\sum_{j=1}^{n} \Phi_{kj} x_j + \sum_{i=1}^{n} \Phi_{ik} x_i \right) \\ &= -\sum_{j=1}^{n} \Phi_{ij} x_j , \end{aligned}$$

so that the required equations of motion become

$$m_i \ddot{x}_i = F_i = -\sum_{j=1}^{n} \Phi_{ij} x_j , \tag{1.218}$$

where m_i is the mass associated with the degree of freedom i. The left-hand side gives mass times acceleration, and the notation $\ddot{x}_i$ means the second derivative with respect to time.

Introducing obvious vector and matrix notations,

$$\mathbf{x} = \begin{pmatrix} x_1 \\ \cdots \\ x_n \end{pmatrix}, \quad \mathbf{\Phi} = \begin{pmatrix} \Phi_{11} & \cdots & \Phi_{1n} \\ \cdots & \Phi_{ij} & \cdots \\ \Phi_{n1} & \cdots & \Phi_{nn} \end{pmatrix} \quad \text{and} \quad \mathbf{M} = \begin{pmatrix} m_1 & \cdots & 0 \\ \cdots & m_i & \cdots \\ 0 & \cdots & m_n \end{pmatrix},$$

where $\mathbf{M}$ is the diagonal matrix of masses, we can rewrite equations of motion (1.218) as a single matrix equation:

$$\mathbf{M\ddot{x}} = -\mathbf{\Phi x} \,. \tag{1.219}$$

In order to solve this equation, we first perform some matrix manipulations. Let us define matrices

$$\mathbf{M}^{1/2} = \begin{pmatrix} \sqrt{m_1} & \cdots & 0 \\ \cdots & \ddots & \cdots \\ 0 & \cdots & \sqrt{m_n} \end{pmatrix}, \quad \mathbf{M}^{-1/2} = \begin{pmatrix} 1/\sqrt{m_1} & \cdots & 0 \\ \cdots & \ddots & \cdots \\ 0 & \cdots & 1/\sqrt{m_n} \end{pmatrix}.$$

These are defined in such a way that $\mathbf{M}^{1/2}\mathbf{M}^{1/2} = \mathbf{M}$ and $\mathbf{M}^{1/2}\mathbf{M}^{-1/2} = \mathbf{E}$. Now, multiply from the left Eq. (1.219) by $\mathbf{M}^{-1/2}$ and insert the unit matrix $\mathbf{E} = \mathbf{M}^{-1/2}\mathbf{M}^{1/2}$ between $\mathbf{\Phi}$ and $\mathbf{x}$ in the right-hand side:

$$\mathbf{M}^{-1/2}\mathbf{M\ddot{x}} = -\left(\mathbf{M}^{-1/2}\mathbf{\Phi}\mathbf{M}^{-1/2}\right)\mathbf{M}^{1/2}\mathbf{x} \implies \left(\mathbf{M}^{1/2}\mathbf{\ddot{x}}\right) = -\mathbf{D}\left(\mathbf{M}^{1/2}\mathbf{x}\right) \,,$$

where

$$\mathbf{D} = \mathbf{M}^{-1/2}\mathbf{\Phi}\mathbf{M}^{-1/2} \tag{1.220}$$

is called the *dynamical matrix* of atomic vibrations. Note that the dynamical matrix is symmetric, its elements are given by

$$d_{ij} = \frac{1}{\sqrt{m_i m_j}}\left(\frac{\partial^2 V}{\partial x_i \partial x_j}\right)_{\mathbf{x}=0} \,.$$

Finally, introducing a vector $\mathbf{u} = \mathbf{M}^{1/2}\mathbf{x}$, we obtain

$$\mathbf{\ddot{u}} = -\mathbf{Du} \,, \tag{1.221}$$

that is the form considered at the end of Sect. 1.19.2.

This is a system of linear second-order differential equations. Since the motion is oscillatory, we anticipate that the displacements $u_i(t) \sim e^{i\omega t}$ in time. This also follows from the one-dimensional analogue of this equation, $\ddot{u} = -\omega^2 u$, which solution is $u(t) \sim e^{i\omega t}$. Therefore, we substitute into Eq. (1.219) a trial solution of the form $\mathbf{u}(t) = \mathbf{v}e^{i\omega t}$ with some unknown scalar ω and a constant vector $\mathbf{v}$ (see the end of Sect. 1.19.2). Since $\mathbf{\ddot{u}} = (i\omega)^2 \mathbf{v}e^{i\omega t} = -\omega^2 \mathbf{v}e^{i\omega t}$, we obtain the equation:

$$\mathbf{Dv} = \omega^2 \mathbf{v} \,. \tag{1.222}$$

This is the central result. It shows that the vibrational problem can be cast into an eigenvector/eigenvalue problem. Squares of the vibrational frequencies $\lambda^{(\alpha)} = \omega_\alpha^2$ appear as αth eigenvalue, and the corresponding eigenvector $\mathbf{v}^{(\alpha)}$ is directly related to atomic displacements via $\mathbf{x}^{(\alpha)} = \mathbf{M}^{-1/2}\mathbf{v}^{(\alpha)}$ which are called *normal modes*. These are collective (synchronised) displacements of many atoms of the system.

Since the matrix $\mathbf{D}$ is symmetric, the modal matrix of it, in which the vectors $\mathbf{v}^{(\alpha)} = \left(v_i^{(\alpha)}\right)$ are placed as its columns, is orthogonal. In turn, this means that its rows (or columns) form orthonormal sets of vectors. Let us adopt α as the column index and i as the row index of the modal matrix. If these conditions are written explicitly in components, the following identities are obtained

$$\sum_i v_i^{(\alpha)} v_i^{(\beta)} = \delta_{\alpha\beta} \quad \text{or} \quad \left(\mathbf{v}^{(\alpha)}\right)^T \mathbf{v}^{(\beta)} = \delta_{\alpha\beta} \text{ (orthonormality)}, \tag{1.223}$$

$$\sum_\alpha v_i^{(\alpha)} v_j^{(\alpha)} = \delta_{ij} \quad \text{or} \quad \sum_\alpha \mathbf{v}^{(\alpha)} \cdot \left(\mathbf{v}^{(\alpha)}\right)^T = \mathbf{E} \text{ (completeness)}. \tag{1.224}$$

The first of these corresponds to orthonormality of the vectors $\mathbf{v}^{(\alpha)}$ (due to orthonormality of the columns of the modal matrix), while the second to their completeness (orthonormality of the rows).

Further, from the theory of symmetric matrices (Sect. 1.8.4) we know that all eigenvalues are real and the eigenvectors $\mathbf{v}^{(\alpha)}$ form a linearly independent set. There are exactly n elementary solutions, $\alpha = 1, \ldots, n$, and their linear combination forms a *general solution* of Eqs. (1.219):

$$\mathbf{x}(t) = \frac{1}{2}\mathbf{M}^{-1/2} \sum_{\alpha=1}^n \mathbf{v}^{(\alpha)} \left(C_\alpha e^{i\omega_\alpha t} + C_\alpha^* e^{-i\omega_\alpha t}\right), \tag{1.225}$$

where the factor of one half was introduced for convenience. Since the matrix $\mathbf{D}$ is real, its eigenvectors $\mathbf{v}^{(\alpha)} = \left(v_i^{(\alpha)}\right)$ are also real. However, the coefficients C_α must be complex to ensure that the vector $\mathbf{x}(t)$ is real:

$$\mathbf{x}(t) = \sum_{\alpha=1}^n \mathbf{M}^{-1/2}\mathbf{v}^{(\alpha)} \text{Re} \left(C_\alpha e^{i\omega_\alpha t}\right) = \sum_{\alpha=1}^n \mathbf{M}^{-1/2}\mathbf{v}^{(\alpha)} g_\alpha(t), \tag{1.226}$$

where

$$g_\alpha(t) = \text{Re} \left(C_\alpha e^{i\omega_\alpha t}\right) = a_\alpha \cos(\omega_\alpha t) + b_\alpha \sin(\omega_\alpha t)$$

with a_α and b_α being two real arbitrary constants. The forms (1.225) and (1.226) chosen above ensures real solutions. The C_α (or a pair of a_α and b_α) are found from the initial ($t = 0$) conditions for atomic displacements and their velocities. There are $2n$ such conditions with n being the number of degrees of freedom, and we have exactly $2n$ real constants to determine (there are n complex constants C_α or n real pairs of a_α and b_α).

As we mentioned above, the eigenvalues ω_α^2 are always real which is guaranteed by the fact that the dynamical matrix $\mathbf{D}$ is symmetric. However, there is no guarantee that they are *positive*. If all eigenvalues (the frequencies squared) are positive, $\omega_\alpha^2 > 0$, then the frequencies are real and (can be chosen) positive, and the vibrational system is in a *stable* mechanical equilibrium. If there is at least one negative eigenvalue, $\omega_\alpha^2 < 0$, then the frequency $\omega_\alpha = \pm i \, |\omega_\alpha|$ is pure imaginary and the corresponding normal mode is no longer sinusoidal: $\mathbf{x}^{(\alpha)}(t) \sim e^{\mp |\omega_\alpha| t}$. This means that the system is *not* stable in this particular atomic configuration and will eventually transform to a different atomic arrangement (e.g. a molecule may dissociate, i.e. break into several parts).

These conclusions about stability can be also directly illustrated on the potential energy itself. As should be clear from Sect. 1.13, the potential energy (1.217) is a *quadratic form* of atomic displacements, which can be brought into a diagonal form (i.e. *diagonalised*) in terms of the normal modes:

$$
V - V_0 = \frac{1}{2}\mathbf{x}^T \boldsymbol{\Phi} \mathbf{x} = \frac{1}{2}\sum_{\alpha,\beta}\left(\mathbf{M}^{-1/2}\mathbf{v}^{(\alpha)}g_\alpha(t)\right)^T \boldsymbol{\Phi}\left(\mathbf{M}^{-1/2}\mathbf{v}^{(\beta)}g_\beta(t)\right)
$$

$$
= \frac{1}{2}\sum_{\alpha,\beta}g_\alpha(t)g_\beta(t)\left(\mathbf{v}^{(\alpha)}\right)^T\left(\mathbf{M}^{-1/2}\boldsymbol{\Phi}\mathbf{M}^{-1/2}\right)\mathbf{v}^{(\beta)}
$$

$$
= \frac{1}{2}\sum_{\alpha,\beta}g_\alpha(t)g_\beta(t)\left(\mathbf{v}^{(\alpha)}\right)^T\underbrace{\mathbf{D}\mathbf{v}^{(\beta)}}_{\omega_\beta^2\mathbf{v}^{(\beta)}} = \frac{1}{2}\sum_{\alpha,\beta}g_\alpha(t)g_\beta(t)\omega_\beta^2\left[\left(\mathbf{v}^{(\alpha)}\right)^T\mathbf{v}^{(\beta)}\right],
$$

where at the last step we made use of the fact that the vector $\mathbf{v}^{(\beta)}$ is an eigenvector of $\mathbf{D}$ with the eigenvalue ω_β^2, so that $\mathbf{D}\mathbf{v}^{(\beta)} = \omega_\beta^2\mathbf{v}^{(\beta)}$. Because of the orthogonality condition, Eq. (1.223), the expression in the square brackets above is equal to $\delta_{\alpha\beta}$, so that we finally obtain

$$
V = V_0 + \frac{1}{2}\mathbf{x}^\dagger \boldsymbol{\Phi} \mathbf{x} = V_0 + \frac{1}{2}\sum_{\alpha=1}^{n}\omega_\alpha^2 g_\alpha^2(t) . \tag{1.227}
$$

One can clearly see that if all eigenfrequencies are real, i.e. $\omega_\alpha^2 > 0$, then the current equilibrium state is indeed stable, i.e. the quadratic form, the potential energy (1.217), is positive definite since it can only increase due to non-zero atomic displacements. If, however, at least one of ω_α is complex, then $\omega_\alpha^2 < 0$ and the current state is not stable as there must be a displacement which would take the potential energy to a value smaller than V_0.

Problem 1.174 Consider a system of particles connected by springs and performing vibrations around their equilibrium positions. If particles coordinates and velocities are combined into vectors $\mathbf{q}$ and $\dot{\mathbf{q}}$, respectively, then the kinetic and potential energies of the system can be written as

$$E_K = \frac{1}{2}\dot{\mathbf{q}}^T \mathbf{K}\dot{\mathbf{q}} \quad \text{and} \quad E_P = \frac{1}{2}\mathbf{q}^T \mathbf{V}\mathbf{q} , \tag{1.228}$$

where $\mathbf{K}$ and $\mathbf{V}$ are symmetric square matrices. Show, considering the energy conservation condition, $E_K + E_P = \text{Const}$, that the motion of the particles is described by the matrix equation $\mathbf{K}\ddot{\mathbf{q}} + \mathbf{V}\mathbf{q} = 0$. Then, assuming an oscillatory motion of frequency ω, i.e. $\mathbf{q}(t) = \mathbf{x}e^{i\omega t}$, show that the oscillation frequencies of the system normal modes are determined by the equation $\left|\mathbf{V} - \omega^2 \mathbf{K}\right| = 0$.

Problem 1.175 Consider a linear symmetric triatomic molecule A–B–A with masses m, μm and m, respectively, see Fig. 1.11. If x_1, x_2 and x_3 are displacements of the atoms from their equilibrium positions along the molecular axis, then one can write the following expressions for the kinetic and potential energies of the system:

$$E_K = \frac{m}{2}\left(\dot{x}_1^2 + \mu\dot{x}_2^2 + \dot{x}_3^2\right) \quad \text{and} \quad E_P = \frac{k}{2}\left[(x_2 - x_1)^2 + (x_3 - x_2)^2\right] ,$$

where k is the elastic constant corresponding to interactions between neighbouring atoms. Assuming the vector $\mathbf{q}$ contains x_1, x_2 and x_3 as its components, rewrite E_K and E_P in the matrix form and hence obtain explicit expressions for the matrices $\mathbf{V}$ and $\mathbf{K}$ of Eq. (1.228). Show that the frequencies of normal vibrations of the molecule are

$$\omega_1 = \sqrt{\frac{k}{m}} , \quad \omega_2 = 0 \quad \text{and} \quad \omega_3 = \sqrt{\frac{k}{m}\left(1 + \frac{2}{\mu}\right)} . \tag{1.229}$$

Show that the corresponding eigenvectors for each mode are $\mathbf{q}_1 = (1, 0, -1)^T$, $\mathbf{q}_2 = (1, 1, 1)^T$ and $\mathbf{q}_3 = (1, -2/\mu, 1)^T$. Sketch them. What motion does the zero frequency mode correspond to?

Problem 1.176 Here we shall solve the previous problem differently. Using the condition that the centre of mass of the molecule is at rest at the origin, eliminate x_2 and thus rewrite both E_K and E_P in the matrix form as

$$E_K = \frac{m}{2}\dot{\mathbf{X}}^T \mathbf{K}\dot{\mathbf{X}} \quad \text{and} \quad E_P = \frac{k}{2}\mathbf{X}^T \mathbf{\Phi}\mathbf{X} ,$$

where

$$\mathbf{X} = \begin{pmatrix} x_1 \\ x_3 \end{pmatrix} \quad \text{and} \quad \dot{\mathbf{X}} = \begin{pmatrix} \dot{x}_1 \\ \dot{x}_3 \end{pmatrix} .$$

Obtain eigenvalues and eigenvectors of the 2×2 matrix $\mathbf{\Phi}$ and hence find the orthogonal transformation $\mathbf{U}$ that diagonalises $\mathbf{\Phi}$. Express new coordinates $\mathbf{Y} = \begin{pmatrix} y_1 \\ y_3 \end{pmatrix} = \mathbf{U}^\dagger \mathbf{X}$ via the old ones, $\mathbf{X}$. Demonstrate explicitly that the new coordinates (y_1, y_3) are no longer coupled in $\mathbf{V}$. Show then that the same orthogonal transformation diagonalises the matrix $\mathbf{K}$ of the kinetic energy as well. Show that the total energy $E = E_K + E_P$ of the molecule in the new coordinates is the sum of the energies of two independent harmonic oscillators. Hence, determine the two oscillation frequencies of the molecule ω_1 and ω_3. Make sure they are the same as in Eq. (1.229).

Problem 1.177 Consider a double compound pendulum (its equations of motion are derived in Problem 9.12). The two strings have the same length and the upper mass is equal three times the lower one. The strings make angles α_1 and α_2 with the vertical axis, and satisfy the following DEs:

$$4\ddot{\alpha}_1 + \ddot{\alpha}_2 = -4\omega_0^2 \alpha_1 \text{ and } \ddot{\alpha}_1 + \ddot{\alpha}_2 = -\omega_0^2 \alpha_2 .$$

(a) Write the two equations in the matrix form $\mathbf{D\ddot{X}} = \mathbf{CX}$ and then rearrange into $\mathbf{\ddot{X}} = \mathbf{AX}$. (b) Using the trial solutions of the form $\mathbf{X}(t) = \mathbf{Y}e^{i\omega t}$, demonstrate that $\mathbf{Y}$ and ω are obtained from the eigenproblem $\mathbf{AY} = -\omega^2 \mathbf{Y}$. (c) Find eigenvectors and eigenvalues of $\mathbf{A}$. (d) Write the general solution $\mathbf{X}(t)$. There should be four real arbitrary constants. (e) Then find the particular solution corresponding to the initial condition at which the whole pendulum was displaced by an angle α_0 and then released. [Answer:

$$\alpha_1(t) = \frac{3\alpha_0}{4} \cos(\omega_1 t) + \frac{\alpha_0}{4} \cos(\omega_2 t) , \quad \alpha_2(t) = \frac{3\alpha_0}{2} \cos(\omega_1 t) - \frac{\alpha_0}{2} \cos(\omega_2 t) ,$$

where $\omega_1 = \sqrt{\frac{2}{3}}\omega_0$ *and* $\omega_2 = \sqrt{2}\omega_0$.]

There is another, rather formal, method of solving vibrational Eq. (1.221) that enables one to obtain the whole general solution algebraically. It is based on the spectral decomposition (1.107) of the dynamical matrix $\mathbf{D}$ via its eigenvectors $\mathbf{v}^{(\alpha)}$ and eigenvalues ω_α^2 and is considered in Problem 1.156.

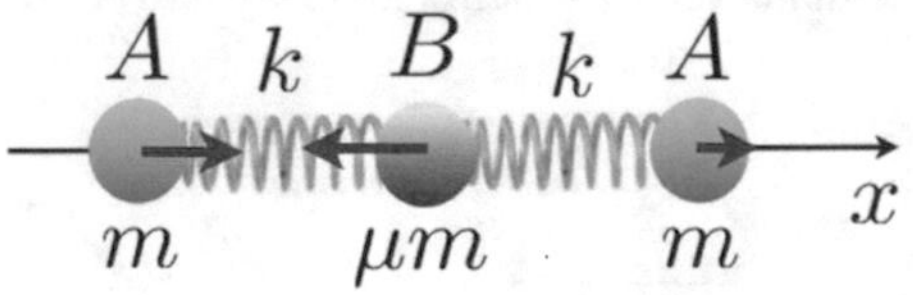

Fig. 1.11 Three atoms connected with springs

1.22.4   Least Square Method and MD Simulations

1.22.4.1   The Least Square Method and Pseudo-Inverse

Consider the following problem: given some 'observation' data forming a vector $\mathbf{y} = (y_i)$ and a known 'theoretical' linear dependence between the data and the values of the 'sources' $\mathbf{x} = (x_j)$ via $\mathbf{y} = \mathbf{A}\mathbf{x}$, where $\mathbf{A} = (a_{ij})$ is the known matrix, determine $\mathbf{x}$. In other words, we would like to solve the system of linear algebraic equations

$$\begin{cases} a_{11}x_1 + a_{12}x_2 + \cdots + a_{1n}x_n & = y_1 \\ a_{21}x_1 + a_{22}x_2 + \cdots + a_{2n}x_n & = y_2 \\ \quad \cdots & \quad \cdots \\ a_{m1}x_1 + a_{m2}x_2 + \cdots + a_{mn}x_n & = y_m \end{cases} . \tag{1.230}$$

Note that here m, the number of equations, could be either larger or smaller than the number n of the unknown $\{x_i\}$, and hence the system, strictly speaking, may either have no solution (if $m > n$, i.e. the system is over-defined) or have an infinite number of solutions (if $m < n$, the system is under-defined). In the case of $m = n$ we may either have a unique solution (if $\det \mathbf{A} \neq 0$), or have an infinite number of them (if $\det \mathbf{A} = 0$).

If the 'exact' solution does not exist or there are too many of them, the best strategy of solving these equations would be to find the' best' one that minimises the error function

$$Q(\mathbf{x}) = \sum_{i=1}^{m} \left(y_i - \sum_{j=1}^{n} a_{ij}x_j \right)^2 .$$

This could also be written as the length $|\mathbf{y} - \mathbf{A}\mathbf{x}|$ of the vector $\mathbf{y} - \mathbf{A}\mathbf{x}$ squared,[24] i.e. $Q(\mathbf{x}) = |\mathbf{y} - \mathbf{A}\mathbf{x}|^2$. We would like to determine the best vector $\mathbf{x} = (x_j)$ that minimises Q. Let us denote that vector $\mathbf{x}^0 = (x_i^0)$. We shall show now that such a vector exists and it is unique subject to an additional condition of its length $|\mathbf{x}^0|$ being minimal as well.

Let $\mathbf{x}$ be an arbitrary vector, and let $\mathbf{x}_1$ be a vector that minimises $Q(\mathbf{x})$. Without loss of generality, we can write this vector as

$$\mathbf{x}_1 = \mathbf{x}^0 + \mathbf{g}, \quad \text{where } \mathbf{x}^0 = \mathbf{A}^+\mathbf{y}, \quad \mathbf{g} = \left(\mathbf{E} - \mathbf{A}^+\mathbf{A}\right)\mathbf{w} \tag{1.231}$$

with the vector $\mathbf{w}$ being arbitrary and $\mathbf{A}^+$ being the pseudo-inverse of $\mathbf{A}$ (Sect. 1.20).

---

[24] This is the square of the so-called Euclidean norm $\|\mathbf{z}\|_E = \sqrt{\mathbf{z}^\dagger \mathbf{z}}$ of the vector $\mathbf{z}$, i.e. $Q(\mathbf{x}) = \|\mathbf{y} - \mathbf{A}\mathbf{x}\|_E^2$,

Consider the error function in more detail:

$$Q(\mathbf{x}) = (\mathbf{y} - \mathbf{Ax})^\dagger (\mathbf{y} - \mathbf{Ax})$$
$$= \left[(\mathbf{y} - \mathbf{Ax}_1) + \mathbf{A}(\mathbf{x}_1 - \mathbf{x}) \right]^\dagger \left[(\mathbf{y} - \mathbf{Ax}_1) + \mathbf{A}(\mathbf{x}_1 - \mathbf{x}) \right]$$
$$= |\mathbf{y} - \mathbf{Ax}_1|^2 + |\mathbf{A}(\mathbf{x}_1 - \mathbf{x})|^2 + \left[(\mathbf{Ax}_1 - \mathbf{Ax})^\dagger (\mathbf{y} - \mathbf{Ax}_1) + \text{h.c.} \right],$$

where 'h.c.' corresponds to adding the Hermitian conjugate term. Firstly, we can simplify it by noticing that

$$\mathbf{Ax}_1 = \mathbf{Ax}^0 + \mathbf{Ag} = \mathbf{Ax}^0 + \underbrace{\left(\mathbf{A} - \mathbf{AA}^+\mathbf{A} \right)}_{=0} \mathbf{w} = \mathbf{Ax}^0,$$

leading to

$$|\mathbf{y} - \mathbf{Ax}|^2 = \left| \mathbf{y} - \mathbf{Ax}^0 \right|^2 + \left| \mathbf{A}(\mathbf{x}^0 - \mathbf{x}) \right|^2 + \left[(\mathbf{Ax}^0 - \mathbf{Ax})^\dagger (\mathbf{y} - \mathbf{Ax}^0) + \text{h.c.} \right].$$

Let us show that both terms within the square brackets are zero. We have for the first one:

$$\left(\mathbf{Ax}^0 - \mathbf{Ax} \right)^\dagger \left(\mathbf{y} - \mathbf{Ax}^0 \right) = \left(\mathbf{AA}^+\mathbf{y} - \mathbf{Ax} \right)^\dagger \left(\mathbf{y} - \mathbf{AA}^+\mathbf{y} \right)$$
$$= \left((\mathbf{A}^+\mathbf{y})^\dagger \mathbf{A}^\dagger - \mathbf{x}^\dagger \mathbf{A}^\dagger \right) \left(\mathbf{y} - \mathbf{AA}^+\mathbf{y} \right)$$
$$= \left((\mathbf{A}^+\mathbf{y})^\dagger - \mathbf{x}^\dagger \right) \mathbf{A}^\dagger \left(\mathbf{E} - \mathbf{AA}^+ \right) \mathbf{y} = \mathbf{0},$$

since $\mathbf{A}^\dagger \left(\mathbf{E} - \mathbf{AA}^+ \right) = \mathbf{A}^\dagger - \mathbf{A}^\dagger\mathbf{AA}^+ = \mathbf{0}$ according to Eq. (1.204). Of course, the other term, which is the Hermitian conjugate of this one, is zero as well. We thus obtain that

$$Q(\mathbf{x}) = |\mathbf{y} - \mathbf{Ax}|^2 = \left| \mathbf{y} - \mathbf{Ax}^0 \right|^2 + \left| \mathbf{A}(\mathbf{x}^0 - \mathbf{x}) \right|^2 \geq \left| \mathbf{y} - \mathbf{Ax}^0 \right|^2 = Q(\mathbf{x}^0),$$

i.e. the error function reaches minimum at $\mathbf{x} = \mathbf{x}^0$. Note, however, that since

$$Q(\mathbf{x}_1) = |\mathbf{y} - \mathbf{Ax}_1|^2 = \left| \mathbf{y} - \mathbf{Ax}^0 \right|^2 \equiv Q(\mathbf{x}^0),$$

the same minimum is reached by any vector $\mathbf{x}_1$ from Eq. (1.231) with arbitrary $\mathbf{w}$.

It is possible, however, to pick up the unique solution if we require, in addition, that the vector $\mathbf{x}_1$ of the solution would have the smallest length. The length of the vector $\mathbf{x}_1$ squared is

$$|\mathbf{x}_1|^2 = \mathbf{x}_1^\dagger \mathbf{x}_1 = \left(\mathbf{x}^0 + \mathbf{g} \right)^\dagger \left(\mathbf{x}^0 + \mathbf{g} \right) = \left| \mathbf{x}^0 \right|^2 + |\mathbf{g}|^2 + \left[\left(\mathbf{x}^0 \right)^\dagger \mathbf{g} + \text{h.c.} \right].$$

It is easy to see that the expression in the square brackets is zero; indeed, the first term there,

$$\left(\mathbf{x}^0\right)^\dagger \mathbf{g} = \left(\mathbf{A}^+\mathbf{y}\right)^\dagger \left(\mathbf{E} - \mathbf{A}^+\mathbf{A}\right)\mathbf{w} = \mathbf{y}^\dagger \left(\mathbf{A}^+\right)^\dagger \left(\mathbf{E} - \mathbf{A}^+\mathbf{A}\right)\mathbf{w}$$
$$= \mathbf{y}^\dagger \underbrace{\left(\left(\mathbf{A}^+\right)^\dagger - \left(\mathbf{A}^+\right)^\dagger \mathbf{A}^+\mathbf{A}\right)}_{=0}\mathbf{w} = \mathbf{0},$$

where we have made use of Eq. (1.205) in the last step. Therefore,

$$|\mathbf{x}_1|^2 = \left|\mathbf{x}^0\right|^2 + |\mathbf{g}|^2 \geq \left|\mathbf{x}^0\right|^2 .$$

Hence, the unique solution, $\mathbf{x}^0$, has the minimum possible length that is reached by taking $\mathbf{w} = \mathbf{0}$ in $\mathbf{x}_1$.

Sometimes in applications one has to consider a certain generalisation of the above consideration. Indeed, above we have established the unique solution $\mathbf{x} = \mathbf{A}^+\mathbf{y}$ for the vector $\mathbf{x}$ that appears in the equation $\mathbf{y} = \mathbf{A}\mathbf{x}$, and the solution was based on minimising the error function $Q(\mathbf{x}) = |\mathbf{y} - \mathbf{A}\mathbf{x}|^2$. Suppose, instead we would like to solve the *matrix* equation $\mathbf{Y} = \mathbf{A}\mathbf{X}$ with respect to the $m \times p$ matrix $\mathbf{X} = \left(x_{ij}\right)$, where $\mathbf{A} = (a_{ki})$ is a $n \times m$ matrix and $\mathbf{Y} = \left(y_{kj}\right)$ is a $n \times p$ matrix, both known matrices. Since 'the length of a matrix' cannot be defined, a different definition of the error function is required in this case. It is appropriate to choose here, as the relevant error function, the sum of errors due to all matrix elements of the matrix $\mathbf{Y} - \mathbf{A}\mathbf{X}$, i.e.[25]

$$Q(\mathbf{X}) = \sum_{j=1}^{p} \sum_{k=1}^{n} \left| y_{kj} - \sum_{i=1}^{m} a_{ki} x_{ij} \right|^2 = \sum_{j=1}^{p} \left| \mathbf{y}_j - \mathbf{A}\mathbf{x}_j \right|^2 , \qquad (1.232)$$

where $\mathbf{y}_j = \left(y_{kj}\right)$ is the jth column of the matrix $\mathbf{Y} = \left(\mathbf{y}_1 \ \mathbf{y}_2 \ \ldots \ \mathbf{y}_p\right)$ and $\mathbf{x}_j = \left(x_{ij}\right)$ is the jth column of the matrix $\mathbf{X} = \left(\mathbf{x}_1 \ \mathbf{x}_2 \ \ldots \ \mathbf{x}_p\right)$. It is seen that each column $\mathbf{x}_j$ of the matrix $\mathbf{X}$ is determined independently as the form (1.232) is a sum of the error functions $Q_j\left(\mathbf{x}_j\right) = \left|\mathbf{y}_j - \mathbf{A}\mathbf{x}_j\right|^2$ associated with the columns of $\mathbf{X}$. But we already know that the minimum of each $Q_j(\mathbf{x}_j)$ is achieved at $\mathbf{x}_j^0 = \mathbf{A}^+\mathbf{y}_j$, i.e.

$$x_{ij}^0 = \sum_{k=1}^{n} \left(\mathbf{A}^+\right)_{ik} y_{kj} , \quad \text{for all } i = 1, \ldots, m \text{ and } j = 1, \ldots, p .$$

Then it immediately follows that the full solution must be $\mathbf{X}^0 = \mathbf{A}^+\mathbf{Y}$. If $\mathbf{A}$ was a square matrix, then from $\mathbf{A}\mathbf{X} = \mathbf{Y}$ we would obtain $\mathbf{X} = \mathbf{A}^{-1}\mathbf{Y}$. Hence, $\mathbf{A}^+$ serves as a generalisation of the inverse of a matrix for the general case of non-square (and singular) matrices.

In the application considered in the next subsection we shall need to determine the matrix $\mathbf{A}$ from the equation $\mathbf{A}\mathbf{X} = \mathbf{Y}$ with known matrices $\mathbf{X}$ and $\mathbf{Y}$. It is easy to see,

---

[25] In mathematics the square root of the sum $\sum_{i,j} \left|a_{ij}\right|^2 = \text{Tr}\left(\mathbf{A}^\dagger\mathbf{A}\right)$ of all matrix elements a_{ij} of the matrix $\mathbf{A}$ is called Frobenius norm $\|\mathbf{A}\|_F$; it is a generalisation of the Euclidean norm $\|\mathbf{z}\|_E = \sqrt{\mathbf{z}^\dagger\mathbf{z}}$ defined for the vector $\mathbf{z}$ and used earlier as the error function.

however, that this problem can be converted to the one that we have just considered by taking the Hermitian conjugate of this equation, $\mathbf{X}^\dagger \mathbf{A}^\dagger = \mathbf{Y}^\dagger$, and hence obtaining for the 'best' matrix $\mathbf{A}$ the solution:

$$\mathbf{A}^\dagger = \left(\mathbf{X}^\dagger\right)^+ \mathbf{Y}^\dagger = \left(\mathbf{X}^+\right)^\dagger \mathbf{Y}^\dagger \implies \mathbf{A} = \mathbf{Y}\mathbf{X}^+. \tag{1.233}$$

A use have been made here of Eq. (1.202).

1.22.4.2 Determining Vibrations from Molecular Dynamics Simulations

As follows from Eq. (1.217), the potential energy, e.g. of a vibrating molecule, can be (up to an irrelevant constant) written as $E = \frac{1}{2}\mathbf{u}^T \mathbf{\Phi} \mathbf{u}$, where $\mathbf{u} = (u_i)$ is the vector-column of atomic displacements (i numerates all participating degrees of freedom) and $\mathbf{\Phi} = (\Phi_{ij})$ is the force constant matrix. Suppose, that one would like to determine this matrix, but the direct calculation of the second derivatives of the potential energy with respect to the atomic displacements is either not possible or difficult; it may also be that the equilibrium geometry of the molecule is not yet known. At the same time, for any atomic arrangement, it is possible to calculate atomic forces $\mathbf{f} = (f_i)$ acting on each degree of freedom. In that case one can run on a computer the so-called molecular dynamics (MD) simulations in which atomic positions $\mathbf{r} = (r_i)$ are changed in time t by solving discretised Newtonian equations of motion $m_i \ddot{r}_i = f_i$ (all i). That means that atomic positions $\mathbf{r} = (r_i)$ and velocities $\mathbf{v} = (v_i) = (\dot{r}_i)$ are recalculated over a very large number of small time steps Δt. One starts from particular initial positions and velocities, $(\mathbf{r}(0), \mathbf{v}(0)) = \{r_i(0), v_i(0)\}$, then their values are updated after propagating the atoms over time Δt to $\{r_i(\Delta t), v_i(\Delta t)\}$ using the current forces $(f_i(\Delta t))$. Using the new (updated) positions, the forces are recalculated, and the positions and velocities updated again, $\{r_i(2\Delta t), v_i(2\Delta t)\}$. This procedure is repeated many times until the required simulation time t_{final} is achieved. In many cases a huge number of MD time steps can be performed (millions).

One advantage of this procedure is that after a certain number of MD steps the atoms in the molecule would start vibrating around the true equilibrium geometry specified by the vector $\mathbf{r}^0 = (r_i^0)$, i.e. their positions $r_i = r_i^0 + u_i$ would only change by small atomic displacements $\mathbf{u} = (u_i)$ from the equilibrium positions. When this stage in the MD simulations is reached, it is possible, by continuing the MD run, to determine numerically the force constant matrix $\mathbf{\Phi}$. It is essential that the whole procedure does not require any *a priory* knowledge of the equilibrium geometry at all.[26]

---

[26] If the reader is interested to learn more about the details of this method, especially in application to crystalline solids, they need to look at the original literature, namely: (i) O. Hellman, I. A. Abrikosov, and S. I. Simak, Lattice dynamics of anharmonic solids from first principles.—Phys. Rev. **B 84**, 180301(R) (2011), and (ii) Olle Hellman, Peter Steneteg, I. A. Abrikosov, and S. I. Simak, Temperature dependent effective potential method for accurate free energy calculations of solids.—Phys. Rev. **B 87**, 104111 (2013).

The idea is based on recognising that over many MD steps one accumulates the vectors $\mathbf{f}^{(k)} = \left(f_i^{(k)} \right)$ of atomic forces due to atomic positions $\mathbf{r}^{(k)} = \left(r_i^{(k)} \right) = \mathbf{r}^0 + \mathbf{u}^{(k)}$ at each MD step k, and there could be a very large number of forces-displacements pairs available (due to a very large number of MD steps). On the other hand, the forces are related to displacements via $\mathbf{f} = -\mathbf{\Phi}\mathbf{u}$. Hence, considering all available forces $\left\{ \mathbf{f}^{(k)} \right\}$ and corresponding to them atomic displacements $\left\{ \mathbf{u}^{(k)} \right\}$ for all MD steps, one may determine the force constant matrix $\mathbf{\Phi}$ by minimising the error function

$$Q(\mathbf{\Phi}) = \sum_k \left| \mathbf{f}^{(k)} + \mathbf{\Phi}\mathbf{u}^{(k)} \right|^2 = \sum_k \left(\mathbf{f}^{(k)} + \mathbf{\Phi}\mathbf{u}^{(k)} \right)^2 ,$$

where we sum over all MD steps and used the fact that all quantities are real. Note that a relatively small number of elements of $\mathbf{\Phi}$ (this is a symmetric $n \times n$ matrix) is to be determined from a very large number of known forces and displacements, so that the problem is over-defined and the exact determination of $\mathbf{\Phi}$ is impossible. However, this is precisely the same problem that we have solved at the end of the previous subsection. Hence, the matrix $\mathbf{\Phi}$ can be determined as the 'best' solution of the matrix equation $-\mathbf{\Phi}\mathbf{U} = \mathbf{F}$, where $\mathbf{F} = \left(\mathbf{f}^{(1)} \, \mathbf{f}^{(2)} \ldots \mathbf{f}^{(N)} \right)$ is the matrix all forces accumulated after N MD steps with the forces from each MD step forming its columns, and $\mathbf{U} = \left(\mathbf{u}^{(1)} \, \mathbf{u}^{(2)} \ldots \mathbf{u}^{(N)} \right)$ is similarly constructed matrix of the atomic displacements. We finally obtain from Eq. (1.233) that the force constant matrix can be determined via

$$\mathbf{\Phi} = -\mathbf{F}\mathbf{U}^+ = -\left(\mathbf{f}^{(1)} \, \mathbf{f}^{(2)} \ldots \mathbf{f}^{(N)} \right) \left(\mathbf{u}^{(1)} \, \mathbf{u}^{(2)} \ldots \mathbf{u}^{(N)} \right)^+ .$$

Hence, for the numerical calculation of the force constant matrix, it is required to form the matrix $\mathbf{U}$ of atomic displacements, calculate its pseudo-inverse and then multiply it with the matrix of the atomic forces.

1.22.5 Vibrations of Atoms in an Infinite Chain. A Point Defect

Consider a one-dimensional chain of identical atoms of mass m connected with springs of elastic constant k, see Fig. 1.12a. At equilibrium, all atoms are equidistant with the distance a between them. At non-zero temperature atoms vibrate around their equilibrium positions.

Let x_n be the instantaneous displacement of atom n from its equilibrium position. Then the distance between atoms n and $n-1$ will be $\Delta_- = a + x_n - x_{n-1}$, while the distance between atom n and its another neighbour $n+1$ is $\Delta_+ = a + x_{n+1} - x_n$. Correspondingly, the force acting on atom n is composed of the force $F_+ = k(\Delta_+ - a) = k(x_{n+1} - x_n)$ due to its right neighbour (number $n+1$), and of the

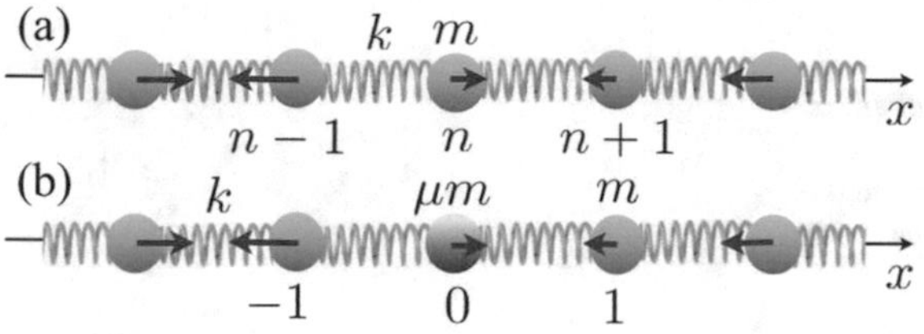

1.22 Examples in Physics

Fig. 1.12 a An infinite chain of identical atoms of mass m connected with identical springs; the atoms are numbered by an integer index n running between $-\infty$ and $+\infty$. **b** The same chain, but the atom with $n = 0$ was replaced with the one having a different mass μm

force $F_- = -k (\Delta_- - a) = -k (x_n - x_{n-1})$ due to its left neighbour (number $n - 1$). Therefore, we can write down an equation of motion for atom n as follows:

$$m\ddot{x}_n = F_+ + F_- = k (x_{n-1} - 2x_n + x_{n+1}) ,$$

where $n = -\infty, \ldots, -1, 0, 1, \ldots, \infty$. Introducing a vector $\mathbf{X} = (x_n)$ of atomic displacements, these equations are rewritten in the matrix form $\ddot{\mathbf{X}} = -\mathbf{DX}$, i.e. in more detail:

$$
\frac{d^2}{dt^2}
\begin{pmatrix}
\cdots \\
x_{n-1} \\
x_n \\
x_{n+1} \\
\cdots
\end{pmatrix}
= -
\begin{pmatrix}
\ddots & \ddots & \ddots & & \\
& -\lambda & 2\lambda & -\lambda & \\
& & -\lambda & 2\lambda & -\lambda \\
& & & -\lambda & 2\lambda & -\lambda \\
& & & & \ddots & \ddots & \ddots
\end{pmatrix}
\begin{pmatrix}
\cdots \\
x_{n-1} \\
x_n \\
x_{n+1} \\
\cdots
\end{pmatrix},
\tag{1.234}
$$

where $\lambda = k/m$. The infinite dimension matrix $\mathbf{D}$ has a tridiagonal form, i.e. it has non-zero elements only on the diagonal itself as well as one element to the left and one to the right. The matrix $\mathbf{D}$ is symmetric with the elements $d_{ij} = 2\lambda\delta_{ij} - \lambda (\delta_{i,j+1} + \delta_{i,j-1})$. We notice also that d_{ij} depends only on the difference of indices $i - j$, but not on both indices themselves. This is due to periodicity of the system at equilibrium. Hence, we can write d_{ij} simply as d_{i-j}, where $d_n = \lambda (2\delta_{n0} - \delta_{n1} - \delta_{n,-1})$.

Now we shall try to solve the equations $\ddot{\mathbf{X}} = -\mathbf{DX}$. To do this, we shall introduce the so-called periodic boundary conditions: we shall say that the chain repeats itself after a very large number N of atoms, i.e. $x_{n+N} = x_n$ for any n between 0 and $N - 1$. This can be imagined in such a way that atom N would coincide with atom 0, i.e. the chain of N atoms is connected to itself forming a ring as depicted in Fig. 1.13. This trick allows us to form a set of N (which is very-very large, but finite) set of Eqs. (1.234) which we shall now attempt to solve.

Using the method of the previous section, we shall attempt a substitution $\mathbf{X}(t) = \mathbf{Y}e^{i\omega t}$, which results in the eigenvector–eigenvalue problem

Fig. 1.13 Boundary
conditions for a
one-dimensional chain of
atoms: N atoms can be
thought of as connected on
themselves forming a ring

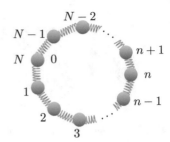

$$\mathbf{DY} = \omega^2 \mathbf{Y} \quad \text{or} \quad \sum_j d_{nj} y_j = \omega^2 y_n . \tag{1.235}$$

To find ω and $\mathbf{Y}$, we shall perform a discrete Fourier Transform of the atomic displacements (see Problem 1.39 and Eq. (1.42)) to diagonalise the matrix $\mathbf{D}$. We have

$$\mathbf{DY} = \omega^2 \mathbf{Y} \implies \left(\mathbf{U}^\dagger \mathbf{DU}\right)\left(\mathbf{U}^\dagger \mathbf{Y}\right) = \omega^2 \left(\mathbf{U}^\dagger \mathbf{Y}\right) \implies \mathcal{D} \mathcal{Y} = \omega^2 \mathcal{Y} . \tag{1.236}$$

In other words, we apply a unitary transformation matrix $\mathbf{U} = \left(u_{nj}\right)$, where $u_{nj} = N^{-1/2} \exp\left(-i2\pi nj/N\right)$, such that elements of the matrix $\mathcal{D}$ become

$$(\mathcal{D})_{js} = \left(\mathbf{U}^\dagger \mathbf{DU}\right)_{js} = \sum_{l,n=0}^{N-1} u_{lj}^* d_{ln} u_{ns} = \frac{1}{N} \sum_{ln} d_{l-n} e^{i2\pi lj/N} e^{-i2\pi ns/N} .$$

Since d_{l-n} depends only on the difference of indices, we can introduce a new index $p = l - n$ to replace n, which yields

$$
\begin{aligned}
(\mathcal{D})_{js} &= \frac{1}{N} \sum_{lp} d_p e^{i2\pi lj/N} e^{-i2\pi(l-p)s/N} \\
&= \left(\sum_{p=0}^{N-1} d_p e^{i2\pi sp/N}\right)\left(\frac{1}{N} \sum_{l=0}^{N-1} e^{i2\pi l(j-s)/N}\right) = \tilde{d}_s \delta_{js} .
\end{aligned}
\tag{1.237}
$$

Here the Kronecker symbol appeared since in the second bracket we simply have $\left(\mathbf{U}^\dagger \mathbf{U}\right)_{js}$ written explicitly (the matrix $\mathbf{U}$ is unitary and hence $\mathbf{U}^\dagger \mathbf{U} = \mathbf{E}$). We have also introduced a new quantity

$$
\begin{aligned}
\tilde{d}_s &= \sum_{p=0}^{N-1} d_p e^{i2\pi sp/N} = \sum_{p=0}^{N-1} \lambda \left(2\delta_{p0} - \delta_{p1} - \delta_{p,-1}\right) e^{i2\pi sp/N} \\
&= \lambda \left[2 - \left(e^{i2\pi s/N} + e^{-i2\pi s/N}\right)\right] = 2\lambda \left(1 - \cos\frac{2\pi s}{N}\right) .
\end{aligned}
\tag{1.238}
$$

Note that, because of the ring structure, the value of $p = -1$ (even though formally not contained in the sum) actually corresponds to the atom $N - 1$, and hence need to be accounted for. We see from Eq. (1.237) that the matrix $\mathcal{D}$ became diagonal after the similarity transformation, and hence we can immediately get the required eigenvalues appearing in (1.236) as

$$\omega^2 = 2\lambda \left(1 - \cos \frac{2\pi s}{N}\right) = \frac{2k}{m} \left(1 - \cos \frac{2\pi s}{N}\right) = \frac{2k}{m} \left(1 - \cos qa\right)$$

or

$$\omega(q) = \sqrt{\frac{4k}{m}} \left|\sin \frac{qa}{2}\right| .$$

There are N eigenvalues corresponding to N possible values of the index $s = 0, 1, \ldots, N - 1$; however, it is more convenient to introduce $q = 2\pi s/aN$ (which is the wavevector) instead of s to label different solutions. It changes between 0 (when $s = 0$) and $2\pi/a$ (when $s = N - 1 \simeq N$ as $N \gg 1$). Since the chain is a ring, we can alternatively consider the values of q between $-\pi/a$ and π/a. This interval is called the Brillouin zone. The wavevector almost continuously changes within the Brillouin zone between these two values as N is very large, i.e. the nearest values of q differ only by $\Delta q = 2\pi/Na = 2\pi/L$, where $L = Na$ is the length of the ring. And the vibrational frequencies $\omega(q)$ change between zero and the value of $\sqrt{4k/m}$ at the Brillouin zone boundaries (at $q = \pm\pi/a$). The dependence of the oscillation frequency $\omega(q)$ of the chain on the wavevector q is called the dispersion relation.

Once we obtained all N eigenvalues, $\omega(q)$, we can calculate the corresponding to them eigenvectors. The simplest choice of orthogonal eigenvectors[27] is to consider the corresponding to $\omega(q)$ eigenvector $\mathcal{Y}^{(q)}$ as a vector with all components equal to zero apart from the sth one which is equal to 1 (here $q = 2\pi s/aN$ and s are directly related):

$$\mathcal{Y}^{(q)} = (\ldots, 0, 1, 0, \ldots)^T , \quad \text{i.e.} \quad \left(\mathcal{Y}^{(q)}\right)_j = \delta_{js} ,$$

where $q = \frac{2\pi s}{L} = s\Delta q$. Then, for $\mathbf{Y}^{(q)} = \mathbf{U}\mathcal{Y}^{(q)}$ we have

$$\left(\mathbf{Y}^{(q)}\right)_n = \sum_{j=0}^{N-1} \frac{1}{\sqrt{N}} e^{-i2\pi nj/N} \left(\mathcal{Y}^{(q)}\right)_j = \frac{1}{\sqrt{N}} e^{-i2\pi ns/N} = \frac{1}{\sqrt{N}} e^{-iqna} ,$$

which immediately gives us the required solution:

$$\left(\mathbf{X}^{(q)}\right)_n = \left(\mathbf{Y}^{(q)}\right)_n e^{i\omega(q)t} = \frac{1}{\sqrt{N}} e^{-iqna} e^{i\omega(q)t} . \tag{1.239}$$

---

[27] As the matrix $\mathcal{D}$ is diagonal and hence Hermitian, this choice is always possible.

The problem which we have just solved corresponds to a periodic chain: all atoms in the chain are identical and repeat themselves after a 'translation' by the distance a (the distance between atoms). In practice, one is sometimes concerned with solving a much more difficult problem of defective systems which have no periodicity. However, before studying such systems, it is instructive first to calculate a special auxiliary object, called the phonon Green's function. It is defined as a resolvent of the dynamical matrix ($\mathbf{D}$ in our case):

$$\mathbf{G}(z) = (z\mathbf{E} - \mathbf{D})^{-1} = \sum_s \frac{\mathbf{y}_s \mathbf{y}_s^\dagger}{z - \omega_s^2} , \tag{1.240}$$

where $\mathbf{y}_s$ (which is the same as $\mathbf{Y}^{(q)}$) and ω_s^2 are the s-th (or qth) eigenvector and eigenvalue of the matrix $\mathbf{D}$, i.e. $\mathbf{D}\mathbf{y}_s = \omega_s^2 \mathbf{y}_s$, and z is a (generally) complex number (see also Eq. (1.113)); z is sometimes called complex 'energy'. For the periodic chain the index s counts different solutions of the eigenproblem, but we can use q for that instead. Using the above results, we write for the elements of the matrix $\mathbf{G}(z)$:

$$g_{nj}(z) = \sum_q \frac{\left(\mathbf{Y}^{(q)}\right)_n \left(\mathbf{Y}^{(q)}\right)_j^*}{z - \omega^2(q)} = \frac{1}{N} \sum_q \frac{1}{z - \omega^2(q)} e^{-iq(n-j)a} . \tag{1.241}$$

Equipped with Green's function of the ideal (perfect) chain, we can now consider a more difficult problem of a defective chain. As the simplest example, let us have a look at the chain in which a single 0th atom of mass m was replaced with an isotope of different mass μm, as in Fig. 1.12b. Since the isotope is chemically identical, the same spring constants can be used as for the ideal chain. The same equations of motion

$$m\ddot{x}_n = k\left(x_{n-1} - 2x_n + x_{n+1}\right) , \quad n = -\infty, \ldots, -1, 1, \ldots, \infty ,$$

can be written for all atoms apart from the one with $n = 0$ (the isotope), for which we have instead:

$$\mu m \ddot{x}_0 = k\left(x_{-1} - 2x_0 + x_1\right) \implies m\ddot{x}_0 = k\left(x_{-1} - 2x_0 + x_1\right) + m\left(1 - \mu\right)\ddot{x}_0 .$$

Therefore, all linear differential equations we have to solve can now be written as

$$m\ddot{x}_n = k\left(x_{n-1} - 2x_n + x_{n+1}\right) + m\left(1 - \mu\right)\ddot{x}_0 \delta_{n0} , \quad n = 0, \pm 1, \ldots, \pm\infty .$$

They differ from the equations for the perfect chain by the second term in the right-hand side which plays the role of a perturbation. Introducing now the same notations we used for the periodic chain, we make the substitution $x_n(t) = y_n(t)e^{i\omega t}$, which enable us to rewrite our equations as follows:

$$\omega^2 y_n = \sum_{j=0}^{N-1} d_{n-j} y_j + (1 - \mu)\, \omega^2 y_0 \delta_{n0}$$

$$\Longrightarrow \quad \omega^2 \mathbf{Y} = \mathbf{D}\mathbf{Y} + \mathbf{W}\mathbf{Y} \tag{1.242}$$

$$\Longrightarrow \quad \left(\omega^2 \mathbf{E} - \mathbf{D} - \mathbf{W}\right)\mathbf{Y} = \mathbf{0}\,,$$

where the 'perturbation' matrix $\mathbf{W}$ was introduced, which has a single non-zero element $W_{00}(\omega) = (1 - \mu)\,\omega^2$. Note that it depends on the frequency explicitly. To solve the above equation, we shall rewrite it for a general complex z (i.e. we replace ω^2 with z) as follows:

$$(z\mathbf{E} - \mathbf{D} - \mathbf{W}(z))\,\mathbf{Y} = \mathbf{0} \quad \Longrightarrow \quad (z\mathbf{E} - \mathbf{D})\,(\mathbf{E} - \mathbf{G}(z)\mathbf{W}(z))\,\mathbf{Y} = \mathbf{0}\,, \tag{1.243}$$

where we have made use of Green's function $\mathbf{G}(z) = (z\mathbf{E} - \mathbf{D})^{-1}$ of the perfect chain. Note that the values of z above correspond to the solutions of the equation

$$|z\mathbf{E} - \mathbf{D} - \mathbf{W}(z)| = 0$$

and hence the determinant of the matrix $z\mathbf{E} - \mathbf{D}$ cannot be zero at these values of z. This allowed us to introduce the matrix $\mathbf{G}(z)$ in Eq. (1.243).

Non-trivial solutions of this equation appear as roots of the equation

$$|(z\mathbf{E} - \mathbf{D})\,(\mathbf{E} - \mathbf{G}(z)\mathbf{W}(z))| = |z\mathbf{E} - \mathbf{D}|\ |\mathbf{E} - \mathbf{G}(z)\mathbf{W}(z)| = 0$$

$$\Longrightarrow \quad |\mathbf{E} - \mathbf{G}(z)\mathbf{W}(z)| = 0\,. \tag{1.244}$$

To solve Eq. (1.244), we introduce a block structure in our matrices by splitting all sites of the chain into two regions: the zeroth region corresponds to the single site 0, while the first region - to all other sites. This gives

$$\mathbf{W} = \begin{pmatrix} W_{00} & \mathbf{0}_{01} \\ \mathbf{0}_{10} & \mathbf{0}_{11} \end{pmatrix}$$

and

$$\mathbf{G}(z) = \begin{pmatrix} g_{00}(z) & \mathbf{G}_{01}(z) \\ \mathbf{G}_{10}(z) & \mathbf{G}_{11}(z) \end{pmatrix}$$

$$\Longrightarrow \quad \mathbf{E} - \mathbf{G}(z)\mathbf{W} = \begin{pmatrix} 1 - g_{00}(z) W_{00}(z) & \mathbf{0}_{01} \\ \mathbf{G}_{10}(z) W_{00}(z) & \mathbf{E}_{11} \end{pmatrix}$$

(note that the 0th region contains just a single lattice site). This is a left triangular matrix. Therefore, additional frequencies as solutions of the equation $|\mathbf{E} - \mathbf{G}\mathbf{W}| = 0$ are obtained by solving the equation (the off-diagonal terms do not contribute to the determinant):

$$|1 - g_{00}(z) W_{00}(z)| = 0 \implies 1 = g_{00}(z) W_{00}(z)$$

$$\implies \frac{1}{(1-\mu)\,\omega^2} = \frac{1}{N} \sum_q \frac{1}{\omega^2 - \omega^2(q)} , \qquad (1.245)$$

where explicit expression for Green's function on the defect site (1.241) (when $n = j = 0$) has been used. The sum in the right-hand side can be turned into an integral (since $N \to \infty$ and hence $\Delta q \to 0$)

$$\frac{1}{N} \sum_q \frac{1}{\omega^2 - \omega^2(q)} = \frac{1}{N\Delta q} \sum_q \frac{\Delta q}{\omega^2 - \omega^2(q)} \implies \frac{a}{2\pi} \int_{-\pi/a}^{\pi/a} \frac{dq}{\omega^2 - \frac{4k}{m} \sin^2 \frac{qa}{2}} ,$$

and hence calculated. Equating this to the left-hand side of Eq. (1.245) allows calculation of all solutions for the frequencies of the defective chain. It should give perturbed 'bulk' solutions close to those of the perfect chain, plus additional solutions may also appear when μ becomes sufficiently different than one. A patient reader should be able to perform the q-integration analytically and obtain a transcendental equation for ω.

Problem 1.178 Show that the above integral is equal to (cf. Eqs. (2.63), (2.119) or (2.120))

$$\frac{a}{2\pi} \int_{-\pi/a}^{\pi/a} \frac{dq}{\omega^2 - \frac{4k}{m} \sin^2 \frac{qa}{2}} = \frac{1}{\omega \sqrt{\omega^2 - 4k/m}} ,$$

and hence find the frequency corresponding to the local vibration of a lighter atom ($\mu < 1$) as

$$\omega_{loc} = \sqrt{\frac{4k}{m\mu\,(2-\mu)}} ,$$

which is positioned above the perfect chain frequencies $0 < \omega \le \sqrt{4k/m}$ (since $0 < \mu\,(2-\mu) < 1$ as can be checked, e.g. by plotting this parabola). This vibration corresponds to a local mode associated with oscillations of atoms in the vicinity of the defect. Explain why there is no extra solution for a heavier atom ($\mu > 1$). [Hint: to find the integral, use the substitution $t = \tan \frac{x}{2}$, where x is the argument of the sine function in the integrand.]

1.22.6   States of an Electron in a Solid

Consider a three-dimensional atomic system, e.g. a solid. The solid we are considering does not need to be periodic; it could be a disordered or defective system. We would like to obtain energy levels ϵ_λ electrons can occupy in this material. To find the energy levels, one needs to solve the Schrödinger equation for the electrons:

$$-\frac{\hbar^2}{2m}\Delta\psi_\lambda(\mathbf{r}) + V(\mathbf{r})\psi_\lambda(\mathbf{r}) = \epsilon_\lambda\psi_\lambda(\mathbf{r}) \ , \qquad (1.246)$$

where ψ_λ is the wavefunction of the electron occupying the state λ with energy ϵ_λ, the first term in the left-hand side corresponds to the kinetic, while the second to the potential energy of the electron with $V(\mathbf{r})$ being the corresponding lattice potential that the electrons experience in the solid, and m is the electron mass. It is convenient to introduce the Hamiltonian *operator* $\widehat{H}$ which is defined in such a way that its action on any function $\varphi(\mathbf{r})$ standing on the right of it results in the following action:

$$\widehat{H}\varphi(\mathbf{r}) = \left[-\frac{\hbar^2}{2m}\Delta + V(\mathbf{r})\right]\varphi(\mathbf{r}) = -\frac{\hbar^2}{2m}\Delta\varphi(\mathbf{r}) + V(\mathbf{r})\varphi(\mathbf{r}) \ .$$

Then, the Schrödinger equation (1.246) takes on a very simple form:

$$\widehat{H}\psi_\lambda(\mathbf{r}) = \epsilon_\lambda\psi_\lambda(\mathbf{r}) \ . \qquad (1.247)$$

To solve the differential equation (1.247), it is convenient to turn it into a matrix form. To this end, we shall expand the electronic wavefunction ψ_λ in terms of localised on atoms orbitals $\chi_A(\mathbf{r})$ (called atomic orbitals),

$$\psi_\lambda(\mathbf{r}) = \sum_A c_A(\lambda)\chi_A(\mathbf{r}) \ . \qquad (1.248)$$

For the sake of simplicity we shall assume that there is only one orbital, $\chi_A(\mathbf{r})$, placed on each atom A. Here the summation is performed with respect to all atoms (i.e. all atomic orbitals). This method is called the *linear combination of atomic orbitals* (LCAO) method. As an example of a possible system we show in Fig. 1.14 a fragment of a two-dimensional (e.g. a surface or a slab) system in which two species of atoms are distributed at random at regular lattice sites. On each atom of the system with position vector $\mathbf{R}_A$ we have placed an orbital $\chi_A(\mathbf{r}) = \chi(\mathbf{r} - \mathbf{R}_A)$; note that in this example all orbitals have an identical shape given by the function $\chi(\mathbf{r})$; only their positions are different.

We shall also assume that the orbitals are normalised to unity and that the orbitals on different atoms do not overlap, i.e. they are orthogonal to each other; in other words,

$$\int \chi_A(\mathbf{r})\chi_B(\mathbf{r})d\mathbf{r} = \delta_{AB} \ , \qquad (1.249)$$

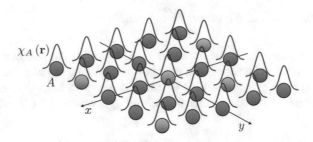

Fig. 1.14 A possible model of an alloy of a two-dimensional solid: we have a regular arrangement of identical atoms (brown) except for some random lattice sites which are occupied by a different species (cyan). At each lattice site a single localised atomic orbital $\chi_A(\mathbf{r})$ is positioned (blue)

where the integration is done with respect to the whole volume of the system, and δ_{AB} is the Kronecker symbol (equal to one when $A = B$, otherwise it is equal to zero).

Before we move any further, we need one important result related to expanding a given function in terms of other functions. This question is considered in more detail in Sect. I.7.2 on functional series. What is essential for us here is that we may *assume* that the atomic orbitals centred on *all* atoms of the system form a set of functions which is very close to a *complete set* of functions, i.e. that any 'good' function can be expanded in terms of them:

$$f(\mathbf{r}) = \sum_A f_A \chi_A(\mathbf{r}) \ .$$

The expansion coefficients f_A are obtained, as this is usually done, by multiplying both sides of the above equation with some function $\chi_B(\mathbf{r})$, integrating over the whole space and making use of the orthogonality condition (1.249). This gives

$$\int f(\mathbf{r})\chi_B(\mathbf{r})d\mathbf{r} = \sum_A f_A \int \chi_B(\mathbf{r})\chi_A(\mathbf{r})d\mathbf{r}$$

$$\Longrightarrow \quad \int f(\mathbf{r})\chi_B(\mathbf{r})d\mathbf{r} = \sum_A f_A \delta_{AB}$$

$$\Longrightarrow \quad \int f(\mathbf{r})\chi_B(\mathbf{r})d\mathbf{r} = f_B \quad \text{or} \quad f_B = \langle f \,|\chi_B\rangle \ ,$$

where we have used notations that are similar to the Dirac's notation for the matrix element, something which is frequently used in quantum mechanics.

Now, if we consider $\widehat{H}\chi_B(\mathbf{r})$ as the function $f(\mathbf{r})$, then the above expressions are rewritten as

$$\widehat{H}\chi_B(\mathbf{r}) = \sum_A f_A \chi_A(\mathbf{r}) \quad \text{with} \quad f_A = \int \chi_A(\mathbf{r})\widehat{H}\chi_B(\mathbf{r})d\mathbf{r} = H_{AB} \ , \qquad (1.250)$$

where the numbers $H_{AB} = \langle \chi_A | \widehat{H} | \chi_B \rangle$ form elements of the Hamiltonian matrix $\mathbf{H} = (H_{BA})$, and yet again notations similar to the Dirac's notations for the matrix elements were applied for convenience.

Now we are ready to continue working on our problem. Substituting the LCAO expansion (1.248) into the Schrödinger equation (1.247), we obtain

$$\sum_A c_A(\lambda) \widehat{H} \chi_A(\mathbf{r}) = \epsilon_\lambda \sum_A c_A(\lambda) \chi_A(\mathbf{r}) .$$

Multiplying both sides of the equation by $\chi_B(\mathbf{r})$ and integrating over the whole volume and using the orthogonality of the atomic orbitals (1.249), we obtain

$$\sum_A H_{BA} c_A(\lambda) = \epsilon_\lambda c_B(\lambda) , \tag{1.251}$$

which presents yet another example of the eigenvalue–eigenvector problem,

$$\mathbf{HC}_\lambda = \epsilon_\lambda \mathbf{C}_\lambda , \tag{1.252}$$

where $\mathbf{C}_\lambda = (c_A(\lambda))$ is the vector-column of all LCAO coefficients corresponding to the state λ. Hence, to determine eigenvalues ϵ_λ and the LCAO coefficients $\mathbf{C}_\lambda$, one has to find eigenvalues and eigenvectors of the Hamiltonian matrix $\mathbf{H} = (H_{AB})$. Note that since $\mathbf{H}$ is a symmetric matrix, the eigenvalues are guaranteed to be real.

For a general disordered infinite system it is rather difficult to solve the eigen-problem since the Hamiltonian matrix has an infinite dimension. On the other hand, one can exploit a local character of the interaction between atoms which can be mathematically formulated as a condition stating, for instance, that only matrix elements H_{AB} between nearest neighbours A and B are non-zero. Then, instead of solving the eigenproblem (1.251) or (1.252) directly, one can consider a *resolvent* $\mathbf{G}(z) = (z\mathbf{E} - \mathbf{H})^{-1}$ of the Hamiltonian matrix $\mathbf{H}$ (z is a complex number) whose singularity points z on the real axis (called poles) give the required energies.

To proceed, we shall use the (slightly modified) Lanczos method (Sect. 1.16). We start by choosing the first orbital $g_1(\mathbf{r})$ as the atomic orbital $\chi_0(\mathbf{r})$ centred at the zero lattice site. We assume, of course, that χ_0 is properly normalised: $\langle g_1 | g_1 \rangle = 1$. Then, the second orbital is chosen using

$$\beta_1 g_2(\mathbf{r}) = \widehat{H} g_1(\mathbf{r}) - \alpha_1 g_1(\mathbf{r}) ,$$
$$\alpha_1 = \langle g_1 | \widehat{H} | g_1 \rangle \quad \text{and} \quad \beta_1 = \langle g_2 | \widehat{H} | g_1 \rangle = \langle g_1 | \widehat{H} | g_2 \rangle . \tag{1.253}$$

The only difference with the Lanczos procedure introduced in Sect. 1.16 is that we work here with functions and the operator $\widehat{H}$ instead of vectors and a matrix. However, as we recall from Sect. 1.1.2, there is a very close analogy between vectors and functions; it is easy to see that the operator $\widehat{H}$ plays here the role of such a matrix.

By virtue of the above construction, the function g_2 must be orthogonal to g_1 and normalised, i.e. $\langle g_2 | g_1 \rangle = 0$ and $\langle g_2 | g_2 \rangle = 1$. It is essential to realise at this stage that g_2, constructed as above, is actually a linear combination of the atomic orbitals. Indeed, according to Eq. (1.250),

$$\widehat{H} g_1(\mathbf{r}) = \widehat{H} \chi_0(\mathbf{r}) = \sum_B H_{0B} \chi_B(\mathbf{r}) \, ,$$

with the expansion coefficients $H_{0B} = \langle \chi_0 | \widehat{H} | \chi_B \rangle$. Since there is only 'interaction' between nearest sites, the expansion above is limited only to the nearest neighbours. Therefore, the orbital $g_2(\mathbf{r})$ from (1.253) is a linear combination of atomic orbitals from the zero and all neighbouring atomic sites only. Let us keep this in mind.

The construction of the consecutive orbitals g_3, g_4, and so on goes along the general recipe of Eqs. (1.144)–(1.146):

$$\beta_k g_{k+1} = \widehat{H} g_k - \alpha_k g_k - \beta_{k-1} g_{k-1} \, ,$$
$$\alpha_k = \langle g_k | \widehat{H} | g_k \rangle \quad \text{and} \quad \beta_{k-1} = \langle g_k | \widehat{H} | g_{k-1} \rangle = \langle g_{k-1} | \widehat{H} | g_k \rangle \, , \tag{1.254}$$

where $k = 2, 3, \ldots$. Constructed in this way, the orbitals $g_k(\mathbf{r})$ are orthogonal to each other and normalised, i.e. $\langle g_l | g_k \rangle = \delta_{kl}$. Then the Hamiltonian matrix in the new orbitals, $T_{lk} = \langle g_l | \widehat{H} | g_k \rangle$, would have a special structure: the only matrix elements which are non-zero are those which relate either the same ($k = l$) or the neighbouring orbitals ($k = l \pm 1$) during the procedure, i.e. $T_{lk} = 0$ as long as $|k - l| > 1$. In other words, the matrix $\mathbf{T}$ is tridiagonal.

Each orbital constructed using the described Lanczos algorithm is a linear combination of atomic orbitals centred around the chosen zero lattice site. Indeed, as we have already pointed out, g_2 contains χ_0 and the orbitals χ_A from the sites nearest to the zero site; g_3 would contain orbitals used in g_2 and the orbitals nearest to them; g_4 would contain all orbitals used in expanding g_3 plus all orbitals χ_A nearest to them, and so on. As more and more orbitals are built, the larger and larger region around the zero lattice site is used for expanding them. At every stage of this procedure a finite number of orbitals is built which at each stage form an orthonormal set.

Let us formally continue the construction procedure until we build as many orbitals as the number of atoms in the system, i.e. we shall formally have the same number of new orbitals in the set as there are atomic orbitals in the initial set. As each new orbital is a linear combination of the original atomic orbitals also forming an orthonormal set, the new set of orbitals is related to the initial set by a unitary transformation matrix $\mathbf{U}$:

$$g_k = \sum_A U_{kA} \chi_A \, , \quad \text{where} \quad \mathbf{U}^{-1} = \mathbf{U}^T \, .$$

Reversely,

$$\chi_A = \sum_k U_{kA} g_k \, .$$

Let us now look at the resolvent matrix $\mathbf{G}(z) = (z\mathbf{E} - \mathbf{H})^{-1}$. Let us express the Hamiltonian matrix $\mathbf{H} = (H_{AB})$ via matrix elements T_{kl} of the Hamiltonian operator $\widehat{H}$ written with respect to the new orbitals:

$$H_{AB} = \langle \chi_A | \widehat{H} | \chi_B \rangle = \sum_{kl} U_{kA} U_{lB} \langle g_k | \widehat{H} | g_l \rangle$$

$$= \sum_{kl} U_{kA} U_{lB} T_{kl} = \sum_{kl} \left(\mathbf{U}^T \right)_{Ak} T_{kl} U_{lB} .$$

In other words, $\mathbf{H} = \mathbf{U}^T \mathbf{T} \mathbf{U}$, where the matrix $\mathbf{T}$ is tridiagonal as explained above. Then, employing the fact that $\mathbf{U}^T \mathbf{U} = \mathbf{E}$, we can write

$$\mathbf{G}(z) = \left(z\mathbf{U}^T\mathbf{U} - \mathbf{U}^T\mathbf{T}\mathbf{U} \right)^{-1} = \left[\mathbf{U}^T \left(z\mathbf{E} - \mathbf{T} \right) \mathbf{U} \right]^{-1}$$

$$= \mathbf{U}^T \left(z\mathbf{E} - \mathbf{T} \right)^{-1} \mathbf{U} .$$

This identity shows that the pole structure of the resolvent matrix is fully contained in the resolvent $\mathcal{G}(z) = (z\mathbf{E} - \mathbf{T})^{-1}$ of the T matrix which is written in terms of the new (Lanczos's) basis set. But the matrix $z\mathbf{E} - \mathbf{T}$ is tridiagonal, and hence the first diagonal element of its inverse, $\mathcal{G}_{11}(z)$, can easily be calculated as a function of z via a continued fraction, see Sect. 1.3.5 and Problem 1.80:

$$\mathcal{G}_{11}(z) = \frac{1|}{|\gamma_1} - \frac{\beta_1^2|}{|\gamma_2} - \frac{\beta_2^2|}{|\gamma_3} - \frac{\beta_3^2|}{|\gamma_4} - \cdots - \frac{\beta_{n-1}^2|}{|\gamma_n} - \cdots ,$$

where $\gamma_k = z - \alpha_k$ are the diagonal elements of the matrix $z\mathbf{E} - \mathbf{T}$, while $-T_{k,k+1} = -T_{k+1,k} = -\beta_k$ are its non-zero off-diagonal elements. As the size of our system is infinite, the continued fraction is infinite as well.

However, calculations usually show that some sort of convergence for the coefficients $\alpha_k \to \alpha_\infty$ and $\beta_k \to \beta_\infty$ may be reached at some (possibly large) step l of the construction procedure. If the convergence is indeed reached, the fraction can be 'summed up' to infinity exactly. Indeed, in this case the fraction looks like this:

$$\mathcal{G}_{11}(z) = \frac{1|}{|\gamma_1} - \frac{\beta_1^2|}{|\gamma_2} - \frac{\beta_2^2|}{|\gamma_3} - \frac{\beta_3^2|}{|\gamma_4} - \cdots - \frac{\beta_{l-2}^2|}{|\gamma_{l-1}}$$

$$- \frac{\beta_l^2|}{|\gamma_\infty} - \frac{\beta_\infty^2|}{|\gamma_\infty} - \frac{\beta_\infty^2|}{|\gamma_\infty} - \frac{\beta_\infty^2|}{|\gamma_\infty} - \cdots ,$$

(1.255)

where $\gamma_\infty = z - \alpha_\infty$ and after the step l all terms in the fraction are identical. If we denote this part of the fraction as S_∞, then it can be calculated:

$$S_\infty = \gamma_\infty - \cfrac{\beta_\infty^2}{\gamma_\infty - \cfrac{\beta_\infty^2}{\gamma_\infty - \cfrac{\beta_\infty^2}{\gamma_\infty - \cdots}}} = \gamma_\infty - \frac{\beta_\infty^2}{S_\infty} = z - \alpha_\infty - \frac{\beta_\infty^2}{S_\infty} ,$$

which yields a quadratic equation for the sum $S_\infty(z)$. Once the infinite tail of the fraction containing identical terms is known, these terms can be all replaced by the infinite sum S_∞ leading to a finite continued fraction

$$\mathcal{G}_{11}(z) = \frac{1|}{|\gamma_1} - \frac{\beta_1^2|}{|\gamma_2} - \frac{\beta_2^2|}{|\gamma_3} - \frac{\beta_3^2|}{|\gamma_4} - \cdots - \frac{\beta_{l-2}^2|}{|\gamma_{l-1}} - \frac{\beta_l^2|}{S_\infty(z)} , \tag{1.256}$$

which can now be calculated exactly.

We have discussed here only the main ideas of Lanczos method. It exploits the 'localised' nature of interactions in the systems, and as such is frequently used, e.g. in the theory of electronic states of disordered system as well as in other fields as it effectively allows considering finite fragments of realistic systems taking explicit account of their specific geometry.

1.22.7   Time Propagation of a Wavefunction

Consider an n-level quantum system described by the Hamiltonian matrix $\mathbf{H} = \left(H_{ij}\right)$. If at the initial time $t = 0$ the wavefunction of the system was a vector-column

$$\mathbf{\Psi}_0 = \begin{pmatrix} \psi_1^0 \\ \cdots \\ \psi_k^0 \\ \cdots \\ \psi_n^0 \end{pmatrix} ,$$

then its evolution in time can be described by the Schrödinger equation

$$i\hbar\frac{d\mathbf{\Psi}_t}{dt} = \mathbf{H}\mathbf{\Psi}_t . \tag{1.257}$$

In the simplest case of time-independent Hamiltonian, a closed formal solution of this equation can be found in a form of an exponential function of the matrix $\mathbf{H}$. Indeed, let λ_k and $\mathbf{x}_k$ are the eigenvalues and eigenvectors of $\mathbf{H}$ (note that since $\mathbf{H}$ is Hermitian, the eigenvalues are guaranteed to be real; however, the eigenvectors may be complex as the matrix $\mathbf{H}$ may in general be complex). It is also guaranteed that the eigenvectors form an orthonormal set, i.e. $\mathbf{x}_k^\dagger\mathbf{x}_j = \delta_{kj}$.

Then, using the spectral theorem, we can write the Hamiltonian via its eigenvectors and eigenvalues and then substitute into the Schrödinger equation (1.257) to obtain

$$\mathbf{H} = \sum_k \lambda_k\mathbf{x}_k\mathbf{x}_k^\dagger \implies i\hbar\frac{d\mathbf{\Psi}_t}{dt} = \sum_k \lambda_k\mathbf{x}_k\mathbf{x}_k^\dagger\mathbf{\Psi}_t . \tag{1.258}$$

Expanding the state vector $\boldsymbol{\Psi}_t$ in terms of the eigenvectors of $\mathbf{H}$, i.e. $\boldsymbol{\Psi}_t = \sum_j \alpha_j(t)\mathbf{x}_j$, we find upon multiplication from the left with $\mathbf{x}_k^\dagger$, that the expansion coefficients $\alpha_j(t) = \mathbf{x}_j^\dagger \boldsymbol{\Psi}_t$. Substituting the obtained expansion of $\boldsymbol{\Psi}_t$ into (1.258), we have

$$i\hbar \sum_k \frac{d\alpha_k}{dt}\mathbf{x}_k = \sum_{kj} \lambda_k \alpha_j \mathbf{x}_k \left(\mathbf{x}_k^\dagger \mathbf{x}_j \right) \quad \Longrightarrow \quad i\hbar \sum_k \frac{d\alpha_k}{dt}\mathbf{x}_k = \sum_k \lambda_k \alpha_k \mathbf{x}_k \ .$$

$$(1.259)$$

In the above manipulation we used the associativity in multiplying matrices (vectors in this particular case) and also the fact that the eigenvectors of $\mathbf{H}$ are orthonormal; because of the latter the double sum in the right-hand side was transformed into a single sum. Finally, multiply (1.259) by $\mathbf{x}_j^\dagger$ from the left again and use the orthonormality property. This way a simple differential equation for the coefficients $\alpha_j(t)$ follows which is trivially solved:

$$i\hbar \frac{d\alpha_j}{dt} = \lambda_j \alpha_j \quad \Longrightarrow \quad \alpha_j(t) = \alpha_j(0)e^{-i\lambda_j t/\hbar} \ ,$$

where $\alpha_j(0) = \mathbf{x}_j^\dagger \boldsymbol{\Psi}_0$ are the initial expansion coefficients. Therefore, the wavefunction at time t becomes

$$\boldsymbol{\Psi}_t = \sum_j \alpha_j(t)\mathbf{x}_j = \sum_j \alpha_j(0)e^{-i\lambda_j t/\hbar}\mathbf{x}_j = \sum_j e^{-i\lambda_j t/\hbar}\mathbf{x}_j \alpha_j(0)$$

$$= \left(\sum_j e^{-i\lambda_j t/\hbar}\mathbf{x}_j \mathbf{x}_j^\dagger \right) \boldsymbol{\Psi}_0 \ .$$

The sum in the right-hand side in the round brackets is recognised to be the spectral representation of the exponential function of the Hamiltonian matrix $\mathbf{H}$:

$$\mathbf{U}_{t0} = e^{-i\mathbf{H}t/\hbar} \ .$$

The matrix $\mathbf{U}_{t0}$ is called *propagator* in quantum mechanics as it propagates the wavefunction $\boldsymbol{\Psi}_0$ from $t = 0$ to $\boldsymbol{\Psi}_t = \mathbf{U}_{t0}\boldsymbol{\Psi}_0$ at any finite value of $t > 0$. It satisfies simple properties which we shall leave to the reader to prove as a problem.

Problem 1.179 Show that

$$\mathbf{U}_{tt'} = \mathbf{U}_{t\tau}\mathbf{U}_{\tau t'} \ , \quad \mathbf{U}_{tt'}^\dagger = \mathbf{U}_{t't} \quad \text{and} \quad \mathbf{U}_{tt'}^{-1} = \mathbf{U}_{t't} \ ,$$

where

$$\mathbf{U}_{tt'} = e^{-i\mathbf{H}(t-t')/\hbar} \ .$$

Problem 1.180 *(Rabi problem)* Consider a two-level quantum system with stationary states $\psi_1 = \begin{pmatrix} 1 \\ 0 \end{pmatrix}$ and $\psi_2 = \begin{pmatrix} 0 \\ 1 \end{pmatrix}$, which are normalised and orthogonal to each other. These states are eigenvectors of the Hamiltonian

$$\mathbf{H}_0 = \begin{pmatrix} \epsilon_1 & 0 \\ 0 & \epsilon_2 \end{pmatrix}$$

with eigenvalues ϵ_1 and ϵ_2, respectively. Suppose, that at time $t = 0$ when the system was in state ψ_1 it was subjected to a perturbation

$$\mathbf{V} = \begin{pmatrix} V_1 & v \\ v^* & V_2 \end{pmatrix} .$$

The state of the system $\psi(t)$ at time t can be considered as a linear combination

$$\psi(t) = C_1(t)\psi_1 + C_2(t)\psi_2 ,$$

written in terms of the two stationary states. By substituting $\psi(t)$ into the time-dependent Schrödinger equation $i\hbar\dot{\psi} = (\mathbf{H}_0 + \mathbf{V})\,\psi$ and considering explicitly both components, show that the coefficients $C_1(t)$ and $C_2(t)$ satisfy the following system of two ordinary differential equations:

$$i\hbar\dot{C}_1 = (\epsilon_1 + V_1)\,C_1 + vC_2 \quad \text{and} \quad i\hbar\dot{C}_2 = (\epsilon_2 + V_2)\,C_2 + v^*C_1 .$$

Introducing a vector $\mathbf{C} = \begin{pmatrix} C_1 \\ C_2 \end{pmatrix}$, rewrite these equations in a matrix form $i\hbar\dot{\mathbf{C}} = \mathbf{DC}$, where $\mathbf{D}$ is a 2×2 Hermitian matrix.

By using a trial solution $\mathbf{C}(t) = e^{\alpha t}\mathbf{X}$, where $\mathbf{X} = \begin{pmatrix} x_1 \\ x_2 \end{pmatrix}$ is a constant vector, show that both $\mathbf{X}$ and α can be obtained by solving an eigenproblem $\mathbf{DX} = \lambda\mathbf{X}$ for the matrix $\mathbf{D}$, where $\lambda = i\hbar\alpha$.

Show that eigenvalues of $\mathbf{D}$ are (cf. Problem 1.114)

$$\lambda_\pm = \frac{1}{2}\left[(\nu_1 + \nu_2) \pm \sqrt{(\nu_1 - \nu_2)^2 + 4\,|v|^2}\right], \quad \text{where} \quad \nu_i = \epsilon_i + V_i , \quad i = 1, 2 ,$$

while the corresponding eigenvectors can for instance be chosen as

$$\mathbf{X}_+ = \begin{pmatrix} v \\ \kappa - \nu/2 \end{pmatrix} \quad \text{and} \quad \mathbf{X}_- = \begin{pmatrix} -v \\ \kappa + \nu/2 \end{pmatrix} ,$$

where $\nu = \nu_1 - \nu_2$ and $\kappa = \frac{1}{2}\sqrt{(\nu_1 - \nu_2)^2 + 4\,|v|^2}$.

> Introducing two arbitrary constants A_+ and A_-, write the general solution of the equation $i\hbar\dot{\mathbf{C}} = \mathbf{D}\mathbf{C}$. Then, applying the initial condition that $\mathbf{C}(0) = \psi_1$, determine these two constants. Finally, show that the probabilities $P_1(t) = |C_1(t)|^2$ and $P_2(t) = |C_2(t)|^2$ of finding the system in the states ψ_1 or ψ_2 at time $t > 0$ are given by the following equations:
>
> $$P_1(t) = 1 - \frac{4\,|v|^2}{v^2 + 4\,|v|^2}\sin^2\frac{\kappa t}{\hbar} \quad \text{and} \quad P_2(t) = \frac{4\,|v|^2}{v^2 + 4\,|v|^2}\sin^2\frac{\kappa t}{\hbar}\,.$$
>
> Note that the probabilities oscillate; however, $P_1 + P_2 = 1$ at any time.

1.22.8 Slater Determinants and Fermionic Many-Body Wavefunctions

The notion of a determinant appears in many areas of physics. The most notable examples of the appearance of determinants can probably be met in electronic many-body theory. The first one comes from the Wicks theorem which states that the n-particle Green's function of non-interacting fermions can be expressed as a determinant constructed of one particle non-interacting Green's functions; this lies at the heart of the Feynman diagrammatic technique of the many-body theory. The second one is at the foundation of the second quantisation for fermions: it states that the exact expansion of a many-fermion wavefunction into a basis set of one-particle orbitals (functions that depend only a single spatial variable) has a very specific form as an expansion over the so-called Slater determinants constructed out of these orbitals. It would not be possible to review here the former application as this would require building up the whole many-body theory; the interested reader should study this in specialised textbooks. However, the latter application of determinants is not very difficult to discuss and this is what we shall do in this section.

Assume that we have a basis set $\{\phi_n(x)\}$ of one-particle functions (for brevity, we shall call them 'orbitals'). Here: x denotes all variables a single particle (e.g. an electron) will depend upon; this is not so important for us here.[28] The basis set satisfies two conditions. The first one is orthonormality[29]:

$$\langle\phi_n|\phi_{n'}\rangle = \int dx\, \phi_n^*(x)\phi_{n'}(x) = \delta_{nn'}\,.$$

---

[28] Basically, these would be the spatial position $\mathbf{r} = (x, y, z)$ and the spin σ.

[29] Strictly speaking, it is needed merely for simplicity and can easily be avoided; however, this leads to unnecessary complications of technical nature in formulating the method that does not bring anything of importance; hence, we shall assume that condition to illustrate the main idea as the algebra is greatly simplified.

The second condition

$$\sum_n \phi_n(x)\phi_n^*(x') = \delta(x - x')$$

is of principal nature: it is called 'completeness' and basically states that any 'good' function $f(x)$ can be expanded with respect to these functions,

$$f(x) = \sum_n c_n \phi_n(x), \tag{1.260}$$

so that the series in the right-hand side at point x converges to the value of $f(x)$ in the left-hand side. The expansion (1.260) represents the so-called generalised functional Fourier series and some aspects of it are considered in Sect. I.7.2 and 3.9.3.

In quantum mechanics, the same expansion (1.260) could be used to represent a wavefunction $\psi(x)$ of a system containing a single electron; if there are N such electrons, the wavefunction $\psi(x_1, \ldots, x_N)$ becomes a function of N variables $(x_1, \ldots, x_N)$ and an appropriate generalisation is required. We can easily construct such a generalisation by performing the Fourier expansion with respect to each variable step by step. Let us first expand the wavefunction with respect to its first variable:

$$\psi(x_1, \ldots, x_N) = \sum_{n_1} C_{n_1}(x_2, \ldots, x_N)\phi_{n_1}(x_1).$$

Of course, we expect that the expansion coefficients would still be functions of the rest of the variables. Therefore, we can next expand the coefficients into the generalised Fourier series with respect to the variable x_2,

$$C_{n_1}(x_2, \ldots, x_N) = \sum_{n_2} C_{n_1 n_2}(x_3, \ldots, x_N)\phi_{n_2}(x_2),$$

where the new expansion coefficients have been introduced that depend on the rest of the variables. Hence, at that stage, we can write an expansion of the wavefunction as a double Fourier series:

$$\psi(x_1, \ldots, x_N) = \sum_{n_1}\sum_{n_2} C_{n_1 n_2}(x_3, \ldots, x_N)\phi_{n_1}(x_1)\phi_{n_2}(x_2).$$

This process can be continued; obviously, at the very last N-st step the expansion coefficients would be just numbers:

$$\psi(x_1, \ldots, x_N) = \sum_{n_1}\sum_{n_2}\cdots\sum_{n_N} C_{n_1 n_2 \cdots n_N}\phi_{n_1}(x_1)\phi_{n_2}(x_2)\cdots\phi_{n_N}(x_N). \tag{1.261}$$

The expansion coefficients can be easily found due to the orthogonality of the orbitals. In order to do this, we multiply both sides of the expansion above with

$\phi_{m_1}^*(x_1)\phi_{m_2}^*(x_2)\cdots\phi_{m_N}^*(x_N)$ with some fixed indices $m_1,\ldots,m_N$, and integrate over all the variables. The result is

$$C_{m_1 m_2 \cdots m_N} = \int \psi(x_1, x_2, \ldots)\phi_{m_1}^*(x_1)\phi_{m_2}^*(x_2)\cdots\phi_{m_N}^*(x_N)\,dx_1 dx_2\cdots dx_N.$$
(1.262)

Problem 1.181 Prove formula (1.262).

So far, our consideration has been rather general and applicable to *any* function of N variables. Now, let us recall that we actually would like to do is to consider an expansion of the wavefunction describing N electrons. Electrons are fermions, and hence not any function of N variables may serve as their wavefunction. Only functions that are antisymmetric (Sect 1.3.3) with respect to their variables can be chosen. That means that a permutation of any two variables in $\psi(x_1,\ldots,x_N)$ should lead to a change of sign,

$$\psi(\ldots,x_i,\ldots,x_j,\ldots) = -\psi(\ldots,x_j,\ldots,x_i,\ldots).$$

That condition of antisymmetricity of the electron wavefunction enables one to significantly simplify the expansion (1.261). To show this, we first note that in this expansion each term contains products of orbitals from the set $\{\phi_n(x)\}$ for all N variables. Amongst these orbitals in some of the terms we may find identical ones, i.e. the orbitals with identical indices n_i, n_j, etc. They will all have however different arguments. To understand what happens, it is instructive to consider a specific case. For instance, let us have a look at the case of $N = 3$ and orbitals $\phi_1(x)$, $\phi_3(x)$ and $\phi_9(x)$. Then, amongst an infinite number of terms in the sum (1.261), there will only be precisely 6 terms containing just these three orbitals:

$$\begin{aligned}
\psi(x_1, x_2, x_3) &= [C_{139}\phi_1(x_1)\phi_3(x_2)\phi_9(x_3) + C_{193}\phi_1(x_1)\phi_9(x_2)\phi_3(x_3) \\
&\quad + C_{319}\phi_3(x_1)\phi_1(x_2)\phi_9(x_3) + C_{391}\phi_3(x_1)\phi_9(x_2)\phi_1(x_3) \\
&\quad + C_{913}\phi_9(x_1)\phi_1(x_2)\phi_3(x_3) + C_{931}\phi_9(x_1)\phi_3(x_2)\phi_1(x_3)] + \ldots \\
&= [C_{139}\phi_1(x_1)\phi_3(x_2)\phi_9(x_3) + C_{193}\phi_1(x_1)\phi_3(x_3)\phi_9(x_2) \\
&\quad + C_{319}\phi_1(x_2)\phi_3(x_1)\phi_9(x_3) + C_{391}\phi_1(x_3)\phi_3(x_1)\phi_9(x_2) \\
&\quad + C_{913}\phi_1(x_2)\phi_3(x_3)\phi_9(x_1) + C_{931}\phi_1(x_3)\phi_3(x_2)\phi_9(x_1)] + \ldots,
\end{aligned}$$
(1.263)

where dots correspond to other terms in the sum containing other sets of orbitals. In the first equality the orbitals $\phi_n(x)$ in each of the six terms are ordered with respect to their arguments so that their indices are permuted (and there are $3! = 6$ permutations possible), while in the second they are ordered with respect to the indices of the orbitals, i.e. in the second equality the arguments of the orbitals are permuted. Of course, both representations are completely identical as only differing by the permutations of orbitals in each term. The whole expansion (1.261) contains the sum of such sets of terms, each such set corresponding to all possible selection of indices and consisting of precisely 6 terms, similar to the ones given above.

What we are about to demonstrate is that, because the wavefunction of fermions is antisymmetric, the coefficients $C_{n_1 n_2 \cdots n_N}$ that have at least two identical indices can be set to zero, and the coefficients that contain all indices different and differ only in the order in which they appear (as in the example above) and differ from each other only by a sign. For instance, returning back to our example, $C_{139} = -C_{193} = -C_{319} = C_{391} = C_{913} = -C_{931}$. However, for instance, $C_{113} = C_{111} = C_{339} = 0$.

To prove this, let us consider all terms in the sum that contain a product of the same N orbitals $\phi_{n_1}(x)$, $\phi_{n_2}(x)$, etc., $\phi_{n_N}(x)$, i.e. the ones that bear indices n_1, n_2, etc., n_N. These terms have C-coefficients with the same indices, however, these will all come in different orders; there will be $N!$ such terms for each set of indices (for instance, in the case of $N = 3$ we shall have $3! = 6$ terms with the coefficients $C_{n_1 n_2 n_3}$, $C_{n_2 n_1 n_3}$, $C_{n_1 n_3 n_2}$, $C_{n_3 n_2 n_1}$, $C_{n_2 n_3 n_1}$, and $C_{n_3 n_1 n_2}$). Let us call this particular sequence of indices $n = (n_1, n_2, \ldots, n_N)$. All $N!$ terms in the expansion of ψ with the same indices (but permuted arguments) we shall call a *set* corresponding to the sequence of indices n.

In these $N!$ terms, orbitals in the product within each term will differ from any other term either by a permutation of their arguments or their indices, both viewpoints are equally applicable. Between the two such views we shall now choose the one in which all orbitals $\phi_n(x)$ in the product are ordered by their arguments, as shown in the first equality in Eq. (1.263). Then the products of the functions within a given set would differ by the order of their indices n_1, n_2, etc. from the given set n of indices. Note that some of the orbitals in the sets may be the same (have equal indices).

Consider now two particular terms in the sum within the same set of $N!$ terms with indices

$$n = (n_1, n_2, \ldots) \text{ and } m = (m_1, m_2, \ldots).$$

Note that the sequence m contains the same indices as the sequence n, but they are permuted, i.e. arranged in a different order. Formally, this can be written via $n = \widehat{P}m$, where $\widehat{P}$ is the corresponding permutation

$$\widehat{P} = \begin{pmatrix} m_1 & m_2 & \ldots & m_N \\ n_1 & n_2 & \ldots & n_N \end{pmatrix} \tag{1.264}$$

that indicates that in order to get the sequence n from m one needs to perform the following replacements: $m_1 \to n_1$, $m_2 \to n_2$, and so on.

We can write for one of the coefficients:

$$C_{m_1 m_2 \cdots} = \int \psi(x_1, x_2, \ldots) \phi_{m_1}^*(x_1) \phi_{m_2}^*(x_2) \cdots dx_1 dx_2 \cdots. \tag{1.265}$$

Let us perform a change of variables here using the inverse permutation

$$\widehat{P}^{-1} = \begin{pmatrix} n_1 & n_2 & \dots & n_N \\ m_1 & m_2 & \dots & m_N \end{pmatrix}$$

of the original variables,

$$(y_1, y_2, \dots, y_N) = \widehat{P}^{-1}(x_1, x_2, \dots x_N),$$

or more concisely, $Y = \widehat{P}^{-1}X$, such that the product $\phi^*_{m_1}(x_1)\phi^*_{m_2}(x_2)\cdots$ of orbitals in the integrand becomes (after appropriately changing their order) $\phi^*_{n_1}(y_1)\phi^*_{n_2}(y_2)\cdots$ $\phi^*_{n_N}(y_N)$, exactly as in $C_{n_1 n_2 \dots}$. We now have the same order of orbitals as in $C_{n_1 n_2 \dots}$ but using other variables $y_1, y_2, \dots$. Since $Y = \widehat{P}^{-1}X$, then $X = \widehat{P}Y$. The variables X in ψ should also be changed into Y inside the integral (1.265) during this change of the variables. Since it is a fermionic wavefunction, it changes sign upon a single permutation of two variables. Hence, a general permutation may consist of many such elementary permutations,

$$\psi(x_1, \dots, x_N) = \psi\left(\widehat{P}(y_1, \dots, y_N)\right) = \epsilon_P \psi(y_1, \dots, y_N),$$

where $\epsilon_P = (-1)^P$ is either plus or minus one depending on whether an even or odd number of elementary permutations[30] P is needed. This P is called parity of the permutation. We see that

$$C_{m_1 m_2 \dots} = \int \epsilon_P \psi(y_1, \dots, y_N)\phi^*_{n_1}(y_1)\phi^*_{n_2}(y_2)\cdots dy_1 dy_2 \cdots = \epsilon_P C_{n_1 n_2 \dots} \quad (1.266)$$

Hence, the expansion coefficients for the same set of orbitals $\{\phi_n(x)\}$, but placed in a different order within the same set, may differ only by a trivial sign factor of $\epsilon_P = (-1)^P$.

As an illustration for the above transformation, let us show that

$$C_{321} = \int \psi(x_1, x_2, x_3)\phi^*_3(x_1)\phi^*_2(x_2)\phi^*_1(x_3)\,dx_1 dx_2 dx_3 = -C_{123}.$$

That would require renaming the variables by means of the permutation

$$\widehat{P} = \begin{pmatrix} 1 & 2 & 3 \\ 3 & 2 & 1 \end{pmatrix},$$

i.e. $x_1 = y_3$, $x_2 = y_2$ and $x_3 = y_1$. The transformation for the indices of the orbitals,

---

[30] An elementary permutation involves only two numbers to be swapped.

$$\widehat{P}^{-1} = \begin{pmatrix} 3 & 2 & 1 \\ 1 & 2 & 3 \end{pmatrix}$$

is the inverse to it. This yields[31]

$$
\begin{aligned}
C_{321} &= \int \psi(y_3, y_2, y_1)\phi_3^*(y_3)\phi_2^*(y_2)\phi_1^*(y_1)\, dy_3 dy_2 dy_1 \\
&= \int \underbrace{\psi(y_3, y_2, y_1)}_{-\psi(y_1, y_2, y_3)} \phi_1^*(y_1)\phi_2^*(y_2)\phi_3^*(y_3)\, dy_1 dy_2 dy_3 \\
&= -\int \psi(y_1, y_2, y_3)\phi_1^*(y_1)\phi_2^*(y_2)\phi_3^*(y_3)\, dy_1 dy_2 dy_3 = -C_{123}\,,
\end{aligned}
$$

as required.

If we collect all $N!$ terms in ψ from the given set together, and sum them all up, we shall get

$$
\sum_{N! \text{ terms of the set } n=(n_1,\dots,n_N)} C_{n_1 n_2 \cdots} \phi_{n_1}(x_1)\phi_{n_2}(x_2)\cdots
$$

$$
= C_{(n_1 n_2 \cdots n_N)} \left[\sum_{P \in S_N} \epsilon_P \widehat{P} \phi_{n_1}(x_1)\phi_{n_2}(x_2)\cdots\phi_{n_N}(x_N) \right],
$$

$$(1.267)$$

where $C_{(n_1 n_2 \cdots n_N)}$ is a coefficient of one particular term in the expansion with the orbitals arranged in a certain order, the most convenient choice being $n_1 < n_2 < \cdots < n_N$, and $\widehat{P}$ acts on the indices $n_1, n_2, \dots, n_N$ of the orbitals in their product in each term inside the square brackets. Under the sum sign, S_N means that we sum over all $N!$ permutations[32] of the given set of N indices $\{n_1, n_2, \dots n_N\}$. Returning back to our $N = 3$ example, the sum of 6 terms in (1.263) can now we simplified into:

$$
\begin{aligned}
\psi(x_1, x_2, x_3) &= C_{139}\, [\phi_1(x_1)\phi_3(x_2)\phi_9(x_3) - \phi_1(x_1)\phi_9(x_2)\phi_3(x_3) \\
&\quad - \phi_3(x_1)\phi_1(x_2)\phi_9(x_3) + \phi_3(x_1)\phi_9(x_2)\phi_1(x_3) \\
&\quad + \phi_9(x_1)\phi_1(x_2)\phi_3(x_3) - \phi_9(x_1)\phi_3(x_2)\phi_1(x_3)] + \dots \\
&= C_{139} \begin{vmatrix} \phi_1(x_1) & \phi_3(x_1) & \phi_9(x_1) \\ \phi_1(x_2) & \phi_3(x_2) & \phi_9(x_2) \\ \phi_1(x_3) & \phi_3(x_3) & \phi_9(x_3) \end{vmatrix} + \dots,
\end{aligned}
$$

since the terms inside the square brackets can easily be seen to form a determinant.

Now it could be appreciated that the expansion in the square brackets in (1.267) is recognised to be a $N \times N$ determinant:

---

[31] The Jacobian of the transformation is obviously 1 as we simply rename the variables.

[32] These permutations form a group S_N called the group of permutations.

$$\Phi_{n_1, n_2, \ldots, n_N}(x_1, x_2, \ldots, x_N) = \frac{1}{\sqrt{N!}} \begin{vmatrix} \phi_{n_1}(x_1) & \phi_{n_2}(x_1) & \phi_{n_3}(x_1) & \ldots \\ \phi_{n_1}(x_2) & \phi_{n_2}(x_2) & \phi_{n_3}(x_2) & \ldots \\ \ldots & \ldots & \ldots & \ldots \end{vmatrix}$$

$$= \frac{1}{\sqrt{N!}} \sum_{P \in S_N} \epsilon_P \widehat{P} \phi_{n_1}(x_1) \phi_{n_2}(x_2) \cdots \phi_{n_N}(x_N) \equiv \Phi_{set},$$

(1.268)

and we agree that from now on the indices in the determinants are ordered as $n_1 < n_2 < \cdots < n_N$. This particular many-body wavefunction is called a Slater determinant and we have also added a normalisation factor to it for convenience.[33] Hence, all terms in the sum (1.261) with the same collection of orbitals (forming a set) amount to a single coefficient $C_{(n_1 n_2 \cdots n_N)}$ multiplied by the determinant Eq. (1.268).

It is seen now that if there are identical functions in the set $\{\phi_n(x)\}$ used in the determinant (1.268), the determinant contains identical columns and thus is equal to zero[34]; hence, the summation (1.261) contains only terms with all N orbitals being different, drawn from the infinite set $\{\phi_n(x)\}$ of orbitals; alternatively, we may simply agree to set all the coefficients $C_{n_1 n_2 \cdots n_N}$, if they contain identical indices, to zero.

Therefore, the exact expansion (1.261) of the N-electron wavefunction can now be rewritten more compactly as

$$\psi(x_1, \ldots, x_N) = \sum_{sets} C_{set} \Phi_{set}(x_1, \ldots, x_N),$$

(1.269)

where we sum over all different *sets* of N *different* indices $n_1, n_2, \ldots$. For instance, when $N = 3$ and limiting to only 5 orbitals, the possible sets are $(1, 1, 1, 0, 0)$, $(1, 1, 0, 1, 0)$, $(0, 1, 1, 1, 0)$, and so on, where we indicated by 1 if the orbital is used and by 0 if it is not, out of the existing five of them; there will be $\binom{5}{3} = \frac{5!}{3!2!} = 10$ such sets and hence only 10 terms in the expansion of $\psi(x_1, x_2, x_3)$. Note that the original generalised Fourier expansion contained $5^3 = 125$ terms.

The obtained formula (1.269) for the wavefunction can now be considered as an expansion of the N-electron wavefunction into a many-body basis set served by the Slater determinants Φ_{set}. It can be shown that these many-body basis functions are orthogonal (if contains a different set of indices) and normalised to one.

All the information each such determinant contains are the numbers of orbitals it consists from, i.e. every Slater determinant can simply be written by indicating its orbitals. Let us have all orbitals ordered; then a given Slater determinant can equivalently be shown by putting 1 at the appropriate positions for the orbitals that are present in it and zeros for orbitals that are not, and show these within a construction called a ket: $|\ldots 010 \ldots 0110 \ldots\rangle$. For instance, the Slater determinant Φ_{123} for a $N = 3$ electron system can be denoted as $|11100 \ldots\rangle$, $\Phi_{235} = |0110100 \ldots\rangle$ and so

---

[33] We shall not derive this normalisation factor here although this is not a very difficult calculation.

[34] This corresponds to the Pauli exclusion principle stating that two electrons cannot occupy the same state.

on. Note that in the ket only zeros and ones can appear as the same orbital cannot be present more than ones as otherwise the ket vector will be equal to zero (the same Pauli exclusion principle). These simple notations are at the heart of the second quantisation formalism.

The reader may wonder what the wavefunction expansion in the case of the bosons would be? For bosons, the wavefunction does not change sign (is symmetric) when two of its variables are permuted. In this case we arrive at the expansion via objects called permanents which are given (up to an appropriate normalisation factor $\mathcal{N}$) by the following formula:

$$
\Phi_{n_1, n_2, \ldots, n_N}(x_1, x_2, \ldots, x_N) = \mathcal{N} \begin{vmatrix} \phi_{n_1}(x_1) & \phi_{n_2}(x_1) & \phi_{n_3}(x_1) & \ldots \\ \phi_{n_1}(x_2) & \phi_{n_2}(x_2) & \phi_{n_3}(x_2) & \ldots \\ \ldots & \ldots & \ldots & \ldots \end{vmatrix}_+
$$

$$
= \mathcal{N} \sum_{P \in S_N} \widehat{P} \phi_{n_1}(x_1) \phi_{n_2}(x_2) \cdots \phi_{n_N}(x_N) .
$$

(1.270)

As you can see, the obtained expression is very similar to that for the fermions Eq. (1.268); the difference is in the absence of the sign factor ϵ_P. Otherwise, the permanent (denoted with the same symbol as the determinant, but with the plus subscript, see above) contains exactly the same terms. This time, however, the same states can be occupied by any number of bosons, i.e. the permanent is not equal to zero if some of its indices are equal, e.g. the kets may look like $|0120342\ldots\rangle$.

1.22.9 Dirac Equation

Here we shall illustrate matrices by considering the relativistic quantum equation for an electron with spin 1/2 that was proposed by Dirac in 1928. The driving idea was to suggest a quantum equation that would satisfy certain properties: (i) it would lead to a positively defined probability density; (ii) it would suggest a corresponding expression for the probability current density, so that both the probability and current densities are related by the continuity equation, (iii) the equation is relativistically invariant (would have the same form under various transformations of the space-time continuum) and, finally, (iv) the electronic energy in the state with a certain momentum would have a known relativistic form. In a non-relativistic quantum mechanics one introduces a scalar wavefunction ψ that satisfies a quantum equation (the Shrödinger equation) that is a partial differential equation with the first-order derivative with respect to time and the second-order partial derivatives. Since in the relativistic case the spatial coordinates and time enter on the same footing (as components of the so-called four-vector), both types of the partial derivatives in the relativistic case must be of the same order. Further, in order to preserve the superposition principle characteristic for quantum mechanical description, Dirac suggested the equation to be linear, i.e. spatial and time derivatives are to be of the first order only. The crucial (and non-trivial) proposal was that the relativistic description of an electron must be based not on a scalar wavefunction, but rather on n of them, i.e. on a a multi-component wavefunction, called *spinors*,

$$\Psi(\mathbf{x}, t) = \begin{pmatrix} \psi_1(\mathbf{x}, t) \\ \psi_2(\mathbf{x}, t) \\ \dots \\ \psi_n(\mathbf{x}, t) \end{pmatrix},$$

consisting of n scalar components $\{\psi_i(\mathbf{x}, t)\}$ each depending on the spatial coordinates $\mathbf{x} = (x_1, x_2, x_3)$ and time t. These components satisfy a set of n first-order partial differential equations of the most general form that we shall write below in the matrix form:

$$\frac{1}{c}\frac{\partial}{\partial t}\Psi + \sum_k \alpha_k \frac{\partial}{\partial x_k}\Psi + \frac{imc}{\hbar}\beta\Psi = \mathbf{0}, \tag{1.271}$$

where $\alpha_k = \left(\alpha_{\mu\nu}^{(k)}\right)$ are the three $n \times n$ matrices ($k = 1, 2, 3$), $\beta = (\mathbf{fi\text{-}\cdot})$ is another $n \times n$ matrix, c speed of light, $\hbar$ Planck constant, and m electron mass. We shall use Greek indices to indicate the components of the alpha and beta matrices that correspond to the spinor components, while Roman indices will be reserved for the Cartesian components. The four matrices introduced above are at this point still arbitrary and of yet unknown dimension n; they must be determined such that to satisfy the announced above properties. Note that the matrix equation (1.271) corresponds to n scalar equations

$$\frac{1}{c}\frac{\partial \psi_\mu}{\partial t} + \sum_k \sum_\nu \alpha_{\mu\nu}^{(k)} \frac{\partial \psi_\nu}{\partial x_k} + \frac{imc}{\hbar} \sum_\nu \beta_{\mu\nu}\psi_\nu = 0, \quad \mu = 1, 2, \ldots, n. \tag{1.272}$$

We shall start from the conditions that the probability density $\rho(\mathbf{r}, t)$ to be positively defined and also be related to some probability of current density $\mathbf{j}(\mathbf{r}, t)$ defined from the continuity equation

$$\text{div}\,\mathbf{j} + \frac{\partial \rho}{\partial t} = 0. \tag{1.273}$$

Since in non-relativistic quantum mechanics the probability density is defined as $|\psi|^2 = \psi^*\psi$, it is reasonable to define the probability density in the relativistic case as

$$\rho(\mathbf{x}, t) = e\Psi^\dagger(\mathbf{x}, t)\Psi(\mathbf{x}, t) = e\sum_\mu \psi_\mu^*(\mathbf{x}, t)\psi_\mu(\mathbf{x}, t), \tag{1.274}$$

with e being the electron charge. This definition guarantees the positiveness of the density. Let us calculate the time derivative of the density:

$$\frac{\partial \rho}{\partial t} = e\sum_\mu \left(\frac{\partial \psi_\mu^*}{\partial t}\psi_\mu + \psi_\mu^*\frac{\partial \psi_\mu}{\partial t}\right).$$

Employing Eq. (1.272) for the time derivative of every component, after some simple algebra we obtain

$$\frac{\partial \rho}{\partial t} = -ec \sum_k \left\{ \sum_{\mu\nu} \frac{\partial \psi_\mu^*}{\partial x_k} \alpha_{\nu\mu}^{(k)*} \psi_\nu + \psi_\mu^* \alpha_{\mu\nu}^{(k)} \frac{\partial \psi_\nu}{\partial x_k} \right\}$$

$$+ \frac{imc}{\hbar} \sum_{\mu\nu} \psi_\nu^* \left(\beta_{\nu\mu}^* - \beta_{\mu\nu} \right) \psi_\mu .$$

If we now assume that the matrix β is Hermitian, $\textbf{fi}^* = \textbf{fi}$, the last term will vanish. In order to make the first term in the right-hand side to be the divergence of a vector, we must also require that each of the three matrices α_k are Hermitian as well, $\alpha_{\mu\nu}^{(k)*} = \alpha_{\nu\mu}^{(k)}$. Then, we can combine the two terms in the curly brackets as

$$\{\ldots\} = \sum_{\mu\nu} \left(\frac{\partial \psi_\mu^*}{\partial x_k} \alpha_{\mu\nu}^{(k)} \psi_\nu + \psi_\mu^* \alpha_{\mu\nu}^{(k)} \frac{\partial \psi_\nu}{\partial x_k} \right) = \frac{\partial}{\partial x_k} \left(\sum_{\mu\nu} \psi_\mu^* \alpha_{\mu\nu}^{(k)} \psi_\nu \right) ,$$

leading finally to the equation:

$$\frac{\partial \rho}{\partial t} = -\sum_k \frac{\partial}{\partial x_k} \left(ec \sum_{\mu\nu} \psi_\mu^* \alpha_{\mu\nu}^{(k)} \psi_\nu \right) \equiv -\text{div} \left(ec \sum_{\mu\nu} \psi_\mu^* \boldsymbol{\alpha}_{\mu\nu} \psi_\nu \right) ,$$

i.e. the current density k-component becomes ($k = 1, 2, 3$)

$$j_k(\mathbf{x}, t) = ec \sum_{\mu\nu} \psi_\mu^* \alpha_{\mu\nu}^{(k)} \psi_\nu \equiv ec \boldsymbol{\Psi}^\dagger \boldsymbol{\alpha}_k \boldsymbol{\Psi} . \tag{1.275}$$

Here we have a matrix product of a vector-row $\boldsymbol{\Psi}^\dagger$, a square matrix $\boldsymbol{\alpha}_k$ that is the k-component of the matrix-vector

$$\boldsymbol{\alpha} = \{\boldsymbol{\alpha}_1, \boldsymbol{\alpha}_2, \boldsymbol{\alpha}_3\} = \left\{ \left(\alpha_{\mu\nu}^{(1)} \right), \left(\alpha_{\mu\nu}^{(2)} \right), \left(\alpha_{\mu\nu}^{(3)} \right) \right\} ,$$

and of the vector-column $\boldsymbol{\Psi}$, yielding the scalar j_k. That expression can also be formally rewritten for the whole vector of the current as $\mathbf{j} = ec\boldsymbol{\Psi}^\dagger \boldsymbol{\alpha} \boldsymbol{\Psi}$, where the vector character of it is provided by the vector-matrix $\boldsymbol{\alpha}$.

Up to this point the matrices α_k and β are not yet defined; we only know that they must be Hermitian. In order to find appropriate expressions for them, let us formally rewrite Eq. (1.272) in a more familiar form of the time-dependent Schrödinger equation as

$$i\hbar \frac{\partial}{\partial t} \boldsymbol{\Psi} = \mathbf{H}_D \boldsymbol{\Psi} ,$$

where the Hamiltonian matrix

$$\mathbf{H}_D = c\boldsymbol{\alpha} \cdot \mathbf{p} + mc^2\beta \tag{1.276}$$

contains the dot product (indicated by the dot for clarity) of the vector-matrix $\boldsymbol{\alpha}$ and of the momentum operator $\mathbf{p} = -i\hbar\nabla$ of the electron. The mechanical energy of a relativistic particle of the rest mass m and momentum $\mathbf{p}$ is $E_p = \sqrt{c^2\mathbf{p}^2 + m^2c^4}$. Hence, it is reasonable to require that $\mathbf{H}_D^2 = c^2\mathbf{p}^2 + m^2c^4$. This is because $\mathbf{H}_D$ contains the momentum linearly, but the energy must be proportional to its square. Let us see if this condition would enable us to impose the necessary constraints on the possible form of the unknown matrices α_k and β. We write

$$\begin{aligned}\mathbf{H}_D^2 &= \left(c\boldsymbol{\alpha} \cdot \mathbf{p} + mc^2\beta\right)\left(c\boldsymbol{\alpha} \cdot \mathbf{p} + mc^2\beta\right) \\ &= c^2\left(\boldsymbol{\alpha} \cdot \mathbf{p}\right)\left(\boldsymbol{\alpha} \cdot \mathbf{p}\right) + mc^3\left[\beta\left(\boldsymbol{\alpha} \cdot \mathbf{p}\right) + \left(\boldsymbol{\alpha} \cdot \mathbf{p}\right)\beta\right] + m^2c^4\beta^2.\end{aligned}$$

Let us write down explicitly the first two terms:

$$\begin{aligned}c^2\left(\boldsymbol{\alpha} \cdot \mathbf{p}\right)\left(\boldsymbol{\alpha} \cdot \mathbf{p}\right) &= c^2\sum_{kk'}\alpha_k\alpha_{k'}p_k p_{k'} \\ &= c^2\sum_{k}\alpha_k^2 p_k^2 + c^2\sum_{k<k'}\left(\alpha_k\alpha_{k'} + \alpha_{k'}\alpha_k\right)p_k p_{k'}\end{aligned}$$

(in the second term we have collected all off-diagonal terms in the double sum over all values of $k < k'$ exploiting the fact that for $k \neq k'$ the momentum operators commute[35]) and

$$mc^3\left[\beta\left(\boldsymbol{\alpha} \cdot \mathbf{p}\right) + \left(\boldsymbol{\alpha} \cdot \mathbf{p}\right)\beta\right] = mc^3\sum_{k}\left(\beta\alpha_k + \alpha_k\beta\right)p_k.$$

Now it is easy to see that the required expression for the square of the Hamiltonian is obtained if we impose the following conditions:

$$\beta^2 = \mathbf{E}, \quad \beta\alpha_k + \alpha_k\beta = \mathbf{0}, \quad \alpha_k\alpha_{k'} + \alpha_{k'}\alpha_k = 2\delta_{kk'}. \tag{1.277}$$

The last condition is equivalent to stating that α_1^2, α_2^2 and α_3^2 are the identity matrices $\mathbf{E}$ and $\alpha_k\alpha_{k'} + \alpha_{k'}\alpha_k$ is the zero matrix $\mathbf{0}$ for any $k \neq k'$, i.e. they anti-commute. Note that matrices α_k also anti-commute with β.

---

[35] Indeed, $p_k p_{k'} = (-i\hbar)^2\frac{\partial^2}{\partial x_k \partial x_{k'}} = p_{k'} p_k$ as the mixed derivative does not depend on the order.

Problem 1.182 Another alternative argument that could be put forward to derive the same conditions for the matrices α_k and β is to claim that each of the components ψ_ν of the vector $\mathbf{\Psi}$ satisfies the second order Klein–Gordon quantum equation

$$\left[\frac{1}{c^2}\frac{\partial^2}{\partial t^2} - \sum_k \frac{\partial^2}{\partial x_k^2} + \left(\frac{mc}{\hbar}\right)^2\right]\psi_\nu = 0, \quad \nu = 1, 2, \ldots, n.$$

Note that the operator in the second term inside the square brackets is the Laplacian Δ. Act with the operator

$$\frac{1}{c}\frac{\partial}{\partial t} - \sum_{k'} \alpha_{k'}\frac{\partial}{\partial x_{k'}} - \frac{imc}{\hbar}\beta$$

from the left on both sides of Eq. (1.271) and derive conditions (1.277).

This equation was developed for π and K mesons by O. Klein in 1926 and somewhat later by A. V. Fock and W. Gordon.

Even though we established (somewhat restrictive) conditions for the unknown matrices α_k and β, their dimension, and hence the number of components n in the state vector $\mathbf{\Psi}$, is still unknown. To investigate the possible values of n, consider the second condition (1.277) which we shall rewrite as $\beta\alpha_k = -\alpha_k\beta = (-\mathbf{E})\alpha_k\beta$, where $-\mathbf{E}$ is the minus identity matrix. Let us take the determinant of both sides of this equation recalling that the determinant of a product of matrices is equal to the product of their determinants, We shall immediately have that

$$\det(-\mathbf{E}) = (-1)^n = 1.$$

Hence, the dimension n must be even. If $n = 2$, then the four matrices α_k and β must be Pauli matrices of Eq. (1.22)

$$\sigma_1 = \begin{pmatrix} 0 & 1 \\ 1 & 0 \end{pmatrix}, \quad \sigma_2 = \begin{pmatrix} 0 & -i \\ i & 0 \end{pmatrix} \quad \text{and} \quad \sigma_3 = \begin{pmatrix} 1 & 0 \\ 0 & -1 \end{pmatrix}, \tag{1.278}$$

and the two-dimensional identity matrix $\beta = \mathbf{E}$ (or their linear combinations), since any two-dimensional matrix can be expressed via them uniquely, see Problem 1.26. However, the identity matrix commutes with any of the Pauli matrix, while the conditions require anti-commutation of β and any of the α_k. Hence, it is not possible to satisfy conditions (1.277) in two dimensions. The next possible dimension is $n = 4$, and it was established that the electron is correctly (i.e. in agreement with experiment) is described by this dimension. Hence, we shall accept that $n = 4$ from now on.

One possible choice of the 4×4 matrices satisfying all conditions is

$$\alpha_k = \begin{pmatrix} 0 & \sigma_k \\ \sigma_k & 0 \end{pmatrix}, \quad k = 1, 2, 3; \quad \beta = \begin{pmatrix} \mathbf{E}_2 & 0 \\ 0 & -\mathbf{E}_2 \end{pmatrix}$$

with $\mathbf{E}_2$ being a two-dimensional identity matrix. We have written these matrices in the block 2×2 form with each block being itself a 2×2 matrix. In more detailed form:

$$\alpha_1 = \begin{pmatrix} 0 & 0 & 0 & 1 \\ 0 & 0 & 1 & 0 \\ 0 & 1 & 0 & 0 \\ 1 & 0 & 0 & 0 \end{pmatrix}, \quad \alpha_2 = \begin{pmatrix} 0 & 0 & 0 & -i \\ 0 & 0 & i & 0 \\ 0 & -i & 0 & 0 \\ i & 0 & 0 & 0 \end{pmatrix}, \quad \alpha_3 = \begin{pmatrix} 0 & 0 & 1 & 0 \\ 0 & 0 & 0 & -1 \\ 1 & 0 & 0 & 0 \\ 0 & -1 & 0 & 0 \end{pmatrix},$$

and

$$\beta = \begin{pmatrix} 1 & 0 & 0 & 0 \\ 0 & 1 & 0 & 0 \\ 0 & 0 & -1 & 0 \\ 0 & 0 & 0 & -1 \end{pmatrix}.$$

Problem 1.183 Demonstrate by direct calculation that these matrices indeed satisfy the required conditions (1.277).

The proposed concrete form for the matrices is not unique as illustrated by the following problem.

Problem 1.184 Consider an arbitrary non-singular 4×4 matrix $\mathbf{S}$. Show that the matrices

$$\alpha'_k = \mathbf{S}\alpha_k\mathbf{S}^{-1} \ (k = 1, 2, 3) \quad \text{and} \quad \beta' = \mathbf{S}\beta\mathbf{S}^{-1}$$

satisfy the same conditions. Moreover, show that the physical properties of the electron also do not change if these matrices are used in the Dirac equation. In other words, show that the solution $\mathbf{\Psi}'$ of the Dirac equation with the primed matrices is related to its solution with the unprimed ones via $\mathbf{\Psi}' = \mathbf{S}\mathbf{\Psi}$, and that the probability and the current densities remain invariant:

$$\rho' = \left(\mathbf{\Psi}'\right)^{\dagger}\mathbf{\Psi}' \equiv e\mathbf{\Psi}^{\dagger}\mathbf{\Psi} = \rho,$$

$$\mathbf{j}' = ec\left(\mathbf{\Psi}'\right)^{\dagger}\alpha'\mathbf{\Psi}' \equiv ec\mathbf{\Psi}^{\dagger}\alpha\mathbf{\Psi} = \mathbf{j}.$$

Next, we shall rewrite the Dirac equation in the relativistically invariant form. To this end, we first generalise the equation and include the electro-magnetic (EM) field with the vector potential $\mathbf{A}(\mathbf{r}, t)$ and the scalar potential $A_0(\mathbf{r}, t)$. The required generalisation is performed by the following substitutions:

$$H_D \longmapsto H_D + eA_0, \quad \mathbf{p} \longmapsto \mathbf{p} - \frac{e}{c}\mathbf{A},$$

leading to the equation

$$i\hbar\frac{\partial}{\partial t}\mathbf{\Psi} = \left[c\boldsymbol{\alpha} \cdot \left(\mathbf{p} - \frac{e}{c}\mathbf{A}\right) + eA_0 + mc^2\beta\right]\mathbf{\Psi}. \qquad (1.279)$$

Next, we shall introduce the following 4-vectors (these will be denoted by their components and written non-bold): space-time $(x_\lambda) = (\mathbf{x}, ict) = (x_1, x_2, x_3, x_4)$ with $x_4 = ict$; four-vector potential of the EM field $(A_\lambda) = (\mathbf{A}, iA_0)$; the momentum vector operator

$$(p_\lambda) = \left(\mathbf{p}, -i\hbar\frac{\partial}{\partial x_4}\right) = \left(-i\hbar\nabla, -i\hbar\frac{\partial}{\partial x_4}\right) = \left(-i\hbar\nabla, -\frac{\hbar}{c}\frac{\partial}{\partial t}\right).$$

Next, we shall introduce four 4×4 matrices:

$$\gamma_k = -i\beta\alpha_k = i\begin{pmatrix} \mathbf{0} & -\sigma_k \\ \sigma_k & \mathbf{0} \end{pmatrix}, \quad k = 1, 2, 3,$$

and $\gamma_4 = \beta$.

Problem 1.185 Show that the gamma matrices satisfy

$$\gamma_\lambda\gamma_{\lambda'} + \gamma_{\lambda'}\gamma_\lambda = 2\mathbf{E}_4\delta_{\lambda\lambda'} \quad (\lambda, \lambda' = 1, 2, 3, 4).$$

Here $\mathbf{E}_4$ is the 4D identity matrix.

Problem 1.186 Show that the Dirac Eq. (1.279) can be rewritten as

$$\left[\sum_\lambda \gamma_\lambda\left(p_\lambda - \frac{e}{c}A_\lambda\right) - imc\right]\mathbf{\Psi} = \mathbf{0}.$$

It can be shown that this equation is relativistically invariant.

Problem 1.187 Show that the continuity Eq. (1.273) can also be rewritten as the zero 4D divergence,

$$\sum_\lambda \frac{\partial j_\lambda}{\partial x_\lambda} = 0,$$

where the four-vector current is defined via $(j_\lambda) = (\mathbf{j}, ic\rho)$.

1.22.10 Classical 1D Ising Model

Here we shall present an extremely elegant derivation of the free energy of a chain of magnetic atoms, the so-called 1D Ising model. Consider a 1D chain of N equidistant atoms whose magnetic moments (spins) S_n (where $n = 1, 2, \ldots, N$) can only take two values: up ($S_n = +1$) or down ($S_n = -1$). We shall assume periodic boundary conditions whereby the first atom neighbours atoms 2 and N, i.e. the atomic chain forms a ring as shown in Fig. 1.15. Formally, we can write that $S_{N+1} = S_1$. Neighbouring spins n and $n + 1$ interact via $-J S_n S_{n+1}$, where the parameter J is called the exchange interaction. The Hamiltonian of the chain

$$H = -J \sum_{n=1}^{N} S_n S_{n+1} - h \sum_{n=1}^{N} S_n, \tag{1.280}$$

where the first term accounts for the exchange interaction between all neighbouring pairs of spins and the second term is due to the interaction of spins with the magnetic field h. In order to calculate the free energy $F = -\beta^{-1} \ln Z$ of the chain, where $\beta = 1/(k_B T)$ is the inverse temperature (k_B is the Boltzmann constant), one has to calculate the partition function Z that accounts for all possible states the system of spins can take

$$Z = \sum_{\text{all states}} e^{-\beta H}.$$

Fig. 1.15 1D Ising chain of atoms in periodic boundary conditions. Each atom possesses a spin (shown in red) that is directed either up or down

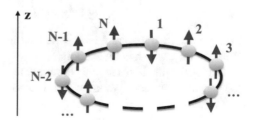

We can count all possible states by trying all possible combinations of spins: $S_1 = \pm 1$, $S_2 = \pm 1$, etc. Correspondingly, we can write

$$Z = \sum_{S_1 = \pm 1} \sum_{S_2 = \pm 1} \cdots \sum_{S_N = \pm 1} e^{-\beta H} .$$

This expression looks rather complicated. However, we shall demonstrate now that it can actually be calculated analytically using elements of the matrix algebra.

Indeed, let us split our Hamiltonian as follows:

$$H = \sum_{n=1}^{N} \left[-J S_n S_{n+1} - \frac{h}{2} (S_n + S_{n+1}) \right] = \sum_{n=1}^{N} H_{n,n+1} ,$$

and then rewrite the partition function by 'assigning' each term in the sum in H to a single sum in Z:

$$Z = \sum_{S_1 = \pm 1} e^{-\beta H_{1,2}} \sum_{S_2 = \pm 1} e^{-\beta H_{2,3}} \cdots \sum_{S_N = \pm 1} e^{-\beta H_{N,1}} .$$

With this formal division of the Hamiltonian, the partition function can be represented as a sum of products of simple terms:

$$Z = \sum_{\{S_n = \pm 1\}} T_{S_1 S_2} T_{S_2 S_3} \cdots T_{S_{N-1} S_N} T_{S_N S_1} , \tag{1.281}$$

where summation is performed over all possible values of the spins and we introduced a convenient notation:

$$T_{S_n S_{n+1}} = \exp \left(-\beta H_{n,n+1} \right) = \exp \left[\beta J S_n S_{n+1} + \frac{\beta h}{2} (S_n + S_{n+1}) \right] . \tag{1.282}$$

The notations just made are very suggestive offering a trick that would enable us to conclude our calculation. Indeed, if we now look at expression (1.281) and imagine that elements (1.282) form a 2×2 matrix

$$\mathbf{T} = \begin{pmatrix} T_{1,1} & T_{1,-1} \\ T_{-1,1} & T_{-1,-1} \end{pmatrix} = \begin{pmatrix} e^{\beta(J+h)} & e^{-\beta J} \\ e^{-\beta J} & e^{\beta(J-h)} \end{pmatrix} ,$$

then it becomes apparent that Z contains the trace of a product of N such matrices, $Z = \mathrm{Tr} \left(\mathbf{T}^N \right)$, which is quite cool!

Equipped with this extremely remarkable observation, we are now able to calculate the partition function. We know that the trace of a matrix is equal to the sum of its eigenvalues. If $\lambda_\pm$ are eigenvalues of $\mathbf{T}$, then the eigenvalues of the matrix $\mathbf{T}^N$ are $\lambda_\pm^N$. Hence, we are able to write a very simple result:

$$Z = \lambda_+^N + \lambda_-^N .$$

So, to conclude this derivation, we only need to calculate the eigenvalues of the matrix $\mathbf{T}$. We shall leave it as a problem for the reader.

Problem 1.188 Show that

$$\lambda_\pm = e^{\beta J} \left[\cosh (\beta h) \pm \sqrt{\sinh^2 (\beta h) + e^{-4\beta J}} \right] .$$

Once the partition function is known as a function of both temperature T and the magnetic field h, thermodynamic properties of the chain can be investigated analytically.

Chapter 2
Complex Numbers and Functions

We introduced complex numbers in Sect. I.1.11 and I.3.4 of the first volume.[1] There
we just defined the numbers themselves, but did not go any further. In fact, since the
introduction of complex numbers a number of centuries ago, the theory based on them
has been substantially developed into an extended analysis of complex functions
defined on the complex plane. The mathematical tool thus created represents an
extremely powerful device for solving practical problems ranging from calculating
real integrals to solving partial differential equations.

The purpose of this chapter is to consider in detail this elegant formalism. We shall
start by returning to complex numbers and the complex plane, then consider functions
on the complex plane, their differentiation and integration, complex functional series,
analytic continuation, residues, Frobenius method for solving ordinary differential
equations, and, finally, some applications in physics will conclude this chapter.

2.1 Representation of Complex Numbers

We shall start by introducing various representations of complex numbers. Some of
the material of this section the reader may find familiar because of the material we
covered in Sect. I.3.4 of the first volume; still, it is convenient to repeat here some
of the arguments to have the presentation more self-contained. Let us recall that a
complex number $z = x + iy$ can be written in the trigonometric form via its length
$r = \sqrt{x^2 + y^2} = |z|$ and phase ϕ (also called the argument of the complex number z
and denoted $\phi = \arg z$) as

---

[1] In the following, references to the first volume of this course (L. Kantorovich, Mathematics for
natural scientists: fundamentals and basics, 2nd Edition, Springer, 2022) will be made by appending
the Roman number I in front of the reference, e.g. Sect. I.1.8 or Eq. (I.5.18) refer to Sect. 1.8 and
Eq. (5.18) of the first volume, respectively.

© The Author(s), under exclusive license to Springer Nature Switzerland AG 2024   257
L. Kantorovich, *Mathematics for Natural Scientists II*, Undergraduate Lecture
Notes in Physics, https://doi.org/10.1007/978-3-031-46320-4_2

$$z = x + iy = r \left(\cos \phi + i \sin \phi \right) \ . \tag{2.1}$$

It is essential that the phase ϕ is defined up to an arbitrary integer number of 2π, i.e. the same complex number is obtained for the phase $\text{Arg} \, z = \phi + 2\pi n$, where $n = 0, \pm 1, \pm 2, \ldots$. This fact has profound consequences for functions of complex variables as we shall see below many times!

A complex number can be shown as a point with coordinates (x, y) on the 2D plane $x - y$, called the complex plane $\mathcal{C}$. It is also convenient to show a complex number $z = x + iy$ as a vector connecting the centre of the coordinate system with the point (x, y). Then r corresponds to the length of the vector, while the phase (up to a multiple of 2π)

$$\phi = \arctan \frac{y}{x} \tag{2.2}$$

is equal to the angle the vector makes with the x axis. By going around the $z = 0$ point (i.e. around the point with coordinates $(0, 0)$, the centre of the coordinate system) the phase acquires either 2π or -2π, depending on whether the direction of traverse is anti-clockwise or clockwise, respectively.

We know that one can perform algebraic operations with complex numbers respecting usual algebraic rules and an additional condition that $i^2 = -1$. Using known properties of the trigonometric functions, Sect. I.2.3.8, we shall now prove useful formulae for the product and division of complex numbers. Consider a product of two complex numbers z_1 and z_2 which are specified by their lengths r_1 and r_2 and their phases ϕ_1 and ϕ_2. Then, their product

$$
\begin{aligned}
z_1 z_2 &= r_1 \left(\cos \phi_1 + i \sin \phi_1 \right) r_2 \left(\cos \phi_2 + i \sin \phi_2 \right) \\
&= r_1 r_2 \left(\cos \phi_1 \cos \phi_2 + i \sin \phi_1 \cos \phi_2 + i \cos \phi_1 \sin \phi_2 + i^2 \sin \phi_1 \sin \phi_2 \right) \\
&= r_1 r_2 \left[\left(\cos \phi_1 \cos \phi_2 - \sin \phi_1 \sin \phi_2 \right) + i \left(\sin \phi_1 \cos \phi_2 + \cos \phi_1 \sin \phi_2 \right) \right] \\
&= r_1 r_2 \left[\cos \left(\phi_1 + \phi_2 \right) + i \sin \left(\phi_1 + \phi_2 \right) \right] \ ,
\end{aligned}
$$
$$\tag{2.3}$$

i.e. the product $z_3 = z_1 z_2$ is the complex with the length $r_3 = r_1 r_2$ and the phase $\phi_3 = \phi_1 + \phi_2$.

Problem 2.1 Show that if two complex numbers z_1 and z_2 are specified by the lengths r_1 and r_2 and the phases ϕ_1 and ϕ_2, then their division $z_3 = z_1/z_2$ is characterised by $r_3 = r_1/r_2$ and the phase $\phi_3 = \phi_1 - \phi_2$, i.e.

$$\frac{z_1}{z_2} = \frac{r_1}{r_2} \left[\cos \left(\phi_1 - \phi_2 \right) + i \sin \left(\phi_1 - \phi_2 \right) \right] \ . \tag{2.4}$$

Problem 2.2 Using the above derived trigonometric representations for a product and ratio of two complex numbers, conclude that:

$$|z_1 z_2| = |z_1| \, |z_2| \quad \text{and} \quad \left| \frac{z_1}{z_2} \right| = \frac{|z_1|}{|z_2|} \, .$$

Problem 2.3 Derive the following result for the power of a complex number by repeatedly using Eq. (2.3):

$$
\begin{aligned}
z^n &= r^n \left[\cos{(n\phi)} + i \sin{(n\phi)} \right] , \\
z^{-n} &= r^{-n} \left[\cos{(n\phi)} - i \sin{(n\phi)} \right] ,
\end{aligned}
\tag{2.5}
$$

where n is a positive integer. In particular, for positive n we have the formula due to de Moivre's:

$$(\cos{\phi} + i \sin{\phi})^n = \cos{(n\phi)} + i \sin{(n\phi)} \tag{2.6}$$

Problem 2.4 Also prove de Moivre's formula by induction.

This formula can actually be used for expressing cosine or sine of the angle $n\phi$ with integer n via cosine and sine of the angle ϕ (angle reduction). Indeed, by taking $n = 2$ and opening brackets in (2.6), we have:

$$\cos^2{\phi} + 2i \cos{\phi} \sin{\phi} - \sin^2{\phi} = \cos{(2\phi)} + i \sin{(2\phi)} \, .$$

The real part in the left-hand side should be equal to the real part in the right, and similarly for the imaginary parts, which give us two identities:

$$\cos{(2\phi)} = \cos^2{\phi} - \sin^2{\phi} \quad \text{and} \quad \sin{(2\phi)} = 2 \sin{\phi} \cos{\phi} \, ,$$

which we are well familiar with already.

Problem 2.5 Similarly, show that (cf. Eqs. (I.2.56)–(I.2.59)):

$$\cos{(3\phi)} = \cos^3{\phi} - 3 \sin^2{\phi} \cos{\phi} \, ;$$

$$\sin{3\phi} = 3 \sin{\phi} \cos^2{\phi} - \sin^3{\phi} \, ;$$

$$\cos{(4\phi)} = \cos^4{\phi} - 6 \sin^2{\phi} \cos^2{\phi} + \sin^4{\phi} \, ;$$

$$\sin{4\phi} = 4 \sin{\phi} \cos^3{\phi} - 4 \sin^3{\phi} \cos{\phi} \, .$$

Problem 2.6 Now obtain a general result applying the binomial expansion to the left-hand side of (2.6):

$$\cos(2p\phi) = \sum_{k=0}^{p} \binom{2p}{2k} (-1)^k \sin^{2k}\phi \cos^{2(p-k)}\phi \, ,$$

$$\sin(2p\phi) = \sum_{k=0}^{p-1} \binom{2p}{2k+1} (-1)^k \sin^{2k+1}\phi \cos^{2(p-k)-1}\phi \, ,$$

$$\cos((2p+1)\phi) = \sum_{k=0}^{p} \binom{2p+1}{2k} (-1)^k \sin^{2k}\phi \cos^{2(p-k)+1}\phi \, ,$$

$$\sin((2p+1)\phi) = \sum_{k=0}^{p} \binom{2p+1}{2k+1} (-1)^k \sin^{2k+1}\phi \cos^{2(p-k)}\phi \, .$$

Check that these equations reproduce the previous results of Problem 2.5.

Next we can re-derive the sum of cosine and sine functions we calculated previously, Eqs. (I.2.63) and (I.2.64), using the trigonometric representation of a complex number. In fact, we shall be even able to generalise these formulae a little. To this end, let us note that the formula (I.1.143) for the geometric progression is valid for complex q as well as we did not use in the derivation there that q was necessarily real (recall that the complex numbers are governed by the same algebraic rules as the real numbers). Then, we can write:

$$\sum_{k=0}^{n} z^k = \frac{1 - z^{n+1}}{1 - z}$$

with

$$z = r(\cos x + i \sin x) \quad \text{and} \quad z^k = r^k [\cos(kx) + i \sin(kx)] \, .$$

Therefore, the real part of the sum (denoted as $\mathrm{Re}(\ldots)$)

$$\mathrm{Re}\left(\sum_{k=0}^{n} z^k\right) = \mathrm{Re}\sum_{k=0}^{n} r^k [\cos(kx) + i \sin(kx)] = \sum_{k=0}^{n} r^k \cos(kx)$$

$$\text{must be the same as} \quad \mathrm{Re}\frac{1 - z^{n+1}}{1 - z} \, ,$$

while the imaginary part (denoted Im $(\ldots)$)

$$\mathrm{Im}\left(\sum_{k=0}^{n} z^k\right) = \mathrm{Im}\sum_{k=0}^{n} r^k \left[\cos(kx) + i\sin(kx)\right] = \sum_{k=0}^{n} r^k \sin(kx)$$

must be the same as $\quad \mathrm{Im}\dfrac{1 - z^{n+1}}{1 - z}$,

Therefore, by taking the real and imaginary parts of the expression of the sum of the geometric progression, we should be able to work out the sums of $r^k \cos(kx)$ and $r^k \sin(kx)$. The calculation is based on the fact that the product $zz^* = x^2 + y^2$ is real, where $x = \mathrm{Re}(z)$ and $y = \mathrm{Im}(z)$. Therefore, we multiply and divide the expression above by the complex conjugate of the denominator:

$$\frac{1 - z^{n+1}}{1 - z} = \frac{1 - z^{n+1}}{1 - z} \cdot \frac{1 - z^*}{1 - z^*}.$$

With this trick the denominator becomes real, so that only the numerator is complex:

$$(1 - z)(1 - z^*) = \left[(1 - r\cos x) - ir\sin x\right]\left[(1 - r\cos x) + ir\sin x\right]$$
$$= (1 - r\cos x)^2 + r^2\sin^2 x = 1 - 2r\cos x + r^2 .$$

Problem 2.7 Calculate the numerator to show that its real and imaginary parts are, correspondingly:

$$\mathrm{Re}\left[\left(1 - z^{n+1}\right)\left(1 - z^*\right)\right] = 1 - r\cos x + r^{n+2}\cos(nx) - r^{n+1}\cos((n+1)x) ,$$
$$\mathrm{Im}\left[\left(1 - z^{n+1}\right)\left(1 - z^*\right)\right] = r\sin x + r^{n+2}\sin(nx) - r^{n+1}\sin((n+1)x) .$$

Therefore, we finally obtain:

$$\sum_{k=0}^{n} r^k \cos(kx) = \frac{1 - r\cos x + r^{n+2}\cos(nx) - r^{n+1}\cos((n+1)x)}{1 - 2r\cos x + r^2} , \qquad (2.7)$$

$$\sum_{k=0}^{n} r^k \sin(kx) = \frac{r\sin x + r^{n+2}\sin(nx) - r^{n+1}\sin((n+1)x)}{1 - 2r\cos x + r^2} . \qquad (2.8)$$

Assuming that $0 < r < 1$, we can also calculate the infinite sums by taking the $n \to \infty$ limit (note that $r^n \to 0$):

$$\sum_{k=0}^{\infty} r^k \cos(kx) = \frac{1 - r \cos x}{1 - 2r \cos x + r^2} \,, \tag{2.9}$$

$$\sum_{k=0}^{\infty} r^k \sin(kx) = \frac{r \sin x}{1 - 2r \cos x + r^2} \,. \tag{2.10}$$

Problem 2.8 Prove that the series (2.9) and (2.10) converge only for $0 < r < 1$.

Problem 2.9 Check that by setting $r = 1$ in Eqs. (2.7) and (2.8), we recover Eqs. (I.2.63) and (I.2.64).

2.2 Functions on a Complex Plane

2.2.1 Regions in the Complex Plane and Mapping

Before introducing complex functions, we have to make some definitions and consider *regions* (or *domains*) in the complex plain $\mathcal{C}$ as these play crucial role in what follows.

In the world of real numbers we consider intervals like $a < x < b, a \le x < b$ or $a \le x \le b$, where the first one excludes the two boundary points a and b, while the other two include either one of them or both. Similarly, in the (two-dimensional) complex plane $\mathcal{C}$ we consider *open domains* (*regions*) where the boundary line (or boundary lines) are not included, or *closed domains* (*regions*) where they are included. To define what we mean by this, let us first define the circular ϵ-vicinity (or *neighbourhood*) of a point z_0 on the complex plane as a set of points z in it which satisfy the condition

$$|z - z_0| < \epsilon \,. \tag{2.11}$$

Here $|z - z_0| = \sqrt{(x - x_0)^2 + (y - y_0)^2}$ corresponds to the distance on the complex plane between the two points, z and z_0. Therefore, Eq. (2.11) selects all the points lying inside a circle of radius ϵ drawn with the point z_0 at its centre, see Fig. 2.1a. Note that the points at the boundary of the circle are strictly *not* selected because of the less (rather than less and equal) sign in Eq. (2.11). If we now consider a region D in the complex plane, see Fig. 2.1b, then three situations can be envisaged: an internal point z_1 of D, a boundary point z_2 lying on the boundary L of D (shown by the solid line), and finally a point z_3 which is outside D. For any internal point one can always find such its ϵ-vicinity (or such value of ϵ) that *all* its points belong to D;

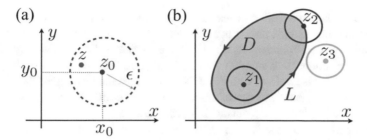

Fig. 2.1 **a** The ϵ-vicinity (neighbourhood) of the point z_0 includes all points z from C which lie inside the circle of radius ϵ centred at the point $z_0 = (x_0, y_0)$. **b** For any internal point z_1 of a region D one can always find its ϵ-vicinity such that it lies entirely inside D. This is however, not the case for the point z_2 lying on the boundary L of the region D: only some of the points of its any ϵ-vicinity lie inside D. For the point z_3 lying outside D, one can always find such its vicinity that lies entirely outside D

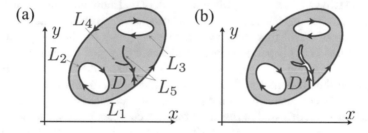

Fig. 2.2 Regions in the complex plane and their boundaries. **a** Region D has five boundaries $L_1 - L_5$, where L_1 is its outer boundary, L_2 and L_3 correspond to the boundaries of internal ovals taken out, and L_4 and L_5 correspond to the cuts made in D. The cuts are shown more clearly in **b**. Note the directions of traverse: their "positive" direction is always chosen such that the region D is on the left

if a point is on the boundary L of D, then for *any* ϵ-vicinity there will be points inside and outside of D including some boundary points, i.e. not all points in the ϵ-vicinity would belong to D; finally, if a point lies outside D, then one can always find such its ϵ-vicinity that all its points lie outside D, i.e. the entire ϵ-vicinity is outside D.

If a region in the complex plane has a single boundary (a single continuous line, either a smooth or a broken one[2]) as the one shown in Fig. 2.1b, it is called *simply connected*. However, regions in the complex plane may have a more complex structure. For instance, they may have not just one but several boundaries (i.e. several closed boundary lines). This happens when some regions of points on C are excluded and/or cuts made through accepted regions, as is schematically shown in Fig. 2.2. Indeed, the two white regions (with boundaries L_2 and L_3) which are cut off from the shaded region add two more boundary lines to it, so that now the shaded region has three boundary lines: L_1, L_2 and L_3. Regions which have more than one con-

---

[2] In other words, a closed piecewise line (i.e. consisting of smooth pieces which connect to each other).

tinuous boundary we shall call *multiply connected* or *non-simply connected*. One may say that our shaded region has two *holes*. Indicating explicitly the number of boundaries (closed boundary lines) existing in D as $k = 1, 2, \ldots$, we say that D is $(k - 1)$−fold-connected; simply connected regions have $k = 1$. Hence, the shaded region in Fig. 2.2 is twofold -connected. Note that the cuts by lines L_4 and L_5 in this figure do *not* change the number of boundary lines as it must be clear from Fig. 2.2b: they simply contribute to the outside boundary line of D.

When generalising the idea of integration, we shall need to define a *direction* along the boundary of a region. We shall consider the direction "positive" when we traverse the boundary in such a way that the region D is on the left (compare the Green's and Stock's theorems in Sects. I.6.4.3 and I.6.5.5). If the opposite direction is chosen with the points of D being on the right, the integrals will change their sign, so that this direction will then be called "negative".

Functions $f(z)$ on the complex plane C can be defined in a way similar to real functions of two variables $g(x, y)$ when each pair of real numbers x and y corresponds a complex number w. Hence, there is an important difference with the former case: if in the case of a real function $g = g(x, y)$ every point on the $x - y$ plane with the coordinates (x, y) is put into correspondence with a *real* number g, in the case of a complex function $w = f(z)$ we define the correspondence between the point $z = (x, y)$ in the 2D complex plane and a point $w = (u, v)$ on the same 2D plane, i.e. $w = u + iv$ contains two real numbers, not one as in the case of real functions $g(x, y)$. Therefore, this kind of functions map the complex plane onto itself. Hence. in this sense, there is more similarity with a real function of a single variable where the 1D space (the real x axis) is mapped onto itself. Hence, the mapping $w = f(z)$ is equivalent to two real functions of two real variables each:

$$ w = f(z) \quad \Longleftrightarrow \quad u = u(x, y) \quad \text{and} \quad v = v(x, y) . \tag{2.12} $$

When considering such a mapping, more complications may arise. For instance, we may have that for the same value of z several values of w are possible. This results in a multi-valued function as schematically shown in Fig. 2.3b. If only a single value of w exists for each value of z, but still several different values of z may result in the same w, the function $w = f(z)$ is called single-valued, see Fig. 2.3a. If any single z within some region D corresponds to a single value of w, and *vice versa*, we then have a mapping with one-to-one correspondence, Fig. 2.3c. We shall see that in many practical situations we shall be trying to find such regions of values of z and w so that the one-to-one mapping between them would be possible.

We note that this is not something really specific for functions of complex variables. Indeed, we know that a square root of a real positive number has two values: negative and positive, i.e. the function $y = \sqrt{x}$ has in fact two values, $\pm\sqrt{x}$, for the same value of x (assuming that by $\sqrt{x}$ we mean only the positive value of the root); therefore, in the world of real numbers we do face a similar problem. It is overcome by taking only the positive root, i.e. by assuming that $y = \sqrt{x}$ is always positive. In the case of complex functions defined on the complex plane this is an issue of fundamental importance and we shall encounter this over and over again. How this

Fig. 2.3 The function $w = f(z)$ performs a mapping from one complex plane C of z (left panels) to another one C of w (right panels): **a** the function is single-valued; **b** multi-valued; **c** mapping is a one-to-one correspondence, so that the inverse function $f^{-1}(w) = z$ can be defined

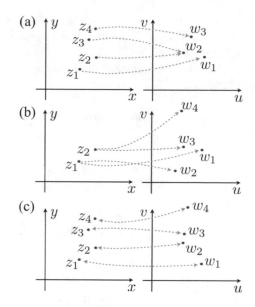

problem is tackled is similar to choosing a single root in the world of real numbers and will be considered later on in detail.

Suppose now that we have a one-to-one mapping between two complex domains of z and w. Then, it must be possible to define the inverse function $w = f^{-1}(z)$ such that $z = f(w)$. For two single-valued functions $w = f(z)$ and $q = \lambda(w)$ one can also introduce their superposition $q = \lambda(f(z))$. In particular, if $w = f(z)$ performs one-to-one mapping, then $f(f^{-1}(z)) = z$ and $f^{-1}(f(z)) = z$.

Next, we have to define the limit of a function in C. The limit $\lim_{z \to z_0} f(z)$ is defined similarly to that of a real function of two variables, i.e.

$$\lim_{z \to z_0} f(z) = \lim_{x \to x_0 , \, y \to y_0} f(z) = \lim_{x \to x_0 , \, y \to y_0} [u(x, y) + iv(x, y)]$$

$$= \lim_{x \to x_0 , \, y \to y_0} u(x, y) + i \lim_{x \to x_0 , \, y \to y_0} v(x, y) .$$

Alternatively, using the $\epsilon - \delta$ language, $\lim_{z \to z_0} f(z) = F$ if for any $\epsilon > 0$ one can always find such $\delta > 0$ that for any z satisfying $0 < |z - z_0| < \delta$ follows $|f(z) - F| = |w - F| < \epsilon$. In other words, points in the δ-vicinity of z_0 (excluding the point z_0 as the function $f(z)$ may not exist there) are mapped onto points of an ϵ-vicinity of F in the w plane. No matter how small the latter vicinity is, one can always find the corresponding vicinity of the point z_0 to accomplish the mapping.

It is clear that the limit exists if and only if the two limits for the functions $u(x, y)$ and $v(x, y)$ exist. The function $f(z)$ is *continuous*, if

$$\lim_{z \to z_0} f(z) = f(z_0) ,$$

similarly to the real functions. Clearly, both functions $u(x, y)$ and $v(x, y)$ are to be continuous for this to happen. It is essential that the limit must not depend on the path $z \to z_0$ in $\mathcal{C}$.

The usual properties of the limits apply then, similarly to the calculus of real functions:

$$\lim_{z \to z_0} (f(z) + g(z)) = \lim_{z \to z_0} f(z) + \lim_{z \to z_0} g(z) ,$$

$$\lim_{z \to z_0} (f(z)g(z)) = \left(\lim_{z \to z_0} f(z) \right) \left(\lim_{z \to z_0} g(z) \right) ,$$

$$\lim_{z \to z_0} \frac{f(z)}{g(z)} = \frac{\lim_{z \to z_0} f(z)}{\lim_{z \to z_0} g(z)} \quad (\text{if } g(z_0) \neq 0) ,$$

and

$$\lim_{z \to z_0} g(f(z)) = g \left(\lim_{z \to z_0} f(z) \right) .$$

It is also possible to show that if a closed region D is considered, the function $f(z)$ will be bounded, i.e. there exists such positive F that $|f(z)| \leq F$ for any z from D.

2.2.2 Differentiation. Analytic Functions

Now we are ready to consider differentiation of $f(z)$ with respect to z. We can define the derivative of $f(z)$ similarly to real functions as

$$f'(z) = \lim_{\Delta z \to 0} \frac{f(z + \Delta z) - f(z)}{\Delta z} . \tag{2.13}$$

It is essential that the limit must *not* depend on the path in which the complex number Δz approaches zero, Fig. 2.4, as otherwise this definition would not have any sense. This condition for a function $f(z)$ to be (complex) differentiable, puts

Fig. 2.4 For the derivative $f'(z)$ to exist, the limit in Eq. (2.13) must not depend on the path in which $\Delta z \to 0$

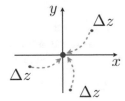

certain limitations on the function $f(z) = w = u(x, y) + iv(x, y)$ itself, which are formulated by the following Theorem.

Theorem 2.1 (Cauchy and Riemann) *In order for the function $f(z) = u + iv$ to be (complex) differentiable at $z = (x, y)$, where both $u(x, y)$ and $v(x, y)$ are differentiable around (x, y), it is necessary and sufficient that the following conditions are satisfied at this point:*

$$\frac{\partial u}{\partial x} = \frac{\partial v}{\partial y} \quad and \quad \frac{\partial u}{\partial y} = -\frac{\partial v}{\partial x}. \tag{2.14}$$

Proof We shall first prove the *necessary condition*. This means that we assume that $f'(z)$ exists and need to show that the conditions (2.14) are satisfied. Indeed, if the derivative exists, it means that it does not depend on the direction in which $\Delta z \to 0$. Therefore, let us take the limit (2.13) by approaching zero along the x axis, i.e. by considering $\Delta x \to 0$ and $\Delta y = 0$. We have:

$$
\begin{aligned}
f'(z) &= \lim_{\Delta x \to 0} \frac{[u(x + \Delta x, y) + iv(x + \Delta x, y)] - [u(x, y) + iv(x, y)]}{\Delta x} \\
&= \lim_{\Delta x \to 0} \frac{u(x + \Delta x, y) - u(x, y)}{\Delta x} + i \lim_{\Delta x \to 0} \frac{v(x + \Delta x, y) - v(x, y)}{\Delta x} \\
&= \frac{\partial u}{\partial x} + i \frac{\partial v}{\partial x},
\end{aligned}
$$

since along this path $\Delta z = \Delta x$. Alternatively, we may approach zero along the imaginary axis by taking $\Delta z = i\Delta y$ with $\Delta x = 0$. This must give the same complex number $f'(z)$:

$$
\begin{aligned}
f'(z) &= \lim_{\Delta y \to 0} \frac{[u(x, y + \Delta y) + iv(x, y + \Delta y)] - [u(x, y) + iv(x, y)]}{i\Delta y} \\
&= \frac{1}{i} \left[\lim_{\Delta y \to 0} \frac{u(x, y + \Delta y) - u(x, y)}{\Delta y} + i \lim_{\Delta y \to 0} \frac{v(x, y + \Delta y) - v(x, y)}{\Delta y} \right] \\
&= -i \frac{\partial u}{\partial y} + \frac{\partial v}{\partial y}.
\end{aligned}
$$

The two expressions must be identical as the derivative must not depend on the direction. Therefore, comparing the real and imaginary parts of the two expressions above, we obtain the required conditions (2.14).

Next we prove *sufficiency*: we are given the conditions, and we have to prove that the limit (2.13) exists. Assuming the functions $u(x, y)$ and $v(x, y)$ are differentiable, we can write (see Sect. I.5.3):

$$\Delta u = u\,(x + \Delta x, y + \Delta y) - u(x, y) = \frac{\partial u}{\partial x}\Delta x + \frac{\partial u}{\partial y}\Delta y + \alpha_1 \Delta x + \alpha_2 \Delta y$$

$$= \frac{\partial u}{\partial x}\Delta x - \frac{\partial v}{\partial x}\Delta y + \alpha_1 \Delta x + \alpha_2 \Delta y \;,$$

$$\Delta v = v\,(x + \Delta x, y + \Delta y) - v(x, y) = \frac{\partial v}{\partial x}\Delta x + \frac{\partial v}{\partial y}\Delta y + \beta_1 \Delta x + \beta_2 \Delta y$$

$$= \frac{\partial v}{\partial x}\Delta x + \frac{\partial u}{\partial x}\Delta y + \beta_1 \Delta x + \beta_2 \Delta y \;,$$

where α_1, α_2, β_1 and β_2 tend to zero if Δx and Δy tend to zero (i.e. $\Delta z \to 0$), and in the second passage in both cases we have made use of the conditions (2.14) by replacing all partial derivatives with respect to y with those with respect to x. Therefore, we can consider the difference of the function in Eq. (2.13):

$$
\begin{aligned}
\Delta f = f\,(z + \Delta z) - f(z) &= \Delta u + i\Delta v \\
&= \left(\frac{\partial u}{\partial x} + i\frac{\partial v}{\partial x}\right)\Delta x + \left(-\frac{\partial v}{\partial x} + i\frac{\partial u}{\partial x}\right)\Delta y + \gamma_1 \Delta x + \gamma_2 \Delta y \\
&= \left(\frac{\partial u}{\partial x} + i\frac{\partial v}{\partial x}\right)(\Delta x + i\Delta y) + \gamma_1 \Delta x + \gamma_2 \Delta y \\
&= \left(\frac{\partial u}{\partial x} + i\frac{\partial v}{\partial x}\right)\Delta z + \gamma_1 \Delta x + \gamma_2 \Delta y \;,
\end{aligned}
\tag{2.15}
$$

where $\gamma_1 = \alpha_1 + i\beta_1$ and $\gamma_2 = \alpha_2 + i\beta_2$ both tend to zero as $\Delta z \to 0$. Therefore,

$$\frac{f\,(z + \Delta z) - f(z)}{\Delta z} = \frac{\partial u}{\partial x} + i\frac{\partial v}{\partial x} + \gamma_1 \frac{\Delta x}{\Delta z} + \gamma_2 \frac{\Delta y}{\Delta z} \;. \tag{2.16}$$

It is easy to see that the fractions $\Delta x/\Delta z$ and $\Delta y/\Delta z$ are limited. To show this, let us write the complex number $\Delta z = \Delta r\,(\cos\phi + i\sin\phi)$ in the trigonometric form, then:

$$
\begin{aligned}
\frac{\Delta x}{\Delta z} &= \frac{\Delta r\cos\phi}{\Delta r\,(\cos\phi + i\sin\phi)} = \frac{\cos\phi}{\cos\phi + i\sin\phi} = \frac{\cos\phi\,(\cos\phi - i\sin\phi)}{(\cos\phi + i\sin\phi)(\cos\phi - i\sin\phi)} \\
&= \frac{\cos\phi\,(\cos\phi - i\sin\phi)}{\cos^2\phi + \sin^2\phi} = \cos\phi\,(\cos\phi - i\sin\phi) \;,
\end{aligned}
$$

and hence

$$\left|\frac{\Delta x}{\Delta z}\right| = |\cos\phi\,(\cos\phi - i\sin\phi)| \le |\cos\phi - i\sin\phi| = \sqrt{\cos^2\phi + \sin^2\phi} = 1 \;,$$

since the complex number $\cos\phi - i\sin\phi$ lies on a circle of the unit radius in the complex plane; similarly for the $\Delta y/\Delta z$. Therefore, when taking the limit $\Delta z \to 0$

in (2.16), we can ignore these fractions and consider only the limit of γ_1 and γ_2 which both tend to zero. Therefore, finally:

$$\lim_{\Delta z \to 0} \frac{f(z + \Delta z) - f(z)}{\Delta z} = \frac{\partial u}{\partial x} + i \frac{\partial v}{\partial x} ,$$

which is a well defined expression (since both u and v are differentiable), i.e. the derivative $f'(z)$ exists. **Q.E.D.**

Using the conditions (2.14) and the fact that the derivative should not depend on the direction in which $\Delta z \to 0$, we can write down several alternative expressions for the derivative:

$$f'(z) = \frac{\partial u}{\partial x} + i \frac{\partial v}{\partial x} = -i \frac{\partial u}{\partial y} + \frac{\partial v}{\partial y} = \frac{\partial v}{\partial y} + i \frac{\partial v}{\partial x} = \frac{\partial u}{\partial x} - i \frac{\partial u}{\partial y} . \qquad (2.17)$$

Since the derivative is basically based on the partial derivatives of the functions $u(x, y)$ and $v(x, y)$, all properties of the derivatives of real functions are carried over into here:

$$(f + g)' = f' + g' , \quad (fg)' = f'g + fg' \quad \text{and} \quad f(g(z))' = \frac{df}{dg}\frac{dg}{dz} .$$

If $z = f(w)$ does the unique (one-to-one) mapping, then the inverse function $w = f^{-1}(z)$ exists. Then, similarly to the real calculus,

$$\frac{dw}{dz} = \frac{d}{dz} f^{-1}(z) = \frac{1}{dz/dw} = \frac{1}{f'(w)}\bigg|_{w = f^{-1}(z)}$$

for the derivative of the inverse function.

It is said that if $f'(z)$ exists for any z belonging to some region D in the complex plane $\mathcal{C}$, it is *analytic* (or *holomorphic*) there. Hence, the complex-valued function is analytic (holomorphic) in a region D if it is complex differentiable at any point of that region. Analytic functions must only depend on z; they cannot depend on both z and $z*$.

As an example, consider the square function $w = z^2 = (x + iy)^2 = (x^2 - y^2) + i2xy$. First, let us show that it is analytic everywhere in $\mathcal{C}$. Indeed, for this function $u = x^2 - y^2$ and $v = 2xy$, and hence

$$\frac{\partial u}{\partial x} = 2x , \quad \frac{\partial v}{\partial y} = 2x , \quad \frac{\partial u}{\partial y} = -2y , \quad \frac{\partial v}{\partial x} = 2y ,$$

and the Cauchy-Riemann conditions (2.14) are indeed satisfied as can be easily checked. Correspondingly,

$$\left(z^2\right)' = \frac{\partial u}{\partial x} + i\frac{\partial v}{\partial x} = 2x + i2y = 2(x + iy) = 2z .$$

Problem 2.10 Show that the functions z^3 and z^4 are analytic everywhere, and hence, using any form of Eq. (2.17), show that $\left(z^3\right)' = 3z^2$ and $\left(2z^4 + z\right)' = 8z^3 + 1$.

Problem 2.11 Generalise these results for the general natural power function $w = z^n$ by showing that $(z^n)' = nz^{n-1}$. [Hint: *use the binomial expansion for* $z^n = (x + iy)^n$.]

Problem 2.12 When proving Theorem 2.1 we considered the derivative $f'(z)$ taken along the x and iy directions. Prove that the derivative $f'(z)$ of the analytic function $f(z)$ would not change if the limit is taken along an arbitrary direction $\mathbf{l}$ in the complex plane. [Hint: *define a unit vector* $\mathbf{l}$ *by the angle* α *it makes with the x axis; then* $\Delta y = \Delta x \tan \alpha$.]

Problem 2.13 Prove that if the function $f(z) = u(x, y) + iv(x, y)$ is analytic, then its real and imaginary parts are harmonic functions, i.e. each of them satisfies the two-dimensional Laplace equation:

$$\frac{\partial^2 u}{\partial x^2} + \frac{\partial^2 u}{\partial y^2} = 0 \quad \text{and} \quad \frac{\partial^2 v}{\partial x^2} + \frac{\partial^2 v}{\partial y^2} = 0 . \tag{2.18}$$

[Hint: *differentiate the conditions* (2.14).]

Because of the Cauchy-Riemann conditions (2.14) the whole function $f(z)$ is in fact nearly fully determined (up to a complex constant) by its either real or imaginary part. We shall illustrate this by the following example.

Example 2.1 ▶ Let the real part of $f(z)$ be $u(x, y) = xy$. Let us determine the imaginary part of it. Use the first condition:

$$\frac{\partial u}{\partial x} = y \quad \Longrightarrow \quad \frac{\partial v}{\partial y} = y \quad \Longrightarrow \quad v(x, y) = \int y\,dy = \frac{1}{2}y^2 + \lambda(x) .$$

Here we wrote v as a $y-$integral since $\partial v/\partial y = y$; this follows from the general property of the indefinite integrals. Also, the integration is performed only over y, i.e. keeping x constant, and $\lambda(x)$ appears as an arbitrary function of x, i.e. $\lambda(x)$ is yet to be determined. The above result however fully determines the dependence of $v(x, y)$ on y. To find the function $\lambda(x)$, we employ the second condition (2.14):

$$\frac{\partial u}{\partial y} = x \implies \frac{\partial v}{\partial x} = -x = \frac{d\lambda}{dx}$$

$$\implies \lambda(x) = -\int x dx = -\frac{1}{2}x^2 + C \,,$$

where C is a real constant. Hence, we finally obtain:

$$v(x, y) = \frac{1}{2}y^2 - \frac{1}{2}x^2 + C$$

$$\implies f(z) = u + iv = xy + i\left(\frac{1}{2}y^2 - \frac{1}{2}x^2 + C\right) = xy + i\frac{y^2 - x^2}{2} + iC \,.$$

It is easy to see now that $f(z) = -iz^2/2 + iC$. Indeed,

$$-\frac{i}{2}z^2 + iC = -\frac{i}{2}(x + iy)^2 + iC = xy - i\frac{1}{2}\left(x^2 - y^2\right) + iC \,,$$

which is precisely our function. This example shows that if $u(x, y)$ is known, then the corresponding imaginary part $v(x, y)$ can indeed be found. It also illustrates that the final function must be expressible entirely via z as $f(z)$. Not in all cases this is possible; some $u(x, y)$ (or $v(x, y)$) do not correspond to any $f(z)$. For instance, if $u(x, y) = x^3 - 3x^2y$, then we first obtain:

$$\frac{\partial u}{\partial x} = 3x^2 - 6xy \implies \frac{\partial v}{\partial y} = 3x^2 - 6xy$$

$$\implies v(x, y) = \int \left(3x^2 - 6xy\right) dy$$

$$= 3x^2y - 3xy^2 + \lambda(x) \,;$$

however, the second conditions yields:

$$\frac{\partial u}{\partial y} = -3x^2 \implies \frac{\partial v}{\partial x} = 3x^2$$

$$\implies 3x^2 = \frac{\partial}{\partial x}\left(3x^2y - 3xy^2 + \lambda(x)\right)$$

$$= 6xy - 3y^2 + \frac{d\lambda}{dx} \,,$$

which does not give an equation for $\lambda(x)$ since terms with y do not cancel out. This means that there is no function $f(z)$ having the real part equal to $x^3 - 3x^2y$.

Problem 2.14 Find the analytic function $f = u + iv$ if $u = x^3 - 3xy^2 + 1$.
[Answer: $v = 3x^2y - y^3 + C$ and $f(z) = z^3 + 1 + iC$.]

Problem 2.15 Find the analytic function $f = u + iv$ if $v = -y/\left(x^2 + y^2\right)$.
[Answer: $u = x/\left(x^2 + y^2\right) + C$ and $f(z) = 1/z + C$.]

Problem 2.16 In this problem we shall derive the Cauchy-Riemann conditions in polar coordinates. Let us consider the functions $u(x, y)$ and $v(x, y)$ specified via (r, ϕ) instead: $u = u(r, \phi)$ and $v = v(r, \phi)$, where $x = r \cos \phi$ and $y = r \sin \phi$. By differentiating the latter equations with respect to x and y, four algebraic equations are obtained for the partial derivatives of r and ϕ with respect to x and y. Show that

$$\frac{\partial r}{\partial x} = \cos \phi, \quad \frac{\partial r}{\partial y} = \sin \phi, \quad \frac{\partial \phi}{\partial x} = -\frac{\sin \phi}{r} \quad \text{and} \quad \frac{\partial \phi}{\partial y} = \frac{\cos \phi}{r}.$$

Then, considering the functions $u = u(x, y)$ and $v = v(x, y)$ as composite functions of r and ϕ, relate the partial derivatives of u and v with respect to x and y to their partial derivatives with respect to r and ϕ, and hence show that the Cauchy-Riemann conditions read as follows:

$$\frac{\partial u}{\partial r} \cos \phi - \frac{\partial u}{\partial \phi} \frac{\sin \phi}{r} = \frac{\partial v}{\partial r} \sin \phi + \frac{\partial v}{\partial \phi} \frac{\cos \phi}{r},$$

$$\frac{\partial u}{\partial r} \sin \phi + \frac{\partial u}{\partial \phi} \frac{\cos \phi}{r} = -\frac{\partial v}{\partial r} \cos \phi + \frac{\partial v}{\partial \phi} \frac{\sin \phi}{r}.$$

Finally, manipulate these equations to show that

$$\frac{\partial u}{\partial r} = \frac{1}{r} \frac{\partial v}{\partial \phi} \quad \text{and} \quad \frac{\partial v}{\partial r} = -\frac{1}{r} \frac{\partial u}{\partial \phi}.$$

2.3   Main Elementary Functions

Having introduced main definitions and concepts related to analytic functions, we are now ready to consider main elementary functions to see if it is possible to define our usual functions on the complex plane.

2.3.1 Integer Power Function

Consider $w = z^n$ with n being a positive integer. If $|z| = r$ and $arg\,(z) = \phi$, then we know from Eq. (2.5) that $|z^n| = r^n$ and $arg\,(z^n) = n\phi$. So, the power function is single-valued.

It is however easy to see that the mapping here is given by that depicted in Fig. 2.3a where two points, z_2 and z_3, go over into a single point w_2 of the function. Indeed, let us see if there exist two *different* points z_1 and z_2 (given by their absolute values r_1 and r_2 and the phases ϕ_1 and ϕ_2) which give the same power, i.e. for which $z_1^n = z_2^n$. Naively, we may write two equations: $r_1^n = r_2^n$ and $n\phi_1 = n\phi_2$. The first one yields simply $r_1 = r_2$. However, the second equation is not quite correct as the phase may differ by an integer number of 2π, and hence is to be replaced by $n\phi_1 = n\phi_2 + 2\pi k$ with k being any integer (including zero); therefore, $\phi_1 = \phi_2 + 2\pi k/n$. Hence, the points on a circle of radius $r_1 = r_2$ which have difference in phases of $2\pi k/n$ transform by the power function $w = z^n$ into the same point w in C. Correspondingly, if we only consider a sector of points with the phases satisfying the inequality

$$\frac{2\pi}{n}(k-1) < \phi < \frac{2\pi}{n}k \,, \tag{2.19}$$

then they would all transform by $w = z^n$ into different points, i.e. there will be a one-to-one mapping of that sector into the complex plane. Here $k = 1, \ldots, n$, i.e. the complex plane is divided into n such sectors as shown in Fig. 2.5 for the case of $n = 8$. If we consider any points z inside of any of these sectors, they will provide a unique mapping by means of the power function $w = z^n$.

Using the binomial expansion, we have proven in problem 2.11 that this function is analytic. In fact, a much simpler prove exists. Indeed, $w = u + iv = (x + iy)^n$, with $u = \mathrm{Re}\,(x + iy)^n$ and $v = \mathrm{Im}\,(x + iy)^n$. Therefore,

$$\frac{\partial u}{\partial x} = \mathrm{Re}\left[\frac{\partial}{\partial x}(x + iy)^n\right] = \mathrm{Re}\left[n(x + iy)^{n-1}\right]$$

$$= \mathrm{Re}\left[n(a + ib)\right] = na \,,$$

Fig. 2.5 In the case of $n = 8$ there will be eight sectors with the angles $2\pi/8 = \pi/4$ (or 45°). The z points inside any of the sectors provide a one-to-one mapping with the function $w = z^n$

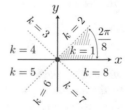

$$\frac{\partial v}{\partial y} = \mathrm{Im}\left[\frac{\partial}{\partial y}(x+iy)^n\right] = \mathrm{Im}\left[ni\,(x+iy)^{n-1}\right]$$

$$= \mathrm{Im}\left[in\,(a+ib)\right] = \mathrm{Im}\,(ina - nb) = na\,,$$

i.e. both are the same. Above, a and b are real and imaginary parts of the complex number $(x+iy)^{n-1}$.

Problem 2.17 Show that the second condition (2.14) is also satisfied.

Correspondingly, as the power function is analytic, we can calculate its derivative using any of the expressions in (2.17); for instance, using the first one (derivative with respect to x), we get

$$\left(z^n\right)' = \frac{\partial}{\partial x}(x+iy)^n = n\,(x+iy)^{n-1} = nz^{n-1}\,,$$

i.e. the same result as in the real calculus, i.e. as if the derivative was taken directly with respect to z.

2.3.2 Integer Root Function

The function $w = \sqrt[n]{z}$ is defined as an inverse to the $n-$power function, i.e. $z = w^n$. We have seen above that n different values of w differing only by $2\pi/n$ in their phase correspond to the same value of z. This means that for the given $z = r\,(\cos\phi + i\sin\phi)$ there are n values of w (roots), which are defined via $z = w^n$, i.e. we have for $w = \rho\,(\cos\psi + i\sin\psi)$ the absolute value ρ and the phase ψ satisfying: $\rho^n = r$ and $n\psi = \phi + 2\pi k$, i.e.

$$\left|\sqrt[n]{z}\right| = \rho = \sqrt[n]{r}\quad\text{and}\quad arg\left(\sqrt[n]{z}\right) = \psi = \frac{\phi+2\pi k}{n} = \frac{\phi}{n} + \frac{2\pi}{n}k\,,\tag{2.20}$$

$$\text{where}\quad k = 0, 1, 2, \ldots, n-1\,,$$

which means

$$\sqrt[n]{z} = \sqrt[n]{r}\,(\cos\psi_k + i\sin\psi_k)\,,\quad \psi_k = \frac{\phi}{n} + \frac{2\pi}{n}k\,,\tag{2.21}$$

where the index k numbers all the roots. Note that there are only n different values of the integer k possible, e.g. the ones given above; all additional values repeat the same roots. So, we conclude, that the $n-$root of any complex number z (except of $z = 0$) has n roots. The point $z = 0$ is indeed somewhat special as it only has a single root for any n. We shall see in a moment the special meaning of this particular point.

Fig. 2.6 Two square roots of a negative real $x = -|x|$

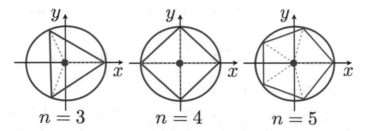

Fig. 2.7 Roots $\sqrt[n]{1}$ for $n = 3, 4, 5$ form vertices of regular polygons inserted inside the circle of unit radius

Example 2.2 ▶ As an example, consider the case of $n = 2$; we should have two roots:

$$\left(\sqrt{z}\right)_1 = \sqrt{r}\left(\cos\frac{\phi}{2} + i\sin\frac{\phi}{2}\right),$$

$$\left(\sqrt{z}\right)_2 = \sqrt{r}\left[\cos\left(\frac{\phi}{2} + \pi\right) + i\sin\left(\frac{\phi}{2} + \pi\right)\right]$$

$$= \sqrt{r}\left(-\cos\frac{\phi}{2} - i\sin\frac{\phi}{2}\right) = -\left(\sqrt{z}\right)_1.$$

In particular, if z is real positive ($\phi = 0$ and $z = x > 0$), the two roots are $\sqrt{x}$ and $-\sqrt{x}$, as one would expect. If z is real negative, then $\phi = \pi$, and the two roots become

$$\left(\sqrt{z}\right)_1 = \sqrt{|x|}\left(\cos\frac{\pi}{2} + i\sin\frac{\pi}{2}\right) = i\sqrt{|x|} \quad \text{and} \quad \left(\sqrt{z}\right)_2 = -i\sqrt{|x|}.$$

These are shown as dots in Fig. 2.6.

As an another example, consider all n−roots of 1. Here $r = 1$ and $\phi = 0$, so the n−roots of 1 would all have the absolute value equal to one, and the phases $\psi_k = 2\pi k/n$. The roots form vertices of a regular n−polygon fit into a circle of unit radius as shown for $n = 3, 4, 5$ in Fig. 2.7.

In a general case of calculating all roots of a number $z = x + iy$, it is necessary to calculate r and ϕ first, and then work out all the roots. For instance, consider

$\sqrt[3]{1 + 2i}$. Here $r = \sqrt{1^2 + 2^2} = \sqrt{5}$ and $\phi = \arctan(2/1) = \arctan 2$. Therefore, the three roots are:

$$z_k = \sqrt[3]{5}\left[\cos\phi_k + i\sin\phi_k\right] \quad \text{with} \quad \phi_k = \frac{1}{3}\arctan 2 + \frac{2\pi}{3}k, \quad k = 0, 1, 2.$$

As an example of an application, let us obtain all roots of a quadratic equation $x^2 + 2x + 4 = 0$. Using the general expression for the roots of the quadratic equation, we can write:

$$x_{1,2} = -1 \pm \sqrt{1 - 4} = -1 \pm \sqrt{-3} = -1 \pm i\sqrt{3}.$$

Here the $\pm$ sign takes care of the two values of the square root. ◀

Problem 2.18 Prove that the sum of all roots of 1 is equal to zero. [*Hint: use Eqs.* (2.7) *and* (2.8) *.*]

Problem 2.19 Obtain all roots of the following quadratic equations:

$$x^2 + x + 1 = 0; \quad x^2 + 2x + i = 0; \quad 2x^2 - 3x + 3 = 0.$$

[Answers: $-1/2 \pm i\sqrt{3}/2$; $-1 \pm 2^{1/4}\left[\cos(\pi/8) - i\sin(\pi/8)\right]$; $\left(3 \pm i\sqrt{15}\right)/4$.]

Problem 2.20 Obtain all roots of the equation $z^4 + a^4 = 0$. [*Answer: the four roots are* $\pm(\pm 1 + i)\,a/\sqrt{2}$]

Problem 2.21 Show that all four roots of the equation ($g > 1$)

$$x^4 + 2\left(2g^2 - 1\right)x^2 + 1 = 0$$

are $\pm i\sqrt{2g^2 - 1 \pm 2g\sqrt{g^2 - 1}}$.

Now let us consider in more detail how different roots of the same complex number z are related to each other. Let us start from the square root $w = \sqrt{z}$. If $|z| = r$ and $arg(z) = \phi$, then the first root w_1 is given by $|w_1| = \sqrt{r}$ and $arg(w_1) = \phi/2$. If we change ϕ of z within the limits of 0 and 2π, excluding the boundary values themselves, i.e. $0 < \phi < 2\pi$, then the argument of w_1 would change (see Eq. (2.21)) within the limits of 0 and π, i.e. $0 < arg(w_1) < \pi$. However, if we consider the second root w_2 of the function $w = \sqrt{z}$, then over the same range of z values the phase of the root w_2 would vary within the interval $\pi < arg(w_2) < 2\pi$. This is schematically shown in Fig. 2.8: when z is passed along the closed loop shown in (a) which is not crossing the positive part of the x axis, the first root w_1 traverses only over the upper loop in (b), while the second root w_2 over the lower

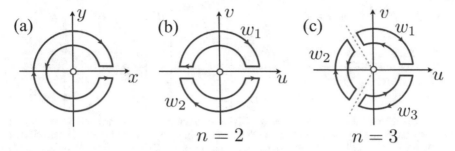

Fig. 2.8 When a contour shown in **a** is passed by the variable z, different regions in the complex plane are passed by the function $w = u + iv = \sqrt[n]{z}$ as shown in **b** and **c** for the cases of $n = 2$ and $n = 3$

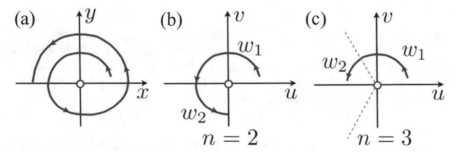

Fig. 2.9 As z changes along the path in **a**, the value of the root $w = \sqrt[n]{z}$ goes from w_1 domain to the w_2 domain as shown for the cases of $n = 2$ (**b**) and $n = 3$ (**c**)

part. Similarly, in the case of the function $w = \sqrt[3]{z}$ the arguments of the three roots lie within the intervals, correspondingly, $0 < arg(w_1) < 2\pi/3, 2\pi/3 < arg(w_2) < 4\pi/3$ and $4\pi/3 < arg(w_3) < 2\pi$. Therefore, if we imagine taking z along the same contour shown in Fig. 2.8a, the three roots would traverse along the three paths shown in (c) of the same Figure. The root function $w = \sqrt[n]{z}$ under this condition $(0 < arg(z) < 2\pi)$ is clearly single valued and we can choose any of the roots.

Therefore, if z does not cross the positive part of the x axis from below, i.e. the $z = 0$ point is not completely circled, each of the roots remains stable within their respective regions. Let us now imagine that we take a contour in which the positive part of the x axis is crossed from below, i.e. the $z = 0$ is fully circled ($arg(z)$ goes beyond 2π) as is shown in Fig. 2.9a. In this case, if we initially start from the root w_1, its phase $arg(w_1)$ goes beyond its range (π in the case of $n = 2$ and $2\pi/3$ in the case of $n = 3$) as shown in Fig. 2.9b, c, and thus the root function takes on the next value w_2.

Looking at this slightly differently, consider a point $z_0 = x > 0$ on the positive part of the x axis (i.e. with $|z_0| = x$ and $arg(z) = 0$). Its first square root $w_1 = \sqrt{z_0} = +\sqrt{x}$ since $\left|\sqrt{z_0}\right| = \sqrt{x}$ and $arg\left(\sqrt{z_0}\right) = 0$. After the complete circle we arrive at the point z_1 where 2π is added to the argument of z_0; the first square root becomes $\left|\sqrt{z_1}\right| = \sqrt{x}$ and $arg\left(\sqrt{z_1}\right) = 2\pi/2 = \pi$. However, z_1 is obviously the

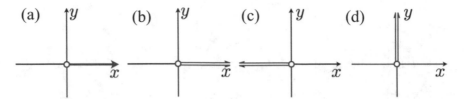

Fig. 2.10 To avoid multi-valued character of the n−root function, a "cut" is to be made from the $z = 0$ point to infinity in order to restrict the phase of z in such a way that the maximum change of $arg(z)$ would not exceed 2π. For example, this can be done either by cutting along the positive part of the x axis as shown in **a** (in **b** the cut is shown clearer), in which case $0 < arg(z) < 2\pi$, or along its negative part as in **c** in which case $-\pi < arg(z) < \pi$, or along e.g. the upper part of the imaginary axis as in **d** when $-3\pi/2 < arg(z) < \pi/2$

same point on the diagram as z_0! So, the function $w = \sqrt{z}$ jumps on the positive part of the x axis to its second instance w_2 and hence w_1 becomes discontinuous. Hence the root function is multi-valued. In the case of the square root, two complete revolutions will bring the root function back to w_1; in the case of $\sqrt[3]{z}$ three revolutions are necessary. Generally, in the case of $w = \sqrt[n]{z}$ each full rotation of z around the $z = 0$ point will take the root w_k to w_{k+1}, and exactly n rotations will return w_n back to w_1.

This complication happens because the $z = 0$ point is so special: it is called the *branch point*. If we limit the argument of z to stay within a 2π interval between $2\pi k$ and $2\pi(k + 1)$, or, more precisely, $2\pi k \leq arg(z) < 2\pi(k + 1)$), then any of the roots, which is specified by the integer $k = 0, 1, \ldots, n - 1$, would correspond to a well defined single-valued function. Each such function is related to different *branches* of the root function. These are related to each other by an extra phase of $2\pi/n$, so that $arg(w_{k-1}) = arg(w_k) + 2\pi/n$. To facilitate this limitation on the possible values of the phase of z, a *cut* is made from the $z = 0$ point to infinity to disallow crossing the cut when traversing the complex plane putting natural limits to the possible change of the phase. An example of such a cut is shown in Fig. 2.10a, b. In this case $0 < arg(z) < 2\pi$. Note however, that the domain of arguments of z depends on where the cut is made. In Fig. 2.10c another possibility is shown when the argument of z varies between $-\pi$ and π; for the cut made along the upper part of the imaginary y axis, Fig. 2.10d, we have instead $-3\pi/2 < arg(z) < \pi/2$.

Some functions may have more than one brunch point to be defined. Consider for instance $w = \sqrt{z^2 - 1} = \sqrt{z - 1}\sqrt{z + 1}$. It has two brunch points, $z_1 = 1$ and $z_2 = -1$ due to each individual square root function. If we take a general point z, then it can be represented either with respect to the one or the other brunch point as must be clearer from the drawing made in Fig. 2.11a:

$$z - 1 = x_1 + iy_1 = r_1 \left(\cos \phi_1 + i \sin \phi_1 \right)$$

or

$$z + 1 = x_2 + iy_2 = r_2 \left(\cos \phi_2 + i \sin \phi_2 \right).$$

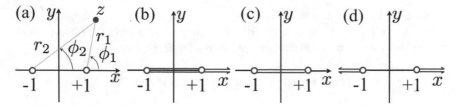

Fig. 2.11 **a** The function $w = \sqrt{z^2 - 1}$ has two brunch points: $z = \pm 1$. A general point z in the complex plane can be represented in two ways: with respect to either the brunch point $z = 1$ (using r_1 and ϕ_1) or $z = -1$ (using r_2 and ϕ_2). Here r_1 and r_2 are the corresponding distances to the brunch points. **b** One brunch cut is drawn from $z = -1$ to infinity along the positive x direction, while the other brunch cut is drawn in the same direction but from the point $z = 1$. **c** The previous case is equivalent to having the brunch cut drawn only between the points $z = -1$ and $z = 1$; as in **b** $w(z)$ appears to be continuous everywhere outside the cut (see text). **d** Another possible choice of the two brunch cuts which leads to a different function $w(z)$

The brunch cuts can be made in various ways, as the idea is to limit the phase of z such that the function $w(z)$ be single-valued. Two such other possibilities for $w = \sqrt{z^2 - 1}$ are shown in Fig. 2.11b, d. Consider first the construction shown in (b) where the two cuts are drawn in the positive direction of the x axis from both points (so that the cuts overlap in the region $x > 1$). In this case the angles ϕ_1 and ϕ_2 change within the same intervals: $0 < \phi_1 < 2\pi$ and $0 < \phi_2 < 2\pi$. To understand the behaviour of $w(z) = \sqrt{z^2 - 1}$, it is sufficient to calculate it on both sides of the x axis only, i.e. above ($y = +0$) and below ($y = -0$) it. We have to consider three cases: (i) $x < -1$, (ii) $-1 < x < 1$ and (iii) $x > 1$. Generally, for any z we can write

$$
\begin{aligned}
w &= \sqrt{z - 1}\sqrt{z + 1} \\
&= \sqrt{r_1 r_2}\, (\cos \phi_1 + i \sin \phi_1)\, (\cos \phi_2 + i \sin \phi_2) \\
&= \sqrt{r_1 r_2}\, [\cos (\phi_1 + \phi_2) + i \sin (\phi_1 + \phi_2)] \\
&= \sqrt{r_1 r_2} \left[\cos \frac{\phi_1 + \phi_2}{2} + i \sin \frac{\phi_1 + \phi_2}{2} \right],
\end{aligned}
$$

i.e.

$$
arg(w) = \frac{\phi_1 + \phi_2}{2}.
$$

For $x > 1$ we have $r_1 = x - 1$ and $r_2 = x + 1$. Then, on the upper side of the cut $\phi_1 = \phi_2 = 0$ and hence $w = \sqrt{r_1 r_2} = \sqrt{x^2 - 1}$. On the lower side $\phi_1 = \phi_2 = 2\pi$ (we have to make a complete circle in the anti-clockwise direction to reach the lower side whereby accumulating in both cases the phase of 2π) and hence $arg(w) = 2\pi$ which yields $w = \sqrt{x^2 - 1}$, which is exactly the same value. Therefore, for any $x > 1$ our function is continuous as crossing the x axis does not change it.

Next, let us consider $x < -1$. Here $r_1 = 1 - x$ and $r_2 = -1 - x$; on the upper side of the x axis $\phi_1 = \phi_2 = \pi$ (a half circle rotation is necessary resulting in the phase change of π only) and thus $arg(w) = \pi$ as well, so that

$$w = \sqrt{(1 - x)(-1 - x)}(\cos \pi + i \sin \pi) = -\sqrt{x^2 - 1}.$$

On the lower side the angles ϕ_1 and ϕ_2 are the same, leading to the same function. We conclude, that $w(z)$ is also continuous at $x < -1$.

Finally, let us consider the interval $-1 < x < 1$ between the two brunch points, where $r_1 = 1 - x$ and $r_2 = x + 1$. On the upper side of the cut $\phi_1 = \pi$ (a half circle rotation) but $\phi_2 = 0$, yielding $arg(w) = \pi/2$ and hence the function there

$$w = \sqrt{(1 - x)(x + 1)}\left(\cos \frac{\pi}{2} + i \sin \frac{\pi}{2}\right) = i\sqrt{1 - x^2}.$$

On the lower side $\phi_1 = \pi$ (a half circle rotation), $\phi_2 = 2\pi$ (a full circle rotation) and $arg(w) = 3\pi/2$, giving

$$w = \sqrt{(1 - x)(x + 1)}\left(\cos \frac{3\pi}{2} + i \sin \frac{3\pi}{2}\right) = -i\sqrt{1 - x^2}.$$

Hence, the function jumps across the cut between the points $(x = -1, y = 0)$ and $(x = 1, y = 0)$, i.e. it is *discontinuous only* across the interval $-1 < x < 1$. There-fore, the two cuts drawn in Fig. 2.11b can be equivalently drawn as a *single* cut between the two points only, Fig. 2.11c. Note however, that the actual cut goes from the point $(x = -1, y = 0)$ to the right end of the x axis, so that both ϕ_1 and ϕ_2 change between 0 and 2π. Indeed, the short cut between the two branch points is actually misleading, as it implies $-\pi < \phi_1 < \pi$ and $0 < \phi_2 < 2\pi$ leading to an undesired behaviour of our function, as can be easily checked.

Problem 2.22 Show that the function $w = \sqrt{z^2 - 1}$ becomes discontinuous for $x > 1$ and $x < -1$ and continuous for $-1 < x < 1$, if we assume that $-\pi < \phi_1 < \pi$ and $0 < \phi_2 < 2\pi$.

Problem 2.23 Show that the function $w = \sqrt{z^2 - 1}$ is continuous for $-1 < x < 1$ and discontinuous everywhere else on the x axis if the two cuts are chosen as in Fig. 2.11d.

Problem 2.24 Analyze the behaviour of the function $w = \sqrt{z^2 + 1}$ on both sides of the imaginary y axis. Consider cuts similar to the choice of Fig. 2.11b, d. Note that the brunch points in this case are at $z = \pm i$. You may find the drawing in Fig. 2.12 useful. *[Answer: for the cut $-1 < y < 1$ (or, which is the same, two cuts $y > -1$ and $y > 1$) the function w is continuous for $y > 1$ and $y < -1$, but is discontinuous across $-1 < y < 1$; in the case of the cuts $y > 1$ and $y < -1$, it is the other way round.]*

Fig. 2.12 For the analysis of the function $w = \sqrt{z^2 + 1}$ which has its brunch points at $\pm i$

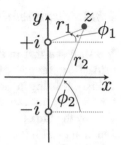

Problem 2.25 Consider the function $w = \sqrt[3]{(z+1)^2 (z-1)}$. Show that in the case of the cut made as in Fig. 2.11b, c it is discontinuous for $-1 < x < 1$, while for the cuts as in Fig. 2.11d this happens at $x > 1$ and $x < -1$ instead.

Consider now one branch of the n−root function, $w = \sqrt[n]{z}$ (we omit the index k for simplicity). Let us show that it is analytic. In order to check the Cauchy-Riemann conditions (2.14), we have to calculate the partial derivatives of u and v with respect to x and y. Here w and z are related via

$$z = w^n \quad \Longrightarrow \quad x + iy = (u + iv)^n .$$

Differentiating with respect to x and y both sides of this identity, we obtain two equations:

$$1 = n (u + iv)^{n-1} \left(\frac{\partial u}{\partial x} + i \frac{\partial v}{\partial x} \right) \quad \text{and} \quad i = n (u + iv)^{n-1} \left(\frac{\partial u}{\partial y} + i \frac{\partial v}{\partial y} \right) ,$$

so that

$$\frac{\partial u}{\partial y} + i \frac{\partial v}{\partial y} = \frac{i}{n (u + iv)^{n-1}} = i \left(\frac{\partial u}{\partial x} + i \frac{\partial v}{\partial x} \right) .$$

Comparing the real and imaginary parts on both sides, we immediately see that the necessary conditions are indeed satisfied, i.e. the root function is analytic. Therefore, its derivative can be calculated using general rules of the inverse function:

$$\frac{dw}{dz} = \left(\sqrt[n]{z} \right)' = \frac{1}{dz/dw} = \frac{1}{(w^n)'_w} = \frac{1}{nw^{n-1}} = \frac{w}{nw^n} = \frac{\sqrt[n]{z}}{nz} = \frac{1}{n} z^{1/n-1} ,$$

which is exactly the same result as for the real n−root function.

Thus we see that we can differentiate both z^n and $z^{1/n}$ with respect to z directly using usual rules; there is no need to split $z = x + iy$ and use any of the formulae (2.17) which followed from the Cauchy-Riemann conditions.

2.3.3 Exponential and Hyperbolic Functions

We shall see that in the complex calculus the trigonometric functions are directly related to the exponential function, something which would be completely impossible to imagine being in the world of real functions! To see this we need first to *define* what we mean by the exponential function on the complex plane $\mathcal{C}$. To this end, we shall employ the famous results of the real calculus, Eq. (I.2.85), linking the exponential function to a limit of a numerical sequence, and will use it as our definition:

$$e^z = e^{x+iy} = \lim_{n\to\infty} \left(1 + \frac{z}{n}\right)^n = \lim_{n\to\infty} \left(1 + \frac{x+iy}{n}\right)^n$$
$$= \lim_{n\to\infty} \left[\left(1 + \frac{x}{n}\right) + i\frac{y}{n}\right]^n = \lim_{n\to\infty} w^n \; . \tag{2.22}$$

Here

$$w = 1 + \frac{z}{n} = \left(1 + \frac{x}{n}\right) + i\frac{y}{n}$$

is a complex number with the phase

$$\phi = \arctan \frac{y/n}{1 + x/n} = \arctan \frac{y}{x+n}$$

and the absolute value $|w|$, where

$$|w|^2 = \left(1 + \frac{x}{n}\right)^2 + \frac{y^2}{n^2} = \left(1 + \frac{2x}{n}\right) + \frac{x^2 + y^2}{n^2} \; .$$

Correspondingly, the absolute value of the complex number $(1 + z/n)^n = w^n$ is

$$|w^n| = (|w|^2)^{n/2} = \left[\left(1 + \frac{2x}{n}\right) + \frac{x^2 + y^2}{n^2}\right]^{n/2} , \tag{2.23}$$

while its phase will be

$$\psi = arg\left[\left(1 + \frac{z}{n}\right)^n\right] = n \cdot arg\,(w) = n\phi = n\arctan\frac{y}{x+n} \; . \tag{2.24}$$

Now we need to take the limit $n \to \infty$ in both expressions (2.23) and (2.24) above. Concerning the phase ψ, the limit is calculated using the L'H?pital's rule and the answer is y (see Problem I.3.54(j)), while when calculating the limit of the absolute value in Eq. (2.23), we can neglect the second term inside the square brackets which is inversely proportional to n^2 as it tends to zero much faster than the first term:

$$\lim_{n \to \infty} \left[\left(1 + \frac{2x}{n} \right) + \frac{x^2 + y^2}{n^2} \right]^{n/2} = \lim_{n \to \infty} \left(1 + \frac{2x}{n} \right)^{n/2}$$

$$= \lim_{n/2 \to \infty} \left(1 + \frac{x}{n/2} \right)^{n/2} = e^x ,$$

where we have used again Eq. (I.2.85).

Problem 2.26 To justify the last step, consider the limit of the logarithm of the above expression,

$$\lim_{n \to \infty} \frac{n}{2} \ln \left[\left(1 + \frac{2x}{n} \right) \left(1 + \frac{x^2 + y^2}{n(n + 2x)} \right) \right] .$$

Split the logarithm into the sum of two terms, then show that the limit of the first term (due to the first multiplier within the square brackets) is x, while the limit of the second term, after using the L'Hôpital's rule, tends to zero as $1/n$.

Thus, the complex number which we call e^z has the absolute value equal to e^x and the phase y, i.e. we obtain the fundamental formula:

$$e^z = e^{x+iy} = e^x \left(\cos y + i \sin y \right) . \tag{2.25}$$

Let us analyse this result. Firstly, if the imaginary part $y = 0$, the exponential turns into the real exponential e^x, i.e. the complex exponential is exactly the real one if z is real. Next, as follows from the two problems below, this function satisfies usual properties of the exponential function of the real functions, and is analytic everywhere.

Problem 2.27 Show that it follows from (2.25) that the following identities are satisfied by the exponential function:

$$e^{z_1} e^{z_2} = e^{z_1 + z_2} \; ; \quad e^{-z} = \frac{1}{e^z} , \quad \left(e^z \right)^n = e^{nz} ,$$

$$\left(e^z \right)^{-n} = e^{-nz} \quad \text{and} \quad \frac{e^{z_1}}{e^{z_2}} = e^{z_1 - z_2} , \tag{2.26}$$

where z, z_1, z_2 are complex numbers and n is an integer.

Problem 2.28 Show that

$$\left(e^z \right)^* = \left(e^{x+iy} \right)^* = e^{x-iy} = e^{z^*} . \tag{2.27}$$

Problem 2.29 Show, using again Eq. (2.25), that the exponential function is analytic everywhere.

Finally, we calculate its derivative: if $e^z = u + iv$ with $u = e^x \cos y$ and $v = e^x \sin y$, then

$$\left(e^z\right)' = \frac{\partial u}{\partial x} + i\frac{\partial v}{\partial x} = e^x \cos y + ie^x \sin y = e^x \left(\cos y + i \sin y\right) = e^z \, ,$$

which again is the familiar result from the world of real functions. So, the exponential function can be differentiated directly with respect to z as the latter was real.

If we set $x = 0$ in Eq. (2.25), we shall obtain:

$$e^{iy} = \cos y + i \sin y \quad \text{and} \quad e^{-iy} = \cos y - i \sin y \, , \tag{2.28}$$

where Eq. (2.27) was used for the second formula. These two identities were derived by Euler and bear his name. Using the exponential function we can write any complex number z with the absolute value r and the phase ϕ simply as

$$z = r \left(\cos \phi + i \sin \phi\right) = re^{i\phi} \, , \tag{2.29}$$

which is called the exponential form of the complex number z. It becomes particularly transparent from this simple form and the properties of the complex exponential function derived above, why upon multiplication of two complex numbers their phases sum up while upon division the phases subtract.

Problem 2.30 Write the following complex numbers in the exponential form:

$$z = \pm 1 \, ; \quad z = 1 \pm i \, ; \quad z = \pm i \, ; \quad z = 1 \pm \sqrt{3}i \, .$$

[Answers: $1, e^{i\pi}; \sqrt{2}e^{\pm i\pi/4}; e^{\pm i\pi/2}; 2e^{\pm i\pi/3}.]

Problem 2.31 Write all roots of $\sqrt[5]{\pm 1}$ in the exponential form.

Problem 2.32 Show that $\left|e^{ia}\right| = 1$, where a is a real number.

Problem 2.33 Show that all roots of the quadratic equation $x^2 + (2 + i)x + 4i = 0$ can be written as $x_{1,2} = -1 - i/2 \pm \left(17^{1/4}\sqrt{3}/2\right)e^{-i\phi/2}$, where $\phi = $ arctan 4.

Problem 2.34 Prove that the sum of all n−roots of 1 is equal to zero,

$$\sum_{k=1}^{n} \left(\sqrt[n]{1}\right)_k = 0 ,$$

by representing the roots in the exponential form and then calculating the sum (cf. Problem 2.18).

From Eq. (2.28) we can express both the cosine and sine functions via the exponential functions:

$$\cos y = \frac{1}{2}\left(e^{iy} + y^{-iy}\right) \quad \text{and} \quad \sin y = \frac{1}{2i}\left(e^{iy} - e^{-iy}\right) . \qquad (2.30)$$

It is seen that the trigonometric functions are indeed closely related to the complex exponential function. These relations are called Euler's identities as well.

Expressions (2.28) or (2.30) (that are easy to remember) can be used for quickly deriving various trigonometric identities (that are not so easy to remember!). For instance, let us prove the double angle formula for the sine function:

$$\sin(2\alpha) = \frac{1}{2i}\left(e^{i2\alpha} - e^{-i2\alpha}\right) = \frac{1}{2i}\left[\left(e^{i\alpha}\right)^2 - \left(e^{-i\alpha}\right)^2\right]$$

$$= \frac{1}{2i}\left(e^{i\alpha} + e^{-i\alpha}\right)\left(e^{i\alpha} - e^{-i\alpha}\right) = 2\left[\frac{1}{2}\left(e^{i\alpha} + e^{-i\alpha}\right)\right]\left[\frac{1}{2i}\left(e^{i\alpha} - e^{-i\alpha}\right)\right]$$

$$= 2\cos\alpha\sin\alpha .$$

Problem 2.35 Derive your favourite trigonometric identities of Sect. I.2.3.8 using Euler's formulae.

Problem 2.36 Prove the formula (p is an integer):

$$\int_{-1}^{1} \frac{t^{2p}}{\sqrt{1 - t^2}} dt = \frac{\pi}{2^{2p}} \frac{(2p)!}{(p!)^2} \qquad (2.31)$$

by following these steps: (i) make the substitution $t = \sin\phi$; (ii) use the Euler's formula (2.30) to express the sine function via complex exponentials; (iii) use the Binomial formula and perform integration over ϕ; (iv) note that only a single term in the sum will give a non-zero contribution.

Because the exponential function of the purely imaginary argument is related directly to the sine and cosine functions, it is periodic (below k is an integer):

$$e^{z+i2\pi k} = e^z e^{i2\pi k} = e^z \left[\cos\left(2\pi k\right) + i \sin\left(2\pi k\right)\right] = e^z ,$$

since the cosine is equal to one and the sine to zero. In other words, $e^{i2\pi k} = 1$ for any integer k. This also means that the exponential function of any two complex numbers z_1 and z_2 related via $z_1 = z_2 + i2\pi k$ is the same: $e^{z_1} = e^{z_2}$. Therefore, if one considers horizontal stripes $2\pi k \le \text{Im}\,(z) < 2\pi(k+1)$ for any fixed integer k, then there will be one-to-one correspondence between z and e^z. This is essential to define the inverse of the exponential function which, as we shall see in the next section, is the logarithm. This situation is similar to the integer power function we considered previously where it was necessary to restrict the phase of z to define the inverse (the $n-$root) function.

The hyperbolic functions are defined identically to their real variables counterparts:

$$\sinh z = \frac{1}{2}\left(e^z - e^{-z}\right) \quad \text{and} \quad \cosh z = \frac{1}{2}\left(e^z + e^{-z}\right) . \tag{2.32}$$

Problem 2.37 Prove that

$$\cosh^2 z - \sinh^2 z = 1 ; \quad \cosh\left(-z\right) = \cosh\left(z\right) ; \quad \sinh\left(-z\right) = -\sinh\left(z\right) .$$

Problem 2.38 Using the Euler's formulae (2.28), sum the appropriate geometric progression to re-derive Eqs. (2.9) and (2.10).

2.3.4 Logarithm

The logarithm is defined as an inverse function to the exponential function, i.e. if we write $w = \ln z$, that means $e^w = z$. Because of the properties (2.26) of the exponential function which are exactly the same as for the real exponential function, the logarithm on the complex plane of a product or ratio of two complex numbers is equal to the sum or difference of their logarithms:

$$\ln\left(z_1 z_2\right) = \ln z_1 + \ln z_2 \quad \text{and} \quad \ln\frac{z_1}{z_2} = \ln z_1 - \ln z_2 .$$

For any complex number z we can write:

$$z = re^{i\phi} = e^{\ln r} e^{i\phi} = e^{\ln r + i\phi}$$
$$\implies \quad \ln z = \ln r + i\phi = \ln|z| + i\,arg(z) . \tag{2.33}$$

Since the phase $arg(z)$ of any z from C is defined up to $2\pi k$ with any integer k, the logarithm of z is a multi-valued function. If ϕ is one particular phase of z, then

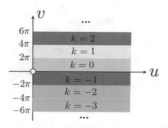

Fig. 2.13 The logarithmic function $w = u + iv = \ln z$ is defined in stripes of the vertical width 2π satisfying (for the given branch k) the inequality $2\pi k \leq v < 2\pi (k + 1)$. Each stripe corresponds to a particular branch of the logarithmic function

$$\ln z = \ln |z| + i\,(\phi + 2\pi k)\ ,\quad k = 0, \pm 1, \pm 2, \ldots\ . \tag{2.34}$$

To choose a single branch of the logarithm, we fix the value of the integer k, and then $\ln z$ will remain within a stripe $2\pi k \leq \mathrm{Im}\,(\ln z) < 2\pi (k + 1)$, see Fig. 2.13. For instance, by choosing $k = 0$ we select the stripe between $0i$ and $2\pi i$. This is the *principal brunch* of the logarithm corresponding to the cut made along the positive part of the x axis shown in Fig. 2.10b as in this case the phase of the logarithm changes only between 0 and 2π.

For instance, consider $\ln (-1)$. This logarithm is not defined in real numbers. However, on $\mathcal{C}$ this quantity has perfect meaning: since $-1 = e^{i\pi}$ then $\ln (-1) = \ln 1 + i\,(\pi + 2\pi k) = i\,(\pi + 2\pi k)$. Here different values of k correspond to the values of $\ln (-1)$ on different branches of the logarithmic function.

Problem 2.39 Present the following expressions in the form $u + iv$ using the kth branch of the logarithm:

$$\ln i\ ;\quad \ln (1 \pm i)\ ;\quad \ln\left(\sqrt{3} \pm i\right)\ ;\quad \ln\left(1 \pm i\sqrt{3}\right)\ ;\quad \ln (1 + 5i)\ .$$

[Answers: $i\pi\,(2k + 1/2)$; $(\ln 2)\,/2 + i\pi\,(2k \pm 1/4)$; $\ln 2 + i\pi\,(2k \pm 1/6)$; $\ln 2 + i\pi\,(2k \pm 1/3)$; $(\ln 26)\,/2 + i\,(\arctan 5 + 2\pi k)$.]

Now let us try to understand if there are any limitations on the domain of allowed z values in order for its logarithm to remain within the particular chosen stripe (and hence to correspond to a single-valued function). In Fig. 2.14a we take a closed path which does not contain the point $z = 0$ within it, with the horizontal parts being on the upper and lower sides of the positive part of the x axis (in the Figure they are shown slightly off the x axis for clarity); then the logarithmic function $w = \ln z$ goes along the particular path shown in (b) for each particular branch k. The vertical parts in the paths in (b) correspond to the circle paths in (a) when only the phase of z changes, while the horizontal parts in (b) correspond to the horizontal paths in (a) when only $|z|$ changes. The situation is different, however, if the point $z = 0$ lies inside the contour as shown in Fig. 2.15. In this case $w = \ln z$ passes through the

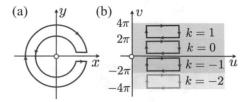

Fig. 2.14 When z goes around a closed contour **a** which avoids the point $z = 0$, the function $w = \ln z = u + iv$ remains within a particular branch making a rectangular path there (**b**)

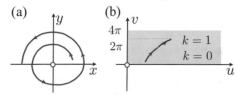

Fig. 2.15 If the z-contour shown in **a** contains the point $z = 0$ inside, then the path taken by the logarithm shown in **b** goes from one logarithm branch into another continuously

current stripe and goes over into the next one by accumulating an extra phase 2π, i.e. revolving around $z = 0$ takes the logarithmic function from one of its branches to the next one. As in the case of the n−root function, this problem is avoided by taking a branch cut from the branch point $z = 0$ which would limit the phase of z between 0 and 2π, Fig. 2.10b.

Above, when choosing the brunches, we assumed that the brunch cut is made along the positive part of the x axis. Another direction of the cut will change the function. For instance, the cut made along the negative part of the x axis shown in Fig. 2.10c restricts the phase of z being between $-\pi$ and π. Hence, the same point on the complex plane will have a different imaginary part of the logarithm. Indeed, the points $z_1 = re^{i3\pi/2}$ and $z_2 = re^{-i\pi/2}$ are equivalent, but the former point (when $0 < arg(z_1) < 2\pi$) corresponds to the cut made along the positive, while the latter (when $-\pi < arg(z_2) < \pi$) along the negative directions of the x axis. Correspondingly, the values of the logarithm in each case are by $i2\pi$ different: $\ln z_1 = \ln r + i3\pi/2$ and $\ln z_2 = \ln r - i\pi/2$.

Considering a particular branch, we can easily establish that the logarithmic function is analytic, and we can calculate its derivative.

Problem 2.40 Show that the logarithmic function $w = \ln z = u + iv$, where $u = \ln \sqrt{x^2 + y^2}$ and $v = \arctan (y/x)$, is analytic.

Problem 2.41 Show that the principal branch of the function ($\omega > 0$)

$$f(z) = \ln \frac{z - \omega}{z + \omega}$$

near the real axis ($z = x + i\delta$ with $\delta \to \pm 0$) is given by

$$f(z) = \ln \left| \frac{x - \omega}{x + \omega} \right| \pm i\pi\chi_\omega(x) \, ,$$

where the plus sign corresponds to $\delta > 0$ (i.e. just above the real axis), while the minus sign to $\delta < 0$ (just below), and $\chi_\omega(x) = 1$ if $-\omega < x < \omega$ and zero otherwise. [*Hint: since we have here two logarithmic functions, $\ln(z - \omega)$ and $\ln(z + \omega)$, there will be two brunch points at $\pm\omega$ on the real axis; then it is convenient to choose the cuts $x > -\omega$ and $x > \omega$ (which is equivalent to the cut $-\omega < x < \omega$).*]

The derivative of the logarithm is obtained in the usual way as it is an inverse function to the exponential one:

$$(\ln z)' = \frac{dw}{dz} = \frac{1}{dz/dw} = \frac{1}{(e^w)'} = \frac{1}{e^w} = \frac{1}{z} \, . \tag{2.35}$$

Again, the formula looks the same as for the real logarithm.

2.3.5 Trigonometric Functions

Sine and cosine functions of a complex variable are defined from the Euler-like equations (2.30) generalised to any complex number z, i.e.

$$\sin z = \frac{1}{2i} \left(e^{iz} - e^{-iz} \right) \quad \text{and} \quad \cos z = \frac{1}{2} \left(e^{iz} + e^{-iz} \right) \, . \tag{2.36}$$

As an example, let us calculate:

$$\begin{aligned}
\sin \left(2i + \frac{\pi}{4} \right) &= \frac{1}{2i} \left[e^{i(2i+\pi/4)} - e^{-i(2i+\pi/4)} \right] = \frac{1}{2i} \left[e^{-2+i\pi/4} - e^{2-i\pi/4} \right] \\
&= \frac{-i}{2} \left[\frac{1}{e^2} e^{i\pi/4} - e^2 e^{-i\pi/4} \right] \\
&= -\frac{i}{2} \left[\frac{1}{e^2} \left(\cos \frac{\pi}{4} + i \sin \frac{\pi}{4} \right) - e^2 \left(\cos \frac{\pi}{4} - i \sin \frac{\pi}{4} \right) \right] \\
&= -\frac{i}{2\sqrt{2}} \left[\frac{1}{e^2} (1+i) - e^2 (1-i) \right] \\
&= \frac{1 + e^4}{2\sqrt{2}e^2} - i \frac{1 - e^4}{2\sqrt{2}e^2} = \frac{1}{\sqrt{2}} (\cosh 2 + i \sinh 2) \, .
\end{aligned}$$

The trigonometric functions satisfy all the usual properties of the sine and cosine functions of the real variable. First of all, if z is real, these definitions become Euler's formulae and hence give us the usual sine and cosine. Then we see that for a complex z the sine is an odd while cosine an even function, e.g.

$$\sin(-z) = \frac{1}{2i} \left(e^{-iz} - e^{iz} \right) = -\frac{1}{2i} \left(e^{iz} - e^{-iz} \right) = -\sin z \ .$$

Similarly one can establish various trigonometric identities for the sine and cosine, and the manipulations are similar to those in Problem 2.35. For instance, consider

$$\sin^2 z + \cos^2 z = \left[\frac{1}{2i} \left(e^{iz} - e^{-iz} \right) \right]^2 + \left[\frac{1}{2} \left(e^{iz} + e^{-iz} \right) \right]^2$$

$$= -\frac{1}{4} \left[e^{2iz} - 2 + e^{-2iz} \right] + \frac{1}{4} \left[e^{2iz} + 2 + e^{-2iz} \right]$$

$$= \frac{1}{2} + \frac{1}{2} = 1 \ ,$$

as expected. It is also obvious that since the sine and cosine functions are composed as a linear combination of the analytic exponential functions, they are analytic. Finally, their derivatives are given by the same formulae as for the real variable sine and cosine.

Problem 2.42 Express the complex numbers below in the form of $u + iv$:

$$\sin i \ ; \quad \cos i \ ; \quad \sin \frac{1+i}{1-i} \ ; \quad \cos \left(\frac{2i}{2+i} - \frac{3i}{3+i} \right) \ .$$

[Answers: $i \sinh(1)$; $\cosh(1)$; $i \sinh(1)$; $\cos \frac{1}{10} \cosh \frac{1}{10} + i \sin \frac{1}{10} \sinh \frac{1}{10}$.]

Problem 2.43 Prove, that generally:

$$\sin(x + iy) = \sin x \, \cosh y + i \cos x \, \sinh y \ ,$$

$$\cos(x + iy) = \cos x \, \cosh y - i \sin x \, \sinh y \ .$$

Problem 2.44 Prove that

$$\cos(2z) = \cos^2 z - \sin^2 z \ ;$$

$$\sin(4z) = 4 \sin z \cos^3 z - 4 \sin^3 z \cos z \ ;$$

$$\sin (z_1 \pm z_2) = \sin z_1 \cos z_2 \pm \cos z_1 \sin z_2 \; ;$$

$$(\sin z)' = \cos z \; ; \quad (\cos z)' = - \sin z \; .$$

Problem 2.45 Prove directly (by checking derivatives of their real and imaginary parts) that both sine and cosine functions are analytic.

Problem 2.46 Prove the following identities using the definitions of the corresponding functions:

$$\cos (iz) = \cosh z \; ; \quad \sin (iz) = i \sinh z \; ;$$
$$\sinh (iz) = i \sin z \; ; \quad \cosh (iz) = \cos z \; . \tag{2.37}$$

Problem 2.47 Prove that

$$\coth x - i \cot y = \frac{\sin (y - ix)}{\sin y \sinh x} \; .$$

Problem 2.48 Show that zeros of $\cos z$ are given by $z = \pi/2 + \pi k$, while zeros of $\sin z$ are given by $z = \pi k$ with $k = 0, \pm 1, \pm 2, \ldots$.

The last point which needs investigating is to determine which z points give the same values for the sine and cosine functions. This is required for selecting such domains of z in $\mathcal{C}$ where the trigonometric functions are single-valued and hence where their inverse functions can be defined.

Let us start from the sine function:

$$\sin z_1 = \sin z_2 \quad \Longrightarrow \quad e^{iz_1} - e^{-iz_1} = e^{iz_2} - e^{-iz_2} \; .$$

If $g_1 = e^{iz_1}$ and $g_2 = e^{iz_2}$, then the last equation can be rewritten as

$$g_1 - \frac{1}{g_1} = g_2 - \frac{1}{g_2} \quad \Longrightarrow \quad (g_1 - g_2) \left(1 + \frac{1}{g_1 g_2} \right) = 0$$
$$\Longrightarrow \quad \left(e^{iz_1} - e^{iz_2} \right) \left(1 + e^{-i(z_1 + z_2)} \right) = 0 \; .$$

Therefore, we obtain that either

$$e^{iz_1} = e^{iz_2} \quad \Longrightarrow \quad z_1 = z_2 + 2\pi k \quad \Longrightarrow \quad \{x_1 = x_2 + 2\pi k \; , \; y_1 = y_2\} \; ,$$

or/and

$$e^{i(z_1 + z_2)} = -1 \quad \Longrightarrow \quad e^{i(z_1 + z_2)} = e^{i\pi} \quad \Longrightarrow \quad z_1 + z_2 = \pi + 2\pi k$$
$$\Longrightarrow \quad \{x_1 + x_2 = \pi (2k + 1) \; , \; y_1 = y_2\} \; .$$

The first expression reflects the periodicity of the sine function along the real axis with the period of 2π; note that this is entirely independent of the imaginary part of z. This gives us vertical stripes (along the y axis) of the width 2π within which the sine function is single-valued. The second condition is trickier. It is readily seen that if z is contained inside the vertical stripe $-\pi/2 < \mathrm{Re}(z) < \pi/2$, then no additional solutions (or relationships between x_1 and x_2) come out of this extra condition. Indeed, it is sufficient to consider the case of $k = 0$ because of the mentioned periodicity. Then, we have the condition $x_1 + x_2 = \pi$. If both x_1 and x_2 are positive, this identity will never be satisfied for both x_1 and x_2 lying between 0 (including) and $\pi/2$ (excluding). Similarly, if both x_1 and x_2 were negative, then this condition will not be satisfied if both x_1 and x_2 lie between $-\pi/2$ and 0. Finally, if x_1 and x_2 are of different sign, then the condition $x_1 + x_2 = \pi$ is not satisfied at all if both of them are contained between $-\pi/2$ and $\pi/2$. Basically, the conclusion is that no identical values of the sine function are found if z is contained inside the vertical stripe $-\pi/2 < \mathrm{Re}(z) < \pi/2$, as required.

Problem 2.49 Show similarly that the equation $\cos z_1 = \cos z_2$ has the solution of either $x_1 = x_2 + 2\pi k$ (k is an integer) and/or $x_1 + x_2 = 2\pi k$. Therefore, one may choose the vertical stripe $0 \le \mathrm{Re}(z) < \pi$ to avoid identical values of the cosine function.

Since the cosine and sine functions were generalised from their definitions given for real variables, it makes perfect sense to define the tangent and cotangent functions accordingly:

$$\tan z = \frac{\sin z}{\cos z} = -i\frac{e^{iz} - e^{-iz}}{e^{iz} + e^{iz}} \quad \text{and} \quad \cot z = \frac{\cos z}{\sin z} = i\frac{e^{iz} + e^{-iz}}{e^{iz} - e^{iz}}. \tag{2.38}$$

2.3.6 Inverse Trigonometric Functions

We shall start from $w = \arcsin z$. It is defined as an inverse to $z = \sin w$. If we solve the last equation with respect to w, we should be able to obtain an explicit expression for the arcsine function. We therefore have:

$$z = \sin w \implies 2iz = e^{iw} - e^{-iw}$$

$$\implies 2iz = p - \frac{1}{p} \implies p^2 - 2izp - 1 = 0,$$

where $p = e^{iw}$. Solving the quadratic equation with respect to p, we obtain:

$$p = iz + \sqrt{(iz)^2 + 1} = iz + \sqrt{1 - z^2} \implies e^{iw} = iz + \sqrt{1 - z^2}$$
$$\implies \quad w = \arcsin z = -i \ln\left(iz + \sqrt{1 - z^2}\right). \tag{2.39}$$

Since the logarithm is a multi-valued function, so is the arcsine. Also, here we do not need to write $\pm$ before the root since here the root is understood as a multi-valued function.

Problem 2.50 Show that arccosine function is related to the logarithm as follows:

$$\arccos z = -i \ln\left(z + \sqrt{z^2 - 1}\right). \tag{2.40}$$

Problem 2.51 Prove the following identities (k-integer):

$$\arcsin z + \arccos z = \frac{\pi}{2} + 2\pi k ; \quad \arcsin(-z) + \arcsin z = 2\pi k .$$

In both cases all values of the functions are implied.

Using these representations, inverse trigonometric functions are expressed via the logarithm and hence easily calculated as is illustrated by the following example:

$$w = \arcsin\left(\frac{1+i}{1-i}\right) = \arcsin\left(\frac{2i}{2}\right) = \arcsin(i) = -i \ln\left(i^2 + \sqrt{1 - i^2}\right)$$
$$= -i \ln\left(-1 \pm \sqrt{2}\right).$$

Here the $\pm$ sign comes from the two possible values of the square root. When we choose the "+" sign, then $z = -1 + \sqrt{2} > 0$ and its phase is zero, so that $w_1 = -i \ln\left(-1 + \sqrt{2}\right) + 2\pi k$. If we choose the "−" sign for the value of the root, with the phase π attached to it, then we arrive at $w_2 = -i \ln\left(1 + \sqrt{2}\right) + (\pi + 2\pi k)$. In this way all possible values of the complex number w are obtained.

Next, arctangent $w = \arctan z$ is defined in such a way that $\tan w = z$, and similarly for the arccotangent. Their derivation we leave as an exercise.

Problem 2.52 Derive the following expressions:

$$\arctan z = \frac{1}{2i} \ln \frac{1+iz}{1-iz} \quad \text{and} \quad \operatorname{arccot} z = \frac{1}{2i} \ln \frac{z+i}{z-i} \; ; \qquad (2.41)$$

$$\arctan z + \operatorname{arccot} z = \frac{\pi}{2} + \pi k \; ,$$

where k is an integer as usual.

All these functions are multi-valued.

2.3.7 General Power Function

General power function $w = z^c$ with $c = \alpha + i\beta$ being a general complex number is defined as follows:

$$w = z^c = e^{c \ln z} \; . \qquad (2.42)$$

Here the logarithmic function is understood as a multi-valued one. To see this explicitly, we recall that $\ln z = \ln r + i\,(2\pi k + \phi)$, where $z = re^{i\phi}$ and k is an integer. Then,

$$w = z^c = e^{R+i\Psi} \; ,$$

where

$$R = \operatorname{Re}\,(c \ln z) = \operatorname{Re}\,[(\alpha + i\beta)\,(\ln r + i2\pi k + i\phi)]$$
$$= \alpha \ln r - (2\pi k + \phi)\,\beta \; ,$$

$$arg\,(z^c) = \Psi = \operatorname{Im}\,(c \ln z) = \beta \ln r + \alpha\,(2\pi k + \phi) \; .$$

These are the general formulae.

Consider now specifically the case of $\beta = 0$, i.e. the power $c = \alpha$ being real. Then $R = \alpha \ln r$ and $\Psi = \alpha\,(2\pi k + \phi)$, i.e.

$$z^\alpha = e^{\alpha \ln r} e^{i\alpha(2\pi k + \phi)} = r^\alpha e^{i\alpha(2\pi k + \phi)} \; .$$

If $\alpha = n$ is an integer, then

$$z^n = r^n e^{in(2\pi k + \phi)} = r^n e^{in\phi} \; ,$$

i.e. we obtain our previous single-valued result of Sect. 2.3.1. Similarly, in the $n-$root case, i.e. when $\alpha = 1/n$ with n being an integer, we obtain

$$z^{1/n} = r^{1/n} e^{i(2\pi k + \phi)/n} = \sqrt[n]{r} \exp\left(i\frac{2\pi}{n}k + i\frac{\phi}{n}\right) ,$$

which is the same result as we obtained earlier in Sect. 2.3.2. Further, if we consider now a rational power $\alpha = n/m$ with both n and m being integers ($m \neq 0$), then

$$z^{n/m} = r^{n/m} \exp\left(i\frac{2\pi n}{m}k + i\frac{\phi n}{m}\right) = \left[r^{1/m} \exp\left(i\frac{2\pi}{m}k + i\frac{\phi}{m}\right)\right]^n$$

and it coincides with the function $\left(\sqrt[m]{z}\right)^n$, as expected. Hence the definition (2.42) indeed generalises our previous definitions of the power function.

Now we shall consider some examples of calculating the general power function. Let us calculate $i^i = \exp(i \ln i)$. Since $\ln i = i(\pi/2 + 2\pi k)$, then $i^i = \exp(-\pi/2 - 2\pi k)$, i.e. the result is a real number which however depends on the branch (the value of k) used to define the logarithm.

Problem 2.53 Calculate the following complex numbers:

$$(1+i)^{1+i} \; ; \quad \left(\frac{1+i\sqrt{3}}{2}\right)^{2i} \; ; \quad 10^{2i+5} \; ; \quad (x+i)^\alpha \; ,$$

where in the last case both x and α are real. [Answers: $\sqrt{2} \exp(-\pi/4 - 2\pi k) \exp\left[i\left(\pi/4 + \ln\sqrt{2}\right)\right]$; $\exp(-2\pi/3 - 4\pi k)$; $10^5 \exp(i2\ln 10)$; $\left(1+x^2\right)^{\alpha/2} \exp[i\alpha(\phi + 2\pi k)]$, where $\phi = \arctan(1/x)$.]

2.4 Integration in the Complex Plane

2.4.1 Definition

Integration of a function $f(z)$ of a complex variable z can be defined similarly to the line integrals considered at length in Sect. I.6.4: we define a line (called *contour*) L in the complex plane $\mathcal{C}$, on which the function $f(z)$ is defined, Fig. 2.16. The line is divided into n portions by points $z_0, z_1, \ldots, z_n$ and within each neighbouring pair of points (z_k, z_{k+1}) an intermediate point ξ_k lying on L is chosen. Then we define the integral of $f(z)$ on the contour as the limit of the sum:

Fig. 2.16 The directed contour L in the complex plane is divided into n portions by $n + 1$ points $z_0, z_1, \ldots, z_n$

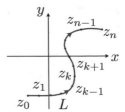

$$\int_L f(z)dz = \lim_{\Delta \to 0} \sum_{k=0}^{n-1} f(\xi_k)(z_{k+1} - z_k) \ , \qquad (2.43)$$

where Δ is the maximum distance $|z_{k+1} - z_k| = \sqrt{(x_{k+1} - x_k)^2 + (y_{k+1} - y_k)^2}$ between any of the two adjacent points on the curve. The limit of $\Delta \to 0$ means that the distances between any two adjacent points become smaller and smaller in the limit (and correspondingly the number of division points $n \to \infty$). It is clear that if the limit exists, then the choice of the internal points ξ_k is not important.

We observe that this definition is very close to the definition of the line integral of a vector field (of the 2nd kind), Sect. I.6.4.2. Indeed, let $f(z) = u(x, y) + iv(x, y)$, $z_k = x_k + iy_k$ and $\xi_k = \alpha_k + i\beta_k$, then

$$\sum_{k=0}^{n-1} f(\xi_k)(z_{k+1} - z_k) = \sum_{k=0}^{n-1} (u_k + iv_k)\left[(x_{k+1} - x_k) + i(y_{k+1} - y_k)\right]$$

$$= \sum_{k=0}^{n-1} \left[u_k(x_{k+1} - x_k) - v_k(y_{k+1} - y_k)\right] + i\sum_{k=0}^{n-1} \left[v_k(x_{k+1} - x_k) + u_k(y_{k+1} - y_k)\right]$$

$$= \sum_{k=0}^{n-1} \left[u_k \Delta x_k - v_k \Delta y_k\right] + i\sum_{k=0}^{n-1} \left[v_k \Delta x_k + u_k \Delta y_k\right] \ ,$$

where $u_k = u(\alpha_k, \beta_k)$ and $v_k = v(\alpha_k, \beta_k)$. In the limit of $\max\{\Delta x_k\} \to 0$ and $\max\{\Delta y_k\} \to 0$, the two sums tend to 2D line integrals, i.e. one can writes:

$$\int_L f(z)dz = \int_L u\,dx - v\,dy + i\int_L v\,dx + u\,dy \ . \qquad (2.44)$$

This result shows that the problem of calculating the integral on the complex plane can in fact be directly related, if needed, to calculating two real line integrals on the $x - y$ plane. If these two integrals exist (i.e. $u(x, y)$ and $v(x, y)$ are piece-wise continuous and their absolute values are limited), then the complex integral also exists and is well defined.

Fig. 2.17 A circle of radius
R centred at point
$z_0 = x_0 + iy_0$

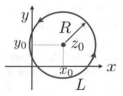

In practice the complex integrals are calculated by using a parametric representation of the contour L. Let $x = x(u)$ and $y = y(u)$ (or $z = x(u) + iy(u) = z(u)$)
define the curve L via a parameter u. Then $dx = x'(u)du$ and $dy = y'(u)du$, so that
we obtain:

$$\int_L f(z)dz = \int_L \left[(ux' - vy') + i \left(vx' + uy' \right) \right] du$$

$$= \int_L (u + iv) \left(x' + iy' \right) du = \int_L f(z)z'(u)du .$$

(2.45)

Example 2.3 ▶ As an example, let us integrate the function $f(z) = 1/(z - z_0)$
around a circle of radius R centred at the point $z_0 = x_0 + iy_0$ in the anti-clockwise
direction, see Fig. 2.17. Here the parameter u can be chosen as a polar angle ϕ since
the points z on the circle can be easily related to ϕ via

$$z(\phi) = z_0 + Re^{i\phi} .$$

(2.46)

Indeed, if the circle was centred at the origin, then we would have $x(\phi) = R \cos \phi$
and $y(\phi) = R \sin \phi$, i.e.

$$z(\phi) = R(\cos \phi + i \sin \phi) = Re^{i\phi} ;$$

however, once the circle is shifted by z_0, we add z_0, which is exactly Eq. (2.46).
Therefore, we get:

$$\int_{circle} \frac{dz}{z - z_0} = \left| \begin{array}{c} z = z_0 + Re^{i\phi} \\ dz = z'(\phi)d\phi = iRe^{i\phi}d\phi \end{array} \right|$$

$$= \int_0^{2\pi} \frac{iRe^{i\phi}d\phi}{Re^{i\phi}} = i \int_0^{2\pi} d\phi = 2\pi i .$$

Note the direction of integration: it is anti-clockwise as shown in the Figure, hence
ϕ changes from 0 to 2π.◀

Note that the result does not depend on the radius R of the circle. We shall better
understand this fact later on in Sect. 2.4.2.

From the relationship (2.44) connecting the complex integral with the real line
integrals, it is clear that the properties of the line integrals are immediately transferred
to the complex ones: (i) the integral changes sign if the direction of integration is

reversed; (ii) the integral of a sum of functions is equal to the sum of the integrals taken with the individual functions; (iii) the integral over a composite curve L consisting of separate pieces (e.g. a piece-wise curve $L = L_1 + L_2 + \ldots$) is equal to the sum of integrals taken over each piece L_1, L_2, etc. The absolute value of the integral is also limited from above if the function $f(z)$ is limited everywhere on the line L. Indeed, if $|f(z)| \leq F = \max\{f(z)\}$ and the integral is the limit of the sum, we can use the inequality

$$\left| \sum_i z_i \right| \leq \sum_i |z_i| \,,$$

see, e.g., Eq. (I.1.193) for it,

$$\left| \int_L f(z)dz \right| \leq \int_L |f(z)dz| = \int_L |f(z)|\,|dz| \leq F \int_L |dz| = Fl \,, \qquad (2.47)$$

where

$$l = \int_L |dz| = \int_L \sqrt{dx^2 + dy^2} = \int_L \sqrt{x'(u)^2 + y'(u)^2}\,du$$

is the length of the curve L, specified with the parameter u, compare with Eq. (I.6.33).

Problem 2.54 Show that $\int_L dz/z^2 = 2/R$, where L is the upper semicircle of radius R centred at the origin which is traversed from the positive direction of the x axis to the negative one. At the same time, demonstrate that the same result is obtained when the integral is taken along the lower part of the semicircle traversed in the negative x direction (i.e. connecting the same initial and final points).

2.4.2 Integration of Analytic Functions

As in the case of real line integrals, the contour integral $\int_L f(z)dz$ would depend on the function $f(z)$ and on the curve L. Recall, however, that for functions in 2D satisfying a certain condition $\partial P/\partial y = \partial Q/\partial x$, Eq. (I.6.45), the line integral $\int_L P dx + Q dy$ does not depend on the path, but only on the initial and final points, see Theorem I.6.5. It appears that exactly the same is true for complex integrals provided that the function $f(z)$ is analytic. This is stated by the following Theorem.

Theorem 2.2 (due to Cauchy) *If the function $f(z)$ is analytic in some simply connected region D of C (recall that this was also an essential condition in Theorem I.6.5 dealing with real line integrals) and has a continuous derivative everywhere in D (in fact, this particular condition can be avoided, although this would make the proof more complex), then for any contour L lying in D and starting and ending at the points $z_A = A(x_A, y_A)$ and $z_B = B(x_B, y_B)$, the integral $\int_L f(z)dz$ would have the same value, i.e. the integral does not depend on the actual path, but only on the initial and final points.*

Proof Indeed, consider both real integrals in formula (2.44) and let us check if these two integrals satisfy conditions of Theorem I.6.5. The region D is simply connected, we hence only need to check whether the condition (I.6.45) is satisfied in each case. In the first integral we have $P = u$ and $Q = -v$, and hence the required condition (I.6.45) corresponds to $\partial u/\partial y = -\partial v/\partial x$, which is the second Cauchy-Riemann condition (2.14). Next, in the second integral in (2.44) we instead have $P = v$ and $Q = u$, so that the required condition becomes $\partial v/\partial y = \partial u/\partial x$, which is exactly the first Cauchy-Riemann condition (2.14). So, the conditions for the integral to be independent on the path appear to be exactly the same as the conditions for $f(z)$ to be analytic, as required. **Q.E.D.**

Problem 2.55 Prove that the integral over any closed contour $\oint_L f(z)dz$ taken inside D is zero:

$$\oint_L f(z)dz = 0 . \qquad (2.48)$$

This is a corollary to the Cauchy Theorem 2.2. The inverse statement is also valid as is demonstrated by the following Theorem.

Theorem 2.3 (by Morera) *If $f(z)$ is continuous in a simply connected region D and the integral $\oint_L f(z)dz = 0$ over any closed contour in D, then $f(z)$ is analytic in D.*

Proof Since the integral over *any* closed contour is equal to zero, the integral

$$\int_{z_0}^{z} f(p)dp = \int_L udx - vdy + i \int_L vdx + udy$$

does not depend on the path L connecting the two points, z_0 and z, but only on the points themselves. There are two line integrals $\int_L Pdx + Qdy$ above, each not depending on the path. Then, from Theorem I.6.5 for line integrals, it follows that in each case $\partial P/\partial y = \partial Q/\partial x$. Applying this condition to each of the two integrals, we immediately obtain the Cauchy-Riemann conditions (2.14) for the functions $u(x, y)$ and $v(x, y)$, which means that indeed the function $f(z)$ is analytic. **Q.E.D.**

The usefulness of these Theorems can be illustrated on the frequently met integral

$$I(a) = \int_{-\infty}^{\infty} e^{-\beta(x+ia)^2} dx = \int_{-\infty+ia}^{\infty+ia} e^{-\beta z^2} dz , \qquad (2.49)$$

where a is a real number (for definiteness, we assume that $a \geq 0$). The integration here is performed along the horizontal line in the complex plane between the points $z_{\pm} = \pm\infty + ia$ crossing the imaginary axis at the number ia. The integrand $f(z) = e^{-\beta z^2}$ does not have any singularities, so any region in C is simply connected. Hence, the integration line can be replaced by (it is often said "deformed into") a three-piece contour connecting the same initial and final points, as shown in Fig. 2.19 in purple. The integrals along the vertical pieces, where $x = \pm\infty$, are equal to zero. Indeed, consider the integral over the right vertical piece for some finite $x = R$, the $R \to \infty$ limit is assumed at the end. There $z = R + iy$ and hence

$$f(z) = e^{-\beta(R+iy)^2} = e^{-\beta(R^2-y^2)} e^{-i2\beta Ry}$$

and hence the integral can be estimated:

$$\left| \int_0^a e^{-\beta(R+iy)^2} i\, dy \right| \leq \int_0^a \left| e^{-\beta(R^2-y^2)} e^{-i2\beta Ry} \right| dy$$

$$= \int_0^a \left| e^{-\beta(R^2-y^2)} \right| dy = e^{-\beta R^2} \int_0^a e^{\beta y^2} dy$$

$$= M e^{-\beta R^2} ,$$

where M is some positive finite number corresponding to the value of the integral $\int_0^a e^{\beta y^2} dy$. It is seen from here that since $e^{-\beta R^2} \to 0$ as $R \to \infty$, the integral tends to zero. Similarly it is shown that the integral over the vertical piece at $x = -R \to -\infty$ is also zero. Therefore, it is only necessary to perform integration over the horizontal x axis between $-\infty$ and $+\infty$. Effectively it appears that the original horizontal contour at $z = ia$ can be shifted down (up if $a < 0$) to coincide with the x axis, in which case the integration is easily performed as explained in Sect. 4.2:

$$I(a) = \int_{-\infty}^{\infty} e^{-\beta(x+ia)^2} dx = \int_{-\infty}^{\infty} e^{-\beta x^2} dx = \sqrt{\frac{\pi}{\beta}} . \qquad (2.50)$$

So, the integral (2.49) does not actually depend on the value of a.

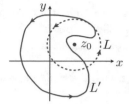

Fig. 2.18 The contour L has the point z_0 inside, while the contour L' avoids this point

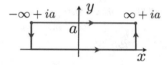

Fig. 2.19 The integration between points $(\pm\infty, ia)$ along the blue horizontal line $z = x + ia$ $(-\infty < x < \infty)$ can alternatively be performed along a different contour consisting of three straight pieces (in purple): a vertical down piece at $x = -\infty$, then horizontal line along the x axis, and finally again a vertical piece to connect up with the final point

Problem 2.56 Show that

$$\int_{-\infty}^{\infty} e^{-\alpha x^2 + i\beta x} dx = e^{-\beta^2/4\alpha} \sqrt{\frac{\pi}{\alpha}}. \tag{2.51}$$

The importance of the region D being simply connected can be illustrated by our Example 2.3: the contour there is taken around the point z_0 at which the function $f(z) = 1/(z - z_0)$ is singular. Because of the singularity, the region inside the circular contour is not simply connected, because it is needed to cut off the singular point. The latter can be done by drawing a small circle around z_0 and removing all points inside that circle. Thus, the region where the integration contour passes has a hole and hence two boundaries: one is the circle itself and another is related to the small circle used to cut off the point z_0. That is why the integral is not zero. However, if the integral was taken around any contour which does not have the point z_0 inside it, e.g. the contour L' shown in Fig. 2.18, then the integral would be zero according to the Cauchy Theorem 2.2.

Since the integral of an analytic function only depends on the starting and ending points of the contour, we may also indicate this explicitly:

$$\int_L f(z)dz = \int_{z_0}^{z} f(z)dz.$$

This notation now looks indeed like the one used for a real one-dimensional definite integral, and this similarity is even stronger established because of the following two theorems which provide us with a formula that is practically exactly the same as the main formula of the integral calculus of Sect. I.4.3.

Theorem 2.4 *If $f(z)$ is an analytic function in a simply connected region D, then*

$$F(z) = \int_{z_0}^{z} f(z')\, dz' , \qquad (2.52)$$

considered as a function of the upper limit , is an analytic function. Note that according to Theorem 2.2 the actual path is not important as long as it lies fully inside the simply connected region D.

Proof Let us write the integral explicitly via real and imaginary parts of the function $f(z) = u + iv$, see Eq. (2.44). However, since we know that both line integrals do not depend on the choice of the path, we can run them using the special path $(x_0, y_0) \to (x, y_0) \to (x, y)$, i.e. we first move along the x and then along the y axis (cf. Sect. I.6.4.4 and especially Fig. I.6.32, as, indeed, we have done this before!). In this case the real $U(x, y)$ and imaginary $V(x, y)$ parts of $F(z)$ are, respectively:

$$U(x, y) = \operatorname{Re} F(z) = \int_L u\,dx - v\,dy = \int_{x_0}^{x} u(\xi, y_0)\, d\xi - \int_{y_0}^{y} v(x, \xi)\, d\xi ,$$

$$V(x, y) = \operatorname{Im} F(z) = \int_L v\,dx + u\,dy = \int_{x_0}^{x} v(\xi, y_0)\, d\xi + \int_{y_0}^{y} u(x, \xi)\, d\xi .$$

Now, let us calculate all the partial derivatives to check if $F(z)$ is analytic. We start with $\partial U / \partial x$:

$$\frac{\partial U}{\partial x} = u(x, y_0) - \int_{y_0}^{y} \frac{\partial v(x, \xi)}{\partial x} d\xi = u(x, y_0) + \int_{y_0}^{y} \frac{\partial u(x, \xi)}{\partial \xi} d\xi \qquad (2.53)$$
$$= u(x, y_0) + [u(x, y) - u(x, y_0)] = u(x, y) ,$$

where we replaced $\partial v/\partial x$ with $-\partial u/\partial y$ (using ξ for the second variable) because the function $f(z)$ is analytic and hence satisfies the conditions (2.14). A similar calculation yields:

$$\frac{\partial V}{\partial x} = v(x, y_0) + \int_{y_0}^{y} \frac{\partial u(x, \xi)}{\partial x} d\xi = v(x, y_0) + \int_{y_0}^{y} \frac{\partial v(x, \xi)}{\partial \xi} d\xi \qquad (2.54)$$
$$= v(x, y_0) + [v(x, y) - v(x, y_0)] = v(x, y) ,$$

while the $y-$derivatives are straightforward:

$$\frac{\partial U}{\partial y} = -v\,(x, y) \quad \text{and} \quad \frac{\partial V}{\partial y} = u\,(x, y) \ .$$

Hence, it is immediately seen that

$$\frac{\partial U}{\partial y} = -\frac{\partial V}{\partial x} \quad \text{and} \quad \frac{\partial V}{\partial y} = \frac{\partial U}{\partial x} \ ,$$

i.e. the Cauchy-Riemann conditions (2.14) are indeed satisfied for $F(z)$, as required.
Q.E.D.

Theorem 2.5 *If $f(z)$ is an analytic function in a simply connected region D and $F(z)$ is given by Eq. (2.52), then $F'(z) = f(z)$.*

Proof Since we have proven in the previous theorem that the function $F(z)$ is analytic, its derivative $F'(z)$ does not depend on the direction in which it is taken. If we take it, say, along the x axis, then, as follows from Eqs. (2.53) and (2.54),

$$F'(z) = (U + iV)' = \frac{\partial U}{\partial x} + i\frac{\partial V}{\partial x} = u(x, y) + iv(x, y) = f(z)\,,$$

as required. **Q.E.D.**

Similarly to the case of real integrals, we can establish a simple formula for calculating complex integrals. Indeed, it is easy to see that different functions $F(z)$, all satisfying the relation $F'(z) = f(z)$, may only differ by a constant. Indeed, suppose there are two such functions, $F_1(z)$ and $F_2(z)$, i.e. $F_1'(z) = F_2'(z) = f(z)$. Consider $\mathcal{F} = F_1 - F_2$ which has zero derivative: $\mathcal{F}' = F_1' - F_2' = f - f = 0$. If $\mathcal{F}(z)$ was a real function, then it would be obvious that it is then a constant. In our case $\mathcal{F} = U + iV$ is in general complex, consisting of two real functions, and hence a proper consideration is needed. Because the derivative can be calculated along any direction, we can write for the real, U, and imaginary, V, parts of the function $\mathcal{F}$ the following equations:

$$\frac{d\mathcal{F}}{dz} = \frac{\partial U}{\partial x} + i\frac{\partial V}{\partial x} = 0 \quad \Longrightarrow \quad \frac{\partial U}{\partial x} = 0 \ \text{and} \ \frac{\partial V}{\partial x} = 0\,,$$

and

$$\frac{d\mathcal{F}}{dz} = \frac{\partial U}{\partial y} + i\frac{\partial V}{\partial y} = 0 \quad \Longrightarrow \quad \frac{\partial U}{\partial y} = 0 \ \text{and} \ \frac{\partial V}{\partial y} = 0\,,$$

from which it is clear that U and V can only be constants, i.e. $\mathcal{F}(z) = C$, where C is a complex number. This means that the two functions F_1 and F_2 may only differ by a complex constant, and therefore, one can write:

$$\int_{z_0}^{z_1} f(z)dz = F(z_1) + C$$

with the constant C defined immediately by setting $z_1 = z_0$. Indeed, in this case the integral is zero and hence $C = -F(z_0)$, which finally gives:

$$\int_{z_0}^{z_1} f(z)dz = F(z_1) - F(z_0) \ , \tag{2.55}$$

which does indeed coincide with the main result of real integral calculus Eq. (I.4.44) (the Newton-Leibnitz formula). The function $F(z)$ may also be called an indefinite integral. This result enables calculation of complex integrals using methods identical to those used in real calculus, such as integration by parts, change of variables, etc. Many formulae of real calculus for simple integrals can also be directly applied here. Indeed, since expressions for derivatives of all elementary functions in $\mathcal{C}$ coincide with those of the functions of a real variable, we can immediately write (assuming the functions in question are defined in a simply connected region):

$$\int e^z dz = e^z + C \ , \quad \int \sin z \, dz = -\cos z + C \ , \quad \int \cos z \, dz = \sin z + C \ ,$$

and so on.

Problem 2.57 Consider the integral $I = \int_A^B z^2 dz$ between points $A(1,0)$ and $B(0,1)$ using several methods: (i) along the straight line AB; (ii) along the quarter of a circle connecting the two points; (iii) going first from A to the centre $O(0,0)$, and then from O to B; (iv) using directly Eq. (2.55) and finding the appropriate function $F(z)$ for which $F' = z^2$. [*Answer: in all cases $I = -(1+i)/3$ and $F(z) = z^3/3$.*]

Problem 2.58 Show by an explicit calculation that for any positive or negative integer $n \neq -1$ the integral $\oint_L (z - z_0)^n dz = 0$, where L is a circle centred at z_0.

The Cauchy theorem above was proven for simply connected regions. We can now generalise this result to multiply connected regions as e.g. the ones shown in Fig. 2.2. To this end, let us consider a region D shown in Fig. 2.20a which has two holes in it. If we calculate the closed-loop integral $\oint_L f(z)dz$ for some analytic function $f(z)$, it would not be zero since $f(z)$ is not analytic where the holes are and hence our region is not simply connected. This is perfectly illustrated by the Example 2.3 where a non-zero value for the integral around the singularity z_0 was found. Therefore, in those cases the Cauchy theorem has to be modified.

The required generalisation can be easily made by constructing an additional path which goes around all the "forbidden" regions as shown in Fig. 2.20b. In this case we make two cuts to transform our region into a simply connected one; then the integral

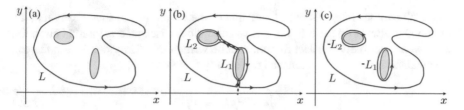

Fig. 2.20 **a** Contour L is taken around two "forbidden" regions shown as yellow with red boundaries. **b** The contour L is deformed such that it goes round each of the forbidden regions with sub-contours L_1 and L_2, both traversed in the clockwise direction in such a way that the "allowed" region is always on the left; the red dashed lines indicate the brunch cuts made to turn the region into a simply connected one; **c** the contours $-L_1$ and $-L_2$ are taken in the opposite direction so that they traverse the "forbidden" regions anti-clockwise

will be zero over the whole closed loop:

$$\int_L f(z)dz + \int_{L_1} f(z)dz + \int_{L_2} f(z)dz + \int_{\Delta L} f(z)dz = 0 \,,$$

where L_1 is taken around the first "forbidden" region, while L_2 around the second, and ΔL corresponds to two connecting lines traversing in the opposite directions when connecting L with L_1 and L_1 with L_2. Since we can arbitrarily deform the contour inside the simply connected region without changing the value of the integral (which is zero), we can make sure that the connecting lines in ΔL are passed very close to each other on both sides of each cut; since they are passed in the opposite directions, they cancel themselves out, and hence their contribution will be zero. Therefore, we can write

$$\int_L f(z)dz = -\int_{L_1} f(z)dz - \int_{L_2} f(z)dz$$
$$= \int_{-L_1} f(z)dz + \int_{-L_2} f(z)dz = g_1 + g_2 \,, \quad (2.56)$$

where g_1 and g_2 are the closed-loop integrals around each of the holes passed in the opposite (anti-clockwise) direction as shown in Fig. 2.20c.

Hence, if a loop L encloses several "forbidden" regions, where $f(z)$ is not analytic, as in Fig. 2.20a, then

$$\oint_L f(z)dz = \sum_k \oint_{-L_k} f(z)dz \,, \quad (2.57)$$

where the sum is taken over all "forbidden" regions falling inside L, and in all cases the integrals are taken in the anti-clockwise direction. One can also write the above formula in an alternative form:

$$\oint_L f(z)dz + \sum_k \oint_{L_k} f(z)dz = 0 \,, \quad (2.58)$$

where all contour integrals over L and any of the L_k are run in such a way that the region D is always on the left (i.e. L is run anti-clockwise and any internal ones, L_k, clockwise). Formally this last formula can be written in a form identical to the one we obtained for a simply connected region, Eq. (2.48):

$$\oint_{L^*} f(z)dz = 0 , \tag{2.59}$$

where the loop L^* is understood as composed of the loop L itself and all the internal loops L_k which surround any of the "forbidden" regions which fall inside L. All loops are taken in such a way that the region D is on the left.

It is clear that if the loop L goes around the kth hole many times, each time the value g_k of the corresponding loop-integral in Eq. (2.56) is added on the right-hand side, in which case

$$\oint_L f(z)dz = \sum_k n_k g_k = \sum_k n_k \oint_{-L_k} f(z)dz , \tag{2.60}$$

where n_k is the number of times the kth hole is traversed. These numbers n_k may also be negative if the traverse is made in the clockwise direction, or zero if no traverse is made at all around the given hole (which happens when the hole is outside L). The values g_k do not depend on the loop shape as within the simply connected region the loop can be arbitrarily deformed, i.e. g_k is the "property" of the function $f(z)$. Thus, it is seen from Eq. (2.60) that the value of the integral with the contour L taken inside a multiply connected region with the "forbidden" regions inside L, may take many values, i.e. it is inherently multi-valued. Formulae (2.59) and (2.60) are known as a Cauchy theorem for a multiply connected region.

Example 2.4 ▶To illustrate this very point, it is instructive to consider a contour integral of $f(z) = 1/z$ between two points $z_0 \neq 0$ and $z \neq 0$. We expect that for any path connecting the points z_0 and z and *not* looping around the $z = 0$ point, as e.g. is the path L shown in Fig. 2.21 by the solid line, the integral is related to the logarithm:

$$\int_{z_0}^{z} \frac{dz'}{z'} = \ln z - \ln z_0 \tag{2.61}$$

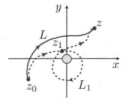

Fig. 2.21 When integrating $f(z) = 1/z$ between z_0 and z, one can choose a direct path L between the two points, or a dashed-line path L_1 which makes a loop around the brunch point $z = 0$

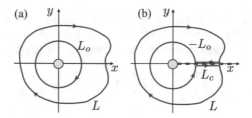

Fig. 2.22 The loop integrals around L and a circle loop L_o in **a** are the same since the two loops can be connected via two lines L_c going in the opposite directions on both sides of the brunch cut (red dashed line) as shown in **b**

(since $(\ln z)' = 1/z$). However, if the path loops around the brunch point $z = 0$ along the way, as does the path L_1 shown on the same Figure by the dashed line, then the result must be different. Indeed, the path L_1 can be split into two parts: the first one, $z_0 \to z_1 \to z$, which goes directly between the initial and ending points, and the second one, which is the loop itself (passed in the clockwise direction). The first part should give the same result (2.61) as for L as the path $z_0 \to z_1 \to z$ can be obtained by deforming L all the time remaining within the simply connected region (this can be done by making a brunch cut going e.g. from $z = 0$ along the positive x direction as shown in Fig. 2.10). Concerning the loop integral around $z = 0$, this can also be arbitrarily deformed, the result will not depend on the actual shape. In Fig. 2.22a two loops are shown: an arbitrary shaped loop L and a circle L_o. We change the direction on the circle and run a brunch cut along the positive direction of the x axis as in Fig. 2.22b; then we connect two loops by straight horizontal lines L_c running on both sides of the cut. They run in the opposite directions and hence do not contribute to the integral. However, since the whole contour $L + L_c - L_o$ lies entirely in the simply connected region, the Cauchy theorem applies, and hence the result must be zero. Considering that the path L_c does not contribute, we have that

$$\int_L \frac{dz'}{z'} + \int_{-L_o} \frac{dz'}{z'} = 0 \implies \int_L \frac{dz'}{z'} = -\int_{-L_o} \frac{dz'}{z'} = \int_{L_o} \frac{dz'}{z'} ,$$

i.e. the two loop integrals in Fig. 2.22a are indeed the same. This means that the integral over the loop in Fig. 2.21 can be replaced with the one where the contour is a circle of any radius. We have already looked at this problem in Example 2.3 for some z_0 and found the value of $2\pi i$ for the value of the integral taken over a single loop going in the anti-clockwise direction; incidentally, we found that the result indeed does not depend on R (as it should as changing the radius would simply correspond to a deformation of the contour). Hence, for the contour L_1 shown in Fig. 2.21, the result will be

$$\int_{z_0}^{z} \frac{dz'}{z'} = \ln z - \ln z_0 - 2\pi i ,$$

where $2\pi i$, which is the contribution from the contour L_0, appeared with the minus sign due to the clockwise direction of the traverse in L_0. Obviously, we can loop the brunch point in either direction and many times, so that the general result for *any* contour is

$$\int_{z_0}^{z} \frac{dz'}{z'} = \ln z - \ln z_0 + i2\pi k \;, \tag{2.62}$$

where $k = 0, \pm 1, \pm 2, \ldots$. We see that the integral is indeed equal to the multi-valued logarithmic function, compare with Eq. (2.34), and the different branches of the logarithm are related directly to the contour chosen. ◄

Problem 2.59 Using the substitution $t = \tan(x/2)$ from Sect. I.4.4.4, show that for $g > 1$

$$\int_{-\pi/2}^{\pi/2} \frac{dx}{g - \sin x} = \int_{-\pi/2}^{\pi/2} \frac{dx}{g + \sin x} = \frac{\pi}{\sqrt{g^2 - 1}} \;. \tag{2.63}$$

[Hint: after the change of the variable $x \to t$, the denominator becomes a square polynomial whose two roots are complex numbers; decompose the fraction into simpler ones and perform integration using a particular branch (sheet) of the logarithm; take the limits and collect all four logarithmic functions together.]

Analytic functions still have more very interesting properties. We shall prove now a famous result that the value of an analytic function $f(z)$ at some point z of a multiply connected region D is determined by its values on any closed contour surrounding the point z; in particular, this could be the boundary of region D. For a multiply connected region this boundary includes both the external and all the internal loops surrounding the "forbidden" regions.

Theorem 2.6 (due to Cauchy) *Let the function $f(z)$ be analytic inside some multiply connected region D. Choose a point z inside D. Then for any contour L surrounding the point z, we have*

$$f(z) = \frac{1}{2\pi i} \oint_{L^*} \frac{f(p)dp}{p - z} \;, \tag{2.64}$$

where L^* contains the loop L and all the internal loops $\{L_k\}$ which surround any holes ("forbidden" regions) lying inside L. Note the direction of the traverse of the outside loop L and any of the internal loops: the "allowed" points of D should always remain on the left.

Proof A function $g(p) = f(p)/(p - z)$ is analytic everywhere in a multiply connected region D except at the point z itself; hence, we can surround z by a small circle C_r of sufficiently small radius r and cut the circle off the region D whereby constructing an additional "forbidden" region in D. Note that the loop C_r is to be traversed in the clockwise direction, see e.g. Fig. 2.20b, keeping points of D always on the left. Let L be a closed loop which contains the point z and may be some "forbidden" regions inside it. Then, as follows from the Cauchy theorem for multiply connected regions, Eq. (2.59),

$$\oint_{L^*} \frac{f(p)}{p - z} dp + \oint_{C_r} \frac{f(p)}{p - z} dp = 0 \,,$$

where L^* is a composite loop consisting of L and all the internal loops $\{L_k\}$ surrounding the "forbidden" regions inside L. Therefore:

$$\oint_{L^*} \frac{f(p)}{p - z} dp = \oint_{-C_r} \frac{f(p)}{p - z} dp = \oint_{-C_r} \frac{f(p) - f(z)}{p - z} dp + f(z) \oint_{-C_r} \frac{dp}{p - z} \,, \tag{2.65}$$

where both integrals in the right-hand side are now taken in the anti-clockwise direction. The second integral we have calculated in Example 2.3 where we found that it is equal to $2\pi i$. The first integral is equal to zero. Indeed, it can be estimated using the inequality (2.47) as

$$\left| \oint_{circle} \frac{f(p) - f(z)}{p - z} dp \right| \leq \max_{circle} \left| \frac{f(p) - f(z)}{p - z} \right| 2\pi r$$

$$= 2\pi r \max_{circle} \left| \frac{f\left(z + r e^{i\phi}\right) - f(z)}{r e^{i\phi}} \right| = 2\pi \max_{circle} \left| \frac{f\left(z + r e^{i\phi}\right) - f(z)}{e^{i\phi}} \right|$$

$$= 2\pi \max_{circle} \left| f\left(z + r e^{i\phi}\right) - f(z) \right| \,.$$

Here we have made use of the fact that on the circle $p = z + re^{i\phi}$. The circle can be continuously deformed without affecting the value of the integral. In particular, we can make it as small as we want. Taking therefore the limit of $r \to 0$, the difference $\left| f\left(z + r e^{i\phi}\right) - f(z) \right|$ tends to zero, and hence the above estimate shows that the first circle integral in the right-hand side of (2.65) tends to zero. Therefore, from (2.65) follows the result we set out to prove. **Q.E.D.**

If we formally differentiate both sides of Eq. (2.64) with respect to z, we get a similar result for the derivative of $f(z)$:

$$f'(z) = \frac{1}{2\pi i} \oint_{L^*} \frac{f(p)}{(p - z)^2} dp \,. \tag{2.66}$$

At this point we have to be careful as the operation of differentiation is not rigorously justified since the integrand in the Cauchy formula (2.64) is singular at $p = z$. We shall show now that Eq. (2.66) is nevertheless still valid.

To this end, consider a function $f(\alpha, z)$ which depends parametrically on a (generally complex) α, and let us discuss the limit of α tending to α_0 (cf. Sects. I.6.1.3 and I.7.2.1). We shall say that $f(\alpha, z)$ converges *uniformly* to $F(z)$ within some region D when $\alpha \to \alpha_0$, if for any $\epsilon > 0$ there exists $\delta = \delta(\epsilon) > 0$ such that $|\alpha - \alpha_0| < \delta$ implies $|f(\alpha, z) - F(z)| < \epsilon$ for all z. It is essential for the uniform convergence to exist, that δ only depends on ϵ, but not on z. Next, we define a function

$$G(\alpha) = \int_{L^*} f(\alpha, z) g(z) dz , \qquad (2.67)$$

where, as above, L^* is a contour in the complex plane consisting of a closed loop L and all the internal loops $\{L_k\}$ going around any of the "forbidden" regions inside L. The function $g(z)$ above can be arbitrary; we only require that the integral of its absolute value, $\int_{L^*} |g(z)| \, dz$, exists.

Theorem 2.7 *If $f(\alpha, z)$ converges uniformly to $F(z)$ on the contour L^* when $\alpha \to \alpha_0$, and $g(z)$ is limited there, $|g(z)| < M$, then*

$$G(\alpha) = \int_{L^*} f(\alpha, z) g(z) dz \;\to\; \int_{L^*} F(z) g(z) dz .$$

In other words, one may take the limit sign inside the integral:

$$\lim_{\alpha \to \alpha_0} \int_{L^*} f(\alpha, z) g(z) dz = \int_{L^*} \left[\lim_{\alpha \to \alpha_0} f(\alpha, z) \right] g(z) dz . \qquad (2.68)$$

Proof Since the function $f(\alpha, z)$ converges uniformly with respect to α, then for any $\epsilon > 0$ one can find $\delta > 0$, not depending on z, such that $|\alpha - \alpha_0| < \delta$ implies $|f(\alpha, z) - F(z)| < \epsilon$. Therefore, considering the limit of the integral $\int_{L^*} f(\alpha, z) g(z) dz$, we can write down an estimate:

$$\left| \int_{L^*} f(\alpha, z) g(z) dz - \int_{L^*} F(z) g(z) dz \right| = \left| \int_{L^*} [f(\alpha, z) - F(z)] g(z) dz \right|$$

$$< \epsilon \int_{L^*} |g(z)| \, dz < \epsilon M \int_{L^*} dz = \epsilon M l^* = \epsilon' ,$$

where l^* is the length of the whole contour L^* (including all the internal parts). The above inequality proves the property (2.68). **Q.E.D.**

The Cauchy Theorem 2.6 enables us to present an analytic function at an internal point z via its values on a contour surrounding it. With the help of formula (2.68) just proven, we can extend the theorem to the derivatives of $f(z)$. We shall now show how to present the derivatives of an analytic $f(z)$ via its values on a contour

L^* surrounding the point z. Indeed, the first derivative is the limit of the expression $(f(z + \Delta z) - f(z))/\Delta z$, which, with the help of Eq. (2.64), can be written as:

$$
\begin{aligned}
\frac{f(z + \Delta z) - f(z)}{\Delta z} &= \frac{1}{2\pi i \Delta z} \left[\oint_{L^*} \frac{f(p)}{p - z - \Delta z} dp - \oint_{L^*} \frac{f(p)}{p - z} dp \right] \\
&= \frac{1}{2\pi i} \oint_{L^*} f(p) \frac{1}{\Delta z} \left(\frac{1}{p - z - \Delta z} - \frac{1}{p - z} \right) dp \quad (2.69) \\
&= \frac{1}{2\pi i} \oint_{L^*} \frac{1}{p - z - \Delta z} \frac{f(p)}{p - z} dp .
\end{aligned}
$$

The function $1/(p - z - \Delta z)$ converges uniformly to $1/(p - z)$ when $\Delta z \to 0$. Indeed, let d be the shortest distance from the point z to the contour L^* (recall that z lies inside L^*). Then, $|p - z| \geq d$ and, if $|\Delta z| < \delta$, then for any p from L^* we can also write

$$
|p - z - \Delta z| > |p - z| - |\Delta z| > d - \delta = d_1 .
$$

For small enough δ one can always ensure that $d_1 > 0$. Note that this estimate is valid for any p on L^* and d_1 is a constant, i.e. it does not depends on p. Then, to justify the uniform convergence $1/(p - z - \Delta z) \to 1/(p - z)$, we have to estimate the difference:

$$
\left| \frac{1}{p - z - \Delta z} - \frac{1}{p - z} \right| = \left| \frac{\Delta z}{(p - z - \Delta z)(p - z)} \right| < \frac{\delta}{d d_1} = \epsilon .
$$

It is seen that the estimate is valid for any p from L^*, so that δ depends only on ϵ but not on p, and this proves the uniform convergence required. Therefore, Eq. (2.68) is applicable, and we can take the limit $\Delta z \to 0$ inside the integral (2.69), yielding Eq. (2.66).

Problem 2.60 Prove the general result for the n-th derivative:

$$
f^{(n)}(z) = \frac{n!}{2\pi i} \oint_{L^*} \frac{f(p)}{(p - z)^{n+1}} dp . \quad (2.70)
$$

[Hint : use induction.]

This result shows that an analytic function $f(z)$ has derivatives of any order which are also analytic functions.

To avoid cumbersome notations, when using the Cauchy theorem we shall write L instead of L^* in the following, assuming that all the internal contours $\{L_k\}$ are included as well if there are "forbidden" regions inside L.

As a simple illustration of the Cauchy theorem, consider the closed-contour integral $\oint_L (p - a)^n \, dp$, where the point $z = a$ is somewhere inside the contour L. Since the function is analytic in the whole complex plane, this integral should be equal to zero. This does not contradict the Cauchy Theorem 2.6. Indeed, in this case $f(p) = (p - a)^{n+1}$ and, according to the Cauchy Theorem, the integral must be equal to $2\pi i f(a) = 0$. Finally, let us choose a circle of radius R around the point a as the contour L. In this case the integral can be calculated explicitly using $p = a + Re^{i\phi}$:

$$\oint_L (p - a)^n dp = \int_0^{2\pi} \left(Re^{i\phi}\right)^n Rie^{i\phi} d\phi = \frac{i R^{n+1}}{i(n+1)} \left(e^{i(n+1)\phi}\right)_0^{2\pi} = 0,$$

as required.

Problem 2.61 In this Problem we shall derive a representation of a Binomial coefficient via an integral in the complex plane:

$$\binom{n}{m} = \frac{1}{2\pi i} \oint_L \frac{(1+p)^n}{p^{m+1}} dp, \quad 0 \le m \le n, \tag{2.71}$$

where L is any closed anticlockwise contour containing the point $z = 0$ inside it. To prove this formula, expand $(1 + p)^n$ in the Binomial expansion, this leads to integrals $F_k = \oint p^k dp$ with the integers k ranging between $-m - 1 < 0$ and $n - m - 1$. Argue that $F_k = 0$ for $k \ge 0$. Next, by deforming the contour L into a circle of the unit radius (why can we do it?) and performing the integration explicitly, show that for any negative $k \ne -1$ the integral is also zero, while for $k = -1$ we have $F_{-1} = 2\pi i$. From this the result we seek follows immediately.

Problem 2.62 Consider a function $f(z) = (z - a)^n \phi(z)$ that has a repeated root at $z = a$ of order n, i.e. $\phi(a) \ne 0$. Show that a closed-contour integral of $f'(z)/f(z)$ taken around the point a and containing only this root is equal to the root order n:

$$\frac{1}{2\pi i} \oint_L \frac{f'(p)}{f(p)} dp = n.$$

Next, consider $f(z)$ within a certain closed contour L, and suppose that it has m roots $a_1, a_2, \ldots, a_m$ there of the orders $n_1, n_2, \ldots, n_m$, respectively. Show that in this case

$$\frac{1}{2\pi i} \oint_L \frac{f'(p)}{f(p)} dp = n_1 + n_2 + \cdots + n_m,$$

i.e. the integral is equal to the total number of roots of $f(z)$ (including their repetitions) that fall inside the contour L.

Problem 2.63 We shall calculate the integral

$$\int_{-\infty}^{\infty} \frac{dx}{x \pm i\epsilon} = \mp i\pi , \quad \text{where } \epsilon > 0 ,\qquad (2.72)$$

using the following method. Calculate first the integral with the plus sign, and for that we shall consider a contour shown in Fig. 2.23a consisting of the portion of the real axis between $-R$ and R, and a semicircle C_R of radius R in the upper half of the complex plane. The integral over the contour,

$$\oint \frac{dz}{z + i\epsilon} = \int_{-R}^{R} \frac{dx}{x + i\epsilon} + \int_{C_R} \frac{dz}{z + i\epsilon} = 0 ,$$

since the integrand $f(z) = 1/(z + i\epsilon)$ has a singularity (indicated by a small orange circle in the Figure) outside the integration area. Using the substitution $z = Re^{i\phi}$, show that the integral over the semicircle in the $R \to \infty$ limit is equal to $i\pi$. Since the integral over dx in the same limit becomes the integral we seek to calculate, we obtain the required result. Similarly, using the contour shown in panel (b) of the same Figure, calculate the integral with the minus.

Problem 2.64 Consider now the integrals of the previous Problem using the Cauchy formula (2.64) and the function $f(p) = 1$ and $z = \pm i\epsilon$. Using the appropriate contour in Fig. 2.23, that contains z inside it, and the values of the integral along C_R from the previous Problem, confirm Eq. (2.72).

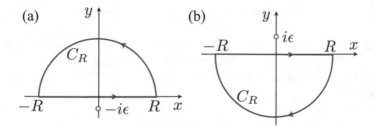

Fig. 2.23 Integration contours used in Problem 2.63

2.4.3 Fundamental Theorem of Algebra

We stated in Sects. I.1.4 and I.2.3.1 of Vol. I that a polynomial $P_n(x)$ of order n always has exactly n roots. However, our proof was based on an assumption that a polynomial of any order always has at least one root. Here we shall lift this assumption and prove this statement, and this would conclude the proof of the fundamental theorem of algebra.

The proof basically follows from the Cauchy formula (2.66) for the derivative $f'(z)$ of a function $f(z)$ that is analytical in the whole complex plane. What we about to prove is that if $|f(z)| \leq M$, i.e. the function is bounded from above and below in the *whole* complex plane, then it must be a constant. Indeed, consider a closed contour integral for the derivative taken along a circle C_R of radius R and centred at z:

$$f'(z) = \frac{1}{2\pi i} \oint_{C_R} \frac{f(p)}{(p-z)^2} dp = \frac{1}{2\pi i R^2} \oint_{C_R} e^{-2i\phi} f(p) dp ,$$

since on the circle $p = z + Re^{i\phi}$. Let us estimate the derivative:

$$|f'(z)| \leq \frac{1}{2\pi R^2} \oint_{C_R} \left| e^{-2i\phi} f(p) \right| dp$$

$$\leq \frac{1}{2\pi R^2} \oint_{C_R} |f(p)| dp = \frac{M}{2\pi R^2} \oint_{C_R} dp = \frac{M 2\pi R}{2\pi R^2} = \frac{M}{R} .$$

By taking the circle of bigger and bigger radius, i.e. by taking the $R \to \infty$ limit, we obtain that $|f'(z)| \leq 0$, i..e. $f'(z) = 0$, from which the made statement immediately follows: the function $f(z)$ is a constant in the whole complex plane if it is bounded there.

Now we can consider a n-th order polynomial

$$P_n(z) = a_0 + a_1 z + a_2 + \cdots + a_n z^n .$$

We would like to prove that it has at least one root, i.e. there exists a point z_1 at which $P_n(z_1) = 0$. We shall prove this by contradiction assuming that this function does not have roots, i.e. it is not equal to zero at all on the whole complex plane. Then, the function

$$\frac{1}{P_n(z)} = \frac{1}{z^n \left(a_n + a_{n-1}/z + a_{n-2}/z^2 + \cdots + a_0/z^n \right)}$$

is well defined in the whole complex plane as its denominator is never zero. Moreover, considering a circle of radius R and centred at zero, $z = Re^{i\phi}$, we can see that the reciprocal of the polynomial tends to zero as $R \to \infty$:

$$\frac{1}{P_n(z)}\Bigg|_{z=Re^{i\phi}} = \frac{e^{-in\phi}}{R^n}\left(a_n + \frac{a_{n-1}}{R}e^{-i\phi} + \frac{a_{n-2}}{R^2}e^{-2i\phi} + \cdots + \frac{a_0}{R^n}e^{-in\phi}\right)^{-1}$$

$$\rightarrow \frac{e^{-in\phi}a_n}{R^n} \rightarrow 0$$

Hence, this function is limited in the complex plane. However, from the statement proven above it then follows that this function must be a constant. Moreover, as the function tends to zero at large $|z|$, it must be the zero constant. Obviously, this cannot be true as the function does change with z according to its inverse polynomial form. Therefore, our assumption is wrong and $P_n(z)$ must have at least one root.

2.4.4 Fresnel Integral

The definite integral of a complex exponential, which we shall consider here, plays a fundamental role in quantum theory based on path integrals.[3] The integral we shall consider,

$$F_\pm = \int_0^\infty e^{\pm it^2}dt \, ,$$

is trivially related to the $x \rightarrow \infty$ limit of the so-called Fresnel integrals

$$S(x) = \int_0^x \sin(t^2)dt \quad \text{and} \quad C(x) = \int_0^x \cos(t^2)dt \tag{2.73}$$

that are also frequently used, e.g., in optics.[4] Both functions cannot be presented for a general x via elementary functions, but could be calculated numerically (e.g., by expanding into the Taylor series). Their graphs are shown in Fig. 2.24.

We shall perform our calculation in the complex plane by considering the closed contours shown in Fig. 2.25. We shall start from F_- and use the contour in panel (a). The contour consists of three pieces: the real axis line from $x = 0$ to $x = R$, the circular arch L_1 of radius R, centred at the origin and of the angle $\pi/4$, and the final straight line L_2 returning back to the origin. The value of R is finite, but we aim at tenting it to infinity in the end. The closed contour integral

$$\oint e^{-z^2}dz = \int_0^R e^{-x^2}dx + \int_{L_1} e^{-z^2}dz + \int_{L_2} e^{-z^2}dz$$

---

[3] The most comprehensive book on the topic, a real "bible" of path integrals is: H. Kleinert, *Path Integrals In Quantum Mechanics, Statistics, Polymer Physics, And Financial Markets*, World Scientific, 4th Edition, 2006.

[4] Specifically, in the theory of near-field Fresnel diffraction phenomena.

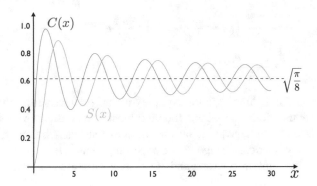

Fig. 2.24 Fresnel functions of Eq. (2.73)

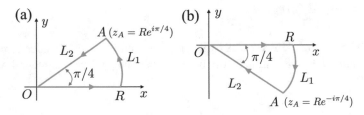

Fig. 2.25 Contours in the complex plane used for the calculation of the Fresnel integrals **a** F_- and **b** F_+

is equal to zero because the integrand $f(z) = e^{-z^2}$ has no singularities in the region bounded by the contour and hence the Cauchy theorem (2.59) applies. The first integral in the $R \to \infty$ limit is the Gaussian integral whose value is $\sqrt{\pi}/2.$, see Eq. (I.6.16). Next, consider the integrals along L_1 and L_2. On L_1 we have $z = Re^{i\phi}$ with the angle ϕ changing between 0 and $\pi/4$. Performing the change of the variables, $z \to \phi$, we have

$$\int_{L_1} e^{-z^2} dz = \int_0^{\pi/4} e^{-R^2(\cos(2\phi)+i\sin(2\phi))} i\, Re^{i\phi} d\phi.$$

Let us estimate this integral in the $R \to \infty$ limit. Firstly, we can write:

$$\left| \int_0^{\pi/4} e^{-R^2(\cos(2\phi)+i\sin(2\phi))} i\, Re^{i\phi} d\phi \right| \leq \int_0^{\pi/4} \left| e^{-R^2(\cos(2\phi)+i\sin(2\phi))} i\, Re^{i\phi} \right| d\phi$$

$$= R \int_0^{\pi/4} e^{-R^2 \cos(2\phi)} d\phi.$$

As at $\phi = \pi/4$ we have $\cos(2\phi) = 0$, the exponent goes to one and hence the behaviour of this integral in the limit is not clear. Therefore, we shall split the integral into two: between 0 and some angle $\alpha < \pi/4$, and for the rest of the interval. In the first one,

$$R \int_0^\alpha e^{-R^2 \cos(2\phi)} d\phi \le R \int_0^\alpha e^{-R^2 \cos(2\alpha)} d\phi = R\alpha \, e^{-R^2 \cos(2\alpha)} ,$$

where we have used the fact that for $0 \le \phi \le \alpha$ we obviously have $\cos(2\phi) \ge \cos(2\alpha)$. The obtained estimate goes to zero as $R \to \infty$. For the second part of the integral, when $\alpha < \phi \le \pi/4$, we can use the fact that $\sin(2\phi) \ge \sin(2\alpha)$, hence

$$R \int_\alpha^{\pi/4} e^{-R^2 \cos(2\phi)} d\phi \le R \int_\alpha^{\pi/4} \frac{\sin(2\phi)}{\sin(2\alpha)} e^{-R^2 \cos(2\phi)} d\phi$$

$$= \left| \begin{matrix} t = \cos(2\phi) \\ dt = -2\sin(2\phi)d\phi \end{matrix} \right|$$

$$= \frac{R}{2\sin(2\alpha)} \int_0^{\cos(2\alpha)} e^{-R^2 t} dt$$

$$= \frac{1}{2R\sin(2\alpha)} \left[1 - e^{-R^2 \cos(2\alpha)} \right] ,$$

which also goes to zero in the $R \to \infty$ limit. Hence, the whole integral over the arch L_1 in the $R \to \infty$ limit goes to zero.

We are left to consider the integral over the line L_2. There $z = te^{i\pi/4}$ with t changing between R and 0:

$$\int_{L_2} e^{z^2} dz = \int_R^0 \exp\left(-t^2 e^{i\pi/2}\right) e^{i\pi/4} dt = -e^{i\pi/4} \int_0^R e^{-it^2} dt .$$

Combining all the results, we obtain

$$\lim_{R \to \infty} \left[\int_0^R e^{-x^2} dx - e^{i\pi/4} \int_0^R e^{-it^2} dt \right] = 0$$

or

$$\int_0^\infty e^{-it^2} dt = e^{-i\pi/4} \int_0^\infty e^{-x^2} dx = \frac{\sqrt{\pi}}{2} e^{-i\pi/4} . \tag{2.74}$$

Problem 2.65 Using the contour shown in Fig. 2.25b, show that

$$\int_0^\infty e^{it^2}dt = \frac{\sqrt{\pi}}{2}e^{i\pi/4}. \qquad (2.75)$$

Problem 2.66 Deduce that

$$\int_0^\infty \cos(t^2)dt = \int_0^\infty \sin(t^2)dt = \sqrt{\frac{\pi}{8}}.$$

2.5 Complex Functional Series

2.5.1 Numerical Series

Similarly to the case of the real calculus, one can consider infinite numerical series

$$z_1 + z_2 + \ldots = \sum_{k=1}^\infty z_k \qquad (2.76)$$

on the complex plane. The series is said to converge to z, if for any $\epsilon > 0$ one can find a positive integer N such that for any $n \geq N$ the partial sum of the series, $S_n = \sum_{k=1}^n z_k$, differs from z by no more than ϵ, i.e. the following inequality holds: $|S_n - z| < \epsilon$. If such N cannot be found, the series is said to diverge.

It is helpful to recognise that a complex numerical series consists of two real series. Since each term $z_k = x_k + iy_k$ consists of real and imaginary parts, we can write:

$$\sum_{k=1}^\infty z_k = \sum_{k=1}^\infty x_k + i\sum_{k=1}^\infty y_k.$$

Therefore, the series (2.76) converges to $z = x + iy$, if and only if the two real series in the right-hand side of the above equation converge to x and y, respectively. This fact allows transferring most of the theorems we proved for real series (Sect. I.7.1) to the complex numerical series.

Especially useful for us here is the notion of *absolute convergence* introduced in Sect. I.7.1.4 for real numerical series with terms which may be either positive or negative. We proved there that a general series necessarily converges if the series consisting of the absolute values of the terms of the original series converges. The same type of statement is valid for complex series as well which is formulated in the following Theorem.

Theorem 2.8 (Sufficient condition) *If the series*

$$p = |z_1| + |z_2| + |z_3| + \ldots = \sum_{k=1}^{\infty} |z_k| = \sum_{k=1}^{\infty} \sqrt{x_k^2 + y_k^2} , \qquad (2.77)$$

constructed from absolute values of the terms of the original series, converges, then so does the original series.

Proof Indeed, since $|x_k| \le \sqrt{x_k^2 + y_k^2}$ and similarly $|y_k| \le \sqrt{x_k^2 + y_k^2}$ for any k, then the series $\sum_{k=1}^{\infty} |x_k|$ and $\sum_{k=1}^{\infty} |y_k|$ will both converge (and converge absolutely) as long as the series (2.77) converges (see Theorem I.7.6). Then, since the real and imaginary series both individually converge and converge absolutely, so are the original real and imaginary series, $\sum_{k=1}^{\infty} x_k$ and $\sum_{k=1}^{\infty} y_k$, and hence the series (2.76). **Q.E.D.**

Similarly to absolutely converging real series, absolutely converging complex series can be summed up, subtracted from and/or multiplied with each other; their sum also does not depend on the order of terms in the series.

The root and ratio tests for the convergence of the series are also valid. Although the proof of the root test remains essentially the same (see Theorem I.7.8), the ratio test proven in Theorem I.7.7 requires some modification due to a different nature of the absolute value $|z|$ of a complex number. We shall therefore sketch the proof of the ratio test here again to adopt it specifically for complex series.

Theorem 2.9 (The ratio test) *The series (2.76) converges absolutely if*

$$\lambda = \lim_{n \to \infty} \left| \frac{z_{n+1}}{z_n} \right| < 1 ,$$

while it diverges if $\lambda > 1$.

Proof Note that λ is a positive number. Since the limit exists, then for any $\epsilon > 0$ one can always find a number N such that any $n \ge N$ implies

$$\left| \frac{z_{n+1}}{z_n} - \lambda \right| < \epsilon \quad \Longrightarrow \quad |z_{n+1} - \lambda z_n| < \epsilon |z_n| . \qquad (2.78)$$

From the inequality $|a - b| \le |a| + |b|$ (valid even for complex a and b) follows that $|c - b| \ge |c| - |b|$ (where $c = a + b$). Therefore,

$$|z_{n+1} - \lambda z_n| \geq |z_{n+1}| - \lambda |z_n| \ ,$$

which, when combined with (2.78), gives

$$|z_{n+1}| - \lambda |z_n| < \epsilon |z_n| \implies |z_{n+1}| < (\epsilon + \lambda) |z_n| \ .$$

Repeated use of this inequality results in an estimate:

$$|z_{n+1}| < (\epsilon + \lambda) |z_n| < (\epsilon + \lambda)^2 |z_{n-1}| < \ldots < (\epsilon + \lambda)^n |z_1| \ .$$

Therefore, the series (2.79) converges absolutely for such values of z for which the geometric progression $\sum_n q^n$ with $q = \epsilon + \lambda$ converges:

$$\sum_{n=1}^{\infty} |z_n| = \sum_{n=0}^{\infty} |z_{n+1}| < |z_1| \sum_{n=0}^{\infty} (\lambda + \epsilon)^n \ ,$$

which is the case only if $0 < q < 1$. If $\lambda < 1$, one can always find a positive ϵ such that $\epsilon + \lambda < 1$, and hence the series (2.76) converges.

Consider now the case of $\lambda > 1$. In this case it is convenient to consider the ratio z_n / z_{n+1} which has a definite limit of $\zeta = 1/\lambda < 1$. Similar argument to the one given in the previous case then leads to an inequality:

$$\left| \frac{z_n}{z_{n+1}} - \zeta \right| < \epsilon \implies |z_n| - \zeta |z_{n+1}| \leq |z_n - \zeta z_{n+1}| < \epsilon |z_{n+1}| \ ,$$

which yields

$$|z_{n+1}| > \frac{1}{\epsilon + \zeta} |z_n| = \frac{\lambda}{1 + \epsilon \lambda} |z_n| > \left(\frac{\lambda}{1 + \epsilon \lambda} \right)^2 |z_{n-1}| > \ldots > \left(\frac{\lambda}{1 + \epsilon \lambda} \right)^n |z_1| \ .$$

Since $\lambda > 1$, one can always find a positive $\epsilon < (\lambda - 1)/\lambda$ such that $q = \lambda/(1 + \epsilon \lambda) > 1$. However, since $q^n \to \infty$ when $n \to \infty$, $|z_{n+1}| \to \infty$ as well, and hence the necessary condition for the convergence of the series, provided by Theorem I.7.4, is not satisfied, i.e. the series (2.76) indeed diverges. **Q.E.D.**

As in the case of the real calculus, nothing can be said about the convergence of the series if $\lambda = 1$.

Problem 2.67 Prove that the geometric progression $S = \sum_{k=0}^{\infty} q^k$ (where q is a complex number) converges absolutely if $|q| < 1$ and diverges if $|q| > 1$. Then show that the sum of the series is still formally given by exactly the same expression, $S = 1/(1 - q)$, as in the real case. [*Hint: derive a recurrence relation for the partial sum, S_N, and then take the limit $N \to \infty$.*]

2.5.2 General Functional Series

In this section we shall generalise some of the results of Chap. I.7 to complex functions. Most of the results obtained in Chap. I.7 are valid in this cases as well, although there are some differences. We shall mostly be interested in *uniform convergence* here (cf. Sect. I.7.2.1).

We shall start by considering a *functional sequence* $f_1(z)$, $f_2(z)$, $f_3(z)$, etc. We know that the sequence $\{f_n(z)\}$ converges uniformly to $f(z)$ if for any $\epsilon > 0$ one can find a number $N = N(\epsilon)$ such that any $n \geq N$ implies $|f_n(z) - f(z)| < \epsilon$ for any z. We stress again, that it is essential that the number N depends exclusively on ϵ, not on z, i.e. the same value of N applies to all z from region D where all the functions are defined; that is why the convergence is called uniform.

Next, consider an infinite functional series

$$f_1(z) + f_2(z) + \ldots = \sum_{n=1}^{\infty} f_n(z) . \tag{2.79}$$

The series (2.79) is said to converge uniformly to $f(z)$ if the functional sequence of its partial sums

$$S_N(z) = \sum_{n=1}^{N} f_n(z) \tag{2.80}$$

converges uniformly when $N \to \infty$. Most of the theorems of Sect. I.7.2.1 are valid here as well. In particular, if the series converges, its n-th term tends to zero as $n \to \infty$ (cf. Theorem I.7.4). Next, if the series converges uniformly to $f(z)$ and the functions $\{f_n(z)\}$ are continuous, then $f(z)$ is continuous as well, which means that (cf. Theorems I.7.16 and I.7.17)

$$\lim_{z \to z_0} \sum_{n=1}^{\infty} f_n(z) = \sum_{n=1}^{\infty} \lim_{z \to z_0} f_n(z) = f(z_0) .$$

Further, one can integrate a uniformly converging series (2.79) term-by-term, i.e. (cf. Theorem I.7.18) for any contour L within region D:

$$\int_L \sum_{n=1}^{\infty} f_n(z)dz = \sum_{n=1}^{\infty} \int_L f_n(z)dz .$$

Also the convergence test due to Weierstrass (Theorem I.7.15) is also valid: if each element of the series $f_n(z)$ beyond some number N (i.e. for all $n > N$) satisfies $|f_n(z)| \leq \alpha_n$ and the series $\sum_n \alpha_n$ converges, then the series (2.79) converges uniformly. Proofs of all these statements are almost identical to those given in Chapter I.7, so we do not need to repeat them here.

There are also some additional Theorems specific for the complex functions which we shall now discuss.

Theorem 2.10 *If the series (2.79) converges uniformly to* $f(z)$*, and all functions* $\{f_n(z)\}$ *are analytic in a simply connected region* D*, then* $f(z)$ *is also analytic in* D*.*

Proof Indeed, since the series converges uniformly for all z from D, we can integrate the series term-by-term, i.e. one can write:

$$\oint_L f(z)dz = \sum_{n=1}^{\infty} \oint_L f_n(z)dz \ ,$$

where L is an arbitrary closed contour in D. Since the functions $f_n(z)$ are analytic, the closed contour integral of any of them is equal to zero (see Problem 2.55). Therefore, the closed contour integral of $f(z)$, from the above equation, is also zero. But this means, according to Theorem 2.3, that $f(z)$ is analytic. **Q.E.D.**

The next Theorem states that the uniformly converging functional series (2.79) can be differentiated any number of times. The situation is much more restrictive in the real calculus (Theorem I.7.19).

Theorem 2.11 (*due to Weierstrass*) *If the series (2.79) converges uniformly to* $f(z)$ *in* D*, it can be differentiated any number of times.*

Proof Consider a closed loop L in D, and let us pick up a point z inside L and a point p on L. Then, since the series (2.79) converges uniformly to $f(z)$ for any z including points p on the contour L, we can write:

$$f(p) = \sum_{n=1}^{\infty} f_n(p) \ .$$

Next, we multiply both sides of this equation by $k!/\left[2\pi i \, (p-z)^{k+1}\right]$ with some positive integer k and integrating over L (note that the integration can be done term-by-term in the right-hand side as the series converges uniformly), we obtain:

$$\frac{k!}{2\pi i} \oint_L \frac{f(p)dp}{(p-z)^{k+1}} = \sum_{n=1}^{\infty} \frac{k!}{2\pi i} \oint_L \frac{f_n(p)dp}{(p-z)^{k+1}} \ .$$

According to the previous theorem, $f(z)$ is analytic. Therefore, we can use formula (2.70) in both sides, which yields:

$$f^{(k)}(p) = \sum_{n=1}^{\infty} f_n^{k)}(p) \, ,$$

which is exactly the result we wanted to prove. **Q.E.D.**

2.5.3 Power Series

The series

$$\sum_{k=0}^{\infty} c_k \, (z - a)^k \, , \tag{2.81}$$

in which functions $f_k(z)$ are powers of $z - a$ (where a is also complex) and c_k are some complex coefficients, is called a power series in the complex plane $\mathcal{C}$. Practically all the results of the real calculus we considered before are transferred (with some modifications) to the complex power series.

We shall start by stating again the Abel's theorem I.7.24, which we shall reformulate for the case of the complex power series here.

Theorem 2.12 (due to Abel) *If the power series (2.81) converges at some point $z_0 \neq a$, see Fig. 2.26a, then it converges absolutely within the circle $|z - a| < r$, where $r = |z_0 - a|$; moreover, it converges uniformly for any z within a circle $|z - a| < \lambda r$, where $0 < \lambda < 1$.*

Proof Since the series (2.81) converges at z_0, it is required by the necessary condition of convergence, that $c_k \, (z_0 - a)^k$ tends to zero as $k \to \infty$. Hence, its general element

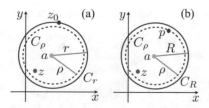

Fig. 2.26 a Point z_0 lies on a circle C_r of radius $r = |z_0 - a|$ with the centre at point a, and the point z is strictly *inside* a circle C_ρ of the radius $\rho = \theta r < r$. **b** Point p is on the circle C_ρ, which has the radius $\rho < R$

$c_k (z_0 - a)^k$ must be bounded, i.e. $\left| c_k (z_0 - a)^k \right| < M$, where M is some positive number. Then, for any z inside the circle of radius r, i.e. within a circle C_ρ with radius $\rho = \theta r$ with $0 < \theta < 1$, we can write:

$$\left| c_k (z - a)^k \right| = \left| c_k (z_0 - a)^k \right| \left| \frac{(z - a)^k}{(z_0 - a)^k} \right| = \left| c_k (z_0 - a)^k \right| \left| \frac{z - a}{z_0 - a} \right|^k < M \zeta^k ,$$

where $\zeta = \left| (z - a) / (z_0 - a) \right| < \rho/r < 1$. Hence, the absolute value of each term of our series is bounded by the elements of the converging geometric progression, $M \zeta^k$, with $0 < \zeta < 1$, and hence, according to the corresponding analog of Weierstrass theorem I.7.15, the series converges absolutely and uniformly within the circle C_ρ.
Q.E.D.

Problem 2.68 Prove by contradiction that if the series (2.81) diverges at some $z_0 \neq a$, then it diverges for any z lying outside the circle C_r of radius $r = |z_0 - a|$.

Problem 2.69 Prove that if it is known that the series (2.81) converges at some $z_0 \neq a$ and diverges at some z_1 (obviously, $|z_1 - a| > |z_0 - a|$), then there exists a positive $R > 0$ such that the series diverges outside the circle C_R, i.e. for any z satisfying $|z - a| > R$, and absolutely converges inside C_R, i.e. for any z satisfying $|z - a| < R$.

The number R is called the *radius of convergence* of the series (cf. Sect. I.7.3.1). It follows now from Theorem 2.10 that the series (2.81) is an analytic function inside the circle C_R of its radius of convergence R. This in turn means that it can be differentiated and integrated term-by-term any number of times. The series obtained this way would have the same radius of convergence. The radius of convergence can be determined from either ratio or root tests via the following formulae (cf. Sect. I.7.3.1):

$$R = \lim_{n \to \infty} \left| \frac{c_n}{c_{n+1}} \right| \quad \text{and/or} \quad \frac{1}{R} = \text{Sup}_{n \to \infty} \sqrt[n]{|c_n|} , \qquad (2.82)$$

where in the latter case the maximum value of the root in the limit is implied.

Problem 2.70 Determine the region of convergence of the power series with the coefficients $c_k = 3^k / k$ around the point $a = i$. [*Answer:* $|z - i| < 1/3$.]

Problem 2.71 Determine the region of convergence of the power series with the coefficients $c_k = 1 / \left(2^k \sqrt{k} \right)$ around the point $a = 0$. Does the series converge at the points: $z = i$ and $z = 3 - i$? [*Answer:* $|z| < 2$; *yes; no.*]

Next, let us consider a function $f(z)$ which is analytic in some region D. We choose a point a inside D and draw the largest possible circle C_R of radius R with the centre at a and lying inside D; all points z inside the circle satisfy the inequality $|z - a| < R$. Choose then another circle C_ρ, also with the centre at a, which encloses the point z and is inside the circle C_R, and let p be some point on C_ρ, so that $|p - z| = \rho$, see Fig. 2.26b. The complex number $q = (z - a) / (p - a)$, which absolute value

$$\lambda = |q| = \left| \frac{z - a}{p - a} \right| = \frac{|z - a|}{\rho} < 1$$

(as the points z and a lie inside C_ρ, while p is on it), may be used to form an infinite geometric progression

$$\sum_{k=0}^{\infty} q^k = \sum_{k=0}^{\infty} \left(\frac{z - a}{p - a} \right)^k = \frac{1}{1 - q}.$$

It converges absolutely to $1 / (1 - q)$ for any p on the circle C_ρ. Moreover, it also converges uniformly with respect to p. Indeed, the absolute value of its k-th term, λ^k, where $\lambda = |z - a| / \rho$, does not depend on p (ρ is the distance from a to the circle C_ρ, which is the same for all points p on it), and the geometric progression $\sum_{k=0}^{\infty} \lambda^k$ converges absolutely since $0 < \lambda < 1$. Therefore, we obtain

$$
\frac{1}{p - z} = \frac{1}{(p - a) - (z - a)} = \frac{1}{p - a} \frac{1}{1 - \frac{z-a}{p-a}} = \frac{1}{p - a} \frac{1}{1 - q}
$$
$$
= \frac{1}{p - a} \sum_{k=0}^{n} q^k = \sum_{k=0}^{n} \frac{(z - a)^k}{(p - a)^{k+1}},
$$

(2.83)

where the series on the right converges uniformly for all p on the circle C_ρ. Therefore, it can be integrated term-by-term. Multiplying both sides of Eq. (2.83) by $f(p)/2\pi i$ and integrating over the circle C_ρ, we get:

$$
\frac{1}{2\pi i} \oint_{C_\rho} \frac{f(p)}{p - z} dp = \sum_{k=0}^{n} \frac{1}{2\pi i} \oint_{C_\rho} \frac{f(p) (z - a)^k}{(p - a)^{k+1}} dp
$$
$$
= \sum_{k=0}^{n} (z - a)^k \left[\frac{1}{2\pi i} \oint_{C_\rho} \frac{f(p) dp}{(p - a)^{k+1}} \right].
$$

(2.84)

Using now formulae (2.64) and (2.70) for the left- and right-hand sides, respectively, and recalling that $f(p)$ is analytic on C_ρ as it is inside C_R, we see that the left-hand side is equal to $f(p)$ and in the right-hand side we have the k-th derivative of $f(p)$. Hence, we finally obtain:

$$f(z) = \sum_{k=0}^{\infty} c_k (z-a)^k \ , \quad \text{where} \quad c_k = \frac{f^{(k)}(a)}{k!} = \frac{1}{2\pi i} \oint_{C_\rho} \frac{f(p)dp}{(p-a)^{k+1}} \ , \quad (2.85)$$

which is the final result. Note that the expansion converges uniformly since it was obtained by a term-by-term integration of the uniformly converging geometric progression; moreover, the series is an analytic function (Theorem 2.10).

The formula for the series above looks exactly the same as the Taylor's formula for real functions (see Sect. I.7.3.3). Hence, since the formulae for differentiation of all elementary functions on the complex plane are identical to those in the real case, the Taylor's expansions for the elementary functions also look identical. For instance, we can immediately write the following expansions around $a = 0$:

$$e^z = 1 + z + \frac{z^2}{2!} + \ldots = \sum_{n=0}^{\infty} \frac{z^n}{n!} \ ; \quad (2.86)$$

$$\sin z = z - \frac{z^3}{3!} + \frac{z^5}{5!} - \frac{z^7}{7!} + \ldots + \frac{(-1)^{n+1} z^{2n-1}}{(2n-1)!} + \ldots$$
$$= \sum_{n=1}^{\infty} \frac{(-1)^{n-1} z^{2n-1}}{(2n-1)!} \ ; \quad (2.87)$$

$$\cos z = 1 - \frac{z^2}{2!} + \frac{z^4}{4!} - \ldots + \frac{(-1)^n z^{2n}}{(2n)!} + \ldots$$
$$= \sum_{n=0}^{\infty} \frac{(-1)^n z^{2n}}{(2n)!} \ ; \quad (2.88)$$

$$\ln(1+z) = z - \frac{z^2}{2} + \frac{z^3}{3} - \ldots + (-1)^{n+1} \frac{z^n}{n} + \ldots$$
$$= \sum_{n=1}^{\infty} (-1)^{n+1} \frac{z^n}{n} \ ; \quad (2.89)$$

$$(1+z)^\alpha = 1 + \alpha z + \frac{\alpha(\alpha-1)}{2} z^2 + \ldots + D_n^\alpha z^n + \ldots = \sum_{n=0}^{\infty} D_n^\alpha z^n \ , \quad (2.90)$$

where α is generally complex and the "generalised binomial" coefficients D_n^α are given by Eq. (I.3.94):

$$D_n^\alpha = \binom{\alpha}{n} = \frac{\alpha(\alpha-1)(\alpha-2)\cdots(\alpha-n+1)}{n!} \ . \quad (2.91)$$

The latter two expansions are written for a single-valued brunches of the functions which correspond to the values of 0 and 1 of the functions at the point $z = 0$, respectively.

Problem 2.72 Show that the radius of convergence of the series (2.86)–(2.88) is $R = \infty$ (i.e. they converge for all z).

Problem 2.73 Show that the radius of convergence of the series (2.89) and (2.90) is $R = 1$ (i.e. they converge for $|z| < 1$).

Problem 2.74 Derive an analog of the Taylor's formula (I.3.84) for complex functions,

$$f(z) = \sum_{k=0}^{n} \frac{(z-a)^k}{k!} f^{(k)}(a) + R_{n+1} \tag{2.92}$$

with the remainder term

$$R_{n+1} = \frac{(z-a)^{n+1}}{2\pi i} \oint_{C_p} \frac{f(p)dp}{(p-z)(p-a)^{n+1}}, \tag{2.93}$$

starting from a finite geometric progression

$$S_n = \sum_{k=0}^{n} q^k = \frac{q^{n+1} - 1}{q - 1} \quad \text{with} \quad q = \frac{z-a}{p-a},$$

proving first that

$$\frac{1}{p-z} = \frac{1}{p-a} \sum_{k=0}^{n} \left(\frac{z-a}{p-a}\right)^k + \frac{(z-a)^{n+1}}{(p-z)(p-a)^{n+1}}$$

and then applying the method we used above when deriving the Taylor's series.

Problem 2.75 Consider the Taylor's series of e^{ix} with x real, separate out the even and odd powers of x and hence prove the Euler's formulae (2.28).

Exactly in the same way as in Sect. I.7.3.3 it can be shown that any power series of an analytic function $f(z)$ coincides with its Taylor's series, i.e. the Taylor's series is unique.

2.5.4   The Laurent Series

The Taylor's series is useful in expanding the function $f(z)$ within a circle $|z - a| < R$ where $f(z)$ is analytic. However, if $f(z)$ is not analytic at the point $z = a$, Taylor's expansion around this point cannot be applied. As was shown by Laurent[5] it is still possible to expand $f(z)$ around the point $z = a$, but in this case the series would not only contain terms $(z - a)^k$ with positive powers k, but also terms with negative k as well, i.e. this, the so-called *Laurent series,* would have the general form:

$$f(z) = \sum_{k=-\infty}^{\infty} c_k (z - a)^k \ . \tag{2.94}$$

For some functions the series may contain a finite number of terms in the part of the series with positive or negative k. This is determined by the character of the point $z = a$ where $f(z)$ has a singularity. We shall postpone considering this particular aspect in more detail until later, but now let us derive the Laurent series.

Consider $f(z)$ which is analytic inside a circle of radius R around the point a, except the point a itself, i.e. the function is analytic in a ring $0 < |z - a| < R$, see Fig. 2.27. Take now a point z inside the ring formed by the circle C_r, surrounding the point $z = a$, and a larger circle C_ρ which encloses the point z but still remains inside the circle C_R of radius R, i.e. $0 < r < \rho < R$. The value of the function $f(z)$ at the point z can then be expressed employing the generalisation (2.64) of the Cauchy theorem, yielding

$$f(z) = \frac{1}{2\pi i} \oint_{C_\rho} \frac{f(p)}{p - z} dp - \frac{1}{2\pi i} \oint_{C_r} \frac{f(p)}{p - z} dp \ . \tag{2.95}$$

Both integrals are taken in the anti-clockwise direction. Note that in the first integral over C_ρ the points p are further away from a than z, i.e. $|p - a| > |z - a|$, and hence $1/(p - z)$ can be expanded into a geometric progression (2.83) leading to the Taylor's series for the first integral in (2.95), i.e.

$$\frac{1}{2\pi i} \oint_{C_\rho} \frac{f(p)}{p - z} dp = \sum_{k=0}^{\infty} c_k (z - a)^k \ , \quad \text{where} \quad c_k = \frac{1}{2\pi i} \oint_{C_\rho} \frac{f(p) dp}{(p - a)^{k+1}} \ , \tag{2.96}$$

see Eq. (2.85). Note that the coefficients c_k here cannot be written via $f^{(k)}(a)$ as the latter does not exist ($f(z)$ is not analytic at $z = a$).

In the second integral in Eq. (2.95) points p lie closer to a then z, i.e. $|p - a| < |z - a|$. In this case we can expand with respect to $q = (p - a)/(z - a)$ (so that $|q| < 1$), i.e.

---

[5] Karl Weierstrass discovered the series two years before Laurent, but published his results more than 50 years later.

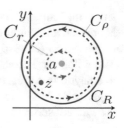

Fig. 2.27 We surround the singularity point $z = a$ by two circles: C_r of radius r and a larger one C_ρ of radius ρ, where $r < \rho < R$ and R is the radius of a circle where $f(z)$ is analytic (except for the point $z = a$ itself)

$$\frac{1}{p-z} = \frac{1}{(p-a)-(z-a)} = -\frac{1}{z-a}\frac{1}{1-\frac{p-a}{z-a}} = -\frac{1}{z-a}\frac{1}{1-q}$$

$$= -\frac{1}{z-a}\sum_{k=0}^{\infty} q^k = -\sum_{k=1}^{\infty} \frac{(p-a)^{k-1}}{(z-a)^k} \,,$$

where in the last step we changed the summation index in the sum, so that now it starts from $k = 1$. This leads to the following formula for the second integral:

$$-\frac{1}{2\pi i}\oint_{C_r}\frac{f(p)}{p-z}dp = \sum_{k=1}^{\infty} c_{-k}(z-a)^{-k} \,, \quad \text{where} \quad c_{-k} = \frac{1}{2\pi i}\oint_{C_r}(p-a)^{k-1}f(p)dp \,.$$

Here the sum runs over negative powers of $(z - a)$ using the positive index k; if we change the index $k \rightarrow -k$, so that the summation index k runs over all negative integer numbers between -1 and $-\infty$, then we obtain for the second integral in (2.95) instead:

$$-\frac{1}{2\pi i}\oint_{C_r}\frac{f(p)}{p-z}dp = \sum_{k=-\infty}^{-1} c_k(z-a)^k \,, \quad \text{where} \quad c_k = \frac{1}{2\pi i}\oint_{C_r}\frac{f(p)}{(p-a)^{k+1}}dp \,.$$

$$(2.97)$$

The latter form now looks extremely similar to the expansion (2.96) for the first integral which allows combining both results into a single formula:

$$f(z) = \sum_{k=-\infty}^{\infty} c_k(z-a)^k \,, \quad \text{where} \quad c_k = \frac{1}{2\pi i}\oint_C\frac{f(p)}{(p-a)^{k+1}}dp \,, \quad (2.98)$$

which is called the Laurent series. Note that here C is any closed contour lying between C_r and C_ρ. Indeed, the loop C_ρ in the formula for c_k with positive k can be deformed into C as long as C remains inside the ring formed by C_r and C_ρ, and for negative k the same can be done with the loop C_r which can be freely deformed into C.

The part of the series containing negative powers of $(z - a)$ is called the *principal part* of the Laurent series. Also note that since both parts of the Laurent series (for positive and negative k) were based on the uniformly converging geometric progressions, the Laurent series also converges uniformly inside the ring $0 < |z - a| < R$. Moreover, the Laurent series represents an analytic function in the ring as consisting of two series (corresponding to negative and positive k), each of which is analytic. The following Theorem establishes uniqueness of the series and shows that if $f(z)$ is analytic in a ring $r < |z - a| < R$, then its expansion (2.94) over positive and negative powers of $(z - a)$ is unique and hence is given by Eq. (2.98), i.e. it must be the Laurent series.

Theorem 2.13 *Consider the series (2.94) that converges to $f(z)$ within a ring $r < |z - a| < R$. Then its expansion over positive and negative powers of $(z - a)$ is analytic, unique and hence coincides with its Laurent expansion.*

Proof Indeed, consider the expansion (2.94). Its part with $k \geq 0$ converges inside the circle $|z - a| < R$, while its part with $k \leq -1$ converges for any z satisfying $|z - a| > r$. Indeed, using the ratio test for the positive part, we have

$$\lim_{k \to \infty} \left| \frac{c_{k+1}}{c_k} \right| |z - a| = |z - a| \lim_{k \to \infty} \left| \frac{c_{k+1}}{c_k} \right| < 1$$

$$\implies \quad |z - a| < \lim_{k \to \infty} \left| \frac{c_k}{c_{k+1}} \right| = R \ ,$$

while for the negative part

$$\lim_{k \to -\infty} \left| \frac{c_{k-1}}{c_k} \right| |z - a|^{-1} = |z - a|^{-1} \lim_{k \to -\infty} \left| \frac{c_{k-1}}{c_k} \right| < 1$$

$$\implies \quad |z - a| > \lim_{k \to -\infty} \left| \frac{c_{k-1}}{c_k} \right| = r \ .$$

Either series converges uniformly; the proof for the positive part of the series is identical to that given by the Abel's Theorem 2.12; for the negative part the proof has to be slightly modified and we shall sketch it here (using temporarily k as positive for convenience). Consider a point z_0 inside the ring such that $|z_0 - a| < |z - a|$. The series converges at z_0 and hence each its term must be bounded: $\left| c_{-k} (z_0 - a)^{-k} \right| < M$. Then,

$$\left| c_{-k} (z - a)^{-k} \right| = \left| c_{-k} (z_0 - a)^{-k} \right| \left| \frac{(z - a)^{-k}}{(z_0 - a)^{-k}} \right| < M \left| \frac{(z_0 - a)^k}{(z - a)^k} \right|$$

$$= M \left| \frac{z_0 - a}{z - a} \right|^k = M \zeta^k \ ,$$

where $\zeta = |z_0 - a| / |z - a| < 1$. Since the absolute value of each element of our series is bounded by the elements ζ^k of the converging geometric progression, the series converges absolutely and uniformly because of the Weierstrass test (Theorem I.7.15).

Now, since the expansion (2.94) converges uniformly, it can be integrated term-by-term. Let us multiply both side of this equation by $(z - a)^n f(z)/2\pi i$ with some fixed integer n and integrate in the anti-clockwise direction over a circle C with the point a in its centre:

$$\frac{1}{2\pi i} \oint_C (z - a)^n f(z)dz = \sum_{k=-\infty}^{\infty} c_k \frac{1}{2\pi i} \oint_C (z - a)^{n+k} dz .$$

The integral in the right-hand side is equal to zero for any $n + k \neq -1$ (see problem 2.58), and is equal to $2\pi i$ for $n + k = -1$ (Example 2.3). Therefore, in the sum in the right-hand side only the single term with $k = -n - 1$ survives, and we obtain:

$$\frac{1}{2\pi i} \oint_C (z - a)^n f(z)dz = c_{-n-1} \implies c_n = \frac{1}{2\pi i} \oint_C \frac{f(z)}{(z - a)^{n+1}} dz ,$$

which is exactly the same as in the Laurent series, see Eq. (2.98) (of course, the circle can be deformed into any contour lying inside the ring). This proves the second part of the Theorem. **Q.E.D.**

Formula (2.98) can be used to find the Laurent expansion of any function $f(z)$ which is analytic in a ring. This requires calculating closed-loop contour integrals for the expansion coefficients c_k via Eq. (2.98). However, in the cases of $f(z) = Q_n(z)/P_m(z)$, where $Q_n(z)$ and $P_m(z)$ are polynomials of the powers n and m, respectively, simpler methods can be used based on a geometric progression.

Example 2.5 ▶ Let us find the Laurent series for the function $f(z) = 1/ (z^2 - 3z + 2)$ around $a = 0$.

The quadratic polynomial in the denominator has two roots at $z_1 = 1$ and $z_2 = 2$, i.e.

$$f(z) = \frac{1}{(z - 1)(z - 2)} = \frac{1}{z - 2} - \frac{1}{z - 1} . \tag{2.99}$$

Since singularities are at $z = 1$ and $z = 2$, we will have to consider three circular regions: $A\,\{0 \leq |z| < 1\}$, $B\,\{1 < |z| < 2\}$ and $C\,\{|z| > 2\}$, see Fig. 2.28a, where each of the two fractions can be considered separately. In region A we can expand directly with respect to z as $|z| < 1$:

$$\frac{1}{z - 1} = -\frac{1}{1 - z} = -\sum_{k=0}^{\infty} z^k \quad \text{and} \quad \frac{1}{z - 2} = -\frac{1}{2} \frac{1}{1 - z/2} = -\frac{1}{2} \sum_{k=0}^{\infty} \left(\frac{z}{2}\right)^k ,$$

and hence in this interval

$$\frac{1}{(z-1)(z-2)} = \sum_{k=0}^{\infty} \left(-\frac{1}{2^{k+1}} + 1\right) z^k \, ,$$

i.e. it is basically represented by the Taylor's series. Region B is a ring $1 < |z| < 2$ and hence we should expect the negative part of the Laurent expansion be presented as well. And, indeed, since $|z| > 1$, we cannot expand the fraction $1/(z-1)$ in terms of z, but rather should be able to do it in terms of $1/z$:

$$\frac{1}{z-1} = \frac{1}{z}\frac{1}{1-1/z} = \frac{1}{z}\sum_{k=0}^{\infty}\left(\frac{1}{z}\right)^k = \sum_{k=0}^{\infty} z^{-k-1} = \sum_{k=1}^{\infty} z^{-k} \, .$$

On the other hand, since $|z| < 2$, the same expansion as above can be used for $1/(z-2)$. Therefore, in this ring region

$$\frac{1}{(z-1)(z-2)} = -\sum_{k=1}^{\infty} z^{-k} - \frac{1}{2}\sum_{k=0}^{\infty}\left(\frac{z}{2}\right)^k \, ,$$

i.e. it contains both negative and positive parts. Finally, in region C we have $|z| > 2$ and hence for both fractions we have to expand in terms of $1/z$. The expansion of $1/(z-1)$ stays the same as for the previous region, while for the other fraction

$$\frac{1}{z-2} = \frac{1}{z}\frac{1}{1-2/z} = \frac{1}{z}\sum_{k=0}^{\infty}\left(\frac{2}{z}\right)^k = \sum_{k=0}^{\infty} 2^k z^{-k-1} = \sum_{k=1}^{\infty} 2^{k-1} z^{-k} \, ,$$

so that finally within C

$$\frac{1}{(z-1)(z-2)} = -\sum_{k=1}^{\infty} z^{-k} + \sum_{k=1}^{\infty} 2^{k-1} z^{-k} = \sum_{k=1}^{\infty} \left(2^{k-1} - 1\right) z^{-k} \, ,$$

i.e. the expansion contains only the negative part of the Laurent series. ◄

Example 2.6 ► In this example we shall expand the same function around $a = 3$ instead.

We again have three regions A, B and C as depicted in Fig. 2.28b. Region A corresponds to a circle centred at $z = 3$ and with the radius 1, i.e. $|z - 3| < 1$, up to the nearest singularity at $z = 2$; region B forms a ring between the two singularities, i.e. $1 < |z - 3| < 2$, while the third region C corresponds to region $|z - 3| > 2$. Let us construct the Laurent series for region B:

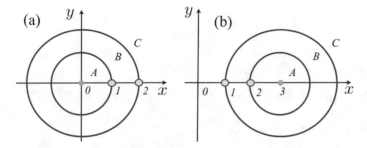

Fig. 2.28 For the expansion into the Laurent series of $f(z) = 1/(z^2 - 3z + 2)$ around **a** $a = 0$ and **b** $a = 3$. Blue circles separate three regions A, B and C in each case

$$\frac{1}{z-2} = \frac{1}{(z-3)+1} = \frac{1}{z-3}\frac{1}{1+\frac{1}{z-3}} = \sum_{k=0}^{\infty}\frac{(-1)^k}{(z-3)^{k+1}}$$

$$= -\sum_{k=1}^{\infty}\frac{(-1)^k}{(z-3)^k},$$

and

$$\frac{1}{z-1} = \frac{1}{(z-3)+2} = \frac{1}{2}\frac{1}{1+\frac{z-3}{2}} = \frac{1}{2}\sum_{k=0}^{\infty}(-1)^k\frac{(z-3)^k}{2^k}$$

$$= \sum_{k=0}^{\infty}(-1)^k\frac{(z-3)^k}{2^{k+1}},$$

and hence we finally obtain

$$\frac{1}{(z-1)(z-2)} = -\sum_{k=1}^{\infty}\frac{(-1)^k}{(z-3)^k} - \sum_{k=0}^{\infty}(-1)^k\frac{(z-3)^k}{2^{k+1}}.$$

Here, when expanding the first fraction, $1/(z-2)$, we have in the denominator $(z-3)+1$, where $|z-3| > 1$, and hence we must expand using inverse powers. In the second case of the fraction $1/(z-1)$, the denominator becomes $(z-3)+2$ with $|z-3| < 2$, and hence we can expand with respect to $(z-3)/2$ which results in terms with positive powers. ◄

Example 2.7 ► Let us expand into the Laurent series the function $f(z) = \sin(1/z)$ around $a = 0$. ◄

Here we have a single region $|z| > 0$. Since the Taylor's expansion for the sine function (2.87) converges for all values of z, we can just use this expansion with respect to $p = 1/z$ to get:

$$\sin\left(\frac{1}{z}\right) = \sum_{n=1}^{\infty} \frac{(-1)^{n-1}}{(2n-1)!} z^{-2n+1} .$$

This series contains only negative powers of z. ▶

Problem 2.76 Expand $f(z) = \exp(1/z)$ into the Laurent series around $a = 0$.

Problem 2.77 Show that the Laurent expansion of $f(z) = 1/\left(z^2 - (i+3)z + 3i\right)$ around $a = 4i$ is:

$$f(z) = \sum_{k=0}^{\infty} \left[\alpha_k (z - 4i)^k + \frac{\beta_k}{(z - 4i)^{k+1}} \right] ,$$

where the coefficients are:

$$\alpha_k = \frac{3+i}{10} \left[\left(\frac{i}{3}\right)^{k+1} - \left(\frac{3+4i}{25}\right)^{k+1} \right] , \quad \beta_k = 0 \quad \text{for} \quad |z - 4i| < 3 ;$$

$$\alpha_k = -\frac{3+i}{10} \left(\frac{3+4i}{25}\right)^{k+1} , \quad \beta_k = -\frac{3+i}{10}(-3i)^k \quad \text{for} \quad 3 < |z - 4i| < 5 ;$$

$$\alpha_k = 0, \quad \beta_k = \frac{3+i}{10} \left[(3 - 4i)^k - (-3i)^k \right] \quad \text{for} \quad |z - 4i| > 5 .$$

[Hint: the roots of the square polynomial in the denominator are 3 and i.]

Problem 2.78 Expand $f(z) = 1/\left(z^2 + 1\right)$ into the Laurent series around $a = 0$ using the method discussed above (by decomposing the fraction). Then check your result by expanding the original fraction directly using $u = z^2$.

2.5.5 Zeros and Singularities of Functions

Singularities of a function $f(z)$ are the points where it is not analytic; as we shall discuss here, the singularities are closely related to the Laurent series of $f(z)$.

Firstly, we shall consider the so-called *isolated singularities*. The point $z = a$ is an isolated singularity, if one can always find its such neighbourhood where $f(z)$ is analytic apart from the singularity point itself, i.e. one can always find $r > 0$

such that in the ring $0 < |z - a| < r$ the function $f(z)$ has no other singularities. For instance, the function $f(z) = 1/\left(z^2 - 3z + 2\right)$ has two isolated singularities at $z = 1$ and $z = 2$, since one can always draw a circle of the radius $r < 1$ around each of these points and find $f(z)$ to be analytic everywhere in those circles apart from the points $z = 1$ and $z = 2$ themselves. Isolated singularities, in turn, may be of three categories:

Removable singularity

The point $z = a$ is removable if $f(z)$ is not analytic there although its limit, $\lim_{z \to a} f(z)$, exists. In this case one can define $f(z)$ at $z = a$ via its limit making $f(z)$ analytic at this point as well, i.e. the singularity can be "removed". Since $f(z)$ has a well defined limit at $z \to a$, its Laurent expansion cannot contain negative power terms; therefore, $f(z)$ must be represented on the ring $0 < |z - a| < R$ (with some R) by the Taylor's series (2.85) with $f(a)$ being defined as its zero power coefficient c_0 (as all other terms tend to zero in the limit).

Example 2.8 ▶ Function $f(z) = \sin z / z$ has $z = 0$ as an isolated removable singularity.

Indeed, we can expand the sine functions for $|z| > 0$ in the Taylor's series to see that the singularity is removed:

$$\frac{\sin z}{z} = \frac{1}{z} \sum_{n=1}^{\infty} \frac{(-1)^{n-1}}{(2n-1)!} z^{2n-1} = \frac{1}{z}\left(z - \frac{z^3}{3!} + \ldots\right) = 1 - \frac{z^2}{3!} + \ldots \, ,$$

and hence it has a well defined limit at $z = 0$. Therefore, we can redefine the function via

$$f(z) = \begin{cases} \sin z / z \, , & \text{if } z \neq 0 \\ 1 \, , & \text{if } z = 0 \end{cases} \, ,$$

which now makes it analytic everywhere. ◀

Poles

The point $z = a$ is called a *pole* if $\lim_{z \to a} f(z) = \infty$ (either positive or negative infinity). In this case the Laurent series contains a finite number of negative power terms, i.e. it has the form:

$$f(z) = \sum_{k=-n}^{\infty} c_k (z-a)^k = \frac{1}{(z-a)^n} \sum_{k=0}^{\infty} c_{-n+k} (z-a)^k = \frac{\varphi(z)}{(z-a)^n} \, , \qquad (2.100)$$

where $\varphi(z)$ has only positive terms in its expansion, i.e. it is expandable into the Taylor's series, and hence is well defined in the neighbourhood of $z = a$ including the point $z = a$ itself. Therefore, the origin of the singularity (and of the infinite limit of $f(z)$ when $z \to a$) is due to the factor $1/(z-a)^n$.

Above, $-n$ corresponds to the largest negative power term in the expansion (2.100). If $n = 1$, i.e. the expansion starts from the term $c_{-1}/(z-a)$, the pole is called *simple*. Otherwise, if its starts from $c_n/(z-a)^n$, the pole is said to be of the order n. It is easy to see that

$$(z-a)^n f(z) = \varphi(z) = \sum_{k=0}^{\infty} c_{-n+k}(z-a)^k$$

has a well defined limit at $z \to a$ equal to c_{-n}, which must be neither zero nor infinity. Therefore, by taking such a limit it is possible to determine the order of the pole:

Order of pole n is when $\lim_{z \to a}(z-a)^n f(z)$ is neither zero nor infinity. (2.101)

Example 2.9 ▶ *The function*

$$f(z) = \frac{z-3}{z^3 + (1-2i)z^2 - (1+2i)z - 1} = \frac{z-3}{(z-i)^2(z+1)}$$

has two poles. The point $z = i$ is the pole of order 2, while $z = -1$ is a simple pole. Indeed, applying the criterion, we have for the former pole:

$$\lim_{z \to i}(z-i)^n f(z) = \lim_{z \to i} \frac{(z-3)(z-i)^{n-2}}{z+1} = \frac{i-3}{i+1}\lim_{z \to i}(z-i)^{n-2}.$$

The limit is not zero or infinity only if $n = 2$ (in which case the limit is $(i-3)/(i+1) = -1+2i$). Similarly, for the other pole:

$$\lim_{z \to -1}(z+1)^n f(z) = \lim_{z \to -1} \frac{(z-3)(z+1)^{n-1}}{(z-i)^2} = \frac{-1-3}{(-1-i)^2}\lim_{z \to -1}(z+1)^{n-1},$$

which gives $n = 1$; for any other values of n the limit is either zero (when $n > 1$) or infinity ($n < 1$).◀

Poles are closely related to *zeros* of complex functions. If the function $f(z)$ is not singular at $z = a$ and its Taylor's expansion around this point is missing the very first (constant) term, i.e. the coefficient $c_0 = 0$ in Eq. (2.85), then $f(a) = 0$. But more first terms in the Taylor's expansion may be missing for some functions, and this would characterise the rate with which $f(z)$ tends to zero as $z \to a$. More precisely, if the Taylor's expansion of $f(z)$ starts from the n-th term, i.e.

$$f(z) = \sum_{k=n}^{\infty} c_n(z-a)^k = (z-a)^n \sum_{k=0}^{\infty} c_{n+k}(z-a)^k = (z-a)^n \varphi(z), \quad (2.102)$$

where $\varphi(z)$ has all terms in its Taylor's expansion, then it is said that the point $z = a$ is a zero of order n of $f(z)$. If $n = 1$, then it is called a *simple zero*. The function $f(z) = \sin z$ has a simple zero at $z = 0$ since its Taylor's expansion starts from the linear term.

By looking at Eqs. (2.100) and (2.102), we can see that if $f(z)$ has a pole of order n at $z = a$, then the same point is a zero of the same order of the function $1/f(z) = (z - a)^n / \varphi(z)$ since $1/\varphi(z)$ tends to a finite limit at $z = a$. Inversely, if $f(z)$ has a zero of order n at $z = a$, then $1/f(z) = (z - a)^{-n} / \varphi(z)$ has a pole of the same order at the same point.

Essential singularity

The point $z = a$ is called an *essential singularity* of $f(z)$ if the limit $\lim_{z \to a} f(z)$ does not exist. This means that by taking different sequences of points $\{z_k\}$ on the complex plane which converge to $z = a$, different limits of $f(z)$ are obtained, i.e. $f(z)$ at $z = a$ is basically not defined; in fact,[6] one can always find a sequence of numbers converging to $z = a$ which results in *any* limit of $\lim_{z \to a} f(z)$. The Laurent series around $z = a$ must have an infinite number of negative power terms (as otherwise we arrive at the two previously considered cases).

Example 2.10 ▶ The function $f(z) = \sin(1/z)$ has an essential singularity at $z = 0$.

Indeed, the Laurent series has the complete principal part, i.e. the negative part has all terms (Example 2.7). If we tend z to zero over a sequence of points on the real axis, $z = x$, then the limit $\lim_{x \to 0}(1/x)$ is not defined as the function oscillates rapidly as $x \to 0$, although it remains bounded by ± 1. If, however, we take the limit along the imaginary axis, $z = iy$, then

$$\sin\left(\frac{1}{iy}\right) = \frac{1}{2i}\left(e^{i/iy} - e^{-i/iy}\right) = -\frac{i}{2}\left(e^{1/y} - e^{-1/y}\right) .$$

If $y \to +0$ (i.e., from above), then $e^{1/y} \to \infty$, $e^{-1/y} \to 0$ and hence $\sin(1/iy) \to -i\infty$. If, however, $y \to -0$ (that is, from below), then $\sin(1/iy) = +i\infty$. This discussion illustrates well that $z = 0$ is an essential singularity of $\sin(1/z)$. ◀

Problem 2.79 Show by taking different sequences of points that $z = 0$ is an essential singularity of the function $f(z) = \exp(1/z)$.

---

[6] This statement was proven independently by Julian Sochocki, Felice Casorati and Karl Weierstrass.

Problem 2.80 Show that $z = 0$ is a removable singularity of $f(z) = \left(1 - e^{z^2}\right) / z^2$ and redefine the function appropriately at this point.

Problem 2.81 Determine all poles of the function $f(z) = e^z / \left(z^2 + 1\right)$ and state their orders.

Problem 2.82 The same for the function $f(z) = \sin z / z^3$.

Problem 2.83 The same for the function $f(z) = z \left(z^2 + i\right)^{-1} (z - 3)^5 (z + 3)^{-3}$.

Holomorphic and meromorphic functions

Functions $f(z)$ which do not have singularities are called *holomorphic* or single-valued analytic functions. For instance, these are sine, cosine and exponential functions, polynomials, as well as their simple combinations not involving division. *Meromorphic functions* only have isolated poles. It follows from Eq. (2.100) that any meromorphic function is the ratio of two holomorphic functions.

Non-isolated singularities

Two more types of singularities may also be present. Firstly, a singularity may be not isolated. Consider the function $f(z) = \left(e^{1/z} + 1\right)^{-1}$. Its denominator is equal to zero, $e^{1/z} + 1 = 0$, when

$$\frac{1}{z} = \ln(-1) = \ln e^{i\pi} = i\pi + i2\pi k \quad \text{with} \quad k = 0, \pm 1, \pm 2, \ldots .$$

Therefore, at points $z_k = (i\pi + i2\pi k)^{-1}$ the function $f(z)$ is singular. However, in the limit of $k \to \infty$ the points z_k form a very dense sequence near $z = 0$, i.e. the singularities are not isolated.

The other types of points where a function is not analytic are brunch points and points on brunch cuts. The points on the brunch cuts form a continuous set and hence are also not isolated.

2.6   Analytic Continuation

It is often necessary to define a function outside the domain of its natural definition. For instance, consider a real function $f(x) = \sin x / \left(x^2 + 1\right)$. It is defined on the real x axis. If we now consider the function $f(z) = \sin z / \left(z^2 + 1\right)$ with z being complex, how would it relate to $f(x)$? Obviously, the two functions coincide on the real axis (when $z = x$), but it should be clarified what happens outside that axis. We need this understanding for instance when calculating real integrals by extending the integrands into the complex plane, a method very often used in practice.

It is easy to see that if a function $f(x)$ is continuous together with its first derivative $f'(x)$, then the complex function $f(z) = f(x + iy)$ (obtained by replacing x with z in its definition) is analytic. Indeed, $f(z)$ is a complex function, i.e. $f(z) = f(x + iy) = u(x, y) + iv(x, y)$, and hence we can write:

$$\frac{\partial f}{\partial x} = \frac{df}{dz}\frac{\partial z}{\partial x} = \frac{df}{dz} \equiv u'_x + iv'_x \,, \quad \frac{\partial f}{\partial y} = \frac{df}{dz}\frac{\partial z}{\partial y} = \frac{df}{dz}i \equiv u'_y + iv'_y \,.$$

Therefore,

$$\frac{df}{dz} = u'_x + iv'_x \quad \text{and also} \quad \frac{df}{dz} = -i\left(u'_y + iv'_y\right) = v'_y - iu'_y \,.$$

Equating real and imaginary parts of the two expressions, we obtain $u'_x = v'_y$ and $v'_x = -u'_y$, which are the Cauchy-Riemann conditions (2.14). Hence, $f(z)$ is indeed analytic.

Problem 2.84 Consider a complex function $f(x) = u(x) + iv(x)$, composed of two real functions $u(x)$ and $v(x)$. Prove that the complex function $f(z) = u(z) + iv(z)$ is analytic.

Another example: consider a function $f(z) = \ln(1 + z) / (1 + z)$ in the complex plane, where we take the principal brunch of the logarithm (with zero imaginary part on the real axis, when $x > 0$); the brunch point is at $z = -1$. This function is analytic everywhere except for the point $z = -1$. For $|z| < 1$ it can be expanded into the Taylor's series by e.g. multiplying the corresponding expansions of $\ln(1 + z)$ and $(1 + z)^{-1}$ (since both converge absolutely and this is legitimate to do):

$$\begin{aligned} f(z) &= \frac{\ln(1 + z)}{1 + z} = \left(z - \frac{z^2}{2} + \frac{z^3}{3} - \ldots\right)\left(1 - z + z^2 - z^3 + \ldots\right) \\ &= z - \frac{3}{2}z^2 + \frac{11}{6}z^3 - \ldots \,. \end{aligned} \tag{2.103}$$

This series converges within the circle $|z| < 1$, and it is not defined outside it including the circle itself, i.e. for $|z| \geq 1$ the series diverges.

Let us now expand the function around some general point $z = a$ (where $a \neq -1$):

$$\ln(1 + z) = \ln[(1 + a)(1 + g)] = \ln(1 + a) + g - \frac{g^2}{2} + \ldots \,,$$

$$(1 + z)^{-1} = [(1 + a)(1 + g)]^{-1} = \frac{1}{1 + a}\left(1 - g + g^2 - \ldots\right) \,,$$

where $g = (z - a) / (1 + a)$, so that we can define a function

Fig. 2.29 Circles of
convergence of $f(z)$ given
by Eq. (2.104) for $a = 0$
(domain A), $a = 1$ (domain
B) and $a = 3$ (domain C).
The brunch cut is made to
the left from $z = -1$ (the red
line)

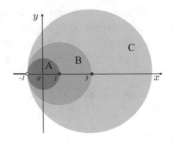

$$f_a(z) = \left(\ln (1+a) + g - \frac{g^2}{2} + \dots \right) \left[\frac{1}{1+a} \left(1 - g + g^2 - \dots \right) \right]$$

$$= \frac{\ln (1+a)}{1+a} + \frac{1 - \ln (1+a)}{(1+a)^2} (z-a) - \frac{3 - 2 \ln (1+a)}{2 (1+a)^3} (z-a)^2 - \dots .$$

$$(2.104)$$

This is a particular expansion of $f(z)$ obtained with respect to the point a, and it converges for $|g| = |(z-a)/(1+a)| < 1$, i.e. in the circle $|z - a| < |1 + a|$, which is centred at the point $z = a$ and has the radius of $R = |1 + a|$. At $a = 0$ the latter expansion reduces to (2.103).

The convergence of the series (2.104) is compared for different values of a in Fig. 2.29. Expanding $f(z)$ around $a = 0$ results in a function $f_0(z)$ which is only defined in the domain A; the function $f_1(z)$ obtained with $a = 1$ is defined in a larger domain which also completely includes A; taking a bigger value of $a = 3$ results in an even bigger domain C which goes beyond the two previous ones. We may say that the expansion (2.104) for $a = 1$ defines our function $f(z)$ in the part of the domain B which goes beyond A, i.e. $f_0(z)$ is said to be continued beyond its domain into a larger one by means of $f_1(z)$. Similarly, $f_2(z)$ defines $f(z)$ in the rest of C which is beyond B. In other words, we may now define *one* function

$$f(z) = \begin{cases} f_0(z) , & z \text{ from } A \\ f_1(z) , & z \text{ from that part of } B \text{ which is outside } A \\ f_2(z) , & z \text{ from that part of } C \text{ which is outside } B \end{cases} .$$

This process can be continued so that $f(z)$ would be defined in an even larger domain in the complex plane.

Of course, in our particular case we know $f(z)$ in the whole complex plane $\mathcal{C}$ via the logarithm from the very beginning, so this exercise seems to be useless. However, it serves to illustrate the general idea and can be used in practice when the function $f(z)$ is actually not known. For instance, when solving a differential equation via a power series (see Sects. I.8.4 and 2.8), expanding about different points $z = a$ allows obtaining different expansions with overlapping circles of convergence. Hence, a single function can be defined in the united domains as a result of the procedure described above.

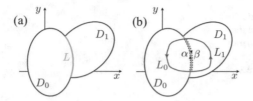

Fig. 2.30 a Tho regions D_0 and D_1 overlap at a line L. **b** A contour is considered which runs across both regions and consists of two parts: L_0 lying in D_0 and L_1 lying fully in D_1. The paths α and β lie exactly on the common line L and are passed in the opposite directions

This operation of defining a function $f(z)$ in a bigger domain is called *analytic continuation*. The appearance of the word "continuous" is not accidental since the resulting function will be analytic as long as its components are. This is proven by the following Theorem.

Theorem 2.14 *Consider two functions $f_0(z)$ and $f_1(z)$ which are analytic in domains D_0 and D_1, respectively. Let the two domains have only a common line L as shown in Fig. 2.30a, and let the two functions be equal on the line, i.e. $f_0(z) = f_1(z)$ for any z on L. If the two functions are also continuous on L, then the combined function*

$$f(z) = \begin{cases} f_0(z), z \text{ in } D_0 \\ f_1(z), z \text{ on } L \\ f_1(z), z \text{ in } D_1 \end{cases} \quad or \quad f(z) = \begin{cases} f_0(z), z \text{ in } D_0 \\ f_0(z), z \text{ on } L \\ f_1(z), z \text{ in } D_1 \end{cases}, \quad (2.105)$$

which can be considered as an analytic continuation of $f_0(z)$ from D_0 into D_1 (or of $f_1(z)$ from D_1 into D_0) is analytic in the whole domain $D_0 + D_1$.

Proof Consider a contour which starts in D_0, then crosses L, goes into D_1 and finally returns back to D_0. It consists of two parts: L_0 which lies fully inside D_0, and of the other part, L_1, which is in D_1, as shown in Fig. 2.30b. We can close L_0 with a line α which lies on the common line L for the two regions; similarly, L_1 can be closed with the line $\beta = -\alpha$, also lying on L and passed in the opposite direction to that of α. Since the two lines α and β are the same, but passed in the opposite directions,

$$\int_\alpha f(z)dz + \int_\beta f(z)dz = \int_\alpha f_0(z)dz + \int_\beta f_1(z)dz$$

$$= \int_\alpha f_0(z)dz + \int_{-\alpha} f_0(z)dz = 0,$$

Fig. 2.31 Two regions D_0
and D_1 overlap over some
domain Δ, i.e. they share
some internal points apart
from the common line L

as $f(z) = f_0(z) = f_1(z)$ everywhere on L. Therefore, The contour integral of $f(z)$
over the path $L_0 + L_1$ can be written as

$$
\oint_{L_0+L_1} f(z)dz = \int_{L_0} f(z)dz + \int_{L_1} f(z)dz = \int_{L_0} f_0(z)dz
$$
$$
+ \int_{L_1} f_1(z)dz + \left(\int_{\alpha} f_0(z)dz + \int_{\beta} f_1(z)dz \right)
$$
$$
= \oint_{L_0+\alpha} f_0(z)dz + \oint_{L_1+\beta} f_1(z)dz = 0 + 0 = 0 \,,
$$

since the loop $L_0 + \alpha$ lies fully in D_0 and hence the closed-contour integral of the
analytic function $f_0(z)$ is equal to zero; similarly, the integral of $f_1(z)$ over the closed
loop $L_1 + \beta$ is zero as well. Therefore, a closed-loop integral of the function (2.105)
anywhere in the entire region $D_0 + D_1$ is zero. This finally means, according to the
Morera's Theorem 2.3, that the function $f(z)$ is analytic. Note that the continuity
of both functions on L is required for the Cauchy Theorem 2.2 we used here and
formula (2.48) as L is a part of the boundary of the two regions. **Q.E.D.**

Above we assumed that the two regions overlap only along a line. In this case
the continuation defines a single-valued function $f(z)$. If the two regions overlap in
their internal parts, i.e. in the subregion Δ shown in Fig. 2.31, then a continuation is
not longer unique as $f_0(z)$ may be quite different to $f_1(z)$ in Δ, and hence in Δ the
function $f(z)$ becomes multi-valued.

2.7 Residues

2.7.1 Definition

Consider a domain D where a function $f(z)$ is analytic everywhere except for an
isolated singularity $z = a$. This could be either a pole or an essential singularity.
Surround the point a with a circle C passed in the anti-clockwise direction and
expand $f(z)$ in the Laurent series around a. The c_{-1} coefficient of the series is called
the *residue* of $f(z)$ at point $z = a$, and is denoted Res $f(a)$ or Res $[f(z);\ a]$. From
(2.98) we deduce that

$$
\text{Res } f(a) = c_{-1} = \frac{1}{2\pi i} \oint_C f(p)dp \,. \tag{2.106}
$$

Obviously the residue at the removable singularity is zero, as the expansion of $f(z)$ around this point does not have the negative (principal) part and hence $c_{-1} = 0$.

Let us derive a convenient expression for calculating the residue of a function at a pole. If the point a is a pole of order n, then the Laurent expansion (2.100) starts from the term $c_{-n}(z-a)^{-n}$, and hence one can write:

$$(z-a)^n f(z) = \sum_{k=-n}^{\infty} c_k (z-a)^{k+n} = c_{-n} + c_{-n+1}(z-a)^1$$
$$+ c_{-n+2}(z-a)^2 + \ldots + c_{-1}(z-a)^{n-1}$$
$$+ c_0(z-a)^n + \ldots .$$

Differentiate this expression $n-1$ times:

$$\frac{d^{n-1}}{dz^{n-1}}\left[(z-a)^n f(z)\right] = (n-1)!c_{-1} + n!c_0(z-a)^1$$
$$+ \frac{(n+1)!}{2!}c_1(z-a)^2 + \ldots ,$$

where all terms preceding the term with c_{-1} vanish upon differentiation. Then, taking the limit $z \to a$, the terms standing behind the c_{-1} term disappear as well, and we finally obtain a very useful expression:

$$c_{-1} = \text{Res}[f(a); a] = \frac{1}{(n-1)!} \lim_{z \to a} \frac{d^{n-1}}{dz^{n-1}}\left[(z-a)^n f(z)\right] . \qquad (2.107)$$

This formula can be used in practice to find the residues. For simple poles this formula can be manipulated into simpler forms. For $n = 1$

$$\text{Res}[f(a); a] = \lim_{z \to a}\left[(z-a)f(z)\right] . \qquad (2.108)$$

Another useful result is obtained when $f(z)$ is represented by a ratio $A(z)/B(z)$ of two functions, where $A(a) \neq 0$ and $B(z)$ has a zero of order one (a simple zero) at the point a, i.e.

$$B(z) = (z-a)B_1(z) = (z-a)[b_1 + b_2(z-a) + \ldots] \quad \text{with} \quad B_1(a) = b_1 .$$

Since $B'(z) = B_1(z) + (z-a)B_1'(z)$ with $B_1'(a) = b_2 \neq 0$ and $B'(a) = B_1(a)$, we obtain in this particular case a simpler formula:

$$\text{Res}\left[\frac{A(z)}{B(z)}; a\right] = \lim_{z \to a}\frac{A(z)}{B_1(z)} = \frac{A(a)}{B_1(a)} = \frac{A(a)}{B'(a)} . \qquad (2.109)$$

Problem 2.85 Show that if the point a is a zero of order 2 of the function $B(z)$, then the above formula is modified:

$$\text{Res}\left[\frac{A(z)}{B(z)}\,;\,a\right] = \frac{2A'(a)}{B''(a)} - \frac{2A(a)B'''(a)}{3\,[B''(a)]^2}\,. \qquad (2.110)$$

Example 2.11 ► Let us find all residues of the function

$$f(z) = \frac{\cos z}{z\,(z-i)^2}\,.$$

There are obviously two of them to find as there are only two singularities: at $z = 0$ and $z = i$. Consider first $z = 0$. Since the cosine function and $(z - i)^2$ behave well there, the function can be presented as $f(z) = \varphi_1(z)/z$ with $\varphi_1(z) = \cos z/(z - i)^2$ being a function which has a finite non-zero limit at $z = 0$. Therefore, $z = 0$ is a simple pole. This also follows from the fact that the limit of $\lim_{z\to 0}[zf(z)]$ is non-zero and finite, see Eq. (2.101). Therefore, from Eq. (2.109) we obtain:

$$\text{Res}\left[\frac{\cos z}{z\,(z-i)^2}\,;\,0\right] = \frac{\cos 0/\,(0 - i)^2}{(z)'} = \frac{1/\,(-1)}{1} = -1\,.$$

The same result comes out from the general formula (2.107) as well:

$$\text{Res}\left[\frac{\cos z}{z\,(z-i)^2}\,;\,0\right] = \frac{1}{0!}\lim_{z\to 0}\frac{\cos z}{(z-i)^2} = \frac{\cos 0}{(0-i)^2} = -1\,.$$

The Laurent expansion would also give the same $c_{-1} = -1$ coefficient. Indeed, expanding around $z = 0$, we have:

$$\cos z = 1 - \frac{z^2}{2} + \ldots\,,$$

$$\frac{1}{(z-i)^2} = \frac{-1}{(1+iz)^2} = -(1 - 2iz + \ldots)\,,$$

giving $c_{-1} = -1$, as expected:

$$\frac{\cos z}{z\,(z-i)^2} = \frac{1}{z}\left(1 - \frac{z^2}{2} + \ldots\right)(-1 + 2iz + \ldots)$$

$$= \frac{1}{z}(-1 + 2iz + \ldots) = -z^{-1} + 2iz^0 + \ldots\,.$$

Now let us consider another pole $z = i$. Since the cosine and $1/z$ behave well at $z = i$, we deal here with the pole of order 2. The criterion (2.101) confirms this: the limit of $(z - i) f(z)$ at $z \to i$ does not exist, while the limit of $(z - i)^2 f(z)$ is finite and non-zero (and equal to $\cos i / i = -i \cos i$). Hence we can use directly formula (2.107) for $n = 2$:

$$\mathrm{Res}\,[f(i); i] = \frac{1}{1!} \lim_{z \to i} \frac{d}{dz} \left[(z - i)^2 \frac{\cos z}{z\,(z - i)^2} \right] = \lim_{z \to i} \frac{d}{dz} \left[\frac{\cos z}{z} \right]$$

$$= \lim_{z \to i} \left(\frac{-\sin z}{z} - \frac{\cos z}{z^2} \right) = i \sin i + \cos i = e^{i^2} = e^{-1}\,.$$

The same result is of course obtained when using the Laurent method:

$$\cos z = \cos i - (z - i) \sin i + \dots$$

and

$$\frac{1}{z} = \frac{1}{(z - i) + i} = \frac{1}{i}\frac{1}{1 + (z - i)/i} = \frac{1}{i}\left(1 - \frac{z - i}{i} + \dots\right)$$

$$= -i + (z - i) + \dots\,,$$

so that we obtain

$$\frac{\cos z}{z\,(z - i)^2} = \frac{1}{(z - i)^2}\,[\cos i - (z - i)\sin i + \dots][-i + (z - i) + \dots]$$

$$= \frac{1}{(z - i)^2}\left[-i\cos i + (\cos i + i \sin i)\,(z - i)^1 + \dots\right]\,,$$

$$= -\frac{i\cos i}{(z - i)^2} + \frac{\cos i + i \sin i}{z - i} + \dots$$

and the same value for the $c_{-1} = \cos i + i \sin i = e^{-1}$ coefficient is calculated. ◄

Problem 2.86 Identify all singularities the following functions have, and then calculate the corresponding residues there:

$$(a)\ \frac{e^{iz}}{z^2 + 1}\ ;\ (b)\ \tan z\ ;\ (c)\ \frac{z^2 - 1}{z^2 - iz + 6}\,.$$

[Answers: (a) $\mathrm{Res}\,f(i) = -i/2e$, $\mathrm{Res}\ f(-i) = ie/2$; (b) $\mathrm{Res}\ f\,(\pi/2 + \pi k) = -1$ for any integer k; (c) $\mathrm{Res}\ f(-2i) = -i$, $\mathrm{Res}\ f(3i) = 2i$.]

Most applications of the residues are based on the following Theorem:

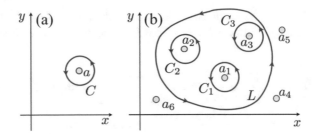

Fig. 2.32 **a** An isolated singularity a is traversed by a circular contour C. **b** Three isolated singularities a_1, a_2 and a_3 fall inside the contour L, while three other singularities are outside the contour. According to the generalised Cauchy theorem (2.57), the integral over L is equal to the sum of three contour integrals around the points a_1, a_2 and a_3

Theorem 2.15 (due to Cauchy) *Consider a function $f(z)$ which is analytic everywhere in some region D except at isolated singularities $a_1, a_2, \ldots, a_n$. Then for any closed contour L inside D (passed, as usual, in the anti-clockwise direction)*

$$\oint_L f(z)dz = 2\pi i \sum_{i \ (inside \ L)} Res \ f(a_i) \ , \qquad (2.111)$$

where the sum is taken over all isolated singularities which lie inside L. (Note that the function $f(z)$ does not need to be analytic on L (which could be e.g. a boundary of D), but need to be continuous there).

Proof First, let us consider a single isolated singularity $z = a$. We surround it by a circle C as shown in Fig. 2.32a to be passed in the anti-clockwise direction as shown. Then, according to the definition of the residue, Eq. (2.106), we have:

$$\oint_C f(z)dz = 2\pi i \, \text{Res} \, f(a) \ .$$

Next, consider a larger contour L which is run (in the same direction) around several such poles as shown in Fig. 2.32b. Then, according to the generalised form of the Cauchy theorem, Eq. (2.57), we can write:

$$\oint_L f(z)dz = \sum_i \oint_{C_i} f(z)dz = 2\pi i \sum_i \text{Res} \, f(a_i) \ ,$$

where the sum is run over all poles which lie inside the contour L. At the last step we again used the definition (2.106) of the residue. This is the result we have set out to prove. **Q.E.D.**

Note that in the above formula (2.111) only poles and essential singularities matter, as the residue of a removable singularity is zero. Therefore, removable singularities can be ignored, and this is perfectly in line with the fact that any function $f(z)$ can be made analytic at the removable singularity by defining its value there with the corresponding limit as was explained in Sect. 2.5.5.

2.7.2 Functional Series: An Example

As our first illustration of the usefulness of the residues, let us calculate a peculiar functional series:

$$f(x) = \sum_{n=0}^{\infty} \binom{2n}{n} x^n . \tag{2.112}$$

The binomial coefficient $\binom{2n}{n}$ can be related to an integral over a closed contour L, see Eq. (2.71), so that we can write:

$$f(x) = \sum_{n=0}^{\infty} \frac{1}{2\pi i} \oint_L \left[\frac{x(1+p)^2}{p} \right]^n \frac{dp}{p} .$$

We shall choose L as the unit circle C around $p = 0$. Next we shall interchange the summation and integration (the justification of this step will be slightly postponed):

$$f(x) = \frac{1}{2\pi i} \oint_C \frac{dp}{p} \left(\sum_{n=0}^{\infty} \left[\frac{x(1+p)^2}{p} \right]^n \right) . \tag{2.113}$$

The series within the round brackets represents a geometric progression

$$S(x, p) = \sum_{n=0}^{\infty} \left[\frac{x(1+p)^2}{p} \right]^n$$

that converges to

$$S(x, p) = \frac{1}{1 - \frac{x(1+p)^2}{p}} = -\frac{p}{xp^2 - (1 - 2x)p + x} . \tag{2.114}$$

On the unit circle $p = e^{i\phi}$ and it can easily be concluded that the convergence condition

$$\left| \frac{x(1+p)^2}{p} \right| = \left| x(1+p)^2 \right| = \left| x \left(1 + e^{i\phi}\right)^2 \right| = \left| x e^{i\phi} \left(e^{-i\phi/2} + e^{i\phi/2}\right)^2 \right|$$

$$= \left| x \left(2\cos\frac{\phi}{2} \right)^2 \right| = 4 |x| \left| \cos^2\frac{\phi}{2} \right| < 4 |x| < 1$$

is satisfied for $|x| < 1/4$. So, we shall only consider $f(x)$ on the interval $-1/4 < x < 1/4$.

Let us investigate whether within this interval the functional series $S(x, p)$ converges uniformly with respect to $p = e^{i\phi}$; this would justify the interchange between the integration and summation made above. We can estimate the sum:

$$|S(x, p)| \le \sum_{n=0}^{\infty} \left| \frac{x(1+p)^2}{p} \right|^n = |x| \sum_{n=0}^{\infty} \left| \left(1 + e^{i\phi}\right)^2 \right|$$

$$= \sum_{n=0}^{\infty} \left(4|x| \cos^2\frac{\phi}{2} \right)^n \le \sum_{n=0}^{\infty} (4|x|)^n = \frac{1}{1 - 4|x|},$$

since $4|x| < 1$. We see that the series $S(x, p)$ can be estimated with a convergent series that does not depend on p on the unit circle and hence converges uniformly via the Weierstrass theorem (see Sect. 2.5.2).

Finally, replacing the sum $S(x, p)$ with its value (2.114) in the integral (2.113), we obtain the relationship between the function $f(x)$ and the contour integral:

$$f(x) = -\frac{1}{2\pi i} \oint_C \frac{dp}{xp^2 - (1 - 2x)p + x} . \tag{2.115}$$

This is a remarkable result since the contour integral can be calculated analytically. Let us first investigate if the roots of the square polynomial in the denominator are within the unit circle or not. Note that we can immediately exclude the point $x = 0$ as obviously $f(0) = 1$.

Problem 2.87 Show that $f(0) = 1$.

Problem 2.88 Show that only the root

$$p_- = \frac{1}{2x} \left[1 - 2x - \sqrt{1 - 4x} \right]$$

of the polynomial $P_2(p) = xp^2 - (1 - 2x)p + x$ for $|x| < 1/4$ (excluding $x = 0$) lies within the unit circle.

Problem 2.89 Using the residue at p_-, calculate the contour integral (2.115) to show that $f(x) = (1 - 4x)^{-1/2}$.

Incidentally, this formula also describes the point $x = 0$ even though it was excluded from our consideration. Hence, the obtained expression for $f(x)$ is valid for all x between $-1/4$ and $1/4$.

2.7.3 Applications of Residues in Calculating Real Axis Integrals

Closed-loop contour integrals are calculated immediately using the residue Theorem 2.15. For instance, any contour L going around a point z_0 which is a simple pole of the function $f(z) = 1/(z - z_0)$, yields

$$\oint_L \frac{dz}{z - z_0} = 2\pi i \operatorname{Res}\left[\frac{1}{z - z_0}; z_0\right] = 2\pi i \, ,$$

which is exactly the same result as in Example 2.3.

More interesting, however, are applications of the residues in calculating definite integrals of real calculus taken along the real axis x. These are based on closing the integration line running along the real x axis in the complex plane and using the appropriate analytic continuation of the function $f(x) \rightarrow f_1(z)$ (in many cases $f_1(z) = f(z)$). It is best to illustrate the method using various examples.

Example 2.12 ▶ Let us calculate the following integral:

$$I = \int_{-\infty}^{\infty} \frac{dx}{1 + x^2} \, . \tag{2.116}$$

To perform the calculation, we consider the following integral on the complex plane:

$$I_R = \oint_L \frac{dz}{1 + z^2} = \int_{-R}^{R} \frac{dz}{1 + z^2} + \int_{C_R} \frac{dz}{1 + z^2} = I_{horiz} + I_{circle} \, ,$$

where the closed contour L shown in Fig. 2.33 consists of a horizontal part from $-R$ to R running along the positive direction of the x axis, and a semicircle C_R which is run in the anti-clockwise direction as shown. This contour is closed and hence the value of the integral can be calculated using the residue Theorem, formula (2.111),

Fig. 2.33 The contour used to calculate the integral (2.116)

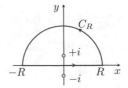

i.e. it is equal to $2\pi i$ times the residue of $f(z) = 1/(z^2 + 1)$ at the point $z = +i$. We only need to consider this pole as the other one ($z = -i$) is outside L. The residue is easily calculated to be:

$$\mathrm{Res}\left[\frac{1}{z^2 + 1} ; i\right] = \mathrm{Res}\left[\frac{1}{(z+i)(z-i)} ; i\right] = \frac{1}{i+i} = \frac{1}{2i} = -\frac{i}{2} ,$$

so that $I_R = 2\pi i \, (-i/2) = \pi$. The integral I_{horiz} is a part of the integral I we need, and the latter can be obtained by taking $R \to \infty$ limit, $I = \lim_{R\to\infty} I_{horiz}$. However, we still need to evaluate the integral over the semicircle I_{circle}. We shall show now that it tends to zero as $R \to \infty$. Indeed, let us estimate the function $f(z) = 1/(z^2 + 1)$ on the circle C_R. There $z = Re^{i\phi}$ (with $0 \leq \phi \leq \pi$), so that

$$\left|\frac{1}{1 + z^2}\right| = \frac{1}{|1 + z^2|} = \frac{1}{|1 + R^2 e^{2i\phi}|} \leq \frac{1}{R^2 - 1} ,$$

where we made use of the fact that

$$|a + b| \geq |a| - |b| > 0 \quad \text{for} \quad |a| > |b| \tag{2.117}$$

(this is follows from the inequality $|c - b| \leq |c| + |b|$ by setting $c = a + b$). Hence, according to the inequality (2.47),

$$|I_{C_R}| = \left|\int_{C_R} \frac{dz}{z^2 + 1}\right| \leq \frac{1}{R^2 - 1} \left|\int_{C_R} dz\right| = \frac{1}{R^2 - 1} \left|\int_0^\pi Re^{i\phi} d\phi\right|$$

$$\leq \frac{R}{R^2 - 1} \int_0^\pi \left|ie^{i\phi}\right| d\phi = \frac{R}{R^2 - 1} \int_0^\pi d\phi = \frac{\pi R}{R^2 - 1} ,$$

and the absolute value of the integral tends indeed to zero as $R \to \infty$, which means that $I_{C_R} \to 0$ in this limit. Therefore,

$$I = \lim_{R\to\infty} I_{horiz} = \lim_{R\to\infty} \left(I_R - I_{C_R}\right) = \pi - \lim_{R\to\infty} I_{C_R} = \pi ,$$

which is our final result. This result can be checked independently as the integral can also be calculated directly:

$$I = \int_{-\infty}^{\infty} \frac{dx}{1 + x^2} = \arctan(+\infty) - \arctan(-\infty) = \frac{\pi}{2} - \left(-\frac{\pi}{2}\right) = \pi ,$$

which is the same. ◄

In a similar manner one can calculate integrals of a more general type:

$$I = \int_{-\infty}^{\infty} \frac{Q_m(x)}{P_n(x)} dx , \tag{2.118}$$

where $Q_m(x)$ and $P_n(x)$ are polynomials of the orders m and n, respectively, and it is assumed that $P_n(x)$ does *not* have real zeros, i.e. its poles do *not* lie on the x axis. First, we note that the convergence of this integral at $\pm\infty$ requires $m + 1 < n$. Then, each polynomial can be expressed via its zeroes as a product:

$$Q_m(x) = q_m x^m + q_{m-1} x^{m-1} + \cdots + q_1 x + q_0 = q_m (x - a_1)(x - a_2)\ldots(x - a_m)$$

$$= q_m \prod_{k=1}^{m} (x - a_k) \, ,$$

$$P_n(x) = p_n x^n + p_{n-1} x^{n-1} + \cdots + p_1 x + p_0 = p_n (x - b_1)(x - b_2)\ldots(x - b_n)$$

$$= p_n \prod_{l=1}^{n} (x - b_l) \, ,$$

where $\{a_k\}$ are zeroes of $Q_m(x)$, while $\{b_k\}$ are zeroes of $P_n(x)$. Again, we consider the contour shown in Fig. 2.33 and hence need to investigate the integral over the semicircle. On it we have:

$$|z - a_k| = \left|Re^{i\phi} - a_k\right| \le \left|Re^{i\phi}\right| + |a_k| = R + |a_k|$$

and also

$$|z - b_k| = \left|Re^{i\phi} - b_k\right| \ge \left|Re^{i\phi}\right| - |b_k| = R - |b_k|$$

$$\implies \quad \frac{1}{|z - b_k|} \le \frac{1}{R - |b_k|} \, ,$$

which allows us to estimate $f(z) = Q_m(z)/P_n(z)$ as follows:

$$\left|\frac{Q_m(z)}{P_n(z)}\right| = \left|\frac{q_m}{p_n}\right| \frac{\left|\prod_{k=1}^{m} (z - a_k)\right|}{\left|\prod_{l=1}^{n} (z - b_l)\right|} \le \left|\frac{q_m}{p_n}\right| \frac{\prod_{k=1}^{m} (R + |a_k|)}{\prod_{l=1}^{n} (R - |b_k|)} = \left|\frac{q_m}{p_n}\right| \frac{A_m(R)}{B_n(R)} \, ,$$

where $A_m(R)$ and $B_n(R)$ are polynomials in R. Hence, the value of the semicircle integral can be estimated as

$$|I_{C_R}| = \left|\int_{C_R} \frac{Q_m(z)}{P_n(z)} dz\right| \le \left|\frac{q_m}{p_n}\right| \frac{A_m(R)}{B_n(R)} \pi R \, .$$

It is clearly seen that the expression in the right-hand side of the inequality tends to zero as $R \to \infty$ if $m + 1 < n$, and therefore, the $I_{C_R} \to 0$. This means that

$$I = \lim_{R \to \infty} \left(I_R - I_{C_R}\right) = \lim_{R \to \infty} I_R - \lim_{R \to \infty} I_{C_R} = \lim_{R \to \infty} I_R$$

$$= 2\pi i \sum_k \text{Res}\left[\frac{Q_m(z)}{P_n(z)} ; b_k\right] \, ,$$

where the sum is taken over all zeroes lying in the upper half of the complex plane.

Problem 2.90 Calculate the following integrals: (assuming $a, b > 0$):

(a) $\displaystyle\int_0^\infty \frac{dx}{\left(x^2+a^2\right)^2}$; (b) $\displaystyle\int_{-\infty}^\infty \frac{dx}{\left(x^2+a^2\right)\left(x^2+b^2\right)}$; (c) $\displaystyle\int_{-\infty}^\infty \frac{dx}{x^2-2x+2}$;

(d) $\displaystyle\int_{-\infty}^\infty \frac{dx}{\left(x^2-2x+2\right)^2}$; (e) $\displaystyle\int_{-\infty}^\infty \frac{dx}{\left(x^2+a^2\right)^3}$.

[Answers: (a) $\pi/4a^3$; (b) $\pi/[ab(a+b)]$; (c) π; (d) $\pi/2$; (e) $3\pi/\left(8a^5\right)$.]

Problem 2.91 Prove the formula

$$\oint \frac{z\,dz}{z^4+2\left(2g^2-1\right)z^2+1} = \frac{\pi i}{2g\sqrt{g^2-1}}, \qquad (2.119)$$

where $g > 1$ and the contour is the circle of unit radius with the centre at the origin and passed in the anti-clockwise direction.

A wide class of integrals of the type

$$I = \int_0^{2\pi} R\left(\sin\phi, \cos\phi\right) d\phi$$

can also be calculated using the residue theorem. Here $R(x, y)$ is a rational function of x and y. The trick here is to notice that the integration over ϕ between 0 and 2π may be related to the integration in the anti-clockwise direction around the circle C_1 of unit radius. Indeed, on the circle $z = e^{i\phi}$, $dz = ie^{i\phi}d\phi = izd\phi$ and also

$$\sin\phi = \frac{1}{2i}\left(e^{i\phi}-e^{-i\phi}\right) = \frac{1}{2i}\left(z-\frac{1}{z}\right)$$

and

$$\cos\phi = \frac{1}{2}\left(e^{i\phi}+e^{-i\phi}\right) = \frac{1}{2}\left(z+\frac{1}{z}\right),$$

so that

$$I = \oint_{C_1} R\left(\frac{1}{2i}\left(z-\frac{1}{z}\right), \frac{1}{2}\left(z+\frac{1}{z}\right)\right)\frac{dz}{iz},$$

i.e. the integration is indeed related now to that on the circle. Therefore, according to the residue theorem, the result is equal to the sum of residues of all poles inside the unit circle C_1, times $2\pi i$.

Example 2.13 ▶ *Consider the integral*

$$I = \int_0^{2\pi} \frac{dx}{2 + \cos x}.$$

Using the transformations described above, we get:

$$I = \oint_{C_1} \frac{dz}{zi\left[2 + \frac{1}{2}(z + 1/z)\right]} = -2i \oint_{C_1} \frac{dz}{z^2 + 4z + 1}.$$

The square polynomial has two roots: $z_{\pm} = -2 \pm \sqrt{3}$ and only the one with the plus sign is inside the circle of unit radius. Therefore, we only need to calculate the residue at the simple pole $z_+ = -2 + \sqrt{3}$, which yields:

$$I = -2i \oint_{C_1} \frac{dz}{(z - z_+)(z - z_-)} = -2i\,(2\pi i)\,\frac{1}{z_+ - z_-} = \frac{4\pi}{2\sqrt{3}} = \frac{2\pi}{\sqrt{3}}.$$

◀

Problem 2.92 Calculate the integrals $\int_0^{2\pi} f(\phi)d\phi$ for the following functions $f(\phi)$:

(a) $\dfrac{1}{1 + \sin^2 \phi}$; (b) $\dfrac{1}{(1 + \sin^2 \phi)^2}$; (c) $\dfrac{1}{1 + \cos^2 \phi}$;

(d) $\dfrac{\cos^2 \phi}{(1 + \sin^2 \phi)^2}$; (e) $\dfrac{\cos^2 \phi}{2 - \cos \phi + \cos^2 \phi}$.

[Answers: (a) $\pi\sqrt{2}$; (b) $3\pi/2\sqrt{2}$; (c) $\pi\sqrt{2}$; (d) $\pi/\sqrt{2}$; (e) $\pi\left(2 - \sqrt{13/14 + 8\sqrt{2}/7}\right)$.]

Problem 2.93 Show that for $g > 1$

$$\int_{-\pi/2}^{\pi/2} \frac{dx}{g^2 - \sin^2 x} = \frac{\pi}{g\sqrt{g^2 - 1}}. \tag{2.120}$$

Problem 2.94 Consider the integral of Problem 2.36 again. Make the substitution $t = \sin \phi$, then replace ϕ with $z = e^{i\phi}$ and show that the integration can be extended over the whole unit circle. Finally, perform the integration using the method of residues. You will also need the result (I.3.69).

A considerable class of other real-axis integrals have the form

$$I = \int_{-\infty}^{\infty} f(x) e^{i\lambda x} dx \tag{2.121}$$

with some real λ; alternatively, cosine or sine functions may appear instead of the exponential function. We shall come across this type of integrals e.g. in considering the Fourier transform in Chap. 5. The idea of their calculation is also based on closing the contour either above the real axis (as we have done so far) or below. Before we formulate a rather general result known as Jordan's lemma, let us consider an example.

Example 2.14 ▶ *Consider the following integral:*

$$I_\lambda = \int_{-\infty}^{\infty} \frac{e^{i\lambda x} dx}{x^2 + a^2} , \quad \text{where} \quad \lambda > 0 .$$

We use the function $f(z) = e^{i\lambda z} / \left(z^2 + a^2\right)$ and the same contour as in Fig. 2.33 can be employed. Then the contribution from the semicircle part C_R appears to be equal to zero in the limit $R \to \infty$ Indeed, on the upper semicircle $z = Re^{i\phi} = R(\cos \phi + i \sin \phi)$ and $0 < \phi < \pi$, so that $\sin \phi > 0$, and hence

$$e^{i\lambda z} = e^{i\lambda R \cos \phi} e^{-\lambda R \sin \phi} \to 0 \quad \text{when} \quad R \to \infty .$$

Notice that the function $1/\left(z^2 + a^2\right)$ remains bounded on the semicircle and in fact goes to zero as well with $R \to \infty$ (since $z = Re^{i\phi}$ on C_R). Even though the length of the semicircle grows proportionally to R, this is overpowered by the decrease of the function $1/\left(z^2 + a^2\right)$ and the exponential function. Therefore, only the horizontal part contributes with a simple pole at $z = ia$ to be considered, and we immediately obtain:

$$I_\lambda + \int_{C_R} f(z) dz = 2\pi i \operatorname{Res}\left[f(z); ia\right]$$

$$\implies \quad I_\lambda = 2\pi i \operatorname{Res}\left[f(z); ia\right] = 2\pi i \left(\frac{e^{i\lambda z}}{2z}\right)_{z=ia}$$

$$= 2\pi i \frac{e^{-\lambda a}}{2ia} = \frac{\pi}{a} e^{-\lambda a} .$$

◀

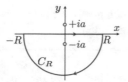

Fig. 2.34 The horizontal path between $-R$ and R has been closed in the lower part of the complex plane by a semicircle C_R. Note that the whole closed contour (horizontal path and C_R) is traversed in the clockwise direction

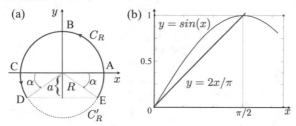

Fig. 2.35 For the proof of Jordan's lemma: **a** when $\lambda > 0$, then the contour C_R can be used which consists of an upper semicircle ABC and two (optional) arches CD and EA; when $\lambda < 0$, then instead the rest of the circle C'_R is to be used. Note that a is kept constant in the $R \to \infty$ limit, when. the angle $\alpha = \arcsin(a/R) \to 0$. **b** the sine function is concave on the interval $0 < x < \pi/2$, i.e. it lies everywhere higher than the straight line $y = 2x/\pi$

How will the result change if λ was negative? We cannot use the same contour as above since

$$e^{i\lambda z} = e^{i\lambda R \cos\phi} e^{-\lambda R \sin\phi} \to +\infty \quad \text{when} \quad R \to \infty$$

in this case. What is needed is to make $\sin\phi < 0$ on the semicircle C_R, then $e^{i\lambda z}$ would still tend to zero in the $R \to \infty$ limit and the contribution from C_R will vanish. This can be achieved simply by closing the horizontal path with a semicircle in the *lower* part of the complex plane, Fig. 2.34.

Problem 2.95 Show that in this case the integral is equal to $(\pi/a) e^{\lambda a}$. *[Hint: note that this time the contour is traversed in the clockwise direction, so that the closed contour integral is equal to a minus residue at $z = -ia$ times $2\pi i$.]*

In the above example we have used a rather intuitive approach. Fortunately, the reasoning behind it can be put on a more rigorous footing. In fact, it appears that if $f(z)$ tends to zero as $z \to \infty$ (i.e. along any direction from $z = 0$), then the integral over the semicircle C_R *always* tends to zero. This latter statement is formulated in the following Lemma where a slightly more general contour C_R is considered than in the example we have just discussed since this is useful for some applications: C_R

may also have a lower part which goes by a below the x axis, i.e. Im $z > -a$ for all z on C_R, where $0 \le a < R$, Fig. 2.35a.

Lemma 2.1 (due to Jordan) *Consider an incomplete circle C_R of radius R as shown in Fig. 2.35a. We shall also need the rest of the same circle, C'_R, shown by the dashed line on the same Figure. Consider the limit of $R \to \infty$ with a being fixed. If $f(z) \to 0$ uniformly with respect to z on C_R and C'_R in the limit, then for $\lambda > 0$*

$$\lim_{R \to \infty} \int_{C_R} f(z)e^{i\lambda z} dz = 0 , \tag{2.122}$$

while for $\lambda < 0$

$$\lim_{R \to \infty} \int_{C'_R} f(z)e^{i\lambda z} dz = 0 . \tag{2.123}$$

Proof Let us start from the case of positive λ, and consider the semicircle part ABC of C_R. Since $f(z)$ converges uniformly to zero as $R \to \infty$, we can say that for any z on the semicircle $|f(z)| < M_R$, where $M_R \to 0$ as $R \to \infty$. Note that due to the uniform convergence M_R does not depend on z, it only depends on R. Next, on the semicircle

$$z = Re^{i\phi} = R(\cos\phi + i\sin\phi) \implies e^{i\lambda z} = e^{i\lambda R \cos\phi} e^{-\lambda R \sin\phi} ,$$

and hence we can estimate the integral as follows ($dz = iRe^{i\phi} d\phi$):

$$\left| \int_{semicircle} f(z)e^{i\lambda z} dz \right| \le M_R \left| \int_{semicircle} e^{i\lambda z} dz \right|$$

$$= M_R \left| \int_0^\pi e^{i\lambda R \cos\phi} e^{-\lambda R \sin\phi} Rie^{i\phi} d\phi \right|$$

$$\le RM_R \int_0^\pi \left| e^{i\lambda R \cos\phi} e^{-\lambda R \sin\phi} ie^{i\phi} \right| d\phi$$

$$= RM_R \int_0^\pi e^{-\lambda R \sin\phi} d\phi = 2RM_R \int_0^{\pi/2} e^{-\lambda R \sin\phi} d\phi .$$

In the last step we were able to use the symmetry of the sine function about $\phi = \pi/2$ and hence consider the integration interval only between 0 and $\pi/2$ (with the corresponding factor of two appearing in front). The last integral can be estimated with the help of the inequality $\sin\phi \ge 2\phi/\pi$ valid for $0 \le \phi \le \pi/2$ (see Fig. 2.35b), yielding

$$\left| \int_{semicircle} f(z)e^{i\lambda z} dz \right| \le 2RM_R \int_0^{\pi/2} e^{-(2\lambda R/\pi)\phi} d\phi = \frac{\pi M_R}{\lambda} \left(1 - e^{-\lambda R} \right) .$$

In the limit of $R \to \infty$ the right-hand side tends to zero because of M_R.

What is left to consider is the effect of the arches CD and AE of C_R, which lie below the x axis. Apply the above method to estimate the integral on the AE first:

$$\left| \int_{AE} f(z) e^{i\lambda z} dz \right| \le M_R \left| \int_{AE} e^{i\lambda z} dz \right| \le M_R R \int_{-\alpha}^{0} e^{-\lambda R \sin \phi} d\phi .$$

For the angles ϕ lying between $-\alpha$ and zero we have $R \sin \phi \ge -a$, so that the last integral can be estimated as:

$$M_R R \int_{-\alpha}^{0} e^{-\lambda R \sin \phi} d\phi \le M_R R \int_{-\alpha}^{0} e^{\lambda a} d\phi = M_R R e^{\lambda a} \alpha$$

$$= M_R e^{\lambda a} R \arcsin \frac{a}{R} .$$

It is easy to see that $R \arcsin (a/R) \to a$ as $R \to \infty$ (use, e.g., the L'Hôpital's rule, Sect. I.3.10), so that again because of M_R the contribution from the arch AE tends to zero in the $R \to \infty$ limit.

Concerning the CD part, there the inequality $R \sin \phi \ge -a$ also holds, and hence the estimate for the exponent made above is valid as well, i.e. the same zero result is obtained.

Let us now consider the case of $\lambda < 0$. In this case we shall consider the incomplete circle contour C'_R instead. Then, the integral is estimated as follows (notice the opposite direction of the traverse this time!):

$$\left| \int_{C'_R} f(z) e^{i\lambda z} dz \right| \le R M_R \left| \int_{-\alpha}^{-\pi+\alpha} e^{-\lambda R \sin \phi} d\phi \right| = R M_R \int_{\alpha}^{\pi-\alpha} e^{\lambda R \sin \psi} d\psi$$

$$= 2 R M_R \int_{\alpha}^{\pi/2} e^{\lambda R \sin \psi} d\psi ,$$

where we have made the substitution $\psi = -\phi$. Since ψ lies between the limits $0 < \alpha < \psi < \pi/2$, the inequality $\sin \psi \ge 2\psi/\pi$ is valid and, since $\lambda < 0$, we can estimate the integral in the same way as above:

$$\left| 2 R M_R \int_{\alpha}^{\pi/2} e^{\lambda R \sin \psi} d\psi \right| = 2 R M_R \left| \int_{\alpha}^{\pi/2} e^{-|\lambda| R \sin \psi} d\psi \right|$$

$$\le 2 R M_R \left| \int_{\alpha}^{\pi/2} e^{-(2|\lambda|R/\pi)\psi} d\psi \right| = \frac{\pi M_R}{|\lambda|} \left(e^{-2R|\lambda|\alpha/\pi} - e^{-R|\lambda|} \right) .$$

It follows then that this expression tends to zero in the $R \to \infty$ limit. The Lemma is now fully proven. **Q.E.D.**

So, when $\lambda > 0$ one has to consider closing a horizontal path in the upper part of the complex plane, while for $\lambda < 0$ in the lower part; we did exactly the same in the last example.

Problem 2.96 Prove the following formulae:

$$\int_{-\infty}^{\infty} \frac{\cos(\lambda x)\,dx}{x^2 + a^2} = \frac{\pi}{a} e^{-|\lambda|a} \; ;$$

$$\int_{-\infty}^{\infty} \frac{\cos(\lambda x)\,dx}{\left(x^2 + a^2\right)^2} = \frac{\pi}{2a^2} \left(|\lambda| + \frac{1}{a}\right) e^{-|\lambda|a} \; ;$$

$$\int_{-\infty}^{\infty} \frac{e^{i\lambda x}\,dx}{\left(x^2 + a^2\right)\left(x^2 + b^2\right)} = \frac{\pi}{b^2 - a^2} \left(\frac{e^{-|\lambda|a}}{a} - \frac{e^{-|\lambda|b}}{b}\right) \; .$$

[Hint: in either of the first two problems replace the cosine with $e^{i|\lambda|z}$, work out the integral over the real axis and then take the real part of it. (In fact, the same integral with $\cos(\lambda x)$ replaced by $\sin(\lambda x)$ is zero because the integrand is an odd function. In the third problem consider both cases of $\lambda > 0$ and $\lambda < 0$.]

Problem 2.97 Prove the following formulae ($\lambda > 0$):

$$\int_{-\infty}^{\infty} \frac{x \sin(\lambda x)\,dx}{x^2 + a^2} = \pi e^{-\lambda a} \; ;$$

$$\int_{-\infty}^{\infty} \frac{x^2 \cos(\lambda x)\,dx}{\left(x^2 + a^2\right)\left(x^2 + b^2\right)} = \frac{\pi}{a^2 - b^2} \left(a e^{-\lambda a} - b e^{-\lambda b}\right) \; ;$$

$$\int_{-\infty}^{\infty} \frac{x^2 \cos(\lambda x)\,dx}{\left(x^2 + a^2\right)^2} = \frac{\pi}{2} \left(\frac{1}{a} - \lambda\right) e^{-\lambda a} \; ;$$

$$\int_{-\infty}^{\infty} \frac{x \sin(\lambda x)\,dx}{x^4 + a^4} = \frac{\pi}{a^2} e^{-\lambda a/\sqrt{2}} \sin \frac{\lambda a}{\sqrt{2}} \; .$$

Problem 2.98 Prove the following integral representations of the Heaviside function :

$$H(t) = \pm \int_{-\infty}^{\infty} \frac{d\omega}{2\pi i} \frac{e^{\pm i\omega t}}{\omega \mp i\delta} = \begin{cases} 1, & t > 0 \\ 0, & t < 0 \end{cases}, \qquad (2.124)$$

where $\delta \to +0$.

Note that, apart from the notation $H(x)$ for the Heaviside function, notation $\theta(t)$ is also frequently used.

Let us now consider several examples in which the poles of the functions $f(z)$ in the complex plane lie exactly on the real axis.

Example 2.15 ▶ *Consider the integral*

$$I_\lambda = \int_{-\infty}^{\infty} \frac{e^{i\lambda x}}{x} dx , \qquad (2.125)$$

where $\lambda > 0$. The contour cannot be taken exactly as in Fig. 2.33 in this case since it goes directly through the point $z = 0$. The idea is to bypass the point with a little semicircle C_ρ as is shown in Fig. 2.36 and then take the limit $\rho \to 0$. Therefore, the contour contains a large semicircle C_R which contribution tends to zero as $R \to \infty$ due to Jordan's Lemma, then a small semicircle C_ρ whose contribution we shall need to investigate, and two horizontal parts $-R < x < -\rho$ and $\rho < x < R$. The horizontal parts upon taking the limits $R \to \infty$ and $\rho \to 0$ converge to the integration over the whole real axis as required in I_λ. Moreover, since the function $f(z) = e^{i\lambda z}/z$ does not have any poles inside the contour (the only pole is $z = 0$ which we avoid), we can write using obvious notations:

$$\int_{-R}^{-\rho} + \int_{C_\rho} + \int_{\rho}^{R} + \int_{C_R} = 0 \qquad (2.126)$$

for any $\rho > 0$ and $R > \rho$. Consider now the integral over C_ρ. There $z = \rho e^{i\phi}$, so that

$$\int_{C_\rho} \frac{e^{i\lambda z}}{z} dz = \int_{\pi}^{0} \frac{e^{i\lambda \rho(\cos\phi + i\sin\phi)}}{\rho e^{i\phi}} \rho i e^{i\phi} d\phi = -i \int_{0}^{\pi} e^{i\lambda \rho(\cos\phi + i\sin\phi)} d\phi .$$

Fig. 2.36 The integration contour used in calculating the integral of Eq. (2.125)

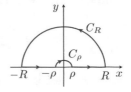

Calculating this integral exactly would be a challenge; however, this is not required, as we only need it calculated in the limit of $\rho \to 0$. This evidently gives simply $-i\pi$ as the exponent in the limit is replaced by one. Therefore, the C_ρ integral in the limit tends simply to $-i\pi$. The two horizontal integrals in (2.126) give in the limit the integral over the whole real axis apart from the point $x = 0$ itself (in the sense of Cauchy principal value, Sect. I.4.5.4), the C_R integral is zero in the limit as was already mentioned, and hence we finally obtain:

$$\mathcal{P} \int_{-\infty}^{\infty} \frac{e^{i\lambda x}}{x} dx = - \int_{C_\rho} = -(-i\pi) = i\pi \ .$$

Noting that $e^{i\lambda x} = \cos(\lambda x) + i \sin(\lambda x)$ and that the integral with the cosine function is zero due to symmetry (the integration limits are symmetric but the integrand is an odd function), we arrive at the following famous result:

$$\int_0^{\infty} \frac{\sin(\lambda x)}{x} dx = \frac{\pi}{2} \ . \tag{2.127}$$

Note that the integrand is even and hence we were able to replace the integration over the whole real axis by twice the integration over its positive part. Most importantly, however, we were able to remove the sign $\mathcal{P}$ of the principal value here since the function $\sin(\lambda x)/x$ is well defined at $x = 0$ (and is equal to λ). ◄

Problem 2.99 Show that

$$\chi_1(\omega) = \int_{-\infty}^{\infty} \frac{\sin t}{\pi t} e^{i\omega t} dt = \begin{cases} 1, & \text{if } -1 < \omega < 1 \\ 0, & \text{otherwise} \end{cases} . \tag{2.128}$$

[*Hint: write the sine via an exponential function and consider all three cases for ω. You may also find the result of the above Example useful.*]

In many other cases of the integrals $\int f(x)dx$ over the real axis the singularity of $f(x)$ within the integration interval cannot be removed and hence integrals may only exist in the sense of the Cauchy principal value. We shall now demonstrate that the method of residues may be quite useful in evaluating some of these integrals.

Example 2.16 ► *The integral ($\lambda > 0$)*

$$\int_{-\infty}^{\infty} f(x)dx \quad \text{with} \quad f(x) = \frac{e^{i\lambda x}}{(x-a)(x^2+b^2)}$$

is not well defined in the usual sense since the integrand has a singularity at $x = a$ lying on the real axis; however, as we shall see, its Cauchy principal value,

Fig. 2.37 The contour taken when calculating the integral (2.129)

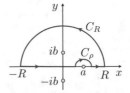

$$I = \mathcal{P} \int_{-\infty}^{\infty} f(x)dx = \lim_{\epsilon \to 0} \left[\int_{-\infty}^{a-\epsilon} f(x)dx + \int_{a+\epsilon}^{\infty} f(x)dx \right], \qquad (2.129)$$

is well defined. To calculate it, we shall take the contour as shown in Fig. 2.37: this contour is similar to the one shown in Fig. 2.36, but the little semicircle C_ρ goes around the singularity at $z = a$. There is only one simple pole at $z = ib$ which is inside the contour, so using the simplified notations we can write:

$$\int_{-R}^{a-\rho} + \int_{C_\rho} + \int_{a+\rho}^{R} + \int_{C_R} = 2\pi i \operatorname{Res}\left[f(z); ib\right] = 2\pi i \frac{e^{-\lambda b}}{(ib-a)(2ib)}$$

$$= -\frac{\pi}{b} \frac{a+ib}{a^2+b^2} e^{-\lambda b}. \qquad (2.130)$$

The integral over the large semicircle C_R is equal to zero in the $R \to \infty$ limit due to the Jordan's Lemma; the integral over the small semicircle is however not zero as is clear from a direct calculation ($z = a + \rho e^{i\phi}$ on C_ρ, and $dz = \rho i e^{i\phi} d\phi$):

$$\int_{C_\rho} = \int_{\pi}^{0} \frac{\exp\left[i\lambda\left(a + \rho e^{i\phi}\right)\right]}{\rho e^{i\phi}\left[\left(a + \rho e^{i\phi}\right)^2 + b^2\right]} \rho i e^{i\phi} d\phi = -i \int_{0}^{\pi} \frac{\exp\left[i\lambda\left(a + \rho e^{i\phi}\right)\right]}{\left[\left(a + \rho e^{i\phi}\right)^2 + b^2\right]} d\phi$$

$$\to -i \int_{0}^{\pi} \frac{e^{i\lambda a}}{a^2 + b^2} d\phi = -\frac{i\pi e^{i\lambda a}}{a^2 + b^2}$$

in the $\rho \to 0$ limit. The two integrals over the parts of the x axis in (2.130) in the limits $R \to \infty$ and $\rho \to 0$ yield exactly the principal value integral (2.129), therefore, we obtain:

$$I = -\frac{\pi}{b} \frac{a+ib}{a^2+b^2} e^{-\lambda b} + \frac{i\pi e^{i\lambda a}}{a^2 + b^2} = \frac{\pi}{a^2 + b^2} \left[-\frac{a+ib}{b} e^{-\lambda b} + i e^{i\lambda a} \right].$$

Separating the real and imaginary parts on both sides, we obtain two useful integrals:

$$\mathcal{P} \int_{-\infty}^{\infty} \frac{\cos(\lambda x)\, dx}{(x-a)(x^2+b^2)} = -\frac{\pi}{a^2+b^2} \left[\frac{a}{b} e^{-\lambda b} + \sin(\lambda a) \right],$$

$$\mathcal{P}\int_{-\infty}^{\infty}\frac{\sin\left(\lambda x\right)dx}{\left(x-a\right)\left(x^2+b^2\right)}=\frac{\pi}{a^2+b^2}\left[-e^{-\lambda b}+\cos\left(\lambda a\right)\right] .$$

◀

Problem 2.100 Prove that

$$\mathcal{P}\int_{-\infty}^{\infty}\frac{xdx}{x^3-a^3}=\frac{\pi}{\sqrt{3}a} .$$

[Hint: *do not forget to take account of a simple pole due to a root of the denominator and prove that the integral over the large semicircle C_R tends to zero as $R \to \infty$.]*

Finally, we shall consider several examples on calculating real integrals containing functions which, when continued into the complex plane, become multi-valued. We shall only consider the logarithmic and the general power functions here.

Example 2.17 ▶ *We shall start from the following integral containing the logarithmic function:*

$$I=\int_0^{\infty}\frac{\ln x}{\left(x^2+a^2\right)^2}dx .$$

We shall adopt the principal brunch of the logarithmic function with $\ln\left(re^{i\phi}\right)=\ln r+i\phi$ and $0<\phi<2\pi$, the brunch cut being chosen along the positive part of the x axis. The contour is the same as the one shown in Fig. 2.36. The four integrals are to be considered which appear in the left-hand side of the Cauchy theorem:

$$
\begin{aligned}
\int_{-R}^{-\rho}+\int_{C_\rho}+\int_{\rho}^{R}+\int_{C_R} &= 2\pi i \operatorname{Res}\left[\frac{\ln z}{\left(z^2+a^2\right)^2};ia\right] \\
&= 2\pi i\left[\frac{d}{dz}\frac{\ln z}{\left(z+ia\right)^2}\right]_{z=ia} \\
&= 2\pi i\left[\frac{1/z}{\left(z+ia\right)^2}-\frac{2\ln z}{\left(z+ia\right)^3}\right]_{z=ia} \\
&= \frac{\pi}{2a^3}\left(\ln a-1\right)+i\frac{\pi^2}{4a^3} .
\end{aligned}
\tag{2.131}
$$

Let us first consider the integral over the small circle C_ρ, where $z = \rho e^{i\phi}$:

$$\int_{C_\rho} = \int_\pi^0 \frac{\ln\rho + i\phi}{\left(\rho^2 e^{i2\phi} + a^2\right)^2} \rho i e^{i\phi} d\phi$$

$$= \int_\pi^0 \frac{e^{i\phi} i \rho \ln\rho}{\left(\rho^2 e^{i2\phi} + a^2\right)^2} d\phi - \int_\pi^0 \frac{\rho\phi e^{i\phi}}{\left(\rho^2 e^{i2\phi} + a^2\right)^2} d\phi .$$

Both integrals tend to zero in the $\rho \to 0$ limit. We shall demonstrate this using two methods: the first one is based on taking the limit inside the integrals; the other method will be based on making estimates of the integrals. Using the first method, the first integral behaves as $\rho \ln \rho$ for small ρ, which tends to zero when $\rho \to 0$; the second integral behaves as ρ and hence also goes to zero.

The same result is obtained using the second method which may be thought of as more rigorous. Indeed, for any ϕ the following inequality holds:

$$|\ln z| = |\ln\rho + i\phi| = \sqrt{\ln^2\rho + \phi^2} = \sqrt{\ln^2\frac{1}{\rho} + \phi^2} \le 2\ln\frac{1}{\rho} ,$$

when ρ becomes sufficiently small.[7] In addition,[8]

$$\left|\frac{1}{\rho^2 e^{i2\phi} + a^2}\right| \le \frac{1}{a^2 - \rho^2} ,$$

so that we can estimate the integral by a product of the semicircle length, $\pi\rho$, and the maximum value of the integrand, see Eq. (2.47):

$$\left|\int_{C_\rho} \frac{\ln z}{\left(z^2 + a^2\right)^2} dz\right| \le \left(\max_{C_\rho}\left|\frac{\ln z}{\left(z^2 + a^2\right)^2}\right|\right)\pi\rho \le \pi\rho\frac{2\ln(1/\rho)}{\left(a^2 - \rho^2\right)^2} ,$$

where the quantity on the right-hand side can easily be seen going to zero when $\rho \to 0$, and so is the integral. We see that a more rigorous approach fully supports the simple reasoning we developed above and hence in most cases our simple method can be used without hesitation.

Now we shall consider the integral over the large semicircle C_R. Either of the arguments used above proves that it goes to zero when $R \to \infty$. Indeed, using our (more rigorous) second method, we can write:

$$\left|\int_{C_R} \frac{\ln z}{\left(z^2 + a^2\right)^2} dz\right| \le \left(\max_{C_R}\left|\frac{\ln z}{\left(z^2 + a^2\right)^2}\right|\right)\pi R \le \pi R\frac{2\ln R}{\left(R^2 - a^2\right)^2} ,$$

---

[7] Taking square of the both sides, we have $3\ln^2(1/\rho) \ge \phi^2$ or $\ln(1/\rho) \ge \phi/\sqrt{3}$, which is fulfilled starting from some large number $1/\rho$ as the logarithm monotoniusly icreases. Note that ϕ is limited to the value of 2π.

[8] Since $|a + b| \ge a - b$, when $a > b > 0$.

and the expression on the right-hand side goes to zero as $\ln R / R^3$. The same result is obtained by using the first (simpler) method.

So, we conclude, there is no contribution coming from the semicircle parts of the contour. Let us now consider the horizontal parts. On the part $-R < x < -\rho$ the phase of the logarithm is π, so that $\ln z = \ln |x| + i\pi = \ln (-x) + i\pi$, and we can write:

$$
\int_{-R}^{-\rho} \frac{\ln (-x) + i\pi}{(x^2 + a^2)^2} dx = \left| \begin{array}{c} t = -x \\ dt = -dx \end{array} \right| = \int_{\rho}^{R} \frac{\ln t + i\pi}{(t^2 + a^2)^2} dt
$$

$$
\rightarrow \int_0^\infty \frac{\ln t}{(t^2 + a^2)^2} dt + i\pi \int_0^\infty \frac{dt}{(t^2 + a^2)^2} = I + i\pi J
$$

in the $R \rightarrow \infty$ and $\rho \rightarrow 0$ limits. Above, J is the second integral which does not contain the logarithm (and which we calculated above in Problem 2.90(a)). On the horizontal part from ρ to R the phase of the logarithm is zero and the integral converges exactly to I. Combining all parts of Eq. (2.131) together, we obtain:

$$
2I + i\pi J = \frac{\pi}{2a^3} (\ln a - 1) + i \frac{\pi^2}{4a^3} ,
$$

which, by equating the real and imaginary parts on both sides, gives the required result:

$$
I = \int_0^\infty \frac{\ln x}{(x^2 + a^2)^2} dx = \frac{\pi}{4a^3} (\ln a - 1) ,
$$

whereas for J we obtain exactly the same result $\pi / 4a^3$ as in Problem 2.90(a). ◄

Problem 2.101 Prove that

$$
\int_0^\infty \frac{\ln x\, dx}{x^2 + a^2} = \frac{\pi \ln a}{2a} ;
$$

$$
\int_0^\infty \frac{\ln x}{(x^2 + a^2)(x^2 + b^2)} dx = \frac{\pi}{2(a^2 - b^2)} \left(\frac{\ln b}{b} - \frac{\ln a}{a} \right) .
$$

Note that when calculating the last integral, as a by-product, you will get a particular case of the last example in Problem 2.96.

Example 2.18 ► *As our second example, let us consider the integral*

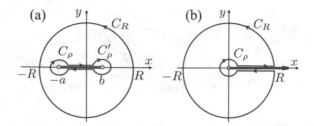

Fig. 2.38 a The contour taken for the calculation of the integral (2.132) in the complex plane. The contour consists of a closed internal part and a closed external large circle C_R. The cut has been made effectively between points $z = -a$ and $z = b$. **b** The contour to be taken when solving Problem 2.102; it goes all the way around the cut taken along the positive part of the x axis

$$I = \int_{-a}^{b} \frac{dx}{\sqrt[3]{(x+a)^2 (b-x)}}, \quad \text{where} \quad a < 0, \quad b > 0. \tag{2.132}$$

To calculate this integral, we consider a slightly different integrand $F(z) = 1/f(z)$ with $f(z) = \sqrt[3]{(z+a)^2 (z-b)}$. How to deal with $f(z)$ of this kind we have already discussed in Problem 2.25. So, we shall consider the brunch cut going to the right from the point $z = -a$ along the x axis as in Fig. 2.11b (where in the Figure $a = -1$ and $b = 1$). We select the branch of the root-three function for which its phase is between 0 and $2\pi/3$, i.e. we restrict the phase of any z to $0 < arg(z) < 2\pi$. As $f(z)$ is only discontinuous between the points $z = -a$ and $z = b$ on the x axis, we choose the contour as shown in Fig. 2.38a. It consists of two parts: an external circle C_R of large radius R (which we shall eventually take to infinity) and of a closed path going around the brunch cut. In turn, this latter part is made of two circles, C_ρ (around $z = -a$) and C'_ρ (around $z = b$), and two horizontal parts, one going along the upper side of the cut, the other along the lower. Let ρ_1 and ρ_2 be the radii of the two small circles C'_ρ and C_ρ, respectively. According to the Cauchy theorem for multiply connected regions, the sum of the integrals going around these two contours is zero as there are no poles between them:

$$\int_{C_R} + \int_{C'_\rho} + \underbrace{\int_{b-\rho_1}^{-a+\rho_2}}_{\text{lower}} + \int_{C_\rho} + \underbrace{\int_{-a+\rho_2}^{b-\rho_1}}_{\text{upper}} = 0. \tag{2.133}$$

On the large circle $z = Re^{i\phi}$, the function $f(z)$ is continuous, and hence we can calculate the integral as follows:

$$\int_{C_R} \frac{dz}{\sqrt[3]{(z+a)^2\,(z-b)}} = \int_0^{2\pi} \frac{Rie^{i\phi}d\phi}{\sqrt[3]{\left(Re^{i\phi}+a\right)^2\left(Re^{i\phi}-b\right)}}$$

$$\to \int_0^{2\pi} \frac{Rie^{i\phi}d\phi}{\sqrt[3]{\left(R^2e^{2i\phi}\right)\left(Re^{i\phi}\right)}} = \int_0^{2\pi} \frac{Rie^{i\phi}d\phi}{Re^{i\phi}} = 2\pi i$$

after taking the limit of $R \to \infty$.

Next, we shall consider the two small circle integrals. In the C'_ρ we have $z = b + \rho_1 e^{i\phi}$ with $0 < \phi < 2\pi$ and $dz = \rho_1 i e^{i\phi}d\phi$, and hence the integral consists of two integrations: one for ϕ between π and 0 (the upper semicircle), and another between 2π and π (the lower semicircle). For either of these integrals, we can write:

$$\int_{C'_\rho} = \int \frac{\rho_1 i e^{i\phi}d\phi}{\sqrt[3]{\left(b+a+\rho_1 e^{i\phi}\right)^2 \rho_1 e^{i\phi}}}$$

$$\to \int \frac{\rho_1 i e^{i\phi}d\phi}{\sqrt[3]{(b+a)^2 \rho_1 e^{i\phi}}} = \frac{i\rho_1^{2/3}}{\sqrt[3]{(a+b)^2}} \int \frac{e^{i\phi}d\phi}{e^{i\phi/3}} \;,$$

which tends to zero as $\rho_1^{2/3}$ when $\rho_1 \to 0$. So, the integral over C'_ρ tends to zero. Similarly for the other small circle integral ($z = -a + \rho_2 e^{i\phi}$ with $0 < \phi < 2\pi$):

$$\int_{C_\rho} = \int_{2\pi}^0 \frac{\rho_2 i e^{i\phi}d\phi}{\sqrt[3]{\left(\rho_2 e^{i\phi}\right)^2\left(-b-a+\rho_2 e^{i\phi}\right)}}$$

$$\to \int_{2\pi}^0 \frac{\rho_2 i e^{i\phi}d\phi}{\rho_2^{2/3} e^{i2\phi/3}\sqrt[3]{-(b+a)}} = -\frac{i\rho_2^{1/3}}{\sqrt[3]{-(a+b)}} \int_0^{2\pi} e^{i\phi/3}d\phi \;,$$

which tends to zero as $\rho_2^{1/3}$. So, both small circle integrals give no contribution.

What is left to consider are the two horizontal parts of the internal contour. Generally $z = -a + \rho_2 e^{i\phi_2}$ or $z = b + \rho_1 e^{i\phi_1}$ (see Fig. 2.11a for the definitions of the two angles ϕ_1 and ϕ_2 in a similar case), and hence $f(z) = \sqrt[3]{\rho_2^2 \rho_1} e^{i(2\phi_2+\phi_1)/3}$. On the upper side of the cut between the two points $\phi_2 = 0$, $\phi_1 = \pi$, $\rho_2 = x + a$ and $\rho_1 = b - x$, so that

$$f(z) = \sqrt[3]{(x+a)^2\,(b-x)}\,e^{i\pi/3}$$

and the corresponding integral over the upper side of the cut becomes

$$\int_{-a+\rho_2}^{b-\rho_1} = \int_{-a+\rho_2}^{b-\rho_1} \frac{e^{-i\pi/3}dx}{\sqrt[3]{(x+a)^2(b-x)}}$$
$$\rightarrow \int_{-a}^{b} \frac{e^{-i\pi/3}dx}{\sqrt[3]{(x+a)^2(b-x)}} = e^{-i\pi/3}I .$$

On the lower side of the cut $\phi_1 = \pi$, $\phi_2 = 2\pi$, $\rho_2 = x + a$ and $\rho_1 = b - x$, so that

$$f(z) = \sqrt[3]{(x+a)^2(b-x)}e^{i(4\pi+\pi)/3} = \sqrt[3]{(x+a)^2(b-x)}e^{-i\pi/3} ,$$

giving for the corresponding horizontal part the contribution of

$$\int_{b-\rho_1}^{-a+\rho_2} = \int_{b-\rho_1}^{-a+\rho_2} \frac{e^{i\pi/3}dx}{\sqrt[3]{(x+a)^2(b-x)}}$$
$$\rightarrow -\int_{-a}^{b} \frac{e^{i\pi/3}dx}{\sqrt[3]{(x+a)^2(b-x)}} = -e^{i\pi/3}I .$$

Combining now all contributions in Eq. (2.133), we obtain:

$$2\pi i - e^{i\pi/3}I + e^{-i\pi/3}I = 0$$
$$\implies I = \frac{2\pi i}{e^{i\pi/3} - e^{-i\pi/3}} = \frac{2\pi i}{2i\sin(\pi/3)} = \frac{\pi}{\sin(\pi/3)} = \frac{2\pi}{\sqrt{3}} .$$

Interestingly, the result does not depend on the values of a and b. This can easily be understood also by making a substitution $t = (2x + a - b) / (a + b)$. ◄

Problem 2.102 Using the contour shown in Fig. 2.38b, prove the following results ($a > 0$):

(a) $\displaystyle\int_0^\infty \frac{x^\lambda dx}{x + a} = -\frac{\pi a^\lambda}{\sin(\pi\lambda)}$, where $-1 < \lambda < 0$;

(b) $\displaystyle\int_0^\infty \frac{\sqrt{x}}{x^2 + a^2}dx = \frac{\pi}{\sqrt{2a}}$; (c) $\displaystyle\int_0^\infty \frac{\sqrt{x}\ln x}{x^2 + a^2}dx = \frac{\pi}{\sqrt{2a}}\left(\frac{\pi}{2} + \ln a\right)$.

Make sure that the integral in (b) comes out correctly when solving for the integral in (c).

Problem 2.103 Consider the same integrals using the contour shown in Fig. 2.36.

2.8   Linear Differential Equations: Series Solutions

In Sect. I.8.4 we considered, without proof, a method of solving a second order linear differential equation (DE)

$$y''(x) + p(x)y'(x) + q(x)y(x) = 0 \tag{2.134}$$

with variable functions $p(x)$ and $q(x)$ based on a generalised series expansion,

$$y(x) = \sum_{r=0}^{\infty} c_r \, (x - x_0)^{r+s} \, , \tag{2.135}$$

about a point $x = a$, where c_r are expansion coefficients, $s = 0$ when x_0 is an ordinary point and $s \neq 0$ (and is a non-integer) when x_0 is a regular singular point. Neither proof of the method or the actual meaning of the different types of the singular points were provided, and only a brief outline was given for special cases which may arise. In this section we shall present a more detailed account of this method and provide some essential proofs, derivations and explanations. Our consideration will be based on a more general equation

$$y''(z) + p(z)y'(z) + q(z)y(z) = 0 \, , \tag{2.136}$$

considered on the complex plane, i.e. the variable functions $p(z)$ and $q(z)$ and the solution itself, $y(z)$, are treated as functions of a complex variable z. What we shall be interested in here is addressing the following question: under which circumstances one can apply a generalised series expansion,

$$y(z) = \sum_{r=0}^{\infty} c_r \, (z - z_0)^{r+s} \, , \tag{2.137}$$

in order to solve the DE (2.136), and what is a practical criterion one may apply in order to determine if this can be done. We shall not be considering the question of convergence of the series.

We shall first consider the case of the point z_0 being an ordinary point, i.e. when the functions $p(z)$ and $q(z)$ are analytic functions within some circle of radius R centred at z_0, i.e. for all z satisfying $|z - z_0| < R$. Therefore, they can be expanded in the Taylor series around z_0:

$$p(z) = p_0 + p_1 u + p_2 u^2 + \dots \quad \text{and} \quad q(z) = q_0 + q_1 u + q_2 u^2 + \dots \, ,$$

where $u = z - z_0$. Suppose, we would like to obtain a solution which satisfies initial conditions $y(z_0) = \alpha$ and $y'(z_0) = \beta$. Then, substituting $z \to z_0$ into Eq. (2.136), we immediately get $y''(z_0) = -p_0\beta - q_0\alpha$. Differentiating (2.136) and substituting $z \to z_0$ again, we get

$$y'''(z_0) = -(q_0 + p_1)\beta - q_1\alpha - p_0 y''(z_0)$$
$$= -\alpha(q_1 - p_0 q_0) - \beta(p_1 - p_0^2 + q_0) \ .$$

This process can be continued. It is clear, that any derivative $y^{(n)}(z_0)$ of the solution can be calculated this way and expressed via the initial conditions and the expansion coefficients of the functions $p(z)$ and $q(z)$. In other words, the solution $y(z)$ in this case can be sought in the form of the Taylor expansion

$$y(z) = \alpha + \beta u + y_2 u^2 + \dots \ .$$

The unknown coefficients y_2, y_3, etc. are obtained by substituting the expansion into the DE, collecting coefficients to the same powers of u and setting them to zero. This is exactly the method we used in Sect. I.8.4. It can be shown that the series obtained in this way converges uniformly to the solution of the DE within the same region $|z - z_0| < R$ where the functions $p(z)$ and $q(z)$ are analytic.

Consider now a much more interesting (and rather non-trivial) case of the functions $p(z)$ and $q(z)$ having a singularity at z_0. In this case these functions of coefficients of the DE are in general expanded into infinite[9] Laurent series (Sect. 2.5.4):

$$p(z) = \sum_{k=-\infty}^{\infty} p_k u^k \quad \text{and} \quad q(z) = \sum_{k=-\infty}^{\infty} q_k u^k \ .$$

The point z_0 becomes a branch point. This means that if we consider two linearly independent solutions $y_1(z)$ and $y_2(z)$ of the DE (we assume that they both exist), then, when passing around z_0 along a closed contour starting from a point z, they arrive at some values $y_1^+(z)$ and $y_2^+(z)$ when arriving back to z, which are *not* the same as the starting points, i.e. $y_1^+(z) \neq y_1(z)$ and $y_2^+(z) \neq y_2(z)$ (the superscript "+" hereafter indicates the value of a function after completing the full closed contour around z_0). However, since y_1 and y_2 are the two linearly independent solutions, the new values must be some linear combinations of the same functions y_1 and y_2, i.e.

$$y_1^+ = a_{11} y_1 + a_{12} y_2 \quad \text{and} \quad y_2^+ = a_{21} y_1 + a_{22} y_2 \tag{2.138}$$

or

$$\begin{pmatrix} y_1^+ \\ y_2^+ \end{pmatrix} = A \begin{pmatrix} y_1 \\ y_2 \end{pmatrix}, \quad \text{where} \quad A = \begin{pmatrix} a_{11} & a_{12} \\ a_{21} & a_{22} \end{pmatrix} \ . \tag{2.139}$$

---

[9] Either of the series may contain a finite number of terms in its fundamental part which is a particular case of the infinite series.

Problem 2.104 Prove by contradiction that $a_{11}a_{22} - a_{12}a_{21} \neq 0$, as otherwise the two solutions y_1 and y_2, which are assumed to be linearly independent, appear to be linearly dependent (i.e. $y_2 = \gamma y_1$ with some complex number γ), which is impossible by the assumption.

One can consider various nonequivalent linear combinations of y_1 and y_2 as new solutions. Let us find a special linear combination $\mathcal{Y} = b_1 y_1 + b_2 y_2$ which transforms upon traversing around the point z_0 into itself with a complex multiplier λ, i.e. that

$$\mathcal{Y}^+ = \lambda \mathcal{Y} \implies b_1 y_1^+ + b_2 y_2^+ = \lambda (b_1 y_1 + b_2 y_2) \ .$$

Substituting here Eq. (2.138) for y_1^+ and y_2^+,

$$b_1 (a_{11} y_1 + a_{12} y_2) + b_2 (a_{21} y_1 + a_{22} y_2) = \lambda (b_1 y_1 + b_2 y_2) \ ,$$

and comparing the coefficients to y_1 and y_2 in both sides, one arrives at an eigenproblem, which gives both the eigenvalue λ and the corresponding vector $\mathbf{B} = \begin{pmatrix} b_1 \\ b_2 \end{pmatrix}$:

$$\mathbf{A}^T \mathbf{B} = \lambda \mathbf{B} \ . \tag{2.140}$$

Solving the equation

$$\begin{vmatrix} a_{11} - \lambda & a_{21} \\ a_{12} & a_{22} - \lambda \end{vmatrix} = 0$$

results in up to two values of λ and hence in up to two linearly independent eigenvectors $\mathbf{B}$. Consider now two possible cases: the two values λ_1 and λ_2 are different, or are the same.

In the former case we obtain two solutions $\mathcal{Y}_1(z)$ and $\mathcal{Y}_2(z)$ which satisfy $\mathcal{Y}_1^+(z) = \lambda_1 \mathcal{Y}_1(z)$ and $\mathcal{Y}_2^+(z) = \lambda_2 \mathcal{Y}_2(z)$. They are linearly independent since $\lambda_1 \neq \lambda_2$ (Theorem 1.18).

Problem 2.105 Prove this by contradiction, i.e. assuming that $\mathcal{Y}_2(z) = \gamma \mathcal{Y}_1(z)$ with some complex number γ.

It is also easy to see that the eigenvalues are the property of the DE itself; they do not depend on the particular linear combinations taken as the linearly independent solutions. Indeed, consider two new functions w_1 and w_2 defined by the transformation

$$\mathbf{W} = \begin{pmatrix} w_1 \\ w_2 \end{pmatrix} = \begin{pmatrix} c_{11} & c_{12} \\ c_{21} & c_{22} \end{pmatrix} \begin{pmatrix} y_1 \\ y_2 \end{pmatrix} = \mathbf{CY} .$$

Upon a traverse around z_0 the vector function $\mathbf{W}$ goes[10] into $\mathbf{W}^+ = \mathbf{CY}^+ = \mathbf{CAY} = \mathbf{CAC}^{-1}\mathbf{W}$, i.e. the role of the matrix $\mathbf{A}$ in the eigenproblem (2.140) plays the matrix $\mathbf{CAC}^{-1}$ obtained using a similarity transformation with the matrix $\mathbf{C}$ (see Sect. 1.9.1). However, similar matrices share their eigenvalues (Theorem 1.23 in Sect. 1.9.1), this proves the above made statement.

Next, we consider a function

$$f(z) = (z - z_0)^{s_1} = e^{s_1 \ln(z - z_0)}$$

with $s_1 = (\ln \lambda_1) / 2\pi i$. Any value of the logarithm can be assumed for $\ln \lambda_1$, i.e. it is defined up to $2\pi m i$ with an arbitrary integer number m. In turn, this means that the number s_1 is defined up to an arbitrary integer m. After the traverse around z_0 the logarithm $\ln(z - z_0)$ acquires the phase 2π, i.e. the function $f(z)$ turns into

$$f^+(z) = e^{s_1 [\ln(z - z_0) + 2\pi i]} = e^{s_1 2\pi i} e^{s_1 \ln(z - z_0)} = e^{\ln \lambda_1} e^{s_1 \ln(z - z_0)} = \lambda_1 f(z) ,$$

i.e. $f(z)$ behaves similarly to $\mathcal{Y}_1(z)$. Hence the function $\mathcal{Y}_1(z)/f(z) = \mathcal{Y}_1(z)/(z - z_0)^{s_1}$ does not change when going around the point z_0,

$$\left(\frac{\mathcal{Y}_1(z)}{f(z)} \right)^+ = \frac{\lambda_1 \mathcal{Y}_1(z)}{\lambda_1 f(z)} = \frac{\mathcal{Y}_1(z)}{f(z)} ,$$

i.e. it is single-valued around the vicinity of this point, except may be at the point z_0 itself. Therefore, it can be expanded in a Laurent series, i.e. one can write

$$\mathcal{Y}_1(z) = (z - z_0)^{s_1} \sum_{k=-\infty}^{\infty} \gamma_k (z - z_0)^k . \tag{2.141}$$

Similarly, the other solution

$$\mathcal{Y}_2(z) = (z - z_0)^{s_2} \sum_{k=-\infty}^{\infty} \zeta_k (z - z_0)^k \tag{2.142}$$

has a similar form but with the $s_2 = (\ln \lambda_2) / 2\pi i$ instead (the coefficients in the Laurent series are most likely also different). Recall that the complex numbers s_1 and s_2 are defined up to an arbitrary integers m_1 and m_2, respectively. However, this does not change the general form of the series (2.141) or (2.142) as the Laurent series in each case contains all possible powers of $(z - z_0)$.

---

[10] Note that the plus superscript here does not correspond to the pseudo-inverse of Sect. 1.20.

Consider now the case of equal eigenvalues $\lambda_1 = \lambda_2$ of the matrix $\mathbf{A}$ (or $\mathbf{A}^T$).
Note that the corresponding numbers s_1 and s_2 may in fact differ by an integer. In
this case $\mathcal{Y}_1$ satisfies $\mathcal{Y}_1^+ = \lambda_1 \mathcal{Y}_1$, while the second possible linearly independent
solution $\mathcal{Y}_2$ must generally satisfy $\mathcal{Y}_2^+ = a_{21}\mathcal{Y}_1 + a_{22}\mathcal{Y}_2$. The corresponding matrix
$\mathbf{A}$ in this case in the basis of $\mathcal{Y}_1$ and $\mathcal{Y}_2$ (we have learned that the eigenvalues are
invariant with respect to the choice of the basis) would have $a_{11} = \lambda_1$ and $a_{12} = 0$;
therefore, the secular equation for the eigenvalues λ is of the form

$$\begin{vmatrix} \lambda_1 - \lambda & a_{21} \\ 0 & a_{22} - \lambda \end{vmatrix} = 0 \; .$$

The condition that the second root of this equation coincides with the first, λ_1, means
that $a_{22} = \lambda_1$, so that the second solutions $\mathcal{Y}_2$ must in fact satisfy the condition
$\mathcal{Y}_2^+ = a_{21}\mathcal{Y}_1 + \lambda_1\mathcal{Y}_2$.

Consider now the function

$$f(z) = \frac{\mathcal{Y}_2}{\mathcal{Y}_1} - \frac{a_{21}}{2\pi i \lambda_1} \ln (z - z_0) \; . \tag{2.143}$$

Upon a complete traverse around z_0 it turns into

$$\begin{aligned} f^+(z) &= \frac{\mathcal{Y}_2^+}{\mathcal{Y}_1^+} - \frac{a_{21}}{2\pi i \lambda_1} [\ln (z - z_0) + 2\pi i] \\ &= \frac{a_{21}\mathcal{Y}_1 + \lambda_1\mathcal{Y}_2}{\lambda_1\mathcal{Y}_1} - \frac{a_{21}}{2\pi i \lambda_1} [\ln (z - z_0) + 2\pi i] \\ &= \frac{\mathcal{Y}_2}{\mathcal{Y}_1} + \frac{a_{21}}{\lambda_1} - \frac{a_{21}}{2\pi i \lambda_1} [\ln (z - z_0) + 2\pi i] \\ &= \frac{\mathcal{Y}_2}{\mathcal{Y}_1} - \frac{a_{21}}{2\pi i \lambda_1} \ln (z - z_0) = f(z) \; , \end{aligned}$$

i.e. the function $f(z)$ is single-valued inside some circle with the centre at z_0 accept
may be at the point z_0 itself, and therefore can be represented by a Laurent series.
This means that the second solution, according to Eq. (2.143), must have the form:

$$\begin{aligned} \mathcal{Y}_2(z) &= \zeta \ln (z - z_0) \, \mathcal{Y}_1(z) + f(z)\mathcal{Y}_1(z) \\ &= \zeta \ln (z - z_0) \, \mathcal{Y}_1(z) + \mathcal{Y}_1(z) \sum_{k=-\infty}^{\infty} \xi_k \, (z - z_0)^k \; , \end{aligned} \tag{2.144}$$

where $\zeta = a_{21}/2\pi i \lambda_1$ is some complex number. Thus, the second solution should
contain a logarithmic term.

So far we did not make any assumptions concerning the nature of the point z_0; it
could have been either a pole (sometimes also called a *regular singular point*, Sect.
I.8.4) or an essential singularity (an *irregular singular point*), see Sect. 2.5.5. Let us

consider specifically the case when the point z_0 is a pole and hence each Laurent series contains a finite number of terms $(z - z_0)^k$ with negative powers k. But then we can add appropriate integers m_1 and m_2 to s_1 and s_2, respectively, to ensure that the Laurent series in either of the cases considered above do not contain the negative powers at all, and this brings us to the method presented in Sect. I.8.4. What is only left to understand is how one can determine from the DE itself whether the point z_0 is regular or irregular.

To this end, let us assume that we know two linearly independent solutions $y_1(z)$ and $y_2(z)$ of the DE (2.136), i.e.

$$y_1'' + py_1' + qy_1 = 0 \quad \text{and} \quad y_2'' + py_2' + qy_2 = 0 .$$

Problem 2.106 Solve the above equations with respect to the functions $p(z)$ and $q(z)$ to show that they can be expressed via the solutions y_1 and y_2 as follows:

$$p(z) = -\frac{W'(z)}{W(z)} \quad \text{and} \quad q(z) = -\frac{y_1''(z)}{y_1(z)} - p(z)\frac{y_1'(z)}{y_1(z)} , \qquad (2.145)$$

where $W(z) = y_1 y_2' - y_1' y_2$ is the Wronskian (see Sect. 1.6.1).

We shall now consider the necessary criterion for the first case, when $\lambda_1 \neq \lambda_2$. Since z_0 is a pole by our assumption, the two solutions can be written as

$$y_1 = u^{s_1} P_1(u) \quad \text{and} \quad y_2 = u^{s_2} P_2(u) , \qquad (2.146)$$

where $u = z - z_0$ and $P_1(u)$ and $P_2(u)$ are two functions which are well defined (analytic) at $u = 0$ and some vicinity around it; they both can be represented by Taylor expansions around $u = 0$ with a non-zero free term, i.e. $P_1(0) \neq 0$ and $P_2(0) \neq 0$. This is because the behaviour of y_1 or y_2 around $u = 0$ has already been described by the corresponding power terms u^{s_1} and u^{s_2}, and starting from a zero free term in the expansions of the functions P_1 and P_2 would simply modify the exponents s_1 and s_2 by an integer. We are using here symbolic notations whereby $P_n(u)$ represents a Taylor expansion with a non-zero free term and with the index n numbering various such functions we shall encounter in what follows.

The idea is to obtain a general form of the functions $p(z)$ and $q(z)$ from (2.145) based on the known general form of the solutions in (2.146). To this end, we need to calculate derivatives of the solutions and then the Wronskian and its derivative. Since a derivative of a function $P_1(u)$ is also some expansion $P_3(u)$, we can write:

$$y_1' = s_1 u^{s_1-1} P_1(u) + u^{s_1} P_3(u) = u^{s_1-1} [s_1 P_1(u) + u P_3(u)] = u^{s_1-1} P_4(u)$$

and similarly for $y_2' = u^{s_2-1} P_5(u)$, so that

$$W = u^{s_1+s_2-1} P_6(u) - u^{s_1-1+s_2} P_7(u) = u^{s_1+s_2-1} P_8(u)$$

and

$$W' = (s_1 + s_2 - 1)\, u^{s_1+s_2-2} P_8(u) + u^{s_1+s_2-1} P_9(u) = u^{s_1+s_2-2} P_{10}(u)\,.$$

Therefore, a general form of the function $p(z)$ must then be:

$$p(z) = -\frac{W'(z)}{W(z)} = -\frac{u^{s_1+s_2-2} P_{10}(u)}{u^{s_1+s_2-1} P_8(u)} = -\frac{P_{11}(u)}{u} = -\frac{P_{11}(z-z_0)}{z-z_0}\,,$$

where $P_{11}(z)$ is some well-behaving function of z that is expandable in the Taylor series around z_0. Next, using the same method as above, we get $y_1'' = u^{s_1-2} P_{12}(u)$. Hence, for $q(z)$ we obtain:

$$q(z) = -\frac{y_1''(z)}{y_1(z)} - p(z)\frac{y_1'(z)}{y_1(z)} = -\frac{u^{s_1-2} P_5(u)}{u^{s_1} P_1(u)} + \frac{P_{11}(u)}{u}\frac{u^{s_1-1} P_4(u)}{u^{s_1} P_1(u)}$$

$$= \frac{P_{13}(u)}{u^2} = \frac{P_{13}(z-z_0)}{(z-z_0)^2}$$

with $P_{13}(z)$ being expandable into a Taylor series around z_0. Therefore, the necessary conditions for the point z_0 to be a regular singular point is that $p(z)$, if singular, has a pole of the first order, while $q(z)$, if singular, has either first or second order pole. Of course, one of these functions may be regular at z_0 (i.e. it may have a finite limit there), but then the other should be singular. The simple criteria for determining if these conditions are satisfied are then based on calculating the limits

$$\lim_{z\to z_0} (z - z_0)\, p(z) \quad \text{and} \quad \lim_{z\to z_0} (z - z_0)^2\, q(z)\,.$$

If both limits result in finite numbers (including zeros), then the point z_0 is a regular singular point. This is exactly the criterion we used in Sect. I.8.4 without proper proof.

Problem 2.107 Use a similar reasoning to show that a general form of the functions $p(z)$ and $q(z)$ still remains the same if the two independent solutions have the form (2.141) and (2.144). [Hint: write $y_2 = y_1 (\zeta \ln u + P_2(u))$ and keep explicitly the terms by the logarithm when calculating W; in this way the logarithm containing terms will cancel out.]

Problem 2.108 Show that the associated Legendre equation

$$\left(1 - x^2\right) y'' - 2xy' + \left(\mu - \frac{m^2}{1 - x^2}\right) y = 0 \,,$$

where μ and m are some parameters, has *regular singular points* at $x = +1$ and $x = -1$. We shall study this DE in more detail in Sect. 4.5.

It can be shown that the stated above conditions for the function $p(z)$ and $q(z)$ are also sufficient for the point z_0 to be a regular singular point.

2.9 Selected Applications in Physics

2.9.1 Dispersion Relations

In many physical problems one encounters complex functions of a real variable x (e.g. of the frequency or energy ω) containing a real and imaginary parts, $f(x) = u(x) + iv(x)$. We know from Problem 2.84 that when $f(x)$ is continued analytically into the complex plane, $x \to z = x + iy$, then the new function $f(z)$ becomes analytic. Let us assume that the function $f(z)$ converges uniformly to zero as $|z| \to \infty$. It is easy to show that if $f(x) = P_n(x)/Q_m(x)$ is a rational function with polynomials $P_n(x)$ and $Q_m(x)$ of orders n and m, respectively, the uniform convergence to zero, $f(x) \to 0$ when $x \to \pm\infty$, immediately follows if $m > n$. Indeed, let us estimate $f(z)$ for $z = Re^{i\phi}$ in the $R \to \infty$ limit. We have using Eq. (2.117):

$$\begin{aligned}
|f(z)| &= \left|\frac{P_n(z)}{Q_m(z)}\right| = \frac{\left|\prod_{i=1}^{n} (z - a_i)\right|}{\left|\prod_{j=1}^{m} (z - b_j)\right|} = \frac{\prod_{i=1}^{n} |z - a_i|}{\prod_{j=1}^{m} |z - b_j|} \\
&\leq \frac{\prod_{i=1}^{n} (|z| + |a_i|)}{\prod_{j=1}^{m} (|z| - |b_j|)} = \frac{\prod_{i=1}^{n} (R + |a_i|)}{\prod_{j=1}^{m} (R - |b_j|)} \,,
\end{aligned}$$

(2.147)

where a_i and b_i are (generally complex) zeros of the two polynomials. The expression in the right-hand side of (2.147) tends to zero as $R \to \infty$ since the polynomial in the denominator is of a higher order ($m > n$) with respect to R than in the numerator. This convergence to zero does not depend on the phase ϕ and hence is uniform. Note, that it is not at all obvious that the same can be proven for a general function $f(x)$; however, in many physical applications $f(x)$ is some rational function of x.

Therefore, let us consider an analytic function $f(z)$ which tends to zero uniformly when $|z| \to \infty$. We shall also assume that $f(z)$ does not have singularities in the upper half of the complex plane including the real axis itself. Then, if x_0 is some real number,

$$\oint_L \frac{f(z)dz}{z - x_0} = 0 \,,$$

according to the Cauchy formula (2.48), where L is an arbitrary loop running in the upper half plane. The loop may include the real axis, but must avoid the point x_0. Consider now a particular loop shown in Fig. 2.39a, where we assume taking the limits $R \to \infty$ and $\rho \to 0$. We can write:

$$\int_{-R}^{x_0 - \rho} + \int_{C_\rho} + \int_{x_0 + \rho}^{R} + \int_{C_R} = 0 \,, \tag{2.148}$$

where the integral over the large semicircle C_R tends to zero due to uniform convergence of $f(z)$ to zero as $|z| = R \to \infty$. Indeed, we can write $\left| f\left(Re^{i\phi}\right) \right| \le M_R$ with $M_R \to 0$ when $R \to \infty$. Hence,

$$\left| \int_{C_R} \frac{f(z)dz}{z - x_0} \right| = \left| \int_0^\pi \frac{f\left(Re^{i\phi}\right) Rie^{i\phi}}{Re^{i\phi} - x_0} d\phi \right| \le \int_0^\pi \frac{\left| f\left(Re^{i\phi}\right) \right| R}{\left| Re^{i\phi} - x_0 \right|} d\phi$$

$$\le \frac{R}{R - x_0} \int_0^\pi \left| f\left(Re^{i\phi}\right) \right| d\phi \le \frac{\pi M_R}{1 - x_0/R} \,.$$

It is obvious now that the estimate in the right-hand side tends to zero as $R \to \infty$. Now, let us calculate the integral over the small semicircle where $z = x_0 + \rho e^{i\phi}$. We have:

$$\int_{C_\rho} \frac{f(z)dz}{z - x_0} = \int_\pi^0 \frac{f\left(x_0 + \rho e^{i\phi}\right) \rho i e^{i\phi}}{\rho e^{i\phi}} d\phi = -i \int_0^\pi f\left(x_0 + \rho e^{i\phi}\right) d\phi$$

$$\implies -if(x_0) \int_0^\pi d\phi = -i\pi f(x_0)$$

in the $\rho \to 0$ limit. The two integrals along the real x axis in Eq. (2.148) combine into the Cauchy principal value integral

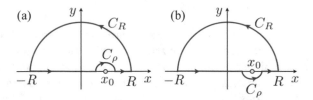

Fig. 2.39 Two possible contours which can be used to prove the dispersion relations (2.150). Here C_R is a semicircle of radius $R \to \infty$, while C_ρ is a semicircle of radius $\rho \to 0$ around the point x_0

$$\int_{-R}^{x_0-\rho} + \int_{x_0+\rho}^{\infty} \;\rightarrow\; \mathcal{P}\int_{-\infty}^{\infty} \frac{f(x)}{x-x_0}dx$$

in the limits of $R \to \infty$ and $\rho \to 0$. Therefore, combining all terms together, we obtain:

$$f(x_0) = -\frac{i}{\pi}\mathcal{P}\int_{-\infty}^{\infty} \frac{f(x)}{x-x_0}dx \ . \tag{2.149}$$

Remarkably, only the values of the function on the real axis enter this expression! Normally in applications this formula is presented in a different form in which real and imaginary parts of the function $f(x)$ are used:

$$\mathrm{Re}\, f(x_0) = \frac{1}{\pi}\mathcal{P}\int_{-\infty}^{\infty} \frac{\mathrm{Im}\, f(x)}{x-x_0}dx \quad \text{and} \quad \mathrm{Im}\, f(x_0) = -\frac{1}{\pi}\mathcal{P}\int_{-\infty}^{\infty} \frac{\mathrm{Re}\, f(x)}{x-x_0}dx \ . \tag{2.150}$$

These relationships were discovered by H. A. Kramers and R. Kronig in relation to real and imaginary parts of the dielectric constant of a material. However, the physical significance of these relations is much wider. For instance, they may relate imaginary and real parts of the time Fourier transform (to be considered in Chap. 5) of a response function $\chi(t-\tau)$ of an observable $G(t)$ of a physical system subjected to an external perturbation $F(t)$ (e.g. a field):

$$G(t) = \int_{-\infty}^{t} \chi(t-\tau)\, F(\tau)d\tau \ . \tag{2.151}$$

Note that the integral here is taken up to the current time t due to *causality*: the observable $G(t)$ can only depend on the values of the field at previous times, it cannot depend on the future times $\tau > t$. An example of Eq. (2.151) could be, for instance, the relationship between the displacement vector $\mathbf{D}(t)$ and the electric field $\mathbf{E}(t)$ in electromagnetism, in which case the response function is the time dependent dielectric function $\epsilon(t-\tau)$.

It can be shown using the Fourier transform method that because of the causality alone (i.e. the response function $\chi(t) = 0$ for $t < 0$), the Kramers-Kronig relations can be derived independently. Note that the imaginary part of the response function is responsible for energy dissipation in a physical system, while the real part of $\chi(t-\tau)$ corresponds to a driving force.

Problem 2.109 Prove formula (2.149) using the contour shown in Fig. 2.39b.

Problem 2.110 Consider specifically a response function $\chi(\omega) = \chi_1(\omega) + i\chi_2(\omega)$ with χ_1 and χ_2 being its real and imaginary parts and $\omega > 0$ a real frequency. For physical systems $\chi(-\omega)^* = \chi(\omega)$ which guarantees that the physical observables remain real. Show that in this case the Kramers-Kronig relations read:

$$\chi_1(\omega_0) = \frac{2}{\pi} \mathcal{P} \int_0^\infty \frac{\omega \chi_2(\omega)}{\omega^2 - \omega_0^2} d\omega \quad \text{and} \quad \chi_2(\omega_0) = -\frac{2\omega_0}{\pi} \mathcal{P} \int_0^\infty \frac{\chi_1(\omega)}{\omega^2 - \omega_0^2} d\omega .$$
$$(2.152)$$

Problem 2.111 Show that

$$\widetilde{K}(\omega) = \int_0^\infty \frac{\sin t}{\pi t} e^{i\omega t} dt = \frac{1}{2}\chi_1(\omega) - \frac{i}{2\pi} \ln \left| \frac{\omega - 1}{\omega + 1} \right| ,$$

where $\chi_1(\omega) = 1$ if $-1 < \omega < 1$ and 0 otherwise. Use the following method:
(i) relate first the real part of the function $\widetilde{K}(\omega)$ to $\chi_1(\omega)$ of Eq. (2.128)
expressed via a similar integral; (ii) then, using the Kramers-Kronig relations,
determine the imaginary part of $\widetilde{K}(\omega)$.

2.9.2 Propagation of Electro-Magnetic Waves in a Material

Let us first discuss a little what is a wave. Consider the following function of the
coordinate x and time t:

$$\Psi(x, t) = \Psi_0 \sin(kx - \omega t) . \qquad (2.153)$$

If we choose a fixed value of the x, the function $\Psi(x, t)$ oscillates at that point with
the frequency ω between the values $\pm\Psi_0$. Consider now a particular value of Ψ at time
t (x is still fixed). The full oscillation cycle corresponds to the time $t + T$ after which
the function returns back to the chosen value. Obviously, this must correspond to the
ωT being equal exactly to the 2π which is the period of the sine function. Therefore,
the minimum period of oscillations is given by $T = 2\pi/\omega$.

Let us now make a snapshot of the function $\Psi(x, t)$ at *different* values of x but
at the given (*fixed*) time t. It is also a sinusoidal function shown in blue in Fig. 2.40.
After a passage of small time Δt the function becomes

$$\Psi(x, t + \Delta t) = \Psi_0 \sin[kx - \omega(t + \Delta t)] ,$$

and its corresponding snapshot is shown in red in Fig. 2.40. We notice that the whole
shape is shifted to the *right* by some distance Δx. This can be calculated by e.g. considering the shift of the maximum of the sine function. Indeed, at time t the maximum
of one of the peaks is when $kx_m - \omega t = \pi/2$. At time $t + \Delta t$ the maximum must be
at the point $x_m + \Delta x_m$, where $k(x_m + \Delta x_m) - \omega(t + \Delta t) = \pi/2$. From these two
equations we immediately get that $\Delta x_m = \omega \Delta t / k$. Therefore, the function (2.153)

Fig. 2.40 The oscillation of a media along the direction x shown at two times t (blue) and $t + \Delta t$ (red)

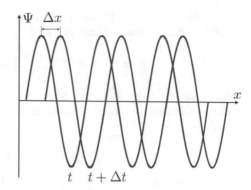

"moves" to the right with the velocity $v_{phase} = \Delta x_m / \Delta t = \omega / k$. Over the time of the period T the function would move by the distance $\lambda = v_{phase} T = \omega T / k = 2\pi / k$.

We see that the specific construction (2.153) corresponds to a wave. It could be e.g. a sound wave: for the given position x the function Ψ oscillates in time (particles of air vibrate in time at the given point), but at each given time the function Ψ at different positions x forms a sinusoidal shape (the displacement of air particles at the given time changes sinusoidally with x). This sinusoidal shape, if considered as a function of time, moves undistorted in the positive x direction with the velocity v_{phase}. This would be the velocity at which the sound propagates: the sound (oscillation of air particles) created at a particular point x_0 (the source) would propagate along the direction x with the velocity v_{phase}. The latter is called phase velocity, ω the frequency, k the wavevector and the distance λ the wave passes over the time T the wavelength.

Since the wave depends only on a single coordinate x, points in space with different y and z but the same x form a plane perpendicular to the x axis and passing through that value of x. All y and z points lying on this plane would have exactly the same properties, i.e. they oscillate in phase with each other; this wave is actually a *plane wave* since its front is a plane. A point source creates a spherical wave with the front being a sphere, but at large distances from the source such waves can approximately be considered as plane waves as the curvature of a sphere of a very large radius is very small.

The function Ψ satisfies a simple *partial* differential equation:

$$\frac{\partial^2 \Psi}{\partial x^2} = \frac{1}{v_{phase}^2} \frac{\partial^2 \Psi}{\partial t^2} , \qquad (2.154)$$

which is called (one-dimensional) wave equation. We have already come across wave equations in Sect. I.6.9.1, but these were more general three-dimensional equations corresponding to waves propagating in 3D space.[11]

---

[11] We shall consider the 1D wave equation in more detail in Sect. 8.2.

Of course, the cosine function in place of the sine function above can serve perfectly well as a one-dimensional wave. Moreover, their linear combination

$$\Psi\,(x,t) = A\sin\,(kx - \omega t) + B\cos\,(kx - \omega t) \tag{2.155}$$

would also describe a wave. Its shape (which moves undistorted with time) is still perfectly sinusoidal as is demonstrated by the following simple manipulation (where $\varphi = kx - \omega t$):

$$
\begin{aligned}
A\sin\varphi + B\cos\varphi &= \sqrt{A^2 + B^2}\left(\frac{A}{\sqrt{A^2 + B^2}}\sin\varphi + \frac{B}{\sqrt{A^2 + B^2}}\cos\varphi\right)\\
&= \sqrt{A^2 + B^2}\,(\cos\theta\sin\varphi + \sin\theta\cos\varphi)\\
&= \sqrt{A^2 + B^2}\,\sin\,(\varphi + \theta)\ ,
\end{aligned}
$$

where $\tan\theta = B/A$. And of course the form (2.155) still satisfies the same wave equation (2.154). At the same time, we also know that both sine and cosine functions can also be written via complex exponentials by virtue of the Euler's formulae (2.30). Therefore, instead of Eq. (2.155) we can also write

$$\Psi\,(x,t) = C_1 e^{i(kx-\omega t)} + C_2 e^{-i(kx-\omega t)}\ , \tag{2.156}$$

where C_1 and C_2 are complex numbers ensuring that Ψ is real. Obviously, since the second exponential is a complex conjugate of the first, the Ψ will only be real if $C_2 = C_1^*$. It is readily checked that the function (2.156) still satisfies the same wave equation (2.154). Although complex exponential functions are perfectly equivalent to the real sine and cosine functions, they are much easier to deal with in practice, and we shall illustrate their use now in a number of simple examples from physics.

Consider an isotropic conducting media with the dielectric constant ϵ and conductivity σ. For simplicity, we shall assume the media is non-magnetic and without free charges. Maxwell equations in such a media differ slightly from those in the vacuum we considered in Sect. I.6.9.1 and have the following form:

$$
\begin{aligned}
\operatorname{curl}\mathbf{H} &= \frac{4\pi}{c}\mathbf{j} + \frac{1}{c}\frac{\partial \mathbf{D}}{\partial t}\ , \quad \operatorname{div}\mathbf{H} = 0\ ,\\
\operatorname{curl}\mathbf{E} &= -\frac{1}{c}\frac{\partial \mathbf{H}}{\partial t}\ , \quad \operatorname{div}\mathbf{E} = 0\ ,
\end{aligned}
\tag{2.157}
$$

where $\mathbf{D} = \epsilon\mathbf{E}$, while the current $\mathbf{j} = \sigma\mathbf{E}$ is related to the electric field via conductivity σ of the media.

Let us establish a wave solution of these equations. To this end, we shall rearrange them by applying the curl to both sides of the first equation (cf. Sect. I.6.9.1):

$$\operatorname{curl}\operatorname{curl}\mathbf{H} = \frac{4\pi\sigma}{c}\operatorname{curl}\mathbf{E} + \frac{\epsilon}{c}\frac{\partial}{\partial t}\operatorname{curl}\mathbf{E}\ .$$

Using the fact that

$$\text{curl curl } \mathbf{H} = \text{grad div } \mathbf{H} - \Delta\mathbf{H} = -\Delta\mathbf{H} ,$$

where we have made use of the second Maxwell equation in (2.157), and the third
equation (2.157) relating curl $\mathbf{E}$ to $\mathbf{H}$, we obtain a closed equation for $\mathbf{H}$:

$$\Delta\mathbf{H} = \frac{4\pi\sigma}{c^2}\frac{\partial\mathbf{H}}{\partial t} + \frac{\epsilon}{c^2}\frac{\partial^2\mathbf{H}}{\partial t^2} . \tag{2.158}$$

A similar calculation starting form the third equation in (2.157) results in an identical
closed equation for $\mathbf{E}$:

$$\Delta\mathbf{E} = \frac{4\pi\sigma}{c^2}\frac{\partial\mathbf{E}}{\partial t} + \frac{\epsilon}{c^2}\frac{\partial^2\mathbf{E}}{\partial t^2} . \tag{2.159}$$

Consider now a plane wave propagating along the z direction:

$$\mathbf{E}(z, t) = \mathbf{E}_0 e^{i(kz-\omega t)} , \tag{2.160}$$

and similarly for the $\mathbf{H}(z, t)$. Substituting this trial solution into Eq. (2.159) should
give us the unknown dependence of the wavevector k on the frequency ω. We obtain:

$$k^2 = \frac{\omega^2}{c^2}\left(\epsilon + i\frac{4\pi\sigma}{\omega}\right) \implies k(\omega) = \frac{\omega}{c}\sqrt{\epsilon + i\frac{4\pi\sigma}{\omega}} = \frac{\omega}{c}(n + i\kappa) . \tag{2.161}$$

We see that the wavevector k is complex. This makes perfect sense as it corresponds
to the propagating waves attenuating into the media. Indeed, using the obtained value
of k in Eq. (2.160), we obtain:

$$\mathbf{E}(z, t) = \mathbf{E}_0 e^{-\zeta z} e^{i(k_0 z-\omega t)} ,$$

where $k_0 = \omega n/c$ and $\zeta = \omega\kappa/c$. It is seen that the amplitude of the wave, $\mathbf{E}_0 e^{-\zeta z}$,
decays in the media as the energy of the wave is spent on accelerating conducting
electrons in it.

Problem 2.112 Show that the solution of the equation (2.161) with respect to
real and imaginary parts of the wavevector reads:

$$n = \sqrt{\frac{1}{2}\left[\sqrt{\epsilon^2 + \frac{16\pi^2\sigma^2}{\omega^2}} + \epsilon\right]} \quad \text{and} \quad \kappa = \sqrt{\frac{1}{2}\left[\sqrt{\epsilon^2 + \frac{16\pi^2\sigma^2}{\omega^2}} - \epsilon\right]} . \tag{2.162}$$

Problem 2.113 Assume that the conductivity σ of the media is small as compared to ϵ, i.e. $\sigma/\omega \ll \epsilon$. Then show that in this case

$$n \simeq \sqrt{\epsilon} \quad \text{and} \quad \kappa \simeq \frac{2\pi\sigma}{\omega\sqrt{\epsilon}} \,,$$

and hence $\kappa \ll n$. This means that absorption of energy in such a media is very small.

As our second example, let us consider how an alternating current is distributed inside a conductor. For simplicity, we shall consider the conductor occupying the half space $z \geq 0$ and the fields are quasi-stationary, i.e. they vary slowly over the dimensions of the system. In this case it can be shown that the displacement current, $\mathbf{j}_{dislp} = c^{-1}\partial \mathbf{D}/\partial t$, can be dropped in the Maxwell's equations (2.157) as compared to the actual current $\mathbf{j}$ and hence the equations for the fields (2.158) and (2.159) take on a simpler form:

$$\Delta \mathbf{H} = \frac{4\pi\sigma}{c^2}\frac{\partial \mathbf{H}}{\partial t} \quad \text{and} \quad \Delta \mathbf{E} = \frac{4\pi\sigma}{c^2}\frac{\partial \mathbf{E}}{\partial t} \,, \tag{2.163}$$

We assume that the current flows along the x axis:

$$j_x = j(z)e^{i\omega t} \,, \quad j_y = 0 \,, \quad j_z = 0 \,.$$

Because of the continuity equation (I.6.104), $\operatorname{div}\mathbf{j} = 0$ (there are no free charges) and hence $\partial j_x/\partial x = 0$, i.e. the current cannot depend on x, it can only depend on z (if we assume, in addition, that there is no y dependence either). This seems a natural assumption for the conductor which extends indefinitely in the positive and negative y directions: the current may only depend on the distance z from the boundary of the conductor, and hence the function $j(z)$ characterises the distribution of the current with respect to that distance. To find this distribution, we note that the current and the electric field are proportional, $\mathbf{j} = \sigma\mathbf{E}$, and hence only the x component of the field $\mathbf{E}$ remains, $E_x(z,t) = E(z)e^{i\omega t}$, where $j(z) = \sigma E(z)$. Substituting this trial solution into the second equation (2.163) for the field $\mathbf{E}$, we obtain:

$$\frac{\partial^2 E(z)}{\partial z^2} = i\frac{4\pi\sigma\omega}{c^2}E(z) \,. \tag{2.164}$$

Problem 2.114 Assuming an exponential solution, $E(z) \sim e^{\lambda z}$, show that the two possible values of the exponent λ are:

$$\lambda_{1,2} = \pm \frac{1+i}{c} \sqrt{2\pi\sigma\omega} = \pm \frac{1+i}{\delta} , \qquad (2.165)$$

where we have introduced a parameter δ which has a meaning of the decay length (see below).

Hence, the general solution of equation (2.164) is

$$E(z) = A e^{z/\delta} e^{iz/\delta} + B e^{-z/\delta} e^{-iz/\delta} .$$

Physically, it is impossible that the field (and, hence, the current) would increase indefinitely with z; therefore, the first term above leading to such an unphysical behavior must be omitted ($A = 0$), and we obtain:

$$E_x(z,t) = B e^{-z/\delta} e^{-iz/\delta} e^{i\omega t} , \quad E_y = E_z = 0 . \qquad (2.166)$$

The current $j_x(z,t) = \sigma E_x(z,t)$ behaves similarly. It appears then that the field and the current decay exponentially into the conductor remaining non-zero within only a thin layer near its boundary, with the width of the layer being of the order of $\delta = c/\sqrt{2\pi\sigma\omega}$. The width decays as $\omega^{-1/2}$ with the frequency ω, i.e. the effect is more pronounced at high frequencies. This is called the *skin effect*. As $\omega \to 0$ the "skin" width $\delta \to \infty$, i.e. there is no skin effect anymore: the direct (not alternating) current is distributed uniformly over the conductor.

Problem 2.115 Using the third equation in (2.157), show that the magnetic field is:

$$H_x = 0, \quad H_y(z,t) = \frac{(1-i)c}{\omega\delta} E_x(z,t) , \quad H_z = 0 ,$$

i.e. the magnetic field also decays exponentially into the body of the conductor remaining perpendicular to the electric field. Numerical estimates show that at high frequencies ω the magnetic field is much larger than the electric field within the skin layer.

2.9.3   Electron Tunnelling in Quantum Mechanics

In quantum mechanics the behaviour of an electron in an one-dimensional potential $V(x)$ is described by its (generally) complex wavefunction $\psi(x)$ which yields a probability $dP(x) = |\psi(x)|^2\, dx$ for an electron to be found during a measurement between x and $x + dx$. The wavefunction satisfies the Schrödinger equation

$$-\frac{\hbar^2}{2m}\psi''(x) + V(x)\psi(x) = E\psi(x) , \qquad (2.167)$$

where $\hbar$ is the Planck constant (divided by 2π), m electron mass and E is the electron energy.

Here we shall consider a number of problems, where an electron behaves as a wave which propagates in space with a particular momentum and may be reflected from a potential barrier; at the same time, we shall also see that under certain circumstances the wave may also penetrate the barrier and hence transmit through it, which is a purely quantum effect called *tunnelling*.

Consider first the case of an electron moving in a free space where $V(x) = 0$. The solution of the differential equation (2.167) can be sought in a form of an exponential, $\psi(x) \sim e^{ikx}$, which upon substitution yields $E = \hbar^2 k^2/2m$, the energy of a free electron. The energy is positive and continuous. The wave function of the electron corresponds to two possible values of the wave vector k (in fact, it is just a number in the one-dimensional case), which are $k_{\pm} = \pm\sqrt{2mE}/\hbar = \pm k$. Correspondingly, the wave function can be written as a linear combination

$$\psi(x) = Ae^{ikx} + Be^{-ikx} ,$$

where A and B are complex constants. The first term in $\psi(x)$ above describes the electron moving along the positive x direction, whereas the second term—in the negative one.

Let us now consider a more interesting case of an electron hitting a wall, as shown in Fig. 2.41a. We shall consider a solution of the Schrödinger equation corresponding to the electron propagating to the right from $x = -\infty$. In this case in the region $x < 0$ (the left region) the wavefunction of the electron

$$\psi_L(x) = Ae^{ikx} + Be^{-ikx} , \qquad (2.168)$$

where $k = \sqrt{2mE}/\hbar$. The solution of the Schrödinger equation for $x > 0$ (the right region) is obtained exactly in the same way, but in this case

$$\psi_R(x) = Ce^{igx} + De^{-igx} ,$$

where $g = \sqrt{2m\,(E - V_0)}/\hbar$. We shall consider the most interesting case of the electron energy $E < V_0$. In a classical set up of the problem the electron would not be

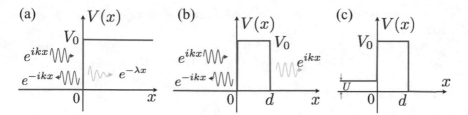

Fig. 2.41 a The potential $V(x)$ makes a step of the height V_0 at $x = 0$, so that the electron wave $\psi = e^{ikx}$ propagating towards the step (to the right), although partially penetrating into the step, will mostly reflect from it propagating to the left as $\psi \sim e^{-ikx}$; **b** in the case of the potential barrier of the height V_0 and width d the electron wave propagates through the barrier as $\psi \sim e^{ikx}$, although a partial reflection from the barrier is also taking place; **c** the same case as (**b**), but in this case a bias $\phi < 0$ is applied to the left electrode, so that the potential experienced by the electrons $U = e\phi > 0$ ($e < 0$ is the electron charge) on the left becomes higher than on the right causing a current (a net flow of electrons to the right)

able to penetrate into the right region with this energy; however, quantum mechanics allows for some penetration. Indeed, in this case $g = i\sqrt{2m\,(V_0 - E)}/\hbar = i\kappa$ becomes purely imaginary and hence the wavefunction in the barrier region becomes

$$\psi_R(x) = Ce^{-\kappa x} + De^{\kappa x} .$$

Since $\kappa > 0$, the second term must be dropped as it leads to an infinite increase of the wavefunction at large x which is unphysical (recall that $|\psi(x)|^2$ represents the probability density). Therefore,

$$\psi_R(x) = Ce^{-\kappa x} . \tag{2.169}$$

To find the amplitudes in Eqs. (2.168) and (2.169), we exploit the fact that the wavefunction $\psi(x)$ and its first derivative must be continuous across the whole x axis. This means that the following conditions must be satisfied at any point $x = x_b$:

$$\psi(x_b - 0) = \psi(x_b + 0) \quad \text{and} \quad \left.\frac{d}{dx}\psi(x)\right|_{x=x_b-0} = \left.\frac{d}{dx}\psi(x)\right|_{x=x_b+0} , \tag{2.170}$$

where $+0$ and -0 correspond to the limits of $x \to x_b$ from the right and left, respectively. In our case the potential makes a jump at $x_b = 0$ and hence this particular point has to be specifically considered; the wavefunction is already continuous anywhere else. Therefore, the above conditions translate into the following:

$$\psi_L(0) = \psi_R(0) \quad \text{and} \quad \left.\frac{d}{dx}\psi_L(x)\right|_{x=0} = \left.\frac{d}{dx}\psi_R(x)\right|_{x=0} .$$

Using these conditions, we immediately obtain two simple equations for the amplitudes:

$$A + B = C \quad \text{and} \quad ik(A - B) = -\kappa C ,$$

solution of which yields:

$$\frac{B}{A} = \frac{ik + \kappa}{ik - \kappa} = \frac{i\sqrt{2mE} + \sqrt{2m(V_0 - E)}}{i\sqrt{2mE} - \sqrt{2m(V_0 - E)}} ,$$

$$\frac{C}{A} = \frac{2ik}{ik - \kappa} = \frac{2i\sqrt{2mE}}{i\sqrt{2mE} - \sqrt{2m(V_0 - E)}}$$

(2.171)

(only *relative* amplitudes can in fact be determined). The extend of reflection is determined by the energy dependent reflection coefficient

$$R(E) = \left|\frac{B}{A}\right|^2 = \left|\frac{ik + \kappa}{ik - \kappa}\right|^2 = \frac{\kappa^2 + k^2}{\kappa^2 + k^2} = 1 .$$

We see that in this case all of the incoming wave gets reflected from the step; however, this coefficient is actually misleading as some of the wave gets transmitted into the step through the wall: the probability density $P(x) = |\psi_R(x)|^2 = |C|^2 e^{-2\kappa x}$ is non-zero within a small region of the width $\Delta x \sim 1/2\kappa$ behind the wall of the step, although it decays exponentially inside the step. This behaviour is entirely due to quantum nature of the electron. Since no electron is to be determined at $x = +\infty$, the transmission through the step is zero in this case. Note that $\Delta x \to 0$ when $V_0 \to \infty$, i.e. the region behind the wall ($x > 0$) is forbidden even for quantum electrons when the step is infinitely high.

A very interesting case from the practical point of view is a step of a finite width shown in Fig. 2.41b. In this case two solutions of the same energy E of the Schrödinger equation exist: one corresponding to the wave propagating from left to right, and one in the opposite direction. Let us first consider the wave propagating to the right. Three regions are to be identified: $x < 0$ (left, L), $0 < x < d$ (central, C) and $x > d$ (right, R). The corresponding solutions of the Schrödinger equation for all these regions are:

$$\psi_L(x) = e^{ikx} + Re^{-ikx} , \quad \psi_C(x) = Ce^{iqx} + De^{-iqx}$$

$$\text{and} \quad \psi_R(x) = Te^{ikx} ,$$

(2.172)

where $k = \sqrt{2mE}/\hbar$ and $q = \sqrt{2m(E - V_0)}/\hbar$. We have set the amplitude of the wave incoming to the barrier (in ψ_L) to one as the amplitudes can only be determined relatively to this one anyway; also note that only a single term exists for the right wave, ψ_R, as we are considering the solution propagating to the right.

Problem 2.116 Using the continuity conditions (2.170) at $x = 0$ and $x = d$, show that:

$$R = \frac{\left(q^2 - k^2\right)\left(e^{-2iqd} - 1\right)}{(q-k)^2 - (q+k)^2 \, e^{-2iqd}}, \quad T = \frac{-4kq \, e^{-i(k+q)d}}{(q-k)^2 - (q+k)^2 \, e^{-2iqd}},$$

$$C = \frac{-2k\,(q+k)\,e^{-2iqd}}{(q-k)^2 - (q+k)^2 \, e^{-2iqd}} \quad \text{and} \quad D = \frac{-2k\,(q-k)}{(q-k)^2 - (q+k)^2 \, e^{-2iqd}}.$$

Problem 2.117 The above solution corresponds to any energy E of the electron. Let us now specifically consider two possible cases: $0 < E < V_0$ and $E > V_0$. Show that for $E > V_0$ the amplitudes read:

$$R = \frac{-i\left(q^2 - k^2\right)\sin\left(qd\right)}{i\left(q^2 + k^2\right)\sin\left(qd\right) - 2kq\cos\left(qd\right)},$$

$$T = \frac{-2kq\,e^{-ikd}}{i\left(q^2 + k^2\right)\sin\left(qd\right) - 2kq\cos\left(qd\right)},$$

$$C = \frac{-k\,(q+k)\,e^{-iqd}}{i\left(q^2 + k^2\right)\sin\left(qd\right) - 2kq\cos\left(qd\right)},$$

$$D = \frac{-k\,(q-k)\,e^{iqd}}{i\left(q^2 + k^2\right)\sin\left(qd\right) - 2kq\cos\left(qd\right)},$$

while in the case of $0 < E < V_0$ (when q becomes imaginary) they are:

$$R = \frac{\left(\kappa^2 + k^2\right)\sinh\left(\kappa d\right)}{\left(k^2 - \kappa^2\right)\sinh\left(\kappa d\right) + 2ik\kappa\cosh\left(\kappa d\right)},$$

$$T = \frac{2ik\kappa\,e^{-ikd}}{\left(k^2 - \kappa^2\right)\sinh\left(\kappa d\right) + 2ik\kappa\cosh\left(\kappa d\right)},$$

$$C = \frac{k\,(i\kappa + k)\,e^{\kappa d}}{\left(k^2 - \kappa^2\right)\sinh(\kappa d) + 2ik\kappa\cosh(\kappa d)} ,$$

$$D = \frac{k\,(i\kappa - k)\,e^{-\kappa d}}{\left(k^2 - \kappa^2\right)\sinh(\kappa d) + 2ik\kappa\cosh(\kappa d)} ,$$

where $\kappa = \sqrt{2m\,(V_0 - E)}/\hbar$.

Problem 2.118 The current can be obtained using an expression:

$$j\,[\psi] = \frac{ie\hbar}{2m}\left(\frac{d\psi^*}{dx}\psi - \psi^*\frac{d\psi}{dx}\right) , \qquad (2.173)$$

where $j[\psi]$ states explicitly that the current depends on the wavefunction ψ to be used (it is said that the current is a *functional* of the wavefunction; we shall consider this notion in more detail in Chap. 9) and e is the (negative) electron charge. The current should not depend on the x, i.e. it is expected that it is constant across the whole system. Show that the currents calculated using $\psi_L(x)$ (i.e. in the left region of Fig. 2.41b) and $\psi_R(x)$ (in the right region) are, respectively:

$$j_L = \frac{e\hbar k}{m}\left(1 - |R|^2\right) \quad \text{and} \quad j_R = \frac{e\hbar k}{m}\,|T|^2 . \qquad (2.174)$$

Then demonstrate by a direct calculation based on the expressions for the amplitudes given in Problem 2.116 that

$$|T|^2 = 1 - |R|^2 .$$

This identity guarantees that the current on the left and on the right due to an electron of any energy E is the same.

Problem 2.119 The reflection coefficient r is defined as the ratio of the current (which is basically a flux of electrons) of the reflected wave, $j\left[Re^{-ikx}\right]$ (here we assume $j[\psi]$ calculated specifically for $\psi = Re^{-ikx}$), to the current due to the incoming one, $j\left[e^{ikx}\right]$, while the transmission coefficient t is defined as the ratio of the transmitted, $j\left[Te^{ikx}\right]$, and the incoming waves. Show that

$$r = \frac{j\left[Re^{-ikx}\right]}{j\left[e^{ikx}\right]} = |R|^2 \quad \text{and} \quad t = \frac{j\left[Te^{ikx}\right]}{j\left[e^{ikx}\right]} = |T|^2 \ .$$

Considering specifically the two possible cases for the electron energy, show that:

$$r(E) = \frac{\left(q^2 - k^2\right)^2}{\left(q^2 + k^2\right)^2 + 4k^2q^2 \cot^2(qd)} \ , \quad t(E) = \left[1 + \left(\frac{k^2 - q^2}{2kq}\right)^2 \sin^2(qd)\right]^{-1} \ ,$$

when $E > V_0$, and

$$r(E) = \left[1 + \frac{4k^2\kappa^2}{\left(k^2 + \kappa^2\right)^2 \sinh^2(\kappa d)}\right]^{-1} \ , \quad t(E) = \left[1 + \left(\frac{k^2 + \kappa^2}{2k\kappa}\right)^2 \sinh^2(\kappa d)\right]^{-1} \ ,$$

when $0 < E < V_0$, where the energy dependence comes from that of k and κ as given above.

Problem 2.120 Consider now the other solution of the Schrödinger equation corresponding to the propagation of an electron of the same energy E from right to left. In this case the solutions for the wavefunctions in the three regions we are seeking are:

$$\psi_L(x) = Te^{-ikx} \ , \quad \psi_C(x) = Ce^{-iqx} + De^{iqx}$$
$$\text{and} \quad \psi_R(x) = e^{-ikx} + Re^{ikx} \ , \tag{2.175}$$

where the same k and q are used. Show that the expressions for the reflection and transmission coefficients for this solution are exactly the same as for the left-to-right propagating solution, and hence the corresponding currents as given by Eq. (2.174) coincide exactly as well.

Therefore, the net current due to electrons traveling from the left and from the right, which is the difference of the two, is zero. This result is to be expected as the situation is totally symmetric: there is no difference between the two sides ($x < 0$ and $x > d$) of the system. In order for the current to flow, it is necessary to distort this balance, and this can be achieved by applying a bias ϕ to the system which would result in the electrons on the right and left to experience a different potential $V(x)$. The simplest case illustrating this situation is shown in Fig. 2.41c where, because of the bias $\phi < 0$ applied to the left electrode, the potential $V(x) = U = e\phi > 0$ exists in the left region ($x < 0$), while $V = 0$ in the right one ($x > d$). Recall that e is the (negative) electron charge.

390   2   Complex Numbers and Functions

Problem 2.121 In this Problem we shall consider this situation explicitly. Using the method we have been using above, write down the left-to-right propagating solution of the Schrödinger equation:

$$\psi_L^{\to}(x) = e^{ikx} + R^{\to}e^{-ikx} , \quad \psi_C^{\to}(x) = C^{\to}e^{iqx} + D^{\to}e^{-iqx}$$
$$\text{and} \quad \psi_R^{\to}(x) = T^{\to}e^{ipx} , \tag{2.176}$$

where $k = \sqrt{2m\,(E-U)}/\hbar$, $q = \sqrt{2m\,(E-V_0)}/\hbar$ and $p = \sqrt{2mE}/\hbar$, and then determine the amplitudes $R^{\to}$ and $T^{\to}$; you should get this:

$$R^{\to} = \frac{q\,(k-p) + i\,(q^2 - pk)\tan(qd)}{q\,(k+p) - i\,(q^2 + pk)\tan(qd)} \quad \text{and}$$
$$T^{\to} = \frac{2kqe^{-ipd}}{q\,(k+p)\cos(qd) - i\,(q^2 + pk)\sin(qd)} . \tag{2.177}$$

These formulae are valid for any energy E; specifically, $q = i\kappa$ when $E < V_0$ and the sine and cosine functions are understood in the proper sense as functions of the complex variable.

The most interesting case to consider is when electrons are with energies $U < E < V_0$ which corresponds to the tunnelling. Then, show that for these energies:

$$t^{\to} = |T^{\to}|^2 = \frac{4k^2\kappa^2}{\left(\kappa^2 - pk\right)^2 \sinh^2(\kappa d) + \kappa^2\,(p+k)^2 \cosh^2(\kappa d)} ,$$

$$r^{\to} = |R^{\to}|^2 = \frac{\left(\kappa^2 + pk\right)^2 \sinh^2(\kappa d) + \kappa^2\,(p-k)^2 \cosh^2(\kappa d)}{\left(\kappa^2 - pk\right)^2 \sinh^2(\kappa d) + \kappa^2\,(p+k)^2 \cosh^2(\kappa d)} .$$

Using the complete wavefunctions in the left and right regions, show that the corresponding currents are:

$$j_L^{\to} = \frac{e\hbar k}{m}\left(1 - |R^{\to}|^2\right) = \frac{e\hbar k}{m}\left(1 - r^{\to}\right) \quad \text{and} \quad j_R^{\to} = \frac{e\hbar p}{m}|T^{\to}|^2 = \frac{e\hbar p}{m}t^{\to} .$$

Show that $j_L^{\to} = j_R^{\to}$. Next, consider the reverse current, from right to left. Repeating the calculations, show that in this case (for any E):

$$R^{\leftarrow} = e^{-2ipd} \frac{q\,(p-k) + i\left(q^2 - pk\right)\tan\left(qd\right)}{q\,(k+p) - i\left(q^2 + pk\right)\tan\left(qd\right)} \quad \text{and}$$

$$T^{\leftarrow} = \frac{2pqe^{-ipd}}{q\,(k+p)\cos\left(qd\right) - i\left(q^2 + pk\right)\sin\left(qd\right)}. \tag{2.178}$$

Therefore, considering again the energies between U and V_0, show that

$$t^{\leftarrow} = |T^{\leftarrow}|^2 = \frac{p^2}{k^2}t^{\rightarrow} \quad \text{and} \quad r^{\leftarrow} = |R^{\leftarrow}|^2 = r^{\rightarrow}.$$

Hence, calculate the currents as measured in the left, $j_L = j_L^{\rightarrow} + j_L^{\leftarrow}$, and right, $j_R = j_R^{\rightarrow} + j_R^{\leftarrow}$, electrodes. Demonstrate that both are the same (as it should be) and given by:

$$j = \frac{e\hbar}{m} \frac{4\kappa^2 pk\,(k-p)}{\left(\kappa^2 - pk\right)^2 \sinh^2\left(\kappa d\right) + \kappa^2\,(p+k)^2 \cosh^2\left(\kappa d\right)}.$$

It is explicitly seen now that if no bias is applied, $U = 0$, $k = p$ and hence $j = 0$.

Using atomic units (a.u., in which $\hbar^2/m = 1$ and $e = -1$) and explicit expressions given above for k, p and κ, investigate the dependence of the current j on the electron energy E (from 0 to say 0.1 a.u.), depth d (between 1 and say 10 a.u.) and height V_0 (up to 0.1 a.u.) of the barrier. Also plot the dependence of the current on the applied bias $\phi = -U$ (the so-called current-voltage characteristics of the device we have constructed).

The problem we have just considered is a very simplified model for a very real and enormously successful experimental tool called Scanning Tunnelling Microscopy (STM), for invention of which Gerd Binnig and Heinrich Rohrer received the Nobel Prize in Physics in 1986. Schematically STM is shown in Fig. 2.42a. An atomically sharp conducting tip is brought very close to a conducting surface and the bias is applied between the two. As the distance between the tip apex atom (the atom which is at the tip end, the closest to the surface) and the surface atoms reaches less than $\simeq 10$ Å, a small current (in the range of pA to nA) is measured due to tunnelling electrons: the vacuum gap between the two surfaces serves as a potential energy barrier we have just considered in Problem 2.121. Most importantly, the current depends on the lateral position of the tip with respect to the atomic structure of the surface. This allows obtaining in many cases *atomic resolution* when scanning the sample with the STM. Nowadays the STM is widely used in surface physics and chemistry as it allows not only imaging atoms and adsorbed molecules on crystal surfaces, but also move them (this is called manipulation) and perform local chemical reactions (e.g. breaking a molecule into parts or fusing parts together). In Fig. 2.42b a real STM image of

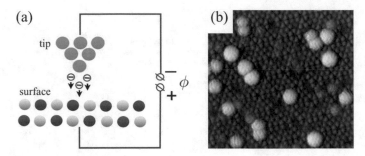

Fig. 2.42 a A sketch of the Scanning Tunnelling Microscope (STM): an atomically sharp tip is placed above a sample surface 4–10 Å from it. When a bias ϕ is applied to the tip and sample, electrons tunnel between the two surfaces. **b** STM image of a Si(111) 7 × 7 reconstructed surface with a small coverage of C_{60} molecules on top of it; bias voltage of -1.8 V and the tunnel current of 0.2 nA were used (reproduced with permission from J. I. Pacual *et al.* - Chem. Phys. Lett. **321** (2000) 78)

C_{60} molecules adsorbed on top of the Si(111) reconstructed surface is shown. You can clearly see atoms of the surface as well as the molecules themselves as much bigger circles. Atoms on the surface form a nearly perfect arrangement, however, many defects are visible. The molecules are seen as various types of bigger balls with some internal structure (the so-called sub-molecular resolution): carbon atoms of the molecules can be distinguished as tiny features on the big circles. Analysing this substructure, it is possible to figure out the orientation of the molecules on the surface. Different sizes of the balls correspond most likely to different adsorption positions of the molecules.

2.9.4 Propagation of a Quantum State

Consider an n-level quantum system described by the $n \times n$ Hamiltonian matrix $\mathbf{H} = (H_{kj})$. If $\mathbf{\Psi}_0 = (\psi_k)$ is a vector-column of the system wavefunction at time t, then at time $t' > t$ its wavefunction can be formally written as $\mathbf{\Psi}_{t'} = \mathbf{U}_{t't} \mathbf{\Psi}_t$, where $\mathbf{U}_{t't}$ is a matrix to be calculated. Due to its nature it is called the *propagation* matrix as it describes a propagation of the system state vector $\mathbf{\Psi}_t$ from the initial time t to the final time t'. Let us calculate the propagation matrix in the case when the Hamiltonian matrix $\mathbf{H}$ does not depend on time.

We shall start by solving approximately the time dependent Schrödinger equation

$$i\hbar \frac{d\mathbf{\Psi}_t}{dt} = \mathbf{H}\mathbf{\Psi}_t \tag{2.179}$$

for a very small propagation time Δt. In this case the time derivative of the wavefunction vector can be approximated as

$$\frac{d\mathbf{\Psi}_t}{dt} \simeq \frac{\mathbf{\Psi}_{t+\Delta t} - \mathbf{\Psi}_t}{\Delta t} \implies i\hbar \frac{\mathbf{\Psi}_{t+\Delta t} - \mathbf{\Psi}_t}{\Delta t} \simeq \mathbf{H}\mathbf{\Psi}_t$$

$$\implies \mathbf{\Psi}_{t+\Delta t} \simeq \left(1 + \frac{\Delta t}{i\hbar}\mathbf{H}\right)\mathbf{\Psi}_t .$$

Applying the obtained relationship m times ($t \to t + \Delta t \to t + 2\Delta t \to \cdots \to t + m\Delta t$), the wavefunction will be propagated by the time $m\Delta t$ as follows:

$$\mathbf{\Psi}_{t+m\Delta t} \simeq \left(1 + \frac{\Delta t}{i\hbar}\mathbf{H}\right)\mathbf{\Psi}_{t+(m-1)\Delta t} \simeq \left(1 + \frac{\Delta t}{i\hbar}\mathbf{\Psi}_t\right)^2 \mathbf{\Psi}_{t+(m-2)\Delta t}$$

$$\simeq \cdots \simeq \left(1 + \frac{\Delta t}{i\hbar}\mathbf{H}\right)^m \mathbf{\Psi}_t . \tag{2.180}$$

Now, let λ_k and $\mathbf{x}_k$ be the eigenvalues and eigenvectors of the matrix $\mathbf{H}$. The latter must be Hermitian on the physical grounds, $\mathbf{H}^\dagger = \mathbf{H}$, and hence the eigenvalues are real and eigenvectors can always be chosen in such a way that they comprise an orthonormal set, $\mathbf{x}_k^\dagger \mathbf{x}_j = \delta_{kj}$. Expand $\mathbf{\Psi}_t$ in terms of the eigenvectors, $\mathbf{\Psi}_t = \sum_k \alpha_k \mathbf{x}_k$. Multiplying both sides of this equation by $\mathbf{x}_k^\dagger$ from the left and using the orthonormality relation between the eigenvectors, the coefficients α_k can be found as $\alpha_k = \mathbf{x}_k^\dagger \mathbf{\Psi}_t$. Hence, one can then rewrite Eq. (2.180) as:

$$\mathbf{\Psi}_{t+m\Delta t} \simeq \left(1 + \frac{\Delta t}{i\hbar}\mathbf{H}\right)^m \sum_k \alpha_k \mathbf{x}_k = \sum_k \alpha_k \left(1 + \frac{\Delta t}{i\hbar}\mathbf{H}\right)^m \mathbf{x}_k$$

$$= \sum_k \alpha_k \left(1 + \frac{\Delta t}{i\hbar}\lambda_k\right)^m \mathbf{x}_k .$$

At the last step we used the spectral theorem, Eq. (1.109), stating that a function of a matrix, acting on its eigenvector, can be replaced with the same function calculated at the corresponding eigenvalue. Let $t' = t + m\Delta t$ be the finite time after the propagation. Then $\Delta t = (t' - t)/m$ and we obtain:

$$\mathbf{\Psi}_{t+m\Delta t} = \mathbf{\Psi}_{t'} \simeq \sum_k \left(\mathbf{x}_k^\dagger \mathbf{\Psi}_t\right) \left[1 + \frac{\lambda_k (t' - t)/i\hbar}{m}\right]^m \mathbf{x}_k .$$

The above formula becomes exact in the limit of $\Delta t \to 0$ or, which is equivalent, when $m \to \infty$. As we already know (see Sect. 2.3.3), that limit is equal to the exponential function:

$$\Psi_{t'} = \sum_k \left(\mathbf{x}_k^\dagger \Psi_t\right) \left\{ \lim_{m \to \infty} \left[1 + \frac{\lambda_k \left(t' - t\right)/i\hbar}{m} \right]^m \right\} \mathbf{x}_k$$

$$= \sum_k \left(\mathbf{x}_k^\dagger \Psi_t\right) e^{\lambda_k (t'-t)/i\hbar} \mathbf{x}_k = \left[\sum_k e^{\lambda_k (t'-t)/i\hbar} \mathbf{x}_k \mathbf{x}_k^\dagger \right] \Psi_t \ .$$

In the square brackets we recognise the spectral theorem expansion of the matrix $e^{\mathbf{H}(t'-t)/i\hbar}$ which serves then as the required propagation matrix:

$$\mathbf{U}_{t't} = e^{\mathbf{H}(t'-t)/i\hbar} = e^{-i\mathbf{H}(t'-t)/\hbar} \ .$$

Note that this result could have been directly obtained following Sect. 1.19.2; the above procedure followed an alternative route.

2.9.5 Matsubara Sums

There is a class of infinite summations that can be calculated using complex calculus. They appear in theoretical physics when dealing with the so-called Matsubara Green's functions. These, the so-called *Matsubara sums,* look like this:

$$S(\tau) = \frac{1}{\beta\hbar} \sum_{n=-\infty}^{\infty} D(i\omega_n) e^{i\omega_n \tau} \tag{2.181}$$

where $D(i\omega_n)$ are expansion coefficients that somehow explicitly depend on $i\omega_n$, and

$$\omega_n = \begin{cases} \frac{2\pi n}{\hbar\beta} & \text{for bosons} \\ \frac{(2n+1)\pi}{\hbar\beta} & \text{for fermions} \end{cases} \tag{2.182}$$

are the so-called Matsubara frequencies. They have different values depending on which type of particles (bosons or fermions) is considered. This is not important for us here, so we shall assume, for definiteness, that we consider fermions and hence accept the second line in the equation above. Also, β is the inverse temperature and τ the so-called "imaginary" time although it is a real quantity between 0 and $\beta\hbar$, i.e. $0 < \tau \le \beta\hbar$.

In various applications the function $D(i\omega_n)$ has a different form, and it is required to perform the summation to determine the function $S(\tau)$. There is a very general procedure for calculating such sums that rests heavily on the mathematical foundations we have developed in this chapter.

We shall start by continuing the function $D(i\omega_n)$ into the complex plane $z \in \mathcal{C}$ via the replacement $i\omega_n \to z$ in the expression for $D(i\omega_n)$. This results in a function $D(z)$ defined in $\mathcal{C}$. That function may have poles in the complex plane and/or have

Fig. 2.43 The circle C_R in the complex plane $\mathcal{C}$, traversed in the anti-clockwise direction, in the limit of its radius $R \to \infty$ encompasses all poles (if exist) of the function $D(z)$ (shown as purple circles) and all the Matsubara frequencies $i\omega_n$ that are arranged evenly along the imaginary axis (shown red)

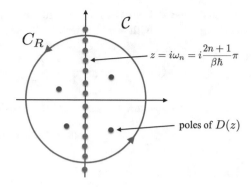

a branch cut, e.g., along the real axis. We shall derive simple general formulae for performing such summations for these two cases.

It is convenient first to define a function

$$\tilde{n}_F(z) = \left[e^{\beta\hbar z} + 1\right]^{-1}.$$

What is remarkable about it is that it has poles exactly at Matsubara frequencies: indeed, $e^{\beta\hbar z} + 1 = 0$ when $\beta\hbar z = i(2n+1)\pi$, and hence $z = i\frac{\pi(2n+1)}{\beta\hbar} = i\omega_n$ for any integer $n = 0, \pm 1, \pm 2, \ldots$. The idea of performing the summation in Eq. (2.181) is to consider such a contour C in $\mathcal{C}$ that comprises all the poles $\{z = i\omega_n\}$; then the sum will emerge from the residue theorem at these poles in the contour integral $\oint_C \tilde{n}_F(z)D(z)e^{z\tau}dz$.

Let us initially consider a simpler case of the function $D(z)$ having only poles at $z_1, \ldots, z_N$ in $\mathcal{C}$, but otherwise analytical everywhere including the real axis.

Consider a circle C_R of radius R centred at the origin, as shown in Fig. 2.43.

For a sufficiently large, but still finite, radius R the contour C_R will include all poles of $D(z)$ (if present) and some of the poles $i\omega_n$ of $\tilde{n}_F(z)$ for which $|\omega_n| < R$. In the limit of $R \to \infty$ the contour C_R will include all the poles. Therefore, in that limit

$$\oint_{C_R} \tilde{n}_F(z)D(z)e^{z\tau}dz = 2\pi i \sum_{i=1}^{N} \text{Res}\left[D(z); z_i\right]\tilde{n}_F(z_i)e^{z_i\tau}$$

$$+ 2\pi i \sum_{n=-\infty}^{\infty} \text{Res}\left[\tilde{n}_F(z); i\omega_n\right] D(i\omega_n)e^{i\omega_n\tau},$$

where the first term in the right hand side accounts for the poles of $D(z)$, while the second term takes care of the poles of the function $\tilde{n}_F(z)$. Assuming that the function $D(z)$ is limited in $\mathcal{C}$ (i.e. $|D(z)| \leq M < +\infty$), it is easy to see that the integral $\oint_{C_R}$ tends to zero in the $R \to \infty$ limit. Indeed, the contour C_R can be considered as consisting of two semicircles: the right one for which $\text{Re}\, z > 0$ and the left one for which Re $z < 0$. Let us estimate the function $\tilde{n}_F(z)e^{z\tau}$ on the *right semicircle* of C_R, where $z = Re^{i\phi}$ with $-\pi/2 < \phi < \pi/2$ (Re $z > 0$)

$$\tilde{n}_F(z)e^{z\tau} = \frac{e^{z\tau}}{e^{\hbar\beta z} + 1} = \frac{e^{z(\tau-\hbar\beta)}}{1 + e^{-\hbar\beta z}} \xrightarrow{R\to\infty} e^{z(\tau-\hbar\beta)} \to 0\,,$$

since $0 < \tau < \beta\hbar$ and the final zero limit results from an obvious estimate:

$$\left| e^{z(\tau-\hbar\beta)} \right| = \left| e^{-(\hbar\beta-\tau)(\text{Re } z + i \text{ Im } z)} \right| \le e^{-(\hbar\beta-\tau)\text{Re } z} \to 0\,.$$

For the *left semicircle*, Re $z < 0$,

$$\tilde{n}_F(z)e^{z\tau} = \frac{e^{z\tau}}{e^{\hbar\beta z} + 1} = \frac{e^{z\tau}}{e^{\hbar\beta(\text{Re } z + i \text{ Im } z)} + 1}$$

$$\xrightarrow{R\to\infty} e^{z\tau} = e^{\tau(\text{Re } z + i \text{ Im } z)} \to 0$$

as $R \to \infty$ since Re $z < 0$. Therefore, the closed contour integral $\oint_{C_R} \to 0$ and we obtain:

$$\sum_{n=-\infty}^{\infty} \text{Res}\,[\tilde{n}_F(z); i\omega_n]\, D(i\omega_n)e^{i\omega_n\tau} = -\sum_{i=1}^{N} \text{Res}\,[D(z); z_i]\, \tilde{n}_F(z_i)e^{z_i\tau}$$

Note that in the discussion above concerning the circle integral we assumed that $\tau \ne 0$. If $\tau = 0$, then the circle integral will also be zero in the $R \to \infty$ limit if $D(z) \to 0$ faster than $1/z$ (e.g. $D(z) \sim z^{-2}$).

We still need to calculate the residue Res $[\tilde{n}_F(z), z = i\omega_n]$ at the Matsubara frequency to relate the sum over n above with the Matsubara sum.

Problem 2.122   Using the fact that

$$\lim_{z\to a} \frac{z-a}{f(z)} = \frac{1}{f'(a)}$$

for any function $f(z)$ such that $f(a) = 0$, show that

$$\text{Res}\,[\tilde{n}_F(z); i\omega_n] = \lim_{z\to i\omega_n} \tilde{n}_F(z)(z - i\omega_n) = -\frac{1}{\beta\hbar}\,.$$

Therefore,

$$\sum_{n=-\infty}^{\infty} \text{Res}\,[\tilde{n}_F(z); i\omega_n]\, D(i\omega_n)e^{i\omega_n\tau} = -\frac{1}{\beta\hbar} \sum_{n=-\infty}^{\infty} D(i\omega_n)e^{i\omega_n\tau}$$

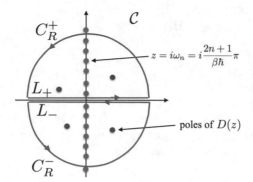

Fig. 2.44 The semicircle C_R^+, traversed in the anti-clockwise direction, and the horizontal line L_+ in the limit $R \to \infty$ encompass poles (if exist) of the function $D(z)$ (purple circles) and the Matsubara frequencies $i\omega_n$ (red) in the upper half of the complex plane. The semicircle C_R^-, also traversed in the anti-clockwise direction, and the horizontal line L_- surround all poles of $D(z)$ and the Matsubara frequencies in the lower half of $\mathcal{C}$

and we finally obtain:

$$\frac{1}{\beta\hbar} \sum_{n=-\infty}^{\infty} D(i\omega_n)e^{i\omega_n\tau} = \sum_i \tilde{n}_F(z_i)\text{Res}[D(z); z_i]e^{z_i\tau} . \tag{2.183}$$

Hence, the infinite sum over Matsubara frequencies is easily calculated as a finite sum over all poles of the function $D(z)$ in $\mathcal{C}$.

Consider now the case when the function $D(z)$ has a branch cut along the real axis. In this case we introduce two semicircle contours C_R^+ and C_R^-, both of the radius R, as illustrated in the Fig. 2.44. The semicircles of the two contours are connected by horizontal lines L_+ and L_- just above and below the real axis, respectively.

We have in the $R \to \infty$ limit:

$$\oint_{C_R^+ + L_+} \tilde{n}_F(z)D(z)e^{z\tau}dz + \oint_{C_R^- + L_-} \tilde{n}_F(z)D(z)e^{z\tau}dz$$

$$= 2\pi i \sum_{i=1}^{N} \text{Res}[D(z); z_i]\tilde{n}_F(z_i)e^{z_i\tau} + 2\pi i \sum_{n=-\infty}^{\infty} \text{Res}[\tilde{n}_F(z); i\omega_n] D(i\omega_n)e^{i\omega_n\tau} .$$

As in the previous case, the integrals over the semicircles tend to zero, $\int_{C_R^+} + \int_{C_R^-} \to 0$, as $R \to \infty$. On L_+ we have $z = \omega + i\delta \, (\delta \to +0)$, and on L_- we have $z = \omega - i\delta$, so that the integration over z on $L_\pm$ can be carried out over the real axis ω:

$$\int_{L_+} + \int_{L_-} = \int_{-\infty}^{\infty} d\omega\, \tilde{n}_F(\omega) D(\omega + i\delta) e^{\omega\tau} + \int_{+\infty}^{-\infty} d\omega\, \tilde{n}_F(\omega) D(\omega - i\delta) e^{\omega\tau}$$

$$= \int_{-\infty}^{\infty} d\omega\, \tilde{n}_F(\omega) \left[D(\omega^+) - D(\omega^-) \right] e^{\omega\tau},$$

where we need to keep the infinitesimal part $i\delta$ only in the function $D(z)$, and $\omega^\pm = \omega \pm i\delta$ as in other cases (the function $\tilde{n}_F(\omega^\pm)$ and the exponential) the limit $\delta \to 0$ can be taken straight away. Other terms are calculated exactly in the same way as in the previous case. We finally obtain:

$$\frac{1}{\beta\hbar} \sum_{n=-\infty}^{\infty} D(i\omega_n) e^{i\omega_n\tau} = \sum_{i=1}^{N} \tilde{n}_F(z_i) \text{Res}[D(z); z_i] e^{z_i\tau}$$

$$- \frac{1}{2\pi i} \int_{-\infty}^{\infty} d\omega\, \tilde{n}_F(\omega)[D(\omega^+) - D(\omega^-)] e^{\omega\tau}.$$

(2.184)

Problem 2.123 Show that

$$\frac{1}{\beta\hbar} \sum_n \frac{1}{i\omega_n - \xi} e^{i\omega_n\tau} = \frac{e^{\xi\tau}}{e^{\beta\hbar\xi} + 1},$$

where $\xi \in \mathcal{C}$.

Problem 2.124 Consider the function

$$D(i\omega_n) = \begin{cases} (i\omega_n - \xi_+)^{-1} & , \text{ if } n \geq 0 \\ (i\omega_n - \xi_-)^{-1} & , \text{ if } n < 0 \end{cases}$$

that makes a jump across the real axis. Here ξ_+ lies somewhere in the upper half of the complex plane and ξ_- in the lower half. Show that in this case

$$\frac{1}{\beta\hbar} \sum_n D(i\omega_n) e^{i\omega_n\tau} = \frac{e^{\xi_+\tau}}{e^{\beta\hbar\xi_+} + 1} + \frac{e^{\xi_-\tau}}{e^{\beta\hbar\xi_-} + 1}$$

$$+ i\frac{\xi_+ - \xi_-}{2\pi} \int_{-\infty}^{\infty} \frac{\tilde{n}_F(\omega)}{(\omega - \xi_+)(\omega - \xi_-)} e^{\omega\tau} d\omega.$$

Problem 2.125 Show that in the case of bosons (the first line in Eq. (2.182)) the formulae (2.183) and (2.184) are still valid if the function $\tilde{n}_F(z)$ is replaced with $\tilde{n}_B(z) = \left[e^{\beta\hbar z} - 1 \right]^{-1}$.

Interestingly, the function $\tilde{n}_F(z)$ is very close to the Fermi-Dirac distribution for fermions, while the function $\tilde{n}_B(z)$ is practically identical to the Bose-Einstein distribution for bosons.

2.9.6 A Beautiful Counting Problem

Here we shall consider a mathematical problem that unlikely has any practical signif-icance. However, its solution is extremely beautiful and hence it is worth reproducing here,[12] especially as it also nicely illustrates the usefulness of the complex calculus.

Let us ask ourselves a question: how many sets of numbers one can select from the integers between 1 and 2000 such that the sum of the numbers in each set be divisible by 5? This looks like a simple problem but it is not as in the sets you may have as many different numbers as you like.

Consider first a simple example of a sequence of five consecutive numbers 1, 2, 3, 4, and 5. Let us construct from these all possible sets in which sum of numbers is divisible by 5. Overall, many sets can be constructed here (see Fig. 2.45 that lists all possible sets), but only some of them satisfy the required condition: (i) 5 sets with a single number, only one (the set $\{5\}$) satisfies our condition; (ii) $\binom{5}{2} = 10$ sets of two numbers, only two sets comply, $\{1, 4\}$ and $\{2, 3\}$; (iii) $\binom{5}{3} = 10$ sets of three numbers, and again only two sets have their numbers summing up to a number divisible by 5, these are $\{1, 4, 5\}$ and $\{2, 3, 5\}$; (iv) there are 5 sets of four numbers, only one set $\{1, 2, 3, 4\}$ is to be selected, and, finally, (v) there is just one set of 5 numbers $\{1, 2, 3, 4, 5\}$, which must also be selected. Including the empty set $\{\}$ as well as containing the sum of 0 (which we consider as divisible by 5 as well), we arrive at $1 + 1 + 2 + 2 + 1 + 1 = 8$ successful sets out of $1 + 5 + 10 + 10 + 5 + 1 = 32$ sets overall.

Of course it would be impossible to repeat this exercise for a sequence of consec-utive 2000 integers, so a clever solution is needed. The idea comes from an auxiliary function[13]

$$f(x) = \prod_{n=1}^{2000}(1 + x^n) = (1 + x)(1 + x^2)\cdots(1 + x^{2000}). \tag{2.185}$$

---

[12] This problem is from the book "T. Andreescu and Z. Feng, "102 Combinatorial Problems. From the Training of the USA IMO team", Birkhauser, Boston, 2003 (Chap. 2, problem 10). There is also an absolutely fantastic and enjoyable presentation of the problem and its solution at https://www.youtube.com/watch?v=bOXCLR3Wric, from which I learned about it.

[13] It may also be called "generating" function as it "generates" the required number of sets.

Fig. 2.45 32 sets of integer numbers between 1 and 5. In 8 sets highlighted in red the sum of their numbers is divisible by 5

No numbers (empty): {}

1 number in each set (5 sets):   {1}{2}{3}{4}{5}

2 numbers in each set (10 sets):
{1,2}{1,3}{1,4}{1,5}{2,3}{2,4}{2,5}{3,4}{3,5}{4,5}

3 numbers in each set (10 sets):
{1,2,3}{1,2,4}{1,2,5}{1,3,4}{1,3,5}{1,4,5}{2,3,4}{2,3,5}{2,4,5}{3,4,5}

4 numbers in each set (5 sets):
{2,3,4,5}{1,3,4,5}{1,2,4,5}{1,2,3,5}{1,2,3,4}

5 numbers in each set (1 set): {1,2,3,4,5}

The trick is that if we multiply these factors, then we shall have a polynomial in x,

$$f(x) = \sum_{n=0}^{M} c_n x^n \qquad (2.186)$$

of the order

$$M = 1 + 2 + 3 + \cdots + 2000 = \frac{1}{2}(1 + 2000)\, 2000 = 2001000 \,.$$

There will be many terms contributing to a particular term x^n, each such a term is obtained by summing up the powers of x in the appropriate factors in Eq. (2.185) when we multiply them to make up the power n; many such combinations may exist and these are given by the expansion coefficients c_n. To illustrate this point, consider simply the polynomial for $M = 6$ as an example:

$$(1 + x^1)(1 + x^2)(1 + x^3) = 1 + x^1 + x^2 + \underbrace{x^{1+2} + x^3}_{c_3=2} + x^{1+3} + x^{2+3} + x^{1+2+3} \,.$$

It is seen that x^3 appears twice due to only two sets $\{1, 2\}$ and $\{3\}$ possible that sum up to 3; all other terms appear only once. This is what you would expect from the original sequence 1, 2, 3.

Therefore, it becomes clear that the coefficients c_n in Eq. (2.186) give the number of all possible sets of different integers selected from the original sequence whose sum is precisely n. Hence, if we would like to know the number of all sets in which the number's sum is divisible by 5 (including the empty set), we need to find the sum

$$S = c_0 + c_5 + c_{10} + \cdots + c_{2000} \,. \qquad (2.187)$$

We have included the coefficient c_0 as well here, it corresponds to the empty set.

This is the first nice trick. Here comes the second one. To determine this quantity S, let us consider all five solutions of the equation $z^5 = 1$. These are given by five roots of one:

Table 2.1 Table of powers of the fifth roots of one

	z_α^1	z_α^2	z_α^3	z_α^4
z_0	z_0	z_0	z_0	z_0
z_1	z_1	z_2	z_3	z_4
z_2	z_2	z_4	z_1	z_3
z_3	z_3	z_1	z_4	z_2
z_4	z_4	z_3	z_2	z_1

$$z_0 = 1, \quad z_1 = e^{2\pi i/5}, \quad z_2 = e^{4\pi i/5}, \quad z_3 = e^{6\pi i/5}, \quad z_4 = e^{8\pi i/5}.$$

The powers of these roots are easily calculated and are given in Table 2.1. What is worth noticing is that powers z_α^n of the roots $\alpha = 1, 2, 3, 4$ with $n = 1, 2, 3, 4$ give the same roots z_α, but in different order. If we consider a higher power $n = 5k + l$ of a root with $k = 1, 2, \ldots$ and $l = 0, 1, 2, 3, 4$, then clearly

$$z_\alpha^{5k+l} = \left(z_\alpha^5\right)^k z_\alpha^l = z_\alpha^l,$$

which is some other root $z_{\alpha'}$. Hence, any integer power of a fifth root of one is one of the fifth roots of one, and if we calculate the powers of a given root one after another, we have the roots of one repeating themselves periodically, each root appearing exactly once within each such a period. For instance, for the root z_2, we have:

$$\overbrace{z_2, \; z_2^2, \; z_2^3, \; z_2^4, \; z_2^5,}^{\text{period}} \; \overbrace{z_2^6, \; z_2^7, \; z_2^8, \; z_2^9, \; z_2^{10},}^{\text{period}} \; \overbrace{z_2^{11}, \; z_2^{12}, \; z_2^{13}, \; z_2^{14}, \; z_2^{15},}^{\text{period}} \; z_2^{16}, \ldots$$

$$\underbrace{z_4 \quad z_1 \quad z_3 \quad z_0}_{} \quad z_2 \quad z_4 \quad z_1 \quad z_3 \quad z_0 \quad z_2 \quad z_4 \quad z_1 \quad z_3 \quad z_0 \quad z_2$$

This is a very useful observation that we shall find essential in the following.

Another important observation comes from the fact that the sum of the five roots amounts to zero: $z_0 + z_1 + z_2 + z_3 + z_4 = 0$ (Problems 2.18 and 2.34).

Next, we shall calculate $f(z_\alpha)$ of a particular root z_α by rearranging the sum so that we would sum by five consecutive terms:

$$f(z_\alpha) = \sum_{n=0}^{400} c_{5n} z_\alpha^{5n} + \sum_{n=0}^{399} \left[c_{5n+1} z_\alpha^{5n+1} + c_{5n+2} z_\alpha^{5n+2} + c_{5n+3} z_\alpha^{5n+3} + c_{5n+4} z_\alpha^{5n+4}\right]$$

$$= \sum_{n=0}^{400} c_{5n} \underbrace{z_\alpha^{5n}}_{=1} + \sum_{n=0}^{399} \underbrace{z_\alpha^{5n}}_{=1} \left[c_{5n+1} z_\alpha^1 + c_{5n+2} z_\alpha^2 + c_{5n+3} z_\alpha^3 + c_{5n+4} z_\alpha^4\right]$$

$$= \sum_{n=0}^{400} c_{5n} + \sum_{n=0}^{399} \left[c_{5n+1} z_\alpha^1 + c_{5n+2} z_\alpha^2 + c_{5n+3} z_\alpha^3 + c_{5n+4} z_\alpha^4\right],$$

and then consider adding together these expressions for all five roots of one:

$$\sum_{\alpha=0}^{4} f(z_\alpha) = 5 \sum_{n=0}^{400} c_{5n} + \sum_{n=0}^{399} \left[c_{5n+1} \sum_{\alpha=0}^{4} z_\alpha^1 + c_{5n+2} \sum_{\alpha=0}^{4} z_\alpha^2 + c_{5n+3} \sum_{\alpha=0}^{4} z_\alpha^3 + c_{5n+4} \sum_{\alpha=0}^{4} z_\alpha^4 \right].$$

The sum of the roots serving as a coefficient to c_{5n+1} is zero; the sum of the roots squared standing as a coefficient to c_{5n+2} is also zero, since the squares of the roots are the same roots but repeated in a different order; the same is true for the other two terms. Hence,

$$\sum_{n=0}^{400} c_{5n} = \frac{1}{5} \sum_{\alpha=0}^{4} f(z_\alpha), \tag{2.188}$$

which is exactly what we are to calculate, S, cf. Eq. (2.187) Hence, we conclude, that what is required is to calculate the function $f(x)$ and all five roots of one and sum up the results.

We shall first calculate $f(z_\alpha)$ for $\alpha = 0$. In this case $z_0 = 1$ and hence

$$f(z_0) = \prod_{n=1}^{2000} (1 + z_0^n) = \prod_{n=1}^{2000} (1 + 1) = 2^{2000}.$$

Next, let us consider the other four roots z_α corresponding to $\alpha \neq 0$. To this end, we use the definition of the function (2.185) again, bit this time we make up the whole product from the portions of five consecutive terms:

$$f(z_\alpha) = \prod_{n=0}^{399} \left[\left(1 + z_\alpha^{5n+1}\right) \left(1 + z_\alpha^{5n+2}\right) \left(1 + z_\alpha^{5n+3}\right) \left(1 + z_\alpha^{5n+4}\right) \left(1 + z_\alpha^{5n+5}\right) \right]$$

$$= \prod_{n=0}^{399} \left[\left(1 + z_\alpha^1\right) \left(1 + z_\alpha^2\right) \left(1 + z_\alpha^3\right) \left(1 + z_\alpha^4\right) \left(1 + z_\alpha^5\right) \right].$$

The powers of the given root form all the roots without repetitions, so the product of five terms behind the product sign are all exactly the same and we have $2000/5 = 400$ such products:

$$f(z_\alpha) = \left[(1 + z_0)(1 + z_1)(1 + z_2)(1 + z_3)(1 + z_4) \right]^{400}, \quad \alpha \neq 0. \tag{2.189}$$

Notice, that this result is valid for any $\alpha = 1, 2, 3, 4$ and it does not depend on the particular root.

In principle, the product inside the square brackets can be calculated by brute force, simply by substituting the expressions of the roots. However, there exists a much more elegant way! As our next trick, consider the polynomial $P_5(z) = z^5 - 1$. The roots of this polynomial are exactly all five roots of one, and hence we can formally write:

$$P_5(z) = (z - z_0)(z - z_1)(z - z_2)(z - z_3)(z - z_4) \equiv z^5 - 1.$$

What we actually need in Eq. (2.189) is

$$P_5(-1) = -(1 + z_0)(1 + z_1)(1 + z_2)(1 + z_3)(1 + z_4),$$

so that the required product is simply equal to $-P_5(-1) = -(-1)^5 + 1 = 2$. Hence, for any $\alpha \neq 0$ we get $f(z_\alpha) = 2^{400}$. Therefore, we finally obtain from Eq. (2.188):

$$S = \frac{1}{5}\left(2^{2000} + 4 \cdot 2^{400}\right).$$

Problem 2.126 Following this method show that the number of sets in which the sum of their numbers is divisible by 5 and selected from $5N$ consecutive integer numbers $1, 2, \ldots, 5N$ is given by $S = \frac{1}{5}\left[2^{5N} + 4 \cdot 2^N\right]$.

And indeed, in our first example at the very beginning of this section, we had $N = 1$, so that that formula gives $S = 8$, as required; in our long example $N = 400$ and we again recover the correct result.

Problem 2.127 Consider a sequence of all integer numbers between 1 and $4N$. Show that the total number of sets in which sum of their numbers is divisible by 4 is $S = 2^{4N}/4$.

Problem 2.128 Consider a general problem now: in a sequence of all integer numbers between 1 and pN (p is an integer), show that the number of sets in which the sum of their numbers is divisible by p is:

$$S = \frac{1}{p}\begin{cases} 2^{pN} & \text{, if } p \text{ is even} \\ \left[2^{pN} + (p-1)2^N\right] & \text{, if } p \text{ is odd} \end{cases}.$$

2.9.7 Landau-Zener Problem

Lev Landau, Clarence Zener, Ernst Stueckelberg and Ettore Majorana independently considered a transition in a quantum two-level system in which the diagonal elements of its Hamiltonian depend linearly on time. It was discovered that if the system is prepared at time $t = -\infty$ in one of the two states, an analytical formula can be derived for the probability of that state to be occupied at $t = +\infty$. We shall follow here a very clever derivation by Alberto G. Rojo[14] because it is really a very beautiful maths that in addition illustrates nicely the power of the complex calculus.

Consider a two-level system with the matrix time-dependent Hamiltonian

$$\mathbf{H}(t) = \begin{pmatrix} \lambda t & \Gamma \\ \Gamma & -\lambda t \end{pmatrix}, \tag{2.190}$$

in which its diagonal elements (energy levels of the two-level system), being initially (at $t = -\infty$) infinitely far away from each other, cross at $t = 0$ and then separate out indefinitely again to be infinitely away from each other in the future; Γ is the off-diagonal element of the Hamiltonian coupling the two states. The state vector $\psi(t) = \begin{pmatrix} a(t) \\ b(t) \end{pmatrix}$ of the quantum system satisfies the Schödinger equation

$$i\hbar \frac{\partial}{\partial t} \begin{pmatrix} a(t) \\ b(t) \end{pmatrix} = \mathbf{H}(t) \begin{pmatrix} a(t) \\ b(t) \end{pmatrix}, \tag{2.191}$$

that we would like to solve. We shall assume that in the remote past the system was in the lowest energy state a, i.e. $a(-\infty) = 1$ and $b(-\infty) = 0$. Note that the eigenstates of the Hamiltonian at each time t are $\epsilon_{\pm}(t) = \pm\sqrt{\Gamma^2 + \lambda^2 t^2}$. These eigenstates are shown in Fig. 2.46. We assume that initially the system was prepared in its lowest energy state (with the minus) at $t = -\infty$. If there was no coupling between the two states, $\Gamma = 0$, the system would at $t = +\infty$ remain at the same eigenstate, which would correspond in this case to the upper energy state. The non-zero coupling leads to a mixture of the states near $t = 0$, but at $t = \pm\infty$ the eigenstates remain largely unaffected. Therefore, the first eigenstate, corresponding to the component $a(t)$ of $\psi(t)$, is the lower one at $t = -\infty$ and the upper one at $t = +\infty$ (the orange curves in Fig. 2.46), while the second eigenstate that is the upper state initially and the lower state in the remote future corresponds to the second component $b(t)$ of $\psi(t)$ and is shown by the blue curves in the Figure. If the system evolved adiabatically slowly, then it would remain in the same state all the time, i.e. being started on the orange curve it would remain staying on it. The question we ask is what is the probability for that to happen in reality when the system is evolved in time by means of the Schödinger equation of quantum mechanics.

To approach the problem, we shall benefit from the method developed in Sect. 1.19.6 and shall split H into two parts:

$$\mathbf{H}_0(t) = \begin{pmatrix} \lambda t & 0 \\ 0 & -\lambda t \end{pmatrix} \text{ and } \mathbf{V} = \begin{pmatrix} 0 & \Gamma \\ \Gamma & 0 \end{pmatrix}.$$

---

[14] Alberto G. Rojo, Matrix exponential solution of the Landau-Zener problem, arXiv:1004.2914 (2010). In our treatment a number of errors/typos of the original derivation are also corrected.

Fig. 2.46 Evolution of
eigenstates $\epsilon_{1,2}$ of the
Landau-Zener model in time

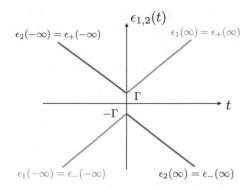

The Schödinger equation with $\mathbf{H}_0$ (instead of $\mathbf{H}$) can be easily solved as the two
levels become completely decoupled: $i\hbar\dot{a}_0(t) = \lambda t a_0(t)$ and $i\hbar\dot{b}_0(t) = -\lambda t b_0(t)$.
Solving two independent DEs, we obtain that

$$a_0(t) \sim \exp\left(-i\frac{\lambda t^2}{2\hbar}\right) \quad \text{and} \quad b_0(t) \sim \exp\left(i\frac{\lambda t^2}{2\hbar}\right).$$

Hence, the state vector of the decoupled system evolves in time according to

$$\begin{pmatrix} a_0(t) \\ b_0(t) \end{pmatrix} = \mathbf{U}(t)\begin{pmatrix} a_0(-\infty) \\ b_0(-\infty) \end{pmatrix},$$

where

$$\mathbf{U}(t) = \begin{pmatrix} e^{-i\lambda t^2/2\hbar} & 0 \\ 0 & e^{i\lambda t^2/2\hbar} \end{pmatrix}, \tag{2.192}$$

is the corresponding evolution operator satisfying $\mathbf{U}^\dagger = \mathbf{U}^* = \mathbf{U}^{-1}$ and the equation

$$i\hbar\dot{\mathbf{U}}(t) = \mathbf{H}_0\mathbf{U}(t). \tag{2.193}$$

Having this in mind, we shall seek the solution of the full equation with $\mathbf{H} = \mathbf{H}_0 + \mathbf{V}$
using the substitution

$$\psi(t) \equiv \begin{pmatrix} a(t) \\ b(t) \end{pmatrix} = \mathbf{U}(t)\begin{pmatrix} a_1(t) \\ b_1(t) \end{pmatrix} = \mathbf{U}(t)\psi_1(t). \tag{2.194}$$

Substituting this into Eq. (2.191), we obtain:

$$i\hbar\dot{\mathbf{U}}\psi_1(t) + \mathbf{U}\,i\hbar\frac{\partial\psi_1(t)}{\partial t} = \mathbf{H}_0\mathbf{U}\psi_1(t) + \mathbf{V}\mathbf{U}\psi_1(t),$$

which, after cancelling out the first terms on both sides because of Eq. (2.193), and multiplying from the left with $\mathbf{U}^\dagger$, results in:

$$i\hbar\frac{\partial\psi_1(t)}{\partial t} = \mathbf{W}(t)\psi_1(t) \quad \text{or} \quad i\hbar\frac{\partial}{\partial t}\begin{pmatrix} a_1(t) \\ b_1(t) \end{pmatrix} = \mathbf{W}(t)\begin{pmatrix} a_1(t) \\ b_1(t) \end{pmatrix}, \quad (2.195)$$

where, *cf.* Eq. (1.194),

$$\mathbf{W}(t) = \mathbf{U}^{-1}\mathbf{V}\mathbf{U} = \mathbf{U}^\dagger\mathbf{V}\mathbf{U} = \begin{pmatrix} 0 & \Gamma e^{i\mu t^2} \\ \Gamma e^{-i\mu t^2} & 0 \end{pmatrix} \quad (2.196)$$

is the operator responsible for the time evolution of the new state vector $\psi_1(t)$, and $\mu = \lambda/\hbar$ is a shorthand.

Let us determine the initial conditions for $\psi_1(t)$ at $t = -\infty$. We have:

$$\psi_1(-\infty) = \begin{pmatrix} a_1(-\infty) \\ b_1(-\infty) \end{pmatrix} = \mathbf{U}^\dagger(-\infty)\psi(-\infty) = \mathbf{U}^\dagger(-\infty)\begin{pmatrix} 1 \\ 0 \end{pmatrix}$$
$$= \begin{pmatrix} a(-\infty) \\ 0 \end{pmatrix},$$

where $a(-\infty) = \left(e^{i\mu t^2/2}\right)_{t=-\infty} = e^{i\phi}$ is a phase factor corresponding to the infinitely remote past. Of course, it is ill-defined at $-\infty$, but what is important for us here is only that $|a(-\infty)| = 1$, which seems to be the case.

We shall next solve Eq. (2.195) by iterations following the method of Sect. 1.19.6:

$$\begin{pmatrix} a_1(t) \\ b_1(t) \end{pmatrix} = \left[\mathbf{E} + \sum_{m=1}^{\infty} \left(-\frac{i}{\hbar}\right)^m \right.$$
$$\left. \times \int_{-\infty}^{t} dt_1 \dots \int_{-\infty}^{t_{m-1}} dt_n \mathbf{W}(t_1) \dots \mathbf{W}(t_m) \right] \begin{pmatrix} a(-\infty) \\ 0 \end{pmatrix}.$$

The order of the matrices $\mathbf{W}$ in the formula above is important as they do not commute at different times, i.e. $\mathbf{W}(t_k)\mathbf{W}(t_{k'}) \neq \mathbf{W}(t_{k'})\mathbf{W}(t_k)$ (check!). Also, since $\mathbf{W}$ only has off-diagonal elements, there is a very peculiar structure of the product of the matrices. Indeed, let us consider a few initial cases corresponding to $m = 2, 3, 4$:

$$\mathbf{W}(t_1)\mathbf{W}(t_2) = \begin{pmatrix} \Gamma^2 e^{i\mu t_1^2} e^{-i\mu t_2^2} & 0 \\ 0 & \Gamma^2 e^{-i\mu t_1^2} e^{i\mu t_2^2} \end{pmatrix},$$

$$\mathbf{W}(t_1)\mathbf{W}(t_2)\mathbf{W}(t_3) = \begin{pmatrix} 0 & \Gamma^3 e^{i\mu t_1^2} e^{-i\mu t_2^2} e^{i\mu t_3^2} \\ \Gamma^3 e^{-i\mu t_1^2} e^{i\mu t_2^2} e^{-i\mu t_3^2} & 0 \end{pmatrix},$$

$$\mathbf{W}(t_1)\mathbf{W}(t_2)\mathbf{W}(t_3)\mathbf{W}(t_4) = \begin{pmatrix} \Gamma^4 e^{i\mu t_1^2} e^{-i\mu t_2^2} e^{i\mu t_3^2} e^{-i\mu t_4^2} & 0 \\ 0 & \Gamma^4 e^{-i\mu t_1^2} e^{i\mu t_2^2} e^{-i\mu t_3^2} e^{i\mu t_4^2} \end{pmatrix}.$$

It is easy to see that the products of an even number of $\mathbf{W}$ matrices result in a matrix with only diagonal elements, while the product of an odd number of $\mathbf{W}$ matrices always yields a matrix only with off-diagonal elements. This is important, since if we multiply a matrix with only off-diagonal elements with the initial state vector $\begin{pmatrix} a(-\infty) \\ 0 \end{pmatrix}$, it will result in the zero contribution for the component $a_1(t)$ that we are interested in, while the products of an even number of $\mathbf{W}$ matrices contribute directly to that component, hence only even terms with $m = 2n$ need to be considered in the sum:

$$\frac{a_1(\infty)}{a(-\infty)} = 1 - \left(\frac{\Gamma}{\hbar}\right)^2 \int_{-\infty}^{\infty} dt_1 \int_{-\infty}^{t_1} dt_2 \, e^{i\mu t_1^2} e^{-i\mu t_2^2}$$

$$+ \left(\frac{\Gamma}{\hbar}\right)^4 \int_{-\infty}^{\infty} dt_1 \int_{-\infty}^{t_1} dt_2 \int_{-\infty}^{t_2} dt_3 \int_{-\infty}^{t_3} dt_4 \, e^{i\mu t_1^2} e^{-i\mu t_2^2} e^{i\mu t_3^2} e^{-i\mu t_4^2}$$

$$+ \dots$$

$$= 1 + \sum_{n=1}^{\infty} \left(-\frac{\Gamma^2}{\hbar^2}\right)^n \int_{-\infty}^{\infty} dt_1 \dots \int_{-\infty}^{t_{2n-1}} dt_{2n} \, e^{i\mu t_1^2} e^{-i\mu t_2^2} \dots e^{i\mu t_{2n-1}^2} e^{-i\mu t_{2n}^2}.$$

We shall next change variables $t_k \to x_k = \sqrt{\mu} t_k$, which enables us to write:

$$\frac{a_1(\infty)}{a(-\infty)} = \sum_{n=0}^{\infty} (-\gamma)^n T_{2n}, \tag{2.197}$$

where $\gamma = \Gamma^2/(\mu\hbar^2) = \Gamma^2/(\lambda\hbar)$ and

$$T_{2n} = \int_{-\infty}^{\infty} dx_1 \int_{-\infty}^{x_1} dx_2 \dots$$

$$\dots \int_{-\infty}^{x_{2n-2}} dx_{2n-1} \int_{-\infty}^{x_{2n-1}} dx_{2n} \, e^{ix_1^2} e^{-ix_2^2} \dots e^{ix_{2n-1}^2} e^{-ix_{2n}^2}. \tag{2.198}$$

In this multiple integral the upper limits of the integrals, starting from the second one, correspond to the previous integration variable which makes the calculation non-trivial. This kind of integration could have been possible to perform analytically, however, would the integrand to contain a product of identical function. This is illustrated by the following Problem.

Problem 2.129 Consider $(n = 1, 2, 3, \dots)$

$$F_n(t) = \int^t dt_1 \int^{t_1} dt_2 \dots \int^{t_{n-1}} dt_n \, f(t_1) f(t_2) \dots f(t_n)$$

with all the integrals having the same bottom limit. Noting that $F_1'(t) = f(t)$ and F_n at the bottom limit equals to zero, so that

$$F_2(t) = \int^t dt_1 \, f(t) F_1(t_1) = \int^{F_1(t)} F_1 d F_1 = \frac{1}{2} F_1(t)^2 \,,$$

show using induction that

$$F_n(t) = \frac{1}{n!} F_1(t)^n = \frac{1}{n!} \left(\int^t dt_1 \, f(t_1) \right)^n .$$

Unfortunately, the exponential functions in the integrand in Eq. (2.198) have alternating signs and hence the functions are not the same, and we cannot use the beautiful result of Problem 2.129. However, a few very nice tricks can help.

First of all, we can formally change all the upper limits in T_{2n} to $+\infty$ by introducing the Heaviside functions as follows:

$$T_{2n} = \int_{-\infty}^{\infty} dx_1 \int_{-\infty}^{\infty} dx_2 \dots \int_{-\infty}^{\infty} dx_{2n-1} \int_{-\infty}^{\infty} dx_{2n}$$
$$\times \, e^{ix_1^2} H(x_1 - x_2) e^{-ix_2^2} H(x_2 - x_3) \dots e^{ix_{2n-1}^2} H(x_{2n-1} - x_{2n}) e^{-ix_{2n}^2} .$$

Next, we shall use the integral representation of the Heaviside function (2.124):

$$H(x) = \frac{1}{2\pi i} \int_{-\infty}^{\infty} \frac{d\omega}{\omega^-} e^{i\omega x} \,,$$

where $\omega^- = \omega - i\delta$. Replacing all the $2n - 1$ Heaviside functions in T_{2n} by their integral representations, we obtain:

$$T_{2n} = \frac{1}{(2\pi i)^{2n-1}} \int \frac{d\omega_1 \dots d\omega_{2n-1}}{\omega_1^- \dots \omega_{2n-1}^-}$$
$$\times \int_{-\infty}^{\infty} dx_1 \, e^{ix_1^2 + i\omega_1 x_1} \int_{-\infty}^{\infty} dx_2 \, e^{-ix_2^2 + i(\omega_2 - \omega_1)x_2} \dots$$
$$\times \int_{-\infty}^{\infty} dx_{2n-1} \, e^{ix_{2n-1}^2 + i(\omega_{2n-1} - \omega_{2n-2})x_{2n-1}}$$
$$\times \int_{-\infty}^{\infty} dx_{2n} \, e^{-ix_{2n}^2 - i\omega_{2n-1} x_{2n}} .$$

The x-integrations can be performed exactly by relating them to the Fresnel integrals (2.74) and (2.75):

$$\int_{-\infty}^{\infty} dx\, e^{ix^2+i\omega x} = \int_{-\infty}^{\infty} dx\, e^{i(x+\omega/2)^2} e^{-i\omega^2/4} = \left(\int_{-\infty}^{\infty} dx\, e^{ix^2}\right) e^{-i\omega^2/4}$$
$$= \sqrt{\pi}\, e^{-i\omega^2/4} e^{i\pi/4}$$

and

$$\int_{-\infty}^{\infty} dx\, e^{-ix^2+i\omega x} = \sqrt{\pi}\, e^{i\omega^2/4} e^{-i\pi/4} .$$

Two observations can be made: if in the exponent in the integrand we have $+ix^2$, then the exponent with $i\omega^2/4$ has the minus sign, while the exponent with $i\pi/4$ has the plus sign; inversely, if the exponent in the integrand is $-ix^2$, then the signs in the two exponents after the integration over x change to the plus and the minus, respectively. Substitution of the results of the x-integrations into the expression for T_{2n} leads then to the following consequences: because we have all the exponents $e^{\pm ix^2}$ appearing in pairs, the terms like $e^{\pm i\omega_k^2/4}$ and $e^{\pm i\pi/4}$ cancel out. Therefore, we obtain:

$$T_{2n} = \frac{\pi^n}{(2\pi i)^{2n-1}} \int_{-\infty}^{\infty} \frac{d\omega_1 \dots d\omega_{2n-1}}{\omega_1^- \dots \omega_{2n-1}^-} e^{i\omega_2(\omega_3-\omega_1)/2} e^{i\omega_4(\omega_5-\omega_3)/2}$$
$$\dots \times e^{i\omega_{2n-2}(\omega_{2n-1}-\omega_{2n-3})/2} .$$

Let us integrate only over the ω_{2k} with *even* indices ($k = 1, 2, \dots, m-1$), this gives in each case

$$\frac{1}{2\pi i} \int_{-\infty}^{\infty} \frac{d\omega_{2k}}{\omega_{2k}^-} e^{i\omega_{2k}(\omega_{2k+1}-\omega_{2k-1})/2} = H\left(\frac{\omega_{2k+1}-\omega_{2k-1}}{2}\right)$$
$$\equiv H(\omega_{2k+1}-\omega_{2k-1})$$

as it is again the Heaviside function, Eq. (2.124). After this calculation, we have the following product of the Heaviside functions in the integrand:

$$H(\omega_{2n-1}-\omega_{2n-3})\, H(\omega_{2n-3}-\omega_{2n-5}) \dots H(\omega_5-\omega_3)\, H(\omega_3-\omega_1) ,$$

that corresponds to the following order of the integration variables,

$$\omega_{2n-1} > \omega_{2n-3} > \omega_{2n-5} > \dots > \omega_5 > \omega_3 > \omega_1 ,$$

in the remaining integrations over ω_{2k-1} with the odd indices ($k = 1, 2, \dots, n$). Correspondingly we are able to rewrite T_{2n} as the following sequence of coupled integrals over the remaining odd ω-variables:

$$T_{2n} = \frac{\pi^n}{(2\pi i)^n} \int_{-\infty}^{\infty} \frac{d\omega_{2n-1}}{\omega_{2n-1}^-} \int_{-\infty}^{\omega_{2n-1}^-} \frac{d\omega_{2n-3}}{\omega_{2n-3}^-} \cdots \int_{-\infty}^{\omega_5} \frac{d\omega_3}{\omega_3^-} \int_{-\infty}^{\omega_3} \frac{d\omega_1}{\omega_1^-} .$$

This time we do have a sequence of coupled integrals with identical functions! Therefore, we can use the result of Problem 2.129 to obtain:

$$T_{2n} = \frac{\pi^n}{(2\pi i)^n} \frac{1}{n!} \left(\int_{-\infty}^{\infty} \frac{d w}{w - i\delta} \right)^n = \frac{\pi^n}{(2\pi i)^n} \frac{(i\pi)^n}{n!} = \frac{1}{n!} \left(\frac{\pi}{2} \right)^n ,$$

where we have used formula (2.72) for the single ω-integral. Concluding,

$$\frac{a_1(\infty)}{a(-\infty)} = \sum_{n=0}^{\infty} (-\gamma)^n \frac{1}{n!} \left(\frac{\pi}{2} \right)^n = \sum_{n=0}^{\infty} \frac{(-\gamma\pi/2)^n}{n!} = e^{-\gamma\pi/2}$$

$$= \exp\left(-\frac{\Gamma^2 \pi}{2\lambda\hbar} \right) .$$

The component $a(t)$ we are actually interested in is related to $a_1(t)$ via a phase factor through Eq. (2.194):

$$a(t) = (\mathbf{U}(t))_{11}\, a_1(t) = e^{-i\lambda t^2/2\hbar} a_1(t) ,$$

the phase factor is however unimportant as having the absolute value of one. Correspondingly, the probability to find the system in the state a in the remote future is

$$P_a(\infty) = |a(\infty)|^2 = |a_1(\infty)|^2 = \exp\left(-\frac{\Gamma^2 \pi}{\lambda\hbar} \right) |a(-\infty)|^2$$

$$= \exp\left(-\frac{\Gamma^2 \pi}{\lambda\hbar} \right) .$$

The obtained expression has 0a clear physical meaning, and having obtained the exact solution we are able to answer the questions posted at the beginning. The probability to remain in the initial state (the orange curves in Fig. 2.46) is 1, as expected, if there is no coupling between the two states, $\Gamma = 0$. However, if the coupling exists, then such a probability becomes exponentially small with the square of the coupling in the exponent, i.e. the system would most likely to appear in the other state, the probability of this being $P_b(\infty) = 1 - P_a(\infty)$: the system would "switch" to the lower blue curve in Fig. 2.46 in the remote future.

I hope every reader would agree that this indeed has been a very beautiful and intricate calculation! And I sincerely hope you have enjoyed various tricks that have been used to obtain it!

Chapter 3
Fourier Series

In Sect. I.7.2[1] of the first book we investigated in detail how an arbitrary function $f(x)$ can be expanded into a functional series in terms of functions $a_n(x)$ with $n = 0, 1, 2, \ldots$, i.e.

$$f(x) = f_0 a_0(x) + f_1 a_1(x) + \cdots = \sum_{n=0}^{\infty} f_n a_n(x) . \tag{3.1}$$

There are many applications in which one has to solve ordinary or partial differential equation with respect to functions which are *periodic*:

$$f(x + T) = f(x) . \tag{3.2}$$

Here $T = 2l$ is the period and l is the corresponding half-period. Expanding the function $f(x)$ we seek in an appropriate series with respect to functions $a_n(x)$, which are also periodic with the same period, may help enormously when solving differential equations. These expansions are called Fourier series. The Fourier series also provides a foundation for the so-called *Fourier transform*, to be considered in Chap. 5, which is an analogous formalism for *non-periodic functions*.

In this chapter we shall consider a specific case when a set of functions $\{a_n(x)\}$ consists of *orthogonal trigonometric functions*. Only briefly we shall dwell on a more general case. The definition and the properties of the Fourier series will be thoroughly considered. The chapter will conclude with some applications of the method in physics.

---

[1] In the following, references to the first volume of this course (L. Kantorovich, Mathematics for natural scientists: fundamentals and basics, 2nd Edition, Springer, 2022) will be made by appending the Roman number I in front of the reference, e.g. Sect. I.1.8 or Eq. (I.5.18) refer to Sect. 1.8 and Eq. (5.18) of the first volume, respectively.

© The Author(s), under exclusive license to Springer Nature Switzerland AG 2024
L. Kantorovich, *Mathematics for Natural Scientists II*, Undergraduate Lecture
Notes in Physics, https://doi.org/10.1007/978-3-031-46320-4_3

3.1   Trigonometric Series: An Intuitive Approach

Consider functions

$$\phi_n(x) = \cos \frac{n\pi x}{l} \text{ and } \psi_n(x) = \sin \frac{n\pi x}{l} , \tag{3.3}$$

where $n = 0, 1, 2, \ldots$. As can easily be seen, they have the same periodicity T for *any* n, i.e.

$$\cos \frac{n\pi(x + 2l)}{l} = \cos \left(\frac{n\pi x}{l} + 2\pi n \right) = \cos \frac{n\pi x}{l}$$

and

$$\sin \frac{n\pi(x + 2l)}{l} = \sin \left(\frac{n\pi x}{l} + 2\pi n \right) = \sin \frac{n\pi x}{l} .$$

What we would like to do is to understand whether it is possible to express $f(x)$ as a linear combination of all these functions for all possible values of n from 0 to ∞. We shall start our discussion by showing that the functions (3.3) have a very simple and important property. Namely, they satisfy the following identities for any $n \neq m$:

$$\int_{-l}^{l} \phi_n(x)\phi_m(x)dx = 0 \text{ and } \int_{-l}^{l} \psi_n(x)\psi_m(x)dx = 0 , \tag{3.4}$$

and also for any n and m (including the case of $n = m$):

$$\int_{-l}^{l} \phi_n(x)\psi_m(x)dx = 0 . \tag{3.5}$$

Indeed, let us first prove Eq. (3.4). Using trigonometric identities from Sect. I.2.3.8, one can write

$$\phi_n(x)\phi_m(x) = \cos \frac{n\pi x}{l} \cos \frac{m\pi x}{l}$$

$$= \frac{1}{2} \left\{ \cos \frac{(n + m)\pi x}{l} + \cos \frac{(n - m)\pi x}{l} \right\} .$$

Note that for any different n and m, the integer numbers $k = n \pm m$ are never equal to zero. But then for any $k \neq 0$:

$$\int_{-l}^{l} \cos \frac{k\pi x}{l} dx = \frac{l}{k\pi} \sin \frac{k\pi x}{l} \Big|_{-l}^{l} = \frac{l}{k\pi} [\sin k\pi - \sin (-k\pi)] = 0 ,$$

so that the first integral in Eq. (3.4) is zero.

Problem 3.1 Prove the second identity in Eq. (3.4).

Problem 3.2 Prove Eq. (3.5).

Consider now similar integrals between two identical functions:

$$\int_{-l}^{l} \phi_n(x)^2 dx = \int_{-l}^{l} \cos^2\left(\frac{n\pi x}{l}\right) dx = \frac{1}{2}\int_{-l}^{l}\left\{1 + \cos\left(\frac{2n\pi x}{l}\right)\right\} dx$$

$$= \frac{1}{2}\int_{-l}^{l} dx = l\,,$$

$$\int_{-l}^{l} \psi_n(x)^2 dx = \int_{-l}^{l} \sin^2\left(\frac{n\pi x}{l}\right) dx = \frac{1}{2}\int_{-l}^{l}\left\{1 - \cos\left(\frac{2n\pi x}{l}\right)\right\} dx$$

$$= \frac{1}{2}\int_{-l}^{l} dx = l\,.$$

The relations we have found can now be conveniently rewritten using the Kronecker symbol δ_{nm} which, we recall, is by definition equal to zero if $n \neq m$ and to unity if $n = m$. Then we can write

$$\int_{-l}^{l} \phi_n(x)\phi_m(x)dx = l\delta_{nm} \text{ and } \int_{-l}^{l} \psi_n(x)\psi_m(x)dx = l\delta_{nm}\,. \tag{3.6}$$

Thus, we find that the integral between $-l$ and l (the interval of periodicity of $f(x)$) of a product of any two *different* functions taken from the set $\{\phi_n, \psi_n\}$ of Eq. (3.3) is always equal to zero. We conclude then that these functions are *orthogonal*, i.e. they form an *orthogonal set* of functions. We shall have a more detailed look at an expansion via orthogonal set of functions in Sect. 3.9.3.

Problem 3.3 *Prove that Eq. (3.6) are valid also for the integral limits $-l + c$ and $l + c$ with c being any real number*. This means that the integrals can be taken between any two limits x_1 and x_2, as long as the difference between them $x_2 - x_1 = 2l = T$ is equal to the period.

Let us now return to our function $f(x)$ that is periodic with the period of $T = 2l$, and let us *assume* that it can be represented as a linear combination of all functions of the set $\{\phi_n, \psi_n\}$, i.e. as an infinite functional series

$$f(x) = \frac{a_0}{2} + \sum_{n=1}^{\infty} \{a_n \phi_n(x) + b_n \psi_n(x)\}$$

$$= \frac{a_0}{2} + \sum_{n=1}^{\infty} \left\{ a_n \cos \frac{n\pi x}{l} + b_n \sin \frac{n\pi x}{l} \right\} \,.$$

(3.7)

Note that $\psi_0(x) = 0$ and thus has been dropped; also, $\phi_0(x) = 1$ and hence the a_0 term has been separated from the sum with its coefficient chosen for convenience as $a_0/2$. Note that $f(x)$ in the left-hand side of the expansion above, as we shall see later on in Sect. 3.2, may not coincide for *all* values of x with the value which the series in the right-hand side converges to; in other words, the two functions, $f(x)$ and the series itself, may differ at specific points. We shall have a proper look at this particular point and the legitimacy of this expansion later on in Sect. 3.9.

Now, what we would like to do is to determine the coefficients a_0, a_1, a_2, a_3, $\ldots$ and b_1, b_2, b_3, $\ldots$, *assuming* that such an expansion exists. To this end, let us also *assume* that we can integrate both sides of Eq. (3.7) from $-l$ to l term by term (of course, as we know well from Sect. I.7.2, this is not always possible). Thus, we have

$$\int_{-l}^{l} f(x)dx = \int_{-l}^{l} \frac{a_0}{2}dx + \sum_{n=1}^{\infty} \left\{ a_n \int_{-l}^{l} \phi_n(x)dx + b_n \int_{-l}^{l} \psi_n(x)dx \right\} \,.$$

The integrals in the right-hand side for $n \geq 1$ in the curly brackets are all equal to zero (this can be checked either by a direct calculation, or also from the fact that these integrals can be considered as orthogonality integrals (3.6) with $\phi_0(x) = 1$), so that the $(a_0/2)\, 2l = a_0 l$ is obtained in the right-hand side, yielding

$$a_0 = \frac{1}{l} \int_{-l}^{l} f(x)dx$$

(3.8)

for the a_0 coefficient. To obtain other coefficients a_n for $n \neq 0$, we first multiply both sides of Eq. (3.7) by $\phi_m(x)$ with some *fixed* value of $m \neq 0$ and then integrate from $-l$ to l:

$$\int_{-l}^{l} f(x)\phi_m(x)dx = \frac{a_0}{2} \int_{-l}^{l} \phi_m(x)dx$$

$$+ \sum_{n=1}^{\infty} \left\{ a_n \int_{-l}^{l} \phi_n(x)\phi_m(x)dx + b_n \int_{-l}^{l} \psi_n(x)\phi_m(x)dx \right\} \,.$$

The first term in the right-hand side is zero since $m \neq 0$, similarly to the above. In the same fashion, due to Eq. (3.5), the second integral in the curly brackets is also equal to zero, and we are left with

$$\int_{-l}^{l} f(x)\phi_m(x)dx = \sum_{n=1}^{\infty} a_n \int_{-l}^{l} \phi_n(x)\phi_m(x)dx .$$

Now in the right-hand side we have an infinite sum of terms containing the same integrals as in Eq. (3.6) which are all equal to zero except for the single one in which $n = m$, i.e. only a single term in the sum above survives

$$\int_{-l}^{l} f(x)\phi_m(x)dx = \sum_{n=1}^{\infty} a_n \int_{-l}^{l} \phi_n(x)\phi_m(x)dx = \sum_{n=1}^{\infty} a_n l \delta_{nm}$$
$$= a_m l ,$$

which gives immediately

$$a_m = \frac{1}{l} \int_{-l}^{l} f(x)\phi_m(x)dx = \frac{1}{l} \int_{-l}^{l} f(x) \cos \frac{m\pi x}{l} dx . \tag{3.9}$$

Note that a_0 of Eq. (3.8) can also formally be obtained from Eq. (3.9) although the latter was, strictly speaking, obtained for non-zero values of m only. This became possible because of the factor of $1/2$ introduced earlier in Eq. (3.7), which now justifies its convenience. Therefore, Eq. (3.9) gives *all* a_n coefficients. Note however that in practical calculations it is frequently required to consider the coefficients a_0 and a_n for $n > 0$ separately.

Problem 3.4 Prove that the coefficients b_n are obtained from

$$b_m = \frac{1}{l} \int_{-l}^{l} f(x)\psi_m(x)dx = \frac{1}{l} \int_{-l}^{l} f(x) \sin \frac{m\pi x}{l} dx . \tag{3.10}$$

Thus, formulae (3.8)–(3.10) solve the problem: if the function $f(x)$ is known, then we can calculate all the coefficients in its expansion of Eq. (3.7). The coefficients a_n and b_n are called *Fourier coefficients*, and the infinite series (3.7) *Fourier series*.

Example 3.1 ▶ Consider a periodic function with the period of 2π specified in the following way: $f(x) = x$ within the interval $-\pi < x < \pi$, and then repeated like this to the left and to the right. This function, when periodically repeated, jumps between its values of $\pm\pi$ at the points $\pm k\pi$ for any integer $k = 1, 3, 5, \ldots$. Calculate the expansion coefficients a_n and b_n and thus write the corresponding Fourier series.

Solution: In this case $l = \pi$, and formulae (3.8)–(3.10) for the coefficients a_n and b_n are rewritten as

$$a_m = \frac{1}{\pi} \int_{-\pi}^{\pi} x \cos(mx) \, dx = 0, \quad m = 0, 1, 2, \ldots ,$$

and

$$b_m = \frac{1}{\pi} \int_{-\pi}^{\pi} x \sin(mx)\, dx = \frac{1}{\pi} \left\{ x \frac{-\cos(mx)}{m} \bigg|_{-\pi}^{\pi} + \frac{1}{m} \int_{-\pi}^{\pi} \cos(mx)\, dx \right\}$$

$$= \frac{1}{\pi} \left\{ -\pi \frac{\cos(m\pi)}{m} - \pi \frac{\cos(-m\pi)}{m} + \underbrace{\frac{1}{m^2} \sin(mx) \bigg|_{-\pi}^{\pi}}_{=0} \right\}$$

$$= -\frac{2}{m}(-1)^m = \frac{2}{m}(-1)^{m+1}, \quad m \neq 0$$

(the integration by parts was used). Note that $a_m = 0$ for any m because the function under the integral for a_m is an odd function and we integrate over a symmetric interval. Thus, in this particular example, the Fourier series consists only of sine functions:

$$f(x) = \sum_{m=1}^{\infty} \frac{2}{m}(-1)^{m+1} \sin(mx) . \tag{3.11}$$

The convergence of the series is demonstrated in Fig. 3.1: the first n terms in the series are only accounted for, i.e. the functions

$$f_n(x) = \sum_{m=1}^{n} \frac{2}{m}(-1)^{m+1} \sin(mx)$$

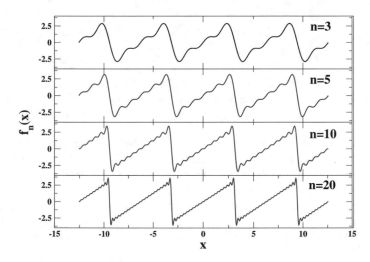

Fig. 3.1 Graphs of $f_n(x)$ corresponding to the first n terms in the series of Eq. (3.11)

for several choices of the upper limit $n = 3, 5, 10, 20$ are plotted. It can be seen that the series converges very quickly to the exact function between $-\pi$ and π. Beyond this interval the function is *periodically repeated*. Note also that the largest error in representing the function $f(x)$ by the series appears at the points $\pm k\pi$ (with $k = 1, 3, 5, \ldots$) where $f(x)$ jumps between the values $\pm\pi$. ◄

The actual integration limits from $-l$ to $+l$ in the above formulae were chosen only for simplicity; in fact, due to periodicity of $f(x)$, $\cos(m\pi x/l)$ and $\sin(m\pi x/l)$, one can use any limits differing by $2l$, i.e. from $-l + c$ to $l + c$ for any value of c (Problem 3.3). For instance, in some cases, it is convenient to use the interval from 0 to $T = 2l$.

The Fourier expansion can be handy in summing up infinite numerical series as illustrated by the following example:

Example 3.2 ► Show that

$$S = 1 - \frac{1}{3} + \frac{1}{5} - \frac{1}{7} + \cdots = \frac{\pi}{4} . \tag{3.12}$$

Solution: Consider the series (3.11) generated in the previous Example 3.1 for $f(x) = x$, $-\pi < x < \pi$, and set there $x = \pi/2$:

$$\frac{\pi}{2} = \sum_{m=1}^{\infty} \frac{2}{m} (-1)^{m+1} \sin\left(\frac{m\pi}{2}\right) .$$

The sine functions are non-zero only for odd values of $m = 1, 3, 5, \ldots$, so that we obtain (convince yourself in what follows by calculating the first few terms explicitly):

$$\frac{\pi}{2} = 2\left(1 - \frac{1}{3} + \frac{1}{5} - \frac{1}{7} + \ldots\right) = 2S ,$$

so that $S = \pi/4$ as required. ◄

Problem 3.5 Show that if the function $f(x) = f(-x)$ is even, the Fourier series is simplified as follows:

$$f(x) = \frac{a_0}{2} + \sum_{n=1}^{\infty} a_n \cos\frac{\pi n x}{l} \quad \text{with} \quad a_n = \frac{2}{l} \int_0^l f(x) \cos\frac{\pi n x}{l} dx . \tag{3.13}$$

This is called the cosine Fourier series.

Problem 3.6 Show that if the function $f(x) = -f(-x)$ is odd, the Fourier
series is simplified as follows:

$$f(x) = \sum_{n=1}^{\infty} b_n \sin \frac{\pi n x}{l} \quad \text{with} \quad b_n = \frac{2}{l} \int_0^l f(x) \sin \frac{\pi n x}{l} dx . \qquad (3.14)$$

This is called the sine Fourier series.

The Fourier series for a periodic function $f(x)$ can be written in an alternative
form which may better illuminate its meaning. Indeed, the expression in the curly
brackets of Eq. (3.7) can be rewritten as a single sine function:

$$a_n \cos \frac{n \pi x}{l} + b_n \sin \frac{n \pi x}{l} = A_n \sin (2 \pi \nu_n t + \phi_n) ,$$

where $A_n = \sqrt{a_n^2 + b_n^2}$ is an amplitude and ϕ_n a phase satisfying

$$\sin \phi_n = \frac{a_n}{\sqrt{a_n^2 + b_n^2}} \quad \text{and} \quad \cos \phi_n = \frac{b_n}{\sqrt{a_n^2 + b_n^2}} ,$$

and $\nu_n = n/T = n/ (2l)$ is the frequency. Correspondingly, the Fourier series (3.7)
is transformed into

$$f(x) = \frac{a_0}{2} + \sum_{n=1}^{\infty} A_n \sin (2 \pi \nu_n x + \phi_n) . \qquad (3.15)$$

This simple result means that any periodic function can be represented as a sum of
sinusoidal functions with discrete frequencies $\nu_1 = \nu_0$, $\nu_2 = 2\nu_0$, $\nu_3 = 3\nu_0$, etc., so
that $\nu_n = n\nu_0$, where $\nu_0 = 1/T$ is the smallest (the so-called fundamental) frequency,
and $n = 1, 2, 3, \ldots$.

For instance, let us consider some signal $f(t)$ in a device or an electric circuit
which is periodic in time t with the period of T. Then, we see from Eq. (3.15)
that such a signal can be *synthesised* by forming a *linear superposition* of simple
harmonic signals having *discrete frequencies* $\nu_n = n\nu_0$ and amplitudes A_n. As we
have seen from the examples given above, in practice very often only a finite number
of lowest frequencies may be sufficient to faithfully represent the signal, with the
largest error appearing at the points where the original signal $f(t)$ experiences jumps
or changes most rapidly. It also follows from the above equations that when expanding
a complicated signal $f(t)$ there is a finite number of harmonics with relatively large
amplitudes, then $f(t)$ could be reasonably well represented only by these.

3.2 Dirichlet Conditions

The discussion above was not rigorous: firstly, we assumed that the expansion (3.7) exists and, secondly, we integrated the infinite expansion term by term, which cannot always be done. A more rigorous formulation of the problem is this: we are given a function $f(x)$ specified in the interval^2 $-l < x < l$. We then form an infinite series (3.7),

$$\frac{a_0}{2} + \sum_{n=1}^{\infty} \left\{ a_n \cos \frac{n\pi x}{l} + b_n \sin \frac{n\pi x}{l} \right\} ,$$

with the coefficients calculated via Eqs. (3.8)–(3.10).

We ask if for any $-l < x < l$ the series converges to some function $f_{FS}(x)$, and if it does, would the resulting function be exactly the same as $f(x)$, i.e. is it true that $f_{FS}(x) = f(x)$ for all values of x? Also, are there any limitations on the function $f(x)$ itself for this to be true? The answers to all these questions are not trivial and the corresponding rigorous discussion is given later on in Sect. 3.9. Here we shall simply formulate the final result established by Dirichlet.

We first give some definitions (see also Sect. I.2.4.3). Let the function $f(x)$ be discontinuous at $x = x_0$, see Fig. 3.2, but has well-defined limits $x \to x_0$ from the left and from the right of x_0 (see Sect. I.2.4.1.2), i.e.

$$f(x_0 + 0) = \lim_{\delta \to 0} f(x_0 + \delta) \quad \text{and} \quad f(x_0 - 0) = \lim_{\delta \to 0} f(x_0 - \delta) ,$$

with $\delta > 0$ in both cases. It is then said that $f(x)$ has a *discontinuity of the first kind* at the point x_0. Then, the function $f(x)$ is said to be *piecewise continuous* in the interval $a < x < b$, if it has a finite number $n < \infty$ of discontinuities of the first kind there, but otherwise is continuous everywhere, i.e. it is continuous between any two adjacent points of discontinuity.

The function $f(x) = x, -\pi < x < \pi$, of Example 3.1, when periodically repeated, represents an example of such a function: in any finite interval crossing points $\pm \pi, \pm 3\pi, \ldots$, it has a finite number of discontinuities; however, at each discontinuity finite limits exist from both sides. For instance, consider the point of discontinuity $x = \pi$. Just on the left of it $f(x) = x$ and hence the limit from the left $f(\pi - 0) = \pi$, while on the right of it $f(\pi + 0) = -\pi$ due to periodicity of $f(x)$. Thus, at $x = \pi$ we have a discontinuity of the first kind.

Then, the following *Dirichlet theorem* addresses the fundamental questions about the expansion of the function $f(x)$ into the Fourier series:

---

^{2} Or, which is the same, which is periodic with the period of $T = 2l$; the 'main' or 'irreducible' part of the function, which is periodically repeated, can start anywhere, one choice is between $-l$ and l, the other between 0 and $2l$, etc.

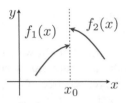

Fig. 3.2 The function $f(x)$ is equal to $f_1(x)$ for $x < x_0$ and to $f_2(x)$ for $x > x_0$, but is discontinuous (makes a 'jump') at x_0. However, finite limits exist at both sides of the 'jump' corresponding to the two different values of the function on both sides of the point $x = x_0$: on the left, $f(x - 0) = \lim_{x \to x_0} f_1(x) = f_1(x_0)$, while on the right $f(x + 0) = \lim_{x \to x_0} f_2(x) = f_2(x_0)$, where $f_1(x_0) \neq f_2(x_0)$

Theorem 3.1 *If $f(x)$ is piecewise continuous in the interval $-l < x < l$ and has the period of $T = 2l$, then the Fourier series*

$$f_{FS}(x) = \frac{a_0}{2} + \sum_{n=1}^{\infty} \left\{ a_n \cos \frac{n\pi x}{l} + b_n \sin \frac{n\pi x}{l} \right\} \qquad (3.16)$$

converges to $f(x)$ at any point x where $f(x)$ is continuous, while it converges to the mean value

$$f_{FS}(x_0) = \frac{1}{2} [f(x_0 - 0) + f(x_0 + 0)] \qquad (3.17)$$

at the points $x = x_0$ of discontinuity.

The proof of this theorem is quite remarkable and will be given in Sect. 3.9 using additional assumptions. Functions $f(x)$ satisfying conditions of the Dirichlet theorem are said to satisfy *Dirichlet conditions*.

As an example, consider again the function $f(x) = x$, $-\pi < x < \pi$, with the period 2π (Example 3.1) whose Fourier series is given by Eq. (3.11):

$$f_{FS}(x) = \sum_{m=1}^{\infty} \frac{2}{m} (-1)^{m+1} \sin(mx) \ .$$

What values does $f_{FS}(x)$ converge to at the points $x = -\pi$, 0, $\pi/2$, π, $3\pi/2$? To answer these questions, we need to check if the function makes a jump at these points. If it does, then we should consider the left and right limits of the function at the point of the jump and calculate the average (the mean value); if there is no jump, the Fourier series converges at this point to the value of the function itself. $f(x)$ is continuous at $x = 0$, $\pi/2$, $3\pi/2$ and thus $f_{FS}(0) = f(0) = 0$, $f_{FS}(\pi/2) = f(\pi/2) = \pi/2$ and

$f_{FS}(3\pi/2) = f(3\pi/2) = f(3\pi/2 - 2\pi) = f(-\pi/2) = -\pi/2$ (by employing $T = 2\pi$ periodicity of $f(x)$, we have moved its argument $3\pi/2$ back into the interval $-\pi < x < \pi$), while at $x = -\pi$ the function $f(x)$ has the discontinuity of the first kind with the limits on both sides equal to $+\pi$ (from the left) and $-\pi$ (right), respectively, so that the mean is zero, i.e. $f_{FS}(-\pi) = 0$. This is also clearly seen in Fig. 3.1.

Problem 3.7 The function $f(x) = 2\cos^2 2x + \sin 2x + 1$ is defined for $-\pi \le x \le \pi$. Show its non-zero Fourier coefficients are $a_0 = 4$ and $a_4 = b_2 = 1$.

Problem 3.8 Consider the periodic function $f(x)$ with the period of 2π, one period of which is specified as $f(x) = x^2$, $-\pi < x < \pi$. Sketch the function in the interval $-3\pi < x < 3\pi$. Is this function continuous everywhere? Show that the Fourier series of this function is

$$f(x) = \frac{\pi^2}{3} + \sum_{n=1}^{\infty} \frac{4(-1)^n}{n^2} \cos(nx) . \tag{3.18}$$

Would the series converge everywhere to x^2?

Problem 3.9 Show that the Fourier series of the function $f(x)$ with the period of 2 defined as $x + 1$ for $-1 \le x < 0$ and $-x + 1$ for $0 \le x < 1$ is

$$f(x) = \frac{1}{2} + \sum_{n=1}^{\infty} \frac{2}{(\pi n)^2} \left[1 - (-1)^n\right] \cos(\pi n x) .$$

Sketch the function and explain why the series converges to 1 at $x = 0$?

Problem 3.10 Show that the sine/cosine Fourier series expansion of the function $f(x) = 1$ for $0 < x < \pi$ and $f(x) = 0$ for $-\pi < x < 0$ is

$$f(x) = \frac{1}{2} + \frac{1}{\pi} \sum_{n=1}^{\infty} \frac{1 - (-1)^n}{n} \sin(nx) . \tag{3.19}$$

What does the series converge to at $x = 0$?

Problem 3.11 Consider the step function

$$g(x) = \begin{cases} 1, & 0 < x < \pi \\ -1, & -\pi < x < 0 \end{cases} . \tag{3.20}$$

Relating $g(x)$ to $f(x)$ defined in the previous Problem, show that the Fourier series for $g(x)$ is

$$g(x) = \frac{4}{\pi}\left[\sin x + \frac{1}{3}\sin 3x + \frac{1}{5}\sin 5x + \dots\right]$$

$$= \frac{4}{\pi}\sum_{n=1}^{\infty}\frac{1}{2n-1}\sin(2n-1)x . \tag{3.21}$$

Check this result by calculating the Fourier coefficients directly.

Problem 3.12 Use the series (3.18) to show that

$$P_2 = 1 - \frac{1}{2^2} + \frac{1}{3^2} - \frac{1}{4^2} + \dots = \sum_{n=1}^{\infty}\frac{(-1)^{n+1}}{n^2} = \frac{\pi^2}{12} . \tag{3.22}$$

Problem 3.13 Use the Fourier series (3.19) to show that the numerical series

$$S = 1 - \frac{1}{3} + \frac{1}{5} - \dots = \frac{\pi}{4} . \tag{3.23}$$

Problem 3.14 Show that the Fourier series expansion of the function $f(x) = x^4$ defined on the interval $-\pi < x < \pi$ and then periodically repeated is

$$f(x) = \frac{\pi^4}{5} + \sum_{n=1}^{\infty}\frac{8\pi^2(-1)^n}{n^2}\left(1 - \frac{6}{\pi^2 n^2}\right)\cos(nx) . \tag{3.24}$$

Problem 3.15 Using the expansion (3.24), show that

$$P_4 = 1 - \frac{1}{2^4} + \frac{1}{3^4} - \frac{1}{4^4} + \dots = \sum_{n=1}^{\infty}\frac{(-1)^{n+1}}{n^4} = \frac{7\pi^4}{720} . \tag{3.25}$$

3.3   Gibbs–Wilbraham Phenomenon

An interesting observation the curious reader may make by looking at Fig. 3.1: no matter how closely $f_{FS}(x)$ approaches the function $f(x)$ (recall that the latter is defined as $f(x) = x$ within the interval $-\pi < x < \pi$ and then periodically repeated) by increasing the number of terms n in the Fourier series (3.11), the partial FS at the points of discontinuity $\pm\pi, \pm 3\pi, \dots$ always overshoots or undershoots the expected values of $+\pi$ and $-\pi$, respectively. In fact, it appears to be a very general phenomenon: even though the Fourier series converges (in the limit of an infinite

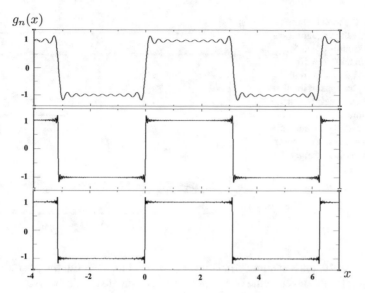

Fig. 3.3 Partial FS of the square wave function calculated with different number of terms: $n = 10$ (top), $n = 50$ (middle), and $n = 100$ (bottom)

number of terms in it) to the actual function $f(x)$ at any point of continuity and to the mean value $\frac{1}{2} \left[f(x_-) + f(x_+) \right]$ at the points of discontinuity, the FS approaching $x_+ = x + 0$ from the right, Fig. 3.2, overshoots the actual value of the function by approximately 9% of the actual jump, while the FS approaching $x_- = x - 0$ from the left undershoots by the same amount. This so-called Gibbs phenomenon was first discovered by Henry Wilbraham in 1848 and then rediscovered by J. Willard Gibbs in 1899.

We shall illustrate this effect by an example of a square wave function (3.20) given by the FS (3.21) as in this case everything can be worked out analytically. In Fig. 3.3 the partial FSs of the square wave function,

$$ g_n(x) = \frac{4}{\pi} \left[\sin x + \frac{1}{3} \sin 3x + \frac{1}{5} \sin 5x + \cdots + \frac{1}{2n-1} \sin(2n-1)x \right] $$

are shown for several values of n. We shall specifically concentrate on the discontinuity jump the function makes at $x = 0$ between ± 1. The close-ups of these partial sums around $x = 0$ are shown in Fig. 3.4. It can be seen that the partial FS makes an overshoot to the right of the jump point that is considerably shifted from the point of the jump if only 10 terms in the partial sum are kept. The more terms are retained, the peak to the right of the jump point is shifted closer to the jump point, but its height does not change noticeably. The same can be said about the undershoot of the curves to the left of the jump point.

This finite sum $g_n(x)$ can actually be calculated analytically and presented as an integral; however, this is not required for our purposes. Instead, anticipating the fact

Fig. 3.4 Close-ups of the partial FS of the square wave function around $x = 0$ for $n = 10$ (black), $n = 50$ (blue), and $n = 100$ (red). Note the positions of the first maximum to the right and the first minimum to the left of $x = 0$ in each of the curves: these move closer and closer to the jump discontinuity point as n increases

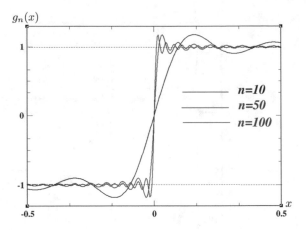

that $g_n(x)$ overshoots at some point above $x = 0$, let us work out the position of that maximum for the given value of n. For this, we calculate the derivative

$$g_n'(x) = \frac{4}{\pi}[\cos x + \cos 3x + \cos 5x + \cdots + \cos(2n - 1)x]$$

$$= \frac{4}{\pi}\sum_{k=1}^{n}\cos(2k - 1)x\,.$$

Problem 3.16 Subtracting summations Eq. (I.2.64) for $\Delta = x$ and $2n$ terms from the same summation for $\Delta = 2x$ but with n terms, show that

$$g_n'(x) = \frac{2\sin(2nx)}{\pi\sin x}\,.$$

Setting $g_n'(x) = 0$, we find that the nearest root above 0 appears at $x = \pi/2n$.

Problem 3.17 Show, by working out the sign of the second derivative of $g_n(x)$ at $x = \pi/2n$, that this point corresponds to the maximum of $g_n(x)$.

The actual value of the partial sum at this value of x,

$$g_n(\pi/2n) = \frac{4}{\pi}\sum_{k=1}^{n}\frac{1}{2k - 1}\sin\frac{(2k - 1)\pi}{2n}\,,$$

can be related to the Riemann sum of the definite integral of the sinc function

$$\mathrm{sinc}(x) = \frac{\sin x}{x} \qquad (3.26)$$

between 0 and π if we divide up the interval $0 \le x \le \pi$ into n equal intervals of width $\Delta = \pi/n$ and choose the middle points $x_k = (2k - 1)\Delta/2$ for the values of the function:

$$\int_0^\pi \frac{\sin x}{x} dx \approx \Delta \sum_{k=1}^n \frac{\sin \frac{(2k-1)\Delta}{2}}{\frac{(2k-1)\Delta}{2}} = 2 \sum_{k=1}^n \frac{\sin \frac{(2k-1)\Delta}{2}}{2k - 1} = \frac{\pi}{2} g_n (\pi/2n) \ ,$$

i.e. it can be seen is exactly the required value! In the limit of $n \to \infty$, the relationship becomes exact:

$$\lim_{n \to \infty} g_n (\pi/2n) = \frac{2}{\pi} \int_0^\pi \frac{\sin x}{x} dx \ .$$

The integral of the sinc function can be calculated numerically and is equal to $1.85194\ldots$. Hence, the overshoot is $(2/\pi) \times 1.85194 - 1 = 1.17898\ldots - 1 = 0.17898\ldots$, which is about 9% of the total jump of 2 of our function at $x = 0$.

Problem 3.18 Repeat the argument by considering the point $x_{-1} = -\pi/2n$ on the left of $x = 0$ and show that there is a minimum and that the partial sum undershoots there by the same amount.

Note that the phenomenon does not contradict the Dirichlet theorem: indeed, as the number of terms in the partial FS $n \to \infty$, the points at which the partial sum is maximum (minimum) tend to the corresponding point of discontinuity; hence, the FS converges to the actual function away from the discontinuity points, while at those points the FS overshoots (undershoots) on both sides of it in equal measure leading still to the correct mean value at the discontinuity itself.

3.4 Integration and Differentiation of the Fourier Series

The Fourier series is an example of a functional series and hence it can be integrated term by term if it converges uniformly (Sect. I.7.2). However, the uniform convergence is only a sufficient criterion. We shall prove now that irrespective of whether the uniform convergence exists or not, the Fourier series can still be integrated term by term any number of times.

To prove this, it is convenient to consider an expansion of $f(x)$ with $a_0 = 0$; it is always possible by considering the function $f(x) - a_0/2 \to f(x)$ whose Fourier expansion would have no constant term, i.e. for this function $a_0 = 0$. Then, we consider an auxiliary function:

$$F(x) = \int_{-l}^{x} f(t)dt \ . \tag{3.27}$$

It is equal to zero at $x = -l$, but it is also zero at $x = l$, since $F(l) = a_0 l = 0$, see expression (3.8) for a_0. These conditions ensure the required periodicity of $F(x)$ with the same period of $2l$ as for $f(x)$:

$$\begin{aligned} F(x + 2l) &= \int_{-l}^{x+2l} f(t)dt = \int_{-l}^{x} f(t)dt + \int_{x}^{x+2l} f(t)dt \\ &= \int_{-l}^{x} f(t)dt + \underbrace{\int_{0}^{2l} f(t)dt}_{=0} = \int_{-l}^{x} f(t)dt = F(x) \ . \end{aligned}$$

Further, if $f(x)$ satisfies the Dirichlet conditions (is piecewise continuous), its integral, $F(x)$, will as well and with the same periodicity. Therefore, $F(x)$ can also formally be expanded into a Fourier series:

$$F(x) = \frac{A_0}{2} + \sum_{n=1}^{\infty} \left[A_n \cos \frac{\pi n x}{l} + B_n \sin \frac{\pi n x}{l} \right] , \tag{3.28}$$

where

$$\begin{aligned} A_0 &= \frac{1}{l} \int_{-l}^{l} F(x)dx = \frac{1}{l} \left[x F(x)|_{-l}^{l} - \int_{-l}^{l} x f(x)dx \right] \\ &= -\frac{1}{l} \int_{-l}^{l} x f(x)dx \ , \end{aligned}$$

and for any $n \geq 1$:

$$\begin{aligned} A_n &= \frac{1}{l} \int_{-l}^{l} F(x) \cos \frac{\pi n x}{l} dx \\ &= \frac{1}{l} \left[F(x) \frac{l}{\pi n} \sin \frac{\pi n x}{l} \Big|_{-l}^{l} - \frac{l}{\pi n} \int_{-l}^{l} f(x) \sin \frac{\pi n x}{l} dx \right] \\ &= -\frac{1}{\pi n} \int_{-l}^{l} f(x) \sin \frac{\pi n x}{l} dx = -\frac{l}{\pi n} b_n \ , \end{aligned}$$

$$\begin{aligned} B_n &= \frac{1}{l} \int_{-l}^{l} F(x) \sin \frac{\pi n x}{l} dx \\ &= \frac{1}{l} \left[-F(x) \frac{l}{\pi n} \cos \frac{\pi n x}{l} \Big|_{-l}^{l} + \frac{l}{\pi n} \int_{-l}^{l} f(x) \cos \frac{\pi n x}{l} dx \right] \\ &= \frac{1}{\pi n} \int_{-l}^{l} f(x) \cos \frac{\pi n x}{l} dx = \frac{l}{\pi n} a_n \ , \end{aligned}$$

where we have used integration by parts in each case and the fact that $F'(x) = f(x)$ and $F(\pm l) = 0$. Above, a_n and b_n are the corresponding Fourier coefficients to the cosines and sines in the Fourier expansion of $f(x)$. Therefore, we can write the expansion of $F(x)$ as follows:

$$
\begin{aligned}
F(x) &= \int_{-l}^{x} f(t)dt \\
&= -\frac{1}{2l} \int_{-l}^{l} tf(t)dt + \sum_{n=1}^{\infty} \left[-\frac{l}{\pi n} b_n \cos \frac{\pi n x}{l} + \frac{l}{\pi n} a_n \sin \frac{\pi n x}{l} \right] .
\end{aligned}
\tag{3.29}
$$

Further, at $x = -l$ we should have zero in the left-hand side, which means that

$$
0 = -\frac{1}{2l} \int_{-l}^{l} tf(t)dt - \sum_{n=1}^{\infty} \frac{l}{\pi n} b_n \cos(\pi n)
$$

or

$$
-\frac{1}{2l} \int_{-l}^{l} tf(t)dt = \sum_{n=1}^{\infty} \frac{l(-1)^n}{\pi n} b_n ,
$$

so as a result, the expansion (3.29) takes on the form:

$$
\begin{aligned}
F(x) &= \int_{-l}^{x} f(t)dt \\
&= \sum_{n=1}^{\infty} \left\{ -\frac{l}{\pi n} b_n \left[\cos \frac{\pi n x}{l} - (-1)^n \right] + \frac{l}{\pi n} a_n \sin \frac{\pi n x}{l} \right\} .
\end{aligned}
\tag{3.30}
$$

It is seen that the function $F(x)$ does not jump at $x = \pm l$ where it is equal to zero.

Now let us look at the Fourier expansion of $f(t)$ (replacing x with t) with $a_0 = 0$, e.g. Eq. (3.7), and formally integrate it term by term between $-l$ and x. It is easy to see that exactly the same expressions are obtained on both sides as above. This proves that integration of the Fourier series is legitimate.

Now, consider a function $f(x)$ for which $a_0 \neq 0$ in its Fourier expansion. The above treatment is valid for $\phi(x) = f(x) - a_0/2$ and the corresponding function in the left-hand side of the integrated Fourier series expansion in this case is the function

$$
F(x) = \int_{-l}^{x} \left[f(t) - \frac{a_0}{2} \right] dt = \int_{l}^{x} f(t)dt - \frac{a_0}{2} (x + l) .
$$

The integral in the right-hand side corresponds to integrating term by term a general Fourier expansion of $f(x)$ disregarding the a_0 term. However, there is an additional term $a_0(x + l)/2$; however, it also appears when integrating the constant $a_0/2$ term of the original expansion of $f(x)$. This discussion fully demonstrates that one can always integrate any Fourier series term by term.

We mention that, in fact, the bottom limit when integrating the series of $f(x)$ is not so important. Indeed, assume that one needs to integrate a series between a and x, where $x > a > -l$. Then, two series can be produced, one integrated between $-l$ and x, and another between $-l$ and a. Subtracting the two series from each other term by term (assuming absolute convergence, Sect. I.7.1.5), the required series is obtained.

Note also that the convergence of the series only improves after integration. This is because, when integrating $\cos(\pi nx/l)$ or $\sin(\pi nx/l)$ with respect to x, an additional factor of $1/n$ arises that can only accelerate the convergence of the Fourier series.

Example 3.3 Obtain the Fourier series for the function $f(x) = x^2$, $-\pi < x < \pi$, that is periodic with the period of 2π.

Solution: This can be obtained by integrating term by term the series (3.11) for $f_1(x) = x$ from 0 to some $0 < x < \pi$:

$$\int_0^x f_1(x_1)dx_1 = \sum_{m=1}^{\infty} \frac{2}{m}(-1)^{m+1} \int_0^x \sin(mx_1)dx_1 .$$

Since $f_1(x_1) = x_1$ in the interval under consideration, the integral in the left-hand side gives $x^2/2$. Integrating the sine functions in the right-hand side, we obtain

$$\frac{x^2}{2} = \sum_{m=1}^{\infty} \frac{2}{m}(-1)^{m+1} \frac{1}{m}[-\cos(mx) + 1]$$

$$= -\sum_{m=1}^{\infty} \frac{2(-1)^{m+1}}{m^2} \cos(mx) + 2\sum_{m=1}^{\infty} \frac{(-1)^{m+1}}{m^2} .$$

The numerical series (the last term) can be shown (using the direct method for $f(x) = x^2$, i.e. expanding it into the Fourier series, see Problems 3.8 and 3.12) to be equal to $\pi^2/12$,

$$\sum_{m=1}^{\infty} \frac{(-1)^{m+1}}{m^2} = \frac{\pi^2}{12} , \tag{3.31}$$

so that we finally obtain

$$x^2 = \frac{\pi^2}{3} - 4\sum_{m=1}^{\infty} \frac{(-1)^{m+1}}{m^2} \cos(mx) . \tag{3.32}$$

The convergence of this series with different number of terms n in the sum ($m \le n$) is pretty remarkable as is demonstrated in Fig. 3.5. ◄

Fig. 3.5 The partial Fourier series of $f(x) = x^2$, see Eq. (3.32), containing $n = 2, 3$ and 10 terms

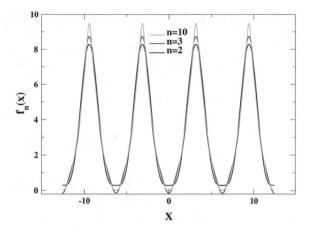

The situation with term-by-term differentiation of the Fourier series is more complex since each differentiation of either $\cos(\pi n x / l)$ or $\sin(\pi n x / l)$ brings in an extra n in the sum which results is slower convergence or even divergence. For example, if we formally differentiate term-by-term formula (3.11) for $f(x) = x$, $-\pi < x < \pi$, we obtain

$$1 = \sum_{m=1}^{\infty} 2(-1)^{m+1} \cos(mx) \,,$$

which contains the diverging series. There are much more severe restrictions on the function $f(x)$ that would enable its Fourier series be differentiable term by term, so this procedure should be performed with caution and proper prior investigation.

3.5 Parseval's Theorem

Consider a periodic function $f(x)$ with the period of $2l$ satisfying Dirichlet conditions, i.e. $f(x)$ has a finite number of discontinuities of the first kind in the interval $-l < x < l$. Thus, it can be expanded in the Fourier series (3.16). If we multiply this expansion by itself, a Fourier series of the square of the function, $f^2(x)$, is obtained. Next, we integrate both sides from $-l$ to l (we know by now that the term-by-term integration of the Fourier series is always permissible), which gives

$$\int_{-l}^{l} f_{FS}^2(x)dx = \int_{-l}^{l} \left\{ \frac{a_0}{2} + \sum_{n=1}^{\infty} \left[a_n \cos \frac{n\pi x}{l} + b_n \sin \frac{n\pi x}{l} \right] \right\}$$

$$\times \left\{ \frac{a_0}{2} + \sum_{m=1}^{\infty} \left[a_m \cos \frac{m\pi x}{l} + b_m \sin \frac{m\pi x}{l} \right] \right\} dx$$

$$= \frac{a_0^2}{4} 2l + 2\frac{a_0}{2} \sum_{n=1}^{\infty} \int_{-l}^{l} \left\{ a_n \cos \frac{n\pi x}{l} + b_n \sin \frac{n\pi x}{l} \right\} dx$$

$$+ \sum_{n,m=1}^{\infty} \int_{-l}^{l} \left\{ a_n \cos \frac{n\pi x}{l} + b_n \sin \frac{n\pi x}{l} \right\} \left\{ a_m \cos \frac{m\pi x}{l} + b_m \sin \frac{m\pi x}{l} \right\} dx .$$

In the left-hand side we used $f_{FS}(x)$ instead of $f(x)$ to stress the fact that these two functions may not coincide at the points of discontinuity of $f(x)$. However, the integral in the left-hand side can be replaced by the integral of $f^2(x)$ if $f(x)$ is continuous everywhere; if it has discontinuities, this can also be done by splitting the integral into a sum of integrals over each region of continuity of $f(x)$, where $f_{FS}(x) = f(x)$. Hence, the integral of $f_{FS}^2(x)$ can be replaced by the integral of $f^2(x)$ in a very general case. Further, in the right-hand side of the above equation any integral in the second term is zero (integrals of cosine and sine functions there can be treated as the orthogonality integrals with the $n = 0$ cosine function which is equal to one). Also, due to the orthogonality of the sine and cosine functions, Eqs. (3.5) and (3.6), in the term with the double sum only integrals with equal indices $n = m$ are non-zero if taken between two cosine or two sine functions; all other integrals are zero. Hence, we obtain the following simple result:

$$\frac{1}{l} \int_{-l}^{l} f^2(x)dx = \frac{a_0^2}{2} + \sum_{n=1}^{\infty} \left(a_n^2 + b_n^2 \right) . \qquad (3.33)$$

This equation is called *Parseval's equality* or *theo*. It can be used for calculating infinite numerical series.

Example 3.4 ▶ Write Parseval's equality for the series (3.11) of $f(x) = x$, $-\pi < x < \pi$, and then sum up the infinite numerical series:

$$1 + \frac{1}{2^2} + \frac{1}{3^2} + \frac{1}{4^2} \cdots .$$

Solution: The integral in the left-hand side of Parseval's equality (3.33) is simply ($l = \pi$ here):

$$\frac{1}{\pi} \int_{-\pi}^{\pi} f^2(x)dx = \frac{1}{\pi} \int_{-\pi}^{\pi} x^2 dx = \frac{1}{\pi} \frac{x^3}{3} \Big|_{-\pi}^{\pi} = \frac{2\pi^2}{3} .$$

In the right-hand side of Eq. (3.33), we have $b_n = 2(-1)^{n+1}/n$ and $a_n = 0$, see Eq. (3.11), i.e.

$$\frac{0^2}{2} + \sum_{n=1}^{\infty} \left\{ 0^2 + \left[\frac{2}{n}(-1)^{n+1} \right]^2 \right\} = \sum_{n=1}^{\infty} \frac{4}{n^2} = 4 \sum_{n=1}^{\infty} \frac{1}{n^2}$$

$$\implies \quad 4 \sum_{n=1}^{\infty} \frac{1}{n^2} = \frac{2\pi^2}{3} ,$$

or

$$S_2 = \sum_{n=1}^{\infty} \frac{1}{n^2} = 1 + \frac{1}{2^2} + \frac{1}{3^2} + \cdots = \frac{\pi^2}{6} , \qquad (3.34)$$

as required. ◄

Problem 3.19 Prove a generalisation of Parseval's theorem,

$$\frac{1}{l} \int_{-l}^{l} f(x)g(x)dx = \frac{a_0 a_0'}{2} + \sum_{n=1}^{\infty} \left(a_n a_n' + b_n b_n' \right) , \qquad (3.35)$$

called Plancherel's theorem. Here $f(x)$ and $g(x)$ are two functions of the same period $2l$, both satisfying the Dirichlet conditions, and $\{a_n, b_n\}$ and $\{a_n', b_n'\}$ are their Fourier coefficients, respectively.

Problem 3.20 Use Parseval's theorem applied to the series (3.19) to show that

$$1 + \frac{1}{3^2} + \frac{1}{5^2} + \ldots = \frac{\pi^2}{8} .$$

Problem 3.21 Applying Parseval's theorem to the series (3.32), show that

$$S_4 = 1 + \frac{1}{2^4} + \frac{1}{3^4} + \ldots = \frac{\pi^4}{90} . \qquad (3.36)$$

Problem 3.22 Show that the Fourier series expansion of $f(x) = x^3 - 4x$, where $-2 \leq x < 2$, is

$$f(x) = \frac{96}{\pi^3} \sum_{n=1}^{\infty} \frac{(-1)^n}{n^3} \sin \frac{\pi n x}{2} .$$

Then apply Parseval's theorem to show that

$$S_6 = 1 + \frac{1}{2^6} + \frac{1}{3^6} + \ldots = \frac{\pi^6}{945} . \qquad (3.37)$$

3.6 Summing Inverse Powers of Integers

Above, using Parceval's theorem, we have derived formulae for the infinite sum

$$S_{2p} = \sum_{k=1}^{\infty} \frac{1}{k^{2p}} \tag{3.38}$$

of inverse even powers of all integer numbers for specific values of p. Indeed, S_2 is given by Eq. (3.34), S_4 and S_6 by Eqs. (3.36) and (3.37), respectively. Remarkably, it is possible to derive a closed expression for S_{2p} for any integer value of p using the Fourier series expansion of a specific cosine function. We shall start from the following problem:

Problem 3.23 Show that the Fourier series expansion of $\cos(xt/\pi)$ with respect to x defined in the interval $-\pi < x < \pi$ and then periodically repeated, is

$$\cos\left(\frac{xt}{\pi}\right) = \frac{\sin t}{t}\left[1 - \sum_{k=1}^{\infty}(-1)^k \frac{2t^2}{(\pi k)^2 - t^2}\cos(kx)\right]. \tag{3.39}$$

Problem 3.24 By choosing appropriate values of x and t in the above formula, prove the following expansion of the inverse sinusoidal function:

$$\frac{1}{\sin \pi z} = \frac{1}{\pi z}\left[1 - \sum_{k=1}^{\infty}(-1)^k \frac{2z^2}{k^2 - z^2}\right].$$

Our next step is to apply the value of $x = \pi$ to the expansion (3.39). Since the function $\cos(xt/\pi)$ is continuous at the point $x = \pi$, the Fourier series in the right-hand side of Eq. (3.39) converges to the function in the left-hand side. Therefore, we can write

$$\cos t = \frac{\sin t}{t}\left[1 - \sum_{k=1}^{\infty}\frac{2t^2}{(\pi k)^2 - t^2}\right] \implies t\cot(t) = 1 - \sum_{k=1}^{\infty}\frac{2t^2}{(\pi k)^2 - t^2}. \tag{3.40}$$

Next we shall manipulate the fraction under the sum as follows:

$$\frac{2t^2}{(\pi k)^2 - t^2} = 2\left(\frac{t}{\pi k}\right)^2 \frac{1}{1 - (t/\pi k)^2} = 2\left(\frac{t}{\pi k}\right)^2 \sum_{n=0}^{\infty}\left(\frac{t}{\pi k}\right)^{2n},$$

where in the last step we have expanded $\left[1 - (t/\pi k)^2\right]^{-1}$ in the power series (in fact, it is a geometrical progression). This can be done for $|t| < \pi k$. Using this result in Eq. (3.40), we obtain

$$t \cot(t) = 1 - 2 \sum_{k=1}^{\infty} \sum_{n=0}^{\infty} \left(\frac{t}{\pi k}\right)^{2(n+1)} = 1 - 2 \sum_{k=1}^{\infty} \sum_{m=1}^{\infty} \left(\frac{t}{\pi k}\right)^{2m}$$

or

$$t \cot(t) = 1 - 2 \sum_{m=1}^{\infty} S_{2m} \left(\frac{t}{\pi}\right)^{2m}. \tag{3.41}$$

This is the desired result: it shows that the values of S_m can be obtained by expanding into the Maclaurin series the function $t \cot(t)$. Indeed, a simple calculation gives

$$t \cot(t) = 1 - \frac{t^2}{3} - \frac{t^4}{45} - \frac{2t^6}{945} - \frac{t^8}{4725} - \frac{2t^{10}}{93555} - \cdots.$$

Comparing the coefficients of the above expansion with those in Eq. (3.41), we readily obtain $S_2 = \pi^2/6$, $S_4 = \pi^4/90$ and $S_6 = \pi^6/945$, our previous results. Using the last two terms in the expansion of $t \cot(t)$, we also obtain new expressions for the series of the inverse powers: $S_8 = \pi^8/9450$ and $S_{10} = \pi^{10}/93555$.

It is probably surprising, although true, that the sums S_{2p} can also be related to Bernoulli numbers that are related to the sums of powers of integer numbers (see Vol. I, Sect. 1.12 and 7.3.4). The Bernoulli numbers B_k are defined as the coefficients in the expansion:

$$\frac{t}{e^t - 1} = \sum_{k=0}^{\infty} B_k \frac{t^k}{k!}. \tag{3.42}$$

Therefore, to establish the relationship between S_{2m} and Bernoulli numbers, we need to relate the two functions, $t \cot(t)$ and $t/(e^t - 1)$. This can be done by considering $t \cot(t)$ at a purely imaginary $t = -ix/2$. Indeed, in this case:

$$-\frac{ix}{2} \cot\left(-\frac{ix}{2}\right) = -\frac{ix \cos\left(-\frac{ix}{2}\right)}{2 \sin\left(-\frac{ix}{2}\right)} = -\frac{ix \frac{1}{2}\left(e^{-i^2 x/2} + e^{i^2 x/2}\right)}{2 \frac{1}{2i}\left(e^{-i^2 x/2} - e^{i^2 x/2}\right)}$$

$$= \frac{x e^{x/2} + e^{-x/2}}{2 e^{x/2} - e^{-x/2}} = \frac{x e^x + 1}{2 e^x - 1} = \frac{x}{2} + \frac{x}{e^x - 1},$$

so that

$$\frac{x}{e^x - 1} = -\frac{x}{2} + \left[-\frac{ix}{2} \cot\left(-\frac{ix}{2}\right)\right] = -\frac{x}{2} + \left[1 - 2 \sum_{m=1}^{\infty} S_{2m} \left(\frac{-ix}{2\pi}\right)^{2m}\right]$$

$$= 1 - \frac{x}{2} - \sum_{m=1}^{\infty} \frac{2(-1)^m}{(2\pi)^{2m}} S_{2m} x^{2m}.$$

On the other hand, the same expansion is provided by Eq. (3.42), and this finally yields the relationship we seek:

$$S_{2m} = B_{2m} \frac{(2\pi)^{2m}}{2(2m)!} (-1)^{m+1}. \tag{3.43}$$

The expansion for the generating function $x/(e^x - 1)$ contains only even powers of x, confirming that Bernoulli coefficients B_{2m+1} with odd indices are equal to zero (apart from $B_1 = -1/2$). Since (see Vol. I, Sect. 1.14) $B_0 = 1$, $B_1 = -1/2$, $B_2 = 1/6$, $B_4 = -1/30$, $B_6 = 1/42$, etc., the same expressions as above for the first few sums S_{2p} of the inverse powers are obtained.

Problem 3.25 Show that the infinite series of alternating reciprocal integers ($m \geq 1$ is a positive integer),

$$P_{2m} = \sum_{n=0}^{\infty} \frac{(-1)^n}{n^{2m}} = 1 - \frac{1}{2^{2m}} + \frac{1}{3^{2m}} - \frac{1}{4^{2m}} + \cdots,$$

is related to the similar series S_{2p} with all positive terms as follows:

$$P_{2m} = \sum_{n=0}^{\infty} \frac{(-1)^n}{n^{2m}} = \left(1 - \frac{1}{2^{2m-1}}\right) S_{2m}. \tag{3.44}$$

This means that such a series is also directly related to Bernoulli numbers. Show, in particular, that for $m = 1$ we obtain our previous result (3.31), while for $m = 2$ we get Eq. (3.25). Finally, show that

$$P_6 = \sum_{n=0}^{\infty} \frac{(-1)^n}{n^6} = \frac{31\pi^6}{30240}. \tag{3.45}$$

As a by-product of the above discussion we are able to write explicitly the power expansion for the cotangent as

$$\cot t = \frac{1}{t} - 2 \sum_{m=1}^{\infty} S_{2m} \frac{t^{2m-1}}{\pi^{2m}} = \sum_{m=0}^{\infty} B_{2m} \frac{(-1)^m 2^{2m}}{(2m)!} t^{2m-1}. \tag{3.46}$$

Problem 3.26 Using the identity

$$\tan t = \cot t - 2\cot(2t),$$

obtain the following power expansion for the tangent function:

$$\tan t = \sum_{m=0}^{\infty} B_{2m} \frac{(-1)^m 2^{2m} \left(1 - 2^{2m}\right)}{(2m)!} t^{2m-1} . \tag{3.47}$$

3.7 Complex (Exponential) form of the Fourier Series

It is also possible to formulate the Fourier series (3.7) of a function $f(x)$ that is periodic with the period of $T = 2l$ in a different form based on complex exponential functions:

$$f(x) = \sum_{n=-\infty}^{\infty} c_n e^{in\pi x/l} = \sum_{n=-\infty}^{\infty} c_n \chi_n(x) . \tag{3.48}$$

Note that here the index n runs over all possible negative and positive integer values (including zero), not just over zero and positive values as in the case of the cosine/sine Fourier series expansion. It will become clearer later on why this is necessary.

The Fourier coefficients c_n can be obtained in the same way as for the sine/cosine series by noting that the functions $\chi_n(x) = \exp(in\pi x/l)$ also form an *orthogonal set*. Indeed, if integers n and m are different, $n \neq m$, then

$$\int_{-l}^{l} \chi_n^*(x)\chi_m(x)dx = \int_{-l}^{l} e^{-in\pi x/l} e^{im\pi x/l} dx = \int_{-l}^{l} e^{-i(n-m)\pi x/l} dx$$

$$= \frac{-l}{i(n-m)\pi} \left. e^{-i(n-m)\pi x/l} \right|_{-l}^{l}$$

$$= \frac{-l}{i(n-m)\pi} \left(e^{-i(n-m)\pi} - e^{i(n-m)\pi} \right)$$

$$= \frac{-l}{i(n-m)\pi} \{-2i \sin[\pi(n-m)]\} = 0 .$$

Note that when formulating the orthogonality condition above, we took one of the functions in the integrand, $\chi_n(x)$, as complex conjugate. If $n = m$, however, then

$$\int_{-l}^{l} \chi_n^*(x)\chi_n(x)dx = \int_{-l}^{l} |\chi_n(x)|^2 dx = \int_{-l}^{l} dx = 2l ,$$

so that we can generally write

$$\int_{-l}^{l} \chi_n^*(x)\chi_m(x)dx = 2l\delta_{nm} , \tag{3.49}$$

where δ_{nm} is the Kronecker symbol. Note again that one of the functions is *complex conjugate* in the above equation.

Thus, *assuming* that $f(x)$ can be expanded into the functional series in terms of the functions $\chi_n(x)$, we can find the expansion coefficients c_n exactly in the same way as for the sine/cosine series: multiply both sides of Eq. (3.48) by $\chi_m^*(x)$ with a fixed index m, and integrate from $-l$ to l on both sides:

$$\int_{-l}^{l} f(x)\chi_m^*(x)dx = \sum_{n=-\infty}^{\infty} c_n \int_{-l}^{l} \chi_n(x)\chi_m^*(x)dx$$

$$= \sum_{n=-\infty}^{\infty} c_n 2l\delta_{nm} = c_m 2l ,$$

which finally gives for any m:

$$c_m = \frac{1}{2l} \int_{-l}^{l} f(x)\chi_m^*(x)dx . \tag{3.50}$$

Note that we never assumed here that the function $f(x)$ is real. So, it could also be complex.

The same expressions (3.48) and (3.50) can also be derived directly from the sine/cosine Fourier series. This exercise helps to understand, firstly, that this new form of the Fourier series is *exactly equivalent* to the previous one based on the sine and cosine functions; secondly, we can see explicitly why both positive and negative values of n are needed. To accomplish this, we start from Eq. (3.16) and replace sine and cosine with complex exponentials by means of Euler's formulae (2.30):

$$\sin x = \frac{1}{2i} \left(e^{ix} - e^{-ix} \right) \quad \text{and} \quad \cos x = \frac{1}{2} \left(e^{ix} + e^{-ix} \right) ,$$

yielding

$$\begin{aligned} f(x) &= \frac{a_0}{2} + \sum_{n=1}^{\infty} \left\{ a_n \frac{1}{2} \left(e^{in\pi x/l} + e^{-in\pi x/l} \right) + b_n \frac{1}{2i} \left(e^{in\pi x/l} - e^{-in\pi x/l} \right) \right\} \\ &= \frac{a_0}{2} + \sum_{n=1}^{\infty} \left\{ a_n \frac{1}{2} e^{in\pi x/l} + b_n \frac{1}{2i} e^{in\pi x/l} \right\} \\ &\quad + \sum_{n=1}^{\infty} \left\{ a_n \frac{1}{2} e^{-in\pi x/l} - b_n \frac{1}{2i} e^{-in\pi x/l} \right\} \\ &= \frac{a_0}{2} + \frac{1}{2} \sum_{n=1}^{\infty} \left(a_n + \frac{1}{i} b_n \right) e^{in\pi x/l} + \frac{1}{2} \sum_{n=1}^{\infty} \left(a_n - \frac{1}{i} b_n \right) e^{-in\pi x/l} . \end{aligned} \tag{3.51}$$

Now look at the formulae (3.9) and (3.10) for a_n and b_n:

$$a_n = \frac{1}{l} \int_{-l}^{l} f(x) \cos \frac{n\pi x}{l} dx \quad \text{and} \quad b_n = \frac{1}{l} \int_{-l}^{l} f(x) \sin \frac{n\pi x}{l} dx .$$

Although these expressions were obtained for positive values of n only, they can *formally* be extended for any values of n including negative and zero values. Then, we observe that $a_{-n} = a_n$, $b_{-n} = -b_n$ and $b_0 = 0$. These expressions allow us to rewrite the second sum in (3.51) as a sum over all *negative* integers n:

$$\frac{1}{2} \sum_{n=1}^{\infty} \left(a_n - \frac{1}{i} b_n \right) e^{-in\pi x/l} = \frac{1}{2} \sum_{n=-1}^{-\infty} \left(a_{-n} - \frac{1}{i} b_{-n} \right) e^{in\pi x/l}$$

$$= \frac{1}{2} \sum_{n=-1}^{-\infty} \left(a_n + \frac{1}{i} b_n \right) e^{in\pi x/l} .$$

We see that this sum looks now exactly the same as the first sum in (3.51) in which n is positive, so that we can combine the two into a single sum in which n takes on all integer values from $-\infty$ to $+\infty$ except for $n = 0$:

$$f(x) = \frac{a_0}{2} + \frac{1}{2} \sum_{n=-\infty, n \neq 0}^{\infty} (a_n - i b_n) e^{in\pi x/l} .$$

Introducing the coefficients

$$c_n = \frac{1}{2} (a_n - i b_n) = \frac{1}{2} \left[\frac{1}{l} \int_{-l}^{l} f(x) \cos \frac{n\pi x}{l} dx - \frac{i}{l} \int_{-l}^{l} f(x) \sin \frac{n\pi x}{l} dx \right]$$

$$= \frac{1}{2l} \int_{-l}^{l} f(x) \left(\cos \frac{n\pi x}{l} - i \sin \frac{n\pi x}{l} \right) dx = \frac{1}{2l} \int_{-l}^{l} f(x) e^{-in\pi x/l} dx ,$$

and noting that $b_0 = 0$ and hence the $a_0/2$ term can also be formally incorporated into the sum as $c_0 = a_0/2$, we can finally rewrite the above expansion in the form of Eq. (3.48). The obtained equations are the same as Eqs. (3.48) and (3.50), respectively, but derived differently. Thus, the two forms of the Fourier series are completely equivalent to each other. The exponential (complex) form looks simpler and thus is easier to remember. It is always possible, using Euler's formula, to obtain any of the forms as illustrated by the following example.

Example 3.5 ▶ Obtain the complex (exponential) form of the Fourier series for $f(x) = x$, $-\pi < x < \pi$ as in Example 3.1.

Solution: We start by calculating the Fourier coefficients c_n from Eq. (3.50) using $l = \pi$. For $n \neq 0$,

$$
\begin{aligned}
c_n &= \frac{1}{2\pi}\int_{-\pi}^{\pi} x e^{-in\pi x/\pi}\,dx = \frac{1}{2\pi}\int_{-\pi}^{\pi} x e^{-inx}\,dx \\
&= \frac{1}{2\pi}\left\{ x\,\frac{1}{-in}e^{-inx}\Big|_{-\pi}^{\pi} + \frac{1}{in}\int_{-\pi}^{\pi} e^{-inx}\,dx \right\} \\
&= \frac{1}{2\pi}\left\{ \frac{\pi}{-in}\left(e^{-in\pi}+e^{in\pi}\right) - \frac{1}{(in)^2}\left(e^{-in\pi}-e^{in\pi}\right) \right\} \\
&= \frac{1}{2\pi}\left\{ \frac{2\pi}{-in}\cos(n\pi) + \frac{2i}{(in)^2}\sin(n\pi) \right\} = -\frac{1}{in}\cos(n\pi) \\
&= \frac{(-1)^{n+1}}{in}\,,
\end{aligned}
$$

and, when $n = 0$,

$$
c_0 = \frac{1}{2\pi}\int_{-\pi}^{\pi} x\,dx = \frac{1}{2\pi}\frac{x^2}{2}\Big|_{-\pi}^{\pi} = 0\,,
$$

so that the Fourier series is

$$
f(x) = \sum_{n=-\infty, n\neq 0}^{\infty} \frac{(-1)^{n+1}}{in} e^{inx}\,. \blacktriangleleft \tag{3.52}
$$

Example 3.6 ▶ Show that the above expansion is equivalent to the series (3.11).

Solution: Since $\exp(inx) = \cos(nx) + i\sin(nx)$, we get by splitting the sum into two with negative and positive summation indices:

$$
\begin{aligned}
f(x) &= \sum_{n=1}^{\infty} \frac{(-1)^{n+1}}{in} e^{inx} + \sum_{n=-1}^{-\infty} \frac{(-1)^{n+1}}{in} e^{inx} \\
&= \sum_{n=1}^{\infty} \frac{(-1)^{n+1}}{in} e^{inx} + \sum_{n=1}^{\infty} \frac{(-1)^{-n+1}}{-in} e^{-inx}\,,
\end{aligned}
$$

where in the second sum we replaced the summation index $n \to -n$, so that the new index would run from 1 to $+\infty$ as in the other sum. Combining the two sums together, and noting that $(-1)^{-n+1} = (-1)^{n+1}$, we get

$$
\begin{aligned}
f(x) &= \sum_{n=1}^{\infty} \frac{(-1)^{n+1}}{in}\left(e^{inx} - e^{-inx}\right) = \sum_{n=1}^{\infty} \frac{(-1)^{n+1}}{in} 2i\sin(nx) \\
&= \sum_{n=1}^{\infty} \frac{2(-1)^{n+1}}{n}\sin(nx)\,,
\end{aligned}
$$

which is exactly the same as in Eq. (3.11) which was obtained using the sine/cosine formulae for the Fourier series. ◀

Problem 3.27 Show that the complex (exponential) Fourier series of the function

$$f(x) = \begin{cases} 0, & -\pi < x < 0 \\ 1, & 0 < x < \pi/2 \\ 0, & \pi/2 < x < \pi \end{cases}$$

is

$$f(x) = \frac{1}{4} + \sum_{n=-\infty, n\neq 0}^{\infty} \frac{1}{2\pi i n} \left[1 - e^{-in\pi/2} \right] e^{inx} . \tag{3.53}$$

Problem 3.28 Use $x = 0$ in the series of the previous problem to obtain the sum of the numerical series (3.23).

Problem 3.29 Show that the expansion of the function

$$f(x) = \begin{cases} \sin x, & 0 \leq x < \pi \\ 0, & -\pi < x < 0 \end{cases}$$

into the complex (exponential) Fourier series is

$$f(x) = \frac{1}{4i} \left(e^{ix} - e^{-ix} \right) + \sum_{n=-\infty, n\neq\pm 1}^{\infty} \frac{1 + (-1)^n}{2\pi(1 - n^2)} e^{inx} .$$

[Hint: *You should consider the c_n coefficients for $n = \pm 1$ separately.*]

Problem 3.30 Derive the sine/cosine Fourier series (3.16) directly from its exponential form (3.48).

Problem 3.31 Derive Parseval's theorem ($f(x)$ is a real function)

$$\frac{1}{2l} \int_{-l}^{l} f^2(x)dx = \sum_{n=-\infty}^{\infty} |c_n|^2 \tag{3.54}$$

for the (complex) exponential Fourier series.

Problem 3.32 Derive the more general Plancherel's theorem

$$\frac{1}{2l} \int_{-l}^{l} f(x)g^*(x)dx = \sum_{n=-\infty}^{\infty} c_n d_n^* \tag{3.55}$$

for two (generally complex) functions $f(x)$ and $g(x)$ of the same period of $2l$, with the corresponding (exponential) Fourier coefficients c_n and d_n.

Problem 3.33 Applying Parseval's theorem (3.54) to the Fourier series of Problem 3.29, show that

$$\sum_{n=0}^{\infty} \frac{1}{\left(4n^2 - 1\right)^2} = 1 + \frac{1}{3^2} + \frac{1}{15^2} + \cdots = \frac{\pi^2}{16} + \frac{1}{2} \, .$$

Problem 3.34 In theoretical many-body physics one frequently uses the so-called Matsubara Green's function $G(\tau_1, \tau_2) = G(\tau_1 - \tau_2)$. The function $G(\tau)$ is defined on the interval $-\beta \le \tau \le \beta$, where $\beta = 1/k_B T$ is the inverse temperature. For *bosons* the values of G for positive and negative arguments are related via the following relationship: $G(\tau) = G(\tau - \beta)$, where $\beta \ge \tau \ge 0$. Show that this function may be expanded into the following Fourier series:

$$G(\tau) = \frac{1}{\beta} \sum_n e^{-i\omega_n \tau} G(\omega_n) \, ,$$

where $\omega_n = \pi n/\beta$ are the so-called Matsubara frequencies, the summation is performed only over *even* (positive, negative, zero) integer values of n, and

$$G(\omega_n) = \int_0^{\beta} e^{i\omega_n \tau} G(\tau) d\tau$$

are the corresponding coefficients.

Problem 3.35 Show that exactly the same expansion exists also for *fermions* when the values of G for positive and negative arguments are related via $G(\tau) = -G(\tau - \beta)$, where $\beta \ge \tau \ge 0$. The only difference is that the summation is run only over *odd* values of n.

3.8 Application to Differential Equations

The Fourier method is frequently used to obtain solutions of differential equations in the form of the Fourier series. Then this series can be either summed up analytically or, if this does not work out, numerically on a computer using a final number of terms in the sum. As an example, consider a harmonic oscillator subject to a periodic excitation (external) force $f(t)$:

$$\ddot{y} + \omega_0^2 y = f(t) \, . \tag{3.56}$$

We assume that the force $f(t)$ is a periodic function with the period $T = 2\pi/\omega$ (frequency ω). Let ω_0 be the fundamental frequency of the harmonic oscillator. Here $\ddot{y}$ is a double derivative of $y(t)$ with respect to time t. We would like to obtain the particular integral of this differential equation to learn about the response of the oscillator to the external force $f(t)$.

Using the Fourier series method, it is possible to write down the solution of (3.56) for a general $f(t)$. Indeed, since $f(t)$ is periodic, we can expand it into the complex Fourier series (we use $l = \pi/\omega$ so that $T = 2l$):

$$f(t) = \sum_{n=-\infty}^{\infty} f_n e^{in\pi t\omega/\pi} = \sum_{n=-\infty}^{\infty} f_n e^{in\omega t} \qquad (3.57)$$

with

$$f_n = \frac{1}{T} \int_0^T f(t) e^{-in\omega t} dt = \frac{\omega}{2\pi} \int_0^{2\pi/\omega} f(t) e^{-in\omega t} dt , \qquad (3.58)$$

where the integration was shifted to the interval $0 < t < T = 2\pi/\omega$.

To obtain $y(t)$ that satisfies the differential equation above, we recognise from Eq. (3.56) that the function $Y(t) = \ddot{y}(t) + \omega_0^2 y(t)$ must also be periodic with the same periodicity as the external force; consequently, the function $y(t)$ sought for must also be such. Hence, we can expand it into a Fourier series as well:

$$y(t) = \sum_{n=-\infty}^{\infty} y_n e^{in\omega t} . \qquad (3.59)$$

Substituting Eqs. (3.57) and (3.59) into Eq. (3.56), we obtain[3]:

$$\sum_{n=-\infty}^{\infty} \left[(in\omega)^2 + \omega_0^2 \right] y_n e^{in\omega t} = \sum_{n=-\infty}^{\infty} f_n e^{in\omega t} ,$$

or

$$\sum_{n=-\infty}^{\infty} \left\{ \left[-(n\omega)^2 + \omega_0^2 \right] y_n - f_n \right\} e^{in\omega t} = 0 . \qquad (3.60)$$

This equation is satisfied for all values of t if and only if all coefficients of $\exp(in\omega t)$ are equal to zero simultaneously for all values of n:

$$\left[-(n\omega)^2 + \omega_0^2 \right] y_n - f_n = 0 . \qquad (3.61)$$

Indeed, upon multiplying both sides of Eq. (3.60) by $\exp(-im\omega t)$ with some fixed value of m and integrating between 0 and T, we get that only the $n = m$ term is

---

[3] Since the Fourier series for both $y(t)$ and $f(t)$ converge, the series for $\ddot{y}(t)$ must converge as well, i.e. the second derivative of the Fourier series of $y(t)$ must be well defined.

left in the left-hand side of Eq. (3.60) due to orthogonality of the functions $\chi_n(t) =$ $\exp{(in\omega t)}$:

$$\sum_{n=-\infty}^{\infty} \left\{ \left[-(n\omega)^2 + \omega_0^2 \right] y_n - f_n \right\} \int_0^T e^{in\omega t} e^{-im\omega t} dt$$

$$= \sum_{n=-\infty}^{\infty} \left\{ \left[-(n\omega)^2 + \omega_0^2 \right] y_n - f_n \right\} \delta_{nm} T$$

$$= T \left\{ \left[-(m\omega)^2 + \omega_0^2 \right] y_m - f_m \right\} = 0 ,$$

which immediately leads to Eq. (3.61). Alternatively, we may argue that functions $\chi_n(t) = \exp{(in\omega t)}$ with different m are linearly independent, so that all coefficients to them in the linear combination in Eq. (3.60) must be zero. Thus, we get the unknown Fourier coefficients y_n of the solution $y(t)$ as

$$y_n = \frac{f_n}{\omega_0^2 - (n\omega)^2} \tag{3.62}$$

and hence the whole solution reads

$$y(t) = \sum_{n=-\infty}^{\infty} \frac{f_n}{\omega_0^2 - (n\omega)^2} e^{in\omega t} . \tag{3.63}$$

We see that the harmonics of $f(t)$ with frequencies $n\omega$ are greatly enhanced in the solution if they come close to the fundamental frequency ω_0 of the harmonic oscillator (resonance).

3.9   A More Rigorous Approach to the Fourier Series

3.9.1   Convergence 'on Average'

Before actually giving the rigorous formulation for the Fourier series, we note that in Eq. (3.7) the function $f(x)$ is expanded into a set of *linearly independent* functions $\{\phi_n(x)\}$ and $\{\psi_n(x)\}$, see Sect. 1.1.2 for the definition of linear independence.

It is easy to see that cosine and sine functions $\{\phi_n(x)\}$ and $\{\psi_n(x)\}$ of Eq. (3.3) are linearly independent. Indeed, let us construct a linear combination of all functions with unknown coefficients α_i and β_i and set it to zero:

$$\sum_{i=0}^{\infty} [\alpha_i \phi_i(x) + \beta_i \psi_i(x)] = 0 . \tag{3.64}$$

Next we multiply out both sides by $\phi_j(x)$ with some fixed j and integrate over x between $-l$ and l:

$$\sum_{i=0}^{\infty} \left[\alpha_i \int_{-l}^{l} \phi_i(x)\phi_j(x)dx + \beta_i \int_{-l}^{l} \psi_i(x)\phi_j(x)dx \right] = 0 \,.$$

Due to the orthogonality of the functions, see Eqs. (3.5) and (3.6), all the integrals between any $\psi_i(x)$ and $\phi_j(x)$ will be equal to zero, while the integrals involving $\phi_i(x)$ $(i = 0, 1, 2, \ldots)$ and $\phi_j(x)$ give the Kronecker symbol δ_{ij}, i.e. only one term in the sum with the value of $i = j$ will survive:

$$\sum_{i=0}^{\infty} \alpha_i \int_{-l}^{l} \phi_i(x)\phi_j(x)dx = \sum_{i=0}^{\infty} \alpha_j \delta_{ij} l = \alpha_j l = 0 \quad \Longrightarrow \quad \alpha_j = 0 \,.$$

By taking different values of j, we find that any of the coefficients α_1, α_2, etc. is equal to zero. Similarly, by multiplying both sides of Eq. (3.64) on $\psi_j(x)$ with fixed j and integrating over x between $-l$ and l, we find $\beta_j = 0$. Since j was chosen arbitrarily, all the coefficients β_1, β_2, etc. are equal to zero. Hence, the sine and cosine functions $\sin(\pi n x/l)$ and $\cos(\pi n x/l)$ for all $n = 0, 1, 2, \ldots$ are linearly independent.

Intuitively, the function $f(x)$ contains an 'infinite' amount of 'information' since the set of x values is continuous; therefore, when we expand it into an infinite set of linearly independent functions, it naively seems that we provide an adequate amount of information for it, and hence the expansion must be meaningful. Of course, this is not yet sufficient for the $f(x)$ to be expandable: the set of functions $\{\phi_n(x)\}$ and $\{\psi_n(x)\}$ must also be *complete* to represent $f(x)$ adequately. We shall elaborate on this point later on.

In order to get some feeling of how the Fourier series, calculated using formulae (3.9) and (3.10) for the a_n and b_n coefficients, converges to the function $f(x)$, let us consider a general linear combination

$$f_N(x) = \frac{\alpha_0}{2} + \sum_{n=1}^{N} (\alpha_n \phi_n(x) + \beta_n \psi_n(x)) \tag{3.65}$$

of the same type as in the Fourier series (3.7), but with *arbitrary* coefficients α_n and β_n. In addition, the sum above is constructed out of only the first N functions $\phi_n(x)$ and $\psi_n(x)$ of the Fourier series.

Theorem 3.2 *The expansion (3.65) converges on average to the function $f(x)$ for any N if the coefficients α_n and β_n of the linear combination coincide with the corresponding Fourier coefficients a_n and b_n defined by Eqs. (3.9) and (3.10), i.e. when $\alpha_n = a_n$ and $\beta_n = b_n$ for any $n = 0, 1, \ldots, N$. By 'average' convergence we mean the minimum of the* mean square error *function*

$$\delta_N = \frac{1}{l} \int_{-l}^{l} [f(x) - f_N(x)]^2 \, dx . \tag{3.66}$$

Note that $\delta_N \geq 0$, i.e. cannot be negative.

Proof Expanding the square in the integrand of the error function (3.66), we get three terms:

$$\delta_N = \frac{1}{l} \int_{-l}^{l} f^2(x)dx - \frac{2}{l} \int_{-l}^{l} f(x)f_N(x)dx + \frac{1}{l} \int_{-l}^{l} f_N^2(x)dx . \tag{3.67}$$

Substituting the expansion (3.65) into each of these terms, they can be calculated. We use the orthogonality of the functions $\{\phi_n(x)\}$ and $\{\psi_n(x)\}$ to calculate first the last term in Eq. (3.67):

$$\frac{1}{l} \int_{-l}^{l} f_N^2(x)dx = \frac{\alpha_0^2}{2} + \frac{2}{l} \sum_{n=1}^{N} \frac{\alpha_0}{2} \int_{-l}^{l} (\alpha_n \phi_n(x) + \beta_n \psi_n(x)) \, dx$$

$$+ \frac{1}{l} \sum_{n,m=1}^{N} \int_{-l}^{l} (\alpha_n \phi_n(x) + \beta_n \psi_n(x)) (\alpha_m \phi_m(x) + \beta_m \psi_m(x)) \, dx .$$

The second term in the right-hand side is zero since the integrals of either ϕ_n or ψ_n are zero for any n. In the third term, only integrals between two ϕ_n or two ψ_n functions *with equal indices* survive, and thus we obtain

$$\frac{1}{l} \int_{-l}^{l} f_N^2(x)dx = \frac{\alpha_0^2}{2} + \sum_{n=1}^{N} (\alpha_n^2 + \beta_n^2) , \tag{3.68}$$

i.e. an identity similar to Parseval's theorem, Sect. 3.5. The second integral in Eq. (3.67) can be treated along the same lines:

$$\frac{2}{l}\int_{-l}^{l}f(x)f_N(x)dx = \frac{\alpha_0}{l}\int_{-l}^{l}f(x)dx$$

$$+\frac{2}{l}\sum_{n=1}^{N}\left(\alpha_n\int_{-l}^{l}f(x)\phi_n(x)dx + \beta_n\int_{-l}^{l}f(x)\psi_n(x)dx\right).$$

Using Eqs. (3.9) and (3.10) for the Fourier coefficients of $f(x)$, we can rewrite the above expression in a simplified form:

$$\frac{2}{l}\int_{-l}^{l}f(x)f_N(x)dx = \alpha_0 a_0 + 2\sum_{n=1}^{N}(\alpha_n a_n + \beta_n b_n). \tag{3.69}$$

Thus, collecting all terms, we get for the error:

$$\delta_N = \frac{1}{l}\int_{-l}^{l}f^2(x)dx - \alpha_0 a_0 - 2\sum_{n=1}^{N}(\alpha_n a_n + \beta_n b_n) + \frac{\alpha_0^2}{2} + \sum_{n=1}^{N}(\alpha_n^2 + \beta_n^2)$$

$$= \frac{1}{l}\int_{-l}^{l}f^2(x)dx + \left(\frac{\alpha_0^2}{2} - \alpha_0 a_0\right) + \sum_{n=1}^{N}\left[(\alpha_n^2 - 2\alpha_n a_n) + (\beta_n^2 - 2\beta_n b_n)\right]$$

$$= \frac{1}{l}\int_{-l}^{l}f^2(x)dx + \left[\frac{1}{2}(\alpha_0 - a_0)^2 - \frac{a_0^2}{2}\right]$$

$$+ \sum_{n=1}^{N}\left[(\alpha_n - a_n)^2 - a_n^2 + (\beta_n - b_n)^2 - b_n^2\right].$$

$$\tag{3.70}$$

It is seen that the minimum of δ_N with respect to the coefficients α_n and β_n of the trial expansion (3.65) is achieved at $\alpha_n = a_n$ and $\beta_n = b_n$, i.e. when the expansion (3.65) coincides with the partial Fourier series containing the first N terms. **Q.E.D.**

Theorem 3.3 *The Fourier coefficients a_n and b_n defined by Eqs. (3.9) and (3.10) tend to zero as $n \to \infty$, i.e.*

$$\lim_{n\to\infty}\int_{-l}^{l}f(x)\sin\frac{\pi n x}{l}dx = 0 \quad and \quad \lim_{n\to\infty}\int_{-l}^{l}f(x)\cos\frac{\pi n x}{l}dx = 0,$$

$$\tag{3.71}$$

if the function $f(x)$ is continuous everywhere within the interval $-l \le x \le l$, apart from points where it has discontinuities of the first kind.

Proof The conditions imposed on the function $f(x)$ in the formulation of the Theorem are needed for all the integrals we wrote above to exist. Then, the minimum error δ_N is obtained from Eq. (3.70) by putting $\alpha_n = a_n$ and $\beta_n = b_n$:

$$\delta_N = \frac{1}{l} \int_{-l}^{l} f^2(x)dx - \frac{a_0^2}{2} - \sum_{n=1}^{N} \left(a_n^2 + b_n^2\right) .$$ (3.72)

Note that the values of the coefficients α_n and β_n do not depend on the value of N; for instance, if N is increased by one, $N \to N+1$, two new coefficients are added to the expansion (3.65), α_{N+1} and β_{N+1}, however, the values of the previous coefficients remain the same. At the same time, the error (3.72) becomes $\delta_{N+1} = \delta_N - a_{N+1}^2 - b_{N+1}^2$, i.e. it gets two extra negative terms and hence can only become smaller since the error is always not negative by construction, i.e. $\delta_N \geq 0$. As the number of terms N in the expansion is increased, the error gets smaller and smaller. Therefore, from Eq. (3.72), given the fact that $\delta_N \geq 0$, we conclude that

$$\frac{a_0^2}{2} + \sum_{n=1}^{N} \left(a_n^2 + b_n^2\right) \leq \frac{1}{l} \int_{-l}^{l} f^2(x)dx .$$ (3.73)

As N is increased, the sum in the left-hand side is getting larger, but will always remain smaller than the positive value of the integral in the right-hand side. This means that the infinite series $\sum_{n=1}^{\infty} \left(a_n^2 + b_n^2\right)$ is absolutely convergent, and we can replace N with ∞:

$$\frac{a_0^2}{2} + \sum_{n=1}^{\infty} \left(a_n^2 + b_n^2\right) \leq \frac{1}{l} \int_{-l}^{l} f^2(x)dx .$$ (3.74)

Thus, the infinite series in the left-hand side above is bound from above. Since the series converges, the terms of it, $a_n^2 + b_n^2$, must tend to zero as $n \to \infty$ (Theorem I.7.4), i.e. each of the coefficients a_n and b_n must tend separately to zero as $n \to \infty$. **Q.E.D.**

As a simple corollary to this theorem, we notice that the integration limits in Eq. (3.71) can be arbitrary; in particular, these could only cover a part of the interval $-l \leq x \leq l$. Indeed, if the interval $a \leq x \leq b$ is given, which lies inside the interval $-l \leq x \leq l$, one can always define a new function which is equal to the original function $f(x)$ inside $a \leq x \leq b$ and is zero in the remaining part of the periodicity interval $-l \leq x \leq l$. The new function would still be integrable, and hence the above theorem would hold for it as well. This proves the made statement which we shall use below.

It can then be shown that the error $\delta_N \to 0$ as $N \to \infty$. This means that actually we have the equal sign in the above equation (3.74):

$$\frac{a_0^2}{2} + \sum_{n=1}^{\infty} \left(a_n^2 + b_n^2\right) = \frac{1}{l} \int_{-l}^{l} f^2(x)dx.$$ (3.75)

This is the familiar *Parseval's equality*, Eq. (3.33). To prove this, however, we have to perform a very careful analysis of the partial sum of the Fourier series which is the subject of the next subsection.

3.9.2 A More Rigorous Approach to the Fourier Series: Dirichlet Theorem

Consider a partial sum of N terms in the Fourier series,

$$S_N(x) = \frac{a_0}{2} + \sum_{n=1}^{N} \left(a_n \cos \frac{\pi n x}{l} + b_n \sin \frac{\pi n x}{l} \right) , \qquad (3.76)$$

where the coefficients a_n and b_n are given precisely by Eqs. (3.9) and (3.10). Using these expressions in (3.76), we get

$$
\begin{aligned}
S_N(x) &= \frac{1}{2l} \int_{-l}^{l} f(t) dt + \sum_{n=1}^{N} \left\{ \left(\frac{1}{l} \int_{-l}^{l} f(t) \cos \frac{\pi n t}{l} dt \right) \cos \frac{\pi n x}{l} \right. \\
&\quad + \left. \left(\frac{1}{l} \int_{-l}^{l} f(t) \sin \frac{\pi n t}{l} dt \right) \sin \frac{\pi n x}{l} \right\} \\
&= \frac{1}{l} \int_{-l}^{l} f(t) \left\{ \frac{1}{2} + \sum_{n=1}^{N} \left[\cos \frac{\pi n t}{l} \cos \frac{\pi n x}{l} + \sin \frac{\pi n t}{l} \sin \frac{\pi n x}{l} \right] \right\} dt \\
&= \frac{1}{l} \int_{-l}^{l} f(t) \left\{ \frac{1}{2} + \sum_{n=1}^{N} \cos \frac{\pi n (t - x)}{l} \right\} dt ,
\end{aligned}
$$

where we have used a well-known trigonometric identity for the expression in the square brackets, as well as the fact that we are dealing with a sum of a finite number of terms, and hence the summation and integration signs are interchangeable. The sum of cosine functions we have already calculated before, Eq. (I.2.64), so that

$$
\begin{aligned}
\frac{1}{2} + \sum_{n=1}^{N} \cos \frac{\pi n (t - x)}{l} &= \frac{1}{2 \sin \frac{\pi(t-x)}{2l}} \sin \left(\frac{\pi (t - x)}{l} \left(N + \frac{1}{2} \right) \right) \\
&= \frac{\sin \left(\frac{\pi}{2l} (t - x) (2N + 1) \right)}{2 \sin \frac{\pi(t-x)}{2l}} .
\end{aligned}
$$

Correspondingly, we can write for the partial sum an expression:

$$S_N(x) = \frac{1}{l} \int_{-l}^{l} f(t) \frac{\sin\left(\frac{\pi}{2l}(t-x)(2N+1)\right)}{2 \sin \frac{\pi(t-x)}{2l}} dt$$

$$= \frac{1}{l} \int_{x-l}^{x+l} f(t) \frac{\sin\left(\frac{\pi}{2l}(t-x)(2N+1)\right)}{2 \sin \frac{\pi(t-x)}{2l}} dt ,$$

(3.77)

where at the last step we shifted the integration interval by x. We can do it, firstly, as the function $f(t)$ is periodic, and, secondly, the ratio of both sine functions is also periodic with respect to t with the same period of $2l$ (check it!). Next, we split the integration region into two intervals: from $x - l$ to x, and from x to $x + l$, and shall make a different change of the variable $t \to p$ in each integral as shown below:

$$\frac{1}{l} \int_{x-l}^{x} f(t) \frac{\sin\left(\frac{\pi}{2l}(t-x)(2N+1)\right)}{2 \sin \frac{\pi(t-x)}{2l}} dt = \begin{vmatrix} t = x - 2p \\ dt = -2dp \end{vmatrix}$$

$$= \frac{1}{l} \int_{0}^{l/2} f(x-2p) \frac{\sin\left(\frac{\pi}{l}(2N+1)p\right)}{\sin\left(\frac{\pi}{l}p\right)} dp ,$$

$$\frac{1}{l} \int_{x}^{x+l} f(t) \frac{\sin\left(\frac{\pi}{2l}(t-x)(2N+1)\right)}{2 \sin \frac{\pi(t-x)}{2l}} dt = \begin{vmatrix} t = x + 2p \\ dt = 2dp \end{vmatrix}$$

$$= \frac{1}{l} \int_{0}^{l/2} f(x+2p) \frac{\sin\left(\frac{\pi}{l}(2N+1)p\right)}{\sin\left(\frac{\pi}{l}p\right)} dp ,$$

i.e. we can write

$$S_N(x) = \frac{1}{l} \int_{0}^{l/2} f(x-2p) \frac{\sin\left(\frac{\pi}{l}(2N+1)p\right)}{\sin\left(\frac{\pi}{l}p\right)} dp$$

$$+ \frac{1}{l} \int_{0}^{l/2} f(x+2p) \frac{\sin\left(\frac{\pi}{l}(2N+1)p\right)}{\sin\left(\frac{\pi}{l}p\right)} dp .$$

(3.78)

Next, we note that if $f(x) = 1$, then the Fourier series consists of a single term which is $a_0/2 = 1$, i.e. $a_0 = 2$, since a_n and b_n for $n \geq 1$ are all equal to zero (due to orthogonality of the corresponding sine and cosine functions, Eq. (3.3), with the cosine function $\phi_0(x) = 1$ corresponding to $n = 0$). Therefore, $S_N(x) = 1$ in this case for any N, both contributions in Eq. (3.78) become identical, and we can write

$$1 = \frac{2}{l} \int_{0}^{l/2} \frac{\sin\left(\frac{\pi}{l}(2N+1)p\right)}{\sin\left(\frac{\pi}{l}p\right)} dp .$$

Multiply this identity on both sides by $[f(x+\delta) + f(x-\delta)]/2$ (with some small $\delta > 0$) and subtract from (3.78) to obtain

$$S_N(x) - \frac{1}{2} [f(x - \delta) + f(x + \delta)]$$

$$= \frac{1}{l} \int_0^{l/2} [f(x - 2p) - f(x - \delta)] \frac{\sin\left(\frac{\pi}{l}(2N+1)p\right)}{\sin\left(\frac{\pi}{l}p\right)} dp$$

$$+ \frac{1}{l} \int_0^{l/2} [f(x + 2p) - f(x + \delta)] \frac{\sin\left(\frac{\pi}{l}(2N+1)p\right)}{\sin\left(\frac{\pi}{l}p\right)} dp$$

$$= \frac{1}{l} \int_0^{l/2} \Psi_1(p) \sin\left(\frac{\pi}{l}(2N+1)p\right) dp + \frac{1}{l} \int_0^{l/2} \Psi_2(p) \sin\left(\frac{\pi}{l}(2N+1)p\right) dp \; ,$$

$$(3.79)$$

where

$$\Psi_1(p) = \frac{[f(x - 2p) - f(x - \delta)]}{\sin\left(\frac{\pi}{l}p\right)} \; , \qquad (3.80)$$

$$\Psi_2(p) = \frac{[f(x + 2p) - f(x + \delta)]}{\sin\left(\frac{\pi}{l}p\right)} \; . \qquad (3.81)$$

The integrals in the right-hand side of Eq. (3.79) are of the form we considered in Theorem 3.3 and the corollary to it. They are supposed to tend to zero as $N \to \infty$ (where $n = 2N + 1$) if the functions $\Psi_1(p)$ and $\Psi_2(p)$ are continuous with respect to their variable p within $-l \le p \le l$, apart may be from some finite number of points of discontinuity of the first kind. If this was true, this theorem would be applicable and then

$$\lim_{N \to \infty} \left\{ S_N(x) - \frac{1}{2} [f(x - \delta) + f(x + \delta)] \right\} = 0 \; .$$

This would be the required result if we assume that $\delta \to 0$, i.e. the Fourier series (which is the limit of $S_N(x)$ for $N \to \infty$) would be equal to the mean value of $f(x)$ at the point x calculated using its left and right limits:

$$\lim_{N \to \infty} S_N(x) = \frac{1}{2} [f(x - \delta) + f(x + \delta)] \; . \qquad (3.82)$$

Hence, what is needed is to analyse the two functions, $\Psi_1(p)$ and $\Psi_2(p)$. We notice that both functions inherit the discontinuities of the first kind of $f(x)$ itself, i.e. they are indeed continuous everywhere apart from possible discontinuities of $f(x)$. The only special point is $p = +0$. Indeed, the $\sin(\pi p/l)$ in the denominator of both functions is zero at $p = 0$, and hence we have to consider this particular point carefully. It is easy to see, however, that both functions have well-defined limits at this point. Indeed, assuming $f(x)$ is differentiable on the left of x (i.e. at $x - 0$), we can apply the Lagrange formula (I.3.88) (recall that p changes between 0 and $l/2$, i.e. is always *positive*):

$$f(x - 2p) = f(x - \delta) + f'(x - \vartheta 2p)(-2p) \quad \text{with} \quad 0 < \vartheta < 1 \; ,$$

and hence

$$
\begin{aligned}
\lim_{p \to +0} \Psi_1(p) &= \lim_{p \to +0} \frac{[f(x - 2p) - f(x - \delta)]}{-2p} \frac{-2p}{\sin\left(\frac{\pi}{l} p\right)} \\
&= \lim_{p \to +0} \frac{[f(x - 2p) - f(x - \delta)]}{-2p} \lim_{p \to +0} \frac{-2p}{\sin\left(\frac{\pi}{l} p\right)} \\
&= \lim_{p \to +0} f'(x - \vartheta 2p) \lim_{p \to +0} \frac{-2p}{\sin\left(\frac{\pi}{l} p\right)} \\
&= f'(x - 0) \lim_{p \to +0} \frac{-2p}{\sin\left(\frac{\pi}{l} p\right)} = -f'(x - 0)\frac{2l}{\pi} \, ,
\end{aligned}
$$

since the limit of $z/\sin z$ when $z \to 0$ is equal to one (e.g. Eq. (I.2.86)). Hence, the limit of $\Psi_1(p)$ at $p \to +0$ is finite and is related to the *left* derivative of $f(x)$. Similarly,

$$
f(x + 2p) = f(x + \delta) + f'(x + \vartheta 2p)\, 2p \quad \text{with} \ \ 0 < \vartheta < 1,
$$

and hence

$$
\begin{aligned}
\lim_{p \to +0} \Psi_2(p) &= \lim_{p \to +0} \frac{[f(x + 2p) - f(x + \delta)]}{2p} \lim_{p \to +0} \frac{2p}{\sin\left(\frac{\pi}{l} p\right)} \\
&= f'(x + 0) \lim_{p \to +0} \frac{2p}{\sin\left(\frac{\pi}{l} p\right)} = f'(x + 0)\frac{2l}{\pi} \, ,
\end{aligned}
$$

it is also finite and is related to the *right* derivative of $f(x)$ (the left and right derivatives at x do not need to be the same). This concludes our proof: the functions $\Psi_1(p)$ and $\Psi_2(p)$ satisfy the conditions of Theorem 3.3 and its corollary, and hence $S_N(x)$ indeed tends to the mean value of $f(x)$ at point x, Eq. (3.82).

In the above proof we have made an assumption concerning the function $f(x)$, specifically, that it has well-defined left and right derivatives (that are not necessarily the same) at *any* point x between $-l$ and l; in fact, this assumption is not necessary and can be lifted leading to the general formulation of the Dirichlet Theorem 3.1; however, this more general proof is much more involved and will not be given here.

3.9.3 Expansion of a Function via Orthogonal Functions

Sine, cosine and complex exponential functions are just few examples of orthogonal sets of functions which can be used to expand 'good' functions in the corresponding functional series. There are many more such examples. Here we shall discuss in some detail this question, a rigorous discussion of this topic goes far beyond this book.

Let us assume that some continuous and generally complex functions $\psi_1(x)$, $\psi_2(x)$, $\psi_3(x)$, etc. form a set $\{\psi_i(x)\}$. We shall call this set of functions *orthonormal* on the interval $a \leq x \leq b$ with weight $w(x) \geq 0$ if for any i and j we have (cf. Sect. 1.1.2)

$$\int_a^b w(x)\psi_i^*(x)\psi_j(x)dx = \delta_{ij} , \qquad (3.83)$$

compare Eqs. (3.6) and (3.49). Next, we consider a generally complex function $f(x)$ with a finite number of discontinuities of the first kind, but continuous everywhere between any two points of discontinuity. Formally, let us *assume*, exactly in the same way as when we investigated the trigonometric Fourier series in Sect. 3.1 that $f(x)$ can be expanded into a functional series in terms of the functions of the set $\{\psi_i(x)\}$, i.e.

$$f(x) = \sum_i c_i \psi_i(x) , \qquad (3.84)$$

where c_i are expansion coefficients. To find them, we assume that the series above can be integrated term by term. Then we multiply both sides of the above equation by $w(x)\psi_j^*(x)$ with some fixed index j and integrate between a and b:

$$\int_a^b w(x)f(x)\psi_j^*(x)dx = \sum_i c_i \int_a^b w(x)\psi_j^*(x)\psi_i(x)dx$$
$$= \sum_i c_i \delta_{ij} = c_j , \qquad (3.85)$$

where we have made use of the orthonormality condition (3.83). Note that the coefficients c_i may be complex. Therefore, *if* the expansion (3.84) exists, the expansion coefficients c_j are to be determined from (3.85).

The expansion (3.84) is called the generalised Fourier expansion and coefficients (3.85) the generalised Fourier coefficients.

Problem 3.36 Assuming the existence of the expansion (3.84) and legitimacy of the integration term by term, prove Parseval's theorem for the generalised Fourier expansion:

$$\int_a^b w(x) |f(x)|^2 \, dx = \sum_i |c_i|^2 . \qquad (3.86)$$

Problem 3.37 Similarly, consider two functions, $f(x)$ and $g(x)$, both expanded into the functional series in terms of the same set $\{\psi_i(x)\}$. Then, prove Plancherel's theorem for the generalised Fourier expansion:

$$\int_a^b w(x) f^*(x) g(x) dx = \sum_j f_i^* g_i , \qquad (3.87)$$

where f_i and g_i are the corresponding generalised Fourier coefficients.

Problem 3.38 Prove that if the functions of the system $\{\psi_i(x)\}$ are orthogonal, then they are linearly independent.

Next, we build an error function

$$\delta_N = \int_a^b w(x) |f(x) - S_N(x)|^2 dx \qquad (3.88)$$

between $f(x)$ and the partial sum

$$S_N(x) = \sum_{i=1}^N f_i \psi_i(x) ,$$

constructed using some *arbitrary* coefficients f_i.

Problem 3.39 Substitute the above expansion for the partial sum into the error function (3.88) and, using explicit expressions for the generalised Fourier coefficients (3.85), show that the error function

$$\delta_N = \int_a^b w(x) |f(x)|^2 dx - \sum_{i=1}^N |c_i|^2 + \sum_{i=1}^N |c_i - f_i|^2 . \qquad (3.89)$$

Since the error function is non-negative, $\delta_N \geq 0$, it is seen from this result, Eq. (3.89), that the error δ_N is minimised if the coefficients f_i and c_i coincide: $f_i = c_i$. In other words, if the generalised Fourier expansion exists, its coefficients must be the generalised Fourier coefficients (3.85). Then the error

$$\delta_N = \int_a^b w(x) |f(x)|^2 \, dx - \sum_{i=1}^{N} |c_i|^2 \geq 0$$

(3.90)

$$\Longrightarrow \quad \int_a^b w(x) |f(x)|^2 \, dx \geq \sum_{i=1}^{N} |c_i|^2 \ .$$

We also notice that the c_i coefficients do not depend on N, and if N is increased, the error δ_N is necessarily reduced remaining non-negative. If the functional series $\sum_i c_i \psi_i(x)$ was equal to $f(x)$ everywhere except may be for a finite number of points with discontinuities of the first kind, then, according to the Parseval's theorem (3.86), we would have the equal sign in (3.90), i.e. the error δ_N in the limit of $N \to \infty$ would tend to zero. Therefore, there is a fundamental connection between Eq. (3.86) and the fact that the generalised Fourier expansion is equal to the original function $f(x)$ at all points apart from some finite number of them where $f(x)$ is discontinuous. Equation (3.86) is called the *completeness* condition, because it is closely related to the fact of whether an arbitrary function can be expanded into the set $\{\psi_i(x)\}$ or not.

Problem 3.40 Prove that if $f(x)$ is orthogonal to all functions of the complete system $\{\psi_i(x)\}$, then $f(x) = 0$. *[Hint: Use the completeness condition (3.86) as well as expressions for the c_i coefficients.]*

We know that orthogonal functions are linearly independent. Consider a subset of functions of a complete set, say with numbers $i = 1, \ldots, N$. If we take any other function from the complete set which is not in the selected subset, i.e. it is $\psi_j(x)$ with $j > N$, then it will be orthogonal to any function of the selected subset. However, obviously, this function is not identically zero. So, if a function is orthogonal to a set of functions, it does not necessarily mean that this function is zero. However, as follows from the last problem, if a function is orthogonal to any function of a *complete set*, then another function orthogonal to any function of the set but not belonging to it must be zero. This means that any function is *linearly dependent* of the functions of the set, which in turn means that it can be expanded in a linear combination of them.

Theorem 3.4 *If the functional series $\sum_i c_i \psi_i(x)$ converges uniformly with c_i being the Fourier coefficients for $f(x)$, and the set $\{\psi_i(x)\}$ is complete, then the series converges to $f(x)$.*

Proof Let us assume that the series converges to another function $g(x)$, i.e.

$$g(x) = \sum_i c_i \psi_i(x) \quad \text{with} \quad c_i = \int_a^b w(x) f(x) \psi_j^*(x) dx.$$

Since the series converges uniformly, $g(x)$ is continuous and we can integrate it term by term. Hence, we multiply both sides of the expression for the series by $w(x)\psi_j^*(x)$ with some fixed j and integrate between a and b:

$$\int_a^b w(x)g(x)\psi_j^*(x)dx = \sum_i c_i \int_a^b w(x)\psi_j^*(x)\psi_i(x)dx$$

$$\implies c_j = \int_a^b w(x)g(x)\psi_j^*(x)dx ,$$

i.e. the coefficients c_j have the same expressions both in terms of $f(x)$ and $g(x)$, i.e.

$$\int_a^b w(x)\left[f(x) - g(x)\right]\psi_j^*(x)dx = 0$$

for any j. The expression above states that the continuous function $h(x) = f(x) - g(x)$ is orthogonal to any of the functions $\psi_j(x)$ of the set. But if the set is complete, then $h(x)$ can only be zero, i.e. $g(x) = f(x)$. **Q.E.D.**

Using the notion of the Dirac delta function (Sect. 4.1), the completeness condition can be formulated directly in terms of the functions $\{\psi_i(x)\}$ of the set and the weight. Indeed, using (3.85), we write

$$f(x) = \sum_i c_i\psi_i(x) = \sum_i \int_a^b w(x)f(x')\psi_i^*(x')dx'\psi_i(x)$$

$$= \int_a^b f(x')\left[w(x)\sum_i \psi_i^*(x')\psi_i(x)\right]dx' .$$

It is now clearly seen, that, according to the filtering theorem for the delta function, Eq. (4.9), the square brackets above should be equal to $\delta(x - x')$, i.e.

$$w(x)\sum_i \psi_i^*(x')\psi_i(x) = \delta(x - x')$$

must hold. This is the required completeness condition.

Note that one may always remove the weight function from the consideration completely by introducing an alternative set of orthogonal functions $\widetilde{\psi}_i(x) = \psi_i(x)\sqrt{w(x)}$. These are orthogonal with the unit weight and for them the completeness condition simply states that

$$\sum_i \widetilde{\psi}_i^*(x')\widetilde{\psi}_i(x) = \delta(x - x') . \tag{3.91}$$

3.10 Investigation of the Convergence of the FS

We have discussed at length various properties of the FS, however, we have not actually investigated how the FS converges. The question we ask here is, for instance, which conditions the function $f(x)$ must satisfy so that its FS would converge uniformly. In fact, we shall show here[4] that the character of the FS convergence is directly related to the smoothness of the function $f(x)$. In particular, we shall find at which conditions the FS of $f(x)$ converges uniformly.

3.10.1 The Second Mean-Value Integral Theorem

We shall first need to establish a useful result for a definite integral that we shall be using later on.

Theorem 3.5 *Consider a continuous function $\phi(x)$ on the interval $a \leq x \leq b$ and a function $f(x)$ on the same interval that is either monotonically increasing or decreasing, and may also have a finite number of discontinuity jumps of the first kind (satisfies Dirichlet conditions). Then, one may write*

$$\int_a^b f(x)\phi(x)dx = f(a^+) \int_a^\xi \phi(x)dx + f(b^-) \int_\xi^b \phi(x)dx . \qquad (3.92)$$

Here $a^+ = \lim_{\delta \to 0}(a + \delta) = a + 0$, $b^- = \lim_{\delta \to 0}(b - \delta) = b - 0$, and ξ is a number in the open interval $a < \xi < b$.

Proof We shall prove this formula first by assuming that $f(x)$ is non-decreasing (i.e. for any $x_2 > x_1$ we have $f(x_2) \geq f(x_1)$) and that $f(a^+) = 0$, in which case we need to show that

$$\int_a^b f(x)\phi(x)dx = f(b^-) \int_\xi^b \phi(x)dx , \quad a < \xi < b . \qquad (3.93)$$

Once we have proven this formula, we shall be able to generalise our result for a more general $f(x)$ stated initially.

---

[4] In this section we closely follow Chapter VI of V. I. Smirnov 'A Course of Higher Mathematics. Advanced Calculus', Vol. II, Pergamon (1964).

Our function $f(x) \geq 0$ because $f(a^+) = 0$ and it is non-decreasing. Let us divide up the whole interval $a \leq x \leq b$ by n points $x_0 \equiv a, x_1, x_2, \ldots, x_n \equiv b$ and consider a sum

$$\sum_{i=1}^{n} f(\xi_i)\phi(\xi_i)(x_i - x_{i-1}),$$

where the point ξ_i is from the interval between x_{i-1} and x_i such that, according to the integral mean-value theorem (Theorem I.4.6),

$$\int_{x_{i-1}}^{x_i} \phi(x)dx = \phi(\xi_i)(x_i - x_{i-1}).$$

Therefore,

$$\sum_{i=1}^{n} f(\xi_i)\phi(\xi_i)(x_i - x_{i-1}) = \sum_{i=1}^{n} f(\xi_i) \int_{x_{i-1}}^{x_i} \phi(x)dx.$$

Each of the integrals in the right-hand side we can equivalently represent as a difference of two integrals with the common upper limit being b:

$$\sum_{i=1}^{n} f(\xi_i)\phi(\xi_i)\Delta_i = \sum_{i=1}^{n} f(\xi_i) \left[\int_{x_{i-1}}^{b} \phi(x)dx - \int_{x_i}^{b} \phi(x)dx \right] = \sum_{i=1}^{n} f(\xi_i) \left[F_{i-1} - F_i \right]$$

$$= \left\{ f(\xi_1)[F_0 - F_1] + f(\xi_2)[F_1 - F_2] + \cdots + f(\xi_n)[F_{n-1} - F_n] \right\}$$

$$= f(\xi_1)F_0 + \sum_{k=2}^{n} \left[f(\xi_k) - f(\xi_{k-1}) \right] F_{k-1}$$

$$= f(\xi_1)F(a) + \sum_{k=2}^{n} \left[f(\xi_k) - f(\xi_{k-1}) \right] F(x_{k-1}),$$

where $\Delta_i = x_i - x_{i-1}$ and $F_k = F(x_k) = \int_{x_k}^{b} \phi(x)dx$ are shortcuts; note that $F_n = 0$. Since the function $\phi(x)$ is continuous, so is the function $F(x)$. Hence, for any value of x, the function $F(x)$ would be bounded by two constants, $m \leq F(x) \leq M$, from below and from above. Therefore, we can make two estimates for the sum under consideration:

$$\sum_{i=1}^{n} f(\xi_i)\phi(\xi_i)(x_i - x_{i-1}) \leq f(\xi_1)M + \sum_{k=2}^{n} \underbrace{\left[f(\xi_k) - f(\xi_{k-1}) \right]}_{\geq 0} M$$

$$= M \left\{ f(\xi_1) + \sum_{k=2}^{n} \left[f(\xi_k) - f(\xi_{k-1}) \right] \right\} = Mf(\xi_n)$$

and

$$\sum_{i=1}^{n} f(\xi_i)\phi(\xi_i)\,(x_i - x_{i-1}) \geq f(\xi_1)m + \underbrace{\sum_{k=2}^{n} \big[f(\xi_k) - f(\xi_{k-1})\big]}_{\geq 0} m$$

$$= m\left\{f(\xi_1) + \sum_{k=2}^{n}\big[f(\xi_k) - f(\xi_{k-1})\big]\right\} = mf(\xi_n)\,,$$

so that

$$mf(\xi_n) \leq \sum_{i=1}^{n} f(\xi_i)\phi(\xi_i)\,(x_i - x_{i-1}) \leq Mf(\xi_n)\,.$$

Let us now make the number of divisions to go to infinity so that the largest interval $\max(x_i - x_{i-1})$ goes to zero. Then the sum above, being an appropriate integral sum, becomes a definite integral, and the last point ξ_n in the limit tends to the 'left edge' b^- of the upper limit b (since ξ_n is strictly smaller than b and remains such as $n \to \infty$):

$$mf(b^-) \leq \int_a^b f(x)\phi(x)dx \leq Mf(b^-)\,.$$

Therefore, the integral can be written as $f(b^-)\mathcal{M}$ with $\mathcal{M}$ being a number between m and M. On the other hand, both m and M are bounding values for the integral $F(x)$; hence, there exists such a value ξ between a and b that $F(\xi) = \mathcal{M}$ as $F(x)$ will pass through all values between m and M (Theorem I.2.19). We have then:

$$\int_a^b f(x)\phi(x)dx = f(b^-)F(\xi) = f(b^-)\int_\xi^b \phi(x)dx\,,$$

which is the required result, Eq. (3.93).

At the next step we shall eliminate the condition $f(a^+) = 0$ by formulating the following Problem:

Problem 3.41 Consider a continuous non-decreasing function $g(x) = f(x) - f(a^+)$. This function satisfies all conditions required for Eq. (3.93) to hold. Show that formula (3.92) then follows for the non-decreasing function $f(x)$.

The obtained formula is also valid for a non-increasing function $f(x)$ which can be easily concluded from the fact that we have already proven its validity for a non-decreasing function $-f(x)$. **Q.E.D.**

3.10.2   First Estimates of the Fourier Coefficients

'

The mean-value formula we obtained above can now be applied to the coefficient

$$a_n = \frac{1}{l} \int_{-l}^{l} f(x) \cos \frac{\pi n x}{l} dx$$

of the FS of $f(x)$. Here the cosine function is continuous, and $f(x)$ satisfies the Dirichlet conditions. It may be neither non-decreasing nor non-increasing; however, the interval $-l \le x \le l$ can always be split into subintervals such that within each of them $f(x)$ be either non-increasing or non-decreasing. Within each such subinterval $\alpha_k \le x \le \beta_k$, we can apply the mean-value formula (3.92) by choosing $\phi(x) = \cos \frac{\pi n x}{l}$:

$$\frac{1}{l} \int_{\alpha_k}^{\beta_k} f(x) \cos \frac{\pi n x}{l} dx = \frac{1}{l} \left[f(\alpha_k^+) \int_{\alpha_k}^{\xi_k} \cos \frac{\pi n x}{l} dx + f(\beta_k^-) \int_{\xi_k}^{\beta_k} \cos \frac{\pi n x}{l} dx \right]$$

$$= \frac{1}{\pi n} \left[f(\alpha_k^+) \left(\sin \frac{\pi n \xi_k}{l} - \sin \frac{\pi n \alpha_k}{l} \right) \right.$$

$$\left. + f(\beta_k^-) \left(\sin \frac{\pi n \beta_k}{l} - \sin \frac{\pi n \xi_k}{l} \right) \right],$$

so that

$$\left| \frac{1}{l} \int_{\alpha_k}^{\beta_k} f(x) \cos \frac{\pi n x}{l} dx \right|$$

$$= \frac{1}{\pi n} \left| f(\alpha_k^+) \left(\sin \frac{\pi n \xi_k}{l} - \sin \frac{\pi n \alpha_k}{l} \right) + f(\beta_k^-) \left(\sin \frac{\pi n \beta_k}{l} - \sin \frac{\pi n \xi_k}{l} \right) \right|$$

$$\le \frac{1}{\pi n} \left\{ \left| f(\alpha_k^+) \left(\sin \frac{\pi n \xi_k}{l} - \sin \frac{\pi n \alpha_k}{l} \right) \right| + \left| f(\beta_k^-) \left(\sin \frac{\pi n \beta_k}{l} - \sin \frac{\pi n \xi_k}{l} \right) \right| \right\}$$

$$\le \frac{1}{\pi n} \left\{ |f(\alpha_k^k)| + |f(\beta_k^-)| \right\} = \frac{M_k}{n},$$

where M_k is some positive constant. Obviously, the same estimate can be made for any other subinterval. Hence,

$$|a_n| = \left| \frac{1}{l} \int_{-l}^{l} f(x) \cos \frac{\pi n x}{l} dx \right| = \left| \sum_k \frac{1}{l} \int_{\alpha_k}^{\beta_k} f(x) \cos \frac{\pi n x}{l} dx \right|$$

$$\le \sum_k \left| \frac{1}{l} \int_{\alpha_k}^{\beta_k} f(x) \cos \frac{\pi n x}{l} dx \right| \le \sum_k \frac{M_k}{n} = \frac{M}{n}.$$

> **Problem 3.42** Repeat the calculation for the FS coefficient b_n and prove that the same estimate can done be for it as well.

So, it follows that FS coefficients can always be estimated as

$$|a_n| \leq \frac{M}{n} \quad \text{and} \quad |b_n| \leq \frac{M}{n}. \tag{3.94}$$

We have proven above (Theorem 3.3) that the FS coefficients tend to zero as $n \to \infty$. Now we can be more precise and state that they tend to zero *at least* as $1/n$.

Consider now a *continuous* periodic function $f(x)$ such that its first derivative $f'(x)$ satisfies the Dirichlet conditions and hence may have jumps. Just to emphasise, $f(x)$ is continuous, so does not have jumps; in particular, $f(-l) = f(l)$. Using integration by parts, we can write

$$
\begin{aligned}
a_n &= \frac{1}{l} \int_{-l}^{l} f(x) \cos \frac{\pi n x}{l} dx \\
&= \frac{1}{\pi n} \left[f(x) \sin \frac{\pi n x}{l} \Big|_{l}^{l} - \int_{-l}^{l} f'(x) \sin \frac{\pi n x}{l} dx \right] \\
&= -\frac{l}{\pi n} \left(\frac{1}{l} \int_{-l}^{l} f'(x) \sin \frac{\pi n x}{l} dx \right).
\end{aligned}
$$

The expression in the round brackets in the right-hand side is the coefficient $b_n^{(1)}$ of the FS of $f'(x)$. We have used the superscript here to indicate that $b_n^{(1)}$ corresponds to the first derivative of $f(x)$. In the next section we shall use notations $a_n^{(k)}$ and $b_n^{(k)}$ for the FS coefficients of the k−th derivative $f^{(k)}(x)$ of $f(x)$. For the coefficient $b_n^{(1)}$ we can apply the above made estimate Eq. (3.94), valid for any FS, which leads to

$$|a_n| \leq \frac{l}{\pi n} \left| b_n^{(1)} \right| \leq \frac{M'}{n^2},$$

where M' is some positive number. Exactly the same calculation leads to $|b_n| \leq M'/n^2$. It follows from these estimates that the FS of such a function converges absolutely and *uniformly* since

$$\left| \sum_{n=1}^{\infty} \left\{ a_n \cos \frac{n\pi x}{l} + b_n \sin \frac{n\pi x}{l} \right\} \right| \leq \sum_{n=1}^{\infty} (|a_n| + |b_n|) \leq M' \sum_{n=1}^{\infty} \frac{1}{n^2}.$$

The infinite sum in the right-hand side converges absolutely, which proves the above made statement.

So, if the function $f(x)$ is continuous and its first derivative $f'(x)$ satisfies the Dirichlet conditions, then its FS converges uniformly. If the function $f(x)$ itself contains discontinuity jumps, its Fourier coefficients are of the order of $1/n$ and the series does not converge uniformly.

The series of $f(x) = x^2$, $-\pi \le x \le \pi$ considered in Problem 3.8 is an example of such a series: $f(x)$ is continuous (no jumps), however, its first derivative does have jumps at $\pm\pi$ with well-defined limits on both sides of them. As a result, the Fourier coefficients of this function given by Eq. (3.18) behave like $1/n^2$, as expected, and hence the series converges absolutely and uniformly. Three successive approximations of the FS of this function are shown in Fig. 3.5. Conversely, the series of $f(x) = x$, $-\pi < x < \pi$ (Example 3.1), experiences jumps at the ends of the interval and its Fourier coefficients behave like $1/n$; the series does not converge uniformly then.

3.10.3 More Detailed Investigation of FS Convergence

In the previous subsection we found that the convergence of the FS could be as slow as $1/n$, but could also be faster, as $1/n^2$. In the latter case the function $f(x)$ must be continuous and only its first derivative may have discontinuities of the first kind. What would then be the character of the convergence of the FS if $f'(x)$ is also continuous? Or, if, together with the function $f(x)$ itself, all its derivatives up to the k-th order (excluding $f^{(k)}(x)$) are continuous and only the $k-$th derivative $f^{(k)}(x)$ has jumps (but still satisfies Dirichlet conditions)? The mentioned conditions the function $f(x)$ must satisfy would tell us a lot about its smoothness.

To answer all these questions and hence relate the convergence of the FS to the smoothness of $f(x)$, we can generalise the method of the previous subsection to a very general case as follows.

Let us assume that the function $f(x)$ has n_0 discontinuity jumps of the first kind at some points

$$x_0^{(0)} \equiv -l, \; x_1^{(0)}, \; x_2^{(0)}, \; \ldots, x_{n_0-1}^{(0)}, \; x_{n_0}^{(0)} \equiv l$$

within the interval $-l \le x \le l$. Note that we have included the end points of the interval as $f(x)$ may jump there as well. Similarly, we can assume that its first derivative $f'(x)$ has also n_1 discontinuities at points (possibly including the ends of the interval)

$$x_0^{(1)} \equiv -l, \; x_1^{(1)}, \; x_2^{(1)}, \; \ldots, x_{n_1-1}^{(1)}, \; x_{n_1}^{(1)} \equiv l \; .$$

Generally, the $k-$th derivative $f^{(k)}(x)$ has n_k discontinuities at points

$$x_0^{(k)} \equiv -l, \; x_1^{(k)}, \; x_2^{(k)}, \; \ldots, x_{n_k-1}^{(k)}, \; x_{n_k}^{(k)} \equiv l \; .$$

We show in Fig. 3.6 an example of a function $f(x)$ that not only has discontinuities itself, but its first derivative has them as well.

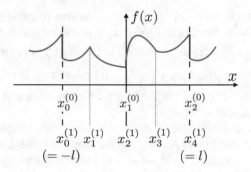

Fig. 3.6 A hypothetical function $f(x)$ defined in the interval $-l < x < l$ and then periodically repeated satisfies Dirichlet conditions with two discontinuities at $x_0^{(0)} = -l$ (and correspondingly at $x_2^{(0)} = l$) and $x_1^{(0)} = 0$. For clarity the values of the jumps are indicated by vertical red lines. It is continuous at points $x_1^{(1)}$ and $x_3^{(1)}$; however, at these two points its first derivative $f'(x)$ experiences discontinuities of the first kind, i.e. satisfies the Dirichlet conditions. Note that $f'(x)$ also experiences discontinuities at points $x_0^{(1)} \equiv x_0^{(0)}$, $x_2^{(1)} \equiv x_1^{(0)}$ and $x_4^{(1)} \equiv x_2^{(0)}$, where $f(x)$ is discontinuous; altogether, there are three discontinuities in $f(x)$ and five in $f'(x)$

Consider now the Fourier coefficient

$$a_n = \frac{1}{l} \int_{-l}^{l} f(x) \cos \frac{\pi n x}{l} dx ,$$

in which we split the integral by the regions of continuity of $f(x)$, which are

$$-l < x < x_1^{(0)}, \quad x_1^{(0)} < x < x_2^{(0)}, \quad \ldots, x_{n_0-1}^{(0)} < x < l .$$

Inside each such interval $f(x)$ is continuous but has discontinuities at its ends. Let us take the integral over the k−th interval $x_k^{(0)} < x < x_{k+1}^{(0)}$ (with $k = 0, 1, \ldots, n_0 - 1$) by parts:

$$\frac{1}{l} \int_{x_k^{(0)}}^{x_{k+1}^{(0)}} f(x) \cos \frac{\pi n x}{l} dx = \frac{1}{\pi n} \left[f(x) \sin \frac{\pi n x}{l} \Big|_{x_k^{(0)}}^{x_{k+1}^{(0)}} - \int_{x_k^{(0)}}^{x_{k+1}^{(0)}} f(x) \sin \frac{\pi n x}{l} dx \right]$$

$$= \frac{1}{\pi n} \left[\left(S_{k+1} f_{k+1}^- - S_k f_k^+ \right) - \int_{x_k^{(0)}}^{x_{k+1}^{(0)}} f(x) \sin \frac{\pi n x}{l} dx \right] ,$$

where

$$S_{k+1} = \sin \frac{\pi n x_{k+1}^{(0)}}{l} , \quad S_k = \sin \frac{\pi n x_k^{(0)}}{l} ,$$

and

$$f_{k+1}^- = f \left(x_{k+1}^{(0)} - 0 \right) , \quad f_k^+ = f \left(x_k^{(0)} + 0 \right) .$$

We have adopted the variables $x_k^{(0)}$ and $x_{k+1}^{(0)}$ for the sine function as it is continuous. However, we had to use the right limit of $f(x)$ at the bottom limit (the left edge of the interval), hence $f\left(x_k^{(0)} + 0\right)$ there, and the left limit of $f(x)$ at the upper limit (the right edge of the interval), hence $f\left(x_{k+1}^{(0)} - 0\right)$ there. This is because the function $f(x)$ has discontinuities exactly at the interval's boundaries and hence we have to use an open interval excluding them.

Let us now sum up these expressions for all our subintervals. In the left-hand side we shall get the full coefficient a_n; in the right-hand side all integrals combine into an integral over the whole interval $-l < x < l$ which is directly related to the b_n coefficients of the FS of $f'(x)$, that is,

$$b_n^{(1)} = \frac{1}{l} \int_{-l}^{l} f'(x) \sin \frac{\pi n x}{l} dx \ ;$$

all the free terms that appeared above after integration by parts can be rearranged. We obtain

$$
\begin{aligned}
a_n &= \frac{1}{\pi n} \left[(S_1 f_1^- - S_0 f_0^+) + (S_2 f_2^- - S_1 f_1^+) + \cdots + \left(S_{n_0} f_{n_0}^- - S_{n_0-1} f_{n_0-1}^+ \right) \right] - \frac{l}{\pi n} b_n^{(1)} \\
&= \frac{1}{\pi n} \left[S_0 \left(f_{n_0}^- - f_0^+ \right) + S_1 \left(f_1^- - f_1^+ \right) + \cdots + S_{n_0-1} \left(f_{n_0-1}^- - f_{n_0-1}^+ \right) \right] - \frac{l}{\pi n} b_n^{(1)} \\
&= -\frac{1}{\pi n} \sum_{k=1}^{n_0} \Delta_k^{(0)} \sin \frac{\pi n x_k^{(0)}}{l} - \frac{l}{\pi n} b_n^{(1)} \\
&= -\frac{A_n^{(0)}}{n} - \frac{l}{\pi n} b_n^{(1)} \ .
\end{aligned}
$$

$$(3.95)$$

On the second line we have used the fact that $S_{n_0} = S_0$ (since $x_0^{(0)} = -l$ and $x_{n_0}^{(0)} = l$, and at both points the sine function is the same and equals to zero[5]). Next, we have introduced the values of the jumps of the function at the points of discontinuity:

$$\Delta_1^{(0)} = f_1^+ - f_1^- = f\left(x_1^{(0)} + 0\right) - f\left(x_1^{(0)} - 0\right) \ ,$$

$$\Delta_2^{(0)} = f_2^+ - f_2^- = f\left(x_2^{(0)} + 0\right) - f\left(x_2^{(0)} - 0\right) \ ,$$

$$\cdots \quad \cdots$$

$$\Delta_{n_0}^{(0)} = f_{n_0}^+ - f_{n_0}^- = f\left(x_{n_0}^{(0)} + 0\right) - f\left(x_{n_0}^{(0)} - 0\right) \ ,$$

where $\Delta_{n_0}^{(0)}$ corresponds to the (possible) jump of the function at the edges of the interval. Finally, we have collected all terms with the values of the jumps and the sinusoidal functions in a parameter $A_n^{(0)}$ which dependence on n is limited (bounded due to the sine functions).

---

[5] Although we keep this term for clarity.

Problem 3.43 Repeat the calculation made above for b_n to show that

$$b_n = \frac{1}{\pi n} \sum_{k=1}^{n_0} \Delta_k^{(0)} \cos \frac{\pi n x_k^{(0)}}{l} + \frac{l}{\pi n} a_n^{(1)} = \frac{B_n^{(0)}}{n} + \frac{l}{\pi n} a_n^{(1)}.$$

Here $a_n^{(1)}$ is the a_n coefficient of the FS of $f'(x)$.

The obtained expressions clearly show that if $f(x)$ has discontinuities, the leading terms in the a_n and b_n coefficients of its FS are of the order of $1/n$ leading to a very slow convergence. If, however, $f(x)$ is itself continuous, then $A_n^{(0)} = B_n^{(0)} = 0$, and the convergence of the FS is determined by the behaviour of the FS coefficients of the derivative $f'(x)$. Then, if it is known that $f'(x)$ has discontinuities, then its FS coefficients $a_n^{(1)}$ and $b_n^{(1)}$ will themselves be of the order of $1/n$ leading to the coefficients of $f(x)$ to behave as $1/n^2$. This is exactly what we have obtained in Sect. 3.10.2.

To be more detailed, we can now proceed further to establish a general behaviour of the FS coefficients by repeating the previous steps for the FS coefficients $a_n^{(1)}$ and $b_n^{(1)}$ of $f'(x)$. Of course, there is no need to duplicate the algebra as the above made analysis for $f(x)$ can be immediately applied to $f'(x)$ by just lifting by one the value of the superscripts:

$$a_n^{(1)} = -\frac{A_n^{(1)}}{n} - \frac{l}{\pi n} b_n^{(2)} \quad \text{and} \quad b_n^{(1)} = \frac{B_n^{(1)}}{n} + \frac{l}{\pi n} a_n^{(2)},$$

where $a_n^{(2)}$ and $b_n^{(2)}$ are the Fourier coefficients for $f''(x)$ and

$$A_n^{(1)} = \frac{1}{\pi} \sum_{k=1}^{n_1} \Delta_k^{(1)} \sin \frac{\pi n x_k^{(1)}}{l} \quad \text{and} \quad B_n^{(1)} = \frac{1}{\pi} \sum_{k=1}^{n_1} \Delta_k^{(1)} \cos \frac{\pi n x_k^{(1)}}{l}$$

with the quantities $\Delta_k^{(1)}$ being values of the jumps of $f'(x)$ at n_1 points. Obviously, this analysis can be repeated further by considering expressions for $a_n^{(2)}$ and $b_n^{(2)}$, which will be related to $a_n^{(3)}$ and $b_n^{(3)}$ of $f^{(3)}(x)$, and so on. Substituting recursively coefficients of higher order derivatives of $f(x)$ to the expression for the Fourier coefficients of the lower order, we in the end obtain the following representations for the Fourier coefficients of $f(x)$ we are interested in:

$$a_n = -\frac{A_n^{(0)}}{n} - \left(\frac{l}{\pi}\right) \frac{B_n^{(1)}}{n^2} + \left(\frac{l}{\pi}\right)^2 \frac{A_n^{(2)}}{n^3} + \left(\frac{l}{\pi}\right)^3 \frac{B_n^{(3)}}{n^4} + \cdots + \left(\frac{l}{\pi}\right)^k \frac{d_n^{(k)}}{n^k},$$

$$b_n = \frac{B_n^{(0)}}{n} - \left(\frac{l}{\pi}\right) \frac{A_n^{(1)}}{n^2} + \left(\frac{l}{\pi}\right)^2 \frac{B_n^{(2)}}{n^3} + \left(\frac{l}{\pi}\right)^3 \frac{A_n^{(3)}}{n^4} + \cdots + \left(\frac{l}{\pi}\right)^k \frac{c_n^{(k)}}{n^k},$$

where $d_n^{(k)}$ and $c_n^{(k)}$ are either $a_n^{(k)}$ or $b_n^{(k)}$ coefficients, depending on the actual value of k. Here $a_n^{(k)}$ or $b_n^{(k)}$ are Fourier series coefficients of $f^{(k)}(x)$. Importantly, the last term in the above expressions is of the order of $1/n^{k+1}$ since the coefficients $a_n^{(k)}$ or $b_n^{(k)}$ are themselves at least of the order of $1/n$.

We have arrived at the important result: if the function $f(x)$ and all its derivatives up to the order $k-1$ are continuous functions, and $f^{(k)}(x)$ satisfies Dirichlet conditions, then all the coefficients $A_n^{(i)}$ and $B_n^{(i)}$ are equal to zero for any $i = 0, 1, \ldots, k-1$, and therefore the expansions of a_n and b_n start from the last term that is of the order of $1/n^{k+1}$. If $k \geq 1$, the coefficients are guaranteed to decay at least as $1/n^2$ resulting in the FS for $f(x)$ to converge absolutely and uniformly.

3.10.4 Improvements of Convergence of the FS

It is clear from the analysis made in the previous subsection that the convergence of a Fourier series could be slow if the function $f(x)$ has discontinuities as in this case the FS coefficients will behave as $1/n$. This will present practical difficulties in calculating the FS as the computational time will be large (many terms need to be retained). Also, such a FS cannot be differentiated since this would result in a diverging series.

If we know the points of discontinuity and their value, there is a very simple fix to this problem. Indeed, one could choose an auxiliary function $f_1(x)$ (e.g. a set of horizontal and/or generally straight lines) that has *the same discontinuities*, so that their difference, $\Delta_1(x) = f(x) - f_1(x)$, would be continuous. Then, $f(x) = f_1(x) + \Delta_1(x)$, where $f_1(x)$ is the selected simple function available analytically and the FS of $\Delta_1(x)$ converging much faster, with the FS coefficients behaving at least as $1/n^2$.

If one needs to improve the convergence of $\Delta_1(x)$, then it is necessary to consider possible jumps of discontinuity in its first derivative. Again, we consider $\Delta_1'(x)$ and chose another auxiliary function $f_2(x)$, e.g. as a set of straight lines (linear polynomials), so that the difference $\Delta_2(x) = \Delta_1'(x) - f_2(x)$ be continuous. As a result, the FS of $\Delta_2(x)$ would contain the coefficients behaving at least as $1/n^2$. Then,

$$\Delta_1'(x) = \Delta_2(x) + f_2(x) \quad \Longrightarrow \quad \Delta_1(x) = \int f_2(x)dx + \int \Delta_2(x)dx ,$$

where the integral of $f_2(x)$ is a well-defined function given as a combination of parabolas, while the last term, the integral of $\Delta_2(x)$, would represent a FS with the coefficients that behave at least as $1/n^3$. Obviously, this procedure can be continued with the result that the function $f(x)$ in question is given as a set of simple analytical functions and a FS that has a much better convergence.

This method is good, however, only if we know (or can somehow deduce from the Fourier coefficients) the points of discontinuity. Usually, these are not known as we do not know what $f(x)$ is; for instance, we may just have its FS expansion as a solution of a differential equation (see Sect. 3.8). In this case another approach (attributed to A. N. Krylov) could be used. The idea is to extract from the coefficients a_n and b_n the parts of the order of $1/n$, $1/n^2$, etc.

To illustrate this method, suppose, a_n contains the leading dependence on n of the form:

$$a_n \sim \frac{n^2 + 3}{n^3 + 1}.$$

The coefficient behaves as $1/n$ at large n leading to a slow convergence. Let us separate out the $1/n$ term:

$$\frac{n^2 + 3}{n^3 + 1} = \frac{A}{n} + c_n,$$

where c_n is at least as $1/n^2$. Multiplying both sides on n, we have

$$n \frac{n^2 + 3}{n^3 + 1} = A + n c_n.$$

By taking $n \to \infty$ limit on both sides and using the fact that $n c_n \to 0$, we obtain

$$A = \lim_{n \to \infty} n \frac{n^2 + 3}{n^3 + 1} = \lim_{n \to \infty} \frac{1 + 3/n^2}{1 + 1/n^3} = 1.$$

Hence, the required remainder

$$c_n = \frac{n^2 + 3}{n^3 + 1} - \frac{1}{n} = \frac{3n - 1}{n(n^3 + 1)}$$

behaves already as $1/n^3$ at large n ensuring better convergence. At the same time, the series with A/n types of terms can be easily calculated or found in the FS tables. For instance,

$$\sum_{n=1}^{\infty} \frac{1}{n} \sin \frac{\pi n x}{l} = \frac{\pi}{2l} \begin{cases} l - x, & 0 < x < 2l \\ 0, & x = 0, 2l \end{cases}, \tag{3.96}$$

$$\sum_{n=1}^{\infty} \frac{(-1)^n}{n} \sin \frac{\pi n x}{l} = -\frac{\pi}{2l} \begin{cases} x, & -l < x < l \\ 0, & x = \pm l \end{cases},$$

i.e. these are precisely simple auxiliary functions we have talked about before. Note that they are periodic, hence, you should be able to apply them in any interval. For instance, if the $1/n$ part of a_n is $\cos(\pi n/2l)/n$, then we can write

$$\sum_{n=1}^{\infty} \frac{\cos(\pi n/2l)}{n} \sin \frac{\pi n x}{l} = \sum_{n=1}^{\infty} \frac{1}{2n} \left[\sin \left(\frac{\pi n}{l} \left(x + \frac{\pi}{2} \right) \right) + \sin \left(\frac{\pi n}{l} \left(x - \frac{\pi}{2} \right) \right) \right]$$

$$= \frac{1}{2} \sum_{n=1}^{\infty} \frac{1}{n} \sin \left(\frac{\pi n}{l} \left(x + \frac{\pi}{2} \right) \right) + \frac{1}{2} \sum_{n=1}^{\infty} \frac{1}{n} \sin \left(\frac{\pi n}{l} \left(x - \frac{\pi}{2} \right) \right),$$

where both series correspond to the FS (3.96) with a shifted argument and hence can be related to it.

Of course, in the same way one can extract the terms of the order of $1/n^2$ if required and continue in this way until the desired convergence is achieved. One would need a series with $1/n^2$. These could be obtained, e.g. by integrating the series above with respect to x between 0 and x. For instance, when integrating the first one, we get

$$\sum_{n=1}^{\infty} \frac{1}{n^2} \cos \frac{\pi n x}{l} = \frac{\pi^2}{12l^2} \left(3x^2 - l^2 \right), \quad -l \le x \le -l.$$

We have made use of Eq. (3.34) here. As expected, the series in the left-hand side is a continuous function as it is the same at $x = \pm l$. Also, at $x = 0$ we recover Eq. (3.22).

In practice, when extracting the terms associated with the 'slow' behaviour, one needs to find simple enough functions whose FS coincides reproduces the required behaviour.

3.11 Applications of Fourier Series in Physics

The method of Fourier series is being used a lot in sciences since it allows presenting a function (e.g. an electric signal or a field), which may have a complicated form, as a linear combination of simple complex exponentials or sine and cosine functions which are much easier to deal with; all the information about the function will then be 'stored' in the expansion coefficients. Therefore, the Fourier series method is also used when solving differential equations, especially partial differential equations, since in this case the problem with respect to all or some of the variables becomes an algebraic one related to finding Fourier coefficients in the expansion of the unknown functions.

Here we shall consider several examples of using Fourier series in condensed matter physics. We shall also introduce notations often used in physics which are slightly different to those used above.

3.11.1 Expansions of Functions Describing Crystal Properties

Consider a periodic solid (a crystal) and a function $f(x, y, z) = f(\mathbf{r})$ which describes one of its properties, e.g. the charge density or the electrostatic potential at point $\mathbf{r} = (x, y, z)$. This function is periodic in several directions, and hence can be expanded into multiple Fourier series. Here we shall consider how this can be done in some detail.

Let us start from a one-dimensional crystal: consider a one-dimensional periodic chain of atoms arranged along the x direction, similar to the one shown in Fig. 3.7. Let $\rho(x)$ be the electron density of the system corresponding to a distribution of electrons on atoms of the chain. The density must repeat the periodicity of the chain itself, i.e. $\rho(x + a) = \rho(x)$ for any point x along the axis, where a ($= 2l$ in our previous notations) is the periodicity, which is the smallest distance between two equivalent atoms. The density is a smooth function of x, continuous everywhere, and hence $\rho(x)$ can be expanded into the Fourier series:

$$\rho(x) = \sum_{n=-\infty}^{\infty} \rho_n e^{i2\pi nx/a} = \sum_n \rho_n e^{ig_n x} = \sum_g \rho_g e^{igx} \ ,$$

where $g_n = n\,(2\pi/a)$ is the one-dimensional reciprocal lattice "vector" g, so that the summation over n can simply be replaced with the sum over all possible such values of g (the last equality), and

$$\rho_n = \rho_g = \frac{1}{a} \int_0^a \rho(x) e^{-ig_n x} dx$$

are the corresponding Fourier coefficients. In the last passage of the expansion formula for $\rho(x)$ we have used simplified notations in which n is dropped; these are also frequently used. The quantity $b = 2\pi/a$ is called a reciprocal lattice vector (of the one-dimensional lattice) as it corresponds to the periodicity of this reciprocal lattice: $g_n = bn$. We can see that the Fourier series in this case can be mapped onto

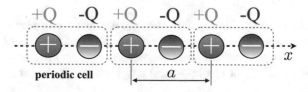

Fig. 3.7 A one-dimensional infinite periodic crystal of alternating point charges $\pm Q$. Here a is the closest distance between equivalent atoms yielding the period. Each unit cell (indicated as 'periodic cell') contains two oppositely charged atoms with charges $Q > 0$ and $-Q < 0$. The total charge of the unit cell is zero and there is an infinite number of such cells running along the positive and negative directions of the x-axis

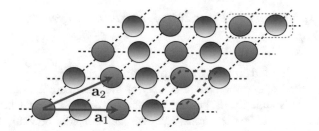

Fig. 3.8 A two-dimensional lattice of atoms of two species. The system is seen periodic across the lattice with the lattice vectors being $\mathbf{a}_1$ and $\mathbf{a}_2$. Two non-equivalent atoms are indicated by a pink dashed line, while the irreducible region in the 2D space associated with the unit cell is shown by a dashed red line

the lattice sites of the (imaginable) reciprocal lattice with the periodicity b, and the single summation over n would take all possible lattice sites g of this lattice into account.

This consideration can be generalised for two- and three-dimensional lattices. Consider first a two-dimensional lattice shown in Fig. 3.8. We introduce two vectors on the plane of the lattice, $\mathbf{a}_1$ and $\mathbf{a}_2$. Then any lattice site can be related to a reference (zero) site via the lattice vector $\mathbf{a} = n_1\mathbf{a}_1 + n_2\mathbf{a}_2$, where n_1 and n_2 are two integers. By taking on all possible negative and positive values of n_1 and n_2 including zero (corresponding to the zero lattice site), the whole infinite two-dimensional lattice can be reproduced. The density in this case $\rho(x, y) = \rho(\mathbf{r})$ is a function of the two-dimensional vector $\mathbf{r} = (x, y)$ or, alternatively, of the two coordinates x and y. However, in general the periodicity of the system does not necessarily follow the Cartesian directions, i.e. the lattice vectors $\mathbf{a}_1$ and $\mathbf{a}_2$ may be directed differently to the x- and y-axes. In this case it is more convenient to consider the density as a function of the so-called *fractional coordinates* r_1 and r_2 instead. These appear if we write $\mathbf{r}$ in terms of the lattice vectors, $\mathbf{r} = r_1\mathbf{a}_1 + r_2\mathbf{a}_2$, with r_1 and r_2 being real numbers between $-\infty$ and $+\infty$. Then the density $\rho(\mathbf{r})$ becomes a function[6] $\rho_1(r_1, r_2)$ of the fractional coordinates r_1 and r_2. The convenience of this representation is in that the density is periodic with respect to both r_1 and r_2 with the period equal to one, i.e.

$$\rho_1(r_1 + 1, r_2) = \rho_1(r_1, r_2 + 1) = \rho_1(r_1, r_2) \ ,$$

since adding the unity to either r_1 or r_2 changes the vector $\mathbf{r}$ in $\rho(\mathbf{r})$ exactly by the lattice vector $\mathbf{a}_1$ or $\mathbf{a}_2$, i.e. in either case an equivalent point in space is obtained. Hence, $\rho_1(r_1, r_2)$ is periodic and can be expanded into the Fourier series. We first consider ρ_1 as a function of r_1 and expand it as above with respect to this variable only:

---

[6] We use ρ_1 here instead of ρ since both functions of their arguments will have a different functional form; however, of course, they give the same value of the density when calculated at the same spatial point, i.e. $\rho_1(r_1, r_2) = \rho(\mathbf{r})$ for $\mathbf{r} = r_1\mathbf{a}_1 + r_2\mathbf{a}_2$.

$$\rho_1(r_1, r_2) = \sum_{n_1=-\infty}^{\infty} \rho_{n_1}(r_2) e^{ik_1 r_1}$$

with

$$\rho_{n_1}(r_2) = \int_{-1/2}^{1/2} \rho_1(r_1, r_2) e^{-ik_1 r_1} dr_1 = \int_0^1 \rho_1(r_1, r_2) e^{-ik_1 r_1} dr_1 \, ,$$

where $k_1 = 2\pi n_1$ and the corresponding Fourier coefficient $\rho_{n_1}(r_2)$ is still an explicit function of r_2 (recall, that the periodicity here is just equal to one). It is easily seen that ρ_1 is indeed periodic with respect to r_1 with the shortest period of 1. Because the density is also periodic with respect to r_2, the Fourier coefficients above must still be periodic with respect to it: $\rho_{n_1}(r_2) = \rho_{n_1}(r_2 + 1)$. Hence these can in turn be expanded in a Fourier series and we arrive at a double Fourier series for the whole density:

$$
\begin{aligned}
\rho_1(r_1, r_2) &= \sum_{n_1=-\infty}^{\infty} \rho_{n_1}(r_2) e^{ik_1 r_1} \\
&= \sum_{n_1=-\infty}^{\infty} \left[\sum_{n_2=-\infty}^{\infty} \rho_{n_1 n_2} e^{ik_2 r_2} \right] e^{ik_1 r_1} \\
&= \sum_{n_1=-\infty}^{\infty} \sum_{n_2=-\infty}^{\infty} \rho_{n_1 n_2} e^{i(k_1 r_1 + k_2 r_2)} \, ,
\end{aligned}
\tag{3.97}
$$

where $k_2 = 2\pi n_2$ and

$$\rho_{n_1 n_2} = \int_0^1 \rho_{n_1}(r_2) e^{-ik_2 r_2} dr_2 = \int_0^1 dr_1 \int_0^1 dr_2 \, \rho_1(r_1, r_2) e^{-i(k_1 r_1 + k_2 r_2)} \, . \tag{3.98}$$

These are the final Fourier coefficients.

The above equations can also be written in a more concise form. To this end, as it is customary done in crystallography and solid-state physics, we introduce reciprocal lattice vectors $\mathbf{b}_1$ and $\mathbf{b}_2$ via $\mathbf{a}_i \cdot \mathbf{b}_j = 2\pi \delta_{ij}$, where $i, j = 1, 2$, and δ_{ij} is the Kronecker delta symbol.

Problem 3.44 Show that the two-dimensional reciprocal lattice vectors are

$$\mathbf{b}_1 = \frac{2\pi}{v_s^2} \left[a_2^2 \mathbf{a}_1 - (\mathbf{a}_1 \cdot \mathbf{a}_2) \mathbf{a}_2 \right] \, , \quad \mathbf{b}_2 = \frac{2\pi}{v_s^2} \left[-(\mathbf{a}_1 \cdot \mathbf{a}_2) \mathbf{a}_1 + a_1^2 \mathbf{a}_2 \right] \, ,$$

where $v_s = |a_{1x} a_{2y} - a_{2x} a_{1y}|$ is the area of the unit cell. [Hint: Expand $\mathbf{b}_1$ and $\mathbf{b}_2$ with respect to $\mathbf{a}_1$ and $\mathbf{a}_2$ and find the expansion coefficients using their definition $\mathbf{a}_i \cdot \mathbf{b}_j = 2\pi \delta_{ij}$.]

Next we define the reciprocal lattice vector $\mathbf{g} = n_1\mathbf{b}_1 + n_2\mathbf{b}_2$ corresponding to the (n_1, n_2) site in the reciprocal lattice. It is easy to see then that

$$\mathbf{g} \cdot \mathbf{r} = (n_1\mathbf{b}_1 + n_2\mathbf{b}_2) \cdot (r_1\mathbf{a}_1 + r_2\mathbf{a}_2) = n_1 r_1 (\mathbf{a}_1 \cdot \mathbf{b}_1) + n_2 r_2 (\mathbf{a}_2 \cdot \mathbf{b}_2)$$
$$= 2\pi (n_1 r_1 + n_2 r_2) = k_1 r_1 + k_2 r_2 \, ,$$

and hence Eqs. (3.97) and (3.98) are simplified into:

$$\rho(\mathbf{r}) = \sum_{\mathbf{g}} \rho_{\mathbf{g}} e^{i\mathbf{g}\cdot\mathbf{r}} \, , \quad \text{where} \quad \rho_{\mathbf{g}} = \frac{1}{v_s} \iint_{cell} \rho(\mathbf{r}) e^{-i\mathbf{g}\cdot\mathbf{r}} d\mathbf{r} \, , \tag{3.99}$$

where we sum over all possible reciprocal lattice vectors (it is in fact a double sum over all possible integer values of n_1 and n_2). Notice that the integration in the definition of the Fourier coefficients $\rho_{\mathbf{g}}$ is performed over the whole area of the unit cell shown in Fig. 3.8, not over a square of side one as before in (3.98), and the factor of $1/v_s$ appeared. This is the result of the change of variables in the double integral which was initially taken over r_1 and r_2. Indeed, $\mathbf{r} = r_1\mathbf{a}_1 + r_2\mathbf{a}_2$, and hence $x = r_1 a_{1x} + r_2 a_{2x}$ and $y = r_1 a_{1y} + r_2 a_{2y}$. Therefore, $d\mathbf{r} = dxdy = |J| dr_1 dr_2$, where the Jacobian

$$J = \frac{\partial(x, y)}{\partial(r_1, r_2)} = \begin{vmatrix} \partial x/\partial r_1 & \partial y/\partial r_1 \\ \partial x/\partial r_2 & \partial y/\partial r_2 \end{vmatrix} = \begin{vmatrix} a_{1x} & a_{1y} \\ a_{2x} & a_{2y} \end{vmatrix}$$
$$= a_{1x}a_{2y} - a_{1y}a_{2x} \quad \Longrightarrow \quad |J| = v_s \, .$$

In the following problems we shall generalise our results to the three-dimensional case.

Problem 3.45 A three-dimensional lattice is specified by three basic direct lattice vectors $\mathbf{a}_1$, $\mathbf{a}_2$ and $\mathbf{a}_3$. The basis vectors of the corresponding reciprocal lattice are defined via $\mathbf{a}_i \cdot \mathbf{b}_j = 2\pi\delta_{ij}$, where $i, j = 1, 2, 3$. Show that

$$\mathbf{b}_1 = \frac{2\pi}{v_c} [\mathbf{a}_2 \times \mathbf{a}_3] \, , \quad \mathbf{b}_2 = \frac{2\pi}{v_c} [\mathbf{a}_3 \times \mathbf{a}_1] \, , \quad \mathbf{b}_3 = \frac{2\pi}{v_c} [\mathbf{a}_1 \times \mathbf{a}_2] \, , \tag{3.100}$$

where $v_c = \|[\mathbf{a}_1, \mathbf{a}_2, \mathbf{a}_3]\|$ is the unit cell volume (see also Sect. I.1.10.1).

Problem 3.46 Correspondingly, show that a function $f(\mathbf{r})$ which is periodic in the direct space with respect to any direct lattice vector $\mathbf{L} = n_1\mathbf{a}_1 + n_2\mathbf{a}_2 + n_3\mathbf{a}_3$ (where $\{n_i\} = (n_1, n_2, n_3)$ are all possible negative and positive integers, including zero, and $\{\mathbf{a}_i\}$ are the unit base vectors) can be expanded in the triple Fourier series

$$f(\mathbf{r}) = \sum_{\mathbf{g}} f_{\mathbf{g}} e^{i\mathbf{g}\cdot\mathbf{r}}, \quad \text{where} \quad f_{\mathbf{g}} = \frac{1}{v_c} \iiint_{cell} f(\mathbf{r}) e^{-i\mathbf{g}\cdot\mathbf{r}} d\mathbf{r}, \quad (3.101)$$

where integration is performed over the volume of the unit cell (i.e. a parallelepiped formed by the three basic direct lattice vectors) and the summation is performed over all possible reciprocal lattice vectors $\mathbf{g} = m_1\mathbf{b}_1 + m_2\mathbf{b}_2 + m_3\mathbf{b}_3$ vectors (i.e. all possible m_1, m_2, m_3).

Problem 3.47 Consider a big macroscopic volume of the solid containing a very large number N of identical unit cells of volume v_c each. Show then that the triple integral in Eq. (3.101) for $f_{\mathbf{g}}$ can in fact be extended to the whole volume $V = N v_c$, i.e.

$$f(\mathbf{r}) = \sum_{\mathbf{g}} f_{\mathbf{g}} e^{i\mathbf{g}\cdot\mathbf{r}}, \quad \text{where} \quad f_{\mathbf{g}} = \frac{1}{V} \iiint_V f(\mathbf{r}) e^{-i\mathbf{g}\cdot\mathbf{r}} d\mathbf{r}. \quad (3.102)$$

This expression for the Fourier coefficients could be sometimes more convenient in considering the so-called thermodynamic limit when $V \to \infty$ since the shape of the unit cell can be ignored.

Problem 3.48 The electrostatic potential $V(\mathbf{r})$ in a three-dimensional periodic crystal caused by its charge density satisfies the Poisson equation of classical electrostatics $\Delta V(\mathbf{r}) = -4\pi\rho(\mathbf{r})$, where $\rho(\mathbf{r})$ is the total charge density in the crystal at point $\mathbf{r}$. Expanding both the potential and the density in the Fourier series show that the solution of this equation is

$$V(\mathbf{r}) = \sum_{\mathbf{g}} \frac{4\pi\rho_{\mathbf{g}}}{\mathbf{g}^2} e^{i\mathbf{g}\cdot\mathbf{r}}. \quad (3.103)$$

This expression has a singularity at $\mathbf{g} = 0$ since the Fourier coefficient of the potential contains division on the square of $\mathbf{g}$. Show that the Fourier component of the charge density, ρ_0, corresponding to the zero reciprocal vector is directly related to the total charge in the unit cell and is then zero for neutral systems. Therefore, the summation in Eq. (3.103) is in fact performed only with respect to non-zero reciprocal lattice vectors, hence the singularity is avoided explicitly.

Problem 3.49 Let $\mathbf{L} = n_1\mathbf{a}_1 + n_2\mathbf{a}_2 + n_3\mathbf{a}_3$ and $\mathbf{g} = m_1\mathbf{b}_1 + m_2\mathbf{b}_2 + m_3\mathbf{b}_3$ be direct and correspondingly reciprocal lattice vectors. Prove the so-called theta-function transformation:

$$\sum_{\mathbf{L}} e^{-|\mathbf{r}-\mathbf{L}|^2 t^2} = \frac{1}{v_c} \sum_{\mathbf{g}} \left(\frac{\sqrt{\pi}}{t}\right)^3 e^{-\mathbf{g}^2/4t^2} e^{i\mathbf{g}\cdot\mathbf{r}} , \qquad (3.104)$$

where t is a real number and $\mathbf{r}$ is a vector. This identity shows that the direct lattice sum in the left-hand side can be expanded in the Fourier series and can thus be equivalently rewritten as a reciprocal lattice sum. *[Hint: First of all, prove that the function in the left-hand side is periodic in $\mathbf{r}$ with respect to an arbitrary lattice vector $\mathbf{L}'$, i.e. adding $\mathbf{L}'$ to $\mathbf{r}$ does not change the function; hence, the left-hand side can be expanded in the Fourier series in terms of $\mathbf{g}$ using Eq. (3.102). Then, when performing triple integration over the whole space, show that each term in the sum makes an identical contribution giving out a factor of N; hence the sum over $\mathbf{L}$ can be removed, while V be replaced with v_c. Next, calculate the triple integral over the whole space (i.e. in the thermodynamic limit).]*

Another proof of the theta-function transformation will be given in Sect. 5.2.6.

3.11.2 Ewald's Formula

The theta-function transformation allows one to obtain a very useful formula for calculating in practice the electrostatic potential of the lattice of point charges (atoms), the so-called Ewald formula. Consider a three-dimensional periodic crystal with point charges q_s in each unit cell, the index s is used to count the charges within the cell; in the zero cell (with the lattice vector $\mathbf{L} = \mathbf{0}$) the positions of charges are given by vectors $\mathbf{X}_s$. Each unit cell is considered charge-neutral, i.e. $\sum_s q_s = 0$. Then, the electrostatic potential at point $\mathbf{r}$ due to all charges of the whole crystal (we only consider $\mathbf{r}$ to be somewhere *between* atoms) will be

$$V(\mathbf{r}) = \sum_{\mathbf{L}} \sum_s \frac{q_s}{|\mathbf{r} - \mathbf{L} - \mathbf{X}_s|} ,$$

since the position of the charge q_s in the unit cell associated with the lattice vector $\mathbf{L}$ is $\mathbf{L} + \mathbf{X}_s$.

This lattice sum converges extremely slowly; in fact, it can be shown that it converges only conditionally (see Sect. I.7.1.5). However, its convergence can be considerably improved using the following trick.[7] Consider the so-called error function

$$\mathrm{erfc}(x) = \frac{2}{\sqrt{\pi}} \int_x^\infty e^{-t^2} dt \ . \tag{3.105}$$

This function tends to zero exponentially fast as $x \to \infty$ (in fact, as e^{-x^2}). Conjugated to this one is another error function

$$\mathrm{erf}(x) = 1 - \mathrm{erfc}(x) = \frac{2}{\sqrt{\pi}} \int_0^x e^{-t^2} dt \ , \tag{3.106}$$

which tends to unity as $x \to \infty$.

Problem 3.50 Prove Eq. (3.106).

Using these two functions, we can rewrite the potential:

$$V(\mathbf{r}) = \sum_{Ls} q_s \frac{\mathrm{erfc}\left(\gamma \left|\mathbf{r} - \mathbf{L} - \mathbf{X}_s\right|\right)}{\left|\mathbf{r} - \mathbf{L} - \mathbf{X}_s\right|} + \sum_s q_s \left[\sum_{\mathbf{L}} \frac{\mathrm{erf}\left(\gamma \left|\mathbf{r} - \mathbf{L} - \mathbf{X}_s\right|\right)}{\left|\mathbf{r} - \mathbf{L} - \mathbf{X}_s\right|} \right] ,$$

where γ is some (so far arbitrary) positive constant (called Ewald's constant). The first sum over the direct lattice $\mathbf{L}$ converges very quickly because of the error function, so only the unit cells around the point $\mathbf{r}$ contribute appreciably. The second sum, however, converges extremely slowly because of the other error function $\mathrm{erf}(x)$ which tends to unity as $x \to \infty$. This is the point where the theta-function transformation proves to be extremely useful. Indeed, the function in the square brackets can be manipulated into the following expression:

$$\sum_{\mathbf{L}} \frac{\mathrm{erf}\left(\gamma \left|\mathbf{r} - \mathbf{L} - \mathbf{X}_s\right|\right)}{\left|\mathbf{r} - \mathbf{L} - \mathbf{X}_s\right|} = \frac{2}{\sqrt{\pi}} \sum_{\mathbf{L}} \frac{1}{\left|\mathbf{r} - \mathbf{L} - \mathbf{X}_s\right|} \int_0^{\gamma|\mathbf{r}-\mathbf{L}-\mathbf{X}_s|} e^{-t^2} dt$$

$$= \left| \begin{array}{l} \lambda = t/\left|\mathbf{r} - \mathbf{L} - \mathbf{X}_s\right| \\ d\lambda = dt/\left|\mathbf{r} - \mathbf{L} - \mathbf{X}_s\right| \end{array} \right| = \frac{2}{\sqrt{\pi}} \int_0^\gamma \left(\sum_{\mathbf{L}} e^{-\lambda^2|\mathbf{r}-\mathbf{L}-\mathbf{X}_s|^2} \right) d\lambda \ .$$

Now, the expression in the round brackets can easily be recognised to be the one for which the theta-function transformation (3.104) can be used:

---

[7] The Ewald method corresponds to a particular regularisation of the conditionally converging series. However, it can be shown (see Sect. I.7.4.2 based on the paper L. Kantorovich and I. Tupitsin - J. Phys. Cond. Matter **11**, 6159 (1999)) that this calculation results in the correct expression for the electrostatic potential in the central part of a large finite sample if the dipole and quadruple moments of the unit cell are equal to zero. Otherwise, an additional macroscopic contribution to the potential is present.

$$\frac{2}{\sqrt{\pi}} \int_0^{\gamma} \left(\sum_{\mathbf{L}} e^{-\lambda^2 |\mathbf{r}-\mathbf{L}-\mathbf{X}_s|^2} \right) d\lambda = \frac{2\pi}{v_c} \int_0^{\gamma} \left(\sum_{\mathbf{g}} \frac{1}{\lambda^3} e^{-\mathbf{g}^2/4\lambda^2} e^{i\mathbf{g}\cdot(\mathbf{r}-\mathbf{X}_s)} \right) d\lambda .$$

The integration over λ is trivially performed using the substitution $x = \mathbf{g}^2/4\lambda^2$, and we obtain

$$\frac{2}{\sqrt{\pi}} \int_0^{\gamma} \left(\sum_{\mathbf{L}} e^{-\lambda^2 |\mathbf{r}-\mathbf{L}-\mathbf{X}_s|^2} \right) d\lambda = \frac{4\pi}{v_c} \sum_{\mathbf{g}} \frac{1}{\mathbf{g}^2} e^{-\mathbf{g}^2/4\gamma^2} e^{i\mathbf{g}\cdot(\mathbf{r}-\mathbf{X}_s)} .$$

Therefore, the potential takes on the final form:

$$V(\mathbf{r}) = \sum_{\mathbf{L}s} q_s \frac{\operatorname{erfc}(\gamma |\mathbf{r} - \mathbf{L} - \mathbf{X}_s|)}{|\mathbf{r} - \mathbf{L} - \mathbf{X}_s|} + \frac{4\pi}{v_c} \sum_{\mathbf{g}} \frac{1}{\mathbf{g}^2} e^{-\mathbf{g}^2/4\gamma^2} \left(\sum_s q_s e^{-i\mathbf{g}\cdot\mathbf{X}_s} \right) e^{i\mathbf{g}\cdot\mathbf{r}} .$$

$$(3.107)$$

This is the required result: because of the exponential function $e^{-\mathbf{g}^2/4\gamma^2}$, the second sum over the reciprocal lattice vectors converges extremely quickly rendering the obtained formula a very efficient instrument in calculating the electrostatic potential of a lattice of point charges. The constant γ regulates the contributions coming from the direct (the first term) and reciprocal (second) lattice sums, and can be optimised to improve the efficiency in numerical calculations. Also note that the $\mathbf{g} = 0$ term in the second sum does not contribute as any physical system is charge-neutral.

In the following two problems it is necessary to apply the theta-function transformation for 1D and 2D cases (see Problem 3.49).

Problem 3.51 Consider a one-dimensional periodic chain of atoms (i.e. a one-dimensional crystal) with periodicity a. Show that in this case the Ewald formula for the electrostatic potential reads

$$V(\mathbf{r}) = \sum_{Ls} q_s \frac{\operatorname{erfc}(\gamma(x - L - X_s))}{|x - L - X_s|} + \frac{1}{a} \sum_{g \neq 0} \operatorname{Ei}\left(\frac{g^2}{4\gamma^2} \right) \left(\sum_s q_s e^{-igX_s} \right) e^{igx} ,$$

$$(3.108)$$

where $g = mb$ is one-dimensional reciprocal vector with $b = 2\pi/a$, $m = \pm 1, \pm 2, \ldots$, and

$$\operatorname{Ei}(x) = \int_x^{\infty} \frac{e^{-t}}{t} dt$$

is the function called exponential integral.

Problem 3.52 Prove that for a two-dimensional solid the Ewald formula reads

$$V(\mathbf{r}) = \sum_{Ls} q_s \frac{\text{erfc}\,(\gamma\,|\mathbf{r} - \mathbf{L} - \mathbf{X}_s|)}{|\mathbf{r} - \mathbf{L} - \mathbf{X}_s|} + \frac{2\pi}{v_s} \sum_{\mathbf{g} \neq 0} \frac{1}{\mathbf{g}^2} \text{erfc}\left(\frac{|\mathbf{g}|}{2\gamma}\right) \left(\sum_s q_s e^{-i\mathbf{g}\cdot\mathbf{X}_s}\right) e^{i\mathbf{g}\cdot\mathbf{r}} ,$$

$$(3.109)$$

where $\mathbf{L}$ *and* $\mathbf{g}$ *are two-dimensional direct and reciprocal lattice vectors.*

3.11.3 Born and von Karman Boundary Conditions

In solid-state physics very often it is necessary to deal with functions $f(\mathbf{r})$ which are not periodic, e.g. an electron wavefunction in a crystal. Even in those cases it is still useful and convenient to employ the Fourier series formalism. The way this is normally done is as follows. We consider, following Born and von Karman, an artificial periodicity in the solid happening on a much larger scale. Namely, we assume that if we move $N_1 \gg 1$ times along the lattice vector $\mathbf{a}_1$, the solid repeats itself, and the same happens if we perform a translation $N_2 \gg 1$ and $N_3 \gg 1$ times along the other two lattice directions (vectors), i.e. we impose an *artificial periodicity* on the function $f(\mathbf{r})$ as follows:

$$f(\mathbf{r}) = f(\mathbf{r} + N_1\mathbf{a}_1) = f(\mathbf{r} + N_2\mathbf{a}_2) = f(\mathbf{r} + N_3\mathbf{a}_3) . \qquad (3.110)$$

This condition is sometimes called a cyclic boundary condition, as explained in Fig. 3.9 for the one-dimensional case: it corresponds to forming a ring of N_1 cells. In the three-dimensional case one may imagine that every two opposite sides of our very big solid are connected with each other like in a torus. This new solid can be imagined as having very large direct lattice vectors $\mathbf{A}_i = N_i\mathbf{a}_i$ $(i = 1, 2, 3)$ and correspondingly very small reciprocal basic lattice vectors $\mathbf{B}_i = \mathbf{b}_i/N_i$ $(i = 1, 2, 3)$. Expanding $f(\mathbf{r})$ now as in Eq. (3.101) we have a sum over all vectors of our new reciprocal lattice:

$$\mathbf{K} = M_1\mathbf{B}_1 + M_2\mathbf{B}_2 + M_3\mathbf{B}_3 = \frac{M_1}{N_1}\mathbf{b}_1 + \frac{M_2}{N_2}\mathbf{b}_2 + \frac{M_3}{N_3}\mathbf{b}_3 ,$$

where numbers M_i take on all possible negative and positive integer values including zero. When each integer M_i $(i = 1, 2, 3)$ becomes equal to an integer number of the corresponding N_i, the vector $\mathbf{K}$ becomes equal to the reciprocal lattice vector $\mathbf{g}$ of the original reciprocal lattice corresponding to the small direct lattice with the basic vectors $\mathbf{a}_i$. For other values of the numbers M_i we can always write: $M_i = m_i N_i + \mu_i$, where m_i is an integer taking all possible values, but the integer μ_i changes only between 0 and $N_i - 1$. Then,

$$\mathbf{K} = (m_1\mathbf{b}_1 + m_2\mathbf{b}_2 + m_3\mathbf{b}_3) + \left(\frac{\mu_1}{N_1}\mathbf{b}_1 + \frac{\mu_2}{N_2}\mathbf{b}_2 + \frac{\mu_3}{N_3}\mathbf{b}_3\right) = \mathbf{g} + \mathbf{k},$$

where the vector $\mathbf{k}$ takes $N_1 N_2 N_3$ values within a parallelepiped with the sides made by the three basic reciprocal lattice vectors $\mathbf{b}_i$ corresponding to the original reciprocal lattice. This parallelepiped, called the (first) Brillouin zone (BZ), is divided by a grid of $N_1 N_2 N_3$ small cells with the sides given by $\mathbf{b}_i/N_i$, and all values of $\mathbf{k}$ correspond to these small cells. Therefore, instead of (3.101) we can write in this case:

$$f(\mathbf{r}) = \sum_{\mathbf{g}}\sum_{\mathbf{k}} f_{\mathbf{g}+\mathbf{k}} e^{i(\mathbf{g}+\mathbf{k})\cdot\mathbf{r}} = \sum_{\mathbf{K}} f_{\mathbf{K}} e^{i\mathbf{K}\cdot\mathbf{r}},$$

where we sum over all reciprocal lattice vectors of the original lattice and all points $\mathbf{k}$ from the first BZ, where $\mathbf{K} = \mathbf{g} + \mathbf{k}$.

The complex exponential functions $\exp(i\mathbf{K}\cdot\mathbf{r})$ with $\mathbf{K} = \mathbf{g} + \mathbf{k}$ are called plane waves, they form the basis of most modern electronic structure calculation methods. It is essential that this basis is complete (see Sect. 3.9.3), i.e. any function $f(\mathbf{r})$ can be expanded in terms of them. In practice the expansion is terminated, and there is a very simple algorithm for doing this. Indeed, plane waves with large reciprocal lattice vectors $\mathbf{K}$ oscillate rapidly in the direct space and hence need to be kept in the expansion only if the function $f(\mathbf{r})$ changes rapidly on a small length scale (e.g. wavefunctions of the electrons oscillate strongly close to atomic nuclei in atoms, molecules and crystals). If, however, $f(\mathbf{r})$ is smooth everywhere in space, then large reciprocal space vectors $\mathbf{K}$ are not needed, i.e. one can simply include in the expansion all vectors whose lengths are smaller than a certain cut-off K_{max}, i.e. $|\mathbf{K}| \le K_{max}$. Special tricks are used to achieve the required smooth behaviour of the valence electron wavefunctions near and far away from the atomic cores by employing special effective core potentials called pseudopotentials.

Fig. 3.9 Born and von Karman periodic boundary conditions for a one-dimensional solid: **a** using an infinite periodic cell containing $N_1 \gg 1$ primitive unit cells, and **b** using a ring consisting of N_1 unit cells

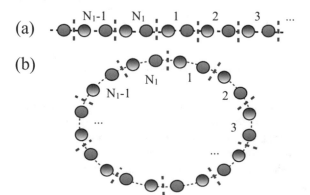

3.11.4 Atomic Force Microscopy

In non-contact atomic force microscopy (NC-AFM) an atomically sharp tip attached to the end of a cantilever beam is made to oscillate above a crystal surface. This is achieved by means of a piezocrystal which the beam is attached to with its fixed end: an alternating voltage of frequency ω (in the order of hundreds of megahertz) applied to the piezocrystal causes the beam to oscillate, see Fig. 3.10. During its oscillations the tip comes very close to the surface, within one-two atomic distances, and this way the surface force F_s acts on the cantilever affecting its oscillations. Depending on whether the tip oscillates above one surface species or another, or between atoms, this force would be different; this variation of the short-range interaction of the tip with the surface due to lateral position of the tip is at the heart of the contrast of this experimental device when it scans the surface.

Modern NC-AFM experiments can achieve a remarkable resolution: in many cases surfaces are resolved atomically. This is done when the scan is performed in vacuum using a special technique called frequency modulation. In this method the cantilever is oscillated at resonance which is established in such a way that there is exactly $-\pi/2$ phase shift between the oscillating cantilever and the sinusoidal voltage applied to the piezo (the driving signal). During the scan this resonance is established automatically by sophisticated electronics, so that for any lateral position of the tip above the surface the tip always oscillates at resonance. In constant (resonance) frequency mode the oscillation frequency is kept constant during the scan, this is achieved by moving the surface vertically by another piezocrystal attached to the sample surface. The values of the vertical movements of the sample Δz for each lateral position of the tip (x, y) form the image of the surface, $\Delta z(x, y)$.

Let us consider the oscillating cantilever theoretically. Let h be the average distance between the tip and surface, which is maintained by a constant external force F_{ext} holding the oscillating cantilever. During the scan this distance changes (via

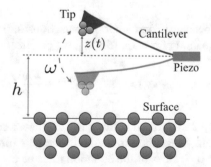

Fig. 3.10 Schematic of NC-AFM: a macroscopic cantilever with a sharp tip at its end, terminated by a single atom, oscillates above a surface with frequency ω. The average distance between the tip and the surface is h, while the deviation of the tip from that average distance is given by the function $z(t)$. The total distance of the tip from the surface is $Z(t) = h + z(t)$

appropriate vertical displacements of the sample) to maintain the constant frequency shift. Then we denote by $z(t)$ the vertical deviation of the tip during its oscillations from that distance. This means that the total distance between the tip and surface at each time t is $Z(t) = h + z(t)$, see Fig. 3.10. If $F_s(Z)$ is the force acting on the tip from the surface, $F_{piezo} = A_0 \cos(\omega t)$ is the excitation (or driving) signal due to the piezo which holds the cantilever, k cantilever elastic constant, $\omega_0 = \sqrt{k/m}$ the cantilever fundamental frequency and m is its effective mass, then the equation of motion of the cantilever can be written as follows:

$$m\ddot{z}(t) + \gamma m\dot{z}(t) + kz(t) = F_s(h + z) + A_0 \cos(\omega t) + F_{ext}. \tag{3.111}$$

Here we introduced also a friction force with γ being a friction constant since there must always be friction in the system. If the cantilever is excited with a signal of frequency ω, after some transient time a steady-state solution will be established as the particular integral of the DE. This solution will also be periodic with the same frequency ω. We would like to obtain such a solution. Since the oscillation is periodic, it can be expanded into a Fourier series (it is convenient to use the exponential form):

$$z(t) = \sum_{n=-\infty}^{\infty} z_n e^{i2\pi nt/T} = \sum_n z_n e^{i\omega nt} = z_0 + \sum_{n\neq 0} z_n e^{i\omega nt}, \tag{3.112}$$

where $T = 2\pi/\omega$ is the period of oscillations. Correspondingly, the surface force $F_s(h + z(t))$ will also be a periodic function and can also be expanded in a similar way:

$$F_s(h + z(t)) = \sum_n F_n e^{i\omega nt}, \tag{3.113}$$

where

$$F_n = \frac{1}{T} \int_0^T F_s(h + z(t)) e^{-i\omega nt} dt \tag{3.114}$$

are constant (time independent) Fourier coefficients. Substituting expansions (3.113) and (3.112) into the equation of motion (3.111), we obtain

$$\sum_n z_n e^{i\omega nt} \left(-\omega^2 n^2 + i\omega\gamma n + \omega_0^2\right) = \sum_n \frac{F_n}{m} e^{i\omega nt} + \frac{A_0}{2m} \left(e^{i\omega t} + e^{-i\omega t}\right) + \frac{F_{ext}}{m}.$$

Note that the constant external force holding the cantilever contributes only to the $n = 0$ Fourier coefficient. Since the exponential functions form a complete set, the coefficients by them in both sides of the equation must be equal to each other. This way we can establish the following equations for the unknown amplitudes z_n:

$$z_0 = \frac{F_0 + F_{ext}}{m\omega_0^2},$$

$$z_1\left(-\omega^2 + i\omega\gamma + \omega_0^2\right) = \frac{1}{m}\left(F_1 + \frac{1}{2}A_0\right),$$

$$z_{-1}\left(-\omega^2 - i\omega\gamma + \omega_0^2\right) = \frac{1}{m}\left(F_{-1} + \frac{1}{2}A_0\right),$$

$$z_n\left(-\omega^2 n^2 + i\omega\gamma n + \omega_0^2\right) = \frac{F_n}{m}, \quad n = \pm 2, \pm 3, \dots .$$

(3.115)

The external force ensures that there is no constant displacement of the cantilever due to oscillations, i.e. $F_{ext} = -F_0$ and hence $z_0 = 0$. The rest of the obtained equations are to be solved self-consistently since the coefficients F_n depend on all amplitudes z_n, see Eq. (3.114): using some initial values for the amplitudes, one calculates the constants F_n; these are then used to update the amplitudes z_n by solving the above equations; new amplitudes are used to recalculate the forces F_n, and so on until convergence. In practice, the Fourier series is terminated, so that only the first N terms are considered, $0 \le n \le N$.

In many cases only the first two terms in the Fourier series (i.e. for $n = \pm 1$) can be retained. Let us establish an expression for the resonance frequency in this case. As was explained above, the resonance is established by the $-\pi/2$ phase shift between the excitation signal, $A_0 \cos(\omega t)$, and the tip oscillation. This means that in this approximation

$$z(t) \simeq A \cos\left(\omega t - \frac{\pi}{2}\right) = A \sin(\omega t) = \frac{A}{2i}\left(e^{i\omega t} - e^{-i\omega t}\right),$$

i.e. $z_1 = A/2i = -iA/2$ and $z_{-1} = iA/2$. Here A is the cantilever oscillation amplitude.

Problem 3.53 Substituting these values of $z_{\pm 1}$ into Eq. (3.115), show that the resonance frequency ω must satisfy the following equation:

$$\left(\frac{\omega}{\omega_0}\right)^2 = 1 - \frac{1}{\pi k A}\int_0^{2\pi} F_s\left(h + A\sin\phi\right)\sin\phi\, d\phi .$$

(3.116)

This formula can be used for calculating the resonance frequency of the tip oscillations for the given lateral position of the tip above the surface for which the tip-surface force F_s is known.

3.11.5   Applications in Quantum Mechanics

Expansion of a function in terms of other functions, which form a complete orthonormal set, is at the heart of quantum mechanics. For simplicity, we shall only consider here a one-dimensional system (e.g. an electron in an external potential moving along the x-axis). Then the state of the electron in quantum mechanics is described by its wavefunction $\psi(x, t)$, which is a function of the electron position x and time t. The time evolution of the wavefunction depends on its value at $t = 0$ (initial conditions) and how the potential $V(x, t)$ acting on it changes in time. If the potential is constant in time, $V = V_0(x)$, then the electron may accept just one state from an infinite set of *stationary states* each described by the wavefunction $\psi_n(x)$ and energy ϵ_n, where the index $n = 0, 1, 2, 3, \ldots$ numbers the stationary states in order of increasing energy, i.e. $\epsilon_{n+1} \geq \epsilon_n$. Any state is legitimate, but most likely the state $n = 0$ with the lowest energy will be occupied, unless a higher energy state with $n \geq 1$ was specifically prepared. If the potential $V_0(x)$ is known, one can solve the Schrödinger equation and determine all the stationary energies and wavefunctions, at least in principle.

The situation, however, becomes more complex if the potential changes in time. Consider as an example the uniform potential (i.e. it does not depend on x) shown in Fig. 3.11a. Initially the potential is equal V_0 and is constant in time. Then between times t_1 and t_2 the potential changes in time somehow, after which it returns back to its initial value. For instance, it may be that within the finite time interval $t_1 \leq t \leq t_2$ an external field was acting on our electron. If initially the electron was prepared to occupy exclusively the ground state, $\psi_0(x)$, after the action of the potential at $t \geq t_1$ its wavefunction $\psi(x, t)$ will depend on time and each stationary state will then become occupied with some probability, Fig. 3.11b. If we expand $\psi(x, t)$ in terms of the complete set of stationary states, $\{\psi_n(x)\}$, i.e. if we write down the corresponding generalised Fourier series,

$$\psi(x, t) = \sum_{n=0}^{\infty} c_n(t)\psi_n(x) ,$$

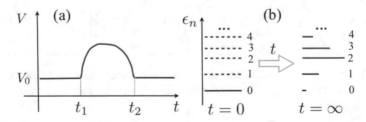

Fig. 3.11   a A spatially uniform external potential $V(t)$ acting on an electron as a function of time, t. **b** Occupation of energy levels ϵ_n by the electron. Left: initially, only the ground state $n = 0$ was occupied. Right: at $t = \infty$, several states are occupied with different probabilities (shown by the length of the horizontal lines representing the states), with the excited state $n = 2$ being the most probable

then, according to the rules of quantum mechanics, the probability at time t to find our electron in state ψ_n is $P_n(t) = |c_n(t)|^2$, i.e. it is given by the module square of the generalised Fourier coefficient. Of course, since the electron should occupy a state with certainty,

$$\sum_n P_n(t) = \sum_n |c_n(t)|^2 = 1 \, .$$

Problem 3.54 Assuming that the stationary states functions $\psi_n(x)$ form an orthonormal set, show that the above condition corresponds to the correct normalisation of the wavefunction at any time: $\int \psi(x, t)^* \psi(x, t) dx = 1$.

If the measurement is done long after the action of the external potential so that we can assume that all the relaxation processes have ceased in the system, then there will be a stationary distribution $P_n(\infty)$ of finding the electron in the state n as schematically shown in Fig. 3.11b.

To calculate the probability, we can use formula (3.85) for the Fourier coefficients:

$$P_n(t) = \left| \int \psi_n^*(x)\psi(x, t)dx \right|^2 \, .$$

Of course, this is just one example; in fact, quantum mechanics is basically built on the idea of expanding states into the Fourier series with respect to some stationary states.

Another application of the functional series in quantum mechanics is related to modern theories of quantum chemistry and condensed matter physics where a many-electron problem is solved. The same Schrödinger equation is solved in both these cases, but essentially different expansion of the electron wavefunctions in basis functions (the basis set) is employed in each case. If in the condensed matter physics plane waves are used representing an orthogonal and complete basis the convergence of which can be easily controlled, in quantum chemistry localised on atoms functions are used which mimic atomic-like orbitals. This way much smaller sets of basis functions are needed to achieve a reasonable precision, however, it is much more difficult to achieve convergence with respect to the basis set as functions localised on different atoms do not represent an orthogonal and complete basis set, so that including too many of them on each atom may result in over-completeness and hence instabilities in numerical calculations.

3.11.6   Free Electron Gas

In Sommerfeld theory of electron gas it is assumed that electrons in a metal can occupy any energy $\epsilon \geq 0$ with the probability given by the Fermi-Dirac distribution

$$f(\epsilon) = \left[e^{\beta(\epsilon - \mu)} + 1 \right]^{-1}, \tag{3.117}$$

where μ is the chemical potential of the electron gas, $\beta = 1/k_B T$ the inverse temperature, and k_B Boltzmann constant. The Fermi–Dirac distribution coincides with the step (Heaviside) function $H(\mu - \epsilon)$ at zero temperature being 1 for $\epsilon < \mu$, 0 for $\epsilon > \mu$ and $1/2$ for $\epsilon = \mu$; at non-zero temperatures the sharp drop of $f(\epsilon)$ at the value of the chemical potential is smoothed out.

The value of the chemical potential depends on temperature, even though this dependence is rather weak. To establish this dependence, one can use the conservation condition for the number of electrons n_e per unit volume:

$$n_e = \int_0^\infty D(\epsilon) f(\epsilon) d\epsilon, \tag{3.118}$$

where $D(\epsilon)$ is the so-called density of states (DOS) of the electron gas giving the number of states for the electrons to occupy per unit energy and volume.

Even though the DOS for the free election gas $D(\epsilon) \propto \sqrt{\epsilon}$ is a rather simple function, this integral cannot be calculated analytically. There are also other quantities related to the electron gas that require calculation of a similar integral. For instance, to calculate the heat capacity of the electrons, one has to differentiate over temperature the electron energy (per unit volume)

$$U = \int_0^\infty \epsilon D(\epsilon) f(\epsilon) d\epsilon. \tag{3.119}$$

Another example is the so-called Pauli paramagnetism: when calculating the response of the electron gas on the applied external magnetic field B one has to calculate the net magnetic moment in the system due to electrons with spin up and down responding differently to the field. Indeed, electrons with spin up move up in energy, while electrons with spin down move down in energy which results in spin polarisation: there are more electrons with spin down than with spin up after some of the electrons flip their spin as the highest energy for both spins is established at the same value of the chemical potential μ as shown in Fig. 3.12.

The total magnetic moment (per unit volume) $M = M_\uparrow + M_\downarrow$ is obtained as a sum of the magnetic moments of both spins:

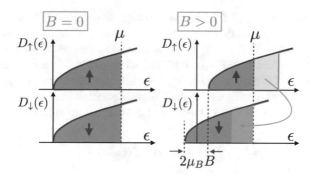

Fig. 3.12 Filled states of the electrons in a metal at zero (left) and non-zero (right) magnetic field B. In both cases electrons of both spins fill up all states up to the common chemical potential μ (this will depend on B though). If at $B = 0$ there are the same numbers of the electrons of both spins, at $B > 0$ electrons of spin up move up in energy by $\mu_B B$, electrons of spin down move down in energy by the same amount, and in order to provide a common value of the chemical potential μ in this case, some of the spin-up electrons (grey area) flip their spin (shown by an arrow). As a result, there are more electrons of spin down than of spin up resulting in a non-zero net magnetisation of the metal

$$M_\uparrow = \int_0^\infty (-\mu_B) D_\uparrow(\epsilon) f(\epsilon + \mu_B B) d\epsilon\,,$$
$$M_\downarrow = \int_0^\infty (+\mu_B) D_\downarrow(\epsilon) f(\epsilon - \mu_B B) d\epsilon\,, \tag{3.120}$$

where $D_\uparrow(\epsilon) = D_\downarrow(\epsilon) = \frac{1}{2}D(\epsilon)$ are the DOS of the electrons with opposite spins, μ_B is the Bohr magneton, and spin-up electrons have their magnetic moment equal to $-\mu_B$, while spin-down electrons equal to $+\mu_B$. Differentiating the net moment M with respect to the magnetic field B yields Pauli susceptibility.

Excited by these examples of applications, let us discuss a general integral containing the Fermi–Dirac distribution,

$$I = \int_0^\infty G(\epsilon) f(\epsilon) d\epsilon\,, \tag{3.121}$$

where $G(\epsilon)$ is some function that may increase to infinity when $\epsilon \to \infty$, but we assume—to ensure convergence of the integral at the upper limit—that this increase is much slower than the slope with which the Fermi–Dirac function goes to zero there (note that at large energies $f(\epsilon) \sim e^{-\beta(\epsilon - \mu)}$ goes to zero exponentially).

Our initial step is to make a change of variables in the integral: let us denote $x = \beta(\epsilon - \mu)$, then

$$I = \frac{1}{\beta} \int_{-\beta\mu}^\infty G\left(\frac{x}{\beta} + \mu\right) \frac{dx}{e^x + 1}\,.$$

For all practical purposes of T being not too large as compared to the room temperature ($T = 300$ K), $\beta\mu$ is rather large (e.g. at room temperature this quantity may be a few hundreds, while at $T = 10^3$ K it could be still many tens). Hence, the bottom limit in the integral can actually be replaced with $-\infty$, while the function $G(x)$ is additionally defined at its negative argument x to be either zero or going to zero sufficiently fast to ensure convergence of the integral (note that the Fermi–Dirac function $f(x) = (e^x + 1)^{-1}$ tends to 1 at $x \to -\infty$). Hence, we replace I with the integral

$$I_1 = \int_{-\infty}^{\infty} G(\epsilon) f(\epsilon) d\epsilon \,, \tag{3.122}$$

which we initially take by parts using an auxiliary function

$$A(\epsilon) = \int_{-\infty}^{\epsilon} G(x) dx \,. \tag{3.123}$$

We write

$$I_1 = \underbrace{A(\epsilon) f(\epsilon)|_{-\infty}^{\infty}}_{=0} + \int_{-\infty}^{\infty} A(\epsilon) \left(-\frac{df}{d\epsilon} \right) d\epsilon = \int_{-\infty}^{\infty} A(\epsilon) \left(-\frac{df}{d\epsilon} \right) d\epsilon \,.$$

The free term is zero since $A(-\infty) = 0$ and $f(\infty) = 0$ (in the latter case $f(\epsilon)$ tends to zero faster than $A(\epsilon)$ may go to infinity). Next, we shall expand the auxiliary function $A(\epsilon)$ in the Taylor series around $\epsilon = \mu$:

$$A(\epsilon) = A(\mu) + \sum_{n=1}^{\infty} \frac{A^{(n)}(\mu)}{n!} (\epsilon - \mu)^n \,,$$

which yields

$$I_1 = A(\mu) \underbrace{\int_{-\infty}^{\infty} \left(-\frac{df}{d\epsilon} \right) d\epsilon}_{f(-\infty)-f(\infty)=1} + \sum_{n=1}^{\infty} \frac{A^{(n)}(\mu)}{n!} \int_{-\infty}^{\infty} (\epsilon - \mu)^n \left(-\frac{df}{d\epsilon} \right) d\epsilon$$

$$= A(\mu) + \sum_{n=1}^{\infty} \frac{A^{(n)}(\mu)}{n!} \int_{-\infty}^{\infty} (\epsilon - \mu)^n \left(-\frac{d}{d\epsilon} \frac{1}{e^{\beta(\epsilon-\mu)} + 1} \right) d\epsilon \,. \tag{3.124}$$

Making the familiar substitution, $x = \beta(\epsilon - \mu)$, the integral corresponding to the n-th term in the sum becomes

$$\frac{1}{\beta^n} \int_{-\infty}^{\infty} x^n \left(-\frac{d}{dx} \frac{1}{e^x + 1} \right) dx = \frac{1}{\beta^n} \int_{-\infty}^{\infty} x^n \left[\frac{e^{-x}}{(1 + e^{-x})^2} \right] dx \,.$$

The function in the square brackets, as can easily be checked, is an even function of x, hence the integral only survives for even values of n, in which cases one can integrate from 0 to ∞ and attach a factor of two:

$$I_1 = A(\mu) + \sum_{m=1}^{\infty} \frac{2A^{(2m)}(\mu)}{(2m)!\beta^{2m}} \int_0^{\infty} \frac{x^{2m}e^{-x}}{(1+e^{-x})^2} dx \,.$$

Since $e^{-x} < 1$ for all $x > 0$, we make use of the appropriate Maclaurin expansion (Sect. I.7.3.3)

$$\frac{1}{(1+t)^2} = \sum_{k=0}^{\infty} \binom{-2}{k} t^k = \sum_{k=0}^{\infty} (-1)^k (k+1) t^k \,,$$

that converges for any t between -1 and 1. Hence,

$$\int_0^{\infty} \frac{x^{2m}e^{-x}}{(1+e^{-x})^2} dx = \sum_{k=0}^{\infty} (-1)^k (k+1) \int_0^{\infty} x^{2m} e^{-(k+1)x} dx$$

$$= \sum_{k=0}^{\infty} \frac{(-1)^k (2m)!}{(k+1)^{2m}} \,,$$

since the integral of an integer power of x and of the exponential can be calculated by successive integrations by parts (see Problem I.4.46).[8] Correspondingly,

$$I_1 = A(\mu) + \sum_{m=1}^{\infty} \frac{2A^{(2m)}(\mu)}{(2m)!\beta^{2m}} \left(\sum_{k=0}^{\infty} \frac{(-1)^k (2m)!}{(k+1)^{2m}} \right)$$

$$= A(\mu) + 2 \sum_{m=1}^{\infty} \frac{A^{(2m)}(\mu)}{\beta^{2m}} \left(\sum_{k=1}^{\infty} \frac{(-1)^{k+1}}{k^{2m}} \right) \,.$$

It is seen that the expression in the round brackets is the series of the alternating reciprocal integers considered in Problem 3.25. Keeping only the first terms in I_1 and making use of Eqs. (3.31), (3.25) and (3.45), we can write

---

[8] The integral $\int_0^{\infty} t^{2m} e^{-t} dt$ is equal to the so-called Gamma function $\Gamma(2m+1) = (2m)!$, see Sect. 4.2.1 and especially Eq. (4.35).

$$I_1 = A(\mu) + \frac{\pi^2}{6\beta^2} A^{(2)}(\mu) + \frac{7\pi^4}{360} A^{(4)}(\mu) + \frac{31\pi^6}{15120} A^{(6)}(\mu) + \dots$$

$$= \int_{-\infty}^{\mu} G(\epsilon)d\epsilon + \frac{1}{6}\left(\frac{\pi}{\beta}\right)^2 G'(\mu) + \frac{7}{360}\left(\frac{\pi}{\beta}\right)^4 G^{(3)}(\mu)$$

$$+ \frac{31}{15120}\left(\frac{\pi}{\beta}\right)^6 G^{(5)}(\mu) + \dots .$$

Using this result, e.g. in Eq. (3.118), and keeping a few terms in the expansion, one can approximately determine the chemical potential μ as a function of β (and hence temperature). Also, one can obtain after that, by using a few terms in the expansion and making use of the temperature dependence of the chemical potential, an approximate expression for the (Pauli) magnetisation of a metal as a function of temperature from Eq. (3.120).

Chapter 4
Special Functions

In this chapter[1], we shall consider various special functions which have found prominent and widespread applications in physics and engineering. Most of these functions
cannot be expressed via a finite combination of elementary functions, and are solutions of non-trivial differential equations. Their careful consideration is our main
objective here.

We shall start from the so-called *Dirac delta function*, which corresponds to the
class of generalised functions, then move on to gamma and beta functions; then
consider in great detail various orthogonal polynomials, hypergeometric functions,
associated Legendre and Bessel functions. The chapter is concluded with differential
equations which frequently appear when solving physics and engineering problems,
where the considered special functions naturally emerge as their solutions.

4.1 Dirac Delta Function

Consider a charge q which is spread along the x-axis. The distribution of the charge
q along x is characterised by the distribution function $\rho(x)$, called charge density,
which is the charge per unit length. Integrating the density over the whole 1D space
would give q,

$$\int_{-\infty}^{\infty} \rho(x)dx = q \, .$$

If the charge q is smeared out along x smoothly, the density $\rho(x)$ would be a smooth
function of a single variable x. In physics, however, it is frequently needed to describe

---

[1] In the following, references to the first volume of this course (L. Kantorovich, Mathematics for
natural scientists: fundamentals and basics, 2nd Edition, Springer, 2022) will be made by appending
the Roman number I in front of the reference, e.g. Sect. I.1.8 or Eq. (I.5.18) refer to Sect. 1.8 and
Eq. (5.18) of the first volume, respectively.

© The Author(s), under exclusive license to Springer Nature Switzerland AG 2024   487
L. Kantorovich, *Mathematics for Natural Scientists II*, Undergraduate Lecture
Notes in Physics, https://doi.org/10.1007/978-3-031-46320-4_4

point charges. A point charge is localised 'at a single point', i.e. there is a *finite* charge at a single value of x only, beyond this point there is no charge at all. How can one in this case specify the corresponding charge density? Obviously, there is a problem. Consider a charge q placed at $x = 0$. Since the charge is assumed to be point-like, the density must be equal to zero everywhere outside the charge, i.e. where $x \neq 0$, and equal to some constant A where the charge is, i.e. at $x = 0$. It is immediately obvious that the constant A cannot be finite. Indeed, integration of the density must recover the total charge q. However, we obtain zero when integrating A since the function $\rho(x) = A$ only within an immediate vicinity of the point $x = 0$, i.e. when $-\epsilon < x < \epsilon$ with $\epsilon \to 0$, but is zero outside it. Integrating such a density gives zero:

$$\int_{-\infty}^{\infty} \rho(x)dx = \lim_{\epsilon \to 0} \int_{-\epsilon}^{\epsilon} Adx = \lim_{\epsilon \to 0} 2A\epsilon = 0 .$$

This paradoxical situation may be resolved in the following way: one hopes that the limit above may *not* be equal to zero if the constant A was infinity as, in this case, the limit would be of the $0 \cdot \infty$ type and hence (one hopes) might be finite. This means that the charge density of a point charge must have a very unusual form: $\rho(x) = 0$ for any $x \neq 0$ and $\rho(0) = +\infty$, i.e. it is zero everywhere apart from the single point $x = 0$ where it is infinite; such a density has an infinitely high zero-width spike at $x = 0$.

This is indeed a very unusual function we never come across before: all functions we have encountered so far were smooth with at most finite number of discontinuities; we have never had a function which jumps infinitely high up on the left of a single point ($x = 0$) and immediately jumps back on the right side of it.

How this kind of a function can be defined mathematically? We shall show below that this function can be defined in a very special way as a limit of an infinite sequence of well-defined functions.

Consider a function $\delta_n(x)$ which has the form of a rectangular impulse for any value of $n = 1, 2, 3, \ldots$ (see Fig. 4.1):

$$\delta_n(x) = \begin{cases} n, & \text{if } -1/2n \leq x \leq 1/2n \\ 0, & \text{if } |x| > 1/2n \end{cases} . \tag{4.1}$$

Fig. 4.1 First elements of the rectangular delta sequence $\delta_n(x)$ for $n = 1, \ldots, 4$. One can see that each next element of the sequence is narrower and higher, but the area under it remains the same and equal to unity

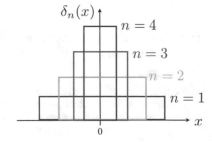

This function is constructed in such a way that the area under it is equal to one for *any* value of n, i.e. the definite integral

$$\int_{-\infty}^{\infty} \delta_n(x)\,dx = \int_{-1/2n}^{1/2n} n\,dx = n\frac{2}{2n} = 1 \qquad (4.2)$$

for any n. When n is increased, the graph of $\delta_n(x)$ becomes narrower and more peaked; at the same time, the area under the curve remains the same and equal to one. In the limit of $n \to \infty$ we shall arrive at a function which has an infinite height and at the same time an infinitely small width—exactly what we need! This idealised 'impulse' function, denoted $\delta(x)$, which is defined in the limit of $n \to \infty$, was first introduced by Dirac and bears his name:

$$\lim_{n \to \infty} \delta_n(x) = \delta(x) . \qquad (4.3)$$

By definition, it has the following properties:

$$\delta(x) = \begin{cases} \infty, & \text{if } x = 0 \\ 0, & \text{if } x \neq 0 \end{cases} \text{ and } \int_{-\infty}^{\infty} \delta(x)\,dx = 1 . \qquad (4.4)$$

It is helpful to think of $\delta(x)$ as being an infinitely sharp impulse function of unit area centred at $x = 0$.

It is important to stress that the function defined in this way is *not* an ordinary function; it can only be understood and defined using a sequence of well-defined functions $\delta_n(x), n = 1, 2, 3...$, called a *delta sequence*. This type of a function belongs to the class of *generalised functions*.

We shall now consider one of the most important properties of this function which is called the *filtering theorem*. Let $f(x)$ be a well-defined function which can be expanded in a Taylor series around $x = 0$. Consider the integral

$$\int_{-\infty}^{\infty} \delta_n(x) f(x)\,dx = n\int_{-1/2n}^{1/2n} f(x)\,dx . \qquad (4.5)$$

For large n the integration interval $-1/2n \leq x \leq 1/2n$ becomes very narrow and hence $f(x)$ can be represented well by the Maclaurin series about the point $x = 0$:

$$f(x) = f(0) + f'(0)x + \frac{f''(0)}{2}x^2 + \ldots + \frac{f^{(k)}(0)}{k!}x^k + \ldots , \qquad (4.6)$$

so that the integral (4.5) becomes

$$\int_{-\infty}^{\infty} \delta_n(x)f(x)dx = f(0) + \sum_{k=1}^{\infty} \frac{f^{(k)}(0)}{k!} \int_{-1/2n}^{1/2n} x^k dx$$

$$= f(0) + \sum_{k=1}^{\infty} \frac{f^{(k)}(0)}{(k+1)!} \frac{1 + (-1)^k}{(2n)^{k+1}}\ .$$

It can be readily seen that at large values of n the integral is mainly determined by $f(0)$, the other terms (in the sum) are very small, the largest one being of the order of $1/n^2$. In fact, by taking the $n \to \infty$ limit, we obtain

$$\lim_{n \to \infty} \left[\int_{-\infty}^{\infty} \delta_n(x)f(x)dx \right] = f(0)\ . \tag{4.7}$$

Formally, we can take the limit under the integral sign which would turn $\delta_n(x)$ into $\delta(x)$, and hence arrive at the following formal result:

$$\int_{-\infty}^{\infty} \delta(x)f(x)dx = f(0)\ . \tag{4.8}$$

One may say that the integration of $f(x)$ with $\delta(x)$ has *filtered* out the value $f(0)$ of the function $f(x)$; this value corresponds to the point $x = 0$ where the delta function is peaked. This result must be easy to understand: since the delta function is infinitely narrow around $x = 0$, only a single value of $f(x)$ at this point, $f(0)$, can be kept in the product $f(x)\delta(x)$ under the integral. Basically, within that infinitesimally narrow interval of x, the function $f(x)$ may be considered as a constant equal to $f(0)$. Therefore, it can be taken out of the integral which then appears simply as the integral of the delta function alone which is equal to unity because of Eq. (4.4).

We shall now discuss some obvious generalisations. Instead of considering the delta sequence centred at $x = 0$, one may define a sequence $\{\delta_n(x - a),\ n = 1, 2, 3, \ldots\}$, centred on the point $x = a$. This would lead us to the delta function $\delta(x - a)$ and the corresponding generalisation of the filtering theorem:

$$\int_{-\infty}^{\infty} \delta(x - a)f(x)dx = f(a)\ . \tag{4.9}$$

Here the delta function $\delta(x - a)$ peaks at $x = a$ and in the integral it filtered out the value of the function $f(x)$ at this point.

Problem 4.1 Derive the above result by making the substitution $t = x - a$ in the integral.

In fact, more complex arguments of the delta function may be considered in the same way by making an appropriate substitution. For instance,

$$\int_{-10}^{5} e^{-x}\delta(2x+1)\,dx = \left| \begin{array}{l} t = 2x+1 \\ dt = 2dx \end{array} \right| = \int_{-19}^{11} \left(\frac{1}{2} e^{-(t-1)/2} \right) \delta(t)\,dt$$

$$= \left(\frac{1}{2} e^{-(t-1)/2} \right)_{t=0} = \frac{1}{2} e^{1/2} = \frac{\sqrt{e}}{2}.$$

Note that the exact numerical values of the boundaries in the integral are unimportant: as long as the singularity of the delta function happens within them, one can use the filtering theorem irrespective of the exact values of the boundaries; otherwise, i.e. when the singularity is outside the boundaries, the integral is simply equal to zero.

Problem 4.2 Prove the following identities:

$$\int_{-\infty}^{1} \delta(x)\frac{\sin x}{x}\,dx = 1 \; ; \quad \int_{-1}^{1} \frac{1+x}{1+2x^2}\delta(2x-4)\,dx = 0 \; ;$$

$$\int_{-1}^{\infty} \frac{1+x}{1+2x^2}\delta(2x-4)\,dx = \frac{1}{6}.$$

Using the filtering theorem, various useful properties of the delta function may be established. First of all, consider $\delta(-x)$. We have for any function $f(x)$ which is smooth around $x = 0$:

$$\int_{-\infty}^{\infty} \delta(-x)f(x)\,dx = -\int_{+\infty}^{-\infty} \delta(t)f(-t)\,dt = \int_{-\infty}^{\infty} \delta(t)f(-t)\,dt$$

$$= f(-0) = f(0),$$

i.e. $\delta(-x)$ does the same job as $\delta(x)$ and hence we formally may write

$$\delta(-x) = \delta(x). \tag{4.10}$$

Because of this property which tells us that the delta function is even (something one would easily accept considering its definition), we may also write the following useful results:

$$\int_{-\infty}^{0} \delta(x)\,dx = \int_{0}^{\infty} \delta(x)\,dx = \frac{1}{2},$$

and hence

$$\int_{-\infty}^{0} \delta(x)f(x)\,dx = \int_{0}^{\infty} \delta(x)f(x)\,dx = \frac{f(0)}{2}. \tag{4.11}$$

Problem 4.3 Prove the following other properties of the delta function:

$$x\delta(x) = 0 \; ;$$

$$\delta(ax + b) = \frac{1}{|a|}\delta\left(x - \frac{b}{a}\right) . \qquad (4.12)$$

Note that in the latter case both positive and negative values of a are to be considered separately.

Often the argument of the delta function is a more complex function than the linear one considered so far. Moreover, it may become equal to zero at more than one point yielding more than one singularity points for the delta function. As a consequence, the filtering theorem in these cases must be modified. As an example, consider the integral

$$I = \int_{-\infty}^{\infty} f(x)\delta\left(x^2 - a^2\right) dx .$$

It contains $\delta\left(x^2 - a^2\right)$ which has two impulses: one at $x = -a$ and another at $x = +a$ (here $a > 0$). We split the integral into two: one performed around the point $x = -a$ and another around $x = a$:

$$I = \int_{-a-\epsilon}^{-a+\epsilon} \delta\left(x^2 - a^2\right) f(x)dx + \int_{a-\epsilon}^{a+\epsilon} \delta\left(x^2 - a^2\right) f(x)dx = I_- + I_+ ,$$

where $0 < \epsilon < a$. In the first integral I_- we change the variable x into $t = x^2 - a^2$, so that $x = -\sqrt{t + a^2}$ and $dx = -dt/\left(2\sqrt{t + a^2}\right)$; here the minus sign is essential as the point $t = 0$ where $\delta(t)$ is peaked must correctly correspond to $x = -a$ where $\delta\left(x^2 - a^2\right)$ is peaked. Performing the integration, we then obtain:

$$I_- = \int_{2\epsilon a + \epsilon^2}^{-2\epsilon a + \epsilon^2} \delta(t)f\left(-\sqrt{t + a^2}\right) \frac{-dt}{2\sqrt{t + a^2}}$$

$$= \int_{-2\epsilon a + \epsilon^2}^{2\epsilon a + \epsilon^2} \delta(t)f\left(-\sqrt{t + a^2}\right) \frac{dt}{2\sqrt{t + a^2}} = \frac{f\left(-\sqrt{a^2}\right)}{2\sqrt{a^2}}$$

$$= \frac{f(-|a|)}{2|a|} .$$

In the second integral I_+ the same substitution is made, however, in this case $x = +\sqrt{t + a^2}$ and $dx = +dt/2\sqrt{t + a^2}$, while the integration results in $f(|a|)/|a|$. We therefore obtain that

$$\int_{-\infty}^{\infty} f(x)\delta\left(x^2 - a^2\right) dx = \frac{1}{2\,|a|}\left[f\left(-\,|a|\right) + f\left(|a|\right)\right] = \frac{1}{2\,|a|}\left[f\left(a\right) + f\left(-a\right)\right].$$

The same result is obtained if we formally accept the following identity:

$$\delta\left(x^2 - a^2\right) = \frac{1}{2\,|a|}\left[\delta(x - a) + \delta(x + a)\right]. \tag{4.13}$$

Problem 4.4 Generalise this result to an arbitrary continuous differentiable function $\varphi(x)$ inside the delta function:

$$\int_{-\infty}^{\infty} \delta(\varphi(x)) f(x) dx = \sum_{i=1}^{n} \frac{f(x_i)}{|\varphi'(x_i)|} \quad \Longrightarrow \quad \delta(\varphi(x)) = \sum_{i=1}^{n} \frac{\delta(x - x_i)}{|\varphi'(x_i)|},$$
$$\tag{4.14}$$

where $x_1, \ldots, x_n$ are n roots of the equation $\varphi(x) = 0$.

The rectangular sequence considered above to define the delta function is not unique; there are many other sequences one can build to define the delta function as the limit. If we consider any bell-like function $\phi(x)$ defined for $-\infty < x < \infty$ and of unit area which tends to zero at $x \to \pm\infty$, as in Fig. 4.2, then one can construct the corresponding delta sequence using the recipe $\delta_n(x) = n\phi(nx)$, which in the limit $n \to \infty$ corresponds to the delta function. Indeed, $\phi(nx)$ tends to become narrower with increasing n and, at the same time, the pre-factor n makes the function $\delta_n(x)$ more peaked as n gets bigger; at the same time, the area under the curve remains the same for any n,

$$\int_{-\infty}^{\infty} \delta_n(x) dx = \int_{-\infty}^{\infty} n\phi(nx) dx = \left|\begin{array}{c} t = nx \\ dt = ndx \end{array}\right| = \int_{-\infty}^{\infty} \phi(t) dt = 1,$$

as required. Also, it is clear that because the function $\delta_n(x)$ defined above gets narrower for larger n, the filtering theorem (4.8) or (4.9) is also valid.

Indeed, consider the integral

$$\int_{-\infty}^{\infty} f(x)\delta_n(x) dx = \int_{-\infty}^{\infty} f(x) n\phi(nx) dx = \int_{-\infty}^{\infty} f\left(\frac{t}{n}\right) \phi(t) dt.$$

Expanding the function $f(t/n)$ in the Maclaurin series, we have:

$$\int_{-\infty}^{\infty} f(x)\delta_n(x) dx = f(0) \int_{-\infty}^{\infty} \phi(t) dt + \frac{f'(0)}{n} \int_{-\infty}^{\infty} t\phi(t) dt$$
$$+ \frac{f''(0)}{2n^2} \int_{-\infty}^{\infty} t^2\phi(t) dt + \cdots.$$

Fig. 4.2 The graph of a typical bell-like function $\phi(x)$ which may serve as the one generating the corresponding delta sequence

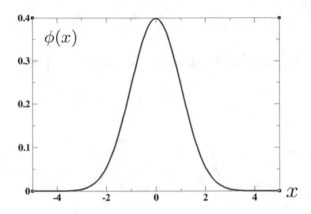

Since the function $\phi(t)$ is of unit area, the first term is simply $f(0)$, while all the other terms tend to zero in the $n \to \infty$ limit, providing that for any positive integer k the integral $\int_{-\infty}^{\infty} t^k \phi(t)\, dt$ converges.

For example, the following functions can be used to generate the delta sequences:

$$\phi_1(x) = \frac{1}{\sqrt{2\pi}} e^{-x^2/2} \;, \tag{4.15}$$

$$\phi_2(x) = \frac{1}{\pi} \cdot \frac{1}{1+x^2} \;, \tag{4.16}$$

$$\phi_3(x) = \frac{\sin x}{\pi x} = \frac{1}{2\pi} \int_{-1}^{1} e^{ikx} dk \tag{4.17}$$

(the last passage in the equality is checked by direct integration). All these functions satisfy the required conditions and have the desired shape as in Fig. 4.2. Therefore, these functions generate the following delta sequences:

$$\delta_n^{(1)}(x) = \frac{n}{\sqrt{2\pi}} e^{-n^2 x^2/2} \;, \tag{4.18}$$

$$\delta_n^{(2)}(x) = \frac{n}{\pi} \cdot \frac{1}{1+n^2 x^2} \;. \tag{4.19}$$

The third sequence,

$$\delta_n^{(3)}(x) = \frac{n}{2\pi} \int_{-1}^{1} e^{iknx} dk = \left| \begin{matrix} t = nk \\ dt = ndk \end{matrix} \right| = \frac{1}{2\pi} \int_{-n}^{n} e^{ixt} dt \;,$$

in the $n \to \infty$ limit tends to the integral

$$\delta_n^{(3)}(x) \quad \Longrightarrow \quad \delta(x) = \frac{1}{2\pi} \int_{-\infty}^{\infty} e^{ixt} dt \,, \tag{4.20}$$

which is a very frequently used integral representation of the delta function. We shall come across it several times in this book.

Various representations of the delta function can be used in deriving useful formulae. We shall illustrate this point here by deriving the Sokhotski–Plemelj formula:

$$\int_{-\infty}^{\infty} \frac{f(x)}{x - x_0 \pm i\delta} dx = \mathcal{P} \int_{-\infty}^{\infty} \frac{f(x)}{x - x_0} dx \mp i\pi f(x_0) \,, \tag{4.21}$$

where $\delta \to +0$, $\mathcal{P}$ is the symbol of Cauchy principal value (Sect. I.4.5.4) and $f(x)$ is some continuous function on the real axis. This formula can be written symbolically as:

$$\frac{1}{x - x_0 \pm i\delta} = \mathcal{P} \frac{1}{x - x_0} \mp i\pi \delta(x - x_0) \,. \tag{4.22}$$

To prove it, let us multiply the denominator and the numerator of the integrand in the original integral by $(x - x_0) \mp i\delta$:

$$\int_{-\infty}^{\infty} \frac{f(x)}{x - x_0 \pm i\delta} dx = \int_{-\infty}^{\infty} \frac{f(x)(x - x_0 \mp i\delta)}{(x - x_0)^2 + \delta^2} dx$$

$$= \int_{-\infty}^{\infty} \frac{x - x_0}{(x - x_0)^2 + \delta^2} f(x) dx \mp i \int_{-\infty}^{\infty} \frac{\delta}{(x - x_0)^2 + \delta^2} f(x) dx. \tag{4.23}$$

In the first term we have to exclude the point x_0 from the integration since otherwise the integral diverges in the $\delta \to 0$ limit:

$$\int_{-\infty}^{\infty} \frac{x - x_0}{(x - x_0)^2 + \delta^2} f(x) dx = \int_{-\infty}^{x_0 - \epsilon} \frac{x - x_0}{(x - x_0)^2 + \delta^2} f(x) dx + \int_{x_0 + \epsilon}^{\infty} \frac{x - x_0}{(x - x_0)^2 + \delta^2} f(x) dx$$

$$\Longrightarrow \int_{-\infty}^{x_0 - \epsilon} \frac{x - x_0}{(x - x_0)^2} f(x) dx + \int_{x_0 + \epsilon}^{\infty} \frac{x - x_0}{(x - x_0)^2} f(x) dx$$

$$\Longrightarrow \mathcal{P} \int_{-\infty}^{\infty} \frac{f(x)}{x - x_0} dx$$

in the $\delta \to 0$ limit. Above, $\epsilon \to +0$. To transform the second term in Eq. (4.23), we notice that an expression $\delta/(\delta^2 + x^2)$ tends to $\pi \delta(x)$ in the $\delta \to 0$ limit, i.e.

$$\lim_{\delta \to 0} \frac{1}{\pi} \frac{\delta}{\delta^2 + x^2} = \delta(x) \,. \tag{4.24}$$

Indeed, this follows from the representation (4.19) of the delta function upon the substitution $n \to 1/\delta$. Therefore, the second term in Eq. (4.23), upon the application of the filtering theorem for the delta function, becomes $\mp i\pi f(x_0)$. This proves completely formulae (4.21) and (4.22).

Problem 4.5 Show that formula (4.22) is consistent with Eq. (2.72).

Problem 4.6 By splitting the integral (4.20) into two by the zero point, show that this formula can also be equivalently written as follows:

$$\delta(x) = \frac{1}{\pi} \int_0^\infty \cos(xt)\, dt . \qquad (4.25)$$

Problem 4.7 Calculate the following integrals:

$$(a) \int_{-10}^{20} \delta(5x+6)e^{-10(x+1)^2} dx ; \quad (b) \int_{-\infty}^{\infty} \delta(-5x+6)e^{-10(x+1)^2} dx ;$$

$$(c) \int_{-\infty}^{\infty} e^{-2x} [\delta(2x+1) + 5H(x+1)]\, dx .$$

Here $H(x)$ is the Heaviside unit step function defined in Sect. I.2.1. [*Answer : (a)* $\exp(-2/5)/5$; *(b)* $\exp(-242/5)/5$; *(c)* $e(1+5e)/2$.]

The Heaviside unit step function $H(x)$ is directly related to the delta function. Indeed, $H(x)$ is constant anywhere apart from the point $x = 0$ and hence its derivative is equal to zero for $x \neq 0$; it is not defined at $x = 0$. We might expect, however, that since it makes a finite jump at this point, its derivative will be equal to infinity at $x = 0$, so we might anticipate that

$$H'(x) = \delta(x) . \qquad (4.26)$$

This guess can be supported by a simple calculation that shows that the integral of the derivative of the Heaviside function is equal to unity:

$$\int_{-\infty}^{\infty} H'(x)dx = H(\infty) - H(-\infty) = 1 - 0 = 1 ,$$

as is required for the delta function. Moreover, the integral:

$$\int_{-\infty}^{x} \delta(t)dt = H(x) \quad \text{for any} \quad x \neq 0 , \qquad (4.27)$$

since for $x < 0$ it is zero as the spike due to singularity of the delta function appears outside the integral limits, while for $x > 0$ we obviously have 1 as the singularity falls inside the limits. The case of $x = 0$ results in fact in the value of the integral equal to $1/2$ due to the fact that the delta function is even, i.e. it follows that $H(0) = 1/2$ is consistent with Eq. (4.27). If we now differentiate both sides of this equation with respect to x, we obtain (4.26). The derivative function $H'(x)$ belongs to the class of generalised functions and is equal to $\delta(x)$ in this sense.

Problem 4.8 By using integration by parts, prove the integral identity

$$\int_{-\infty}^{\infty} f(x)\delta'(x)dx = -f'(0),$$

which can be used as a definition of the impulse function $\delta'(x) = \frac{d}{dx}\delta(x)$.

Problem 4.9 By using repeatedly integration by parts or the method of induction, prove that generally

$$\int_{-\infty}^{\infty} f(x)\delta^{(n)}(x)dx = (-1)^n f^{(n)}(0). \tag{4.28}$$

This identity can be used to define higher derivatives of $\delta(x)$.

Problem 4.10 Prove that

$$\iint_{-\infty}^{\infty} x^n y^m e^{ixy} dx dy = \delta_{nm} 2\pi i^n n!.$$

[Hint: *first, relate, e.g. the x-integral $\int x^n e^{ixy} dx$ to n derivatives over y of the delta function $\delta(y)$, the latter arises via its representation* (4.20); *secondly, calculate the y-integral using Eq.* (4.28) *or by the repetitive integration by parts; note that both the delta function and its derivatives are equal to zero at $\pm\infty$.*] Next, show that

$$\iint_{-\infty}^{\infty} x^m y^n e^{2i(y-a)(x-b)} dx dy = \pi \sum_{k=0}^{\min(m,n)} \frac{m!n!}{k!(m-k)!(n-k)!} \left(\frac{i}{2ab}\right)^k a^n b^m.$$

Problem 4.11 By calculating the integral $\int_{-\infty}^{\infty} f(x)H'(x)dx$ by parts, prove that the derivative of the Heaviside function, $H'(x)$, is equal to the Dirac delta function $\delta(x)$ in a sense that $H'(x)$ works as a filter for $f(x)$, exactly as the delta function.

Problem 4.12 Consider a system of n unit mass particles which oscillate around their equilibrium positions (e.g. a crystal). The displacement of a particle p in direction $\alpha = x, y, z$ is denoted $x_{p\alpha}(t) = x_i(t)$, where $i = (p\alpha)$ is the joined index associated with the corresponding degree of freedom. If we define a vector $\mathbf{X} = (x_i)$ of all displacements of the particles from their equilibrium positions, then this vector would satisfy the following equation of motion:

$$\ddot{\mathbf{X}} + \Gamma \dot{\mathbf{X}} + \mathbf{D}\mathbf{X} = \mathbf{\Phi}, \tag{4.29}$$

where $\mathbf{D}$ is a (symmetric) dynamical matrix of the system associated with atoms vibrating around their equilibrium positions. The second term on the left-hand side corresponds to a friction acting on each degree of freedom, with $\Gamma > 0$ being the corresponding (scalar) friction constant (i.e. all degrees of freedom experience an identical friction). The force $\mathbf{\Phi}(t) = (\varphi_i(t))$ on the right-hand side is a vector of stochastic (random) forces $\varphi_i(t)$ acting on each degree of freedom i. The statistical average, shown by the angle brackets $\langle \ldots \rangle$, of a product of these forces calculated at different times satisfies the following equation:

$$\langle \varphi_i(t)\varphi_j(t') \rangle = 2\Gamma k_B T \delta_{ij} \delta(t - t'). \tag{4.30}$$

This expression shows that stochastic forces are not correlated in time: indeed, if $t \neq t'$, then the delta function is zero meaning there are no correlations. Only the forces at the same times $t = t'$ correlate (when the delta function is not zero). This also means that the system of oscillators does not possess any memory as the past ($t' < t$) does not influence the future at time t due to lack of correlation between the forces at different times. Also forces corresponding to different degrees of freedom are not correlated with each other due to δ_{ij}. The appearance of temperature T on the right-hand side of Eq. (4.30) is not accidental: this ensures that the so-called fluctuation–dissipation theorem is obeyed. This is necessary to satisfy the equipartition theorem of statistical mechanics as we shall see later on in this problem.

(i) Let $\mathbf{e}_\lambda$ and ω_λ^2 be eigenvectors and eigenvalues of the dynamical matrix $\mathbf{D}$. By writing $\mathbf{X}$ as a linear combination of all eigenvectors, $\mathbf{X}(t) = \sum_\lambda \xi_\lambda(t)\mathbf{e}_\lambda$, show that each scalar coordinate $\xi_\lambda(t)$ satisfies the following DE:

$$\ddot{\xi}_\lambda + \Gamma \dot{\xi}_\lambda + \omega_\lambda^2 \xi_\lambda = \psi_\lambda, \tag{4.31}$$

where $\psi_\lambda = \mathbf{e}_\lambda^T \mathbf{\Phi}$ is the scalar product which can also be interpreted as a projection of the vector $\mathbf{\Phi}$ onto the λ-th normal coordinate.

(ii) Show that the general solution of Eq. (4.31) contains, apart from a decaying (with time) component (the transient), a particular integral

$$\xi_\lambda(t) = \frac{1}{\overline{\omega}_\lambda} \int_{-\infty}^{t} \psi_\lambda(\tau) e^{-\Gamma(t-\tau)/2} \sin\left[\overline{\omega}_\lambda (t-\tau)\right] d\tau,$$

where $\overline{\omega}_\lambda = \sqrt{\omega_\lambda^2 - \Gamma^2/4}$ (assumed real).

(iii) Correspondingly, show that the solution of Eq. (4.29) which survives at long times can be written in the matrix form as follows:

$$\mathbf{X}(t) = \int_{-\infty}^{t} e^{-\Gamma(t-\tau)/2} \frac{\sin\left[\sqrt{\mathbf{G}}\,(t-\tau)\right]}{\sqrt{\mathbf{G}}} \boldsymbol{\Phi}(\tau) d\tau,$$

while the velocity vector, $\mathbf{V}(t) = \dot{\mathbf{X}}(t)$ reads:

$$\mathbf{V}(t) = \int_{-\infty}^{t} e^{-\Gamma(t-\tau)/2} \left\{ -\frac{\Gamma}{2} \frac{\sin\left[\sqrt{\mathbf{G}}\,(t-\tau)\right]}{\sqrt{\mathbf{G}}} + \cos\left[\sqrt{\mathbf{G}}\,(t-\tau)\right] \right\} \boldsymbol{\Phi}(\tau) d\tau.$$

Here the matrix $\mathbf{G} = \mathbf{D} - \left(\Gamma^2/4\right) \mathbf{E}$, where $\mathbf{E}$ is the unit matrix.

Problem 4.13 (Continuation of the previous problem) (iv) Show that the half of the same time velocity–velocity autocorrelation function at long times, i.e. the half of the average of the square of the velocity is

$$\frac{1}{2} \left\langle \mathbf{V}(t) \mathbf{V}^T(t) \right\rangle = \frac{k_B T}{2} \left(\delta_{ij}\right) = \frac{k_B T}{2} \mathbf{E},$$

i.e. $\frac{1}{2} k_B T$ of the kinetic energy (recall that our particles are of unit mass) is associated with a single degree of freedom, while the same time displacement–displacement autocorrelation function is

$$\left\langle \mathbf{X}(t) \mathbf{X}^T(t) \right\rangle = k_B T \mathbf{D}^{-1}.$$

These are manifestations of the equipartition theorem for the vibrational potential energy. Indeed, if $\mathbf{D}$ was diagonal, $\mathbf{D} = \left(\delta_{ij} \chi_i\right)$, with χ_i being an elastic constant of the oscillator associated with the i-th degree of freedom, the diagonal element $j = i$ of the latter correlation function would be given by

$$\left\langle \mathbf{X}(t) \mathbf{X}^T(t) \right\rangle_{ii} = \frac{k_B T}{\chi_i} \implies \frac{\chi_i \left\langle x_i^2(t) \right\rangle}{2} = \frac{1}{2} k_B T,$$

as expected. These expressions justify the particular choice (4.30) for the random forces which ensure the manifestation of the equipartition theorem.

Problem 4.14 Consider N point particles at positions $\mathbf{r}_i(t)$ and with charges q_i that are allowed to move around in some volume, $i = 1, \ldots, N$. The charge density and the current due to the particles are given by

$$\rho(\mathbf{r}, t) = \sum_{i=1}^{N} q_i \delta\left(\mathbf{r} - \mathbf{r}_i(t)\right) \text{ and } \mathbf{j}(\mathbf{r}, t) = \sum_{i=1}^{N} q_i \dot{\mathbf{r}}_i \delta\left(\mathbf{r} - \mathbf{r}_i(t)\right),$$

respectively. By differentiating $\rho(\mathbf{r}, t)$ with respect to time t, derive the continuity equation:

$$\frac{\partial \rho}{\partial t} + \operatorname{div} \mathbf{j} = 0.$$

Here the three-dimensional delta function $\delta(\mathbf{r})$ is defined as a product of three one-dimensional delta functions, $\delta(x)\delta(y)\delta(z)$.

Problem 4.15 In this problem we shall also illustrate the 3D delta function $\delta(\mathbf{r}) = \delta(x)\delta(y)\delta(z)$ by proving that

$$\delta(\mathbf{r}) = -\Delta \frac{1}{4\pi r}, \tag{4.32}$$

where Δ is the Laplacian and $r = |\mathbf{r}|$. Follow the following steps:

(i) show that outside the centre of the coordinate system, i.e. for $r \neq 0$, we have $\Delta \frac{1}{r} = 0$; note that at $r = 0$ this calculation is ill-defined;

(ii) next, show that $\Delta \frac{1}{r}$ can also be formally written as div $\mathbf{E}$, where $\mathbf{E} = -\mathbf{r}/r^3$;

(iii) then, consider a sphere of radius R with its centre placed at the centre of the coordinate system $\mathbf{r} = \mathbf{0}$; use the divergence theorem to demonstrate that a volume integral of $-\Delta \frac{1}{4\pi r}$ over that sphere is equal to 1; argue that this result must also be valid for any volume as long as it contains the centre of the coordinate system in it;

(iv) hence, conclude that Eq. (4.32) makes sense since we have the expected normalisation:

$$\iiint \delta(\mathbf{r}) d\mathbf{r} = \iiint \delta(x)\delta(y)\delta(z) dx dy dz$$

$$= \int \delta(x) dx \int \delta(y) dy \int \delta(z) dz = 1;$$

(v) next, take an arbitrary function $f(\mathbf{r})$ that is finite at $\mathbf{r} = \mathbf{0}$, and consider the volume integral

$$\iiint f(\mathbf{r}) \left(\Delta \frac{1}{r} \right) dV = \iiint f(\mathbf{r}) \operatorname{div} \mathbf{E} \, dV \, .$$

Argue, that one can take an arbitrary volume containing the centre of the coordinate system in it; therefore, choose a sphere of radius $R \to 0$;

(vi) using a formula for $\operatorname{div}(f(\mathbf{r})\mathbf{E})$, represent the above integral as a sum of two: one containing the divergence, and another a volume integral containing the vector $\mathbf{F} = \operatorname{grad} f(\mathbf{r})$;

(vii) convert the volume integral with the divergence into a surface integral over the surface of the sphere and calculate it in the limit $R \to 0$; note that in this limit the expression containing the arbitrary function $f(\mathbf{r})$ can be taken out of the integral at $\mathbf{r} = \mathbf{0}$; then show that this integral is equal to $-4\pi f(\mathbf{0})$;

(viii) finally, calculate the remaining volume integral taken over the sphere with the radius $R \to 0$ to show that it tends to zero; this calculation is convenient to do in spherical coordinates; note also that the dot product $\mathbf{r} \cdot \mathbf{F}$ can be replaced with $r F_r$, where F_r is a projection of the vector field $\mathbf{F}$ on the direction $\mathbf{r}$; as we are interested in the limit when the sphere radius tends to zero, the function F_r can be taken out of the integral at $\mathbf{r} = \mathbf{0}$;

(ix) therefore, conclude that

$$\iiint f(\mathbf{r}) \left(-\Delta \frac{1}{4\pi r} \right) dV = f(\mathbf{0}) \, ,$$

as required.

4.2 The Gamma Function

4.2.1 Definition and Main Properties

We define the gamma function, $\Gamma(z)$, as the following integral in which its argument, generally a complex number z, appears as a parameter:

$$\Gamma(z) = \int_0^\infty t^{z-1} e^{-t} dt \, . \tag{4.33}$$

We initially assume that $\operatorname{Re} z > 0$ as this integral converges in the right half of the complex plane. Indeed, because of e^{-t}, the convergence at $t \to \infty$ is achieved. However, the integral may diverge at $t = 0$ since for small t one puts $e^{-t} \simeq 1$ arriving at the integral $\int_0^\epsilon t^{z-1} dt$ (with some $0 < \epsilon \ll 1$) which diverges for $z = 0$

(logarithmically). Therefore, the question of convergence of the integral is not at all straightforward. In order to prove the convergence of the integral for $\operatorname{Re} z > 0$, it is wise to split the integral into two parts corresponding to intervals $0 \le t \le 1$ and $1 \le t < \infty$:

$$\Gamma(z) = \Gamma_1(z) + \Gamma_2(z)\,, \quad \Gamma_1(z) = \int_0^1 t^{z-1} e^{-t} dt \quad \text{and} \quad \Gamma_2(z) = \int_1^\infty t^{z-1} e^{-t} dt\,.$$

Consider first $\Gamma_2(z)$. For a vertical stripe $0 < \operatorname{Re} z \le x_{max}$, we write $z = x + iy$, and then the integral can be estimated as

$$\begin{aligned}
|\Gamma_2(z)| &\le \int_1^\infty \left| t^{z-1} e^{-t} \right| dt = \int_1^\infty e^{-t} \left| e^{(x-1)\ln t + iy\ln t} \right| dt \\
&= \int_1^\infty e^{-t} \left| e^{(x-1)\ln t} \right| \left| e^{iy\ln t} \right| dt = \int_1^\infty e^{-t} \left| e^{(x-1)\ln t} \right| dt \\
&\le \int_1^\infty e^{-t} e^{(x_{max}-1)\ln t} dt = \int_1^\infty t^{x_{max}-1} e^{-t} dt\,.
\end{aligned}$$

In writing the third line above we have made use of the fact that $\ln t > 0$ when $t > 1$. Since the integral on the right-hand side converges (its convergence at $t \to \infty$ is obvious because of the exponential function e^{-t}), then $\Gamma_2(z)$ converges as well. Moreover, since the estimate above is valid for any z within the stripe, it converges uniformly. Since the value of x_{max} was chosen arbitrarily, the integral $\Gamma_2(z)$ converges everywhere to the right of the imaginary axis, i.e. for any $\operatorname{Re} z > 0$, and is analytic there.

Consider now $\Gamma_1(z)$ which is obviously analytic for $\operatorname{Re}(z-1) > 0$ (or when $x = \operatorname{Re} z > 1$). Within the vertical stripe $0 < x_{min} < \operatorname{Re} z < 1$ in the complex plane, the integral can be estimated as follows (note that here $\ln t < 0$):

$$\begin{aligned}
|\Gamma_1(z)| &\le \int_0^1 \left| t^{z-1} e^{-t} \right| dt = \int_0^1 e^{-t} \left| e^{(x-1)\ln t} \right| dt \\
&\le \int_0^1 e^{-t} e^{(x_{min}-1)\ln t} dt = \int_0^1 t^{x_{min}-1} e^{-t} dt \le \int_0^1 t^{x_{min}-1} dt \\
&= \left. \frac{t^{x_{min}}}{x_{min}} \right|_0^1 = \frac{1}{x_{min}}\,.
\end{aligned}$$

This means that $\Gamma_1(z)$ converges and converges uniformly. Hence, $\Gamma(z) = \Gamma_1(z) + \Gamma_2(z)$ is analytic everywhere to the right of the imaginary axis, $\operatorname{Re} z > 0$.

The function $\Gamma(z)$ satisfies a simple recurrence relation. Indeed, let us calculate the integral for $\Gamma(z+1)$ by parts:

$$\Gamma(z+1) = \int_0^\infty t^z e^{-t} dt = \left[-t^z e^{-t}\right]_0^\infty + z \int_0^\infty t^{z-1} e^{-t} dt$$
$$= z \int_0^\infty t^{z-1} e^{-t} dt = z\Gamma(z),$$
(4.34)

where the free term above is zero both at $t = 0$ (for Re z) and $t = \infty$.

Problem 4.16 Prove that in the case of z being a positive integer, $z = n$, the gamma function is equal to the factorial function:

$$\Gamma(n+1) = n!.$$
(4.35)

[*Hint: to see this, first check that* $\Gamma(1) = 1$ *and then apply induction.*]

Problem 4.17 Using induction and the fact that $\Gamma(1/2) = \sqrt{\pi}$ (see below), prove that

$$\Gamma\left(n+\frac{1}{2}\right) = \frac{1 \cdot 3 \cdot 5 \cdot \ldots \cdot (2n-1)}{2^n}\sqrt{\pi} = \frac{(2n-1)!!}{2^n}\sqrt{\pi} = \frac{(2n)!}{2^{2n}n!}\sqrt{\pi},$$
(4.36)

where the double factorial $(2n-1)!!$ corresponds to a product of all odd integers between 1 and $2n-1$.

Next we shall calculate $\Gamma(1/2)$. To do this, we first write its definition::

$$\Gamma\left(\frac{1}{2}\right) = \int_0^\infty t^{-1/2} e^{-t} dt.$$

Then, we make a substitution $t = x^2$, which yields:

$$\Gamma\left(\frac{1}{2}\right) = 2\int_0^\infty e^{-x^2} dx = \int_{-\infty}^\infty e^{-x^2} dx,$$

where we have also made use of the factor of two in front of the integral to extend the bottom limit to the $-\infty$. The really clever trick that we shall apply next is to consider the square of that expression and use x in the first integral and y in the second:

$$\left[\Gamma\left(\frac{1}{2}\right)\right]^2 = \int_{-\infty}^\infty e^{-x^2} dx \int_{-\infty}^\infty e^{-y^2} dy$$
$$= \int_{-\infty}^\infty \int_{-\infty}^\infty e^{-(x^2+y^2)} dx dy.$$

Now it must be clear why we have used x and y for the integration variables: $\left[\Gamma\left(\frac{1}{2}\right)\right]^2$ can now be viewed as a double integral over the entire $x - y$ plane! Hence, we can be calculate it by going into polar coordinates $x = r\cos\phi$ and $y = r\sin\phi$ (with $dxdy = rdrd\phi$), which gives:

$$\left[\Gamma\left(\frac{1}{2}\right)\right]^2 = \int_0^{2\pi} d\phi \int_0^\infty e^{-r^2} r\,dr = 2\pi \int_0^\infty e^{-r^2} r\,dr$$

$$= \pi \int_0^\infty e^{-t} dt = \pi \implies \Gamma\left(\frac{1}{2}\right) = \sqrt{\pi}.$$

(4.37)

Hence, we can also write

$$\Gamma\left(\frac{1}{2}\right) = \int_{-\infty}^{+\infty} e^{-x^2} dx = \sqrt{\pi}.$$

(4.38)

This is called the Gaussian integral.

Many integrals encountered in statistical mechanics are of the form:

$$I_n(a) = \int_{-\infty}^{+\infty} t^n e^{-\alpha t^2} dt,$$

(4.39)

where α is a positive constant and n a positive integer. For odd values of n the integrand is an odd function and the integral is obviously equal to zero. For even values of n the integral $I_n(\alpha)$ is directly related to the gamma function of half-integer argument as is demonstrated by the following problem.

Problem 4.18 Using a new variable $x = \alpha t^2$ in the integral of Eq. (4.39), show explicitly that for even n

$$I_n(a) = \alpha^{-(n+1)/2} \Gamma\left(\frac{n+1}{2}\right).$$

(4.40)

Problem 4.19 Prove the following formulae:

$$\int_{-\infty}^\infty t^2 e^{-\alpha t^2} dt = \frac{1}{2\alpha}\sqrt{\frac{\pi}{\alpha}}; \quad \int_{-\infty}^\infty t^4 e^{-\alpha t^2} = \frac{3\sqrt{\pi}}{4\alpha^{5/2}};$$

$$\int_{-\infty}^\infty t^8 e^{-\alpha t^2} = \frac{105\sqrt{\pi}}{16\alpha^{9/2}};$$

$$\int_{-\infty}^\infty \left(1+2x^2\right) e^{2x-\alpha x^2} dx = \sqrt{\frac{\pi}{\alpha}}\left(1+\frac{1}{\alpha}+\frac{2}{\alpha^2}\right) e^{1/\alpha}.$$

Problem 4.20 Prove that

$$\int_{-\infty}^{+\infty} e^{-ax^2 \pm bx} dx = \sqrt{\frac{\pi}{a}} e^{b^2/4a}. \tag{4.41}$$

Problem 4.21 In physics (as well as in many other disciplines) it is often necessary to consider the so-called Gaussian (named after Johann Carl Friedrich Gauss) function, or the Gaussian,

$$G(x) = \frac{1}{\sigma\sqrt{2\pi}} e^{-(x-x_0)^2/2\sigma^2}, \tag{4.42}$$

centred at the point x_0 and with dispersion σ which characterises the width of the function. An example of a Gaussian for $x_0 = 0$ and $\sigma = 1$ is depicted in Fig. 4.2. Show that the full width at half maximum (FWHM) for the Gaussian is equal to $2\sigma\sqrt{2\ln 2}$. Then, prove that $G(x)$ is correctly normalised to unity. Finally, calculate the first two momenta of the Gaussian:

$$\int_{-\infty}^{\infty} x G(x) dx = x_0 \quad \text{and} \quad \int_{-\infty}^{\infty} x^2 G(x) dx = \sigma^2 + x_0^2.$$

Above, the gamma function $\Gamma(z)$ was defined in the right half of the complex plane, to the right of the imaginary axis. The recursion relation (4.34) can be used to analytically continue (Sect. 2.6) $\Gamma(z)$ to the left half of the complex plane as well, where $\text{Re}\, z < 0$. Indeed, let us apply the recurrence relation to $\Gamma(z+n)$ consecutively $n - 1$ times (here $n = 1, 2, \ldots$):

$$\Gamma(z + n + 1) = (z + n)\,\Gamma(z + n) = \ldots$$
$$= (z + n)(z + n - 1) \cdots (z + 1)\, z\Gamma(z).$$

Solving for $\Gamma(z)$, we obtain

$$\Gamma(z) = \frac{\Gamma(z + n + 1)}{(z + n)(z + n - 1) \cdots (z + 1)\, z}. \tag{4.43}$$

This formula can be used for calculating the gamma function for $\text{Re}\, z \leq 0$. Indeed, using in the above formula for $n = 0$, we can write $\Gamma(z) = \Gamma(z+1)/z$, relating $\Gamma(z)$ within the vertical stripe $-1 < \text{Re}\, z \leq 0$ to the values of the gamma function $\Gamma(z+1)$ with $0 < \text{Re}\,(z+1) \leq 1$, where it is well defined. Similarly, by choosing different values of n one can define the gamma function for the corresponding vertical stripes of the width one to the left of the imaginary axis.

This formula also clearly shows that the gamma function in the left half of the complex plane will have poles at $z = -n$, where $n = 0, 1, 2, \ldots$, i.e. the gamma function defined this way is analytic in the whole complex plane apart from the points $z = 0, -1, -2, -3, \ldots$, where it has singularities. These are simple poles, however, since (see Sect. 2.5.5) only the limit

$$\lim_{z \to -n} (z + n) \, \Gamma \, (z) = \lim_{z \to -n} \frac{\Gamma(z + n + 1)}{(z + n - 1) \cdots (z + 1) z}$$

$$= \frac{\Gamma \, (1)}{(-1) \, (-2) \cdots (-n)} = \frac{(-1)^n}{n!}$$

is finite (recall that $\Gamma(1) = 1$); the limits of $(z + n)^k \, \Gamma(z)$ for any $k > 1$ are equal to zero. The limit above (see Sect. 2.7.1) also gives the residue at the pole $z = -n$, which is $(-1)^n / n!$.

There is also a simple identity involving the gamma function which we shall now derive. Using the integral representation (4.33) and assuming a real z between 0 and 1, let us consider the product $\Gamma(z)\Gamma(1 - z)$. We shall employ a similar trick to that we have used before when deriving, e.g. Eq. (4.37). We write

$$\Gamma \, (z) \, \Gamma \, (1 - z) = \left(\int_0^\infty t_1^{z-1} e^{-t_1} dt_1 \right) \left(\int_0^\infty t_2^{-z} e^{-t_2} dt_2 \right).$$

In the first integral we make the substitution $x^2 = t_1$, while in the second integral the substitution will be $y^2 = t_2$. This brings us to a double integral

$$\Gamma \, (z) \, \Gamma \, (1 - z) = 4 \int_0^\infty \int_0^\infty e^{-(x^2 + y^2)} \left(\frac{x}{y} \right)^{2z-1} dx dy,$$

in which we next use the polar coordinates $x = r \cos \phi$ and $y = r \sin \phi$. This gives (note that we only integrate over a quarter of the $x - y$ plane and hence $0 \le \phi \le \pi/2$):

$$\Gamma \, (z) \, \Gamma \, (1 - z) = 4 \int_0^\infty e^{-r^2} r dr \int_0^{\pi/2} (\cot \phi)^{2z-1} d\phi$$

$$= 2 \int_0^{\pi/2} (\cot \phi)^{2z-1} d\phi.$$

At the final step, we have made the substitution $\sqrt{t} = \cot \phi$, $d\phi = -dt / \left[2\sqrt{t} \, (1 + t) \right]$, which transforms the integral into the form which can be handled:

$$\Gamma \, (z) \, \Gamma \, (1 - z) = \int_0^\infty \frac{t^{z-1}}{1 + t} dt = \frac{\pi}{\sin(\pi z)}, \tag{4.44}$$

where at the last step we have used the result we obtained earlier in Problem 2.102 for the integral on the right-hand side. This result was derived for $0 < z < 1$. However, it can be analytically continued to the whole complex plane and hence this formula is valid for any z (apart from $z = -1, -2, \ldots$ where in both sides we have infinity).

Since $\Gamma(1-z) = -z\Gamma(-z)$ because of the recurrence relation the gamma function satisfies, the obtained identity can also be written as

$$\Gamma(z)\Gamma(-z) = -\frac{\pi}{z\sin(\pi z)}.\qquad(4.45)$$

This result allows deriving an interesting formula in which the gamma function is expressed via an infinite product. Let us derive it ignoring some mathematical subtleties. Consider a sequence $(n = 1, 2, \ldots)$ of functions

$$\Gamma_n(z) = \int_0^n \left(1 - \frac{t}{n}\right)^n t^{z-1} dt = \left|\begin{matrix} \tau = t/n \\ dt = nd\tau \end{matrix}\right|$$
$$= n^z \int_0^1 (1-\tau)^n \tau^{z-1} d\tau$$

for $z > 0$. This sequence converges to the gamma function in the limit of $n \to \infty$ because the sequence $(1 - t/n)^n$ converges to e^{-t}. The integral above can be calculated by repeated integration by parts.

Problem 4.22 Performing n integrations by parts, show that

$$\Gamma_n(z) = n^z \frac{n(n-1)(n-2)\cdots 1}{z(z+1)(z+2)\cdots(z+n-1)} \int_0^1 \tau^{z+n-1} d\tau$$
$$= \frac{n!}{z(z+1)(z+2)\cdots(z+n)} n^z.$$

Since $n^z = \exp(z\ln n)$, the above result can be rewritten as:

$$\frac{1}{\Gamma_n(z)} = z\frac{z+1}{1}\frac{z+2}{2}\cdots\frac{z+n}{n} e^{-z\ln n} = ze^{-z\ln n}\prod_{k=1}^n \frac{z+k}{k}$$
$$= ze^{-z\ln n}\prod_{k=1}^n \left(1 + \frac{z}{k}\right).$$

Next, we shall multiply the right-hand side of the above formula by

$$1 = e^{z(1+1/2+1/3+\ldots+1/n)} e^{-z(1+1/2+1/3+\ldots+1/n)}$$
$$= e^{z(1+1/2+1/3+\ldots+1/n)} \prod_{k=1}^n e^{-z/k},$$

giving:

$$\frac{1}{\Gamma_n(z)} = ze^{z(1+1/2+1/3+\ldots+1/n-\ln n)} \prod_{k=1}^{n}\left(1+\frac{z}{k}\right)e^{-z/k} .$$

Since $\Gamma(z) = \lim_{n\to\infty}\Gamma_n(z)$, we can take the $n \to \infty$ limit in the above formula which yields:

$$\frac{1}{\Gamma(z)} = z\exp\left[z\lim_{n\to\infty}\left(1+\frac{1}{2}+\frac{1}{3}+\ldots+\frac{1}{n}-\ln n\right)\right]$$
$$\times \prod_{k=1}^{\infty}\left(1+\frac{z}{k}\right)e^{-z/k} .$$

Here the finite product becomes an infinite one, and the limit in the exponent is nothing but the Euler–Mascheroni constant $\gamma = 0.5772\ldots$ we introduced in Sect. I.7.1.3. Therefore, we finally obtain a representation of the gamma function via an infinite product as follows:

$$\frac{1}{\Gamma(z)} = ze^{\gamma z}\prod_{k=1}^{\infty}\left(1+\frac{z}{k}\right)e^{-z/k} . \tag{4.46}$$

Problem 4.23 Using this product representation in Eq. (4.45), derive the following representation of the sine function via an infinite product:

$$\sin(\pi z) = \pi z \prod_{k=1}^{\infty}\left(1-\frac{z^2}{k^2}\right) . \tag{4.47}$$

Problem 4.24 Taking the logarithms of both sides of (4.47) and differentiating, show that

$$\cot(\pi z) = \frac{1}{\pi}\sum_{k=-\infty}^{\infty}\frac{1}{k+z} . \tag{4.48}$$

We shall need this beautiful result at the end of this chapter in Sect. 4.7.5.

4.2.2 Beta Function

Another class of integrals, the so-called beta function, defined as

$$B(\alpha,\beta) = \int_{0}^{1} t^{\alpha-1}(1-t)^{\beta-1}\,dt , \tag{4.49}$$

can be directly expressed via the gamma function.

Problem 4.25 Similarly to the way we calculated above $\Gamma(1/2)$, consider the integral

$$I = \left(\int_0^\infty e^{-x^2} x^{2\alpha-1} dx \right) \left(\int_0^\infty e^{-y^2} y^{2\beta-1} dy \right)$$

using two methods: (i) firstly, show that each integral in the brackets above is related to the gamma function, so that $I = \frac{1}{4}\Gamma(\alpha)\Gamma(\beta)$; (ii) secondly, combine the two integrals together into a double integral and then change into the polar coordinates $(x, y) \to (r, \varphi)$; relate the r-integral to $\Gamma(\alpha + \beta)$ by means of an appropriate substitution, while the φ-integral can be manipulated into $\frac{1}{2}B(\alpha, \beta)$ by means of the substitution $t = \cos^2 \varphi$. Hence show that

$$B(\alpha, \beta) = \frac{\Gamma(\alpha)\Gamma(\beta)}{\Gamma(\alpha + \beta)}. \tag{4.50}$$

Problem 4.26 Using the substitution $t \to x = -1 + 1/t$ in Eq. (4.49), derive another integral representation of the beta function:

$$B(\alpha, \beta) = \int_0^\infty \frac{x^{\beta-1} dx}{(x+1)^{\alpha+\beta}} = \int_0^\infty \frac{x^{\alpha-1} dx}{(x+1)^{\alpha+\beta}}. \tag{4.51}$$

The second form (with $x^{\alpha-1}$) follows from the symmetry of the beta function (4.50).

Problem 4.27 Show that

$$\int_{-1}^1 \left(1 - x^2\right)^n dx = 2^{2n+1} \frac{(n!)^2}{(2n+1)!}. \tag{4.52}$$

[Hint: using a new variable t via $x = 1 - 2t$, express the integral via the beta function $B(n+1, n+1)$.]

4.3 Orthogonal Polynomials

Orthogonal polynomials appear in a wide range of important physical problems as solutions of the corresponding equations of mathematical physics or in functional series used to expand their solutions. We have already encountered some of the polynomials in Sect. 1.1.3 where they were constructed using the Gram–Schmidt procedure applied to powers of x. Here we consider polynomials in more detail starting from the Legendre polynomials; at the end of this section, we shall build a general theory of the polynomials and will hence discuss other polynomials as well.

4.3.1   Legendre Polynomials

4.3.1.1   Generating Function

As will be clear later on in Sect. 4.3.2, it is always possible to generate all polynomials using the so-called *generating functions*. Legendre polynomials $P_n(x)$ of the variable x can be generated by the Taylor's expansion with respect to an auxiliary variable t of the generating function

$$G(x,t) = \frac{1}{\sqrt{1 - 2xt + t^2}} = \sum_{n=0}^{\infty} P_n(x)t^n . \tag{4.53}$$

Here $-1 < x < 1$, otherwise the square root is complex. Indeed, the function under the square root,

$$f(t) = 1 - 2xt + t^2 = (t - x)^2 + \left(1 - x^2\right) , \tag{4.54}$$

is a parabola, see Fig. 4.3. It is positive for all values of t *only* if $1 - x^2 > 0$, i. e. when $-1 < x < 1$.

Let us expand the generating function $G(x, t)$ explicitly into the Taylor's series with respect to t and thus calculate several first polynomials:

$$G(x,t) = \frac{1}{\sqrt{1 - 2xt + t^2}} = \sum_{n=0}^{\infty} \frac{t^n}{n!} \left[\frac{\partial^n G(x,t)}{\partial t^n} \right]_{t=0}$$

$$= G(x,0) + \left. \frac{\partial G}{\partial t} \right|_{t=0} t + \frac{1}{2} \left. \frac{\partial^2 G}{\partial t^2} \right|_{t=0} t^2 + \dots$$

$$= 1 + \left. \frac{x - t}{\left(1 - 2xt + t^2\right)^{3/2}} \right|_{t=0} t$$

$$+ \frac{1}{2} \left[\frac{-1}{\left(1 - 2xt + t^2\right)^{3/2}} - \frac{3}{2} \frac{(x - t)(2t - 2x)}{\left(1 - 2xt + t^2\right)^{5/2}} \right]_{t=0} t^2 + \dots$$

$$= 1 + xt + \frac{1}{2} \left(3x^2 - 1\right) t^2 + \dots .$$

Fig. 4.3 Function $f(t)$ of
Eq. (4.54)

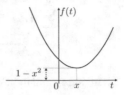

Therefore, comparing this expansion with the definition of the Legendre polynomials
(4.53), we conclude that

$$P_0(x) = 1; \quad P_1(x) = x; \quad P_2(x) = \frac{1}{2}\left(3x^2 - 1\right). \tag{4.55}$$

This procedure of direct expansion of the generating function can be continued;
however, the calculation becomes increasingly tedious.

A more convenient and much simpler method is based on generating recurrence
relations for the polynomials which enables one to calculate them consecutively
starting from the very first one, $P_0 = 1$. To this end, we shall first differentiate both
sides of Eq. (4.53) with respect to t:

$$\frac{\partial G}{\partial t} = \sum_{n=0}^{\infty} n P_n(x) t^{n-1}. \tag{4.56}$$

Using the explicit expression of $G(x, t)$, we can also write

$$\frac{\partial G}{\partial t} = \frac{\partial}{\partial t} \frac{1}{\sqrt{1 - 2xt + t^2}} = \frac{x - t}{\left(1 - 2xt + t^2\right)^{3/2}}$$

$$= \frac{x - t}{1 - 2xt + t^2} G(x, t) = \frac{x - t}{1 - 2xt + t^2} \sum_{n=0}^{\infty} P_n(x) t^n. \tag{4.57}$$

The two expressions should be identical. Therefore, we must have:

$$\left(1 - 2xt + t^2\right) \sum_{n=0}^{\infty} n P_n(x) t^{n-1} = (x - t) \sum_{n=0}^{\infty} P_n(x) t^n,$$

or

$$\sum_{n=0}^{\infty} n P_n t^{n-1} - \sum_{n=0}^{\infty} 2xn P_n t^n + \sum_{n=0}^{\infty} n P_n t^{n+1} = \sum_{n=0}^{\infty} x P_n t^n - \sum_{n=0}^{\infty} P_n t^{n+1}.$$

Collecting sums with similar powers of t, we get

$$\sum_{n=0}^{\infty} n P_n t^{n-1} - \sum_{n=0}^{\infty} (2n+1) x P_n t^n + \sum_{n=0}^{\infty} (n+1) P_n t^{n+1} = 0 .$$

$$\underbrace{\phantom{\sum_{n=0}^{\infty} n P_n t^{n-1}}}_{n \to n+1} \qquad\qquad \underbrace{\phantom{+ \sum_{n=0}^{\infty} (n+1) P_n t^{n+1}}}_{n+1 \to n}$$

Changing summation indices as indicated above, we arrive at:

$$\sum_{n=-1}^{\infty} (n+1) P_{n+1} t^n - \sum_{n=0}^{\infty} (2n+1) x P_n t^n + \sum_{n=1}^{\infty} n P_{n-1} t^n = 0 .$$

Several simplifications are possible: the $n = -1$ term in the first sum does not contribute because of the factor of $(n+1)$, and hence the summation can start from $n = 0$; in the third term we can add the $n = 0$ term since it is zero anyway because of the pre-factor n in front of P_{n-1}. Then, all three summations now run from $n = 0$, have the same power of t and hence can be combined into one:

$$\sum_{n=0}^{\infty} \left[(n+1) P_{n+1} - x(2n+1) P_n + n P_{n-1} \right] t^n = 0 .$$

Since this expression must be valid for any t, each and every coefficient to t^n should be equal to zero[2]:

$$(n+1) P_{n+1} + n P_{n-1} = (2n+1) x P_n , \quad \text{where} \quad n \geq 1 . \qquad (4.58)$$

Note that this recurrence relation is formally valid even for $n = 0$ as well if we postulate that $P_{-1} = 0$, in which case we simply get $P_1(x) = x P_0(x) = x$, the result we already knew.

This recurrent relation can be used to generate the functions $P_n(x)$. Indeed, using $P_0 = 1$ and $P_1 = x$, we have using $n = 1$ in the recurrence relation that $2P_2 + P_0 = 3x P_1$, which gives

$$P_2 = \frac{1}{2} (3x P_1 - P_0) = \frac{1}{2} \left(3x^2 - 1 \right) ,$$

i.e. the same expression as obtained above using the direct method. All higher order functions (corresponding to larger values of n) are obtained in exactly the same way, i.e. using a very simple algebra.

---

[2] Functions t^n with different powers n are linearly independent.

Problem 4.28 Show using the direct method (the Taylor's expansion) that:

$$P_3 = \frac{1}{2}(5x^3 - 3x);$$

$$P_4(x) = \frac{1}{8}(35x^4 - 30x^2 + 3); \qquad (4.59)$$

$$P_5(x) = \frac{1}{8}\left(63x^5 - 70x^3 + 15x\right).$$

Problem 4.29 Confirm these results by repeating the calculation using the recurrence relations.

Problem 4.30 Prove that $P_n(x)$ is a polynomial of order n. [Hint: use the recurrence relation (4.58) and induction.]

A number of other recurrence relations relating the function P_n with different values of n can also be established. Above we differentiated both sides of Eq. (4.53) with respect to t. This time, we shall differentiate it with respect to x. On the one hand,

$$\frac{\partial G}{\partial x} = \sum_{n=0}^{\infty} P_n'(x)t^n,$$

while on the other,

$$\frac{\partial G}{\partial x} = \frac{t}{\left(1 - 2xt + t^2\right)^{3/2}} = \frac{t}{1 - 2xt + t^2}G(x,t)$$

$$= \frac{t}{1 - 2xt + t^2}\sum_{n=0}^{\infty} P_n t^n.$$

The two expressions must be equal, so

$$\left(1 - 2xt + t^2\right)\sum_{n=0}^{\infty} P_n' t^n = \sum_{n=0}^{\infty} P_n t^{n+1},$$

or

$$\underbrace{\sum_{n=0}^{\infty} P_n' t^n}_{n \to n+1} - 2x\sum_{n=0}^{\infty} P_n' t^{n+1} + \underbrace{\sum_{n=0}^{\infty} P_n' t^{n+2}}_{n+1 \to n} = \sum_{n=0}^{\infty} P_n t^{n+1},$$

which, after the corresponding index substitutions indicated above, transforms into:

$$\sum_{n=0}^{\infty} P'_{n+1} t^{n+1} - \sum_{n=0}^{\infty} \left(2x P'_n + P_n\right) t^{n+1} + \sum_{n=1}^{\infty} P'_{n-1} t^{n+1} = 0 .$$

Note that the $n = -1$ term in the first sum does not contribute since $P'_0 = 0$ and hence was omitted. Separating out the $n = 0$ terms in the first and second sums and collecting other terms together, we obtain

$$\left[P'_1 t - \left(2x P'_0 + P_0\right) t\right] + \sum_{n=1}^{\infty} \left[P'_{n+1} - 2x P'_n - P_n + P'_{n-1}\right] t^{n+1} = 0 .$$

The expression in the square brackets is in fact zero if we recall that $P_0 = 1$ and $P_1 = x$. Thus, we immediately obtain a different recurrence relation:

$$P'_{n+1} - 2x P'_n + P'_{n-1} = P_n , \quad n \geq 1 . \tag{4.60}$$

Note that the two recurrence relations we have derived contain the Legendre polynomials with *three* consecutive indices. Other identities can also be obtained via additional differentiations as described explicitly below. In doing this, we aim at obtaining such identities which relate only Legendre polynomials with *two* consecutive indices. Using these we shall derive a differential equation the functions $P_n(x)$ must satisfy.

To this end, we first differentiate Eq. (4.58) with respect to x:

$$(n + 1) P'_{n+1} + n P'_{n-1} = (2n + 1)x P'_n + (2n + 1) P_n . \tag{4.61}$$

Solve (4.61) with respect to $x P'_n$ and substitute into (4.60); after straightforward algebra, we obtain

$$(2n + 1) P_n = P'_{n+1} - P'_{n-1} . \tag{4.62}$$

Solving (4.62) with respect to P'_{n+1} and substituting into Eq. (4.60), gives:

$$P'_{n-1} = x P'_n - n P_n , \tag{4.63}$$

while solving (4.62) with respect to P'_{n-1} and substituting into Eq. (4.60), results in:

$$P'_{n+1} = (n + 1) P_n + x P'_n . \tag{4.64}$$

If we make the index change $n + 1 \to n$ in the last expression, we obtain

$$P'_n = n P_{n-1} + x P'_{n-1} . \tag{4.65}$$

Now we have both P'_{n-1} and P'_{n+1} expressed via P_n and P'_n by means of Eqs. (4.63) and (4.64), respectively. This should allow us to formulate a differential equation for the polynomials $P_n(x)$. Differentiating Eq. (4.63), we can write

$$P''_{n-1} = x P''_n + P'_n - n P'_n .\tag{4.66}$$

Differentiating (4.65), we obtain

$$P''_n = n P'_{n-1} + x P''_{n-1} + P'_{n-1} = (n+1) P'_{n-1} + x P''_{n-1} .\tag{4.67}$$

Substituting P'_{n-1} and P''_{n-1} from (4.63) and (4.66), respectively, into (4.67), gives an equation containing functions P_n with the same index n:

$$\left(1 - x^2\right) P''_n - 2x P'_n(x) + n(n+1) P_n(x) = 0 ,\tag{4.68}$$

which is called *Legendre's differential equation*.

Using the generating function $G(x, t)$, one can easily establish some general properties of the polynomials and calculate them at the ends and the centre of the $-1 \le x \le 1$ interval.

Problem 4.31 Expanding directly the generating function into the Taylor's series for $x = \pm 1$ and comparing this expansion with the definition of the polynomials, Eq. (4.53), show that

$$P_n(1) = 1 \quad \text{and} \quad P_n(-1) = (-1)^n .\tag{4.69}$$

Problem 4.32 Using the fact that $G(-x, t) = G(x, -t)$, prove that

$$P_n(-x) = (-1)^n P_n(x) ,\tag{4.70}$$

i.e. polynomials are even functions (contain only even powers of x) for even n, while polynomials with odd n are odd (contain only odd powers of x). In particular, $P_n(0) = 0$ for odd n. In other words, the polynomials $P_n(x)$ for odd n do not contain constant terms, e.g. $P_3(x) = \frac{1}{2}\left(5x^3 - 3x\right)$.

Problem 4.33 In this problem we shall calculate $P_{2n}(0)$ using the method of the generating function. First show that at $x = 0$ the Taylor's expansion of the generating function is

$$G(0, t) = \sum_{n=0}^{\infty} (-1)^n \frac{(2n)!}{(n!)^2 2^{2n}} t^{2n} .$$

Then, alternatively, $G(0, t)$ must be equal to the series $\sum_{n=0}^{\infty} P_n(0) t^n$. Rewriting this latter series as an expansion with respect to t^2 (why this can only be done for even n?), show that

$$P_{2n}(0) = (-1)^n \frac{(2n)!}{2^{2n}(n!)^2} .\tag{4.71}$$

4.3.1.2 Orthogonality and Normalisation

The Legendre polynomials with different indices n are orthogonal to each other with the weight one. To show this, we first rewrite the differential equation (4.68) in the following equivalent form:

$$\frac{d}{dx}\left[(1 - x^2)\, P_n'\right] + n(n + 1)P_n = 0\,. \tag{4.72}$$

Multiplying it with $P_m(x)$ with $m \neq n$ and integrating between -1 and 1, gives:

$$\int_{-1}^{1} P_m \frac{d}{dx}\left[(1 - x^2)\, P_n'\right] dx + n(n + 1) \int_{-1}^{1} P_n P_m dx = 0\,.$$

The first integral is calculated by parts:

$$\underbrace{P_m(x)\, (1 - x^2)\, P_n'(x)\Big|_{-1}^{1}}_{=0} - \int_{-1}^{1} (1 - x^2)\, P_n'(x) P_m'(x) dx$$

$$+ n(n + 1) \int_{-1}^{1} P_n(x) P_m(x) dx = 0\,,$$

or

$$n(n + 1) \int_{-1}^{1} P_n(x) P_m(x) dx = \int_{-1}^{1} (1 - x^2)\, P_n' P_m' dx\,. \tag{4.73}$$

Alternatively, we can start from Eq. (4.72) written for $P_m(x)$, then multiply it by $P_n(x)$ with some $n \neq m$ and integrate; this way we would obtain the same result as above but with n and m interchanged:

$$m(m + 1) \int_{-1}^{1} P_m(x) P_n(x) dx = \int_{-1}^{1} (1 - x^2)\, P_m' P_n' dx\,.$$

Subtracting one equation from another yields:

$$[n(n + 1) - m(m + 1)] \int_{-1}^{1} P_n(x) P_m(x) dx = 0$$

$$\implies \int_{-1}^{1} P_n(x) P_m(x) dx = 0\,, \tag{4.74}$$

since $n \neq m$. So, it is seen that indeed $P_n(x)$ and $P_m(x)$ are *orthogonal*.

Consider now the special case of equal n and m by calculating

$$\int_{-1}^{1} G^2(x,t)dx = \sum_{n=0}^{\infty}\sum_{m=0}^{\infty} t^{n+m}\int_{-1}^{1} P_n(x)P_m(x)dx$$

$$= \sum_{n=0}^{\infty} t^{2n}\int_{-1}^{1} P_n^2(x)dx \; , \tag{4.75}$$

where use has been made of the already established orthogonality of the Legendre polynomials, so that only $n = m$ terms survive in the above double sum and hence only a single sum remains. On the other hand, the left-hand side of the above equation can be calculated directly using the known expression for the generating function where we, without loss of generality, may assume that $-1 < t < 1$:

$$\int_{-1}^{1} G^2(x,t)dx = \int_{-1}^{1}\frac{dx}{1-2xt+t^2} = -\frac{1}{2t}\int_{-1}^{1}\frac{dx}{x-\frac{t^2+1}{2t}}$$

$$= -\frac{1}{2t}\ln\left|x-\frac{t^2+1}{2t}\right|_{-1}^{1} = -\frac{1}{2t}\ln\left|\frac{1-\frac{t^2+1}{2t}}{-1-\frac{t^2+1}{2t}}\right|$$

$$= -\frac{1}{2t}\ln\left|\frac{2t-t^2-1}{-2t-t^2-1}\right| = -\frac{1}{2t}\ln\left|\frac{t^2-2t+1}{t^2+2t+1}\right|$$

$$= -\frac{1}{2t}\ln\left|\frac{(t-1)^2}{(t+1)^2}\right| = -\frac{1}{t}\ln\left|\frac{1-t}{1+t}\right|$$

$$= -\frac{1}{t}\left[\ln(1-t)-\ln(1+t)\right] \; .$$

Using the Taylor expansion for the logarithms (recall that $-1 < t < 1$ and hence the expansions converge),

$$\ln(1+t) = \sum_{k=1}^{\infty}(-1)^{k+1}\frac{t^k}{k} \quad \text{and} \quad \ln(1-t) = -\sum_{k=1}^{\infty}\frac{t^k}{k} \; ,$$

we can manipulate the difference of the two logarithms into

$$\ln(1-t)-\ln(1+t) = \sum_{k=1}^{\infty}\frac{t^k}{k}\left[-1-(-1)^{k+1}\right]$$

$$= -\sum_{k=1}^{\infty}\left[1+(-1)^{k+1}\right]\frac{t^k}{k} \; .$$

Only terms with the odd summation indices k survive, hence we can replace the summation index according to the recipe $k \to 2n+1$, yielding

$$\ln(1 - t) - \ln(1 + t) = -\sum_{n=0}^{\infty} 2 \frac{t^{2n+1}}{2n + 1} ,$$

and thus obtain that

$$\int_{-1}^{1} G^2(x, t)dx = -\frac{1}{t}[\ln(1 - t) - \ln(1 + t)] = \sum_{n=0}^{\infty} \frac{2}{2n + 1} t^{2n} .$$

Comparing this expansion with that in Eq. (4.75), we conclude that

$$\int_{-1}^{1} P_n^2(x)dx = \frac{2}{2n + 1} . \tag{4.76}$$

Orthogonality (4.74) and normalisation (4.76) properties can be combined into a single equation:

$$\int_{-1}^{1} P_n(x) P_m(x)dx = \frac{2}{2n + 1} \delta_{mn} , \tag{4.77}$$

where δ_{nm} is the Kronecker delta symbol.

Problem 4.34 Show, using explicit calculation of the integrals, that $P_3(x)$ is orthogonal to $P_4(x)$ (their expressions are given in Eq. (4.59)) and that $P_3(x)$ is properly normalised, i.e.

$$\int_{-1}^{1} P_3(x) P_4(x)dx = 0 \quad \text{and} \quad \int_{-1}^{1} P_3^2(x)dx = \frac{2}{7} .$$

Problem 4.35 Prove the following identities:

$$\int_{-1}^{1} P_n(x)dx = 2\delta_{n0} \quad \text{and} \quad \int_{-1}^{1} x P_n(x)dx = \frac{2}{3}\delta_{n1} .$$

Since all polynomials contain different powers of x, they are all linearly independent. It can also be shown that they form a complete set (see Sect. 4.3.2.6), and hence a function $f(x)$ defined on the interval $-1 \le x \le 1$ can be expanded in them:

$$f(x) = \sum_{n=0}^{\infty} a_n P_n(x) . \tag{4.78}$$

Multiplying both sides of this equation by $P_m(x)$ with some fixed value of m, integrating between -1 and 1 and using the orthonormality condition (4.77), we get

$$\int_{-1}^{1} f(x) P_m(x) dx = \sum_{n=0}^{\infty} a_n \int_{1}^{1} P_n(x) P_m(x) dx$$

$$\implies \int_{-1}^{1} f(x) P_m(x) dx = \sum_{n=0}^{\infty} a_n \frac{2}{2n+1} \delta_{nm} = \frac{2}{2m+1} a_m ,$$

from which the following expression for the expansion coefficient follows:

$$a_n = \frac{2n+1}{2} \int_{-1}^{1} f(x) P_n(x) dx. \tag{4.79}$$

Example 4.1 ▶ Let us expand the Dirac delta function in Legendre polynomials:

$$\delta(x) = \sum_{n=0}^{\infty} a_n P_n(x) .$$

Using Eq. (4.79), we get

$$a_n = \frac{2n+1}{2} \int_{-1}^{1} \delta(x) P_n(x) dx = \frac{2n+1}{2} P_n(0)$$

$$\implies \delta(x) = \sum_{n=0}^{\infty} \frac{2n+1}{2} P_n(0) P_n(x).$$

◀

Problem 4.36 Expand the function $f(x) = 1 + x + 2x^2$ into a series with respect to the Legendre polynomials:

$$1 + x + 2x^2 = \sum_{n=0}^{2} c_n P_n(x) .$$

Why from the start only polynomials up to the order two need to be considered? Using explicit expressions for the several first Legendre polynomials, verify your expansion. *[Answer: $c_0 = 5/3$, $c_1 = 1$ and $c_2 = 4/3$.]*

Problem 4.37 Expand $f(x) = x^3$ via the appropriate Legendre polynomials.
[Answer: $x^3 = \frac{3}{5}P_1(x) + \frac{2}{5}P_3(x)$.]

Problem 4.38 Hence, show that

$$\int_{-1}^{1} P_n(x)x^3 dx = \frac{2}{5}\delta_{n1} + \frac{4}{35}\delta_{n3} .$$

4.3.1.3 Rodrigues Formula

We shall now prove the so-called Rodrigues formula which allows writing the Legendre polynomial $P_n(x)$ with general n in an explicit and compact form:

$$P_n(x) = \frac{1}{2^n n!} \frac{d^n}{dx^n} (x^2 - 1)^n . \tag{4.80}$$

To prove it, we consider an auxiliary function $\vartheta(x) = (x^2 - 1)^n$, which satisfies the equation (check!):

$$(x^2 - 1)\frac{d\vartheta}{dx} = 2xn\vartheta(x) . \tag{4.81}$$

We shall now differentiate this equation $n + 1$ times using Leibniz formula (I.3.71). On the left-hand side, we get

$$
\begin{aligned}
\text{LHS} &= \frac{d^{n+1}}{dx^{n+1}}\left[(x^2 - 1)\frac{d\vartheta}{dx}\right] = \frac{d^{n+1}}{dx^{n+1}}\left[(x^2 - 1)\vartheta^{(1)}\right] \\
&= \sum_{k=0}^{n+1} \binom{n+1}{k} (x^2 - 1)^{(k)} \vartheta^{(n-k+2)} \\
&= \binom{n+1}{0}(x^2 - 1)\vartheta^{(n+2)} + \binom{n+1}{1}2x\vartheta^{(n+1)} + \binom{n+1}{2}2\vartheta^{(n)} \\
&= (x^2 - 1)\vartheta^{(n+2)} + (n+1)2x\vartheta^{(n+1)} + \frac{n(n+1)}{2}2\vartheta^{(n)} ,
\end{aligned}
\tag{4.82}
$$

since only $k = 0, 1, 2$ terms give non-zero contributions. On the other hand, the right-hand side of (4.81) after $n + 1$ differentiations results in:

$$2n \frac{d^{n+1}}{dx^{n+1}} (x\vartheta) = 2n \sum_{k=0}^{n+1} \binom{n+1}{k} x^{(k)} \vartheta^{(n+1-k)}$$

$$= 2n \binom{n+1}{0} x\vartheta^{(n+1)} + 2n \binom{n+1}{1} \vartheta^{(n)}$$

$$= 2nx\vartheta^{(n+1)} + 2n(n+1)\vartheta^{(n)} \ .$$

Since both expressions should be identical, we obtain

$$\left(x^2 - 1\right) \vartheta^{(n+2)} + 2(n+1)x\vartheta^{(n+1)}$$

$$+ n(n+1)\vartheta^{(n)} = 2nx\vartheta^{(n+1)} + 2n(n+1)\vartheta^{(n)} \ ,$$

or

$$\left(1 - x^2\right) \vartheta^{(n+2)} - 2x\vartheta^{(n+1)} + n(n+1)\vartheta^{(n)} = 0$$

$$\implies \left(1 - x^2\right) U'' - 2xU' + n(n+1)U = 0 \ ,$$

which is the familiar Legendre equation (4.68) for the function $U(x) = \vartheta^{(n)}$.

Since the function $U(x)$ satisfies the correct differential equation for $P_n(x)$, it must be equal to $P_n(x)$ up to an unknown constant factor. To find this factor, and hence prove the Rodrigues formula (4.80), we shall calculate $U(1)$ and compare it with $P_n(1)$ which we know is equal to one:

$$U(x) = \vartheta^{(n)} = \frac{d^n}{dx^n} \left(x^2 - 1\right)^n = \frac{d^n}{dx^n} \left[(x+1)^n (x-1)^n\right] \ .$$

Use Leibniz formula again:

$$U(x) = \sum_{k=0}^{n} \binom{n}{k} \left[(x+1)^n\right]^{(k)} \left[(x-1)^n\right]^{(n-k)} \ ,$$

where

$$\left[(x+1)^n\right]^{(k)} = n(n-1)...(n-k+1)(x+1)^{n-k} = \frac{n!}{(n-k)!}(x+1)^{n-k} \ ,$$

and similarly

$$\left[(x-1)^n\right]^{(k)} = \frac{n!}{(n-k)!}(x-1)^{n-k} \implies \left[(x-1)^n\right]^{(n-k)} = \frac{n!}{k!}(x-1)^k \ .$$

Hence, we obtain

$$U(x) = \sum_{k=0}^{n} \binom{n}{k} \frac{n!}{(n-k)!} \frac{n!}{k!}(x+1)^{n-k}(x-1)^k \ .$$

Because of $(x - 1)^k$, at $x = 1$ only one term with $k = 0$ is left:

$$U(1) = \binom{n}{0} \frac{n! \, n!}{n! \, 0!} (x + 1)^n \bigg|_{x=1} = n! 2^n \, ,$$

and this proves the normalisation factor in Eq. (4.80). The Rodrigues formula is proven completely.

Problem 4.39 Using Rodrigues formula, derive polynomials $P_n(x)$ for $n = 0, \ldots, 5$. [Hint: to perform multiple differentiations, it is convenient first to expand $(x^2 - 1)^n$ using the binomial formula and then differentiate each power of x separately.]

4.3.2   General Theory of Orthogonal Polynomials

4.3.2.1   General Properties

There are several functions–polynomials defined in a similar way to Legendre polynomials. Apparently, they all can be considered on the same footing in a rather general way. We shall briefly discuss this theory here for the reader to appreciate that all orthogonal polynomials have the same foundation.

Let $Q_n(x)$ be a real $n-$order polynomial with respect to a real variable x. We assume that polynomials of different orders are *orthogonal* to each other on the interval $a \leq x \leq b$ (a and b could be either finite or infinite, e.g. a could be $-\infty$, while b could be either finite or $+\infty$) in the following sense:

$$(Q_n, Q_m) = \int_a^b w(x) Q_n(x) Q_m(x) dx = 0 \quad \text{for any} \quad n \neq m \, , \tag{4.83}$$

where $w(x) \geq 0$ is a *weight function* (see Sects. 1.1.2 and 3.9.3). In this section, we shall call the expression (f, g) defined by Eq. (4.83) an overlap integral between two functions $f(x)$ and $g(x)$. Note that the overlap integral is fully defined if the weight function $w(x)$ is given. For the moment we shall not assume any particular form of the weight function, but later on several forms of it will be considered.

Theorem 4.1 *Any polynomial $H_n(x)$ of order n can be presented as a linear combination of Q_m polynomials with $m = 0, 1, \ldots, n$, i.e. higher order polynomials are not required:*

$$H_n(x) = \sum_{k=0}^{n} c_{nk} Q_k(x) . \qquad (4.84)$$

Proof: The $Q_k(x)$ polynomial contains only powers of x from zero to k. Then, collecting all polynomials $Q_0(x), Q_1(x), \ldots, Q_n(x)$ into a vector-column Q, one can write this statement in a compact form using the following formal matrix equation:

$$
\begin{pmatrix} Q_0(x) \\ Q_1(x) \\ Q_2(x) \\ \vdots \\ Q_n(x) \end{pmatrix}
=
\begin{pmatrix}
a_{11} & & & & \\
a_{21} & a_{22} & & & \\
a_{31} & a_{32} & a_{33} & & \\
\vdots & \vdots & \vdots & \ddots & \\
a_{n1} & a_{n2} & a_{n3} & \cdots & a_{nn}
\end{pmatrix}
\begin{pmatrix} 1 \\ x \\ x^2 \\ \vdots \\ x^n \end{pmatrix} ,
$$

which can also be written simply as $Q = AX$, where A is the left triangular matrix of the coefficients a_{ij}, and X is the vector-column of powers of x. Correspondingly, $X = A^{-1}Q$. It is known (see Problem 1.34 in Sect. 1.2.3) that the inverse of a triangular matrix has the same structure as the matrix itself, i.e. A^{-1} is also left triangular:

$$
\begin{pmatrix} 1 \\ x \\ x^2 \\ \vdots \\ x^n \end{pmatrix}
=
\begin{pmatrix}
b_{11} & & & & \\
b_{21} & b_{22} & & & \\
b_{31} & b_{32} & b_{33} & & \\
\vdots & \vdots & \vdots & \ddots & \\
b_{n1} & b_{n2} & b_{n3} & \cdots & b_{nn}
\end{pmatrix}
\begin{pmatrix} Q_0(x) \\ Q_1(x) \\ Q_2(x) \\ \vdots \\ Q_n(x) \end{pmatrix} ,
$$

where b_{ij} are elements of the matrix A^{-1}. In other words, the $k-$th power of x is expanded only in polynomials $Q_0, Q_1, \ldots, Q_k$. Since the polynomial $H_n(x)$ contains only powers of x from 0 to n, it is expanded exclusively in polynomials Q_k with $k \leq n$, as required. **Q.E.D.**

A simple corollary to this theorem is that any polynomial $H_n(x)$ of order n is orthogonal (in the sense of the definition (4.83)) to any of the orthogonal polynomials $Q_k(x)$ with $k > n$. Indeed, $H_n(x)$ can be expanded in terms of the orthogonal polynomials as shown in Eq. (4.84). Therefore, the overlap integral

$$(H_n, Q_k) = \sum_{l=0}^{n} c_{nl} (Q_l, Q_k) = \sum_{l=0}^{n} c_{nl} \delta_{kl} .$$

Fig. 4.4 Polynomials $Q_3(x)$, $Q_4(x)$ and $Q_5(x)$ cross the x-axis three, four and five times, respectively, and hence change their sign the same number of times

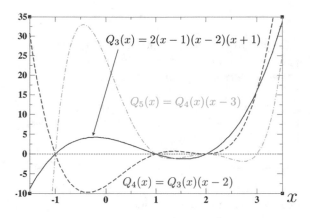

It seen from here that the overlap integral $(H_n, Q_k) \neq 0$ only if the summation index l may accept the value of k, but this can only happen if $k \leq n$. This proves the above made statement: $(H_n, Q_k) = 0$ for any $k > n$.

Theorem 4.2 *The polynomial $Q_n(x)$ has exactly n roots on the interval $a \leq x \leq b$, i.e. all roots are* simple *(non-degenerate).*

Proof: Let us assume that $Q_n(x)$ changes its sign k times on the interval, and that $0 \leq k < n$ (strictly). This means that the function $Q_n(x)$ must cross the x-axis at some k points $x_1, x_2, \ldots, x_k$ lying within the same interval, see an example in Fig. 4.4. Consider then a polynomial $H_k(x) = (x - x_1)(x - x_2) \cdots (x - x_k)$ which also changes its sign k times on the same interval and has its roots at the same points. Correspondingly, the product $F(x) = Q_n(x)H_k(x)$ does not change its sign at all, and hence the integral

$$\int_a^b w(x)F(x)dx = \int_a^b w(x)H_k(x)Q_n(x)dx \neq 0$$

(recall, that $w(x) \geq 0$). This result, however, contradicts the fact proven above that Q_n is orthogonal to any polynomial of a lower order than itself. It follows therefore that k must be equal to n as only in this case the contradiction is eliminated. Since a polynomial of order n may have no more than n distinct roots (the total number of roots, including repetitions, is n), we conclude that all n roots of $Q_n(x)$ must be distinct (hence, are simple).**Q.E.D.**

It follows then that the orthogonal polynomials $\{Q_n(x)\}$ must be somewhat special as they possess a special property that each of them has exactly as many simple (distinct) roots within the interval of orthogonality as its order.

Table 4.1 Approximate roots x_i $(i = 1, \ldots, n)$ of a first few Legendre polynomials $P_n(x)$

n	x_1	x_2	x_3	x_4	x_5
1	0.0	–	–	–	–
2	−0.57735	0.57735	–	–	–
3	−0.7746	0.0	0.7746	–	–
4	−0.86114	−0.34	0.34	0.86114	–
5	−0.90618	−0.53847	0.0	0.53847	0.90618

To illustrate the point about the roots, we give the roots of a few first Legendre polynomials in Table 4.1. It is seen that all roots are distinct and indeed fall within the interval between -1 and 1 in which the polynomials are defined.

4.3.2.2 Recurrence Relation

Next we shall derive a recurrence relation relating three consecutive polynomials. To this end, we shall consider a polynomial $x Q_n(x)$ which is of order $n + 1$. It can be expanded into polynomials Q_k with $k = 0, 1, \ldots, n + 1$ as follows:

$$x Q_n = \sum_{k=0}^{n+1} h_{n,k} Q_k , \quad \text{with} \quad h_{n,k} = \frac{1}{D_{k0}} \int_a^b w(x) x Q_n(x) Q_k(x) dx = \frac{(x Q_n, Q_k)}{(Q_k, Q_k)} ,$$

$$(4.85)$$

where

$$D_{k0} = \int_a^b w(x) Q_k^2(x) dx = (Q_k, Q_k) \tag{4.86}$$

is the corresponding normalisation given by the overlap integral of Q_k with itself. The above expression for the expansion coefficient $h_{n,k}$ is obtained by multiplying both sides of the first equation above by Q_m with some fixed m $(0 \le m \le n + 1)$, integrating over x between a and b and using the orthogonality relation between the polynomials Q_k.

The expression for $h_{n,k}$ is proportional to the overlap integral $(x Q_n, Q_k)$. This overlap integral can however be considered alternatively as the overlap $(Q_n, x Q_k)$ between Q_n and the polynomial $H_{k+1}(x) = x Q_k(x)$ of the order $k + 1$. The latter polynomial can be expanded in Q_m polynomials with m ranging between 0 and $k + 1$. Hence, $h_{n,k} \ne 0$ only if $n \le k + 1$ or $k \ge n - 1$. However, when expanding $x Q_n$ in terms of Q_k in Eq. (4.85), we have the summation index $k \le n + 1$. Combining the last two inequalities, it is seen that in the sum in Eq. (4.85) only three terms will survive for which k is either $n - 1$, n or $n + 1$. In other words, we may write

$$x Q_n = h_{n,n-1} Q_{n-1} + h_{n,n} Q_n + h_{n,n+1} Q_{n+1} , \tag{4.87}$$

which is the required recurrence relation relating three consecutive polynomials: Q_{n-1}, Q_n and Q_{n+1}.

It is actually possible to express the expansion coefficients in the above recurrence relation via the coefficients $a_k^{(n)}$ of the polynomials Q_n themselves. Let

$$Q_n(x) = a_n^{(n)} x^n + a_{n-1}^{(n)} x^{n-1} + \ldots + a_0^{(n)} . \tag{4.88}$$

By comparing the coefficients to the x^{n+1} in (4.87), we get $a_n^{(n)} = h_{n,n+1} a_{n+1}^{(n+1)}$ which gives us the last coefficient in Eq. (4.87) as $h_{n,n+1} = a_n^{(n)} / a_{n+1}^{(n+1)}$. From the definition of $h_{n,k}$ in (4.85) it also follows that

$$
h_{n,n+1} = \frac{(xQ_n, Q_{n+1})}{(Q_{n+1}, Q_{n+1})} = \frac{(Q_n, xQ_{n+1})}{(Q_{n+1}, Q_{n+1})} = \frac{(Q_n, Q_n)}{(Q_{n+1}, Q_{n+1})} \frac{(Q_n, xQ_{n+1})}{(Q_n, Q_n)}
$$
$$
= \frac{(Q_n, Q_n)}{(Q_{n+1}, Q_{n+1})} \frac{(xQ_{n+1}, Q_n)}{(Q_n, Q_n)} = \frac{D_{n,0}}{D_{n+1,0}} h_{n+1,n} ,
$$

so that

$$
h_{n+1,n} = \frac{D_{n+1,0}}{D_{n,0}} h_{n,n+1} = \frac{D_{n+1,0}}{D_{n,0}} \frac{a_n^{(n)}}{a_{n+1}^{(n+1)}} .
$$

This identity must be valid for any n. Therefore, using $n - 1$ instead of n in it, we can write

$$
h_{n,n-1} = \frac{D_{n,0}}{D_{n-1,0}} \frac{a_{n-1}^{(n-1)}}{a_n^{(n)}} .
$$

This gives us the first coefficient in the expansion (4.87). It is now left to calculate the second coefficient $h_{n,n}$, which can be done by comparing the coefficients to x^n in both sides of Eq. (4.87):

$$
a_{n-1}^{(n)} = h_{n,n} a_n^{(n)} + h_{n,n+1} a_n^{(n+1)}
$$
$$
\Longrightarrow \quad h_{n,n} = \frac{a_{n-1}^{(n)}}{a_n^{(n)}} - h_{n,n+1} \frac{a_n^{(n+1)}}{a_n^{(n)}} = \frac{a_{n-1}^{(n)}}{a_n^{(n)}} - \frac{a_n^{(n+1)}}{a_{n+1}^{(n+1)}} .
$$

After collecting all the coefficients we have just found, the recurrence relation (4.87) takes on the following form:

$$
xQ_n = \frac{D_{n,0}}{D_{n-1,0}} \frac{a_{n-1}^{(n-1)}}{a_n^{(n)}} Q_{n-1} + \left(\frac{a_{n-1}^{(n)}}{a_n^{(n)}} - \frac{a_n^{(n+1)}}{a_{n+1}^{(n+1)}} \right) Q_n + \frac{a_n^{(n)}}{a_{n+1}^{(n+1)}} Q_{n+1} . \tag{4.89}
$$

Therefore, the knowledge of the normalisation constants and the coefficients to the first two highest powers of x in the polynomials allows one to establish the recurrence relations between them.

4.3.2.3 Differential Equation

Here we shall derive the differential equation which the polynomials $Q_n(x)$ should satisfy. However, at this point more information is to be given concerning the weight function $w(x)$. As it is customary done, we shall assume that it satisfies the following first-order DE:

$$\frac{w'(x)}{w(x)} = \frac{\alpha(x)}{\sigma(x)} , \qquad (4.90)$$

where $\alpha(x) = \alpha_0 + \alpha_1 x$ and $\sigma(x) = \sigma_0 + \sigma_1 x + \sigma_2 x^2$ are polynomials of at most the first and second orders, respectively. For the moment we shall not assume any particular values for the coefficients of $\alpha(x)$ and $\sigma(x)$; several particular cases will be specifically considered below. We shall also assume that at the end of the interval $a \le x \le b$ the following boundary conditions are satisfied:

$$\sigma(x)w(x)|_{x=a,b} = 0 . \qquad (4.91)$$

It will become apparent later on why it is convenient that these are obeyed.

Next we shall consider the following integral which we take by parts:

$$I = \int_a^b x^k \left(w\sigma Q_n'\right)' dx = x^k \left(w\sigma Q_n'\right)\big|_a^b - k \int_a^b \left(x^{k-1} w\sigma\right) Q_n' dx .$$

The first (free) term on the right-hand side is equal to zero because of the boundary condition (4.91); to calculate the integral on the right-hand side, we use the integration by parts again:

$$I = -k \left(x^{k-1} w\sigma\right) Q_n\big|_a^b + k \int_a^b \left(x^{k-1} w\sigma\right)' Q_n dx .$$

Again, due to the boundary condition, the first term on the right-hand side is zero, and hence

$$I = k \int_a^b \left(x^{k-1} w\sigma\right)' Q_n dx$$

$$= k \int_a^b \left[(k-1) x^{k-2} w\sigma + x^{k-1} w'\sigma + x^{k-1} w\sigma'\right] Q_n dx .$$

The second term in the square brackets can be rearranged into $x^{k-1} w'\sigma = x^{k-1} w\alpha$ because of the DE (4.90) the weight function satisfies. Then we can finally write

$$I = k \int_a^b w \left[(k-1) x^{k-2} \sigma + x^{k-1}\alpha + x^{k-1}\sigma'\right] Q_n dx .$$

Now the whole expression in the square brackets is a polynomial of the order k (note that σ' is the first-order polynomial), and therefore $I = 0$ for any $k < n$ as any polynomial of order less than n is orthogonal to Q_n.

On the other hand, the integral I can be written directly as

$$I = \int_a^b x^k \left(w\sigma Q_n' \right)' dx = \int_a^b x^k \left(w'\sigma Q_n' + w\sigma' Q_n' + w\sigma Q_n'' \right) dx$$

$$= \int_a^b x^k \left(w\alpha Q_n' + w\sigma' Q_n' + w\sigma Q_n'' \right) dx = \int_a^b w x^k \left[\left(\alpha + \sigma' \right) Q_n' + \sigma Q_n'' \right] dx ,$$

where we have used (4.90) again. Here the expression in the square brackets is some polynomial H_n of order n. Since we know that $I = 0$ for any $k < n$, the polynomial H_n must be proportional to Q_n, i.e. we must have:

$$\left(\alpha + \sigma' \right) Q_n' + \sigma Q_n'' = \gamma_n Q_n \quad \Longrightarrow \quad \sigma Q_n'' + \left(\alpha + \sigma' \right) Q_n' - \gamma_n Q_n = 0 , \quad (4.92)$$

which is the desired DE. The constant γ_n can be expressed via the coefficients of the polynomials $\alpha(x)$ and $\sigma(x)$. Indeed, comparing the coefficients to x^n in the DE above, we obtain

$$n(n-1)\sigma_2 a_n^{(n)} + (\alpha_1 + 2\sigma_2) n a_n^{(n)} - \gamma_n a_n^{(n)} = 0 \quad \Longrightarrow \quad \gamma_n = n\left[(n+1)\sigma_2 + \alpha_1 \right] . \tag{4.93}$$

Hence, the functions $\alpha(x)$ and $\sigma(x)$, which define the weight function $w(x)$ via Eq. (4.90), also precisely determine the DE for the orthogonal polynomials. This DE, which we shall rewrite in a simpler form as

$$\sigma Q_n'' + \tau Q_n' + \lambda_n Q_n = 0 , \tag{4.94}$$

has a form in which the coefficient $\sigma(x)$ to the second derivative of Q_n is a polynomial of no more than the second order, the coefficient $\tau(x) = \alpha + \sigma'$ to the first derivative of Q_n is a polynomial of no more than the first order, while the pre-factor to the function Q_n itself is a constant

$$\lambda_n = -\gamma_n = -n\left[(n+1)\sigma_2 + \alpha_1 \right] = -n\left[\frac{1}{2}(n-1)\sigma'' + \tau' \right] , \tag{4.95}$$

since $\alpha_1 = \alpha' = \left(\tau - \sigma' \right)' = \tau' - \sigma''$. The same expression for the constants λ_n, also called *eigenvalues* of the DE, will be derived in the next section using a very different approach.

4.3.2.4 Rodrigues Formula

There exists a compact general formula for the polynomials $Q_n(x)$ corresponding to a particular weight function $w(x)$ which also bears the name of Olinde Rodrigues[3]:

$$Q_n(x) = \frac{C_n}{w(x)} \frac{d^n}{dx^n} \left[w(x)\sigma^n(x) \right] , \qquad (4.96)$$

where C_n is a constant determined by the chosen normalisation of the polynomials.

To prove Eq. (4.96) we shall use a method similar to the one we employed in Sect. 4.3.1.3 for Legendre polynomials. Consider a function $v(x) = w(x)\sigma^n(x)$ that appears under n differentiations in the Rodrigues formula above. It satisfies a DE. To derive it, consider:

$$v'(x) = w'(x)\sigma^n(x) + w(x)n\sigma^{n-1}(x)\sigma'(x) = \sigma^{n-1}(x)w(x)\left[\alpha(x) + n\sigma'(x)\right] ,$$

where we expressed $\sigma(x)w'(x)$ from Eq. (4.90). Multiplying both sides of the above equation by $\sigma(x)$, we obtain

$$\sigma(x)v'(x) = \left[\alpha(x) + n\sigma'(x)\right] v(x) .$$

In this equation $\sigma(x)$ and $\alpha(z) + n\sigma'(x)$ are the second- and first-order polynomials, respectively. Therefore, we can differentiate both sides of this equation $n + 1$ times using the Leibniz formula (I.3.71) and a finite number of terms will be obtained on the right- and the left-hand sides:

$$\text{LHS} = \left(\sigma v'\right)^{(n+1)} = \binom{n+1}{0} \sigma v^{(n+2)} + \binom{n+1}{1} \sigma' v^{(n+1)}$$

$$+ \binom{n+1}{2} \sigma'' v^{(n)}$$

$$= \sigma v^{(n+2)} + (n+1)\,\sigma' v^{(n+1)} + \frac{1}{2}n\,(n+1)\,\sigma'' v^{(n)} ,$$

$$\text{RHS} = \left[\left(\alpha + n\sigma'\right) v\right]^{(n+1)} = \left(\alpha + n\sigma'\right) v^{(n+1)} + (n+1)\left(\alpha' + n\sigma''\right) v^{(n)} .$$

Since the two expressions must be equal, we obtain after small rearrangements:

$$\sigma v^{(n+2)} + \left(\sigma' - \alpha\right) v^{(n+1)} - (n+1)\left(\alpha' + \frac{n}{2}\sigma''\right) v^{(n)} = 0 .$$

To obtain a DE for Q_n we notice that, according to the Rodrigues formula (4.96) we are set to prove here, $v^{(n)}$ is supposed to be proportional to wQ_n. Therefore, we have

---

[3] He actually derived it only for Legendre polynomials in 1816.

to replace $v^{(n)}$ with $w Q_n$ in the above DE for the final rearrangement (and ignoring the constant pre-factor between them):

$$\sigma \left(w Q_n\right)'' + \left(\sigma' - \alpha\right)\left(w Q_n\right)' - (n + 1)\left(\alpha' + \frac{n}{2}\sigma''\right) w Q_n = 0 \ .$$

Problem 4.40 Performing the necessary differentiations in the above equation (e.g. using the Leibniz formula) and using repeatedly equation (4.90) to express derivatives of the weight function via itself, $w' = w\alpha/\sigma$ and

$$w'' = \left(w\frac{\alpha}{\sigma}\right)' = w'\frac{\alpha}{\sigma} + w\frac{\alpha'}{\sigma} - w\frac{\alpha\sigma'}{\sigma^2} = \frac{w}{\sigma}\left[\alpha' + \frac{\alpha}{\sigma}(\alpha - \sigma')\right] ,$$

show that DE (4.92) for $Q_n(x)$ is obtained precisely with the same pre-factor (4.93) to Q_n.

This proves the Rodrigues formula in the very general case.

4.3.2.5   Generating Function

We know from the very beginning that one can define a generating function $G(x, t)$ for Legendre polynomials, Eq. (4.53), which can then be used as a starting point in deriving all their properties. In the current section, we have taken a different approach by deriving the polynomials from the weight function. Still, it is important to show that a generating function can be constructed in the very general case as well. This can easily be shown using the Rodrigues formula.

Indeed, consider the function of two variables:

$$G(x, t) = \sum_{n=0}^{\infty} \frac{Q_n(x)}{C_n} \frac{t^n}{n!} \ , \tag{4.97}$$

where C_n is the constant pre-factor in the Rodrigues formula, Eq. (4.96). Using the Rodrigues formula, we first rewrite $G(x, t)$:

$$G(x, t) = \frac{1}{w(x)} \sum_{n=0}^{\infty} \frac{t^n}{n!} \frac{d^n}{dx^n}\left(w\sigma^n\right) \ , \tag{4.98}$$

and then use the Cauchy formula (2.70) for the n-th derivative:

$$G(x, t) = \frac{1}{w(x)} \sum_{n=0}^{\infty} \frac{t^n}{n!} \frac{n!}{2\pi i} \oint_L \frac{w(z)\sigma^n(z)}{(z-x)^{n+1}} dz$$

$$= \frac{1}{w(x)} \frac{1}{2\pi i} \oint_L \frac{w(z)}{z-x} \sum_{n=0}^{\infty} \left(\frac{\sigma(z)t}{z-x} \right)^n dz ,$$

where L is some contour in the complex plane that surrounds the point x; it is to be chosen such that the function $w(z)$, when analytically continued into the complex plane from the real axis, is analytic inside L including L itself ($\sigma(z) = \sigma_0 + \sigma_1 z + \sigma_2 z^2$ is obviously analytic everywhere). Assuming that $|\sigma(z)t / (z-x)| < 1$ for any z on L (this can always be achieved by choosing t sufficiently small), we can sum up the geometric progression inside the integral to obtain:

$$G(x, t) = \frac{1}{w(x)} \frac{1}{2\pi i} \oint_L \frac{w(z)}{z-x} \frac{1}{1 - \frac{\sigma(z)t}{z-x}} dz = \frac{1}{w(x)} \frac{1}{2\pi i} \oint_L \frac{w(z)}{z-x-\sigma(z)t} dz .$$

The function $f(z) = z - x - \sigma(z)t$ in the denominator in the contour integral is a polynomial in z which could be either of the first or second order. Consider first the latter case which requires for $\sigma(x)$ to be a second-order polynomial. In this case $f(z)$ has two roots:

$$z_{\pm} = \frac{1}{2\sigma_2 t} \left[1 - t\sigma_1 \pm \sqrt{(1 - t\sigma_1)^2 - 4t\sigma_2 (t\sigma_0 + x)} \right] .$$

At this point we need to understand where the roots are for small enough t. To this end, let us expand the square root in terms of t up to the first order:

$$\sqrt{(1 - t\sigma_1)^2 - 4t\sigma_2 (t\sigma_0 + x)} = 1 - (\sigma_1 + 2\sigma_2 x) t + \dots .$$

It is easy to see that the pole $z_+ = 1/ (\sigma_2 t) - x - \sigma_1/\sigma_2 + \dots$ can be made as remote from x as is wished for by choosing t appropriately small, while $z_- = x + \dots$ is very close to x for small t (the points stand for the terms which vanish at $t = 0$). Therefore, choosing the contour such that only the pole z_- near $z = x$ lies inside it, we can calculate the contour integral above using the single residue at that pole only:

$$G(x, t) = \frac{1}{w(x)} \text{Res} \left[\frac{w(z)}{z - x - \sigma(z)t}; z_- \right] = \frac{1}{w(x)} \left[\frac{w(z)}{[z - x - \sigma(z)t]'} \right]_{z=z_-}$$

$$= \frac{1}{w(x)} \frac{w(z_-)}{1 - t\sigma'(z_-)} ,$$

(4.99)

which is the required expression for the generating function.

If $\sigma(x)$ is a constant or a first-order polynomial, then there is only one root of $f(z)$ which is easily seen to be always close to x for small t. Hence, the formula derived above is valid formally in this case as well, where z_- is the root in question.

4.3.2.6   Expansion of Functions in Orthogonal Polynomials

We have already discussed general theory of functional series in Sect. I.7.2. We
also stated in Sect. 4.3.1.2 that 'good' functions $f(x)$ can be expanded in Legendre
polynomials since these form a complete set. This is actually true for any orthogonal
polynomials. This is because the orthogonal polynomials $Q_n(x), n = 0, 1, 2, \ldots$, as
it can be shown, form a *closed* set which is the necessary condition for them to form
a *complete* set.

To explain what that means, consider functions $f(x)$ for which the integral

$$\int_a^b f^2(x)w(x)dx < +\infty .$$

If for any of such functions the conditions

$$\int_a^b f(x)w(x)Q_n(x)dx = 0 \quad \text{for any} \quad n = 0, 1, 2, 3, \ldots$$

are   satisfied,   that   implies   that $f(x) = 0$ and   hence   the   functions
$\{Q_n(x), \ n = 0, 1, 2, \ldots\}$ form a closed set. This is analogous to a statement that
if a vector in an n-dimensional space is orthogonal to every basis vector of that
space, then this vector is a zero vector, i.e. the collection of basis vectors is complete.

The point about the orthogonal polynomials is that any family of them (i.e. for the
given $\alpha(x)$, $\sigma(x)$, see the following subsection) forms such a closed set, and hence
any function $f(x)$ for which integrals

$$\int_a^b f^2(x)w(x)dx \quad \text{and} \quad \int_a^b \left[f'(x)\right]^2 w(x)\sigma(x)dx$$

converge, can be expanded in a functional series with respect to them:

$$f(x) = \sum_{n=0}^{\infty} f_n Q_n(x) \implies f_n = \frac{(f, Q_n)}{(Q_n, Q_n)} = \frac{1}{(Q_n, Q_n)} \int_a^b w(x)f(x)Q_n(x)dx .$$

$$(4.100)$$

We shall leave this statement without proof.

4.3.2.7   Classical Orthogonal Polynomials

It is seen that orthogonal polynomials are uniquely determined by the functions $\alpha(x)$
and $\sigma(x)$, and hence a particular choice of these functions would generate a particular
family of orthogonal polynomials. We shall investigate this point in more detail here
and construct from the general equations worked out above the particular ones for
Legendre, Hermite, Chebyshev, Jacobi and Laguerre polynomials.

As our first example, let us choose $\alpha(x) = 0$ ($\alpha_0 = \alpha_1 = 0$) and $\sigma(x) = 1 - x^2$ ($\sigma_0 = 1, \sigma_1 = 0$ and $\sigma_2 = -1$), and the interval $-1 \leq x \leq 1$. In this case Eq. (4.90) for the weight function reads simply $w'(x) = 0$ and hence $w(x)$ is a constant. We shall set $w = 1$. Note that the boundary condition (4.91) is also satisfied due to $\sigma(\pm 1) = 0$ by construction. The DE for the polynomials has the general form (4.92) with the constant pre-factor to Q_n being $\gamma_n = -n(n+1)$ (see Eq. (4.93)). Correspondingly, the DE becomes the same as Eq. (4.68) for the Legendre polynomials. Hence, this particular choice of the interval and of the weight function leads to the Legendre polynomials. Using the unit weight and $\sigma = 1 - x^2$, we immediately obtain from the general result (4.96) the corresponding Rodrigues formula $Q_n(x) = C_n \left[\left(1 - x^2\right)^n \right]^{(n)}$. Comparing this with our previous result, Eq. (4.80), gives $C_n = (-1)^n / (2^n n!)$ for the pre-factor. To derive the generating function, we need first to find one specific root of the equation

$$z - x - t\sigma(z) = tz^2 - x + (z - t) = 0 ,$$

which is close to x for small t. This is easily found to be

$$z_- = -\frac{1}{2t} \left(1 - \sqrt{1 + 4t^2 + 4tx} \right)$$

and hence the generating function, Eq. (4.99), becomes

$$G(x, t) = \frac{1}{1 + 2tz_-} = \frac{1}{\sqrt{1 + 4tx + 4t^2}} ,$$

which by definition (4.97) is also equal to

$$G(x, t) = \sum_{n=0}^{\infty} Q_n(x) (-2)^n n! \frac{t^n}{n!} = \sum_{n=0}^{\infty} Q_n(x) (-2t)^n .$$

By replacing $-2t \to t$ in the last two formulae, we obtain the usual definition (4.53) of the generating function for the Legendre polynomials.

Calculations for other classical polynomials are performed in the same way. These are considered in the following Problems.

Problem 4.41 (Hermite polynomials $H_n(x)$) Consider on the interval $-\infty < x < \infty$ functions $\alpha(x) = -2x$ and $\sigma(x) = 1$. Show that the weight function in this case can be chosen as $w(x) = e^{-x^2}$. Verify that the boundary conditions (4.91) are satisfied. Show next that the DE, the Rodrigues formula and the generating function for the polynomials are:

$$H_n'' - 2x H_n' + 2n H_n = 0 , \tag{4.101}$$

$$H_n(x) = (-1)^n \, e^{x^2} \frac{d^n}{dx^n} \left(e^{-x^2} \right) \tag{4.102}$$

$$G(x, t) = e^{-t^2 - 2xt} \, . \tag{4.103}$$

This corresponds to the customary chosen pre-factor $C_n = (-1)^n$. Use the definition of the generating function (4.97) and replace there $t \to -t$ to finally obtain:

$$e^{-t^2 + 2xt} = \sum_{n=0}^{\infty} \frac{1}{n!} H_n(x) t^n \, , \tag{4.104}$$

which is the expression that is sometimes used to define the Hermite polynomials.

Problem 4.42 Verify, using the method of the generating function and the Rodrigues formula that several first Hermite polynomials are:

$$H_0 = 1 \, ; \quad H_1 = 2x \, ; \quad H_2 = 4x^2 - 2 \, ; \quad H_3 = 8x^3 - 12x \, ;$$
$$H_4 = 16x^4 - 48x^2 + 12 \, . \tag{4.105}$$

Problem 4.43 (Laguerre polynomials $L_n(x)$) Consider on the interval $0 \leq x < \infty$ functions $\alpha(x) = -x$ and $\sigma(x) = x$. Show that the weight function is $w(x) = e^{-x}$ and the boundary conditions are satisfied. Choosing $C_n = 1/n!$, demonstrate that the DE, the Rodrigues formula and the generating function for the polynomials are:

$$x L_n'' + (1 - x) \, L_n' + n L_n = 0 \, , \tag{4.106}$$

$$L_n(x) = \frac{e^x}{n!} \frac{d^n}{dx^n} \left(e^{-x} x^n \right) \tag{4.107}$$

$$G(x, t) = \frac{e^{-xt/(1-t)}}{1 - t} = \sum_{n=0}^{\infty} L_n(x) t^n \, . \tag{4.108}$$

Problem 4.44 Verify, using the method of the generating function and the Rodrigues formula that several first Laguerre polynomials are:

$$L_0 = 1 \, ; \quad L_1 = 1 - x \, ; \quad L_2 = \frac{1}{2} \left(x^2 - 4x + 2 \right) \, ;$$
$$L_3 = \frac{1}{6} \left(-x^3 + 9x^2 - 18x + 6 \right) \, ; \quad L_4 = \frac{1}{24} \left(x^4 - 16x^3 + 72x^2 - 96x + 24 \right) \, . \tag{4.109}$$

Problem 4.45 (Generalised Laguerre polynomials $L_n^{(\lambda)}(x)$) The only difference with the previous case is that $\alpha(x) = \lambda - x$. Show that, in this case, $w(x) = x^\lambda e^{-x}$, the boundary conditions are satisfied for $\lambda > -1$, and by choosing $C_n = 1/n!$, we get

$$x L_n^{(\lambda)''} + (\lambda + 1 - x) L_n^{(\lambda)'} + n L_n^{(\lambda)} = 0, \qquad (4.110)$$

$$L_n^{(\lambda)}(x) = \frac{x^{-\lambda} e^x}{n!} \frac{d^n}{dx^n} \left(e^{-x} x^{\lambda+n} \right) \qquad (4.111)$$

$$G(x, t) = \frac{e^{-xt/(1-t)}}{(1-t)^{\lambda+1}} = \sum_{n=0}^{\infty} L_n^{(\lambda)}(x) t^n. \qquad (4.112)$$

Problem 4.46 (Chebyshev polynomials $T_n(x)$) Consider on the interval $-1 \leq x \leq 1$ functions $\alpha(x) = x$ and $\sigma(x) = 1 - x^2$. Show that the weight function in this case can be chosen as $w(x) = \left(1 - x^2\right)^{-1/2}$. Verify the boundary conditions (4.91) and show that the DE and the Rodrigues formula are:

$$\left(1 - x^2\right) T_n'' - x T_n' + n^2 T_n = 0, \qquad (4.113)$$

$$T_n(x) = \frac{(-2)^n n!}{(2n)!} \sqrt{1 - x^2} \frac{d^n}{dx^n} \left(1 - x^2\right)^{n-1/2}, \qquad (4.114)$$

where the factor C_n is chosen as it is customary done.

Problem 4.47 Verify, using the Rodrigues formula that several first Chebyshev polynomials are:

$$T_0 = 1; \quad T_1 = x; \quad T_2 = 2x^2 - 1; \quad T_3 = 4x^3 - 3x;$$
$$T_4 = 8x^4 - 8x^2 + 1. \qquad (4.115)$$

Problem 4.48 (Jacobi polynomials, $P_n^{(\lambda,\mu)}(x)$) Legendre and Chebyshev polynomials are particular cases of a more general class of polynomials due to Jacobi. They are generated using $\sigma(x) = 1 - x^2$ and $\alpha(x) = (\mu - \lambda) - (\mu + \lambda) x$ on the interval $-1 \leq x \leq 1$, where μ and λ are some real numbers. Show that for the Jacobi polynomials $P_n^{(\lambda,\mu)}(x)$ the weight function $w(x) = (1 + x)^\mu (1 - x)^\lambda$, the boundary conditions at $x = \pm 1$ are satisfied for $\mu > -1$ and $\lambda > -1$, and the corresponding DE and the Rodrigues formula read:

$$\left(1-x^2\right)\frac{d^2}{dx^2}P_n^{(\lambda,\mu)} + [(\mu-\lambda)-(\mu+\lambda+2)x]\frac{d}{dx}P_n^{(\lambda,\mu)}$$
$$+ n\left(\mu+\lambda+n+1\right)P_n^{(\lambda,\mu)} = 0 , \tag{4.116}$$

$$P_n^{(\lambda,\mu)} = \frac{(-1)^n}{2^n n!}(1-x)^{-\lambda}(1+x)^{-\mu}\frac{d^n}{dx^n}\left[(1-x)^{n+\lambda}(1+x)^{n+\mu}\right]. \tag{4.117}$$

Here the factor C_n is chosen appropriately as shown to simplify the normalisation conditions as in the cases of other polynomials.

It is seen from the above formulae that the Legendre polynomials can indeed be obtained from Jacobi ones at $\lambda=\mu=0$, i.e. $P_n(x)=P_n^{(0,0)}(x)$. Chebyshev polynomials, $T_n(x)$, follow by choosing $\lambda=\mu=-1/2$, i.e. $T_n(x)\sim P_n^{(-1/2,-1/2)}(x)$. In fact, it can be shown that Jacobi, Hermite and Laguerre polynomials cover all possible cases of orthogonal polynomials.

It is possible to derive an explicit expression for the Jacobi polynomials (4.117). This is done by applying the Leibniz formula (I.3.71) when differentiating n times the product of $(1-x)^{n+\lambda}$ and $(1+x)^{n+\mu}$,

$$\frac{d^n}{dx^n}\left[(1-x)^{n+\lambda}(1+x)^{n+\mu}\right] = \sum_{k=0}^{n}\binom{n}{k}\left[(1-x)^{n+\lambda}\right]^{(k)}\left[(1+x)^{n+\mu}\right]^{(n-k)}.$$

Using the fact that

$$\left[(1-x)^{n+\lambda}\right]^{(k)} = (-1)^k(n+\lambda)(n+\lambda-1)\ldots(n+\lambda-k+1)(1-x)^{n+\lambda-k}$$
$$= (-1)^k k!\frac{(n+\lambda)\ldots(n+\lambda-k+1)}{k!}(1-x)^{n+\lambda-k}$$
$$= (-1)^k k!\binom{n+\lambda}{k}(1-x)^{n+\lambda-k} ,$$

where $\binom{n+\lambda}{k}$ are generalised binomial coefficients (I.3.94), and similarly

$$\left[(1+x)^{n+\mu}\right]^{(n-k)} = (n-k)!\binom{n+\mu}{n-k}(1+x)^{\mu+k} ,$$

we obtain

$$P_n^{(\lambda,\mu)}(x) = \frac{1}{2^n}\sum_{k=0}^{n}\binom{\lambda+n}{k}\binom{\mu+n}{n-k}(x-1)^{n-k}(x+1)^k , \tag{4.118}$$

from which it is evident that this is indeed a polynomial for any real values of λ and μ. Note that the obtained expression is formally valid for any λ and μ including negative integer values for which some values of k in the sum are cut off. This would happen automatically because of the numerators in the generalised binomial coefficients. The obtained expression can be used for writing down an explicit formula for e.g. Legendre polynomials (when $\lambda = \mu = 0$).

4.4 Differential Equation of Generalised Hypergeometric Type

In the previous section, we considered orthogonal polynomials $Q_n(x)$ and investigated their various properties. In particular, we found that they satisfy the DE (4.92). In this section, we shall give a brief introduction into a theory which considers solutions of a more general DE of the following type:

$$y''(z) + \frac{\beta(z)}{\sigma(z)} y'(z) + \frac{\xi(z)}{\sigma^2(z)} y(z) = 0 , \qquad (4.119)$$

where $\beta(z)$ is a polynomial of up to the first order, while $\sigma(z)$ and $\xi(z)$ are polynomials of the order not higher than two. This equation is called a *generalised equation of the hypergeometric type*.

For generality we shall consider solutions in the complex plane, i.e. z in the DE above is complex. This type of equation is frequently encountered when solving partial differential equations of mathematical physics using the method of separation of variables. We shall consider a number of examples which would emphasise this point in Sect. 4.7. Here, however, we shall simply try to investigate solutions of the above equation. More specifically, we shall investigate under which conditions its solutions on the real axis (when $z = x$) are bound (limited) within a particular interval of x between a and b; note that the latter boundaries could be also $-\infty$ and/or $+\infty$. This consideration is very important when obtaining physically meaningful solutions because in physics we normally expect the solutions not to be infinite in the spatial region of interest.

4.4.1 Transformation to a Standard Form

Equation (4.119) can be transformed into a simpler form which is easier to investigate. This can be accomplished using a transformation of the unknown function $y \rightarrow u$ using some (yet arbitrary) function $\phi(z)$ via $y(z) = \phi(z)u(z)$. Performing differentiation of $y(z)$, the following DE for the new function $u(z)$ is obtained:

$$u'' + \left(\frac{2\phi'}{\phi} + \frac{\beta}{\sigma} \right) u' + \left(\frac{\phi''}{\phi} + \frac{\beta}{\sigma} \frac{\phi'}{\phi} + \frac{\xi}{\sigma^2} \right) u = 0 \,. \qquad (4.120)$$

It appears that a significant simplification is possible if the function $\phi(z)$ satisfies the following first-order DE:

$$\frac{\phi'}{\phi} = \frac{\zeta(z)}{\sigma(z)} \,, \qquad (4.121)$$

where $\zeta(z)$ is an unknown polynomial of the first order which we shall try to select in order to perform the required transformation. Since

$$\left(\frac{\phi'}{\phi} \right)' = \frac{\phi''}{\phi} - \left(\frac{\phi'}{\phi} \right)^2 \implies \frac{\phi''}{\phi} = \left(\frac{\phi'}{\phi} \right)' + \left(\frac{\phi'}{\phi} \right)^2 = \left(\frac{\zeta}{\sigma} \right)' + \left(\frac{\zeta}{\sigma} \right)^2 \,,$$

we obtain instead of (4.120):

$$u''(z) + \frac{\tau(z)}{\sigma(z)} u'(z) + \frac{\overline{\xi}(z)}{\sigma^2(z)} u(z) = 0 \,, \qquad (4.122)$$

where

$$\begin{aligned} \tau(z) &= \beta(z) + 2\zeta(z) \,, \\ \overline{\xi}(z) &= \xi(z) + \zeta^2(z) + \zeta(z) \left[\beta(z) - \sigma'(z) \right] + \zeta'(z)\sigma(z) \end{aligned} \qquad (4.123)$$

are the first- and the second-order polynomials, respectively. The unknown polynomial $\zeta(z)$ is now selected in a specific way so that $\overline{\xi} = \lambda\sigma$ with some constant λ. This must be possible as the polynomial $\overline{\xi}(z)$ in (4.123) only depends on the unknown first-order polynomial $\zeta(z) = \zeta_0 + \zeta_1 z$ (with two unknown parameters ζ_0 and ζ_1) and $\sigma(z)$, both polynomials on the left- and right-hand sides of the equation $\overline{\xi} = \lambda\sigma$ are of the second order, so that equating the coefficients to z^0, z^1 and z^2 on both sides in this equation should give three algebraic equations for λ, ζ_0 and ζ_1.

Although this procedure would formally allow us to transform Eq. (4.122) into the required standard form (compare with Eq. (4.94) for classical orthogonal polynomials),

$$\sigma(z)u'' + \tau(z)u' + \lambda u = 0 \,, \qquad (4.124)$$

it is inconvenient in practice. A better method exists which is as follows. Equating $\overline{\xi}$ from (4.123) to $\lambda\sigma$, we obtain a quadratic equation

$$\zeta^2 + \left(\beta - \sigma' \right) \zeta + \left[\xi + \left(\zeta' - \lambda \right) \sigma \right] = 0$$

with respect to ζ which has two solutions:

$$\zeta = -\frac{1}{2}(\beta - \sigma') \pm \sqrt{\frac{1}{4}(\beta - \sigma')^2 - [\xi + (\zeta' - \lambda)\sigma]}$$

$$= -\frac{1}{2}(\beta - \sigma') \pm \sqrt{\frac{1}{4}(\beta - \sigma')^2 - \xi + k\sigma}, \qquad (4.125)$$

where

$$k = \lambda - \zeta' \qquad (4.126)$$

is a constant (recall, that by our assumption the function $\zeta(z)$ is a first-order polynomial and hence its derivative is a constant). In order for the function $\zeta(z)$ be a first-order polynomial, both terms on the right-hand side of Eq. (4.125) have to be polynomials of up to the first order. The free term $-(\beta - \sigma')/2$ is already a first-order polynomial, but we also have to ensure that the square root is a first-order polynomial as well. The expression under the square root, as it can easily be seen, is a second-order polynomial. However, the square root of it may still be an irrational function. The square root is going to be a first-order polynomial if and only if the second-order polynomial under the root is the exact square of a first-order polynomial. Then the square root is a rational function which is a first-order polynomial.

Equipped with this idea, we write the expression inside the square root explicitly as a quadratic polynomial, $R_2(z) = a_0 + a_1 z + a_2 z^2$ (with the coefficients a_0, a_1 and a_2 which can be expressed via the corresponding coefficients of the polynomials $\xi(z), \sigma(z)$ and $\beta(z)$), and make it up to the complete square:

$$R_2 = a_2 \left(z + \frac{a_1}{2a_2}\right)^2 + D, \quad D = a_0 - \frac{a_1^2}{4a_2}.$$

This polynomial is going to be an exact square if and only if the constant term is zero: $D = 0$. This procedure gives possible values for the constant k. There could be more than one solution. Once k is known, we find the complete function $\zeta(z)$ from (4.125), and hence by solving equation (4.121) obtain the transformation function $\phi(z)$. The pre-factors $\tau(z)$ and λ of the new form of the DE (4.124) are then obtained from (4.123) and (4.126).

Example 4.2 ▶ As an example, consider the transformation of the following DE

$$y'' + \frac{z+1}{z}y' + \frac{(z+1)^2}{z^2}y = 0$$

into the form (4.124). Here $\sigma = z$, $\beta = z + 1$ and $\xi = (z+1)^2$, compare with Eq. (4.119). The polynomial inside the square root in (4.125) becomes

$$R_2 = -\frac{3}{4}\left(z - \frac{2}{3}k + \frac{4}{3}\right)^2 + D \quad \text{with} \quad D = -1 + \frac{(k-2)^2}{3},$$

which gives $k = 2 \pm \sqrt{3}$ as a solution of $D = 0$. Let us choose here the plus sign for definiteness. We also choose the plus sign before the square root in Eq. (4.125). This yields

$$\zeta(z) = -i - \frac{1}{2}\left(1 - i\sqrt{3}\right)z .$$

Once we know this function, we can find

$$\lambda = k + \zeta' = \left(\frac{3}{2} + \sqrt{3}\right) + i\frac{\sqrt{3}}{2}$$

and determine the transformation function $\phi(z)$ satisfying Eq. (4.121):

$$\frac{\phi'}{\phi} = \frac{\zeta}{z} = -\frac{i}{z} - \frac{1 - i\sqrt{3}}{2} \quad \Longrightarrow \quad \ln\phi = -i\ln z - \frac{1 - i\sqrt{3}}{2}z$$

$$\Longrightarrow \quad \phi(z) = z^{-i}\exp\left[-\frac{z}{2}\left(1 - i\sqrt{3}\right)\right] .$$

Finally, we calculate the pre-factor $\tau(z)$ from (4.123) yielding $\tau = (1 - 2i) + i\sqrt{3}z$. Now the initial DE accepts a new form (4.124) for $u(z)$ with the above values of λ and $\tau(z)$. Once the solution $u(z)$ of the new equation is obtained, the required function $y(z)$ can be found via $y(z) = \phi(z)u(z)$. ◀

We shall come across several more examples of this transformation below.

4.4.2   Solutions of the Standard Equation

In this section, we shall study (mostly polynomial) solutions of the DE (4.124) which is called the DE of a *hypergeometric type*.

Let $u(z)$ be a solution of such an equation. It is easy to find the DE which is satisfied by the n-th derivative of u, i.e. by the function $g_n(z) = u^{(n)}(z)$. Indeed, using Leibniz formula (I.3.71), one can differentiate the DE (4.124) n times recalling that $\sigma(z)$ and $\tau(z)$ are polynomials of the second and first orders, respectively. We have:

$$\left(\sigma u''\right)^{(n)} = \sigma u^{(n+2)} + n\sigma' u^{(n+1)} + \frac{1}{2}n\left(n - 1\right)\sigma'' u^{(n)} ,$$

$$\left(\tau u'\right)^{(n)} = \tau u^{(n+1)} + n\tau' u^{(n)} ,$$

which results in the DE for g_n:

$$\sigma g_n'' + \tau_n g_n' + \mu_n g_n = 0 , \tag{4.127}$$

where

$$\tau_n(z) = n\sigma'(z) + \tau(z) \quad \text{and} \quad \mu_n = \lambda + n\tau' + \frac{1}{2}n(n-1)\sigma'' \tag{4.128}$$

are a first-order polynomial and a constant.

In particular, if $\mu_n = 0$ for some integer n, then one of the solutions of the DE (4.127), $\sigma g_n'' + \tau_n g_n' = 0$, is a constant. If $g_n = u^{(n)}(z)$ is a constant, then surely this in turn means that $u(z)$ must be an n-th-order polynomial. We see from here immediately that if the constant λ takes on one of the following *eigenvalues*, corresponding to $\mu_n = 0$:

$$\lambda_n = -n\tau' - \frac{1}{2}n(n-1)\sigma'' \,, \tag{4.129}$$

where $n = 0, 1, 2, \ldots$, then one of the solutions of the DE (4.124) is a polynomial (recall that the DE (4.127) is of the second order and hence there must be two linearly independent solutions, so this is only one of them). This is exactly the same expression as the one we obtained earlier, see Eq. (4.95). Of course, this is not accidental considering that Eqs. (4.124) and (4.94) have an identical form.

Now let us consider one such solution, $u_n(x)$, corresponding to the value of λ_n given by Eq. (4.129) with some n. It satisfies the DE (4.124). If we differentiate this equation m times (where $m \leq n$), then according to the above discussion the function $v_m(x) = u_n^{(m)}(x)$ should satisfy the DE (4.127) where n is replaced by m, i.e.

$$\sigma v_m'' + \tau_m v_m' + \mu_m v_m = 0 \,, \tag{4.130}$$

where

$$\tau_m(x) = m\sigma'(x) + \tau(z) \tag{4.131}$$

and

$$\mu_m = \lambda_n + m\tau' + \frac{1}{2}m(m-1)\sigma'' = -(n-m)\left[\tau' + \frac{n+m-1}{2}\sigma''\right] \,. \tag{4.132}$$

It is explicitly seen from this that when $m = n$ we have $\mu_n = 0$, as it should be.

We shall now rewrite DEs (4.124) and (4.130) in the *self-adjoint form* in which the terms with the second and first-order derivatives are combined into a single expression. To this end, let us multiply (4.124) and (4.130) by some functions $w(z)$ and $w_n(z)$, respectively, so that the two DEs would take on the self-adjoint form each:

$$\left(\sigma w u'\right)' + \lambda w u = 0 \quad \text{and} \quad \left(\sigma w_m v_m'\right)' + \mu_m w_m v_m = 0 \,, \tag{4.133}$$

where $u(x) = u_n(x)$ is the initial polynomial.

Problem 4.49 Show that functions $w(z)$ and $w_m(z)$ must satisfy the following DEs:

$$(\sigma w)' = \tau w \quad \text{and} \quad (\sigma w_m)' = \tau_m w_m \; . \tag{4.134}$$

Problem 4.50 Using expression (4.128) for $\tau_m(z)$, manipulate the equation for w_m into $w_m'/w_m = w'/w + m\sigma'/\sigma$ and hence show that

$$w_m(z) = \sigma^m(z) w(z) \; , \quad m = 0, 1, 2, \ldots, n \tag{4.135}$$

(when integrating, an arbitrary constant was set to zero here which corresponds to the pre-factor of one in the relation above).

The above equations should help deriving an explicit expression for the polynomials $u_n(z)$ corresponding to $\mu_n = 0$ for a particular (positive integer) value of n. Note that $\mu_m \neq 0$ for $m < n$. Recalling that $v_m = u_n^{(m)}(z)$ is the m-th derivative of the polynomial $u_n(z)$, we can write $v_{m+1} = u_n^{(m+1)} = v_m'$. Also, from (4.135), we have $w_{m+1} = \sigma^{m+1} w = \sigma w_m$, and therefore, $(w_{m+1} v_{m+1})' = (\sigma w_m v_m')'$. By virtue of the second DE in (4.133), this expression should also be equal to $-\mu_m w_m v_m$, i.e. we obtain the recurrence relation:

$$w_m v_m = -\frac{1}{\mu_m} (w_{m+1} v_{m+1})' \; . \tag{4.136}$$

Applying this relation $n - m$ times consecutively, one obtains:

$$
\begin{aligned}
w_m v_m &= -\frac{1}{\mu_m} (w_{m+1} v_{m+1})' = \left(-\frac{1}{\mu_m}\right)\left(-\frac{1}{\mu_{m+1}}\right)(w_{m+2} v_{m+2})'' \\
&= \left(-\frac{1}{\mu_m}\right)\left(-\frac{1}{\mu_{m+1}}\right) \cdots \left(-\frac{1}{\mu_{n-1}}\right)(w_n v_n)^{(n-m)} \\
&= \left[\prod_{j=m}^{n-1}(-\mu_j)\right]^{-1}(w_n v_n)^{(n-m)} \; .
\end{aligned}
$$

Expressing v_m from the left-hand side and realising that $v_n = u_n^{(n)}(z)$ is a constant, we obtain

$$v_m = u_n^{(m)}(z) = \frac{C_{nm}}{w_m(z)}(w_n(z))^{(n-m)} \; , \tag{4.137}$$

where

$$C_{nm} = u_n^{(n)} \left[\prod_{j=m}^{n-1} (-\mu_j) \right]^{-1} = u_n^{(n)} \left[\prod_{j=0}^{n-1} (-\mu_j) \right]^{-1} \left[\prod_{j=0}^{m-1} (-\mu_j) \right]$$

$$= u_n^{(n)} \frac{M_m}{M_n} \tag{4.138}$$

is some constant pre-factor, and

$$M_l = \prod_{j=0}^{l-1} (-\mu_j) = \prod_{j=0}^{l-1} (n-j) \left[\tau' + \frac{n+j-1}{2} \sigma'' \right]$$

$$= \frac{n!}{(n-l)!} \prod_{j=0}^{l-1} \left[\tau' + \frac{n+j-1}{2} \sigma'' \right] . \tag{4.139}$$

By definition it is convenient to assume that $M_0 = 1$. Formula (4.137) demonstrates that the polynomials and their derivatives are closely related.

In particular, when $m = 0$, one obtains an explicit (up to a constant pre-factor) expression for the polynomial solution of the DE (4.124) we have been looking for:

$$u_n(z) = \frac{C_{n0}}{w(z)} (w_n(z))^{(n)} = \frac{C_{n0}}{w(z)} \left(\sigma^n(z) w(z) \right)^{(n)} , \tag{4.140}$$

where we have made use of Eq. (4.135). This is the familiar Rodrigues formula (4.96). The pre-factor

$$C_{n0} = u_n^{(n)} \frac{M_0}{M_n} = \frac{u_n^{(n)}}{M_n}$$

contains M_n, for which we can write down an explicit expression from (4.139), and a constant pre-factor $u_n^{(n)}$ which is set by the normalisation of the polynomials (this will be discussed later on).

It is seen that we have recovered some of our previous results of Sect. 4.3.2 using a rather different approach which started from the DE itself.

It is easy to show that *on the real axis* the polynomial functions $u_n(x)$ corresponding to different values of λ_n in Eq. (4.124) are orthogonal if the following condition is satisfied at the boundary points $x = a$ and $x = b$ of the interval:

$$\sigma(z) w(z) z^l \big|_{z=a,b} = 0 , \tag{4.141}$$

where $l = 0, 1, 2, \ldots$ This is proven in the following Problems:

Problem 4.51 Using the method developed in Sect. 4.3.1.2 where we proved orthogonality of Legendre polynomials, show that *on the real axis* two solutions $u_n(x)$ and $u_m(x)$ of the DE (4.124) with λ_n and λ_m, respectively, are orthogonal:

$$\int_a^b w(x)u_n(x)u_m(x)dx = 0 , \quad n \neq m . \tag{4.142}$$

[Hint: instead of Eq. (4.124) use its self-adjoint form (the first equation (4.133)) and then note that either $u_n(x)$ or $u_m(x)$ consists of a sum of powers of x.]

Problem 4.52 Similarly, consider the DE (4.130) for the m-th derivative, $v_m(x) = u_n^{(m)}(x)$, of the polynomial $u_n(x)$, which obviously is also a polynomial. Show that these are also orthogonal with respect to the weight function $w_m(x) = \sigma^m(x)w(x)$ for different n and the same m:

$$\int_a^b w(x)\sigma^m(x)u_n^{(m)}(x)u_l^{(m)}(x)dx = 0 , \quad l \neq n . \tag{4.143}$$

[Hint: also use the self-adjoint form for the DE, the second equation in (4.133).]

Let us now derive a relationship between the normalisation integral (4.86) for the polynomials,

$$D_{n0} = \int_a^b w(x)u_n^2(x)dx , \tag{4.144}$$

and a similar integral defined for their m-th derivative:

$$D_{nm} = \int_a^b w_m(x)\left[u_n^{(m)}(x)\right]^2 dx = \int_a^b w(x)\sigma^m(x)\left[u_n^{(m)}(x)\right]^2 dx . \tag{4.145}$$

This can be done by considering the second DE in Eq. (4.133) for the function $v_m(x) = u_n^{(m)}(x)$ which we shall rewrite using (4.136) as

$$(w_{m+1}v_{m+1})' + \mu_m w_m v_m = 0 .$$

Multiplying both sides of it by $v_m(x)$, integrating between a and b and applying integration by parts for the derivative term we obtain

$$(w_{m+1}v_{m+1}v_m)\big|_a^b - \int_a^b w_{m+1}v_{m+1}v_m'dx + \mu_m \int_a^b w_m v_m^2 dx = 0 .$$

The first term is zero due to the boundary condition (4.141). In the second term $v'_m(x) = v_{m+1}(x)$, and hence we immediately obtain a recurrence relation: $D_{n,m+1} = \mu_m D_{nm}$. Repeatedly applying this relation, D_{nm} can be directly related to D_{n0}:

$$D_{nm} = \mu_{m-1} D_{n,m-1} = \mu_{m-1} \mu_{m-2} D_{n,m-2} = \ldots = \left(\prod_{l=0}^{m-1} \mu_l \right) D_{n0} \tag{4.146}$$

$$= (-1)^m M_m D_{n0} \, ,$$

which is the required relationship. The quantity M_m was defined earlier by Eq. (4.139). Recall that μ_m are given by Eq. (4.132). We shall employ this identity in the next section to calculate the normalisation integral for the associated Legendre functions.

Another useful application of the above result is in calculating the normalisation D_{n0} of the polynomials. Indeed, setting $m = n$ in Eq. (4.146) and noticing that

$$D_{nn} = \int_a^b w(x) \sigma^n(x) \left(u_n^{(n)} \right)^2 dx = \left(u_n^{(n)} \right)^2 \int_a^b w(x) \sigma^n(x) dx \, ,$$

we obtain

$$D_{n0} = (-1)^n \frac{\left(u_n^{(n)} \right)^2}{M_n} \int_a^b w(x) \sigma^n(x) dx$$

$$= (-1)^n \frac{\left(a_n^{(n)} n! \right)^2}{M_n} \int_a^b w(x) \sigma^n(x) dx \, . \tag{4.147}$$

Here we have made use of the fact that $u_n^{(n)}$ is a constant and hence can be taken out of the integral. Also, this constant can trivially be related to the coefficient $a_n^{(n)}$ to the highest power x^n in the polynomial $u_n(x)$ as $u_n^{(n)} = a_n^{(n)} n!$, leading finally to the above relationship for the normalisation integral. Therefore, what is required for the calculation of the normalisation D_{n0} is the knowledge of the highest power coefficient $a_n^{(n)}$ and the value of the integral of $w\sigma^n$. These can be obtained in each particular case of the polynomials as is done in the next section.

We recall from Sect. 4.4.1 that when transforming the original DE (4.119) into the standard form (4.124), several cases for choosing the constant k and the polynomial $\zeta(x)$ might be possible. The obtained above result for the normalisation constants may help in narrowing down that uncertainty. Indeed, consider the case of $n = m = 1$. Then, from (4.146) follows that $D_{11} = \mu_0 D_{10}$, where $\mu_0 = \mu_0|_{n=1} = -\tau'$, see Eq. (4.132). Note that both quantities, D_{11} and D_{10}, must be positive, see Eqs. (4.144) and (4.145). Therefore, τ' must be *negative*. Let us remember this result. This is a necessary condition which can be employed when choosing particular signs for k and the first-order polynomial $\zeta(x)$ when applying the transformation method of Sect. 4.4.1.

So far we have discussed mostly orthogonal polynomials as solutions of the hypergeometric type Eq. (4.124). Polynomial solutions correspond to a particular values of λ given by the eigenvalues of Eq. (4.129). It is also possible to construct solutions of such an equation for other values of λ. Although we are not going to do this here as it goes way beyond this course, we state without proof a very important fact that complete solution of the original generalised equation of the hypergeometric type (4.119) on the real axis corresponding to other values of λ than those given by Eq. (4.129) *are not bound* within the interval $a \leq x \leq b$; only solutions corresponding to the eigenvalues λ_n from (4.129), i.e. orthogonal polynomials, result in bound solutions of the original equation (4.124). In other words, only such solutions are everywhere finite in the interval of their definition, any other solutions will indefinitely increase (decrease) within that interval or at its boundaries. As was mentioned at the beginning of this section, this is extremely essential in solving physical problems when quantities of interest can only take on finite values.

4.4.3 Classical Orthogonal Polynomials

Here we shall revisit Jacobi, Hermite and Laguerre polynomials using the general theory developed above. We shall derive their explicit recurrence relations and establish their normalisation.

We shall first consider the question of normalisation. As an example, let us look first at the Legendre polynomials. Their DE is given by Eq. (4.68), it is already in the standard form with $\sigma(x) = 1 - x^2$, $\tau(x) = -2x$ and $\lambda_n = n(n+1)$. We also know that the weight function in this case is $w(x) = 1$. Hence from Eq. (4.139) we can calculate

$$M_l = \frac{n!}{(n-l)!} \prod_{j=0}^{l-1} \left[-2 + \frac{n+j-1}{2}(-2) \right] = \frac{(-1)^l n!}{(n-l)!} \prod_{j=0}^{l-1} (n+j+1)$$

$$= \frac{(-1)^l n! (n+l)!}{(n-l)! n!} = \frac{(-1)^l (n+l)!}{(n-l)!} ,$$

and hence $M_n = (-1)^n (2n)!$. We also need the integral in Eq. (4.147), $\int_{-1}^{1} (1 - x^2)^n \, dx$, which has been calculated before, see Eq. (4.52). The final ingredient is the coefficient $a_n^{(n)}$ to the x^n term in the polynomial. To find it, consider the Rodrigues formula (4.80) in which we shall expand $(x^2 - 1)^n$ using the binomial formula and differentiate each term n times:

$$\frac{1}{2^n n!} \left[(x^2 - 1)^n \right]^{(n)} = \frac{1}{2^n n!} \sum_{k=0}^{n} \binom{n}{k} (-1)^{n-k} (x^{2k})^{(n)} .$$

The term with the highest power of x (the term with x^n) arises when $k = n$, and the required coefficient is

$$a_n^{(n)} = \frac{1}{2^n n!} \binom{n}{n} (2n) (2n - 1) \ldots (2n - n + 1) = \frac{1}{2^n n!} (2n) (2n - 1) \ldots (n + 1)$$

$$= \frac{1}{2^n n!} \frac{(2n)!}{n!} ,$$

so that

$$a_n^{(n)} = \frac{(2n)!}{2^n (n!)^2} . \tag{4.148}$$

Collecting all our findings in Eq. (4.147), we obtain the final result,

$$D_{n0} = (-1)^n \frac{1}{(-1)^n (2n)!} \left[\frac{(2n)!}{2^n (n!)^2} n! \right]^2 \frac{2^{2n+1} (n!)^2}{(2n + 1)!} = \frac{2}{2n + 1} ,$$

the result we already know, see Eq. (4.76).

Problem 4.53 Consider Hermite polynomials (4.102) satisfying the DE (4.101) and defined on the whole x-axis. In this case $\sigma = 1$, $\tau = -2x$ and $w = e^{-x^2}$. Convince yourself by repeated differentiation of e^{-x^2} in the Rodrigues formula, that the term with the highest power of x arises from differentiating n times the exponential function only, i.e.

$$a_n^{(n)} = 2^n . \tag{4.149}$$

Then prove that, in this case,

$$D_{n0} = \sqrt{\pi} 2^n n! . \tag{4.150}$$

Problem 4.54 The generalised Laguerre polynomials $L_n^\lambda(x)$ defined for $x \geq 0$ are specified by the Rodrigues formula (4.111); they satisfy the DE (4.110). In this case: $\sigma = x$, $\tau = 1 + \lambda - x$ and $w = x^\lambda e^{-x}$. Show by repeated differentiation of $x^{n+\lambda} e^{-x}$ in the Rodrigues formula that the term with the highest power of x arises when differentiating the exponential function only, and hence in this case

$$a_n^{(n)} = \frac{(-1)^n}{n!} . \tag{4.151}$$

Correspondingly, verify that the normalisation is

$$D_{n0} = \frac{\Gamma (n + \lambda + 1)}{n!} . \tag{4.152}$$

Problem 4.55 This result is needed for the problem that follows next. Consider the Taylor expansion of $(1 + x)^{\alpha+\beta}$ using Eq. (I.7.52). By comparing it to the product of the expansions of $(1 + x)^\alpha$ and $(1 + x)^\beta$, establish the following identity related to generalised binomial coefficients:

$$\sum_{m=1}^{n} \binom{\alpha}{m} \binom{\beta}{n - m} = \binom{\alpha + \beta}{n} . \tag{4.153}$$

Problem 4.56 Consider the Jacobi polynomials $P_n^{(\lambda,\mu)}(x)$ given by Rodrigues formula (4.117), defined on the interval $-1 \le x \le 1$ and satisfying the DE (4.116). The coefficient to the highest power of x (which is x^n) can be obtained by considering the limit

$$a_n^{(n)} = \lim_{x \to \infty} \left[x^{-n} P_n^{(\lambda,\mu)}(x) \right] . \tag{4.154}$$

Use this formula in conjunction with the general expression (4.118) for the Jacobi polynomials and formula (4.153) for the generalised binomial coefficients, to find that

$$a_n^{(n)} = \frac{1}{2^n} \binom{2n + \lambda + \mu}{n} = \frac{\Gamma(2n + \mu + \lambda + 1)}{2^n n! \, \Gamma(n + \mu + \lambda + 1)} . \tag{4.155}$$

The last passage is verified by repeatedly applying the recurrence relation (4.34) to the gamma function $\Gamma(2n + \mu + \lambda + 1)$.

Problem 4.57 In the case of Jacobi polynomials $\sigma = 1 - x^2$, $\tau = \mu - \lambda - (\mu + \lambda + 2) x$ and the weight function $w = (1 + x)^\mu (1 - x)^\lambda$. Show that, in this case, the normalisation of the polynomials is given by:

$$D_{n0} = \frac{2^{\lambda+\mu+1}}{n!} \frac{\Gamma(n + \lambda + 1) \Gamma(n + \mu + 1)}{(2n + \lambda + \mu + 1) \Gamma(n + \lambda + \mu + 1)} . \tag{4.156}$$

[Hint: when calculating the integral appearing in Eq. (4.147), relate it to the beta function using the method employed in deriving Eq. (4.52), and then relate it to the gamma function.]

The last point of the general theory which remains to be considered systematically concerns the derivation of the recurrence relation for all classical polynomials. We have considered in detail various recurrence relations for Legendre polynomials in Sect. 4.3.1.1 (see Eq. (4.58) in particular), and a general formula for any orthogonal polynomials have been derived in Sect. 4.3.2.2. It follows from Eq. (4.89) derived in the latter section that in order to set up explicitly the recurrence relation between any three consecutive polynomials, one needs to know both the coefficients $a_n^{(n)}$ and $a_{n-1}^{(n)}$, and the normalisation D_{n0}. We have calculated the former and the latter above in this section; however, we still need to calculate the coefficient $a_{n-1}^{(n)}$ to the power x^{n-1} in the polynomial.

This task can be accomplished easily if we notice that

$$u_n(x) = a_n^{(n)} x^n + a_{n-1}^{(n)} x^{n-1} + \ldots \quad \Longrightarrow \quad u_n^{(n-1)} = \left(a_n^{(n)} n!\right) x + a_{n-1}^{(n)} (n-1)! \, .$$

On the other hand, application of Eq. (4.137) for $m = n - 1$ yields:

$$v_{n-1} = u_n^{(n-1)}(x) = \frac{u_n^{(n)}}{w_{n-1}(z)} \frac{M_{n-1}}{M_n} (w_n(x))^{(1)} = a_n^{(n)} n! \frac{M_{n-1}}{M_n} \frac{w_n'}{w_{n-1}} \, ,$$

where we have made use also of Eq. (4.138). However, because of the second equation in (4.134) and of Eq. (4.131), we can write

$$\frac{w_n'}{w_{n-1}} = \frac{w_n}{w_{n-1}} \frac{w_n'}{w_n} = \frac{w_n}{w_{n-1}} \frac{\tau_n - \sigma'}{\sigma} = \frac{w\sigma^n}{w\sigma^{n-1}} \frac{\tau_n - \sigma'}{\sigma}$$
$$= \tau_n - \sigma' = n\sigma' + \tau - \sigma' = (n-1)\sigma' + \tau \, .$$

Also, from (4.139),

$$\frac{M_{n-1}}{M_n} = \frac{1}{-\mu_{n-1}} = \frac{1}{\tau' + (n-1)\sigma''} \, ,$$

so that we finally obtain:

$$\left(a_n^{(n)} n!\right) x + a_{n-1}^{(n)} (n-1)! = \left(a_n^{(n)} n!\right) \frac{\tau + (n-1)\sigma'}{\tau' + (n-1)\sigma''}$$

$$\Longrightarrow \quad \frac{a_{n-1}^{(n)}}{a_n^{(n)}} = n \left[\frac{\tau + (n-1)\sigma'}{\tau' + (n-1)\sigma''} - x \right] . \tag{4.157}$$

Note that the first term in the square brackets above is the first-order polynomial which must start from x and cancel out the second term giving the required constant ratio of the two coefficients.

As an example, let us consider the case of Laguerre polynomials $L_n(x) = L_n^0(x)$. For them $\sigma = x$ and $\tau = 1 - x$, so that $a_{n-1}^{(n)}/a_n^{(n)} = -n^2$. According to recurrence relation (4.89), we also need the ratio $a_n^{(n+1)}/a_{n+1}^{(n+1)}$, which is obtained from the previous one by the substitution $n \rightarrow n + 1$. Then, from (4.152) we have $D_{n0} = \Gamma(n+1)/n! = 1$ and according to (4.151) we know the expressions for the coefficients $a_n^{(n)}$, $a_{n-1}^{(n-1)}$ and $a_{n+1}^{(n+1)}$. Hence the recurrence relation (4.89) reads in this case:

$$x L_n = -n L_{n-1} + (2n+1) L_n - (n+1) L_{n+1} , \qquad (4.158)$$

since $a_{n-1}^{(n)}/a_n^{(n)} = -n^2$, $a_n^{(n+1)}/a_{n+1}^{(n+1)} = -(n+1)^2$, $a_{n-1}^{(n-1)}/a_n^{(n)} = -n$ and $a_n^{(n)}/a_{n+1}^{(n+1)} = -(n+1)$.

Problem 4.58 Prove that for the generalised Laguerre polynomials, $L_n^\lambda(x)$, the recurrence relation reads:

$$x L_n^\lambda = -(n+\lambda) L_{n-1}^\lambda + (2n+\lambda+1) L_n^\lambda - (n+1) L_{n+1}^\lambda . \qquad (4.159)$$

Problem 4.59 Prove that the recurrence relation for the Hermite polynomials is:

$$2x H_n = 2n H_{n-1} + H_{n+1} . \qquad (4.160)$$

Problem 4.60 Prove that the recurrence relation for the Legendre polynomials is given by Eq. (4.58).

Problem 4.61 Prove that the recurrence relation for the Jacobi polynomials is:

$$x P_n^{(\lambda,\mu)} = \frac{2(n+\lambda)(n+\mu)}{(2n+\lambda+\mu+1)(2n+\lambda+\mu)} P_{n-1}^{(\lambda,\mu)} + \frac{\mu^2-\lambda^2}{(2n+\lambda+\mu)(2n+\lambda+\mu+2)} P_n^{(\lambda,\mu)}$$
$$+ \frac{2(n+1)(n+\lambda+\mu+1)}{(2n+\lambda+\mu+2)(2n+\lambda+\mu+1)} P_{n+1}^{(\lambda,\mu)} . \qquad (4.161)$$

4.5 Associated Legendre Function

Here we shall consider special functions which have a solid significance in many applications in physics. They are related to Legendre polynomials and are called *associated Legendre functions*. We shall see in Sect. 4.5.3 that these functions appear naturally while solving the Laplace equation in spherical coordinates; similar partial differential equations appear in various fields of physics, notably in quantum mechanics (e.g. Sect. 4.7.2), electrostatics (Sect. 4.7.7), etc.

4.5.1 Bound Solutions of the Associated Legendre Equation

Consider the so-called *associated Legendre equation*

$$\left(1 - x^2\right)\Theta'' - 2x\Theta' + \left(\mu - \frac{m^2}{1 - x^2}\right)\Theta = 0 , \tag{4.162}$$

where $m = 0, \pm 1, \pm 2, \ldots$ We are interested in only such solutions of the above equation that are bound within the interval $-1 \le x \le 1$. Whether or not there are such solutions of the DE, would certainly depend on the values of the parameter μ. Therefore, one task here is to establish if such particular values of μ exist that ensure the solutions are bound, and if they do, what are those eigenvalues. Finally, we would like to obtain the bound solutions explicitly.

To accomplish this goal, we shall use the method developed above in Sect. 4.4. Using notations from this section, we notice that this equation is of the gener-alised hypergeometric type (4.119) with $\sigma(x) = 1 - x^2$, $\beta(x) = -2x$ and $\xi(x) = \mu\left(1 - x^2\right) - m^2$. It can be transformed into the standard form (4.124) by means of the transformation $\Theta(x) = \phi(x)u(x)$ with the transformation function $\phi(x)$ satisfy-ing Eq. (4.121) where the first-order polynomial $\zeta(x)$ is determined from Eq. (4.125). The polynomial $\zeta(x)$ has the form (note that $\beta - \sigma' = 0$ in our case)

$$\zeta(x) = \pm\sqrt{(k - \mu)\left(1 - x^2\right) + m^2} , \tag{4.163}$$

where $k = \lambda - \zeta'$, see Eq. (4.126). Here λ, $\sigma(x)$ and $\tau(x) = \beta(x) + 2\zeta(x)$ (see Eq. (4.123)) enter the DE (4.124) for $u(x)$. We now need to find such values of k that would guarantee $\zeta(x)$ from (4.163) to be a first-order polynomial. It is easily seen that two cases are only possible: (i) $k = \mu$, in which case $\zeta(x) = \pm m$, and (ii) $k - \mu = -m^2$, in which case $\zeta(x) = \pm mx$. We therefore have four possibilities, some of them would provide us with the required solution.

Let us specifically consider the case of $m \ge 0$ and choose $\zeta(x) = -mx$ which yields $\tau(x) = -2x - 2mx = -2(m + 1)x$. This choice guarantees that $\tau' < 0$ as required (see the end of Sect. 4.4.2). This case corresponds to $k - \mu = -m^2$. The transformation function $\phi(x)$, satisfying the DE $\phi'/\phi = \zeta/\sigma = -mx/\left(1 - x^2\right)$ from (4.121), is immediately found to be (up to an insignificant multiplier)

$$\phi(x) = \left(1 - x^2\right)^{m/2} . \tag{4.164}$$

Next,

$$\lambda = k + \zeta' = k - m = \mu - m(m + 1) . \tag{4.165}$$

The DE in this case,

$$\left(1 - x^2\right)u'' - 2(m + 1)xu' + \lambda u = 0 ,$$

has nontrivial bound solutions only if the parameter λ takes on the (eigen)values from Eq. (4.129), i.e.

$$\lambda_n = -n\tau' + \frac{1}{2}n(n-1)\sigma'' = -n\left[-2(m+1)\right] + \frac{1}{2}n(n-1)(-2)$$

$$= 2n(m+1) + n(n-1),$$

where $n = 0, 1, 2, \ldots$ is a positive integer which can be used to number different solutions $u(x) \rightarrow u_n(x)$ that are polynomials of order n. Consequently, the required values of μ become, from Eq. (4.165),

$$\mu \implies \mu_n = \lambda_n + m(m+1) = l(l+1) \quad \text{with} \quad l = n + m . \qquad (4.166)$$

As $n \geq 0$, we should have $l \geq m \geq 0$.

Next, we shall find the weight function $w(x)$. It satisfies the first DE given in (4.134), i.e.

$$(\sigma w)' = \tau w \implies \frac{w'}{w} = \frac{-2mx}{1 - x^2} \implies w(x) = \left(1 - x^2\right)^m . \qquad (4.167)$$

Therefore, the (eigen)function $u_n(x)$ corresponding to the eigenvalue λ_n (or, equivalently, to μ_n) is given by the Rodrigues formula (4.140)

$$u_n(x) = \frac{C_n}{w(x)} \left[w(x)\sigma^n(x)\right]^{(n)} = C_n \left(1 - x^2\right)^{-m} \left[\left(1 - x^2\right)^{n+m}\right]^{(n)}$$

$$= C_n \left(1 - x^2\right)^{-m} \left[\left(1 - x^2\right)^l\right]^{(l-m)},$$

with C_n being a normalisation constant. We shall determine it later. Historically, instead of n the number $l = 0, 1, 2, \ldots$ is normally used to number all possible bound solutions, $u_n(x) \rightarrow u_l(x)$, and for each value of l we shall only have a limited number of the values of m, namely, $m = 0, 1, \ldots, l$ since, as was determined above, $m \leq l$.

The above findings enable us to finally write the bound solutions of the original DE (4.162) as follows:

$$\Theta_{lm}(x) = \phi(x)u_l(x) = C_{lm} \left(1 - x^2\right)^{-m/2} \left[\left(1 - x^2\right)^l\right]^{(l-m)}, \qquad (4.168)$$

where in writing the constant factor C_{lm} and the solution itself, $\Theta_{lm}(x)$, we indicated specifically that they would not only depend on l, but also on the value of m. The obtained functions are called associated Legendre functions and denoted $P_l^{-m}(x)$ because they have a direct relation to the Legendre polynomials $P_l(x)$ as we shall see in a moment. Choosing appropriately the proportionality constant in accord with tradition, we write

$$P_l^{-m}(x) = \frac{1}{2^l l!} \left(1 - x^2\right)^{-m/2} \frac{d^{l-m}}{dx^{l-m}} \left(1 - x^2\right)^l . \tag{4.169}$$

It may seem that because of the pre-factor $\left(1 - x^2\right)^{-m/2}$ this function is infinite (not bound) at the boundary points $x = \pm 1$; however, this is not the case as $\left(1 - x^2\right)^l$ is differentiated $l - m$ times yielding the final power of $1 - x^2$ being larger than $-m/2$.

We can also see this point more clearly by directly relating these functions with the Jacobi polynomials (that are obviously bound at $x = \pm 1$). This can be done in the following way:

$$P_l^{-m}(x) = (-1)^{l+m} \frac{(l-m)!}{2^m l!} \left(1 - x^2\right)^{m/2} P_{l-m}^{(m,m)}(x) , \tag{4.170}$$

as can be easily checked by comparing (4.169) with the Rodrigues formula (4.117) for the Jacobi polynomials: obviously, as $P_{l-m}^{(m,m)}(x)$ is a polynomials and $(1 - x^2)^{m/2}$ is raised in a positive power $m/2$, the function $P_l^{-m}(x)$ is finite at $x = \pm 1$.

Replacing formally $m \to -m$ in the above formula, another form of the associated Legendre functions is obtained:

$$P_l^m(x) = \frac{1}{2^l l!} \left(1 - x^2\right)^{m/2} \frac{d^{l+m}}{dx^{l+m}} \left(1 - x^2\right)^l = (-1)^l \left(1 - x^2\right)^{m/2} \frac{d^m}{dx^m} P_l(x) , \tag{4.171}$$

which shows the mentioned relationship with the Legendre polynomials. Above we have made use of the Rodrigues formula (4.80) for these polynomials. In particular, $P_l^0(x) = (-1)^l P_l(x)$. This other form is also bound everywhere within $-1 \leq x \leq 1$ and is also related to the Jacobi polynomials:

$$P_l^m(x) = (-1)^{l+m} \frac{2^m (l+m)!}{l!} \left(1 - x^2\right)^{-m/2} P_{l+m}^{(-m,-m)}(x) . \tag{4.172}$$

We need to show though that $P_l^m(x)$ (where still $m \geq 0$) is a solution of the associated Legendre equation. In fact, we shall show that by proving that the two functions, $P_l^{-m}(x)$ and $P_l^m(x)$ (where $m \geq 0$), are directly proportional to each other.

To demonstrate this fact, let us derive an explicit expression for the function $P_l^m(x)$ inspired by some tricks we used when deriving expression (4.118) for the Jacobi polynomials. Writing $\left(1 - x^2\right)^l$ as $(1 - x)^l (1 + x)^l$ and performing differentiation with the help of the Leibniz formula (I.3.71), we obtain

$$P_l^m(x) = \frac{1}{2^l l!} \left(1 - x^2\right)^{m/2} \sum_{r=0}^{l+m} \binom{l+m}{r} \left[(1 - x)^l\right]^{(r)} \left[(1 + x)^l\right]^{(l+m-r)} ,$$

where

$$\left[(1 - x)^l\right]^{(r)} = (-1)^r \frac{l!}{(l-r)!} (1 - x)^{l-r} \quad \text{for} \quad r \leq l ,$$

and zero otherwise, which limits the summation index from above, and

$$\left[(1+x)^l\right]^{(l+m-r)} = \frac{l!}{(r-m)!}(1+x)^{r-m},$$

which is only non-zero for $r \geq m$ and hence it limits the values of r from below. Therefore, finally:

$$P_l^m(x) = \frac{l!\,(l+m)!}{2^l}\left(1-x^2\right)^{m/2}$$

$$\times \sum_{r=m}^{l} \frac{(-1)^r}{r!\,(l+m-r)!\,(l-r)!\,(r-m)!}(1-x)^{l-r}(1+x)^{r-m}. \tag{4.173}$$

Problem 4.62 Perform a similar calculation for $P_l^{-m}(x)$ of Eq. (4.169) to obtain:

$$P_l^{-m}(x) = \frac{l!\,(l-m)!}{2^l}\left(1-x^2\right)^{m/2}$$

$$\times \sum_{r=0}^{l-m} \frac{(-1)^r}{r!\,(l-m-r)!\,(l-r)!\,(r+m)!}(1-x)^{l-r-m}(1+x)^r.$$

$$\tag{4.174}$$

Now, changing the summation index $r \to r+m$, rearrange the sum and then derive the following relationship between the two representations of the associated Legendre function:

$$P_l^{-m}(x) = (-1)^m \frac{(l-m)!}{(l+m)!} P_l^m(x), \quad l \geq m \geq 0. \tag{4.175}$$

We see that the two functions are proportional to each other and hence are solutions of the same DE for the same value of m.

Problem 4.63 Derive several first associated Legendre functions:

$$P_1^1 = -\sqrt{1-x^2}; \quad P_2^1 = -3x\sqrt{1-x^2}; \quad P_2^2 = 3\left(1-x^2\right);$$

$$P_3^1 = -\frac{3}{2}\left(5x^2-1\right)\sqrt{1-x^2}; \quad P_3^2 = 15x\left(1-x^2\right); \quad P_3^3 = -15\left(1-x^2\right)^{3/2};$$

$$P_4^1(x) = -\frac{5}{2}\sqrt{1-x^2}\left(7x^3-3x\right); \quad P_4^2 = \frac{15}{2}\left(7x^2-1\right)\left(1-x^2\right);$$

$$P_4^3 = -105x\left(1-x^2\right)^{3/2}; \quad P_4^4 = 105\left(1-x^2\right)^2.$$

These are polynomials only for even values of m.

Problem 4.64 Consider the associated Legendre equation (4.162) with $\mu = l(l+1)$. By making the substitution $\Theta(x) = \left(1 - x^2\right)^{m/2} W(x)$, derive the following DE for the function $W(x)$:

$$\left(1 - x^2\right) W'' - 2(m+1)x W' + (l-m)(l+m+1)W = 0 .$$

On the other hand, show that by differentiating m times the Legendre equation (4.68) for $P_l(x)$, the same equation is obtained, i.e. $W(x) = [P_l(x)]^{(m)}$, which is basically formula (4.171) for $P_l^m(x)$.

It should now be obvious that one can equivalently use either $P_l^m(x)$ or $P_l^{-m}(x)$ as a solution of the associated Legendre equation. It is customary to use the function $P_l^m(x)$ with the non-negative value of m.

Note that $P_l^m(x)$ is *not* the only solution of the differential equation (4.162). However, the other solution is infinite at $x \to \pm 1$ and thus is *not* acceptable for many physical problems. Hence it will not be considered here.

4.5.2 Orthonormality of Associated Legendre Functions

Here we shall consider some properties of the functions $P_l^m(x)$. Firstly, let us show that the functions $P_l^m(x)$ and $P_{l'}^m(x)$ are orthogonal for $l \neq l'$ and the same m:

$$I_{ll'} = \int_{-1}^{1} P_l^m(x) P_{l'}^m(x) dx = 0 , \quad l \neq l' . \tag{4.176}$$

Indeed, because of the relationship (4.172) between the associated Legendre functions and the Jacobi polynomials, we can write

$$I_{ll'} = (-1)^{l+l'} \frac{4^m (l+m)! \, (l'+m)!}{l! l'!} \int_{-1}^{1} \left(1 - x^2\right)^{-m} P_{l+m}^{(-m,-m)}(x) P_{l'+m}^{(-m,-m)}(x) dx .$$

But the integral is nothing but the orthogonality condition (4.142) written for the Jacobi polynomials $P_n^{(-m,-m)}(x)$, for which the weight function is $w(x) = (1 - x)^{-1}(1 + x)^{-m} = \left(1 - x^2\right)^{-m}$ (see Problem 4.48). Therefore, the integral is equal to zero.

Problem 4.65 Derive the orthogonality condition for the associated Legendre functions exploiting the relationship (4.171) between them and the Legendre polynomials and the orthogonality condition (4.143) for the derivatives of the polynomials.

Next, let us derive the normalisation integral for the associated Legendre functions:

$$D_{lm} = \int_{-1}^{1} \left[P_l^m(x) \right]^2 dx = \int_{-1}^{1} \left(1 - x^2 \right)^m \left[P_l^{(m)}(x) \right]^2 dx .$$

The functions $P_l^{(m)}(x)$ are the m-th derivatives of the Legendre polynomials. The latter are characterised by the unit weight function $w(x) = 1$, $\tau(x) = -2x$ and $\sigma(x) = 1 - x^2$. The weight $w_m(x)$ associated with the m-th derivative of $P_l(x)$, according to Eq. (4.135), must be $w_m(x) = \sigma^m(x)w(x) = \left(1 - x^2 \right)^m$. Therefore, the integral above is the normalisation integral (4.145) for the $m-$the derivative of the Legendre polynomials $P_l(x)$, and hence we can directly use our result (4.146) to relate it to the normalisation D_{l0} of the functions $P_l(x)$ themselves:

$$D_{lm} = \left(\prod_{r=0}^{m-1} \mu_r \right) D_{l0} .$$

For Legendre polynomials $D_{l0} = 2/(2l + 1)$, see Eq. (4.76), and, according to (4.132),

$$\mu_r = -(l - r)\left[\tau' + \frac{1}{2}(l + r - 1)\sigma'' \right] = 2(l - r) + (l - r)(l + r - 1)$$

$$= (l - r)(l + r + 1) ,$$

where $\tau = -2x$ and $\sigma = 1 - x^2$ correspond to the DE (4.68) for $P_l(x)$. Therefore,

$$D_{lm} = \left(\prod_{r=0}^{m-1} \mu_r \right) \frac{2}{2l + 1} = \frac{2}{2l + 1} \prod_{r=0}^{m-1} (l - r)(l + r + 1)$$

$$= \frac{2}{2l + 1} \underbrace{\left[\prod_{r=0}^{m-1} (l - r) \right]}_{l!/(l-m)!} \underbrace{\left[\prod_{r=0}^{m-1} (l + r + 1) \right]}_{(l+m)!/l!} = \frac{2}{2l + 1} \frac{(l + m)!}{(l - m)!} .$$

This results allows us to write the orthonormality condition for the associated Legendre functions as:

$$I_{ll'} = \int_{-1}^{1} P_l^m(x) P_{l'}^m(x) dx = \frac{2}{2l+1} \frac{(l+m)!}{(l-m)!} \delta_{ll'} . \qquad (4.177)$$

It is convenient to redefine the solutions $\Theta_{lm}(x)$ of the DE (4.162) in such a way that their normalisation would be equal to one:

$$\int_{-1}^{1} \Theta_{lm}(x) \Theta_{l'm}(x) dx = \delta_{ll'} . \qquad (4.178)$$

It is readily seen to be done, e.g. by choosing

$$\Theta_{lm}(x) = (-1)^l \sqrt{\frac{2l+1}{2} \frac{(l-m)!}{(l+m)!}} P_l^m(x)$$

$$= \frac{(-1)^l}{2^l l!} \sqrt{\frac{2l+1}{2} \frac{(l-m)!}{(l+m)!}} \left(1-x^2\right)^{m/2} \frac{d^{l+m}}{dx^{l+m}} \left(1-x^2\right)^l . \qquad (4.179)$$

Note that $m \geq 0$ here.

4.5.3 Laplace Equation in Spherical Coordinates

4.5.3.1 Separation of Variables

The associated Legendre functions we have encountered above form the main component of the so-called spherical functions or spherical harmonics which appear in a wide class of physical problems where partial differential equations containing the Laplacian are solved in spherical coordinates. For instance this happens when considering central field problems of quantum mechanics (Sect. 4.7). Therefore, it is essential to introduce these functions. It is natural to do this by considering the simplest problem of the Laplace equation in spherical coordinates.

Consider the Laplace equation $\Delta \psi = 0$. Solutions of such an equation are called *harmonic functions*. We shall obtain these by considering the Laplace equation in spherical coordinates (r, θ, ϕ). We shall learn in Sect. 7.11 that the Laplacian of $\psi(r, \theta, \phi)$ in the spherical coordinates can be written as:

$$\frac{1}{r^2} \frac{\partial}{\partial r} \left(r^2 \frac{\partial \psi}{\partial r} \right) + \frac{1}{r^2 \sin \theta} \frac{\partial}{\partial \theta} \left(\sin \theta \frac{\partial \psi}{\partial \theta} \right) + \frac{1}{r^2 \sin^2 \theta} \frac{\partial^2 \psi}{\partial \phi^2} = 0 . \qquad (4.180)$$

First of all, we note that $1/r^2$ appears in all terms and hence can be cancelled. Next, we shall attempt[4] to separate the variables in Eq. (4.180). For that, we shall be looking

---

[4] We shall basically use the method described in more detail in Sect. 8.2.5.

for solutions of this partial differential equation (PDE) which are in the form of a
product of three functions (a product solution),

$$\psi(r, \theta, \phi) = R(r)\Theta(\theta)\Phi(\phi) , \tag{4.181}$$

each depending on its own variable. Substituting this product solution into the PDE
and dividing through by $\psi = R\Theta\Phi$, we obtain

$$\frac{1}{R}\frac{d}{dr}\left(r^2\frac{dR}{dr}\right) + \frac{1}{\Theta \sin\theta}\frac{d}{d\theta}\left(\sin\theta\frac{d\Theta}{d\theta}\right) + \frac{1}{\sin^2\theta}\left[\frac{1}{\Phi}\frac{d^2\Phi}{d\phi^2}\right] = 0 .$$

It is seen that the part depending on the angle ϕ is 'localised': nowhere else in the
equation there is any dependence on the angle ϕ. Hence one can solve the above
equation with respect to this part:

$$-\frac{1}{\Phi}\frac{d^2\Phi}{d\phi^2} = \sin^2\theta\frac{1}{R}\frac{d}{dr}\left(r^2\frac{dR}{dr}\right) + \frac{\sin\theta}{\Theta}\frac{d}{d\theta}\left(\sin\theta\frac{d\Theta}{d\theta}\right) . \tag{4.182}$$

The left-hand side of (4.182) depends only on ϕ, while the right-hand side only on
the other two variables (r, θ). This is only possible if both sides are equal to the
same constant; let us call it the *separation constant* λ. Hence, we can write the above
equation equivalently as *two* equations:

$$-\frac{1}{\Phi}\frac{d^2\Phi}{d\phi^2} = \lambda \implies \frac{d^2\Phi}{d\phi^2} + \lambda\Phi = 0 , \tag{4.183}$$

and

$$\lambda = \sin^2\theta\frac{1}{R}\frac{d}{dr}\left(r^2\frac{dR}{dr}\right) + \frac{\sin\theta}{\Theta}\frac{d}{d\theta}\left(\sin\theta\frac{d\Theta}{d\theta}\right) .$$

It is seen that the variable ϕ was 'separated' from the other two variables. Now we
need to separate the variables r and θ in the equation above. This is easily done by
dividing through both sides on $\sin^2\theta$:

$$\frac{1}{R}\frac{d}{dr}\left(r^2\frac{dR}{dr}\right) = -\frac{1}{\Theta \sin\theta}\frac{d}{d\theta}\left(\sin\theta\frac{d\Theta}{d\theta}\right) + \frac{\lambda}{\sin^2\theta} . \tag{4.184}$$

In Eq. (4.184) all terms depending on r are collected on the left, while the right-hand
side depends only on θ. It follows, therefore, that both sides must be equal to the
same (separation) constant which this time we shall call μ. This way we arrive at the
following two final equations:

$$\frac{d}{dr}\left(r^2\frac{dR}{dr}\right) - \mu R = 0 , \tag{4.185}$$

$$\frac{1}{\sin\theta}\frac{d}{d\theta}\left(\sin\theta\frac{d\Theta}{d\theta}\right) + \left[\mu - \frac{\lambda}{\sin^2\theta}\right]\Theta = 0 , \tag{4.186}$$

clearly demonstrating that our method of separating the variables succeeded. We obtained three *ordinary* differential equations which should be solved: Eq. (4.185) for the radial function $R(r)$ and Eqs. (4.183) and (4.186) for the angular part provided by the functions $\Phi(\phi)$ and $\Theta(\theta)$, respectively. However, we also have two unknown separation constants, μ and λ, which we have to find.

4.5.3.2 Solution of the Φ Equation

To find λ, we note that the function $\Phi(\phi)$ must be periodic with respect to its argument with the period of 2π due to the geometric meaning of this angle. Indeed, if we increase ϕ by 2π, we come back to the *same point* in space, and hence we must require that the solution cannot change because of the transformation $\phi \to \phi + 2\pi$. Therefore, $\psi(r, \theta, \phi + 2\pi) = \psi(r, \theta, \phi)$, which is ensured if $\Phi(\phi + 2\pi) = \Phi(\phi)$, i.e. if $\Phi(\phi)$ is a periodic function with the period of 2π.

Let us consider Eq. (4.183) for different values of λ to verify which ones would ensure such periodicity. When $\lambda = m^2 > 0$, the solution of (4.183) is

$$\Phi(\phi) = A\sin(m\phi) + B\cos(m\phi) ; \tag{4.187}$$

when $\lambda = 0$, we find $\Phi(\phi) = A\phi + B$, and for $\lambda = -p^2 < 0$ the solution is

$$\Phi(\phi) = A\sinh(p\phi) + B\cosh(p\phi) .$$

In all these solutions A and B are arbitrary constants. It is readily seen that the required periodicity of $\Phi(\phi)$ is only possible in the first case when $\Phi(\phi)$ is a sum of sine and cosine functions. But even in that case, this may only happen if the values of $m = 0, \pm 1, \pm 2, \dots$ are integers. Hence, the *eigenvalues* are $\lambda = m^2$ and the corresponding *eigenfunctions* are given by Eq. (4.187), i.e. these are either $\Phi_m = \sin(m\phi)$ or $\Phi_m = \cos(m\phi)$ (or their arbitrary linear combinations). In fact, it is very convenient to use the *complex* eigenfunctions $\Phi_m = e^{im\phi}$ and $\Phi_m = e^{-im\phi}$ instead. This is particularly useful for *quantum-mechanical* applications.

Using the exponential form of the eigenfunctions, it is especially easy to see that they satisfy

$$\int_0^{2\pi} \Phi_m(\phi)^* \Phi_{m'}(\phi) d\phi = 2\pi \delta_{mm'} , \tag{4.188}$$

i.e. the eigenfunctions are orthogonal. They will be orthonormal if a factor of $1/\sqrt{2\pi}$ is attached to them, $\Phi_m(\phi) = \frac{1}{\sqrt{2\pi}} e^{\pm im\phi}$; this is what will be assuming in the following.

4.5.3.3 Solution of the Θ Equation

Having equipped with the knowledge of the separation constant $\lambda = m^2$ being necessarily non-negative, we found while solving the equation for $\Phi(\phi)$, we can rewrite equation (4.186) for $\Theta(\theta)$ as follows:

$$\sin\theta \frac{d}{d\theta}\left(\sin\theta \frac{d\Theta}{d\theta}\right) + \left[\mu\sin^2\theta - m^2\right]\Theta = 0 . \tag{4.189}$$

In order to simplify this equation, we shall change the variable θ using the substitution $x = \cos\theta$, where $-1 \le x \le 1$. For any function $f(\theta)$,

$$\frac{df}{d\theta} = \frac{df}{dx}\frac{dx}{d\theta} = -\sin\theta\frac{df}{dx} \quad\Longrightarrow\quad \sin\theta\frac{df}{d\theta} = -\left(1-x^2\right)\frac{df}{dx} ,$$

so that the first term in the DE above can be obtained by the operator replacement

$$\sin\theta\frac{d}{d\theta} \quad\Longrightarrow\quad -\left(1-x^2\right)\frac{d}{dx}$$

that is applied twice, i.e. the term in question becomes:

$$\sin\theta\frac{d}{d\theta}\left(\sin\theta\frac{d\Theta}{d\theta}\right) = \sin\theta\frac{d}{d\theta}\left[-\left(1-x^2\right)\frac{d\Theta}{dx}\right]$$
$$= -\left(1-x^2\right)\frac{d}{dx}\left[-\left(1-x^2\right)\frac{d\Theta}{dx}\right] ,$$

which results in the following DE for $\Theta(x)$ in terms of the new variable:

$$\left(1-x^2\right)\frac{d}{dx}\left[\left(1-x^2\right)\frac{d\Theta}{dx}\right] + \left[\mu\left(1-x^2\right) - m^2\right]\Theta = 0 .$$

After performing the single differentiation in the first term and dividing through by $(1-x^2)$, we finally get:

$$\left(1-x^2\right)\frac{d^2\Theta}{dx^2} - 2x\frac{d\Theta}{dx} + \left(\mu - \frac{m^2}{1-x^2}\right)\Theta = 0 , \tag{4.190}$$

which is precisely the same equation as the one we considered in detail in Sect. 4.5. Therefore, physically acceptable solutions of this DE which are bound everywhere within the interval $-1 \le x \le 1$ are given by the associated Legendre functions $\Theta_{lm}(x)$, Eq. (4.179), with $\mu = l(l+1)$, where $l = 0, 1, 2, \ldots$ (see Eq. (4.166)). With these values of μ this is the *only* solution which is finite along the z-axis ($\theta = 0$ for $x = 1$ and $\theta = \pi$ for $x = -1$).

A simple illustration of this requirement for the possible eigenvalues, $\mu = l(l+1)$, can be given based on solving the DE (4.190) using the Frobenius method around the point $x = 0$. We have shown in Sect. I.8.4.2 in Example I.8.9 when considering series solutions of the Legendre equation (4.68) (i.e. for $m = 0$) that a polynomial solution is only possible when $\mu = l(l+1)$ as, in this case, one of the series of the general solution $y(x) = C_1 y_1(x) + C_2 y_2(x)$, either $y_1(x)$ or $y_2(x)$, terminates and hence is guaranteed to remain finite at the boundary points $x = \pm 1$; the other series solution diverges at these points, and hence should be rejected. The bound solution coincides with the Legendre polynomials which are a particular case of the associated Legendre functions when $m = 0$.

4.5.3.4 Spherical Harmonics

The above consideration brings us to the point when we can finally write done the complete solution of the *angular* part of the Laplace equation in spherical polar coordinates. These solutions are all possible products of the individual solutions of the equations for $\Theta(\theta)$ and $\Phi(\phi)$, see Eq. (4.181). These products, $\Theta_{lm}(x)\Phi_m(\phi) = \Theta_{lm}(\cos\theta)\Phi_m(\phi)$, denoted as $Y_l^m(\theta, \phi)$ or $Y_{lm}(\theta, \phi)$, are widely known as s*pherical harmonics*. For $m \geq 0$ these functions are defined by[5]

$$Y_l^m(\theta, \phi) = (-1)^m \sqrt{\frac{2l+1}{4\pi}\frac{(l-m)!}{(l+m)!}} P_l^m(\cos\theta)e^{im\phi} . \qquad (4.191)$$

Here the associated Legendre function is considered only for positive values of m, i.e. $0 \leq m \leq l$. At the same time, two possible functions $\Phi_m(\phi) = e^{\pm im\phi}$ exist for $m > 0$. That means that altogether $2l+1$ possible values of the index m can be considered between $-l$ and l for each $l = 0, 1, 2, 3, \ldots$, i.e. including the negative values as well, and hence $2l+1$ spherical harmonics can also be defined for each l. The spherical harmonics for negative values $m = -1, -2, \ldots, -l$ are defined such that

$$Y_l^{-m}(\theta, \phi) = (-1)^m Y_l^m(\theta, \phi)^* . \qquad (4.192)$$

This definition is in sync with Eq. (4.175) as changing $m \to -m$ in the expression for $P_l^m(x)$ results in an appropriate change of the pre-factor including the sign factor of $(-1)^m$. In other words, with the definition (4.192) one can use Eq. (4.191) not only for positive and zero, but also for the negative values of m.

The spherical harmonics are normalised so that

$$\int_0^{2\pi} d\phi \int_0^{\pi} \left(Y_l^m(\theta, \phi)\right)^* Y_j^i(\theta, \phi) \sin\theta d\theta = \delta_{lj}\delta_{mi} , \qquad (4.193)$$

---

[5] Various sign factors, such as e.g. the $(-1)^m$ factor we have in Eq. (4.191), can be also found in the literature.

which is easily checked by making the substitution $x = \cos\theta$ and using orthonormality of the associated Legendre functions $\Theta_{lm}(x)$ and that of $\Phi_m(\phi) = e^{im\phi}/\sqrt{2\pi}$. Here the integration is performed over the so-called solid angle $d\Omega = \sin\theta d\theta d\varphi$ (which integrates to 4π over the whole sphere).

Problem 4.66 Show that a first few spherical harmonics are:

$$Y_0^0 = \sqrt{\frac{1}{4\pi}} \;;$$

$$Y_1^0 = \sqrt{\frac{3}{4\pi}} \cos\theta \;; \quad Y_1^1 = \sqrt{\frac{3}{8\pi}} \sin\theta e^{i\varphi} \;; \quad Y_1^{-1} = -\sqrt{\frac{3}{8\pi}} \sin\theta e^{-i\phi} \;;$$

$$\tag{4.194}$$

$$\begin{cases} Y_2^1 = \sqrt{\frac{15}{8\pi}} \sin\theta \cos\theta e^{i\phi} \;; \quad Y_2^{-1} = -\sqrt{\frac{15}{8\pi}} \sin\theta \cos\theta e^{-i\phi} \;; \\[2mm] Y_2^0 = \sqrt{\frac{5}{16\pi}} (3\cos^2\theta - 1) \;; \\[2mm] Y_2^2 = \sqrt{\frac{15}{32\pi}} \sin^2\theta e^{2i\phi} \;; \quad Y_2^{-2} = \sqrt{\frac{15}{32\pi}} \sin^2\theta e^{-2i\phi} \;. \end{cases} \tag{4.195}$$

So far we have defined complex spherical harmonics. *Real spherical harmonics* can also be defined by mixing (for $m \neq 0$) Y_l^m and Y_l^{-m}. Assuming that $m > 0$, we obtain two real functions:

$$S_l^m(\theta, \phi) = \frac{1}{\sqrt{2}} \left[Y_l^m + (-1)^m Y_l^{-m} \right] = \frac{1}{\sqrt{2}} \left(Y_l^m + Y_l^{m*} \right)$$

$$= (-1)^m \sqrt{\frac{2l+1}{2\pi} \frac{(l-m)!}{(l+m)!}} P_l^m(\cos\theta) \cos(m\phi) \;, \tag{4.196}$$

$$S_l^{-m}(\theta, \phi) = \frac{1}{\sqrt{2}i} \left[Y_l^m - (-1)^m Y_l^{-m} \right] = \frac{1}{\sqrt{2}i} \left(Y_l^m - Y_l^{m*} \right)$$

$$= (-1)^m \sqrt{\frac{2l+1}{2\pi} \frac{(l-m)!}{(l+m)!}} P_l^m(\cos\theta) \sin(m\phi) \;. \tag{4.197}$$

Note that Y_l^0 is already real and hence is kept the same: $S_l^0(\theta, \phi) = Y_l^0(\theta, \phi)$. Here $m > 0$.

Problem 4.67 Show that a first few real spherical harmonics for $l = 1, 2$ are:

$$S_1^0 = Y_1^0 = \sqrt{\frac{3}{4\pi}} n_z \; ; \quad S_1^1 = \sqrt{\frac{3}{4\pi}} n_x \; ; \quad S_1^{-1} = \sqrt{\frac{3}{4\pi}} n_y \; ; \qquad (4.198)$$

$$\begin{cases} S_2^1 = \sqrt{\dfrac{15}{4\pi}} n_x n_z \; ; \quad S_2^{-1} = \sqrt{\dfrac{15}{4\pi}} n_y n_z \; ; \\[2ex] S_2^0 = Y_2^0 = \sqrt{\dfrac{5}{16\pi}} (3n_z^2 - 1) \; ; \\[2ex] S_2^2 = \sqrt{\dfrac{15}{16\pi}} (n_x^2 - n_y^2) \; ; \quad S_2^{-2} = \sqrt{\dfrac{15}{4\pi}} n_x n_y \; , \end{cases} \qquad (4.199)$$

where $n_x = \sin\theta \cos\phi$, $n_y = \sin\theta \sin\phi$ and $n_z = \cos\theta$ are the components of the unit vector $\mathbf{n} = \mathbf{r}/r$.

Spherical harmonics Y_l^m are useful in expressing angular dependence of functions of vectors, $f(\mathbf{r})$. Indeed, it can be explicitly shown that any 'good' function $f(\mathbf{r}) = f(r, \theta, \phi)$ can be expanded into a functional series with respect to all spherical harmonics:

$$f(r, \theta, \phi) = \sum_{l=0}^{\infty} \sum_{m=-l}^{l} f_{lm}(r) Y_l^m(\theta, \phi) \; , \qquad (4.200)$$

where by virtue of the orthogonality of the spherical functions, Eq. (4.193), the expansion coefficients are given by

$$f_{lm}(r) = \int_0^{2\pi} d\phi \int_0^{\pi} \left(Y_n^m(\theta, \phi) \right)^* f(r, \theta, \phi) \sin\theta d\theta \; . \qquad (4.201)$$

Note that the expansion coefficients only depend on the length $r = |\mathbf{r}|$ of the vector $\mathbf{r}$.

We finish this section with another important result, which we shall also leave without proof. If there are two unit vectors $\mathbf{n}_1$ and $\mathbf{n}_2$ with the angle $\theta = (\mathbf{n}_1, \mathbf{n}_2)$ between them, then

$$P_l(\cos\theta) = \frac{4\pi}{2l+1} \sum_{m=-l}^{l} Y_l^m(\theta_1, \phi_1) Y_l^m(\theta_2, \phi_2)^* \; , \qquad (4.202)$$

where and angles (θ_1, ϕ_1) and (θ_2, ϕ_2) correspond to the orientation of the first and the second vectors, respectively, in the spherical coordinates.

Problem 4.68 Show that formula (4.202) remains invariant upon replacement of the complex harmonics Y_l^m with the real ones:

$$P_l(\cos\theta) = \frac{4\pi}{2l+1} \sum_{m=-l}^{l} S_l^m(\theta_1, \phi_1) S_l^m(\theta_2, \phi_2) \,. \tag{4.203}$$

4.6   Bessel Equation

4.6.1   Bessel Differential Equation and its Solutions

These functions frequently appear when solving partial differential equations of mathematical physics in polar or cylindrical coordinates. They correspond to solutions of the following DE:

$$x^2 y'' + x y' + \left(x^2 - \nu^2\right) y = 0 \,, \tag{4.204}$$

where ν is a real parameter.

We shall solve this DE using the Frobenius method of Sect. I.8.4 (see also Sect. 2.8). An example of the simplest DE corresponding to $\nu = 0$ was also considered in Problem I.8.70. We shall consider the most general case of arbitrary ν in the following two Problems.

Problem 4.69 Consider the generalised power series

$$y(x) = \sum_{r=0}^{\infty} c_r x^{r+s} \,,$$

where s is a number yet to be determined. Substitute this expansion into the DE (4.204) assuming that c_0 is arbitrary. Show then that $c_r = 0$ for any odd values of r, while for its even values

$$c_{r+2} = -\frac{c_r}{(r+s+2)^2 - \nu^2} \,, \tag{4.205}$$

where s can only take on two values, $s = \pm\nu$, giving rise to two independent solutions. Recall from Sect. I.8.4 that when the difference of two s values is an integer (which is 2ν in our case), only a single solution may be obtained by this method. Therefore, let us initially assume that 2ν is not an integer. Show then that several first terms in the expansion of the two solutions are:

$$y_1(x) = c_0 x^\nu \left[1 - \frac{x^2}{2\,(2\nu + 2)} + \frac{x^4}{2 \cdot 4 \cdot (2\nu + 2)\,(2\nu + 4)} \right.$$
$$\left. - \frac{x^6}{2 \cdot 4 \cdot 6\,(2\nu + 2)\,(2\nu + 4)\,(2\nu + 6)} + \cdots \right], \qquad (4.206)$$

$$y_2(x) = c_0 x^{-\nu} \left[1 - \frac{x^2}{2\,(-2\nu + 2)} + \frac{x^4}{2 \cdot 4 \cdot (-2\nu + 2)\,(-2\nu + 4)} \right.$$
$$\left. - \frac{x^6}{2 \cdot 4 \cdot 6\,(-2\nu + 2)\,(-2\nu + 4)\,(-2\nu + 6)} + \cdots \right]; \qquad (4.207)$$

In fact, the second independent solution is obtained from the first one simply by the substitution $\nu \to -\nu$.

Problem 4.70 The recurrence relation for the c_r coefficients (4.205) can be used to obtain a general expression for them. Prove, using the method of mathematical induction or by a repeated application of the recurrence relation that for $s = \nu$ one can write

$$y_1(x) = c_0 x^\nu \left[1 + \sum_{r=1}^{\infty} \frac{(-1)^r}{r!\,(\nu + 1)\,(\nu + 2) \cdots (\nu + r)} \left(\frac{x}{2} \right)^{2r} \right], \quad (4.208)$$

while the other solution corresponding to $s = -\nu$ is obtained formally by replacing $\nu \to -\nu$.

The above expansion can be written in a much more compact form if, as it is usually done, c_0 is chosen as $c_0 = 2^{-\nu} / \Gamma(\nu + 1)$. In this case the first solution is called *Bessel function of order ν of the first kind* and denoted $J_\nu(x)$. Using the recurrence relation (4.34) for the gamma function, we can write

$$\Gamma(\nu + 1)\,(\nu + 1)\,(\nu + 2) \cdots (\nu + r) = \Gamma(\nu + r + 1) \ ,$$

and hence from (4.208) we finally obtain:

$$J_\nu(x) = \sum_{r=0}^{\infty} \frac{(-1)^r}{r!\,\Gamma(r + \nu + 1)} \left(\frac{x}{2} \right)^{2r + \nu} . \qquad (4.209)$$

Note that the very first term in the expansion (4.208) is now nicely incorporated into the summation. The second solution of the DE corresponding to $-\nu$, is obtained by replacing $\nu \to -\nu$ in the first one:

$$J_{-\nu}(x) = \sum_{r=0}^{\infty} \frac{(-1)^r}{r!\,\Gamma(r - \nu + 1)} \left(\frac{x}{2} \right)^{2r - \nu} . \qquad (4.210)$$

It is also the Bessel function of the first kind. Both functions for non-integer ν are linearly independent and hence their linear combination, $y(x) = C_1 J_\nu(x) + C_2 J_{-\nu}(x)$, is a general solution of the DE (4.204).

Now we shall consider the case when 2ν is an integer. This is possible in either of the following two cases: (i) ν is an integer; (ii) ν is a half of an odd integer (obviously, half of an even integer is an integer and hence this is the first case). Consider these two cases separately.

If $\nu = n$ is a *positive integer*, the solution $J_n(x)$ is perfectly valid as the first independent solution. The function $J_{-n}(x)$ contains the gamma function $\Gamma(r - n + 1)$ which is equal to infinity for negative integer values, i.e. when $r - n + 1 = 0, -1, -2, -3, \ldots$ or simply when $r \leq n - 1$ and is an integer. Since the gamma function is in the numerator, contributions of these values of the index r in the sum, i.e. of $r = 0, 1, \ldots, n - 1$, are equal to zero and the sum can be started from $r = n$:

$$
\begin{aligned}
J_{-n}(x) &= \sum_{r=n}^{\infty} \frac{(-1)^r (x/2)^{2r-n}}{r! \Gamma(r-n+1)} = \left| \begin{array}{c} \text{change summation index} \\ r \to k = r - n \end{array} \right| \\
&= \sum_{k=0}^{\infty} \frac{(-1)^{k+n} (x/2)^{2k+n}}{(k+n)! \Gamma(k+1)} = (-1)^n \sum_{k=0}^{\infty} \frac{(-1)^k (x/2)^{2k+n}}{k! \Gamma(k+n+1)} \qquad (4.211) \\
&= (-1)^n J_n(x) ,
\end{aligned}
$$

where on the second line we have repetitively used the recurrence relation for the gamma function to demonstrate that

$$
\begin{aligned}
\Gamma(k+n+1) &= (k+n)(k+n-1)\ldots(k+1)\Gamma(k+1) \\
&= \frac{(k+n)!}{k!}\Gamma(k+1) .
\end{aligned}
$$

This derivation shows that the Bessel function of a negative integer index is directly proportional to the one with the corresponding positive integer index, i.e. $J_{-n}(x)$ is linearly dependent on $J_n(x)$ and hence cannot be taken as the second independent solution of the DE. In this case, according to general theory of Sects I.8.4 and 2.8, one has to look for the second solution in the form containing a logarithmic function, Eq. (I.8.100) and Problem I.8.70:

$$
K_n(x) = J_n(x) \ln x + g(x) , \quad \text{where} \quad g(x) = x^{-n} \sum_{r=0}^{\infty} c_r x^r . \qquad (4.212)
$$

This function, $K_n(x)$, will be the second independent solution of the Bessel DE in the case of $\nu = n$ being a positive integer (including $\nu = 0$). It diverges at $x = 0$. The functions $J_n(x)$ and $K_n(x)$ are called Bessel functions of the first and second kind, respectively.

Problem 4.71 By substituting $K_n(x)$ into the DE (4.204) and using the fact that $J_n(x)$ is already its solution, show that the following DE is obtained for the function $g(x)$:

$$x^2 g'' + x g' + \left(x^2 - n^2\right) g = -2x J_n'(x) . \qquad (4.213)$$

Since $J_n(x)$ is represented by a Taylor's expansion, expansion coefficients of $g(x)$ in (4.212) can be obtained from the above equation by comparing coefficients to the same powers of x. An example of such a calculation is presented in Problem I.8.70.

Problem 4.72 The modified Bessel function of the first kind $I_\nu(x)$ is related to the Bessel function $J_\nu(x)$ via $I_\nu(x) = i^{-\nu} J_\nu(ix)$. Show that $I_\nu(x)$ satisfies the following differential equation:

$$x^2 y'' + x y' - \left(x^2 + \nu^2\right) y = 0 . \qquad (4.214)$$

4.6.2 Half-Integer Bessel Functions

Next, let us consider the case when $\nu = (2n + 1)/2 = n + 1/2$ is a half-integer. In this case, from Eq. (4.209),

$$J_{n+1/2}(x) = \sum_{r=0}^{\infty} \frac{(-1)^r}{r! \, \Gamma\left(r + n + \frac{3}{2}\right)} \left(\frac{x}{2}\right)^{2r+n+1/2} \qquad (4.215)$$

and

$$J_{-n-1/2}(x) = \sum_{r=0}^{\infty} \frac{(-1)^r}{r! \, \Gamma\left(r - n + \frac{1}{2}\right)} \left(\frac{x}{2}\right)^{2r-n-1/2} . \qquad (4.216)$$

These expressions can be manipulated into a combination of elementary functions. However, it would be much easier to do so after we worked out the appropriate recurrence relations for the Bessel function which will be done in the next section. Here we shall only derive explicit expressions for the case of $\nu = \pm 1/2$, i.e. for $J_{\pm 1/2}(x)$. Since

$$\Gamma\left(r + \frac{3}{2}\right) = \left(r + \frac{1}{2}\right)\Gamma\left(r + \frac{1}{2}\right) = \frac{(2r+1)!}{2^{2r+1}r!}\sqrt{\pi}$$

according to Eq. (4.36), we can write

$$J_{1/2}(x) = \sum_{r=0}^{\infty} \frac{(-1)^r \, 2^{2r+1} r!}{r! \, (2r+1)! \sqrt{\pi}} \left(\frac{x}{2}\right)^{2r+1/2} = \sqrt{\frac{2}{\pi x}} \left[\sum_{r=0}^{\infty} \frac{(-1)^r}{(2r+1)!} x^{2r+1}\right]$$

$$= \sqrt{\frac{2}{\pi x}} \sin x \,,$$

(4.217)

where a use has been made of the Taylor's expansion of the sine function (which is the expression within the square brackets, see Eq. (I.7.50)).

Problem 4.73 Show similarly that

$$J_{-1/2}(x) = \sqrt{\frac{2}{\pi x}} \left[\sum_{r=0}^{\infty} \frac{(-1)^r}{(2r)!} x^{2r}\right] = \sqrt{\frac{2}{\pi x}} \cos x \,. \qquad (4.218)$$

Expression (4.216) is well defined and hence can be used as the second linearly independent solution of the Bessel DE, i.e. a general solution in the case of a half-integer ν reads $y(x) = C_1 J_{n+1/2}(x) + C_2 J_{-n-1/2}(x)$.

Interestingly, the half-integer Bessel functions, $J_{n+1/2}(x)$, are the only ones which are bound on the real axis. We can investigate this by applying our general approach for investigating DEs of hypergeometric type developed in Sect. 4.4.

Problem 4.74 Here we shall transform the Bessel DE (4.204) into the standard form (4.124). Show that, in this case, $k = \pm 2i\nu$ and $\zeta(x) = \pm (ix \pm \nu)$. Choosing $k = 2i\nu$ and $\zeta(x) = -(ix + \nu)$, show that the DE in the standard form is characterised by $\sigma(x) = x$, $\tau(x) = -2\nu + 1 - 2ix$ and $\lambda = i(2\nu - 1)$. Next, on the other hand, comparing this with expression (4.129) for the eigenvalues, show that, to guarantee polynomial solutions of the transformed equation (4.124) for the auxiliary function $u(x) = J_\nu(x)/\phi(x)$ (which are bound in arbitrary finite interval of x including the point $x = 0$), the values of ν must be positive half-integer numbers, i.e. they must satisfy $\nu = n + 1/2$, where n is a *positive* integer (including zero).

For instance, $n = 0$ corresponds to $\nu = 1/2$, which gives $J_{1/2}(x) \sim \sin x / \sqrt{x} = \sqrt{x} \, (\sin x / x)$ which is obviously finite at $x = 0$ and tends to zero at $x = \pm\infty$. On the other hand, the function $J_{-1/2}(x) \sim \cos x / \sqrt{x}$ tends to infinity when $x \to 0$ and is therefore not finite there. The possible values of ν we found in the above problem guarantee that all functions $J_{n+1/2}(x)$ with any $n = 0, 1, 2, \ldots$ are bound for all values of x. Moreover, as was said, it can be shown that these are the only Bessel functions which possess this property.

4.6.3 Recurrence Relations for Bessel Functions

Starting from the general expression (4.209) for $J_\nu(x)$, consider the derivative

$$\frac{d}{dx}\frac{J_\nu}{x^\nu} = \sum_{r=0}^{\infty} \frac{(-1)^r}{r!\,\Gamma\,(\nu+r+1)\,2^{2r+\nu}}\,(x^{2r})' = \sum_{r=1}^{\infty} \frac{(-1)^r\,2rx^{2r-1}}{r!\,\Gamma\,(\nu+r+1)\,2^{2r+\nu}}$$

$$= \sum_{r=1}^{\infty} \frac{(-1)^r}{(r-1)!\,\Gamma\,(\nu+r+1)\,2^\nu}\,\left(\frac{x}{2}\right)^{2r-1}.$$

Note that summation starts from $r=1$ after differentiation. Changing the summation index $r-1 \to r$, we can rewrite the above result as

$$\frac{d}{dx}\frac{J_\nu}{x^\nu} = -\frac{1}{x^\nu}\left[\sum_{r=0}^{\infty} \frac{(-1)^r}{r!\,\Gamma\,(\nu+r+2)}\,\left(\frac{x}{2}\right)^{2r+\nu+1}\right] = -\frac{1}{x^\nu}J_{\nu+1},$$

which can be rearranged:

$$\left(\frac{d}{xdx}\right)\frac{J_\nu(x)}{x^\nu} = -\frac{J_{\nu+1}(x)}{x^{\nu+1}}. \qquad (4.219)$$

Thus, applying the operator $\frac{d}{xdx}$ to J_ν/x^ν gives (up to a sign) the Bessel function with $\nu \to \nu+1$. Applying this recurrence relation n times, we will have:

$$\left(\frac{d}{xdx}\right)^n \frac{J_\nu(x)}{x^\nu} = -\left(\frac{d}{xdx}\right)^{n-1}\frac{J_{\nu+1}(x)}{x^{\nu+1}} = (-1)^2\left(\frac{d}{xdx}\right)^{n-2}\frac{J_{\nu+2}(x)}{x^{\nu+2}}$$

$$= \ldots = (-1)^n \frac{J_{\nu+n}(x)}{x^{\nu+n}}. \qquad (4.220)$$

Problem 4.75 Prove the other two recurrence relations using a similar method:

$$\left(\frac{d}{xdx}\right)x^\nu J_\nu(x) = x^{\nu-1}J_{\nu-1}(x), \qquad (4.221)$$

$$\left(\frac{d}{xdx}\right)^n x^\nu J_\nu(x) = x^{\nu-n}J_{\nu-n}(x). \qquad (4.222)$$

Therefore, it is readily seen that if the first set of relations, (4.219) and (4.220), serves to increase the order of the Bessel function, the second set of relations, (4.221) and (4.222), reduces it. This property can be employed to derive recurrence relations between Bessel functions of different orders. Indeed, from (4.219) we have, performing differentiation:

$$J_\nu' = -J_{\nu+1} + \frac{\nu}{x} J_\nu ,$$
(4.223)

while similarly from (4.221) ones obtains

$$J_\nu' = J_{\nu-1} - \frac{\nu}{x} J_\nu .$$
(4.224)

Combining these, we can obtain other two recurrence relations:

$$J_{\nu-1} + J_{\nu+1} = \frac{2\nu}{x} J_\nu \quad \text{and} \quad 2J_\nu' = J_{\nu-1} - J_{\nu+1} .$$
(4.225)

Recurrence relations (4.220) allow for an explicit calculation of the Bessel functions of a positive half integer. Indeed, since we already know $J_{1/2}(x)$, we can write

$$J_{n+1/2}(x) = (-1)^n x^{n+1/2} \left(\frac{d}{xdx} \right)^n \frac{J_{1/2}(x)}{x^{1/2}} = (-1)^n x^{n+1/2} \sqrt{\frac{2}{\pi}} \left(\frac{d}{xdx} \right)^n \frac{\sin x}{x} .$$
(4.226)

Problem 4.76 Similarly prove, using (4.222), that

$$J_{-n-1/2}(x) = x^{n+1/2} \sqrt{\frac{2}{\pi}} \left(\frac{d}{xdx} \right)^n \frac{\cos x}{x} .$$
(4.227)

Problem 4.77 Show using the explicit formulae (4.226) and (4.227), that

$$J_{3/2}(x) = \sqrt{\frac{2}{\pi x}} \left(-\cos x + \frac{\sin x}{x} \right) , \quad J_{-3/2}(x) = -\sqrt{\frac{2}{\pi x}} \left(\sin x + \frac{\cos x}{x} \right) .$$

Then, using the first of the recurrence relations (4.225) demonstrate that

$$J_{5/2}(x) = \sqrt{\frac{2}{\pi x}} \left[\left(\frac{3}{x^2} - 1 \right) \sin x - \frac{3}{x} \cos x \right]$$

and

$$J_{-5/2}(x) = \sqrt{\frac{2}{\pi x}} \left[\frac{3}{x} \sin x + \left(\frac{3}{x^2} - 1 \right) \cos x \right] .$$

Verify the above expressions for $J_{\pm 5/2}(x)$ by applying directly expressions (4.226) and (4.227).

It is seen now that the functions $J_{-n-1/2}(x)$ diverge at $x = 0$. At the same time, $J_{n+1/2}(x)$ behaves well around the point $x = 0$. This can be readily seen e.g. by expanding $\sin x/x$ in (4.226) into the Taylor's series. It also follows directly from (4.215).

4.6.4 Generating Function and Integral Representation for Bessel Functions

Here we shall derive some formulae for integer-index Bessel functions which are frequently found useful in applications. We shall start by expanding the following exponential function into the Laurent series:

$$e^{\frac{x}{2}(u-1/u)} = \sum_{k=-\infty}^{\infty} c_k(x)u^k \ . \tag{4.228}$$

To calculate the coefficients $c_k(x)$ in the expansion of the exponential function in (4.228), we write the latter as a product of two exponential functions, $e^{xu/2}$ and $e^{-x/2u}$, which we both expand into the Taylor series:

$$e^{xu/2} = \sum_{n=0}^{\infty} \frac{1}{n!}\left(\frac{x}{2}\right)^n u^n \quad \text{and} \quad e^{-x/2u} = \sum_{m=0}^{\infty} \frac{(-1)^m}{m!}\left(\frac{x}{2}\right)^m u^{-m} \ .$$

Multiplying both expansions, we get:

$$e^{\frac{x}{2}(u-1/u)} = \sum_{n=0}^{\infty}\sum_{m=0}^{\infty} A_{n,m} u^{n-m} \quad \text{with} \quad A_{n,m} = \frac{(-1)^m}{n!m!}\left(\frac{x}{2}\right)^{n+m} \ .$$

Both summations can be combined by introducing a single summation index $k = n - m$. Then, one has:

$$e^{\frac{x}{2}(u-1/u)} = \sum_{k=0}^{\infty}\left(A_{k,0} + A_{k+1,1} + A_{k+2,2} + \ldots\right)u^k$$

$$+ \sum_{k=-1}^{-\infty}\left(A_{0,-k} + A_{1,-k+1} + A_{2,-k+2} + \ldots\right)u^k$$

$$= \sum_{k=0}^{\infty}\left(A_{k,0} + A_{k+1,1} + A_{k+2,2} + \ldots\right)u^k$$

$$+ \sum_{k=1}^{\infty}\left(A_{0,k} + A_{1,k+1} + A_{2,k+2} + \ldots\right)u^{-k}$$

$$= \sum_{k=0}^{\infty}\left(\sum_{n=0}^{\infty} A_{k+n,n}\right)u^k + \sum_{k=1}^{\infty}\left(\sum_{n=0}^{\infty} A_{n,k+n}\right)u^{-k} \ .$$

It follows from here that $(k > 0)$

$$c_k(x) = \sum_{n=0}^{\infty} A_{k+n,n} = \sum_{n=0}^{\infty} \frac{(-1)^n}{n!\,(k+n)!} \left(\frac{x}{2}\right)^{2n+k} , \tag{4.229}$$

$$c_{-k}(x) = \sum_{n=0}^{\infty} A_{n,k+n} = \sum_{n=0}^{\infty} \frac{(-1)^{k+n}}{n!\,(k+n)!} \left(\frac{x}{2}\right)^{2n+k} . \tag{4.230}$$

Comparing Eqs. (4.229) and (4.209), we immediately recognise in $c_k(x)$ the Bessel function $J_k(x)$ of a positive index $k = 0, 1, 2, \ldots$. Also, comparing Eqs. (4.229) and (4.230), one can see that $c_{-k}(x) = (-1)^k c_k(x)$. Exactly the same relationship exists between $J_{-n}(x)$ and $J_n(x)$, see Eq. (4.211). Therefore, the coefficients (4.230) correspond to $J_{-k}(z)$. In other words, expansion (4.228) can be now written as

$$e^{\frac{x}{2}(u-1/u)} = \sum_{k=-\infty}^{\infty} J_k(x)u^k , \tag{4.231}$$

so that the exponential function on the left-hand side serves as the generating function for the Bessel functions of integer indices.

On the other hand, the coefficients of the Laurent expansion are given by the contour integral of Eq. (2.98), so that we can write

$$J_k(x) = \frac{1}{2\pi i} \oint_C e^{\frac{x}{2}(p-1/p)} p^{-k-1} dp , \tag{4.232}$$

where the contour C is taken anywhere in the complex plane around the $p = 0$ point. This formula is valid for both negative and positive integer values of the index k.

Problem 4.78 By taking the contour C to be a circle of unit radius centred at the origin, $p = e^{i\phi}$, show that the Bessel function can also be represented as a definite integral:

$$J_k(x) = \frac{1}{2\pi} \int_0^{2\pi} e^{i(x\sin\phi - k\phi)} d\phi . \tag{4.233}$$

Then show that alternatively this formula can also be rewritten as

$$J_k(x) = \frac{1}{2\pi} \int_0^{2\pi} \cos(x\sin\phi - k\phi)\, d\phi . \tag{4.234}$$

[Hint: *by shifting the integration range to* $-\pi \le \phi \le \pi$ *(which can always be done as the integrand is a periodic function of* ϕ *with the period of* 2π*); argue that the integral of the sine function does not contribute.*]

Problem 4.79 Prove the following formula:

$$e^{iz \sin \phi} = \sum_{n=-\infty}^{\infty} J_n(z) e^{in\phi} .$$

Problem 4.80 Consider now specifically $J_1(z)$. Shift the integration interval to $-\pi/2 \le \phi \le 3\pi/2$, split the integral into two, one for $-\pi/2 \le \phi \le \pi/2$, and another for $\pi/2 \le \phi \le 3\pi/2$, change the variable $\phi \to \pi - \phi$ in the second integral, and then combine the two before making the final change of variable $t = \sin \phi$. Hence, show that

$$J_1(x) = -\frac{i}{\pi} \int_{-1}^{1} \frac{t e^{ixt}}{\sqrt{1-t^2}} dt . \qquad (4.235)$$

The modified function of the first kind, see Problem 4.72, is then

$$I_1(x) = i^{-1} J_1(ix) = -\frac{1}{\pi} \int_{-1}^{1} \frac{t e^{-xt}}{\sqrt{1-t^2}} dt . \qquad (4.236)$$

4.6.5 Orthogonality and Functional Series Expansion

In some physical applications one has to solve partial differential equations of mathematical physics in cylindrical coordinates. In those cases it is possible to seek the solution as a functional series expansion with respect to Bessel functions using the method of separation of variables to be considered in detail in Chap. 8. We shall only briefly discuss this point later on in this section. However, one essential issue of this method is to be able to find coefficients of such an expansion, and for that it is extremely convenient if the functions used in the expansion are orthogonal. It appears that Bessel functions of the index $\nu > -1$ do possess such a property in some specific sense (i.e. not with respect to integration of two functions with different indices as e.g. is the case for orthogonal polynomials), which is nevertheless exactly what is needed for using the method of separation of variables in cylindrical coordinates. Hence it is worth discussing it in detail.

Consider the function $J_\nu(\alpha z)$ with α being some real parameter and z generally a complex number.

Problem 4.81 Show that $J_\nu(\alpha z)$ satisfies the following DE:

$$\frac{d^2 J_\nu(\alpha z)}{dz^2} + \frac{1}{z}\frac{d J_\nu(\alpha z)}{dz} + \left(\alpha^2 - \frac{\nu^2}{z^2}\right) J_\nu(\alpha z) = 0 , \qquad (4.237)$$

which can also be manipulated into the following self-adjoint form:

$$\frac{d}{dz}\left(z\frac{d J_\nu(\alpha z)}{dz}\right) + \left(\alpha^2 z - \frac{\nu^2}{z}\right) J_\nu(\alpha z) = 0 . \qquad (4.238)$$

Similarly to the method developed in Sect. 4.3.1.2, let us multiply both sides of the equation above by $J_\nu(\beta z)$ with some real β and integrate between 0 and some $l > 0$:

$$\int_0^l \frac{d}{dz}\left(z\frac{d J_\nu(az)}{dz}\right) J_\nu(\beta z)\, dz + \int_0^l \left(\alpha^2 z - \frac{\nu^2}{z}\right) J_\nu(\alpha z) J_\nu(\beta z)\, dz = 0 .$$

Taking the first integral by parts, we obtain

$$z\frac{d J_\nu(\alpha z)}{dz} J_\nu(\beta z)\bigg|_0^l - \int_0^l z\frac{d J_\nu(\alpha z)}{dz}\frac{d J_\nu(\beta z)}{dz}\, dz$$
$$+ \int_0^l \left(\alpha^2 z - \frac{\nu^2}{z}\right) J_\nu(\alpha z) J_\nu(\beta z)\, dz = 0 . \qquad (4.239)$$

In the same fashion, we start from (4.238) written for $J_\nu(\beta z)$, multiply it by $J_\nu(\alpha z)$ and integrate between 0 and l; this procedure readily yields the same equation as above but with α and β swapped:

$$z\frac{d J_\nu(\beta z)}{dz} J_\nu(\alpha z)\bigg|_0^l - \int_0^l z\frac{d J_\nu(\beta z)}{dz}\frac{d J_\nu(\alpha z)}{dz}\, dz$$
$$+ \int_0^l \left(\beta^2 z - \frac{\nu^2}{z}\right) J_\nu(\beta z) J_\nu(\alpha z)\, dz = 0 .$$

Subtracting one equation from the other, we get

$$z\left[\frac{d J_\nu(\alpha z)}{dz} J_\nu(\beta z) - \frac{d J_\nu(\beta z)}{dz} J_\nu(\alpha z)\right]\bigg|_0^l$$
$$+ \left(\alpha^2 - \beta^2\right)\int_0^l z J_\nu(\alpha z) J_\nu(\beta z)\, dz = 0 . \qquad (4.240)$$

Let us now carefully analyse the first term above near $z = 0$ (the bottom limit). The leading term in the Bessel function expansion, Eq. (4.209), is $J_\nu(\alpha z) \sim z^\nu$, while its derivative $\frac{d}{dz} J_\nu(\alpha z) \sim z^{\nu-1}$. Correspondingly, it may seem that the whole first term above has the behaviour $z[\ldots] \sim z^{2\nu}$ around $z = 0$. A more careful consideration, however, results in $z[\ldots] \sim z^{2\nu+2}$ as is demonstrated by the following problem.

Problem 4.82 Show using the explicit expansion of $J_\nu(\alpha z)$ from Eq. (4.209) that

$$z\left[\frac{dJ_\nu(\alpha z)}{dz} J_\nu(\beta z) - \frac{dJ_\nu(\beta z)}{dz} J_\nu(\alpha z)\right]$$

$$= \sum_{k=0}^{\infty}\sum_{n=0}^{\infty} \frac{2(k-n)(-1)^{k+n}}{k!n!\,\Gamma(\nu+k+1)\,\Gamma(\nu+n+1)}\left(\frac{\alpha z}{2}\right)^{\nu+2k}\left(\frac{\beta z}{2}\right)^{\nu+2n}.$$

$$\tag{4.241}$$

The above result shows that the terms with $k = n$ can be dropped; therefore, the first term in the expansion starts not from $n = k = 0$ as we thought before, but from $k, n = 0, 1$ and $k, n = 1, 0$, which give the $z^{2\nu+2}$ type of behaviour for the first (leading) term in Eq. (4.241). Therefore, the free term in (4.240) is well defined at $z = 0$ if $2\nu + 2 > 0$ or $\nu > -1$.

We still need to check the convergence of the two integrals in (4.239). Let us start from considering the first one: it contains two derivatives of the Bessel function which behave as $z^{\nu-1}$ near $z = 0$ each, i.e. the integrand is $\sim z^{2\nu-1}$ near its bottom limit. We know that the integral $\int_0^l z^{2\nu-1}dz$ converges, if $2\nu - 1 > -1$ which gives $\nu > -1$ again. Similarly for the second integral: its first term behaves as $z^{2\nu+1}$ around $z = 0$, while the second one as $z^{2\nu-1}$, both converge for $\nu > -1$.

This analysis proves that our consideration is valid for any $\nu > -1$. Because the free term (4.241) in Eq. (4.240), prior to taking the limits, behaves as $z^{2\nu+2} = z^{2(\nu+1)}$, it is equal to zero when $z = 0$ since $\nu + 1 > 0$. Therefore, the bottom limit applied to the free term can be dropped and we obtain instead of (4.240):

$$l\left[\frac{dJ_\nu(\alpha z)}{dz} J_\nu(\beta z) - \frac{dJ_\nu(\beta z)}{dz} J_\nu(\alpha z)\right]_{z=l}$$

$$+ \left(\alpha^2 - \beta^2\right)\int_0^l z J_\nu(\alpha z) J_\nu(\beta z)\, dz = 0.$$

$$\tag{4.242}$$

So far we have not specified the values of the real constants α or β. Now we are going to be more specific. Consider solutions of the equation

$$J_\nu(z) = 0 \tag{4.243}$$

with respect to (generally complex) z; these are the roots of the Bessel function. It can be shown that this equation cannot have complex roots at all, but there is an infinite number of real roots $\pm x_1, \pm x_2$, etc. Let us choose the values of α and β such that αl and βl be two distinct roots ($\alpha \neq \beta$). Then, $J_\nu(\alpha l) = 0$ and $J_\nu(\beta l) = 0$, and the first term in (4.242) becomes zero, so that we immediately obtain:

$$\int_0^l z J_\nu(\alpha z) J_\nu(\beta z)\, dz = 0 \qquad (4.244)$$

for $\alpha \neq \beta$. This is the required orthogonality condition for the Bessel functions: they appear to be orthogonal not with respect to their index ν, but rather with respect to the 'scaling' of their arguments. Note the weight z in the orthogonality condition above.

It is also possible to choose αl and βl as roots of the equation $J_\nu'(z) = 0$; it is readily seen that, in this case, the first term in Eq. (4.242) is also zero and we again arrive at the orthogonality condition (4.244).

Problem 4.83 Prove that αl and βl can also be chosen as roots of the equation $C_1 J_\nu(z) + C_2 z J_\nu'(z) = 0$, where C_1 and C_2 are arbitrary real constants. This condition generalises the two conditions given above as it represents their linear combination.

As was mentioned, under certain conditions a properly integrable function $f(x)$ can be expanded into Bessel functions with a desired index ν:

$$f(x) = \sum_i f_i J_\nu(\alpha_i x) \ , \qquad (4.245)$$

where summation is made over all roots $x_i = \alpha_i l$ of the equation $J_\nu(x) = 0$; note that the roots depend on the index ν chosen, although this is not explicitly indicated. Using the orthogonality of the Bessel functions, we immediately obtain for the expansion coefficients the following formula:

$$f_i = \frac{\int_0^l x f(x) J_\nu(\alpha_i x)\, dx}{\int_0^l x J_\nu(\alpha_i x)^2\, dx} \ . \qquad (4.246)$$

It is only left to discuss how to calculate the normalisation integral standing in the denominator of the above equation. To do this, we first solve for the integral in Eq. (4.242):

$$\int_0^l x J_\nu(\alpha x) J_\nu(\beta x)\, dx = \frac{-l}{\alpha^2 - \beta^2} \left[\frac{d J_\nu(\alpha x)}{dx} J_\nu(\beta x) - \frac{d J_\nu(\beta x)}{dx} J_\nu(\alpha x) \right]_{x=l} ,$$

and then consider the limit $\beta \to \alpha$. This is because, on the left-hand side under this limit, we would have exactly the normalisation integral we need. Note that $x = \alpha l$ is one of the roots of the equation $J_\nu(x) = 0$. Therefore, on the right-hand side, the second term inside the square brackets can be dropped and the expression is simplified:

$$\int_0^l x J_\nu (\alpha x)^2 \, dx = \lim_{\beta \to \alpha} \frac{-l}{\alpha^2 - \beta^2} \left[\frac{d J_\nu (\alpha x)}{dx} J_\nu (\beta x) \right]_{x=l} .$$

Since under the limit $J_\nu (\beta l) \to J_\nu (\alpha l) = 0$, we have to deal with the $0/0$ uncertainty. Using the L'Hôpital's rule, we have:

$$\begin{aligned} \int_0^l x J_\nu (\alpha x)^2 \, dx &= \lim_{\beta \to \alpha} \frac{-l}{-2\beta} \left[\frac{d J_\nu (\alpha x)}{dx} \frac{d J_\nu (\beta x)}{d\beta} \right]_{x=l} \\ &= \frac{l}{2\alpha} \lim_{\beta \to \alpha} \left(\alpha \frac{d J_\nu (u)}{du} \right)_{u=\alpha l} \left(l \frac{d J_\nu (u)}{du} \right)_{u=\beta l} \\ &= \frac{l^2}{2} \left(\frac{d J_\nu (u)}{du} \right)_{u=\alpha l}^2 . \end{aligned} \quad (4.247)$$

This is the required result. It can also be rearranged using Eq. (4.223). Using there $x = \alpha l$ and recalling that $J_\nu (\alpha l) = 0$, we obtain that $[d J_\nu (u) / du]_{u=\alpha l} = -J_{\nu+1} (\alpha l)$, which yields

$$\int_0^l x J_\nu (\alpha x)^2 \, dx = \frac{l^2}{2} J_{\nu+1} (\alpha l)^2 . \quad (4.248)$$

Problem 4.84 Perform a similar calculation assuming that αl is a root of the equation $J_\nu'(x) = 0$. Show that, in this case,

$$\int_0^l x J_\nu (\alpha x)^2 \, dx = \frac{1}{2} \left(l^2 - \frac{\nu^2}{\alpha^2} \right) J_\nu (\alpha l)^2 . \quad (4.249)$$

[Hint: express the second derivative of $J_\nu(u)$ from the Bessel DE.]

4.7 Selected Applications in Physics

4.7.1 Schrödinger Equation for a Harmonic Oscillator

Consider a quantum particle of mass m moving in a quadratic potential $V(x) = kx^2/2$. In classical mechanics a particle in this potential would oscillate with the frequency $\omega = \sqrt{k/m}$. In quantum mechanics stationary states of the particle are obtained by solving the Schrödinger equation:

$$-\frac{\hbar^2}{2m}\frac{d^2\psi(x)}{dx^2} + \frac{kx^2}{2}\psi(x) = E\psi(x) \,, \tag{4.250}$$

where E is the energy of a stationary state and $\psi(x)$ is the corresponding wave-function, $\hbar = h/2\pi$ is the Planck constant; the probability $dP(x)$ to find the particle between x and $x + dx$ in this particular state is given by $|\psi(x)|^2 dx$. Because the particle must be somewhere, the probability to find it anywhere on the x-axis must be equal to one:

$$\int_{-\infty}^{+\infty} P(x)dx = \int_{-\infty}^{+\infty} |\psi(x)|^2 dx = 1 \,, \tag{4.251}$$

which gives a normalisation condition for the wavefunction. What we shall try to do now is to find the energies and the wavefunctions of the harmonic oscillator.

Let us introduce a new variable $u = x\sqrt{m\omega/\hbar}$ instead of x. Then, $\psi'_x = \psi'_u u'_x = \sqrt{m\omega/\hbar}\,\psi'_u$, $\psi''_{xx} = (m\omega/\hbar)\,\psi''_{uu}$ and the DE is transformed into:

$$\psi'' + \left(\frac{2E}{\hbar\omega} - u^2\right)\psi = 0 \,. \tag{4.252}$$

This DE is of the hypergeometric type (4.119) with $\sigma(u) = 1$, $\beta(u) = 0$ and $\xi(u) = 2E/\hbar\omega - u^2$, and can be transformed into the standard form (4.124) using the method we developed in Sect. 4.4.1.

Problem 4.85 Show using this method that the transformation from $\psi(u)$ to a new function $g(u)$ via $\psi(u) = \phi(u)g(u)$ can be accomplished by the transformation function $\phi(u) = e^{-u^2/2}$, which corresponds to the polynomial (4.125) being $\zeta(u) = -u$ (the sign here was chosen such that $\tau'(u)$, which is a constant, be negative, see the end of Sect. 4.4.2), while the corresponding eigenvalues $\lambda \to \lambda_n = 2n$ with $n = 0, 1, 2, \ldots$. Hence demonstrate that the transformed DE for the new function $g(u)$,

$$g'' - 2ug' + \lambda_n g = 0 \,, \tag{4.253}$$

coincides with the DE for the Hermite polynomials (4.101), where $\lambda_n = 2E/\hbar\omega - 1$.

Therefore, we obtain for the energies of the oscillator:

$$\frac{2E_n}{\hbar\omega} - 1 = 2n \implies E_n = \hbar\omega\left(n + \frac{1}{2}\right) \,, \quad n = 0, 1, 2, 3, \ldots \,. \tag{4.254}$$

Hence, stationary states of the harmonic oscillator are characterised by a discrete set of equidistant energies (differing exactly by a single oscillation quanta $\hbar\omega$), and their wavefunctions are proportional to Hermite polynomials:

$$\psi_n(u) = C_n e^{-u^2/2} H_n(u) \quad \Longrightarrow \quad \psi_n(x) = C_n e^{-m\omega x^2/2\hbar} H_n\left(\sqrt{\frac{m\omega}{\hbar}}x\right),$$

where the normalisation constant is to be defined from Eq. (4.251):

$$\int_{-\infty}^{\infty} \psi_n^2(x)dx = C_n^2 \int_{-\infty}^{\infty} e^{-m\omega x^2/\hbar} \left[H_n\left(\sqrt{\frac{m\omega}{\hbar}}x\right)\right]^2 dx$$

$$= C_n^2 \sqrt{\frac{\hbar}{m\omega}} \int_{-\infty}^{\infty} e^{-u^2} H_n(u)^2 du = 1.$$

The last integral on the right-hand side corresponds to the normalisation of the Hermite polynomials, see Eq. (4.150), and therefore we finally obtain:

$$C_n = \left(\frac{m\omega}{\pi\hbar}\right)^{1/4} \frac{2^{-n/2}}{\sqrt{n!}}. \tag{4.255}$$

The ground state of the quantum oscillator, $\psi_0(x)$, has a non-zero energy $E = \hbar\omega/2$, called the *zero point energy*; excited states of the oscillator, $\psi_n(x)$ with $n \geq 1$, are obtained by adding n quanta $\hbar\omega$ to the energy of the ground state. A single quanta of oscillation is called in solid state physics a '*phonon*'. It is therefore said that there are no phonons in the ground state, but there are n phonons, each carrying the same energy of $\hbar\omega$, in the n-th excited state.

4.7.2 Schrödinger Equation for the Hydrogen Atom

Consider an electron of charge $-e$ (the elementary charge e is assumed positive here) moving in the Coulomb field $V(\mathbf{r}) = -Ze^2/r$ of a nucleus of a positive charge Ze placed in the centre of coordinates. The nucleus potential $V(\mathbf{r}) = V(r)$ depends only on the distance $r = |\mathbf{r}|$ between the nucleus, placed in the centre of the coordinate system, and the electron position described by the vector $\mathbf{r}$. The Schrödinger equation in this case for the stationary states of the electron with energy E has the form:

$$-\frac{\hbar^2}{2m}\Delta\psi(\mathbf{r}) + V(r)\psi(\mathbf{r}) = E\psi(\mathbf{r}), \tag{4.256}$$

where $\psi(\mathbf{r})$ is the electron wavefunction and m is its mass, $\hbar = h/2\pi$ is the Planck constant. Writing the Laplacian in the spherical coordinates (see Sect. 7.11, cf. Sect. 4.5.3), after simple rearrangements, we obtain

$$\frac{1}{r^2}\frac{\partial}{\partial r}\left(r^2\frac{\partial\psi}{\partial r}\right) + \frac{1}{r^2\sin\theta}\frac{\partial}{\partial\theta}\left(\sin\theta\frac{\partial\psi}{\partial\theta}\right) + \frac{1}{r^2\sin^2\theta}\frac{\partial^2\psi}{\partial\phi^2} = -\frac{2m}{\hbar^2}\left(E + \frac{Ze^2}{r}\right)\psi.$$

Problem 4.86 Using the method of separation of variables identical to the one we used in Sect. 4.5.3, show that the solution of the above equation can be written as $\psi(\mathbf{r}) = R(r)\Theta(\theta)\Phi(\phi)$, where $\Phi(\phi) = e^{\pm im\phi}$ with a positive integer m (including zero) and the function $\Theta(\theta)$ satisfies Eq. (4.189) with $\mu = l(l+1)$. In other words, the angular part of the wavefunction is identical to the angular solution of the Laplace equation we obtained in Sect. 4.5.3, i.e. it is basically given by the spherical functions $Y_{lm}(\theta, \phi)$ with $l = 0, 1, \ldots$ and $m = -l, \ldots, 0, \ldots, l$. At the same time, the equation for the function $R(r)$ reads:

$$R'' + \frac{2}{r}R' + \left[\frac{2m}{\hbar^2}\left(E + \frac{Ze^2}{r}\right) - \frac{l(l+1)}{r^2}\right]R = 0. \qquad (4.257)$$

Let us solve the above equation for the radial part $R(r)$ of the wavefunction for the case of *bound states*, i.e. the states which energy is negative: $E < 0$. These states correspond to a discrete spectrum of the atom, i.e. there will be an energy gap between any two states. There are also continuum states when $E > 0$ (no gap), we shall not consider them here.

It is convenient to start by introducing a new variable $\rho = r/a$, where $a = \hbar/\sqrt{-8Em}$. Then, $R' \to R'/a$ and $R'' \to R''/a^2$, and Eq. (4.257) is rewritten as:

$$R'' + \frac{2}{\rho}R' + \left[-\frac{1}{4} + \frac{\epsilon}{\rho} - \frac{l(l+1)}{\rho^2}\right]R = 0, \qquad (4.258)$$

where $\epsilon = Ze^2\sqrt{-m/2\hbar^2 E}$.

The obtained DE is of the generalised hypergeometric type (4.119) with $\sigma(\rho) = \rho$, $\beta = 2$ and $\xi(\rho) = -\rho^2/4 + \epsilon\rho - l(l+1)$. In the general method of Sect. 4.4.1, we express $R(\rho)$ via another function $u(\rho)$ using $R(\rho) = \phi(\rho)u(\rho)$, where the auxiliary function $\phi(\rho)$ is chosen such that the new function $u(\rho)$ satisfies a DE in the standard form (4.124). When applying the method developed in this section, several choices must be made concerning the sign in defining the constant k and the $\zeta(\rho)$—polynomial (4.125), which in our case is as follows:

$$\zeta(\rho) = -\frac{1}{2} \pm \sqrt{\frac{1}{4} + k\rho - \xi(\rho)} = -\frac{1}{2} \pm \sqrt{\left(l + \frac{1}{2}\right)^2 + (k - \epsilon)\rho + \frac{\rho^2}{4}}.$$

Recall that the constant k is chosen such that the expression under the square root becomes a full square.

Problem 4.87 Show using the method developed in Sect. 4.4.1, that the constant k has to be chosen as $k = \epsilon \pm (l + 1/2)$. Then, if one chooses the *minus* sign in the expression for the k and hence the polynomial $\zeta(\rho) = -\rho/2 + l$, then the DE for the function $u(\rho)$,

$$\rho u'' + (2l + 2 - \rho)u' + (\epsilon - l - 1)u = 0 , \qquad (4.259)$$

takes on indeed the form of Eq. (4.124) with $\sigma = \rho$, $\tau(\rho) = 2l + 2 - \rho$ and $\lambda = \epsilon - l - 1$. Note the minus sign before ρ in $\zeta(\rho) = -\rho/2 + l$; this ensures that the derivative $\tau' = -\epsilon$ is negative. Finally, show that, in this case, the transformation function $\phi(\rho) = \rho^l e^{-\rho/2}$, i.e. the radial part is expressed in terms of the function $u(\rho)$ of Eq. (4.259) as

$$R(\rho) = \rho^l e^{-\rho/2} u(\rho) . \qquad (4.260)$$

Note that this solution is finite at $\rho = 0$ and hence is physically acceptable.

Problem 4.88 Show that if instead the constant k was chosen as $k = \epsilon + (l + 1/2)$, then $\zeta(\rho) = -\rho/2 - l - 1$ and the transformation between $R(\rho)$ and $u(\rho)$ is given by $R = \rho^{-l-1} e^{-\rho/2} u$. This solution is physically unacceptable as being singular at $\rho = 0$.

Problem 4.89 Consider the solution of Eq. (4.258) near $\rho = 0$ in terms of a power series: $R(\rho) = r_1 \rho^s + r_2 \rho^{s+1} + \ldots$. Show that the power s of the leading term can only be either l or $-l - 1$.

Problem 4.90 Similarly, consider the solution of Eq. (4.258) for large ρ by dropping terms containing $1/\rho$. Show that, in this case, $R(\rho) \sim e^{-\rho/2}$ (the other solution, $R(\rho) \sim e^{\rho/2}$, is physically unacceptable as goes to infinity in the $\rho \to \infty$ limit).

So we have two possible choices for the constant k and hence two possible transformations $R(\rho) \to u(\rho)$, namely, either $R = \rho^l e^{-\epsilon\rho} u$ or $R = \rho^{-l-1} e^{-\epsilon\rho} u$. As we expect that the solution of the DE for $u(\rho)$ is to be bound on the whole interval $0 \le \rho < \infty$ (by choosing the appropriate eigenvalues λ), then only the first choice can be accepted; the second one is unacceptable as in this case $R(\rho)$ diverges at $\rho = 0$ for any $l = 0, 1, 2, \ldots$. Hence, we conclude that $R(\rho)$ is to be chosen from Eq. (4.260) with $u(\rho)$ satisfying the DE (4.259). This DE has a bound solution if and only if the coefficient to the u term in DE (4.259), i.e. $\lambda = \epsilon - l - 1$, is given by Eq. (4.129) with integer $n = 0, 1, 2, \ldots$. Since in our case $\sigma'' = 0$ and $\tau' = (2l + 2 - \rho)' = -1$, we then have:

$$\lambda = \epsilon - l - 1 \equiv -n\tau' = n \quad \Longrightarrow \quad \epsilon = n + l + 1$$

$$\Longrightarrow \quad E = -\frac{mZ^2 e^4}{2\hbar^2 n_r^2} \, , \qquad (4.261)$$

where $n_r = n + l + 1$ is called the *main quantum number*. For the given value of l, the main quantum number takes on values $n_r = l + 1, l + 2, \ldots$. In particular, for $l = 0$ (the so-called s states) we have $n_r = 1, 2, \ldots$ which correspond to $1s$, $2s$, etc. states of the H atom electron; for $l = 1$ (the p states) we have $n_r = 2, 3, \ldots$ corresponding to $2p$, $3p$, etc. states, and so on. We see that the allowed energy levels are indeed discrete and they converge at large quantum numbers to zero from below. As n_r is getting bigger, the gap between energy levels gets smaller tending to zero as $n_r \to \infty$.

Replacing $\lambda = \epsilon - l - 1$ in Eq. (4.259) with $\lambda = n$, we obtain the DE for generalised Laguerre polynomials (4.110) $L_n^{2l+1}(\rho) = L_{n_r - l - 1}^{2l+1}(\rho)$. In other words, the radial part of the wavefunction of an electron in a hydrogen-like atom is given by

$$R_{n_r l}(\rho) \sim \rho^l e^{-\rho/2} L_{n_r - l - 1}^{2l+1}(\rho) \, .$$

Normalisation of these functions can now be obtained using Eq. (4.152).

4.7.3   A Free Electron in a Cylindrical Ptential Well

Consider eigenstates of an electron in a cylindrical potential well pictured in Fig. 4.5. We assume a constant potential V_0 acting on the electron, and that the electron is confined in the cylinder, i.e. its wavefunction ψ must be strictly zero at all surfaces of the cylinder. We have to solve the following Schrödinger equation,

$$-\frac{\hbar^2}{2m} \Delta \psi(\mathbf{r}) + V_0 \psi(\mathbf{r}) = E\psi(\mathbf{r}) \, , \qquad (4.262)$$

that give us the stationary states of the electron with energies E. The symmetry of the problem inspire us to use cylindrical coordinates, $\psi(\mathbf{r}) \to \psi(r, \phi, z)$, and hence

Fig. 4.5 A free electron is confined in a vertical cylinder of hight H and radius a

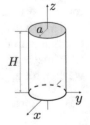

it is convenient to rewrite the Laplacian Δ in these coordinates (that transformation will be proven in Sect. 7.11, see Eq. (7.109)):

$$-\frac{\hbar^2}{2m}\left[\frac{1}{r}\frac{\partial}{\partial r}\left(r\frac{\partial\psi}{\partial r}\right) + \frac{1}{r^2}\frac{\partial^2\psi}{\partial\phi^2} + \frac{\partial^2\psi}{\partial z^2}\right] + (V_0 - E)\psi = 0.\tag{4.263}$$

Using the method of separation of variables similar to that of Sect. 4.5.3, we seek the solution as a product of functions each depending on a single variable: $\psi(\mathbf{r}) = R(r)\Phi(\phi)Z(z)$.

Problem 4.91 Substitute the product solution into Eq.(4.263) to obtain the modified equation appropriate for the separation of variables:

$$\frac{(rR')'}{rR} + \frac{1}{r^2}\left(\frac{\Phi''}{\Phi}\right) + \left(\frac{Z''}{Z} - \frac{2m}{\hbar^2}(V_0 - E)\right) = 0.\tag{4.264}$$

Argue, that, to ensure periodicity with respect to the angle ϕ, the quantity Φ''/Φ must be equal to the minus square of an integer m, and hence Φ is either $\Phi_m(\phi) = \cos(m\phi)$ or $\Phi_{-m}(\phi) = \sin(m\phi)$, assuming $m \geq 0$. Correspondingly, the z dependence comes from solving the ODE

$$Z''(z) + \xi Z(z) = 0,\tag{4.265}$$

where

$$\xi = -\frac{2m}{\hbar^2}(V_0 - E) + \frac{\hbar^2}{2m}\lambda\tag{4.266}$$

and λ is a separation parameter, yet to be determined, while the r dependence is obtained by solving

$$r^2 R'' + rR' + \left(\lambda r^2 - m^2\right)R = 0.\tag{4.267}$$

By making the substitution $x = \sqrt{\lambda}r$, Eq. (4.267) is transformed into

$$x^2 R'' + xR' + \left(x^2 - m^2\right)R = 0\tag{4.268}$$

for the function $R(x) = R(\sqrt{\lambda}r)$. It is seen that this is the Bessel equation (4.204), hence, its solutions are given by the Bessel functions of the first, $J_m(x) = J_m(\sqrt{\lambda}r)$, and second, $K_m(x) = K_m(\sqrt{\lambda}r)$, kind, where $m = 0, 1, 2, 3, \ldots$. Since the Bessel function of the second kind diverges at $r = 0$, this solution is physically unacceptable and has to be dropped. Hence, $R(x) \to R_m(r) = J_m(\sqrt{\lambda}r)$ with $m = 0, 1, 2, \ldots$.

The constant λ is still undetermined. It can be obtained from the boundary conditions that at the side surface of the cylinder the wavefunction must be zero. To ensure this, we can set $R(a) = 0$. This means that $J_m(\sqrt{\lambda}a) = 0$, giving all possible values of the constant λ, for each particular value of the index m. Using the index α to number roots $\mu_\alpha^{(m)}$ of the Bessel function, $J_m(\mu_\alpha^{(m)}) = 0$, we obtain $\lambda \to \lambda_{m\alpha} = \left(\mu_\alpha^{(m)}/a\right)^2$, where $\alpha = 1, 2, 3 \ldots$.

The function $Z(z)$ is determined by solving Eq. (4.265).

Problem 4.92 Show that the only solutions of Eq. (4.265) satisfying the zero boundary conditions at the upper and lower surfaces of the cylinder, $Z(0) = Z(H) = 0$, correspond to the positive value of the constant $\xi = p^2$ and are given by

$$Z(z) \to Z_n(z) = \sin\frac{\pi n z}{H}, \quad p \to p_n = \frac{\pi n}{H}, \quad n = 1, 2, 3, \ldots .$$

Hence, different solutions $Z_n(z)$ are characterised by the integer n that takes on all possible positive values (excluding zero).

So, the complete solution of the Schrödinger equation in our case is given by the wavefunction

$$\psi_{\pm m, \alpha, n}(r, \phi, z) \sim J_m\left(\mu_\alpha^{(m)}r/a\right) \Phi_{\pm m}(\phi) Z_n(z),$$

that is characterised by three quantum numbers: $\pm m$, α and n. The corresponding energies $E_{\pm m, \alpha, n}$ of the electron are obtained from Eq. (4.266), where $\xi = p_n^2$, and the known value $\lambda_{m\alpha} = \left(\mu_\alpha^{(m)}/a\right)^2$ of the constant λ:

$$-\frac{2m}{\hbar^2}(V_0 - E_{\pm m, \alpha, n}) + \frac{\hbar^2}{2m}\lambda_{m\alpha} = p_n^2 \implies E_{\pm m, \alpha, n} = V_0 + \frac{\hbar^2}{2m}\left[\left(\frac{\pi n}{H}\right)^2 + \left(\frac{\mu_\alpha^{(m)}}{a}\right)^2\right].$$

The energy does not depend on the plus or minus sign of m, it is degenerate for the states characterised by m and $-m$ (where $m > 0$). Several first roots of the Bessel function for several initial values of m are given[6] in Table 4.2.

Since the roots of the Bessel function quickly increase with α, the first few lowest states of the electron confined in the cylinder are obtained by using the lowest values of n, m and α.

---

[6] These were obtained using the Mathematica software (specifically, by applying the command N[BesselJZero[m,α]]).

Table 4.2 The lowest five zeros $\mu_\alpha^{(m)}$ of the Bessel function $J_m(x)$ for $m = 0, 1, 2, 3$ (up to the five significant digits)

m	α	$\mu_\alpha^{(m)}$	m	α	$\mu_\alpha^{(m)}$
0	1	2.40483	2	1	5.13562
0	2	5.52008	2	2	8.41724
0	3	8.65373	2	3	11.6198
0	4	11.7915	2	4	14.796
0	5	14.9309	2	5	17.9598
1	1	3.83171	3	1	6.38016
1	2	7.01559	3	2	9.76102
1	3	10.1735	3	3	13.0152
1	4	13.3237	3	4	16.2235
1	5	16.4706	3	5	19.4094

4.7.4 Stirling's Formula and Phase Transitions

Statistical mechanics deals with very large numbers N of particles which may be indistinguishable. In these cases it is often needed to calculate the factorial $N!$ of very large numbers. There exists an elegant approximation to $N!$ for very large integer numbers N, the so-called Stirling's approximation,

$$N! \approx N^N e^{-N} \sqrt{2\pi N} \,, \tag{4.269}$$

which we shall derive here first. Then we shall show some applications of this result.

Since $\Gamma(N+1) = N!$, we have to investigate the Gamma function for large integer N. We shall consider a more general problem of a real number z, where $|z| \gg 1$. Let us start by making the following transformation in $\Gamma(z+1)$:

$$
\begin{aligned}
\Gamma(z+1) &= \int_0^\infty t^z e^{-t} dt = \int_0^\infty (zx)^z e^{-zx} z dx \\
&= z^{z+1} e^{-z} \int_0^\infty \exp\left[-z(x - \ln x - 1)\right] dx \,,
\end{aligned} \tag{4.270}
$$

where the substitution $t = zx$ has been made.

Now, the function $f(x) = x - \ln x - 1$ appearing in the exponential, see Fig. 4.6, has a minimum at $x = 1$ with the value of $f(1) = 0$. Thus the function $e^{-zf(x)}$ has a maximum at $x = 1$ and this maximum becomes very sharply peaked when $|z|$ is large, see the same figure. This observation allows us to evaluate the integral in

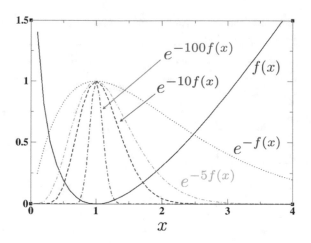

Fig. 4.6 Functions
$f(x) = x - \ln x - 1$ (black)
and $\exp(-zf(x))$ (other
colours) for $z = 1, 5, 10$, and
100

Eq. (4.270) approximately.[7] Indeed, two things can be done. Firstly, the bottom limit of the integral can be extended to $-\infty$ as the exponential function in the integral is practically zero for $x < 0$. Secondly, we can expand $f(x)$ about $x = 1$ and retain only the leading order term. That is, if we write $x = 1 + y$, we obtain

$$f(x) = f(1 + y) = 1 + y - \ln(1 + y) - 1 = y - \ln(1 + y),$$

which for small y becomes

$$f(x) = y - \left(y - \frac{y^2}{2} + \ldots \right) = \frac{y^2}{2} + \ldots$$

after expanding the logarithm in the Taylor series about $y = 0$. Consequently, we obtain

$$\Gamma(z + 1) \approx z^{z+1} e^{-z} \int_{-\infty}^{+\infty} e^{-zy^2/2} dy = \sqrt{2\pi z} z^z e^{-z}.$$

As a consequence of the Stirling's approximation, consider the logarithm of $N!$ for large values of N. Letting $z = N$ in the Stirling's formula, one obtains:

$$\ln(N!) \approx N \ln N - N. \tag{4.271}$$

In most applications the terms we dropped here are negligible as compared to the two terms retained.

As an example of an application of the Stirling's formula (4.271), we shall consider a paramagnetic–ferromagnetic phase transition in a simple three-dimensional lattice within a rather simple Bragg–Williams theory. Let us consider a lattice of atoms

---

[7] In fact, what we are about to derive corresponds to the leading terms of the so-called asymptotic expansion of the gamma function for $|z| \gg 1$.

assuming that each atom may have its magnetic moment directed either 'up' or 'down'. We shall also assume that only magnetic moments of nearest atoms interact via the so-called exchange interaction with z being the number of nearest neighbours for each lattice site. Let $n_\downarrow$ and $n_\uparrow$ be average densities of atoms with the moments directed up and down, respectively. We shall also introduce the order parameter $m = (n_\uparrow - n_\downarrow)/n$, which is the relative magnetisation of the solid and $n = n_\uparrow + n_\downarrow$ is the total number of atoms in the unit volume. Therefore, the densities of the atoms having moments up and down are given, respectively, by $n_\uparrow = n(1+m)/2$ and $n_\downarrow = n - n_\uparrow = n(1-m)/2$ via the order parameter, as is easily checked.

Our goal is to calculate the free energy density $F = U - TS$ of this lattice, where U is the internal energy density, T temperature and S the entropy density. We start by calculating the internal energy U. Let $N_{\uparrow\uparrow}$, $N_{\downarrow\downarrow}$ and $N_{\uparrow\downarrow}$ be the numbers of different pairs (per unit volume) of the *nearest* moments which are aligned, respectively, parallel up, parallel down or antiparallel to each other. It is more energetically favourable for the moments to be aligned in parallel, the energy, in this case, is $-J_0$; otherwise, the energy is increased by $2J_0$, i.e. it is equal to $+J_0$. Then on average one can write the following expression for the internal energy:

$$U = -J_0 \left(N_{\uparrow\uparrow} + N_{\downarrow\downarrow}\right) + J_0 N_{\uparrow\downarrow} - \mu_B H \left(n_\uparrow - n_\downarrow\right), \qquad (4.272)$$

where the last term corresponds to an additional contribution due to the applied magnetic field H along the direction of the moments, μ_B being Bohr magneton. To calculate the number density of pairs, we note that, for the given lattice site, the probability of finding a nearest site with the moment up is $p_\uparrow = n_\uparrow/n$. For each site with the moment up (whose density is $n_\uparrow$) there will be $zp_\uparrow$ nearest neighbours with the same direction of the moment, i.e.

$$N_{\uparrow\uparrow} = \frac{1}{2} n_\uparrow z p_\uparrow = \frac{z n_\uparrow^2}{2n} = \frac{1}{8} z n (1+m)^2,$$

where the factor of one half is required to avoid double counting of pairs. Similarly, one can calculate densities of pairs of atoms with other arrangements of their moments:

$$N_{\downarrow\downarrow} = \frac{1}{2} n_\downarrow z p_\downarrow = \frac{1}{8} z n (1-m)^2 \quad \text{and} \quad N_{\uparrow\downarrow} = n_\uparrow z p_\downarrow = \frac{1}{4} z n \left(1 - m^2\right).$$

Note that the factor of one half is missing in the last case. This is because, in this case, there is no double counting: we counted atoms with the moment up surrounded by those with the moment down.

The obtained expressions for the density of pairs yields for the internal energy density (4.272):

$$U = -\frac{1}{2} J_0 z n m^2 - m \mu_B n H. \qquad (4.273)$$

This formula expresses the internal energy entirely via the order parameter m.

The entropy density can be worked out from the well-known expression $S = k_B \ln W$, where k_B is the Boltzmann constant and W is the number of possibilities in which one can allocate $n_\uparrow$ moments up and $n_\downarrow$ moments down on a lattice with n sites. Obviously, $W = \binom{n}{n_\downarrow} = \binom{n}{n_\uparrow}$, and hence

$$S = k_B \ln \binom{n}{n_\uparrow} = k_B \ln \frac{n!}{n_\uparrow! \, (n - n_\uparrow)!}$$
$$= k_B \left[\ln(n!) - \ln(n_\uparrow!) - \ln((n - n_\uparrow)!) \right] .$$

To calculate the factorials above, we use the Stirling's formula (4.271) which gives:

$$
\begin{aligned}
S &\simeq k_B \left[(n \ln n - n) - (n_\uparrow \ln n_\uparrow - n_\uparrow) - ((n - n_\uparrow) \ln(n - n_\uparrow) - (n - n_\uparrow)) \right] \\
&= k_B \left[n \ln n - n_\uparrow \ln n_\uparrow - (n - n_\uparrow) \ln(n - n_\uparrow) \right] \\
&= k_B \left[n \ln n - n_\uparrow \ln n_\uparrow - n_\downarrow \ln n_\downarrow \right] \\
&= k_B n \left[\ln 2 - \frac{1 + m}{2} \ln(1 + m) - \frac{1 - m}{2} \ln(1 - m) \right] .
\end{aligned}
$$

(4.274)

A small algebraic manipulation is required when going from the third to the fourth line.

As we now have both components of the free energy, U and S, we can combine them to obtain the free energy density as

$$
\begin{aligned}
F = U - TS = &-\frac{1}{2} J_0 z n m^2 - m \mu_B n H \\
&- n k_B T \left[\ln 2 - \frac{1 + m}{2} \ln(1 + m) - \frac{1 - m}{2} \ln(1 - m) \right] .
\end{aligned}
$$

(4.275)

To calculate the magnetisation $M = n \mu_B m$ (per unit volume), we need to find the minimum of the free energy for the given value of H and T which should correspond to the stable phase at these parameters:

$$\frac{\partial F}{\partial m} = 0 \quad \Longrightarrow \quad -n k_B T_c m - \mu_B n H + \frac{1}{2} n k_B T \ln \frac{1 + m}{1 - m} = 0 , \qquad (4.276)$$

where T_c is defined by the identity $k_B T_c = J_0 z$. The above equation can be rearranged in the following way. If we denote $x = (\mu_B H + k_B T_c m) / k_B T$, then

$$\ln \frac{1 + m}{1 - m} = 2x \quad \Longrightarrow \quad m = \frac{e^{2x} - 1}{e^{2x} + 1} = \frac{e^x - e^{-x}}{e^x + e^{-x}} = \tanh(x)$$
$$= \tanh \left[\frac{1}{k_B T} (\mu_B H + k_B T_c m) \right] .$$

Fig. 4.7 Graphical solution of Eq. (4.277) in the case of zero magnetic field (for simplicity, we have used the units in which $n\mu_B = 1$)

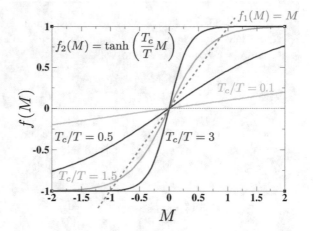

Therefore, the magnetisation is given by:

$$M = n\mu_B m = n\mu_B \tanh\left[\frac{\mu_B H}{k_B T} + \frac{T_c}{Tn\mu_B}M\right]. \qquad (4.277)$$

This is a transcendental equation for M which can be solved, e.g. graphically, by plotting $f_1(M) \sim M$ and $f_2(M) = \tanh(a + \lambda M)$ for different values of $\lambda = T_c/(Tn\mu_B)$ and $a = \mu_B H/k_B T$ (or the magnetic field H) and looking for intersection(s). Consider the simplest case of zero magnetic field ($a = 0$) which is shown in Fig. 4.7. Then it can be seen that Eq. (4.276) has a single solution $M = 0$ if $T > T_c$; however, for $T < T_c$ two additional solutions are possible corresponding to two symmetric minima in F, both giving non-zero magnetisation determined by Eq. (4.277). These correspond to a spontaneous magnetisation and the ferromagnetic phase. Correspondingly, T_c is a critical temperature at which the transition from the non-magnetic to ferromagnetic phase happens, i.e. it is the phase transition temperature. Note that the solution $M = 0$ also exists for $T < T_c$, but it corresponds to a higher free energy F and hence has to be omitted.

The situation is a bit more complex for a non-zero magnetic field: the two minima are no longer equivalent, and the system prefers to align the magnetic moment along the field. With decreasing T the magnetisation tends to the saturation limit at $m = \pm 1$.

Using an essentially identical argument, one can also consider statistics of the order–disorder phase transition in binary alloys.

590   4   Special Functions

Problem 4.93 The probability for a particle diffusing on a 1D lattice of equidistant lattice points to find itself at the site n after N steps is given by the formula (see Problem 1.196 in Vol. I):

$$P_N(n) = \frac{N!}{\left(\frac{1}{2}(N-n)\right)! \left(\frac{1}{2}(N+n)\right)!} p^{(N+n)/2} q^{(N-n)/2} , \qquad (4.278)$$

where p is the probability for the particle to jump one site to the right and $q = 1 - p$ to jump one site to the left. Show that after many steps (assuming $N \gg n \gg 1$) the probability becomes Gaussian,

$$\ln P_N(n) \approx C - \frac{n^2}{2N} \implies P_N(n) \propto e^{-n^2/2N} ,$$

with respect to n (above, C does not depend on n).

4.7.5   Band Structure of a Solid

It is well known from elementary quantum mechanics that electrons in a periodic lattice form energy bands, i.e. energies of the electrons can take values within some continuous intervals of energy called *bands*, and there could be gaps between such intervals (called energy gaps). We shall now discuss a famous model due to R. de L. Kronig and W. G. Penney which demonstrates the formation of bands in an extremely clear way.

Consider a one-dimensional metal in which positive ions are arranged periodically along the x-axis with a distance a between them as shown in Fig. 4.8a. Positive ions create a potential $V(x)$ for the electrons which is also periodic, i.e. $V(x + na) = V(x)$ for any integer n. This potential is also schematically shown in Fig. 4.8a. An

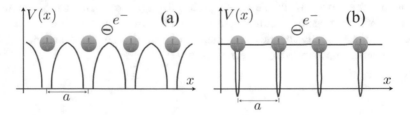

Fig. 4.8 One dimensional lattice of positive ion cores produces a potential $V(x)$ for the electrons (in blue); the potential goes to $-\infty$ at the positions of the nuclei. **a** A realistic potential; **b** a model potential in which each nucleus creates a delta function like potential

essential feature of $V(x)$, apart from its periodicity, is that it goes to $-\infty$ at the nuclei themselves, the minus sign is due to attraction between negative electrons and positive nuclei.

What we would like to do is to calculate the energy levels of the electrons in this solid by solving the corresponding Schrödinger equation:

$$-\frac{\hbar^2}{2m}\frac{d^2\psi_k(x)}{dx^2} + V(x)\psi_k(x) = E_k\psi_k(x) \,, \qquad (4.279)$$

where k is the electron momentum used to distinguish different states, and $\psi_k(x)$ and E_k are the corresponding electron wavefunction and energy. Because the lattice is periodic, each electron has a well defined momentum k serving as a good quantum number. Also, the wavefunction satisfies the so-called Bloch theorem, whereby $\psi_k(x) = e^{ikx}u_k(x)$, where $u_k(x)$ is a periodic function with the same period a as the lattice potential.

Following Kronig and Penney, we shall consider an extremely simplified nuclear potential which is simply a sum of delta functions placed at the nuclei:

$$V(x) = Aa \sum_{l=-\infty}^{\infty} \delta(x - la) \,, \qquad (4.280)$$

where $A < 0$ is some parameter. Here we sum over all atoms whose positions are given by la. So in this model negative singularities at atomic cores are correctly described, however, the potential everywhere between cores is zero instead of going smoothly between two minus infinity values at the nearest cores, compare Fig. 4.8a, b.

Since the functions $V(x)$ and $u_k(x)$ are periodic, we can expand them in an (exponential) Fourier series:

$$V(x) = \sum_g V_g e^{igx} \quad \text{and} \quad u_k(x) = \sum_g u_g(k)e^{igx} \,, \qquad (4.281)$$

where $g = (2\pi/a)\,j$ with $j = 0, \pm 1, \pm 2, \dots$ is the one-dimensional reciprocal 'lattice vector', and the summations are performed over all such possible values of g (in fact, over all integer j). The Fourier image V_g of the potential is given by

$$V_g = \frac{1}{a}\int_{-a/2}^{a/2} V(x)e^{-igx}dx = A\sum_l \int_{-a/2}^{a/2}\delta(x-la)e^{-igx}dx$$

$$= A\sum_l \delta_{l0}\left(e^{-igx}\right)_{x=0} = A \,.$$

Note that only the term $l = 0$ contributed to the integral. Thus, the delta-function lattice potential has all its Fourier components equal to each other. Now use the

Bloch theorem and then substitute the expansions (4.281) of the potential and of the wavefunction into the Schrödinger equation (4.279):

$$\sum_g \left[\frac{\hbar^2}{2m}(k+g)^2 + A \sum_{g'} e^{ig'x} \right] u_g(k) e^{i(k+g)x} = E_k \sum_g u_g(k) e^{i(k+g)x} . \quad (4.282)$$

Problem 4.94 Using the definition of the reciprocal one-dimensional vector g, prove that the plane waves e^{igx} are orthonormal on the lattice, i.e. they satisfy the following identity:

$$\frac{1}{a} \int_0^a e^{i(g-g')x} dx = \delta_{gg'} . \quad (4.283)$$

[Hint: consider separately the case of $g = g'$ and $g \neq g'$.]

Problem 4.95 By multiplying both sides of Eq. (4.282) with e^{-ig_1x} and integrating over x between 0 and a, obtain the following equations for the wavefunction Fourier components $u_g(k)$:

$$\left[\frac{\hbar^2}{2m}(k+g)^2 - E_k \right] u_g(k) + A \sum_{g'} u_{g-g'}(k) = 0 . \quad (4.284)$$

Note that the second term on the left-hand side can actually be written simply as $\sum_{g'} u_{g'}(k)$ since the summation there is performed over all reciprocal lattice vectors and hence the vector $g - g'$ will take all possible values anyway.

This equation can be solved exactly to obtain the energies E_k. The trick is to introduce a constant $\lambda = \sum_{g'} u_{g'}(k)$. Indeed, from (4.284)

$$u_g(k) = -\frac{A\lambda}{\frac{\hbar^2}{2m}(k+g)^2 - E_k} .$$

Summing both sides with respect to all g, we shall recognise λ on the left-hand side. Hence, cancelling on λ, we obtain the following exact equation for the energies:

$$-\frac{\hbar^2}{2mA} = \sum_g \frac{1}{(k+g)^2 - \frac{2mE_k}{\hbar^2}} . \quad (4.285)$$

The sum on the right-hand side is calculated analytically. Indeed, using the identity

$$\frac{1}{a^2 - b^2} = \frac{1}{2b}\left(\frac{1}{a-b} - \frac{1}{a+b} \right) ,$$

Eq. (4.285) is manipulated into the following form:

$$\frac{\pi \hbar}{Aa}\sqrt{\frac{8E_k}{m}} = \sum_{j=-\infty}^{\infty} \frac{1}{j+(d+c)/\pi} - \sum_{j=-\infty}^{\infty} \frac{1}{j+(d-c)/\pi} ,$$

where $d = ka/2$ and $c = (a/2\hbar)\sqrt{2mE_k}$. Next, we use Eq. (4.48) for the cotangent function as well as the trigonometric identity (I.2.66):

$$\frac{\pi \hbar}{Aa}\sqrt{\frac{8E_k}{m}} = \pi\left[\cot(d-c) - \cot(d+c)\right] = -\pi\frac{2\sin(2c)}{\cos(2d)-\cos(2c)} .$$

This finally gives:

$$\cos ka = \cos Ka + P\frac{\sin Ka}{Ka}, \tag{4.286}$$

where $P = mAa^2/\hbar^2$ and $K = 2c/a = \hbar^{-1}\sqrt{2mE_k}$, i.e. $E_k = \hbar^2 K^2/2m$. The obtained equation (4.286) fully solves the problem as it gives possible values of K for any k, i.e. the dependence $K(k)$, which in turn yields E_k for each k. Because of the cosine function on the left-hand side, the equation has real solutions only if the right-hand side of the equation is between -1 and 1. This condition restricts possible values of the wavevector K and hence of the energies E_k, and therefore results in bands. The function

$$f(Ka) = \cos(Ka) + P\frac{\sin(Ka)}{Ka}$$

for $P = -20$ is plotted in Fig. 4.9. The bands, i.e, the regions of allowed values of Ka, are coloured green in the figure. Solving Eq. (4.286) with respect to $K = K(k)$ for each given value of k allows calculating dispersion of energies E_k in each band.

Fig. 4.9 Plot of the function $f(z) = \cos z - 20\sin z/z$ versus $z = Ka$. The allowed regions of the parameter Ka for which the function lies between -1 and 1 are indicated by green regions. The first four energy bands are numbered

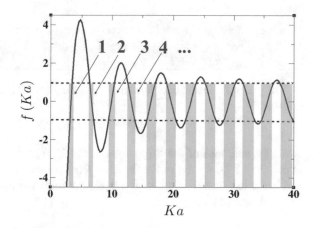

4.7.6 Oscillations of a Circular Membrane

Consider oscillations in a circular membrane of radius a which is fixed at its rim. The mathematical description of this problem is based on the theory of Bessel functions. The partial DE which needs to be solved in this case, assuming the membrane is positioned within the $x - y$ plane, has the form (see Chap. 8 for a more detailed discussion on partial DEs of mathematical physics):

$$\frac{1}{c^2} \frac{\partial^2 u}{\partial t^2} = \Delta u \implies \frac{1}{c^2} \frac{\partial^2 u}{\partial t^2} = \frac{1}{r} \frac{\partial}{\partial r} \left(r \frac{\partial u}{\partial r} \right) + \frac{1}{r^2} \frac{\partial^2 u}{\partial \phi^2}, \qquad (4.287)$$

where $u(r, \phi, t)$ is the vertical displacement of the membrane (along the z-axis) written in polar coordinates (r, ϕ), and c is the sound velocity. Correspondingly, we wrote the Laplacian on the right-hand side in the polar coordinates and discarded the $z-$dependent term, see Eq. (7.109).

To solve this equation, we apply again the method of separation of variables that has already been mentioned several times and to be discussed in more detail in Chap. 8.

Problem 4.96 Seeking the solution in the factorised form, $u(r, \phi, t) = R(r)\Phi(\phi)T(t)$, substitute it into the DE and show, assuming an oscillatory in time solution, that the three functions satisfy the following ordinary DEs:

$$T'' + \lambda^2 c^2 T = 0, \qquad (4.288)$$

$$\Phi'' + n^2 \Phi = 0, \qquad (4.289)$$

$$R'' + \frac{1}{r} R' + \left(\lambda^2 - \frac{n^2}{r^2} \right) R = 0, \qquad (4.290)$$

where $\lambda^2 > 0$ and n^2 are the corresponding separation constants. Then argue that the number n must be an integer (cf. discussion, e.g. in Sect. 4.5.3).

The general solution of equation (4.288) for $T(t)$ is given by (A and B are arbitrary constants)

$$T(t) = A \cos (\lambda c t) + B \cos (\lambda c t),$$

while the solutions of the $\Phi(\phi)$ equation (4.289) are periodic functions $e^{\pm in\phi}$. The equation (4.290) for the radial function $R(r)$ coincides with the Bessel DE (4.237), and hence its general solutions is a linear combination of the Bessel functions of the first and second kind:

$$R(r) = C_1 J_n (\lambda r) + C_2 K_n (\lambda r).$$

Since the K_n function diverges at $r = 0$, it has to be discarded as we expect the solution to be everywhere finite including the centre of the membrane. Hence, $R(r) = J_n(\lambda r)$ must be the only acceptable solution.

To find the appropriate values of the separation constant λ, we have to make sure that the solution we are seeking behaves appropriately at the boundary of the membrane, the so-called boundary conditions. Since the membrane is fixed at the boundary $r = a$, there is no vertical displacement on the rim of the membrane, and hence we have to set $R(a) = 0$. This means that $J_n(\lambda a) = 0$, and this equation may serve as the equation for the determination of the unknown separation constant λ. If $\mu_i^{(n)}$ are roots (numbered by the index $i = 1, 2, \ldots$) of the Bessel function for the given value of n, i.e. $J_n\left(\mu_i^{(n)}\right) = 0$, then $\lambda \Longrightarrow \lambda_i^{(n)} = \mu_i^{(n)}/a$. We know from Sect. 4.6.5 that there is an infinite number of roots $\mu_i^{(n)}$ of the equation $J_n(x) = 0$, and hence there is an infinite number of the possible values of the separation constant $\lambda_i^{(n)} = \mu_i^{(n)}/a$. Therefore, for any n and i (the root number) the products

$$\left[A_{ni} \cos\left(\lambda_i^{(n)}ct\right) + B_{ni} \sin\left(\lambda_i^{(n)}ct\right)\right] e^{\pm in\phi} J_n\left(\lambda_i^{(n)}r\right)$$

satisfy the partial DE (4.287) and the boundary conditions. Here A_{ni} and B_{ni} are arbitrary constants. By taking their linear combination, we can construct the general solution of the problem since the DE is linear:

$$
\begin{aligned}
u(r, \phi, t) = \sum_{n=0}^{\infty} \sum_{i=1}^{\infty} &\left\{\left[A_{ni} \cos\left(\lambda_i^{(n)}ct\right) + B_{ni} \sin\left(\lambda_i^{(n)}ct\right)\right] e^{in\phi}\right. \\
&\left. + \left[A_{ni}^* \cos\left(\lambda_i^{(n)}ct\right) + B_{ni}^* \sin\left(\lambda_i^{(n)}ct\right)\right] e^{-in\phi}\right\} J_n\left(\lambda_i^{(n)}r\right),
\end{aligned}
\tag{4.291}
$$

where the constants A_{ni} and B_{ni} are assumed to be generally complex. Note that the arbitrary constants by the sine and cosine functions of the $e^{-in\phi}$ part are complex conjugates of those we used with $e^{in\phi}$; this particular construction ensures that the function $u(r, \phi, t)$ is real. Also note that the summation over i corresponds to all the roots (zeros) of the Bessel function for the particular value of n in the first sum.

Formally, the above formula can be rewritten in a simpler form by extending the sum over n to run from $-\infty$ to ∞ and defining $A_{-ni} = A_{ni}^*$, $B_{-ni} = B_{ni}^*$ and $\lambda_i^{(-n)} = \lambda_i^{(n)}$:

$$u(r, \phi, t) = \sum_{n=-\infty}^{\infty} \sum_{i=1}^{\infty} \left[A_{ni} \cos\left(c\lambda_i^{(n)}t\right) + B_{ni} \sin\left(c\lambda_i^{(n)}t\right)\right] e^{in\phi} J_{|n|}\left(\lambda_i^{(n)}r\right).$$

$$\tag{4.292}$$

To determine the constants, we have to apply the initial conditions. In fact, the problem can be solved for very general initial conditions:

$$u(r, \phi, t)|_{t=0} = f(r, \phi) \quad \text{and} \quad \left.\frac{\partial u(r, \phi, t)}{\partial t}\right|_{t=0} = \varphi(r, \phi). \tag{4.293}$$

Applying $t = 0$ to the function (4.292) and its time derivative, these conditions are transformed into the following two equations:

$$u|_{t=0} = \sum_{n=-\infty}^{\infty} \sum_{i=1}^{\infty} A_{ni} e^{in\phi} J_{|n|}\left(\lambda_i^{(n)} r\right) = f(r, \phi) , \qquad (4.294)$$

$$\frac{\partial u}{\partial t}\bigg|_{t=0} = \sum_{n=-\infty}^{\infty} \sum_{i=1}^{\infty} c\lambda_i^{(n)} B_{ni} e^{in\phi} J_{|n|}\left(\lambda_i^{(n)} r\right) = \varphi(r, \phi) . \qquad (4.295)$$

The complex coefficients A_{ni} and B_{ni} contain two indices and correspondingly are under double sum. They can be found then in two steps.

Problem 4.97 Using the orthogonality of the exponential functions,

$$\int_0^{2\pi} e^{i(n-m)\phi} d\phi = 2\pi \delta_{nm} ,$$

multiply both sides of the above equations (4.294) and (4.295) by $e^{-im\phi}$ and then integrate over the angle ϕ to show that

$$\frac{1}{2\pi} \int_0^{2\pi} f(r, \phi) e^{-in\phi} d\phi = \sum_{i=1}^{\infty} A_{ni} J_{|n|}\left(\lambda_i^{(n)} r\right) , \qquad (4.296)$$

$$\frac{1}{2\pi} \int_0^{2\pi} \varphi(r, \phi) e^{-in\phi} d\phi = \sum_{i=1}^{\infty} c\lambda_i^{(n)} B_{ni} J_{|n|}\left(\lambda_i^{(n)} r\right) . \qquad (4.297)$$

At the second step, we recall that the Bessel functions $J_{|n|}\left(\lambda_i^{(n)} r\right)$ also form an orthogonal set, see Eq. (4.244), so that coefficients in the expansion of any function in Eq. (4.245) can be found from Eq. (4.246). Therefore, we can finally write

$$A_{ni} = \frac{1}{2\pi J_{ni}} \int_0^a r dr \int_0^{2\pi} d\phi e^{-in\phi} f(r, \phi) J_{|n|}\left(\lambda_i^{(n)} r\right) , \qquad (4.298)$$

$$B_{ni} = \frac{1}{2\pi c\lambda_i^{(n)} J_{ni}} \int_0^a r dr \int_0^{2\pi} d\phi e^{-in\phi} f(r, \phi) J_{|n|}\left(\lambda_i^{(n)} r\right) , \qquad (4.299)$$

where

$$J_{ni} = \int_0^a r\left[J_{|n|}\left(\lambda_i^{(n)} r\right)\right]^2 dr \qquad (4.300)$$

is the normalisation integral which we calculated in Eq. (4.247). These equations fully solve our problem.

In a similar way one can solve heat transport problem in cylindrical coordinates, as well as vibrational problems. In both cases Bessel functions appear.

4.7.7 Multipole Expansion of the Electrostatic Potential

Consider an electrostatic potential at point $\mathbf{r}$ created by some charge density distribution $\rho(\mathbf{r})$. A little volume $d\mathbf{r}'$ inside the distribution, see Fig. 4.10, contains the charge $dq = \rho(\mathbf{r}')\, d\mathbf{r}'$ which is a source of the electrostatic potential $dq/\left|\mathbf{r} - \mathbf{r}'\right|$ at the 'observation' point $\mathbf{r}$. The latter point may be inside the distribution as shown in the figure, or outside it. The total potential at the point $\mathbf{r}$ is given by the integral

$$U(\mathbf{r}) = \int_V \frac{\rho(\mathbf{r}')}{|\mathbf{r} - \mathbf{r}'|}\, d\mathbf{r}' \, , \qquad (4.301)$$

taken over the whole volume V where the charge distribution is non-zero. Note that if we have a collection of point charges q_i located at points $\mathbf{r}_i$, then the charge density due to a single charge q_i is given by $q_i \delta(\mathbf{r} - \mathbf{r}_i)$, where

$$\delta(\mathbf{r} - \mathbf{r}_i) = \delta(x - x_i)\, \delta(y - y_i)\, \delta(z - z_i)$$

is a three-dimensional Dirac delta function. Indeed, this definition makes perfect sense: it is only non-zero at the position $\mathbf{r}_i$ where the charge is and integrates exactly to q_i:

$$\int q_i \delta(\mathbf{r} - \mathbf{r}_i)\, d\mathbf{r} = q_i \int \delta(x - x_i)\, dx \int \delta(y - y_i)\, dy \int \delta(z - z_i)\, dz = q_i \, .$$

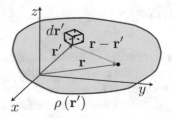

Fig. 4.10 A small volume $d\mathbf{r}'$ within a charge density distribution $\rho(\mathbf{r}')$ creates an electrostatic potential at point $\mathbf{r}$. Note that the 'observation' point $\mathbf{r}$ may in general be either inside (as in the figure) or outside the distribution

The filtering theorem is also valid for this delta function as the volume integral can always be split into a sequence of three one-dimensional integrals, and the filtering theorem can be applied to each of them one after another:

$$
\begin{aligned}
\int f\left(\mathbf{r}'\right) \delta\left(\mathbf{r} - \mathbf{r}'\right) d\mathbf{r}' &= \int \delta\left(x - x'\right) dx' \int \delta\left(y - y'\right) dy' \int f\left(x', y', z'\right) \delta\left(z - z'\right) dz' \\
&= \int \delta\left(x - x'\right) dx' \int \delta\left(y - y'\right) f\left(x', y', z\right) dy' \\
&= \int \delta\left(x - x'\right) f\left(x', y, z\right) dx' = f\left(x, y, z\right) = f\left(\mathbf{r}\right) ,
\end{aligned}
$$

where $f(\mathbf{r})$ is an arbitrary continuous function.

Therefore, for a collection of point charges

$$
\rho(\mathbf{r}) = \sum_i q_i \delta\left(\mathbf{r} - \mathbf{r}_i\right) ,
$$

and the potential becomes

$$
U_{charges} = \int_V \frac{\rho\left(\mathbf{r}'\right)}{|\mathbf{r} - \mathbf{r}'|} d\mathbf{r}' = \sum_i q_i \int_V \frac{\delta\left(\mathbf{r}' - \mathbf{r}_i\right)}{|\mathbf{r} - \mathbf{r}'|} d\mathbf{r}' = \sum_i \frac{q_i}{|\mathbf{r} - \mathbf{r}_i|} ,
$$

as expected.

So, Formula (4.301) is general: it can be used both for continuous and discrete charge distributions. We shall now derive the so-called multipole expansion of the potential $U(\mathbf{r})$. Let the angle between the vectors $\mathbf{r}$ and $\mathbf{r}'$ be θ. Then, using the cosine theorem, we can write

$$
\left|\mathbf{r} - \mathbf{r}'\right| = \sqrt{r^2 + r'^2 - 2rr' \cos \theta} .
$$

The volume integration in (4.301) is convenient to perform in spherical coordinates (r, θ, ϕ):

$$
U(\mathbf{r}) = \int_0^\infty r'^2 dr' \int d\Omega \frac{\rho\left(\mathbf{r}'\right)}{\sqrt{r^2 + r'^2 - 2rr' \cos \theta}} ,
$$

where $d\Omega = \sin \theta d\theta d\phi$ is an element of the solid angle (which integrates to 4π); this is just a shorter, fully equivalent, way of writing the integration in spherical coordinates. The integration over r' is formally performed up to infinity; in practice, it is terminated if the charge distribution occupies a finite volume. Now, we shall divide the r' integration into two parts: $0 \leq r' \leq r$ and $r \leq r' < \infty$, and then expand the integrand into a power series exploiting the generating function (4.53) for the Legendre polynomials. In the case of $r'/r < 1$, we write

$$\frac{1}{\sqrt{r^2 + r'^2 - 2rr'\cos\theta}} = \frac{1}{r}\frac{1}{\sqrt{1 + (r'/r)^2 - 2(r'/r)\cos\theta}}$$

$$= \frac{1}{r}\sum_{l=0}^{\infty} P_l(\cos\theta)\left(\frac{r'}{r}\right)^l,$$

while, for the case of $r/r' < 1$, we similarly have

$$\frac{1}{\sqrt{r^2 + r'^2 - 2rr'\cos\theta}} = \frac{1}{r'}\frac{1}{\sqrt{1 + (r/r')^2 - 2(r/r')\cos\theta}}$$

$$= \frac{1}{r'}\sum_{l=0}^{\infty} P_l(\cos\theta)\left(\frac{r}{r'}\right)^l.$$

Next we shall expand the Legendre polynomial into spherical functions using Eq. (4.202) which yields the following: expression for the potential:

$$U(\mathbf{r}) = \sum_{l=0}^{\infty}\sum_{m=-l}^{l}\frac{4\pi}{2l+1}Y_l^m(\theta,\phi)^*\left[\frac{1}{r^{l+1}}\int_0^r (r')^{l+2}\, dr'\int d\Omega\,\rho(\mathbf{r}')\,Y_l^m(\theta',\phi')\right.$$

$$\left. + r^l\int_r^{\infty}(r')^{-l+1}\, dr'\int d\Omega\,\rho(\mathbf{r}')\,Y_l^m(\theta',\phi')\right],$$

(4.302)

which is the required general formula. Here $\mathbf{r} = (r, \theta, \phi)$ and $\mathbf{r}' = (r', \theta', \phi')$ are the two points (vectors) in the spherical coordinates. Note that complex spherical harmonics were used here; however, since the expansion formula (4.202) for the Legendre polynomials is invariant under the substitution of the complex harmonics Y_l^m with the real ones, $Y_l^m \rightarrow S_l^m$ (see Problem 4.68), one can also use the real harmonics in the formula above.

We shall now consider a particular case of a potential outside a *confined* charge distribution. In this case, we assume that there exists such $a = \max_V(r')$ that $\rho(\mathbf{r}') = 0$ for $r' > a$. Then the potential outside the charge distribution (i.e. for $r > a$) would only contain the first integral of Eq. (4.302) in which the r' integration is done between 0 and a:

$$U(\mathbf{r}) = \sum_{lm}\frac{4\pi}{2l+1}Y_l^m(\theta,\phi)^*\frac{1}{r^{l+1}}\int_0^a (r')^{l+2}\, dr'\int d\Omega\,\rho(\mathbf{r}')\,Y_l^m(\theta',\phi')$$

(4.303)

$$= \sum_{lm}\sqrt{\frac{4\pi}{2l+1}}Y_l^m(\theta,\phi)^*\frac{Q_{lm}}{r^{l+1}},$$

where we introduced *multipole moments* of the charge distribution:

$$Q_{lm} = \sqrt{\frac{4\pi}{2l+1}} \int_0^a (r')^{l+2} dr' \int d\Omega\, \rho\left(\mathbf{r}'\right) Y_l^m\left(\theta', \phi'\right)$$

$$= \sqrt{\frac{4\pi}{2l+1}} \int_V \rho\left(\mathbf{r}'\right) Y_l^m\left(\theta', \phi'\right) \left(r'\right)^l d\mathbf{r}' \tag{4.304}$$

with $d\mathbf{r}' = \left(r'\right)^2 dr' d\Omega$. The moments Q_{lm} depend entirely on the charge distribution and are therefore its intrinsic property.

Expression (4.303) is the required multipole expansion. It shows that the potential outside the charge distribution can be expanded into a series with respect to inverse powers of the distance measured from the 'centre' of the distribution (where the centre of the coordinate system was chosen):

$$U(\mathbf{r}) = \frac{A}{r} + \frac{B}{r^2} + \frac{C}{r^3} + \ldots = U_1(\mathbf{r}) + U_2(\mathbf{r}) + U_3(\mathbf{r}) + \ldots\ .$$

We expect that A must be the total charge of the distribution, B is related to its dipole moment, C to its quadrupole moment, and so on. For instance, consider the $l = m = 0$ term:

$$Q_{00} = \sqrt{\frac{4\pi}{1}} \int_V \rho\left(\mathbf{r}'\right) \frac{1}{\sqrt{4\pi}} d\mathbf{r}' = \int_V \rho\left(\mathbf{r}'\right) d\mathbf{r}'\ ,$$

which is the total charge Q of the distribution, and the corresponding term in the potential, $U_0(\mathbf{r}) = Q/r$, does indeed have the correct form. We have used here the fact that $Y_0^0 = 1/\sqrt{4\pi}$ (see Sect. 4.5.3.4).

Problem 4.98 Similarly, show that the $l = 1$ term in the expansion (4.303) corresponds to the dipole term:

$$U_1(\mathbf{r}) = \frac{\mathbf{P} \cdot \mathbf{n}}{r^2}\ ,$$

where $\mathbf{n} = \mathbf{r}/r$ is the unit vector in the direction $\mathbf{r}$, and

$$\mathbf{P} = \int_V \rho\left(\mathbf{r}'\right) \mathbf{r}' d\mathbf{r}'$$

is the dipole moment of the charge distribution. [Hint: use real spherical harmonics and explicit expressions for them from Eq. (4.198).]

Problem 4.99 Also show that the $l = 2$ term is associated with the quadrupole contribution

$$U_2(\mathbf{r}) = \frac{1}{2r^3} \sum_{\alpha,\beta=1}^{3} D_{\alpha\beta} n_\alpha n_\beta \,,$$

where summation is performed over three Cartesian components and

$$D_{\alpha\beta} = \int_V \rho\left(\mathbf{r}'\right) \left(3r'_\alpha r'_\beta - \delta_{\alpha\beta} r'^2\right) d\mathbf{r}'$$

is the quadrupole moment matrix. *[Hint: use real spherical harmonics (4.199) and the fact that $D_{xx} + D_{yy} + D_{zz} = 0$.]*

Note that the quadrupole matrix is symmetric and contains only five independent elements (since the diagonal elements sum up to zero); this is not surprising as there are only five spherical harmonics of $l = 2$.

4.8   Van Hove Singularities

One of the important characteristics of the spectra of electrons in a crystal[8] is their *density of states* (DOS), $D(\epsilon)$. The DOS shows the number of electronic states $dN(\epsilon) = D(\epsilon)d\epsilon$ contained between the energies ϵ and $\epsilon + d\epsilon$.

Let us consider an electron gas in a metal where electronic energies $\epsilon_\mathbf{k}$ are numbered by the so-called wavevector $\mathbf{k}$. We shall consider a single continuous stretch of these energies called a band. Then the DOS of this band can be written as a volume integral

$$D(\epsilon) = \frac{V}{(2\pi)^3} \int \delta\left(\epsilon - \epsilon_\mathbf{k}\right) d\mathbf{k} \,, \tag{4.305}$$

where the 3D integration is performed over all vectors $\mathbf{k}$ (they form the primitive cell in the so-called reciprocal space although this is not essential for us here) and V is the crystal volume. It appears (due to Leon van Hove) that the DOS has a special form in the vicinity of special points in which the gradient of the energy, $\nabla \epsilon_\mathbf{k}$ (calculated in the $\mathbf{k}$ space), is equal to zero. These are called van Hove singularities. We shall briefly consider them here.

Let us start from the contribution to the DOS due to the region in the $\mathbf{k}$ space corresponding to the top of the band. If the top of the band happens at the wavevector $\mathbf{k}_t$, then the energies in the immediate vicinity of this point can be approximated as

---

[8] Vibrations in a crystal can also be characterised in the same way; however, here we shall only consider electrons.

$$\epsilon_{\mathbf{k}} = \epsilon_t - \sum_{\alpha,\beta=1}^{3} A_{\alpha\beta} \left(k_\alpha - k_{t\alpha}\right) \left(k_\beta - k_{t\beta}\right),$$

where $\epsilon_t = \epsilon_{\mathbf{k}_t}$ is the energy at the top of the band and the coefficients $A_{\alpha\beta}$ form a 3×3 positively definite matrix $\mathbf{A}$ (all its eigenvalues are positive). Since we are going to integrate over $\mathbf{k}$, we can always choose the coordinate system defining the $\mathbf{k}$ vectors such that the matrix $\mathbf{A}$ be diagonal. Moreover, one can always rescale the $\mathbf{k}$ vectors components as well and shift the coordinate system by $\mathbf{k}_t$. Hence, it is always possible to ensure that the energy has a simpler form:

$$\epsilon_{\mathbf{k}} = \epsilon_t - A\mathbf{k}^2, \quad A > 0.$$

To see how the DOS behaves near the top of the band, we need to consider the integral

$$D(\epsilon) = \frac{V}{(2\pi)^3} \int \delta\left(\epsilon - \epsilon_t + A\mathbf{k}^2\right) d\mathbf{k}.$$

Using spherical coordinates, and integrating over the angles (the delta function only depends on the length k of the vector $\mathbf{k}$), we obtain

$$D(\epsilon) = \frac{4\pi V}{(2\pi)^3} \int \delta\left(\epsilon - \epsilon_t + Ak^2\right) k^2 dk = \frac{V}{2\pi^2 A} \int \delta\left(k^2 - a^2\right) k^2 dk,$$

where $a = \sqrt{(\epsilon_t - \epsilon)/A}$. Next we shall use Eq. (4.13) for the delta function and recognise that $k > 0$ (and hence only the delta function $\delta(k - a)$ will contribute), leading us to

$$D(\epsilon) = \frac{V}{2\pi^2 A} \int \frac{1}{2a} [\delta(k - a) + \delta(k + a)] k^2 dk \sim \frac{1}{a} \int \delta(k - a) k^2 dk$$

$$= \frac{1}{a} a^2 \sim a \sim \sqrt{\epsilon_t - \epsilon}.$$

Note that for energies ϵ close to the top of the band, the value of a is small and hence the corresponding value of $k = a$ can always be found when integrating over the vicinity of $\mathbf{k}_t$. Hence, near the top of the band, the DOS behaves like $B\sqrt{\epsilon_t - \epsilon}$, where B is a positive constant, see Fig. 4.11a.

Problem 4.100 Show that DOS near the bottom of the band at ϵ_b behaves like $C\sqrt{\epsilon - \epsilon_b}$, see Fig. 4.11b.

We shall next consider a more interesting case of a saddle point $\mathbf{k}_s$ of energy ϵ_s, in the vicinity of which the energies are (after an appropriate choice of the coordinate system)

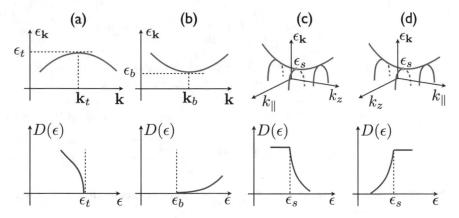

Fig. 4.11 Four types of van Hove singularities: **a** top and **b** bottom of a band, and **c, d** two types of the saddle point

$$\epsilon_{\mathbf{k}} = \epsilon_s - A\left(k_x^2 + k_y^2\right) + Bk_z^2, \quad A > 0 \text{ and } B > 0. \tag{4.306}$$

(This is indeed a saddle as the energy in the $\mathbf{k}$ space has the maximum along the k_x and k_y directions and the minimum in the k_z direction.)

In this case, one has to consider separately two cases: the energies on the left of the saddle point energy, $\epsilon < \epsilon_s$, and on the right, $\epsilon > \epsilon_s$. Let us start from the former case. Using the cylindrical coordinates $(k_\parallel, \phi, k_z)$, integrating over the angles, dropping an unimportant pre-factor, and assuming the $k_\parallel$ integration first, we obtain

$$D(\epsilon) \sim \int_{-\delta_z}^{\delta_z} dk_z \int_0^\delta k_\parallel \delta\left(k_\parallel^2 - a^2\right) dk_\parallel,$$

where $a = \sqrt{\left(\epsilon_s - \epsilon + Bk_z^2\right)/A}$ is assumed to be small ensuring both integrations to give a non-zero result, and for clarity, we indicated the integration limits (with δ and δ_z being sufficiently small). Performing the integration using the same method as above, we obtain a constant.

Consider now the other case when $\epsilon > \epsilon_s$. Here it is more convenient to perform the k_z integration first:

$$D(\epsilon) \sim \int_0^\delta k_\parallel dk_\parallel \int_{-\delta_z}^{\delta_z} \delta\left(k_z^2 - a^2\right) dk_z,$$

where this time $a = \sqrt{\left(\epsilon - \epsilon_s + Ak_\parallel^2\right)/B}$. In this case, both delta functions $\delta(k_z \pm a)$ contribute, and we obtain

$$D(\epsilon) \sim \int_0^{\delta} \frac{k_{\parallel}}{2a} dk_{\parallel} \int_{-\delta_z}^{\delta_z} \left[\delta \left(k_z - a \right) + \delta \left(k_z + a \right) \right] dk_z$$

$$= \int_0^{\delta} \frac{k_{\parallel}}{a} dk_{\parallel} \sim \int_0^{\delta} \frac{k_{\parallel}}{\sqrt{\epsilon - \epsilon_s + A k_{\parallel}^2}} dk_{\parallel} \, .$$

This integral is easily calculated using the substitution $t = \epsilon - \epsilon_s + A k_{\parallel}^2$, and we obtain

$$D(\epsilon) \sim \sqrt{\delta^2 + \lambda} - \sqrt{\lambda}, \quad \lambda = \sqrt{\frac{\epsilon - \epsilon_s}{A}} \, .$$

The quantity λ is to be considered small, and hence can be dropped in the first term. As a result, we obtain $D(\epsilon) \sim \delta - A^{-1/2} \sqrt{\epsilon - \epsilon_s}$.

Hence, we conclude that in the case of the saddle point (4.306) the DOS in the vicinity of it must be:

$$D(\epsilon) = \begin{cases} D(\epsilon_s) & , \quad \epsilon < \epsilon_s \\ D(\epsilon_s) - \gamma \sqrt{\epsilon - \epsilon_s} & , \quad \epsilon > \epsilon_s \end{cases} ,$$

where $\gamma > 0$ is a constant. This case is schematically illustrated in Fig. 4.11c.

Problem 4.101 Show that if the saddle point is given via (after an appropriate choice of the coordinate system)

$$\epsilon_{\mathbf{k}} = \epsilon_s + A \left(k_x^2 + k_y^2 \right) - B k_z^2, \quad A > 0 \text{ and } B > 0 \, ,$$

the DOS in the vicinity of ϵ_s is (see Fig. 4.11d)

$$D(\epsilon) = \begin{cases} D(\epsilon_s) - \gamma \sqrt{\epsilon - \epsilon_s} & , \quad \epsilon < \epsilon_s \\ D(\epsilon_s) & , \quad \epsilon > \epsilon_s \end{cases} .$$

Chapter 5
Fourier Transform

We know from Chap. 3 that any piecewise continuous *periodic* function $f(x)$ can be expanded into a Fourier series.[1] One may ask if a similar expansion can be constructed for a function which is *not* periodic? The purpose of this chapter is to address this very question. We shall show that for non-periodic functions $f(x)$ an analogous representation of the function exists, but in this case the Fourier series is replaced by an integral, called Fourier integral. This development enables us to go even further and introduce a concept of an integral transform. The definition and various properties of the Fourier integral and Fourier transform are given, with various applications in physics appearing at the end of the chapter.

5.1 The Fourier Integral

5.1.1 Intuitive Approach

This will be done using a *Fourier integral*. We begin by considering an arbitrary piecewise smooth *non-periodic* function $f(x)$, e.g. the one shown in Fig. 5.1. The driving idea here is to cut out a central piece of $f(x)$ between two symmetric points $-T/2$ and $T/2$ (T is a positive number), and then repeat this piece of $f(x)$ periodically in both directions whereby introducing an associated *periodic* function $f_T(x)$, with a fundamental period T, shown in Fig. 5.2:

$$f_T(x) = f(x) \quad \text{for} \quad \frac{-T}{2} \leq x < \frac{T}{2} , \tag{5.1}$$

---

[1] In the following, references to the first volume of this course (L. Kantorovich, Mathematics for natural scientists: fundamentals and basics, 2nd Edition, Springer, 2022) will be made by appending the Roman number I in front of the reference, e.g. Sect. I.1.8 or Eq. (I.5.18) refer to Sect. 1.8 and Eq. (5.18) of the first volume, respectively.

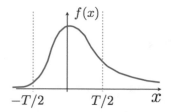

Fig. 5.1 An example of a non-periodic function $f(x)$

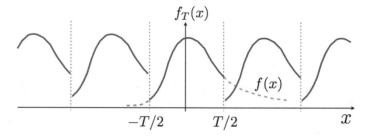

Fig. 5.2 The non-periodic function $f(x)$ shown in Fig. 5.1 is compared with its periodic (with the period T) approximation $f_T(x)$ (the solid line). The part of $f(x)$ which was cut out in $f_T(x)$ is shown with the dashed line

while the expression $f_T(x + T) = f_T(x)$ extends this approximation to the entire x axis. It is clear from Eq. (5.1) and Fig. 5.2 that the periodic function $f_T(x)$ will provide an exact representation of $f(x)$ within the interval $-T/2 \le x < T/2$; beyond this interval these two functions are obviously different. However, the two functions should become identical in the limit of $T \to \infty$, i.e. we should have

$$f(x) = \lim_{T \to \infty} f_T(x) \quad \text{for} \quad -\infty < x < \infty . \tag{5.2}$$

The function $f_T(x)$ is periodic and hence can be expanded into the Fourier series. Using the exponential form of the series, Sect. 3.7, we can write

$$f_T(x) = \sum_{n=-\infty}^{\infty} F_T(\nu_n)e^{i2\pi\nu_n x} \Delta\nu , \tag{5.3}$$

where $\nu_n = n/T$, $\Delta\nu = \nu_{n+1} - \nu_n = 1/T$, and the Fourier coefficients are

$$F_T(\nu_n) = \int_{-T/2}^{T/2} f_T(x)e^{-i2\pi\nu_n x}dx . \tag{5.4}$$

In the $T \to \infty$ limit $\Delta\nu = 1/T \to 0$ and the summation in (5.3) becomes an integral sum, i.e. the sum tends to a definite integral in the right-hand side, while $f_T(x) \to f(x)$ in the left-hand side:

$$f(x) = \int_{-\infty}^{\infty} F(\nu) e^{i2\pi\nu x} d\nu , \tag{5.5}$$

where $F(\nu) = \lim_{T\to\infty} F_T(\nu)$. At the same time, in the $T \to \infty$ limit the integral in the right-hand side of the Fourier coefficients (5.4) extends to the whole x axis, while in the integrand $f_T(x) \to f(x)$, i.e. we can also write

$$F(\nu) = \int_{-\infty}^{\infty} f(x) e^{-i2\pi\nu x} dx . \tag{5.6}$$

Combining formulae (5.5) and (5.6), one obtains

$$\begin{aligned} f(x) &= \int_{-\infty}^{\infty} \left[\int_{-\infty}^{\infty} f(t) e^{-i2\pi\nu t} dt \right] e^{i2\pi\nu x} d\nu \\ &= \int_{-\infty}^{\infty} d\nu \int_{-\infty}^{\infty} dt \, f(t) e^{-i2\pi\nu(t-x)} . \end{aligned} \tag{5.7}$$

This very important result is known as the Fourier integral representation for the *non-periodic* function $f(x)$. After calculating the integral over t (if possible), we shall arrive at a representation of $f(x)$ via a single integral over ν, Eq. (5.5). Therefore, it is seen from (5.7) that if we wish to represent a *non-periodic* function $f(x)$ over the infinite range $-\infty < x < \infty$, then the Fourier *sum* over the *discrete* frequency spectrum $\{\nu_n; n = 0, \pm 1, \pm 2, ...\}$ must be replaced by an *integral* over a *continuous* frequency spectrum ν.

The discussion which led us to Eqs. (5.5)–(5.7) was mostly based on intuition; it was not very rigorous. It can be proved that the Fourier integral representation (5.7) is valid for all *piecewise smooth* non-periodic functions $f(x)$, provided that the integral $\int_{-\infty}^{\infty} |f(x)| dx$ exists. If the function $f(x)$ has a *discontinuity* at $x = x_1$ then the Fourier integral (5.7) will give the *mean value* of $f(x)$ at $x = x_1$, i.e. $\frac{1}{2}[f(x_1 - 0) + f(x_1 + 0)]$. But before we provide a more rigorous discussion of the Fourier integral, let us first consider its other alternative forms.

5.1.2 Alternative Forms of the Fourier Integral

Let us split the integral over ν in Eq. (5.7) into two: one between $-\infty$ and 0, and another between 0 and $+\infty$. In the first integral we shall then make a change of variables $\nu \to -\nu$. This gives

$$\begin{aligned} f(x) &= \int_{+\infty}^{0} (-d\nu) \int_{-\infty}^{\infty} f(t) e^{i2\pi\nu(t-x)} dt + \int_{0}^{\infty} d\nu \int_{-\infty}^{\infty} f(t) e^{-i2\pi\nu(t-x)} dt \\ &= \int_{0}^{\infty} d\nu \int_{-\infty}^{\infty} f(t) \left[e^{-i2\pi\nu(t-x)} + e^{i2\pi\nu(t-x)} \right] dt . \end{aligned} \tag{5.8}$$

The expression in the square brackets is recognised to be the two times cosine, and we finally obtain the so-called *trigonometric* form of the Fourier integral:

$$f(x) = 2 \int_0^\infty d\nu \int_{-\infty}^\infty f(t) \cos[2\pi\nu(t - x)]\, dt \ . \tag{5.9}$$

Problem 5.1 Show that if $f(x)$ is an *even* function $f_e(x)$, then Eq. (5.7) is simplified further:

$$f_e(x) = 4 \int_0^\infty \left[\int_0^\infty f_e(t) \cos(2\pi\nu t)dt\right] \cos(2\pi\nu x)d\nu \ , \tag{5.10}$$

which is known as the Fourier cosine integral for $f_e(x)$.

Problem 5.2 Similarly, show that for an *odd* function $f_o(x)$ we have

$$f_o(x) = 4 \int_0^\infty \left[\int_0^\infty f_o(t) \sin(2\pi\nu t)dt\right] \sin(2\pi\nu x)d\nu \ , \tag{5.11}$$

which is called Fourier sine integral.

Problem 5.3 It is also possible to use the Fourier cosine and sine integrals to represent a *general* function $f(x)$ which is only defined on the half of the real axis, e.g. for $0 \le x < \infty$ (i.e. $f(x) = 0$ outside this range). Show that in this case the following formula is valid:

$$f(x) = 2 \int_0^\infty d\nu \int_0^\infty dt\, f(t) \cos(2\pi\nu(x - t)) \ . \tag{5.12}$$

As an example, let us obtain the Fourier integral representation of the function $\Pi(x)$ which is equal to 1 for $-1 \le x \le 1$ and zero otherwise. In this particular case the function $\Pi(x)$ is *even*, so that we can write the Fourier integral in the form of Eq. (5.10). Hence we find

$$\Pi(x) = 4 \int_0^\infty \left[\int_0^1 \cos(2\pi\nu t)dt\right] \cos(2\pi\nu x)d\nu$$
$$= 4 \int_0^\infty \frac{\sin(2\pi\nu)}{2\pi\nu} \cos(2\pi\nu x)d\nu \ . \tag{5.13}$$

This is the required integral representation of the function $\Pi(x)$. When $x = \pm1$, where the original function $\Pi(x)$ makes a jump, the Fourier integral (5.13) gives the mean value of $\frac{1}{2}$.

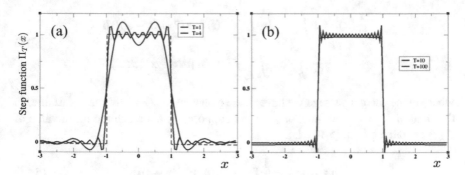

Fig. 5.3 Fourier integral for $\Pi(x)$ calculated using the spectroscopic range $0 \leq \nu \leq T$ for **a** $T = 1$ and $T = 4$, and **b** $T = 10$ and $T = 100$. The original function $\Pi(x)$ is shown in (**a**) by the dashed line

---

Problem 5.4 Show this using Eq. (2.127). Also show that at $x = 0$ the integral is equal to one, as required.

Let us try to understand how this integral representation actually works. To this end let us consider an approximation

$$\Pi_T(x) = 4 \int_0^T \frac{\sin(2\pi\nu)}{2\pi\nu} \cos(2\pi\nu x) d\nu$$

for the integral (5.13), where integration is performed not up to infinity, but up to some finite real number T. The results of a numerical integration for $T = 1, 4, 10$ and 100 shown in Fig. 5.3 clearly demonstrate convergence of the function $\Pi_T(x)$ to the exact function $\Pi(t)$ as the value of T (the upper limit in the integral) is increased.

5.1.3 A More Rigorous Derivation of Fourier Integral

The discussion which led us to Eqs. (5.7) or (5.9), which are absolutely equivalent, was mostly intuitive. A more rigorous derivation leading to either of the forms of the Fourier integral requires a more careful analysis. Let us aim to prove specifically Eq. (5.9). This basically entails proving the following formula:

$$2 \int_0^\infty d\nu \int_{-\infty}^\infty f(t) \cos\left[2\pi\nu(t - x)\right] dt = \frac{1}{2} [f(x - 0) + f(x + 0)] . \quad (5.14)$$

To this end, let us consider the following function

$$g_\lambda(x) = 2 \int_0^{\lambda/2\pi} d\nu \int_{-\infty}^{\infty} f(t) \cos\left[2\pi\nu(t-x)\right] dt$$

$$= \frac{1}{\pi} \int_0^\lambda d\nu \int_{-\infty}^{\infty} f(t) \cos\left[\nu(t-x)\right] dt \,, \tag{5.15}$$

where in the second passage we have made a change $2\pi\nu \to \nu$ of the ν variable. In the limit of $\lambda \to \infty$ the function $g_\lambda(x)$ is supposed to approach the mean value of $f(x)$ as stated by Eq. (5.14), i.e.

$$\lim_{\lambda \to \infty} g_\lambda(x) = \frac{1}{2} \left[f(x-0) + f(x+0) \right] \,. \tag{5.16}$$

This is the result we aim to prove.

We assume that $f(x)$ satisfies Dirichlet conditions and that the integral $\int_{-\infty}^{\infty} |f(t)|$ $dt < \infty$, i.e. it exists. Then, one notices that the t integral in (5.15) converges absolutely and uniformly for any ν since it can be estimated by the integral which (by our assumption) converges

$$\left| \int_{-\infty}^{\infty} f(t) \cos\left[\nu(t-x)\right] dt \right| \le \int_{-\infty}^{\infty} |f(t) \cos\left[\nu(t-x)\right]| \, dt$$

$$= \int_{-\infty}^{\infty} |f(t)| \, |\cos\left[\nu(t-x)\right]| \, dt \le \int_{-\infty}^{\infty} |f(t)| \, dt \,.$$

Therefore, one can swap the integrals in Eq. (5.15) and integrate over ν to obtain

$$g_\lambda(x) = \frac{1}{\pi} \int_{-\infty}^{\infty} f(t) \frac{\sin\left[\lambda(t-x)\right]}{t-x} dt \,.$$

We split this integral into two by the point x; in the integral between $-\infty$ and x we shall then change the variable $x - t \to t$, and in the integral between x and $+\infty$ we shall apply the substitution $t - x \to t$. This yields

$$g_\lambda(x) = \frac{1}{\pi} \int_0^\infty f(x-t) \frac{\sin(\lambda t)}{t} dt + \frac{1}{\pi} \int_0^\infty f(x+t) \frac{\sin(\lambda t)}{t} dt \,. \tag{5.17}$$

Now we have to analyse both these integrals in the limit of $\lambda \to \infty$. We shall start from the first integral in which we shall change the variable $t \to z = \lambda t$:

$$\frac{1}{\pi} \int_0^\infty f(x-t) \frac{\sin(\lambda t)}{t} dt = \frac{1}{\pi} \int_0^\infty f(x - z/\lambda) \frac{\sin z}{z} dz \,.$$

In the $\lambda \to \infty$ limit the function $f(x - z/\lambda)$ inside the integral goes to $f(x-0)$ and can be taken out of the integral. The remaining integral is well-known,

$$\int_0^\infty \frac{\sin z}{z} dz = \frac{\pi}{2} \,,$$

see Eq. (2.127), and this leads to the required result:

$$\frac{1}{\pi} \int_0^\infty f(x-t) \frac{\sin(\lambda t)}{t} dt \to \frac{1}{2} f(x-0).$$

Similarly, for the second integral:

$$\frac{1}{\pi} \int_0^\infty f(x+t) \frac{\sin(\lambda t)}{t} dt = \frac{1}{\pi} \int_0^\infty f(x+z/\lambda) \frac{\sin z}{z} dz \to \frac{1}{2} f(x+0).$$

Both these limits lead to the desired result of

$$\lim_{\lambda \to \infty} g_\lambda(x) = \frac{1}{2} [f(x-0) + f(x+0)].$$

The given calculation seems simple but requires justification as we have an integral depending on the parameter λ and have taken the liberty of taking the limit with respect to this parameter inside the integral. This can only be done if the integral converges uniformly with respect to λ. This can easily be established only for the part of the integral between some $a > 1$ and ∞:

$$\left| \int_a^\infty f(x-t) \frac{\sin(\lambda t)}{t} dt \right| \le \int_a^\infty |f(x-t)| \left| \frac{\sin(\lambda t)}{t} \right| dt < \int_a^\infty |f(x-t)| \, dt$$

$$\le \int_{-\infty}^\infty |f(x-t)| \, dt = \int_{-\infty}^\infty |f(t)| \, dt < \infty.$$

We have used above the fact that for the integration variable $t \ge a > 1$ we can write

$$\left| \frac{\sin(\lambda t)}{t} \right| \le \frac{1}{t} < 1.$$

Unfortunately, the difficulty lies in the vicinity of the zero bottom limit of the original integral where care is needed. Hence, we shall use a different approach.

Consider first the integral between $a > 0$ and ∞, where we shall make a change of the variable $t \to z = \lambda t$:

$$\lim_{\lambda \to \infty} \frac{1}{\pi} \int_a^\infty f(x-t) \frac{\sin(\lambda t)}{t} dt = \lim_{\lambda \to \infty} \frac{1}{\pi} \int_{a\lambda}^\infty f(x-z/\lambda) \frac{\sin z}{z} dz = 0.$$

(5.18)

This is because the bottom limit goes to $+\infty$ as well.

Hence, it is left to consider the original integral between 0 and $a > 0$. As the choice for the value of a in the part of the integral we have already considered is arbitrary (as long as $a > 0$), we may now be more specific and require a to be such that the function $f(x-t)$ be monotonous (for the given x) within the interval $0 \le t \le a$, i.e.

it is either monotonously increasing or decreasing. This will then enable us to apply
the mean value formula (3.92):

$$\int_0^a \psi(t)\phi(t)dt = \psi(+0)\int_0^\xi \phi(t)dt + \psi(a-)\int_\xi^a \phi(t)dt \,,$$

where $\psi(t) = f(x - t), \phi(t) = \sin(\lambda t)/t$ and $0 < \xi < a$. Then, in the integrals in the
right-hand side we shall make the same change of the variable as before, $t \to z = \lambda t$.
We have

$$\int_0^a f(x - t)\frac{\sin(\lambda t)}{t}dt = f(x - 0)\int_0^\xi \frac{\sin(\lambda t)}{t}dt + f(x - a)\int_\xi^a \frac{\sin(\lambda t)}{t}dt$$

$$= f(x - 0)\int_0^{\xi\lambda} \frac{\sin z}{z}dz + f(x - a)\int_{\xi\lambda}^{a\lambda} \frac{\sin z}{z}dz \,.$$

The upper limit in the first integral in the right-hand side goes to ∞ giving the value
of $\pi/2$ for the integral; in the second integral both limits go to infinity in the $\lambda \to \infty$
limit making the integral to go to zero. Therefore, the first integral in Eq. (5.17) does
indeed go to the required limit of $\frac{1}{2}f(x - 0)$.

Problem 5.5 Show that the second integral in Eq. (5.17) goes to $\frac{1}{2}f(x + 0)$.

This finalises the proof of formula (5.16) and hence of Eq. (5.14).

5.2   Fourier Transform

5.2.1   General Idea

If the application of Eq. (5.7) is split into two consecutive steps, then a very useful
and widely utilised device is created called Fourier transform. The Fourier transform
of a function $f(x)$, defined on the whole real axis x and satisfying the Dirichlet
conditions, is defined as the integral (5.6):

$$F(\nu) = \int_{-\infty}^{\infty} f(x)e^{-i2\pi\nu x}dx \,. \tag{5.19}$$

We know from the previous section that the integral $\int_{-\infty}^{\infty} |f(x)|\, dx$ should exist. From
(5.19) it is seen that the function $f(x)$ is *transformed* by a process of integration into
a *spectral* function $F(\nu)$. If we introduce a *functional operator* $\mathcal{F}$ acting on $f(x)$
which converts $f(x) \mapsto F(\nu)$, then we can recast (5.19) in the form:

$$F(\nu) = \mathcal{F}[f(x)] = \int_{-\infty}^{\infty} f(x)e^{-i2\pi\nu x}dx \,. \tag{5.20}$$

Note, in particular, that if x is time and $f(t)$ is a signal of some kind, then ν is a frequency, i.e. in this case the transformation is performed into the frequency domain and the Fourier transform $F(\nu) = \mathcal{F}[f(t)]$ essentially gives a *spectral analysis* of the *signal* $f(t)$.

It is also possible to convert $F(\nu)$ back into $f(x)$, i.e. to perform the inverse transformation $F(\nu) \mapsto f(x)$, by using the formal relation $f(x) = \mathcal{F}^{-1}[F(°)]$, where $\mathcal{F}^{-1}$ denotes the *inverse* functional operator. Fortunately, an explicit formula for this inversion procedure is readily provided by the Fourier integral (5.7). Indeed, the expression in the square brackets there, according to Eq. (5.19), is exactly $F(\nu)$, so that we find

$$f(x) = \mathcal{F}^{-1}[F(\nu)] = \int_{-\infty}^{\infty} F(\nu)e^{i2\pi\nu x}d\nu . \tag{5.21}$$

This result is called the inverse Fourier transform of $F(\nu)$. For a time dependent signal $f(t)$ the inverse transform shows how the signal can be *synthesised* from its frequency spectrum given by $F(\nu)$.

Problem 5.6 Show that for an *even* function $f_e(x)$ the Fourier transform pair (i.e. both direct and inverse transformations) can be expressed in the following alternative form:

$$F(\nu) = \mathcal{F}[f_e(x)] = 2 \int_0^{\infty} f_e(x)\cos(2\pi\nu x)dx , \tag{5.22}$$

$$f_e(x) = \mathcal{F}^{-1}[F(\nu)] = 2 \int_0^{\infty} F(\nu)\cos(2\pi\nu x)d\nu . \tag{5.23}$$

It should be noted that the Fourier transform $F(\nu)$ of a *real* even function $f_e(x)$ is also a *real* function.

Problem 5.7 Similarly, show that for an *odd* function $f_o(t)$ we find

$$F(\nu) = \mathcal{F}[f_o(x)] = -2i \int_0^{\infty} f_o(x)\sin(2\pi\nu x)dx . \tag{5.24}$$

$$f_o(x) = \mathcal{F}^{-1}[F(\nu)] = +2i \int_0^{\infty} F(\nu)\sin(2\pi\nu x)d\nu . \tag{5.25}$$

Hence, the Fourier transform $F(\nu)$ of a *real* odd function is a purely imaginary function.

Problem 5.8 Show that the Fourier transform $F(\nu)$ of a real function, $f(x)^* = f(x)$, satisfies the following identity:

$$F(\nu)^* = F(-\nu) .$$

Fig. 5.4 The Fourier
transforms $F_n(\nu)$ of the
functions $\delta_n(x)$ of the delta
sequence for selected values
of n

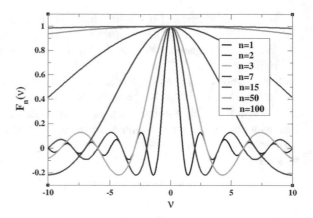

As an example, let us determine the Fourier transforms $F_n(\nu)$ of a set of the unit
impulse functions

$$\delta_n(x) = \begin{cases} n\,, & \text{for } |x| \le 1/2n \\ 0\,, & \text{for } |x| > 1/2n \end{cases}, \tag{5.26}$$

where $n = 1, 2, \dots$. As is explained in Sect. 4.1, the sequence of functions, $\delta_n(x)$,
tends to the Dirac delta function $\delta(x)$ in the $n \to \infty$ limit. Therefore, the Fourier
transforms $\mathcal{F}\,[\delta_n(x)] = F_n(\nu)$ of the functions in this sequence form another sequence
$F_n(\nu)$ which in the same limit tends to the Fourier transform $F(\nu) = \lim_{n\to\infty} F_n(\nu)$
of the delta function itself. We shall also see that as a by-product of this calculation
we would be able to obtain an integral representation for $\delta(x)$. And indeed, since the
function $\delta_n(t)$ is even, we can use Eq. (5.22) which gives

$$F_n(\nu) = \mathcal{F}\,[\delta_n(t)] = 2n \int_0^{1/2n} \cos(2\pi\nu t)dt = \frac{\sin(\pi\nu/n)}{\pi\nu/n}\,. \tag{5.27}$$

The functions $F_n(\nu)$ for selected values of n from 1 to 100 are shown in Fig. 5.4. It
is clearly seen from the definition (5.26) of the $\delta_n(x)$ that as n increases the width
of the peak $\Delta x = 1/n$ of it becomes *smaller*; at the same time, the width $\Delta\nu$ of the
central peak of $F_n(\nu)$ becomes *larger*. Indeed, $\Delta\nu$ may be determined by the twice
the smallest root (excluding zero) of the function $\sin(\pi\nu/n)$, i.e. $\Delta\nu = 2n$, i.e. for
any n we find that $\Delta\nu\Delta x \simeq 2$ (a some sort of the 'uncertainty principle' of quantum
mechanics).

In the $n \to \infty$ limit, we obtain

$$\lim_{n\to\infty} \mathcal{F}\,[\delta_n(x)] = \lim_{n\to\infty} \frac{\sin(\pi\nu/n)}{(\pi\nu/n)} = \lim_{t\to 0} \frac{\sin t}{t} = 1\,. \tag{5.28}$$

This is nicely confirmed by the graphs of the Fourier transform of $\delta_n(x)$ for large
values of n shown in Fig. 5.4, where one can appreciate that the function $F_n(\nu)$

tends more or more to the horizontal line $F_n(\nu) = 1$ with the increase of n. Thus, the Fourier transform of the Dirac delta function, to be obtained in the limit, is

$$\mathcal{F}[\delta(x)] = \int_{-\infty}^{\infty} \delta(x)e^{-i2\pi\nu x}dx = 1 . \qquad (5.29)$$

Note that the integral above also perfectly conforms to the delta function filtering theorem, see Sect. 4.1.

Now, the application of the inverse transform (5.21) to the result (5.29) gives the important *formal* integral representation of the delta function:

$$\delta(x) = \int_{-\infty}^{\infty} \mathcal{F}[\delta(x)] e^{i2\pi\nu x}d\nu = \int_{-\infty}^{\infty} e^{i2\pi\nu x}d\nu = \frac{1}{2\pi} \int_{-\infty}^{\infty} e^{ikx}dk . \qquad (5.30)$$

By changing the variable $k \to -k$ in the integral, another integral representation is obtained with the minus sign in the exponent. Both forms, of course, are fully equivalent. The same expressions for the Dirac delta function are also derived in Sect. 4.1 using a different method, based on general properties of the delta function.

The appearance of the integral representation of the delta function in the theory of Fourier transform is not accidental either. Indeed, let us return to the Fourier integral (5.7) which we shall rewrite this time slightly differently by interchanging the order of the integrals:

$$f(x) = \int_{\infty}^{\infty} \left[\int_{-\infty}^{\infty} e^{-i2\pi\nu(t-x)}d\nu \right] f(t)dt . \qquad (5.31)$$

Comparing the expression in the square brackets with Eq. (5.30) we immediately recognise the delta function $\delta(t-x)$. Therefore, the expression above becomes

$$f(x) = \int_{-\infty}^{\infty} \delta(t-x) f(t)dt .$$

The integral in the right-hand side above gives $f(x)$ due to the filtering theorem for the delta function, i.e. the same function as in the left-hand side, as expected! Note, however, that the above consideration would only be allowed if the $f(t)$ was continuous at the point x. If it is not, i.e. $f(x-0) \neq f(x+0)$, then we have to split the integral into two:

$$f(x) = \int_{-\infty}^{x} \delta(t-x) f(t)dt + \int_{x}^{\infty} \delta(t-x) f(t)dt .$$

Changing the variables $t \to t - x$ in both integrals, we get

$$f(x) = \int_{-\infty}^{0} \delta(t) f(x+t)dt + \int_{0}^{\infty} \delta(t) f(x+t)dt .$$

Applying now the filtering theorem to each of the integrals, we have to introduce the
factor of $1/2$ since the integration is performed over the half of the real axis. Then the
first integral gives $f(x - 0)/2$, while the second $f(x + 0)/2$, i.e. the mean value, as
it must be!

Problem 5.9 Show that

$$\mathcal{F}\left[\frac{1}{2}\left(\delta(x - a) + \delta(x + a)\right)\right] = \cos(2\pi a \nu) \; .$$

Problem 5.10 Show that

$$\mathcal{F}[H(1 - x) - H(-1 - x)] = \frac{\sin(2\pi\nu)}{\pi\nu} \; ,$$

where $H(x)$ is the Heaviside function.

Problem 5.11 Show that

$$\mathcal{F}\left[xe^{-\alpha|x|}\right] = \frac{-8\pi\alpha i\nu}{\left(\alpha^2 + 4\pi^2\nu^2\right)^2} \; , \quad \text{while} \quad \mathcal{F}\left[|x|\,e^{-\alpha|x|}\right] = \frac{2\left(\alpha^2 - 4\pi^2\nu^2\right)}{\left(\alpha^2 + 4\pi^2\nu^2\right)^2} \; ,$$

where $\alpha > 0$.

Problem 5.12 Show that the Fourier transform of the function $f(t)$, which
is equal to -1 for $-1 \le t < 0, +1$ for $0 \le t \le 1$ and zero otherwise, is given
by

$$F(\nu) = -\frac{2i \sin^2 \pi\nu}{\pi\nu} \; .$$

Then, using the inverse Fourier transform show that

$$f(x) = \frac{8}{\pi} \int_0^\infty \frac{1}{t} \sin^2 t \, \sin(tx) \, \cos(tx) \, dt \; .$$

Finally, choose an appropriate value of x to show that the integral

$$\int_0^\infty \frac{\sin^3 t \, \cos t}{t} \, dt = \frac{\pi}{16} \; .$$

Problem 5.13 Show that the Fourier transform of the function $f(t) = e^{-\alpha|x|}$ (where $\alpha > 0$) is given by

$$F(\nu) = \frac{2\alpha}{\alpha^2 + (2\pi\nu)^2} \ .$$

Consequently, prove the following integral representation of it:

$$e^{-\alpha|x|} = \frac{2\alpha}{\pi} \int_0^\infty \frac{\cos(ux)}{\alpha^2 + u^2} du \ .$$

Using this result, show that the integral

$$\int_0^\infty \frac{du}{1 + u^2} = \frac{\pi}{2} \ .$$

Check this result by calculating the integral directly using a different method. *[Hint: look up for the integral representation of the arctangent.]*

Problem 5.14 Prove the so-called modulation theorem:

$$\mathcal{F}[f(x)\cos(2\pi\nu_0 x)] = \frac{1}{2}[F(\nu + \nu_0) + F(\nu - \nu_0)] \ ,$$

where $F(\nu) = \mathcal{F}[f(x)]$.

Problem 5.15 Prove that the Fourier transform of $f(x) = \sin x$ for $|x| \le \pi/2$ and zero otherwise is

$$F(\nu) = \frac{4\pi\nu i}{(2\pi\nu)^2 - 1} \cos\left(\pi^2\nu\right) \ .$$

Using this result and the inverse Fourier transform, show that

$$\int_0^\infty \frac{x\sin(\pi x)}{1 - x^2} dx = \frac{\pi}{2} \ .$$

Verify that the integrand is well defined at $x = 1$.

Problem 5.16 Show that the Fourier transform of the Gaussian delta sequence (where $n = 1, 2, ...$)

$$\delta_n(x) = \frac{n}{\sqrt{2\pi}} e^{-n^2 x^2 / 2} \tag{5.32}$$

is

$$F_n(\nu) = e^{-2(\pi/n)^2 \nu^2} . \tag{5.33}$$

[Hint: you may find formula (2.51) useful.] Show next that if Δx and $\Delta \nu$ are the widths of $\delta_n(x)$ and $F_n(\nu)$, respectively, at their half height, then their product $\Delta x \Delta \nu = 4 \ln 2 / \pi \simeq 0.88$ and does not depend on n.

Again, several interesting observations can be made for the functions (5.32) and their Fourier transforms (5.33). Indeed, the functions $\delta_n(x)$ defined above represent another example of the delta sequence (see Sect. 4.1): as n is increased, the Gaussian function $\delta_n(x)$ becomes thinner and more picked, while the full area under the curve remains equal to unity. Its Fourier transform $F_n(\nu)$ is also a Gaussian; however, with the increase of n the Gaussian $F_n(\nu)$ gets more and more spread out tending to the value of unity in the limit, $\lim_{n \to \infty} F_n(\nu) = 1$, which agrees with the result for the rectangular delta sequence (5.26) considered above.

So, we see yet again that if a function in the direct $x-$space gets thinner, its Fourier transform in the $\nu-$space gets fatter. The opposite is also true which follows from the symmetry of the direct and inverse transforms, Eqs. (5.19) and (5.21).

5.2.2   *Fourier Transform of Derivatives*

One of the important strengths of the Fourier transform stems from the fact that it frequently helps solving linear differential equations (DE), and several examples of this application will be given in the following sections. The idea is to first transform the equation into the $\nu-$space by performing the Fourier transform of all terms in it. In the transformed equation the unknown function $f(x)$ appears as its transform (or 'image') $F(\nu)$, and, as will be shown in this subsection, derivatives of $f(x)$ in the DE are transformed into algebraic expressions linear in $F(\nu)$ rendering the DE in the $\nu-$space to become *algebraic* with respect to $F(\nu)$. Then, at the second step the inverse transform $F(\nu) \mapsto f(x)$ is performed. If successful, the solution is obtained in the analytical form. Otherwise, the solution is obtained as the spectral integral over ν which may be calculated numerically.

However, an application of this scheme may be drastically simplified if certain general *rules* of working with the Fourier transform were established first. Several such rules we shall derive here and also in the next subsection. These may be useful in performing either direct or inverse Fourier transform in practice.

We start by considering the calculation of the Fourier transform of the derivative $f'(x)$ of $f(x)$. It is given by

$$\mathcal{F}\left[f'(x)\right] = \int_{-\infty}^{\infty} f'(x)e^{-i2\pi\nu x}dx .$$ (5.34)

If we integrate the right-hand side of (5.34) by parts, we find

$$\mathcal{F}\left[f'(x)\right] = f(x)e^{-i2\pi\nu x}\Big|_{-\infty}^{\infty} + i2\pi\nu \int_{-\infty}^{\infty} f(x)e^{-i2\pi\nu x}dx .$$ (5.35)

Since we have to assume that the integral $\int_{-\infty}^{\infty} |f(x)|dx$ exists, it is necessary that $\lim_{x\to\pm\infty} f(x) = 0$. Hence, the free term in the right-hand side can be dropped and we obtain a simple result:

$$\mathcal{F}\left[f'(x)\right] = i2\pi\nu F(\nu) ,$$ (5.36)

where $F(\nu) = \mathcal{F}[f(x)]$.

Problem 5.17 Using the method of induction, generalise this result for a derivative of any order m:

$$\mathcal{F}\left[f^{(m)}(x)\right] = (i2\pi\nu)^m F(\nu) ,$$ (5.37)

provided that $\lim_{x\to\infty} f^{(j)}(x) = 0$ for any $j = 0, 1, 2, ..., m - 1$.

Problem 5.18 The n-th moment of a function $f(x)$ is defined by the expression:

$$\mu_n = \int_{-\infty}^{\infty} x^n f(x)dx .$$

Show that if $F(\nu) = \mathcal{F}[f(x)]$, then $\mu_n = F^{(n)}(0)/(2\pi i)^n$.

It is seen that, as mentioned above, the Fourier transform of any derivative of a function is proportional to the Fourier transform of the function itself. This guarantees that any linear DE will transform into a linear algebraic equation with respect to the image $F(\nu)$, and this can always be solved. This is the main reason for the Fourier transform to be so widely used in physics and engineering since very often one faces a problem of solving linear ordinary and partial differential equations.

5.2.3   Fourier Transform of an Integral

It is possible to calculate the FT of a definite integral of a function:

$$g(x) = \int_0^x f(x')dx' .$$

Let $G(\nu)$ and $F(\nu)$ be the Fourier transforms of the functions $g(x)$ and $f(x)$, respectively. We know from the previous section that $\mathcal{F}[g'(x)] = i2\pi\nu G(\nu)$. But $g'(x) = f(x)$. Hence,

$$i2\pi\nu G(\nu) = F(\nu) \implies G(\nu) = \frac{F(\nu)}{i2\pi\nu} , \tag{5.38}$$

i.e.

$$\mathcal{F}\left[\int_0^x f(x')dx'\right] = \frac{1}{i2\pi\nu}\mathcal{F}[f(x)] . \tag{5.39}$$

Hence, the FT of the integral is directly related to the FT of the integrand. Correspondingly,

$$\mathcal{F}^{-1}\left[\frac{F(\nu)}{\nu}\right] = 2\pi i \, \mathcal{F}^{-1}[G(\nu)] = 2\pi i \, g(x) = 2\pi i \int_0^x f(x')dx' . \tag{5.40}$$

Problem 5.19   Derive Eq. (5.39) directly using the definition of the FT. Note that to ensure convergence of the FT, we must assume that $g(\pm\infty) = 0$.

Problem 5.20   Show that

$$\mathcal{F}^{-1}\left[F(\nu)\frac{\sin(\nu\lambda)}{\nu}\right] = \pi \int_{x-\lambda/2\pi}^{x+\lambda/2\pi} f(x')dx' , \tag{5.41}$$

where $\mathcal{F}[f(x)] = F(\nu)$.

5.2.4   Convolution Theorem

Consider two functions $f(x)$ and $g(x)$ with Fourier transforms $F(\nu)$ and $G(\nu)$, respectively. The *convolution* of $f(x)$ and $g(x)$ is *defined* as the following function of x:

$$f(x) * g(x) = \int_{-\infty}^{\infty} f(t)g(x - t)dt . \tag{5.42}$$

It is readily verified that

$$f * g = g * f .$$

Let us calculate the Fourier transform of this new function. Our goal is to express it via the Fourier transforms $F(\nu)$ and $G(\nu)$ of the two functions involved.

To this end, let us substitute into Eq. (5.42) the inverse Fourier transform of $g(x - t)$,

$$g(x - t) = \int_{-\infty}^{\infty} G(\nu)e^{i2\pi\nu(x-t)}d\nu , \qquad (5.43)$$

and then change the order of integrations:

$$\begin{aligned} f(x) * g(x) &= \int_{-\infty}^{\infty} G(\nu)e^{i2\pi\nu x}d\nu \left[\int_{-\infty}^{\infty} f(t)e^{-i2\pi\nu t}dt\right] \\ &= \int_{-\infty}^{\infty} F(\nu)G(\nu)e^{i2\pi\nu x}d\nu , \end{aligned} \qquad (5.44)$$

since the integral within the square brackets above is nothing but the Fourier transform $F(\nu)$ of $f(x)$. The result we have just obtained corresponds to the inverse Fourier transform of the convolution, i.e. $f(t) * g(t) = \mathcal{F}^{-1}[F(\nu)G(\nu)]$, and hence we can finally write

$$\mathcal{F}[f(x) * g(x)] = F(\nu)G(\nu) . \qquad (5.45)$$

Formula (5.45) is known as the *convolution theorem*. We see that the Fourier transform of a convolution $f(x) * g(x)$ is simply equal to the *product* of the Fourier transform of $f(x)$ and $g(x)$. Inversely, if one needs to calculate the inverse Fourier transform of a product of two functions $F(\nu)G(\nu)$, then it is equal to the convolution integral (5.42). This may be useful in performing the inverse transform. Indeed, if a function $T(\nu)$ to be inverted back into the x-space looks too complex for performing such a transformation, it might sometimes be possible to split it into a product of two functions $F(\nu)$ and $G(\nu)$ in such a way, that their individual inversion (into $f(x)$ and $g(x)$, respectively) can be done. Then, the inversion of the whole function $T(\nu) = F(\nu)G(\nu)$ will be given by the integral (5.42), which can then be either attempted analytically or numerically thereby solving the problem, at least in principle.

As the simplest example, let us determine the convolution of the Dirac delta function $\delta(x - b)$ with some function $f(x)$. From the definition (5.42) of the convolution, we find that

$$\delta(x - b) * f(x) = \int_{-\infty}^{\infty} \delta(t - b)f(x - t)dt = f(x - b) , \qquad (5.46)$$

where in the last passage we have applied the filtering theorem for the Dirac delta function, Sect. 4.1. Considering now the ν-space, the Fourier image of $\delta(x - b)$ is

$$\mathcal{F}\left[\delta\left(x-b\right)\right]=\int_{-\infty}^{\infty}\delta\left(x-b\right)e^{-i2\pi\nu x}dx=e^{-i2\pi\nu b}\ ,$$

while the Fourier transform of $f(x)$ is $F(\nu)$. Therefore, the Fourier transform of their convolution is simply $F(\nu)e^{-i2\pi\nu b}$. On the other hand, calculating directly the Fourier transform of their convolution given by Eq. (5.46), we obtain

$$\mathcal{F}\left[f(x-b)\right]=\int_{-\infty}^{\infty}f\left(x-b\right)e^{-i2\pi\nu x}dx=\begin{vmatrix}t=x-b\\ dt=dx\end{vmatrix}$$

$$=e^{-i2\pi\nu b}\int_{-\infty}^{\infty}f\left(t\right)e^{-i2\pi\nu t}dt=e^{-i2\pi\nu b}F(\nu)\ ,$$

which is exactly the same result.

Problem 5.21 Derive formula (5.45) directly by calculating the transform of the convolution from the definition of the Fourier transform:

$$\mathcal{F}\left[f(x)*g(x)\right]=\int_{-\infty}^{\infty}\left[f(x)*g(x)\right]e^{i2\pi\nu x}dx\ .$$

Problem 5.22 The function $f(t)$ is defined as $e^{-\alpha t}$ for $t\geq 0$ and zero otherwise ($\alpha>0$). Show by direct calculation of the convolution integral, considering separately the cases of positive and negative t, that the convolution of this function with itself is $g(t)=f(t)*f(t)=tf(t)$.

Problem 5.23 Find the Fourier transforms of the functions $f(t)$ and $g(t)$ defined in the previous problem directly and using the convolution theorem. [Answer: $\mathcal{F}\left[f(t)\right]=F(\nu)=(\alpha+i2\pi\nu)^{-1}$ and $\mathcal{F}\left[g(t)\right]==F(\nu)^2(\alpha+i2\pi\nu)^{-2}$.]

5.2.5   Parseval's Theorem

Due to close relationship between the Fourier series and transform, it is no surprise that there exists a theorem similar to the corresponding Parseval's theorem we proved in Sects. 3.5 and 3.7 in the context of the Fourier series.

Here we shall formulate this statement in its most general form as Plancherel's theorem. It states that, if $F(\nu)$ and $G(\nu)$ are the Fourier transforms of the functions $f(x)$ and $g(x)$ respectively, then

$$\int_{-\infty}^{\infty}F(\nu)^*G(\nu)d\nu=\int_{-\infty}^{\infty}f(x)^*g(x)dx\ . \qquad (5.47)$$

To prove, we use the definition of the Fourier transforms in the left-hand side:

$$\int_{-\infty}^{\infty} F(\nu)^* G(\nu) d\nu = \int_{-\infty}^{\infty} \left(\int_{-\infty}^{\infty} f(x) e^{-2\pi i \nu x} dx \right)^* \left(\int_{-\infty}^{\infty} g(y) e^{-2\pi i \nu y} dy \right) d\nu$$

$$= \int_{-\infty}^{\infty} \int_{-\infty}^{\infty} \left[\int_{-\infty}^{\infty} e^{2\pi i \nu (x-y)} d\nu \right] f(x)^* g(y) dx dy$$

$$= \int_{-\infty}^{\infty} \int_{-\infty}^{\infty} \delta(x-y) f(x)^* g(y) dx dy$$

$$= \int_{-\infty}^{\infty} f(x)^* \left[\int_{-\infty}^{\infty} \delta(x-y) g(y) dy \right] dx = \int_{-\infty}^{\infty} f^*(x) g(x) dx \, ,$$

where in the second line we recognised the integral representation (5.30) for the delta function $\delta(x - y)$ within the expression in the square brackets, while on the third line we have used the filtering theorem for the delta function.

In particular, if $f(x) = g(x)$, then we obtain Parseval's theorem:

$$\int_{-\infty}^{\infty} |f(x)|^2 dx = \int_{-\infty}^{\infty} |F(\nu)|^2 d\nu \, . \tag{5.48}$$

Problem 5.24 Using the function from Problem 5.13 and Parseval's theorem, show that

$$\int_0^{\infty} \frac{dx}{(1+x^2)^2} = \frac{\pi}{4} \, .$$

Problem 5.25 Consider a function $f(x) = \cos x$ for $-\pi/2 \le x \le \pi/2$ and zero otherwise. Show that its Fourier transform

$$F(\nu) = -\frac{2\cos\left(\pi^2 \nu\right)}{(2\pi\nu)^2 - 1} \, .$$

Hence, using the equal functions Parseval's theorem, show that

$$\int_0^{\infty} \frac{\cos^2(\pi x/2)}{(x^2 - 1)^2} dx = \frac{\pi^2}{8} \, .$$

5.2.6 Poisson Summation Formula

In some physics applications a simple formula is used, which shows an interesting interconnection between the Fourier series and Fourier integral, called Poisson summation formula.

To derive it, consider an arbitrary function $f(x)$ and let us construct an infinite sum

$$\phi(x) = \sum_{n=-\infty}^{\infty} f(n+x) \, .$$

Problem 5.26 Prove that $\phi(x) = \phi(x + T)$ is a periodic function with the period of $T = 1$.

Since the function $\phi(x)$ is periodic, we can expand it into the appropriate Fourier series

$$\phi(x) = \sum_{k=-\infty}^{\infty} \phi_k e^{i2\pi kx}, \quad \phi_k = \int_0^1 e^{-i2\pi kx} \phi(x) dx .$$

Let us calculate the Fourier coefficients using the explicit expression of the function $\phi(x)$:

$$\phi_k = \int_0^1 e^{-i2\pi kx} \left[\sum_{n=-\infty}^{\infty} f(n+x) \right] dx = \sum_n \int_0^1 e^{-i2\pi kx} f(x+n) dx$$

$$= \sum_n \int_n^{n+1} e^{-i2\pi k(t-n)} f(t) dt = \sum_n \underbrace{e^{i2\pi kn}}_{=1} \int_n^{n+1} e^{-i2\pi kt} f(t) dt$$

$$= \sum_{n=-\infty}^{\infty} \int_n^{n+1} e^{-i2\pi kt} f(t) dt .$$

Note that we have exchanged summation and integration on the first line; this requires the uniform convergence of the sum $\sum_n f(x+n)$ with respect to x. We shall assume that this condition is satisfied for our function $f(x)$.

It is seen that each term in the sum covers the values of the integration variable over the interval $n \leq t \leq n+1$, so that by summing over n we shall cover the whole real axis. Hence,

$$\phi_k = \int_{-\infty}^{\infty} e^{-i2\pi kt} f(t) dt .$$

This is exactly the Fourier transform $F(k)$ of $f(x)$ calculated at the integer k, i,.e., $\phi_k = F(k)$. Therefore,

$$\sum_{n=-\infty}^{\infty} f(n+x) = \sum_{k=-\infty}^{\infty} F(k) e^{i2\pi kx} .$$

In particular, by setting $x = 0$ we arrive at the celebrated Poisson summation formula

$$\sum_{n=-\infty}^{\infty} f(n) = \sum_{k=-\infty}^{\infty} F(k) . \tag{5.49}$$

The theta-function transformation Eq. (3.104) we derived previously and used in obtaining Ewald's formula in Sect. 3.11.2 is just one example of the usefulness of the Poisson summation formula. We have derived it differently there by directly applying the Fourier series, but it is instructive to see that the same result can also be obtained using the Poisson summation formula. We shall first consider a simple case of the so-called simple cubic lattice in which the direct basic lattice vectors are

$$
\mathbf{a}_1 = a \begin{pmatrix} 1 \\ 0 \\ 0 \end{pmatrix}, \quad \mathbf{a}_2 = a \begin{pmatrix} 0 \\ 1 \\ 0 \end{pmatrix}, \quad \mathbf{a}_3 = a \begin{pmatrix} 0 \\ 0 \\ 1 \end{pmatrix}.
$$

These vectors define the lattice translation vector $\mathbf{L} = n_1\mathbf{a}_1 + n_2\mathbf{a}_2 + n_3\mathbf{a}_3$, where the integers n_i ($i = 1, 2, 3$) may take all possible values between $\pm\infty$. By trying these values, the whole lattice is built up by these lattice translations.

Consider the lattice summation ($\mathbf{r}$ is a vector, t parameter)

$$
\sum_{\mathbf{L}} e^{-t^2|\mathbf{L}-\mathbf{r}|} = \sum_{n_1,n_2,n_3} \exp\left[-t^2 \sum_{i=1}^{3} a^2 (n_i - r_i)^2\right]
$$

$$
= \prod_{i=1}^{3} \left\{ \sum_{n_i=-\infty}^{\infty} \exp\left[-t^2 a^2 (n_i - r_i)^2\right] \right\},
$$

where we have also expanded the vector $\mathbf{r} = r_1\mathbf{a}_1 + r_2\mathbf{a}_2 + r_3\mathbf{a}_3$ in terms of the basic elementary translations, with the coefficients r_i being real numbers. The expression in the curly brackets is an infinite summation over n_i to which we can apply the Poisson summation formula:

$$
\sum_{n_i=-\infty}^{\infty} \exp\left[-t^2 a^2 (n_i - r_i)^2\right] = \sum_{m_i=-\infty}^{\infty} F(m_i), \tag{5.50}
$$

where

$$
F(\nu_i) = \int_{-\infty}^{\infty} e^{-i2\pi\nu_i x} e^{-(ta)^2(x-r_i)^2} dx
$$

is the Fourier transform of the exponential function $f(x) = \exp\left[-t^2 a^2 (x - r_i)^2\right]$ in the left-hand side of (5.50) obtained by replacing $m_i \mapsto \nu_i$.

Problem 5.27 Show that

$$
\int_{-\infty}^{\infty} e^{-i2\pi\nu_i x} e^{-(ta)^2(x-r_i)^2} dx = \frac{\sqrt{\pi}}{at} e^{-(\pi\nu_i/at)^2} e^{-i2\pi\nu_i r_i}. \tag{5.51}
$$

Therefore, using the Fourier transforms and replacing $\nu_i \mapsto m_i$ in them, we obtain

$$\sum_{\mathbf{L}} e^{-t^2 |\mathbf{L}-\mathbf{r}|} = \prod_{i=1}^{3} \left\{ \sum_{m_i=-\infty}^{\infty} \frac{\sqrt{\pi}}{at} e^{-(\pi m_i/at)^2} e^{-i2\pi m_i r_i} \right\}$$

$$= \left(\frac{\sqrt{\pi}}{at} \right)^3 \sum_{m_1,m_2,m_3} e^{-i2\pi(m_1 r_1 + m_2 r_2 + m_3 r_3)} e^{-(\pi/at)^2 (m_1^2 + m_2^2 + m_3^2)} .$$

It is easy to see that what we have just obtained coincides precisely with the right-hand side of Eq. (3.104) in this case of the particular lattice type. Indeed, in the case of the simple cubic lattice the volume of the primitive unit cell $v_c = |\mathbf{a}_1 \cdot (\mathbf{a}_2 \times \mathbf{a}_3)| = a^3$ (it is a cube with side a), the basic reciprocal lattice vectors $\mathbf{b}_i$ ($i = 1, 2, 3$), defined via $\mathbf{b}_i \cdot \mathbf{a}_j = 2\pi \delta_{ij}$, are

$$\mathbf{b}_1 = \frac{2\pi}{a} \begin{pmatrix} 1 \\ 0 \\ 0 \end{pmatrix}, \quad \mathbf{b}_2 = \frac{2\pi}{a} \begin{pmatrix} 0 \\ 1 \\ 0 \end{pmatrix}, \quad \mathbf{b}_3 = \frac{2\pi}{a} \begin{pmatrix} 0 \\ 0 \\ 1 \end{pmatrix},$$

and hence an arbitrary reciprocal lattice vector $\mathbf{g} = m_1 \mathbf{b}_1 + m_2 \mathbf{b}_2 + m_3 \mathbf{b}_3$. Therefore,

$$2\pi (m_1 r_1 + m_2 r_2 + m_3 r_3) = \mathbf{g} \cdot \mathbf{r},$$

$$\left(\frac{\pi}{at} \right)^2 (m_1^2 + m_2^2 + m_3^2) = \frac{\mathbf{g}^2}{4t^2},$$

leading immediately to the required result as summing over all possible integers m_1, m_2 and m_3 corresponds to summing over all possible reciprocal lattice vectors $\mathbf{g}$.

The only point that needs to be checked is that the infinite series

$$\phi(x) = \sum_{n=-\infty}^{\infty} f(n+x) = \sum_{n=-\infty}^{\infty} e^{-\alpha(n+x-r)^2} \quad \text{with} \quad \alpha > 0$$

converges uniformly with respect to x (we have introduced obvious simplifications to the unessential parameters in the formula above). To prove the uniform convergence, we recall that $\phi(x)$ is periodic with the period of 1, so that we can limit ourselves with the interval $0 \le x < 1$. Then, we can write using the Weierstrass test (Theorem I.6.1) that in this interval,

$$\left| \sum_{n=-\infty}^{\infty} e^{-\alpha(n+x-r)^2} \right| \le \sum_{n=-\infty}^{\infty} \left| e^{-\alpha(n+x-r)^2} \right| \le \sum_{n=-\infty}^{\infty} \underbrace{e^{-\alpha(n-r)^2}}_{c_n},$$

that is the converging series since (the ratio test)

$$\lim_{n\to\infty} \frac{c_{n+1}}{c_n} = \lim_{n\to\infty} e^{-\alpha[(n+1-r)^2 - (n-r)^2]} = \lim_{n\to\infty} e^{-\alpha[2n-2r+1]} = 0.$$

In order to consider a general case of an arbitrary lattice type given by basic direct, $\mathbf{a}_i$, and related reciprocal, $\mathbf{b}_i$, lattice vectors, we need first to generalise the Poisson formula for the triple sum:

$$\sum_{n_1,n_2,n_3} f(n_1, n_2, n_3) = \sum_{m_1,m_2,m_3} F(m_1, m_2, m_3), \qquad (5.52)$$

where

$$F(\nu_1, \nu_2, \nu_3) = \int dx_1 dx_2 dx_3 \, e^{-i2\pi(x_1\nu_1 + x_2\nu_2 + x_3\nu_3)} f(x_1, x_2, x_3) \qquad (5.53)$$

is the triple Fourier transform (more on these in Sect. 5.3.1).

Problem 5.28 Prove formulae (5.52) and (5.53).

Problem 5.29 In this problem we shall derive Eq. (3.104) for a general lattice type. To start with, we must calculate the triple Fourier transform $F(\nu_1, \nu_2, \nu_3)$ of the function

$$f(x_1, x_2, x_3) = e^{-t^2 |x_1 \mathbf{a}_1 + x_2 \mathbf{a}_2 + x_3 \mathbf{a}_3 - \mathbf{r}|^2}$$

that is obtained from the original one,

$$e^{-t^2 |\mathbf{L} - \mathbf{r}|} = e^{-t^2 |n_1 \mathbf{a}_1 + n_2 \mathbf{a}_2 + n_3 \mathbf{a}_3 - \mathbf{r}|^2},$$

by replacing $n_i \mapsto x_i$. To perform the integration, introduce new variables given by the three components of the vector

$$\mathbf{R} = x_1 \mathbf{a}_1 + x_2 \mathbf{a}_2 + x_3 \mathbf{a}_3 - \mathbf{r}. \qquad (5.54)$$

Show that the Jacobian of the transformation $(x_1, x_2, x_3) \to (R_1, R_2, R_3)$ is $J = 1/v_c$, where $v_c = |\mathbf{a}_1 \cdot (\mathbf{a}_2 \times \mathbf{a}_3)|$ is the unit cell volume. The old variables $\{x_i\}$ are expressed via the new ones $\{R_j\}$ by dot-multiplying both sides of Eq. (5.54) with the reciprocal lattice vectors $\mathbf{b}_i$. The triple integral (5.53) in the new variables factorises into three integrals of the type (5.51) leading to

$$F(\nu_1, \nu_2, \nu_3) = \frac{1}{v_c} e^{-i\mathbf{G}\cdot\mathbf{r}} e^{-G^2/4t^2} \left(\frac{\sqrt{\pi}}{t} \right)^3,$$

where $\mathbf{G} = \nu_1 \mathbf{b}_1 + \nu_2 \mathbf{b}_2 + \nu_3 \mathbf{b}_3$ is a vector of the reciprocal space. Replacing $\nu_i \mapsto m_i$ in $F(\nu_1, \nu_2, \nu_3)$, the general reciprocal vector $\mathbf{G}$ goes over into the reciprocal lattice vector $\mathbf{g} = m_1 \mathbf{b}_1 + m_2 \mathbf{b}_2 + m_3 \mathbf{b}_3$, and hence the right-hand side of Eq. (5.52) becomes identical to the right-hand side of Eq. (3.104).

5.3   Applications of the Fourier Transform in Physics

5.3.1   Various Notations and Multiple Fourier Transform

In physics the Fourier transform is used both for functions depending on time t and spatial coordinates x, y and z. There are several ways in which the Fourier transform may be written. Consider first the time Fourier transforms. If $f(t)$ is some function satisfying the necessary conditions for the transform to exist, then its direct and inverse transforms can be written, for instance, as

$$F(\omega) = \int_{-\infty}^{\infty} f(t)e^{-i\omega t}\,dt \quad \text{and} \quad f(t) = \int_{-\infty}^{\infty} F(\omega)e^{i\omega t}\frac{d\omega}{2\pi}\,, \tag{5.55}$$

where $\omega = 2\pi\nu$ is a new frequency and the integration variable in the inverse Fourier transform. This is just another way of writing the Fourier transform. Indeed, substituting $F(\omega)$ from the first equation into the integrand of the second, we obtain the Fourier integral:

$$f(t) = \int_{-\infty}^{\infty}\left[\int_{-\infty}^{\infty} f(\tau)e^{-i\omega\tau}\,d\tau\right]e^{i\omega t}\frac{d\omega}{2\pi} = \int_{-\infty}^{\infty} f(\tau)\left[\int_{-\infty}^{\infty} e^{i\omega(t-\tau)}\frac{d\omega}{2\pi}\right]d\tau$$

$$= \int_{-\infty}^{\infty} f(\tau)\delta(t-\tau)\,d\tau = f(t)\,,$$

i.e. the expected result. The integral representation (5.30) of the delta function was used here.

Note that the $1/2\pi$ multiplier may appear instead in the equation for the direct transform, or it may also be shared in both expressions symmetrically:

$$F(\omega) = \int_{-\infty}^{\infty} f(t)e^{-i\omega t}\frac{dt}{\sqrt{2\pi}} \quad \text{and} \quad f(t) = \int_{-\infty}^{\infty} F(\omega)e^{i\omega t}\frac{d\omega}{\sqrt{2\pi}}\,. \tag{5.56}$$

The mentioned Fourier transforms are related to each other by a trivial constant pre-factor.

It is also important to mention the sign in the exponential functions in the two (direct and inverse) transforms: either plus or minus can equivalently be used in the direct transform; the important point is that then the opposite sign is to be used in the inverse one. Of course, the images (or transforms) of the function $f(t)$ calculated using one or the other formulae would differ. However, since at the end of the day one always goes back to the direct space using the inverse transform, the final result will always be the same no matter which particular definition has been used.

Functions depending on spatial coordinates may also be Fourier transformed. Consider first a function $f(x)$ depending only on the coordinate x. In this case, using one of the forms above, one can write for both transforms:

$$F(k_x) = \int_{-\infty}^{\infty} f(x)e^{-ik_x x}dx \quad \text{and} \quad f(x) = \int_{-\infty}^{\infty} F(k_x)e^{ik_x x}\frac{dk_x}{2\pi} . \qquad (5.57)$$

Here k_x plays the role of the wavevector.

If a function depends on more than one variable, one can transform it with respect to some or all of its variables. Consider, for instance, a function depending on all three spatial variables, $f(x, y, z)$ (e.g. a temperature distribution in a sample). We first calculate its transform with respect to the x variable:

$$F(k_x, y, z) = \int_{-\infty}^{\infty} f(x, y, z)e^{-ik_x x}dx \quad \text{and} \quad f(x, y, z) = \int_{-\infty}^{\infty} F(k_x, y, z)e^{ik_x x}\frac{dk_x}{2\pi} .$$

Then, the function $F(k_x, y, z)$ can be transformed with respect to its y variable:

$$F(k_x, k_y, z) = \int_{-\infty}^{\infty} F(k_x, y, z)e^{-ik_y y}dy \quad \text{and} \quad F(k_x, y, z) = \int_{-\infty}^{\infty} F(k_x, k_y, z)e^{ik_y y}\frac{dk_y}{2\pi} .$$

Finally, we can also perform a similar operation with respect to z:

$$F(k_x, k_y, k_z) = \int_{-\infty}^{\infty} F(k_x, k_y, z)e^{-ik_z z}dz \quad \text{and} \quad F(k_x, k_y, z) = \int_{-\infty}^{\infty} F(k_x, k_y, k_z)e^{ik_z z}\frac{dk_z}{2\pi} .$$

Combining all these expressions together, one writes

$$\begin{aligned} F(k_x, k_y, k_z) &= \int_{-\infty}^{\infty} F(k_x, k_y, z)e^{-ik_z z}dz = \int_{-\infty}^{\infty}\left\{\int_{-\infty}^{\infty} F(k_x, y, z)e^{-ik_y y}dy\right\}e^{-ik_z z}dz \\ &= \int_{-\infty}^{\infty}\left\{\int_{-\infty}^{\infty}\left[\int_{-\infty}^{\infty} f(x, y, z)e^{-ik_x x}dx\right]e^{-ik_y y}dy\right\}e^{-ik_z z}dz \\ &= \int\int\int f(x, y, z)e^{-i(k_x x+k_y y+k_z z)}dxdydz , \end{aligned}$$

where the triple (volume) integral is performed over the whole infinite space. Our result can be further simplified by introducing vectors $\mathbf{r} = (x, y, z)$ and $\mathbf{k} = (k_x, k_y, k_z)$. Then, we finally obtain the direct Fourier transform of the function $f(\mathbf{r})$ specified in 3D space:

$$F(\mathbf{k}) = \int f(\mathbf{r})e^{-i\mathbf{k}\cdot\mathbf{r}}d\mathbf{r} , \qquad (5.58)$$

where $d\mathbf{r}$ corresponds to the volume element $dxdydz$, and the integration is performed over the whole space.

Note that the Fourier transform of a constant function equal to one is given by the 3D delta function:

$$\begin{aligned} \int e^{-i\mathbf{k}\cdot\mathbf{r}}d\mathbf{r} &= \int e^{-ik_x x}dx \int e^{-ik_y y}dy \int e^{-ik_z z}dz \\ &= (2\pi)^3\,\delta(x)\,\delta(y)\,\delta(z) = (2\pi)^3\,\delta(\mathbf{r}) . \end{aligned} \qquad (5.59)$$

Problem 5.30 Show that the inverse transform reads

$$f(\mathbf{r}) = \frac{1}{(2\pi)^3} \int F(\mathbf{k}) e^{i\mathbf{k}\cdot\mathbf{r}} d\mathbf{k} , \qquad (5.60)$$

where the integration is performed over the whole (reciprocal) space spanned by the wave vector $\mathbf{k}$.

Problem 5.31 Assuming that the Fourier transform of a function $f(\mathbf{r})$ is defined as in Eq. (5.58), prove the following identities

$$\mathcal{F}[\nabla f(\mathbf{r})] = i\mathbf{k}F(\mathbf{k}) \quad \text{and} \quad \mathcal{F}[\Delta f(\mathbf{r})] = -k^2 F(\mathbf{k}) ,$$

where $k = |\mathbf{k}|$. Obtain these results in two ways: (i) act with operators ∇ and Δ, respectively, on both sides of the inverse Fourier transform (5.60); (ii) calculate the Fourier transform of $\nabla f(\mathbf{r})$ and $\Delta f(\mathbf{r})$ directly by means of Eq. (5.58) (using integration by parts).

Problem 5.32 Consider a vector function $\mathbf{g}(\mathbf{r})$. Its Fourier transform $\mathbf{G}(\mathbf{k})$ is also some vector function in the $\mathbf{k}$ space defined as above for each Cartesian component of $\mathbf{g}(\mathbf{r})$. Then show that

$$\mathcal{F}\left[\nabla \times \mathbf{g}(\mathbf{r})\right] = -i\mathbf{k} \times \mathbf{G}(\mathbf{k}) .$$

Problem 5.33 Let $f(t)$ be some function of time and its Fourier transforms be given by Eqs. (5.55). Show that

$$\mathcal{F}\left[\frac{\partial f}{\partial t}\right] = i\omega F(\omega) \quad \text{and} \quad \mathcal{F}\left[\frac{\partial^2 f}{\partial t^2}\right] = -\omega^2 F(\omega) .$$

Problem 5.34 Also, show that

$$\mathcal{F}^{-1}[F(\omega)\cos(\omega t_0)] = \frac{1}{2}[f(t-t_0) + f(t+t_0)] \qquad (5.61)$$

and

$$\mathcal{F}^{-1}\left[F(\omega)\frac{\sin(\omega t_0)}{\omega}\right] = \frac{1}{2}\int_{t-t_0}^{t+t_0} f(\tau)d\tau . \qquad (5.62)$$

Problem 5.35 If a function of interest depends on both spatial and time variables, e.g. $f(x, t)$, then the Fourier transform can be performed over both variables. Show that in this case one may write for the direct and inverse transforms:

$$F(k, \omega) = \int_{-\infty}^{\infty} dx \int_{-\infty}^{\infty} dt \, f(x, t) e^{-i(kx - \omega t)} \tag{5.63}$$

and

$$f(x, t) = \int_{-\infty}^{\infty} \frac{dk}{2\pi} \int \frac{d\omega}{2\pi} F(k, \omega) e^{i(kx - \omega t)} . \tag{5.64}$$

Here opposite signs in the exponent before x and t were used as is customarily done in the physics literature.

Problem 5.36 Calculate the Fourier transform

$$\varphi_{\mathbf{k}} = \int \frac{1}{r} e^{-i\mathbf{k} \cdot \mathbf{r}} d\mathbf{r}$$

of the function $1/r$, where $r = |\mathbf{r}|$ is the length of the vector $\mathbf{r}$. First of all, using the spherical coordinates (r, θ, ϕ) and choosing the direction of the vector $\mathbf{k}$ along the z axis (the integral cannot depend on its direction anyway), integrate over the angles. When doing the r integral, attach a factor $e^{-\epsilon r}$ to the integrand which would ensure the convergence of the integral (assuming $\epsilon \to 0$), and then perform the integration to get $4\pi / (k^2 + \epsilon^2)$. Therefore, in the $\epsilon \to 0$ limit $\varphi_{\mathbf{k}} = 4\pi / k^2$. The obtained result corresponds to the (inverse) Fourier transform for the Coulomb potential:

$$\frac{1}{|\mathbf{r} - \mathbf{r}'|} = \int \frac{d\mathbf{k}}{(2\pi)^3} \frac{4\pi}{k^2} e^{i\mathbf{k} \cdot (\mathbf{r} - \mathbf{r}')} . \tag{5.65}$$

Problem 5.37 Act with the Laplacian operator Δ on both sides of Eq. (5.65) and use Eq. (5.59) to prove that

$$\Delta \frac{1}{r} = -4\pi \delta(\mathbf{r}) . \tag{5.66}$$

A different proof of this relationship was given in Problem 4.15; yet another one will be given below in Sect. 5.3.3.

5.3.2 Retarded Potentials

Here we shall show that a particular solution of the Maxwell equations has a form of a retarded potential. As is shown in electrodynamics, all four Maxwell equations for the fields can be recast as only two equations for the corresponding scalar $\varphi\,(\mathbf{r}, t)$ and vector $\mathbf{A}\,(\mathbf{r}, t)$ potentials:

$$\Delta\varphi - \frac{1}{c^2}\frac{\partial^2\varphi}{\partial t^2} = -4\pi\rho \quad \text{and} \quad \Delta\mathbf{A} - \frac{1}{c^2}\frac{\partial^2\mathbf{A}}{\partial t^2} = -\frac{4\pi}{c}\mathbf{J}\,. \tag{5.67}$$

These are the so-called d'Alembert equations. Here $\rho\,(\mathbf{r}, t)$ and $\mathbf{J}\,(\mathbf{r}, t)$ are the charge and current densities, respectively.

We shall derive here a special solution of this equations corresponding to the following problem. Suppose, the potentials are known at $t \leq 0$ where the charges (responsible for the non-zero charge density ρ) are stationary. Then, starting from $t = 0$ the charges start moving causing time and space dependence of the 'sources' $\rho\,(\mathbf{r}, t)$ and $\mathbf{J}\,(\mathbf{r}, t)$. Assuming next that the latter are known, we would like to determine the changes in the potentials due to this movement of charges. As we are concerned only with the changes here, we may assume that the potentials satisfy zero initial conditions:

$$\varphi\,(\mathbf{r}, t = 0) = \left.\frac{\partial\varphi\,(\mathbf{r}, t)}{\partial t}\right|_{t=0} = 0\,, \tag{5.68}$$

and similarly for the vector potential. We also assume that the charges occupy at any time a finite portion of space, so that the potentials tend to zero infinitely far away from the charges. In fact, both φ and $\mathbf{A}$ must tend to zero at least as $1/r$, where $r = |\mathbf{r}|$. This latter condition is essential to make sure that the corresponding Fourier integrals exist.

Let us consider the equation for the scalar potential first. We expand both the potential and the charge density in the corresponding Fourier integrals with respect to the spatial variables only:

$$\varphi\,(\mathbf{r}, t) = \int \frac{d\mathbf{k}}{(2\pi)^3}\varphi_{\mathbf{k}}(t)e^{i\mathbf{k}\cdot\mathbf{r}} \quad \text{and} \quad \rho\,(\mathbf{r}, t) = \int \frac{d\mathbf{k}}{(2\pi)^3}\rho_{\mathbf{k}}(t)e^{i\mathbf{k}\cdot\mathbf{r}}\,. \tag{5.69}$$

Next we apply the Fourier transform to both sides of the differential equation for φ in (5.67). The action of the Laplacian corresponds to the sum of the second derivatives with respect to each of the three spatial variables. Therefore, according to Problem 5.31,

$$\mathcal{F}\,[\Delta\varphi\,(\mathbf{r}, t)] = -k^2\varphi_{\mathbf{k}}(t)$$

(here and in the following $k = |\mathbf{k}|$). For other terms the transformation is trivial:

$$\mathcal{F}\left[-\frac{1}{c^2}\frac{\partial^2\varphi}{\partial t^2}\right] = -\frac{1}{c^2}\frac{\partial^2\varphi_{\mathbf{k}}(t)}{\partial t^2} \quad \text{and} \quad \mathcal{F}\,[-4\pi\rho\,(\mathbf{r}, t)] = -4\pi\rho_{\mathbf{k}}(t)\,.$$

Therefore, after the transformation into the Fourier space (or $\mathbf{k}$-space), we obtain the following time differential equation for the image $\varphi_{\mathbf{k}}(t)$ of the unknown potential:

$$-k^2 \varphi_{\mathbf{k}} - \frac{1}{c^2} \frac{d^2 \varphi_{\mathbf{k}}}{dt^2} = -4\pi\rho_{\mathbf{k}} \quad \Longrightarrow \quad \frac{d^2 \varphi_{\mathbf{k}}}{dt^2} + (kc)^2 \, \varphi_{\mathbf{k}} = 4\pi c^2 \rho_{\mathbf{k}} \, . \qquad (5.70)$$

This is a linear second-order inhomogeneous differential equation which can be solved using e.g. the method of variation of parameters, considered in detail in Sect. I.8.2.3.2.

Problem 5.38 Show that the general solution of this equation is

$$\varphi_{\mathbf{k}}(t) = C_1 e^{ikct} + C_2 e^{-ikct} + \frac{4\pi c}{k} \int_0^t \rho_{\mathbf{k}}(\tau) \sin\left[kc\,(t-\tau)\right] d\tau \, .$$

From the initial conditions, it follows that

$$\varphi_{\mathbf{k}}(0) = \int d\mathbf{r}\, \varphi(\mathbf{r}, 0) e^{-i\mathbf{k}\cdot\mathbf{r}} = 0$$

and

$$\left. \frac{d\varphi_{\mathbf{k}}}{dt} \right|_{t=0} = \int d\mathbf{r} \left[\frac{\partial\varphi(\mathbf{r}, t)}{\partial t} \right]_{t=0} e^{-i\mathbf{k}\cdot\mathbf{r}} = 0 \, ,$$

so that the arbitrary constants above should be zero. Indeed, $\varphi_{\mathbf{k}}(0) = C_1 + C_2 = 0$, while

$$\frac{d\varphi_{\mathbf{k}}}{dt} = ikc \left(C_1 e^{ikct} - C_2 e^{ikct}\right) + \frac{4\pi c}{k} \left\{\rho_{\mathbf{k}}(\tau) \sin\left[kc\,(t-\tau)\right]\right\}_{\tau=t}$$

$$+ 4\pi c^2 \int_0^t \rho_{\mathbf{k}}(\tau) \cos\left[kc\,(t-\tau)\right] d\tau$$

$$= ikc \left(C_1 e^{ikct} - C_2 e^{-ikct}\right) + 4\pi c^2 \int_0^t \rho_{\mathbf{k}}(\tau) \cos\left[kc\,(t-\tau)\right] d\tau \, ,$$

which after setting $t = 0$ yields

$$\left. \frac{d\varphi_{\mathbf{k}}}{dt} \right|_{t=0} = ikc\,(C_1 - C_2) = 0 \quad \Longrightarrow \quad C_1 = C_2 \, ,$$

resulting in $C_1 = C_2 = 0$. Therefore,

$$\varphi_{\mathbf{k}}(t) = \frac{4\pi c}{k} \int_0^t \rho_{\mathbf{k}}(\tau) \sin\left[kc\,(t-\tau)\right] d\tau$$

$$= \frac{4\pi c}{k} \int_0^t d\tau \, \sin\left[kc\,(t-\tau)\right] \left(\int d\mathbf{r}'\, \rho(\mathbf{r}', \tau) e^{-i\mathbf{k}\cdot\mathbf{r}'}\right) ,$$

where in the second passage we have replaced the Fourier transform of the density by the density itself using the Fourier transform of it. Substituting now the image of the potential we have just found into the inverse Fourier transform of it, the first Eq. (5.69), we obtain after slight rearrangements:

$$\varphi(\mathbf{r}, t) = \frac{c}{2\pi^2} \int d\mathbf{r}' \int_0^t d\tau\, \rho(\mathbf{r}', \tau) \left[\int \frac{d\mathbf{k}}{k} e^{i\mathbf{k}\cdot(\mathbf{r}-\mathbf{r}')} \sin(kc(t-\tau)) \right] . \quad (5.71)$$

The 3D integral over $\mathbf{k}$ within the square brackets may only depend on the length R of the vector $\mathbf{R} = \mathbf{r} - \mathbf{r}'$ since $\mathbf{R}$ appears only in the dot product with the vector $\mathbf{k}$ and we integrate over the whole reciprocal $\mathbf{k}$ space. Therefore, this integral can be simplified by directing $\mathbf{R}$ along the z axis and then using spherical coordinates.

Problem 5.39 Using this method, show that integration over the spherical angles of $\mathbf{k}$ yields

$$\int \frac{d\mathbf{k}}{k} e^{i\mathbf{k}\cdot\mathbf{R}} \sin(\xi k) = \frac{2\pi}{R} \left[\int_0^\infty \cos(k(\xi - R))\, dk - \int_0^\infty \cos(k(\xi + R))\, dk \right] ,$$

where $\xi = c(t - \tau)$.

In these integrals one can recognise Dirac delta functions, see Eq. (4.25), so that

$$\int \frac{d\mathbf{k}}{k} e^{i\mathbf{k}\cdot\mathbf{R}} \sin[kc(t-\tau)] = \frac{2\pi^2}{R} [\delta(c(t-\tau) - R) - \delta(c(t-\tau) + R)] .$$

This simple result allows manipulating Eq. (5.71) further:

$$\varphi(\mathbf{r}, t) = c \int \frac{d\mathbf{r}'}{|\mathbf{r}-\mathbf{r}'|} \int_0^t d\tau\, \rho(\mathbf{r}', \tau) [\delta(c(t-\tau) - R) - \delta(c(t-\tau) + R)] .$$

The two delta functions give rise to two integrals over τ. We shall consider them separately. The first one, after changing the variable $\tau \to \lambda = c(t-\tau) - R$, results in

$$\int_0^t d\tau\, \rho(\mathbf{r}', \tau)\, \delta(c(t-\tau) - R) = \frac{1}{c} \int_{-R}^{ct-R} \rho\left(\mathbf{r}', t - \frac{R+\lambda}{c}\right) \delta(\lambda)\, d\lambda .$$

If the point $\lambda = 0$ does not lie within the limits, the integral is equal to zero. Since $R > 0$, the integral is zero if $ct - R < 0$, i.e. if $R > ct$ or $t < R/c$. If, however, $t > R/c$ or $R < ct$, then the integral results in $(1/c)\rho(\mathbf{r}', t - R/c)$ by virtue of the filtering theorem for the delta function. The second delta function does not contribute to the

τ integral in this case since after the change of variables $\tau \to \lambda = c\,(t - \tau) + R$ the limits of λ become $ct + R$ and $+R$ which are both positive, and hence exclude the point $\lambda = 0$. Therefore, only the first delta function contributes for $t > R/c$ and we finally obtain

$$\varphi\,(\mathbf{r}, t) = \int \frac{d\mathbf{r}'}{|\mathbf{r} - \mathbf{r}'|} H\left(t - \frac{|\mathbf{r} - \mathbf{r}'|}{c}\right) \rho\left(\mathbf{r}', t - \frac{|\mathbf{r} - \mathbf{r}'|}{c}\right), \qquad (5.72)$$

where $H(x)$ is the Heaviside function. The Heaviside function indicates that if there is a charge density in some region around $\mathbf{r}' = 0$, then the effect of this charge will only be felt at points $\mathbf{r}$ at times $t > R/c \simeq r/c$. In other words, the field produced by the moving charges is not felt immediately at any point $\mathbf{r}$ in space; the effect of the charges will propagate to the point $\mathbf{r}$ over some time and will only be felt there when $t \geq r/c$. Also note that for $t = 0$ the Heaviside function is exactly equal to zero and hence is the potential - in full accordance with our initial conditions.

The second equation in (5.67) for the vector potential can be solved in the same way. In fact, these are three scalar equations for the three components of $\mathbf{A}$, each being practically identical to the one we have just solved. Therefore, the solution for each component of $\mathbf{A}$ is obtained by replacing ρ with J_α/c (where $\alpha = x, y, z$) in formula (5.72), and then combining all three contributions together:

$$\mathbf{A}\,(\mathbf{r}, t) = \frac{1}{c} \int \frac{d\mathbf{r}'}{|\mathbf{r} - \mathbf{r}'|} H\left(t - \frac{|\mathbf{r} - \mathbf{r}'|}{c}\right) \mathbf{J}\left(\mathbf{r}', t - \frac{|\mathbf{r} - \mathbf{r}'|}{c}\right). \qquad (5.73)$$

This is a retarded solution as well. Hence, either of the two potentials at time t is determined by the densities ρ or $\mathbf{J}$ calculated at time $t - |\mathbf{r} - \mathbf{r}'|/c$ in the past. In other words, the effect of the densities (the 'sources') from point $\mathbf{r}'$ is felt at the 'observation' point $\mathbf{r}$ only after time $|\mathbf{r} - \mathbf{r}'|/c$ which is required for the light to travel the distance $|\mathbf{r} - \mathbf{r}'|$ between these two points.

The other term which contains the delta function $\delta\,(c\,(t - \tau) + R)$ (which we dropped) corresponds to the advanced part of the general solution. This term is zero for the boundary conditions we have chosen. In the case of general initial and boundary conditions both potentials contribute together with the free term (the one which contains the constants C_1 and C_2).

Problem 5.40 Using the Fourier transform method, show that the solution of the Poisson equation

$$\Delta\varphi(\mathbf{r}) = -4\pi\rho(\mathbf{r})$$

of classical electrostatics is

$$\varphi(\mathbf{r}) = \int \frac{\rho(\mathbf{r}')}{|\mathbf{r} - \mathbf{r}'|} d\mathbf{r}' . \tag{5.74}$$

Of course, this famous Coulomb formula can be obtained directly from Eq. (5.72) for stationary charges (i.e. for the charge density which does not depend on time), but it is a good exercise to repeat the steps of the method of this subsection in this much simpler case.

Problem 5.41 Consider an infinitely long string along the x-axis that is pulled up at $x = 0$ and then let go oscillating in the $x - y$ plane. The initial shape of the string is described by the function $y(x, t)|_{t=0} = he^{-|x|/a}$ (h and a being positive constants). Show by using the Fourier transform method that the solution of the equation of motion for the string (the wave equation),

$$\frac{1}{v^2}\frac{\partial^2 y}{\partial t^2} = \frac{\partial^2 y}{\partial x^2} ,$$

where v is the wave velocity, can be written as

$$y(x, t) = \frac{ha}{\pi} \int_{-\infty}^{\infty} \frac{\cos(vkt)}{1 + k^2 a^2} e^{ikx} dk .$$

Problem 5.42 Diffusion of particles immersed in a liquid flowing in the x direction with constant velocity μ is described by the following partial differential equation:

$$D\frac{\partial^2 n}{\partial x^2} - \mu\frac{\partial n}{\partial x} = \frac{\partial n}{\partial t} ,$$

where D is the diffusion coefficient and $n(x, t)$ is the density of the particles. Perform the Fourier transform of this equation with respect to x and solve the obtained ordinary differential equation assuming that initially at $t = 0$ all particles were located at some point x_0, i.e. $n(x_0, 0) = n_0\delta(x - x_0)$. Finally, performing the inverse Fourier transform of the obtained solution, show that

$$n(x, t) = \frac{n_0}{\sqrt{4\pi Dt}} e^{-(x-x_0-\mu t)^2/4Dt} .$$

Describe what happens to the particles distribution over time.

5.3.3 *Green's Function of a Differential Equation*

The so-called Green's functions allow establishing a general formula for a particular integral of an inhomogeneous differential equation as a *convolution* with the function in the right-hand side. And the Fourier transform method has become an essential tool in finding Green's functions for each particular equation. Let us consider this question in a bit more detail.

Consider an inhomogeneous differential equation for a function $y(x)$:

$$\mathcal{L}_x y(x) = f(x) , \tag{5.75}$$

where $\mathcal{L}_x$ is some operator acting on $y(x)$. For instance, the forced damped harmonic oscillator of the unit mass is described by the equation

$$\frac{d^2 y}{dt^2} + 2\gamma\omega_0 \frac{dy}{dt} + \omega_0^2 y = f(t) , \tag{5.76}$$

where ω_0 is its fundamental frequency, γ is related to the friction, and m mass, while $f(t)$ is an external excitation signal, and hence the operator

$$\mathcal{L}_t = \frac{d^2}{dt^2} + 2\gamma\omega_0 \frac{d}{dt} + \omega_0^2 .$$

Suppose that we can find a solution $G(x)$ of the auxiliary equation

$$\mathcal{L}_x G(x) = \delta(x) , \tag{5.77}$$

where the operator $\mathcal{L}_x$ acts on the variable x, and in the right-hand side we have a Dirac delta function. The function $G(x)$ is called Green's function of the differential equation (5.75) or its fundamental solution. It is easy to see then that a particular solution of our differential equation for any function $f(x)$ in the right-hand side of the differential equation can be written as its convolution with Green's function:

$$y(x) = \int_a^b G(x - x') f(x') \, dx' , \tag{5.78}$$

where the integration is performed over the interval $a \leq x \leq b$ of interest, where the function $f(x)$ is defined. Indeed, by acting with the operator $\mathcal{L}_x$ on both sides, we obtain

$$\mathcal{L}_x y(x) = \int_a^b \left[\mathcal{L}_x G(x - x') \right] f(x') \, dx' = \int_a^b \left[\mathcal{L}_{x - x'} G(x - x') \right] f(x') \, dx'$$

$$= \int_a^b \delta(x - x') f(x') \, dx' = f(x) ,$$

where in the last passage we have used the filtering theorem for the delta function. Noe also that before that we have replaced the operator $\mathcal{L}_x$ with $\mathcal{L}_{x-x'}$; this can always be done if the operator $\mathcal{L}_x$ contains just differentiations. This calculation proves that formula (5.78) does indeed provide a particular solution of our equation (5.75) for an arbitrary right-hand side $f(x)$.

Let us now consider some examples, both for ordinary and partial differential equations.

We shall start by looking for Green's function of the damped (small friction) harmonic oscillator equation. Green's function $G(t)$ satisfies the differential equation (5.76) with $f(t)$ replaced by $\delta(t)$. We shall use the Fourier transform method. If we define[2]

$$G(\omega) = \mathcal{F}[G(t)] = \int_{-\infty}^{\infty} G(t)e^{i\omega t}\,dt \ ,$$

then the equation

$$\frac{d^2 G}{dt^2} + 2\gamma\omega_0\frac{dG}{dt} + \omega_0^2 G = \delta(t)$$

for Green's function transforms into

$$-\omega^2 G(\omega) - 2i\gamma\omega_0\omega G(\omega) + \omega_0^2 G(\omega) = 1 \ ,$$

yielding immediately

$$G(\omega) = -\frac{1}{\omega^2 + 2i\gamma\omega_0\omega - \omega_0^2} \ .$$

Therefore, using the inverse Fourier transform, we can write for Green's function we are interested in:

$$G(t) = \int_{-\infty}^{\infty} G(\omega)e^{-i\omega t}\frac{d\omega}{2\pi} = -\frac{1}{2\pi}\int_{-\infty}^{\infty} \frac{e^{-i\omega t}}{\omega^2 + 2i\gamma\omega_0\omega - \omega_0^2}\,d\omega \ .$$

The $\omega-$integral is most easily taken in the complex z plane. Indeed, there are two poles here,

$$\omega_\pm = \omega_0\left(-i\gamma \pm \sqrt{1 - \gamma^2}\right) = \pm\varpi - i\omega_0\gamma \ ,$$

which are solutions of the quadratic equation $\omega^2 + 2i\gamma\omega_0\omega - \omega_0^2 = 0$. Here the frequency $\varpi = \omega_0\sqrt{1 - \gamma^2}$ is positive (and real) as $\gamma < 1$ for a weakly damped oscillator. Both poles lie in the lower part of the complex plane. We may use the contour which closes the horizontal axis with a semicircle of radius $R \to \infty$ either in the upper or lower part of the complex plane. To choose the contour, we have to

---

[2] For a change, a different sign is used in the exponent here as compared to Eq. (5.55); as long as the opposite sign is to be used in the inverse Fourier transform, this particular selection of signs results in the same final result.

consider two cases: $t > 0$ and $t < 0$. In the case of positive times, the exponent $-i\omega t \to -izt = -i(x+iy)t = -ixt + yt$, so that the integral $\int_{C_R}$ over the semicircle of radius R of the contour will tend to zero for $y < 0$, i.e. we have to choose the semicircle in the lower part of the complex plane, and both poles would contribute. Therefore,

$$G(t) = -\frac{1}{2\pi} \oint_C \frac{e^{-izt}}{z^2 + 2i\gamma\omega_0 z - \omega_0^2} dz$$

$$= \frac{2\pi i}{2\pi} \left\{ \mathrm{Res}\left[\frac{e^{-izt}}{z^2 + 2i\gamma\omega_0 z - \omega_0^2}, \omega_+\right] + \mathrm{Res}\left[\frac{e^{-izt}}{z^2 + 2i\gamma\omega_0 z - \omega_0^2}, \omega_-\right] \right\}$$

$$= i \left\{ \frac{e^{-i\omega_+ t}}{2\omega_+ + 2i\gamma\omega_0} + \frac{e^{-i\omega_- t}}{2\omega_- + 2i\gamma\omega_0} \right\} = i \left\{ \frac{e^{-i\omega_+ t}}{2\varpi} + \frac{e^{-i\omega_- t}}{-2\varpi} \right\}$$

$$= \frac{\sin(\varpi t)}{\varpi} e^{-\gamma\omega_0 t} .$$

Note that the contour runs around the poles in the clockwise direction bringing an additional minus sign.

In the case of $t < 0$, one has to enclose the horizontal axis in the upper part of the complex plane to ensure the integral $\int_{C_R}$ over the semicircle of radius R in the $R \to \infty$ limit goes to zero; however, there are no poles, and the result is zero. Therefore, the Heaviside function $H(t)$ appears in Green's function in front of the above expression:

$$G(t) = H(t)\frac{\sin(\varpi t)}{\varpi} e^{-\gamma\omega_0 t} .$$

Problem 5.43 Show that Green's function corresponding to the operator $\mathcal{L}_t = \gamma + d/dt$ is $G(t) = H(t)e^{-\gamma t}$, where $H(t)$ is the Heaviside function.

Problem 5.44 Show that if

$$\mathcal{L}_t = \left(\gamma + \frac{d}{dt}\right)^2 = \frac{d^2}{dt^2} + 2\gamma\frac{d}{dt} + \gamma^2 ,$$

then $G(t) = H(t)te^{-\gamma t}$.

Now let us briefly touch upon the issue of calculating Green's functions for partial differential equations. Consider, as an example, the Poisson equation,

$$\Delta G(\mathbf{r}) = \delta(\mathbf{r}) . \tag{5.79}$$

In this case the boundary conditions are important, as many Green's functions exist for the same equation depending on these. We shall consider the case when Green's function vanishes at infinity, i.e. $G(\mathbf{r}) \to 0$ when $r = |\mathbf{r}| \to \infty$. In this case we can apply to $G(\mathbf{r})$ the 3D Fourier transform,

$$G(\mathbf{k}) = \int e^{-i\mathbf{k}\cdot\mathbf{r}} G(\mathbf{r}) d\mathbf{r} \quad \text{and} \quad G(\mathbf{r}) = \int \frac{d\mathbf{k}}{(2\pi)^3} e^{i\mathbf{k}\cdot\mathbf{r}} G(\mathbf{k}) .$$

In the $\mathbf{k}$−space the differential equation reads simply $-k^2 G(\mathbf{k}) = 1$, so that $G(\mathbf{k}) = -1/k^2$. Correspondingly, Green's function $G(\mathbf{r})$ can be calculated using the inverse Fourier transform taken in spherical coordinates and with the vector $\mathbf{r}$ directed along the z axis:

$$G(\mathbf{r}) = -\frac{1}{(2\pi)^3} \int e^{i\mathbf{k}\cdot\mathbf{r}} \frac{d\mathbf{k}}{k^2} = -\frac{1}{(2\pi)^3} \int_0^\infty dk \int_0^\pi e^{ikr\cos\vartheta} \sin\vartheta d\vartheta \int_0^{2\pi} d\phi$$

$$= -\frac{1}{(2\pi)^2} \int_0^\infty \frac{e^{ikr} - e^{-ikr}}{ikr} dk = -\frac{1}{2\pi^2 r} \int_0^\infty \frac{\sin p}{p} dp = -\frac{1}{4\pi r} ,$$

where we have used Eq. (2.127). Hence, we see that $\Delta \left(-\frac{1}{4\pi r}\right) = \delta(\mathbf{r})$, the formula for the 3D delta function we have already encountered previously (see Problems 4.15 and 5.37). Using this Green's function, one can easily write a solution of the inhomogeneous Poisson equation $\Delta\varphi(\mathbf{r}) = -4\pi\rho(\mathbf{r})$ as a convolution:

$$\varphi(\mathbf{r}) = \int G(\mathbf{r} - \mathbf{r}') \left[-4\pi\rho(\mathbf{r}')\right] d\mathbf{r}' = \int \rho(\mathbf{r}') \frac{d\mathbf{r}'}{|\mathbf{r} - \mathbf{r}'|} ,$$

which is exactly the result (5.74) we obtained previously.

Problem 5.45 Consider Green's function for the so-called Helmholtz partial differential equation. Green's function is defined via

$$\Delta G(\mathbf{r}) + p^2 G(\mathbf{r}) = \delta(\mathbf{r}) . \tag{5.80}$$

First of all, using the Fourier transform method, show that

$$G(\mathbf{r}) = \frac{1}{ir(2\pi)^2} \int_{-\infty}^\infty \frac{k e^{-ikr} dk}{p^2 - k^2} .$$

Then take the k−integral using two methods: (i) replace $p \to p + i\epsilon$ and then, performing the integration in the complex plane, show that in the $\epsilon \to +0$ limit $G(\mathbf{r}) = e^{ipr}/(4\pi r)$; (ii) similarly, by replacing $p \to p - i\epsilon$ show that in this case $G(\mathbf{r}) = e^{-ipr}/(4\pi r)$.

Finally, we shall consider the diffusion equation

$$\Delta\rho(\mathbf{r}, t) = \frac{1}{\kappa}\frac{\partial\rho(\mathbf{r}, t)}{\partial t} , \tag{5.81}$$

and correspondingly yet another definition of Green's function. Here $\rho(\mathbf{r}, t)$ describes the probability density of a particle to be found at point $\mathbf{r}$ at time t, and κ is the diffusion constant. First of all, let us determine the solution of this equation corresponding to the initial condition that the particle was in the centre of the coordinate system at time $t = 0$. In this case $\rho(\mathbf{r}, 0) = \delta(\mathbf{r})$, since integrating this density over the whole space gives the total probability equal to unity, as required. Again, we shall use the Fourier transform method. Define

$$\rho(\mathbf{k}, t) = \int e^{-i\mathbf{k}\cdot\mathbf{r}}\rho(\mathbf{r}, t)d\mathbf{r} \quad \text{and} \quad \rho(\mathbf{r}, t) = \int \frac{d\mathbf{k}}{(2\pi)^3}e^{i\mathbf{k}\cdot\mathbf{r}}\rho(\mathbf{k}, t) ,$$

so that the differential equation in the $\mathbf{k}$−space becomes

$$-k^2\rho(\mathbf{k}, t) = \frac{1}{\kappa}\frac{\partial\rho(\mathbf{k}, t)}{\partial t} \quad \Longrightarrow \quad \rho(\mathbf{k}, t) = Ce^{-k^2\kappa t} .$$

The arbitrary constant C is found using the initial condition:

$$\rho(\mathbf{k}, 0) = \int e^{-i\mathbf{k}\cdot\mathbf{r}}\rho(\mathbf{r}, 0)d\mathbf{r} = \int e^{-i\mathbf{k}\cdot\mathbf{r}}\delta(\mathbf{r})d\mathbf{r} = 1 ,$$

yielding $C = 1$. Hence, performing the inverse transform, we write

$$\rho(\mathbf{r}, t) = \int \frac{d\mathbf{k}}{(2\pi)^3}e^{i\mathbf{k}\cdot\mathbf{r}}e^{-k^2\kappa t} = I_x I_y I_z ,$$

where, for instance,

$$I_x = \int_{-\infty}^{\infty}\frac{dk_x}{2\pi}e^{ik_x x}e^{-(\kappa t)k_x^2} = \frac{1}{\sqrt{4\pi\kappa t}}e^{-x^2/4\kappa t} ,$$

where we have used Eq. (2.51). This finally gives

$$\rho(\mathbf{r}, t) = \frac{1}{(4\pi\kappa t)^{3/2}}e^{-r^2/4\kappa t} . \tag{5.82}$$

This particular solution may also be called Green's function $G(\mathbf{r}, t)$ of the diffusion equation, $\rho(\mathbf{r}, t) \Longrightarrow G(\mathbf{r}, t)$, corresponding to the initial condition $G(\mathbf{r}, 0) = \delta(\mathbf{r})$. Indeed, a particular solution of the diffusion equation at time t for *arbitrary* initial distribution $\rho(\mathbf{r}, 0)$ can be written as a convolution with that Green's function:

$$\rho(\mathbf{r}, t) = \int G\left(\mathbf{r} - \mathbf{r}', t\right) \rho\left(\mathbf{r}', 0\right) d\mathbf{r}' , \tag{5.83}$$

where $G\left(\mathbf{r} - \mathbf{r}', 0\right) = \delta\left(\mathbf{r} - \mathbf{r}'\right)$ by construction. Indeed, acting with the operator $\mathcal{L}_{\mathbf{r}} = \Delta - \kappa^{-1}\left(\partial/\partial t\right)$ on both sides of the above formula and taking into account that the Laplacian acts on the variable $\mathbf{r}$ (and not on the integration variable $\mathbf{r}'$) and hence $\mathcal{L}_{\mathbf{r}} G\left(\mathbf{r} - \mathbf{r}', t\right) = 0$, we obtain immediately that also $\mathcal{L}_{\mathbf{r}} \rho(\mathbf{r}, t) = 0$ for any time $t > 0$, i.e. it satisfies the diffusion equation. At the same time, at $t = 0$

$$\rho(\mathbf{r}, t)|_{t=0} = \int G\left(\mathbf{r} - \mathbf{r}', 0\right) \rho\left(\mathbf{r}', 0\right) d\mathbf{r}' = \int \delta\left(\mathbf{r} - \mathbf{r}'\right) \rho\left(\mathbf{r}', 0\right) d\mathbf{r}' = \rho(\mathbf{r}, 0) ,$$

i.e. it also satisfies the initial conditions, as required.

Problem 5.46   Prove that the density (5.83) remains properly normalised for any $t \geq 0$, i.e.

$$\int \rho(\mathbf{r}, t)\, d\mathbf{r} = \int \rho(\mathbf{r}, 0)\, d\mathbf{r} .$$

[Hint: demonstrate by direct integration that $\int G(\mathbf{r}, t)d\mathbf{r} = 1$.]

Note that the above Green's function, if multiplied with the Heaviside unit step function $H(t)$, also satisfies the corresponding *inhomogeneous* diffusion equation containing a product of two delta functions, both in $\mathbf{r}$ and t, in the right-hand side, as shown in the following Problem 5.48.

Problem 5.47 Calculate Green's function $G(x, t)$ of the one-dimensional Schrödinger equation for a free electron:

$$\left(\frac{\hbar^2}{2m} \frac{\partial^2}{\partial x^2} + i\hbar \frac{\partial}{\partial t} \right) G(x, t) = i\hbar\delta(x)\,\delta(t) \ . \qquad (5.84)$$

Firstly, solve for the Fourier transform $G(k, \omega)$ of Green's function. Then apply the inverse Fourier transform to write $G(x, t)$ as a double integral over ω and k. Perform first the ω-integration in the complex plane using the contour which contains the horizontal part shifted upwards by $i\epsilon$ and enclosed by a semicircle from below, so that the pole (which is on the real axis) appears inside the contour. This gives in the $\epsilon \to +0$ limit:

$$G(x, t) = H(t) \frac{1}{2\pi} e^{imx^2/2\hbar t} \int_{-\infty}^{\infty} e^{-i(\hbar t/2m)\lambda^2} d\lambda \ ,$$

where, as usual, $H(t)$ is the Heaviside function. The λ-integral can be taken in the complex plane, it is the Fresnel integral we considered before, Eq. (2.74). Note that the same result can also be obtained using the following *regularisation*:

$$\int_{-\infty}^{\infty} e^{-i\alpha\lambda^2} d\lambda \ \Rightarrow \ \lim_{\epsilon \to +0} \int_{-\infty}^{\infty} e^{-(i\alpha+\epsilon)\lambda^2} d\lambda = \lim_{\epsilon \to +0} \sqrt{\frac{\pi}{i\alpha + \epsilon}} = \sqrt{\frac{\pi}{i\alpha}} \ ,$$

where $\alpha = \hbar t/2m$. Hence, show that

$$G(x, t) = H(t) \sqrt{\frac{m}{2\pi\hbar t i}} e^{imx^2/2\hbar t} \ .$$

This formula plays central role in the theory of path integrals developed by R. Feynman, which serves as an alternative formulation of quantum mechanics.

Note that formally the manipulations made in the last problem can also be supported by noticing that the equation for Green's function (5.84) can formally be considered identical to the diffusion equation (5.85) by choosing $\kappa = i\hbar/2m$ in the latter. Then, also formally, Green's function for the Schrödinger equation can be obtained directly from the 1D version ($p = 1$) of Eq. (5.86), using the corresponding substitution.

Problem 5.48 Show that the solution of the differential equation

$$- \Delta G_1(\mathbf{r}, t) + \frac{1}{\kappa} \frac{\partial G_1(\mathbf{r}, t)}{\partial t} = \frac{1}{\kappa} \delta(\mathbf{r}) \delta(t) \qquad (5.85)$$

is $G_1(\mathbf{r}, t) = H(t) G(\mathbf{r}, t)$, where $G(\mathbf{r}, t)$ is Green's function given by Eq. (5.82). *[Hint: use the corresponding generalisation of the Fourier transformation (5.63) written both for all three spatial and one temporal variables; then, when calculating the inverse Fourier transform, calculate first the ω integral in the complex plane, then perform a direct integration in the $\mathbf{k}$ space.]*

Problem 5.49 Consider the inhomogeneous diffusion equation of the previous problem in $p = 1, 2, 3$ dimensions. Show that in this case

$$G(\mathbf{r}, t) = \frac{H(t)}{(4\pi \kappa t)^{p/2}} e^{-r^2/4\kappa t}. \qquad (5.86)$$

5.3.4 Time Correlation Functions

Time correlation functions play an important role in physics. When a particular observable is considered, e.g. a position $\mathbf{r}(t)$ or velocity $\mathbf{v}(t) = \dot{\mathbf{r}}(t)$ of a particle in a liquid at time t, this quantity is subjected to random fluctuations due to interaction with the environment (other particles); in other words, physical quantities of a subsystem will fluctuate even if the whole system is in thermodynamic equilibrium. At the same time, the values of the given observable taken at different times are correlated, i.e. there is a 'memory' in the system whereby fluctuations of the observable at time t somehow depend on its values at previous times $t' < t$. The quantitative measure of this kind of memory is provided by the so-called correlation functions. If $f(t)$ is a particular quantity of interest measured at time t, e.g. the velocity along the x direction of a particular particle in the system, then the correlation function is defined as

$$K_{ff}(\tau, t) = \langle \Delta f(t + \tau) \Delta f(t) \rangle, \qquad (5.87)$$

where $\Delta f(t) = f(t) - \langle f(t) \rangle$ corresponds to the deviation of $f(t)$ from its average $\langle f(t) \rangle$ and the angle brackets correspond to the averaging over the statistical ensemble. Here the average $\langle A(t) \rangle$ is understood as a mean value of the function $A(t)$ (in general, of coordinates and velocities) over a thermodynamic ensemble of independent systems (that have the same number and type of particles interacting via the same potentials), $\langle A(t) \rangle = \sum_i w_i A_i(t)$, where the sum is taken over all systems in the ensemble i, w_i is the probability to find the system i in the ensemble, and $A_i(t)$ is the value of the function $A(t)$ reached by the particular system i. For instance, if we would like to calculate the velocity autocorrelation function $K_{vv}(\tau, t)$ of a Brownian particle in a liquid, then $f(\tau) \longmapsto v(t)$ and an ensemble would consist of different identical systems in which initial (at $t = 0$) velocities and positions of particles in the liquid are different, e.g. drawn from the equilibrium (Gibbs) distribution. In this distribution some states of the liquid particles (i.e. their positions and velocities) are more probable than the others, this is determined by the corresponding distribution of statistical mechanics. Finally, the sum over i in the definition of the average would correspond to an integral over all initial positions and momenta of the particles.

Two limiting cases are worth mentioning. If there are no correlations in the values of the observable $f(t)$ at different times, then the average of the product at two times is equal to the product of the individual averages,

$$K_{ff}(\tau, t) = \langle \Delta f (t + \tau) \rangle \langle \Delta f(t) \rangle \ ,$$

leading to $K_{ff}(\tau) = 0$, since

$$\langle \Delta f \rangle = \langle f - \langle f \rangle \rangle = \langle f \rangle - \langle f \rangle = 0$$

by definition. On the other hand, at large times $\tau \to \infty$ the correlations between the quantities $f(t + \tau)$ and $f(t)$ calculated at remote times must weaken, and hence $K_{ff}(\tau, t) \to 0$ in this limit.

When a system is in thermodynamic equilibrium or in a stationary state, the correlation function should not depend on the time t as only the time difference τ between the two functions $\Delta f (t + \tau)$ and $\Delta f(t)$ actually matters. In other words, translation in time (given by t) should not affect the correlation function which would be a function of τ only. Because of this it is clear that the correlation function must be an even function of time:

$$K_{ff}(-\tau) = \langle \Delta f (t - \tau) \Delta f (t) \rangle = \langle \Delta f (t) \Delta f (t - \tau) \rangle = \langle \Delta f (t + \tau) \Delta f (t) \rangle$$
$$= K_{ff}(\tau) \ ,$$

$$(5.88)$$

where we have added τ to both times at the last step. Note that we also omitted the time t in the argument of the correlation function for simplicity of notations. Also, under these conditions, the average $\langle f(t) \rangle$ does not depend on time t.

In the following we shall measure the quantity $f(t)$ relative to its average value and hence will drop the Δ symbol from the definition of the correlation function.

There is a famous theorem due to Wiener and Khinchin which we shall briefly discuss now. Let us assume that the Fourier transform $f(\omega)$ can be defined for the real stochastic variable $f(t)$. Next, we define the spectral power density as the limit:

$$S(\omega) = \lim_{T \to \infty} S_T(\omega) = \lim_{T \to \infty} \frac{1}{T} \langle |f_T(\omega)|^2 \rangle \ ,$$

$$(5.89)$$

Fig. 5.5 The integration region in Eq. (5.90)

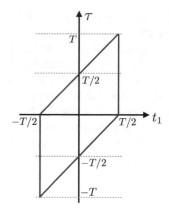

where

$$f_T(\omega) = \int_{-T/2}^{T/2} f(t)e^{-i\omega t}\,dt$$

is the partial Fourier transform, i.e. the Fourier transform of $f(t)$ will then be $f(\omega) = \lim_{T\to\infty} f_T(\omega)$. Consider now $S_T(\omega)$ as defined above:

$$
\begin{aligned}
S_T(\omega) &= \frac{1}{T}\left\langle \left| \int_{-T/2}^{T/2} dt_1\, e^{-i\omega t_1} f(t_1) \right|^2 \right\rangle \\
&= \frac{1}{T}\left\langle \int_{-T/2}^{T/2} dt_1 \int_{-T/2}^{T/2} dt_2\, e^{-i\omega(t_1-t_2)} f(t_1)\, f(t_2) \right\rangle \\
&= \frac{1}{T} \int_{-T/2}^{T/2} dt_1 \int_{-T/2}^{T/2} dt_2\, e^{-i\omega(t_1-t_2)} \langle f(t_1)\, f(t_2) \rangle \\
&= \frac{1}{T} \int_{-T/2}^{T/2} dt_1 \int_{-T/2}^{T/2} dt_2\, e^{-i\omega(t_1-t_2)} K_{ff}(t_1-t_2) = \left| \begin{matrix} t_2 \to \tau = t_1 - t_2 \\ d\tau = -dt_2 \end{matrix} \right| \\
&= \frac{1}{T} \int_{-T/2}^{T/2} dt_1 \int_{t_1-T/2}^{t_1+T/2} d\tau\, e^{-i\omega\tau} K_{ff}(\tau) \ .
\end{aligned}
$$

$$(5.90)$$

The next step consists of interchanging the order of integration: we shall perform integration over t_1 first and over τ second. The integration region on the (t_1, τ) plane is shown in Fig. 5.5. When choosing the τ integration to be performed last, one has to split the integration region into two regions: $-T \leq \tau < 0$ and $0 \leq \tau \leq T$. This yields

$$
\begin{aligned}
S_T(\omega) &= \frac{1}{T} \int_{-T}^{0} e^{-i\omega\tau} K_{ff}(\tau)\, d\tau \int_{-T/2}^{\tau+T/2} dt_1 + \frac{1}{T} \int_{0}^{T} e^{-i\omega\tau} K_{ff}(\tau)\, d\tau \int_{\tau-T/2}^{T/2} dt_1 \\
&= \frac{1}{T} \left[\int_{-T}^{0} K_{ff}(\tau)\,(T+\tau)\, e^{-i\omega\tau} d\tau + \int_{0}^{T} K_{ff}(\tau)\,(T-\tau)\, e^{-i\omega\tau} d\tau \right] \\
&= \frac{1}{T} \int_{-T}^{T} (T - |\tau|)\, e^{-i\omega\tau} K_{ff}(\tau)\, d\tau = \int_{-T}^{T} \left(1 - \frac{|\tau|}{T}\right) e^{-i\omega\tau} K_{ff}(\tau)\, d\tau \ .
\end{aligned}
$$

Taking the $T \to \infty$ limit, we finally obtain

$$
S(\omega) = \int_{-\infty}^{\infty} e^{-i\omega\tau} K_{ff}(\tau)\, d\tau \ , \tag{5.91}
$$

which is the required result. It shows that the spectral power density is in fact the Fourier transform of the correlation function.

As an example of calculating a time correlation function, let us consider a one-dimensional Brownian particle. Its equation of motion reads

$$
\dot{p} = -\gamma p + \xi(t) \ , \tag{5.92}
$$

where $p(t)$ is the particle momentum, γ friction coefficient and $\xi(t)$ a random force. The two forces in the right-hand side are due to random collisions of liquid particles with the Brownian particle: the random force tends to provide energy to the Brownian particle, while the friction force, $-\gamma p$, is responsible for taking any extra energy out, so that on balance the average kinetic energy of the particle,

$$
\frac{1}{2m} \langle p(t)^2 \rangle = \frac{1}{2} k_B T \tag{5.93}
$$

would correspond correctly to the temperature T of the liquid by virtue of the equipartition theorem. Above m is the Brownian particle mass and k_B is Boltzmann's constant.

The random force acting on the particle does not have any memory[3]; correspondingly, we shall assume that its correlation function is proportional to the delta function of the time difference:

$$\langle \xi(t)\, \xi(t') \rangle = \alpha \delta(t - t') \, , \tag{5.94}$$

where $\alpha = 2mk_BT\gamma$. This particular choice of the proportionality constant will become apparent later on. This expression shows that there is no correlation between the random force at different times $t \neq t'$ since $\delta(t - t') = 0$ in this case.

Our goal is to calculate the momentum-momentum correlation function $K_{pp}(\tau) = \langle p(t + \tau)p(t) \rangle$. We shall use the Fourier transform method to perform this calculation. Defining

$$p(\omega) = \int e^{-i\omega t} p(t) dt \quad \text{and} \quad p(t) = \int \frac{d\omega}{2\pi} e^{i\omega t} p(\omega) \, ,$$

and applying the Fourier transform to the both sides of the equation of motion (5.92), we obtain

$$i\omega p(\omega) = -\gamma p(\omega) + \xi(\omega) \quad \Longrightarrow \quad p(\omega) = -i \frac{\xi(\omega)}{\omega - i\gamma} \, ,$$

where $\xi(\omega)$ is the Fourier transform of the random force. The correlation function becomes

$$K_{pp}(\tau) = \int \frac{d\omega d\omega'}{(2\pi)^2} e^{i\omega(t+\tau)} e^{i\omega' t} \langle p(\omega) p(\omega') \rangle$$

$$= - \int \frac{d\omega d\omega'}{(2\pi)^2} e^{i\omega(t+\tau)} e^{i\omega' t} \frac{\langle \xi(\omega) \xi(\omega') \rangle}{(\omega - i\gamma)(\omega' - i\gamma)} \, .$$

To calculate the correlation function of the random force, we shall return back from frequencies to times by virtue of the random force Fourier transform:

$$\langle \xi(\omega) \xi(\omega') \rangle = \int dt \int dt' e^{-i\omega t} e^{-i\omega' t'} \langle \xi(t) \xi(t') \rangle$$

$$= \alpha \int dt \int dt' e^{-i\omega t} e^{-i\omega' t'} \delta(t - t')$$

$$= \alpha \int dt\, e^{-i(\omega + \omega')t} = 2\pi\alpha \delta(\omega + \omega') \, .$$

---

[3] The case with the memory will be considered in Sect. 6.6.2.

Therefore,

$$
\begin{aligned}
K_{pp}(\tau) &= -\int \frac{d\omega\, d\omega'}{(2\pi)^2} e^{i\omega(t+\tau)} e^{i\omega' t} \frac{2\pi\alpha\delta\,(\omega+\omega')}{(\omega-i\gamma)\,(\omega'-i\gamma)} \\
&= -\alpha \int \frac{d\omega}{2\pi} \frac{e^{i\omega\tau}}{(\omega-i\gamma)\,(-\omega-i\gamma)} = \alpha \int \frac{d\omega}{2\pi} \frac{e^{i\omega\tau}}{\omega^2+\gamma^2}\ .
\end{aligned}
$$

The integral is performed in the complex z plane. There are two poles $z = \pm i\gamma$, one in the upper and one in the lower halves of the complex plane. For $\tau > 0$ we enclose the horizontal axis in the upper half where only the pole $z = i\gamma$ contributes, which gives $K_{pp}(\tau) = (\alpha/2\gamma)\, e^{-\gamma\tau}$. For negative times $\tau < 0$ the horizontal axis is enclosed by a large semicircle in the lower part of the complex plane, where the pole $z = -i\gamma$ contributes and an extra minus sign comes from the opposite direction in which the contour is traversed. This gives $K_{pp}(\tau) = (\alpha/2\gamma)\, e^{\gamma\tau}$. Therefore, for any time, we can write

$$
K_{pp}(\tau) = \frac{\alpha}{2\gamma} e^{-\gamma|\tau|} = mk_B T e^{-\gamma|\tau|}\ . \tag{5.95}
$$

Several interesting observations can be made. Firstly, the correlation function does indeed decay (exponentially) with time, i.e. collisions with particles of the liquid destroy any correlation in motion of the Brownian particles on the time scale of $1/\gamma$. Indeed, as one would expect, the stronger the friction, the faster any correlation in the particle motion is destroyed as the Brownian particle undergoes many frequent collisions; if the friction is weak, the particle moves longer without collisions, they happen less frequently and hence the correlation between different times decays weaker. Secondly, at $\tau = 0$ we obtain the average momentum square, $K_{pp}(0) = \langle p(t)p(t)\rangle = \langle p(t)^2\rangle$, which at equilibrium should remain unchanged on average and equal to $mk_B T$ in accordance with Eq. (5.93). And indeed, $K_{pp}(0) = mk_B T$ as required. It is this particular physical condition which fixes the value of the constant α in Eq. (5.94). Finally, the correlation function is indeed even with respect to the time variable τ as required by Eq. (5.88).

Problem 5.50 Consider a one-dimensional damped harmonic oscillator,

$$m\ddot{x} + kx = -\gamma m\dot{x} + \xi(t) ,$$

which experiences a random force $\xi(t)$ satisfying Eq. (5.94) with the same $\alpha = 2mk_BT\gamma$ as above. Show first that after the Fourier transform the particle velocity $v(\omega) = \dot{x}(\omega) = i\omega x(\omega)$ is given by

$$v(\omega) = \frac{-i}{m} \frac{\omega \xi(\omega)}{\left(\omega^2 - \omega_0^2\right) - i\gamma\omega} ,$$

where $\omega_0 = \sqrt{k/m}$ is the fundamental frequency of the oscillator. Then, using this result shows that the velocity autocorrelation function

$$K_{vv}(t) = \langle v(t)v(0) \rangle = \frac{2k_BT\gamma}{m} \int \frac{d\omega}{2\pi} \frac{\omega^2}{\left(\omega^2 - \omega_0^2\right)^2 + (\gamma\omega)^2} e^{i\omega t} .$$

Calculating this integral in the complex plane separately for $t > 0$ and $t < 0$, show that

$$\langle v(t)v(0) \rangle = \frac{k_BT}{m} \left[\cos\left(\sqrt{D}t\right) - \frac{\gamma}{2} \frac{\sin\left(\sqrt{D}|t|\right)}{\sqrt{D}} \right] e^{-\gamma|t|/2} \quad \text{if} \quad \gamma < 2\omega_0$$

and

$$\langle v(t)v(0) \rangle = \frac{k_BT}{m} \left[\cosh\left(\sqrt{-D}t\right) - \frac{\gamma}{2} \frac{\sinh\left(\sqrt{-D}|t|\right)}{\sqrt{-D}} \right] e^{-\gamma|t|/2} \quad \text{if} \quad \gamma > 2\omega_0 .$$

Above, $D = \omega_0^2 - \gamma^2/4$. Note that $\langle v(0)^2 \rangle = k_BT/m$ as required by the equipartition theorem (cf. also with Problem 4.12, where a different method was used).

5.3.5 *Fraunhofer Diffraction*

Consider propagation of light through a wall with a hole in it, an obstacle called aperture. On the other side of the wall a diffraction pattern will be seen because of the wave nature of light. If in the experiment the source of light is placed far away from the aperture so that the coming light can be considered as consisting of plane

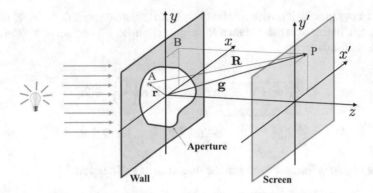

Fig. 5.6 Fraunhofer diffraction: parallel rays of light from a source (on the left) are incident on the wall with an aperture. They are observed at point P on the screen which is far away from the wall. Every point A on the $x - y$ plane within the aperture serves as a source of a spherical wave of light. All such waves from all points within the aperture contribute to the final signal observed at P

waves, and the observation of the diffraction pattern is made on the screen placed also far away from the aperture, this particular diffraction bears the name of Fraunhofer.

Consider an infinitely adsorbing wall (i.e. fully non-radiative) placed in the $x - y$ plane with a hole of an arbitrary shape. The middle of the hole (to be conveniently defined in each particular case) is aligned with the centre of the coordinate system, Fig. 5.6. The positive direction z is chosen towards the observation screen (on the right in the figure). Each point A within the aperture with the vector $\mathbf{r} = (x, y, 0)$ (more precisely, a small surface area $dS = dx dy$ around this point) becomes an independent source of light waves which propagate out of it in such a way that their amplitude

$$dF(\mathbf{g}) \propto \frac{1}{R} e^{-i\omega(t - R/c)} dS ,$$

where $R = |\mathbf{R}|$ is the distance between the point A and the observation point P which position is given by the vector $\mathbf{g}$ drawn from the centre of the coordinate system, c speed of light and ω frequency ($k = \omega/c$ is the wave vector). Note that the amplitude of a spherical wave decays as $1/R$ with distance, and we have also explicitly accounted for in the formula above the retardation effects (see Sect. 5.3.2). In order to calculate the total contribution at point P, one has to integrate over the whole area of the aperture:

$$F(\mathbf{g}) \propto e^{-i\omega t} \int_S \frac{dS}{R} e^{ikR} . \tag{5.96}$$

The observation point is positioned far away from the small aperture, i.e. $r \ll g = |\mathbf{g}|$, see Fig. 5.6. In this case the distance $R = |\mathbf{g} - \mathbf{r}|$ can be worked out using the cosine theorem as follows:

$$R = \sqrt{g^2 + r^2 - 2\mathbf{g} \cdot \mathbf{r}} = g\sqrt{1 + \left(\frac{r}{g}\right)^2 - 2\widehat{\mathbf{g}} \cdot \frac{\mathbf{r}}{g}}$$

$$\simeq g\sqrt{1 - 2\widehat{\mathbf{g}} \cdot \frac{\mathbf{r}}{g}} \simeq g\left(1 + \widehat{\mathbf{g}} \cdot \frac{\mathbf{r}}{g}\right) = g + \widehat{\mathbf{g}} \cdot \mathbf{r} \, ,$$

where $\widehat{\mathbf{g}} = \mathbf{g}/g$ is the unit vector in the direction of $\mathbf{g}$. Therefore,

$$F(\mathbf{g}) \propto e^{-i\omega t} e^{ikg} \int_S \frac{dS}{g + \widehat{\mathbf{g}} \cdot \mathbf{r}} e^{i\mathbf{k}\cdot\mathbf{r}} \propto \int_S \frac{dS}{g} e^{i\mathbf{k}\cdot\mathbf{r}} \propto \int_S e^{i\mathbf{k}\cdot\mathbf{r}} dS \, , \tag{5.97}$$

where $\mathbf{k} = k\widehat{\mathbf{g}}$ and we have made use again of the fact that the observation point is far away from the aperture and hence have removed the $\widehat{\mathbf{g}} \cdot \mathbf{r}$ term in the denominator. We have also omitted the whole pre-factor at the last steps; we shall reinstate it later on.

In order to connect the direction $\mathbf{g}$ (or $\mathbf{k} = k\widehat{\mathbf{g}}$) with the coordinates x' and y' of the observation point P on the screen, we notice that projections of this point on both the screen and on the wall are the same as the two planes are parallel to each other, see additional dotted lines in the figure. In particular, the point B on the wall corresponds to the observation point P on the screen. Therefore, $g_x = x'$ and $g_y = y'$. If by z we denote the distance between the screen and the wall, then $z = g_z \simeq g$, and hence

$$k_x = k\widehat{\mathbf{g}}_x = \frac{k}{g} g_x \simeq \frac{k}{z} x' \quad \text{and} \quad k_y \simeq \frac{k}{z} y' \, , \tag{5.98}$$

yielding $F(\mathbf{g})$ being directly dependent on the coordinates (x', y') on the screen, i.e. it can be written as $F(k_x, k_y)$.

In order to work out the pre-factor to the integral in Eq. (5.97), we first notice that this integral is performed over the aperture area only. It can be extended to the whole $x - y$ plane if we introduce the so-called *aperture function* $\zeta(x, y)$, which is equal to one within the aperture and zero otherwise. Then,

$$F(k_x, k_y) = \lambda \int \zeta(x, y) e^{i(k_x x + k_y y)} dx dy \, , \tag{5.99}$$

i.e. it is seen as a two-dimensional Fourier transform of the aperture function. Here λ is a yet unknown amplitude. We shall now relate it to the total intensity density I_0 (i.e. calculated per unit area) of the incident light waves on the aperture. Indeed, the inverse Fourier transform of the last equation reads

$$\zeta(x, y) = \frac{1}{\lambda} \int F(k_x, k_y) e^{-i(k_x x + k_y y)} \frac{dk_x dk_y}{(2\pi)^2} \, .$$

Both functions, $\lambda\zeta(x, y)$ and $F(k_x, k_y)$ must then be related by Parseval's theorem (5.48) which in our notations reads

$$\lambda^2 \int |\zeta(x, y)|^2 \, dx\,dy = \frac{1}{(2\pi)^2} \int |F(k_x, k_y)|^2 \, dk_x\,dk_y \ .$$

The integral in the left-hand side is equal to the area $S = \int_S dx\,dy$ of the aperture. The integral in the right-hand side gives the total intensity $I_0 S$ of all waves which made it through the aperture. Indeed, the total intensity is given by the module of the amplitude squared, $|F(k_x, k_y)|^2$, integrated over all possible $\mathbf{k}$ vectors; so, if I_0 is the intensity density (i.e. intensity per unit area), then in the right-hand side we have indeed $I_0 S$. Therefore, $\lambda^2 S = I_0 S/(2\pi)^2$, hence $\lambda = \sqrt{I_0}/2\pi$ and we can finally write

$$F(k_x, k_y) = \frac{\sqrt{I_0}}{2\pi} \int \zeta(x, y) \, e^{i(k_x x + k_y y)} dx\,dy \ . \qquad (5.100)$$

This is the final result. The amplitude is proportional to the two-dimensional Fourier transform of the aperture function. The intensity distribution at point $P(x', y')$ on the screen is then given by $dI(x', y') = |F(k_x, k_y)|^2 \, dk_x\,dk_y$, where the relations between the components of the wave vector $\mathbf{k}$ and the coordinates (x', y') on the screen are given by Eq. (5.98).

Problem 5.51 Consider a square aperture with dimensions $h \times h$ separated by the distance z ($z \gg h$) from the screen. Calculate the Fourier transform of the corresponding aperture function and hence show that the intensity of light on the $dx'dy'$ area of the screen is given by

$$dI(x', y') = I_0 \left(\frac{h^2 k}{4\pi z}\right)^2 \text{sinc}^2\left(\frac{kh}{2z}x'\right) \text{sinc}^2\left(\frac{kh}{2z}y'\right) dx'dy' \ , \qquad (5.101)$$

where $\text{sinc}(x) = \sin x / x$ is the so-called sinc function.

Problem 5.52 Consider a circular aperture of radius R. Using polar coordinates for points on the aperture, i.e. $(x, y) \rightarrow (r, \phi)$, and on the screen, i.e. $(x', y') \rightarrow (r', \phi')$, first show that $k_x x + k_y y = (k/z) \, r r' \cos (\phi - \phi')$. Then, integrating in polar coordinates, show that the amplitude in this case

$$F\left(r', \phi'\right) = \sqrt{I_0} \int_0^R r J_0 \left(\frac{krr'}{z}\right) dr ,$$

where $J_0(x)$ is the corresponding Bessel function, Sect. 4.6, and we have made use of a well-known integral representation of it, Eq. (4.233):

$$J_0(x) = \frac{1}{2\pi} \int_0^{2\pi} e^{ix \cos \phi} d\phi .$$

Finally, using the recurrence relation (4.221) for the Bessel functions, integrate over r and show that

$$F\left(r', \phi'\right) = \frac{Rz\sqrt{I_0}}{kr'} J_1 \left(\frac{kR}{z} r'\right) .$$

Correspondingly, the intensity distribution is given by

$$dI\left(r', \phi'\right) = I_0 \left(\frac{Rz}{k}\right)^2 \frac{1}{r'} J_1^2 \left(\frac{kR}{z} r'\right) dr' d\phi'$$

in this case.

Problem 5.53 Consider now an infinite slit of width h running along the y direction. In that case the light arriving at a point (x, y) within the aperture will only be diffracted in the $x - z$ plane. Repeating the above derivation which led us to Eq. (5.100), show that in this essentially one-dimensional case the amplitude on the screen per unit length along the slit is given by the following formula:

$$F\left(k_x\right) = \sqrt{\frac{I_0}{2\pi}} \int e^{ik_x x} dx = h \sqrt{\frac{I_0}{2\pi}} \mathrm{sinc} \left(\frac{kh}{2z} x'\right) , \qquad (5.102)$$

leading to the intensity distribution

$$dI\left(x'\right) = \frac{kh^2 I_0}{2\pi z} \mathrm{sinc}^2 \left(\frac{kh}{2z} x'\right) dx' . \qquad (5.103)$$

Chapter 6
Laplace Transform

In Chap. 5 we considered the Fourier transform (FT) of a function $f(x)$ defined on the whole real axis $-\infty < x < \infty$. Recall that for the FT to exist, the function $f(x)$ must tend to zero at $x \to \pm\infty$ to ensure the convergence of the integral $\int_{-\infty}^{\infty} |f(x)|\, dx$. However, sometimes it is useful to study functions $f(x)$ defined only for $x \geq 0$ which may also increase indefinitely as $x \to +\infty$ (but not faster than an exponential function, see below). For instance, this kind of problems are encountered in physics when dealing with processes in time t when we are interested in the behaviour of some function $f(t)$ at times *after* some 'initial' time (considered as $t = 0$), at which the system was 'prepared' (the initial value problem), e.g. prior to some perturbation acting on it. In such cases another integral transform, named after Laplace, the so-called Laplace transform (LT), has been found very useful in many applications.

In this chapter[1] main properties and applications of the Laplace transform are presented.

6.1 Definition

The idea of why the LT might be useful can be understood if we recall why the FT could be useful. Indeed, the FT was found useful, for instance, in solving differential equations (DE): (i) one first transforms the DE into the Fourier space, i.e. $\{x,\ f(x)\} \to \{\nu,\ F(\nu)\}$; (ii) the DE in the Fourier space (the $\nu-$space) for the 'image' $F(\nu)$ appears to be simpler (e.g. it becomes entirely algebraic with no derivatives) and can then be solved; then, finally, (iii) using the inverse FT, $\{\nu,\ F(\nu)\} \to \{x,\ f(x)\}$, the original function $f(x)$ we are interested in is found. In

---

[1] In the following, references to the first volume of this course (L. Kantorovich, Mathematics for natural scientists: fundamentals and basics, 2nd Edition, Springer, 2022) will be made by appending the Roman number I in front of the reference, e.g. Sect. ,I.1.8 or Eq. (I.5.18) refer to Sect. 1.8 and Eq. (5.18) of the first volume, respectively.

---

other words, to solve the problem, one 'visits' the Fourier space where the problem at hand looks simpler, solves it there, but then 'returns back' to the original x−space by means of the inverse FT. Similarly one operates with the LT: first, the problem of finding the function $f(t)$ is converted into the 'Laplace space' by performing the LT ($t \to p$ and $f(t) \to F(p)$), where p is a complex number; then the 'image' $F(p)$ of the function of interest is found, which is then converted back into the t−space, $F(p) \to f(t)$, by means of the inverse LT. In fact, we shall see later on that the two transforms are closely related to each other.

Consider a (generally) complex function $f(t)$ of the *real* argument $t \geq 0$ which is continuous everywhere apart from some number of discontinuities of the first kind[2]. Moreover, we assume that within each finite interval there could only be a finite number of such discontinuities. Next, for definiteness, we shall set the values of $f(t)$ at negative t to zero: $f(t) = 0$ for $t < 0$. Finally, we shall assume that $f(t)$ may increase with t, but this can happen not faster than for an exponential function $e^{x_0 t}$ with some positive exponent x_0. In other words, we assume that

$$|f(t)| \leq M e^{x_0 t} \quad \text{for any} \quad t \geq 0 , \tag{6.1}$$

where M is some positive constant. An example of such a function is, for instance, the exponential e^{2t}. It goes to infinity when $t \to +\infty$, however, this happens not faster than $e^{x_0 t}$ with any $x_0 > 2$. At the same time, the function $f(t) = e^{2t^2}$ grows much faster than the exponential function $e^{x_0 t}$ with *any* x_0, and hence this class of functions we exclude from our consideration. Note that if $f(t)$ is limited, i.e. $|f(t)| \leq M$, then $x_0 = 0$. The positive number x_0 characterising the exponential growth of $f(t)$ we shall call the growth order parameter of $f(t)$. We shall call the function $f(t)$ the *original*.

We shall then define the Laplace transform (LT) of $f(t)$ by means of the following formula:

$$\mathcal{L}[f(t)] = \int_0^\infty f(t)e^{-pt}dt = F(p) . \tag{6.2}$$

Here $f(t)$ is a function in the 't-space', while its transform, $F(p) = \mathcal{L}[f]$, is the corresponding function in the 'p-space', where p is generally a complex number. The function $F(p)$ in the complex plane will be called the *image* of the original $f(t)$.

Before we investigate the properties of the LT, it is instructive to consider some examples.

The simplest case corresponds to the function $f(t) = 1$. According to the definition of the LT, we can write

$$\mathcal{L}[f] = \int_0^\infty e^{-pt}dt = -\frac{1}{p}e^{-pt}\Big|_0^\infty = \frac{1}{p} .$$

---

[2] Recall that this means that one side limits, not equal to each other, exist on both sides of the discontinuity.

A very important point here is that we set to zero the value of the exponential function at the $t = +\infty$ boundary. This is legitimate only if a part of the complex plane $\mathcal{C}$ is considered for the complex numbers p. Indeed, if we only consider the numbers $p = x + iy$ with the positive real part, $\mathrm{Re}(p) = x > 0$, i.e. the right semi-plane of $\mathcal{C}$, then $e^{-pt} = e^{-xt}e^{-iyt} \to 0$ at $t \to \infty$ and this ensures the convergence of the integral at the upper limit (note that $\left|e^{-iyt}\right| = 1$).

Let us next calculate the LT of $f(t) = e^{-\alpha t}$ with some complex α. According to the definition,

$$\mathcal{L}[f] = \int_0^\infty e^{-(p+\alpha)t}\,dt = \left.\frac{e^{-(p+\alpha)t}}{-(p+\alpha)}\right|_0^\infty = \frac{1}{p+\alpha}\,.$$

To ensure the convergence of the integral at the upper limit we have to consider only such values of p in the complex plane $\mathcal{C}$ which satisfy the following condition: $\mathrm{Re}(p + \alpha) > 0$. Indeed, only in this case $e^{-(p+\alpha)t} \to 0$ as $t \to \infty$. In other words, the LT of the function $e^{-\alpha t}$ exists for any p to the right of the vertical line drawn in $\mathcal{C}$ via the point $-\mathrm{Re}(\alpha)$.

Since the function $f(t)$ enters the integral (6.2) linearly, the LT represents a *linear operator*, i.e. for any two functions $f(t)$ and $g(t)$ satisfying the necessary conditions outlined above,

$$\mathcal{L}[\alpha f(t) + \beta f(t)] = \int_0^\infty [\alpha f(t) + \beta f(t)]e^{-pt}\,dt = \alpha\mathcal{L}[f] + \beta\mathcal{L}[g]\,, \quad (6.3)$$

where α and β are arbitrary complex numbers. The linearity property allows simplifying the calculation of some transforms as illustrated by the example of calculating the LT of the sine and cosine functions $f(t) = \cos\omega t$ and $g(t) = \sin\omega t$ (assuming ω is real). Since

$$f(t) = \cos\omega t = \frac{1}{2}\left(e^{i\omega t} + e^{-i\omega t}\right)\,,$$

we write

$$\mathcal{L}[\cos\omega t] = \frac{1}{2}\left\{\mathcal{L}\left[e^{i\omega t}\right] + \mathcal{L}\left[e^{-i\omega t}\right]\right\} = \frac{1}{2}\left(\frac{1}{p - i\omega} + \frac{1}{p + i\omega}\right)$$

$$= \frac{p}{p^2 + \omega^2}\,.$$

(6.4)

Here we must assume that $\mathrm{Re}\,(p) > 0$.

Problem 6.1 Similarly, show that

$$\mathcal{L}\left[\sin \omega t\right] = \frac{\omega}{p^2 + \omega^2} \; . \tag{6.5}$$

Problem 6.2 Obtain the same formulae for the LT of the sine and cosine functions directly by calculating the LT integral.

Problem 6.3 Prove that for real ω and $\mathrm{Re}(p) > 0$

$$\mathcal{L}\left[t \sin \omega t\right] = \frac{2\omega p}{\left(p^2 + \omega^2\right)^2} \quad \text{and} \quad \mathcal{L}\left[t \cos \omega t\right] = \frac{p^2 - \omega^2}{\left(p^2 + \omega^2\right)^2} \; . \tag{6.6}$$

Problem 6.4 Prove that for $\mathrm{Re}(p) > |\mathrm{Re}(\alpha)|$

$$\mathcal{L}\left[\sinh\left(\alpha t\right)\right] = \frac{\alpha}{p^2 - \alpha^2} \quad \text{and} \quad \mathcal{L}\left[\cosh\left(\alpha t\right)\right] = \frac{p}{p^2 - \alpha^2} \; . \tag{6.7}$$

Problem 6.5 Prove that for real ω and $\mathrm{Re}\left(p\right) > -\mathrm{Re}(\alpha)$

$$\mathcal{L}\left[e^{-\alpha t} \sin \omega t\right] = \frac{\omega}{(p + \alpha)^2 + \omega^2} \quad \text{and} \quad \mathcal{L}\left[e^{-\alpha t} \cos \omega t\right] = \frac{p + \alpha}{(p + \alpha)^2 + \omega^2} \; . \tag{6.8}$$

Problem 6.6 Prove that for real α and β, and for $\mathrm{Re}\left(p\right) > \max\left(\alpha \pm \beta\right)$

$$\mathcal{L}\left[e^{\alpha t} \sinh\left(\beta t\right)\right] = \frac{\beta}{(p - \alpha)^2 - \beta^2} \quad \text{and} \quad \mathcal{L}\left[e^{\alpha t} \cosh\left(\beta t\right)\right] = \frac{p - \alpha}{(p - \alpha)^2 - \beta^2} \; . \tag{6.9}$$

Problem 6.7 It was shown above that $\mathcal{L}\left[t^0\right] = 1/p$. Show that $\mathcal{L}\left[t\right] = 1/p^2$. Then prove by induction that

$$\mathcal{L}\left[t^n\right] = \frac{n!}{p^{n+1}} \; , \quad n = 0, 1, 2, 3 \ldots . \tag{6.10}$$

Problem 6.8 Prove by induction that

$$\mathcal{L}\left[t^n e^{-\alpha t}\right] = \frac{n!}{(p+\alpha)^{n+1}} , \quad n = 0, 1, 2, 3\dots. \tag{6.11}$$

Problem 6.9 In fact, show that the following result is generally valid:

$$\mathcal{L}\left[(-t)^n f(t)\right] = \frac{d^n}{dp^n} F(p) , \tag{6.12}$$

where $\mathcal{L}[f] = F(p)$. *[Hint: the formula follows from the definition (6.2) upon differentiating its both sides n times.]*

Problem 6.10 Using the rule (6.12), re-derive Eq. (6.10).

Problem 6.11 Using the rule (6.12), re-derive Eq. (6.11).

Problem 6.12 Show that

$$\mathcal{L}\left[e^{-\alpha t} \cos^2(\beta t)\right] = \frac{1}{2(p+\alpha)} + \frac{1}{2} \frac{p+\alpha}{(p+\alpha)^2 + 4\beta^2} .$$

Problem 6.13 Show that for $\tau \geq 0$

$$\mathcal{L}[H(t-\tau)] = \frac{1}{p} e^{-p\tau} ,$$

where $H(t)$ is the Heaviside unit step function.

Problem 6.14 Show that

$$\mathcal{L}\left[\frac{e^{\pm it}}{\sqrt{2\pi t}}\right] = \frac{1}{\sqrt{2}} \frac{1}{\sqrt{p \mp i}} . \tag{6.13}$$

[Hint: make the change of variables $t \rightarrow x = \sqrt{t}$ and note that for any α with $\mathrm{Re}(\alpha) > 0$ we have

$$\int_0^\infty e^{-\alpha x^2} dx = \sqrt{\frac{\pi}{4\alpha}} ,$$

which follows from Eq. (4.38).]

As another useful example, let us consider the LT of the Dirac delta function $f(t) = \delta(t - \tau)$ with some positive $\tau > 0$:

$$\mathcal{L}[\delta(x - \tau)] = \int_0^\infty \delta(t - \tau) e^{-pt} dt = e^{-p\tau} . \tag{6.14}$$

In particular, by taking the limit $\tau \to +0$, we conclude that

$$\mathcal{L}[\delta(t)] = 1 . \tag{6.15}$$

6.2 Method of Partial Fractions

If we are faced with a problem of calculating the inverse LT of a rational fraction $R_n(p)/Q_m(p)$ given by the ratio of two polynomials (of orders n and m, respectively, with $n < m$), this calculation can always be done by presenting the fraction as a sum of simpler fractions. The latter can always be done using the method we developed in Sect. I.2.3.2, see Theorem I.2.9, of Vol. I.

There, the Theorem was proven for polynomials with real coefficients and it states that if the denominator $Q_m(p)$ of the rational function has a root a of order k (i.e. it is a repeated root, repeated k times), then in the expansion of the rational function into simpler fractions there will be terms (among others)

$$\frac{A_1}{p - a} + \frac{A_2}{(p - a)^2} + \cdots + \frac{A_k}{(p - a)^k} , \tag{6.16}$$

where A_i are some coefficients. If, however, the root a is complex, then it always comes as a complex conjugate pair a and a^* of two roots, in which case the following terms will appear in expanding the rational fraction:

$$\frac{B_1 p + C_1}{p^2 + sp + q} + \frac{B_2 p + C_2}{(p^2 + sp + q)^2} + \cdots + \frac{B_k p + C_k}{(p^2 + sp + q)^k} , \tag{6.17}$$

where k is the repetition of the pair of the roots, and $p^2 + sp + q$ is the square polynomial which has the pair of the roots a and a^*; yet again, B_i and C_i are numerical coefficients. Note that the given terms represent just a portion of the whole expansion of the original fraction, the portion that is due to the particular real root or a pair of complex conjugate ones.

This kind of reasoning is very useful in calculating the inverse LT of a rational fraction since the inverse LT of each of the terms in the simple fractions given above can actually be calculated without difficulty. Consider first the terms that appear in the expansion (6.16). We know that

$$\mathcal{L}^{-1}\left[\frac{1}{p-a}\right] = e^{at}.$$

Then, according to formula (6.11),

$$\mathcal{L}^{-1}\left[\frac{1}{(p-a)^{n+1}}\right] = \frac{t^n}{n!}e^{at}.$$

Consider now the fractions that appear in Eq. (6.17). In view of Eq. (6.8), it is actually more convenient to rewrite the terms of Eq. (6.17) as

$$\frac{B_1(p+b)+D_1}{(p+b)^2+c} + \frac{B_2(p+b)+D_2}{\left((p+b)^2+c\right)^2} + \cdots + \frac{B_k(p+b)+D_k}{\left((p+b)^2+c\right)^k}, \tag{6.18}$$

where in the denominators we have completed the square[3]:

$$(p+b)^2 + c = p^2 + sp + q \ \text{ with } \ b = \frac{s}{2} \ \text{ and } \ c = q - \frac{s^2}{4}.$$

Then, it follows that

$$\mathcal{L}^{-1}\left[\frac{B_1(p+b)+D_1}{(p+b)^2+c}\right] = B_1\mathcal{L}^{-1}\left[\frac{p+b}{(p+b)^2+c}\right] + D_1\mathcal{L}^{-1}\left[\frac{1}{(p+b)^2+c}\right]$$

$$= B_1 e^{-bt}\cos\left(\sqrt{c}t\right) + \frac{D_1}{\sqrt{c}}e^{-bt}\sin\left(\sqrt{c}t\right).$$

The calculation of the other terms in Eq. (6.18) can also be performed considering derivatives of the first fraction calculated with respect to p and using Eq. (6.12). However, it is much easier to work directly with complex roots of the polynomial $Q_m(p)$. This is because then both roots can be considered as different and the expansion would contain only terms like the ones in Eq. (6.16) for which the inverse LT is trivially calculated. Also, this method is not limited to polynomials with real coefficients and can be used for general rational functions containing complex coefficients as well.

As an example, consider the decomposition of the fraction in Eq. (I.2.19) that we have considered in Vol. I:

$$\frac{3p^2+p-1}{(p-1)^2\left(p^2+1\right)} = \frac{2}{p-1} + \frac{3}{2(p-1)^2} - \frac{4p+1}{2\left(p^2+1\right)}.$$

---

[3] Since both roots form a complex conjugate pair $p_{\pm} = -s/2 \pm i\sqrt{q - s^2/4}$, the constant $c = q - s^2/4$ must definitely be positive.

Then, we shall have:

$$\mathcal{L}^{-1}\left[\frac{3p^2 + p - 1}{(p-1)^2(p^2+1)}\right] = \mathcal{L}^{-1}\left[\frac{2}{p-1}\right] + \mathcal{L}^{-1}\left[\frac{3}{2(p-1)^2}\right] - \mathcal{L}^{-1}\left[\frac{4p+1}{2(p^2+1)}\right]$$

$$= 2e^t + \frac{3}{2}te^t - 2\cos t - \frac{1}{2}\sin t .$$

Another example, in which we shall deal directly with complex roots: consider the fraction expansion of

$$\frac{1}{p^4 - 5p^3 + 7p^2 - 5p + 6} .$$

The polynomial here has four roots: $\pm i$, 2 and 3. Expanding into the partial fractions, we obtain

$$\frac{1}{(p-i)(p+i)(p-2)(p-3)} = \frac{1}{10(p-3)} - \frac{1}{5(p-2)} + \left(\frac{1-i}{20}\right)\frac{1}{p-i} + \left(\frac{1+i}{20}\right)\frac{1}{p+i} ,$$

so that

$$\mathcal{L}^{-1}\left[\frac{1}{(p-i)(p+i)(p-2)(p-3)}\right] = \frac{1}{10}\mathcal{L}^{-1}\left[\frac{1}{p-3}\right] - \frac{1}{5}\mathcal{L}^{-1}\left[\frac{1}{p-2}\right]$$

$$+ \left(\frac{1-i}{20}\right)\mathcal{L}^{-1}\left[\frac{1}{p-i}\right] + \left(\frac{1+i}{20}\right)\mathcal{L}^{-1}\left[\frac{1}{p+i}\right]$$

$$= \frac{1}{10}e^{3t} - \frac{1}{5}e^{2t} + \left(\frac{1-i}{20}\right)e^{it} + \left(\frac{1+i}{20}\right)e^{-it}$$

$$= \frac{1}{10}e^{3t} - \frac{1}{5}e^{2t} + \frac{1}{10}\cos t + \frac{1}{10}\sin t .$$

Problem 6.15 Show that

$$\mathcal{L}^{-1}\left[\frac{2p^2 - 9p + 11}{(p-3)(p^2 - 4p + 5)}\right] = e^{3t} + e^{2t}\cos t ;$$

$$\mathcal{L}^{-1}\left[\frac{p-1}{(p^2 - 4p + 5)^2}\right] = \frac{1}{2}e^{2t}\sin t - \frac{t}{2}e^{2t}(\cos t - \sin t) .$$

6.3 Detailed Consideration of the LT

Here we shall consider the LT in more detail including the main theorems related to it. Various properties of the LT which are needed for the actual use of this method in solving practical problems will be considered in the next section.

6.3.1 Analyticity of the LT

Theorem 6.1 *If $f(t)$ is of an exponential growth, i.e. it goes to infinity not faster than the exponential function $e^{x_0 t}$ with some positive growth order parameter $x_0 > 0$, then the LT $\mathcal{L}[f(t)] = F(p)$ of $f(t)$ exists in the semi-plane $Re(p) > x_0$.*

Proof: If $f(t)$ is of the exponential growth, Eq. (6.1), where the positive number x_0 may be considered as a characteristic exponential of the function $f(t)$, then the LT integral (6.2) converges absolutely. Indeed, consider the absolute value of the LT of $f(t)$ at the point $p = x + iy$:

$$|F(p)| = \left| \int_0^\infty f(t) e^{-pt} dt \right| \leq \int_0^\infty \left| f(t) e^{-pt} \right| dt$$

$$= \int_0^\infty |f(t)| \cdot \left| e^{-(x+iy)t} \right| dt = \int_0^\infty |f(t)| e^{-xt} dt ,$$

so that

$$|F(p)| \leq \int_0^\infty |f(t)| e^{-xt} dt \leq M \int_0^\infty e^{x_0 t} e^{-xt} dt$$

$$= M \int_0^\infty e^{-(x-x_0)t} dt = -\frac{M}{x - x_0} e^{-(x-x_0)t} \Big|_0^\infty .$$

If $x - x_0 = Re(p) - x_0 > 0$, i.e. if $Re(p) > x_0$, then the value of the expression above at the upper limit $t = \infty$ is zero and we obtain

$$|F(p)| = \left| \int_0^\infty f(t) e^{-pt} dt \right| \leq \frac{M}{x - x_0} , \tag{6.19}$$

which means that the LT integral converges absolutely and hence the LT $F(p)$ exists. **Q.E.D.**

Note that the estimate above establishes the *uniform* convergence of the LT integral with respect to the imaginary part of p since $|F(p)|$ is bounded by an expression containing only the real part of p. We shall use this fact later on in Sect. 6.3.3.

It follows from this theorem that $F(p) \to 0$ when $|p| \to \infty$ since the expression in the right-hand side of (6.19) tends to zero in this limit. Note that since $\text{Re}(p) > x_0$, this may only happen for arguments ϕ of $p = |p|\,e^{i\phi}$ satisfying the inequality $-\pi/2 < \phi < \pi/2$ with the points $\pm\pi/2$ strictly excluded.

Theorem 6.2 *If $f(t)$ is of an exponential growth with the growth order parameter $x_0 > 0$, then the LT $\mathcal{L}[f(t)] = F(p)$ of $f(t)$ can be differentiated with respect to p in the complex semi-plane $\text{Re}(p) > x_0$.*

Proof: Consider the derivative of $F(p)$:

$$\frac{dF(p)}{dp} = \frac{d}{dp}\int_0^\infty f(t)e^{-pt}\,dt = \int_0^\infty f(t)\frac{d}{dt}\left(e^{-pt}\right)dt = -\int_0^\infty f(t)te^{-pt}\,dt \ .$$

This integral converges absolutely:

$$\left|\frac{dF}{dp}\right| = \left|\int_0^\infty f(t)te^{-pt}\,dt\right| \le \int_0^\infty |f(t)|\,te^{-xt}\,dt$$

$$\le M\int_0^\infty e^{x_0 t}te^{-xt}\,dt = M\int_0^\infty te^{-(x-x_0)t}\,dt \ .$$

This integral can be calculated by parts in the usual way:

$$\int_0^\infty te^{-(x-x_0)t}\,dt = \left.\frac{t}{-(x-x_0)}e^{-(x-x_0)t}\right|_0^\infty + \frac{1}{x-x_0}\int_0^\infty e^{-(x-x_0)t}\,dt \ .$$

The first term is equal to zero both at $t = 0$ (because of the t explicitly present there) and at $t = \infty$ if the condition $x = \text{Re}(p) > \text{Re}(x_0)$ is satisfied (note that the exponential $e^{-\alpha t}$ with $\alpha > 0$ tends to zero much faster than any power of t when $t \to \infty$). Then, only the integral in the right-hand side remains which is calculated trivially to yield:

$$\int_0^\infty te^{-(x-x_0)t}\,dt = \frac{1}{x-x_0}\int_0^\infty e^{-(x-x_0)t}\,dt = -\left.\frac{1}{(x-x_0)^2}e^{-(x-x_0)t}\right|_0^\infty$$

$$= \frac{1}{(x-x_0)^2} \ ,$$

so that we finally conclude that

$$\left|\frac{dF}{dp}\right| \le M\int_0^\infty te^{-(x-x_0)t}\,dt = \frac{M}{(x-x_0)^2} \ ,$$

i.e. it is indeed finite. This means that the function $F(p)$ can be differentiated. **Q.E.D.**

It also follows from the two theorems that $F(p)$ does not have singularities in the right semi-plane $\mathrm{Re}(p) > x_0$, i.e. all possible poles of $F(p)$ can only be in the complex plane on the left of the vertical line $\mathrm{Re}(p) = x_0$. Therefore, the growth order parameter demonstrating the growth of the function $f(t)$ in the 'direct' $t-$space also determines the analytical properties of the 'image' $F(p)$ in the complex $p-$plane.

Theorem 6.2 is proven to be very useful for obtaining LT of new functions without calculating the LT integral. Indeed, as an example, let us derive again formulae (6.6). Differentiating both sides of the LT of the cosine function, Eq. (6.4), with respect to p, we have

$$\frac{d}{dp}\mathcal{L}[\cos(\omega t)] = \frac{d}{dp}\int_0^\infty e^{-pt}\cos(\omega t)dt = -\int_0^\infty e^{-pt}t\cos(\omega t)dt$$
$$= -\mathcal{L}[t\cos(\omega t)].$$

On the other hand,

$$\frac{d}{dp}\mathcal{L}[\cos(\omega t)] = \frac{d}{dp}\frac{p}{p^2+\omega^2} = -\frac{p^2-\omega^2}{\left(p^2+\omega^2\right)^2}.$$

Equating the two, we reproduce the second identity in Eq. (6.6). The other formula is reproduced similarly.

6.3.2   Relation to the Fourier Transform

It appears that the LT and the FT are closely related. We shall establish here the relationship between the two which would also allow us to derive a direct formula for the inverse LT. A different (and more rigorous) derivation of the inverse LT will be given in the next subsection.

Consider a function $f(t)$ for which the LT exists in the region $\mathrm{Re}(p) > x_0$. If we complement $f(t)$ with an extra exponential factor e^{-at} with some *real* parameter a, we shall arrive at the function $g_a(t) = f(t)e^{-at}$ which for any $a > x_0$ will tend to zero at the $t \to +\infty$ limit:

$$|g_a(t)| = \left|f(t)e^{-at}\right| \le Me^{-(a-x_0)t}.$$

Therefore, the FT $G_a(\nu)$ of $g_a(t)$ can be defined.[4] It will depend on both a and ν:

$$G_a(\nu) = \int_{-\infty}^\infty g_a(t)e^{-i2\pi\nu t}dt = \int_0^\infty f(t)e^{-at}e^{-i2\pi\nu t}dt$$
$$= \int_0^\infty f(t)e^{-(a+i2\pi\nu)t}dt .$$

(6.20)

---

[4] Since $f(t)$ is piecewise continuous, so is $g_a(t)$, and hence the other Dirichlet condition is also satisfied for the FT to exist.

Note that we have replaced the bottom integration limit by zero as $f(t) = 0$ for any negative t. The inverse FT of $G_a(t)$ is then given by

$$\int_{-\infty}^{\infty} G_a(\nu)e^{i2\pi\nu t}\,dt = \begin{cases} g_a(t) = f(t)e^{-at}\,, & t > 0 \\ 0\,, & t < 0 \end{cases}.$$

Multiplying both sides of this equation by e^{at}, we obtain

$$\int_{-\infty}^{\infty} G_a(\nu)e^{(a+i2\pi\nu)t}\,d\nu = \begin{cases} f(t)\,, & t > 0 \\ 0\,, & t < 0 \end{cases}. \tag{6.21}$$

The number $p = a + i2\pi\nu$ is some complex number, so that Eqs. (6.20) and (6.21) can also be alternatively written as

$$G_a(\nu) = \int_0^{\infty} f(t)e^{-pt}\,dt \tag{6.22}$$

and

$$f(t) = \int_{-\infty}^{\infty} G_a(\nu)e^{pt}\,d\nu\,. \tag{6.23}$$

One can recognise in Eq. (6.22) the LT of the function $f(t)$, i.e. $G_a(\nu) = \mathcal{L}[f(t)] = F(p)$. In the other Eq. (6.23) we shall change the integration variable from ν to $p = a + i2\pi\nu$ and will replace $G_a(\nu)$ with $F(p)$. This gives:

$$f(t) = \frac{1}{2\pi i}\int_{a-i\infty}^{a+i\infty} F(p)e^{pt}\,dp\,. \tag{6.24}$$

This formula provides a recipe for the inverse LT. Here a is any positive number such that $a > x_0$. One can see that in order to calculate $f(t)$ from its LT $F(p)$, one has to perform an integration in the complex plane of the function $F(p)e^{pt}$ along the vertical line $\text{Re}(p) = a$ from $-i\infty$ to $+i\infty$.

6.3.3   Inverse Laplace Transform

Here we shall re-derive formula (6.24) using an approach similar to the one used when proving the Fourier integral in Sect. 5.1.3.

Theorem 6.3 *If $f(t)$ satisfies all the necessary conditions for its LT $F(p)$ to exist and $f(t)$ is differentiable (strictly speaking, this condition is not necessary, but we shall assume it to simplify the proof), then the following identity is valid:*

$$f(t) = \lim_{b \to \infty} \frac{1}{2\pi i} \int_{a-ib}^{a+ib} e^{pt} \left[\int_0^\infty f(t_1) e^{-pt_1} dt_1 \right] dp . \qquad (6.25)$$

The limit above essentially means that the integral over p is to be understood as the principal value integral when both limits go to $\pm i\infty$ simultaneously.

Proof: Let us consider the function under the limit above:

$$f_b(t) = \frac{1}{2\pi i} \int_{a-ib}^{a+ib} e^{pt} \left[\int_0^\infty f(t_1) e^{-pt_1} dt_1 \right] dp .$$

At the next step we would like to exchange the order of integrals. This is legitimate if the internal integral (over t_1) converges uniformly with respect to the variable y of the external integral taken over the vertical line $p = a + iy$ in the complex plane. It follows from Theorem 6.1 (see the comment immediately following its proof) that for any $a > x_0$ the integral over t_1 indeed converges absolutely and uniformly with respect to the imaginary part of p. Therefore, the order of the two integrals can be interchanged (cf. Sect. I.6.1.3) enabling one to calculate the p−integral:

$$\begin{aligned}
f_b(t) &= \frac{1}{2\pi i} \int_0^\infty dt_1 \, f(t_1) \left[\int_{a-ib}^{a+ib} e^{p(t-t_1)} dp \right] \\
&= \frac{1}{2\pi i} \int_0^\infty dt_1 \, f(t_1) \frac{e^{(a+ib)(t-t_1)} - e^{(a-ib)(t-t_1)}}{t - t_1} \\
&= \frac{1}{\pi} \int_0^\infty f(t_1) e^{a(t-t_1)} \frac{\sin[b(t-t_1)]}{t - t_1} dt_1 \\
&= \frac{1}{\pi} \int_{-\infty}^\infty f(t_1) e^{a(t-t_1)} \frac{\sin[b(t-t_1)]}{t - t_1} dt_1 ,
\end{aligned}$$

where in the last step we replaced the bottom limit to $-\infty$ as $f(t_1) = 0$ for $t_1 < 0$ by definition. Using the substitution $\tau = t_1 - t$ and introducing the function $g(t) = f(t)e^{-at}$, we obtain

$$\begin{aligned}
f_b(t) &= \frac{1}{\pi} \int_{-\infty}^\infty f(t+\tau) e^{-a\tau} \frac{\sin(b\tau)}{\tau} d\tau = \frac{e^{at}}{\pi} \int_{-\infty}^\infty g(t+\tau) \frac{\sin(b\tau)}{\tau} d\tau \\
&= \frac{e^{at}}{\pi} \int_{-\infty}^\infty \left[\frac{g(t+\tau) - g(t)}{\tau} \right] \sin(b\tau) d\tau + \frac{e^{at}}{\pi} g(t) \int_{-\infty}^\infty \frac{\sin(b\tau)}{\tau} d\tau .
\end{aligned}$$

The last integral, see Eq. (2.127), is equal to π, and hence the whole second term above amounts exactly to $f(t)$. In the first term we introduce the function $\Psi(\tau) = [g(t + \tau) - g(t)]/\tau$ within the square brackets, which yields

$$f_b(t) = f(t) + \frac{e^{at}}{\pi} \int_{-\infty}^{\infty} \Psi(\tau) \sin(b\tau) \, d\tau . \qquad (6.26)$$

The function $\Psi(\tau)$ is piecewise continuous. Indeed, for $\tau \neq 0$ this follows from the corresponding property of $f(t)$. At $\tau = 0$, however,

$$\lim_{\tau \to 0} \Psi(\tau) = \lim_{\tau \to 0} \frac{g(t + \tau) - g(t)}{\tau} = g'(t) ,$$

and is well defined as we assumed that $f(t)$ is differentiable. Hence, $\Psi(\tau)$ is piecewise continuous everywhere.

To show that the integral in Eq. (6.26) tends to zero in the $b \to \infty$ limit, we shall split it up into three using some $A > 0$:

$$\int_{-\infty}^{\infty} \Psi(\tau) \sin(b\tau) \, d\tau = \underbrace{\int_{-\infty}^{-A} \Psi(\tau) \sin(b\tau) \, d\tau}_{I_{-\infty}} + \underbrace{\int_{-A}^{A} \Psi(\tau) \sin(b\tau) \, d\tau}_{I_A} + \underbrace{\int_{A}^{\infty} \Psi(\tau) \sin(b\tau) \, d\tau}_{I_\infty} .$$

Assuming that the function $\Psi(\tau)$ is differentiable within the interval $-B < \tau < B$, we can take the second integral by parts:

$$I_A = \int_{-A}^{A} \Psi(\tau) \sin(b\tau) \, d\tau = -\frac{\cos(b\tau)}{b} \Psi(\tau) \Big|_{-A}^{A} + \frac{1}{b} \int_{-A}^{A} \Psi'(\tau) \cos(b\tau) \, d\tau ,$$

$$(6.27)$$

where both terms tend to zero in the $b \to \infty$ limit. Indeed, it is evident for the free term; in the second term it follows from the fact that the integral is finite for any b:

$$\left| \int_{-A}^{A} \Psi'(\tau) \cos(b\tau) \, d\tau \right| \leq \int_{-A}^{A} \left| \Psi'(\tau) \right| d\tau .$$

In the integrals $I_{\pm\infty}$ we change the variable $\tau \to x = b\tau$. Then,

$$I_{-\infty} = \frac{1}{b} \int_{-\infty}^{-bA} \Psi\left(\frac{x}{b}\right) \sin x \, dx \quad \text{and} \quad I_\infty = \frac{1}{b} \int_{bA}^{\infty} \Psi\left(\frac{x}{b}\right) \sin x \, dx .$$

Both integrals tend to zero in the $b \to \infty$ limit because of the $1/b$ pre-factor and because the bottom and top integral limits approach each other. Note also that in this limit $\Psi\left(\frac{x}{b}\right) \to \Psi(0) = g'(0)$ and is well defined.

Hence, $\lim_{b \to \infty} f_b(t) = f(t)$. As the function in the square brackets in Eq. (6.25) is $F(p)$, the proven result corresponds to the inverse LT of Eq. (6.24). The subtle

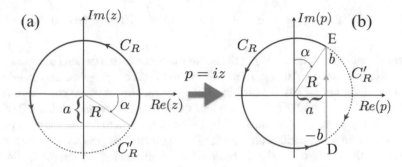

Fig. 6.1 Possible selections of the contour used in calculating the inverse LT integral in the complex p plane. **a** Contours C_R (solid line) and C'_R (dotted line) corresponding to, respectively, $\lambda > 0$ and $\lambda < 0$ in the exponential term $e^{i\lambda z}$ of the standard formulation of Jordan's lemma, and **b** contours C_R (solid) and C'_R (dotted) corresponding to the exponential term e^{pt} of the inverse LT for $t > 0$ and $t < 0$, respectively. The vertical line DE corresponds to the vertical line in Eq. (6.24)

point here is that the result holds in the $b \to \infty$ limit, i.e. the formula is indeed valid in the sense of the principal value. **Q.E.D.**

Formula (6.24) for the inverse LT is very useful in finding the original function $f(t)$ from its 'image' $F(p)$. This may be done by using the methods of residues. Let us discuss this now in some detail. Integration in formula (6.24) for the inverse LT is performed over the vertical line $p = a + iy$, $-\infty < y < \infty$, where $a > x_0$. Consider a part DE of this vertical line in Fig. 6.1b, corresponding to $-b < y < b$. A closed contour can be constructed by attaching to the vertical line an incomplete circle of the radius $R = \sqrt{a^2 + b^2}$ either from the left, C_R (the solid line), or from the right, C'_R (the dotted line). We shall now show that one can use a reformulated Jordan's lemma for expressing the inverse LT integral via residues of the image $F(p)$.

Indeed, let us reformulate Jordan's lemma, Sect. 2.7.3, by making the substitution

$$z = re^{i\phi} \quad \to \quad p = iz = re^{i(\phi + \pi/2)}$$

in Eqs. (2.122) and (2.123). It is seen that each complex number z acquires an additional phase $\pi/2$; this corresponds to the 90° anti-clockwise rotation of the construction we made originally in Fig. 2.35a as illustrated in Fig. 6.1. In particular, upon the substitution the horizontal line $z = x - ia$, $-b < x < b$, in Fig. 6.1a is transformed into the vertical line $p = iz = a + ix$ with $y = x$ satisfying $-b < y < b$, see Fig. 6.1b, while the two parts C_R and C'_R of the circle of radius R in Fig. 6.1a turn into the corresponding parts C_R and C'_R shown in Fig. 6.1b. Replacing $z \to p = iz$ and also $\lambda \to t$ in Eqs. (2.122) and (2.123), we obtain that the contour integrals along the parts C_R and C'_R of the circle shown in Fig. 6.1b for $t > 0$ and $t < 0$, respectively, tend to zero in the $R \to \infty$ limit,

$$\lim_{R\to\infty} \int_{C_R} F(p)e^{pt}dp = 0 \quad \text{if} \quad t{>}0 \quad \text{and} \quad \lim_{R\to\infty} \int_{C'_R} F(p)e^{pt}dp = 0 \quad \text{if} \quad t{<}0 \,,$$

$$(6.28)$$

provided that the function $F(p)$ in the integrand tends to zero when $|p| \to \infty$.

We know that the LT image $F(p) \to 0$ in the limit of $|p| \to \infty$ everywhere on the right of x_0, i.e. for any p satisfying $\mathrm{Re}(p) > x_0$. However, we can perform the analytic continuation of $F(p)$ into the half of the complex plane on the left of the vertical line $p = x_0 + iy$. Then, $F(p)$ is defined everywhere on the complex plane, apart from some singularities on the left of the vertical line $p = x_0 + iy$, and tends to zero on the circle $C_R + C'_R$ in the $R \to \infty$ limit. Hence the lemma is directly applicable to this function.

Consider now the case of $t < 0$ and the closed contour consisting of the vertical line and the part C'_R of the circle, Fig. 6.1b. Since all poles of $F(p)$ are on the left of the vertical part of the contour, the integral over the closed contour is zero. However, according to Jordan's lemma, the integral over the circle part C'_R is zero as well in the $b \to \infty$ limit (and, hence, in the $R \to \infty$ limit). Therefore, the integral over the vertical part is zero as well, i.e. $f(t) = 0$ for $t < 0$, as required.

Let us now consider positive times $t > 0$ and the closed contour formed by the vertical line and the part C_R of the circle. In the $b \to \infty$ limit all poles of $F(p)$ will be enclosed by such a contour, so that

$$\int_{a-ib}^{a+ib} F(p)e^{pt}dp + \int_{C_R} F(p)e^{pt}dp = 2\pi i \sum_k \mathrm{Res}\left[F(p)e^{pt}, p_k\right] \,,$$

where p_k is a pole of $F(p)$ and we sum over all poles. Since the integral over the part of the circle C_R in the $R \to \infty$ is zero according to Jordan's lemma, Eqs. (6.28), we finally obtain

$$f(t) = \lim_{b\to\infty} \frac{1}{2\pi i} \int_{a-ib}^{a+ib} F(p)e^{pt}dp = \sum_k \mathrm{Res}\left[F(p)e^{pt}, p_k\right] \,, \qquad (6.29)$$

where we sum over *all poles* on the left of the vertical line $p = x_0 + iy$.

To illustrate this powerful result, let us first calculate the function $f(t)$ corresponding to the image $F(p) = 1/(p + \alpha)$. We know from the direct calculation that this image corresponds to the exponential function $f(t) = e^{-\alpha t}$. Let us see if the inverse LT formula gives the same result. Choosing the vertical line at $x = a > \mathrm{Re}(\alpha)$, so that the pole at $p = -\alpha$ is positioned on the left of the vertical line, we have

$$f(t) = \mathrm{Res}\left[\frac{e^{pt}}{p+\alpha}, -\alpha\right] = e^{-\alpha t} = \mathcal{L}^{-1}\left[\frac{1}{p+\alpha}\right] \,,$$

i.e. exactly the same result. Consider next the image

$$F(p) = \frac{1}{(p + \alpha)\,(p + \beta)}, \quad \alpha \ne \beta.$$

It has two simple poles, $p_1 = -\alpha$ and $p_2 = -\beta$. Therefore, choosing the vertical line to the right of both poles, the original becomes

$$f(t) = \frac{e^{-\alpha t}}{-\alpha + \beta} + \frac{e^{-\beta t}}{-\beta + \alpha} = \frac{e^{-\alpha t} - e^{-\beta t}}{\beta - \alpha}.$$

The same result can also be obtained by decomposing $F(p)$ into simpler fractions:

$$F(p) = \frac{1}{(p + \alpha)\,(p + \beta)} = \frac{A}{p + \alpha} + \frac{B}{p + \beta}, \quad A = -B = \frac{1}{\beta - \alpha},$$

so that

$$f(t) = \mathcal{L}^{-1}[F(p)] = \frac{1}{\beta - \alpha}\left(\mathcal{L}^{-1}\left[\frac{1}{p + \alpha}\right] - \mathcal{L}^{-1}\left[\frac{1}{p + \beta}\right]\right)$$

$$= \frac{1}{\beta - \alpha}\left(e^{-\alpha t} - e^{-\beta t}\right).$$

This example shows clearly that in some cases it is possible to simplify the expression $F(p)$ for the image by decomposing it into simpler expressions for which the originals $f(t)$ are known.

Problem 6.16 Represent the fraction

$$F(p) = \frac{3p + 2}{3p^2 + 5p - 2}$$

as a sum of partial fractions and then show using the LTs we calculated earlier that the inverse LT of it is $f(t) = \frac{1}{7}\left(4e^{-2t} + 3e^{t/3}\right)$.

Problem 6.17 Similarly, show that

$$\mathcal{L}^{-1}\left[\frac{1 - p}{p^2 + 4p + 13}\right] = e^{-2t}\,(\sin 3t - \cos 3t)\,.$$

Problem 6.18 Use the inverse LT formula (6.24) and the theory or residues to find the original function $f(t)$ from its image:

$$\mathcal{L}^{-1}\left[\frac{p}{p^2 + \omega^2}\right] = \cos\omega t\,; \quad \mathcal{L}^{-1}\left[\frac{\omega}{p^2 + \omega^2}\right] = \sin\omega t\,;$$

$$\mathcal{L}^{-1}\left[\frac{p}{(p+\alpha)\,(p+\beta)}\right] = \frac{1}{\alpha-\beta}\left(\alpha e^{-\alpha t} - \beta e^{-\beta t}\right)\,,\quad \alpha\neq\beta;$$

$$\mathcal{L}^{-1}\left[\frac{a}{p^2-a^2}\right] = \sinh(at)\,;\quad \mathcal{L}^{-1}\left[\frac{p}{p^2-a^2}\right] = \cosh(at)\,;$$

$$\mathcal{L}^{-1}\left[\frac{2\omega p}{\left(p^2+\omega^2\right)^2}\right] = t\sin\omega t\,;\quad \mathcal{L}^{-1}\left[\frac{p^2-\omega^2}{\left(p^2+\omega^2\right)^2}\right] = t\cos\omega t\,;$$

$$\mathcal{L}^{-1}\left[\frac{2ap}{\left(p^2-a^2\right)^2}\right] = t\sinh(at)\,;\quad \mathcal{L}^{-1}\left[\frac{p^2+a^2}{\left(p^2-a^2\right)^2}\right] = t\cosh(at)\,;$$

$$\frac{1}{p\left(p^2+4p+3\right)} = \frac{1}{3} + \frac{1}{6}e^{-3t} - \frac{1}{2}e^{-t}\,;$$

$$\mathcal{L}^{-1}\left[\frac{p+k}{p\,(p+\alpha)\,(p+\beta)}\right] = \frac{k}{\alpha\beta} + \frac{1}{\alpha-\beta}\left(\frac{k-\alpha}{\alpha}e^{-\alpha t} - \frac{k-\beta}{\beta}e^{-\beta t}\right)\,;$$

$$\mathcal{L}^{-1}\left[\frac{ap^2+bp+1}{p\,(p-\alpha)\,(p-\beta)}\right] = \frac{1}{\alpha\beta} + \frac{1}{\alpha-\beta}\left[\left(\frac{1}{\alpha}+a\alpha+b\right)e^{\alpha t} - \left(\frac{1}{\beta}+a\beta+b\right)e^{\beta t}\right]\,.$$

Problem 6.19 Repeat the calculations of the previous problem using the method of decomposition of $F(p)$ into partial fractions.

Multiple valued functions $F(p)$ can also be inverted using formula (6.24). Consider as an example the image $F(p) = p^\lambda$, where $-1 < \lambda < 0$. Note that the fact that λ is negative, guarantees the necessary condition $F(p) \to 0$ at the large circle C_R in the $R \to \infty$ limit. The subtle point here is that we have to modify the contour C_R since a brunch cut from $p=0$ to $p=-\infty$ along the negative part of the real axis is required as shown in Fig. 6.2. Therefore, the residue theorem cannot be

Fig. 6.2 The contour needed for the calculation of the inverse LT of $F(p) = p^\lambda$, $-1 < \lambda < 0$

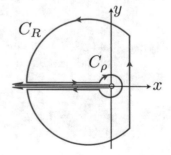

applied immediately as we have to integrate explicitly over each part of the closed contour. On the upper side of the cut $p = re^{i\pi}$, while on the lower $p = re^{-i\pi}$, where $0 < r < \infty$ and $-\pi \le \phi \le \pi$. The function $F(p)$ does not have any poles inside the closed contour, therefore, the sum of integrals over each part of it is zero. The contour consists of the vertical part, of the incomplete circle C_R with $R \to \infty$ (of two parts), a small circle C_ρ (where $\rho \to 0$), and the upper and the lower horizontal parts of the negative x-axis. Therefore,

$$f(t) = \frac{1}{2\pi i} \int_{vertical} = \frac{1}{2\pi i} \left(- \int_{C_R} - \int_{upper} - \int_{lower} - \int_{C_\rho} \right) .$$

The integral over both parts of C_R tends to zero by means of Jordan's lemma[5], so we need to consider the integrals over the small circle and over the upper and lower sides of the brunch cut. The integral over C_ρ tends to zero in the $\rho \to 0$ limit for $\lambda + 1 > 0$:

$$\int_{C_\rho} p^\lambda e^{pt} dp = \left| \begin{array}{l} p = \rho e^{i\phi} \\ dp = i\rho e^{i\phi} d\phi \end{array} \right| = \int_\pi^{-\pi} \rho^\lambda e^{i\lambda\phi} e^{t\rho e^{i\phi}} i\rho e^{i\phi} d\phi \propto \rho^{\lambda+1} \to 0 .$$

On the upper side of the cut

$$\int_\infty^\rho r^\lambda e^{i\lambda\pi} e^{-rt} (-dr) = e^{i\lambda\pi} \int_\rho^\infty r^\lambda e^{-rt} dr = \left| \begin{array}{l} v = tr \\ dv = tdr \end{array} \right|$$

$$= \frac{e^{i\lambda\pi}}{t^{\lambda+1}} \int_{t\rho}^\infty v^\lambda e^{-v} dv \to e^{i\lambda\pi} \frac{\Gamma(\lambda+1)}{t^{\lambda+1}} ,$$

where we have introduced the gamma function, Sect. 4.2, and applied the $\rho \to 0$ limit and hence replaced the lower integration limit with zero. Similarly, on the lower side of the cut:

$$\int_\rho^\infty r^\lambda e^{-i\lambda\pi} e^{-rt} (-dr) = -e^{-i\lambda\pi} \int_\rho^\infty r^\lambda e^{-rt} dr = -e^{-i\lambda\pi} \frac{\Gamma(\lambda+1)}{t^{\lambda+1}} .$$

Therefore, combining all contributions,

$$f(t) = \mathcal{L}^{-1} \left[p^\lambda \right] = \frac{1}{2\pi i} \left(-e^{i\lambda\pi} + e^{-i\lambda\pi} \right) \frac{\Gamma(\lambda+1)}{t^{\lambda+1}}$$

$$= -\frac{\sin(\lambda\pi) \Gamma(\lambda+1)}{\pi \quad t^{\lambda+1}} . \tag{6.30}$$

In particular, for $\lambda = -1/2$ we obtain

---

[5] This follows from the fact that $p^\lambda \to 0$ on *any part* of C_R in the $R \to \infty$ limit.

$$\mathcal{L}^{-1}\left[\frac{1}{\sqrt{p}}\right] = \frac{1}{\pi}\frac{\Gamma\left(1/2\right)}{\sqrt{t}} = \frac{1}{\sqrt{\pi t}} . \tag{6.31}$$

Problem 6.20 Show that

$$\mathcal{L}^{-1}\left[e^{-\alpha\sqrt{p}}\right] = \frac{\alpha}{2\sqrt{\pi t^3}}e^{-\alpha^2/4t} . \tag{6.32}$$

Problem 6.21 Show that

$$\mathcal{L}^{-1}\left[\frac{1}{\sqrt{p}}e^{-\alpha\sqrt{p}}\right] = \frac{1}{\sqrt{\pi t}}e^{-\alpha^2/4t} . \tag{6.33}$$

Some general relationships for the LT can sometimes be established as well. We shall illustrate this by deriving the following useful formula:

$$\mathcal{L}\left[t^{-\alpha}\psi(t)\right] = \frac{1}{\Gamma\left(\alpha\right)}\int_0^\infty x^{\alpha-1}\Psi\left(x+p\right)dx , \tag{6.34}$$

where $0 < \alpha < 1$ and $\Psi(p) = \mathcal{L}\left[\psi(t)\right]$ is the image of $\psi(t)$; $\Psi(p)$ is assumed to vanish when $|p| \to \infty$ in the whole complex plane. We however, assume, for simplicity, that $p > 0$ is a real number.

To prove this formula, we first write $\psi(t)$ using the inverse LT integral:

$$\begin{aligned}
\mathcal{L}\left[t^{-\alpha}\psi(t)\right] &= \int_0^\infty t^{-\alpha}\psi(t)e^{-pt}dt \\
&= \int_0^\infty t^{-\alpha}dt\left[\frac{1}{2\pi i}\int_{a-i\infty}^{a+i\infty}\Psi\left(z\right)e^{-(p-z)t}dz\right] .
\end{aligned} \tag{6.35}$$

Here $\mathrm{Re}\left(z\right) = a < p$, as the vertical line drawn at $\mathrm{Re}\left(z\right) = a$ is always to the left of the region in the complex plane p where the LT is analytic. The function $\Psi\left(z\right)$ decays to zero at $|z| \to \infty$, and hence we can assume that the integral

$$\int_{a-i\infty}^{a+i\infty}\left|\Psi\left(z\right)\right|dz$$

converges. Since $z = a + iy$ with $-\infty < y < \infty$, the integral along the vertical line in Eq. (6.35) converges uniformly with respect to t:

$$\begin{aligned}
\left|\int_{a-i\infty}^{a+i\infty}\Psi\left(z\right)e^{-(p-z)t}dz\right| &= e^{-(p-a)t}\left|\int_{a-i\infty}^{a+i\infty}\Psi\left(z\right)e^{iyt}dz\right| \\
&\leq e^{-(p-a)t}\int_{a-i\infty}^{a+i\infty}\left|\Psi\left(z\right)e^{iyt}\right|dz \leq \int_{a-i\infty}^{a+i\infty}\left|\Psi\left(z\right)\right|dz ,
\end{aligned}$$

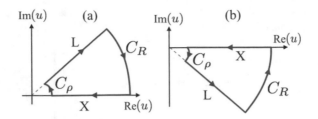

Fig. 6.3 For the proof of the relation of the t integral in Eq. (6.36) to the gamma function integral upon the change of the variable $t \to u = (p - z)\, t = (p - a - iy)\, t$. The two cases correspond to $y < 0$ (**a**) and $y > 0$ (**b**). Note that $p > a$

since $p > a$ and hence $e^{-(p-a)t} < 1$. The demonstrated uniform convergence allows us to exchange the order of integrals in Eq. (6.35) to get

$$\mathcal{L}\left[t^{-\alpha}\psi(t)\right] = \frac{1}{2\pi i} \int_{a-i\infty}^{a+i\infty} \Psi(z)\, dz \left[\int_0^\infty t^{-\alpha} e^{-(p-z)t}\, dt\right]. \tag{6.36}$$

The t integral by means of the change of variable $t \to u = (p - z)\, t$ can be related to the gamma function (4.33). Indeed,

$$\int_0^\infty t^{-\alpha} e^{-(p-z)t}\, dt = \frac{1}{(p-z)^{1-\alpha}} \lim_{t\to\infty} \int_0^{(p-a-iy)t} u^{-\alpha} e^{-u}\, du .$$

A subtle point here is that the u-integration is performed not along the real axis assumed in the definition of the gamma function, but rather along the straight line $u = 0 \longmapsto u = (p - a - iy)\, t$ in the complex plane (note that a real y is fixed). This is line L in Fig. 6.3. To covert this integral into the one along the real axis $\mathrm{Re}(u)$, we introduce a closed contour $L + C_R + X + C_\rho$ shown in the figure. Note that the horizontal line X is being passed in the negative direction as shown. The contour avoids the point $u = 0$ with a circular arc C_ρ of radius $\rho \to 0$. Also, another circular arc C_R of radius $R \to \infty$ connects L with the horizontal line X. Since the integrand $u^{-\alpha} e^{-u}$ does not have poles inside the contour, the sum of the integrals along the whole closed contour is zero. At the same time, it is easy to see that the integral along C_ρ behaves like $\rho^{1-\alpha}$ and hence tends to zero when $\rho \to 0$ (recall that $0 < \alpha < 1$), while the one along C_R behaves like $R^{1-\alpha} e^{-R\cos\phi}$ (where $-\pi/2 < \phi < \pi/2$) and hence goes exponentially to zero as well in the $R \to \infty$ limit. Therefore, $\int_L = -\int_X = \int_{X^+}$, where X^+ is the horizontal line along the real axis $u > 0$ taken in the positive direction. Hence, we arrive at the u-integral in which u changes between 0 and ∞ yielding the gamma function $\Gamma(1 - \alpha)$ as required.

Therefore, Eq. (6.36) can be rewritten as follows:

$$\mathcal{L}\left[t^{-\alpha}\psi(t)\right] = \frac{\Gamma(1-\alpha)}{2\pi i} \int_{a-i\infty}^{a+i\infty} \frac{\Psi(z)}{(p-z)^{1-\alpha}}\, dz . \tag{6.37}$$

Fig. 6.4 The contour used in
calculating the integral in
Eq. (6.37)

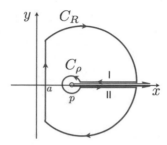

Recall that here $p > a = \mathrm{Re}\,(z)$.

To calculate the vertical line integral above, we use the contour shown in Fig. 6.4. As α is a positive real number, a brunch cut has been made as shown in the figure with the phase ϕ for $z = |z|\,e^{i\phi}$ being zero on the upper and 2π on the lower sides of it. The function $\Psi(z)$ does not have poles to the right of $\mathrm{Re}(z) = a$, so that the integral over the closed contour is equal to zero.

Consider the integral over the upper part of the large circular contour C_R:

$$
\left| \int_{C_R(\mathrm{up})} \frac{\Psi(z)}{(p-z)^{1-\alpha}}dz \right| = \left| \begin{array}{l} z = p + Re^{i\phi} \\ dz = i\,Re^{i\phi}d\phi \end{array} \right| = \left| \int_0^{\phi_{\mathrm{up}}(R)} R^\alpha \Psi(p + Re^{i\phi}) \frac{i e^{i\phi}d\phi}{(-e^{i\phi})^{1-\alpha}} \right|
$$

$$
\leq \int_0^{\phi_{\mathrm{up}}(R)} \left| R^\alpha \Psi(p + Re^{i\phi}) \frac{i e^{i\phi}}{(-e^{i\phi})^{1-\alpha}} \right| d\phi
$$

$$
= \int_0^{\phi_{\mathrm{up}}(R)} R^\alpha \left| \Psi(p + Re^{i\phi}) \right| d\phi .
$$

To make sure that this integral tends to zero as $R \to \infty$, we have to impose on $\Psi(p)$ a slightly more stringent condition that the function $R^\alpha \left| \Psi\left(p + Re^{i\phi}\right) \right|$ also tends to zero in that limit. The same is true for the lower part of the large circular contour. Hence, under this assumption, the integral over the two parts of the large circular contour C_R tends to zero as $R \to \infty$.

Consider the integral over the small circle C_ρ. There $z - p = \rho e^{i\phi}, dz = i\rho e^{i\phi}d\phi$ and hence the integral behaves as $\rho^{\alpha-1}\rho = \rho^\alpha$ and hence also tends to zero as $\rho \to 0$ (recall that $0 < \alpha < 1$). Hence, we have to consider the integrals over the upper I and lower II horizontal lines. On the upper side

$$
\int_I = \int_\infty^p \frac{\Psi(z)}{(p-z)^{1-\alpha}}dz = \left| \begin{array}{l} z - p = s \\ dz = ds \end{array} \right| = -\int_0^\infty \frac{\Psi(s+p)}{(-s)^{1-\alpha}}ds
$$

$$
= -\int_0^\infty (-s)^{\alpha-1}\,\Psi(s+p)\,ds .
$$

The function $(-s)^{\alpha-1}$ originates from $(p-z)^{\alpha-1}$. On the vertical part of the contour the latter function is *positive* when z is on the real axis since there $p > \mathrm{Re}\,(z) = a$

(recall that p is real and positive) and hence $-s = p - z = p - a > 0$. Since $z = a$ is on the left of p, it corresponds to the phase $\phi = \pi$ of $s = |s| e^{i\phi}$ (our change of variables $s = z - p$ is simply a horizontal shift to place the centre of the coordinate system at point p). Therefore, when choosing the correct branch of the function $(-s)^{\alpha-1}$, we have to make sure that the function is positive at the $\phi = \pi$ phase of s. There are basically two possibilities to consider for the pre-factor -1 before the s in $(-s)^{\alpha-1}$: when $-1 = e^{i\pi}$ and $-1 = e^{-i\pi}$. For the former choice (recall that $\phi = \pi$ when $z = a$),

$$(-s)^{\alpha-1} = |s|^{\alpha-1} e^{i(\pi+\phi)(\alpha-1)} = |s|^{\alpha-1} e^{i2\pi(\alpha-1)} = |s|^{\alpha-1} e^{i2\pi\alpha} .$$

It is easy to see that for a general non-integer α it is impossible to guarantee that this number is positive. Indeed, if we, for instance, take $\alpha = 1/2$, then $e^{i2\pi\alpha} = e^{i\pi} = -1$ leading to a negative value. On the other hand, the other choice guarantees the positive value for any α and the phase $\phi = \pi$:

$$(-s)^{\alpha-1} = |s|^{\alpha-1} e^{i(-\pi+\phi)(\alpha-1)} = |s|^{\alpha-1} .$$

Therefore, the correct branch of the function $(-s)^{\alpha-1}$ is chosen by setting $-1 = e^{-i\pi}$.

With this choice in mind, we can now proceed to the calculation of the integral on the upper side of the branch cut. On this side $\phi = 0$ and $s = x - p > 0$. Therefore,

$$(-s)^{\alpha-1} = \left(e^{-i\pi} s\right)^{\alpha-1} = e^{-i(\alpha-1)\pi} s^{\alpha-1} ,$$

so that the integral over the upper side becomes

$$\int_I = -e^{-i(\alpha-1)\pi} \int_0^\infty s^{\alpha-1} \Psi (s + p) \, ds .$$

Similarly, the integral on the lower side,

$$\int_{II} = \int_p^\infty \frac{\Psi (z)}{(p - z)^{1-\alpha}} dz = \int_0^\infty (-s)^{\alpha-1} \Psi (s + p) \, ds ,$$

is evaluated by noting that in this case s has the phase of $\phi = 2\pi$, so that

$$(-s)^{\alpha-1} = \left(e^{-i\pi} e^{i2\pi} s\right)^{\alpha-1} = \left(e^{i\pi} s\right)^{\alpha-1} = e^{i(\alpha-1)\pi} s^{\alpha-1} ,$$

and we obtain

$$\int_{II} = e^{i(\alpha-1)\pi} \int_0^\infty s^{\alpha-1} \Psi (s + p) \, ds .$$

Therefore, the vertical line integral

$$\int_{a-i\infty}^{a+i\infty} \frac{\Psi(z)}{(p-z)^{1-\alpha}} dz = -\int_I - \int_{II} = \left[e^{-i(\alpha-1)\pi} - e^{i(\alpha-1)\pi} \right] \int_0^\infty s^{\alpha-1} \Psi(s+p) \, ds$$

$$= -2i \sin((\alpha-1)\pi) \int_0^\infty s^{\alpha-1} \Psi(s+p) \, ds$$

$$= 2i \sin(\alpha\pi) \int_0^\infty s^{\alpha-1} \Psi(s+p) \, ds \ .$$

$$(6.38)$$

This finally gives

$$\mathcal{L}\left[t^{-\alpha}\psi(t) \right] = \frac{\Gamma(1-\alpha)}{2\pi i} \int_{a-i\infty}^{a+i\infty} \frac{\Psi(z)}{(p-z)^{1-\alpha}} dz$$

$$= \frac{\Gamma(1-\alpha) \sin(\alpha\pi)}{\pi} \int_0^\infty s^{\alpha-1} \Psi(s+p) \, ds \ .$$

Recalling the relationship (4.44), we immediately recover Eq. (6.34), as required.

The obtained result can be illustrated on several examples for the calculated transforms. For instance, if $\psi(t) = 1$, then $\Psi(p) = 1/p$, and we obtain

$$\mathcal{L}\left[t^{-\alpha} \right] = \frac{1}{\Gamma(\alpha)} \int_0^\infty \frac{s^{\alpha-1}}{s+p} ds = \left| \begin{array}{c} s = pt \\ ds = pdt \end{array} \right|$$

$$= \frac{p^{\alpha-1}}{\Gamma(\alpha)} \int_0^\infty \frac{t^{\alpha-1} dt}{t+1} = \frac{p^{\alpha-1}}{\Gamma(\alpha)} B(\alpha, 1-\alpha) = p^{\alpha-1} \Gamma(1-\alpha) \ .$$

Here we used formulae (4.51) and (4.50) for the beta function. For instance, by taking $\alpha = 1/2$, we immediately recover our previous result (6.31).

Problem 6.22 Verify formula (6.34) for $\psi(t) = \delta(t-b)$ with $b > 0$ by showing that both sides give $\mathcal{L}\left[t^{-\alpha}\delta(t-b) \right] = b^{-\alpha} e^{-bp}$.

Problem 6.23 Consider $\alpha = \beta + [\alpha] > 0$, where $[\alpha]$ is the integer part of α and $0 < \beta < 1$ is the rest of it. Demonstrate that

$$\mathcal{L}\left[t^{\alpha}\psi(t) \right] = \frac{(-1)^n}{\Gamma(1-\beta)} \int_0^\infty s^{-\beta} \Psi^{(n)}(s+p) \, ds \ , \qquad (6.39)$$

where $n = [\alpha] + 1$ is a positive integer, and $\Psi^{(n)}(s)$ is the n-th derivative of $\Psi(s)$. [*Hint: repeat the steps which led us to Eq. (6.37) in the previous case, and then showing that*

$$\left(\frac{d}{dp} \right)^n \frac{1}{(p-z)^\beta} = \frac{(-1)^n \, \Gamma(\alpha+1)}{\Gamma(\beta)} \frac{1}{(p-z)^{\alpha+1}} \ ,$$

relate the integral over z to that given in Eq. (6.38).]

6.4 Properties of the Laplace Transform

By calculating the LT of various functions $f(t)$ we would find the correspondences $\mathcal{L}[f(t)] \longleftrightarrow F(p)$. It is convenient to collect these all together in a table, and an example of this is Table 6.1. This table may be found very useful in applications, both at the first step when going into the Laplace p−space from the original time formulation of the problem, and also back from the p−space into the t−space, when the original $f(t)$ is to be obtained from the image $F(p)$. Of course, it is not possible to list all existing functions in such a table; not only because in some cases the LT cannot be obtained analytically; it is simply because the number of possible functions is infinite. That is why establishing *rules* for calculating the LT has been found extremely useful in practical applications of the method as these rules, if applied properly, enable one to use a few entries in the table to calculate the LT (and the inverse LT) of a wide class of functions. In this section, we shall consider most frequently used rules.

6.4.1 Derivatives of Originals and Images

When transforming differential equations into the Laplace p−space, it is necessary to convert derivatives of the original $f(t)$. To this end, let the LT of the function $f(t)$ be $F(p)$, i.e. $\mathcal{L}(f(t)) = F(p)$. The question we ask is whether the LT of the derivative $f'(t)$ can be related to the LT of the function $f(t)$ itself. This question is trivially answered by the direct calculation in which the integration by parts is used:

$$\mathcal{L}\left[f'(t)\right] = \int_0^\infty e^{-pt} \frac{df}{dt} dt = f(t)e^{-pt}\Big|_0^\infty - \int_0^\infty f(t)(-p)e^{-pt} dt$$

$$= -f(0) + p \int_0^\infty e^{-pt} f(t) dt = -f(0) + p\mathcal{L}[f(t)] \,, \tag{6.40}$$

where we have made use of the fact that the upper limit $t = \infty$ in the free term can be omitted. Indeed, since the LT of $f(t)$ exists for any p satisfying $\mathrm{Re}(p) = \mathrm{Re}(x + iy) = x > x_0$ with some growth order parameter x_0, then $|f(t)| \leq Me^{x_0 t}$. Therefore, the free term is limited from above,

$$\left| f(t)e^{-pt} \right| \leq Me^{x_0 t} \left| e^{-(x+iy)t} \right| = Me^{x_0 t} e^{-xt} = Me^{-(x-x_0)t} \,,$$

and hence tends to zero at the $t \to \infty$ limit.

For the second derivative one can use the above formula twice since $f''(t) = g'(t)$ with $g(t) = f'(t)$. Hence,

$$\mathcal{L}\left[f''(t)\right] = \mathcal{L}\left[g'(t)\right] = p\mathcal{L}[g(t)] - g(0) = p\mathcal{L}\left[f'(t)\right] - f'(0)$$

$$= p\left(p\mathcal{L}[f(t)] - f(0)\right) - f'(0) \,,$$

Table 6.1 Laplace transforms of some functions

	$y = f(t),\ t > 0$ $[y = f(t) = 0,\ t < 0]$	$Y = \mathcal{L}(y) = F(p) = \int_0^\infty e^{-pt} f(t)dt$	Range of p in the complex plane
$L1$	1	$\frac{1}{p}$	$\operatorname{Re} p > 0$
$L2$	e^{-at}	$\frac{1}{p+a}$	$\operatorname{Re}(p+a) > 0$
$L3$	$\sin at$	$\frac{a}{p^2+a^2}$	$\operatorname{Re} p > |\operatorname{Im} a|$
$L4$	$\cos at$	$\frac{p}{p^2+a^2}$	$\operatorname{Re} p > |\operatorname{Im} a|$
$L5$	$t^k,\ k > -1$	$\frac{k!}{p^{k+1}}$ or $\frac{\Gamma(k+1)}{p^{k+1}}$	$\operatorname{Re} p > 0$
$L6$	$t^k e^{-at},\ k > -1$	$\frac{k!}{(p+a)^{k+1}}$ or $\frac{\Gamma(k+1)}{(p+a)^{k+1}}$	$\operatorname{Re}(p+a) > 0$
$L7$	$\frac{e^{-at}-e^{-bt}}{b-a}$	$\frac{1}{(p+a)(p+b)}$	$\operatorname{Re}(p+a) > 0$
$L8$	$\frac{ae^{-at}-be^{-bt}}{a-b}$	$\frac{p}{(p+a)(p+b)}$	$\operatorname{Re}(p+b) > 0$
$L9$	$\sinh at$	$\frac{a}{p^2-a^2}$	$\operatorname{Re} p > |\operatorname{Re} a|$
$L10$	$\cosh at$	$\frac{p}{p^2-a^2}$	$\operatorname{Re} p > |\operatorname{Re} a|$
$L11$	$t\sin at$	$\frac{2ap}{(p^2+a^2)^2}$	$\operatorname{Re} p > |\operatorname{Im} a|$
$L12$	$t\cos at$	$\frac{p^2-a^2}{(p^2+a^2)^2}$	$\operatorname{Re} p > |\operatorname{Im} a|$
$L13$	$e^{-at}\sin bt$	$\frac{b}{(p+a)^2+b^2}$	$\operatorname{Re}(p+a) > |\operatorname{Im} b|$
$L14$	$e^{-at}\cos bt$	$\frac{p+a}{(p+a)^2+b^2}$	$\operatorname{Re}(p+a) > |\operatorname{Im} b|$
$L15$	$1 - \cos at$	$\frac{a^2}{p(p^2+a^2)}$	$\operatorname{Re} p > |\operatorname{Im} a|$
$L16$	$at - \sin at$	$\frac{a^3}{p^2(p^2+a^2)}$	$\operatorname{Re} p > |\operatorname{Im} a|$
$L17$	$\sin at - at\cos at$	$\frac{2a^3}{(p^2+a^2)^2}$	$\operatorname{Re} p > |\operatorname{Im} a|$
$L18$	$e^{-at}(1 - at)$	$\frac{p}{(p+a)^2}$	$\operatorname{Re}(p+a) > 0$
$L19$	$\frac{\sin at}{t}$	$\arctan\frac{a}{p}$	$\operatorname{Re} p > |\operatorname{Im} a|$
$L20$	$\frac{1}{t}\sin at\cos bt,\ a > 0,\ b > 0$	$\frac{1}{2}\left(\arctan\frac{a+b}{p} + \arctan\frac{a-b}{p}\right)$	$\operatorname{Re} p > 0$
$L21$	$\frac{e^{-at}-e^{-bt}}{t}$	$\ln\frac{p+b}{p+a}$	$\operatorname{Re}(p+a) > 0$ and $\operatorname{Re}(p+b) > 0$
$L22$	$1 - \operatorname{erf}\left(\frac{a}{2\sqrt{t}}\right),\ a > 0$	$\frac{1}{p}e^{-a\sqrt{p}}$	$\operatorname{Re} p > 0$
$L23$	$J_0(at)$	$\left(p^2 + a^2\right)^{-1/2}$	$\operatorname{Re} p > |\operatorname{Re} a|$, or for real $a \neq 0$, $\operatorname{Re} p \geq 0$
$L24$	$f(t) = \begin{cases} 1, & t > a > 0 \\ 0, & t < a \end{cases}$	$\frac{1}{p}e^{-pa}$	$\operatorname{Re} p > 0$

i.e.

$$\mathcal{L}\left[f''(t)\right] = p^2 \mathcal{L}\left[f(t)\right] - pf(0) - f'(0) .\qquad(6.41)$$

The obtained results (6.40) and (6.41) clearly show that the LT of the first and the second derivatives of a function $f(t)$ are expressed via the LT of the function itself multiplied by p or p^2, respectively, minus a constant. This means that since the differentiation turns into multiplication in the Laplace space, a differential equation would turn into an algebraic one allowing one to find the image $F(p)$ of the solution of the equation. Of course, in this case, the crucial step is the one of inverting the LT and finding the original from its image, a problem which may be nontrivial.

Problem 6.24 In fact, this way it is possible to obtain a general formula for the n-th derivative, $\mathcal{L}\left[f^{(n)}(t)\right]$. Using induction, prove the following result:

$$\mathcal{L}\left[f^{(n)}(t)\right] = p^n F(p) - \sum_{k=0}^{n-1} p^{n-1-k} f^{(k)}(0) , \quad n = 1, 2, 3, \dots ,\qquad(6.42)$$

where $F(p) = \mathcal{L}[f(t)]$ and $f^{(0)}(0) = f(0)$. Note that $f(0)$ and the derivatives $f^{(k)}(0)$ are understood as calculated in the limit $t \to +0$.

So, any differentiation of $f(t)$ always turns into multiplication after the LT.

There is also a simple formula allowing one to calculate the inverse LT of the n-th derivative $F^{(n)}(p)$ of the image, see Eq. (6.12):

$$\mathcal{L}^{-1}\left[F^{(n)}(p)\right] = (-t)^n f(t) .\qquad(6.43)$$

Problem 6.25 Generalise the result (6.40) for the derivative of $f(t)$ for the case when $f(t)$ has a discontinuity of the first kind at the point $t_1 > 0$:

$$\mathcal{L}\left[f'(t)\right] = p\mathcal{L}[f(t)] - f(0) - \left[f\left(t_1^+\right) - f\left(t_1^-\right)\right]e^{-pt_1} ,$$

where $t_1^+ = t_1 + 0$ and $t_1^- = t_1 - 0$. Note that the last extra term which is proportional to the value of the jump of $f(t)$ at t_1 disappears if the function does not jump, and we return to our previous result (6.40). *[Hint: Split the integration in the definition of the LT into two regions by the point t_1 and then take each integral by parts in the same way as when deriving Eq. (6.40).]*

In applying the LT method to solving linear ODEs with variable coefficients of the polynomial form (Sect. 6.5), we shall need a certain generalisation of the above rules to turning into the 'Laplace space' expressions like $t^m f^{(n)}(t)$, where m is a positive integer. Indeed, using Eq. (6.12), we can write

$$\mathcal{L}\left[tf'(t)\right] = -\frac{d}{dp}\mathcal{L}\left[f'(t)\right] = -\frac{d}{dp}\left[-f(0) + pF(p)\right]$$

$$= -F(p) - pF'(p),$$

(6.44)

where $F(p) = \mathcal{L}[f(t)]$. Hence, the pre-factor t to $f'(t)$ resulted in the appearance of the first derivative of the image $F(p)$.

Similarly,

$$\mathcal{L}\left[tf''(t)\right] = -\frac{d}{dp}\mathcal{L}\left[f''(t)\right] = -\frac{d}{dp}\left[p^2 F(p) - pf(0) - f'(0)\right]$$

$$= f(0) - 2pF(p) - p^2 F'(p).$$

It is seen from Eq. (6.42) that the LT of $tf^{(n)}(t)$ becomes a polynomial in p containing no more than the first derivative of the image $F(p)$, for any n.

The order of the derivative of the image increases if the pre-factor to $f^{(n)}(t)$ has the power larger than 1. For instance,

$$\mathcal{L}\left[t^2 f'(t)\right] = (-1)^2 \frac{d^2}{dp^2}\mathcal{L}\left[f'(t)\right] = \frac{d^2}{dp^2}\left[-f(0) + pF(p)\right]$$

$$= 2F'(p) + pF''(p).$$

Problem 6.26   Show that

$$\mathcal{L}\left[t^2 f''(t)\right] = p^2 F''(p) + 4pF'(p) + 2F(p).$$

(6.45)

6.4.2   Shift in Images and Originals

There are two simple properties of the LT related to a shift of either the original or the image which we formulate as a problem for the reader to prove:

Problem 6.27 Using the definition of the LT, prove the following formulae:

$$\mathcal{L}[f(t-\tau)] = e^{-p\tau}\mathcal{L}[f(t)] = e^{-p\tau}F(p);$$ (6.46)

$$\mathcal{L}\left[e^{-\alpha t}f(t)\right] = F(p+\alpha).$$ (6.47)

Note that it is implied (as usual) that $f(t) = 0$ for $t < 0$ in both these equations. In particular, $f(t-\tau) = 0$ for $t < \tau$ (where $\tau > 0$).

Let us illustrate the first identity which may be found useful when calculating the LT of functions which are obtained by shifting on the t-axis a given function. As an example, we shall first work out the LT of a final width step function shown in the left panel of Fig. 6.5:

$$\mathcal{L}[\Pi(t)] = \int_0^T e^{-pt}dt = \frac{1}{p}\left(1 - e^{-pT}\right).$$ (6.48)

Correspondingly, the LT of the function shifted to the right by τ (the right panel of the same Figure) is then

$$\mathcal{L}[\Pi(t-\tau)] = \frac{e^{-p\tau}}{p}\left(1 - e^{-p\tau}\right).$$

Consider now a wave signal which is composed of identical unit step impulses which start at positions $t_k = k(T + \Delta)$, where $k = 0, 1, 2, \ldots$, as shown in Fig. 6.6. The LT of such a function is a sum of contributions from each impulse:

Fig. 6.5 A step $\Pi(t)$ of unit height and width T is defined in the left panel. In the right panel the function is shifted as a whole by τ to the right

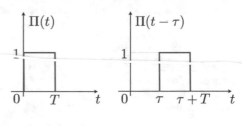

Fig. 6.6 The function $f(t)$ is formed by a sum of an infinite series of identical steps $\Pi(t-t_k)$

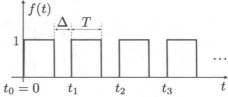

$$\mathcal{L}[f(t)] = \mathcal{L}[\Pi(t)] + \mathcal{L}[\Pi(t - t_1)] + \mathcal{L}[\Pi(t - t_2)] + \ldots$$

$$= \sum_{k=0}^{\infty} \mathcal{L}[\Pi(t - t_k)] = \sum_{k=0}^{\infty} \frac{e^{-pt_k}}{p}\left(1 - e^{-pT}\right)$$

$$= \frac{1}{p}\left(1 - e^{-pT}\right)\sum_{k=0}^{\infty} e^{-p(T+\Delta)k} = \frac{1}{p}\frac{1 - e^{-pT}}{1 - e^{-p(T+\Delta)}}\,.$$

Problem 6.28 Show that the LT of the waveform shown in Fig. 6.7 is given by

$$\mathcal{L}[f(t)] = \frac{\omega}{p^2 + \omega^2}\frac{1 + e^{-pT/2}}{1 - e^{-p(\Delta+T/2)}}\,,$$

where $T = 2\pi/\omega$ and the first waveform is defined on the interval $0 < t < T/2 + \Delta$ as $f(t) = \sin\omega t$ for $0 < t < T/2$ and $f(t) = 0$ for $T/2 < t < T/2 + \Delta$. In the case of $\Delta = T/2$ this waveform corresponds to a half-wave rectifier which removes the negative part of the signal.

Problem 6.29 The same for the waveform shown in Fig. 6.8:

$$\mathcal{L}[f(t)] = \frac{2}{Tp^2}\frac{\left(1 - e^{-pT/2}\right)^2}{1 - e^{-p(\Delta+T)}}\,.$$

Problem 6.30 The same for the waveform shown in Fig. 6.9a:

$$\mathcal{L}[f(t)] = \frac{A}{p}\frac{1}{1 - e^{-pT}}\,.$$

Fig. 6.7 A waveform composed of the periodically repeated positive half-period piece of the sine function

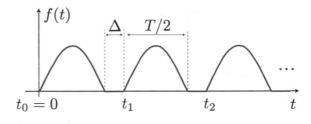

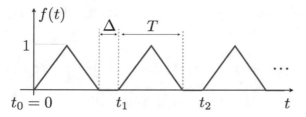

Fig. 6.8 A waveform formed by periodically repeated hat-like signal

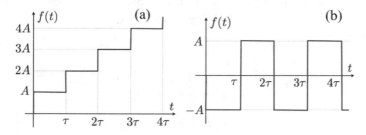

Fig. 6.9 Waveforms to Problems 6.30 (**a**) and 6.31 (**b**)

Problem 6.31 The same for the waveform shown in Fig. 6.9b:

$$\mathcal{L}[f(t)] = -\frac{A}{p}\frac{1-e^{-p\tau}}{1+e^{-p\tau}} \ .$$

6.4.3 Integration of Images and Originals

We saw in Sect. 6.4.1 that the LT of the first derivative of $f(t)$ is obtained essentially by multiplying its image $F(p)$ with p. We shall now see that the LT of the integral of $f(t)$ can be obtained by dividing $F(p)$ with p:

$$\mathcal{L}\left[\int_0^t f(\tau)d\tau\right] = \frac{F(p)}{p} \ . \tag{6.49}$$

Indeed, consider

$$\mathcal{L}\left[\int_0^t f(\tau)d\tau\right] = \int_0^\infty dt\, e^{-pt} \int_0^t d\tau\, f(\tau) \ .$$

Interchanging the order of integration (considering carefully the change of the limits on the $t - \tau$-plane), we obtain

$$\mathcal{L}\left[\int_0^t f(\tau)d\tau\right] = \int_0^\infty d\tau\, f(\tau)\left(\int_\tau^\infty e^{-pt}dt\right) = \int_0^\infty d\tau\, f(\tau)\left(\frac{e^{-p\tau}}{p}\right)$$

$$= \frac{1}{p}\int_0^\infty d\tau\, f(\tau)e^{-p\tau} = \frac{F(p)}{p}\,,$$

as required. When integrating the exponential function e^{-pt} with respect to t, we set to zero the result at $t \to \infty$ which is valid for any $\mathrm{Re}(p) > 0$.

For instance,

$$\mathcal{L}^{-1}\left[\frac{a^2}{p(p^2 + a^2)}\right] = \int_0^t \underbrace{\mathcal{L}^{-1}\left[\frac{a^2}{p^2 + a^2}\right]}_{a\sin(a\tau)}d\tau = a\int_0^t \sin(a\tau)d\tau$$

$$= 1 - \cos(at)\,,$$

which coincides with entry L15 of Table 6.1.

Problem 6.32   Prove that

$$\mathcal{L}^{-1}\left[\frac{1}{p}e^{-\alpha\sqrt{p}}\right] = \mathrm{erfc}\left(\frac{\alpha}{2\sqrt{t}}\right)\,.$$

[*Hint*: note the result of Problem 6.20.]

In the above problem we have introduced a useful function which is called error function:

$$\mathrm{erfc}(x) = \frac{2}{\sqrt{\pi}}\int_x^\infty e^{-t^2}dt\,. \tag{6.50}$$

A function complimentary to this one is

$$\mathrm{erf}(x) = \frac{2}{\sqrt{\pi}}\int_0^x e^{-t^2}dt\,, \tag{6.51}$$

also called the error function. Both functions are related:

$$\mathrm{erf}(x) + \mathrm{erfc}(x) = \frac{2}{\sqrt{\pi}}\int_0^\infty e^{-t^2}dt = 1\,,$$

see Sect. 4.2 for the value of the last integral.

Similarly is proven the 'reciprocal' property:

$$\mathcal{L}^{-1}\left[\int_p^\infty F(s)ds\right] = \frac{f(t)}{t}. \tag{6.52}$$

Here the s-integral is taken along any path in the complex plane connecting the points p and $s_\infty = \infty$, where $|s_\infty| = \infty$ and Re $(s_\infty) > 0$. Indeed,

$$\int_p^\infty F(s)ds = \int_p^\infty ds \int_0^\infty dt\, f(t)e^{-st} = \int_0^\infty dt\, f(t)\left(\int_p^\infty ds\, e^{-st}\right)$$

$$= \int_0^\infty dt\, f(t)\left(\frac{e^{-pt}}{t}\right) = \int_0^\infty dt\, \frac{f(t)}{t}e^{-pt} = \mathcal{L}\left[\frac{f(t)}{t}\right],$$

the desired result. Again, when integrating over s, we set the exponential function e^{-st} to zero at the upper limit of $s_\infty = \infty$ since Re$(s_\infty) > 0$.

Problem 6.33 Consider the function

$$g(t) = \int_0^t e^{-\alpha\tau}d\tau = \frac{1}{\alpha}\left(1 - e^{-\alpha t}\right).$$

Check that its LTs calculated directly and when using Eq. (6.49) do coincide.

Problem 6.34 Prove the following formulae:

$$\mathcal{L}\left[\frac{e^{-\alpha t} - e^{-\beta t}}{t}\right] = \ln\frac{p+\beta}{p+\alpha};$$

$$\mathcal{L}\left[\frac{\sin \omega t}{t}\right] = \frac{\pi}{2} - \arctan\frac{p}{\omega} = \arctan\frac{\omega}{p};$$

$$\mathcal{L}\left[\frac{\cos \omega_1 t - \cos \omega_2 t}{t}\right] = \frac{1}{2}\ln\frac{p^2 + \omega_2^2}{p^2 + \omega_1^2}.$$

Problem 6.35 Prove the following formulae valid for Re$(p) > 0$:

$$\mathcal{L}\left[\int_0^t \frac{\cos \tau}{\sqrt{2\pi\tau}}d\tau\right] = \frac{1}{2\sqrt{2}p}\frac{\sqrt{p+i} + \sqrt{p-i}}{\sqrt{p^2+1}};$$

$$\mathcal{L}\left[\int_0^t \frac{\sin \tau}{\sqrt{2\pi\tau}}d\tau\right] = \frac{-i}{2\sqrt{2}p}\frac{\sqrt{p+i} - \sqrt{p-i}}{\sqrt{p^2+1}}.$$

[Hint: You may find it useful to use Eq. (6.13).]

6.4.4 Convolution Theorem

Let $G(p)$ and $F(p)$ be LTs of the functions $g(t)$ and $f(t)$, respectively:

$$G(p) = \mathcal{L}[g(t)] = \int_0^\infty e^{-pt_1} g(t_1)\, dt_1 \,, \tag{6.53}$$

$$F(p) = \mathcal{L}[f(t)] = \int_0^\infty e^{-pt_2} f(t_2)\, dt_2 \,. \tag{6.54}$$

Consider their product:

$$G(p)F(p) = \int_0^\infty dt_2 \int_0^\infty dt_1\, e^{-p(t_1+t_2)} g(t_1)\, f(t_2) \,.$$

This is a double integral in the $(t_1 - t_2)$-plane. Let us replace the integration variable t_1 with $t = t_1 + t_2$. This yields

$$G(p)F(p) = \int_0^\infty dt_2 \left[\int_{t_2}^\infty dt\, e^{-pt} g(t - t_2)\, f(t_2) \right]. \tag{6.55}$$

At the next step we interchange the order of the integrals. This has to be done with care as the limits will change:

$$
\begin{aligned}
G(p)F(p) &= \int_0^\infty dt \left[\int_0^t dt_2\, e^{-pt} g(t - t_2)\, f(t_2) \right] \\
&= \int_0^\infty dt\, e^{-pt} \left[\int_0^t g(t - t_2)\, f(t_2)\, dt_2 \right].
\end{aligned}
\tag{6.56}
$$

The function

$$h(t) = \int_0^t g(t - \tau)\, f(\tau) d\tau = (g * f)(t) \tag{6.57}$$

is called a *convolution* of functions $g(t)$ and $f(t)$. Note that the convolution is symmetric:

$$\int_0^t g(t - \tau)\, f(\tau) d\tau = \int_0^t f(t - \tau)\, g(\tau) d\tau \quad \Longrightarrow \quad (g * f)(t) = (f * g)(t) \,,$$

which is checked immediately by changing the integration variable. Hence, we have just proven the following identity called the convolution theorem:

$$\mathcal{L}[(f * g)(t)] = F(p)G(p) \,. \tag{6.58}$$

The convolution we introduced here is very similar to the one we defined when considering the FT in Sect. 5.2.4. Since the LT and FT are closely related, there is no surprise that in both cases the convolution theorem has exactly the same form.[6]

Problem 6.36 Use the convolution theorem to prove the integral rule (6.49).

6.5 Solution of Ordinary Differential Equations (ODEs)

One of the main applications of the LT is in solving ordinary differential equations and their systems for specific initial conditions. Here we shall illustrate this point on a number of simple examples. More examples from physics will be given in Sect. 6.6. We shall be using simplified notations from now on: if the original function is $f(t)$, its image will be denoted by the corresponding capital letter, F in this case, with its argument p usually omitted; similarly, if the original is $y(t)$, its image is Y, and so on.

The general scheme for the application of the LT method is sketched in Fig. 6.10: since direct solution of the differential equation may be difficult, one uses the LT to rewrite the equation into a simpler form for the image Y, which is then solved. In particular, as we shall see, a linear DE with constant coefficients turns into a linear algebraic equation for the image Y which can always be solved. Once the image is known, one performs the inverse LT to find the function $y(t)$ of interest. The latter will automatically satisfy the initial conditions.

Example 6.1 ▶ Consider the following problem:

$$y'' + 4y' + 4y = t^2 e^{-2t} \quad \text{with} \quad y(0) = y'(0) = 0 . \tag{6.59}$$

Solution: Take the LT of both sides of the DE:

$$\left[p^2 Y - py(0) - y'(0)\right] + 4\left[pY - y(0)\right] + 4Y = \mathcal{L}\left[t^2 e^{-2t}\right] ,$$

where we made use of Eqs. (6.40) and (6.41) for the derivatives. From Eq. (6.11), the right-hand side is

$$\mathcal{L}\left[t^2 e^{-2t}\right] = \frac{2}{(p+2)^3} ,$$

---

[6] Note, however, that the two definitions of the convolution in the cases of LT and FT are not identical: the limits are 0 and t in the case of the LT, while when we considered the FT the limit were $\pm\infty$, Eq. (5.42).

Fig. 6.10 The working chart for using the LT method when solving differential equations

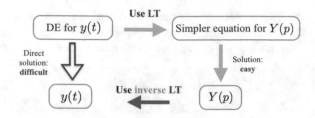

so that, after using the initial conditions, we obtain the following *algebraic* equation for Y:

$$p^2 Y + 4pY + 4Y = \frac{2}{(p+2)^3} \, ,$$

which is trivially solved to yield

$$Y = \frac{2}{(p+2)^3} \cdot \frac{1}{p^2 + 4p + 4} = \frac{2}{(p+2)^5} \implies y(t) = \frac{1}{12} t^4 e^{-2t} \, , \quad (6.60)$$

where we have used Eq. (6.11) again to perform the inverse LT. ◀

This example illustrates the power of the LT method. Normally we would look for a general solution of the corresponding homogeneous equation with two arbitrary constants; then we would try to find a particular integral which satisfies the whole equation with the right-hand side. Finally, we would use the initial conditions to find the two arbitrary constants. Using the LT, the full solution satisfying the initial conditions is obtained in just two steps!

Moreover, it is easy to see that a general solution of the DE with constant coefficients can always be obtained in a form of an integral for an arbitrary function $f(t)$ in the right-hand side of the DE. We have already discussed this point in Sect. 5.3.3 when considering an application of the Fourier transform method for solving DEs. There we introduced the so-called Green's function of the DE. A very similar approach can be introduced within the framework of the LT as well. Indeed, consider a second-order inhomogeneous DE:

$$y'' + a_1 y' + a_2 y = f(t) \, , \quad (6.61)$$

where a_1 and a_2 are some constant coefficients. By applying the LT to both sides, we obtain

$$\left[p^2 Y - py(0) - y'(0) \right] + a_1 \left[pY - y(0) \right] + a_2 Y = F \, ,$$

which yields

$$Y = \frac{1}{p^2 + a_1 p + a_2} F + \frac{(p + a_1) y(0) + y'(0)}{p^2 + a_1 p + a_2} \, . \quad (6.62)$$

Let us introduce a function

$$G(p) = \frac{1}{p^2 + a_1 p + a_2} = \frac{1}{(p - p_+)(p - p_-)} , \qquad (6.63)$$

where p_+ and p_- are the two roots of the corresponding quadratic polynomial in the denominator. This function can serve as the image for the following original:

$$g(t) = \mathcal{L}^{-1}[G(p)] = \frac{1}{p_+ - p_-} \left(e^{p_+ t} - e^{p_- t}\right) \quad \text{or} \quad g(t) = t e^{p_+ t} , \qquad (6.64)$$

depending on whether p_+ and p_- are different or the same (repeated roots)[7]. Then, the solution (6.62) can be written as

$$Y(p) = G(p)F(p) + G(p)\left[(p + a_1)\, y(0) + y'(0)\right] . \qquad (6.65)$$

The second term in the right-hand side corresponds to the solution of the corresponding *homogeneous* DE which is already adapted to the initial conditions; obviously, it corresponds to the complementary solution (use integration in the complex plane, Eq. (6.29), using two poles p_+ and p_-):

$$y_c(t) = \frac{(p_+ + a_1)\, y(0) + y'(0)}{p_+ - p_-} e^{p_+ t} + \frac{(p_- + a_1)\, y(0) + y'(0)}{p_- - p_+} e^{p_- t}$$

(assuming that $p_+ \neq p_-$). The first term in Eq. (6.65), however, corresponds to the particular integral of the *inhomogeneous* DE, and can be written (utilising the convolution theorem) as

$$y_p(t) = \int_0^t g(t - \tau)\, f(\tau)\, d\tau = \int_0^t f(t - \tau)\, g(\tau)\, d\tau . \qquad (6.66)$$

Either form is, of course, valid, as the convolution is symmetric! The function $g(t)$ is called Green's function of the DE (6.61). It satisfies the DE with the delta function $f(t) = \delta(t)$ in the right-hand side (cf. Sect. 5.3.3) and zero initial conditions:

$$g'' + a_1 g' + a_2 g = \delta(t) . \qquad (6.67)$$

Recall that the image of the delta function is just unity, see Eq. (6.15).

Example 6.2 ▶ Consider the following problem:

$$y'' + 3y' + 2y = f(t) \quad \text{with} \quad y(0) = y'(0) = 0 . \qquad (6.68)$$

Our task is to obtain a particular solution of this DE, subject to the given initial conditions, but for an arbitrary function $f(t)$.

---

[7] In the latter case the inverse LT of $G(p) = 1/(p - p_+)^2$ can be obtained either directly or by taking the limit $p_- \to p_+$ in $g(t)$ from the first formula in Eq. (6.64) obtained by assuming $p_+ \neq p_-$.

Solution: Let $\mathcal{L}\left(f(t)\right) = F$ and $\mathcal{L}\left(y(t)\right) = Y$. Then, performing the LT of both sides of the DE, we obtain

$$\left[p^2 Y - py(0) - y'(0)\right] + 3\left[pY - y(0)\right] + 2Y = F .$$

This equation is simplified further by applying our zero initial conditions. Then, the obtained equation is easily solved for the image Y of $y(t)$:

$$p^2 Y + 3pY + 2Y = F \quad \Longrightarrow \quad Y = \frac{1}{p^2 + 3p + 2} F .$$

Here $G(p) = \left(p^2 + 3p + 2\right)^{-1}$ is the image of Green's function. The corresponding to it original is easily calculated, e.g. by decomposing into partial fractions:

$$g(t) = \mathcal{L}^{-1}\left[G(p)\right] = \mathcal{L}^{-1}\left[\frac{1}{p^2 + 3p + 2}\right] = \mathcal{L}^{-1}\left[\frac{1}{p+1}\right] - \mathcal{L}^{-1}\left[\frac{1}{p+2}\right]$$

$$= e^{-t} - e^{-2t} .$$

Correspondingly, the full solution of the DE, because of the convolution theorem, Sect. 6.4.4, is

$$y(t) = \int_0^t \left[e^{-(t-\tau)} - e^{-2(t-\tau)}\right] f(\tau) d\tau \quad \text{or} \quad \int_0^t f\left(t - \tau\right) \left[e^{-\tau} - e^{-2\tau}\right] d\tau .$$

Note that the obtained forms of the solution do correspond to the general result (6.66). This should not be surprising as in this problem we deal with zero initial conditions. ◄

Problem 6.37 Show that the equation $y' + 3y = e^t$, $y(0) = 0$, has the solution $y(t) = \frac{1}{4}\left(e^t - e^{-3t}\right)$.

Problem 6.38 Show that the equation $y' - y = 2e^t$ with the initial condition $y(0) = 3$ has the solution $y(t) = (3 + 2t)\, e^t$.

Problem 6.39 Show that the solution of the following DE,

$$y'' + 9y = \cos 3t , \quad y(0) = 0 , \quad y'(0) = 6 ,$$

is $y(t) = (2 + t/6) \sin 3t$.

Problem 6.40 Show that the function $y(t) = e^{2t}$ is the solution of the equation $y'' + y' - 5y = e^{2t}$ subject to the initial conditions $y(0) = 1$ and $y'(0) = 2$.

Problem 6.41 A response $x(t)$ of an electronic device to an external excitation $f(t)$ (assuming that $f(t) = 0$ for $t < 0$) is described by the differential equation

$$T\frac{dx}{dt} + x = f(t),$$

where T is some positive constant. Using the LT method, show that the general solution of this equation satisfying the initial condition $x(0) = 0$ is

$$x(t) = \frac{1}{T}e^{-t/T}\int_0^T f(\tau)e^{\tau/T}d\tau.$$

Next, consider an excitation in the form of a rectangular impulse $f(t) = 1/\epsilon$ for $0 < t - t_0 < \epsilon$ (with $t_0 > 0$) and zero otherwise. Obtain the $x(t)$ in this case and show that in the $\epsilon \to +0$ limit $x(t) = \frac{1}{T}e^{-(t-t_0)/T}$. Since in this limit the rectangular impulse represents the delta function $\delta(t - t_0)$, the solution $x(t)$ in this case in fact represents Green's function of the differential equation. Hence, show that the same Green's function of the DE can be obtained directly by setting $f(\tau) = \delta(\tau - t_0)$.

Problem 6.42 A harmonic oscillator is initially at rest. Then at $t = 0$ a driving force $f(t)$ is applied. Show that the solution of the corresponding harmonic oscillator equation

$$y'' - 2y' + y = f(t),$$

can be generally written as

$$y(t) = \int_0^t (t - \tau)e^{t-\tau}f(\tau)d\tau.$$

Problem 6.43 As an example of the application of the formalism developed in the previous problem, consider an experiment, in which a finite pulse ($n = 1, 2, 3, \ldots$)

$$f(t) = \Delta_n(t) = \begin{cases} n, & \text{if } -1/2n < t - t_0 < 1/2n \\ 0, & \text{if } \quad |t - t_0| > 1/2n \end{cases}$$

was externally applied to the oscillator around some time $t_0 \gg 0$. Show that the response of the system in each case is

$$y_n(t) = ne^{t-t_0}\left[(t - t_0 - 1)\left(e^{1/2n} - e^{-1/2n}\right) + \frac{1}{2n}\left(e^{1/2n} + e^{-1/2n}\right)\right].$$

By taking the $n \rightarrow \infty$ limit, show that the response

$$y_n(t) \; \rightarrow \; y_\infty(t) = (t - t_0) \, e^{t - t_0} \; .$$

Show that the same result is obtained with $f(t) = \delta(t - t_0)$. Is this coincidence accidental?

Problem 6.44 Show that the equation $y'' + 4y = \sin 2t$ with the initial conditions $y(0) = 10$ and $y'(0) = 0$ has the solution

$$y(t) = 10 \cos 2t + \frac{1}{8} \left(\sin 2t - 2t \cos 2t \right) \; .$$

The LT method simplifies significantly solutions of a *system of ordinary DEs* as well. As an example, consider the following system of two first-order differential equations:

$$\frac{dy}{dt} - 2z = 2 \, , \quad \frac{dz}{dt} + 2y = 0 \, ,$$

which are subject to the zero initial conditions $y(0) = z(0) = 0$ (for simplicity). Applying the LT to both equations and introducing the images $y(t) \rightarrow Y(p)$ and $z(t) \rightarrow Z(p)$, we obtain two algebraic equations for them:

$$pY - 2Z = \frac{2}{p} \quad \text{and} \quad pZ + 2Y = 0 \, ,$$

which are easily solved to yield

$$Y = \frac{p}{4 + p^2} \, , \quad Z = - \frac{4}{p\left(p^2 + 4\right)} \; .$$

Applying the inverse LT, we finally obtain $y(t) = \sin 2t$ and $z(t) = -1 + \cos 2t$. The solution satisfies the equations and the initial conditions. One can also see that there is one more advantage of using the LT if not all but only one particular function is needed. At the intermediate step, when solving for the images in the Laplace space, each solution is obtained independently. Hence, if only one unknown function is needed, it can be obtained and inverted; this brings significant time savings for systems of more than two equations. This situation often appears when solving electrical circuits equations (Sect. 6.6.1) when only the current (or a voltage drop) along a particular circuit element is sought for.

Problem 6.45 Show that the solution of the equations

$$y'' + z'' - z' = 0, \quad y' + z' - 2z = 1 - e^t,$$

which is subject to the initial conditions $y(0) = 0$, $y'(0) = z(0) = z'(0) = 1$ is $y(t) = t$ and $z(t) = e^t$.

Problem 6.46 Show that the solution of the system of two ODEs,

$$y' + z = 2\cos t, \quad z' - y = 1,$$

subject to the initial conditions $y(0) = -1$ and $z(0) = 1$ is $y(t) = t\cos t - 1$ and $z(t) = t\sin t + \cos t$.

Systems of DEs we considered in Sect. 1.22 can also be solved using the LT method. The convenience of the latter method is that the corresponding eigenproblem does not appear. Moreover, we assumed there an exponential trial solution; no need for this assumption either: the whole solution appears naturally if the LT method is applied.

Problem 6.47 In this problem we shall consider again a particle in the magnetic field and obtain a general solution for an arbitrary direction of the magnetic field $\mathbf{B}$ and the initial velocity $\mathbf{v}_0$. Apply the LT method to Eqs. (1.211) to show that

$$\mathbf{v}(t) = \left(\widehat{\mathbf{B}} \cdot \mathbf{v}_0\right)\widehat{\mathbf{B}} + \left[\mathbf{v}_0 - \left(\widehat{\mathbf{B}} \cdot \mathbf{v}_0\right)\widehat{\mathbf{B}}\right]\cos(\omega t) + \left[\mathbf{v}_0 \times \widehat{\mathbf{B}}\right]\sin(\omega t),$$

where $\mathbf{v}_0$ is the initial velocity of the particle and $\widehat{\mathbf{B}} = \mathbf{B}/B$ is the unit vector in the direction of the magnetic field, and $\omega = qB/m$. It is easy to see that in the case of $\mathbf{v}_0 = \left(0, v_\perp, v_\parallel\right)$ and $\widehat{\mathbf{B}} = (0, 0, 1)$ the same result as in Eq. (1.215) is immediately recovered.

So far we have only discussed solving ODEs with constant coefficients. Our consideration at the end of Sect. 6.4.1 sets the grounds for also trying to solve linear ODEs with variable coefficients, specifically, when these coefficients are polynomials in t. We shall illustrate this method on the Cauchy–Euler equation

$$t^2 y'' + t y' - y = 0,$$

that we considered in Problem I.8.38 of Vol. I. It has two elementary solutions, $y_1(t) = t$ and $y_2(t) = t^{-1}$. Using Eqs. (6.44) and (6.45), a second-order differential equation for the image $Y(p)$ is obtained. Indeed, applying the LT,

$$\mathcal{L}\left[t^2 y''\right] + \mathcal{L}\left[ty'\right] - \mathcal{L}[y] = 0,$$

we get

$$\left[p^2 Y'' + 4pY' + 2Y\right] + \left[-pY' - Y\right] - Y = 0,$$

$$pY'' + 3Y' = 0.$$

The latter equation is easily solved by making a substitution $F(p) = Y'(p)$. We obtain $pF' = -3F$, giving $F(p) = C/p^3$, so that

$$Y(p) = \int F(p)dp = C \int \frac{dp}{p^3} = Cp^{-2} + C_1.$$

To ensure the correct behaviour of the image at $|p| \rightarrow \infty$, the constant C_1 must be dropped (see Sect. 6.3.1). Hence, $Y(p) = C/p^2$ and, therefore, $y(t) = Ct$, where C is an arbitrary constant. Therefore, we have obtained the first elementary solution $y_1(t) = t$ of this ODE. The second solution cannot be obtained with the LT method as the LT of the function $1/t$ does not exist (the Laplace integral diverges at $t = 0$).

Some integro-differential equations can also be approached using the LT method. Consider, as an example, the so-called Volterra equation

$$y'(t) = 1 - \int_0^t f(t - \tau)y(\tau)d\tau \tag{6.69}$$

with the initial condition $y(0) = 1$. Due to the fact that the integral in the right-hand side is recognised to be a convolution $f(t) * y(t)$, we can easily take the LT of both sides:

$$pY(p) - 1 = \frac{1}{p} - F(p)Y(p) \implies Y(p) = \frac{p+1}{p\left[p + F(p)\right]},$$

where $F(p) = \mathcal{L}\left[f(t)\right]$. So, the LT of the unknown function can be calculated. What is left is to take the inverse LT, which, of course, might not be that easy; the success of the method depends on the particular form of the kernel $f(t)$.

Problem 6.48 Show that in the special case of the exponential kernel $f(t) = e^{-2t}$ one obtains

$$Y(p) = \frac{p+2}{p(p+1)} \implies y(t) = 2 - e^{-t}.$$

The LT method can be very useful in solving partial differential equations as well. This particular application will be considered in some detail in Sect. 8.6.

6.6 Applications in Physics

6.6.1 *Application of the LT Method in Electronics*

The LT method has been found extremely useful is solving electronic circuits problems. This is because the famous Kirchhoff's laws for the circuits represent a set of integro-differential equations with constant coefficients which turn into a set of algebraic equations for the current images in the Laplace space and hence can be easily solved there. The resulting expressions in most cases represent a rational function of p for which the inverse LT can always be calculated. Therefore, any circuit, however complex, can in principle be solved. We shall illustrate the power of the LT method for solving problems of electrical circuits on a number of simple examples and problems.

We start by analysing the main elements of any electrical circuit. There are four of them, all shown in Fig. 6.11. The precise knowledge of the voltage drop $u(t)$ on each of the elements caused by the flowing current $i(t)$ is required in order to write Kirchhoff's equations. The simplest relationship is for the resistance: $u_R(t) = Ri(t)$. In the case of a conductance, $u_C(t) = q(t)/C$, where $q(t)$ is the charge stored there. Since $i(t) = dq/dt$, we obtain in this case:

$$u_C(t) = \frac{1}{C}\left[\int_0^t i(\tau)d\tau + q_0\right] , \qquad (6.70)$$

where q_0 is the initial charge on the capacitor. In the case of induction,

$$u_L(t) = L\frac{di(t)}{dt} , \qquad (6.71)$$

while in the case of the mutual induction,

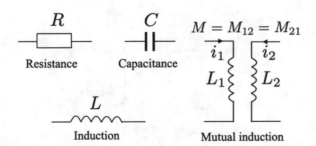

R
Resistance

C
Capacitance

$M = M_{12} = M_{21}$

i_1 i_2

L_1 L_2

L
Induction

Mutual induction

Fig. 6.11 Main elements of electrical circuits: a resistor R, capacitance C, induction L and mutual induction M. In the latter case, the voltage drop in the left mesh due to the current i_2 in the right one is given by $u_{M,1} = M\,di_2/dt$, while the voltage drop in the right mesh due to the current i_1 in the left one is $u_{M,2} = M\,di_1/dt$

$$u_M(t) = M \frac{di(t)}{dt} \, , \tag{6.72}$$

where the directions of the current as shown in Fig. 6.11 are assumed. The first Kirchhoff's law states that for any closed mesh in a circuit the total drop of the voltage across the mesh calculated along a particular direction (see an example in Fig. 6.12) is equal to the applied voltage (zero if no applied voltage is present, i.e. there is no battery attached):

$$\sum_k R_k i_k + \sum_k \frac{1}{C_k} \left[\int_0^t i_k(\tau)d\tau + q_{0k} \right] + \sum_k L_k \frac{di_k}{dt}$$
$$+ \sum_k M_{kk'} \frac{di_{k'}}{dt} = \sum_k v_k(t) \, , \tag{6.73}$$

where we sum over all elements appearing in the given mesh, with $M_{kk'}$ being the mutual induction coefficient between two neighbouring meshes k and k', see Fig. 6.11 (note that $M_{kk'} = M_{k'k}$). The current depending terms (in the left-hand side) are to be added algebraically with the sign defined by the chosen directions of the currents with respect to the chosen positive direction in the mesh (e.g. as indicated in Fig. 6.12), and the voltages $v_k(t)$ in the right-hand side are also to be summed up with the appropriate sign depending on their polarity.

The second Kirchhoff's law states that the sum of all currents through any vertex is zero:

$$\sum_k i_k = 0 \, . \tag{6.74}$$

To solve these equations, we apply the LT method. Since the terms are added algebraically, we can reformulate the rules of constructing Kirchhoff's equations directly in the Laplace space if we consider how each element of the mesh would contribute. If $I(p)$ is the image for a current passing through a resistance R and

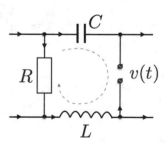

Fig. 6.12 A typical mesh in a circuit containing a set of elements and vertex points where the current goes along several different directions. A dashed line indicates a chosen 'positive' direction needed to select correctly the signs in the first Kirchhoff's law for each element of the mesh

$U_R(p)$ is the corresponding voltage drop, then the contribution due to the resistance would simply be $U_R = RI$. For a capacitance C, the LT of Eq. (6.70) gives

$$U_C = \frac{1}{pC} (I + q_0) , \tag{6.75}$$

while for the induction elements we would have similarly

$$U_L = L (pI - i_0) \quad \text{or} \quad U_M = M (pI - i_0) . \tag{6.76}$$

Here q_0 and i_0 are the initial (at $t = 0$) charge on the capacitor and the current through the induction. Correspondingly, the vertex equation (6.74) has the same form for the images as for the originals, while Eq. (6.73) describing the voltage drop across a mesh is rewritten via images as

$$\sum_k R_k I_k + \sum_k \frac{1}{pC_k} (I_k + q_{0k}) + \sum_k L_k (pI_k - i_{0k})$$
$$+ \sum_k M_{kk'} (pI_{k'} - i_{0k'}) = \sum_k V_k(p) . \tag{6.77}$$

In the case of a constant voltage v_k (a battery or Emf) $V_k(p) = v_k/p$. Here and in the following we use capital letters for images and small letters for the originals.

Example 6.3 ▶ Consider a circuit with Emf v given in Fig. 6.13a. Initially, the switch was opened. Calculate the current after the switch was closed at $t = 0$ and then determine the charge on the capacitor at long times.

Solution: In this case we have a single mesh with zero charge on the capacitor. Then, the first Kirchhoff's equation reads

$$\frac{v}{p} = RI + \frac{1}{Cp}I \quad \Longrightarrow \quad I = \frac{v}{R} \frac{1}{p + \frac{1}{CR}} .$$

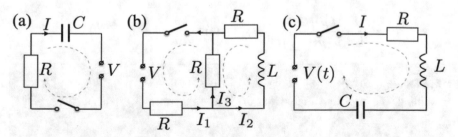

Fig. 6.13 Simple electrical circuits for **a** Example 6.3, **b** Example 6.4 and **c** Problem 6.49. Here voltages and currents are understood to be images of the corresponding quantities in the Laplace space and hence are shown using capital letters

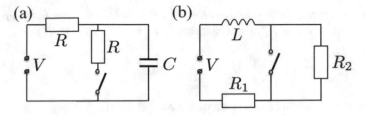

Fig. 6.14 Electrical circuits used in Problems **a** 6.50 and **b** 6.51 and 6.52

Inverting the image, we immediately obtain

$$i(t) = \frac{v}{R} e^{-t/CR} \ .$$

Therefore, the current is smoothly reduces to zero leading to the capacitor being charged up. Its charge is then

$$q_\infty = \int_0^\infty i(t)dt = -vC \ e^{-t/CR}\big|_0^\infty = vC \ .$$

◀

Example 6.4 ▶ Consider a circuit with Emf v shown in Fig. 6.13b. Determine the current passing through the Emf after the switch is closed.

Solution: The chosen directions of the three currents and the positive directions in each mesh are indicated in the figure. Let us write Kirchhoff's equations in the Laplace space following the chosen selections of the directions:

$$RI_1 + RI_3 = \frac{v}{p} \ , \quad LpI_2 + RI_2 - RI_3 = 0 \ , \quad \text{and} \quad I_1 = I_2 + I_3 \ .$$

Solving these algebraic equations with respect to the current I_1, which is the only one of interest for us here, yields

$$I_1 = \frac{v}{2LR} \frac{Lp + 2R}{p\left(p + \frac{3R}{2L}\right)} \quad \Longrightarrow \quad i_1(t) = \frac{v}{6R} \left(4 - e^{-3Rt/2L}\right) \ . \blacktriangleleft$$

Problem 6.49 Show that the current in the circuit shown in Fig. 6.13c containing an arbitrarily variable bias $v(t)$, after the switch was closed, is

$$i(t) = \int_0^t v(t - \tau)g(\tau)d\tau \ ,$$

where

$$g(t) = \frac{1}{L}\left(\frac{\lambda_-}{\lambda_- - \lambda_+}e^{\mu\lambda_- t} - \frac{\lambda_+}{\lambda_- - \lambda_+}e^{\mu\lambda_+ t}\right),$$

$$\lambda_\pm = -1 \pm \sqrt{1 - \frac{4L}{CR^2}}, \quad \mu = \frac{R}{2L}.$$

Problem 6.50 Consider the circuit shown in Fig. 6.14a. Initially the switch was opened and the capacitor gets charged up. Then at $t = 0$ it was closed. Show that the currents through the resistance next to the Emf v and the capacitor, respectively, are given by the equations

$$i_1(t) = \frac{v}{2R}\left(1 - e^{-2t/CR}\right), \quad i_2(t) = -\frac{v}{R}e^{-2t/CR}.$$

Problem 6.51 In the circuit shown in Fig. 6.14b initially the switch was opened. After a sufficiently long time at $t = 0$ the switch was closed. Show that the current through the Emf v is

$$i(t) = \frac{v}{R_1}\left(1 - \frac{R_2}{R_1 + R_2}e^{-R_1 t/L}\right).$$

[*Hint: Note that at $t = 0$ there is a current passing through the outer mesh.*]

Problem 6.52 Consider the same circuit as in the previous problem, but this time initially for a rather long time the switch was closed and then at $t = 0$ it was opened. Show that this time the current flowing through the Emf v will be

$$i(t) = \frac{v}{R_1 + R_2}\left(1 + \frac{R_2}{R_1}e^{(R_1 + R_2)t/L}\right).$$

6.6.2 Harmonic Particle with Memory

In Sect. 5.3.4, we discussed the solution of an equation of motion for a Brownian particle. We established there that if the time correlation function of the random force due to liquid molecules (hitting the particle all the time) satisfies a certain relationship (5.94) with the constant $\alpha = 2mk_BT\gamma$, where m is the particle mass, γ friction coefficient and T temperature, then on average the kinetic energy of the particle satisfies the equipartition theorem (5.93). In our treatment there we did not consider any memory effects assuming that the friction force is proportional to the instantaneous particle velocity.

In this section we shall consider a (still one dimensional) problem[8] of a particle in a harmonic potential well. We shall assume that the friction force will be dependent on the values of the velocity at preceding times; this corresponds to accounting for so-called 'memory' effects. Our aim here is to derive an expression for the distribution function of the particle at long times, both in its position x and velocity $v = \dot{x}$, accepting a certain relationship for the random force autocorrelation function.

Assuming for simplicity that the particle is of unit mass, its equation of motion can be cast in the following form:

$$\ddot{x} + \int_0^t \Gamma\left(t - \tau\right) \dot{x}\left(\tau\right) d\tau + \omega_0^2 x = \xi(t) , \qquad (6.78)$$

where ω_0 is the fundamental frequency associated with the harmonic potential, $\Gamma\left(t\right)$ is the so-called friction kernel (without loss of generality, it can be considered as an even function of time) and $\xi\left(t\right)$ is the random force. The latter force is due to interaction with the surrounding environment with the state of it being uncertain. The friction kernel must be a decaying function of time tending to zero at long times.

Applying the LT to both sides of the equation and using the convolution theorem, we obtain

$$\left[p^2 X(p) - px_0 - v_0\right] + \Gamma(p)\left[pX(p) - x_0\right] + \omega_0^2 X(p) = \xi(p) ,$$

which gives for the LT of the particle position:

$$X(p) = G(p)\left\{\xi(p) + \left[\Gamma(p) + p\right] x_0 + v_0\right\} . \qquad (6.79)$$

Here x_0 and $v_0 = \dot{x}_0$ are initial position and velocity of the particle, and

$$G(p) = \frac{1}{p^2 + p\Gamma(p) + \omega_0^2} . \qquad (6.80)$$

It is easy to see that $G(p)$ is the LT of Green's function $G(t)$ of Eq. (6.78). Indeed, replacing the right-hand side in the equation with the delta function, $\xi(t) \rightarrow \delta(t)$, assuming zero initial conditions, and performing the LT, we obtain exactly the expression (6.80) for $X(p)$ in this case.

Performing the inverse LT of Eq. (6.79), we obtain

$$x(t) = \eta(t)x_0 + G(t)v_0 + \int_0^t G(t - \tau)\xi\left(\tau\right) d\tau , \qquad (6.81)$$

where

---

[8] For a more general discussion, including a derivation of the equations of motion for a multi-dimensional open classical system (the so-called Generalised Langevin Equation), as well as references to earlier literature, see L. Kantorovich, Phys. Rev. B **78**, 094304 (2008).

$$\eta(t) = \mathcal{L}^{-1}\left[G(p)\,(p + \Gamma(p))\right] = \mathcal{L}^{-1}\left[\frac{1}{p} - \frac{\omega_0^2}{p}G(p)\right]$$

$$= 1 - \omega_0^2 \int_0^t G(\tau)d\tau . \tag{6.82}$$

Applying $t = 0$ in the solution (6.81), we deduce that

$$\eta(0) = 1 \quad \text{and} \quad G(0) = 0 . \tag{6.83}$$

Note that the first identity is consistent with the full solution (6.82) for the function $\eta(t)$.

Differentiating now the solution (6.81) with respect to time, we get for the velocity:

$$v(t) = \dot{x}(t) = \dot{\eta}(t)x_0 + \dot{G}(t)v_0 + \int_0^t \dot{G}(t - \tau)\xi(\tau)\,d\tau , \tag{6.84}$$

where the term arising from the differentiation of the integral with respect to the upper limit vanishes due to the second identity in Eq. (6.83). Applying $t = 0$ to the solution (6.84) for the velocity, we can deduce that

$$\dot{\eta}(0) = 0 \quad \text{and} \quad \dot{G}(0) = 1 . \tag{6.85}$$

Note that the first of these identities also follows immediately from Eqs. (6.82) and (6.83).

Equations (6.81) and (6.84) provide us with exact solutions for the position and velocity of the particle in the harmonic well under the influence of the external force $\xi(t)$. However, in our case, this force is random, and hence these 'exact' solutions have little value. Instead, it would be interesting to obtain a statistical information about the position and velocity of the particle at long time $t \to \infty$ when the particle has 'forgotten' its initial state described by x_0 and v_0. To investigate the behaviour of the particle at long times, it is customary considering the appropriate correlation functions. The position autocorrelation function (or *position–position* correlation function)

$$\langle x(t)x(0)\rangle = \langle x(t)x_0\rangle$$

$$= \eta(t)\langle x_0^2\rangle + G(t)\langle v_0 x_0\rangle + \int_0^t G(t - \tau)\langle \xi(\tau)x_0\rangle\,d\tau ,$$

where averaging $\langle \ldots \rangle$ is performed with respect to degrees of freedom of the environment and all possible initial positions and velocities of the particle. The random force $\xi(t)$ is related to the environment and hence is not correlated with the particle initial position, i.e. $\langle \xi(\tau)x_0\rangle = 0$; in addition, the particle initial position and the velocity are totally independent of each other and hence their correlation must be zero, $\langle v_0 x_0\rangle = 0$. Therefore, only the first term survives in the correlation function $\langle x(t)x(0) \rangle$, and hence we conclude that the position autocorrelation function $\langle x(t)x(0) \rangle \propto \eta(t)$. As the correlations with the initial position are expected to die off with time, we deduce that $\eta(t) \to 0$ when $t \to \infty$. Next, consider the *position–velocity* correlation function:

$$\langle x(t)v(0) \rangle = \langle x(t)v_0 \rangle$$

$$= \eta(t) \langle x_0 v_0 \rangle + G(t) \langle v_0^2 \rangle + \int_0^t G(t - \tau) \langle \xi(\tau) v_0 \rangle d\tau .$$

In this case only the second term survives rendering the function $G(t)$ to tend to zero when $t \to \infty$. Similarly, the velocity autocorrelation function

$$\langle v(t)v(0) \rangle = \dot{\eta}(t) \langle x_0 v_0 \rangle + \dot{G}(t) \langle v_0^2 \rangle + \int_0^t \dot{G}(t - \tau) \langle \xi(\tau) v_0 \rangle d\tau ,$$

being proportional to $\dot{G}(t)$ (since the first and the last terms in the right-hand side must be zero), shows that the derivative of Green's function should also tend to zero in the long time limit. So, we have established that $\eta(t)$, $G(t)$ and $\dot{G}(t)$ must decay with time to zero since the particle must forget its 'past' at long enough times. Note that, according to (6.82),

$$\dot{\eta}(t) = -\omega_0^2 G(t) , \tag{6.86}$$

and hence tends to zero at long times together with $G(t)$.

Therefore, if we now consider the function

$$u_1(t) = x(t) - \eta(t)x_0 - G(t)v_0 = \int_0^t G(t - \tau)\xi(\tau) d\tau , \tag{6.87}$$

then it must tend to $x(\infty) = x_\infty$ at long times as both $\eta(t)$ and $G(t)$ tend to zero in this limit. Similarly, we can also introduce another function,

$$u_2(t) = v(t) - \dot{\eta}(t)x_0 - \dot{G}(t)v_0 = \int_0^t \dot{G}(t - \tau)\xi(\tau) d\tau , \tag{6.88}$$

which must tend to the particle velocity $v(\infty) = \dot{x}(\infty) = v_\infty$ at long times.

Next we shall consider the same time correlation function $A_{11} = \langle u_1(t)u_1(t) \rangle$ of the function $u_1(t)$. To calculate it, we shall use the so-called second fluctuation–dissipation theorem according to which

$$\langle \xi(t)\xi(\tau) \rangle = \frac{1}{\beta}\Gamma(t - \tau) , \tag{6.89}$$

where $\beta = 1/k_B T$. This relationship shows that the random forces at the current and previous times are correlated with each other. It is said that the noise provided by this type of the random force is 'coloured', as opposite to the 'white' noise of Eq. (5.94), when such correlations are absent.

Then,

$$
\begin{aligned}
A_{11}(t) = \langle u_1(t)u_1(t) \rangle &= \int_0^t d\tau_1 \int_0^t d\tau_2\, G\,(t - \tau_1)\,\langle \xi\,(\tau_1)\,\xi\,(\tau_2) \rangle\, G\,(t - \tau_2) \\
&= \frac{1}{\beta} \int_0^t d\tau_1 \int_0^t d\tau_2\, G\,(\tau_1)\, \Gamma\,(\tau_1 - \tau_2)\, G\,(\tau_2)\ ,
\end{aligned}
$$
(6.90)

where we used Eq. (6.89) and changed the variables $t - \tau_1 \to \tau_1$ and $t - \tau_2 \to \tau_2$. Two other correlation functions can be introduced in a similar way for which we obtain

$$
A_{12}(t) = \langle u_1(t)u_2(t) \rangle = \frac{1}{\beta} \int_0^t d\tau_1 \int_0^t d\tau_2\, G\,(\tau_1)\, \Gamma\,(\tau_1 - \tau_2)\, \dot{G}\,(\tau_2)\ ,
$$
(6.91)

$$
A_{22}(t) = \langle u_2(t)u_2(t) \rangle = \frac{1}{\beta} \int_0^t d\tau_1 \int_0^t d\tau_2\, \dot{G}\,(\tau_1)\, \Gamma\,(\tau_1 - \tau_2)\, \dot{G}\,(\tau_2)\ .
$$
(6.92)

To calculate these correlation functions, we shall evaluate their time derivatives. Let us start from A_{11} given by Eq. (6.90). Differentiating the right-hand side with respect to t, we find

$$
\dot{A}_{11}(t) = \frac{2}{\beta}G(t) \int_0^t \Gamma\,(t - \tau)\, G(\tau) d\tau = \frac{2}{\beta}G(t)\mathcal{L}^{-1}\left[\Gamma(p)G(p)\right]\ .
$$

Using the definition of the auxiliary function $\eta(t)$ in the Laplace space, see Eq. (6.82), the inverse LT above is easily calculated to yield

$$
\mathcal{L}^{-1}\left[\Gamma(p)G(p)\right] = \mathcal{L}^{-1}\left[(p + \Gamma(p))\,G(p)\right] - \mathcal{L}^{-1}\left[pG(p)\right] = \eta(t) - \dot{G}(t)\ .
$$

In the last passage we have used the fact that $G(0) = 0$ and hence $\mathcal{L}\left[\dot{G}(t)\right] = pG(p)$. Therefore,

$$
\begin{aligned}
\dot{A}_{11}(t) &= \frac{2}{\beta}G(t)\left[\eta(t) - \dot{G}(t)\right] = -\frac{2}{\beta\omega_0^2}\dot{\eta}(t)\eta(t) - \frac{2}{\beta}G(t)\dot{G}(t) \\
&= -\frac{1}{\beta}\frac{d}{dt}\left[\frac{\eta^2(t)}{\omega_0^2} + G^2(t)\right]\ ,
\end{aligned}
$$

where Eq. (6.86) was employed to relate $G(t)$ to $\dot{\eta}(t)$ in the first term. Therefore, integrating, and using the initial condition that $A_{11}(0) = 0$ (it follows from its definition as an average of $u_1(0) = 0$ squared, or directly from the expression (6.90)), we obtain

$$
A_{11}(t) = \frac{1}{\beta\omega_0^2} - \frac{1}{\beta}\left[\frac{\eta^2(t)}{\omega_0^2} + G^2(t)\right]\ .
$$
(6.93)

Problem 6.53 Using a similar reasoning, demonstrate that

$$\dot{A}_{12}(t) = \frac{1}{\beta}\frac{d}{dt}\left[G(t)\eta(t) - G(t)\dot{G}(t)\right] \;\Rightarrow\; A_{12}(t) = \frac{1}{\beta}G(t)\left[\eta(t) - \dot{G}(t)\right] ,$$
(6.94)

$$\dot{A}_{22}(t) = \frac{1}{\beta}\frac{d}{dt}\left[-\omega_0^2 G^2(t) - \dot{G}^2(t)\right] \;\Rightarrow\; A_{22}(t) = \frac{1}{\beta} - \frac{1}{\beta}\left[\omega_0^2 G^2(t) + \dot{G}^2(t)\right] .$$
(6.95)

Hence, at long times $A_{11} \to 1/\beta\omega_0^2$, $A_{12}(t) \to 0$ and $A_{22}(t) \to 1/\beta$ as $\eta(t)$, $G(t)$ and $\dot{G}(t)$ vanish in this limit as discussed above.

The probability distribution function $\mathcal{P}(u_1, u_2)$ gives the probability

$$dW(u_1, u_2) = \mathcal{P}(u_1, u_2)\, du_1 du_2$$

for the particle to be found with the variable u_1 being between u_1 and $u_1 + du_1$, and u_2 being between u_2 and $u_2 + du_2$. From the fact that both $u_1(t)$ and $u_2(t)$ are linear with respect to the noise $\xi(t)$, see Eqs. (6.87) and (6.88), it follows that either of the variables is Gaussian. That basically means that the probability distribution is proportional to the exponential function with the exponent which is quadratic with respect to $u_1(t)$ and $u_2(t)$:

$$\mathcal{P}(u_1, u_2) \propto \exp\left[-\frac{1}{2}\mathbf{Y}^T(t)\mathbf{A}^{-1}(t)\mathbf{Y}(t)\right] ,$$

where

$$\mathbf{Y}(t) = \begin{pmatrix} u_1(t) \\ u_2(t) \end{pmatrix} ,$$

and the inverse of the 2×2 matrix $\mathbf{A}(t)$ can be directly related to the correlation functions between the components of the vector $\mathbf{Y}$, since, see Eq. (7.71),

$$\left(\mathbf{A}^{-1}\right)_{11} = \langle u_1(t)u_1(t)\rangle , \;\; \left(\mathbf{A}^{-1}\right)_{12} = \langle u_1(t)u_2(t)\rangle , \;\; \left(\mathbf{A}^{-1}\right)_{22} = \langle u_2(t)u_2(t)\rangle .$$

We are interested, however, only in the long time limit when $u_1 = u_1(\infty) = x_\infty$, $u_2 = u_2(\infty) = v_\infty$ and

$$\mathbf{Y} = \begin{pmatrix} x_\infty \\ v_\infty \end{pmatrix} ,$$

while the matrix $\mathbf{A}$ tends to

$$\mathbf{A} = \begin{pmatrix} A_{11} & A_{12} \\ A_{21} & A_{22} \end{pmatrix}_{t\to\infty} = \begin{pmatrix} 1/\beta\omega_0^2 & 0 \\ 0 & 1/\beta \end{pmatrix} .$$

Since at long times the matrix $\mathbf{A}$ is diagonal, its inverse is trivially calculated:

$$\mathbf{A}^{-1} = \begin{pmatrix} \beta\omega_0^2 & 0 \\ 0 & \beta \end{pmatrix},$$

so that

$$\mathcal{P}(u_1, u_2) \propto \exp\left[-\beta\left(\frac{\omega_0^2 x_\infty^2}{2} + \frac{v_\infty^2}{2}\right)\right], \qquad (6.96)$$

i.e. at long times the distribution function of the particle tends to the Gibbsian distribution $\mathcal{P} \propto e^{-\beta E}$ containing the total energy $E = v_\infty^2/2 + \omega_0^2 x_\infty^2/2$ of the harmonic oscillator (recall that we set the mass of the vibrating particle to be equal to one here).

Problem 6.54 Determine the normalisation constant in the distribution (6.96), i.e. show that

$$\int_{-\infty}^{\infty} du_1 \int_{-\infty}^{\infty} du_2\, \mathcal{P}(u_1, u_2) = 1 \;\Rightarrow\; \mathcal{P}(u_1, u_2) = \frac{\beta\omega_0}{2\pi} \exp\left[-\beta\left(\frac{u_2^2}{2} + \frac{\omega_0^2 u_1^2}{2}\right)\right].$$

Problem 6.55 Show that the average potential and kinetic energies satisfy the equipartition theorem:

$$\langle U \rangle = \int_{-\infty}^{\infty} du_1 \int_{-\infty}^{\infty} du_2\, \frac{\omega_0^2 u_1^2}{2} \mathcal{P}(u_1, u_2) = \frac{1}{2\beta} = \frac{k_B T}{2},$$

$$\langle K \rangle = \int_{-\infty}^{\infty} du_1 \int_{-\infty}^{\infty} du_2\, \frac{u_2^2}{2} \mathcal{P}(u_1, u_2) = \frac{1}{2\beta} = \frac{k_B T}{2}.$$

[Hint: You may find results of Problem 4.19 useful.]

Problem 6.56 Consider now a problem of the unit mass Brownian particle with memory. There is no conservative force acting on the particle (in the previous case such was the harmonic force), therefore the equation of motion is more conveniently formulated with respect to the particle velocity:

$$\dot{v} + \int_0^t \Gamma(t - \tau) v(\tau)\, d\tau = \xi(t),$$

where the random force satisfies the same condition (6.89) as for the particle in the harmonic well. Show that in this case Green's function $G(t) = \mathcal{L}^{-1}[1/(p + \Gamma(p))]$, the velocity

$$v(t) = G(t)v_0 + \int_0^t G(t - \tau)\xi(\tau)\,d\tau\,,$$

while the same time correlation function for the variable $u(t) = v(t) - G(t)v_0$ is

$$A(t) = \langle u(t)u(t)\rangle = \frac{1}{\beta} - \frac{1}{\beta}G^2(t)\,,$$

which tends to $1/\beta$ at long times. *[Hint: Note that $G(0) = 1$ and relate $\Gamma(p)G(p)$ to $\mathcal{L}[\dot{G}(t)]$.]* Consequently, argue that the probability distribution for the Brownian particle at long times is Maxwellian:

$$\mathcal{P}(v_\infty) = \sqrt{\frac{\beta}{2\pi}}e^{-\beta v_\infty^2/2}\,,$$

so that the equipartition theorem is fulfilled in this case as well: $\langle v_\infty^2/2\rangle = 1/2\beta$.

6.6.3 Probabilities of Hops

Here we shall consider a problem which frequently appears in many areas of physics, chemistry and biology, e.g. in materials modelling, when a system hops between different energy minima (states) during some process (e.g. crystal growth).[9]

Consider a system prepared at time $t = 0$ in some initial state. Over time the system may undergo a transition, with the rate $r_{i_1}^{(1)}$, to a different state i_1 taken from a collection of n_1 possible states. Once the state i_1 is reached, the second transition into state i_2 with the rate $r_{i_2}^{(2)}$ may occur, and so on. Figure 6.15 illustrates this. Over the whole observation time t the system may either remain in the initial state, make just one transition, two, three, etc. The sum of all these events (with up to an infinite number of transitions, or hops) should form a complete system of events, and hence the probabilities of all these events must sum up to unity. We shall derive here these probabilities assuming that the individual rates do not depend on time.

Over time t the system may hop an arbitrary number of times N ranging between 0 and ∞. We shall indicate the hop number by a superscript; for instance, $r_{i_2}^{(3)}$ corresponds to the rate to hop to state i_2 during the third hop from the state reached after the second hop. Summing up all rates at the given $k-$th hop,

---

[9] Here we loosely follow (and in a rather simplified form) a detailed discussion which a reader can find in L. Kantorovich, Phys. Rev. B **75**, 064305 (2007).

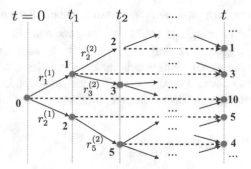

Fig. 6.15 A 'tree' of transitions from the initial state marked as 0 to all other possible states over time t via a final number of hops performed at times t_1, t_2, etc., $t_n = t$. Dashed horizontal lines correspond to the system remaining in the state from which the line starts over all the remaining time, while blue arrows indicate transitions between states, the latter are indicated by filled circles. State numbers are shown by the circles for clarity

$$R^{(k)} = \sum_{i=1}^{n^{(k)}} r_i^{(k)} \, , \tag{6.97}$$

gives the total rate of leaving the state the system was in prior to the k−th hop, and hence it corresponds to the *escape rate* from this state. Above, $n^{(k)}$ is the total number of states available to the system to hop into during the k−th hop.

Now, let the system be in some state at time t' after $(k - 1)$ hops. We can then define the *residence probability* $P_0^{(k)} (t', t)$ for the system to *remain* in the current state until the final time t. In fact, this probability was already calculated in Sect. I.8.6.7 and is given by the formula:

$$P_0^{(k)} (t', t) = e^{-R^{(k)}(t-t')} \, , \tag{6.98}$$

where $R^{(k)}$ is the corresponding escape rate during the k−th hop. Therefore, the probability for the system to remain in the current state over the whole time $t - t'$ (and hence to make no hops at all during the k-th step) is given as $P_0^{(k)}(t', t)$.

Consider now an event whereby the system makes a single hop from the initial state to state i_1 at some time between t_0 and t, and then remains in that state until time t. The probability of this, an essentially one-hop event, is given by the integral:

$$
\begin{aligned}
P_{i_1 0}^{(12)} (t_0, t) &= \int_{t_0}^{t} P_0^{(1)} (t_0, t_1) \left(r_{i_1}^{(1)} dt_1 \right) P_0^{(2)} (t_1, t) \\
&= \int_{t_0}^{t} dt_1 \, P_0^{(1)} (t_0, t_1) \, r_{i_1}^{(1)} P_0^{(2)} (t_1, t) \, .
\end{aligned}
\tag{6.99}
$$

Here the system remains in its initial state up to some time t_1 (where $t_0 < t_1 < t$), the probability of that is $P_0^{(1)} (t_0, t_1)$, then it makes a single hop into state i_1 (the probability of this is $r_{i_1}^{(1)} dt_1$) and then remains in that state all the remaining time, the probability of the latter being $P_0^{(2)}(t_1, t)$. We integrate over all possible times t_1 to obtain the whole probability. The superscript in $P_{i_1 0}^{(12)}$ shows that this event is based on two elementary events: (i) a single hop, indicated by i_1 as a subscript and (ii) no transition thereafter, indicated by 0 next to i_1 in the subscript. Using Eq. (6.98) for the residence probability and performing the integration, we obtain

$$P_{i_1 0}^{(12)}(t_0, t) = \frac{r_{i_1}^{(1)}}{R^{(1)} - R^{(2)}} \left(e^{-R^{(2)}t} - e^{-R^{(1)}t} \right) . \tag{6.100}$$

This expression was obtained assuming different escape rates $R^{(1)} \neq R^{(2)}$. If the escape rates are equal, then one can either perform the integration directly or take the limit of $x = R^{(2)} - R^{(1)} \to 0$ in the above formula. In either case we obtain

$$P_{i_1 0}^{(12)}(t_0, t) = r_{i_1}^{(1)} t e^{-R^{(1)}t} . \tag{6.101}$$

Along the same lines one can calculate the probability to make exactly two hops over time between t_0 and t: initially to state i_1 and then to state i_2:

$$P_{i_1 i_2 0}^{(123)}(t_0, t) = \int_{t_0}^{t} dt_1 \int_{t_1}^{t} dt_2 \, P_0^{(1)}(t_0, t_1) \, r_{i_1}^{(1)} \, P_0^{(2)}(t_1, t_2) \, r_{i_2}^{(2)} \, P_0^{(3)}(t_2, t) . \tag{6.102}$$

Integrating this expression is a bit more tedious, but still simple; assuming that all escape rates are different, we obtain (please, check!)

$$P_{i_1 i_2 0}^{(123)}(t_0, t) = r_{i_1}^{(1)} r_{i_2}^{(2)} \left[\frac{1}{R_{21} R_{31}} e^{-R^{(1)}t} + \frac{1}{R_{12} R_{32}} e^{-R^{(2)}t} + \frac{1}{R_{13} R_{23}} e^{-R^{(3)}t} \right] ,$$
$$\tag{6.103}$$

where it was denoted, for simplicity, $R_{ij} = R^{(i)} - R^{(j)}$. If some of the rates coincide, then a slightly different expression is obtained.

Problem 6.57 Show that the three-hop probability

$$P_{i_1 i_2 i_3 0}^{(1234)}(t_0, t) = r_{i_1}^{(1)} r_{i_2}^{(2)} r_{i_3}^{(3)} \left[\frac{1}{R_{21} R_{31} R_{41}} e^{-R^{(1)}t} + \frac{1}{R_{12} R_{32} R_{42}} e^{-R^{(2)}t} \right.$$
$$\left. + \frac{1}{R_{13} R_{23} R_{43}} e^{-R^{(3)}t} + \frac{1}{R_{14} R_{24} R_{34}} e^{-R^{(4)}t} \right]$$

where all escape rates were assumed to be different.

This type of calculation can be continued, although the calculation of probabilities associated with larger number of hops becomes very cumbersome. Moreover, one has to separately consider all possible cases of rates being different and some of them (or all) the same. The important point is that, as can be shown, the sum of all probabilities for arbitrary number of hops is equal to unity:

$$P_0^{(1)}(t_0, t) + \sum_{i_1=1}^{n^{(1)}} \left\{ P_{i_1 0}^{(12)}(t_0, t) + \sum_{i_2=1}^{n^{(2)}} \left\{ P_{i_1 i_2 0}^{(123)}(t_0, t) + \sum_{i_3=1}^{n^{(3)}} \left\{ P_{i_1 i_2 i_3 0}^{(1234)}(t_0, t) + \ldots \right\} \right\} \right\} = 1 .$$

(6.104)

Using the LT method the calculation of the probabilities can be drastically simplified. Indeed, let us start from writing out an obvious recurrence relation:

$$P_{i_1 i_2 \cdots i_{N-1} 0}^{(12 \cdots N)}(t_0, t) = \int_{t_0}^t P_0^{(1)}(t_0, t_1) \, r_{i_1}^{(1)} dt_1 \, P_{i_2 i_3 \cdots i_{N-1} 0}^{(23 \cdots N)}(t_1, t)$$

$$= \int_{t_0}^t dt_1 \, P_0^{(1)}(t_0, t_1) \, r_{i_1}^{(1)} \, P_{i_2 i_3 \cdots i_{N-1} 0}^{(23 \cdots N)}(t_1, t) ,$$

which states that the N−hop transition can be thought of as a single hop at time t_1 into state i_1 followed by $(N - 1)$ remaining hops into states $i_2 \to i_3 \to \ldots \to i_{N-1}$, after which the system remains at the last state for the rest of the time. Note that the probabilities depend only on the time difference. Setting the initial time t_0 to zero, one can then clearly see that the time integral above represents a convolution integral. Therefore, performing the LT of this expression, we immediately obtain

$$P_{i_1 i_2 \cdots i_{N-1} 0}^{(12 \cdots N)}(p) = \frac{r_{i_1}^{(1)}}{p + R^{(1)}} P_{i_2 i_3 \cdots i_{N-1} 0}^{(23 \cdots N)}(p) ,$$

since $\mathcal{L}\left[P_0^{(1)}(0, t)\right] = 1 / \left(p + R^{(1)}\right)$. Applying this recurrence relation recursively, the probability in the Laplace space can be calculated explicitly:

$$P_{i_1 i_2 \cdots i_{N-1} 0}^{(12 \cdots N)}(p) = \frac{r_{i_1}^{(1)}}{p + R^{(1)}} P_{i_2 i_3 \cdots i_{N-1} 0}^{(23 \cdots N)}(p) = \frac{r_{i_1}^{(1)}}{p + R^{(1)}} \frac{r_{i_2}^{(2)}}{p + R^{(2)}} P_{i_3 i_4 \cdots i_{N-1} 0}^{(34 \cdots N)}(p)$$

$$= \frac{r_{i_1}^{(1)}}{p + R^{(1)}} \frac{r_{i_2}^{(2)}}{p + R^{(2)}} \frac{r_{i_3}^{(3)}}{p + R^{(3)}} P_{i_4 i_5 \cdots i_{N-1} 0}^{(45 \cdots N)}(p) = \cdots$$

$$= \frac{r_{i_1}^{(1)}}{p + R^{(1)}} \cdots \frac{r_{i_{N-1}}^{(N-1)}}{p + R^{(N-1)}} P_0^{(N)}(p)$$

$$= \frac{1}{p + R^{(N)}} \prod_{k=1}^{N-1} \frac{r_{i_k}^{(k)}}{p + R^{(k)}} .$$

(6.105)

To calculate the probability, it is now only required to take the inverse LT. If the escape rates are all different, then we have simple poles on the negative part of the real axis at $-R^{(1)}, -R^{(2)}$, etc., and hence the inverse LT is easily calculated:

$$P_{i_1 i_2 \cdots i_{N-1} 0}^{(12 \cdots N)} (0, t) = \left(\prod_{k=1}^{N-1} r_{i_k}^{(k)} \right) \sum_{k=1}^{N} \mathrm{Res} \left[e^{pt} \prod_{l=1}^{N} \frac{1}{p + R^{(l)}} ; \; p = -R^{(k)} \right]$$

$$= \left(\prod_{k=1}^{N-1} r_{i_k}^{(k)} \right) \sum_{k=1}^{N} e^{-R^{(k)} t} \left(\prod_{l=1 \, (l \neq k)}^{N} \frac{1}{-R^{(k)} + R^{(l)}} \right) .$$

It can be checked immediately that this expression is a generalisation of the previous results obtained above for the cases of N between 1 and 4. If some of the escape rates are the same for different hops, then the poles associated with those terms are no longer simple poles; in this more general case the probability (6.105) can be written slightly differently:

$$P_{i_1 i_2 \cdots i_{N-1} 0}^{(12 \cdots N)} (p) = \left(\prod_{k=1}^{N-1} r_{i_k}^{(k)} \right) \prod_{k} \frac{1}{\left(p + R^{(k)} \right)^{m_k}} ,$$

where the second product runs over all distinct escape rates and m_k is their repetition. Correspondingly, $-R^{(k)}$ becomes the pole of order m_k and one has to use general formula (2.107) for calculating the corresponding residues.

Problem 6.58 Show that if all escape rates are the same and equal to R, then the N-hop probability is

$$P_{i_1 i_2 \cdots i_{N-1} 0}^{(12 \cdots N)} (0, t) = \left(\prod_{k=1}^{N-1} r_{i_k}^{(k)} \right) \frac{t^{N-1}}{(N-1)!} e^{-Rt} .$$

Problem 6.59 The probability $P_N(t)$ of performing N hops (no matter into which states) over time t can be obtained by summing up all possible $(N + 1)$−hops probabilities:

$$P_N(t) = \sum_{i_1=1}^{n^{(1)}} \sum_{i_2=1}^{n^{(2)}} \cdots \sum_{i_N=1}^{n^{(N)}} P_{i_1 i_2 \cdots i_N 0}^{(12 \cdots N+1)} (0, t) .$$

Assuming that the escape rates are all the same and using the result of the previous problem, show that

$$P_N(t) = \frac{(Rt)^N}{N!} e^{-Rt} ,$$

which is the famous Poisson distribution. Then, demonstrate that the sum of all possibilities of performing 0, 1, 2, etc. hops is equal to unity:

$$\sum_{N=0}^{\infty} P_N = 1 \,.$$

6.6.4 Inverse NC-AFM Problem

In Sect. 3.11.4, we introduced Non-contact Atomic Force Microscopy (NC-AFM), shown in Fig. 3.10, a revolutionary technique which is capable of imaging surfaces as well as atoms and molecules on them with sub-molecular, and sometimes even with atomic, resolution. We then derived an expression (3.116) which allows calculating the resonance frequency ω of the oscillating cantilever of this experimental probe from the force acting on the tip due to interaction with the surface. This expression is very useful in comparing theory and experiment as it allows calculating the frequency shift $\Delta\omega = \omega - \omega_0$, the quantity which is directly measured. Here ω_0 is the oscillation frequency of the cantilever far away from the surface. If one can calculate the quantity which is directly measured experimentally, then it would become possible to verify theoretical models. Of course, many such models may need to be tried before a good agreement is reached.

However, one may also ask a different question: how to determine the tip force $F_s(z)$ as a function of the tip-surface distance z from the experimentally measured frequency shift $\Delta\omega(z)$ curve? If the force can be obtained in this way, then it might be easier to choose the theoretical model which is capable of reproducing this particular force $z-$dependence. This problem is basically an inverse problem to the one we solved in Sect. 3.11.4. It requires solving the integral equation (3.116) with respect to the force.

Here we shall obtain a nearly exact solution of this integral equation using the method of LT.[10] We shall start by rewriting the integral equation (3.116) slightly differently:

$$\pi k A \left[1 - \left(\frac{\omega}{\omega_0} \right)^2 \right] = \int_{\pi/2}^{5\pi/2} F_s \left(h_0 + A \sin \phi \right) \sin \phi d\phi \,. \tag{6.106}$$

---

[10] The solution to this problem was first obtained by J. E. Sader and S. P. Jarvis in their highly cited paper [Appl. Phys. Lett. **84**, 1801 (2004)] using the methods of LT and the so-called fractional calculus. We adopted here some ideas of their method, but did not use the fractional calculus at all relying instead on conventional techniques.

Recall that A is the oscillation amplitude and $z = h_0 + A \sin \phi$ corresponds to the tip height above the surface; k is the elastic constant of the cantilever. It was convenient here to shift the integration limits by $\pi/2$. This can always be done as the integrand is periodic with respect to ϕ with the period of 2π. Next, note that within the span of the ϕ values in the integral the tip makes the full oscillation cycle by starting at the height $h_0 + A$ (at $\phi = \pi/2$), then moving to the position $z = h_0 - A$ closest to the surface (at $\phi = 3\pi/2$), and then returning back (retracting) to its initial position of $z = h_0 + A$ at $\phi = 5\pi/2$.

Problem 6.60 Next, we shall split the integral into two: for angles $\pi/2 < \phi < 3\pi/2$ when the tip moves down, and for angles $3\pi/2 < \phi < 5\pi/2$ when it is retracted back up. If $F_\downarrow(z)$ and $F_\uparrow(z)$ are the tip forces for the tip moving down and up, respectively, show, by making a substitution $x = \sin \phi$, that Eq. (6.106) can be rewritten as follows:

$$\int_{-1}^{1} F(z + Ax) \frac{x \, dx}{\sqrt{1 - x^2}} = \Psi(z) , \qquad (6.107)$$

where

$$\Psi(z) = \frac{\pi k A}{2} \left[1 - \frac{\omega^2(z)}{\omega_0^2} \right] \qquad (6.108)$$

is to be considered as a known function of z, and $F(z) = \frac{1}{2} \left[F_\downarrow(z) + F_\uparrow(z) \right]$ is the average (over a cycle) tip force. When deriving formula (6.107), choose the correct sign in working out $dx = \cos \phi \, d\phi = \pm\sqrt{1 - x^2} d\phi$ in each of the two integrals.

Note that the tip force on the way down and up could be different due to a possible atomic reconstruction at the surface (and/or at the tip) when the tip approaches the surface on its way down; this reconstruction sets in and affects the force when the tip is retracted. If such a reconstruction takes place, the tip force experiences a hysteresis over the whole oscillation cycle which results in the energy being dissipated in the junction.

To solve Eq. (6.107), we assume that $F(z)$ is the LT of some function $f(t)$:

$$\int_{-1}^{1} \frac{x \, dx}{\sqrt{1 - x^2}} \left[\int_0^\infty f(t) e^{-(z + Ax)t} dt \right] = \Psi(z) .$$

Our current goal is to find the function $f(t)$. The improper integral over t converges for any x since $z + Ax \geq z - A > 0$ as z is definitely bigger than A (the distance of closest approach $h_0 - A > 0$); moreover, it converges uniformly with respect to x, since

$$\left| \int_0^\infty f(t) e^{-(z + Ax)t} dt \right| \leq \left| \int_0^\infty f(t) e^{-(z - A)t} dt \right| = |F(z - A)| ,$$

so that the two integrals can be swapped:

$$\int_0^\infty f(t) e^{-zt} dt \left[\int_{-1}^1 \frac{x e^{-(At)x}}{\sqrt{1-x^2}} dx \right] = \Psi(z) . \qquad (6.109)$$

The integral in the square brackets can be directly related to the modified Bessel function of the first kind, see Eq. (4.236); it is equal to $-\pi I_1(At)$. This enables us to rewrite Eq. (6.109) as follows:

$$\int_0^\infty f(t) e^{-zt} I_1(At) \, dt = -\frac{1}{\pi} \Psi(z) ,$$

or simply as

$$\mathcal{L} [I_1(At) f(t)] = -\frac{1}{\pi} \Psi(p) , \qquad (6.110)$$

where p is the real positive number; we have changed z to p here as this is the letter we have been using in this chapter as the variable for the LT. Hence, it follows:

$$f(t) = -\frac{1}{\pi} \frac{1}{I_1(At)} \mathcal{L}^{-1} [\Psi(p)] = -\frac{1}{\pi} \frac{\psi(t)}{I_1(At)} , \qquad (6.111)$$

where $\psi(t) = \mathcal{L}^{-1} [\Psi(p)]$. So far, our manipulations have been exact.

To proceed, we shall now apply an approximation to the Bessel function which works really well across a wide range of its variable:

$$\frac{1}{I_1(x)} \simeq e^{-x} \left(\frac{2}{x} + \frac{1}{4\sqrt{x}} + \sqrt{2\pi x} \right) . \qquad (6.112)$$

An amazing agreement of this approximation with the exactly calculated function (4.236) is demonstrated in Fig. 6.16.

By using approximation (6.112) in Eq. (6.111), applying the LT to both sides and recalling that the force $F(p) = \mathcal{L}[f(t)]$, we can now write

$$F(p) = -\frac{1}{\pi} \left\{ \frac{2}{A} \mathcal{L} \left[\frac{e^{-At}}{t} \psi(t) \right] + \frac{1}{4\sqrt{A}} \mathcal{L} \left[\frac{e^{-At}}{\sqrt{t}} \psi(t) \right] + \sqrt{2\pi A} \, \mathcal{L} \left[\sqrt{t} e^{-At} \psi(t) \right] \right\} .$$
$$(6.113)$$

Introducing notations

$$G_1(p) = \mathcal{L} \left[t^{-1} \psi(t) \right] , \quad G_2(p) = \mathcal{L} \left[t^{-1/2} \psi(t) \right] \quad \text{and} \quad G_3(p) = \mathcal{L} \left[t^{1/2} \psi(t) \right] , \qquad (6.114)$$

and using the property (6.47), we obtain

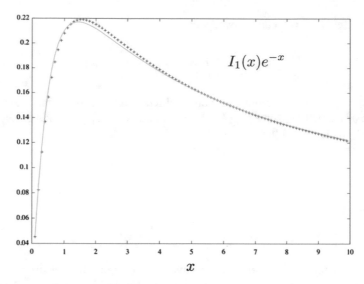

Fig. 6.16 Comparison of the function $I_1(x)$ calculated exactly (by a numerical integration in Eq. (4.236)) and using the approximation of Eq. (6.112). In both cases for convenience the function $I_1(x)e^{-x}$ is actually shown

$$F(p) = -\frac{1}{\pi}\left\{\frac{2}{A}G_1(p+A) + \frac{1}{4\sqrt{A}}G_2(p+A) + \sqrt{2\pi A}G_3(p+A)\right\}.$$
(6.115)

Now we need to calculate all three G functions in Eq. (6.114). The first one is calculated immediately using property (6.52):

$$G_1(p) = \mathcal{L}\left[\frac{\psi(t)}{t}\right] = \int_p^\infty \Psi(z)\,dz = \int_0^\infty \Psi(z+p)\,dz.$$
(6.116)

For the calculation of the second and the third ones, we can use Eqs. (6.34) and (6.39), respectively, with $\alpha = 1/2$. These give

$$G_2(p) = \mathcal{L}\left[t^{-1/2}\psi(t)\right] = \frac{1}{\sqrt{\pi}}\int_0^\infty \frac{\Psi(z+p)}{\sqrt{z}}dz,$$

$$G_3(p) = \mathcal{L}\left[t^{1/2}\psi(t)\right] = -\frac{1}{\sqrt{\pi}}\int_0^\infty \frac{\Psi'(z+p)}{\sqrt{z}}dz,$$

which allow us to obtain the final result:

$$F(p) = -\frac{2}{\pi A}\int_0^\infty dz \left\{\left[1 + \frac{A^{1/2}}{8\sqrt{\pi}}\frac{1}{\sqrt{z}}\right]\Psi(z+p+A) - \frac{A^{3/2}}{\sqrt{2}}\frac{\Psi'(z+p+A)}{\sqrt{z}}\right\}.$$
(6.117)

This formula solves our problem: the function $\Psi(z)$ is measured experimentally by performing the so-called spectroscopy experiments when the oscillating tip gradually approaches the surface and the frequency shift $\Delta\omega(z) = \omega(z) - \omega_0$ is measured as a function of the tip average height z. Then, by performing the numerical integration with the functions $\Psi(z)$ and $\Psi'(z)$ as prescribed by this formula, the force acting on the tip as a function of the tip height p is obtained.

Chapter 7
Curvilinear Coordinates

In many applications physical systems possess symmetry. For instance, the magnetic field of an infinite vertical wire with a current flowing through it has a cylindrical symmetry (i.e. the field depends only on the distance from the wire), while the field radiated by a point source has the characteristic spherical symmetry (i.e. depends only on the distance from the source). In these and many other cases Cartesian coordinates are not the most convenient choice; a special choice of the coordinates (such as e.g. cylindrical or spherical ones for the two examples mentioned above, respectively) may however simplify the problem considerably and hence enable one to obtain a closed solution. In particular, investigation of a large number of physical problems requires solving so-called partial differential equations (PDE). Using the appropriate coordinates in place of the Cartesian ones allow one to obtain simpler forms of these equations (which may e.g. contain a smaller number of variables) that can be easier to solve.

The key objective of this chapter[1] is to present a general theory which allows introduction of such alternative coordinate systems and how general differential operators such as gradient, divergence, curl and the Laplacian can be written in terms of them. Some applications of these so-called *curvilinear coordinates* in solving PDEs will be considered in Sect. 7.13.1 and then in Chap. 8.

7.1 Definition of Curvilinear Coordinates

Instead of the Cartesian coordinate system, we define a different system of coordinates (q_1, q_2, q_3) using a *coordinate transformation* which relates to the new and old coordinates:

---

[1] In the following, references to the first volume of this course (L. Kantorovich, Mathematics for natural scientists: fundamentals and basics, 2nd Edition, Springer, 2022) will be made by appending the Roman number I in front of the reference, e.g. Sect. I.1.8 or Eq. (I.5.18) refer to Sect. 1.8 and Eq. (5.18) of the first volume, respectively.

L. Kantorovich, *Mathematics for Natural Scientists II*, Undergraduate Lecture Notes in Physics, https://doi.org/10.1007/978-3-031-46320-4_7

$$x = x(q_1, q_2, q_3) , \quad y = y(q_1, q_2, q_3) \quad \text{and} \quad z = z(q_1, q_2, q_3) . \tag{7.1}$$

We shall also suppose that Eqs. (7.1) can be solved for each point (x, y, z) with respect to the new coordinates:

$$q_1 = q_1(x, y, z) , \quad q_2 = q_2(x, y, z) \quad \text{and} \quad q_3 = q_3(x, y, z) , \tag{7.2}$$

yielding the corresponding *inverse* relations. In practice, for many such transformations, at *certain* points (x, y, z) the solutions (7.2) are *not* unique, i.e. several (sometimes, infinite) number of coordinates (q_1, q_2, q_3) exist corresponding to the same point in the 3D space. Such points are called *singular points* of the coordinate transformation. The new coordinates (q_1, q_2, q_3) are called *curvilinear coordinates*.

The reader must already be familiar with at least two such curvilinear systems (see Sect. I.1.19.2): the cylindrical and spherical curvilinear coordinate systems. In the case of *cylindrical coordinates* $(q_1, q_2, q_3) = (r, \phi, z)$ the corresponding transformation is given by the following equations:

$$x = r \cos \phi , \quad y = r \sin \phi \quad \text{and} \quad z = z , \tag{7.3}$$

where $0 \le r < \infty, 0 \le \phi < 2\pi$ and $-\infty < z < +\infty$, see Fig. 7.1b. Points along the z axis (when $r = 0$) are all singular in this case: one obtains $x = y = 0$ (any z) for any angle ϕ. By taking square of the first two equations and adding them together, one obtains $r = \sqrt{x^2 + y^2}$, while dividing the first two equations gives $\phi = \arctan(y/x)$. These relations serve as the inverse relations of the transformation. The polar system, Fig. 7.1a, corresponds to the 2D space and is obtained by omitting the z coordinate altogether. In this case only the single point $x = y = 0$ is singular.

For the case of the *spherical coordinates* $(q_1, q_2, q_3) = (r, \theta, \phi)$ we have the transformation relations

$$x = r \sin \theta \cos \phi , \quad y = r \sin \theta \sin \phi \quad \text{and} \quad z = r \cos \theta , \tag{7.4}$$

where $0 \le r < \infty, 0 \le \theta \le \pi$ and $0 \le \phi < 2\pi$, see Fig. 7.1c.

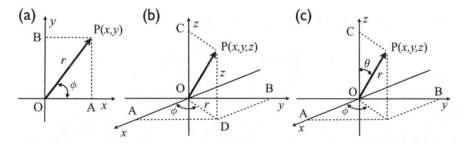

Fig. 7.1 Frequently used curvilinear coordinate systems: **a** polar, **b** cylindrical and **c** spherical

Problem 7.1 *Show that the inverse relations for the spherical coordinate system are*

$$r = \sqrt{x^2 + y^2 + z^2}, \quad \phi = \arctan(y/x) \quad and \quad \theta = \arccos\left(z/\sqrt{x^2 + y^2 + z^2}\right).$$

Which points are singular points in this case?

Hence, transformation relations (7.1) and (7.2) enable one to define a mapping of the Cartesian system onto the curvilinear one. This mapping does not need to be one-to-one correspondence: the same point (x, y, z) may be obtained by several sets of the chosen curvilinear coordinates.

Note that the transformation (7.1) allows one to represent any *scalar field* $G(x, y, z)$ in curvilinear coordinates as

$$G(x(q_1, q_2, q_3), y(q_1, q_2, q_3), z(q_1, q_2, q_3)) = G_{cc}(q_1, q_2, q_3).$$

It is some function G_{cc} of the curvilinear coordinates. For instance, the scalar field $G(x, y, z) = \left(x^2 + y^2 + z^2\right)^2$ is equivalent to the scalar function $G_{cc}(r, \theta, \phi) = r^4$ in the spherical system (or coordinates).

It is seen that once the transformation relations are known, any scalar field can readily be written in the chosen curvilinear coordinates. Next, we need to discuss a general procedure for representing an arbitrary *vector field* $\mathbf{F}(x, y, z)$ in terms of the curvilinear coordinates (q_1, q_2, q_3). This is by far more challenging as the field has a direction and hence we need to understand how to write the field in the vector form which does not rely on Cartesian unit base vectors $\mathbf{i}, \mathbf{j}$ and $\mathbf{k}$.

7.2 Unit Base Vectors

In order to define vector fields in a general curvilinear coordinate system, we have to generalise the definition of the unit base vectors $\mathbf{i}, \mathbf{j}$ and $\mathbf{k}$ of the Cartesian system. To prepare ourselves for the new notions we require, let us first revisit the Cartesian system. If we 'move' along say vector $\mathbf{i}$ starting from some point $P(x, y, z)$, then only the coordinate x will change, the other two coordinates y and z would remain the same; our trajectory will be a straight line parallel to the x axis. Let us call it an x−line. Similarly, 'moving' along $\mathbf{j}$ or $\mathbf{k}$ would only change the coordinates y or z, respectively, leaving the other two coordinates unchanged; this exercise yields y− and z−lines, respectively. We may say that lines parallel to the Cartesian axes and crossing at right angles characterise the Cartesian system.

Conversely, instead of allowing only a single coordinate to change, we can also fix just one coordinate and allow the other two coordinates to change. For instance, by

Fig. 7.2 Coordinate lines
and unit base vectors in a
general curvilinear
coordinate system
(q_1, q_2, q_3)

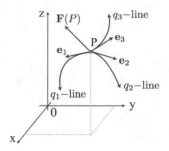

fixing x coordinate of the point $P(x, y, z)$ and making all possible changes in y and z, one constructs a plane passing through the point P which is parallel to the $(y - z)$ plane of the Cartesian system. We can call this a σ_x plane. Similarly, planes σ_z and σ_y parallel to the $(x - y)$ and $(x - z)$ Cartesian planes, respectively, and passing through P correspond to fixing only z or y coordinates, respectively. Any two such planes cross at the corresponding line. Indeed, σ_x and σ_y planes both pass through the point P and cross at the z−line; σ_x and σ_z planes cross at the y−line, while σ_y and σ_z at the x−line, all passing through the point P.

Consider now a general case of a curvilinear system specified by Eqs. (7.1) and (7.2). Choose a point P with the coordinates $\left(q_1^0, q_2^0, q_3^0\right)$. If we constrain two coordinates $q_2(x, y, z) = q_2^0$ and $q_3(x, y, z) = q_3^0$ and allow only one coordinate q_1 to change, we shall 'draw' in space a *coordinate line*, or q_1−line, passing through the point $P\left(q_1^0, q_2^0, q_3^0\right)$. This line may not necessarily be straight as in the Cartesian system; it may be curved. Similarly, one defines q_2− and q_3−lines by allowing only either one of these two coordinates to change. All three lines pass through the point P, but go in different directions as shown in Fig. 7.2. Generally the direction of a coordinate line depends also on the position of the point P, i.e. it might change across space. Note that the directions of the Cartesian coordinate lines remain the same, they do not depend on the position of the point P.

Next, we introduce *coordinate surfaces*. If a single coordinate $q_1(x, y, z) = q_1^0$ is fixed, but the other two coordinates q_2 and q_3 are allowed to change, a surface is constructed given by the equation $q_1(x, y, z) = q_1^0$. We shall call it a σ_1 surface. Similarly, σ_2 and σ_3 surfaces are built by equations $q_2(x, y, z) = q_2^0$ and $q_3(x, y, z) = q_3^0$, respectively. In many frequently used curvilinear coordinate systems at least one of these three surfaces is *not* planar.

All three coordinate surfaces will intersect at a point P. Moreover, any pair of surfaces will intersect at the corresponding coordinate line passing through the point P: σ_1 and σ_3 intersect at the q_2−line, σ_2 and σ_3 and the q_1−line, and so on.

As an example, let us construct coordinate lines and surfaces for the cylindrical coordinate system, Eq. (7.3) and Fig. 7.1b. By changing only r, we draw the r−line which is a ray starting at the z axis and moving outwards perpendicular to it remaining at the height z and making the angle ϕ to the x axis. The coordinate ϕ−line will be a circle of radius r and at the height z, while the coordinate z−line is the vertical line drawn at a distance r from the z axis such that its projection on the $(x - y)$ plane

Fig. 7.3 Coordinate lines
(dashed directed lines) and
unit base vectors (solid
vectors) for the cylindrical
coordinate system at point P

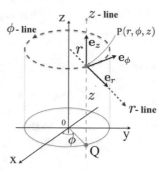

is given by the polar angle ϕ, see Fig. 7.3. The coordinate surfaces for this case are
obtained by fixing a single coordinate: σ_r is a cylinder coaxial with the z axis, σ_ϕ is
a vertical semi-plane hinged to the z axis, and σ_z is a horizontal plane at the height z.
Obviously, σ_r and σ_ϕ intersect at the corresponding z-line, σ_r and σ_z at the ϕ−line,
while σ_ϕ and σ_z at the r−line, as expected.

Next we introduce a set of vectors $\mathbf{e}_1$, $\mathbf{e}_2$ and $\mathbf{e}_3$ at the intersection point P. Each of
these is a unit vector; the direction of $\mathbf{e}_i$ ($i = 1, 2, 3$) is chosen along the tangent to the
corresponding q_i−coordinate line and in such a way that it points in the direction in
which the coordinate q_i increases. These vectors, which are called *unit base vectors*
of the curvilinear coordinate system, enable one to represent an arbitrary vector field
$\mathbf{F}(\mathbf{r})$ at point $P(\mathbf{r}) = P(x, y, z)$ in the form

$$\mathbf{F}(P) = F_1\mathbf{e}_1 + F_2\mathbf{e}_2 + F_3\mathbf{e}_3 \; , \tag{7.5}$$

where F_i ($i = 1, 2, 3$) are scalar functions of the coordinates which can be treated as
components of the vector $\mathbf{F}(P)$ in the given coordinate system. In particular, in the
Cartesian system $\mathbf{e}_1 = \mathbf{i}$, $\mathbf{e}_2 = \mathbf{j}$ and $\mathbf{e}_3 = \mathbf{k}$, and $F_1 = F_x$, $F_2 = F_y$ and $F_3 = F_z$.

The dot product of two vectors written in a particular curvilinear coordinate system
becomes

$$\mathbf{F} \cdot \mathbf{G} = \sum_{i,j=1}^{3} F_i G_j \left(\mathbf{e}_i \cdot \mathbf{e}_j\right) \; ,$$

while the vector product is simply

$$\mathbf{F} \times \mathbf{G} = \sum_{i,j=1}^{3} F_i G_j \left(\mathbf{e}_i \times \mathbf{e}_j\right) \; .$$

Note that in the latter case only off-diagonal elements need to be retained in the
double sum.

If the coordinate system is orthogonal, simple relationships exist between the unit base vectors:

$$\mathbf{e}_i \cdot \mathbf{e}_j = \delta_{ij}$$

and

$$\mathbf{e}_1 \times \mathbf{e}_2 = \mathbf{e}_3 , \quad \mathbf{e}_1 \times \mathbf{e}_3 = -\mathbf{e}_2 , \quad \mathbf{e}_2 \times \mathbf{e}_3 = \mathbf{e}_1 ,$$

which are identical to those existing between the unit base vectors $\mathbf{i}$, $\mathbf{j}$ and $\mathbf{k}$ of the Cartesian system. In that case

$$\mathbf{F} \cdot \mathbf{G} = \sum_{i=1}^{3} F_i G_i ,$$

while in the case of the vector product:

$$\mathbf{F} \times \mathbf{G} = (F_1 G_2 - G_1 F_2) \underbrace{(\mathbf{e}_1 \times \mathbf{e}_2)}_{\mathbf{e}_3} + (F_2 G_3 - G_2 F_3) \underbrace{(\mathbf{e}_2 \times \mathbf{e}_3)}_{\mathbf{e}_1}$$

$$+ (F_1 G_3 - G_1 F_3) \underbrace{(\mathbf{e}_1 \times \mathbf{e}_3)}_{-\mathbf{e}_2} = \begin{vmatrix} \mathbf{e}_1 & \mathbf{e}_2 & \mathbf{e}_3 \\ F_1 & F_2 & F_3 \\ G_1 & G_2 & G_3 \end{vmatrix} , \tag{7.6}$$

as in the Cartesian system.

If a coordinate line curves, the direction of the corresponding unit base vector will vary depending on the position of the point P. This can easily be seen on the example of the cylindrical coordinate system shown in Fig. 7.3: the r-lines are directed away from the z axis parallel to the $(x - y)$ plane; hence, the vector $\mathbf{e}_r$ has the same direction along the line but changes in a different line which have a different value of the angle ϕ. On the other hand, $\mathbf{e}_\phi$ goes around the circle of radius r, i.e. the direction of this unit base vector changes along the line; but it does not depend on z. At the same time, $\mathbf{e}_z = \mathbf{k}$ is always directed along the z axis as the third coordinate z of this system is identical to the corresponding coordinate of the Cartesian system.

Let us now derive explicit expressions for the unit base vectors. We shall relate them to the curvilinear coordinates and the Cartesian base vectors $\mathbf{i}$, $\mathbf{j}$ and $\mathbf{k}$. The latter are convenient for that purpose as they are fixed vectors. Consider the general position vector $\mathbf{r} = x\mathbf{i} + y\mathbf{j} + z\mathbf{k}$ written in the Cartesian system. Because of the transformation equations (7.1), $\mathbf{r} = \mathbf{r}(q_1, q_2, q_3)$, i.e. each Cartesian component of $\mathbf{r}$ depends on the three curvilinear coordinates. Since the vector $\mathbf{e}_i$ is directed along the tangent to the q_i-line and out of the two possible directions we choose the one along which the coordinate q_i increases, the direction of $\mathbf{e}_i$ will be proportional to the partial derivative of the vector $\mathbf{r}$ with respect to this coordinate at the point P, i.e.

$$\mathbf{e}_i \propto \frac{\partial \mathbf{r}}{\partial q_i} = \frac{\partial x}{\partial q_i}\mathbf{i} + \frac{\partial y}{\partial q_i}\mathbf{j} + \frac{\partial z}{\partial q_i}\mathbf{k} , \quad i = 1, 2, 3 . \tag{7.7}$$

A proportionality constant h_i between the derivative $\partial \mathbf{r}/\partial q_i$ and $\mathbf{e}_i$ in the relation $\partial \mathbf{r}/\partial q_i = h_i \mathbf{e}_i$ can be chosen to ensure that the unit base vector is of unit length. This finally allows us to write the required relationships which relate the unit base vectors of an arbitrary curvilinear coordinate system to the Cartesian unit base vectors:

$$\mathbf{e}_i = \frac{1}{h_i} \frac{\partial \mathbf{r}}{\partial q_i} = \frac{1}{h_i} \left(\frac{\partial x}{\partial q_i} \mathbf{i} + \frac{\partial y}{\partial q_i} \mathbf{j} + \frac{\partial z}{\partial q_i} \mathbf{k} \right) , \tag{7.8}$$

where

$$h_i = \left| \frac{\partial \mathbf{r}}{\partial q_i} \right| = \sqrt{ \left(\frac{\partial x}{\partial q_i} \right)^2 + \left(\frac{\partial y}{\partial q_i} \right)^2 + \left(\frac{\partial z}{\partial q_i} \right)^2 } . \tag{7.9}$$

The factors h_i, called *scale factors* and introduced above, are chosen positive to ensure that $\mathbf{e}_i$ is directed along the q_i-line in the direction of increasing q_i. It is readily seen from the above equations that the unit base vectors $\mathbf{e}_i$ are expressed as a linear combination of the Cartesian vectors $\mathbf{i}, \mathbf{j}$ and $\mathbf{k}$ which do not change their direction with the position of the point P. For convenience, these can be collected in three unit vectors $\mathbf{e}_i^0$, $i = 1, 2, 3$. Therefore, a change (if any) experienced by the vectors $\mathbf{e}_i$ is contained in the coefficients m_{ij} of the expansion. The relationship between the new (curvilinear) $\{\mathbf{e}_i\}$ and Cartesian $\{\mathbf{e}_i^0\}$ vectors is conveniently written in the matrix form[2]:

$$\mathbf{e}_i = \sum_{j=1}^{3} m_{ij} \mathbf{e}_j^0 \quad \text{or} \quad \begin{pmatrix} \mathbf{e}_1 \\ \mathbf{e}_2 \\ \mathbf{e}_3 \end{pmatrix} = \mathbf{M} \begin{pmatrix} \mathbf{e}_1^0 \\ \mathbf{e}_2^0 \\ \mathbf{e}_3^0 \end{pmatrix} = \mathbf{M} \begin{pmatrix} \mathbf{i} \\ \mathbf{j} \\ \mathbf{k} \end{pmatrix} , \tag{7.10}$$

where $\mathbf{M} = (m_{ij})$ is the 3×3 matrix of the coefficients,

$$m_{i1} = \frac{1}{h_i} \frac{\partial x}{\partial q_i} , \quad m_{i2} = \frac{1}{h_i} \frac{\partial y}{\partial q_i} , \quad m_{i3} = \frac{1}{h_i} \frac{\partial z}{\partial q_i} .$$

Note that the inverse transformation,

$$\mathbf{e}_i^0 = \sum_{j=1}^{3} \left(\mathbf{M}^{-1} \right)_{ij} \mathbf{e}_j \quad \text{or} \quad \begin{pmatrix} \mathbf{e}_1^0 \\ \mathbf{e}_2^0 \\ \mathbf{e}_3^0 \end{pmatrix} = \mathbf{M}^{-1} \begin{pmatrix} \mathbf{e}_1 \\ \mathbf{e}_2 \\ \mathbf{e}_3 \end{pmatrix} , \tag{7.11}$$

enables one to express the Cartesian vectors in terms of the unit base vectors $\mathbf{e}_i$, if needed.

We know that the unit base vectors of the Cartesian system are mutually orthogonal: $\mathbf{i} \cdot \mathbf{j} = \mathbf{i} \cdot \mathbf{k} = \mathbf{j} \cdot \mathbf{k} = 0$. The Cartesian coordinate system is said to be *orthogonal*.

---

[2] We use bold capital letters for matrices and the corresponding small non-bold letters for their matrix elements.

A curvilinear coordinate system may also be orthogonal. It is called orthogonal if at *each* point P the triple of its unit base vectors remain mutually orthogonal:

$$\mathbf{e}_1 \cdot \mathbf{e}_2 = \mathbf{e}_1 \cdot \mathbf{e}_3 = \mathbf{e}_2 \cdot \mathbf{e}_3 = 0 . \tag{7.12}$$

These relationships hold in spite of the fact that directions of $\mathbf{e}_i$ may vary from point to point. For an orthogonal system the coordinate surfaces through any point P will all intersect at right angles. Indeed, consider for instance the angle between two surfaces σ_1 and σ_2: this angle is the same as the angle between their normal vectors. The normal to σ_1 surface will be given by the unit base vector $\mathbf{e}_1$, while $\mathbf{e}_2$ would serve as the normal of the surface σ_2. Since the two unit base vectors are orthogonal, the two surfaces cross at the right angle.

If the given curvilinear system is orthogonal, then the matrix $\mathbf{M}$ must be orthogonal as well. It is easy to understand as the transformation $\{\mathbf{e}_i^0\} \to \{\mathbf{e}_i\}$, Eq. (7.10), can be considered as a *rotation* in 3D space (Sect. 1.2.5.2): indeed, from the three Cartesian vectors $\{\mathbf{e}_i^0\}$ we obtain a new set of vectors $\{\mathbf{e}_i\}$ in the same space. Hence, if from one set of orthogonal vectors we obtain another orthogonal set, this can only be accomplished by an orthogonal transformation (Sect. 1.2.5). This can also be shown explicitly: if both sets are orthogonal, $\mathbf{e}_i \cdot \mathbf{e}_j = \delta_{ij}$ and $\mathbf{e}_k^{(0)} \cdot \mathbf{e}_{k'}^{(0)} = \delta_{kk'}$, then

$$\mathbf{e}_i \cdot \mathbf{e}_j = \sum_{kk'} m_{ik} m_{jk'} \left(\mathbf{e}_k^{(0)} \cdot \mathbf{e}_{k'}^{(0)} \right) = \sum_{kk'} m_{ik} m_{jk'} \delta_{kk'}$$

$$= \sum_{k} m_{ik} m_{jk} = \sum_{k} m_{ik} \left(\mathbf{M}^T \right)_{kj} = \left(\mathbf{M} \mathbf{M}^T \right)_{ij} .$$

Since the dot product $\mathbf{e}_i \cdot \mathbf{e}_j$ must be equal to δ_{ij}, then $\mathbf{M} \mathbf{M}^T$ is the unity matrix $\mathbf{E}$, i.e. $\mathbf{M}^T = \mathbf{M}^{-1}$, which is the required statement. For orthogonal systems the inverse transformation, Eq. (7.11), is provided by the transposed matrix $\mathbf{M}^T$:

$$\mathbf{e}_i^0 = \sum_{j=1}^{3} \left(\mathbf{M}^T \right)_{ij} \mathbf{e}_j = \sum_{j=1}^{3} m_{ji} \mathbf{e}_j .$$

To illustrate the material presented above it is instructive to consider in detail an example; we shall choose again the case of the cylindrical coordinate system. To express the unit base vectors $\mathbf{e}_r$, $\mathbf{e}_\phi$ and $\mathbf{e}_z$ of this system[3] via the Cartesian unit base vectors, we shall write explicitly the vector $\mathbf{r}$ via the cylindrical coordinates $x = r \cos \phi$, $y = r \sin \phi$, $z = z$ to get

$$\mathbf{r} = \mathbf{r} (r, \phi, z) = x (r, \phi, z) \mathbf{i} + y (r, \phi, z) \mathbf{j} + z (r, \phi, z) \mathbf{k}$$
$$= (r \cos \phi) \mathbf{i} + (r \sin \phi) \mathbf{j} + z \mathbf{k}.$$

---

[3] It is convenient to use the corresponding symbols of the curvilinear coordinates as subscripts for the unit base vectors instead of numbers in each case, and we shall be frequently using this notation.

We now calculate the derivatives of the vector $\mathbf{r}$ and their absolute values as required by Eqs. (7.8) and (7.9):

$$\frac{\partial \mathbf{r}}{\partial r} = (\cos\phi)\,\mathbf{i} + (\sin\phi)\,\mathbf{j} \quad, \quad h_r = \left|\frac{\partial \mathbf{r}}{\partial r}\right| = \left(\cos^2\phi + \sin^2\phi\right)^{1/2} = 1\;; \quad (7.13)$$

$$\frac{\partial \mathbf{r}}{\partial \phi} = (-r\sin\phi)\,\mathbf{i} + (r\cos\phi)\,\mathbf{j}, \quad h_\phi = \left|\frac{\partial \mathbf{r}}{\partial \phi}\right| = \left(r^2\sin^2\phi + r^2\cos^2\phi\right)^{1/2} = r\;;$$
$$(7.14)$$

$$\frac{\partial \mathbf{r}}{\partial z} = \mathbf{k}\,, \quad h_z = \left|\frac{\partial \mathbf{r}}{\partial z}\right| = 1\;. \quad (7.15)$$

Hence, we obtain for the unit base vector the following explicit expressions:

$$\mathbf{e}_r = \cos\phi\,\mathbf{i} + \sin\phi\,\mathbf{j}\,, \quad \mathbf{e}_\phi = -\sin\phi\,\mathbf{i} + \cos\phi\,\mathbf{j}\,, \quad \mathbf{e}_z = \mathbf{k}\,. \quad (7.16)$$

It is easily verified that the vectors are of unit length (by construction) and are all mutually orthogonal: $\mathbf{e}_r \cdot \mathbf{e}_\phi = \mathbf{e}_r \cdot \mathbf{e}_z = \mathbf{e}_\phi \cdot \mathbf{e}_z = 0$. These results show *explicitly* that the cylindrical coordinate system is orthogonal as these conditions were derived for any point P (i.e. any r, ϕ and z). Hence, in the case of the cylindrical system the transformation matrix $\mathbf{M}$ reads

$$\mathbf{M} = \begin{pmatrix} \cos\phi & \sin\phi & 0 \\ -\sin\phi & \cos\phi & 0 \\ 0 & 0 & 1 \end{pmatrix}. \quad (7.17)$$

The matrix $\mathbf{M}$ is orthogonal since its rows (or columns) form an orthonormal set of vectors, as expected; hence

$$\mathbf{M}^{-1} = \mathbf{M}^T = \begin{pmatrix} \cos\phi & -\sin\phi & 0 \\ \sin\phi & \cos\phi & 0 \\ 0 & 0 & 1 \end{pmatrix}. \quad (7.18)$$

In fact, $\mathbf{M}$ is readily seen to be the rotation matrix $R_z(-\phi)$ (rotation by angle $-\phi$ around the z axis, Sect. 1.2.5.2 and Eq. (1.47)). Once the inverse matrix of the transformation $\mathbf{M}^{-1}$ is known, one can write the inverse transformation of the unit base vectors explicitly:

$$\mathbf{i} = \cos\phi\,\mathbf{e}_r - \sin\phi\,\mathbf{e}_\phi\,, \quad \mathbf{j} = \sin\phi\,\mathbf{e}_r + \cos\phi\,\mathbf{e}_\phi\,, \quad \mathbf{k} = \mathbf{e}_z\,.$$

These relationships enable one to rewrite any vector field from the Cartesian to the cylindrical coordinates:

Fig. 7.4 Coordinate lines
and unit base vectors for the
spherical coordinate system

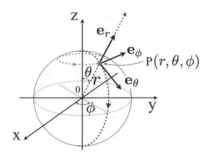

$$\mathbf{F} = F_x\,(\mathbf{r})\,\mathbf{i} + F_y\,(\mathbf{r})\,\mathbf{j} + F_z\,(\mathbf{r})\,\mathbf{k}$$
$$= F_x(\mathbf{r})\left(\cos\phi\,\mathbf{e}_r - \sin\phi\,\mathbf{e}_\phi\right) + F_y(\mathbf{r})\left(\sin\phi\,\mathbf{e}_r + \cos\phi\,\mathbf{e}_\phi\right) + F_z(\mathbf{r})\mathbf{e}_z$$
$$= \left[F_x(\mathbf{r})\cos\phi + F_y(\mathbf{r})\sin\phi\right]\mathbf{e}_r + \left[-F_x(\mathbf{r})\sin\phi + F_y(\mathbf{r})\cos\phi\right]\mathbf{e}_\phi + F_z(\mathbf{r})\mathbf{e}_z\ ,$$

where the three components of the field are explicit functions of the cylindrical
coordinates, e.g. $F_x(\mathbf{r}) = F_x(x, y, z) = F_x\,(r\cos\phi, r\sin\phi, z)$. For instance,

$$\mathbf{F} = e^{-x^2 - y^2}\,(x\mathbf{i} + y\mathbf{j})$$
$$= e^{-r^2}\left[r\cos\phi\,(\cos\phi\,\mathbf{e}_r - \sin\phi\,\mathbf{e}_\phi) + r\sin\phi\,(\sin\phi\,\mathbf{e}_r + \cos\phi\,\mathbf{e}_\phi)\right]$$
$$= e^{-r^2}r\mathbf{e}_r\ .$$

It is seen that the field is radial.

Hence, in order to rewrite a vector field in a curvilinear coordinate system, one
has to express the Cartesian unit base vectors via the unit base vectors of the curvilinear system and replace the Cartesian coordinates via the curvilinear ones using
the corresponding transformation relations.

Problem 7.2 *Using Fig. 7.4, describe the coordinate surfaces and lines for
the spherical coordinate system* (r, θ, ϕ), *Eq. (7.4). Show that the unit base
vectors and the corresponding scale factors for this system are*

$$\mathbf{e}_r = \sin\theta\,(\cos\phi\,\mathbf{i} + \sin\phi\,\mathbf{j}) + \cos\theta\,\mathbf{k}\ , \quad h_r = 1\ ; \qquad (7.19)$$

$$\mathbf{e}_\theta = \cos\theta\,(\cos\phi\,\mathbf{i} + \sin\phi\,\mathbf{j}) - \sin\theta\,\mathbf{k}\ , \quad h_\theta = r\ ; \qquad (7.20)$$

$$\mathbf{e}_\phi = -\sin\phi\,\mathbf{i} + \cos\phi\,\mathbf{j}\ , \quad h_\phi = r\sin\theta\ . \qquad (7.21)$$

*Prove by checking the dot products between unit base vectors that this system
is orthogonal, and show that in this case the transformation matrix*

$$\mathbf{M} = \begin{pmatrix} \sin\theta\cos\phi & \sin\theta\sin\phi & \cos\theta \\ \cos\theta\cos\phi & \cos\theta\sin\phi & -\sin\theta \\ -\sin\phi & \cos\phi & 0 \end{pmatrix}$$

is orthogonal, i.e. $\mathbf{M}^T\mathbf{M} = \mathbf{E}$, where $\mathbf{E}$ is the unit matrix. Also demonstrate by a direct calculation that

$$\mathbf{e}_r \times \mathbf{e}_\theta = \mathbf{e}_\phi , \quad \mathbf{e}_r \times \mathbf{e}_\phi = -\mathbf{e}_\theta \quad and \quad \mathbf{e}_\theta \times \mathbf{e}_\phi = \mathbf{e}_r .$$

Problem 7.3 *The parabolic cylinder coordinate system (u, v, z) is specified by the transformation relations:*

$$x = \frac{1}{2}\left(u^2 - v^2\right), \quad y = uv , \quad z = z . \tag{7.22}$$

Sketch coordinate surfaces and lines for this system. Show that the unit base vectors and the scale factors are

$$\mathbf{e}_u = \frac{1}{h_u}\left(u\mathbf{i} + v\mathbf{j}\right) , \quad h_u = \sqrt{u^2 + v^2} ;$$

$$\mathbf{e}_v = \frac{1}{h_v}\left(-v\mathbf{i} + u\mathbf{j}\right) , \quad h_v = h_u ; \quad \mathbf{e}_z = \mathbf{k} , \quad h_z = 1 .$$

Finally, prove that this system is orthogonal.

Problem 7.4 *The parabolic coordinates (u, v, θ) (where $u \geq 0$, $v \geq 0$ and $0 \leq \theta \leq 2\pi$) are specified by the following transformation equations:*

$$x = uv\cos\theta, \quad y = uv\sin\theta , \quad z = \frac{1}{2}\left(u^2 - v^2\right) . \tag{7.23}$$

Show that in this case the unit base vectors and the scale factors are

$$\mathbf{e}_u = \frac{v}{h_u}\left(\cos\theta\mathbf{i} + \sin\theta\mathbf{j} + \frac{u}{v}\mathbf{k}\right) , \quad h_u = \sqrt{u^2 + v^2} ;$$

$$\mathbf{e}_v = \frac{u}{h_v}\left(\cos\theta\mathbf{i} + \sin\theta\mathbf{j} - \frac{v}{u}\mathbf{k}\right) , \quad h_v = h_u ;$$

$$\mathbf{e}_\theta = -\sin\theta\mathbf{i} + \cos\theta\mathbf{j} , \quad h_\theta = uv .$$

Finally, prove that this system is orthogonal.

Problem 7.5 *Rewrite the vector field*

$$\mathbf{F}(x, y, z) = \frac{x}{\sqrt{x^2 + y^2}}\mathbf{i} + \frac{y}{\sqrt{x^2 + y^2}}\mathbf{j} + z\mathbf{k}$$

in cylindrical coordinates. [*Answer*: $\mathbf{F} = \mathbf{e}_r + z\mathbf{e}_z$.]

7.3 Line Elements and Line Integral

7.3.1 Line Element

If we make a *small* change in the coordinates of a point $P(q_1, q_2, q_3)$, we shall arrive at the point $(q_1 + dq_1, q_2 + dq_2, q_3 + dq_3)$. To the first order the change in the position vector $\mathbf{r}(q_1, q_2, q_3)$ is given by the differential line element

$$d\mathbf{r} = \sum_{i=1}^{3} \frac{\partial \mathbf{r}}{\partial q_i} dq_i = \sum_{i=1}^{3} h_i dq_i \mathbf{e}_i , \tag{7.24}$$

where Eq. (7.8) for the unit base vectors has been used. Note that expression (7.24) for the change of $\mathbf{r}$ is valid for general non-orthogonal curvilinear coordinate systems.

The square of the length ds of the displacement vector $d\mathbf{r}$ is then given by

$$(ds)^2 = d\mathbf{r} \cdot d\mathbf{r} = \sum_{i=1}^{3}\sum_{j=1}^{3} g_{ij} dq_i dq_j , \tag{7.25}$$

where the coefficients

$$g_{ij} = h_i h_j \left(\mathbf{e}_i \cdot \mathbf{e}_j\right) = \frac{\partial \mathbf{r}}{\partial q_i} \cdot \frac{\partial \mathbf{r}}{\partial q_j} \tag{7.26}$$

form a symmetric matrix (since obviously $g_{ij} = g_{ji}$)

$$G = \begin{pmatrix} g_{11} & g_{12} & g_{13} \\ g_{21} & g_{22} & g_{23} \\ g_{31} & g_{32} & g_{33} \end{pmatrix} , \tag{7.27}$$

called a *metric tensor.*

Fig. 7.5 The line elements for a general curvilinear system. The three displacements along each of the q_i−lines form a parallelepiped; if the system is orthogonal, the volume is a cuboid

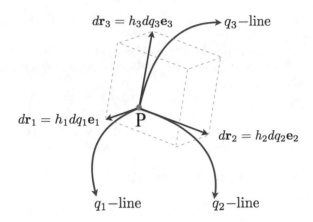

Consider now moving exactly along the q_i−coordinate line through P with the single coordinate q_i changing by dq_i. Since the other curvilinear coordinates are kept constant, then from Eq. (7.24) the change of $\mathbf{r}$ in space is given by the vector

$$d\mathbf{r}_i = h_i dq_i \mathbf{e}_i \ . \tag{7.28}$$

The length of this differential line element is

$$ds_i = |d\mathbf{r}_i| = h_i dq_i \ , \tag{7.29}$$

explaining why h_i is called a *scale factor*. The small displacements $d\mathbf{r}_1$, $d\mathbf{r}_2$ and $d\mathbf{r}_3$ made along each of the coordinate lines are shown in Fig. 7.5.

If the curvilinear system is orthogonal, the metric tensor is diagonal:

$$\mathbf{e}_i \cdot \mathbf{e}_j = \frac{1}{h_i h_j} \frac{\partial \mathbf{r}}{\partial q_i} \cdot \frac{\partial \mathbf{r}}{\partial q_j} = \delta_{ij} \implies g_{ij} = \frac{\partial \mathbf{r}}{\partial q_i} \cdot \frac{\partial \mathbf{r}}{\partial q_j} = h_i^2 \delta_{ij} \tag{7.30}$$

and the metric tensor

$$\mathbf{G} = \begin{pmatrix} h_1^2 & 0 & 0 \\ 0 & h_2^2 & 0 \\ 0 & 0 & h_3^2 \end{pmatrix}$$

has the diagonal form.

For an orthogonal system the three vectors $d\mathbf{r}_i$ are orthogonal independently of the choice of the point P; but this is not generally the case for an arbitrary system: at some points the unit base vectors may be orthogonal, but they will not be orthogonal for all points. One can also see that for an orthogonal system

$$(ds)^2 = h_1^2 (dq_1)^2 + h_2^2 (dq_2)^2 + h_3^2 (dq_3)^2 = \sum_{i=1}^{3} (d\mathbf{r}_i)^2 = \sum_{i=1}^{3} (ds_i)^2 \ , \tag{7.31}$$

which is nothing but a statement of the (three-dimensional) Pythagoras' theorem, see Fig. 7.5.

In special relativity the Minkowski space is defined by four coordinates: $x_0 = ct$, $x_1 = x$, $x_2 = y$, and $x_3 = z$, where c is the speed of light and t time. The metric tensor is adopted as

$$\mathbf{G} = \begin{pmatrix} 1 & 0 & 0 & 0 \\ 0 & -1 & 0 & 0 \\ 0 & 0 & -1 & 0 \\ 0 & 0 & 0 & -1 \end{pmatrix} ,$$

so that the square of the length is

$$ds^2 = (dx_0)^2 - (dx_1)^2 - (dx_2)^2 - (dx_3)^2 = c^2 dt^2 - dx^2 - dy^2 - dz^2 .$$

7.3.2 Line Integrals

In Sect. I.6.4.1 we defined a line integral of a scalar field and calculated a length of a curve specified by parametric equations $x = x(t)$, $y = y(t)$ and $z = z(t)$ (which, e.g. may describe a trajectory of a point particle as a function of time t). We showed there that the length of the curve corresponding to the change of the parameter t from zero (e.g. the total path passed by the particle during time t) is given by the formula (see Eq. (I.6.33) with $f(x, y, z) = 1$):

$$s = \int_0^t \sqrt{x'(t_1)^2 + y'(t_1)^2 + z'(t_1)^2}\, dt_1 . \tag{7.32}$$

We can generalise this result now by considering a space curve specified in curvilinear coordinates by the parametric equations $q_i = q_i(t)$ ($i = 1, 2, 3$). Then $ds = |d\mathbf{r}|$, where $d\mathbf{r}$ given by Eq. (7.24) would correspond to the differential arc length of the curve. The *total* arc length is obtained by integrating over the parameter t:

$$d\mathbf{r} = \sum_{i=1}^{3} \frac{\partial \mathbf{r}}{\partial q_i} dq_i = \left(\sum_{i=1}^{3} \frac{\partial \mathbf{r}}{\partial q_i} \frac{dq_i}{dt} \right) dt ,$$

so that

$$ds = \sqrt{d\mathbf{r} \cdot d\mathbf{r}} = \sqrt{ \sum_{i=1}^{3} \frac{\partial \mathbf{r}}{\partial q_i} \frac{dq_i}{dt} \cdot \sum_{j=1}^{3} \frac{\partial \mathbf{r}}{\partial q_j} \frac{dq_j}{dt} }\, dt$$

$$= \sqrt{ \sum_{i,j=1}^{3} \left(\frac{\partial \mathbf{r}}{\partial q_i} \cdot \frac{\partial \mathbf{r}}{\partial q_j} \right) \frac{dq_i}{dt} \frac{dq_j}{dt} }\, dt = \sqrt{ \sum_{i,j=1}^{3} g_{ij} \frac{dq_i}{dt} \frac{dq_j}{dt} }\, dt ,$$

and hence finally

$$s = \int_0^t \left[\sum_{i,j=1}^{3} g_{ij} \frac{dq_i}{dt} \frac{dq_j}{dt} \right]^{1/2} dt . \tag{7.33}$$

Of course, the same expression is obtained directly from Eq. (7.25) by considering $ds = \sqrt{(ds)^2}$ and $dq_i = (dq_i/dt)\,dt$.

In the case of an orthogonal system this formula is simplified:

$$s = \int_0^t \left[\sum_{i=1}^{3} h_i^2 \left(\frac{dq_i}{dt} \right)^2 \right]^{1/2} dt . \tag{7.34}$$

It is readily seen that in the Cartesian system, when $q_1 = x$, $q_2 = y$ and $q_3 = z$, the previous formula (7.32) is recovered.

This is a general expression for the line integral of the first kind for the length of a curve given parametrically via $q_i = q_i(t)$ ($i = 1, 2, 3$). Correspondingly, the line integral of the first kind for a scalar field $f(\mathbf{r})$ has the form:

$$\int_{s(t_1)}^{s(t_2)} f(\mathbf{r})\,ds = \int_{t_1}^{t_2} f(\mathbf{r}(q_1(t), q_2(t), q_3(t))) \left[\sum_{i,j=1}^{3} g_{ij} \frac{dq_i}{dt} \frac{dq_j}{dt} \right]^{1/2} dt . \tag{7.35}$$

As a simple illustration of the above formula, consider a particle moving around a circular trajectory with angular velocity ω. We use cylindrical coordinates, when r and z are constants, but the angle $\phi(t) = \omega t$ changes with the time t between zero and $T = 2\pi/\omega$ that corresponds to one full circle. Then, the path passed by the particle (note that $g_{rr} = h_r^2 = 1$, $g_{zz} = h_z^2 = 1$, and $g_{\phi\phi} = h_\phi^2 = r^2$, see Eqs. (7.13)–(7.15)) is, according to Eq. (7.34):

$$s = \int_0^T \sqrt{h_r^2 \dot{r}^2 + h_\phi^2 \dot{\phi}^2 + h_z^2 \dot{z}^2}\, dt = \int_0^T \sqrt{h_\phi^2 \dot{\phi}^2}\, dt$$

$$= \int_0^T h_\phi \dot{\phi}\, dt = \int_0^T r\omega\, dt = r\omega T = 2\pi r ,$$

as expected.

Problem 7.6 *Show that the length of a spiral $x = R\cos\phi$, $y = R\sin\phi$ and $z = b\phi$ of radius R with the angle changing as $0 \le \phi \le 2\pi$ (a single revolution) is $s = 2\pi\sqrt{R^2 + b^2}$. Use cylindrical coordinates $(q_1, q_2, q_3) = (r, \phi, z)$, in which case the spiral is specified parametrically as $r(t) = R$, $\phi(t) = t$ and $z(t) = bt$. Note that in the case of a circle of radius R we should put $b = 0$ in which case we obtain $s = 2\pi R$, as it must be.*

Problem 7.7 *Show that* $(ds)^2$ *on the surface of a cylinder of radius R is*

$$(ds)^2 = R^2 (d\phi)^2 + (dz)^2 \ , \tag{7.36}$$

while in the case of a sphere of the same radius,

$$(ds)^2 = R^2 (d\theta)^2 + R^2 \sin^2 \theta \, (d\phi)^2 \ . \tag{7.37}$$

[*Hint*: note that on the surfaces of the cylinder or sphere $dr = 0$.]

Problem 7.8 *Show that* $(ds)^2$ *on the surface of a cone* $z = \sqrt{x^2 + y^2}$*, in polar coordinates, is given by*

$$(ds)^2 = 2 (dr)^2 + r^2 (d\phi)^2 \ , \tag{7.38}$$

while for the paraboloid of revolution $z = x^2 + y^2$ *it is*

$$(ds)^2 = \left(4r^2 + 1\right) (dr)^2 + r^2 (d\phi)^2 \ . \tag{7.39}$$

In a similar manner one can write the line integral of the second kind, $\int \mathbf{F} \cdot d\mathbf{r}$, for the vector field $\mathbf{F}(\mathbf{r})$ via curvilinear coordinates:

$$\int_{\mathbf{r}(t_1)}^{\mathbf{r}(t_2)} \mathbf{F}(\mathbf{r}) \cdot d\mathbf{r} = \int_{t_1}^{t_2} \mathbf{F}\left(\mathbf{r}\left(q_1(t), q_2(t), q_3(t)\right)\right) \cdot \sum_{i=1}^{3} h_i \frac{dq_i}{dt} \mathbf{e}_i \, dt$$

$$= \sum_{i=1}^{3} \int_{t_1}^{t_2} h_i \frac{dq_i}{dt} \left(\mathbf{F} \cdot \mathbf{e}_i\right) dt \ ,$$

where we used $d\mathbf{r}$ from Eq. (7.24). Expanding the vector field $\mathbf{F} = \sum_j F_j \mathbf{e}_j$ into the unit base vectors of the given curvilinear system, we finally obtain

$$\int_{\mathbf{r}(t_1)}^{\mathbf{r}(t_2)} \mathbf{F}(\mathbf{r}) \cdot d\mathbf{r} = \sum_{i,j=1}^{3} \int_{t_1}^{t_2} F_j h_i \frac{dq_i}{dt} \left(\mathbf{e}_j \cdot \mathbf{e}_i\right) dt \ . \tag{7.40}$$

In the case of an orthogonal curvilinear system formula (7.40) for the line integral of the second kind is simplified:

$$\int_{\mathbf{r}(t_1)}^{\mathbf{r}(t_2)} \mathbf{F}(\mathbf{r}) \cdot d\mathbf{r} = \sum_{i=1}^{3} \int_{t_1}^{t_2} F_i h_i \frac{dq_i}{dt} dt \ . \tag{7.41}$$

The formulae derived here for the line integrals are basically identical to the ones we considered previously in Volume I (see Sect. I.6.4). Indeed, let us just consider the Cartesian coordinate system as an example. In this case the scale factors $h_i = 1$ and the metric tensor is the identity matrix, $g_{ij} = \delta_{ij}$. Hence, for the integral of the first kind we have

$$\int_{s(t_1)}^{s(t_2)} f(\mathbf{r})\, ds = \int_{t_1}^{t_2} f(\mathbf{r}(x(t), y(t), z(t)))\, \sqrt{\dot{x}^2 + \dot{y}^2 + \dot{z}^2}\, dt$$

while the formula

$$\int_{\mathbf{r}(t_1)}^{\mathbf{r}(t_2)} \mathbf{F}(\mathbf{r}) \cdot d\mathbf{r} = \sum_{i=1}^{3} \int_{t_1}^{t_2} F_i \frac{dq_i}{dt}\, dt = \sum_{i=1}^{3} \int_{t_1}^{t_2} F_i dq_i$$

$$= \int_{t_1}^{t_2} F_x dx + F_y dy + F_z dz$$

is obtained for the second kind.

As an example of calculating the line integral for a vector field, consider the calculation of a potential $\varphi(\mathbf{r})$ of a conservative field $\mathbf{F}(\mathbf{r}) = \frac{q}{r^3}\mathbf{r}$ (as, e.g. in electrostatics). We know from Sect. I.6.5.6, that the scalar potential of the conservative vector field $\mathbf{F}$ is given by the line integral $\int_A^B \mathbf{F} \cdot d\mathbf{r}$ between two points, A and B, the former one is arbitrary (contributes to a constant associated with the potential) and the latter one corresponds to the vector $\mathbf{r}$. It is convenient in our case to take the point A at infinity. Then:

$$\varphi(\mathbf{r}) = \int_{\infty}^{\mathbf{r}} \frac{q}{r^3}\mathbf{r} \cdot d\mathbf{r} = -q \int_{\mathbf{r}}^{\infty} \frac{\mathbf{r}}{r^3} \cdot d\mathbf{r}.$$

We know that the path connecting the two points is not important. Let us then choose the spherical coordinate system for the calculation. In this system $\mathbf{r} = r\mathbf{e}_r$, $h_r = 1$ (see Problem 7.2) and hence $\mathbf{F} = \frac{q}{r^2}\mathbf{e}_r$, and the integral, via Eq. (7.41), becomes simply

$$\varphi(\mathbf{r}) = -q \int_{r}^{\infty} F_r h_r dr = -q \int_{r}^{\infty} \frac{dr}{r^2} = -\frac{q}{r}.$$

The path between the point $\mathbf{r}$ and an infinitely remote point was taken here along the ray in the direction of the vector $\mathbf{r}$, but it does not need to be. The point is that only the projection of any path on that ray contributes to the integral since $\mathbf{F}$ only has the F_r components (along $\mathbf{e}_r$) being non-zero.

7.4 Surface Normal and Surface Integrals

Consider a continuous one-sided surface S specified parametrically via parameters u and v in some curvilinear coordinates

$$q_i = q_i(u, v), \quad i = 1, 2, 3.$$

First of all, we shall obtain an expression for its normal vector **N** (unnormalised) as defined in Sect. I.6.5.1. This is obtained as the cross product of two tangential vectors to the surface taken along the corresponding u and v lines on the surface:

$$\boldsymbol{\tau}_u = \frac{\partial \mathbf{r}}{\partial u} = \sum_{i=1}^{3} \frac{\partial \mathbf{r}}{\partial q_i} \frac{\partial q_i}{\partial u} = \sum_{i=1}^{3} h_i \mathbf{e}_i \frac{\partial q_i}{\partial u} ,$$

$$\boldsymbol{\tau}_v = \frac{\partial \mathbf{r}}{\partial v} = \sum_{i=1}^{3} \frac{\partial \mathbf{r}}{\partial q_i} \frac{\partial q_i}{\partial v} = \sum_{i=1}^{3} h_i \mathbf{e}_i \frac{\partial q_i}{\partial v}$$

obtained by changing only one of the two parameters at a time. Then,

$$\mathbf{N} = \boldsymbol{\tau}_u \times \boldsymbol{\tau}_v = \sum_{i,j=1}^{3} h_i h_j \frac{\partial q_i}{\partial u} \frac{\partial q_j}{\partial v} \left(\mathbf{e}_i \times \mathbf{e}_j \right) . \tag{7.42}$$

Note that in this double sum the diagonal terms ($i = j$) will not contribute (why?).

The obtained expression can be somewhat simplified for orthogonal curvilinear coordinates. Indeed, in this case the unit base vectors are related to each other in the same way as the Cartesian vectors **i**, **j**, **k**:

$$\mathbf{e}_1 \times \mathbf{e}_2 = \mathbf{e}_3, \quad \mathbf{e}_1 \times \mathbf{e}_3 = -\mathbf{e}_2 \text{ and } \mathbf{e}_2 \times \mathbf{e}_3 = \mathbf{e}_1 .$$

We only need to consider six off-diagonal terms in Eq. (7.42). These can be split into three pairs with particular values of the indices (i, j) and (j, i). We shall combine these recalling that $\mathbf{e}_i \times \mathbf{e}_j = -\mathbf{e}_j \times \mathbf{e}_i$. We then get

$$\mathbf{N} = \sum_{i \neq j} h_i h_j \frac{\partial q_i}{\partial u} \frac{\partial q_j}{\partial v} \left(\mathbf{e}_i \times \mathbf{e}_j \right) = h_1 h_2 \left(\frac{\partial q_1}{\partial u} \frac{\partial q_2}{\partial v} - \frac{\partial q_2}{\partial u} \frac{\partial q_1}{\partial v} \right) \underbrace{(\mathbf{e}_1 \times \mathbf{e}_2)}_{\mathbf{e}_3}$$

$$+ h_1 h_3 \left(\frac{\partial q_1}{\partial u} \frac{\partial q_3}{\partial v} - \frac{\partial q_3}{\partial u} \frac{\partial q_1}{\partial v} \right) \underbrace{(\mathbf{e}_1 \times \mathbf{e}_3)}_{-\mathbf{e}_2} + h_2 h_3 \left(\frac{\partial q_2}{\partial u} \frac{\partial q_3}{\partial v} - \frac{\partial q_3}{\partial u} \frac{\partial q_2}{\partial v} \right) \underbrace{(\mathbf{e}_2 \times \mathbf{e}_3)}_{\mathbf{e}_1}$$

$$= h_1 h_2 \left(\frac{\partial q_1}{\partial u} \frac{\partial q_2}{\partial v} - \frac{\partial q_2}{\partial u} \frac{\partial q_1}{\partial v} \right) \mathbf{e}_3 - h_1 h_3 \left(\frac{\partial q_1}{\partial u} \frac{\partial q_3}{\partial v} - \frac{\partial q_3}{\partial u} \frac{\partial q_1}{\partial v} \right) \mathbf{e}_2$$

$$+ h_2 h_3 \left(\frac{\partial q_2}{\partial u} \frac{\partial q_3}{\partial v} - \frac{\partial q_3}{\partial u} \frac{\partial q_2}{\partial v} \right) \mathbf{e}_1$$

$$= \begin{vmatrix} \mathbf{e}_1 & \mathbf{e}_2 & \mathbf{e}_3 \\ h_1 \partial q_1/\partial u & h_2 \partial q_2/\partial u & h_3 \partial q_3/\partial u \\ h_1 \partial q_1/\partial v & h_2 \partial q_2/\partial v & h_3 \partial q_3/\partial v \end{vmatrix} .$$

$$\tag{7.43}$$

In particular, in the case of an orthogonal Cartesian system $h_i = 1$ and we obtain

$$N = \begin{vmatrix} \mathbf{i} & \mathbf{j} & \mathbf{k} \\ \partial x/\partial u & \partial y/\partial u & \partial z/\partial u \\ \partial x/\partial v & \partial y/\partial v & \partial z/\partial v \end{vmatrix} ,$$

which is exactly our previous expression (I.6.5.2).

Next we shall derive appropriate expressions for surface integrals of the scalar (the first kind) and vector (the second kind, or flux) fields. A surface element dA obtained by changing both u and v by, correspondingly, du and dv, is

$$dA = |\boldsymbol{\tau}_u du \times \boldsymbol{\tau}_v dv| = \left| \sum_{i,j=1}^{3} h_i h_j \frac{\partial q_i}{\partial u} \frac{\partial q_j}{\partial v} (\mathbf{e}_i \times \mathbf{e}_j) \right| du\, dv$$

$$= |\mathbf{N}|\, du\, dv ,$$

so that the surface integral of the first kind for a scalar field $F(\mathbf{r})$ taken along the surface S can be written as

$$\int\int_S F(\mathbf{r}) dA = \int\int_S F(\mathbf{r}) |\mathbf{N}|\, du\, dv$$

$$= \int\int_S F\left(\mathbf{r}\left(q_1(t), q_2(t), q_3(t)\right)\right) \left| \sum_{i,j=1}^{3} h_i h_j \frac{\partial q_i}{\partial u} \frac{\partial q_j}{\partial v} (\mathbf{e}_i \times \mathbf{e}_j) \right| du\, dv .$$

$$(7.44)$$

The flux integral for a vector field $\mathbf{F}(\mathbf{r})$ is derived as follows:

$$\int\int_S \mathbf{F} \cdot d\mathbf{A} = \int\int_S \mathbf{F} \cdot \frac{\mathbf{N}}{|\mathbf{N}|} dA = \int\int_S \mathbf{F} \cdot \mathbf{N}\, du\, dv$$

$$= \int\int_S \mathbf{F} \cdot \sum_{i,j=1}^{3} h_i h_j \frac{\partial q_i}{\partial u} \frac{\partial q_j}{\partial v} (\mathbf{e}_i \times \mathbf{e}_j)\, du\, dv .$$

Writing up again the field $\mathbf{F}$ via the unit base vectors, we finally get

$$\int\int_S \mathbf{F} \cdot d\mathbf{A} = \sum_{i,j,k=1}^{3} \int\int_S F_k h_i h_j \frac{\partial q_i}{\partial u} \frac{\partial q_j}{\partial v} \left[\mathbf{e}_k \cdot (\mathbf{e}_i \times \mathbf{e}_j) \right] du\, dv . \qquad (7.45)$$

As we have explicitly checked that the expression for the normal $\mathbf{N}$ in the Cartesian system coincides with the one we obtained previously, the formulae for the two surface integrals would also coincide with their Cartesian analogues we derived in Vol. I (Sects. I.6.5.3 and I.6.5.4).

As an example, consider the calculation of the normal to a sphere of radius R centred at the origin. Its parametric equation in spherical coordinates (r, θ, ϕ) is $r = R$, $\theta = u$ and $\phi = v$, i.e. basically its angles θ, ϕ serve as the two parameters. For this coordinate system (see Problem 7.2) the scale factors are $h_r = 1$, $h_\theta = r$ and $h_\phi = r \sin \theta$. The system is orthogonal so that we can use Eq. (7.43):

$$\mathbf{N} = \begin{vmatrix} \mathbf{e}_r & \mathbf{e}_\theta & \mathbf{e}_\phi \\ \partial r/\partial\theta & r\,(\partial\theta/\partial\theta) & r \sin\theta\,(\partial\phi/\partial\theta) \\ \partial r/\partial\phi & r\,(\partial\theta/\partial\phi) & r \sin\theta\,(\partial\phi/\partial\phi) \end{vmatrix} = \begin{vmatrix} \mathbf{e}_r & \mathbf{e}_\theta & \mathbf{e}_\phi \\ 0 & r & 0 \\ 0 & 0 & r \sin\theta \end{vmatrix}$$

$$= r^2 \sin\theta \mathbf{e}_r \,,$$

the result to be expected, see Eq. (I.6.58).

Next we shall calculate the surface of the upper hemisphere. In this case $r = R$, $0 \le \theta \le \pi/2$ and $0 \le \phi \le 2\pi$. Using Eq. (7.44) with the normal just calculated, we obtain

$$A = \int_0^{2\pi} d\phi \int_0^{\pi/2} d\theta\, R^2 \sin\theta = 2\pi R^2 \,,$$

as it must be.

7.5 Volume Element and Jacobian in 3D

As we know from Sect. I.6.2.2, curvilinear coordinates could also be useful in performing volume integration, and we derived there the conversion formula for the volume element $dV = dx\,dy\,dz$ between the Cartesian and any other coordinates q_1, q_2, q_3 (we did not call them curvilinear though at the time):

$$dV = |J|\, dq_1 dq_2 dq_3 \,,$$

where $J = \partial(x, y, z)/\partial(q_1, q_2, q_3)$ is the Jacobian of the transformation, Eq. (I.6.24). Exactly the same result can be derived using our new mathematical tools developed above.

Indeed, consider Fig. 7.5 again. A parallelepiped there is formed by three vectors: $d\mathbf{r}_1 = h_1 \mathbf{e}_1 dq_1$, $d\mathbf{r}_2 = h_2 \mathbf{e}_2 dq_2$ and $d\mathbf{r}_3 = h_3 \mathbf{e}_3 dq_3$. These vectors are orthogonal to each other only for an orthogonal curvilinear coordinate system, in which case, obviously, the parallelepiped becomes a cuboid with the volume

$$dV = |d\mathbf{r}_1|\, |d\mathbf{r}_2|\, |d\mathbf{r}_3| = h_1 h_2 h_3 dq_1 dq_2 dq_3 \,,$$

i.e. the Jacobian in this case $J = h_1 h_2 h_3$ is simply given by a product of all three scale factors. In the case of the cylindrical system, when $h_r = h_z = 1$ and $h_\phi = r$, we obtain $J = r$, a well-known result (Sect. I.6.1.4), while in the case of the spherical system, for which $h_r = 1, h_\theta = r$ and $h_\phi = r \sin \theta$, we obtain $J = r^2 \sin \theta$, cf. Sect. I.6.2.2.

In a general case, however, the sides of the parallelepiped in Fig. 7.5 are not orthogonal, and its volume is given by the absolute value of the mixed product of all three vectors:

$$dV = |(d\mathbf{r}_1 \cdot [d\mathbf{r}_2 \times d\mathbf{r}_3])| = |h_1 h_2 h_3 (\mathbf{e}_1 \cdot [\mathbf{e}_2 \times \mathbf{e}_3])| \, dq_1 dq_2 dq_3 .$$

This formula can already be used in practical calculations since it gives a general result for the Jacobian as

$$J = h_1 h_2 h_3 (\mathbf{e}_1 \cdot [\mathbf{e}_2 \times \mathbf{e}_3]) . \tag{7.46}$$

However, it is instructive to demonstrate that this is actually the same result as the one derived previously in Sect. I.6.2.2 where J was expressed directly via partial derivatives. To this end, recall the actual expressions for the unit base vectors, Eq. (7.8); hence, our previous result can be rewritten as

$$dV = \left| \left(\frac{\partial \mathbf{r}}{\partial q_1} \cdot \left[\frac{\partial \mathbf{r}}{\partial q_2} \times \frac{\partial \mathbf{r}}{\partial q_3} \right] \right) \right| dq_1 dq_2 dq_3 = J dq_1 dq_2 dq_3 . \tag{7.47}$$

It is not difficult to see now that the mixed product of the derivatives above is exactly the Jacobian $J = \partial (x, y, z) / \partial (q_1, q_2, q_3)$. Indeed, the mixed product of any three vectors can be written as a determinant (see Sect. I.1.10.1 and Eq. (I.1.108)). Therefore, the Jacobian in Eq. (7.47) can finally be manipulated into

$$J = \begin{vmatrix} \partial x/\partial q_1 & \partial y/\partial q_1 & \partial z/\partial q_1 \\ \partial x/\partial q_2 & \partial y/\partial q_2 & \partial z/\partial q_2 \\ \partial x/\partial q_3 & \partial y/\partial q_3 & \partial z/\partial q_3 \end{vmatrix} = \frac{\partial (x, y, z)}{\partial (q_1, q_2, q_3)} ,$$

as required.

Problem 7.9 *Show that the 3D Fourier transform*

$$F(\mathbf{k}) = \frac{1}{(2\pi)^{3/2}} \int_{-\infty}^{\infty} f(\mathbf{r}) e^{i\mathbf{k}\cdot\mathbf{r}} d\mathbf{r}$$

of the function $f(\mathbf{r}) = (r^2 + \lambda^2)^{-1}$ *can be written as a one-dimensional integral*

$$F(\mathbf{k}) = \frac{1}{k} \sqrt{\frac{2}{\pi}} \int_0^{\infty} \frac{r \sin(kr)}{r^2 + \lambda^2} dr .$$

Problem 7.10 *A charge density of a unit point charge that is positioned at point r_0 is described by the distribution function in the form of the 3D delta function:*

$$\rho(\mathbf{r}) = \delta(\mathbf{r} - \mathbf{r}_0) = \delta(x - x_0)\,\delta(y - y_0)\,\delta(z - z_0)\ .$$

Show that in the spherical coordinates (r, θ, ϕ) this formula takes on the following form:

$$\rho(\mathbf{r}) = \delta(\mathbf{r} - \mathbf{r}_0) = \frac{1}{r^2 \sin\theta}\,\delta(r - r_0)\,\delta(\theta - \theta_0)\,\delta(\phi - \phi_0)\ ,$$

where $\mathbf{r} = (r, \theta, \phi)$ and $\mathbf{r}_0 = (r_0, \theta_0, \phi_0)$.[*Hint*: the simplest method is to ensure that in both coordinate systems, spherical and Cartesian, a volume integral of a general function $f(\mathbf{r})$ multiplied by $\delta(\mathbf{r} - \mathbf{r}_0)$ equals $f(\mathbf{r}_0)$.]

Problem 7.11 *Generalise this result to a general orthogonal curvilinear system:*

$$\delta(\mathbf{r} - \mathbf{r}_0) = \frac{1}{|h_1 h_2 h_3|}\,\delta\left(q_1 - q_1^0\right)\,\delta\left(q_2 - q_2^0\right)\,\delta\left(q_3 - q_3^0\right)\ , \qquad (7.48)$$

where $\mathbf{r} = (q_1, q_2, q_3)$ and $\mathbf{r}_0 = \left(q_1^0, q_2^0, q_3^0\right)$.

It is instructive to derive the above formula for the delta function in an orthogonal curvilinear system also using the exponential delta-sequence (4.18):

$$\delta_{nml}(\mathbf{r} - \mathbf{r}_0) = \delta_n(x - x_0)\,\delta_m(y - y_0)\,\delta_l(z - z_0)$$
$$= \frac{nml}{(2\pi)^{3/2}} \exp\left[-\frac{n^2}{2}(x - x_0)^2 - \frac{m^2}{2}(y - y_0)^2 - \frac{l^2}{2}(z - z_0)^2\right]\ ,$$

where n, m and l are integers. These are meant to go to infinity. The result should not depend on the way these three numbers tend to infinity though; therefore, we shall consider the limit of $n = m = l \to \infty$:

$$\delta_{nnn}(\mathbf{r} - \mathbf{r}_0) = \left(\frac{n}{\sqrt{2\pi}}\right)^3 \exp\left[-\frac{n^2}{2}(\mathbf{r} - \mathbf{r}_0)^2\right]\ .$$

We are interested here in having $\mathbf{r} - \mathbf{r}_0$ being very close to zero. In this case $(\mathbf{r} - \mathbf{r}_0)^2$ is a distance square between two close points with the curvilinear coordinated differing by $\Delta q_1 = q_1 - q_1^0$, $\Delta q_2 = q_2 - q_2^0$ and $\Delta q_3 = q_3 - q_3^0$. The distance squared between these close points is given by Eq. (7.31):

$$(\mathbf{r} - \mathbf{r}_0)^2 = \Delta s^2 = h_1^2 \, (\Delta q_1)^2 + h_2^2 \, (\Delta q_2)^2 + h_3^2 \, (\Delta q_3)^2$$
$$= h_1^2 \left(q_1 - q_1^0 \right)^2 + h_2^2 \left(q_2 - q_2^0 \right)^2 + h_3^2 \left(q_3 - q_3^0 \right)^2 \,,$$

so that

$$\delta_{nnn} \, (\mathbf{r} - \mathbf{r}_0) = \prod_{i=1}^{3} \frac{n}{\sqrt{2\pi}} \exp\left[-\frac{n^2 h_i^2}{2} \left(q_i - q_i^0 \right)^2 \right] \,.$$

Taking the limit of $n \to \infty$ and using the fact that each individual exponent (for the given i) tends to a separate delta function $\delta \left(h_i \left(q_i - q_0^0 \right) \right)$, we obtain

$$\delta \, (\mathbf{r} - \mathbf{r}_0) = \prod_{i=1}^{3} \lim_{n \to \infty} \frac{n}{\sqrt{2\pi}} \exp\left[-\frac{n^2 h_i^2}{2} \left(q_i - q_i^0 \right)^2 \right] = \prod_{i=1}^{3} \delta \left(h_i \left(q_i - q_i^0 \right) \right)$$

$$= \prod_{i=1}^{3} \frac{1}{|h_i|} \delta \left(q_i - q_i^0 \right) \,,$$

which is the same result as the above Eq. (7.48). In the last passage we have used the property (4.12) of the delta function.

7.6 Change of Variables in Multiple Integrals

We have seen in Sects. I.6.1.4 and I.6.2.2 that one has to calculate a Jacobian of transformation when changing the variables in double and triple integrals, respectively. In the previous section we derived an expression for the Jacobian again using the general technique of curvilinear coordinates developed in the preceding sections. However, we have so far only limited ourselves to double and triple integrals. Here we shall prove that there is a straightforward generalisation of these results to any number of multiple integrals.

Consider an n−fold integral

$$I_n = \underbrace{\int \cdots \int}_{n} F \, (x_1, \ldots, x_n) \, dx_1 \ldots dx_n \tag{7.49}$$

of some function $F \, (x_1, \ldots, x_n)$ of n variables. To calculate the integral, we would like to use another set of variables $y_1, y_2, \ldots, y_n$, which are related to the former set via the corresponding transformation relations:

$$x_i = f_i \, (y_1, \ldots, y_n) \,, \quad i = 1, \ldots, n \,. \tag{7.50}$$

We would like to prove that

$$I_n = \underbrace{\int \cdots \int}_{n} F\left(x_1\left(y_1, \ldots, y_n\right), \ldots, x_n\left(y_1, \ldots, y_n\right)\right) \left|J_n\right| dy_1 \ldots dy_n , \quad (7.51)$$

where

$$J_n = \begin{vmatrix} \partial x_1/\partial y_1 & \partial x_2/\partial y_1 & \cdots & \partial x_n/\partial y_1 \\ \partial x_1/\partial y_2 & \partial x_2/\partial y_2 & \cdots & \partial x_n/\partial y_2 \\ \cdots & \cdots & \cdots & \cdots \\ \partial x_1/\partial y_n & \partial x_2/\partial y_n & \cdots & \partial x_n/\partial y_n \end{vmatrix} \qquad (7.52)$$

is the corresponding Jacobian. Note that the absolute value of J is to be taken in Eq. (7.51). To simplify our notations in this section, we shall also be using the following convenient notation for the Jacobian:

$$J_n = \left| \partial x_1/\partial y_k \ \partial x_2/\partial y_k \ \cdots \ \partial x_n/\partial y_k \right| ,$$

where we only write explicitly elements of the k−th row. You must assume that each such a term, e.g. $\partial x_2/\partial y_k$, represents the whole column of such elements with k changing between 1 and n.

We shall prove this formula using induction. The formula is valid for $n = 2, 3$; hence, we assume that it is valid for the $(n-1)$-dimensional integral, and then shall prove that it is also valid for the n-dimensional integral. Consider the original integral (7.49) in which we shall integrate over the variable x_1 at the very end as the last integral:

$$I_n = \int dx_1 \left[\underbrace{\int \cdots \int}_{n-1} F\left(x_1, \ldots, x_n\right) dx_2 \ldots dx_n \right] . \qquad (7.53)$$

When the internal $(n-1)$−dimensional integral over the variables x_2, etc., x_n is calculated, the variable x_1 is fixed. Now, let us solve the first transformation equation in (7.50) (i.e. for $i = 1$) with respect to the new variable y_1; it will be some function of the fixed old variable x_1 and all the other new variables:

$$y_1 = \phi\left(x_1, y_2, \ldots, y_n\right) . \qquad (7.54)$$

Therefore, the transformation relations between the old and new variables can be rewritten by excluding the first variable from either of the sets:

$$x_i = f_i\left(\phi\left(x_1, y_2, \ldots, y_n\right), y_2, \ldots, y_n\right) = g_i\left(x_1, y_2, \ldots, y_n\right) , \quad i = 2, \ldots, n , \qquad (7.55)$$

where g_2, g_3, etc. are the new transformation functions. Hence, the internal integral in (7.53) can be transformed to the new set of variables y_2, etc., y_n by means of the Jacobian J_{n-1}. This is possible to do due to our assumption:

$$I_n = \int dx_1 \underbrace{\int \cdots \int}_{n-1} F(x_1, g_2(x_1, y_2, \ldots, y_n), \ldots, g_n(x_1, y_2, \ldots, y_n)) \, |J_{n-1}| \, dy_2 \ldots dy_n \, ,$$

where

$$J_{n-1} = \begin{vmatrix} \partial g_2/\partial y_2 & \partial g_3/\partial y_2 & \cdots & \partial g_n/\partial y_2 \\ \partial g_2/\partial y_3 & \partial g_3/\partial y_3 & \cdots & \partial g_n/\partial y_3 \\ \cdots & \cdots & \cdots & \cdots \\ \partial g_2/\partial y_n & \partial g_3/\partial y_n & \cdots & \partial g_n/\partial y_n \end{vmatrix} = \left| \partial g_2/\partial y_k \ \partial g_3/\partial y_k \ \cdots \ \partial g_n/\partial y_k \right| \, .$$

(7.56)

Let us calculate the derivatives appearing in $J_{n-1} = \left| \partial g_i/\partial y_j \right|$ explicitly. From Eq. (7.55) for $i = 2, \ldots, n$, we can write

$$\frac{\partial x_i}{\partial y_j} = \frac{\partial g_i}{\partial y_j} = \frac{\partial f_i}{\partial y_j} + \frac{\partial f_i}{\partial \phi} \frac{\partial \phi}{\partial y_j} = \frac{\partial x_i}{\partial y_j} + \frac{\partial x_i}{\partial y_1} \frac{\partial \phi}{\partial y_j} \, .$$

To calculate the derivative $\partial \phi/\partial y_j$, consider the equation

$$x_1 = f_1(y_1, y_2, \ldots, y_n) \, . \tag{7.57}$$

Here x_1 is fixed, and hence the dependence of ϕ (or of y_1, see Eq. (7.54)) is implicit. Differentiating both sides of (7.57) with respect to y_j and keeping in mind that y_1 also depends on y_j via Eq. (7.54), we obtain

$$0 = \frac{\partial f_1}{\partial y_j} + \frac{\partial f_1}{\partial y_1} \frac{\partial \phi}{\partial y_j} \implies \frac{\partial \phi}{\partial y_j} = -\frac{\partial f_1/\partial y_j}{\partial f_1/\partial y_1} = -\frac{\partial x_1/\partial y_j}{\partial x_1/\partial y_1} = -\alpha_j \, ,$$

where α_j is a convenient shorthand. This allows us to write

$$\frac{\partial g_i}{\partial y_j} = \frac{\partial x_i}{\partial y_j} - \alpha_j \frac{\partial x_i}{\partial y_1} \, .$$

Now we are in a position to write the Jacobian J_{n-1} in another form:

$$J_{n-1} = \left| (\partial x_2/\partial y_k) - \alpha_k (\partial x_2/\partial y_1) \ \cdots \ (\partial x_n/\partial y_k) - \alpha_k (\partial x_n/\partial y_1) \right| \, . \tag{7.58}$$

The first column contains a difference of two terms; according to Properties 1.4 and 1.5 of the determinants (Sect. 1.3.2), we can split the first column and rewrite J_{n-1} as two determinants:

$$J_{n-1} = \left| \partial x_2/\partial y_k \; (\partial x_3/\partial y_k) - \alpha_k \, (\partial x_3/\partial y_1) \cdots (\partial x_n/\partial y_k) - \alpha_k \, (\partial x_n/\partial y_1) \right|$$
$$- \frac{\partial x_2}{\partial y_1} \left| \alpha_k \; (\partial x_3/\partial y_k) - \alpha_k \, (\partial x_3/\partial y_1) \cdots (\partial x_n/\partial y_k) - \alpha_k \, (\partial x_n/\partial y_1) \right| .$$

Similarly, we can split the second column in both determinants:

$$J_{n-1} = \left| \partial x_2/\partial y_k \; \partial x_3/\partial y_k \cdots (\partial x_n/\partial y_k) - \alpha_k \, (\partial x_n/\partial y_1) \right|$$
$$- \frac{\partial x_3}{\partial y_1} \left| \partial x_2/\partial y_k \; \alpha_k \cdots (\partial x_n/\partial y_k) - \alpha_k \, (\partial x_n/\partial y_1) \right|$$
$$- \frac{\partial x_2}{\partial y_1} \left| \alpha_k \; \partial x_3/\partial y_k \cdots (\partial x_n/\partial y_k) - \alpha_k \, (\partial x_n/\partial y_1) \right|$$
$$+ \frac{\partial x_2}{\partial y_1} \frac{\partial x_3}{\partial y_1} \left| \alpha_k \; \alpha_k \cdots (\partial x_n/\partial y_k) - \alpha_k \, (\partial x_n/\partial y_1) \right| .$$

The last determinant contains two identical columns and hence is equal to zero (Property 1.3 of determinants from Sect. 1.3.2). It is clear now that if we continue this process and split all columns in J_{n-1} of Eq. (7.58), we shall arrive at a sum of determinants in which the column α_k can only appear once; there will also be one determinant without α_k, which we shall write first:

$$J_{n-1} = \left| \partial x_2/\partial y_k \; \partial x_3/\partial y_k \cdots \partial x_n/\partial y_k \right|$$
$$- \frac{\partial x_2}{\partial y_1} \left| \alpha_k \; \partial x_3/\partial y_k \cdots \partial x_n/\partial y_k \right|$$
$$- \frac{\partial x_3}{\partial y_1} \left| \partial x_2/\partial y_k \; \alpha_k \cdots \partial x_n/\partial y_k \right| \tag{7.59}$$
$$\cdots$$
$$- \frac{\partial x_n}{\partial y_1} \left| \partial x_2/\partial y_k \; \partial x_3/\partial y_k \cdots \alpha_k \right| .$$

So far, we have our integral I_n calculated by first performing $(n-1)$ integrations over the variables $y_2, \ldots, y_n$ and only then over x_1. Now we shall change the order of integration so that the x_1 integration would appear first:

$$I_n = \underbrace{\int \cdots \int}_{n-1} |J_{n-1}| \, dy_2 \ldots dy_n \int F\left(x_1, g_2\left(x_1, y_2, \ldots, y_n\right), \ldots, g_n\left(x_1, y_2, \ldots, y_n\right)\right) dx_1 .$$

Here the x_1 integration is performed between two boundaries corresponding to the fixed values of $y_2, \ldots, y_n$. Since the latter variables are fixed, $dx_1 = (\partial x_1/\partial y_1) \, dy_1$, which gives

$$I_n = \underbrace{\int \cdots \int}_{n} F\left(f_1\left(y_1, \ldots, y_n\right), \ldots, f_n\left(y_1, \ldots, y_n\right)\right) \left| J_{n-1} \frac{\partial x_1}{\partial y_1} \right| dy_1 dy_2 \ldots dy_n \, .$$

(7.60)

What is left to show is that $J_{n-1}\left(\partial x_1/\partial y_1\right) = J_n$. To see this, we take the expression (7.59) for J_{n-1} and multiply it by $\partial x_1/\partial y_1$. In doing so, we shall keep this factor outside the first determinant of J_{n-1}, but in all other determinants in Eq. (7.59) we shall insert this factor inside them into the column containing α_k. We obtain

$$\begin{aligned}
J_{n-1}\frac{\partial x_1}{\partial y_1} = {} & \frac{\partial x_1}{\partial y_1} \left| \partial x_2/\partial y_k \; \partial x_3/\partial y_k \; \cdots \; \partial x_n/\partial y_k \right| \\
& - \frac{\partial x_2}{\partial y_1} \left| \alpha_k \left(\partial x_1/\partial y_1\right) \; \partial x_3/\partial y_k \; \cdots \; \partial x_n/\partial y_k \right| \\
& - \frac{\partial x_3}{\partial y_1} \left| \partial x_2/\partial y_k \; \alpha_k \left(\partial x_1/\partial y_1\right) \; \cdots \; \partial x_n/\partial y_k \right| \\
& \cdots \\
& - \frac{\partial x_n}{\partial y_1} \left| \partial x_2/\partial y_k \; \partial x_3/\partial y_k \; \cdots \; \alpha_k \left(\partial x_1/\partial y_1\right) \right|
\end{aligned}$$

(7.61)

Noticing now that, by definition of α_k we can replace $\alpha_k \left(\partial x_1/\partial y_1\right)$ with $\partial x_1/\partial y_k$, we can write

$$\begin{aligned}
J_{n-1}\frac{\partial x_1}{\partial y_1} = {} & \frac{\partial x_1}{\partial y_1} \left| \partial x_2/\partial y_k \; \partial x_3/\partial y_k \; \cdots \; \partial x_n/\partial y_k \right| \\
& - \frac{\partial x_2}{\partial y_1} \left| \partial x_1/\partial y_k \; \partial x_3/\partial y_k \; \cdots \; \partial x_n/\partial y_k \right| \\
& - \frac{\partial x_3}{\partial y_1} \left| \partial x_2/\partial y_k \; \partial x_1/\partial y_k \; \cdots \; \partial x_n/\partial y_k \right| \\
& \cdots \\
& - \frac{\partial x_n}{\partial y_1} \left| \partial x_2/\partial y_k \; \partial x_3/\partial y_k \; \cdots \; \partial x_1/\partial y_k \right| .
\end{aligned}$$

(7.62)

As the final step, in each of the determinants, apart from the first and the second ones, we move the column with $\partial x_1/\partial y_k$ to the first position; the order of other columns we do not change. This operation requires several pair permutations. If the column with $\partial x_1/\partial y_k$ is at the position of the l-th column (where $l = 1, 2, \ldots, n-1$), this would require $(l-1)$ permutations giving a factor of $(-1)^{l-1}$. This yields

$$J_{n-1} \frac{\partial x_1}{\partial y_1} = \frac{\partial x_1}{\partial y_1} \left| \partial x_2/\partial y_k \; \partial x_3/\partial y_k \; \cdots \; \partial x_n/\partial y_k \right|$$

$$+ (-1)^1 \frac{\partial x_2}{\partial y_1} \left| \partial x_1/\partial y_k \; \partial x_3/\partial y_k \; \cdots \; \partial x_n/\partial y_k \right|$$

$$+ (-1)^2 \frac{\partial x_3}{\partial y_1} \left| \partial x_1/\partial y_k \; \partial x_2/\partial y_k \; \cdots \; \partial x_n/\partial y_k \right| \tag{7.63}$$

$$\cdots$$

$$+ (-1)^{n-1} \frac{\partial x_n}{\partial y_1} \left| \partial x_1/\partial y_k \; \partial x_2/\partial y_k \; \cdots \; \partial x_{n-1}/\partial y_k \right| .$$

In the first $(n-1) \times (n-1)$ determinant the column with $\partial x_1/\partial y_k$ is missing, in the second—the column with $\partial x_2/\partial y_k$, and so on; in the final determinant the column with $\partial x_n/\partial y_k$ is absent. Looking now very carefully, we can verify that this expression is exactly an expansion of J_n in Eq. (7.52) along the first row, i.e. $J_{n-1} (\partial x_1/\partial y_1)$ is indeed J_n, and hence Eq. (7.60) is the required formula (7.51). **Q.E.D.**

7.7   Multi-variable Gaussian

We have already come across, on a number of occasions, the Gaussian function (e.g. in Sect. 4.2.1):

$$G(x) = \frac{1}{\sqrt{2\pi\sigma^2}} \exp\left(-\frac{(x-a)^2}{2\sigma^2}\right)$$

that depends on a single variable x. Here σ and a are parameters: a indicates the values of x at which this function attains the maximum value, and σ is related to the spread (width) of the function at its half height.

In applications it is often necessary to deal with the so-called multi-variable Gaussian functions that depend on $n \geq 2$ variables $x_1, \ldots, x_n$:

$$G(x_1, \ldots, x_n) = \frac{\sqrt{\det \mathbf{A}}}{(2\pi)^{n/2}} \exp\left(-\frac{1}{2} \sum_{i,j=1}^n a_{ij} x_i x_j\right) = \frac{\sqrt{\det \mathbf{A}}}{(2\pi)^{n/2}} \exp\left(-\frac{1}{2}\mathbf{X}^T \mathbf{A} \mathbf{X}\right),$$
$$\tag{7.64}$$

where $\mathbf{X}$ is the vector-column of all the variables and $\mathbf{A} = (a_{ij})$ is a symmetric positively definite (all eigenvalues are positive) real matrix. As we shall see in a moment, this function is normalised to one:

$$\int_{-\infty}^{\infty} dx_1 \ldots \int_{-\infty}^{\infty} dx_n \, G(x_1, \ldots, x_n) = 1 .$$

To show this, we shall consider eigenvectors $\mathbf{e}_\lambda = (e_{\lambda i})$ and eigenvalues a_λ of the matrix $\mathbf{A}$, i.e. $\mathbf{A}\mathbf{e}_\lambda = a_\lambda \mathbf{e}_\lambda$. Using the spectral representation of the matrix $\mathbf{A}$, see Eq. (1.107),

$$\mathbf{A} = \sum_\lambda a_\lambda \mathbf{e}_\lambda \mathbf{e}_\lambda^T ,$$

where the eigenvectors $\mathbf{e}_\lambda$ are orthonormal, the exponent of the Gaussian can be written as a sum of individual scalar quadratic terms (cf. Sect. 1.13):

$$-\frac{1}{2}\mathbf{X}^T \mathbf{A} \mathbf{X} = -\frac{1}{2}\mathbf{X}^T \left(\sum_\lambda a_\lambda \mathbf{e}_\lambda \mathbf{e}_\lambda^T \right) \mathbf{X} = -\frac{1}{2}\sum_\lambda a_\lambda (\mathbf{e}_\lambda \mathbf{X})^T \left(\mathbf{e}_\lambda^T \mathbf{X} \right) = -\frac{1}{2}\sum_\lambda a_\lambda y_\lambda^2 ,$$

where $y_\lambda = \mathbf{e}_\lambda^T \mathbf{X}$ is a scalar given by the dot product of the two vectors: $\mathbf{e}_\lambda$ and $\mathbf{X}$. The equation for y_λ can be solved for $\mathbf{X}$ by making use of the completeness relation (1.104) of the eigenvectors by calculating

$$\underbrace{\sum_\lambda \mathbf{e}_\lambda y_\lambda = \sum_\lambda \mathbf{e}_\lambda \mathbf{e}_\lambda^\dagger}_{\mathbf{E}} \mathbf{X} = \mathbf{X} \text{ or } x_i = \sum_\lambda e_{\lambda i} y_\lambda . \tag{7.65}$$

Now we are ready to justify the choice of the pre-factor in the definition of the Gaussian (7.64).

Problem 7.12 *Prove the following formula for the multi-dimensional Gaussian integral,*

$$I_n = \underbrace{\int_{-\infty}^{\infty} dx_1 \int_{-\infty}^{\infty} dx_2 \cdots \int_{-\infty}^{\infty} dx_n}_{\int d\mathbf{X}} \exp\left(-\frac{1}{2}\sum_{i,j=1}^n a_{ij} x_i x_j \right)$$

$$= \int d\mathbf{X} \exp\left(-\frac{1}{2}\sum_{i,j=1}^n a_{ij} x_i x_j \right) = \int d\mathbf{X} \exp\left(-\frac{1}{2}\mathbf{X}^T \mathbf{A} \mathbf{X} \right) = \frac{(2\pi)^{n/2}}{\sqrt{|\mathbf{A}|}} ,$$

$$\tag{7.66}$$

following these steps: (i) write the quadratic form in the exponent via the sum of the quadratic terms in y_λ as was demonstrated above; (ii) change variables in the integral from $\{x_i\}$ to $\{y_\lambda\}$ and explain why the Jacobian $J = |\partial x_i / \partial y_\lambda|$ is equal to unity; (iii) then, the n−fold integral splits into a product of n independent Gaussian integrals

$$\int_{-\infty}^{\infty} \exp\left(-\frac{1}{2}a_\lambda y_\lambda^2 \right) dy_\lambda = \sqrt{\frac{2\pi}{a_\lambda}} ;$$

(iv) finally, multiply all such contributions to get the required result.

Problem 7.13 *Show that for any square matrix* **D**,

$$\int d\mathbf{X}\,\left(\mathbf{X}^T\mathbf{D}\mathbf{X}\right)G(\mathbf{X}) = Tr\left(\mathbf{D}\mathbf{A}^{-1}\right).$$

Actually, the transformation from the variables $\{x_i\}$ to 'normal coordinates' ($\{y_\lambda\}$) (*cf.* Sect. 1.22.3) makes it clear that the multi-variable Gaussian as actually–in this new coordinates–a product of independent single-variable Gaussians. Indeed, recalling that det $\mathbf{A} = \prod_\lambda a_\lambda$ and replacing the quadratic form in the exponent as before, one obtains

$$G(x_1,\ldots,x_n) = G(\{x_i\}) \;\Rightarrow\; G(\{y_\lambda\}) = \prod_\lambda G_\lambda(y_\lambda), \qquad (7.67)$$

where

$$G_\lambda(y) = \sqrt{\frac{a_\lambda}{2\pi}}\exp\left(-\frac{1}{2}a_\lambda y_\lambda^2\right) \qquad (7.68)$$

is the properly normalised one-dimensional (single variable) Gaussian.

Let us now consider the integral

$$\langle x_i x_j\rangle = \int_{-\infty}^{\infty} dx_1 \ldots \int_{-\infty}^{\infty} dx_n\, x_i x_j G(x_1,\ldots,x_n). \qquad (7.69)$$

This can be interpreted as a pair correlation function if the Gaussian itself represents the appropriate probability density. Using Eq. (7.65), we obtain

$$\langle x_i x_j\rangle = \sum_{\lambda_1\lambda_2} e_{\lambda_1 i}e_{\lambda_2 j}\int_{-\infty}^{\infty} dx_1 \ldots \int_{-\infty}^{\infty} dx_n\, y_{\lambda_1}y_{\lambda_2}G(x_1,\ldots,x_n).$$

After changing the variables from $\{x_i\}$ to $\{y_\lambda\}$ and using Eqs. (7.67) and (7.68), we write

$$\langle x_i x_j\rangle = \sum_{\lambda_1\lambda_2} e_{\lambda_1 i}e_{\lambda_2 j}\int_{-\infty}^{\infty} dy_1 \ldots \int_{-\infty}^{\infty} dy_n\, y_{\lambda_1}y_{\lambda_2}G(y_1,\ldots,y_n)$$

$$= \sum_{\lambda_1\lambda_2} e_{\lambda_1 i}e_{\lambda_2 j}\int_{-\infty}^{\infty} dy_1 \ldots \int_{-\infty}^{\infty} dy_n\, y_{\lambda_1}y_{\lambda_2}\prod_\lambda G_\lambda(y_\lambda)$$

$$= \sum_{\lambda_1\lambda_2} e_{\lambda_1 i}e_{\lambda_2 j}\langle y_{\lambda_1}y_{\lambda_2}\rangle.$$

If $\lambda_1 \neq \lambda_2$, then the multi-dimensional integral $\langle y_{\lambda_1}y_{\lambda_2}\rangle$ factorises into a product of integrals over each of the variables y_λ,

$$\langle y_{\lambda_1} y_{\lambda_2}\rangle = \left[\prod_{\lambda \neq \lambda_1, \lambda_2} \int_{-\infty}^{\infty} dy_\lambda\, G_\lambda(y_\lambda)\right] \underbrace{\int_{-\infty}^{\infty} dy_{\lambda_1}\, y_{\lambda_1} G_{\lambda_1}(y_{\lambda_1})}_{=0} \underbrace{\int_{-\infty}^{\infty} dy_{\lambda_2}\, y_{\lambda_2} G_{\lambda_2}(y_{\lambda_2})}_{=0},$$

where the two integrals over y_{λ_1} and y_{λ_2} give zero due to the Gaussians $G_{\lambda_1}(y)$ and $G_{\lambda_2}(y)$ being even functions. Hence, only the terms with $\lambda_1 = \lambda_2$ remain

$$\langle y_{\lambda_1} y_{\lambda_2}\rangle = \delta_{\lambda_1 \lambda_2} \left[\prod_{\lambda \neq \lambda_1} \int_{-\infty}^{\infty} dy_\lambda\, G_\lambda(y_\lambda)\right] \int_{-\infty}^{\infty} dy_{\lambda_1}\, y_{\lambda_1}^2\, G_{\lambda_1}(y_{\lambda_1}).$$

Here the integrals within the square brackets each give one due to the normalisation of the single-variable Gaussian; the integral over y_{λ_1} can easily be checked to give $1/a_{\lambda_1}$ (see Problem 4.19), i.e.

$$\langle y_{\lambda_1} y_{\lambda_2}\rangle = \delta_{\lambda_1 \lambda_2} \frac{1}{a_{\lambda_1}}, \tag{7.70}$$

and hence,

$$\langle x_i x_j\rangle = \sum_{\lambda_1} \frac{1}{a_{\lambda_1}} e_{\lambda_1 i} e_{\lambda_1 j} = \left(\mathbf{A}^{-1}\right)_{ij}, \tag{7.71}$$

as this is the i, j element of the inverse matrix $\mathbf{A}$ written in its spectral representation. This is a remarkable result showing that the pair correlation function is equal to the appropriate element of the inverse of the matrix $\mathbf{A}$ used in the exponent of the multi-variable Gaussian.

Now we are in a position to discuss a more general result associated with the higher order correlation functions. We shall show that a higher order correlation function of an odd order is zero and the one of an even order is equal to the sum of products of the pair correlation functions made from all possible pairings of the variables. The first part of the statement is simple: if the order of the correlation function is odd, then one of the y_λ integrals would necessarily contain an odd power of this variable and hence would inevitably be zero. In order to prove the second part, it is instructive to start from the simplest case of the fourth order:

$$\begin{aligned}
\langle x_{i_1} x_{i_2} x_{i_3} x_{i_3}\rangle &= \sum_{\lambda_1 \lambda_2 \lambda_3 \lambda_4} e_{\lambda_1 i_1} e_{\lambda_2 i_2} e_{\lambda_3 i_3} e_{\lambda_4 i_4} \int_{-\infty}^{\infty} dy_1 \ldots \int_{-\infty}^{\infty} dy_n\, y_{\lambda_1} y_{\lambda_2} y_{\lambda_3} y_{\lambda_4} \\
&\quad \times \prod_\lambda G_\lambda(y_\lambda) \\
&= \sum_{\lambda_1 \lambda_2 \lambda_3 \lambda_4} e_{\lambda_1 i_1} e_{\lambda_2 i_2} e_{\lambda_3 i_3} e_{\lambda_4 i_4} \langle y_{\lambda_1} y_{\lambda_2} y_{\lambda_3} y_{\lambda_4}\rangle.
\end{aligned}$$

We need to calculate the multi-dimensional integral $\langle y_{\lambda_1} y_{\lambda_2} y_{\lambda_3} y_{\lambda_4} \rangle$ (which is also a correlation function, but written via the y_λ variables). It is clear that this function is only non-zero if the variables before the Gaussian in its integral appear in even powers. Hence, four cases are possible:

- $\lambda_1 = \lambda_2, \lambda_3 = \lambda_4$ and $\lambda_1 \neq \lambda_3$; this means that in this pairing system we shall pair the first two and the last two variables; this can be symbolically indicated by

$$\langle \overline{y_{\lambda_1} y_{\lambda_2}} \, \overline{y_{\lambda_3} y_{\lambda_4}} \rangle \, ;$$

  note that the condition includes two parts: (i) there are two pairs formed by equal indices $\lambda_1 = \lambda_2$ and $\lambda_3 = \lambda_4$, as indicated by the brackets, and (ii) there is an additional condition that the indices in both pairs are different, that is $\lambda_1 \neq \lambda_3$;
- $\lambda_1 = \lambda_3, \lambda_2 = \lambda_4$ and $\lambda_1 \neq \lambda_2$; this pairing system can be indicated as follows:

$$\langle \overline{y_{\lambda_1} y_{\lambda_2} y_{\lambda_3} y_{\lambda_4}} \rangle \, ;$$

- $\lambda_1 = \lambda_4, \lambda_2 = \lambda_3$ and $\lambda_1 \neq \lambda_2$, and this pairing system corresponds to

$$\langle \overline{y_{\lambda_1} y_{\lambda_2} y_{\lambda_3} y_{\lambda_4}} \rangle \, ;$$

- $\lambda_1 = \lambda_2 = \lambda_3 = \lambda_4$. We shall not need to indicate this case symbolically as will be clear in a moment.

All these four cases are mutually exclusive. Clearly, in any other case the correlation $\langle y_{\lambda_1} y_{\lambda_2} y_{\lambda_3} y_{\lambda_4} \rangle = 0$. Let us ignore for the moment the last case (all indices are the same) and concentrate on the first three cases. We then calculate their contribution. Consider for instance one of these terms, say the first one, explicitly:

$$\langle \overline{y_{\lambda_1} y_{\lambda_2}} \, \overline{y_{\lambda_3} y_{\lambda_4}} \rangle = \delta_{\lambda_1 \lambda_2} \delta_{\lambda_3 \lambda_4} \underbrace{\left[\prod_{\lambda \neq \lambda_1, \lambda_3} \int_{-\infty}^{\infty} dy_\lambda G_\lambda(y_\lambda) \right]}_{=1}$$

$$\times \underbrace{\int_{-\infty}^{\infty} dy_{\lambda_1} \, y_{\lambda_1}^2 G_{\lambda_1}(y_{\lambda_1})}_{=1/a_{\lambda_1}} \underbrace{\int_{-\infty}^{\infty} dy_{\lambda_3} \, y_{\lambda_3}^2 G_{\lambda_3}(y_{\lambda_3})}_{=1/a_{\lambda_3}}$$

$$= \delta_{\lambda_1 \lambda_2} \delta_{\lambda_3 \lambda_4} \frac{1}{a_{\lambda_1} a_{\lambda_3}} = \left(\delta_{\lambda_1 \lambda_2} \frac{1}{a_{\lambda_1}} \right) \left(\delta_{\lambda_3 \lambda_4} \frac{1}{a_{\lambda_3}} \right) = \langle y_{\lambda_1} y_{\lambda_2} \rangle \langle y_{\lambda_3} y_{\lambda_4} \rangle \, .$$

Similar results are obtained for the other two cases. It appears that, once we split the y_λ variables into pairs, the final expression for their correlation is simply given by the product of the individual pair correlations.

Let us now consider the sum of the contributions due to the first three cases:

$$\langle y_{\lambda_1} y_{\lambda_2} y_{\lambda_3} y_{\lambda_4} \rangle = \delta_{\lambda_1 \lambda_2} \delta_{\lambda_3 \lambda_4} \langle \overline{y_{\lambda_1} y_{\lambda_2}} \, \overline{y_{\lambda_3} y_{\lambda_4}} \rangle$$
$$+ \delta_{\lambda_1 \lambda_3} \delta_{\lambda_2 \lambda_4} \langle \overline{y_{\lambda_1} y_{\lambda_2} y_{\lambda_3} y_{\lambda_4}} \rangle \qquad (7.72)$$
$$+ \delta_{\lambda_1 \lambda_4} \delta_{\lambda_2 \lambda_3} \langle \overline{y_{\lambda_1} y_{\lambda_2} y_{\lambda_3} y_{\lambda_4}} \rangle$$
$$= \langle y_{\lambda_1} y_{\lambda_2} \rangle \langle y_{\lambda_3} y_{\lambda_4} \rangle + \langle y_{\lambda_1} y_{\lambda_3} \rangle \langle y_{\lambda_2} y_{\lambda_4} \rangle + \langle y_{\lambda_1} y_{\lambda_4} \rangle \langle y_{\lambda_2} y_{\lambda_3} \rangle .$$

It is easy to see, that this expression is actually formally valid for all four cases! Indeed, it is obviously valid for any of the first three cases as these are mutually exclusive. In the fourth case we formally obtain, using Eq. (7.72):

$$\langle y_{\lambda_1} y_{\lambda_1} y_{\lambda_1} y_{\lambda_1} \rangle = 3 \langle y_{\lambda_1}^2 \rangle^2 = \frac{3}{a_{\lambda_1}^2} .$$

However, the same result is obtained explicitly as well by simply calculating this correlation function directly:

$$\langle y_{\lambda_1}^4 \rangle = \underbrace{\left[\prod_{\lambda \neq \lambda_1} \int_{-\infty}^{\infty} dy_\lambda G_\lambda(y_\lambda) \right]}_{=1} \underbrace{\int_{-\infty}^{\infty} dy_{\lambda_1} \, y_{\lambda_1}^4 G_{\lambda_1}(y_{\lambda_1})}_{=3/a_{\lambda_1}^2} = \frac{3}{a_{\lambda_1}^2} .$$

(the integral with $y_{\lambda_1}^4$ is given in Problem 4.19). It is seen then that Eq. (7.72) is indeed formally valid for all four cases, not just for the first three!

Equipped by this example, we should now be able to consider a general case. Consider the correlation function (in y_λ variables) $\langle y_{\lambda_1} \ldots y_{\lambda_n} \rangle$ (n - even). It is not equal to zero if for each index λ_i there is an equal index pair, or that the same index appears an even number of times. If we now list all possible cases, there will be two types of them: (i) all indices are split into pairs of equal indices in such a way that indices in different pairs are different; (ii) there are equal indices in some of the pairs, so that these pairs can be combined into sets of 4,6, etc. equal indices. We state that the correlation function $\langle y_{\lambda_1} \ldots y_{\lambda_n} \rangle$ can be written as a sum of all possible pair systems, i.e. the way in which all n variables are split into pairs,

$$\langle y_{\lambda_1} \ldots y_{\lambda_n} \rangle = \sum_{\text{all systems of pairings}} \langle y_{\lambda_i} y_{\lambda_{i'}} \rangle \langle y_{\lambda_j} y_{\lambda_{j'}} \rangle \langle y_{\lambda_k} y_{\lambda_{k'}} \rangle \ldots , \qquad (7.73)$$

and then this expression would cover the cases of some of the pairs having equal indices automatically. Indeed, let us check that this is the case. Let us select some m pairs of indices; we shall denote them with $\mathcal{M}$. All other indices we shall denote as belonging to $\overline{\mathcal{M}}$. The $2m$ indices from $\mathcal{M}$ would contribute to more than one term in the sum (7.73) as we can construct all possible pairing between them. In fact, there will be exactly (see Problem I.1.181) $\mathcal{N}_{pairs}(m) = (2m)!/(2^m m!)$ such terms involving the same indices but paired differently. In each term in the sum over the system of pairings (7.73) these terms come in a product with other pairings involving indices from $\overline{\mathcal{M}}$. Hence, we can write

$$
\langle y_{\lambda_1} \cdots y_{\lambda_n} \rangle = \underbrace{\left[\sum_{\text{all systems of pairings within } \mathcal{M}} \langle y_{\lambda_i} y_{\lambda_{i'}} \rangle \langle y_{\lambda_j} y_{\lambda_{j'}} \rangle \cdots \right]}_{P_{\mathcal{M}}}
$$

$$
\times \left[\sum_{\text{all systems of pairings within } \overline{\mathcal{M}}} \langle y_{\lambda_l} y_{\lambda_{l'}} \rangle \langle y_{\lambda_p} y_{\lambda_{p'}} \rangle \cdots \right] + \cdots ,
$$

where the dots at the end of the expression in the right-hand side indicate other possible terms (systems of pairings) that involve indices from *both* $\mathcal{M}$ and $\overline{\mathcal{M}}$. What we need to show now is that if all indices within $\mathcal{M}$ are the same and, say, are equal to λ_i, then the direct calculation of the correlation

$$
\langle y_{\lambda_i} y_{\lambda_i} \cdots y_{\lambda_i} \rangle = \langle y_{\lambda_i}^{2m} \rangle
$$

(y_{λ_i} is repeated $2m$ times) gives the same contribution as that of $P_{\mathcal{M}}$. If all $2m$ indices are the same, we have $\mathcal{N}_{pairs}(m)$ identical terms in the sum $P_{\mathcal{M}}$, each containing a product of m pairs and hence each giving $1/a_{\lambda_i}^m$. Therefore, the overall contribution is

$$
P_{\mathcal{M}} = \mathcal{N}_{pairs}(m) \frac{1}{a_{\lambda_i}^m} = \frac{(2m)!}{2^m m!} \frac{1}{a_{\lambda_i}^m} .
$$

On the other hand,

$$
\langle y_{\lambda_i}^{2m} \rangle = \underbrace{\left[\prod_{\lambda \neq \lambda_i} \int_{-\infty}^{\infty} dy_\lambda \, G_\lambda(y_\lambda) \right]}_{=1} \int_{-\infty}^{\infty} dy_{\lambda_i} \, y_{\lambda_i}^{2m} G_{\lambda_i}(y_{\lambda_i}) = \frac{(2m)!}{2^m m!} \frac{1}{a_{\lambda_i}^m} ,
$$

which is exactly the same (the Gaussian integral with $y_{\lambda_i}^{2m}$ is calculated using Eq. (4.40)). Our consideration was completely general as m was arbitrary.

This proves that, when calculating $\langle y_{\lambda_1} \cdots y_{\lambda_n} \rangle$, one can simply write it down as a sum of products of all possible pairings, and this way the cases of more than two indices being equal are covered automatically. Now we should be able to return to our original variables and consider the correlation function in terms of them:

$$\langle x_{i_1} \ldots x_{i_n} \rangle = \sum_{\lambda_1 \ldots \lambda_n} e_{\lambda_1 i_1} \ldots e_{\lambda_n i_n} \langle y_{\lambda_1} \ldots y_{\lambda_n} \rangle$$

$$= \sum_{\lambda_1 \ldots \lambda_n} e_{\lambda_1 i_1} \ldots e_{\lambda_n i_n} \left[\sum_{\text{all systems of pairings}} \langle y_{\lambda_1} y_{\lambda_{1'}} \rangle \langle y_{\lambda_2} y_{\lambda_{2'}} \rangle \ldots \right]$$

$$= \sum_{\text{all systems of pairings}} \langle x_{i_1} x_{i_{1'}} \rangle \langle x_{i_2} x_{i_{2'}} \rangle \ldots ,$$

which is the result we have set out to prove.

In some applications a shifted Gaussian function (distribution) is encountered,

$$G_{shift}(\mathbf{X}) = \mathcal{N} \exp \left(-\frac{1}{2} \mathbf{X}^T \mathbf{A} \mathbf{X} + \mathbf{B}^T \mathbf{X} \right), \tag{7.74}$$

where $\mathbf{X}$ is the vector of all n variables $x_1, \ldots, x_n$, and $\mathbf{B}$ is a vector-column. This distribution, apart from the quadratic term in the exponent, contains also a liner in $\mathbf{X}$ term there. Here

$$\mathcal{N} = \frac{\sqrt{\det \mathbf{A}}}{(2\pi)^{n/2}} \exp \left(-\frac{1}{2} \mathbf{B}^T \mathbf{A}^{-1} \mathbf{B} \right) \tag{7.75}$$

is the appropriate normalisation.

Problem 7.14 *Consider the scalar $-\frac{1}{2}\mathbf{X}^T \mathbf{A}\mathbf{X} + \mathbf{B}^T \mathbf{X}$ in the exponent of the function. Show that one can introduce a new variable vector $\mathbf{Y} = \mathbf{X} + \mathbf{R}$ such that with an appropriate choice of $\mathbf{R}$ the expression becomes quadratic in $\mathbf{Y}$ without the linear term:*

$$-\frac{1}{2}\mathbf{X}^T \mathbf{A}\mathbf{X} + \mathbf{B}^T \mathbf{X} = -\frac{1}{2}\mathbf{Y}^T \mathbf{A}\mathbf{Y} + \frac{1}{2}\mathbf{B}^T \mathbf{A}^{-1}\mathbf{B}. \tag{7.76}$$

Hence demonstrate that the shifted Gaussian can also be written as

$$G_{shift}(\mathbf{X}) = \frac{\sqrt{\det \mathbf{A}}}{(2\pi)^{n/2}} \exp \left[-\frac{1}{2} \left(\mathbf{X} - \mathbf{A}^{-1}\mathbf{B} \right)^T \mathbf{A} \left(\mathbf{X} - \mathbf{A}^{-1}\mathbf{B} \right) \right]. \tag{7.77}$$

Problem 7.15 *Hence, show that $\mathcal{N}$ from Eq. (7.75) does indeed normalise the shifted Gaussian.*

Problem 7.16 *Show that for any square matrix $\mathbf{D}$,*

$$\int d\mathbf{X} \left(\mathbf{X}^T \mathbf{D}\mathbf{X} \right) G_{shift}(\mathbf{X}) = Tr \left(\mathbf{D}\mathbf{A}^{-1} \right) + \mathbf{B}^T \mathbf{A}^{-1} \mathbf{D}\mathbf{A}^{-1}\mathbf{B}.$$

Problem 7.17 *Show that for any vector-column* **C**,

$$\int d\mathbf{X}\,\left(\mathbf{C}^T\mathbf{X}\right)G_{shift}(\mathbf{X}) = \mathbf{C}^T\mathbf{A}^{-1}\mathbf{B}.$$

Problem 7.18 *In bosonic quantum field theory one needs to consider the following multiple Gaussian integrals:*

$$G[\mathbf{C}, \mathbf{C}^*] = \int \exp\left[-\mathbf{Z}^\dagger\mathbf{A}\mathbf{Z} + \mathbf{Z}^\dagger\mathbf{C} + \mathbf{C}^\dagger\mathbf{Z}\right]\prod_{j=1}^{N}\frac{dx_j dy_j}{\pi}$$

$$= \int \exp\left[-\sum_{i,j=1}^{N} z_i^* a_{ij} z_j + \sum_{j=1}^{N}\left(z_j^* c_j + c_j^* z_j\right)\right]\prod_{j=1}^{N}\frac{dx_j dy_j}{\pi}$$

$$= \frac{1}{\det \mathbf{A}}\exp\left(\mathbf{C}^\dagger\mathbf{A}^{-1}\mathbf{C}\right) = \frac{1}{\det \mathbf{A}}\exp\left[\sum_{i,j=1}^{N} c_i^*\left(\mathbf{A}^{-1}\right)_{ij} c_j\right],$$

(7.78)

where $\mathbf{Z} = \left(z_j\right)$ *and* $\mathbf{C} = \left(c_j\right)$ *are complex N-vectors and* $\mathbf{A} = \left(a_{ij}\right)$ *is a Hermitian $N \times N$ matrix with positive eigenvalues, and the integration is performed over the real x_j and imaginary y_j parts of each component $z_j = x_j + i y_j$ of the vector* **Z**. *Prove this formula in the following way: (i) Decompose the matrix* **A** *into its eigenstates,* $\mathbf{A} = \sum_\lambda a_\lambda \mathbf{e}_\lambda \mathbf{e}_\lambda^\dagger$, *and introduce* $z_\lambda = \sum_j e_{\lambda j} z_j$; *(ii) rewrite the expression in the exponential via real, x_λ, and imaginary, y_λ, parts of z_λ; (iii) prove that the Jacobian from the variables $\{x_j, y_j\}$ to $\{x_\lambda, y_\lambda\}$ is equal to one; (iv) this procedure leads to a product of each individual Gaussian integrals for x_λ and y_λ that can be calculated leading to the required result.*

Problem 7.19 *Consider now the pair correlation function*

$$\langle z_a z_b^* \rangle = \int z_a z_b^* \exp\left[-\sum_{i,j=1}^{N} z_i^* a_{ij} z_j\right]\prod_{j=1}^{N}\frac{dx_j dy_j}{\pi},$$

where the same notations as in the previous problem are used. Prove that

$$\langle z_a z_b^* \rangle = \frac{1}{G[\mathbf{0}, \mathbf{0}]}\left(\frac{\partial^2}{\partial c_a^* \partial c_b} G[\mathbf{C}, \mathbf{C}^*]\right)_{\mathbf{C}=0} = \left(\mathbf{A}^{-1}\right)_{ab}.$$

This result represents a single pairing between z_a and z_b^*. Similarly, prove the following expression for the double correlation:

$$\langle z_a z_b z_e^* z_f^* \rangle = \frac{1}{G[0,0]} \left(\frac{\partial^4}{\partial c_a^* \partial c_b^* \partial c_e \partial c_f} G[C, C^*] \right)_{C=0}$$
$$= (A^{-1})_{ae} (A^{-1})_{bf} + (A^{-1})_{af} (A^{-1})_{be} .$$

Note that in the latter result only two systems of 'pairings' are present, that are between z and z^* components; the pairing system between z_a and z_b and correspondingly between z_e^* and z_f^* are missing.

7.8 *N*-Dimensional Sphere

So far we have only limited ourselves to spherical coordinates applied to triple integrals. In fact, one can define a generalisation of these coordinates to $N \geq 3$ dimensions which could be used in taking $N-$dimensional integrals. Here we shall consider, as an example, calculation of a volume of an N-dimensional sphere and a related integral.

In physical applications it is sometimes needed to calculate $N-$dimensional integrals of the following type:

$$I_N = \int \cdots \int f \left(x_1^2 + x_2^2 + \ldots + x_N^2 \right) dx_1 dx_2 \cdots dx_N . \tag{7.79}$$

Here the integration is performed over the whole space. We shall now simplify this $N-$dimensional integral using a kind of heuristic argument.

Indeed, let us first calculate the integral,

$$V_N = \int \cdots \int_{x_1^2 + \ldots + x_N^2 \leq R^2} dx_1 dx_2 \cdots dx_N ,$$

that can be thought of as a volume of an $N-$dimensional 'sphere' of radius R. Indeed, the integration is performed in such a way that the condition

$$x_1^2 + \ldots + x_N^2 \leq R^2$$

is satisfied. It is readily seen that this is indeed a generalisation of the three-dimensional case. Clearly, V_N must be proportional to R^N, so that we write $V_N = C_N R^N$, where C_N is a constant to determine. On the other hand, one may presume that the volume element dV_N can be written as

$$dV_N = dx_1 \ldots dx_N = r^{N-1} dr \, d\Omega_{N-1} \,. \tag{7.80}$$

where $d\Omega_{N-1}$ is the corresponding angular part. Then,

$$V_N = \int_0^R r^{N-1} dr \left[\int \cdots \int d\Omega_{N-1} \right], \tag{7.81}$$

where the $(N-1)$−dimensional integration within the square brackets is performed over all angular dependencies, while the first integration is done with respect to the distance

$$r = \sqrt{x_1^2 + \ldots + x_N^2}$$

to the centre of the coordinate system which can be considered as the 'radial' coordinate of the N−dimensional spherical coordinate system.

We shall show now that for functions $f\left(r^2\right)$ which depend only on the distance r, the precise form of the angular part, $d\Omega_N$, is not important. Indeed, assuming that the angle integral does not depend on r, the r integration can be performed first:

$$V_N = \frac{R^N}{N} \int \cdots \int d\Omega_{N-1} \implies \int \cdots \int d\Omega_{N-1} = N C_N \,. \tag{7.82}$$

To calculate the constant C_N, let us consider the integral (7.79) with $f(x) = e^{-x}$. We can write

$$\int \cdots \int \exp\left[-\left(x_1^2 + \ldots + x_N^2\right) \right] dx_1 \cdots dx_N = \prod_{i=1}^N \int_{-\infty}^{\infty} e^{-x^2} dx$$

$$= \prod_{i=1}^N \sqrt{\pi} = \pi^{N/2} \,.$$

On the other hand, the same integral can be written via the corresponding radial and angular arguments as

$$\int_0^{\infty} e^{-r^2} r^{N-1} dr \left[\int \cdots \int d\Omega_{N-1} \right] = \frac{1}{2} \Gamma \left(\frac{N}{2} \right) \left[\int \cdots \int d\Omega_{N-1} \right],$$

where we have calculated the radial integral explicitly. Since the angular integration gives $N C_N$, see Eq. (7.82), we immediately obtain that

$$\frac{1}{2} \Gamma \left(\frac{N}{2} \right) N C_N = \pi^{N/2} \implies C_N = \frac{2\pi^{N/2}}{N \Gamma \left(\frac{N}{2} \right)} \,. \tag{7.83}$$

This solves the problem of calculating the volume of the N−dimensional sphere:

$$V_N = C_N R^N = \frac{2\pi^{N/2} R^N}{N\Gamma\left(\frac{N}{2}\right)} . \tag{7.84}$$

It is easy to see that the particular values of V_N for $N = 1, 2, 3$ do indeed make perfect sense: $V_1 = 2R$, $V_2 = \pi R^2$ and $V_3 = \frac{4\pi}{3} R^3$.

Incidentally, we can also easily consider the surface area of the $N-$dimensional sphere, $S_{N-1}(R)$. Indeed, since

$$V_N = \int_0^R S_{N-1}(r)dr ,$$

then $dV_N/dR = S_{N-1}(R)$, which yields

$$S_{N-1}(R) = \frac{2\pi^{N/2} R^{N-1}}{\Gamma\left(\frac{N}{2}\right)} . \tag{7.85}$$

Again, in particular cases one obtains expressions we would expect: $S_0 = 2$, $S_1 = 2\pi R$ (the length of a circle of radius R) and $S_2 = 4\pi R^2$ (the surface of a sphere of radius R).

Now we are able to return to the integral (7.79). Noting that the argument of the function f in the integrand is simply r^2, we obtain

$$I_N = \int_0^\infty f\left(r^2\right) r^{N-1} dr \left[\int \cdots \int d\Omega_{N-1}\right] = \frac{2\pi^{N/2}}{\Gamma\left(\frac{N}{2}\right)} \int_0^\infty f\left(r^2\right) r^{N-1} dr , \tag{7.86}$$

where we have used Eqs. (7.82) and (7.83) for the angular integral. This is the final result.

The essential point of our discussion above is based on an assumption, made in Eq. (7.81), that the volume element can be written as in Eq. (7.80). Although this formula is based on the dimensional argument, it is worth deriving it. We shall do this by generalising spherical coordinates to the N-dimensional space:

$$(x_1, x_2, \ldots, x_N) \implies (r, \varphi_1, \varphi_2, \ldots, \varphi_{N-1}) ,$$

where the two systems of coordinates are related to each other via the following transformation equations:

$$x_1 = r \cos \varphi_1 , \quad x_2 = r \sin \varphi_1 \cos \varphi_2 , \quad x_3 = r \sin \varphi_1 \sin \varphi_2 \cos \varphi_3 ,$$

$$\cdots \quad \cdots \quad \cdots$$

$$x_i = r \sin \varphi_1 \sin \varphi_2 \cdots \sin \varphi_{i-1} \cos \varphi_i ,$$

$$\cdots \quad \cdots \quad \cdots$$

$$x_{N-1} = r \sin \varphi_1 \sin \varphi_2 \cdots \sin \varphi_{i-1} \sin \varphi_i \cdots \sin \varphi_{N-2} \cos \varphi_{N-1} \,,$$

$$x_N = r \sin \varphi_1 \sin \varphi_2 \cdots \sin \varphi_{i-1} \sin \varphi_i \cdots \sin \varphi_{N-2} \sin \varphi_{N-1} \,.$$

Here $r \geq 0$ and $0 \leq \varphi_{N-1} \leq 2\pi$, while all other angles range between 0 and π.

The volume elements in the two coordinates systems, the Cartesian and the N-dimensional spherical ones, are related by the Jacobian (Sect. 7.6):

$$
\begin{aligned}
dx_1 dx_2 \cdots dx_N &= \left| \frac{\partial (x_1, x_2, \ldots, x_N)}{\partial (r, \varphi_1, \ldots, \varphi_{N-1})} \right| dr d\varphi_1 \cdots d\varphi_{N-1} \\
&= |J_N| \, dr d\varphi_1 \cdots d\varphi_{N-1} \,.
\end{aligned}
$$

Problem 7.20 *Check that*

$$x_1^2 + x_2^2 + \ldots + x_N^2 = r^2 \,.$$

Problem 7.21 *Prove, by writing down the Jacobian J_N explicitly employing the transformation relations above and using properties of the determinants, that $J_N = r^{N-1} F(\varphi_1, \ldots, \varphi_{N-1})$. This proves Eq. (7.80) with*

$$d\Omega_{N-1} = F\left(\varphi_1, \ldots, \varphi_{N_1}\right) d\varphi_1 \cdots d\varphi_{N-1} \,.$$

A full expression for the Jacobian (i.e. the function F) can also be derived, if needed, by calculating explicitly the Jacobian determinant. This would be needed when calculating integrals in which the integrand, apart from r, depends also on other coordinates (or their combinations).

7.9   Gradient of a Scalar Field

In this and the following sections we shall revisit several important notions of the vector calculus introduced in Sects. I.5.8, I.6.8.1 and I.6.8.2. There we obtained explicit formulae for the gradient, divergence, and curl for the Cartesian system. Our task here is to generalise these for a general curvilinear coordinates system. These would allow us in the next chapter to consider partial differential equations of mathematical physics exploiting symmetry of the problem at hand. Although the derivation can be done for a very general case, we shall limit ourselves here only to orthogonal curvilinear coordinate systems as these are more frequently found in actual applications.

Consider a scalar field $\Psi(P)$ defined at each point $P(x, y, z)$ in a 3D region R. The directional derivative, $d\Psi/dl$, of $\Psi(P)$ was defined in Sect. I.5.8 as the rate of change of the scalar field along a direction specified by the unit vector $\mathbf{l}$. Then, the gradient of $\Psi(P)$, written as grad $\Psi(P)$, was defined as a vector satisfying the relation:

$$(\text{grad } \Psi) \cdot \mathbf{l} = \frac{d\Psi}{dl} . \tag{7.87}$$

Note that the gradient does not depend on the direction $\mathbf{l}$, only on the value of the field at the point P.

Our task now is to obtain an expression for the gradient in a general orthogonal curvilinear coordinate system. In order to do that, we first expand the gradient (which is a vector field) in terms of the unit base vectors of this system:

$$\text{grad } \Psi = \sum_{j=1}^{3} (\text{grad } \Psi)_j \, \mathbf{e}_j . \tag{7.88}$$

Multiplying both sides by $\mathbf{e}_i$ and using the fact that the system is orthogonal, $\mathbf{e}_i \cdot \mathbf{e}_j = \delta_{ij}$, we have

$$(\text{grad } \Psi)_i = (\text{grad } \Psi) \cdot \mathbf{e}_i . \tag{7.89}$$

Comparing this equation with Eq. (7.87), we see that the i-th component of gradΨ is provided by the directional derivative of Ψ along the q_i coordinate line, i.e. along the direction $\mathbf{e}_i$,

$$(\text{grad } \Psi)_i = \frac{d\Psi}{ds_i} , \tag{7.90}$$

where $ds_i = h_i dq_i$ is the corresponding distance in space associated with the change of q_i from q_i to $q_i + dq_i$, see (7.29). Therefore, in order to calculate $(\text{grad } \Psi)_i$, we have to calculate the change $d\Psi$ of Ψ along the direction $\mathbf{e}_i$. In this direction only the coordinate q_i is changing by dq_i, i.e.

$$d\Psi = \frac{\partial \Psi}{\partial q_i} dq_i , \tag{7.91}$$

so that the required directional derivative

$$(\text{grad } \Psi)_i = \frac{d\Psi}{ds_i} = \frac{(\partial \Psi/\partial q_i)\, dq_i}{h_i dq_i} = \frac{1}{h_i} \frac{\partial \Psi}{\partial q_i} . \tag{7.92}$$

Therefore, finally, we can write

$$\text{grad } \Psi = \sum_{i=1}^{3} \frac{1}{h_i} \frac{\partial \Psi}{\partial q_i} \mathbf{e}_i . \tag{7.93}$$

It is easy to see that this expression indeed generalises Eq. (I.5.70) derived for the Cartesian system. In this case $(q_1, q_2, q_3) \rightarrow (x, y, z)$, $(\mathbf{e}_1, \mathbf{e}_2, \mathbf{e}_3) \rightarrow (\mathbf{i}, \mathbf{j}, \mathbf{k})$ and $h_1 = h_2 = h_3 = 1$, so that we immediately recover our previous result. For cylindrical coordinates $(q_1, q_2, q_3) \rightarrow (r, \phi, z)$ and $h_r = h_z = 1$, $h_\phi = r$, so that in this case

$$\text{grad } \Psi = \frac{\partial \Psi}{\partial r}\mathbf{e}_r + \frac{1}{r}\frac{\partial \Psi}{\partial \phi}\mathbf{e}_\phi + \frac{\partial \Psi}{\partial z}\mathbf{e}_z . \tag{7.94}$$

Problem 7.22 *Show that the gradient in the spherical system is*

$$grad \ \Psi = \frac{\partial \Psi}{\partial r}\mathbf{e}_r + \frac{1}{r}\frac{\partial \Psi}{\partial \theta}\mathbf{e}_\theta + \frac{1}{r \sin \theta}\frac{\partial \Psi}{\partial \phi}\mathbf{e}_\phi . \tag{7.95}$$

Problem 7.23 *Using the spherical coordinates, calculate $grad \Psi$ for $\Psi(x, y, z) = 10 \exp\left(-x^2 - y^2 - z^2\right)$ at the point $P\left(r = 1, \theta = \frac{\pi}{2}, \phi = \frac{\pi}{2}\right)$. Check your result by calculating the gradient of the same function directly in Cartesian coordinates. [Answer: $-(20/e)\,\mathbf{e}_r$.]*

Problem 7.24 *Consider a particle of unit mass moving within the $z = 0$ plane in a central field with the potential $U(r) = -\alpha/r$, where r is the distance from the centre. Show that the force field, $\mathbf{F} = -grad \ U$, acting on the particle in this coordinate system is radial, $\mathbf{F} = -\alpha r^{-2}\mathbf{e}_r$.*

7.10 Divergence of a Vector Field

Consider a vector field $\mathbf{F}(P)$. We showed in Sect. I.6.8.1 that the divergence of the vector field at the point P is given by the flux of $\mathbf{F}$ through a closed surface S of volume V containing the point P, divided by that volume,

$$\text{div } \mathbf{F} = \lim_{V \to 0} \frac{1}{V} \oint_S \mathbf{F} \cdot d\mathbf{S} = \lim_{V \to 0} \frac{1}{V} \oint_S \mathbf{F} \cdot \mathbf{n} dS , \tag{7.96}$$

in the limit of $V \to 0$. Note that dS is an area element on S, and $d\mathbf{S} = \mathbf{n} dS$ is the directed area element, where $\mathbf{n}$ is a unit vector normal of S at point P pointing in the outward direction, i.e. outside the volume. An important feature of the definition given by Eq. (7.96) is that it is *intrinsic* since it makes no reference to any particular coordinate system. Therefore, it is exactly what we need to derive an expression for the divergence in a general curvilinear coordinate system.

We have also stressed in Sect. İ.6.8.1 that for the limit in Eq. (7.96) to exist, it should not depend on the *shape* of the volume ensuring that div $\mathbf{F}(P)$ is a well-

Fig. 7.6 For the calculation
of the flux through the faces
$EFGH$ and $IJKL$ of the
closed surface S. Here points
$A\left(q_1 + \frac{1}{2}\delta q_1, q_2, q_3\right)$,
$B\left(q_1, q_2 + \frac{1}{2}\delta q_2, q_3\right)$ and
$C\left(q_1, q_2, q_3 + \frac{1}{2}\delta q_3\right)$ are
positioned in the centre of
the q_1-coordinate surface
$EFGH$, q_2-coordinate
surface $EIJF$ and
q_3-coordinate surface
$KGFJ$, respectively

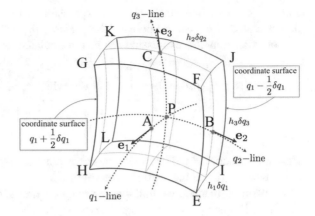

defined *scalar* field which only depends on the value of the vector field $\mathbf{F}$ at point P.
We shall now use this property of the intrinsic definition given above to derive the
corresponding expression for the divergence for an orthogonal curvilinear coordinate
system.

Consider a point P (q_1, q_2, q_3). Let us surround this point with a small curvilinear
box V. Since the limit should not depend on the shape of the box, it is convenient to
choose it to be formed by six-coordinate surfaces corresponding to particular values
of the curvilinear coordinates. Three pairs of surfaces are chosen, by two for each
curvilinear coordinate q_i $(i = 1, 2, 3)$; each pair crosses the corresponding q_i-line
before and after the point P (q_1, q_2, q_3) and corresponds to the q_i coordinate equal to
$q_i \pm \frac{1}{2}\delta q_i$ as is shown in Fig. 7.6, where δq_i is chosen infinitesimally small. Hence,
there are two surfaces crossing q_1, two crossing q_2 and another two crossing q_3. The
vector field $\mathbf{F}$ via its components is

$$\mathbf{F}(q_1, q_2, q_3) = \sum_{i=1}^{3} F_i(q_1, q_2, q_3)\,\mathbf{e}_i \,. \tag{7.97}$$

We first calculate the flux of $\mathbf{F}$ across the coordinate surface $EFGH$ crossing
the coordinate line q_1 at the value $q_1 + \frac{1}{2}\delta q_1$. In an orthogonal system that surface can be considered, to the leading order, rectangular (its sides are orthogonal),
with the sides $HE = GF = h_2\delta q_2$ and $GH = FE = h_3\delta q_3$. Its area therefore is
$dS_1 = h_2 h_3 \delta q_2 \delta q_3$. Also, the outward normal to the surface is $\mathbf{n} = \mathbf{e}_1$ since, again,
we consider an orthogonal system: indeed, the vectors $\mathbf{e}_2$ and $\mathbf{e}_3$ are tangential vectors
to the surface drawn from the point A, and $\mathbf{e}_1$ is orthogonal to them, hence is the
surface normal. Therefore, the flux through this surface

$$d_{1+} = \mathbf{F} \cdot d\mathbf{S}_1 = \mathbf{F} \cdot \mathbf{e}_1 dS_1 = (F_1 h_2 h_3)_{q_1+\delta q_1/2}\,\delta q_2 \delta q_3 \,,$$

since $\mathbf{F} \cdot \mathbf{e}_1 = F_1$, see Eq. (7.97). The expression in the round brackets above is to be calculated at the central point of the surface, $(q_1 + \frac{1}{2}\delta q_1, q_2, q_3)$. Applying the Taylor expansion, this expression can be manipulated, to the first order, to

$$
\begin{aligned}
d_{1+} &= \left\{ (F_1 h_2 h_3)_P + \left[\frac{\partial}{\partial q_1} (F_1 h_2 h_3) \right]_P \frac{\delta q_1}{2} \right\} \delta q_2 \delta q_3 \\
&= (F_1 h_2 h_3)_P \, \delta q_2 \delta q_3 + \left[\frac{\partial}{\partial q_1} (F_1 h_2 h_3) \right]_P \frac{\delta q_1}{2} \delta q_2 \delta q_3 .
\end{aligned}
\tag{7.98}
$$

Here the subscript P indicates that the corresponding expression is to be calculated at the point P (i.e. at the centre of the volume V).

In a similar manner, we find that the *outward* flux across the coordinate surface $IJKL$ (with the q_1 coordinate equal to $q_1 - \frac{1}{2}\delta q_1$) is

$$
d_{1-} = -(F_1 h_2 h_3)_P \, \delta q_2 \delta q_3 + \left[\frac{\partial}{\partial q_1} (F_1 h_2 h_3) \right]_P \frac{\delta q_1}{2} \delta q_2 \delta q_3 ,
\tag{7.99}
$$

the two particular signs are, in part, due to the fact that for this surface the outward normal $\mathbf{n} = -\mathbf{e}_1$ and hence $\mathbf{F} \cdot \mathbf{n} = -F_1$. Thus, the total outward flux across the opposite pair of surfaces $EFGH$ and $IJKL$ is equal to

$$
d_1 = d_{1+} + d_{1-} = \left[\frac{\partial}{\partial q_1} (F_1 h_2 h_3) \right]_P \delta q_1 \delta q_2 \delta q_3 .
\tag{7.100}
$$

The same analysis is repeated for the other two pairs of opposite faces to yield the contributions

$$
d_2 = \left[\frac{\partial}{\partial q_2} (F_2 h_3 h_1) \right]_P \delta q_1 \delta q_2 \delta q_3 \quad \text{and} \quad d_3 = \left[\frac{\partial}{\partial q_3} (F_3 h_1 h_2) \right]_P \delta q_1 \delta q_2 \delta q_3
$$

for the q_2 and q_3 surfaces, respectively. Summing up all three contributions leads to the final value of the flux through the whole surface S as

$$
\left[\frac{\partial}{\partial q_1} (F_1 h_2 h_3) + \frac{\partial}{\partial q_2} (F_2 h_3 h_1) + \frac{\partial}{\partial q_3} (F_3 h_1 h_2) \right] \delta q_1 \delta q_2 \delta q_3 .
\tag{7.101}
$$

The volume V enclosed by S is given by (recall that our system is orthogonal):

$$
\delta V = h_1 h_2 h_3 \delta q_1 \delta q_2 \delta q_3
\tag{7.102}
$$

to the leading order. Finally, dividing the flux (7.101) by the volume (7.102), we finally obtain

$$
\text{div } \mathbf{F} = \frac{1}{h_1 h_2 h_3} \left[\frac{\partial}{\partial q_1} (F_1 h_2 h_3) + \frac{\partial}{\partial q_2} (h_1 F_2 h_3) + \frac{\partial}{\partial q_3} (h_1 h_2 F_3) \right] .
\tag{7.103}
$$

Note that it might seem that we have never applied the limit $V \rightarrow 0$. In reality, we did. Indeed, in the expressions for the fluxes and the volume we only kept leading terms; other terms are proportional to higher powers of δq_i and would have disappeared in the limits $\delta q_i \rightarrow 0$ $(i = 1, 2, 3)$ which correspond to the volume tending to zero. Hence, the result above is the general formula we sought for.

Let us verify if this formula goes over to the result (I.6.88) we had for the Cartesian system when $(q_1, q_2, q_3) \rightarrow (x, y, z)$. In this case all scale factors are equal to one and our result does indeed reduces to the expected one:

$$\text{div } \mathbf{F} = \frac{\partial F_x}{\partial x} + \frac{\partial F_y}{\partial y} + \frac{\partial F_z}{\partial z} .$$

For cylindrical coordinates (r, ϕ, z) the scale factors read $h_r = h_z = 1$ and $h_\phi = r$, so that we find

$$
\begin{aligned}
\text{div } \mathbf{F} &= \frac{1}{r} \left[\frac{\partial}{\partial r} (F_r r) + \frac{\partial}{\partial \phi} (F_\phi) + \frac{\partial}{\partial z} (F_z r) \right] \\
&= \frac{1}{r} \frac{\partial}{\partial r} (r F_r) + \frac{1}{r} \frac{\partial F_\phi}{\partial \phi} + \frac{\partial F_z}{\partial z} .
\end{aligned}
\tag{7.104}
$$

Only the first two terms are to be kept if the two-dimensional polar coordinate system (r, ϕ) is considered.

Problem 7.25 *Show that the divergence of a vector field $\mathbf{F}$ in the spherical system is*

$$
div \, \mathbf{F} = \frac{1}{r^2} \frac{\partial}{\partial r} \left(r^2 F_r \right) + \frac{1}{r \sin \theta} \frac{\partial}{\partial \theta} (F_\theta \sin \theta) + \frac{1}{r \sin \theta} \frac{\partial F_\phi}{\partial \phi} .
\tag{7.105}
$$

Problem 7.26 *Using the spherical system, calculate the divergence of the vector field $\mathbf{F} = r^{-2} \mathbf{e}_r + r \sin \theta \, \mathbf{e}_\theta$ at the point $P \left(r = 1, \theta = \frac{\pi}{2}, \phi = \frac{\pi}{2} \right)$.* [Answer: $2 \cos \theta = 0$]

7.11 Laplacian

We are now in a position to work out an expression for the Laplacian $\Delta \Psi = \text{div grad } \Psi$ of a scalar field in a general curvilinear coordinate system. Indeed, a scalar field Ψ we have can be used to generate a vector field

$$
\mathbf{F} = F_1 \mathbf{e}_1 + F_2 \mathbf{e}_2 + F_3 \mathbf{e}_3 = \text{grad } \Psi = \sum_{i=1}^{3} \frac{1}{h_i} \frac{\partial \Psi}{\partial q_i} \mathbf{e}_i ,
\tag{7.106}
$$

where we have used Eq. (7.93). Next, this vector field can be turned into a scalar field by applying the divergence operation for which we derived a general expression (7.103). We can now combine the two expressions to evaluate $\Delta\Psi$ if we notice that the components of the vector $\mathbf{F}$ above are

$$F_i = \frac{1}{h_i} \frac{\partial \Psi}{\partial q_i} . \tag{7.107}$$

Therefore, substituting (7.107) into (7.103) gives the final expression for the Laplacian sought for

$$\Delta\Psi = \frac{1}{h_1 h_2 h_3} \left[\frac{\partial}{\partial q_1} \left(\frac{h_2 h_3}{h_1} \frac{\partial \Psi}{\partial q_1} \right) + \frac{\partial}{\partial q_2} \left(\frac{h_3 h_1}{h_2} \frac{\partial \Psi}{\partial q_2} \right) + \frac{\partial}{\partial q_3} \left(\frac{h_1 h_2}{h_3} \frac{\partial \Psi}{\partial q_3} \right) \right] . \tag{7.108}$$

It is easy to see that in Cartesian coordinates this expression simplifies to a well-known result

$$\Delta\Psi = \frac{\partial^2 \Psi}{\partial x^2} + \frac{\partial^2 \Psi}{\partial y^2} + \frac{\partial^2 \Psi}{\partial z^2} ,$$

while for cylindrical coordinates (r, ϕ, z) we find that

$$\Delta\Psi = \frac{1}{r} \frac{\partial}{\partial r} \left(r \frac{\partial \Psi}{\partial r} \right) + \frac{1}{r^2} \frac{\partial^2 \Psi}{\partial \phi^2} + \frac{\partial^2 \Psi}{\partial z^2} . \tag{7.109}$$

Correspondingly, in polar coordinates one drops the last term.

Problem 7.27 *Show that the Laplacian in the spherical system* (r, θ, ϕ) *is*

$$\Delta\Psi = \frac{1}{r^2} \frac{\partial}{\partial r} \left(r^2 \frac{\partial \Psi}{\partial r} \right) + \frac{1}{r^2 \sin\theta} \frac{\partial}{\partial \theta} \left(\sin\theta \frac{\partial \Psi}{\partial \theta} \right) + \frac{1}{r^2 \sin^2\theta} \frac{\partial^2 \Psi}{\partial \phi^2} . \tag{7.110}$$

Problem 7.28 *Calculate* $\Delta\Psi$ *in the spherical system for* $\Psi = e^{-r^2} \cos\theta$. [*Answer:* $2e^{-r^2} \left(2r^2 - 3 - r^{-2} \right) \cos\theta$.]

Problem 7.29 *Show that in parabolic coordinates (see Problem 7.4) the Laplacian has the form:*

$$\Delta\Psi = \frac{1}{u^2 + v^2} \left[\frac{1}{u} \frac{\partial}{\partial u} \left(u \frac{\partial \Psi}{\partial u} \right) + \frac{1}{v} \frac{\partial}{\partial v} \left(v \frac{\partial \Psi}{\partial v} \right) \right] + \frac{1}{(uv)^2} \frac{\partial^2 \Psi}{\partial \theta^2} .$$

Problem 7.30 *Using the Laplacian written in the spherical coordinates* (r, θ, ϕ), *show that*

$$\Delta \frac{e^{-\kappa r} - 1}{r} = \frac{\kappa^2}{r} e^{-\kappa r} .$$

Hence, show that the function $\psi\,(\mathbf{r}) = e^{-\kappa r}/r$ *satisfies the differential equation*

$$\left(\Delta - \kappa^2\right) \psi\,(\mathbf{r}) = -4\pi\delta\,(\mathbf{r}) .$$

[*Hint*: you may find Eq. (5.66) useful.] *Verify then that for* $r > 0$ *(strictly) this is a Schrödinger equation for a free electron with energy* $E < 0$, *mass* m *and the wavefunction* $\psi\,(\mathbf{r})$ *if* $\kappa = \frac{1}{\hbar}\sqrt{-2mE}$.

7.12 Curl of a Vector Field

To calculate the curl of a vector field $\mathbf{F}$ in curvilinear coordinates, we shall use the intrinsic definition of the curl given in Sect. I.6.8.2. It states that the curl of $\mathbf{F}$ at point P can be defined in the following way: choose an arbitrary smooth surface S through the point P and draw a contour L around that point within the surface. Then the curl of $\mathbf{F}$ at point P is given by the limit in which the area A enclosed by the contour is tending to zero keeping the point P inside it:

$$\operatorname{curl}\mathbf{F} \cdot \mathbf{n} = \lim_{A \to 0} \frac{1}{A} \oint_L \mathbf{F} \cdot d\mathbf{l} , \tag{7.111}$$

where $\mathbf{n}$ is the normal to the surface at point P and in the line integral the contour is traversed such that the point P is always on the left. The direction of the normal $\mathbf{n}$ is related to the direction of the traverse around the contour L by the right-hand screw rule.

To calculate the curl of $\mathbf{F}$, we note that the limit should not depend on the shape of the region enclosed by the contour L (or the shape of the latter) as long as the point P remains inside it when taking the limit (the curl $\mathbf{F}$ is a well-defined vector field which only depends on $\mathbf{F}$ at the point P). Hence, to perform the calculation, it is convenient to choose the surface S as the q_i coordinate surface for the calculation of the i-th component of the curl. Indeed, with such a choice and for an orthogonal curvilinear system, the normal $\mathbf{n}$ to the surface at point P coincides with the unit base vector $\mathbf{e}_i$. The case of $i = 1$ is illustrated in Fig. 7.7. Hence, with that particular choice of S, from Eq. (7.111),

$$\operatorname{curl}\mathbf{F} \cdot \mathbf{n} = \operatorname{curl}\mathbf{F} \cdot \mathbf{e}_i = (\operatorname{curl}\mathbf{F})_i = \lim_{A \to 0} \frac{1}{A} \oint_L \mathbf{F} \cdot d\mathbf{l} ,$$

Fig. 7.7 For the calculation of the line integral along $ABCD$ serving as the contour L: the q_1 coordinate surface is chosen as S to calculate the component $(\text{curl}\,\mathbf{F})_1$ of the curl. Points $H\left(q_1, q_2 - \frac{1}{2}\delta q_2, q_3\right)$, $F\left(q_1, q_2 + \frac{1}{2}\delta q_2, q_3\right)$, $E\left(q_1, q_2, q_3 - \frac{1}{2}\delta q_3\right)$ and $G\left(q_1, q_2, q_3 + \frac{1}{2}\delta q_3\right)$ in the middle of the sides of the distorted rectangular $ABCD$ are also indicated

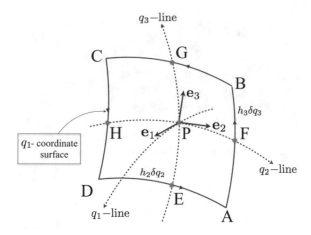

where the vector field

$$\text{curl}\,\mathbf{F} = (\text{curl}\,\mathbf{F})_1\,\mathbf{e}_1 + (\text{curl}\,\mathbf{F})_2\,\mathbf{e}_2 + (\text{curl}\,\mathbf{F})_3\,\mathbf{e}_3 \;.$$

Therefore, to calculate the i-th component of the curl, we can simply choose the q_i coordinate surface passing through the point P. In addition, we can also choose the contour L to go along the other two coordinate lines—see again Fig. 7.7, where in the case of $i = 1$ the contour is a distorted rectangular with its sides running along the q_2 and q_3 lines shifted from their values at the point P by $\pm\frac{1}{2}\delta q_2$ and $\pm\frac{1}{2}\delta q_3$, respectively.

Let us calculate $(\text{curl}\,\mathbf{F})_1$ at P. The surface S is taken as the q_1 coordinate surface with the contour L being $ABCD$ in Fig. 7.7 passed anticlockwise as indicated (note the direction of the normal $\mathbf{e}_1$ to S). The line integral along the contour contains four components which we have to calculate one by one. The line AB corresponds to the q_3 coordinate line through the point $F\left(q_1, q_2 + \frac{1}{2}\delta q_2, q_3\right)$. To the leading order, $\mathbf{F}$ can be considered fixed at its value at the point F, resulting in the contribution

$$\int_{AB} \mathbf{F} \cdot d\mathbf{l} = \int_{AB} \mathbf{F} \cdot \mathbf{e}_3 dl = \int_{AB} F_3 dl = F_3 \int_{AB} dl = (F_3 h_3)_F\,\delta q_3 \;,$$

where $h_3 \delta q_3$ is the length of the line AB. The expression in the round brackets is to be calculated at the point F as indicated above. Similarly, the contribution from the opposite piece CD is

$$\int_{CD} \mathbf{F} \cdot d\mathbf{l} = -\,(F_3 h_3)_H\,\delta q_3 \;,$$

where the minus sign comes from the fact that in this case the direction is opposite to that of $\mathbf{e}_3$, i.e. $d\mathbf{l} = -\mathbf{e}_3 dl$, and the index H means that the expression in the round brackets is to be calculated at point $H\left(q_1, q_2 - \frac{1}{2}\delta q_2, q_3\right)$. The latter differs from F only in its second coordinate. Therefore, the sum of these two contributions is

$$\int_{AB+CD} \mathbf{F} \cdot d\mathbf{l} = (F_3 h_3)_F \, \delta q_3 - (F_3 h_3)_H \, \delta q_3 = \left[(F_3 h_3)_F - (F_3 h_3)_H\right] \delta q_3$$

$$= \left[(F_3 h_3)_{q_2 + \delta q_2/2} - (F_3 h_3)_{q_2 - \delta q_2/2}\right] \delta q_3 .$$

The terms in the expression in the square brackets can be expanded in the Taylor series keeping the first two terms:

$$(F_3 h_3)_{q_2 + \delta q_2/2} = (F_3 h_3)_{q_2} + \left[\frac{\partial}{\partial q_2} (F_3 h_3)\right]_{q_2} \frac{\delta q_2}{2}$$

$$= (F_3 h_3)_P + \left[\frac{\partial}{\partial q_2} (F_3 h_3)\right]_P \frac{\delta q_2}{2} ,$$

$$(F_3 h_3)_{q_2 - \delta q_2/2} = (F_3 h_3)_{q_2} - \left[\frac{\partial}{\partial q_2} (F_3 h_3)\right]_{q_2} \frac{\delta q_2}{2}$$

$$= (F_3 h_3)_P - \left[\frac{\partial}{\partial q_2} (F_3 h_3)\right]_P \frac{\delta q_2}{2} ,$$

so that

$$\int_{AB+CD} \mathbf{F} \cdot d\mathbf{l} = \left[(F_3 h_3)_{q_2 + \delta q_2/2} - (F_3 h_3)_{q_2 - \delta q_2/2}\right] \delta q_3$$

$$= \left[\frac{\partial}{\partial q_2} (F_3 h_3)\right]_P \delta q_2 \delta q_3 .$$

In a similar manner, we find that the line integrals along DA and BC are

$$\int_{DA+BC} \mathbf{F} \cdot d\mathbf{l} = \left[(F_2 h_2)_{q_3 - \delta q_3/2} - (F_2 h_2)_{q_3 + \delta q_3/2}\right] \delta q_2$$

$$= -\left[\frac{\partial}{\partial q_3} (F_2 h_2)\right]_P \delta q_2 \delta q_3 .$$

Hence, we obtain for the whole closed-path line integral:

$$\oint_{ABCD} \mathbf{F} \cdot d\mathbf{l} = \left[\frac{\partial}{\partial q_2} (F_3 h_3) - \frac{\partial}{\partial q_3} (F_2 h_2)\right]_P \delta q_2 \delta q_3 . \qquad (7.112)$$

This expression is valid to the leading order in δq_i.

Next, the area enclosed by $ABCD$, to the leading order, is $A = (h_2 \delta q_2)(h_3 \delta q_3)$. Therefore, dividing the value of the line integral along $ABCD$ by the area A, we obtain

$$(\text{curl}\,\mathbf{F})_1 = \frac{1}{h_2 h_3}\left[\frac{\partial}{\partial q_2}(h_3 F_3) - \frac{\partial}{\partial q_3}(h_2 F_2)\right]. \tag{7.113}$$

This expression is finite; any other terms corresponding to higher orders in δq_i vanish in the $A \to 0$ (or $\delta q_i \to 0$) limit.

The other two components of the curl, namely $(\text{curl}\,\mathbf{F})_2$ and $(\text{curl}\,\mathbf{F})_3$, are obtained in the similar fashion (perform this calculation as an exercise!) giving in the end:

$$\text{curl}\,\mathbf{F} = \frac{\mathbf{e}_1}{h_2 h_3}\left[\frac{\partial}{\partial q_2}(h_3 F_3) - \frac{\partial}{\partial q_3}(h_2 F_2)\right] + \frac{\mathbf{e}_2}{h_3 h_1}\left[\frac{\partial}{\partial q_3}(h_1 F_1) - \frac{\partial}{\partial q_1}(h_3 F_3)\right]$$

$$+ \frac{\mathbf{e}_3}{h_1 h_2}\left[\frac{\partial}{\partial q_1}(h_2 F_2) - \frac{\partial}{\partial q_2}(h_1 F_1)\right]. \tag{7.114}$$

This formula can be expressed more compactly as a determinant:

$$\text{curl}\,\mathbf{F} = \frac{1}{h_1 h_2 h_3}\begin{vmatrix} h_1\mathbf{e}_1 & h_2\mathbf{e}_2 & h_3\mathbf{e}_3 \\ \partial/\partial q_1 & \partial/\partial q_2 & \partial/\partial q_3 \\ h_1 F_1 & h_2 F_2 & h_3 F_3 \end{vmatrix}. \tag{7.115}$$

It must be immediately seen that in the Cartesian coordinates we recover our old Eq. (I.6.82).

Problem 7.31 *Show that the curl in the cylindrical system* (r, ϕ, z) *is*

$$\text{curl}\,\mathbf{F} = \mathbf{e}_r\left(\frac{1}{r}\frac{\partial F_z}{\partial \phi} - \frac{\partial F_\phi}{\partial z}\right) + \mathbf{e}_\phi\left(\frac{\partial F_r}{\partial z} - \frac{\partial F_z}{\partial r}\right) + \frac{\mathbf{e}_z}{r}\left[\frac{\partial}{\partial r}(r F_\phi) - \frac{\partial F_r}{\partial \phi}\right]. \tag{7.116}$$

Problem 7.32 *Show that the curl in the spherical system* (r, θ, ϕ) *is*

$$\text{curl}\,\mathbf{F} = \frac{\mathbf{e}_r}{r \sin\theta}\left[\frac{\partial}{\partial \theta}(F_\phi \sin\theta) - \frac{\partial F_\theta}{\partial \phi}\right] + \frac{\mathbf{e}_\theta}{r \sin\theta}\left[\frac{\partial F_r}{\partial \phi} - \sin\theta\frac{\partial}{\partial r}(r F_\phi)\right]$$

$$+ \frac{\mathbf{e}_\phi}{r}\left[\frac{\partial}{\partial r}(r F_\theta) - \frac{\partial F_r}{\partial \theta}\right]. \tag{7.117}$$

Problem 7.33 *Calculate in the spherical coordinates the curl of the vector field* $\mathbf{F} = r^2\mathbf{e}_r + r^2 \sin\theta\,\mathbf{e}_\theta$. *Show that only the second term in the vector field contributes. Hence, calculate the curl at the point* $P\left(r = 1, \theta = \frac{\pi}{2}, \phi = \frac{\pi}{2}\right)$. *[Answer:* $3r \sin\theta\,\mathbf{e}_\phi = 3\mathbf{e}_\phi$.*]*

7.13 Some Applications in Physics

7.13.1 Partial Differential Equations of Mathematical Physics

Many problems in physics such as heat and mass transport, wave propagation, etc, are described by partial differential equations (PDE). These are equations which are generally very difficult to solve. However, if a physical system has a symmetry, then an appropriate curvilinear coordinate system may help in solving these PDEs. Here we consider some of the well-known PDEs of mathematical physics in cylindrical and spherical coordinate systems and illustrate on simple examples how their solution can be obtained in some cases.

We shall start from the wave equation (Sect. I.6.9.1):

$$\Delta \Psi = \frac{1}{c^2} \frac{\partial^2 \Psi}{\partial t^2} ,$$

where $\Psi (\mathbf{r}, t)$ is a wave field of interest, c is a constant corresponding to the speed of wave propagation. If the system possesses a cylindrical symmetry, then the Laplacian has the form (7.109) and the PDE can be rewritten as

$$\frac{1}{r} \frac{\partial}{\partial r} \left(r \frac{\partial \Psi}{\partial r} \right) + \frac{1}{r^2} \frac{\partial^2 \Psi}{\partial \phi^2} + \frac{\partial^2 \Psi}{\partial z^2} = \frac{1}{c^2} \frac{\partial^2 \Psi}{\partial t^2} . \tag{7.118}$$

Here $\Psi = \Psi (r, \phi, z)$ is the function of the cylindrical coordinates.[4] This equation may seem to be as difficult as the original one, but this is not the case as the variables r and ϕ instead of x and y are more appropriate for that symmetry and hence may help in obtaining the required solution. An example of this was given in Sect. 4.7.6 when we considered an oscillation of the circular membrane. In the latter case the z coordinate was in fact dropped and only polar coordinates were used. The z coordinate can also be dropped (together with the last term in the left-hand side in Eq. (7.118)) if for instance one expects that the solution should not depend on z due to symmetry. For instance, the wave produced by an infinitely long thin source positioned along the z axis would have that kind of symmetry.

Similarly, propagation of a wave from a spherical source should have spherical symmetry. The corresponding wave equation is obtained by taking the Laplacian in the spherical coordinates, Eq. (7.110),

$$\frac{1}{r^2} \frac{\partial}{\partial r} \left(r^2 \frac{\partial \Psi}{\partial r} \right) + \frac{1}{r^2 \sin \theta} \frac{\partial}{\partial \theta} \left(\sin \theta \frac{\partial \Psi}{\partial \theta} \right) + \frac{1}{r^2 \sin^2 \theta} \frac{\partial^2 \Psi}{\partial \phi^2} = \frac{1}{c^2} \frac{\partial^2 \Psi}{\partial t^2} , \tag{7.119}$$

where $\Psi = \Psi (r, \theta, \phi)$.

---

[4] We keep the same notation for it as in the Cartesian coordinates solely for convenience; however, when written via cylindrical coordinates, it will become a different function.

The diffusion or heat transport equations can be considered exactly along the same lines; the only difference is that in the right-hand side instead of the second time derivative we have the first.

As an example, consider a simple electrostatic problem of a potential φ due to a radially symmetric charge distribution $\rho(r)$. We shall assume that $\rho(r) \neq 0$ only for $0 \le r \le R$, i.e. we consider a spherical charged ball of radius R with the charge density which may change with the distance from the centre. In this case the potential φ satisfying the Poisson equation,

$$\Delta\varphi = -4\pi\rho \ ,$$

will only depend on r, and hence we can drop the θ and ϕ terms in the Laplacian, which leads us to a much simpler equation:

$$\frac{1}{r^2}\frac{d}{dr}\left(r^2\frac{d\varphi}{dr}\right) = -4\pi\rho(r) \quad \Longrightarrow \quad \frac{d}{dr}\left(r^2\frac{d\varphi}{dr}\right) = -4\pi r^2\rho(r) \ .$$

This is an ordinary DE which can be easily solved. Integrating both sides between 0 and $r \ge 0$ and assuming that the derivative of the potential at the centre is finite, we get

$$r^2\frac{d\varphi}{dr} = -\int_0^r 4\pi r_1^2 \rho\,(r_1)\,dr_1 \quad \Longrightarrow \quad \frac{d\varphi}{dr} = -\frac{Q(r)}{r^2} \ , \tag{7.120}$$

where

$$Q(r) = \int_0^r 4\pi r_1^2 \rho\,(r_1)\,dr_1 = \int_{Sphere} \rho\,dV \tag{7.121}$$

is the total charge contained in the sphere of radius r. This is because $dV = 4\pi r_1^2 dr_1$ corresponds exactly to the volume of a spherical shell of radius r_1 and width dr_1 (indeed, integrating over the angles $0 \le \theta \le \pi$ and $0 \le \phi \le 2\pi$ of the volume element $dV = r_1^2 \sin\theta d\theta d\phi$ in the spherical system, one obtains this expression for dV).

Consider first the potential outside the sphere, $r \ge R$. In this case $Q(r) = Q(R) = Q_0$ is the total charge of the sphere. Then, integrating equation (7.120) between r and ∞ and setting the potential φ at infinity to zero, we obtain a simple result $\phi(r) = -Q_0/r$. Remarkably, this point charge formula is valid for any distribution of charge inside the sphere provided it remains spherically symmetric.

Consider now the potential inside the sphere, i.e. for $0 \le r \le R$. In this case $Q(r)$ corresponds to the charge inside the sphere of radius r. Integrating Eq. (7.120) between r and R, we find

$$\varphi(r) = \varphi(R) + \int_r^R \frac{Q\,(r_1)\,dr_1}{r_1^2} = -\frac{Q_0}{R} + \int_r^R \frac{Q\,(r_1)\,dr_1}{r_1^2} \ . \tag{7.122}$$

Here we set $\varphi(R) = -Q_0/R$ to ensure continuity of the solution across the boundary of the sphere.

It is immediately seen from this formula, that the potential of a spherical layer of a finite width inside it is constant. Indeed, if $\rho \neq 0$ only for $R_1 \leq r \leq R_2$, then $Q(r) = 0$ for $0 \leq r \leq R_1$, and hence for these distances from the centre

$$\varphi(r) = -\frac{Q_0}{R_2} + \int_{R_1}^{R_2} \frac{Q(r_1)\,dr_1}{r_1^2} \, ,$$

i.e. the potential does not depend on r, it is a constant inside the layer. Correspondingly, the electric field,

$$\mathbf{E} = -\nabla\varphi = -\frac{d\varphi}{dr}\mathbf{e}_r \, ,$$

is zero. Above, Eq. (7.95) was used for the gradient in the spherical coordinates.

Problem 7.34 *Show that the above result can also be written in the form:*

$$\varphi(r) = -\frac{2Q_0}{R} + \frac{Q(r)}{r} + \int_r^R 4\pi r_1 \rho\,(r_1)\,dr_1 \, .$$

[*Hint:* use the explicit definition of $Q(r)$ as an integral in Eq. (*7.122*) and then change the order of integration.]

Problem 7.35 *Consider a uniformly charged sphere of radius R and charge Q_0. Show that in this case inside the sphere*

$$\varphi(r) = -\frac{Q_0}{2R}\left[1 + \left(\frac{r}{R}\right)^2\right] \, .$$

7.13.2 Classical Mechanics of a Particle

Suppose, we would like to solve Newton's equations of motion for an electron in an external field which is conveniently given in some curvilinear coordinates (q_1, q_2, q_3), i.e. the force is $\mathbf{F}(q_1, q_2, q_3)$. Of course, everything can be formulated in the conventional Cartesian system, but that might be extremely (and unnecessarily!) complicated. Instead, if the equation of motion,

$$m\frac{d^2\mathbf{r}}{dt^2} = \mathbf{F} \, , \tag{7.123}$$

is rewritten in the same curvilinear system, it may result in a drastic simplification and even possibly solved.

What we need to do is to transform both sides of Eq. (7.123) into a general curvilinear coordinate system. Since the force is assumed to be already written in this system,

$$\mathbf{F} = \sum_{i=1}^{3} F_i \mathbf{e}_i \, , \tag{7.124}$$

it remains to calculate the acceleration $d^2\mathbf{r}/dt^2$ in these curvilinear coordinates. The starting point is Eq. (7.24) for a displacement $d\mathbf{r}$ of the particle due to small changes $dq_i = \dot{q}_i dt$ of its curvilinear coordinates over the time interval dt, when the particle moved from point P (q_1, q_2, q_3) to point P' $(q_1 + dq_1, q_2 + dq_2, q_3 + dq_3)$:

$$d\mathbf{r} = \sum_i h_i dq_i \mathbf{e}_i = \sum_i h_i \dot{q}_i \mathbf{e}_i dt \, . \tag{7.125}$$

Hence, the velocity is

$$\mathbf{v} = \frac{d\mathbf{r}}{dt} = \sum_i h_i \dot{q}_i \mathbf{e}_i \, , \tag{7.126}$$

where the dot above q_i means its time derivative. The acceleration entering Newton's equations (7.123) is the time derivative of the velocity:

$$\mathbf{a} = \frac{d\mathbf{v}}{dt} = \sum_i \left[\mathbf{e}_i \frac{d}{dt} (h_i \dot{q}_i) + h_i \dot{q}_i \frac{d\mathbf{e}_i}{dt} \right] \, . \tag{7.127}$$

The second term within the brackets is generally not zero as the unit base vectors may change their direction as the particle moves. For instance, if the particle moves along a circular trajectory around the z axis and we use the spherical system, the unit base vectors $\mathbf{e}_r$, $\mathbf{e}_\theta$ and $\mathbf{e}_\phi$ will all change in time.

To calculate the time derivative of the unit base vectors, we recall the general relationship (7.10) between the unit base vectors of the curvilinear system in question and the Cartesian vectors $\mathbf{e}_i^0$. The idea here is that the Cartesian vectors do *not* change in time with the motion of the particle, they always keep their direction. Therefore,

$$\frac{d\mathbf{e}_i}{dt} = \frac{d}{dt} \left(\sum_{j=1}^{3} m_{ij} \mathbf{e}_j^0 \right) = \sum_{j=1}^{3} \frac{dm_{ij}}{dt} \mathbf{e}_j^0 \, .$$

Here elements of the matrix $\mathbf{M} = (m_{ij})$ are some functions of the curvilinear coordinates $(q_1(t), q_2(t), q_3(t))$, and hence the derivatives of m_{ij} are expressed via time derivatives of the coordinates. Expressing back the Cartesian vectors $\mathbf{e}_i^0$ via the unit base vectors $\mathbf{e}_i$ using the inverse matrix $\mathbf{M}^{-1}$, we arrive at

$$\frac{d\mathbf{e}_i}{dt} = \sum_{j=1}^{3} \frac{dm_{ij}}{dt} \mathbf{e}_j^0 = \sum_{j=1}^{3} \frac{dm_{ij}}{dt} \sum_{k=1}^{3} \left(\mathbf{M}^{-1}\right)_{jk} \mathbf{e}_k = \sum_{k=1}^{3} d_{ik}\mathbf{e}_k, \qquad (7.128)$$

where we introduced elements d_{ik} of a matrix $\mathbf{D}$ that are given by

$$d_{ik} = \sum_{j=1}^{3} \frac{dm_{ij}}{dt} \left(\mathbf{M}^{-1}\right)_{jk} \implies \mathbf{D} = \frac{d\mathbf{M}}{dt}\mathbf{M}^{-1}. \qquad (7.129)$$

Substituting Eq. (7.128) into (7.127), we obtain an equation for the acceleration $\mathbf{a}$ in the given curvilinear system expressed via the coordinates and their time derivatives.

Finally, note that if the force is initially given in the Cartesian coordinates,

$$\mathbf{F}(x, y, z) = F_x(x, y, z)\mathbf{i} + F_y(x, y, z)\mathbf{j} + F_z(x, y, z)\mathbf{k}$$
$$= F_x\mathbf{e}_1^0 + F_y\mathbf{e}_2^0 + F_z\mathbf{e}_3^0,$$

it can always be transformed into the preferred curvilinear system using the transformation relations (7.1) and the relationship (7.11) between the Cartesian $\mathbf{e}_i^0$ and the curvilinear $\mathbf{e}_i$ unit base vectors. As a result, the force takes on the form (7.124).

As an example, let us work out explicit equations for the velocity and acceleration for the cylindrical system (r, ϕ, z). We start by writing the velocity (recall that $h_r = 1$, $h_\phi = r$ and $h_z = 1$). Using Eq. (7.126), we immediately obtain

$$\mathbf{v} = \dot{r}\mathbf{e}_r + r\dot{\phi}\mathbf{e}_\phi + \dot{z}\mathbf{e}_z. \qquad (7.130)$$

To calculate the acceleration, we need to work out the elements of the matrix $\mathbf{D}$, Eq. (7.129). Since for the cylindrical system the matrices $\mathbf{M}$ and $\mathbf{M}^{-1}$ are given by Eqs. (7.17) and (7.18), respectively, we obtain

$$\mathbf{D} = \frac{d\mathbf{M}}{dt}\mathbf{M}^{-1} = \left[\frac{d}{dt}\begin{pmatrix} \cos\phi & \sin\phi & 0 \\ -\sin\phi & \cos\phi & 0 \\ 0 & 0 & 1 \end{pmatrix}\right]\begin{pmatrix} \cos\phi & -\sin\phi & 0 \\ \sin\phi & \cos\phi & 0 \\ 0 & 0 & 1 \end{pmatrix}$$

$$= \begin{pmatrix} -\dot{\phi}\sin\phi & \dot{\phi}\cos\phi & 0 \\ -\dot{\phi}\cos\phi & -\dot{\phi}\sin\phi & 0 \\ 0 & 0 & 0 \end{pmatrix}\begin{pmatrix} \cos\phi & -\sin\phi & 0 \\ \sin\phi & \cos\phi & 0 \\ 0 & 0 & 1 \end{pmatrix}$$

$$= \begin{pmatrix} 0 & \dot{\phi} & 0 \\ -\dot{\phi} & 0 & 0 \\ 0 & 0 & 0 \end{pmatrix},$$

so that

$$\frac{d\mathbf{e}_r}{dt} = d_{12}\mathbf{e}_2 = \dot{\phi}\mathbf{e}_\phi, \quad \frac{d\mathbf{e}_\phi}{dt} = d_{21}\mathbf{e}_1 = -\dot{\phi}\mathbf{e}_r \quad \text{and} \quad \frac{d\mathbf{e}_z}{dt} = 0. \qquad (7.131)$$

Therefore, the acceleration (7.127) becomes

$$\mathbf{a} = \left[\mathbf{e}_r \frac{d}{dt}\left(h_r \dot{r}\right) + h_r \dot{r}\frac{d\mathbf{e}_r}{dt}\right] + \left[\mathbf{e}_\phi \frac{d}{dt}\left(h_\phi \dot\phi\right) + h_\phi \dot\phi \frac{d\mathbf{e}_\phi}{dt}\right]$$

$$+ \left[\mathbf{e}_z \frac{d}{dt}\left(h_z \dot{z}\right) + h_z \dot{z}\frac{d\mathbf{e}_z}{dt}\right]$$

$$= \left(\ddot{r}\mathbf{e}_r + \dot{r}\dot{\mathbf{e}}_r\right) + \left[\frac{d}{dt}\left(r\dot\phi\right)\mathbf{e}_\phi + r\dot\phi\dot{\mathbf{e}}_\phi\right] + \ddot{z}\mathbf{e}_z .$$

Using expressions (7.131) for the derivatives of the unit base vectors obtained above, we can write the final expression:

$$\mathbf{a} = \left(\ddot{r} - r\dot\phi^2\right)\mathbf{e}_r + \left(2\dot{r}\dot\phi + r\ddot\phi\right)\mathbf{e}_\phi + \ddot{z}\mathbf{e}_z . \tag{7.132}$$

Let us also calculate the vector of the angular momentum $\mathbf{L} = \mathbf{r}\times\mathbf{p}$ of a particle at position $\mathbf{r}$ and with momentum $\mathbf{p} = m\mathbf{v}$ in the cylindrical coordinate system. Using Eq. (7.6), $\mathbf{L}$ is written via a 3×3 determinant using vectors $\mathbf{r} = r\mathbf{e}_r + z\mathbf{e}_z$ and $\mathbf{v}$ from Eq. (7.130):

$$\mathbf{L} = m\begin{vmatrix} \mathbf{e}_r & \mathbf{e}_\phi & \mathbf{e}_z \\ r & 0 & z \\ \dot{r} & r\dot\phi & \dot{z} \end{vmatrix}$$

$$= m\left[-rz\dot\phi\mathbf{e}_r - (r\dot{z} - z\dot{r})\mathbf{e}_\phi + r^2\dot\phi\mathbf{e}_z\right] .$$

For a particle moving within the $x-y$ plane ($z=0$) we have simply $\mathbf{L} = mr^2\dot\phi\mathbf{e}_z$.

Problem 7.36 *Show that for the spherical system the derivatives of its unit base vectors are*

$$\dot{\mathbf{e}}_r = \dot\theta\mathbf{e}_\theta + \dot\phi\sin\theta\mathbf{e}_\phi , \quad \dot{\mathbf{e}}_\theta = -\dot\theta\mathbf{e}_r + \dot\phi\cos\theta\mathbf{e}_\phi , \quad \dot{\mathbf{e}}_\phi = -\dot\phi\left(\sin\theta\mathbf{e}_r + \cos\theta\mathbf{e}_\theta\right) . \tag{7.133}$$

Correspondingly, the velocity and acceleration in this system are

$$\mathbf{v} = \dot{r}\mathbf{e}_r + r\dot\theta\mathbf{e}_\theta + r\dot\phi\sin\theta\mathbf{e}_\phi , \tag{7.134}$$

$$\mathbf{a} = \left(\ddot{r} - r\dot\theta^2 - r\dot\phi^2\sin^2\theta\right)\mathbf{e}_r + \left(2\dot{r}\dot\theta + r\ddot\theta - r\dot\phi^2\sin\theta\cos\theta\right)\mathbf{e}_\theta$$
$$+ \left(2\dot{r}\dot\phi\sin\theta + r\ddot\phi\sin\theta + 2r\dot\theta\dot\phi\cos\theta\right)\mathbf{e}_\phi . \tag{7.135}$$

Problem 7.37 *Prove the following relations for the parabolic cylinder system, Problem 7.3:*

$$\dot{\mathbf{e}}_u = \frac{u\dot{v} - v\dot{u}}{u^2 + v^2}\mathbf{e}_v , \quad \dot{\mathbf{e}}_v = \frac{-u\dot{v} + v\dot{u}}{u^2 + v^2}\mathbf{e}_u , \quad \dot{\mathbf{e}}_z = 0 .$$

Then show that the velocity and acceleration of a particle in these coordinates are

$$\mathbf{v} = \sqrt{u^2 + v^2}\,(\dot{u}\mathbf{e}_u + \dot{v}\mathbf{e}_v) + \dot{z}\mathbf{e}_z \,,$$

$$\mathbf{a} = \frac{1}{\sqrt{u^2 + v^2}} \left\{ \left[u\left(\dot{u}^2 - \dot{v}^2\right) + \ddot{u}\left(u^2 + v^2\right) + 2v\dot{u}\dot{v} \right] \mathbf{e}_u \right.$$
$$\left. + \left[v\left(\dot{v}^2 - \dot{u}^2\right) + \ddot{v}\left(u^2 + v^2\right) + 2u\dot{u}\dot{v} \right] \mathbf{e}_v \right\} + \ddot{z}\mathbf{e}_z \,.$$

Problem 7.38 *If a particle moves within the $x - y$ plane (i.e. $z = 0$), the spherical coordinate system becomes identical to the polar one. Show then that the equations for the velocity (7.134) and acceleration (7.135) obtained in the spherical system coincide with those for the polar system, Eqs. (7.130) and (7.132).*

Problem 7.39 *Consider a particle of mass m moving under a central force $\mathbf{F} = F_r \mathbf{e}_r$, where $F_r = F(r)$ depends only on the distance r from the centre of the coordinate system.*

(i) Show that the equations of motion, $\mathbf{F} = m\mathbf{a}$, in this case, when projected onto the unit base vectors of the spherical system, have the form:

$$a_r = \ddot{r} - r\dot{\theta}^2 - r\dot{\phi}^2 \sin^2\theta = \frac{F_r}{m} \,, \quad a_\theta = a_\phi = 0 \,, \tag{7.136}$$

where a_r, a_θ and a_ϕ are components of the acceleration given in Eq. (7.135).

(ii) The angular momentum of the particle is defined via $\mathbf{L} = m\,[\mathbf{r} \times \mathbf{v}]$. Show that

$$\mathbf{L} = mr^2 \left(-\dot{\phi}\sin\theta\,\mathbf{e}_\theta + \dot{\theta}\mathbf{e}_\phi \right) \,. \tag{7.137}$$

(iii) Differentiate $\mathbf{L} = m\,[\mathbf{r} \times \mathbf{v}]$ with respect to time and, using equations of motion (7.136), to prove that $\dot{\mathbf{L}} = 0$, i.e. the angular momentum is conserved.

(iv) Choose the coordinate system such that it is oriented with the angular momentum directed along the z axis, i.e. $\mathbf{L} = L\mathbf{k}$. Express $\mathbf{k}$ via the unit base vectors of the spherical system and, comparing with Eq. (7.137), show that $\cos\theta = 0$ (and hence $\theta = \pi/2$ remaining the same during the whole trajectory) and

$$\dot{\phi} = \frac{L}{mr^2} \,. \tag{7.138}$$

Curvilinear Coordinates

777

Curvilinear Coordinates

Therefore, the particle moves within the $x - y$ plane performing a two-dimensional motion. Note that $r = r(t)$ in Eq. (7.138) and the angle $\phi(t)$ does not change its sign along the whole trajectory. For instance if $L > 0$, then $\dot{\phi} > 0$ during the whole motion, i.e. the particle performs a rotation around the centre with its angle ϕ advancing all the time.

(v) Show that the total energy of the particle,

$$E = \frac{m}{2}\left(\dot{r}^2 + r^2\dot{\phi}^2\right) + U(r) = \frac{mv_r^2}{2} + U_{eff}(r) , \qquad (7.139)$$

where $v_r = \dot{r}$ is the radial component of the velocity, and $U_{eff}(r) = U(r) + L^2/2mr^2$ is the so-called effective potential energy of the particle, while $U(r)$ is the potential energy of the field itself, $F(r) = -dU/dr$.

(vi) Therefore, establish the following equation for the distance $r(t)$ to the centre:

$$\ddot{r} - \frac{L^2}{m^2 r^3} = \frac{F(r)}{m} \implies m\ddot{r} = -\frac{dU_{eff}}{dr} . \qquad (7.140)$$

This radial equation corresponds to a one-dimensional motion under the effective potential $U_{eff}(r)$, which is also justified by the energy expression (7.139).

(vii) Then prove that the energy is conserved, $\dot{E} = 0$.

(viii) Choosing r as a new argument in Eq. (7.140) (instead of time) and integrating (cf. Sect. I.8.3), show that the equation for the radial velocity v_r can be written as

$$v_r = \sqrt{\frac{2}{m}\left[E - U_{ff}(r)\right]} . \qquad (7.141)$$

Note that this formula is equivalent to the energy expression (7.139). The above result gives an equation for the velocity v_r as a function of r. Integrating it, obtain an equation for $r(t)$:

$$\int_{r_0}^{r} \frac{dr}{\sqrt{\frac{2}{m}\left[E - U_{eff}(r)\right]}} = t , \qquad (7.142)$$

where $r_0 = r(0)$ is the initial distance to the centre.

(ix) Finally, using the fact that $\dot{r} = \dot{\phi}\,(dr/d\phi)$ and setting $\phi(0) = 0$, obtain the equation for the trajectory $r(\phi)$:

$$\int_{r_0}^{r} \frac{dr}{r^2\sqrt{\frac{2}{m}\left[E - U_{eff}(r)\right]}} = \frac{m}{L}\phi , \qquad (7.143)$$

Fig. 7.8 A boat on a river, Problem 7.40. Its velocity is composed of two components: one is **V** due to the river flow, and another **v**, due to the boat engine, that is directed towards point B along the whole trajectory

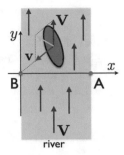

river

Problem 7.40 *A boat crosses a river of width a with a constant speed $v = |\mathbf{v}|$, see Fig. 7.8. Assume that the water in the river flows with a constant velocity V at any point across its width. The boat starts at a point A at one bank of the river and along the whole journey its velocity $\mathbf{v}$ is kept directed towards the point B which is exactly opposite the starting point at the opposite bank. Using the polar system (r, ϕ) with the centre of the coordinate system placed at point B, the x axis pointing towards point A and the y axis chosen along the river flow, show that the trajectory $r(\phi)$ of the boat can be written as*

$$r = \frac{a}{\cos\phi}\left(\frac{1 + \sin\phi}{1 - \sin\phi}\right)^{-v/2V}.$$

7.13.3 Distribution Function of a Set of Particles

Consider a set of n particles of mass m_p $(p = 1, \ldots, n)$ moving with velocities $\mathbf{v}_p = \left(v_{px}, v_{py}, v_{pz}\right)$ each in a box. This could for example be atoms in a liquid or in a solid which in their movement follow Newtonian equations of motion (this type of simulations is called molecular dynamics in material science). It is sometimes useful to know what is the probability distribution of the total kinetic energy of the particles, that is the probability $dP = f(E)dE$ of finding the particles in the system with their kinetic energy between E and $E + dE$.

The total kinetic energy of the system is

$$E_{KE} = \sum_{p=1}^{n}\sum_{\alpha} \frac{m_p v_{p\alpha}^2}{2} = \sum_{i=1}^{3n} \frac{m_i v_i^2}{2},$$

where $\alpha = x, y, z$ corresponds to the Cartesian components of the velocity of each particle, and in the second passage we used the index $i = (p, \alpha)$ to designate all degrees of freedom of the system ($i = 1, \ldots, 3n$) to simplify our notations.

The probability distribution for the degree of freedom 1 to have its velocity between v_1 and $v_1 + dv_1$, for the degree of freedom 2 to have its velocity between v_2 and $v_2 + dv_2$, and so on, is given by Maxwell's distribution:

$$dP_v = \frac{1}{Z} e^{-\beta E_{KE}} dv_1 dv_2 \cdots dv_N ,$$

where $\beta = 1/k_B T$ is the inverse temperature and $N = 3n$ is the total number of degrees of freedom in the whole system. Z is the normalisation constant.

Now, let us find a probability $P_E dE$ for the whole system to have its total kinetic energy to be within the interval between E and $E + dE$. This can be obtained by calculating the integral

$$P_E = \frac{1}{Z} \int_{-\infty}^{\infty} \cdots \int_{-\infty}^{\infty} \delta \left(E - \sum_{i=1}^{N} \frac{m_i v_i^2}{2} \right) \exp \left(-\beta \sum_{i=1}^{N} \frac{m_i v_i^2}{2} \right) dv_1 dv_2 \cdots dv_N .$$

Indeed, because of the delta function, only such combinations of the velocities will be accounted for which correspond to the total kinetic energy equal exactly to E.

Problem 7.41 *The total probability of finding particles velocities somewhere between $-\infty$ and $+\infty$ is one; it is given by summing up all probabilities for all possible values of their velocities, i.e. integrating dP_v with respect to all velocities:*

$$\int_{-\infty}^{\infty} \cdots \int_{-\infty}^{\infty} \frac{1}{Z} \exp \left(-\beta \sum_{i=1}^{N} \frac{m_i v_i^2}{2} \right) dv_1 dv_2 \cdots dv_N = 1 .$$

Perform the integration and show that the normalisation constant

$$Z = \left(\frac{2\pi}{\beta} \right)^{N/2} \prod_i \frac{1}{\sqrt{m_i}} .$$

Then, using results of Sect. 7.8 and the properties of the delta function, show that

$$P_E = \frac{1}{\Gamma\left(\frac{N}{2}\right)} E^{N/2-1} e^{-E} . \tag{7.144}$$

Note that this distribution does not depend on temperature.

Problem 7.42 *Let us derive the distribution P_E in a different way. Consider the probability*

$$P^{\leq}(E) = \frac{1}{Z} \int \cdots \int_{E_{KE} \leq E} \exp\left(-\beta \sum_{i=1}^{N} \frac{m_i v_i^2}{2}\right) dv_1 dv_2 \cdots dv_N$$

of finding the system with its kinetic energy E_{KE} being not bigger than E. Using again ideas borrowed from Sect. 7.8 on the $N-$dimensional sphere, show that

$$P^{\leq}(E) = \frac{2}{\Gamma\left(\frac{3n}{2}\right)} \int_0^{\sqrt{E}} e^{-r^2} r^{N-1} dr \ .$$

Since $P_E dE = P^{\leq}(E + dE) - P^{\leq}(E)$, derive Eq. (7.144) from the above formula for $P^{\leq}(E)$.

Problem 7.43 *Show that the distribution P_E is properly normalised, i.e. $\int_0^{\infty} P_E dE = 1$.*

Chapter 8
Partial Differential Equations of Mathematical Physics

If an unknown function of several variables and its partial derivatives are combined in an equation, the latter is called a *partial differential equation* (PDE). Several times above we have come across such PDEs when dealing with functions of several variables. For instance, in Sect. I.6.7.1[1] we derived the continuity equation for the particle density which contains its partial derivatives with respect to time t and the spatial coordinates x, y and z. In Sect. I.6.9.1 the so-called wave equation

$$\frac{1}{c^2}\frac{\partial^2 \Psi}{\partial t^2} = \frac{\partial^2 \Psi}{\partial x^2} + \frac{\partial^2 \Psi}{\partial y^2} + \frac{\partial^2 \Psi}{\partial z^2} \quad \text{or} \quad \frac{1}{c^2}\frac{\partial^2 \Psi}{\partial t^2} = \Delta \Psi \tag{8.1}$$

was derived for components $\Psi(x, y, z, t)$ of the electric and magnetic fields which they satisfy in free space, while a heat transport equation

$$\frac{1}{D}\frac{\partial \Psi}{\partial t} = \frac{\partial^2 \Psi}{\partial x^2} + \frac{\partial^2 \Psi}{\partial y^2} + \frac{\partial^2 \Psi}{\partial z^2} \quad \text{or} \quad \frac{1}{D}\frac{\partial \Psi}{\partial t} = \Delta \Psi \tag{8.2}$$

was derived in Sect. I.6.9.2 for a distribution of temperature $\Psi(x, y, z, t)$ in a system. An identical equation (but with the constant D having a different physical meaning) is satisfied by the particles density, $\Psi(x, y, z, t)$, in the mass transport (i.e. diffusion) problem. Both Eqs. (8.1) and (8.2) are somewhat similar, but note that the diffusion equation contains the first-order derivative of the unknown function Ψ with respect to time while the second-order time derivative enters the wave equation.

---

[1] In the following, references to the first volume of this course (L. Kantorovich, Mathematics for natural scientists: fundamentals and basics, 2nd Edition, Springer, 2022) will be made by appending the Roman number I in front of the reference, e.g. Sect. I.1.8 or Eq. (I.5.18) refer to Sect. 1.8 and Eq. (5.18) of the first volume, respectively.

© The Author(s), under exclusive license to Springer Nature Switzerland AG 2024
L. Kantorovich, *Mathematics for Natural Scientists II*, Undergraduate Lecture Notes in Physics, https://doi.org/10.1007/978-3-031-46320-4_8

In the *stationary* case when e.g. the distribution of temperature (or density) across the system stopped changing with time, the time derivative $\partial\Psi/\partial t = 0$ and one arrives at the Laplace equation

$$\Delta\Psi = 0 \, . \tag{8.3}$$

For instance, the 1D variant of this equation, $\partial^2\Psi/\partial x^2 = 0$, describes at long times the distribution of temperature in a rod when its both ends are kept at two fixed temperatures. The Laplace equation is also encountered in other physical problems. For instance, it is satisfied by an electrostatic potential in regions of space where there are no charges.

Of course, the wave, diffusion and Laplace equations do not exhaust all possible types of PDEs which are encountered in solving physical problems; however, in this chapter we shall limit ourselves to discussing only these.

8.1 General Consideration

8.1.1 Characterisation of Second-Order PDEs

The PDE is characterised by its *order* (the highest order of the partial derivatives) and whether it is *linear* or not (i.e. whether the unknown function appears only to the first degree anywhere in the equation, either on its own or when differentiated). If an additional function of the variables appears as a separate term in the equation, it is called inhomogeneous; otherwise, the PDE is homogeneous. It follows then, that both diffusion and wave equations are linear, homogeneous and of the second order. At the same time, the PDE

$$\frac{\partial^2\psi}{\partial x^2} + \frac{\partial^2\psi}{\partial y^2} = ae^{\psi}$$

is non-linear, although also of the second order and homogeneous, while the PDE

$$F(x,t,z,t) + \frac{1}{D}\frac{\partial\Psi}{\partial t} = \frac{\partial^2\Psi}{\partial x^2} + \frac{\partial^2\Psi}{\partial y^2} + \frac{\partial^2\Psi}{\partial z^2} \, ,$$

which contains an additional function $F(x,y,z,t)$, is inhomogeneous and linear, and still of the second order.

Let us have a closer look at a general second-order linear PDE with constant coefficients. For simplicity, we shall assume that all the coefficients are real; a generalisation to complex coefficients is actually straightforward. Let the unknown function of n variables $\mathbf{X} = (x_i) = (x_1, x_2, \ldots, x_n)$ be $\Psi(x_1, x_2, \ldots, x_n) = \Psi(\mathbf{X})$. Then the general form of such an equation reads

$$\sum_{i,j=1}^{n} a_{ij} \Psi_{x_i x_j} + \sum_{i=1}^{n} b_i \Psi_{x_i} + c\Psi + f(\mathbf{X}) = 0 , \qquad (8.4)$$

where we have used simplified notations for the partial derivatives: $\Psi_{x_i x_j} = \partial^2 \Psi / \partial x_i \partial x_j$ and $\Psi_{x_i} = \partial \Psi / \partial x_i$. The first term in the above equation contains all second-order derivatives, while the second term all first-order derivatives. Because the mixed second order derivatives are symmetric, $\Psi_{x_i x_j} = \Psi_{x_j x_i}$, the square $n \times n$ matrix $\mathbf{A} = (a_{ij})$ of the coefficients[2] to the second derivatives can always be chosen symmetric, $a_{ij} = a_{ji}$. Indeed, a pair of terms $a_{ij} \Psi_{x_i x_j} + a_{ji} \Psi_{x_j x_i}$ with $a_{ij} \neq a_{ji}$ can always be written as $\widetilde{a}_{ij} \Psi_{x_i x_j} + \widetilde{a}_{ji} \Psi_{x_j x_i}$ with $\widetilde{a}_{ij} = \frac{1}{2}(a_{ij} + a_{ji}) = \widetilde{a}_{ji}$. Note also that one of the coefficients to the second derivative of the unknown function, a_{ij} or a_{ji}, could be zero; still, we can always split it equally into two, i.e. $\widetilde{a}_{ij} = \frac{1}{2} a_{ij} = \widetilde{a}_{ji}$, to have the matrix $\mathbf{A}$ of the coefficients of $\Psi_{x_i x_j}$ in Eq. (8.4) symmetric. The coefficients b_i to the first-order derivatives form an n-dimensional vector $\mathbf{B} = (b_i)$. We assume that the matrix $\mathbf{A} = (a_{ij})$, vector $\mathbf{B}$ and the scalar c are constants, i..e they do not depend on the variables $\mathbf{X}$. The PDE may also contain a general function $f(\mathbf{X})$. As was mentioned above, if this function is not present, the PDE is homogeneous; otherwise, it is inhomogeneous.

We shall now show that the PDE (8.4) can always be transformed into a *canonical form* which does not have mixed second derivatives. Then the transformed PDE can be *characterised* into several types. This is important as the method of solution depends on the type of the PDE as will be clarified in later sections.

The transformation of the PDE (8.4) into the canonical form is made using a change of variables $\mathbf{X} \to \mathbf{Y} = (y_i)$ by means of a *linear transformation*:

$$y_i = \sum_{j=1}^{n} u_{ij} x_j \quad \text{or} \quad \mathbf{Y} = \mathbf{U}\mathbf{X} ,$$

where $\mathbf{U} = (u_{ij})$ is a yet unknown square $n \times n$ transformation matrix. To determine $\mathbf{U}$, let us rewrite our PDE via the new coordinates. We have

$$\Psi_{x_i} = \frac{\partial \Psi}{\partial x_i} = \sum_{k=1}^{n} \frac{\partial \Psi}{\partial y_k} \frac{\partial y_k}{\partial x_i} = \sum_{k=1}^{n} \frac{\partial \Psi}{\partial y_k} u_{ki} = \sum_{k=1}^{n} u_{ki} \Psi_{y_k} ,$$

---

[2] Following our notations of Chap. 1, we shall use bold Capital letters to designate vectors and matrices, and the corresponding small letters to designate their components.

$$\Psi_{x_j x_i} = \frac{\partial}{\partial x_j} \Psi_{x_i} = \frac{\partial}{\partial x_j} \left(\sum_{k=1}^{n} \frac{\partial \Psi}{\partial y_k} u_{ki} \right)$$

$$= \sum_{l=1}^{n} \frac{\partial}{\partial y_l} \left(\sum_{k=1}^{n} \frac{\partial \Psi}{\partial y_k} u_{ki} \right) \frac{\partial y_l}{\partial x_j} = \sum_{l,k=1}^{n} \frac{\partial^2 \Psi}{\partial y_l \partial y_k} u_{lj} u_{ki}$$

$$= \sum_{l,k=1}^{n} u_{ki} \Psi_{y_k y_l} u_{lj} \, .$$

Substituting these derivatives into the PDE (8.4), we obtain

$$\sum_{i,j=1}^{n} a_{ij} \left(\sum_{l,k=1}^{n} u_{ki} \Psi_{y_k y_l} u_{lj} \right) + \sum_{i=1}^{n} b_i \left(\sum_{k=1}^{n} u_{ki} \Psi_{y_k} \right) + c\Psi + f = 0 \, ,$$

or

$$\sum_{l,k=1}^{n} a'_{kl} \Psi_{y_k y_l} + \sum_{k=1}^{n} b'_k \Psi_{y_k} + c\Psi + f_1(\mathbf{Y}) = 0 \, , \tag{8.5}$$

where we introduced a new matrix

$$\mathbf{A}' = \left(a'_{kl} \right) \, , \quad \text{where} \quad a'_{kl} = \sum_{i,j=1}^{n} u_{ki} a_{ij} u_{lj} \quad \text{or} \quad \mathbf{A}' = \mathbf{U} \mathbf{A} \mathbf{U}^T \, , \tag{8.6}$$

a new vector

$$\mathbf{B}' = \left(b'_k \right) \, , \quad \text{where} \quad b'_k = \sum_{i=1}^{n} u_{ki} b_i \quad \text{or} \quad \mathbf{B}' = \mathbf{U} \mathbf{B} \, , \tag{8.7}$$

and a new function $f_1(\mathbf{Y}) = f \left(\mathbf{U}^{-1} \mathbf{Y} \right)$.

To eliminate mixed second-order derivatives of Ψ, one has to choose the transformation matrix $\mathbf{U}$ in such a way that the matrix $\mathbf{A}'$ be diagonal. Since the matrix $\mathbf{A}$ is symmetric, this can always be done by choosing $\mathbf{U}$ to be the modal matrix of $\mathbf{A}$ (see Sect. 1.9.1): if eigenvalues of $\mathbf{A}$ are numbers $\lambda_1, \lambda_2, \ldots, \lambda_n$ (which are real since $\mathbf{A}$ is symmetric, Sect. 1.8.4), with the corresponding eigenvectors $\mathcal{B}_1, \mathcal{B}_2, \ldots, \mathcal{B}_n$ (so that $\mathbf{A}\mathcal{B}_i = \lambda_i \mathcal{B}_i$), then one can define the matrix $\mathbf{U} = (\mathcal{B}_1 \mathcal{B}_2 \cdots \mathcal{B}_n)$ consisting of the eigenvectors of $\mathbf{A}$ as its column. Hence, the matrix $\mathbf{A}' = \mathbf{U} \mathbf{A} \mathbf{U}^T = \left(\delta_{ij} \lambda_i \right)$ has the diagonal form with the eigenvalues of $\mathbf{A}$ standing on its diagonal in the same order as the corresponding eigenvectors in $\mathbf{U}$. Therefore, after this choice, PDE (8.5) transforms into:

$$\sum_{k=1}^{n} \lambda_k \Psi_{y_k y_k} + \sum_{k=1}^{n} b'_k \Psi_{y_k} + c\Psi + f_1(\mathbf{Y}) = 0 \, . \tag{8.8}$$

This PDE does not have mixed derivatives anymore, only the diagonal second-order derivatives $\Psi_{y_k y_k} = \partial^2 \Psi / \partial y_k^2$ are present. This finalises the transformation into the canonical form.

Now we are ready to introduce the characterisation scheme. It is based entirely on the values of the eigenvalues $\{\lambda_k\}$. The original PDE (8.4) is said to be of *elliptic type* if all eigenvalues of **A** are of the same sign (e.g. all positive). It is of *hyperbolic type* if all eigenvalues, *apart from precisely one*, are of the same sign. Finally, the PDE is said to be of *parabolic type* if at least one $\Psi_{y_k y_k}$ term is missing which happens if the corresponding eigenvalue λ_k is zero. For functions of more than three variables ($n \geq 4$) intermediate cases are also possible when more than one eigenvalue λ_k has a different sign to the others, or more than one eigenvalue is zero. We shall not consider those cases here.

It follows from these definitions that the wave equation (8.1) is of the hyperbolic type, the Laplace equation (8.3) is elliptic, while the diffusion equation (8.2) is parabolic. Indeed, in all these three cases the equations already have the canonical form. Then, in the case of the wave equation

$$-\frac{1}{c^2} \frac{\partial^2 \Psi}{\partial t^2} + \frac{\partial^2 \Psi}{\partial x^2} + \frac{\partial^2 \Psi}{\partial y^2} + \frac{\partial^2 \Psi}{\partial z^2} = 0, \qquad (8.9)$$

one coefficient (by the second-order time derivative) is of the opposite sign to the coefficients of the spatial derivatives which means that the PDE is hyperbolic. The heat transport equation does not have the second-order time derivative term and hence is parabolic. Finally, the Laplace equation has all coefficients to the second-order derivatives equal to unity and hence is elliptic.

The PDE (8.8) can be simplified even further: it appears that it is also possible, in some cases, to eliminate the terms with the first derivatives. To do this, we introduce a new function $\Phi(\mathbf{Y})$ via

$$\Psi(\mathbf{Y}) = \exp\left(\sum_{i=1}^{n} \gamma_i y_i\right) \Phi(\mathbf{Y}), \qquad (8.10)$$

where γ_i are new parameters to be determined. These are chosen in such a way as to eliminate terms with first-order derivatives. We have

$$\Psi_{y_k} = \left(\Phi_{y_k} + \gamma_k \Phi\right) \exp\left(\sum_{i=1}^{n} \gamma_i y_i\right),$$

$$\Psi_{y_k y_k} = \left(\Phi_{y_k y_k} + 2\gamma_k \Phi_{y_k} + \gamma_k^2 \Phi\right) \exp\left(\sum_{i=1}^{n} \gamma_i y_i\right),$$

so that Eq. (8.8) is manipulated into the following PDE with respect to the new function $\Phi(\mathbf{Y})$:

$$\sum_{k=1}^{n}\lambda_k\Phi_{y_k y_k} + \sum_{k=1}^{n}\left(2\lambda_k\gamma_k + b_k'\right)\Phi_{y_k}$$

$$+ \left[c + \sum_{k}\left(\lambda_k\gamma_k + b_k'\right)\gamma_k\right]\Phi + f_2(\mathbf{Y}) = 0 , \tag{8.11}$$

where

$$f_2(\mathbf{Y}) = \exp\left(-\sum_{i=1}^{n}\gamma_i y_i\right)f_1(\mathbf{Y}) . \tag{8.12}$$

We now consider a non-zero eigenvalue λ_k. It is seen that choosing $\gamma_k = -b_k'/2\lambda_k$ eliminates the term with Φ_{y_k}. If all eigenvalues are non-zero, then the PDE (8.11) takes on the form:

$$\sum_{k=1}^{n}\lambda_k\Phi_{y_k y_k} + c'\Phi + f_2(\mathbf{Y}) = 0 , \tag{8.13}$$

where

$$c' = c - \sum_{k=1}^{n}\frac{\left(b_k'\right)^2}{4\lambda_k} .$$

Hence, all first-order derivatives disappeared. If some of the eigenvalues $\lambda_k = 0$, then the corresponding linear term Φ_{y_k} cannot be eliminated.

The two-step transformation procedure considered above shows, firstly, that the second-order derivatives term can always be diagonalised (i.e. all mixed derivatives can be eliminated), and, secondly, that the first-order derivatives terms can also be eliminated if the corresponding eigenvalues of the matrix $\mathbf{A}$ of the coefficients to the second derivatives are non-zero (the complete corresponding second order derivative term after the diagonalisation is present).

Problem 8.1 Consider the following PDE for the function $\Psi(x_1, x_2)$:

$$a\Psi_{x_1 x_1} + 2b\Psi_{x_1 x_2} + a\Psi_{x_2 x_2} + b\Psi_{x_1} + c\Psi_{x_2} + \Psi = 0 ,$$

assuming that $a \neq \pm b$. Show that this PDE can be transformed into the following canonical form:

$$(a+b)\Phi_{y_1 y_1} + (a-b)\Phi_{y_2 y_2} + \gamma\Phi = 0 ,$$

where

$$\gamma = 1 - \frac{(b+c)^2}{8(a+b)} - \frac{(b-c)^2}{8(a-b)} ,$$

and the new variables are related to the old ones via

$$\begin{pmatrix} y_1 \\ y_2 \end{pmatrix} = \frac{1}{\sqrt{2}} \begin{pmatrix} 1 & 1 \\ 1 & -1 \end{pmatrix} \begin{pmatrix} x_1 \\ x_2 \end{pmatrix} ,$$

and

$$\Psi(y_1, y_2) = \Phi(y_1, y_2) \exp\left(-\frac{b+c}{2\sqrt{2}(a+b)} y_1 - \frac{b-c}{2\sqrt{2}(a-b)} y_2 \right) .$$

Argue that the PDE is elliptic if and only if $a^2 > b^2$, and it is hyperbolic if $a^2 < b^2$. The PDE cannot be parabolic as this would require $a^2 = b^2$, which is not possible due to the restriction on the coefficients in this problem.

Problem 8.2 Consider a function $\Psi(x, y)$ satisfying the following PDE:

$$a_{11}\Psi_{xx} + 2a_{12}\Psi_{xy} + a_{22}\Psi_{yy} + b_1\Psi_x + b_2\Psi_y + c\Psi = 0 , \qquad (8.14)$$

where the coefficients a_{ij} to the second derivatives are real numbers. Show that this PDE is elliptic if and only if $a_{11}a_{22} > a_{12}^2$, it is parabolic if $a_{11}a_{22} = a_{12}^2$ and is hyperbolic if $a_{11}a_{22} < a_{12}^2$. Verify explicitly that the eigenvalues λ_1 and λ_2 are always real.

Problem 8.3 Consider the PDE:

$$\Psi_{x_1 x_1} + 2\Psi_{x_1 x_2} + \alpha\Psi_{x_2 x_2} = 0 .$$

Show that this PDE is elliptic for $\alpha > 1$, hyperbolic for $\alpha < 1$ and parabolic for $\alpha = 1$.

Problem 8.4 Show that the PDE

$$4\Psi_{x_1 x_2} + \Psi_{x_3 x_3} + \Psi_{x_1} - \Psi_{x_3} + 3\Psi = 0$$

for the function $\Psi(x_1, x_2, x_3)$ can be simplified as follows:

$$2\left(\Phi_{y_1 y_1} - \Phi_{y_2 y_2}\right) + \Phi_{y_3 y_3} + \frac{11}{4}\Phi = 0 ,$$

where the transformation matrix between the original $\mathbf{X}$ and new $\mathbf{Y}$ variables

$$\mathbf{U} = \begin{pmatrix} 1/\sqrt{2} & 1/\sqrt{2} & 0 \\ 1/\sqrt{2} & -1/\sqrt{2} & 0 \\ 0 & 0 & 1 \end{pmatrix}$$

and

$$\Psi(y_1, y_2, y_3) = \Phi(y_1, y_2, y_3) \exp\left(-\frac{1}{4\sqrt{2}}(y_1 - y_2) + \frac{y_3}{2}\right).$$

This manipulation shows that the original PDE is hyperbolic.

Problem 8.5 Show that the PDE

$$2\Psi_{xx} + 4\Psi_{xy} + \Psi_{yy} = 0$$

is hyperbolic.

Problem 8.6 Consider a function $\Psi(x, y)$ satisfying the PDE (8.14). Show that there exists a linear transformation to new variables,

$$\begin{pmatrix} \xi_1 \\ \xi_2 \end{pmatrix} = \mathbf{U} \begin{pmatrix} x \\ y \end{pmatrix} = \begin{pmatrix} u_{11} & u_{12} \\ u_{21} & u_{22} \end{pmatrix} \begin{pmatrix} x \\ y \end{pmatrix},$$

which transforms this PDE into a form containing only the mixed derivative:

$$a'_{12}\Psi_{\xi_1\xi_2} + b'_1\Psi_{\xi_1} + b'_2\Psi_{\xi_2} + c\Psi = 0.$$

Show that in order to eliminate the terms with diagonal double derivatives, $\Psi_{\xi_1\xi_1}$ and $\Psi_{\xi_2\xi_2}$, the ratios u_{12}/u_{11} and u_{22}/u_{21} must be *different* roots of the quadratic equation

$$a_{22}\lambda^2 + 2a_{12}\lambda + a_{11} = 0 \tag{8.15}$$

with respect to λ. Why does the condition on the roots being different guarantee that the determinant of $\mathbf{U}$ is not zero?

Problem 8.7 As a simple application of the previous problem, show that the PDE

$$2\Psi_{x_1x_1} + 3\Psi_{x_1x_2} + \Psi_{x_2x_2} = 0$$

is equivalent to the PDE $\Psi_{y_1y_2} = 0$, where the new variables can be chosen as $y_1 = x_1 - 2x_2$ and $y_2 = x_1 - x_2$.

Problem 8.8 Here we shall return back to Problem 8.6. Specifically, we shall consider a PDE which results in the roots of Eq. (8.15) being equal:

$$\Psi_{x_1 x_1} + 4\Psi_{x_1 x_2} + 4\Psi_{x_2 x_2} = 0 .$$

Show that in this case it is not possible to find a linear transformation, $\mathbf{Y} = \mathbf{UX}$, such that in the new variables the PDE would contain only the single term with the mixed derivative, $\Psi_{y_1 y_2} = 0$. Instead, show that the PDE can be transformed into its canonical form containing (in this case) only a single double derivative, $\Psi_{y_2 y_2} = 0$, where the new variables may be chosen, e.g. as $y_1 = 2x_1 + x_2$ and $y_2 = -x_1 + 2x_2$.

Problem 8.9 A small transverse displacement $\Psi(x, t)$ of an incompressible fluid contained in a flexible tube of negligible viscosity flowing along it in direction x is described by the PDE

$$\Psi_{tt} + A\Psi_{xt} + B\Psi_{xx} = 0 ,$$

where A and B are some constants, and we assume that $A^2 > 4B$. Find the transformation of variables, $(x, t) \rightarrow (x_1, x_2)$ with $x_1 = x + \gamma_1 t$ and $x_2 = x + \gamma_2 t$, in which the PDE reads $\Psi_{x_1 x_2} = 0$. *[Answer:* $\gamma_{1,2} = -\frac{1}{2}\left(A \pm \sqrt{A^2 - 4B}\right).]$

Problem 8.10 Show that the 1D wave equation, $\Psi_{xx} = \frac{1}{c^2}\Psi_{tt}$, is invariant with respect to the (relativistic) Lorentz transformation of coordinates:

$$x' = \gamma(x - vt) , \quad t' = \gamma\left(t - \frac{v}{c^2}x\right) , \quad \gamma = \left(1 - v^2/c^2\right)^{-1/2} ,$$

i.e. in the new (primed) coordinates the equation has an identical form: $\Psi_{x'x'} = \frac{1}{c^2}\Psi_{t't'}$. Here the non-primed variables, (x, t), correspond to the position and time in a laboratory coordinate system, while the primed variables, (x', t'), correspond to a coordinate system moving with velocity v with respect to the laboratory system along the positive direction of the x axis. Next show that if the non-relativistic Galilean transformation,

$$x' = x - vt , \quad t' = t ,$$

is applied, the wave equation does change its form, i.e. it is not invariant with respect to this transformation.

Concluding, we mention that, if desired, one may rescale all or some of the variables $y_k \rightarrow z_k = y_k/\sqrt{|\lambda_k|}$ as an additional step. Of course, this can only be done for those variables y_k for which $\lambda_k \neq 0$; one do not need to do rescaling for those y_k for which λ_k is zero since in these case the second derivative term is missing anyway. This additional transformation leads to the corresponding coefficients to the second derivatives being $\lambda_k/\sqrt{|\lambda_k|} = \pm 1$, i.e. just plus or minus one.

8.1.2 Initial and Boundary Conditions

We know that when solving an ordinary (one-dimensional) DE, we first obtain a general solution which contains one or more arbitrary constants. Then, using initial conditions which provide information about the function we are seeking at a single point $x = x_0$ (the value of the function for the first-order DE or the values of the function and its derivatives at this point in the case of higher order DE), the constant(s) are determined yielding the final particular solution of the DE.

In the case of PDEs there are more than one variable involved, and we also need to know something additional about the function we are seeking in order to determine it. What kind of such additional information is usually available? When solving physics problems in which spatial coordinates are involved, normally one knows the values of the unknown function at a boundary of a region of interest. For instance, in a heat transport problem across a 1D rod we are usually given the temperatures at its both ends and are seeking the temperature distribution at all internal points along the rod in time. This type of additional conditions is called *boundary conditions*. The boundary conditions supply the necessary information on the function of interest associated with its spatial variables. If, as is the case for the wave or diffusion equations, the time is also involved, then usually we know the whole function (and may be its time derivative) at the initial ($t = 0$) time and are interested in determining its evolution at later times. This type of additional conditions is called *initial conditions* and is analogous to the case of an ordinary DE.

As an example, consider oscillations of a string of length L stretched along the x axis between the points $x = 0$ and $x = L$. As we shall see below, the vertical displacement $u(x, t)$ of the point of the string with coordinate x ($0 \le x \le L$) satisfies the wave equation

$$\frac{1}{c^2} \frac{\partial^2 u}{\partial t^2} = \frac{\partial^2 u}{\partial x^2} .$$

Therefore, alongside the initial conditions,

$$u(x, 0) = \phi_1(x) \quad \text{and} \quad \left. \frac{\partial u}{\partial t} \right|_{t=0} = \phi_2(x) , \tag{8.16}$$

giving values of the unknown function and its first-time derivative at all values of the spatial variable x, one has to supply the boundary conditions,

$$u(0, t) = \varphi_1(t) \quad \text{and} \quad u(L, t) = \varphi_2(t) , \tag{8.17}$$

as well. The boundary conditions establish the values of the function at the edge (boundary) points $x = 0$ and $x = L$ of the string at all times. Note that the boundary conditions may include instead a derivative with respect to x at one or both of these points, or some linear combination of the various types of terms.

Also note that since the wave equation is of the second order with respect to time, both the function $u(x, 0)$ and its first-time derivative at $t = 0$ are required to be specified in the initial conditions (8.16). In the case of the diffusion equation where only the first order time derivative is used, only the value of the unknown function at $t = 0$ need to be given; the first derivative cannot be given in addition as this complementary condition may be contradictory. This is because the first-time derivative enters the PDE itself, and hence must be equal, at all times, to the rest of the equation; it cannot be set to a given function.

This idea of the boundary conditions is easily generalised to two- and three-dimensional spatial PDEs where the values of the unknown functions are to be specified at all times at the boundary of a spatial region of interest. For instance, in the case of oscillations of a circular membrane of radius R, one has to specify as the boundary conditions the displacement $u(x, y, t)$ of all boundary points $x^2 + y^2 = R^2$ of the membrane at all times t.

What kind of initial and boundary conditions are necessary to specify in each case to guarantee that a solution of the given PDE can uniquely be found? This depends on the PDE in question so that this highly important question has to be considered individually for each type of the PDE. It will be discussed below only specifically for the wave and the diffusion equations.

8.2 Wave Equation

In Sect. I.6.9 the wave equation was derived for the electric and magnetic fields. However, it is encountered in many other physical problems and hence has a very general physical significance. To stress this point, it is instructive, before discussing methods of solution of the wave PDEs, consider two other problems in which the same wave equation appears: oscillations of a string and sound propagation in a condensed media (liquid or gas). This is what we are going to do in the next two subsections.

8.2.1 One-Dimensional String

Consider a string which lies along the x axis and is subjected to a tension T_0 in the same direction. At equilibrium the string will be stretched along the x axis. If we apply a perpendicular external force F (per unit length) and/or take the string out of its equilibrium position and then release, it will start oscillating vertically, see Fig. 8.1. Let the vertical displacement of a point with coordinate x be $u(x, t)$; note that the vertical displacement is a function of both x and the time, t.

Let us consider a small element AB of the string, with the point A being at x and the point B at $x + dx$, see Fig. 8.1. The total force acting on this element is due to two tensions applied at points A and B which work in the (approximately) opposite

Fig. 8.1 An oscillating string. The tangent direction to the string at point A with the coordinate x is indicated by a vector which makes the angle α with the x axis. Here $F(x)$ is a force (per unit length) applied to the string at point x (it may also depend on time)

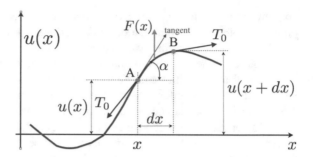

directions, and an (optional) external force $F(x)$, which is applied in the vertical direction. We assume that the oscillations are small (i.e. the vertical displacement $u(x, t)$ is small compared to the string length). This means that the tensions may be assumed to be nearly horizontal (although in the figure vertical components of the forces are greatly amplified for clarity), and cancel each other in this direction (if they did not, then the string would move in the lateral direction!). Therefore, we should only care about the balance of the forces in the vertical direction.

Let $\alpha(x)$ be the angle the tangent line to the string makes with the x axis at the point x, as shown in the figure. Then, the difference in the heights between points B and A can be calculated as

$$u(x + dx) - u(x) = \frac{\partial u}{\partial x}dx = \tan \alpha(x)\, dx \simeq \alpha(x)dx \quad \Longrightarrow \quad \frac{\partial u}{\partial x} \simeq \alpha(x)\, ,$$

because for small oscillations the angle α is small and therefore $\tan \alpha \simeq \sin \alpha \simeq \alpha$. Here the time was omitted for convenience (we consider the difference in height at the given time t). We stress that the angle $\alpha(x)$ depends on the point at which the tangent line is drawn, i.e. it is a function of x.

Then, the vertical component of the tension force acting downwards at the left point A is

$$- T_0 \sin \alpha(x) \simeq -T_0 \alpha(x) = -T_0 \left(\frac{\partial u}{\partial x}\right)_A .$$

On the other hand, the vertical component of the tension applied at the point B is similarly

$$T_0 \sin [\alpha(x + dx)] \simeq T_0 \alpha(x + dx) = T_0 \left(\frac{\partial u}{\partial x}\right)_B ,$$

but this time the partial derivative of the displacement is calculated at the point B. Therefore, the net force acting on the element dx in the vertical direction will be

$$F dx + T_0 \left(\frac{\partial u}{\partial x}\right)_B - T_0 \left(\frac{\partial u}{\partial x}\right)_A = F dx + T_0 \left[\left(\frac{\partial u}{\partial x}\right)_B - \left(\frac{\partial u}{\partial x}\right)_A\right] .$$

The expression in the square brackets gives a change of the function $f(x) = \partial u/\partial x$ between two points B and A which are separated by dx, hence one can write

$$f(x + dx) - f(x) = \frac{\partial f}{\partial x}dx = \frac{\partial^2 u}{\partial x^2}dx \,,$$

so that the total force acting on the element dx of the string in the vertical direction becomes

$$\left(F + T_0\frac{\partial^2 u}{\partial x^2}\right)dx \,.$$

On the other hand, due to Newton's equation of motion, this force should be equal to the product of the mass ρdx of the piece of the string of length dx and its vertical acceleration, $\partial^2 u/\partial t^2$. Therefore, equating the two expressions and cancelling out on dx, the final equation of motion for the string is obtained

$$\frac{1}{c^2}\frac{\partial^2 u}{\partial t^2} = G(x) + \frac{\partial^2 u}{\partial x^2} \,, \tag{8.18}$$

where $G(x) = F(x)/T_0$ and the velocity $c = \sqrt{T_0/\rho}$. As one can see, this PDE is in general inhomogeneous. When the external force F is absent, we arrive at the familiar (homogeneous) wave equation with the velocity c.

8.2.2 Propagation of Sound

Consider a propagation of sound in a gas of density ρ, e.g. in the air. As the gas is perturbed, this creates density fluctuations in it which propagate in space and time. Corresponding to these fluctuations, any little gas volume $dV = d\mathbf{r}$ can be assigned a velocity $\mathbf{v}(\mathbf{r})$. The latter must satisfy the hydrodynamic equation of motion (I.6.156):

$$\frac{\partial \mathbf{v}}{\partial t} + \mathbf{v} \cdot \mathrm{grad}\mathbf{v} = \mathbf{F} - \frac{1}{\rho}\mathrm{grad}\, P \,,$$

where $\mathbf{F}$ is a vector field of external forces (e.g. gravity) and P is the local pressure. We assume that the velocity field $\mathbf{v}(\mathbf{r})$ changes very little in space, and hence we can neglect the $\mathbf{v} \cdot \mathrm{grad}\mathbf{v}$ term in the above equation, leading to

$$\frac{\partial \mathbf{v}}{\partial t} = \mathbf{F} - \frac{1}{\rho}\mathrm{grad}\, P \,. \tag{8.19}$$

This equation has to be supplemented by the continuity equation (I.6.102),

$$\frac{\partial \rho}{\partial t} + \rho\, \mathrm{div}\, \mathbf{v} + \mathbf{v} \cdot \mathrm{grad}\, \rho = 0 \,,$$

in which one can neglect the $\mathbf{v} \cdot \text{grad}\rho$ term containing a product of two small terms: a small velocity and a small variation of the density. Hence,

$$\frac{\partial \rho}{\partial t} + \rho \, \text{div} \, \mathbf{v} = 0 \; .$$

Let ρ_0 be the density of the gas at equilibrium. Then, the density of the gas during the sound propagation (when the system is out of equilibrium) can be written via $\rho = \rho_0 (1 + s)$, where $s(\mathbf{r}) = (\rho - \rho_0)/\rho_0$ is the relative fluctuation of the gas density which is considered much smaller than unity. Then, $d\rho = \rho_0 ds$ and therefore

$$\frac{\partial \rho}{\partial t} = \rho_0 \frac{\partial s}{\partial t} \quad \Longrightarrow \quad \rho_0 \frac{\partial s}{\partial t} + \rho \, \text{div} \, \mathbf{v} = 0 \; . \tag{8.20}$$

Also, in the continuity equation we can replace ρ with ρ_0 since

$$\rho \, \text{div} \, \mathbf{v} = \rho_0 (1 + s) \, \text{div} \, \mathbf{v} = \rho_0 \text{div} \, \mathbf{v} + \rho_0 s \, \text{div} \, \mathbf{v} \simeq \rho_0 \text{div} \, \mathbf{v} \; ,$$

where we dropped the second-order term containing a product of two small terms (s and a small variation of the velocity) to be consistent with the previous approximations. Hence, the continuity equation takes on a simpler form:

$$\frac{\partial s}{\partial t} + \text{div} \, \mathbf{v} = 0 \; . \tag{8.21}$$

Finally, we have to relate the pressure to the density. In an ideal gas the sound propagation is an adiabatic process in which the pressure P is proportional to ρ^γ, where $\gamma = c_P/c_V$ is the ratio of heat capacities for the constant pressure and volume. Therefore, if P_0 is the pressure at equilibrium, then

$$\frac{P}{P_0} = \left(\frac{\rho}{\rho_0}\right)^\gamma \quad \Longrightarrow \quad \frac{P}{P_0} = (1 + s)^\gamma \simeq 1 + \gamma s$$

$$\Longrightarrow \quad \text{grad} \, P = \gamma P_0 \, \text{grad} \, s \; .$$

Substituting this into Eq. (8.19) and replacing ρ with ρ_0 there (which corresponds again to keeping only the first-order terms), we get

$$\frac{\partial \mathbf{v}}{\partial t} = \mathbf{F} - c^2 \text{grad} \, s \; , \tag{8.22}$$

where $c = \sqrt{\gamma P_0/\rho_0}$. Next, we take the divergence of both sides of this equation. The left-hand side then becomes

$$\text{div}\left(\frac{\partial \mathbf{v}}{\partial t}\right) = \frac{\partial}{\partial t} (\text{div} \, \mathbf{v}) = -\frac{\partial^2 s}{\partial t^2} \; ,$$

where we have used Eq. (8.21) at the last step. The divergence of the right-hand side of Eq. (8.22) is worked out as follows:

$$\operatorname{div}\left(\mathbf{F} - c^2 \operatorname{grad} s\right) = \operatorname{div}\mathbf{F} - c^2 \operatorname{div} \operatorname{grad} s = \operatorname{div}\mathbf{F} - c^2 \Delta s \ ,$$

i.e. it contains the Laplacian of s. Equating the left and the right-hand sides, we finally obtain the equation sought for

$$\frac{\partial^2 s}{\partial t^2} = c^2 \Delta s - \operatorname{div}\mathbf{F} \ . \tag{8.23}$$

If the external forces are absent, this equation turns into a familiar wave equation in 3D space with the constant c being the corresponding sound velocity:

$$\frac{1}{c^2}\frac{\partial^2 s}{\partial t^2} = \Delta s \ . \tag{8.24}$$

8.2.3   General Solution of PDE

When solving an ordinary DE, we normally first obtain its general solution, which contains arbitrary constants, and then, by imposing corresponding initial conditions, a particular integral of the DE is obtained so that both the DE and the initial conditions are satisfied. The case of PDEs is more complex as a function with more than one variable is to be sought. However, in most cases one can provide some analogy with the 1D case of an ordinary DE: if in the latter case the general solution contains arbitrary constants, the general solution of a PDE contains *arbitrary functions*. Then, a particular integral of the PDE is obtained by finding the particular functions so that both initial and boundary conditions are satisfied, if present.

To illustrate this point, let us consider a specific problem of string oscillations considered in Sect. 8.2.1. We shall assume that the string is stretched along the x axis and is of an *infinite length*, i.e. $-\infty < x < \infty$. The condition of the string being infinite simplifies the problem considerably as one can completely ignore any boundary conditions. Then, only initial conditions remain.

Therefore, the whole problem can be formulated as follows. The vertical displacement $\Psi(x, t)$ of the point with coordinate x must be a solution of the wave equation:

$$\frac{\partial^2 \Psi}{\partial x^2} = \frac{1}{c^2}\frac{\partial^2 \Psi}{\partial t^2} \ , \tag{8.25}$$

subject to the *initial conditions*

$$\Psi(x, 0) = \phi_1(x) \quad \text{and} \quad \Psi_t(x, 0) = \left.\frac{\partial \Psi}{\partial t}\right|_{t=0} = \phi_2(x) \ . \tag{8.26}$$

The wave equation is of hyperbolic type. As follows from Problem 8.6, in some cases it is possible to find such a linear transformation of the variables $(x, t) \Longrightarrow (\eta, \xi)$ that the PDE would only contain a mixed derivative.

Problem 8.11 Show that the new variables (η, ξ), in which the PDE (8.25) has the form

$$\frac{\partial^2 \Psi}{\partial \eta \partial \xi} = 0 \, , \tag{8.27}$$

are related to the old ones via:

$$\eta = x + ct \quad \text{and} \quad \xi = x - ct \, . \tag{8.28}$$

[Hint: recall Problem 8.6.]

The obtained PDE (8.27) can be easily integrated. Indeed,

$$\frac{\partial^2 \Psi}{\partial \eta \partial \xi} = \frac{\partial}{\partial \eta} \left(\frac{\partial \Psi}{\partial \xi} \right) = 0 \quad \Longrightarrow \quad \frac{\partial \Psi}{\partial \xi} = C \left(\xi \right) \, ,$$

where $C \left(\xi \right)$ is an arbitrary function of the variable ξ. Integrating the latter equation, we obtain

$$\frac{\partial \Psi}{\partial \xi} = C \left(\xi \right) \quad \Longrightarrow \quad \Psi \left(\eta, \xi \right) = \int C \left(\xi \right) d\xi + u \left(\eta \right) = v \left(\xi \right) + u \left(\eta \right) \, ,$$

where $u \left(\eta \right)$ must be another, also arbitrary, function of the other variable η. Above, $v \left(\xi \right)$ is also an arbitrary function of ξ since it is obtained by integrating an arbitrary function $C \left(\xi \right)$. Recalling what the new variables actually are, Eq. (8.28), we immediately arrive at the following *general solution* of the PDE (8.25):

$$\Psi(x, t) = v(x - ct) + u(x + ct) \, . \tag{8.29}$$

So, the general solution appears to be a sum of two *arbitrary functions* of the variables $x \pm ct$. Note that this result is general; in particular, we have not used the fact that the string is of infinite length. However, applying boundary conditions for a string of a final length directly to the general solution (8.29) is non-trivial, so we shall not consider this case here; another method will be considered instead later on in Sect. 8.2.5.

Before applying the initial conditions (8.26) to obtain the formal solution for the particular integral of our wave equation of the infinite string, it is instructive first to illustrate the meaning of the obtained general solution. We start by analysing the function $v \left(x - ct \right)$. It is sketched for two times t_1 and $t_2 > t_1$ as a function of x in Fig. 8.2. It is easy to see that the profile (or shape) of the function remains the same

Fig. 8.2 A sketch of the function $v(x - ct)$ at two times t_1 and $t_2 > t_1$

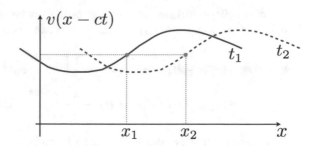

at later times; the whole function simply shifts to the right, i.e. the wave propagates without any distortion of its shape. This can be seen in the following way. Consider a point x_1. At time t_1 the function has some value $v_1 = v(x_1 - xt_1)$. At time $t_2 > t_1$ the function becomes $v(x - ct_2)$; at this later time it will reach the same value v_1 at some point x_2 if the latter satisfies the following condition:

$$x_1 - ct_1 = x_2 - ct_2 \implies x_2 = x_1 + c(t_2 - t_1) ,$$

which immediately shows that the function as the whole shifts to larger values of x since $t_2 > t_1$. This is also shown in Fig. 8.2. It is seen that the function shifts exactly by the distance $\Delta x = c(t_2 - t_1)$ over the time interval $\Delta t = t_2 - t_1$. Therefore, the first part of the solution (8.29), $v(x - ct)$, describes propagation of the wave shape $v(x)$ with velocity c to the right. Similarly it is verified that the second part, $u(x + ct)$, of the solution (8.29) describes the propagation of the wave shape $u(x)$ to the left with the same velocity.

Now, let us find yet unknown functions $u(x)$ and $v(x)$ by satisfying the initial conditions (8.26). Applying the particular form of the general solution (8.29), the initial conditions read

$$v(x) + u(x) = \phi_1(x) \quad \text{and} \quad c\left[-\frac{dv(x)}{dx} + \frac{du(x)}{dx} \right] = \phi_2(x) . \tag{8.30}$$

Integrating the second equation between, say, zero and x, we obtain

$$-[v(x) - v(0)] + [u(x) - u(0)] = \frac{1}{c} \int_0^x \phi_2(\xi) \, d\xi ,$$

which, when combined with the first equation in (8.30), allows solving for both functions:

$$u(x) = \frac{1}{2}\left[A + \phi_1(x) + \frac{1}{c} \int_0^x \phi_2(\xi) \, d\xi \right] ,$$

$$v(x) = \frac{1}{2}\left[\phi_1(x) - A - \frac{1}{c} \int_0^x \phi_2(\xi) \, d\xi \right] ,$$

where $A = u(0) - v(0)$ is a constant. Interestingly, when substituting these functions into the general solution (8.29), the constant A cancels out, and we obtain

$$\Psi(x, t) = \frac{1}{2} [\phi_1(x - ct) + \phi_1(x + ct)] + \frac{1}{c} \left[\int_0^{x+ct} \phi_2(\xi) \, d\xi - \int_0^{x-ct} \phi_2(\xi) \, d\xi \right].$$

$$= \frac{1}{2} [\phi_1(x - ct) + \phi_1(x + ct)] + \frac{1}{2c} \int_{x-ct}^{x+ct} \phi_2(\xi) \, d\xi.$$

(8.31)

This solution is known as the d'Alembert's formula. It gives the full solution of the problem given by Eqs. (8.25) and (8.26) for an infinite string.

As was already mentioned, it is possible to generalise the described method so that solution of the wave equation of a string of a finite length L ($0 \le x \le L$), with boundary conditions explicitly given is possible. However, as was remarked before, the method becomes very complicated and will not be considered here since simpler methods exist. One of such methods, the method due to Fourier, based on separation of variables will be considered in detail in the following sections.

Problem 8.12 Explain why the solution (8.31) conforms to the general form of Eq. (8.29).

Problem 8.13 Show that the general solution of the PDE of Problem 8.7 can be written as

$$\Psi(x_1, x_2) = u(x_1 - 2x_2) + v(x_1 - x_2),$$

where $u(x)$ and $v(x)$ are two arbitrary functions.

Problem 8.14 Show that the general solution of the PDE of Problem 8.8 can be written as

$$\Psi(x_1, x_2) = u(2x_1 + x_2) + v(-x_1 + 2x_2),$$

where $u(x)$ and $v(x)$ are arbitrary functions.

Problem 8.15 Consider propagation of a spherically symmetric wave in 3D space. This problem is described by the wave equation (7.119) in which the dependence of the function Ψ on the angles θ and ϕ is completely ignored:

$$\frac{1}{r^2} \frac{\partial}{\partial r} \left(r^2 \frac{\partial \Psi}{\partial r} \right) = \frac{1}{c^2} \frac{\partial^2 \Psi}{\partial t^2}.$$

(8.32)

By introducing a new function $\Phi(r, t)$ such that $\Psi(r, t) = r^\alpha \Phi(r, t)$, show that by taking $\alpha = -1$ the PDE for Φ will be of the form $\Phi_{rr} = \frac{1}{c^2} \Phi_{tt}$. Correspondingly, the general solution of the problem (8.32) can then be written as

$$\Psi(r,t) = \frac{1}{r}\left[v\left(r - ct\right) + u\left(r + ct\right)\right].$$

This solution describes the propagation of spherical waves from (the first term) and to (the second) the centre of the coordinate system ($r = 0$). The attenuation factor $1/r$ corresponds to a decay of the wave amplitude with the distance r to the centre. Since the energy of the wave front is proportional to the square of the amplitude, the energy decays as $1/r^2$. Since the area of the wave front increases as $4\pi r^2$, the decay of wave's amplitude ensures that wave's total energy is conserved.

Problem 8.16 Similarly, consider a wave propagation in the cylindrical symmetry, Eq. (7.118). In this case the wave equation reads (ignoring the angle ϕ and the coordinate z)

$$\frac{1}{r}\frac{\partial}{\partial r}\left(r\frac{\partial \Psi}{\partial r}\right) = \frac{1}{c^2}\frac{\partial^2 \Psi}{\partial t^2}. \tag{8.33}$$

Show that with an appropriate choice of the power α in the transformation $\Psi(r,t) = r^\alpha \Phi(r,t)$, the wave equation for the new function $\Phi(r,t)$ can be manipulated into the following form:

$$\Phi_{rr} + \frac{1}{4r^2}\Phi = \frac{1}{c^2}\Phi_{tt}.$$

Problem 8.17 Consider Problem 8.9 again. Show that the general solution of the PDE is

$$\Psi(x,t) = u\left(x + \gamma_1 t\right) + v\left(x + \gamma_2 t\right),$$

where $u(x)$ and $v(x)$ are two arbitrary functions. Then, assume that the fluid in the tube was initially at rest and had a small transverse displacement $\Psi = \Psi_0 \cos(kx)$. Show that the subsequent motion of the fluid is given by

$$\Psi(x,t) = \frac{\Psi_0}{\gamma_1 - \gamma_2}\left[\gamma_1 \cos\left(k\left(x + \gamma_2 t\right)\right) - \gamma_2 \cos\left(k\left(x + \gamma_1 t\right)\right)\right].$$

Problem 8.18 Show that the general solution of the PDE

$$(y - z)\frac{\partial f}{\partial x} + 2(z - x)\frac{\partial f}{\partial y} - (x - y)\frac{\partial f}{\partial z} = 0$$

is an arbitrary function of $2x + y - 2z$ and $2x^2 + y^2 - 2z^2$.

8.2.4 Uniqueness of Solution

In the following, we shall study oscillations of a string of a finite length L with the coordinate x being within the interval $0 \le x \le L$. Two types of the boundary conditions at each end of the string we shall consider: when either the function $\Psi(x, t)$ is specified at that value of x (either $x = 0$ or $x = L$), or its x−derivative.

For instance, when the $x = L$ end of the string is free (not fixed), the tension force at this end is zero and hence the x−derivative must be zero, i.e. the boundary condition at this end is $\Psi_x(L, t) = 0$. At the same time, at the other end ($x = 0$) the string is fixed, i.e. at this end the boundary condition must be $\Psi(0, t) = 0$. This simple example shows, that at one end one type of the boundary conditions can be specified, e.g. formulated for the function, while at the other end another, e.g. formulated for its derivative. Altogether, four possibilities of the boundary conditions exist, two of them are distinct. The initial conditions are given in the usual way by Eq. (8.16):

$$\Psi(x, 0) = \phi_1(x) \quad \text{and} \quad \left. \frac{\partial \Psi}{\partial t} \right|_{t=0} = \phi_2(x) . \tag{8.34}$$

What we are interested in here is to demonstrate that by specifying initial and any of the four types of the boundary conditions a unique solution of the wave equation

$$\frac{\partial^2 \Psi}{\partial t^2} = c^2 \frac{\partial^2 \Psi}{\partial x^2} + f(x, t) \tag{8.35}$$

is guaranteed. Here $f(x, t)$ is a function of an 'external force' acting on the string at point x. That force makes the PDE inhomogeneous.

Proving by contradiction, let us assume that two different solutions, $\Psi^{(1)}(x, t)$ and $\Psi^{(2)}(x, t)$, exist which satisfy the same PDE,

$$\Psi_{tt}^{(1)} = c^2 \Psi_{xx}^{(1)} + f(x, t) , \quad \Psi_{tt}^{(2)} = c^2 \Psi_{xx}^{(2)} + f(x, t) , \tag{8.36}$$

and the same initial and boundary conditions. Consider then their difference

$$\Psi(x, t) = \Psi^{(1)}(x, t) - \Psi^{(2)}(x, t) .$$

Obviously, it satisfies the homogeneous wave equation, $\Psi_{tt} = c^2 \Psi_{xx}$ (without the force term f) and the zero initial,

$$\Psi(x, 0) = 0 \quad \text{and} \quad \left. \frac{\partial \Psi}{\partial t} \right|_{t=0} = 0 , \tag{8.37}$$

and boundary,

$$\Psi(0, t) = 0 \quad \text{or} \quad \Psi_x(0, t) = 0 \tag{8.38}$$

and

$$\Psi(L, t) = 0 \quad \text{or} \quad \Psi_x(L, t) = 0 , \tag{8.39}$$

conditions. Next, we shall consider an auxiliary function of time

$$E(t) = \frac{1}{2} \int_0^L \left[(\Psi_x)^2 + \frac{1}{c^2} (\Psi_t)^2 \right] dx . \tag{8.40}$$

Let us differentiate this function with respect to time:

$$\frac{dE}{dt} = \int_0^L \Psi_x \Psi_{xt} dx + \int_0^L \frac{1}{c^2} \Psi_t \Psi_{tt} dx .$$

The first term we shall integrate by parts:

$$\int_0^L \Psi_x (\Psi_{xt} dx) = (\Psi_x \Psi_t)|_{x=L} - (\Psi_x \Psi_t)_{x=0} - \int_0^L \Psi_{xx} \Psi_t dx . \tag{8.41}$$

Consider the free term, $\Psi_x \Psi_t$, which is calculated at $x = 0$. At this end of the string the boundary condition states that either Ψ equals to zero at all times or Ψ_x does. If the latter is true, then we immediately see that the free term is zero. In the former case differentiation of the condition $\Psi = 0$ at $x = 0$ with respect to time gives $\Psi_t = 0$, which again results in the corresponding free term being zero. So, the free term, $(\Psi_x \Psi_t)_{x=0}$, is zero in either case. Similarly, the other free term, calculated at $x = L$ is also zero. So, for all four combinations of the boundary conditions both free terms are zero and hence only the integral term remains in the right-hand side in Eq. (8.41). Therefore,

$$\frac{dE}{dt} = \int_0^L \Psi_t \left(-\Psi_{xx} + \frac{1}{c^2} \Psi_{tt} \right) dx = 0 ,$$

as the expression in the brackets is zero because of the PDE for Ψ itself. Therefore, E must be a constant in time. Calculating this constant at $t = 0$,

$$E(0) = \frac{1}{2} \int_0^L \left[(\Psi_x(x, 0))^2 + \frac{1}{c^2} (\Psi_t(x, 0))^2 \right] dx ,$$

and employing the initial conditions, Eq. (8.37) and the fact that from $\Psi(x, 0) = 0$ immediately follows, upon differentiation, that $\Psi_x(x, 0) = 0$, we must conclude that $E(0) = 0$. In other words, the function $E(t) = 0$ at all times. On the other hand, the function $E(t)$ of Eq. (8.40) consists of the sum of two non-negative terms and hence can only be zero if and only if both these terms are equal to zero at the same time. This means that both $\Psi_x(x, t) = 0$ and $\Psi_t(x, t) = 0$ for any x and t. Since $\Psi_x = 0$, the function $\Psi(x, t)$ cannot depend on x. Similarly, since $\Psi_t = 0$, our function cannot depend on t. In other words, $\Psi(x, t)$ must be simply a constant. However, because of the initial conditions, $\Psi(x, 0) = 0$, so that this constant must be zero, i.e. $\Psi(x, t) = 0$, which contradicts our assumption of two different solutions being possible. Hence, it is wrong.

We have just proved, that specifying initial and any of the four types of the boundary conditions uniquely defines the solution of the problem $\Psi(x, t)$ sought for.

8.2.5   Fourier Method

As was mentioned above, it is possible to devise a method of solution of the wave equation based on the sum of two arbitrary functions, Eq. (8.29); however, it is rather complicated for finite systems when boundary conditions are to be specified. At the same time, a simpler method exists, called the Fourier method, which allows solving a wide range of problems related to the wave equation, and in this section we shall start developing essential ingredients of this method. Before considering a very general case of inhomogeneous wave equation with arbitrary initial and boundary conditions, it is instructive to start from a simple case of the homogeneous equation with trivial zero boundary conditions. Once we develop all essential ideas and techniques for this simpler problem, we shall then be able to consider the general case.

Consider the following 1D problem in which the wave equation,

$$\Psi_{xx} = \frac{1}{c^2} \Psi_{tt} \tag{8.42}$$

is to be solved with general initial,

$$\Psi(x, 0) = \phi_1(x) , \quad \Psi_t(x, 0) = \phi_2(x) , \tag{8.43}$$

and *zero boundary conditions*:

$$\Psi(0, t) = 0 \quad \text{and} \quad \Psi(L, t) = 0 . \tag{8.44}$$

To solve this problem, we shall first seek the solution in a very special form as a *product* of two functions,

$$\Psi(x, t) = X(x)T(t) , \tag{8.45}$$

where $X(x)$ is a function of only one variable x and $T(t)$ is a function of another variable t. This trial solution may seem to be too specific and hence may not serve as a solution of the whole problem (8.42)–(8.44); however, as will be seen below, we shall be able to construct *a linear combination* of such product solutions which will then satisfy our PDE together with the initial and boundary conditions. However, we shall build our solution gradually, step-by-step.

Substituting the product of the two functions, Eq. (8.45), into the PDE (8.42), gives

$$T\frac{d^2X}{dx^2} = \frac{1}{c^2}X\frac{d^2T}{dt^2} \quad \Longrightarrow \quad \frac{1}{X}\frac{d^2X}{dx^2} = \frac{1}{c^2T}\frac{d^2T}{dt^2} . \tag{8.46}$$

To get the form of the equation written to the right of the arrow, we divided both sides of the equation on the left of the arrow by the product XT. What we obtained is quite peculiar. Indeed, the left-hand side of the obtained equation, $\frac{1}{X}\frac{d^2X}{dx^2}$, is a function of x only, while the right-hand side, $\frac{1}{c^2T}\frac{d^2T}{dt^2}$, is a function of t only. One may think that this cannot possibly be true as this 'equality' must hold for all values of x and t. However, there is still exists one and only one possibility here, which would resolve this paradoxical situation: if both functions, $\frac{1}{X}\frac{d^2X}{dx^2}$ and $\frac{1}{c^2T}\frac{d^2T}{dt^2}$, are *constants*. Calling this constant K, we then must have

$$\frac{d^2X}{dx^2} = KX , \tag{8.47}$$

and

$$\frac{d^2T}{dt^2} = c^2KT , \tag{8.48}$$

which are two ordinary DEs for the functions $X(x)$ and $T(t)$, respectively. The constant K is called the *separation constant*. This is because in the right form in Eq. (8.46) the variables x and t has been separated. Therefore, the method we are discussing is called the *method of separation of variables*.

The constant K introduced above is yet unknown. However, available values of it can be determined if we impose the boundary conditions (8.44) on our product solution (8.45). Since in the product the function $T(t)$ does not depend on x, it is clear that the boundary conditions must be applied to the function $X(x)$ only. Hence, we should have

$$X(0) = X(L) = 0 . \tag{8.49}$$

Let us try to solve Eq. (8.47) subject to this boundary conditions to see which values of K are consistent with them. Three cases must be considered[3]: $K < 0$, $K > 0$ and $K = 0$.

1. When $K > 0$, we can write $K = p^2$ with $p > 0$. For this case the solution of (8.47) is a sum of two exponential functions:

$$X(x) = Ae^{px} + Be^{-px} . \tag{8.50}$$

If the boundary conditions (8.49) at $x = 0$ and $x = L$ are applied to Eq. (8.50), one finds

$$A + B = 0 \quad \text{and} \quad Ae^{pL} + Be^{-pL} = 0 .$$

---

[3] Obviously, the constant K must be real.

It is clear that there is only one solution of this system of two simultaneous algebraic equations which is the trivial solution: $A = 0$ *and* $B = 0$. This in turn leads to the *trivial* solution $X(x) = 0$, which results in the trivial trial product solution $\Psi(x, t) = 0$, and it is *not* of any physical interest!

2. When $K = 0$, we find that $d^2 X / dx^2 = 0$, which yields simply $X(x) = Ax + B$. This solution is also of no physical interest, because the boundary conditions at $x = 0$ and $x = L$ can only be satisfied by $A = 0$ and $B = 0$ leading again to the trivial solution.

3. What is left to consider is the last case of $K < 0$. This is more interesting! We write $K = -k^2$ with (without of loss of generality) some positive k (see below). In this case we obtain the harmonic oscillator equation

$$\frac{d^2 X}{dx^2} + k^2 X = 0 ,$$

which solution is $X(x) = A \sin(kx) + B \cos(kx)$. The application of the boundary conditions at $x = 0$ and $x = L$ to this solution results in two algebraic equations:

$$X(0) = B = 0 \quad \text{and} \quad X(L) = A \sin(kL) = 0 . \tag{8.51}$$

When solving these equations, we have to consider al possible cases. Choosing the constant $A = 0$ would again yield the trivial solution; therefore, this equation can only be satisfied if $\sin(kL) = 0$, which gives us a possibility to choose all possible values of k. These obviously are

$$k_n = \frac{n\pi}{L}, \quad n = 1, 2, \dots . \tag{8.52}$$

We obtained not one but an infinite number of possible solutions for k, which we distinguish by the subscript n. Note that $n \neq 0$ as this gives the zero value for k which we know is to be rejected as leading to the trivial solution. Also, negative values of n (and hence negative values of $k_n = k_{-|n|} = -\pi |n| / L = -k_{|n|}$) do not give anything new as these result in the same values of the separation constant $K_n = -k_n^2$ and to the solutions

$$X_{-|n|}(x) = A \sin\left(k_{-|n|} x\right) = -A \sin\left(k_{|n|} x\right) ,$$

which differ only by the sign from the solutions $X_{|n|}(x) = A \sin\left(k_{|n|} x\right)$ associated with the positive n. Hence, the choice of $n = 1, 2, 3, \dots$ in Eq. (8.52); no need to consider negative and zero n.

From the above analysis we see that the boundary conditions can only be satisfied when the separation constant K takes certain *discrete eigenvalues* $K_n = -k_n^2$, where $n = 1, 2, \dots$. This type of situation is of frequent occurrence in the theory of PDEs. In fact, we did come across this situation already in Sect. 4.7.2 when we considered a quantum-mechanical problem of a hydrogen atom.

Associated with the *eigenvalue* k_n we have an *eigenfunction*

$$X_n(x) = A_n \sin \frac{n\pi x}{L} , \quad n = 1, 2, \dots , \tag{8.53}$$

where A_n are some constants; these may be different for different n, so we distinguished them with the subscript n.

Next we have to solve the corresponding differential equation (8.48) for $T(t)$ with $K = -k_n^2$, which is again a harmonic oscillator equation:

$$\frac{d^2 T}{dt^2} + (k_n c)^2 T = 0$$

with the general solution

$$T_n(t) = D_n \sin(\omega_n t) + E_n \cos(\omega_n t) , \tag{8.54}$$

where

$$\omega_n = \frac{\pi c}{L} n , \quad n = 1, 2, \dots . \tag{8.55}$$

We can now collect our functions (8.53) and (8.54) to obtain the desired product solution (8.45):

$$\Psi_n(x, t) = \sin \frac{n\pi x}{L} \left[D_n' \sin(\omega_n t) + E_n' \cos(\omega_n t) \right] . \tag{8.56}$$

Here we combined products $A_n D_n$ and $A_n E_n$ of arbitrary constants into new arbitrary constants D_n' and E_n'. The function above, for any $n = 1, 2, \dots$, satisfies the PDE and the *boundary conditions* by construction. Note that we have obtained not one but an infinite number of such solutions $\Psi_n(x, t)$ labelled by the index n.

This is all good, but neither of the obtained solutions may satisfy the initial conditions (8.43). In order to overcome this difficulty, we remark that our PDE is linear, and hence any linear combination of the solutions will also be a solution. This is called the *superposition principle*. To understand this better, let us first rewrite our PDE in an *operator* form:

$$\Psi_{xx} - \frac{1}{c^2}\Psi_{tt} = 0 \quad \Longrightarrow \quad \left(\frac{\partial^2}{\partial x^2} - \frac{1}{c^2}\frac{\partial^2}{\partial t^2} \right) \Psi = 0 ,$$

or simply $\widehat{L}\Psi = 0$, where

$$\widehat{L} = \frac{\partial^2}{\partial x^2} - \frac{1}{c^2}\frac{\partial^2}{\partial t^2}$$

is the operator expression contained in the round brackets in the PDE above. It is easy to see that this operator is linear, i.e. for any numbers α and β and any two functions $\Phi_1(x, t)$ and $\Phi_2(x, t)$ we have

$$\widehat{L}\left(\alpha\Phi_1 + \beta\Phi_2\right) = \alpha\left(\widehat{L}\Phi_1\right) + \beta\left(\widehat{L}\Phi_2\right) .$$

Indeed, for instance, consider the first part of the operator $\widehat{L}$:

$$\frac{\partial^2}{\partial x^2}\left(\alpha\Phi_1 + \beta\Phi_2\right) = \alpha\frac{\partial^2\Phi_1}{\partial x^2} + \beta\frac{\partial^2\Phi_2}{\partial x^2} ,$$

as required. Therefore, if functions Φ_1 and Φ_2 are solutions of the PDE, i.e. $\widehat{L}\Phi_1 = 0$ and $\widehat{L}\Phi_2 = 0$, then $\widehat{L}\left(\alpha\Phi_1 + \beta\Phi_2\right) = 0$ as well, i.e. their arbitrary linear combination, $\alpha\Phi_1 + \beta\Phi_2$, must also satisfy the PDE. On top of that, if Φ_1 and Φ_2 satisfy the boundary conditions, then their linear combination, $\alpha\Phi_1 + \beta\Phi_2$, will satisfy them as well.

It is now clear how the superposition principle may help in devising a solution which satisfies the PDE, the boundary *and* the initial conditions. We have already built individual solutions (8.56) that each satisfies the PDE and the zero boundary conditions. If we construct now a linear combination of all these solutions with arbitrary coefficients α_n,

$$\Psi(x, t) = \sum_{n=1}^{\infty} \alpha_n \Psi_n(x, t) = \sum_{n=1}^{\infty} [B_n \sin(\omega_n t) + C_n \cos(\omega_n t)] \sin\frac{n\pi x}{L} , \quad (8.57)$$

then this function will satisfy the PDE due to the superposition principle. An essential point now to realise is that it will also satisfy the zero boundary conditions as each term in the sum obeys them! Therefore, this construction satisfies the PDE *and* the boundary conditions giving us enough freedom to obey the initial conditions as well: we can try to find the coefficients B_n and C_n in such a way as to accomplish this last hurdle of the method. Note that the new constants $B_n = \alpha_n D'_n$ and $C_n = \alpha_n E'_n$ are at this point still arbitrary since α_n are arbitrary as well as D'_n and E'_n.

To satisfy the initial conditions, we substitute the linear combination (8.57) into conditions (8.43). This procedure yields

$$\Psi(x, 0) = \sum_{n=1}^{\infty} C_n \sin\frac{n\pi x}{L} = \phi_1(x) , \quad (8.58)$$

$$\Psi_t(x, t)|_{t=0} = \left\{\sum_{n=1}^{\infty} \omega_n [B_n \cos(\omega_n t) - C_n \sin(\omega_n t)] \sin\frac{n\pi x}{L}\right\}_{t=0}$$
$$= \sum_{n=1}^{\infty} B_n \omega_n \sin\frac{n\pi x}{L} = \phi_2(x) . \quad (8.59)$$

Now, what we have just obtained is quite curious: we have just one equation (8.58) for an infinite number of coefficients C_n, and equally another single equation (8.59) for an infinite set of B_n coefficients. How does it help in finding these coefficients?

Well, although we do indeed have just one equation to find an either set of the coefficients, these equations are written for a continuous set of x values between 0 and L. That means, that strictly speaking, we have an infinite number of such equations, and that must be sufficient to find all the coefficients. In fact, what we have just obtained are expansion of functions $\phi_1(x)$ and $\phi_2(x)$ into infinite functional series!

Equipped with this understanding, we now need to devise a practical way of finding the coefficients C_n and B_n. For this we note that we have already come across the functions $\sin(n\pi x/L)$ in Sect. 3.1 when considering Fourier series. We have shown there that these eigenfunctions satisfy the *orthogonality relation* (3.6) with the integration performed from $-L$ to L. So, Eqs. (8.58) and (8.59) are just Fourier series expansions!

Problem 8.19 Prove the orthogonality relation

$$\int_0^L \sin\frac{n\pi x}{L}\sin\frac{m\pi x}{L}dx = \frac{L}{2}\delta_{nm} , \qquad (8.60)$$

where integration is performed for $0 \le x \le L$. Here δ_{nm} denotes the *Kronecker delta* symbol ($\delta_{nm} = 1$ if $n = m$ and $\delta_{nm} = 0$ if $n \ne m$).

Therefore, we can employ a general method developed there (see also a more fundamental discussion in Sect. 3.9.3) to find the coefficients: multiply both sides of Eqs. (8.58) and (8.59) by the function $\sin(m\pi x/L)$ with some fixed $m\ (= 1, 2, \ldots)$, then integrate both sides with respect to x from 0 to L, and use the orthogonality relations (8.60). However, in our case this step is even not needed: as has been stated above, Eqs. (8.58) and (8.59) are just *Fourier sine series* for $\phi_1(x)$ and $\phi_2(x)$, respectively. It follows, therefore, that in both cases the expressions for the coefficients can be borrowed directly from Eq. (3.10):

$$C_n = \frac{2}{L}\int_0^L \phi_1(x)\sin\frac{n\pi x}{L}dx , \qquad (8.61)$$

$$B_n = \frac{2}{\omega_n L}\int_0^L \phi_2(x)\sin\frac{n\pi x}{L}dx , \qquad (8.62)$$

where $n = 1, 2, \ldots$. This result finally solves the entire problem as the solution (8.57) with the coefficients given by Eqs. (8.61) and (8.62) satisfies the PDE *and* both the boundary and initial conditions.

In special cases when the string is initially at equilibrium (the initial displacement is zero, $\phi_1(x) = 0$) and is initially given a kick (so that $\phi_2(x) \ne 0$ at least for some values of x), the coefficients $C_n = 0$ for all n. Conversely, if the string has been initially displaced, $\phi_1(x) \neq 0$, and then released so that initial velocities, $\Psi_t(x, 0)$, are zeros (and hence $\phi_2(x) = 0$), then all the coefficients $B_n = 0$.

Hence, the solution consists of a superposition of oscillations, $\Psi_n(x, t)$, associated with different frequencies ω_n, Eq. (8.55): each point x of the string in the elementary oscillation $\Psi_n(x, t)$ performs a simple harmonic motion with that frequency. At the same time, if we look at the shape of the string due to the same elementary motion at a particular time t, then it is given by the sine function $\sin(n\pi x/L)$. The *elementary* motion $\Psi_n(x, t)$ is called the n-th *normal mode of vibration* of the string, and its frequency of vibration, $\omega_n = n\pi c/L$, is called its n-th harmonic or *normal mode* frequency. The $n = 1$ normal mode is called *fundamental*, with the frequency $\omega_1 = \pi c/L$. Frequencies of all other modes are integer multiples of the fundamental frequency, $\omega_n = n\omega_1$, i.e. ω_n is exactly n times larger.

Note that the fundamental frequency is the lowest sound frequency to be given by the given string. Recalling the expression derived in Sect. 8.2.1 for the wave velocity of the string, $c = \sqrt{T_0/\rho}$, where T_0 is the tension and ρ string density, we see that there are several ways in affecting the frequency

$$\omega_1 = \frac{\pi}{L}\sqrt{\frac{T_0}{\rho}}$$

of the fundamental mode. Indeed, e.g. taking a longer string and/or applying less tension would reduce ω_1, while increasing the tension and/or taking a shorter string would increase the lowest frequency of sound the string can produce. These principles are widely used in various musical instruments such as guitar, violin, piano, etc. For instance, in a six-string guitar by tuning peg heads (or tuning keys) the tension in a string can be adjusted without changing its length, whereby considerably causing the string fundamental frequency to go either up (higher pitch) or down (lower pitch). Also note, that thinner strings in the guitar (they are arranged at a lower position) produce higher pitch than the thicker ones (which are set up at a higher position in the set) since thinner strings are denser (thicker strings are hollow inside). All six strings in a classical guitar are approximately of the same length.

Problem 8.20 Consider a string of length L fixed at both ends. Assume that the string is initially (at $t = 0$) pulled by 0.06 at $x = L/5$ and then released. Show that the corresponding solution of the wave equation is

$$\Psi(x, t) = \sum_{n=1}^{\infty} \left(\frac{3}{4\pi^2 n^2} \sin \frac{\pi n}{5} \right) \sin \frac{\pi n x}{L} \cos \frac{\pi c n t}{L}.$$

Problem 8.21 Consider the propagation of sound in a pipe. The longitudinal oscillations of the air in the pipe are described by the function $\Psi(x, t)$, which obeys the wave equation $\Psi_{tt} = c^2 \Psi_{xx}$ with the initial conditions $\Psi(x, 0) = 0$ and $\Psi_t(x, 0) = v_0$. This corresponds to somebody blowing into one end (at

$x = 0$) of the pipe with velocity v_0 at the initial moment only. The boundary condition at the mouth end of the pipe ($x = 0$) is $\Psi(0, t) = 0$, i.e. the mouth is attached to the pipe all the time and hence there are no vibrations of the air there, while the other end, $x = L$, of the pipe is opened, i.e. the corresponding boundary condition there can be written as $\Psi_x(L, t) = 0$. Separate the variables in the PDE, apply the boundary conditions, and then show that the corresponding eigenvalues and eigenfunctions are $p_n = \frac{\pi}{L}\left(n + \frac{1}{2}\right)$ and $X_n(x) = \sin(p_n x)$, respectively, where $n = 0, 1, 2, 3, \ldots$. Argue why negative values of n yield linearly dependent eigenfunctions and hence can be ignored. Next, show that the eigenfunctions form an orthogonal set:

$$\int_0^L X_n(x) X_m(x)dx = \frac{L}{2}\delta_{nm} .$$

Then, construct the general solution of the equation, apply the initial conditions and show that the solution of the problem reads

$$\Psi(x, t) = \sum_{n=0}^{\infty} \frac{2v_0}{cLp_n^2} \sin(p_n ct) \sin(p_n x) .$$

Problem 8.22 Consider the previous problem again, but assume that the $x = L$ end of the pipe is closed as well. Show that in this case

$$\Psi(x, t) = \sum_{n=1\,(odd)}^{\infty} \frac{4v_0}{cLp_n^2} \sin(p_n ct) \sin(p_n x) ,$$

where the summation is run only over odd values of n and $p_n = \pi n/L$.

8.2.6 Forced Oscillations of the String

Let us now consider a more complex problem of the forced oscillations of the string. In this case the PDE has the form

$$c^2\Psi_{xx} + f(x, t) = \Psi_{tt} , \tag{8.63}$$

where $f(x, t)$ is some function of x and t. We shall again consider zero boundary conditions,

$$\Psi(0, t) = \Psi(L, t) = 0 , \tag{8.64}$$

but general initial conditions,

$$\Psi(x, 0) = \phi_1(x) \quad \text{and} \quad \Psi_t(x, 0) = \phi_2(x) . \tag{8.65}$$

The method of separation of variables is not directly applicable here because of the function $f(x, t)$. Still, as we shall see below, even in the general case of arbitrary $f(x, t)$, it is easy to reformulate the problem in such a way that it can be solved using the method of separation of variables.

Indeed, let us seek the solution $\Psi(x, t)$ as a sum of two functions: $U(x, t)$ and $V(x, t)$. The first one satisfies the following homogeneous problem with zero boundary conditions and the original initial conditions,

$$c^2 U_{xx} = U_{tt} , \quad U(0, t) = U(L, t) = 0 ,$$
$$U(x, 0) = \phi_1(x) \quad \text{and} \quad U_t(x, 0) = \phi_2(x) , \tag{8.66}$$

while the other function satisfies the inhomogeneous PDE with both zero initial and boundary conditions:

$$c^2 V_{xx} + f = V_{tt} , \quad V(0, t) = V(L, t) = 0 , \quad V(x, 0) = V_t(x, 0) = 0 . \tag{8.67}$$

It is easy to see that the function $\Psi = U + V$ satisfies the original problem (8.63)–(8.65). The beauty of this separation into two problems (which is reminiscent of splitting the solution of an inhomogeneous linear DE into a complementary solution and a particular integral) is that the U problem is identical to the one we considered before in Sect. 8.2.5 and hence can be solved by the method of separation of variables. It takes full care of our general initial conditions. Therefore, we only need to consider the second problem related to the function $V(x, t)$ which contains the inhomogeneity. We only need to find just one solution (the particular integral) of the second problem. Recall (Sect. 8.2.4) that the solution of the problem (8.67) is unique.

To find the particular integral $V(x, t)$, we shall use the following trick. This function is specified on a finite interval $0 \le x \le L$ only, and is equal to zero at its ends. However, we are free to extend ('continue') it to the two times larger interval $-L \le x \le L$. It is also convenient to define $V(x, t)$ such that it is odd in the whole interval: $V(-x, t) = -V(x, t)$. Moreover, after that, we can also periodically repeat $V(x, t)$ thus defined over the whole x axis. This makes $V(x, t)$ periodic and hence expandable into the Fourier series (at each t) with the period of $2L$:

$$V(x, t) = \frac{a_0}{2} + \sum_{n=1}^{\infty} \left[W_n(t) \cos \frac{n\pi x}{L} + V_n(t) \sin \frac{n\pi x}{L} \right] ,$$

where (see Sect. 3.1)

$$W_n(t) = \frac{1}{L} \int_{-L}^{L} V(x, t) \cos \frac{n\pi x}{L} dx ,$$

while

$$V_n(t) = \frac{1}{L} \int_{-L}^{L} V(x,t) \sin \frac{n\pi x}{L} dx = \frac{2}{L} \int_{0}^{L} V(x,t) \sin \frac{n\pi x}{L} d .$$

Note that the Fourier coefficients W_n and V_n generally depend on time. Due to the fact that $V(x,t)$ is odd on the whole interval $-L \leq x \leq L$, all the W_n coefficients and a_0 are to be set to zero, so that only sine terms in the Fourier series for $V(x,t)$ remain. This proves that we can seek the solution of the problem (8.67) as a sine Fourier series:

$$V(x,t) = \sum_{n=1}^{\infty} V_n(t) \sin \frac{\pi n x}{L} . \tag{8.68}$$

This expansion satisfies the zero boundary conditions automatically: $V(0,t) = V(L,t) = 0$. Note that if we chose the function $V(x,t)$ to be even in the interval $-L \leq x \leq L$, then the coefficients V_n would be all equal to zero, while the W_n coefficients, including the W_0, would in general be not. Then, the Fourier series for $V(x,t)$ will include in this case only the free and the cosine terms and may not guarantee that at $x = 0$ and $x = L$ the expansion is equal to zero. This would introduce unnecessary complications into the procedure of finding the solution.

By choosing the coefficients-functions $V_n(t)$ appropriately, we hope to satisfy the PDE and the zero initial conditions.

Substituting this expansion into the PDE in (8.67), one obtains

$$-\sum_{n=1}^{\infty} V_n(t) \left(\frac{\pi cn}{L}\right)^2 \sin \frac{\pi n x}{L} + f(x,t) = \sum_{n=1}^{\infty} \ddot{V}_n(t) \sin \frac{\pi n x}{L} ;$$

here $\ddot{V}_n = \partial^2 V_n / \partial t^2$. Multiply both sides by $X_m(x) = \sin(\pi m x/L)$ with some integer m and then integrate with respect to x between 0 and L. Because the eigenfunctions X_n are orthogonal, only the single term with $n = m$ will be left in both sums, yielding:

$$-V_m(t) \left(\frac{\pi cm}{L}\right)^2 \frac{L}{2} + \int_{0}^{L} f(x,t) \sin \frac{\pi m x}{L} dx = \ddot{V}_m(t) \frac{L}{2} ,$$

or

$$\ddot{V}_m + \omega_m^2 V_m = f_m(t) , \tag{8.69}$$

where $\omega_m = \pi cm/L$ and

$$f_m(t) = \frac{2}{L} \int_{0}^{L} f(x,t) \sin \frac{\pi m x}{L} dx . \tag{8.70}$$

It is easy to see that $f_m(t)$ can formally be considered as a Fourier coefficient in the expansion of the function $f(x,t)$ into the sine Fourier series:

$$f(x,t) \longmapsto \sum_{m=1}^{\infty} f_m(t) \sin \frac{\pi m x}{L} \ . \tag{8.71}$$

However, great care is needed here when offering this 'interpretation'. Indeed, although originally the function $f(x,t)$ was defined only for $0 \le x \le L$, it can be 'defined' additionally (and arbitrarily) for $-L < x < 0$ as well. Then the piece thus defined for the interval $-L < x \le L$ can be periodically repeated into the whole x axis, justifying an expansion of $f(x,t)$ into the Fourier series. This, however, will contain all terms including the free and the cosine terms. Since $f(x,t)$ may not in general be equal to zero at the boundaries $x = 0$ and $x = L$, it is impossible to justify the sine Fourier series in Eq. (8.71) for $f(x,t)$. At the same time, we have arrived at Eq. (8.70) for the coefficients $f_m(t)$ without assuming anything about the function $f(x,t)$.

So, what is left to do is to solve DE (8.69) with respect to $V_m(t)$. At this point it is convenient to recall that $V(x,t)$ must satisfy zero initial conditions. This can only be accomplished if for all values of m we have

$$V_m(0) = \frac{d}{dt} V_m(t) \bigg|_{t=0} = 0 \ .$$

Therefore, Eq. (8.69) has to be solved subject to zero initial conditions.

Problem 8.23 Using the method of variation of parameters (Sect. I.8.2.3.2 and Problem 1.158), show that the general solution of the DE

$$y''(x) + \omega^2 y(x) = f(x) \tag{8.72}$$

is

$$y(x) = \left[C_1 + \frac{1}{2i\omega} \int_0^x f(x_1) e^{-i\omega x_1} dx_1 \right] e^{i\omega x} \\ + \left[C_2 - \frac{1}{2i\omega} \int_0^x f(x_1) e^{i\omega x_1} dx_1 \right] e^{-i\omega x} , \tag{8.73}$$

where C_1 and C_2 are arbitrary constants. Next, assuming the zero initial conditions, $y(0) = y'(0) = 0$, show that $C_1 = C_2 = 0$, and hence the final solution can be written simply as (cf. Eq. (1.180)):

$$y(x) = \frac{1}{\omega} \int_0^x f(x_1) \sin [\omega (x - x_1)] dx_1 \ . \tag{8.74}$$

Using formula (8.74) of the problem and Eq. (8.70), we can write down the solution of the DE (8.69) as

$$V_m(t) = \frac{1}{\omega_m} \int_0^t f_m(t_1) \sin[\omega_m(t - t_1)] \, dt_1$$

$$= \frac{2}{\omega_m L} \int_0^t dt_1 \int_0^L dx \, f(x, t_1) \sin[\omega_m(t - t_1)] \sin\frac{\pi m x}{L} . \tag{8.75}$$

Once the Fourier coefficients $V_m(t)$ are defined via the function $f(x, t)$, then the full Fourier series (8.68) is completely defined for the auxiliary function $V(x, t)$ yielding the solution as $\Psi = U + V$. This fully solves our problem.

8.2.7 General Boundary Problem

Now we are prepared to solve the most general boundary problem in which both initial and boundary conditions are arbitrary and the equation is inhomogeneous:

$$c^2 \Psi_{xx} + f(x, t) = \Psi_{tt} , \tag{8.76}$$

$$\Psi(x, 0) = \phi_1(x) , \quad \left.\frac{\partial \Psi}{\partial t}\right|_{t=0} = \phi_2(x) ; \tag{8.77}$$

$$\Psi(0, t) = \varphi_1(t) , \quad \Psi(L, t) = \varphi_2(t) . \tag{8.78}$$

The trick here is to first introduce an auxiliary function, $U(x, t)$, which satisfies the boundary conditions above:

$$U(0, t) = \varphi_1(t) \quad \text{and} \quad U(L, t) = \varphi_2(t) .$$

There could be many choices to accommodate this requirement. The simplest choice seems to be a linear function in x:

$$U(x, t) = \varphi_1(t) + [\varphi_2(t) - \varphi_1(t)]\frac{x}{L} . \tag{8.79}$$

It is easy to see why this is useful: with this choice the function $V(x, t) = \Psi(x, t) - U(x, t)$ satisfies zero boundary conditions. Of course, the PDE for the function $V(x, t)$ would look more complex and the initial conditions will be modified. However, it is easy to see that the problem for this new function can be solved since we have eliminated the main difficulty of having arbitrary boundary conditions. Then, the full solution is obtained via

$$\Psi(x, t) = V(x, t) + U(x, t) .$$

Problem 8.24 Show that the problem for the auxiliary function $V(x, t)$ corresponds to solving an inhomogeneous wave equation with general initial and zero boundary conditions:

$$V_{tt} = c^2 V_{xx} + \widetilde{f}(x, t) ,$$

$$V(x, 0) = \widetilde{\phi}_1(x) , \quad V_t(x, 0) = \widetilde{\phi}_2(x) ; \quad V(0, t) = V(L, t) = 0 ,$$

where

$$\widetilde{f}(x, t) = f(x, t) - \varphi_1''(t) - \frac{x}{L} \left[\varphi_2''(t) - \varphi_1''(t) \right] ,$$

$$\widetilde{\phi}_1(x) = \phi_1(x) - \varphi_1(0) - \frac{x}{L} \left[\varphi_2(0) - \varphi_1(0) \right] ,$$

$$\widetilde{\phi}_2(x) = \phi_2(x) - \varphi_1'(0) - \frac{x}{L} \left[\varphi_2'(0) - \varphi_1'(0) \right] .$$

The full solution of this particular problem was given in Sect. 8.2.6.

Problem 8.25 Consider the following stationary (i.e. fully time independent) problem:
$$c^2 \Psi_{xx} + f(x) = 0 ; \quad \Psi(0) = \varphi_1 , \quad \Psi(L) = \varphi_2 ,$$

where φ_1 and φ_2 are constants. Note that we do not have any initial conditions here; it is assumed that when the system was subjected to some initial conditions with stationary boundary conditions and time independent external force, after a very long time the system would no longer 'remember' the initial conditions, their effect will be washed out completely.

Integrate twice the PDE with respect to x (keeping the arbitrary constants) to show that

$$\Psi(x) = C_1 x + C_2 - \frac{1}{c^2} \int_0^x dx_1 \int_0^{x_1} f(x_2) \, dx_2 .$$

Then show, by applying the boundary conditions, that

$$C_1 = \frac{1}{L} (\varphi_2 - \varphi_1) + \frac{1}{c^2 L} \int_0^L dx_1 \int_0^{x_1} f(x_2) \, dx_2$$

and $C_2 = \varphi_1$.

Problem 8.26 Consider a general problem with arbitrary initial conditions, stationary boundary conditions and time independent function $f(x, t) = f(x)$:

$$c^2 \Psi_{xx} + f(x) = \Psi_{tt} \; ; \quad \Psi(0,t) = \varphi_1 \; , \quad \Psi(L,t) = \varphi_2 \; ,$$

$$\Psi(x,0) = \phi_1(x) \; , \quad \left. \frac{\partial \Psi}{\partial t} \right|_{t=0} = \phi_2(x) \; .$$

Construct the solution as $\Psi(x,t) = U(x) + V(x,t)$, where $U(x)$ is the solution of the corresponding stationary problem from Problem 8.25. Show that the function $V(x,t)$ satisfies the following homogeneous problem with zero boundary conditions:

$$c^2 V_{xx} = V_{tt} \; ; \quad V(0,t) = V(L,t) = 0 \; ,$$

$$V(x,0) = \phi_1(x) - U(x) \; , \quad V_t(x,0) = \phi_2(x) \; .$$

This problem is solved using the Fourier method, Sect. 8.2.5.

8.2.8 Oscillations of a Rectangular Membrane

The method of solving the wave equation for the 1D case considered above can be generalised to the cases of 2D and 3D as well. To illustrate this point, we shall consider here a 2D case of transverse oscillations of a square membrane fixed around its boundary (this corresponds to zero boundary conditions). The membrane is shown in Fig. 8.3; it is stretched in the $x - y$ plane in the intervals $0 \le x \le L$ and $0 \le y \le L$.

Let $\Psi(x, y, t)$ be the transverse displacement of the point (x, y) of the membrane (basically, Ψ is the z coordinate of the point (x, y) of the oscillating membrane). Then, the corresponding wave equation to solve is $\Delta \Psi = \frac{1}{c^2} \Psi_{tt}$, or

$$\frac{\partial^2 \Psi}{\partial x^2} + \frac{\partial^2 \Psi}{\partial y^2} = \frac{1}{c^2} \frac{\partial^2 \Psi}{\partial t^2} \; . \tag{8.80}$$

The membrane is fixed along its perimeter, so that the appropriate boundary conditions are

$$\begin{cases} \Psi(x,0,t) = \Psi(x,L,t) = 0 \; , \; 0 \le x \le L \\ \Psi(0,y,t) = \Psi(L,y,t) = 0 \; , \; 0 \le y \le L \end{cases} . \tag{8.81}$$

Fig. 8.3 2D square membrane

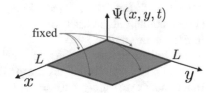

The corresponding initial conditions are assumed to be very general:

$$\Psi(x, y, 0) = \phi_1(x, y) \quad \text{and} \quad \left.\frac{\partial \Psi}{\partial t}\right|_{t=0} = \Psi_t(x, y, 0) = \phi_2(x, y) . \qquad (8.82)$$

To solve this problem, we will first attempt to separate all three variables (x, y, t) by seeking a solution as a product of functions each depending only on a single variable:

$$\Psi(x, y, t) = X(x)Y(y)T(t) . \qquad (8.83)$$

Substituting this trial solution into the PDE (8.80), one gets:

$$X''YT + XY''T = \frac{1}{c^2}XYT'' ,$$

which upon dividing by XYT results in:

$$\frac{X''}{X} + \frac{Y''}{Y} = \frac{1}{c^2}\frac{T''}{T} . \qquad (8.84)$$

Here in the left-hand side, the two terms depend each only on their own variable, x or y, respectively, while the function in the right-hand side depends only on the variable t. This is only possible if each of the expressions, X''/X, Y''/Y and T''/T, are constants. Therefore, we can write

$$\frac{X''}{X} = k_1 , \quad \frac{Y''}{Y} = k_2 , \quad \text{and hence} \quad \frac{1}{c^2}\frac{T''}{T} = k_1 + k_2 , \qquad (8.85)$$

where k_1 and k_2 are two independent *separation constants*. Thus, we have three ordinary DEs for each of the three functions:

$$X'' = k_1 X , \qquad (8.86)$$

$$Y'' = k_2 Y , \qquad (8.87)$$

$$T'' = c^2 (k_1 + k_2) T . \qquad (8.88)$$

The next steps are similar to the 1D case considered above: we first consider equations for $X(x)$ and $Y(y)$ trying to satisfy the boundary conditions; this would give us the permissible values for the separation constants, k_1 and k_2. Once this is done, the DE for $T(t)$ is solved. Finally, a general solution is constructed as a linear combination of all product solutions, with the coefficients to be determined from the initial conditions.

So, following this plan, let us consider the DE for the function $X(x)$, Eq. (8.86). The boundary conditions on the product (8.83) at $x = 0$ and $x = L$ require the function $X(x)$ to satisfy $X(0) = X(L) = 0$. This problem is however fully equivalent to the one we considered in Sect. 8.2.5 when discussing a one-dimensional string. Therefore, it immediately follows that $k_1 = -\rho^2$ must only be negative with ρ taking on the following discrete values:

$$\rho \to \rho_n = \frac{\pi}{L} n , \quad n = 1, 2, \dots ,$$

while the allowed solutions for $X(x)$ are given by the eigenfunctions

$$X_n(x) = \sin\left(\frac{\pi n}{L} x\right) , \tag{8.89}$$

corresponding to the eigenvalues $k_1 = -\rho_n^2$ for any $n = 1, 2, \dots$. We do not need to bother about the constant amplitude (a pre-factor to the sine function) here as it will be absorbed by other constants in $T(t)$ in the product elementary solution; recall, this is exactly what happened in the case of the one-dimensional string. So we simply choose it as one here.

The boundary condition (8.81) applied to (8.87) gives a similar result for k_2, namely: $k_2 = -\rho_m^2$ with $m = 1, 2, \dots$ being *another* positive integer, and the corresponding eigenfunctions are

$$Y_m(y) = \sin\left(\frac{\pi m}{L} y\right) , \quad m = 1, 2, \dots . \tag{8.90}$$

Again, we do not keep the amplitude (pre-factor) to the sine function as it will be absorbed by other constants to appear in $T(t)$.

Next we consider Eq. (8.88) for $T(t)$:

$$T'' + c^2 \left(\rho_n^2 + \rho_m^2\right) T = 0 , \tag{8.91}$$

which contains both ρ_n and ρ_m. It is a harmonic oscillator equation with the solution

$$T_{nm}(t) = A'_{nm} \sin(\omega_{nm} t) + B'_{nm} \cos(\omega_{nm} t) , \tag{8.92}$$

and the frequency

$$\omega_{nm} = c\sqrt{\rho_n^2 + \rho_m^2} = \frac{c\pi}{L}\sqrt{n^2 + m^2} . \tag{8.93}$$

So, in the case of the two-dimensional membrane we have an infinite number of product solutions

$$\Psi_{mn}(x, y, t) = \left[A'_{nm} \sin(\omega_{nm} t) + B'_{nm} \cos(\omega_{nm} t)\right] \sin\left(\frac{\pi n}{L} x\right) \sin\left(\frac{\pi m}{L} y\right) ,$$

which are numbered by *two* integers $n, m = 1, 2, 3, \dots$. Any of these solutions satisfies the PDE *and* the boundary conditions. However, neither may satisfy the initial conditions. As you can see, there was no need to keep the pre-factors in $X_n(x)$ and

$Y_m(y)$ as these would indeed be simply absorbed by the constants A'_{nm} and B'_{nm} already contained in $T_{nm}(t)$.

To satisfy the initial conditions as well, we first construct the *general solution* as a linear combination

$$\Psi(x, y, t) = \sum_{n,m=1}^{\infty} \alpha_{nm} \Psi_{nm}(x, y, t)$$

of all possible elementary solutions:

$$\Psi(x, y, t) = \sum_{n,m=1}^{\infty} [A_{nm} \sin(\omega_{nm}t) + B_{nm} \cos(\omega_{nm}t)] \sin\left(\frac{\pi n}{L}x\right) \sin\left(\frac{\pi m}{L}y\right),$$

$$(8.94)$$

where $A_{nm} = \alpha_{nm} A'_{nm}$ and $B_{nm} = \alpha_{nm} B'_{nm}$ are new arbitrary constants.

The function (8.94), by construction, satisfies the PDE (8.80) and the boundary conditions (8.81). The former is due to the superposition principle; the latter - due to the fact that each elementary product solution obeys the boundary conditions, and hence their linear combination will as well. Now, what is left to do is to choose the constants A_{nm} and B_{nm} in such a way as to satisfy the *initial conditions* (8.82), similarly to the case of the 1D string. To obtain B_{nm}, we apply the initial condition to $\Psi(x, y, t)$, yielding:

$$\Psi(x, y, 0) = \phi_1(x, y) \implies \sum_{n,m=1}^{\infty} B_{nm} \sin\left(\frac{\pi n}{L}x\right) \sin\left(\frac{\pi m}{L}y\right) = \phi_1(x, y).$$

Here $\Psi(x, y, 0)$ is expanded into a double Fourier series with respect to the sine functions, so that the expansion coefficients, B_{nm}, are found from it in the following way. Multiply both sides of the above equation by the product $X_{n'}(x)Y_{m'}(y) = \sin\left(\frac{\pi n'}{L}x\right) \sin\left(\frac{\pi m'}{L}y\right)$ with some fixed positive integers n' and m', and then integrate both sides over x and y between 0 and L. The eigenfunctions $X_n(x)$ for all values of $n = 1, 2, \ldots$ form an orthogonal set, and so do the functions $Y_m(y)$. Therefore, in the left-hand side integration over x leaves only a single $n = n'$ term in the sum over n, while integration over y results in a single $m = m'$ term in the m-sum remaining; in the right-hand side there is a double integral with respect to x and y. Hence, we immediately obtain

$$B_{nm} = \left(\frac{2}{L}\right)^2 \int_0^L dx \int_0^L dy\, \phi_1(x, y) \sin\left(\frac{\pi n}{L}x\right) \sin\left(\frac{\pi m}{L}y\right). \qquad (8.95)$$

To obtain coefficients A_{nm}, we apply the initial condition to the time derivative of $\Psi(x, y, t)$ at $t = 0$. Since

$$\Psi_t(x, y, t) = \frac{\partial \Psi}{\partial t}$$

$$= \sum_{n,m=1}^{\infty} \omega_{nm} [A_{nm} \cos(\omega_{nm} t) - B_{nm} \sin(\omega_{nm} t)] \sin\left(\frac{\pi n}{L} x\right) \sin\left(\frac{\pi m}{L} y\right),$$

the initial condition $\Psi_t(x, y, 0) = \phi_2(x, y)$ yields

$$\sum_{nm} \omega_{nm} A_{nm} \sin\left(\frac{\pi n}{L} x\right) \sin\left(\frac{\pi m}{L} y\right) = \varphi_2(x, y),$$

and, therefore, after a similar reasoning to the one used to derive the above formula for the B_{nm} coefficient, we obtain

$$A_{nm} = \frac{1}{\omega_{nm}} \left(\frac{2}{L}\right)^2 \int_0^L dx \int_0^L dy \, \phi_2(x, y) \sin\left(\frac{\pi n}{L} x\right) \sin\left(\frac{\pi m}{L} y\right). \qquad (8.96)$$

Equations (8.94), (8.95) and (8.96) fully solve the problem as they define such a function $\Psi(x, y, t)$ that satisfies the PDE and both the initial and boundary conditions.

The case of a circular membrane, also treated using the method of separation of variables, but employing the polar coordinates, was considered in detail in Sect. 4.7.6.

8.2.9 General Remarks on the Applicability of the Fourier Method

We have seen above in Sect. 8.2.7 that the Fourier method plays a central role in solving a general problem with non-zero boundary and general initial conditions. Therefore, it is worth repeating again the main steps which lie as a foundation of this method.

The method of separation of variables is only applicable to the problems with zero boundary conditions. For a PDE of any number of dimensions written for any number of coordinates, the construction of the general solution is based on several well-defined steps:

1. Construct a *product solution* as a product of functions each depending just on one variable.
2. Substitute the product solution into the PDE and attempt to separate the variables in the resulting equation. This may be done in steps. The simple separation like the one in Eq. (8.84) may not always be possible; instead, a step-by-step method is to be used. This works like this: using algebraic manipulations, 'localise' all terms depending on a single variable into a single expression that does not contain other variables; hence, this expression must only be a constant. Equate this expression

to the first constant k_1 yielding the first ordinary DE for the variable in question. Once the whole term in PDE is replaced with that constant, the obtained equation would only depend on other variables for which the same procedure is repeated until only the last single variable is left; this will be an ordinary DE containing all separation constants.

3. This procedure gives *separated ordinary* DEs for each of the functions in the elementary product we have started from.

4. Solve the ordinary DEs for the functions in the product using zero boundary conditions. This procedure gives an infinite number of possible solutions for these functions together with possible values for the corresponding separation constants, i.e. a set of *eigenfunctions* and corresponding to them *eigenvalues*. It is now possible to write down a set of *product solutions* Ψ_i (where $i = 1, 2, \ldots$) for the PDE. Each of these solutions satisfies the equation and the zero boundary conditions.

5. Next we use the *superposition principle* to construct the *general solution* of the PDE in the form of the linear combination of all elementary product solutions:

$$\Psi = \sum_{i=1}^{\infty} C_i \Psi_i . \tag{8.97}$$

In the case of non-zero boundary conditions, an auxiliary function U is defined which satisfies the full set of boundary conditions. It is convenient to choose this function as simple as possible. Then, the solution is constructed as a sum of two functions, $\Psi = U + V$, and the corresponding problem for the other auxiliary function, V, is formulated using the original problem. This function would satisfy zero boundary conditions for which the Fourier method is directly applicable.

The most crucial step which decides whether the whole method would work is the procedure of separation of variables itself. No general rules exist to tell us whether a given PDE is *separable* or not, we must just test and see!

For example, consider the following linear homogeneous PDE of the second order:

$$\Delta \Psi + \left(x^2 + y^2\right)^{\alpha} \Psi = 0 , \tag{8.98}$$

where $\Delta = \frac{\partial^2}{\partial^2 x} + \frac{\partial^2}{\partial^2 y}$ is the 2D Laplacian operator. An application of the product solution $\Psi(x, y) = X(x)Y(y)$ in this case gives

$$\left(\frac{1}{X} \frac{d^2 X}{dx^2}\right) + \left(\frac{1}{Y} \frac{d^2 Y}{dy^2}\right) + \left(x^2 + y^2\right)^{\alpha} = 0 . \tag{8.99}$$

We see that in Cartesian coordinates Eq. (8.98) is *not separable*; this is because of the free term $\left(x^2 + y^2\right)^{\alpha}$.

This, however, does not necessarily mean that the PDE is not separable. This is because in some *special coordinate systems* it may appear to be separable! Indeed, suppose now we transform the PDE (8.98) to *polar coordinates* (r, ϕ). This procedure gives

$$\frac{1}{r}\frac{\partial}{\partial r}\left(r\frac{\partial \Psi}{\partial r}\right) + \frac{1}{r^2}\frac{\partial^2 \Psi}{\partial \phi^2} + r^{2\alpha}\Psi = 0 . \qquad (8.100)$$

Now we check if a solution in the form of an elementary product $\Psi(r, \phi) = R(r)\Phi(\phi)$ can lead to a separation of the variables r and ϕ. Substituting this product solution into the transformed Eq. (8.100) and dividing both sides by $R\Phi$, we find

$$\frac{1}{rR}\frac{d}{dr}\left(r\frac{dR}{dr}\right) + \frac{1}{r^2}\left[\frac{1}{\Phi}\frac{d^2\Phi}{d\phi^2}\right] + r^{2\alpha} = 0 .$$

Here the expression in the square brackets is 'localised' in the ϕ variable and hence must be a constant. Let us call it k. Hence, we obtain two ordinary DEs:

$$\Phi'' = k\Phi \quad \text{and} \quad \frac{1}{rR}\left(rR'\right)' + \frac{k}{r^2} + r^{2\alpha} = 0 .$$

We have succeeded in separating our PDE into two ordinary DEs in the polar coordinate system, something which was impossible to do in the Cartesian system. We see that a PDE may be *separable* in one coordinate system, but *not* in another! This shows importance of knowing various curvilinear coordinate systems!

Problem 8.27 Consider a partial differential equation:

$$\frac{\partial^2 \Psi}{\partial x^2} + \frac{\partial^2 \Psi}{\partial y^2} + e^{-\alpha(x^2+y^2)}\Psi = 0 ,$$

written in the Cartesian coordinates. Using $\Psi(x, y) = X(x)Y(y)$, try to separate the variables x, y in the equation. Why do you think it does not work? Choose now the *polar coordinate system* (r, ϕ) and rewrite the equation in this system for $\Psi(x, y) = \Psi_1(r, \phi)$ employing the (r, ϕ)-part of the Laplacian. Demonstrate that in this coordinate system separation of variables is possible, and $\Psi_1(r, \phi) = R(r)\Phi(\phi)$ leads to two ordinary DEs for the functions $R(r)$ and $\Phi(\phi)$:

$$\Phi'' + k\Phi = 0 \quad \text{and} \quad r\left(rR'\right)' + \left(r^2 e^{-r^2} - k\right)R = 0 .$$

Problem 8.28 The method of separation of variables in the case of equations with more than two variables can be done in stages, when each variable is separated individually from all others, i.e. one after the other, until one ends up with a set of ordinary DEs. Consider the Laplace equation in the Cartesian coordinate system:

$$\frac{\partial^2 \Psi}{\partial x^2} + \frac{\partial^2 \Psi}{\partial y^2} + \frac{\partial^2 \Psi}{\partial z^2} = 0 .$$

By substituting $\Psi(x, y, z) = \Phi(x, y)Z(z)$ into the equation, separate the variables (x, y) from z by introducing a separation constant k_1. Obtain two equations: one for $Z(z)$ and another for $\Phi(x, y)$. Consider then the equation for $\Phi(x, y)$ and attempt to separate the variables there by considering $\Phi(x, y) = X(x)Y(y)$ and introducing another separation constant k_2. Find the corresponding DEs for $X(x)$ and $Y(y)$. What is the final form of $\Psi(x, y, z)$? [Answer: *the DEs for the three functions may be written as* $X'' + k_2 X = 0$, $Y'' - (k_1 + k_2) Y = 0$ *and* $Z'' + k_1 Z = 0$.]

8.3 Heat-Conduction Equation

In this section we shall consider parabolic PDEs using, as an important physical example, the heat-conduction equation

$$\frac{\partial \Psi}{\partial t} = \kappa \Delta \Psi + f ,$$

where $\Psi(x, y, z, t)$ is the temperature, κ the thermal diffusivity, $f(x, y, z, t)$ some function due to internal sources, and t time. This PDE has to be supplied with the corresponding initial and boundary conditions.

If the boundary conditions do not depend on time and there are no sources ($f = 0$), the temperature in the system will reach at long times ($t \to \infty$) its stationary distribution $\Psi_\infty(x, y, z)$ which satisfies the Laplace equation:

$$\Delta \Psi = 0 . \tag{8.101}$$

For definiteness, let us consider a 1D heat transport equation

$$\Psi_t = \kappa \Psi_{xx} + f . \tag{8.102}$$

Initial conditions are given simply by

$$\Psi(x, 0) = \phi(x) . \tag{8.103}$$

Note that since our PDE has only the first-time derivative, this condition is sufficient. Various boundary conditions can be supplied. The most simple ones correspond to certain temperatures at the end points of the interval[4]:

---

[4] It can be shown that the functions φ_1 and φ_2 do not need to satisfy the consistency conditions $\varphi_1(0) = \phi(0)$ and $\varphi_2(0) = \phi(L)$.

$$\Psi(0,t) = \varphi_1(t) \quad \text{and} \quad \Psi(L,t) = \varphi_2(t) \ , \tag{8.104}$$

however, other possibilities also exist. For instance, if at $x = 0$ end the heat flux is known, then the above boundary condition at $x = 0$ is replaced by

$$\Psi_x(0,t) = \varphi_1(t) \ .$$

For simplicity, most of our analysis in the coming discussion will correspond to the 1D heat transport equation, although theorems and methods to be considered can be generalised for the 2D and 3D cases as well.

8.3.1 Uniqueness of the Solution

As in the case with the wave equation, one may question whether the above conditions (8.103) and (8.104) which supplement the heat transport equation (8.102) guarantee that the solution exists and it is unique. We shall positively answer the first part of that question later on by showing explicitly how the solution can be constructed. Here we shall prove the second part, that under certain conditions the solution of the heat transport problem is unique. We start from the following theorem.

Theorem 8.1 *If* $\Psi(x,t)$ *is a continuous function of both variables for all values of* x, *i.e. for* $0 \le x \le L$, *and all times up to some time* T, *i.e. for* $0 \le t \le T$, *and* Ψ *satisfies the PDE (without the internal sources term)*

$$\Psi_t = \kappa \Psi_{xx} \tag{8.105}$$

for all internal points $0 < x < L$ *and times* $0 < t \le T$, *then* $\Psi(x,t)$ *reaches its maximum and minimum values either at the initial time* $t = 0$ *and/or at the boundaries* $x = 0$ *and* $x = L$.

Proof Consider the first part of the theorem, stating that Ψ reaches its *maximum* value either at the initial time and/or at the end points. We shall prove that by contradiction assuming that Ψ reaches its maximum value at some *internal* point (x_0, t_0), where $0 < t_0 \le T$ and $0 < x_0 < L$. Let M be the maximum value of the function Ψ at $t = 0$, $x = 0$ and $x = L$. In the combined (x,t) notation these are the points $(x,0)$, $(0,t)$ and (L,t) which we shall call 'end points' for brevity. Then, according to our assumption,

$$\Psi(x_0, t_0) = M + \epsilon \ ,$$

where $\epsilon > 0$. Next, consider an auxiliary function

$$V\,(x,t) = \Psi\,(x,t) + k\,(t_0 - t)\ .$$

Obviously, $V\,(x_0, t_0) = M + \epsilon$. Let us choose the positive constant k such that $kT < \epsilon/2$. Then,

$$k\,(t_0 - t) \le k\,(T - t) \le kT < \frac{\epsilon}{2}\ .$$

Hence, if we now consider any of the end points $(x, 0)$, $(0, t)$ and (L, t), then for them

$$V\,(x,t) = \Psi\,(x,t) + k\,(t_0 - t) \le M + \frac{\epsilon}{2}\,, \tag{8.106}$$

since M is the maximum value of the function at the end points.

Since $V\,(x,t)$ is continuous (because Ψ is), it has to reach its maximum value at some point (x_1, t_1):

$$V\,(x_1, t_1) \ge V\,(x_0, t_0) = M + \epsilon = \left(M + \frac{\epsilon}{2}\right) + \frac{\epsilon}{2}\ .$$

However, because at the end points we have that $V\,(x,t) \le M + \epsilon/2$, see Eq. (8.106), the above inequality cannot be satisfied there. Hence, the point (x_1, t_1) must be an *internal* point satisfying $0 < x_1 < L$ and $0 < t_1 \le T$.

Since this internal point is a maximum of a function of two variables, x and t, we should at least have (see Sects. I.5.10.1 and I.5.10.2)[5]:

$$V_x\,(x_1, t_1) = 0\,, \quad V_t\,(x_1, t_1) = 0 \quad \text{and} \quad V_{xx}\,(x_1, t_1) \le 0\ .$$

However, since $\Psi = V + k\,(t - t_0)$, then

$$\Psi_{xx}\,(x_1, t_1) = V_{xx}\,(x_1, t_1) \le 0 \quad \text{and} \quad \Psi_t\,(x_1, t_1) = V_t\,(x_1, t_1) + k = k > 0\,,$$

and therefore

$$\Psi_t\,(x_1, t_1) - \kappa \Psi_{xx}\,(x_1, t_1) = k - \kappa V_{xx}(x_1, t_1) > 0\,,$$

which means that the PDE is not satisfied at the point (x_1, t_1). We arrived at a contradiction to the conditions of the theorem which states that Ψ does satisfy the equation at any internal point. Therefore, our assumption was wrong.

The case of the minimum, corresponding to the second part of the theorem, is proven by noting that this case can be directly related to the case we have just considered (the maximum) if the function $\Psi' = -\Psi$ is considered instead. **Q.E.D.**

The result we have just obtained has a clear physical meaning: since the internal sources are absent (Eq. (8.105) is homogeneous), the heat cannot be created in the

---

[5] The other sufficient condition (see Eq. (I.5.90)), that $V_{xx} V_{tt} - (V_{xt})^2 > 0$, is not needed here.

system during the heat flow, and hence the temperature cannot exceed either its initial values or values at both boundaries.

This theorem has a number of important consequences of which we only consider one related to the uniqueness of the solution of the heat transport equation with general boundary conditions.

Theorem 8.2 *The problem (8.102)–(8.104) has a unique solution.*

Proof Indeed, assume there are two such solutions, Ψ_1 and Ψ_2, both satisfying the same PDE and the same initial and boundary conditions. Consider then their difference $V = \Psi_1 - \Psi_2$. It satisfies the homogeneous PDE and *zero* initial and boundary conditions. Since we have just shown that the solution of the homogeneous problem can reach its maximum *and* minimum values only at the end points (i.e. at $t = 0$ or at $x = 0$ or $x = L$), where V is zero, it follows that $V(x, t) = 0$ everywhere. Hence, $\Psi_1 = \Psi_2$, our assumption is wrong, there is only one solution to the problem possible. **Q.E.D.**

8.3.2 Fourier Method

Before considering the general problem (8.102)–(8.104), we shall first discuss the simplest case of a one-dimensional problem with zero boundary conditions:

$$\Psi_t = \kappa \Psi_{xx} , \quad \Psi(x, 0) = \phi(x) , \quad \Psi(0, t) = \Psi(L, t) = 0 . \tag{8.107}$$

We shall solve this problem using the method of separation of variables already discussed above. Assuming the product solution, $\Psi(x, t) = X(x)T(t)$, substituting it into the PDE and separating the variables, we obtain

$$XT' = \kappa X''T \quad \Longrightarrow \quad \frac{1}{\kappa}\frac{T'}{T} = \frac{X''}{X} .$$

Since the expressions on both sides of the equal sign depend on different variables, each of the expressions must be a constant. Let us denote this separation constant λ. Hence, two ordinary DEs are obtained for the two functions, $X(x)$ and $T(t)$:

$$X'' - \lambda X = 0 \quad \text{and} \quad T' - \lambda \kappa T = 0 . \tag{8.108}$$

In our discussion on solving a similar problem for the wave equation in Sect. 8.2.5, we arrived at something similar. Our next step there was to solve the equation for the function $X(x)$ first to deduce the possible values for the separation constant λ. The peculiarity of the current problem related to the heat transport equation is that for this job it is convenient to start from solving the equation for the function $T(t)$ instead.

Indeed, the solution reads $T(t) = Ce^{\lambda\kappa t}$, where C is an arbitrary constant. It is clear from this result then that the constant λ cannot be positive. Indeed, in the case of $\lambda > 0$ the temperature in the system may grow indefinitely with time which contradicts physics. Therefore, we conclude immediately, that λ must be non-positive, i.e. negative or equal to zero: $\lambda \leq 0$. We can then write it as $\lambda = -p^2$ with p being non-negative. Hence, the solution for $T(t)$ must be

$$T(t) = Ce^{-\kappa p^2 t} \ . \tag{8.109}$$

Next, we consider the equation for $X(x)$ which now has the form of the harmonic oscillator equation:

$$X'' + p^2 X = 0$$

with the solution

$$X(x) = A \sin(px) + B \cos(px) \ .$$

Applying the boundary conditions $X(0) = X(L) = 0$ yields $B = 0$ and

$$p \ \rightarrow \ p_n = \frac{\pi n}{L} \ , \quad n = 0, 1, 2, \ldots \ . \tag{8.110}$$

Therefore, as in the case of the Fourier method applied to the wave equation, we also have here an infinite discrete set of the permissible values for the separation constant $\lambda_n = -p_n^2$, where p_n, given above, is an eigenvalue corresponding to the eigenfunction

$$X_n(x) = \sin(p_n x) \ . \tag{8.111}$$

Note that we dropped the constant pre-factor here anticipating it to be absorbed by other constants in the end.

Therefore, concluding this part, we find not one but an infinite set of product solutions

$$\Psi_n(x, t) = X_n(t)T_n(t) = e^{-\kappa p_n^2 t} \sin(p_n t) \ , \quad n = 0, 1, 2, \ldots \ ,$$

each satisfies the PDE and the boundary conditions. Actually, since $X_0(x) = 0$, there is no need to consider the $n = 0$ function $\Psi_0(x, t)$; it can be dropped.

Our strategy from this point on should be obvious: the PDE is linear and hence, according to the superposition principle, a general linear combination of the elementary product solutions,

$$\Psi(x, t) = \sum_{n=1}^{\infty} \alpha_n \Psi_n(x, t) = \sum_{n=1}^{\infty} \alpha_n e^{-\kappa p_n^2 t} \sin(p_n t) \ , \tag{8.112}$$

is constructed which also satisfies the PDE and the boundary conditions for any arbitrary constants $\{\alpha_n\}$. Note that we started the summation from $n = 1$. It may however not satisfy the initial conditions. So, we have to find such coefficients α_n which would ensure this happening as well. Applying the initial conditions to the function above,

$$\Psi(x, 0) = \phi(x) \implies \sum_{n=1}^{\infty} \alpha_n \sin(p_n t) = \phi(x),$$

we see that this corresponds to expanding the function $\phi(x)$ into the sine Fourier series. Therefore, the coefficients α_n can be derived without difficulty (see also Sect. 8.2.5):

$$\alpha_n = \frac{2}{L} \int_0^L \phi(x) \sin(p_n x) \, dx . \tag{8.113}$$

Equations (8.112) and (8.113) fully solve our problem.

8.3.3 Stationary Boundary Conditions

The method considered in the previous section is easily generalised to non-zero but *stationary* boundary conditions:

$$\Psi_t = \kappa \Psi_{xx}, \quad \Psi(x, 0) = \phi(x), \quad \Psi(0, t) = \Psi_1 \quad \text{and} \quad \Psi(L, t) = \Psi_2 . \tag{8.114}$$

Here Ψ_1 and Ψ_2 are two temperatures which are maintained constant for $t > 0$ at both ends of our 1D system. This problem corresponds to finding the temperature distribution $\Psi(x, t)$ in a rod of length L which ends are maintained at temperatures Ψ_1 and Ψ_2 and the initial temperature distribution was $\phi(x)$.

The trick to applying the Fourier method to this problem is first to get rid of the non-zero boundary conditions by assuming the *stationary* solution to the problem, i.e. the solution $\Psi_\infty(x)$ which would be established after a very long time (at $t \to \infty$). Obviously, one would expect this to happen as the boundary conditions are kept constant (fixed).

The stationary solution of the heat-conduction equation does not depend on time and hence, after dropping the $\Psi_t = 0$ term in the PDE, in the 1D case the temperature distribution is found to satisfy an ordinary differential equation $\Psi_{xx} = 0$. This is to be expected as $\Psi_{xx} = 0$ equation is the particular case for the Laplace equation (8.101) we already mentioned above as the PDE describing the stationary heat conduction solution.

Solution of the equation $\Psi_{xx} = 0$ is a linear function, $\Psi_\infty(x) = Ax + B$, where A, B are found from the boundary conditions:

$$B = \Psi_1 \quad \text{and} \quad AL + B = \Psi_2 \quad \Longrightarrow \quad A = \frac{1}{L} (\Psi_2 - \Psi_1) \; ,$$

so that the required stationary solution reads

$$\Psi_\infty(x) = \Psi_1 + (\Psi_2 - \Psi_1) \frac{x}{L} \; . \tag{8.115}$$

Thus, the distribution of temperature in a rod kept at different temperatures at both ends, becomes linear at long times.

Once the stationary solution is known, we seek the solution of the whole problem (8.114) by writing $\Psi(x, t) = \Psi_\infty(x) + V(x, t)$. Since $\Psi_\infty(x)$ does not depend on time, but satisfies the boundary conditions, the problem for the auxiliary function $V(x, t)$ is the one with zero boundary conditions:

$$V_t = \kappa V_{xx} \; , \quad V(x, 0) = \phi(x) - \Psi_\infty(x) \quad \text{and} \quad \Psi(0, t) = \Psi(L, t) = 0 \; .$$
$$\tag{8.116}$$

This is the familiar problem (8.107) which can be solved by the Fourier method as was discussed in the previous section. Since $\Psi \to \Psi_\infty$ at long times, we expect that $V \to 0$ in this limit.

Although we have just considered the 1D case, the general principle remains the same for any dimensions: first, find the stationary solution by solving the Laplace equation, $\Delta \Psi_\infty(\mathbf{r}) = 0$, subject to the given boundary conditions, and then formulate the zero boundary conditions problem for the rest of the solution, $V(\mathbf{r}, t) = \Psi(\mathbf{r}, t) - \Psi_\infty(\mathbf{r})$. The latter problem can then be solved using the Fourier method.

As an example of the considered method, we shall discuss a 3D problem of a ball of radius S in a hot bath. The ball initially has the temperature Ψ_0. At $t = 0$ it was placed in a water tank of temperature $\Psi_1 > \Psi_0$. We need to determine the temperature distribution in the ball at all times.

If the tank is much bigger than the ball itself, one can assume that the ball is surrounded by an infinitely large heat bath, and hence the temperature Ψ inside the ball would only depend on time and the distance from ball's centre. In other words, after rewriting the problem in the spherical coordinates (r, θ, φ), we state that Ψ only depends on r and t, there is no dependence on the angles. This in turn means that the 3D heat-conduction equation, $\kappa \Delta \Psi = \Psi_t$, becomes (after writing the Laplacian $\Delta \Psi$ in the spherical coordinates, Sect. 7.11, and dropping the angle terms):

$$\frac{1}{\kappa} \frac{\partial \Psi}{\partial t} = \frac{1}{r^2} \frac{\partial}{\partial r} \left(r^2 \frac{\partial \Psi}{\partial r} \right) \; .$$

This PDE is only slightly different from the 1D heat-conduction equation we have been considering up to now. The initial and boundary conditions are, respectively:

$$\Psi(r, 0) = \Psi_0 \quad \text{and} \quad \Psi(S, t) = \Psi_1 \; .$$

We first have to solve the stationary problem, i.e. the equation:

$$\frac{1}{r^2}\frac{d}{dr}\left(r^2\frac{d\Psi}{dr}\right) = 0 \implies r^2\frac{d\Psi}{dr} = C_1 \implies \Psi(r) = C_1\int\frac{dr}{r^2} = -\frac{C_1}{r} + C_2 \,,$$

where C_1 and C_2. However, since the solution must be finite at all r including the ball centre, $r = 0$, the constant C_1 must be zero. Thus, $\Psi(r) = C_2$ must be constant for all $0 \le r \le S$, the temperature inside the ball should be the same across all of it. It is easy to see that in order to satisfy the boundary condition, $\Psi(S, t) = \Psi_1$, which is valid for all times (including very long ones), one has to take $C_2 = \Psi_1$. Hence, $\Psi_\infty(r) = \Psi_1$. This result is to be expected: at long enough times the whole ball would have eventually the same temperature as the water in the tank around it.

Next, we should solve the full heat-conduction equation subject to the corresponding initial and boundary conditions. Separating out the stationary part of the temperature distribution, $\Psi(r, t) = \Psi_1 + V(r, t)$, we obtain the following problem for the auxiliary function $V(r, t)$:

$$\frac{1}{\kappa}\frac{\partial V}{\partial t} = \frac{1}{r^2}\frac{\partial}{\partial r}\left(r^2\frac{\partial V}{\partial r}\right)\,, \quad V(r, 0) = \Psi_0 - \Psi_1 \text{ and } V(S, t) = 0\,.$$

Next, we seek the solution as an elementary product of two functions, $V(r, t) = R(r)T(t)$, substitute it into the PDE above and separate the variables:

$$\frac{1}{\kappa}RT' = \frac{1}{r^2}T\left(r^2 R'\right)' \implies \frac{1}{\kappa}\frac{T'}{T} = \frac{1}{r^2 R}\left(r^2 R'\right)'\,.$$

The both sides depend on their own variables and hence must be both equal to the same non-positive constant $\lambda = -p^2$. This ensures a non-increasing $T(t)$ in time:

$$\frac{1}{\kappa}\frac{T'}{T} = -p^2 \implies T(t) = Ce^{-\kappa p^2 t}\,.$$

Consider now the equation for $R(r)$:

$$\frac{1}{r^2}\left(r^2 R'\right)' = -p^2 R \implies R'' + \frac{2}{r}R' + p^2 R = 0\,.$$

Problem 8.29 Show that a new function $R_1 = rR$ satisfies the harmonic oscillator equation $R_1'' + p^2 R_1 = 0$.

The ordinary DE for R_1 is easily solved,

$$R_1(r) = A \sin(pr) + B \cos(pr) \, ,$$

yielding the following solution for the R function:

$$R(r) = \frac{R_1}{r} = A \frac{\sin(pr)}{r} + B \frac{\cos(pr)}{r} \, .$$

Similarly to the stationary case considered above, we must set the constant B to zero as the solution must be finite everywhere including the point $r = 0$ (note that $\frac{\sin pr}{r}$ is finite at $r = 0$)[6]. Thus:

$$R(r) = \frac{\sin(pr)}{r} \, ,$$

where we have dropped the arbitrary constant as it will be absorbed by other constants later on when constructing the general solution as the linear combination of all elementary solutions.

Use now the boundary conditions:

$$R(S) = \frac{\sin(pS)}{S} = 0 \quad \Longrightarrow \quad \sin(pS) = 0$$

$$\Longrightarrow \quad p \Rightarrow p_n = \frac{\pi n}{S} \, , \quad n = 1, 2, 3, \ldots \, .$$

We do not need to consider the zero value of n as the eigenfunction $R_n(r) = r^{-1} \sin(p_n r)$ is equal to zero for $n = 0$.

Therefore, by constructing an appropriate linear combination, we obtain the following *general solution*:

$$V(r, t) = \sum_{n=1}^{\infty} \alpha_n \frac{\sin(p_n r)}{r} e^{-\kappa p_n^2 t} \quad \text{with} \quad p_n = \frac{\pi n}{S} \, , \quad n = 1, 2, 3, \ldots \, .$$

Note that n starts from the value of one. We also see that $V \to 0$ at long times as expected.

The linear combination above satisfies our PDE and the boundary conditions. To force it to satisfy the initial conditions as well, we write

$$V(r, 0) = \Psi_0 - \Psi_1 = \sum_{n=1}^{\infty} \alpha_n \frac{\sin(p_n r)}{r} \, .$$

It is convenient first to rewrite this equation in a slightly different form:

$$r(\Psi_0 - \Psi_1) = \sum_{n=1}^{\infty} \alpha_n \sin(p_n r)$$

----

[6] Recall that $\lim_{x \to 0} \frac{\sin x}{x} = 1$.

from which it immediately follows that the coefficients are given by

$$\alpha_n = \frac{2}{S} (\Psi_0 - \Psi_1) \int_0^S r \sin(p_n r) \, dr \, , \tag{8.117}$$

since the functions $\{\sin p_n r\}$ form an orthogonal set:

$$\int_0^S \sin(k_n r) \sin(k_m r) \, dr = \frac{S}{2} \delta_{nm} \, .$$

Problem 8.30 Calculate the integral in Eq. (8.117) to show that

$$\alpha_n = \frac{2S}{\pi n} (-1)^n (\Psi_1 - \Psi_0) \, ,$$

and hence the final solution of the problem is

$$\Psi(r, t) = \Psi_1 + \frac{2S}{\pi r} (\Psi_1 - \Psi_0) \sum_{n=1}^{\infty} \frac{(-1)^n}{n} \sin(p_n r) \, e^{-\kappa p_n^2 t} \, .$$

At $t \to \infty$ it tends to Ψ_1 as expected.

Problem 8.31 Consider the heat flow in a bar of length L, described by the 1D heat transport equation $\kappa \Psi_{xx} = \Psi_t$, where κ is the thermal diffusivity. Initially the distribution of temperature in the bar is $\Psi(x, 0) = \phi(x)$, while the two ends of the bar are maintained at constant temperatures $\Psi(0, t) = \Psi_1$ and $\Psi(L, t) = \Psi_2$. Show that the temperature distribution in the bar is given by

$$\Psi(x, t) = \Psi_1 + \frac{x}{L} (\Psi_2 - \Psi_1) + \sum_{n=1}^{\infty} \alpha_n \sin(p_n x) \, e^{-\kappa p_n^2 t} \, ,$$

where $p_n = \pi n / L$ and

$$\alpha_n = -\frac{2}{\pi n} \left[\Psi_1 - (-1)^n \Psi_2 \right] + \frac{2}{L} \int_0^L \phi(x) \sin(p_n x) \, dx \, .$$

Problem 8.32 Consider a similar problem to the previous one, but assume that at one end the temperature is fixed, $\Psi(0, t) = \Psi_1$, while there is no heat loss at the $x = L$ end, i.e. that $\Psi_x(L, t) = 0$. Show that in this case the general solution of the problem is

$$\Psi(x, t) = \Psi_1 + \sum_{n=0}^{\infty} \alpha_n \sin(p_n x) e^{-\kappa p_n^2 t} ,$$

where $p_n = \frac{\pi}{L}\left(n + \frac{1}{2}\right)$ and

$$\alpha_n = \frac{2}{L} \int_0^L \phi(x) \sin(p_n x)\, dx - \frac{2\Psi_1}{p_n L} .$$

Problem 8.33 Consider the heat flow in a bar of length L with the initial distribution of temperature given by $\Psi(x, 0) = \Psi_0 \sin \frac{3\pi x}{2L}$ and the boundary conditions $\Psi(0, t) = 0$ and $\Psi_x(L, t) = 0$ (cf. the previous problem). Show that the temperature distribution in the bar is

$$\Psi(x, t) = \Psi_0 e^{-\kappa p_1^2 t} \sin(p_1 x) ,$$

where $p_1 = 3\pi/2L$.

8.3.4 Heat Transport with Internal Sources

There is one last specific auxiliary problem we need to study before we would be ready to consider the most general boundary problem for the heat-conduction equation.

So, let us consider the following inhomogeneous problem with stationary boundary conditions:

$$\Psi_t = \kappa \Psi_{xx} + f(x, t) , \quad \Psi(x, 0) = \phi(x) ,$$
$$\Psi(0, t) = \Psi_1 \quad \text{and} \quad \Psi(L, t) = \Psi_2 . \tag{8.118}$$

The solution of this problem follows more or less the same route as the one we used in Sect. 8.2.6 when solving a similar problem for the wave equation. First of all, we split the solution into three parts:

$$\Psi(x, t) = U(x, t) + V(x, t) + \left[\Psi_1 + \frac{x}{L}(\Psi_2 - \Psi_1)\right] ,$$

where the third function (in the square brackets) satisfies the boundary conditions so that the other two functions, V and U, satisfy zero boundary conditions. Next, the function $V(x, t)$ is chosen to satisfy the inhomogeneous equation with zero initial and boundary conditions,

$$V_t = \kappa V_{xx} + f(x, t) , \quad V(x, 0) = 0 , \quad V(0, t) = V(L, t) = 0 , \tag{8.119}$$

while the function $U(x, t)$ satisfies the homogeneous equation with modified initial conditions:

$$U_t = \kappa U_{xx} , \quad U(x, 0) = \widetilde{\phi}(x) , \quad U(0, t) = U(L, t) = 0 , \qquad (8.120)$$

where

$$\widetilde{\phi}(x) = \phi(x) - \Psi_1 - \frac{x}{L}(\Psi_2 - \Psi_1) .$$

This latter problem is solved directly using the Fourier method as explained in the previous section. The problem (8.119) is solved using the method developed in Sect. 8.2.6.

Problem 8.34 Show that the solution of the problem (8.119) is given by the following formulae:

$$V(x, t) = \sum_{n=1}^{\infty} V_n(t) \sin(p_n x) , \quad p_n = \frac{\pi n}{L} ,$$

where

$$V_n(t) = \int_0^t f_n(\tau) e^{-\kappa p_n^2(t-\tau)} d\tau \quad \text{and} \quad f_n(\tau) = \frac{2}{L} \int_0^L f(x, \tau) \sin(p_n x) dx .$$

8.3.5 Solution of the General Boundary Heat-Conduction Problem

Finally, let us consider the most general 1D heat-conduction problem:

$$\Psi_t = \kappa \Psi_{xx} + f(x, t) , \quad \Psi(x, 0) = \phi(x) , $$
$$\Psi(0, t) = \varphi_1(t) \quad \text{and} \quad \Psi(L, t) = \varphi_2(t) . \qquad (8.121)$$

This problem corresponds e.g. to the following situation: a rod which initially had a temperature distribution $\phi(x)$ at $t = 0$ is subjected at its $x = 0$ and $x = L$ ends to some heating according to functions $\varphi_1(t)$ and $\varphi_2(t)$. The solution is obtained using essentially the same method as the one we developed in Sect. 8.2.7 for the wave equation.

Problem 8.35 Show that if we split the function Ψ as

$$\Psi(x,t) = \varphi_1(t) + \frac{x}{L}[\varphi_2(t) - \varphi_1(t)] + V(x,t),$$

where the first two terms are constructed in such a way as to satisfy the boundary conditions, then the problem for the auxiliary function V reads

$$V_t = \kappa V_{xx} + \widetilde{f}(x,t), \quad V(x,0) = \widetilde{\phi}(x), \quad V(0,t) = V(L,t) = 0,$$
(8.122)

where

$$\widetilde{f}(x,t) = f(x,t) - \varphi_1'(t) - \frac{x}{L}\left[\varphi_2'(t) - \varphi_1'(t)\right]$$

and

$$\widetilde{\phi}(x) = \phi(x) - \varphi_1(0) - \frac{x}{L}[\varphi_2(0) - \varphi_1(0)].$$

The problem (8.122) has been solved in the previous section. In fact, there we considered a slightly more difficult problem with non-zero stationary boundary conditions; the problem (8.122) is even easier as it has zero boundary conditions.

8.4 Problems Without Boundary Conditions

Sometimes boundary conditions are absent, only the initial conditions are specified. This kind of problems appears, for instance, when the region in which the solution is sought for is infinite. Even in this case the Fourier method can be used, albeit with some modifications. As an example of this kind of situation, let us consider a heat transport PDE in an infinite one-dimensional region:

$$\Psi_t = \kappa \Psi_{xx}, \quad \Psi(x,0) = \phi(x), \quad -\infty < x < \infty.$$
(8.123)

Here we assume that the function $\phi(x)$ is absolutely integrable, i.e. that

$$\int_{-\infty}^{\infty} |\phi(x)|\,dx = M < \infty.$$
(8.124)

The idea is to apply the separation of variables and build up all possible elementary product solutions satisfying the PDE, and then apply the initial conditions. The calculation, at least initially, goes along the same route as in Sect. 8.3.2. We start by writing $\Psi(x,t) = X(x)T(t)$, next substitute this trial solution into the PDE and separate the variables:

$$\frac{1}{\kappa}\frac{T'}{T} = \frac{X''}{X} = -\lambda^2 ,$$

where $-\lambda^2$ is the separation constant. Note that, as was discussed in Sect. 8.3.2, it has to be non-positive to ensure that the obtained solution has proper physical behaviour. Solving the obtained ordinary DEs,

$$X'' + \lambda^2 X = 0 \quad \text{and} \quad T' = -\kappa \lambda^2 T ,$$

for the functions $T(t)$ and $X(x)$, we can write the obtained elementary solution for the given value of λ as follows:

$$\Psi_\lambda(x, t) = \left[C_1(\lambda)e^{i\lambda x} + C_2(\lambda)e^{-i\lambda x} \right] e^{-\kappa \lambda^2 t} . \tag{8.125}$$

Here the subscript λ in Ψ_λ shows explicitly that the function above corresponds to the particular value of the parameter λ, and C_1 and C_2 which may also depend on λ, i.e. these become some functions of λ.

The obtained solution (8.125) satisfies the PDE, but not yet the initial conditions. Due to the linearity of the PDE, any linear combination of such elementary solutions must also be a solution. However, we do not have any boundary conditions in place to help us to choose the permissible values of λ. Hence, we have to admit that λ can take *any* positive real values, $0 \le \lambda < \infty$, and hence the linear combination (the sum) turns into an integral:

$$\Psi(x, t) = \int_0^\infty \Psi_\lambda(x, t)\, d\lambda = \int_0^\infty \left[C_1(\lambda)e^{i\lambda x} + C_2(\lambda)e^{-i\lambda x} \right] e^{-\kappa \lambda^2 t} d\lambda .$$

Because both positive and negative exponents are present, $e^{\pm i\lambda x}$, there is no need to keep both terms in the square brackets. Instead, we simply take one of the terms and extend the integration over the whole λ axis, i.e. the expression above can be rewritten simply as

$$\Psi(x, t) = \int_{-\infty}^\infty C(\lambda)e^{i\lambda x} e^{-\kappa \lambda^2 t} d\lambda , \tag{8.126}$$

where $C(\lambda)$ is some (generally complex) function of λ.

This function can be found from the initial conditions:

$$\Psi(x, 0) = \phi(x) \quad \Longrightarrow \quad \int_{-\infty}^\infty C(\lambda)e^{i\lambda x} d\lambda = \phi(x) .$$

The obtained expression is nothing but an expansion of the function $\phi(x)$ into the Fourier integral. Therefore, the function $C(\lambda)$ can be found from this equation simply by writing the inverse Fourier transform:

$$C(\lambda) = \frac{1}{2\pi} \int_{-\infty}^{\infty} \phi(x) e^{-i\lambda x} dx .$$

Substituting this expression into Eq. (8.126), we obtain

$$\Psi(x,t) = \frac{1}{2\pi} \int_{-\infty}^{\infty} d\lambda \left[\int_{-\infty}^{\infty} dx_1 \phi(x_1) e^{i\lambda(x-x_1)} \right] e^{-\kappa\lambda^2 t} . \tag{8.127}$$

The integral over x_1 converges uniformly with respect to λ,

$$\left| \int_{-\infty}^{\infty} dx_1 \phi(x_1) e^{i\lambda(x-x_1)} \right| \leq \int_{-\infty}^{\infty} dx_1 \left| \phi(x_1) e^{i\lambda(x-x_1)} \right|$$

$$= \int_{-\infty}^{\infty} dx_1 |\phi(x_1)| = M < \infty ,$$

and hence the order of the integrations in Eq. (8.127) can be interchanged:

$$\Psi(x,t) = \frac{1}{2\pi} \int_{-\infty}^{\infty} dx_1 \phi(x_1) \left[\int_{-\infty}^{\infty} d\lambda \, e^{i\lambda(x-x_1)} e^{-\kappa\lambda^2 t} \right] . \tag{8.128}$$

The integral over λ is calculated explicitly, see Eq. (2.51), and we obtain

$$\Psi(x,t) = \int_{-\infty}^{\infty} dx_1 \, G(x - x_1, t) \phi(x_1) , \tag{8.129}$$

where

$$G(x,t) = \frac{1}{\sqrt{4\pi\kappa t}} e^{-x^2/4\kappa t} .$$

This is the final result. We have already met a 3D analogue of the function $G(x,t)$ in Sect. 5.3.3, see Eq. (5.82). There we solved, using the Fourier transform (FT) method, the heat conduction equation in an infinite 3D space. The reader can appreciate that the separation of variables method developed above brings us to exactly the same result as applying the FT method directly. In fact, this should not come as a complete surprise as both methods are very close in spirit; moreover, we benefited directly from the FT in the derivation performed above.

The function G, called Green's function of the 1D heat-conduction equation, has a very simple physical meaning: it corresponds to the solution of the heat-conduction equation for the initial conditions $\Psi(x, 0) = \delta(x)$ (this follows directly from Eq. (8.129)), when the temperature at the point $x = 0$ was infinite while the temperature at all other points zero, i.e. the initial temperature distribution was at $t = 0$ an infinite spike at $x = 0$. Then the function $G(x, t)$ describes the distribution of the temperature in our 1D system in time: it is easily seen than the spike gets smoothed out in both directions $x \to \pm\infty$ with time, approaching the uniform $\Psi(x, t) = 0$ distribution in a long time limit as $G(x, t) \sim t^{-1/2}$.

8.5 Application of Fourier Method to Laplace Equation

The method of separation of variables can also be very useful in solving the Laplace equation with specific boundary conditions. We discussed in detail the solution of the Laplace equation in spherical coordinates in Sect. 4.5.3. Here we shall briefly illustrate the application of the Fourier method to the Laplace problem on a number of other simple examples.

Consider an infinite cylinder of radius S, the surface of which is kept at a constant temperature $\phi(\varphi)$ which depends on the polar angle φ. We are interested in the stationary distribution of the temperature in the cylinder. This problem is most easily formulated in polar coordinates assuming that the cylinder axis coincides with the z axis (the coordinate z can be ignored altogether as the cylinder is of infinite length):

$$\frac{1}{r}\frac{\partial}{\partial r}\left(r\frac{\partial \Psi}{\partial r}\right) + \frac{1}{r^2}\frac{\partial^2 \Psi}{\partial \varphi^2} = 0\,, \quad \Psi(S, \varphi) = \phi(\varphi)\,, \tag{8.130}$$

where in the left-hand side the (r, φ) part of Laplacian in the cylinder coordinates, Eq. (7.109), has been written.

We start by trying a product solution, $\Psi(r, \varphi) = R(r)\Phi(\varphi)$. Substituting it into the PDE and separating the variables, we find

$$\Phi\left(rR'\right)' + \frac{R}{r}\Phi'' = 0 \quad \Longrightarrow \quad \left(rR'\right)' + \frac{R}{r}\left\{\frac{\Phi''}{\Phi}\right\} = 0\,.$$

The expression inside the curly brackets depends only on the angle φ, hence, it must be a constant λ. Therefore, two ordinary DEs for the functions $R(r)$ and $\Phi(\varphi)$ are obtained

$$r^2 R'' + r R' + \lambda R = 0 \quad \text{and} \quad \Phi'' - \lambda \Phi = 0\,. \tag{8.131}$$

We now have to discuss how the values of the separation constant λ are to be chosen. It is easy to see that $\lambda = -m^2$, where m is any integer including zero, otherwise the function $\Phi(\varphi)$ would not be periodic with the period of 2π (we have already come across this situation before in Sect. 4.5.3 where this point was thoroughly discussed).

Two cases are to be considered: $m = 0$ and $m \neq 0$. In the latter case, the solution for the function Φ is indeed a periodic function,

$$\Phi(\varphi) = A \cos(m\varphi) + B \sin(m\varphi)\,,$$

while the solution of the equation for $R(r)$ is sought using the trial function $R(r) = r^n$, where n is a constant to be determined. Substituting this into the DE for R, we get

$$n(n-1)r^n + nr^n - m^2 r^n = 0 \quad \Longrightarrow \quad n^2 - m^2 = 0$$

$$\Longrightarrow \quad n = \pm m\,.$$

Therefore, the solution for the radial function must be the function

$$R(r) = Cr^{|m|} + \frac{D}{r^{|m|}} \, ,$$

with C and D being some arbitrary constants. The solution must be finite at $r = 0$, and hence the term with the constant D must be omitted, and hence the product solution in the case of $m \neq 0$ is

$$\Psi_m (r, \varphi) = r^{|m|} [A_m \cos (m\varphi) + B_m \sin (m\varphi)] \, , \qquad (8.132)$$

where A_m and B_m are arbitrary constants. Here formally $m = \pm 1, \pm 2, \ldots$, however, negative values of m result in linearly dependent solutions to the ones with the positive ones, so that we can limit ourselves with the positive values of the integer m only.

In the case of $m = 0$ (and hence $\lambda = 0$) the DE for the function R reads

$$r^2 R'' + r R' = 0 \quad \Longrightarrow \quad (r R')' = 0 \quad \Longrightarrow \quad r R' = C_0$$
$$\Longrightarrow \quad R(r) = C_0 \ln r + D_0$$

with C_0 and D_0 being arbitrary constants, while the DE for Φ is simply $\Phi'' = 0$ yielding a linear function, $\Phi (\varphi) = A'_0 + B'_0 \varphi$. As we must have periodicity with respect to the angle, the constant B'_0 must be equal to zero. Also, since the solution must be finite at $r = 0$, the constant C_0 in the solution for $R(r)$ must also be zero as the logarithm has a singularity at $r = 0$. Hence, for $m = 0$ our product solution is simply a constant. Taking a general linear combination of all solutions and absorbing the expansion coefficients in our arbitrary constants, we can write the general solution of the PDE as

$$\Psi (r, \varphi) = \frac{A_0}{2} + \sum_{m=1}^{\infty} r^m [A_m \cos (m\varphi) + B_m \sin (m\varphi)] \, . \qquad (8.133)$$

Here the constant term (which comes from the $m = 0$ contribution discussed above) was conveniently written as $A_0/2$. The reason for that is that when applying the boundary conditions at $r = S$,

$$\frac{A_0}{2} + \sum_{m=1}^{\infty} S^m [A_m \cos (m\varphi) + B_m \sin (m\varphi)] = \phi (\varphi) \, , \qquad (8.134)$$

we arrive exactly at the expansion of the function $\phi (\varphi)$ into the Fourier series with the period of 2π (see Sect. 3.1 and specifically Eq. (3.7) with $l = \pi$), and hence the explicit expressions for the unknown coefficients A_m ($m = 0, 1, 2, \ldots$) and B_m ($m = 1, 2, \ldots$) are obtained directly from Eqs. (3.9) and (3.10), respectively:

$$A_m = \frac{1}{S^m \pi} \int_0^{2\pi} \phi\,(\varphi) \cos\,(m\varphi)\, d\varphi \,, \tag{8.135}$$

$$B_m = \frac{1}{S^m \pi} \int_0^{2\pi} \phi\,(\varphi) \sin\,(m\varphi)\, d\varphi \,. \tag{8.136}$$

The obtained expressions fully solve the problem: the solution is given by the Fourier expansion (8.133) with the expansion coefficients (8.135) and (8.136).

Problem 8.36 Consider a stationary distribution $\Psi\,(r, \varphi)$ of temperature in a hollow cylinder with the internal and external radii being a and b, respectively. The internal surface of the cylinder is maintained at the time-constant temperature distribution $\phi_a(\varphi)$, while the external one at $\phi_b(\varphi)$.

(i) Show first that a general solution of the Laplace equation can be written in this case as follows:

$$\Psi\,(r, \varphi) = C_0 \ln r + D_0 + \sum_{m=-\infty\,(m\neq0)}^{\infty} r^m \left(A_m e^{im\varphi} + B_m e^{-im\varphi}\right) \,.$$

(ii) Applying the boundary conditions, show that the coefficients above are

$$C_0 = \frac{\alpha_0 - \beta_0}{\ln\,(a/b)} \,, \quad D_0 = \alpha_0 - C_0 \ln a \,,$$

$$A_m = \frac{\alpha_m b^{-m} - \beta_m a^{-m}}{(a/b)^m - (b/a)^m} \,, \quad B_m = \frac{\alpha_{-m} b^{-m} - \beta_{-m} a^{-m}}{(a/b)^m - (b/a)^m} \,, \quad m = \pm 1, \pm 2, \ldots \,,$$

where for any $m = 0, \pm 1, \pm 2, \ldots$ we have

$$\alpha_m = \frac{1}{2\pi} \int_0^{2\pi} \phi_a\,(\varphi)\, e^{-im\varphi}\, d\varphi \quad \text{and} \quad \beta_m = \frac{1}{2\pi} \int_0^{2\pi} \phi_b\,(\varphi)\, e^{-im\varphi}\, d\varphi \,.$$

Problem 8.37 Consider a thin circular plate of radius R. One semicircular boundary of it is held at a constant temperature T_1, while the other at T_2. Using polar coordinates and the method of separation of variables, show that the stationary distribution of temperature in the plate is given by

$$\Psi\,(r, \varphi) = \frac{1}{2}\,(T_1 + T_2) + \sum_{m=1\,(odd)}^{\infty} \frac{2\,(T_1 - T_2)}{\pi m} \left(\frac{r}{R}\right)^m \sin\,(m\varphi) \,.$$

Here the summation is performed only over odd values of m.

Problem 8.38 A rectangular plate occupying $0 \leq x \leq L_1$ and $0 \leq y \leq L_2$, is brought into contact with a heat source which maintains its two opposite sides of length L_1 at temperature T_1. Two other sides of the plate (of length L_2) are kept at temperature T_0. The plate is then left for a very long time, so that some stationary distribution of temperature $\Psi(x, y)$ established.

(i) Introduce an auxiliary function $\Psi_1 = \Psi - T_1$ and show, using the method of separation of variables, that

$$\Psi_1(x, y) = \sum_{n=1}^{\infty} \left[A_n e^{p_n x} + B_n e^{-p_n x} \right] \sin(p_n y) , \quad p_n = \frac{\pi n}{L_2} ,$$

satisfies the corresponding PDE and the boundary conditions at the sides $y = 0$ and $y = L_2$.

(ii) Show that the coefficients A_n and B_n satisfy the following equation:

$$A_n e^{p_n x} + B_n e^{-p_n x} = \frac{2}{L_2} \int_0^{L_2} \Psi_1(x, y) \sin(p_n y) \, dy .$$

(iii) Then, by considering the boundary conditions at the other two sides, find the coefficients A_n and B_n, and hence show that the distribution of temperature is given by

$$\Psi(x, y) = T_1 + \frac{4(T_0 - T_1)}{\pi} \sum_{n=1 \, (odd)}^{\infty} \frac{1}{n(1 + \lambda_n)} \left[e^{p_n x} + \lambda_n e^{-p_n x} \right] \sin(p_n y) ,$$

where the summation is run only over odd values of n and $\lambda_n = e^{p_n L_1}$.

8.6 Method of Integral Transforms

Integral transforms (such as Fourier and Laplace) are also frequently used to solve the PDEs. In this section a few examples will be considered to illustrate the usefulness of the integral transforms in solving PDEs.

8.6.1 Fourier Transform

An example of an application of the Fourier transform method for solving d'Alembert PDE (which is an inhomogeneous hyperbolic PDE) was considered in Sect. 5.3.2 and of the heat-conduction equation in Sect. 5.3.3. Here we shall consider more examples.

Consider first a 2D problem of, e.g. a stationary temperature distribution in a semi-infinite $x - y$ plate ($y \geq 0$ and $-\infty < x < \infty$) formulated in Cartesian coordinates:

$$\frac{\partial^2 \psi(x, y)}{\partial x^2} + \frac{\partial^2 \psi(x, y)}{\partial y^2} = 0 , \quad y \geq 0 \text{ and } \psi(x, y = 0) = \phi(x) . \qquad (8.137)$$

We shall also adopt another boundary condition that $\psi(x, y)$ is limited as $y \to \infty$. This would be an obvious choice if the problem in question arises from a real physical problem.

Since the range of y values is limited to $y \geq 0$, but $-\infty < x < \infty$ covers the whole x axis, we can only define the FT with respect to the x variable. Hence, we introduce the $x-$variable FT as

$$\Psi(k, y) = \frac{1}{\sqrt{2\pi}} \int dk \, e^{ikx} \psi(x, y) ,$$

and rewrite the PDE correspondingly:

$$\frac{\partial^2 \Psi(k, y)}{\partial y^2} - k^2 \Psi(k, y) = 0 .$$

This is an ODE with respect to $\Psi(k, y)$ being a function of y (fixed k), that can be easily solved:

$$\psi(k, y) = A(k)e^{ky} + B(k)e^{-ky} .$$

Note that the constants may in fact be functions of k and this has been explicitly indicated. There comes a subtle point: the solution cannot indefinitely increase when y tends to $+\infty$ as otherwise we shall dissatisfy the given boundary condition at $y = \infty$. Therefore, if $k > 0$, then the first exponent need to be dropped by setting $A(k) = 0$; if $k < 0$, then the same fate is to be expected for the other function. Therefore, only a single exponent needs to be kept that decays to zero as $y \to \infty$. Both conditions can be written in a unique way as

$$\Psi(k, y) = C(k)e^{-|k|y} ,$$

where $C(k)$ is a function that needs to be determined.

Of course, one has to have sufficient additional information in the problem in order to determine $C(k)$, and luckily we have the boundary condition at $y = 0$ as well that $\psi(x, 0) = \phi(x)$. Let us apply the FT to both sides of this condition: $\Psi(k, 0) = \Phi(k)$, where $\Phi(k) = \mathcal{F}[\phi(x)]$. Applying this condition to the solution $\Psi(k, y)$ given above, we obtain

$$\Psi(k, y) = \Phi(k)e^{-|k|y} .$$

This is the solution of the entire problem, but obtained in the Fourier space with respect to x. The inverse Fourier transform should give as the full solution of the problem:

$$\psi(x, y) = \mathcal{F}^{-1}[\Psi(k, y)] = \mathcal{F}^{-1}\left[\Phi(k)e^{-|k|y}\right].$$

Problem 8.39   Calculate explicitly the inverse FT to show that

$$\Psi(x, y) = \frac{1}{\pi}\int_{-\infty}^{\infty}\frac{y}{y^2 + (x - x')^2}\phi(x')dx'.$$

Next we shall consider the application of the FT method to a 1D heat transport and wave equation PDEs. These will be formulated as problems for the reader.

Problem 8.40   Using the FT method, show that the solution of the following initial value PDE problem,

$$\frac{1}{\kappa}\frac{\partial\psi(x, t)}{\partial t} = \frac{\partial^2\psi(x, t)}{\partial x^2}, \quad \psi(x, t = 0) = \phi(x)$$

is

$$\psi(x, t) = \int G(x - x', t)\phi(x')dx',$$

where

$$G(x, t) = \frac{1}{\sqrt{4\kappa\pi t}}e^{-x^2/4\kappa t}.$$

The reader may recognise in $G(x, t)$ Green's function of the 1D diffusion problem considered in Sect. 5.3.3, which was obtained there for the delta-function initial condition (see also Sect. 8.4 where it was obtained using the method of separation of variables).

Problem 8.41   Using the FT method, show that the solution of the general initial value 1D wave equation problem,

$$\frac{1}{v^2}\frac{\partial^2\psi(x, t)}{\partial t^2} = \frac{\partial^2\psi(x, t)}{\partial x^2}, \quad \psi(x, t = 0) = \phi_1(x), \quad \left.\frac{\partial\psi(x, t)}{\partial t}\right|_{t=0} = \phi_2(x),$$

is

$$\psi(x,t) = \frac{1}{2}\left[\phi_1(x+vt) + \phi_1(x-vt)\right] + \frac{1}{2v}\int_{x-vt}^{x+vt}\phi_2(x')dx'.$$

[Hint: you may find Eqs. (5.61) and (5.62) useful.]

This is precisely the famous d'Alembert's formula (8.31) that we derived in Sect. 8.2.3 differently.

8.6.2 Laplace Transform

Here we shall show how the Laplace transform (LT) method can also be used in solving PDEs in a general case of arbitrary initial and boundary conditions. Of course, the success depends very much on whether the inverse LT can be performed; however, in many cases exact analytical solutions in the form of integrals and infinite series can be obtained.

We shall illustrate this method by solving a rather general one-dimensional heat-conduction problem described by the following equations:

$$\frac{1}{\kappa}\frac{\partial \psi}{\partial t} = \frac{\partial^2 \psi}{\partial x^2}, \quad \psi(x,0) = \phi(x), \quad \psi(0+,t) = \varphi_1(t), \quad \psi(L-,t) = \varphi_2(t),$$
(8.138)

where $0 \leq x \leq L$, and the notations $0+$ and $L-$ in the boundary conditions mean that these are obtained by taking the limits $x \to 0$ and $x \to L$ from the right and left sides, respectively, in the function $\psi(x,t)$. We need to calculate the distribution $\psi(x,t)$ along the rod at any time $t > 0$. An important variant of this problem may be that of a semi-infinite rod; this can be obtained by taking the limit of $L \to \infty$.

First, we shall consider the case of the rod of a finite length L. Performing the LT of the above equations *with respect to time only*, yields

$$\frac{1}{\kappa}\left[p\Psi(x,p) - \phi(x)\right] = \frac{\partial^2 \Psi(x,p)}{\partial x^2}, \quad \Psi(0+,p) = \Phi_1(p), \quad \Psi(L-,p) = \Phi_2(p),$$
(8.139)

where p is a number in the complex plane, $\Psi(x,p) = \mathcal{L}[\psi(x,t)]$ is the LT of the unknown temperature distribution, while $\Phi_1(p)$ and $\Phi_2(p)$ are LTs of the boundary functions $\varphi_1(t)$ and $\varphi_2(t)$, respectively.

Problem 8.42 Using the method of variation of parameters (Section I.8.2.3.2) shows that the general solution of the DE

$$y''(x) - \lambda^2 y(x) = f(x)$$
(8.140)

is

$$y(x) = \left[C_1 + \frac{1}{2\lambda} \int_0^x f(x_1) e^{-\lambda x_1} dx_1\right] e^{\lambda x}$$
$$+ \left[C_2 - \frac{1}{2\lambda} \int_0^x f(x_1) e^{\lambda x_1} dx_1\right] e^{-\lambda x} ,$$

(8.141)

where C_1 and C_2 are arbitrary constants [cf. *Problem 8.23.*]

When applying the result of this problem to Eq. (8.139), we note: $y(x) \rightarrow \Psi(x, p)$, $\lambda^2 = p/\kappa$ and $f(x) = -\phi(x)/\kappa$. To calculate the two constants (which of course are constants only with respect to x; they may and will depend on p), one has to apply the boundary conditions stated in Eq. (8.139), which gives two equations for calculating them:

$$\Phi_1 = C_1 + C_2 , \tag{8.142}$$

$$\Phi_2 = \left[C_1 + \frac{1}{2\lambda} \int_0^L f(x_1) e^{-\lambda x_1} dx_1\right] e^{\lambda L}$$
$$+ \left[C_2 - \frac{1}{2\lambda} \int_0^L f(x_1) e^{\lambda x_1} dx_1\right] e^{-\lambda L} . \tag{8.143}$$

Problem 8.43 Show that the solution of these equations is

$$C_1(p) = \frac{1}{2 \sinh(\lambda L)} \left[\Phi_2 - \Phi_1 e^{-\lambda L} - \frac{1}{\lambda} \int_0^L f(\xi) \sinh(\lambda(L - \xi)) d\xi\right] ,$$

(8.144)

$$C_2(p) = \frac{1}{2 \sinh(\lambda L)} \left[-\Phi_2 + \Phi_1 e^{\lambda L} + \frac{1}{\lambda} \int_0^L f(\xi) \sinh(\lambda(L - \xi)) d\xi\right] ,$$

(8.145)

where the dependence on p comes from Φ_1, Φ_2 and $\lambda = \sqrt{p/\kappa}$.

Problem 8.44 Then, substituting the expressions for the 'constants' C_1 and C_2 found in the previous problem into the solution (8.141) for $y(x) \rightarrow \Psi(x, p)$, show by splitting the integral $\int_0^L = \int_0^x + \int_x^L$, that the solution of the boundary problem (8.139) can be written as follows:

$$\Psi\,(x,p) = \frac{\sinh\,(\lambda\,(L-x))}{\sinh\,(\lambda L)}\Phi_1(p) + \frac{\sinh\,(\lambda x)}{\sinh\,(\lambda L)}\Phi_2(p)$$

$$-\frac{1}{\kappa\lambda}\int_0^x \phi\,(\xi)\left[\sinh\,(\lambda\,(x-\xi)) - \frac{\sinh\,(\lambda x)}{\sinh\,(\lambda L)}\sinh\,(\lambda\,(L-\xi))\right]d\xi$$

$$+\frac{1}{\kappa\lambda}\int_x^L \phi\,(\xi)\frac{\sinh\,(\lambda x)}{\sinh\,(\lambda L)}\sinh\,(\lambda\,(L-\xi))\,d\xi\,.$$

Finally, using the definition of the sinh function, simplify the expression in the square brackets to obtain the final solution of the boundary problem:

$$\Psi\,(x,p) = \frac{\sinh\,(\lambda\,(L-x))}{\sinh\,(\lambda L)}\Phi_1(p) + \frac{\sinh\,(\lambda x)}{\sinh\,(\lambda L)}\Phi_2(p)$$

$$+\frac{1}{\kappa}\int_0^L \phi\,(\xi)\,K\,(x,p,\xi)\,d\xi\,, \tag{8.146}$$

where

$$K\,(x,p,\xi) = \frac{1}{\lambda\sinh\,(\lambda L)}\begin{cases} \sinh\,(\lambda\xi)\sinh\,(\lambda\,(L-x))\,,\ 0 < \xi < x \\ \sinh\,(\lambda x)\sinh\,(\lambda\,(L-\xi))\,,\ x < \xi < L \end{cases}. \tag{8.147}$$

This is the required solution of the full initial and boundary problem (8.139) in the Laplace space. In order to obtain the final solution $\psi\,(x,t)$, we have to perform the inverse LT. This can only be done in terms of an expansion into an infinite series.

Four functions need to be considered which appear in Eqs. (8.146) and (8.147). The coefficient to the function $\Phi_1(p)$ is the function

$$Y_1(p) = \frac{\sinh\,(\lambda\,(L-x))}{\sinh\,(\lambda L)} = \frac{\sinh\,\big((L-x)\,\sqrt{p/\kappa}\big)}{\sinh\,\big(L\sqrt{p/\kappa}\big)}$$

$$= \frac{e^{(L-x)\sqrt{p/\kappa}} - e^{-(L-x)\sqrt{p/\kappa}}}{e^{L\sqrt{p/\kappa}} - e^{-L\sqrt{p/\kappa}}}$$

$$= \frac{1}{e^{L\sqrt{p/\kappa}}}\frac{e^{(L-x)\sqrt{p/\kappa}} - e^{-(L-x)\sqrt{p/\kappa}}}{1 - e^{-2L\sqrt{p/\kappa}}}$$

$$= \frac{e^{-x\sqrt{p/\kappa}} - e^{-(2L-x)\sqrt{p/\kappa}}}{1 - e^{-2L\sqrt{p/\kappa}}}\,.$$

Expanding the denominator into the geometric progression, we obtain

$$Y_1(p) = \left[e^{-x\sqrt{p/\kappa}} - e^{-(2L-x)\sqrt{p/\kappa}} \right] \sum_{n=0}^{\infty} e^{-2Ln\sqrt{p/\kappa}}$$

$$= \sum_{n=0}^{\infty} \left[e^{-(2Ln+x)\sqrt{p/\kappa}} - e^{-(2L(n+1)-x)\sqrt{p/\kappa}} \right] . \tag{8.148}$$

Note that each exponential term is of the form $e^{-\alpha\sqrt{p}}$ with some positive $\alpha > 0$. This ensures that the inverse LT of each exponent in the series exists and is given by Eq. (6.32). Denoting it by $\eta(\alpha, t) = \mathcal{L}^{-1}\left[e^{-\alpha\sqrt{p}} \right]$, we can write[7]:

$$\mathcal{L}^{-1}[Y_1(p)] = y_1(x, t) = \sum_{n=0}^{\infty} \left[\eta\left(\frac{2Ln + x}{\sqrt{\kappa}}, t \right) - \eta\left(\frac{2L(n+1) - x}{\sqrt{\kappa}}, t \right) \right] . \tag{8.149}$$

Problem 8.45 Using a similar method, show that the other three functions appearing in Eqs. (8.146) and (8.147) can be similarly expressed as infinite series:

$$Y_2(p) = \frac{\sinh(\lambda x)}{\sinh(\lambda L)} = \sum_{n=0}^{\infty} \left[e^{-(L(2n+1)-x)\sqrt{p/\kappa}} - e^{-(L(2n+1)+x)\sqrt{p/\kappa}} \right] ; \tag{8.150}$$

$$K(x, p, \xi) = \frac{\sqrt{\kappa}}{2} \sum_{n=0}^{\infty} \frac{1}{\sqrt{p}} \left[e^{-(2Ln+x-\xi)\sqrt{p/\kappa}} - e^{-(2Ln+x+\xi)\sqrt{p/\kappa}} \right.$$

$$\left. - e^{-(2L(n+1)-x-\xi)\sqrt{p/\kappa}} + e^{-(2L(n+1)-x+\xi)\sqrt{p/\kappa}} \right] , \tag{8.151}$$

if $0 < \xi \le x$, while $K(x, p, \xi)$ for the case of $x \le \xi \le L$ is obtained from the previous expression by the interchange $x \leftrightarrow \xi$.

Then, show that the inverse transform of the coefficient to Φ_2 is

$$y_2(x, t) = \mathcal{L}^{-1} \left[\frac{\sinh(\lambda x)}{\sinh(\lambda L)} \right]$$

$$= \sum_{n=0}^{\infty} \left[\eta\left(\frac{L(2n+1) - x}{\sqrt{\kappa}}, t \right) - \eta\left(\frac{L(2n+1) + x}{\sqrt{\kappa}}, t \right) \right] , \tag{8.152}$$

while the inverse LT of the two parts of the kernel $K(x, p, \xi)$ are as follows: for $0 < \xi \le x$ we have

---

[7] Here we accepted without proof that the operation of LT can be applied to the infinite series term-by-term. This can be shown to be true if the series in the Laplace space converges. This is true in our case as we have a geometric progression.

$$K(x, t, \xi) = \mathcal{L}^{-1}[K(x, p, \xi)] = \mathcal{L}^{-1}\left[\frac{\sinh(\lambda\xi)\sinh(\lambda(L - x))}{\lambda\sinh(\lambda L)}\right]$$

$$= \frac{\sqrt{\kappa}}{2}\sum_{n=0}^{\infty}\left[\chi\left(\frac{2Ln + x - \xi}{\sqrt{\kappa}}, t\right) - \chi\left(\frac{2Ln + x + \xi}{\sqrt{\kappa}}, t\right)\right.$$

$$\left. -\chi\left(\frac{2L(n + 1) - x - \xi}{\sqrt{\kappa}}, t\right) + \chi\left(\frac{2L(n + 1) - x + \xi}{\sqrt{\kappa}}, t\right)\right],$$

$$(8.153)$$

whereas for $x \leq \xi \leq L$ we swap x and ξ in the previous expression:

$$K(x, t, \xi) = \mathcal{L}^{-1}[K(x, p, \xi)] = \mathcal{L}^{-1}\left[\frac{\sinh(\lambda x)\sinh(\lambda(L - \xi))}{\lambda\sinh(\lambda L)}\right]$$

$$= \frac{\sqrt{\kappa}}{2}\sum_{n=0}^{\infty}\left[\chi\left(\frac{2Ln + \xi - x}{\sqrt{\kappa}}, t\right) - \chi\left(\frac{2Ln + \xi + x}{\sqrt{\kappa}}, t\right)\right.$$

$$\left. -\chi\left(\frac{2L(n + 1) - \xi - x}{\sqrt{\kappa}}, t\right) + \chi\left(\frac{2L(n + 1) - \xi + x}{\sqrt{\kappa}}, t\right)\right],$$

$$(8.154)$$

where $\chi(\alpha, t)$ is the inverse LT of the function $e^{-\alpha\sqrt{p}}/\sqrt{p}$ which is given by Eq. (6.33).

The obtained formulae allow us finally to construct the exact solution of the heat conductance problem (8.138). Using the above notations and applying the convolution theorem for the LT (Section 6.4.4), we obtain from Eq. (8.146):

$$\psi(x, t) = \int_0^t [\varphi_1(t - \tau)y_1(x, \tau) + \varphi_2(t - \tau)y_2(x, \tau)]d\tau$$

$$+ \frac{1}{\kappa}\int_0^L \phi(\xi)K(x, t, \xi)d\xi.$$

$$(8.155)$$

This is the required general result. It follows that a general problem for a rod of a finite length with arbitrary initial and boundary conditions can indeed be solved analytically, although the final result is expressed via infinite series and convolution integrals.

Consider now a semi-infinite rod, $L \to \infty$, in which case the above expression is drastically simplified. Indeed, in this case the $x = L$ boundary condition should be replaced by $\psi(\infty, t) = 0$. Correspondingly, in the limit the expressions for the functions (8.148), (8.150) and (8.151) simplify to

$$Y_1(p) = e^{-x\sqrt{p/\kappa}}, \quad Y_2(p) = 0,$$

$$K(x, p, \xi) = \sqrt{\frac{\kappa}{p}} \begin{cases} \exp\left(-x\sqrt{p/\kappa}\right) \sinh\left(\xi\sqrt{p/\kappa}\right) , & 0 < \xi < x \\ \exp\left(-\xi\sqrt{p/\kappa}\right) \sinh\left(x\sqrt{p/\kappa}\right) , & x < \xi < \infty \end{cases} .$$

For instance, in the expression for $Y_1(p)$ only one exponential function with $n = 0$ survives, all others tend to zero in the $L \to \infty$ limit. Similarly, one can establish the expressions given above for the other functions.

Problem 8.46 Show that the final solution for a semi-infinite rod is

$$\psi(x, t) = \int_0^t \varphi_1(t - \tau) \eta\left(\frac{x}{\sqrt{\kappa}}, \tau\right) d\tau + \frac{1}{\kappa} \int_0^\infty \phi(\xi) K_\infty(x, t, \xi) d\xi ,$$

$$(8.156)$$

where

$$K_\infty(x, t, \xi) = \frac{\sqrt{\kappa}}{2} \left[\chi\left(\frac{x-\xi}{\sqrt{\kappa}}, t\right) - \chi\left(\frac{x+\xi}{\sqrt{\kappa}}, t\right) \right]$$

for $0 < \xi \leq \infty$, i.e. for the whole integration interval we have a single function. Note that the solution does not contain the term associated with the zero boundary condition at the infinite end of the rod as $y_2(x, t) = 0$ exactly in the $L \to \infty$ limit.

Problem 8.47 Consider a semi-infinite rod $0 \leq x \leq \infty$ which initially had a uniform temperature T_0. At $t = 0$ the $x = 0$ end of the rod is subjected to a heating according to the law $\varphi_1(t) = \sqrt{t}$. Show that at $t > 0$ the distribution of the temperature in the rod is given by the following formula:

$$\psi(x, t) = \int_0^t \eta\left(\frac{x}{\sqrt{\kappa}}, \tau\right) \sqrt{t - \tau} d\tau + T_0 \left[1 - \mathrm{erfc}\left(\frac{x}{\sqrt{4\kappa t}}\right) \right] .$$

The error function was introduced in Eq. (6.50).

Problem 8.48 A semi-infinite rod was initially at zero temperature. Then from $t = 0$ its $x = 0$ end was kept at temperature T_1. Show that the distribution of temperature in the rod is given by

$$\psi(x, t) = T_1 \mathrm{erfc}\left(\frac{x}{\sqrt{4\kappa t}}\right) .$$

Chapter 9
Calculus of Variations

Often in many problems which are encountered in practice one needs to find a minimum (or a maximum) of a function $f(x)$ with respect to its variable x, i.e. the point x_0 where the value of the function is the smallest (largest) within a certain interval $a \leq x \leq b$. The necessary condition for this to happen is for the point x_0 to satisfy the equation $f'(x_0) = 0$. However, sometimes one may need to solve a much more difficult problem of finding the whole function, e.g. $f(x)$, which results in a minimum (or a maximum) of some scalar quantity L which directly depends on it. We shall call this type of dependence a *functional dependence* and denote it as $L[f]$ using square brackets. Taking some particular function $f = f_1(x)$, one gets a particular numerical value L_1 of this functional, while by taking a different function $f = f_2(x)$ one gets a different value L_2. Therefore, it seems that indeed, by taking all possible choices of functions $f(x)$, various values of the quantity L, called the functional of the function f, are obtained, and it is perfectly legitimate to ask a question of whether it is possible to develop a mathematical method whereby the *optimum* function $f = f_0(x)$ that yields the minimum (maximum) value of that scalar quantity (functional) L can be found.

To understand the concept better, let us consider a simple example. Suppose, we would like to prove that the shortest line connecting two points in the 2D space is the straight line. The length of a line between two points (x_1, y_1) and (x_2, y_2) specified by the equation $y = y(x)$ on the $x - y$ plane is given by the integral (Sect. I.4.7.1[1])

$$L = \int_{x_1}^{x_2} \sqrt{1 + (y')^2} dx . \tag{9.1}$$

---

[1] In the following, references to the first volume of this course (L. Kantorovich, Mathematics for natural scientists: fundamentals and basics, 2nd Edition, Springer, 2022) will be made by appending the Roman number I in front of the reference, e.g. Section I.1.8 or Eq. (I.5.18) refer to Section 1.8 and Eq. (5.18) of the first volume, respectively.

Here the function we seek passes through the two points, i.e. it must satisfy the boundary conditions $y(x_1) = y_1$ and $y(x_2) = y_2$. We see that L directly depends on the function $y(x)$ chosen: by taking different functions $y(x)$ passing through the same two points, different values of L are obtained, i.e. the length of the line between the two points depends directly on how we connect them. In this example the length L depends on the function $y(x)$ via its derivative only. The question we are asking is this: prove that the functional dependence $y(x)$ which yields the shortest line between the two points is the straight line. In other words, we need to *minimise* the functional $L[y(x)]$ with respect to the form (shape) of the function $y(x)$, derive the corresponding (hopefully, differential) equation which would correspond to this condition, and then solve it. We expect that solution is a straight line $y = \alpha x + \beta$ with the appropriate values of the two constants corresponding to the line fixed end points.

Similarly, one may consider a more complex 3D problem of finding the shortest line connecting two points A and B when the line lies completely on a surface given by the equation $G(x, y, z) = 0$. In this case we need to seek the minimum of the functional

$$L[x, y, z] = \int_a^b \sqrt{(x')^2 + (y')^2 + (z')^2}\, dt \tag{9.2}$$

with respect to the three functions $x(t)$, $y(t)$ and $z(t)$ which give us the desired line in the parametric form via t. If we disregard the constraint $G(x, y, z) = 0$, then of course the straight line connecting the points A and B would be the answer. However, this may not be the case if we require that the line is to lie on a surface, e.g. on the surface of a sphere. This kind of variational problems is called problems with constraints. In that case the solution will crucially depend on the constraint specified.

It is the purpose of this chapter to discuss how these problems can be solved in a number of most frequently encountered cases. But before we proceed, we have to define a linear functional. A functional $L[f(x)]$ is *linear* if for any two functions $\phi(x)$ and $\varphi(x)$, one can write

$$L[\alpha\phi(x) + \beta\varphi(x)] = \alpha L[\phi] + \beta L[\varphi]$$

with α and β being arbitrary numbers. For instance, the functional

$$L[f] = \int_a^b \left[x^2 f + (f')^\gamma\right] dx$$

is linear with respect to the function $f(x)$ for $\gamma = 0$ or $\gamma = 1$, while it is not linear for any other values of γ.

Fig. 9.1 The two functions, $f(x)$ and $f_1(x)$, have the same initial and final points, and their difference, $\delta f(x) = f_1(x) - f(x)$, is equal to zero at the end points

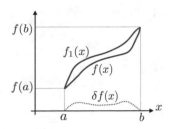

9.1 Functions of a Single Variable

9.1.1 Functionals Involving a Single Function

9.1.1.1 Euler's Equation

Consider the functional dependence, $L[f(x)]$, in the case of a single function of a single variable, $f(x)$, in more detail. Let $L = L[f]$ be the value of the functional corresponding to the function $f(x)$. Using a different function, $f_1(x)$, which satisfies the same boundary conditions, Fig. 9.1, will result in the value $L_1 = L[f_1]$ of the functional. The difference between the two functions, $\delta f(x) = f_1(x) - f(x)$, yields the difference in the values of the functional:

$$\Delta L = L[f_1] - L[f] = L[f + \delta f] - L[f] \ .$$

Next, we define ρ as the maximum deviation of f_1 from f within the interval $a < x < b$, i.e.

$$\rho = \max_{a<x<b} |\delta f| \ .$$

Then, ΔL can be written as

$$\Delta L = \Delta L_1[f, \delta f] + \Delta L_2[f, \delta f] \ ,$$

where the functional ΔL_1 is *linear* with respect to δf, while ΔL_2 is not. In fact, ΔL_1 tends to zero linearly with $\rho \to 0$, while the functional ΔL_2 tends to zero faster than ρ. We can say that $\Delta L_2/\rho \to 0$ as $\rho \to 0$.[2] The first part of the difference, ΔL_1, is called the *variation* of the functional due to change $\delta f(x)$ of the function and is usually denoted δL.

As an example, let us consider the functional

$$L[f] = \int_a^b \left(2f^2 + x^3 f'\right) dx \tag{9.3}$$

---

[2] This is very similar to the differential of a function: $\Delta y = y(x + \Delta x) - y(x)$ is given by a sum of two terms, one being linear in Δx, the other depends on higher powers of Δx, i.e. tends to zero much faster than Δx itself (see Sect. I.3.1).

of a function $f(x)$. In this case,

$$\Delta L = \int_a^b \left\{ 2 \left(f + \delta f \right)^2 + x^3 \left(f' + \delta f' \right) \right\} dx - \int_a^b \left(2 f^2 + x^3 f' \right) dx$$

$$= \Delta L_1 + \Delta L_2 ,$$

where

$$\Delta L_1 = \int_a^b \left[4 f \delta f + x^3 \delta f' \right] dx \quad \text{and} \quad \Delta L_2 = \int_a^b 2 \left(\delta f \right)^2 dx .$$

One can see that ΔL_1 behaves like ρ, while ΔL_2 as ρ^2. Indeed, since $|\delta f| \le \rho$, then the first term in ΔL_1 can be estimated as

$$\left| \int_a^b \left[4 f \delta f \right] dx \right| \le \int_a^b 4 |f| \, |\delta f| \, dx \ll 4 \rho \int_a^b |f| \, dx \sim \rho ,$$

while the second contribution can be estimated by first integrating by parts:

$$\int_a^b x^3 \delta f' dx = \int_a^b x^3 \left(\delta f \right)' dx = \underbrace{x^3 \delta f \big|_a^b}_{=0} - \int_a^b 3 x^2 \delta f \, dx = - \int_a^b 3 x^2 \delta f \, dx ,$$

and hence the module of this term is also proportional to ρ as

$$\left| - \int_a^b 3 x^2 \delta f \, dx \right| \le \int_a^b \left| 3 x^2 \right| \, |\delta f| \, dx \le \rho \int_a^b 3 x^2 \, dx = \rho \left(b^3 - a^3 \right) .$$

Note that we have used here that $\delta f' = (\delta f)'$ and that $\delta f = 0$ at the end points of our interval as the two functions are equal there, Fig. 9.1. Let us now estimate ΔL_2:

$$\left| \int_a^b 2 \left(\delta f \right)^2 dx \right| \le 2 \int_a^b |\delta f|^2 \, dx \le 2 \rho^2 \left(b - a \right) ,$$

and hence $|\Delta L_2 / \rho| \sim \rho \to 0$ as $\rho \to 0$, as required. Hence, the variation of L is $\delta L = \Delta L_1$.

The decomposition of ΔL into the two contributions can be simplified in practice using the following simple method. The trick is to take some *fixed* function δf which is equal to zero at the ends of the interval, and then consider a function

$$L \left(\alpha \right) = L \left[f + \alpha \delta f \right] .$$

Here $\alpha \delta f$ serves as a variation of $f(x)$ and $L(\alpha)$ becomes an ordinary function of a single variable α. By taking small values of α one can make the deviation $\alpha \delta f$ from f to be arbitrarily small for all values of x within the interval $a < x < b$. Then, the division of ΔL into two terms (one being a linear functional and the other of higher order in terms of the deviation) can be made similarly to the way this is done for an ordinary function of a single variable as $L(\alpha)$ is indeed such a function of α. Expanding $L(\alpha)$ into the Taylor series around $\alpha = 0$, one gets

$$\Delta L = L(\alpha) - L(0) = L'(0)\alpha + \frac{1}{2}L''(0)\alpha^2 + \dots . \tag{9.4}$$

Here $L(0) = L[f]$. On the other hand,

$$\begin{aligned} \Delta L &= L[f + \alpha\delta f] - L[f] = \Delta L_1[f, \alpha\delta f] + \Delta L_2[f, \alpha\delta f] \\ &= \alpha \Delta L_1[f, \delta f] + \Delta L_2[f, \alpha\delta f] , \end{aligned} \tag{9.5}$$

where we were able to take α out in the ΔL_1 term since it is a linear functional with respect to its second argument (the variation of the function). Comparing Eqs. (9.4) and (9.5), we conclude that $L'(0)$ must be equal to $\Delta L_1[f, \delta f]$, while the following terms in the expansion (9.4), which behave at least as α^2, will correspond to $\Delta L_2[f, \alpha\delta f]$. Indeed, if $\rho_0 = \max_{a<x<b}|\delta f| > 0$ is the maximum value of the function $\delta f(x)$ within our interval, then $\rho = \max_{a<x<b}|\alpha\delta f| = |\alpha|\rho_0$ is the maximum deviation of $f + \alpha\delta f$ from f, and hence the limit $\rho \to 0$ is equivalent to taking the $\alpha \to 0$ limit. It is clear then that

$$\lim_{\rho\to 0}\frac{|\Delta L_2[f, \alpha\delta f]|}{\rho} = \left|\lim_{\alpha\to 0}\frac{\Delta L_2[f, \alpha\delta f]}{\alpha\rho_0}\right| = \frac{1}{\rho_0}\left|\lim_{\alpha\to 0}\frac{\Delta L_2[f, \alpha\delta f]}{\alpha}\right| = 0 ,$$

as required. Therefore, the differential of $L[f]$ can be defined as

$$\delta L = \Delta L_1[f, \delta f] = L'(0) = L'(\alpha)\big|_{\alpha=0} . \tag{9.6}$$

As an illustration of this method, let us apply this simple result again to the functional (9.3):

$$L(\alpha) = \int_a^b \left[2(f + \alpha\delta f)^2 + x^3\left(f' + \alpha\delta f'\right)\right] dx .$$

Differentiating with respect to α under the integral sign, we get

$$L'(0) = \lim_{\alpha\to 0}\int_a^b \left[4(f + \alpha\delta f)\delta f + x^3\delta f'\right] dx = \int_a^b \left[4f\delta f + x^3\delta f'\right] dx .$$

This is the same result as before.

Problem 9.1 Show that the variation of the functional

$$L[f] = \int_a^b \left[3x^3 f^3 + 2\left(f'\right)^2 \right] dx \quad \text{is} \quad \delta L = \Delta L_1 = \int_a^b \left[9x^3 f^2 \delta f + 4 f' \delta f' \right] dx ,$$

and that the corresponding ΔL_2 term behaves in such a way that ΔL_2 $[f, \alpha \delta f]/\alpha \to 0$ as $\alpha \to 0$.

Next we shall consider the simplest case of the functional $L[f(x)]$ given by an integral

$$L[f] = \int_a^b F\left(x, f, f'\right) dx . \tag{9.7}$$

Here $F\left(x, f, f'\right)$ is some function which we assume is continuous with respect to its three variables and has the necessary continuous derivatives with respect to any of them. Note that F depends on x and, at the same time, is a functional of $f(x)$, depending on the function itself and its first derivative. Let us calculate the variation δL of this functional:

$$L(\alpha) = \int_a^b F\left(x, f + \alpha \delta f, f' + \alpha \delta f'\right) dx$$

$$\implies \quad L'(\alpha)\big|_{\alpha=0} = \int_a^b \left(\frac{dF}{d\alpha}\right)_{\alpha=0} dx = \int_a^b \left[\frac{\partial F}{\partial f}\delta f + \frac{\partial F}{\partial f'}\delta f'\right] dx .$$

There are two contributions, coming from the second and the third arguments of the function F. The contribution to the integral due to the second term in the integrand above (related to the variation of the derivative $\delta f'$) we shall take by parts:

$$\int_a^b \frac{\partial F}{\partial f'}\delta f' dx = \int_a^b \frac{\partial F}{\partial f'}\left(\delta f\right)' dx = \underbrace{\frac{\partial F}{\partial f'}\delta f \bigg|_a^b}_{=0} - \int_a^b \delta f \frac{d}{dx}\left(\frac{\partial F}{\partial f'}\right) dx$$

$$= -\int_a^b \delta f \frac{d}{dx}\left(\frac{\partial F}{\partial f'}\right) dx ,$$

where the free (first) term which appeared after the integration by parts is zero since $\delta f(a) = \delta f(b) = 0$. An important point here is that the x-derivative here is the *total derivative* which takes account not only of the direct dependence of F on x via the first argument, but also its dependence on x via f and f', its second and the third arguments. This is because we have taken the integral over x by parts treating $F_1(x) = \partial F/\partial f'$ as a function of x only, i.e. including its dependence on x via f and f' as well.

Hence, we obtain

$$\delta L = L'(\alpha)\big|_{\alpha=0} = \int_a^b \left[\frac{\partial F}{\partial f} \delta f + \frac{\partial F}{\partial f'} \delta f' \right] dx$$

$$= \int_a^b \left[\frac{\partial F}{\partial f} - \frac{d}{dx} \left(\frac{\partial F}{\partial f'} \right) \right] \delta f \, dx \ . \tag{9.8}$$

Next, we should consider the necessary condition for the functional $L[f]$ be minimum (maximum). Let us assume that the function $f_0(x)$ gives the maximum to the functional. This means that for any function $f(x)$ such that $|f(x) - f_0(x)| < \epsilon$ (ϵ is an arbitrary positive number) follows that $L[f] < L[f_0]$. This is to be satisfied for any x within the interval $a \le x \le b$. The minimum of the functional is defined in a similar manner.

What we are going to do now is to employ the given above definition to establish the necessary condition for the maximum. Choose $f = f_0 + \alpha \delta f$. Then,

$$|f(x) - f_0(x)| = |\alpha \delta f(x)| = |\alpha| \, |\delta f| < |\alpha| \rho_0 \ ,$$

so that one can always choose a sufficiently small α such that $|\alpha| \rho_0 < \epsilon$, where ρ_0 is the largest fluctuation of δf. But if f_0 gives the maximum, then $L[f_0 + \alpha \delta f] < L[f_0]$, which means that $L(\alpha) < L(0)$ for any α satisfying the above condition. Therefore, the value of $\alpha = 0$ corresponds to the maximum of the function $L(\alpha)$ for which the necessary condition is obviously $L'(0) = 0$. Hence, this must be the necessary condition we are looking for

$$L'(0) = L'(\alpha)\big|_{\alpha=0} = \delta L = 0 \ . \tag{9.9}$$

Naturally, the same condition is obtained for the minimum of a functional.

To obtain the appropriate DE for the function $f(x)$ which satisfies this necessary condition, we need the following theorem.

Theorem 9.1 *Let $\eta(x)$ be any function which is zero at the ends of the interval, $\eta(a) = \eta(b) = 0$, and is continuous together with its first derivative within the whole interval. Consider a continuous function $f(x)$. If for any such $\eta(x)$ the integral*

$$\int_a^b f(x)\eta(x)dx = 0 \ ,$$

then $f(x) = 0$.

Proof We shall prove this theorem by contradiction. Assume that there is a point x_0 within our interval $a < x < b$ such that $f(x_0) > 0$. Since $f(x)$ is assumed to be continuous, this means that there must be a vicinity of the point x_0 of some width 2ϵ (with $\epsilon > 0$), where the function $f(x)$ is also positive (see Theorem I.2.17), i.e.

$f(x) > 0$ for any x from some interval $\epsilon_- < x < \epsilon_+$ (with $\epsilon_\pm = x_0 \pm \epsilon$). Next, we can construct a particular function $\eta(x)$ in such a way that

$$\eta(x) = (x - \epsilon_+)^2 (x - \epsilon_-)^2 \qquad (9.10)$$

within that interval and zero otherwise. The value of ϵ can be chosen small enough so that the points $\epsilon_\pm$ both lie inside the original interval $a < x < b$. The function $\eta(x)$ defined in this way is everywhere continuous and is equal to zero at the points a and b. Its first derivative is also continuous (and is equal to zero at the points $x = \epsilon_\pm$ and beyond the interval $\epsilon_- < x < \epsilon_+$). At the same time, the integral

$$\int_a^b f(x)\eta(x)dx = \int_{\epsilon_-}^{\epsilon_+} f(x)\eta(x)dx = \int_{\epsilon_-}^{\epsilon_+} f(x)(x - \epsilon_+)^2 (x - \epsilon_-)^2 \, dx > 0 \, ,$$

since $f(x) > 0$ within the chosen integration limits. This contradicts the condition of the theorem, and hence our assumption was wrong. Note that the integration limits in the integral above were changed to $\epsilon_\pm$ because the function $\eta = 0$ outside it. **Q.E.D.**

This simple theorem allows us to find the DE for the function $f(x)$ which delivers a minimum (or a maximum) to the functional (9.7). Indeed, the variation of this functional is given by Eq. (9.8) containing the arbitrary variation $\delta f(x)$ under the integral. Since the necessary condition for the extrema is that the variation of the functional to be zero, the integral (9.8) must be equal to zero for arbitrary variation δf. Using the proven theorem, this means that the expression in the square brackets in Eq. (9.8) must necessarily be zero, i.e.

$$\frac{\partial F}{\partial f} - \frac{d}{dx}\left(\frac{\partial F}{\partial f'}\right) = 0 \, . \qquad (9.11)$$

This is called Euler's equation. This DE is to be solved subject to the boundary conditions at the ends of the interval: $f(a) = A$ and $f(b) = B$. In practical calculations it is sometimes useful to write the second term in the left-hand side, containing the total derivative with respect to x, explicitly. Indeed, the partial derivative of F with respect to f' depends on x, f and f', with the latter two being also functions of the x, and hence

$$\frac{d}{dx}\left(\frac{\partial F}{\partial f'}\right) = \frac{\partial^2 F}{\partial x \partial f'} + \frac{\partial^2 F}{\partial f \partial f'}f' + \frac{\partial^2 F}{\partial f' \partial f'}f''$$
$$= F_{xf'} + F_{ff'}f' + F_{f'f'}f'' \, .$$

This manipulation allows rewriting Euler's equation (9.11) as follows:

$$F_f - F_{xf'} - F_{ff'}f' - F_{f'f'}f'' = 0 \, . \qquad (9.12)$$

It becomes clear now that this is a second-order DE with respect to the function $f(x)$, and hence its solution $f(x, C_1, C_2)$ will depend on two arbitrary constants C_1 and

C_2 which are to be chosen to satisfy the boundary conditions $f(a, C_1, C_2) = A$ and $f(b, C_1, C_2) = B$. Three cases are possible: (i) there are no solutions and hence the functional (9.7) has no extrema; (ii) there is only one solution and hence only one extremum, and, finally, (iii) there are several solutions, i.e. several extrema exist.

It is instructive at this point to consider several particular cases of the function $F(x, f, f')$ serving as the integrand in the functional (9.7).

Case 1 If $F = F(x, f)$, i.e. this function does not depend on f', then Euler's equation is simply $F_f = 0$. This is an algebraic equation with respect to the function f, and hence the solution does not contain any arbitrary constants. This means that it may not satisfy the boundary conditions.

Case 2 If $F(x, f, f')$ is linear with respect to the derivative of f, i.e.

$$F(x, f, f') = g(x, f) + h(x, f) f' , \tag{9.13}$$

then Euler's equation gives

$$\frac{\partial g}{\partial f} + f' \frac{\partial h}{\partial f} - \frac{dh}{dx} = 0 \implies \frac{\partial g}{\partial f} + f' \frac{\partial h}{\partial f} - \left(\frac{\partial h}{\partial x} + \frac{\partial h}{\partial f} f' \right) = 0$$

$$\implies \frac{\partial g}{\partial f} = \frac{\partial h}{\partial x} . \tag{9.14}$$

This is again an algebraic equation for $f(x)$ and hence f will not contain any arbitrary constants. However, as we shall show presently, this case is really problematic as it appears that the functional does not depend on the function f at all! Indeed, because of the condition (9.14), there is a function $U(x, f)$ such that its exact differential is $dU = g dx + h df$ with $g = \partial U / \partial x$ and $h = \partial U / \partial f$ (see Sect. I.5.5). The total derivative of U with respect to x is

$$\frac{dU}{dx} = \frac{\partial U}{\partial x} + \frac{\partial U}{\partial f} f' = g + h f' ,$$

which coincides exactly with the function F in Eq. (9.13). Therefore, the functional containing it,

$$L[f] = \int_a^b \left(g + h f' \right) dx = \int_a^b \frac{dU}{dx} dx = U(b, f(b)) - U(a, f(a)) ,$$

is simply a constant, it only depends on the points at the ends of the interval! So, no solution in this case is possible at all.

Case 3 Consider now the function F being independent of x and f, i.e. it only depends on the derivative f', i.e. $F = F(f')$. Euler's equation in this case is

$$\frac{d}{dx}\left(\frac{\partial F}{\partial f'}\right) = 0 \quad \Longrightarrow \quad \frac{\partial F}{\partial f'} = C_1 \, ,$$

where C_1 is a constant. This is an algebraic equation with respect to the derivative f', resulting in f' being equal to some number, let us call it α (it is a function of C_1): $f' = \alpha$. The only solution of this DE is the linear function $f(x) = \alpha x + \beta$.

This particular case we shall illustrate by a simple example, mentioned at the beginning of this chapter, of finding the shortest line on the $x - y$ plane connecting two points. The appropriate functional for the line $y(x)$ is given by Eq. (9.1) which depends exclusively on the derivative y'. Hence, Euler's equation immediately gives a linear function $y = \alpha x + \beta$ with the two constants to be obtained from the coordinates of the two points. We have obtained the result to be expected!

Case 4 Finally, let $F = F\left(f, f'\right)$, i.e. the function F does not explicitly depend on the variable x. In this case Euler's equation (9.12) reads

$$F_f - F_{ff'} f' - F_{f'f'} f'' = 0 \, . \tag{9.15}$$

This is the second-order DE which can be integrated once and then transformed into a first-order DE with respect to the function $f(x)$. Do this end, multiply both sides of Eq. (9.15) by f' and add and subtract $F_{f'} f''$:

$$f'\left(F_f - F_{ff'} f' - F_{f'f'} f''\right) + F_{f'} f'' - F_{f'} f'' = 0 \, ,$$

or, after some trivial manipulation,

$$\left(f' F_f + F_{f'} f''\right) - \left[F_{ff'}\left(f'\right)^2 + F_{f'f'} f' f'' + F_{f'} f''\right] = 0 \, . \tag{9.16}$$

The first term within the round brackets is simply the total derivative $\frac{dF}{dx}$. At the same time, the second term in the square brackets is also the total derivative of $f' \frac{\partial F}{\partial f'}$ with respect to x (recall, that F only explicitly depends on f and f'):

$$\frac{d}{dx}\left(f' \frac{\partial F}{\partial f'}\right) = f'' \frac{\partial F}{\partial f'} + f'\left(\frac{\partial^2 F}{\partial f \partial f'} f' + \frac{\partial^2 F}{\partial f' \partial f'} f''\right) \, ,$$

which is the same as in the square brackets in Eq. (9.16). Therefore, Eq. (9.16) can be written as

$$\frac{d}{dx}\left(F - f' \frac{\partial F}{\partial f'}\right) = 0 \quad \Longrightarrow \quad F - f' \frac{\partial F}{\partial f'} = C_1 \, . \tag{9.17}$$

This is the final result: the obtained DE is of the first order, and C_1 is an arbitrary constant. Solving the obtained DE gives another constant, C_2, and this is enough for the two boundary conditions to be satisfied.

Problem 9.2 Consider the functional

$$L = \int_a^b \left[x^2 f + (f')^2 \right] dx \ .$$

Show that its stationary value is given by the function

$$f(x) = \frac{1}{24} x^4 + C_1 x + C_2 \ ,$$

which is a solution of the DE $f'' = \frac{1}{2} x^2$. Here C_1 and C_2 are two constants.

Problem 9.3 Show that the stationary value of the functional

$$L[f] = \int_0^\pi \left[3f^2 + f \sin x + 2 (f')^2 \right] dx$$

is given by the function satisfying the DE

$$f'' - \frac{3}{2} f = \frac{1}{4} \sin x \ .$$

Assuming that the boundary conditions are $f(0) = 0$ and $f(\pi) = 0$ (i.e. $a = 0$ and $b = \pi$), show that $f(x) = -\frac{1}{10} \sin x$. Use this function in the functional to show that its value at this optimum function is equal to $-\pi/40$.

Problem 9.4 Consider the functional

$$L = \int_a^b \left[f + f^2 + e^x - A f f' \right] dx \ .$$

Show that the minimum (maximum) of L is given by a constant function $f(x) = -1/2$ (which may not satisfy the boundary conditions) irrespective of the value of the constant A. Which case does this problem correspond to? [*Hint: could the x dependence in the functional be ignored? Could the term $-A f f'$ be integrated for any $f(x)$ and hence also ignored?*]

Problem 9.5 Show that the stationary value of the functional

$$L = \int_a^b \frac{\sqrt{1 + (f')^2}}{f} dx$$

is given by the function

$$f(x) = \pm\sqrt{C_1 - (x + C_2)^2},$$

where C_1 and C_2 are two arbitrary constants.

9.1.1.2 Geodesic Problem

We shall illustrate the application of the Euler formula (9.11) by solving a classical *geodesic* problem of finding the shortest line between two points $A\,(R, \theta_A, \phi_A)$ and $B\,(R, \theta_B, \phi_B)$ (specified using the spherical coordinates) if the line lies on a sphere of radius R and centred at the origin. A square of the element of the line length, $(ds)^2$, lying on the sphere is given by Eq. (7.37), so that the required element of the arc length is

$$ds = R\sqrt{(d\theta)^2 + \sin^2\theta\,(d\phi)^2}.$$

It is convenient to define the line by the equation $\theta = \theta\,(\phi)$ (with $\theta_A = \theta\,(\phi_A)$ and $\theta_B = \theta\,(\phi_B)$) in which the angle ϕ is used as an independent variable (the line parameter). Then,

$$ds = R\sqrt{\left(\frac{d\theta}{d\phi}\right)^2 + \sin^2\theta}\,d\phi,$$

and hence the line length is given by integrating ds along the line:

$$L = R\int_{\phi_A}^{\phi_B} \sqrt{\left(\frac{d\theta}{d\phi}\right)^2 + \sin^2\theta}\,d\phi.$$

We see that the function F in the Euler formula is

$$F\left(\theta, \theta'\right) = R\sqrt{(\theta')^2 + \sin^2\theta}.$$

It does not depend on the variable ϕ at all, so that the variation of the functional with respect to the unknown function $\theta\,(\phi)$ results in the equation (9.17) with that F. Performing the required differentiation $\partial F/\partial\theta'$, we obtain from (9.17):

$$\sqrt{(\theta')^2 + \sin^2\theta} - \frac{(\theta')^2}{\sqrt{(\theta')^2 + \sin^2\theta}} = \frac{C_1}{R} \quad\Longrightarrow\quad \frac{\sin^2\theta}{\sqrt{(\theta')^2 + \sin^2\theta}} = \frac{1}{\lambda},$$

with $\lambda = R/C_1$. The obtained expression can be solved easily with respect to θ' leading to an ordinary first-order DE in which the variables are separated:

$$\theta' = \sqrt{\lambda^2 \sin^4 \theta - \sin^2 \theta} \quad \Longrightarrow \quad \int \frac{d\theta}{\sqrt{\lambda^2 \sin^4 \theta - \sin^2 \theta}} = \phi + C_2 \,.$$

The integral is calculated using the substitution

$$t = \frac{\cot \theta}{\sqrt{\lambda^2 - 1}} \,, \quad dt = -\frac{1}{\sqrt{\lambda^2 - 1}} \left(1 + \cot^2 \theta\right) d\theta = -\frac{1}{\sqrt{\lambda^2 - 1}} \frac{d\theta}{\sin^2 \theta} \,,$$

and noting that

$$\sqrt{\lambda^2 \sin^4 \theta - \sin^2 \theta} = \sin^2 \theta \sqrt{\lambda^2 - \frac{1}{\sin^2 \theta}} = \sin^2 \theta \sqrt{\lambda^2 - \left(1 + \cot^2 \theta\right)}$$
$$= \sin^2 \theta \sqrt{\lambda^2 - 1}\sqrt{1 - t^2} \,.$$

We obtain

$$-\int \frac{dt}{\sqrt{1 - t^2}} = \phi + C_2 \quad \Longrightarrow \quad \arcsin t = -\phi - C_2 \,,$$

which leads to

$$\sin \left(-C_2 - \phi\right) = t \quad \Longrightarrow \quad \sin \left(C_2 + \phi\right) = -\frac{\cot \theta}{\sqrt{\lambda^2 - 1}} \,,$$

or, writing the sine in the left-hand side via cosine and sine of the angles C_2 and ϕ using the well-known trigonometric identity, and splitting up the cotangent, we obtain

$$\sin C_2 \cos \phi + \cos C_2 \sin \phi = \frac{-\cot \theta}{\sqrt{\lambda^2 - 1}}$$

$$\Longrightarrow \quad \sin C_2 \left(R \sin \theta \cos \phi\right) + \cos C_2 \left(R \sin \theta \sin \phi\right) = \frac{-\left(R \cos \theta\right)}{\sqrt{\lambda^2 - 1}} \,.$$

We recognise in the expressions in the round brackets the coordinates $x = R \sin \theta \cos \phi$, $y = R \sin \theta \sin \phi$ and $z = R \cos \theta$ of the line on the surface of the sphere. Therefore, we obtain

$$\alpha x + \beta y + \gamma z = 0 \,,$$

where $\alpha = \sin C_2$ and $\beta = \cos C_2$ are two related constants, while $\gamma = 1/\sqrt{\lambda^2 - 1}$ is another independent constant, so we have two independent constants in total. We have obtained an equation for a plane passing through the centre of the coordinate system and the two points A and B (which determine the two constants C_2 and γ).

Hence, the shortest line connecting two points on a sphere lies on the intersection of the sphere and this plane.

Problem 9.6 Consider two points A and B on a cylinder of radius R. Show that the shortest line connecting these two points and lying on the cylinder surface is given in the cylindrical coordinates (R, ϕ, z) by $z(\phi) = C_1\phi + C_2$, where the constants C_1 and C_2 are determined from the coordinates of the two points. [*Hint: use formula (7.36).*]

Problem 9.7 Consider the same problem for the cone surface given by the equation $z = \sqrt{x^2 + y^2}$. Using polar coordinates (r, ϕ), show that in this case the optimum curve is given by

$$r(\phi) = \frac{C_1}{\cos\left(\sqrt{2}C_1\phi + C_2\right)} ,$$

where C_1 and C_2 are two constants.

9.1.1.3 A Few Classical Variational Problems

Here we shall formulate as problems two classical problems involving the usage of the Euler equation.

Problem 9.8 (*Classical* catenary *problem*) Consider a curve $y = y(x)$ drawn between two points $A(a, y(a))$ and $B(b, y(b))$ on the $x - y$ plane. If we rotate this line around the x axis (Sect. I.4.7.4), a surface of revolution will be drawn with the surface area given by Eq. (I.4.107):

$$S = 2\pi \int_a^b y\sqrt{1 + (y')^2}\,dx .$$

Show that the curve resulting in the minimal possible surface area is given by solving the differential equation $yy'' = 1 + y'^2$, and is provided by

$$y(x) = C_1 \cosh\frac{x + C_2}{C_1} ,$$

which is called *catenary*. Here C_1 and C_2 are two constants which are determined from the coordinates of the two fixed points. [*Hint: the integral of*

$(x^2 - 1)^{-1/2}$ *is calculated using the substitution* $x = \cosh t$ *leading to the inverse hyperbolic cosine function for the integral.]*

Problem 9.9 *(Classical* brachistochrone *problem)* Consider a slope described by the equation $y = y(x)$ with the x axis running horizontally and the y axis vertically down, Fig. 9.2. An object is released from the top (point A with zero x and y coordinates) which slides down along the slope to reach its bottom at point $B(x_B, y_B)$. The velocity $v(x)$ of the object at its horizontal position x can be determined from the conservation of energy:

$$mgy = \frac{mv^2}{2} \implies v(x) = \sqrt{2gy(x)} \,,$$

where m is the mass of the object and g the gravitational constant (on Earth). If $ds = \sqrt{1 + (y')^2}dx$ is the elementary length of the slide around point x, the time required to cross it is $ds/v(x)$. The total travel time to slide from the top of the slide to its bottom is given by the integral

$$L = \int_0^{x_B} \frac{ds}{v} = \frac{1}{\sqrt{2g}} \int_0^{x_B} \sqrt{\frac{1 + (y')^2}{y}} dx \,.$$

Show first that the Euler equation reads $-1 - (y')^2 = 2yy''$. Next, transform this equation to $y' = \sqrt{(2C_1 - y)/y}$, where C_1 is a constant. Finally, integrate this DE over y by making the substitution

$$y = 2C_1 \sin^2 (t/2) = C_1 (1 - \cos t)$$

to show that

$$x = C_1 (t - \sin t) + C_2 \,.$$

The two equations $x = x(t)$ and $y = y(t)$ give the optimum curve of the slide (i.e. corresponding to the shortest sliding time) in a parameter form. The first of the boundary conditions $x = y = 0$ is satisfied with $t = 0$ and $C_2 = 0$. Demonstrate by plotting $f_1(t) = 1/(1 - \cos t)$ and $f_2(t) = 1/(t - \sin t)$ for $0 \leq t \leq 2\pi$ that they intersect only at a single point ensuring that it is always possible to determine the other constant C_1 from the coordinates of the point B. The curve we obtained is drawn in the $x - y$ plane by a fixed point on a wheel rolling in the x direction; it is called cycloid.

Fig. 9.2 An object of mass
m slides down the slope
$y = y(x)$

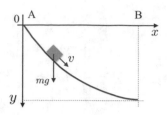

9.1.1.4   The Width of a Domain Wall

In magnetic materials atoms possess localised magnetic moments (also called 'spins')
due to partially filled d or f electronic shells. In ferromagnetic materials in particular
magnetic moments on atoms may spontaneously align in the same direction so that
the whole sample may acquire a large magnetic moment. For relatively large crystalline samples this situation is however highly energetically unfavourable as large
regions of aligned magnetic moments carry significant positive energy. To reduce
the total energy of the system, a compensation mechanism is required. And indeed,
in relatively large magnetic samples, regions of different magnetisation are formed,
called domains, such that overall the magnetisation of the whole sample becomes
very small.

Therefore, in such samples there must be interfacial regions across which the
magnetisation changes from one direction on one side of the interface to an opposite
one on the other. This interfacial regions are called domain walls. In this section we
shall discuss the calculation of the width of such a wall and will try to understand
which factors affect it.

There are two main factors that influence the alignment of magnetic moments
(spins): the so-called exchange interaction between neighbouring spins and a crystal
anisotropy that favours the direction of the spin along a particular crystalline axis.
By assuming this axis to be the z axis, we may say that the second factor favours
alignment of spins along either positive or negative direction of the z axis.

The energy of such a magnetic material can be described by the so-called Heisenberg Hamiltonian

$$E = - \sum_{(i,j) \text{ pairs}} J \mathbf{S}_i \cdot \mathbf{S}_j - \kappa \sum_i S_{iz}^2 , \qquad (9.18)$$

where the first term describes the magnetic interaction between neighbouring spins
$\mathbf{S}_i$ and $\mathbf{S}_j$ with $J > 0$ being its strength, and we sum over all nearest neighbour pairs
(i, j) of atoms, while the second term takes account of the anisotropy contribution
where we sum over all spins. The spins $\mathbf{S}_i$ here are treated as classical vectors. With
the strength J being positive, the first term favours alignment of the spins since an
increase of the value of the dot product $\mathbf{S}_i \cdot \mathbf{S}_j$ leads to a more negative contribution;

Fig. 9.3 The interfacial region (blue spins) between two domains with opposite direction of their spins may be either **a** abrupt, with the extension of just a few interatomic distances, or **b** long, with $N \gg 1$ spins changing their direction over a large distance $L = Na$. The cross section of the domain wall of area A is also indicated in each case

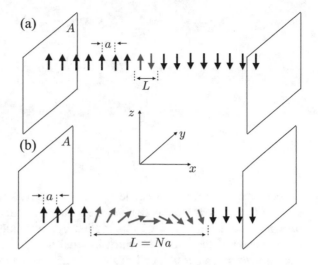

the anisotropy contribution reduces the system energy as well with the amount of such a decrease depending on the magnitude of the projection of the spins, S_{iz}, on the z axis, acting equally in either of the two directions.

For simplicity, we shall assume a very simple cubic structure with an interatomic distance a. Let us direct the x axis across the interface with its centre being at $x = 0$. The spins on the left of the interface point (at $x < 0$) are directed towards the positive z direction (up), and those on the right ($x > 0$) in the negative one (down), see Fig. 9.3. Hence, around the $x = 0$ we should expect an intermediate region in which the spins change their direction. If the ratio J/κ of the two energies is small, one would expect an abrupt change across the interface since in this case the second term is dominant, Fig. 9.3a; if the ratio J/κ is large, then the domain wall is expected to be wide since in this case the first term is dominant favouring a slow rotation of the spins, Fig. 9.3b. Note that we have assumed that the flip of the spins happens within the $x - z$ plane; the other possibility that the spins flip out of this plane is also possible but will not be considered here.

To proceed, we shall assume a slow change of the direction of the spins across the interface; this enables us to approach the problem approximately applying the continuous limit $a \to 0$ when atoms (spins) are assumed to be very close to each other. This approximation is reasonable if the region of the interface contains many atomic sites. Then, the sum over spins can be converted into the integral over x, the direction across the interface. Let us consider the spin interaction term first:

$$-\sum_{(i,j)\text{ pairs}} J\mathbf{S}_i \cdot \mathbf{S}_j = -\frac{A}{a^2}\sum_{k\in x} JS^2 \cos \underbrace{(\theta_k - \theta_{k+1})}_{\Delta\theta_k}$$

$$\approx -\frac{A}{a^2}\sum_{k\in x} JS^2 \left[1 - \frac{1}{2}(\Delta\theta_k)^2\right]$$

$$\implies \frac{A}{2a^2}\sum_{k\in x} JS^2 (\Delta\theta_k)^2 \implies \frac{AJS^2}{2a^2}\sum_{k\in x}\left(\frac{\Delta\theta_k}{\Delta x}\right)^2 (a\Delta x)$$

$$= \frac{AJS^2}{2a}\int_{-\infty}^{\infty}\left(\frac{d\theta(x)}{dx}\right)^2 dx.$$

Here, on the first line we have done two things: firstly, we have replaced the sum over all spin pairs across the junction via the sum over spins along a single x direction (using the summation index k) multiplied by the number of such x–lines in the cross section of the interface, which is equal to A/a^2 (assuming that there is one such line of spins per unit cell of the area a^2); secondly, we have written the dot product of spins via the angle $\Delta\theta_k = \theta_k - \theta_{k+1}$ between the neighbouring spins, where θ_k is the angle the spin vector $\mathbf{S}_k$ makes with the z axis. On the second line we have expanded the cosine keeping only the first two terms assuming small angles $\Delta\theta_k$ between every pair of neighbouring spins. On the third line we, first, dropped the unimportant constant term; then, secondly, we introduced the distance $\Delta x = a$ between the neighbouring spins, which enabled us on the last line to replace the sum over spins along the x direction with the x integral, which is the desired continuum limit. In that limit the direction of the spins along each line across the interface becomes a continuous function of x measured by the angle $\theta(x)$.

The second energy term in the Hamiltonian due to anisotropy can be worked out similarly:

$$-\kappa\sum_i S_{iz}^2 = -\left(\frac{A}{a^2}\right)\kappa\sum_{k\in x}(S\cos\theta_k)^2 = -\kappa\frac{AS^2}{a^2}\sum_{k\in x}\cos^2\theta_k\frac{\Delta x}{a}$$

$$\implies -\kappa\frac{AS^2}{a^3}\int_{-\infty}^{\infty}\cos^2\theta(x)\,dx.$$

Combining both energy contributions, we obtain the following expression for the energy of the interface, per its unit cross-sectional area:

$$\frac{E}{A} = \frac{S^2\kappa}{a^3}\int_{-\infty}^{\infty}\left[\frac{Ja^2}{2\kappa}\left(\frac{d\theta(x)}{dx}\right)^2 - \cos^2\theta(x)\right]dx. \tag{9.19}$$

To find the most energetically favourable configuration of the spins along the x direction, i.e. the dependence $\theta(x)$, one has to minimise this expression for the energy. The integrand here is the functional

$$F\left(\theta(x), \theta'(x)\right) = \frac{Ja^2}{2\kappa}\left(\theta'(x)\right)^2 - \cos^2\theta(x)$$

of both the unknown function and its first derivative. Hence, we can use Euler's equation (9.11) to minimise the energy expression:

$$2\cos\theta(x)\sin\theta(x) = \frac{Ja^2 2}{2\kappa}\theta''(x) \implies \frac{d^2\theta}{dx^2} = \lambda\sin(2\theta), \qquad (9.20)$$

with $\lambda = \kappa/(Ja^2)$.

We have obtained a non-linear second-order differential equation for the function $\theta(x)$. In order to solve it, we have to first establish the required initial conditions. At $x = -\infty$ the spins should be aligned along the z axis, i.e. we must set $\theta(-\infty) = 0$. Also, we suspect that the flip of the spins must happen gradually, so that we can write that $\theta'(-\infty) = 0$.

The solution of the ODE (9.20) is obtained by replacing $\theta'(x) = y(\theta)$ (see Sect. I.8.3.1), which gives

$$y\frac{dy}{d\theta} = \lambda\sin(2\theta) \implies ydy = \lambda\sin(2\theta)d\theta$$

that is easily integrated yielding

$$y^2(\theta) = \lambda\left[1 - \cos(2\theta)\right] + C = 2\lambda\sin^2\theta + C.$$

The constant C must be taken to be zero to ensure $y(0) = \theta'(-\infty) = 0$. Hence, we obtain the first-order differential equation $\theta'(x) = \sqrt{2\lambda}\sin\theta$, that is easily integrated:

$$\int\frac{d\theta}{\sin\theta} = \sqrt{2\lambda}x + C_1,$$

where C_1 is another constant. The integral is taken using the substitution $t = \cos\theta$:

$$\int\frac{d\theta}{\sin\theta} = \int\frac{\sin\theta d\theta}{\sin^2\theta} = \int\frac{-d(\cos\theta)}{1 - \cos^2\theta} = \int\frac{dt}{t^2 - 1}$$

$$= \frac{1}{2}\int\left[\frac{1}{t - 1} - \frac{1}{t + 1}\right]dt = \frac{1}{2}\ln\left|\frac{t - 1}{t + 1}\right|$$

$$= \frac{1}{2}\ln\left|\frac{1 - \cos\theta}{1 + \cos\theta}\right| = \frac{1}{2}\ln\frac{2\sin^2(\theta/2)}{2\cos^2(\theta/2)} = \ln\left|\tan\frac{\theta}{2}\right|,$$

so that we obtain

$$\ln\left|\tan\frac{\theta}{2}\right| = \sqrt{2\lambda}x + C_1 \implies \theta(x) = 2\arctan\left[\exp\left(\sqrt{2\lambda}x + C_1\right)\right].$$

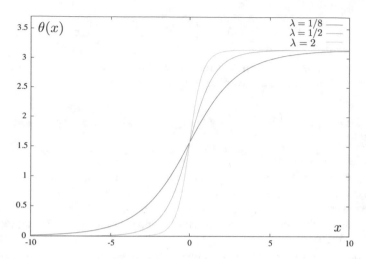

Fig. 9.4 The variation of the angle $\theta(x)$ the spins make with the z axis across the domain wall interface. On the left of the interface the spins are aligned along the z axis ($\theta = 0$), while on the right - in the negative direction ($\theta = \pi$)

The constant C_1 has to be taken to be zero since at $x \to -\infty$ the angle $\theta(-\infty)$ must be equal to 0. Hence, we finally obtain

$$\theta(x) = 2 \arctan \left[\exp \left(\sqrt{2\lambda} x \right) \right].$$

The obtained dependence is plotted in Fig. 9.4 for three values of the constant $\lambda = (\kappa/J)/a^2$. It is seen that as the ratio κ/J becomes larger, the angle θ changes from 0 to π over shorter distances, i.e. the width of the domain wall gets smaller, as expected.

9.1.2   Functionals Involving More Than One Function

Let us now consider a more general case of a functional depending on more than one function:

$$L[f_1, f_2, \ldots, f_n] = \int_a^b F\left(x, f_1, \ldots f_n, f_1', \ldots, f_n'\right) dx . \qquad (9.21)$$

In this case one may consider a partial variation of the functional with respect to the function f_i for a specific i between 1 and n. We then introduce a function

$$L_i(\alpha) = L[f_1, \ldots, f_i + \alpha\delta f_i, \ldots f_n]$$

$$= \int_a^b F\left(x, f_1, \ldots, f_i + \alpha\delta f_i, \ldots, f_n, f_1', \ldots, f_i' + \alpha\delta f_i', \ldots, f_n'\right) dx \,, \tag{9.22}$$

in which only one function, $f_i(x)$, is varied. Then the partial variation of L with respect to $f_i(x)$ is defined as $\delta L_i = L_i'(0)$, and the necessary condition for L to be minimum (maximum) will be

$$\delta L_i = L_i'(0) = 0 \,, \quad i = 1, \ldots, n \,. \tag{9.23}$$

Correspondingly, basically the whole discussion of the previous section can be repeated without change. This results in Euler's equation (9.11) for each i, i.e. one has to solve *a system of DEs* to obtain all the functions:

$$\frac{\partial F}{\partial f_i} - \frac{d}{dx}\left(\frac{\partial F}{\partial f_i'}\right) = 0 \,, \quad i = 1, \ldots, n \,, \tag{9.24}$$

or, if written in the expanded form (9.12),

$$F_{f_i} - F_{xf_i'} - \sum_{j=1}^n \left(F_{f_j f_i'} f_j' + F_{f_j' f_i'} f_j''\right) = 0 \,, \quad i = 1, \ldots, n \,. \tag{9.25}$$

As an example, let us consider again the problem of the shortest line connecting 2 points, $A(x_A, y_A, z_A)$ and $B(x_B, y_B, z_B)$. Suppose, the line is specified parametrically as $x = x(t)$, $y = y(t)$ and $z = z(t)$. Then the line length is given by the line integral

$$L = \int_{t_A}^{t_B} \sqrt{(x')^2 + (y')^2 + (z')^2}\, dt \,.$$

Euler's equations give

$$\frac{d}{dx}\left(\frac{\partial F}{\partial x'}\right) = 0 \quad \Longrightarrow \quad \frac{\partial F}{\partial x'} = \frac{x'}{\sqrt{(x')^2 + (y')^2 + (z')^2}} = \frac{1}{C_1} \,,$$

and similarly for the other two functions:

$$\frac{\partial F}{\partial y'} = \frac{y'}{\sqrt{(x')^2 + (y')^2 + (z')^2}} = \frac{1}{C_2} \,,$$

$$\frac{\partial F}{\partial z'} = \frac{z'}{\sqrt{(x')^2 + (y')^2 + (z')^2}} = \frac{1}{C_3} \,,$$

where C_1, C_2 and C_3 are three independent constants. Square both sides of each equation and rearrange. We get

$$(1 - C_1) \left(x'\right)^2 + \left(y'\right)^2 + \left(z'\right)^2 = 0 \, ,$$

$$\left(x'\right)^2 + (1 - C_2) \left(y'\right)^2 + \left(z'\right)^2 = 0 \, ,$$

$$\left(x'\right)^2 + \left(y'\right)^2 + (1 - C_3) \left(z'\right)^2 = 0 \, .$$

This is a system of three linear algebraic equations with respect to the three squares of the derivatives which has a non-trivial solution only if its determinant is equal to zero. This however would imply some special relationship between the arbitrary constants which is unreasonable. Therefore, we can accept the trivial solution that $\left(x'\right)^2 = \left(y'\right)^2 = \left(z'\right)^2 = 0$. This means that all three functions we are seeking, $x(t)$, $y(t)$ and $z(t)$, are linear functions of t. This corresponds to a straight line connecting the two points as anticipated; the six constants for the three linear functions, $x = \alpha_1 t + \beta_1$, $y = \alpha_2 t + \beta_2$ and $z = \alpha_3 t + \beta_3$, are determined immediately from the six coordinates of the two points.

9.1.3 Functionals Containing Higher Derivatives

Euler's type equations are to be modified if the functional

$$L[f] = \int_a^b F\left(x, f, f', f'', \ldots, f^{(n)}\right) dx \tag{9.26}$$

contains derivatives of the unknown function $f(x)$ up to order $n \geq 2$. Again, we define the function $L(\alpha) = L[f + \alpha \delta f]$ and its variation

$$\delta L = L'(0) = \int_a^b \left[\frac{\partial F}{\partial f} \delta f + \frac{\partial F}{\partial f'} \delta f' + \frac{\partial F}{\partial f''} \delta f'' + \ldots + \frac{\partial F}{\partial f^{(n)}} \delta f^{(n)} \right] dx \, . \tag{9.27}$$

To calculate the variation, however, more than one integration by parts will be required. Consider the contribution due to the $m-$th derivative ($2 \leq m \leq n$). We shall calculate it by parts m times:

$$\int_a^b \frac{\partial F}{\partial f^{(m)}} \delta f^{(m)} dx = \frac{\partial F}{\partial f^{(m)}} \delta f^{(m-1)} \Big|_a^b - \int_a^b \frac{d}{dx}\left(\frac{\partial F}{\partial f^{(m)}}\right) \delta f^{(m-1)} dx$$

$$= \frac{\partial F}{\partial f^{(m)}} \delta f^{(m-1)} \Big|_a^b - \frac{d}{dx}\left(\frac{\partial F}{\partial f^{(m)}}\right) \delta f^{(m-2)} \Big|_a^b + \int_a^b \frac{d^2}{dx^2}\left(\frac{\partial F}{\partial f^{(m)}}\right) \delta f^{(m-2)} dx$$

$$\cdots \qquad \cdots$$

$$= \frac{\partial F}{\partial f^{(m)}} \delta f^{(m-1)} \Big|_a^b - \frac{d}{dx}\left(\frac{\partial F}{\partial f^{(m)}}\right) \delta f^{(m-2)} \Big|_a^b + \ldots +$$

$$+ (-1)^{m-1} \frac{d^{m-1}}{dx^{m-1}}\left(\frac{\partial F}{\partial f^{(m)}}\right) \delta f \Big|_a^b + (-1)^m \int_a^b \frac{d^m}{dx^m}\left(\frac{\partial F}{\partial f^{(m)}}\right) \delta f \, dx \ .$$

To make progress, we have to assume that not only the deviation δf is zero at the boundary points $x = a$ and $x = b$, as we have assumed so far, but also all its derivatives up to the order $(m - 1)$ are equal to zero at the end points, i.e.

$$\delta f'(a) = \delta f''(a) = \ldots = \delta f^{(m-1)}(a) = 0 \ ,$$

and the same for the point $x = b$. Then all free terms in the above integration by parts become zero, and we obtain

$$\int_a^b \frac{\partial F}{\partial f^{(m)}} \delta f^{(m)} dx = (-1)^m \int_a^b \frac{d^m}{dx^m}\left(\frac{\partial F}{\partial f^{(m)}}\right) \delta f \, dx \ .$$

This will happen to each term in Eq. (9.27), which results in the following expression for the variation:

$$\delta L = \int_a^b \left[\frac{\partial F}{\partial f} - \frac{d}{dx}\left(\frac{\partial F}{\partial f'}\right) + \frac{d^2}{dx^2}\left(\frac{\partial F}{\partial f''}\right) + \ldots + (-1)^n \frac{d^n}{dx^n}\left(\frac{\partial F}{\partial f^{(n)}}\right)\right] \delta f \, dx \ . \tag{9.28}$$

To obtain the corresponding Euler's equation from this expression, we need first to modify Theorem 9.1 since when proving it we only assumed that the function $\eta(x)$ is continuous together with its first derivative. Here we have to make sure that $\eta(x)$ in the condition

$$\int_a^b f(x)\eta(x)dx = 0$$

of Theorem 9.1 is continuous together with its all derivatives up to the n-th one and that the function itself and its derivatives are equal to zero at the end points of the interval. This is easy to accomplish if we define $\eta(x)$ within a small interval $\epsilon_- < x < \epsilon_+$ (see the proof of Theorem 9.1) as

$$\eta(x) = (x - \epsilon_+)^{n+1}(x - \epsilon_-)^{n+1} \tag{9.29}$$

instead of the form (9.10) considered previously; outside the interval $\eta = 0$ as before. In addition, $\eta^{(k)} = 0$ for any $k = 1, 2, \ldots, n$ at the end points. This guarantees that $\eta(x)$ and all its derivatives up to the order n are continuous functions of x. Then the proof is repeated almost without change and the theorem remains valid in this case as well.

Hence, setting $\delta L = 0$ in Eq. (9.28) and applying the result of the modified theorem, we obtain the following Euler's equation in this case:

$$\frac{\partial F}{\partial f} - \frac{d}{dx}\left(\frac{\partial F}{\partial f'}\right) + \frac{d^2}{dx^2}\left(\frac{\partial F}{\partial f''}\right) + \ldots + (-1)^n \frac{d^n}{dx^n}\left(\frac{\partial F}{\partial f^{(n)}}\right) = 0 . \quad (9.30)$$

This is the DE of order $2n$: indeed, the derivative $\partial F/\partial f^{(n)}$, which may contain $f^{(n)}$, is differentiated n times with respect to x in the last term in the left-hand side. Hence, the solution $f(x)$ must contain $2n$ arbitrary constants which are obtained from $2n$ boundary conditions, which are that $f, f', \ldots, f^{(n-1)}$ have all well-defined (zero) values at the end points.

9.1.4 Variation with Constraints Given by Zero Functions

In applications it is often required to obtain the function(s) by minimising a functional subject to some additional conditions (or constraints) imposed on the function(s) to be determined. Suppose, we would like to optimise the functional

$$L[f_1, f_2, \ldots, f_n] = \int_a^b F\left(x, f_1, \ldots f_n, f_1', \ldots, f_n'\right) dx , \quad (9.31)$$

subject to additional conditions imposed on the unknown functions. We shall consider in detail the case of the so-called *holonomic* constraints which do not contain derivatives of the functions $f_1, \ldots, f_n$:

$$G_j(x, f_1, \ldots, f_n) = 0 , \quad j = 1, \ldots, k . \quad (9.32)$$

We have $k \geq 1$ such constraints. The case of *non-holonomic* constraints which contain derivatives of the functions as well is much more complicated and will only be briefly mentioned at the end of this section.

Here k, i.e. the number of constrains, must be smaller than n. Indeed, Eqs. (9.32) can, at least formally and within some interval of x, be then solved with respect to k functions $f_1, \ldots, f_k$ if $k < n$; these functions will be related to the rest of the functions, $f_{k+1}, \ldots, f_n$. Substituting these relationships into the original functional (9.31), L would still be a functional of the functions $f_{k+1}, \ldots, f_n$, and hence the optimisation problem would make perfect sense. If $k = n$, then all functions may be determined from the constraints alone and the variational problem would be impossible.

In addition to the constraints themselves, we also require, as usual, that our functions are to be fixed at the end points, $f_i(a) = A_i$ and $f_i(b) = B_i$ (for all $i = 1, \ldots, n$), so that the constraints are satisfied at the end points $x = a$ and $x = b$ as well.

The idea behind solving this problem is to try to convert it into the problem without constraints which we have considered above and know how to deal with. This can be done, as was mentioned above, by formally solving the constraints equations with respect to some k functions and replacing these solutions into the functional L. The latter will then be a functional of the rest of the $n - k$ functions without any additional constraints and hence Euler's equations can be established. This should solve the problem.

9.1.4.1 The Case of Two Functions and of a Single Constraint

To accomplish this programme, it is instructive to consider first the simplest case of the functional

$$L[f_1, f_2] = \int_a^b F\left(x, f_1, f_2, f_1', f_2'\right) dx$$

of just two functions, $f_1 = f_1(x)$ and $f_2 = f_2(x)$, and a single constraint

$$G(x, f_1, f_2) = 0. \tag{9.33}$$

Assume, that we can find f_2 from Eq. (9.33) above; then $f_2 = \varphi(x, f_1)$ will become some function of x and f_1. Substituting this solution into the functional, we obtain an explicit functional of only one function $f_1(x)$:

$$L[f_1] = \int_a^b F\left(x, f_1, \varphi(x, f_1), f_1', \varphi'(x, f_1)\right) dx = \int_a^b \Phi\left(x, f_1, f_1'\right) dx.$$

Note that here $\varphi'(x, f_1)$ is the total derivative of φ with respect to x, i.e.

$$\varphi' = \frac{\partial \varphi}{\partial x} + \frac{\partial \varphi}{\partial f_1} f_1'. \tag{9.34}$$

Also, Φ is the new integrand in the functional which depends only on the first function; we have introduced it for convenience.

The function f_1 (note that f_2 follows from it immediately as it is related to f_1 via $f_2 = \varphi(x, f_1)$) is obtained from Euler's equation

$$\frac{\partial \Phi}{\partial f_1} - \frac{d}{dx}\left(\frac{\partial \Phi}{\partial f_1'}\right) = 0. \tag{9.35}$$

We shall now need to work out all the derivatives in this equation. First of all,

$$
\frac{\partial \Phi}{\partial f_1} = \frac{\partial}{\partial f_1} F\left(x, f_1, \varphi\left(x, f_1\right), f_1', \varphi'\left(x, f_1\right)\right) = \frac{\partial F}{\partial f_1} + \frac{\partial F}{\partial \varphi} \frac{\partial \varphi}{\partial f_1} + \frac{\partial F}{\partial \varphi'} \frac{\partial \varphi'}{\partial f_1}
$$

$$
= \frac{\partial F}{\partial f_1} + \frac{\partial F}{\partial f_2} \frac{\partial \varphi}{\partial f_1} + \frac{\partial F}{\partial f_2'} \frac{\partial \varphi'}{\partial f_1} .
$$

To calculate the derivative $\partial \varphi' / \partial f_1$ needed above, we use the explicit expression (9.34) for φ':

$$
\frac{\partial \varphi'}{\partial f_1} = \frac{\partial}{\partial f_1} \left(\frac{\partial \varphi}{\partial x} + \frac{\partial \varphi}{\partial f_1} f_1' \right) = \frac{\partial^2 \varphi}{\partial x \partial f_1} + \frac{\partial^2 \varphi}{\partial f_1 \partial f_1} f_1' .
$$

Hence,

$$
\frac{\partial \Phi}{\partial f_1} = \frac{\partial F}{\partial f_1} + \frac{\partial F}{\partial f_2} \frac{\partial \varphi}{\partial f_1} + \frac{\partial F}{\partial f_2'} \left(\frac{\partial^2 \varphi}{\partial x \partial f_1} + \frac{\partial^2 \varphi}{\partial f_1 \partial f_1} f_1' \right) . \tag{9.36}
$$

Similarly we calculate the other derivative needed in Eq. (9.35):

$$
\frac{\partial \Phi}{\partial f_1'} = \frac{\partial F}{\partial f_1'} + \frac{\partial F}{\partial f_2'} \frac{\partial \varphi'}{\partial f_1'} .
$$

The second contribution is necessary since the derivative φ', according to Eq. (9.34), also explicitly depends on f_1'. Moreover, from this expression, we immediately get $\partial \varphi' / \partial f_1' = \partial \varphi / \partial f_1$. Therefore,

$$
\frac{\partial \Phi}{\partial f_1'} = \frac{\partial F}{\partial f_1'} + \frac{\partial F}{\partial f_2'} \frac{\partial \varphi}{\partial f_1} . \tag{9.37}
$$

Substituting Eqs. (9.36) and (9.37) into Euler's equation (9.35), we obtain

$$
\frac{\partial F}{\partial f_1} + \frac{\partial F}{\partial f_2} \frac{\partial \varphi}{\partial f_1} + \frac{\partial F}{\partial f_2'} \left(\frac{\partial^2 \varphi}{\partial x \partial f_1} + \frac{\partial^2 \varphi}{\partial f_1 \partial f_1} f_1' \right) = \frac{d}{dx} \left(\frac{\partial F}{\partial f_1'} + \frac{\partial F}{\partial f_2'} \frac{\partial \varphi}{\partial f_1} \right) . \tag{9.38}
$$

The right-hand side requires additional attention:

$$
\frac{d}{dx} \left(\frac{\partial F}{\partial f_1'} + \frac{\partial F}{\partial f_2'} \frac{\partial \varphi}{\partial f_1} \right) = \frac{d}{dx} \left(\frac{\partial F}{\partial f_1'} \right) + \frac{d}{dx} \left(\frac{\partial F}{\partial f_2'} \right) \frac{\partial \varphi}{\partial f_1} + \frac{\partial F}{\partial f_2'} \frac{d}{dx} \left(\frac{\partial \varphi}{\partial f_1} \right) ,
$$

where

$$
\frac{d}{dx} \left(\frac{\partial \varphi}{\partial f_1} \right) = \frac{\partial^2 \varphi}{\partial x \partial f_1} + \frac{\partial^2 \varphi}{\partial f_1 \partial f_1} f_1' .
$$

Substituting these results into Eq. (9.38) and performing necessary cancellations, we obtain

$$\frac{\partial F}{\partial f_1} - \frac{d}{dx}\left(\frac{\partial F}{\partial f_1'}\right) + \frac{\partial \varphi}{\partial f_1}\left[\frac{\partial F}{\partial f_2} - \frac{d}{dx}\left(\frac{\partial F}{\partial f_2'}\right)\right] = 0 . \tag{9.39}$$

This equation is to be complemented by the constraint itself. This can be done in the following way: let us differentiate both sides of the constraint equation,

$$G\left(x, f_1, \varphi\left(x, f_1\right)\right) = 0$$

with respect to f_1:

$$\frac{\partial G}{\partial f_1} + \frac{\partial G}{\partial f_2}\frac{\partial \varphi}{\partial f_1} = 0 , \tag{9.40}$$

from which we can get the derivative of φ with respect to f_1:

$$\frac{\partial \varphi}{\partial f_1} = -\frac{\partial G/\partial f_1}{\partial G/\partial f_2} .$$

This result allows rewriting Euler's equation (9.39) in the following final form:

$$\frac{\partial F}{\partial f_1} - \frac{d}{dx}\left(\frac{\partial F}{\partial f_1'}\right) - \frac{\partial G/\partial f_1}{\partial G/\partial f_2}\left[\frac{\partial F}{\partial f_2} - \frac{d}{dx}\left(\frac{\partial F}{\partial f_2'}\right)\right] = 0 . \tag{9.41}$$

The obtained DE may have solved the problem; however, it looks complicated and hence very difficult to remember. As with the optimisation of functions with constraints (see Sect. I.5.10.4), a simple method exists (proposed by Euler) which can simplify the procedure. Indeed, let us multiply Eq. (9.40) by some function $\lambda(x)$ and then add the obtained equation and Euler's equation (9.39) together. After simple manipulation:

$$\frac{\partial F}{\partial f_1} + \lambda(x)\frac{\partial G}{\partial f_1} - \frac{d}{dx}\left(\frac{\partial F}{\partial f_1'}\right) + \frac{\partial \varphi}{\partial f_1}\left[\frac{\partial F}{\partial f_2} - \frac{d}{dx}\left(\frac{\partial F}{\partial f_2'}\right) + \lambda(x)\frac{\partial G}{\partial f_2}\right] = 0 . \tag{9.42}$$

This result is valid for any function $\lambda(x)$. Let us now make a specific choice of it. Namely, let us choose $\lambda(x)$ in such a way that the expression in the square brackets above be zero:

$$\frac{\partial F}{\partial f_2} - \frac{d}{dx}\left(\frac{\partial F}{\partial f_2'}\right) + \lambda(x)\frac{\partial G}{\partial f_2} = 0 , \tag{9.43}$$

and, therefore,

$$\lambda(x) = -\frac{1}{\partial G/\partial f_2}\left[\frac{\partial F}{\partial f_2} - \frac{d}{dx}\left(\frac{\partial F}{\partial f_2'}\right)\right] . \tag{9.44}$$

Hence, with this choice, what is left from Eq. (9.42) be simply this:

$$\frac{\partial F}{\partial f_1} + \lambda(x)\frac{\partial G}{\partial f_1} - \frac{d}{dx}\left(\frac{\partial F}{\partial f_1'}\right) = 0 . \tag{9.45}$$

It is easy to see that this is exactly equivalent to Eq. (9.41) if we replace here λ with its expression (9.44).

Hence, the two equations (9.43) and (9.45) are fully equivalent to Eq. (9.41). These, however, can be rewritten in a form which is easy to remember. Consider an auxiliary function of x, of both functions and their derivatives:

$$H\left(x, f_1, f_2, f_1', f_2'\right) = F\left(x, f_1, f_2, f_1', f_2'\right) + \lambda(x) G\left(x, f_1, f_2\right) . \tag{9.46}$$

Since G does not depend on the derivatives of f_1 and f_2,

$$\frac{\partial F}{\partial f_1'} = \frac{\partial H}{\partial f_1'} \quad \text{and} \quad \frac{\partial F}{\partial f_2'} = \frac{\partial H}{\partial f_2'} ;$$

other terms in Eqs. (9.43) and (9.45) can also be combined, and we obtain instead the two equations for the function H:

$$\frac{\partial H}{\partial f_1} - \frac{d}{dx}\left(\frac{\partial H}{\partial f_1'}\right) = 0 \quad \text{and} \quad \frac{\partial H}{\partial f_2} - \frac{d}{dx}\left(\frac{\partial H}{\partial f_2'}\right) = 0 . \tag{9.47}$$

These look exactly as Euler's equations of the unconstrained problem formulated for the auxiliary function H, Eq. (9.46), and hence, this formulation seems to be much easier and more attractive. Similarly to the case of optimising functions of many variables subject to constrains, this method is also called the *method of Lagrange multipliers*.

9.1.4.2 The General Case

Once we understood the main idea, we can now apply the same method to the general case of Eqs. (9.31) and (9.32). Suppose, we can solve equations of constrains (9.32) with respect to the first k functions[3]:

$$f_j = \varphi_j\left(x, f_{k+1}, \ldots, f_n\right) , \quad j = 1, \ldots, k . \tag{9.48}$$

We also note for the later that their derivatives are

$$f_j' = \varphi_j' = \frac{\partial \varphi_j}{\partial x} + \sum_{i_1=k+1}^{n} \frac{\partial \varphi_j}{\partial f_{i_1}} f_{i_1}' . \tag{9.49}$$

The functional (9.31) becomes dependent only on the $n - k$ functions f_i with $i = k+1, \ldots, n$,

---

[3] It can be shown that the necessary condition for the constraints (9.32) to be solvable with respect to the functions $f_1, \ldots, f_k$ is that $D \neq 0$, where $D = \frac{\partial(G_1,\ldots,G_k)}{\partial(f_1,\ldots,f_k)}$ is the Jacobian. This point goes deeply into the inverse function theorem in the case of functions of many variables.

$$L\left[f_{k+1}, \ldots, f_n\right] = \int_a^b \Phi\left(x, f_{k+1}, \ldots, f_n\right) dx \,,$$

where we have introduced an auxiliary function

$$\Phi\left(x, f_{k+1}, \ldots, f_n\right) = F\left(x, \varphi_1, \ldots, \varphi_k, f_{k+1}, \ldots, f_n, \varphi_1', \ldots, \varphi_k', f_{k+1}', \ldots, f_n'\right) \,,$$

which yields immediately a set of the corresponding Euler's equations:

$$\frac{\partial \Phi}{\partial f_i} - \frac{d}{dx}\left(\frac{\partial \Phi}{\partial f_i'}\right) = 0 \,, \quad i = k+1, \ldots, n \,. \tag{9.50}$$

It is important to remember that functions φ_j and φ_j' (with $j = 1, \ldots, k$) depend on all the functions f_i with $i = k+1, \ldots, n$. We are now ready to work out the derivatives we encountered in Eq. (9.50) above. Firstly,

$$\frac{\partial \Phi}{\partial f_i} = \frac{\partial F}{\partial f_i} + \sum_{j=1}^{k} \frac{\partial F}{\partial f_j}\frac{\partial \varphi_j}{\partial f_i} + \sum_{j=1}^{k} \frac{\partial F}{\partial f_j'}\frac{\partial \varphi_j'}{\partial f_i}$$

$$= \frac{\partial F}{\partial f_i} + \sum_{j=1}^{k} \frac{\partial F}{\partial f_j}\frac{\partial \varphi_j}{\partial f_i} + \sum_{j=1}^{k} \frac{\partial F}{\partial f_j'}\left(\frac{\partial^2 \varphi_j}{\partial x \partial f_i} + \sum_{i_1=k+1}^{n} \frac{\partial^2 \varphi_j}{\partial f_{i_1} \partial f_i} f_{i_1}'\right) \,,$$

where we have used Eq. (9.49) for the derivative φ_j'. The other derivative we need is

$$\frac{\partial \Phi}{\partial f_i'} = \frac{\partial F}{\partial f_i'} + \sum_{j=1}^{k} \frac{\partial F}{\partial f_j'}\frac{\partial \varphi_j'}{\partial f_i'} = \frac{\partial F}{\partial f_i'} + \sum_{j=1}^{k} \frac{\partial F}{\partial f_j'}\frac{\partial \varphi_j}{\partial f_i} \,.$$

In the last passage we replaced $\partial \varphi_j'/\partial f_i'$ with $\partial \varphi_j/\partial f_i$. This follows from Eq. (9.49) after differentiating it with respect to f_i'. Replacing the above derivatives in Euler's equation (9.50), we get

$$\frac{\partial F}{\partial f_i} + \sum_{j=1}^{k} \frac{\partial F}{\partial f_j}\frac{\partial \varphi_j}{\partial f_i} + \sum_{j=1}^{k} \frac{\partial F}{\partial f_j'}\left(\frac{\partial^2 \varphi_j}{\partial x \partial f_i} + \sum_{i_1=k+1}^{n} \frac{\partial^2 \varphi_j}{\partial f_{i_1} \partial f_i} f_{i_1}'\right)$$

$$- \frac{d}{dx}\left(\frac{\partial F}{\partial f_i'} + \sum_{j=1}^{k} \frac{\partial F}{\partial f_j'}\frac{\partial \varphi_j}{\partial f_i}\right) = 0 \,. \tag{9.51}$$

In the last term we need to calculate the total derivative with respect to x:

$$\frac{d}{dx}\left(\sum_{j=1}^{k} \frac{\partial F}{\partial f_j'}\frac{\partial \varphi_j}{\partial f_i}\right) = \sum_{j=1}^{k} \frac{d}{dx}\left(\frac{\partial F}{\partial f_j'}\right)\frac{\partial \varphi_j}{\partial f_i} + \sum_{j=1}^{k} \frac{\partial F}{\partial f_j'}\frac{d}{dx}\left(\frac{\partial \varphi_j}{\partial f_i}\right) \,,$$

where

$$\frac{d}{dx}\left(\frac{\partial \varphi_j}{\partial f_i}\right) = \frac{\partial^2 \varphi_j}{\partial x \partial f_i} + \sum_{i_1=k+1}^{n} \frac{\partial^2 \varphi_j}{\partial f_i \partial f_{i_1}} f_{i_1}' \; .$$

Using these results in Eq. (9.51) and simplifying, we obtain

$$\frac{\partial F}{\partial f_i} - \frac{d}{dx}\left(\frac{\partial F}{\partial f_i'}\right) + \sum_{j=1}^{k} \frac{\partial \varphi_j}{\partial f_i}\left[\frac{\partial F}{\partial f_j} - \frac{d}{dx}\left(\frac{\partial F}{\partial f_j'}\right)\right] = 0 \; , \quad i = k+1, \ldots, n \; .$$

$$(9.52)$$

The derivatives $\partial \varphi_j / \partial f_i$ can be obtained by differentiating the constraints (9.32) with respect to f_i. Writing the constraints more explicitly as

$$G_j\left(x, \varphi_1 \ldots, \varphi_k, f_{k+1}, \ldots, f_n\right) = 0 \; , \quad j = 1, \ldots, k \; ,$$

we have

$$\frac{\partial G_j}{\partial f_i} + \sum_{j_1=1}^{k} \frac{\partial G_j}{\partial f_{j_1}} \frac{\partial \varphi_{j_1}}{\partial f_i} = 0 \; , \quad i = k+1, \ldots, n \quad \text{and} \quad j = 1, \ldots, k \; . \quad (9.53)$$

The above equations represent (for each fixed index i) a set of k linear algebraic equation with respect to the derivatives $\partial \varphi_{j_1}/\partial f_i$. Introducing the $k \times k$ matrix $\mathbf{D} = \left(D_{j j_1}\right) = \left(\partial G_j / \partial f_{j_1}\right)$ of the derivatives and assuming that its determinant is not equal to zero[4], these equations can be solved using the inverse matrix $\mathbf{D}^{-1}$. This way, at least formally, the derivatives $\partial \varphi_{j_1}/\partial f_i$ can be obtained:

$$\frac{\partial \varphi_{j_1}}{\partial f_i} = - \sum_{j=1}^{k} \left(\mathbf{D}^{-1}\right)_{j_1 j} \frac{\partial G_j}{\partial f_i} \; . \quad (9.54)$$

Now we shall reformulate equations (9.52) by introducing Lagrange multipliers. Interchange the indices $j \longleftrightarrow j_1$ in Eq. (9.53), multiply each by $\lambda_{j_1}(x)$, and add to Eq. (9.52) for each j_1. After grouping the terms appropriately, we obtain

$$\frac{\partial F}{\partial f_i} - \frac{d}{dx}\left(\frac{\partial F}{\partial f_i'}\right) + \sum_{j_1=1}^{k} \lambda_{j_1}(x) \frac{\partial G_{j_1}}{\partial f_i}$$

$$+ \sum_{j=1}^{k} \frac{\partial \varphi_j}{\partial f_i}\left[\frac{\partial F}{\partial f_j} - \frac{d}{dx}\left(\frac{\partial F}{\partial f_j'}\right) + \sum_{j_1=1}^{k} \lambda_{j_1}(x) \frac{\partial G_{j_1}}{\partial f_j}\right] = 0 \; ,$$

$$(9.55)$$

---

[4] The determinant of this matrix represents the Jacobian $D = \frac{\partial(G_1,\ldots,G_k)}{\partial(f_1,\ldots,f_k)}$. As was mentioned in the previous footnote above, we must assume that $D = \det \mathbf{D} \neq 0$.

where $i = k + 1, \ldots, n$. We choose the functions $\lambda_{j_1}(x)$ in such a way as to get the expression in the square brackets equal to zero. This gives two equations:

$$\frac{\partial F}{\partial f_i} - \frac{d}{dx}\left(\frac{\partial F}{\partial f_i'}\right) + \sum_{j_1=1}^{k} \lambda_{j_1}(x)\frac{\partial G_{j_1}}{\partial f_i} = 0 , \quad i = k+1, \ldots, n , \qquad (9.56)$$

and

$$\frac{\partial F}{\partial f_j} - \frac{d}{dx}\left(\frac{\partial F}{\partial f_j'}\right) + \sum_{j_1=1}^{k} \lambda_{j_1}(x)\frac{\partial G_{j_1}}{\partial f_j} = 0 , \quad j = 1, \ldots, k . \qquad (9.57)$$

It can be seen now that if we solve Eq. (9.57) with respect to the Lagrange multipliers and substitute this result into the other equation (9.56), then we shall obtain correctly Eq. (9.52). Indeed, Eq. (9.57) can be rewritten as

$$\sum_{j_1=1}^{k} D_{j_1 j}\lambda_{j_1}(x) = -\left[\frac{\partial F}{\partial f_j} - \frac{d}{dx}\left(\frac{\partial F}{\partial f_j'}\right)\right] .$$

Notice, that here we have the matrix $\mathbf{D}$ transposed. Therefore, when solving for λ_{j_1}, we also transpose the inverse matrix:

$$\lambda_{j_1}(x) = -\sum_{j=1}^{k} (\mathbf{D}^{-1})_{jj_1}\left[\frac{\partial F}{\partial f_j} - \frac{d}{dx}\left(\frac{\partial F}{\partial f_j'}\right)\right] .$$

Replacing λ_{j_1} just obtained in Eq. (9.56), we get

$$\frac{\partial F}{\partial f_i} - \frac{d}{dx}\left(\frac{\partial F}{\partial f_i'}\right) + \sum_{j=1}^{k}\left\{-\sum_{j_1=1}^{k}(\mathbf{D}^{-1})_{jj_1}\frac{\partial G_{j_1}}{\partial f_i}\right\}\left[\frac{\partial F}{\partial f_j} - \frac{d}{dx}\left(\frac{\partial F}{\partial f_j'}\right)\right] = 0 .$$

Making use of Eq. (9.54), we see that the expression in the curly brackets is in fact $\partial \varphi_j/\partial f_i$. Therefore, comparing with the original form (9.52) of Euler's equation, we conclude that it has been recovered. This means that Eqs. (9.56) and (9.57) are completely equivalent to it.

We shall now rewrite these equations in a more convenient form by introducing an auxiliary function

$$H = F + \sum_{j=1}^{k} \lambda_j(x)G_j . \qquad (9.58)$$

Since the constraints do not depend on the derivatives of our functions, we immediately see that Eqs. (9.56) and (9.57) are equivalent to these ones:

$$\frac{\partial H}{\partial f_i} - \frac{d}{dx}\left(\frac{\partial H}{\partial f_i'}\right) = 0 , \quad i = k+1, \ldots, n , \tag{9.59}$$

and

$$\frac{\partial H}{\partial f_j} - \frac{d}{dx}\left(\frac{\partial H}{\partial f_j'}\right) = 0 , \quad j = 1, \ldots, k . \tag{9.60}$$

We conclude that the same equations are obtained for *all* functions f_l with $l = 1, \ldots, n$. In other words, the problem with constraints is completely equivalent to the problem without constraints but applied to the auxiliary function (9.58). This finally proves the required general statement in the case of an arbitrary holonomic constraints.

9.1.4.3   Non-holonomic Constraints

As was mentioned at the beginning of this section, the case of non-holonomic constraints is much more complicated. Indeed, our treatment above was based on an assumption that one can solve equations (9.32) of the constraints with respect to some of the functions and hence express them via the others. This trick made the functional to depend only on these other functions without any additional constraints and hence the usual technique based on Euler's equations would be directly applicable.

It is clear now that if the constraints contain also derivatives of the functions we are seeking, then these equations are differential equations and hence expressing some of the functions via the others may be much more complicated. Still, it can be shown that the method of Lagrange multipliers is still applicable in this case as well.

9.1.5   Variation with Constraints Given by Integrals

Suppose now that our constrains have the form of the integral:

$$\int_a^b G_i\left(x, f_1, \ldots, f_n, f_1', \ldots, f_n'\right) dx = g_i , \quad i = 1, \ldots, k . \tag{9.61}$$

This is sometimes called an *isoparametric problem*. This is because a classical problem of finding a shape of a closed curve on the plane which has a fixed length and which encloses the largest area has this name. We shall demonstrate now that this problem can be reformulated to the one we considered in the previous section when the constraints did not contain integrals. We shall also find that the number of constraints could be arbitrary, e.g. k could be larger or equal to n. Recall, that in the previous case k was necessarily smaller than n.

Let us introduce auxiliary functions

$$y_i(x_1) = \int_a^{x_1} G_i\left(x, f_1, \dots, f_n, f_1', \dots, f_n'\right) dx , \quad i = 1, \dots, k . \tag{9.62}$$

These are functions due to the upper limit in the integral. Since at the upper limit, $x_1 = b$, we should reproduce the constraints themselves, we must have $y_i(b) = g_i$. Obviously, by construction,

$$y_i' - G_i\left(x, f_1, \dots, f_n, f_1', \dots, f_n'\right) = 0 , \quad i = 1, \dots, k .$$

These conditions can serve as constraints without integrals if we agree to consider the functions $y_1, \dots, y_k$ as additional unknown functions of our variational problem. Then, the total number of functions we are seeking increases to $n + k$, with the number of constraints being k. Here k could be either smaller or bigger than n, the problem is well-defined regardless.

According to the discussion of the previous section, to solve our problem we have to consider the auxiliary function

$$H\left(x, f_1, \dots, f_n, f_1', \dots, f_n', y_1', \dots y_k'\right) = F\left(x, f_1, \dots, f_n, f_1', \dots, f_n'\right)$$
$$+ \sum_{i=1}^{k} \lambda_i(x)\left[y_i' - G_i\left(x, f_1, \dots, f_n, f_1', \dots, f_n'\right)\right] , \tag{9.63}$$

which does depend only on the derivatives of the newly added functions $y_i(x)$. Euler's equations in this case are

$$\frac{\partial H}{\partial f_j} - \frac{d}{dx}\left(\frac{\partial H}{\partial f_j'}\right) = 0 , \quad j = 1, \dots, n , \tag{9.64}$$

and

$$\frac{\partial H}{\partial y_i} - \frac{d}{dx}\left(\frac{\partial H}{\partial y_i'}\right) = 0 , \quad i = 1, \dots, k . \tag{9.65}$$

Since the function H does not actually depend on the functions y_i themselves, the second equation gives

$$\frac{\partial H}{\partial y_i'} = \lambda_i(x) = C_i , \quad i = 1, \dots, k ,$$

i.e. the Lagrange multipliers must be constants, i.e. they cannot depend on x. Moreover, when writing Euler's equations (9.64) for the functions f_j, the derivatives of y_i in H of Eq. (9.63) do not contribute, and, therefore, the function H can be built simply as

Fig. 9.5 A rope of length l
fixed between two points
$x = 0$ and $x = x_0$ will
accept some shape $y = y(x)$
due to gravity

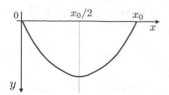

$$H = F + \sum_{i=1}^{k} \lambda_i G_i \tag{9.66}$$

and applied only to finding the functions f_j.

To illustrate the method, let us consider the following problem. Consider a rope suspended between two points $x = 0$ and $x = x_0$ as illustrated in Fig. 9.5. Because of the gravity, the rope will accept a certain shape which minimises the total potential energy of the rope. As with brachistochrone problem (see Problem 9.9), the shape of the rope will be described by a function $y = y(x)$ with the y axis drawn down. The piece $ds = \sqrt{1 + (y')^2}dx$ of the rope accumulates the potential energy $dU = \rho g y ds$, where ρ is the (uniform) linear mass density of the rope and g the gravity constant. Therefore, the total potential energy of the rope due to gravity will be proportional to (we can omit the irrelevant pre-factor ρg):

$$U = \int_0^{x_0} y\sqrt{1 + (y')^2}dx \ . \tag{9.67}$$

We have to minimise U with respect to the shape $y(x)$ of the rope subject to the condition that the length of the rope

$$l = \int_0^{x_0} \sqrt{1 + (y')^2}dx \tag{9.68}$$

is fixed and equal to l. Following the recipe discussed above, we construct the auxiliary function

$$H\left(y, y'\right) = y\sqrt{1 + (y')^2} + \lambda\sqrt{1 + (y')^2} = (y + \lambda)\sqrt{1 + (y')^2} \ .$$

Since it depends only on y and its derivative, Euler's equation yields (see Eq. (9.17)):

$$H - y'\frac{\partial H}{\partial y'} = C_1 \implies (y + \lambda)\sqrt{1 + (y')^2} - (y + \lambda)\frac{(y')^2}{\sqrt{1 + (y')^2}} = C_1 \ .$$

Multiplying both sides by $\sqrt{1 + (y')^2}$ and simplifying, we obtain

$$\sqrt{1+(y')^2} = \frac{y+\lambda}{C_1} \quad \Longrightarrow \quad \frac{dy}{dx} = \pm\frac{1}{C_1}\sqrt{(y+\lambda)^2 - C_1^2},\qquad (9.69)$$

where it is assumed that C_1 is positive.

The curve must be symmetric about the midpoint $x = x_0/2$ and hence at the midpoint y' must be equal to zero. But symmetry condition also enables us to limit ourselves in finding the shape of the curve only within the interval $0 \le x \le x_0/2$. where the first derivative $y' \ge 0$. To satisfy this, we choose the plus sign.

Rearranging the above DE and integrating, we obtain

$$\int_0^y \frac{dy_1}{\sqrt{(y_1+\lambda)^2 - C_1^2}} = \int_0^x \frac{dx_1}{C_1},$$

where the zero bottom limits in both integrals take account of the fact that $(0, 0)$ is the starting point of the line. The integral in the left-hand side by the substitution $y_1 + \lambda = C_1 t$ is manipulated into the inverse hyperbolic cosine integral, the integral in the right-hand side is straightforward, so that we obtain

$$\int_{\lambda/C_1}^{(y+\lambda)/C_1} \frac{dt}{\sqrt{t^2-1}} = \frac{x}{C_1} \quad \Longrightarrow \quad \cosh^{-1}\left(\frac{y+\lambda}{C_1}\right) - \cosh^{-1}\left(\frac{\lambda}{C_1}\right) = \frac{x}{C_1}$$

from where the solution readily follows:

$$\frac{y+\lambda}{C_1} = \cosh\left[\frac{x}{C_1} + \cosh^{-1}\left(\frac{\lambda}{C_1}\right)\right].\qquad (9.70)$$

This is a general solution as it still contains two unknown constants: C_1 and λ. We have already used the condition that the rope starts at the point $(0, 0)$ when choosing the integration limits above; we still need to satisfy the condition that the length of the rope is l and at the midpoint $x_0/2$ the derivative of $y(x)$ is zero. It is convenient first to take account of the latter condition. Using the first equation in (9.69), we find that at the midpoint

$$\left(\frac{y(x)+\lambda}{C_1}\right)_{x=x_0/2} = 1 \quad \Longrightarrow \quad \cosh\left[\frac{x_0}{2C_1} + \cosh^{-1}\left(\frac{\lambda}{C_1}\right)\right] = 1,$$

which in turn means that the expression in the square brackets is zero:

$$\frac{x_0}{2C_1} + \cosh^{-1}\left(\frac{\lambda}{C_1}\right) = 0 \quad \Longrightarrow \quad \lambda = C_1\cosh\left(-\frac{x_0}{2C_1}\right) = C_1\cosh\left(\frac{x_0}{2C_1}\right).$$

This gives λ in terms of C_1. Our solution (9.70) can now be simplified:

$$y = C_1\left[\cosh\left(\frac{x-x_0/2}{C_1}\right) - \cosh\left(\frac{x_0}{2C_1}\right)\right].\qquad (9.71)$$

We now need to satisfy the condition related to the length of the curve to obtain C_1. We have

$$\frac{l}{2} = \int_0^{x_0/2} \sqrt{1 + (y')^2}\, dx = \int_0^{x_0/2} \frac{y + \lambda}{C_1}\, dx$$

$$= \int_0^{x_0/2} \cosh\left(\frac{x - x_0/2}{C_1}\right) dx = C_1 \sinh\left(\frac{x_0}{2C_1}\right).$$

It is easy to verify that this transcendental equation always has a solution for C_1. The shape of the curve (9.71) is fully defined. In fact, this formula is valid for the whole range $0 \leq x \leq x_0$.

Problem 9.10 Consider a line of length l fixed at two points $A(-x_0, 0)$ and $B(x_0, 0)$ (where $l > 2x_0$). Determine the shape $y = y(x)$ of the curve which gives the largest area $A = \int_{-x_0}^{x_0} y(x)\, dx$ under it, i.e. between the curve and the x axis (it may be assumed that $y \geq 0$).

(i) show that the curve which optimises the auxiliary function $H = y + \lambda\sqrt{1 + (y')^2}$ is a circle $(x - a)^2 + (y + b)^2 = \lambda^2$, where a and b are arbitrary constants and λ the Lagrange multiplier;

(ii) prove that $a = 0$, $\lambda = \sqrt{x_0^2 + b^2}$ and b is determined from the equation

$$l = 2\sqrt{x_0^2 + b^2}\, \arctan\frac{x_0}{b}.$$

(iii) Prove that the latter equation always has one solution with respect to b.

This problem also shows that the largest possible area made by a close-looped line of a fixed length l is a circle of radius $R = l/2\pi$, something which can easily be accepted intuitively.

9.2 Functions of Many Variables

So far we have considered the case when the functional L depends on one or more functions (and their derivatives) which depend on a single variable x; consequently, the functional L has the form of a single integral with respect to x. In some applications it is necessary to find the minimum (maximum) of a functional which is written as a multiple integral over some multi-dimensional region and contains an unknown function f (and its partial derivatives) defined in this region, i.e. we have to deal with the function of more than one variable.

For instance, consider a closed curve $\mathcal{L}$ in the 3D space. This curve could be a boundary to many curvilinear surfaces S which may have different surface areas (Sect. I.6.5.2)

$$A = \int \int_{\Sigma} \sqrt{1 + \left(z'_x\right)^2 + \left(z'_y\right)^2} dxdy .$$

Here the surface S is specified as $z = z(x, y)$, and Σ is the projection of S on the $x - y$ plane. It is then legitimate to ask a question of what is the surface that has the minimum possible surface area A. For instance, if the contour $\mathcal{L}$ is planar (e.g. lies in the $x - y$ plane), then the minimum of the area is achieved by the planar surface enclosed by the contour $\mathcal{L}$. However, if $\mathcal{L}$ is not planar, the answer is not so obvious.

We shall first consider the following functional:

$$L[f] = \int \int_D F\left(x, y, f, f'_x, f'_y\right) dxdy . \tag{9.72}$$

It depends on a single function $f = f(x, y)$ of two variables (and its partial derivatives), and the integration is performed over some 2D region D in the $x - y$ plane. We also assume that the function f has definite values on the whole boundary $\mathcal{L}$ of the region D and is continuous everywhere in D together with its first derivatives f'_x and f'_y.

The variation δL in this case is constructed exactly in the same way as for a function of a single variable using the function $L(\alpha) = L[f + \alpha \delta f]$, where δf is a fixed function of two variables such that on the boundary $\mathcal{L}$ it is zero, $\delta f(\mathcal{L}) = \delta f(x, y)|_{\mathcal{L}} = 0$. Then,

$$\delta L = L'(\alpha)|_{\alpha=0} = L'(0) ,$$

as before. To proceed, we need a theorem similar to Theorem 9.1 we proved above for the 1D case.

Theorem 9.2 *Let $\eta(x, y)$ be any function which is zero at the boundary $\mathcal{L}$ of some 2D region D, and is continuous together with its first derivatives η'_x and η'_y within the whole region D including its boundary. If $f(x, y)$ is a continuous function in D and for any such $\eta(x, y)$ the integral*

$$\int \int_D f(x, y)\eta(x, y)dxdy = 0 ,$$

then $f(x, y) = 0$.

Proof This is proven by contradiction as well. If we assume that at some point (x_0, y_0) the function $f(x, y)$ is, say, positive, then it will be such within some vicinity U_ϵ of this point, e.g. in a circle of radius ϵ:

$$f(x, y) > 0 \quad \text{for} \quad (x - x_0)^2 + (y - y_0)^2 < \epsilon^2 .$$

9 Calculus of Variations

Next, we define the function $\eta(x, y)$ in the following way: it is zero at the boundary of the same circle and beyond it, while inside the circle it is defined as

$$\eta(x, y) = \left[(x - x_0)^2 + (y - y_0)^2 - \epsilon^2\right]^2 .$$

It is easy to see that this function is indeed zero at the boundary of the circle and hence is continuous everywhere in D. Its first partial derivatives behave similarly. Indeed, within the circle

$$\eta'_x = 4(x - x_0)\left[(x - x_0)^2 + (y - y_0)^2 - \epsilon^2\right] ,$$

and hence is zero at its boundary; it continues to be zero beyond the circle by construction. Hence, η defined in that manner satisfies the conditions of the theorem. At the same time, the surface integral

$$\int\int_D f(x, y)\eta(x, y)dxdy = \int\int_{U_\epsilon} f(x, y)\eta(x, y)dxdy > 0 ,$$

i.e. it is not zero as both functions are positive within the circle U_ϵ. Hence, our assumption has been proven wrong. **Q.E.D.**

Using this theorem, we can derive an appropriate partial differential equation for the function $f(x, y)$ corresponding to the optimum of the functional (9.72). Indeed, the variation

$$\begin{aligned}
\delta L &= \frac{d}{d\alpha}\left[\int\int_D F\left(x, y, f + \alpha\delta f, (f + \alpha\delta f)'_x, (f + \alpha\delta f)'_y\right)dxdy\right]_{\alpha=0} \\
&= \frac{d}{d\alpha}\left[\int\int_D F\left(x, y, f + \alpha\delta f, f'_x + \alpha(\delta f)'_x, f'_y + \alpha(\delta f)'_y\right)dxdy\right]_{\alpha=0} \\
&= \int\int_D\left[\frac{\partial F}{\partial f}\delta f + \frac{\partial F}{\partial f'_x}(\delta f)'_x + \frac{\partial F}{\partial f'_y}(\delta f)'_y\right]dxdy \\
&= \int\int_D \frac{\partial F}{\partial f}\delta f\, dxdy + \int\int_D\left[\frac{\partial F}{\partial f'_x}(\delta f)'_x + \frac{\partial F}{\partial f'_y}(\delta f)'_y\right]dxdy .
\end{aligned}$$

$$(9.73)$$

To calculate the last double integral, we consider the partial derivative

$$\frac{\partial}{\partial x}\left(\frac{\partial F}{\partial f'_x}\delta f\right) = \frac{\partial}{\partial x}\left(\frac{\partial F}{\partial f'_x}\right)\delta f + \frac{\partial(\delta f)}{\partial x}\frac{\partial F}{\partial f'_x} = \frac{\partial}{\partial x}\left(\frac{\partial F}{\partial f'_x}\right)\delta f + (\delta f)'_x\frac{\partial F}{\partial f'_x} ,$$

$$\Longrightarrow \quad (\delta f)'_x\frac{\partial F}{\partial f'_x} = \frac{\partial}{\partial x}\left(\frac{\partial F}{\partial f'_x}\delta f\right) - \frac{\partial}{\partial x}\left(\frac{\partial F}{\partial f'_x}\right)\delta f .$$

Similarly,

$$(\delta f)'_y \frac{\partial F}{\partial f'_y} = \frac{\partial}{\partial y}\left(\frac{\partial F}{\partial f'_y}\delta f\right) - \frac{\partial}{\partial y}\left(\frac{\partial F}{\partial f'_y}\right)\delta f .$$

Using these expressions to replace the two terms within the square brackets in Eq. (9.73), we obtain

$$\delta L = \int\int_D\left[\frac{\partial F}{\partial f} - \frac{\partial}{\partial x}\left(\frac{\partial F}{\partial f'_x}\right) - \frac{\partial}{\partial y}\left(\frac{\partial F}{\partial f'_y}\right)\right]\delta f \, dxdy$$

$$+ \int\int_D\left[\frac{\partial}{\partial x}\left(\frac{\partial F}{\partial f'_x}\delta f\right) - \frac{\partial}{\partial y}\left(-\frac{\partial F}{\partial f'_y}\delta f\right)\right]dxdy .$$

The second integral is zero. Indeed, it can be handled, e.g. by means of Green's formula, Sect. I.6.4.3, with the following choice of the two functions Q and P:

$$Q(x, y) = \frac{\partial F}{\partial f'_x}\delta f \quad\text{and}\quad P(x, y) = -\frac{\partial F}{\partial f'_y}\delta f .$$

Then, according to Green's formula, the double integral $\int\int_D\left(\frac{\partial Q}{\partial x} - \frac{\partial P}{\partial y}\right)dxdy$ equals the line integral

$$\oint_{\mathcal{L}} Pdx + Qdy = \oint_{\mathcal{L}} \delta f\left(-\frac{\partial F}{\partial f'_y}dx + \frac{\partial F}{\partial f'_x}dy\right)$$

along the boundary $\mathcal{L}$ of D. However, $\delta f = 0$ everywhere on the boundary $\mathcal{L}$, and hence this integral is identically zero as required.

Hence,

$$\delta L = \int\int_D\left[\frac{\partial F}{\partial f} - \frac{\partial}{\partial x}\left(\frac{\partial F}{\partial f'_x}\right) - \frac{\partial}{\partial y}\left(\frac{\partial F}{\partial f'_y}\right)\right]\delta f \, dxdy .$$

At the stationary point $\delta L = 0$. As this is to happen for any variation δf, then based on Theorem 9.2 we arrive at the following equation the function $f(x, y)$ must satisfy

$$\frac{\partial F}{\partial f} - \frac{\partial}{\partial x}\left(\frac{\partial F}{\partial f'_x}\right) - \frac{\partial}{\partial y}\left(\frac{\partial F}{\partial f'_y}\right) = 0 . \tag{9.74}$$

This result was first obtained by Ostrogradsky and bears his name.

A generalisation to functions of more variables can easily be done. First of all, we need the necessary generalisation of Theorem 9.2 to the case of n dimensions ($n \geq 2$). This is done by introducing

$$\eta(x_1, \ldots, x_n) = \left[\sum_{i=1}^{n} (x_i - x_{i0})^2 - \epsilon^2 \right]^2$$

within a small vicinity U_ϵ of the point $(x_{10}, \ldots, x_{n0})$ where we assume that the function $f(x_{10}, \ldots, x_{n0}) > 0$. The rest of the proof remains exactly the same.

Before going to a general n–dimensional case, let us next consider the case of three dimensions. The functional is a volume integral:

$$L = \int \int \int_D F\left(x, y, z, f, f_x', f_y', f_z'\right) dx dy dz \ , \tag{9.75}$$

where $f = f(x, y, z)$ is the function of three variables to be determined as the stationary value of this functional. We assume that this function is continuous together with its necessary derivatives in the 3D region D and is fixed at the (closed) boundary surface S of the volume. As before, the variation

$$\delta L = \int \int \int_D \left[\frac{\partial F}{\partial f} \delta f + \frac{\partial F}{\partial f_x'} (\delta f)_x' + \frac{\partial F}{\partial f_y'} (\delta f)_y' + \frac{\partial F}{\partial f_z'} (\delta f)_z' \right] dx dy dz$$

$$= \int \int \int_D \left[\frac{\partial F}{\partial f} - \frac{\partial}{\partial x} \left(\frac{\partial F}{\partial f_x'} \right) - \frac{\partial}{\partial y} \left(\frac{\partial F}{\partial f_y'} \right) - \frac{\partial}{\partial z} \left(\frac{\partial F}{\partial f_z'} \right) \right] \delta f \, dx dy dz$$

$$+ \int \int \int_D \left[\frac{\partial}{\partial x} \left(\frac{\partial F}{\partial f_x'} \delta f \right) + \frac{\partial}{\partial y} \left(\frac{\partial F}{\partial f_y'} \delta f \right) + \frac{\partial}{\partial z} \left(\frac{\partial F}{\partial f_z'} \delta f \right) \right] dx dy dz \ .$$

This time, to transform the last integral into the integral over the surface boundary S of D, we notice that the integrand there can be thought of as a divergence of the vector field $\mathbf{G}\delta f$, where $\mathbf{G} = \left(G_x, G_y, G_z\right)$ is the vector field with the components

$$G_x = \frac{\partial F}{\partial f_x'}, \quad G_y = \frac{\partial F}{\partial f_y'}, \quad G_z = \frac{\partial F}{\partial f_z'},$$

so that the integral becomes

$$\int \int \int_D \text{div}\,(\mathbf{G}\delta f) \, dx dy dz = \oiint_S \delta f \, \mathbf{G} \cdot d\mathbf{S} \ ,$$

where we have used the Ostrogradsky–Gauss theorem (Sect. I.6.5.7) to transform the volume integral into the integral over its surface boundary. Since everywhere on S the variation $\delta f = 0$, this volume integral is zero. Consequently,

$$\delta L = \int \int \int_D \left[\frac{\partial F}{\partial f} - \frac{\partial}{\partial x} \left(\frac{\partial F}{\partial f_x'} \right) - \frac{\partial}{\partial y} \left(\frac{\partial F}{\partial f_y'} \right) - \frac{\partial}{\partial z} \left(\frac{\partial F}{\partial f_z'} \right) \right] \delta f \, dx dy dz \ ,$$

and the necessary condition $\delta L = 0$ leads to the required generalisation of Eq. (9.74) to three dimensions:

$$\frac{\partial F}{\partial f} - \frac{\partial}{\partial x}\left(\frac{\partial F}{\partial f'_x}\right) - \frac{\partial}{\partial y}\left(\frac{\partial F}{\partial f'_y}\right) - \frac{\partial}{\partial z}\left(\frac{\partial F}{\partial f'_z}\right) = 0 \,. \tag{9.76}$$

The case of any n dimensions ($n \geq 4$) is worked out similarly. What is needed is to transform the n-dimensional integral

$$\underbrace{\int \cdots \int}_{n}\left[\frac{\partial}{\partial x_1}\left(\frac{\partial F}{\partial f'_{x_1}}\delta f\right) + \frac{\partial}{\partial x_2}\left(\frac{\partial F}{\partial f'_{x_2}}\delta f\right) + \ldots + \frac{\partial}{\partial x_n}\left(\frac{\partial F}{\partial f'_{x_n}}\delta f\right)\right] dx_1 dx_2 \ldots dx_n$$

over the whole n-dimensional region into an $(n-1)$-dimensional integral corresponding to its closed 'surface' S. Basically, what we are to do here is to generalise the Ostrogradsky–Gauss theorem to more than 3 dimensions. This step requires performing one 1D integration explicitly. For instance, consider the first term:

$$\underbrace{\int \cdots \int}_{n}\frac{\partial}{\partial x_1}\left(\frac{\partial F}{\partial f'_{x_1}}\delta f\right) dx_1 dx_2 \ldots dx_n$$

$$= \underbrace{\int \cdots \int}_{n-1} dx_2 \ldots dx_n \left[\int_{x_1(x_2,\ldots,x_n)_-}^{x_1(x_2,\ldots,x_n)_+}\frac{\partial}{\partial x_1}\left(\frac{\partial F}{\partial f'_{x_1}}\delta f\right) dx_1\right]$$

$$= \underbrace{\int \cdots \int}_{n-1}\left[\left(\frac{\partial F}{\partial f'_{x_1}}\delta f\right)_{x_1(x_2,\ldots,x_n)_+} - \left(\frac{\partial F}{\partial f'_{x_1}}\delta f\right)_{x_1(x_2,\ldots,x_n)_-}\right] dx_2 \ldots dx_n \,.$$

Here $x_1(x_2,\ldots,x_n)_{\pm}$ are two values of x_1 at the boundary surface S of the n-dimensional region D for the given values $x_2,\ldots,x_n$ of the other variables. Since $\delta f = 0$ on the bounding surface, the surface integral is zero again. Similarly, we repeat for other terms by performing integration over x_2, x_3, etc. for the second, third and so on terms, respectively, to arrive in the end at the following equation:

$$\frac{\partial F}{\partial f} - \sum_{i=1}^{n}\frac{\partial}{\partial x_i}\left(\frac{\partial F}{\partial f'_{x_i}}\right) = 0 \,. \tag{9.77}$$

It is easy to see that in the cases of $n = 1, 2, 3$ we obtain the previously derived equations (9.11), (9.74) and (9.76), respectively.

This results is generalised to the functionals containing more than one function and/or higher derivatives of the functions along the same lines as before.

Problem 9.11 Consider the functional

$$L = \int \int_D F\left(x, y, f, f'_x, f'_y, f''_{xx}, f''_{xy}, f''_{yy}\right) dx dy .$$

Show that in this case the stationary value of the functional is given by the function $f(x, y)$ which satisfies the equation:

$$\frac{\partial F}{\partial f} - \frac{\partial}{\partial x}\left(\frac{\partial F}{\partial f'_x}\right) - \frac{\partial}{\partial y}\left(\frac{\partial F}{\partial f'_y}\right) + \frac{\partial^2}{\partial x^2}\left(\frac{\partial F}{\partial f''_{xx}}\right)$$

$$+ \frac{\partial^2}{\partial x \partial y}\left(\frac{\partial F}{\partial f''_{xy}}\right) + \frac{\partial^2}{\partial y^2}\left(\frac{\partial F}{\partial f''_{yy}}\right) = 0 . \tag{9.78}$$

9.3 Applications in Physics

9.3.1 Mechanics

For a system of N particles of mass m_i (where $i = 1, \ldots, N$) described by their coordinates $\mathbf{r}_i = (x_i, y_i, z_i)$ and velocities $\mathbf{v}_i = \dot{\mathbf{r}}_i = (\dot{x}_i, \dot{y}_i, \dot{z}_i)$ (the dot above symbols here and in what follows corresponds to the time derivative), the usual approach to describe their evolution in time t under forces $\mathbf{F}_i$ is to write Newton's equations of motion, $m_i \ddot{\mathbf{r}}_i = \mathbf{F}_i$, where the vector $\ddot{\mathbf{r}}_i$ is the acceleration of the i-th particle. A general solution of these second-order DEs will contain $6N$ arbitrary constants which can be found from the known initial positions and velocities of the particles.

However, in many practical cases the situation may be much more complex. For instance, in some mechanical problems certain constraints may exist relating Cartesian atomic positions and velocities of particles with each other effectively reducing the actual number of degrees of freedom n. In these and also in other cases one may also choose the so-called generalised coordinates instead of the Cartesian ones (e.g. distances between atoms, angles formed by three atoms, etc.) which are more convenient for the problem at hand. Hence, a more abstract and general formulation of mechanics is desirable that is based on a such general set of independent coordinates chosen to fully describe the system of interest. We shall denote these generalised coordinates q_i ($i = 1, \ldots, n$, where $n \leq 3N$), while their time derivatives will be denoted $\dot{q}_i$ all called generalised velocities.

In order to describe the time evolution of the generalised coordinates, one may use a very general principle of mechanics, called the *principle of least action*, proposed by Hamilton. It states that any mechanical system can be described by a function

$L = L(q_1, \ldots, q_n, \dot{q}_1, \ldots, \dot{q}_n, t)$ called Lagrangian[5] which depends on all atomic coordinates and velocities (and possibly time), and that the true particles trajectories $q_i = q_i(t)$ (amongst all possible trajectories) between times t_1 and t_2 must correspond to the minimum of the action (or action integral)

$$ S = \int_{t_1}^{t_2} L(q_1, \ldots, q_n, \dot{q}_1, \ldots, \dot{q}_n, t)\, dt \ . \tag{9.79} $$

Here it is assumed that the coordinates are fixed at the initial and final times. The Lagrange function is constructed as a difference between system's kinetic, K, and potential, U, energies, $L = K - U$. Note that previously L was denoting the whole functional; here it is denoted S while the integrand is L. These are the notations widely accepted in physics literature for the action and the Lagrangian, respectively, and hence we shall stick to them here.

The corresponding equations of motion are then obtained by optimising the action functional (9.79), which leads to the set of familiar Euler's equations:

$$ \frac{\partial L}{\partial q_i} - \frac{d}{dt}\left(\frac{\partial L}{\partial \dot{q}_i}\right) = 0 , \quad i = 1, \ldots, n \ . \tag{9.80} $$

In physics these are normally called Lagrange equations.

If the system of n degrees of freedom is given with $k < n$ holonomic constraints

$$ G_i(q_1, \ldots, q_n, t) = 0 , \quad i = 1, \ldots k \ , $$

then we have to consider the auxiliary function

$$ H = L + \sum_{i=1}^{k} \lambda_i(t) G_i $$

with $\lambda_i(t)$ being the corresponding Lagrange multipliers, and hence the corresponding Lagrange equations would contain an additional term due to these constraints:

$$ \frac{d}{dt}\left(\frac{\partial L}{\partial \dot{q}_i}\right) = \frac{\partial L}{\partial q_i} + \sum_{j=1}^{k} \lambda_j \frac{\partial G_j}{\partial q_i} , \quad i = 1, \ldots, n \ . \tag{9.81} $$

The terms in the right-hand side serve as forces acting on the coordinate q_i; the first term is a real (physical) force, while the second is an artificial force due to constraints.

As our first example of application of the Lagrange equations, let us consider a single particle of mass m moving under the external potential $U(\mathbf{r})$. Here $\mathbf{r} = (x, y, z)$ are the three Cartesian coordinates describing the system. In this case there is no need for special generalised coordinates, the Cartesian coordinates will do. The Lagrangian

---

[5] After Joseph-Louis Lagrange.

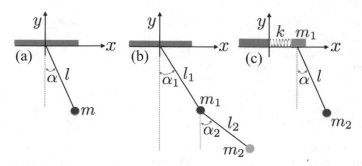

Fig. 9.6 Pendulum problems: **a** a single pendulum; **b** a double pendulum; **c** a pendulum with an oscillating support

is

$$L = \frac{1}{2}m\dot{\mathbf{r}}^2 - U(\mathbf{r}) = \frac{1}{2}m\left(\dot{x}^2 + \dot{y}^2 + \dot{z}^2\right) - U(x, y, z) ,$$

leading to three Lagrange equations for each Cartesian component, e.g. for one of them we have

$$\frac{\partial L}{\partial x} - \frac{d}{dt}\left(\frac{\partial L}{\partial \dot{x}}\right) = 0 \implies -\frac{\partial U}{\partial x} - \frac{d}{dt}(m\dot{x}) = 0$$

$$\implies -\frac{\partial U}{\partial x} = m\ddot{x} ,$$

which is nothing but the second Newton's equation of motion projected on the x axis, since $-\frac{\partial U}{\partial x}$ is the x component of the force, F_x.

In our second example, consider a pendulum of length l with a particle of mass m hanging at its end, Fig. 9.6a. The Cartesian coordinates of the mass are $x = l\sin\alpha$ and $y = -l\cos\alpha$. However, it is clear that only a single coordinate is needed to adequately describe the oscillations of the mass. The most convenient choice here is the angle α of the pendulum with the y (vertical) axis. Hence, choosing α as the required independent generalised coordinate, we have to express both the potential and kinetic energies via it. The potential energy is straightforward: $U = mgy = -mgl\cos\alpha$. The kinetic energy is most easily worked out starting from the Cartesian coordinates:

$$K = \frac{1}{2}m\left(\dot{x}^2 + \dot{y}^2\right) = \frac{1}{2}m\left[(l\dot{\alpha}\cos\alpha)^2 + (l\dot{\alpha}\sin\alpha)^2\right] = \frac{1}{2}ml^2\dot{\alpha}^2 ,$$

so that the Lagrangian becomes

$$L = \frac{1}{2}ml^2\dot{\alpha}^2 + mgl\cos\alpha .$$

The required equation of motion is obtained from the Lagrange equation:

$$\frac{\partial L}{\partial \alpha} - \frac{d}{dt}\left(\frac{\partial L}{\partial \dot{\alpha}}\right) = 0 \quad \Longrightarrow \quad -mgl \sin \alpha - \frac{d}{dt}\left(ml^2 \dot{\alpha}\right) = 0$$

$$\Longrightarrow \quad \ddot{\alpha} + \frac{g}{l} \sin \alpha = 0 \,.$$

This is the same equation as in Sect. I.8.6.4 where we also considered the finite amplitude pendulum. For small oscillations $\sin \alpha \simeq \alpha$ and we arrive at the harmonic oscillator equation with the frequency $\omega = \sqrt{g/l}$.

For a system of particles, we have to sum up over all contributions to get their total kinetic and potential energies.

Problem 9.12 Consider a double pendulum system shown in Fig. 9.6b. By considering two angles, α_1 and α_2, as independent generalised coordinates, write the Cartesian coordinates (x_1, y_1) and (x_2, y_2) of the two masses m_1 and m_2 and hence show that the Lagrangian of the system is

$$L = \frac{1}{2}\left[(m_1 + m_2)\, l_1^2 \dot{\alpha}_1^2 + m_2 l_2^2 \dot{\alpha}_2^2\right] + m_2 l_1 l_2 \cos(\alpha_1 - \alpha_2)\, \dot{\alpha}_1 \dot{\alpha}_2$$

$$+ (m_1 + m_2)\, gl_1 \cos \alpha_1 + m_2 gl_2 \cos \alpha_2 \,.$$

Then, derive equations of motion for the two masses:

$$l_1 \ddot{\alpha}_1 + \frac{m_2}{m_1 + m_2} l_2 \left[\ddot{\alpha}_2 \cos(\alpha_1 - \alpha_2) + \dot{\alpha}_2^2 \sin(\alpha_1 - \alpha_2)\right] = -g \sin \alpha_1 \,,$$

and

$$l_2 \ddot{\alpha}_2 + l_1 \left[\ddot{\alpha}_1 \cos(\alpha_1 - \alpha_2) - \dot{\alpha}_1^2 \sin(\alpha_1 - \alpha_2)\right] = -g \sin \alpha_2 \,.$$

Problem 9.13 Consider a pendulum suspended at a support which is attached by a spring with the elastic constant k to a wall, Fig. 9.6c. Choosing the horizontal coordinate x of the oscillating support and the angle α of the pendulum as independent generalised coordinates, show that the Lagrangian of this system is

$$L = \frac{1}{2}\,(m_1 + m_2)\, \dot{x}^2 + \frac{1}{2} m_2 l^2 \dot{\alpha}^2 + m_2 l \dot{x} \dot{\alpha} \cos \alpha - \frac{1}{2} kx^2 + m_2 gl \cos \alpha \,,$$

while the corresponding equations of motion read

$$l\ddot{\alpha} + \ddot{x} \cos \alpha + g \sin \alpha = 0$$

and

$$\ddot{x} + \frac{m_2 l}{m_1 + m_2}\left(\ddot{\alpha} \cos \alpha - \dot{\alpha}^2 \sin \alpha\right) + \frac{k}{m_1 + m_2} x = 0 \,.$$

Fig. 9.7 A ball on a rotating
rod

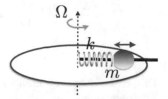

Fig. 9.7 A ball on a rotating
rod

Problem 9.14 A ball of mass m is set on a horizontal rod along which it can
slide without friction; at the same time, the ball is also attached by a spring
with the elastic constant k to a vertical axis, see Fig. 9.7. The rod rotates with
the angular frequency Ω around the vertical axis. Show that the Lagrangian in
this case is

$$L = \frac{1}{2}m\left(\dot{r}^2 + \Omega^2 r^2\right) - \frac{1}{2}k\left(r - r_0\right)^2 ,$$

where r and r_0 are the distances of the mass to the centre during the motion
and at equilibrium, respectively. Then derive the corresponding equations of
motion to show that in the case of $\omega_0 = \sqrt{k/m} > \Omega$ the ball oscillates along
the rod with the frequency $\omega = \sqrt{\omega_0^2 - \Omega^2}$ and the average distance $r_{av} =$
$(\omega_0/\omega)^2 r_0 > r_0$ from the centre. Why the average position of the mass moved
more away from the centre?

The formalism based on the Lagrange equations can be used to establish important conservation quantities of a mechanical system such as energy, linear (or trans-
lational) momentum and angular momentum. These are particular cases of a more
general Noether theorem stating that any symmetry of a physical system has a conservation law associated with it.

First of all, consider the invariance of an isolated system with respect to time[6].
Since the system is conservative, its Lagrangian cannot depend explicitly on time,
i.e. the total time derivative of L should be zero:

$$\frac{dL}{dt} = \sum_{i=1}^{n}\left(\frac{\partial L}{\partial q_i}\dot{q}_i + \frac{\partial L}{\partial \dot{q}_i}\ddot{q}_i\right) = \sum_{i=1}^{n}\left[\frac{d}{dt}\left(\frac{\partial L}{\partial \dot{q}_i}\right)\dot{q}_i + \frac{\partial L}{\partial \dot{q}_i}\ddot{q}_i\right] .$$

Here we replaced the derivative of L with respect to the coordinate q_i with the
time derivative term coming from the Lagrange equations (9.80). Then, we notice
that an expression in the square brackets is nothing but the total time derivative of
$\dot{q}_i\left(\partial L/\partial \dot{q}_i\right)$, so that

---

[6] Our consideration is also valid for systems in an external field which does not depend on time.

$$\frac{dL}{dt} = \sum_{i=1}^{n} \frac{d}{dt}\left(\frac{\partial L}{\partial \dot{q}_i}\dot{q}_i\right) \implies \frac{d}{dt}\left(\sum_{i=1}^{n} \frac{\partial L}{\partial \dot{q}_i}\dot{q}_i - L\right) = 0 .$$

In other words, the total time derivative of the quantity

$$E = \sum_{i=1}^{n} \frac{\partial L}{\partial \dot{q}_i}\dot{q}_i - L , \qquad (9.82)$$

is equal to zero, i.e. it must be time conserved for an isolated system. This quantity is called system energy. Indeed, since the kinetic energy K may depend on both generalised coordinates and their velocities, while the potential energy U only depends on the coordinates, then

$$E = \sum_{i=1}^{n} \frac{\partial (K - U)}{\partial \dot{q}_i}\dot{q}_i - (K - U) = \sum_{i=1}^{n} \frac{\partial K}{\partial \dot{q}_i}\dot{q}_i - K + U .$$

Assuming that K is a quadratic form of the velocities,

$$K = \frac{1}{2} \sum_{i,j=1}^{n} \alpha_{ij}\dot{q}_i\dot{q}_j$$

with the coefficients α_{ij} forming an $n \times n$ matrix[7] (which generally may depend on the coordinates), we can write

$$\sum_{i=1}^{n} \frac{\partial K}{\partial \dot{q}_i}\dot{q}_i = \sum_{i=1}^{n}\left(\sum_{j=1}^{n} \alpha_{ij}\dot{q}_j\right)\dot{q}_i = 2K ,$$

leading to $E = K + U$, which is a usual expression for the energy.

The momentum of the system appears due to its invariance with respect to an arbitrary translation in space. Let us use the Cartesian coordinates of all particles of our system to describe this. Displacing all the positions $\mathbf{r}_i$ of the particles by the same vector $\delta\mathbf{r} = (\delta x, \delta y, \delta z)$ will cause the following change in the Lagrangian:

$$\delta L = \sum_{i=1}^{N}\left(\frac{\partial L}{\partial x_i}\delta x + \frac{\partial L}{\partial y_i}\delta y + \frac{\partial L}{\partial z_i}\delta z\right) = \sum_{i=1}^{N} \frac{\partial L}{\partial \mathbf{r}_i}\cdot\delta\mathbf{r} = \delta\mathbf{r}\cdot\sum_{i=1}^{N} \frac{\partial L}{\partial \mathbf{r}_i} ,$$

where $\partial L/\partial \mathbf{r}_i$ corresponds to a vector with components $(\partial L/\partial x_i, \partial L/\partial y_i, \partial L/\partial z_i)$, the notation frequently used in physics literature. In the formula above we perform summation over all particles in the system, N. Since such a translation cannot change

---

[7] Of course, the matrix can always be chosen symmetric in the quadratic form.

the physical behaviour of the system, we should have $\delta L = 0$. Since the translation $\delta \mathbf{r}$ is arbitrary, one can only satisfy this by having

$$\sum_{i=1}^{N} \frac{\partial L}{\partial x_i} = 0 \implies \sum_{i=1}^{N} \frac{d}{dt}\left(\frac{\partial L}{\partial \dot{x}_i}\right) = 0 \implies \frac{d}{dt}\left(\sum_{i=1}^{N} \frac{\partial L}{\partial \dot{x}_i}\right) = 0 ,$$

and similarly for the y and z components. Here again the Lagrange equations (9.80) were used. Hence the vector $\mathbf{P}$ with the Cartesian components

$$\begin{aligned} P_x &= \sum_{i=1}^{N} \frac{\partial L}{\partial \dot{x}_i} = \sum_{i=1}^{N} \frac{\partial L}{\partial v_{ix}} , \\[2mm] P_y &= \sum_{i=1}^{N} \frac{\partial L}{\partial \dot{y}_i} = \sum_{i=1}^{N} \frac{\partial L}{\partial v_{iy}} , \\[2mm] P_z &= \sum_{i=1}^{N} \frac{\partial L}{\partial \dot{z}_i} = \sum_{i=1}^{N} \frac{\partial L}{\partial v_{iz}} \end{aligned} \qquad (9.83)$$

is to be conserved. This vector is called momentum of the system,

$$\mathbf{P} = \sum_{i=1}^{N} \frac{\partial L}{\partial \mathbf{v}_i} ,$$

while its individual components, $\partial L/\partial v_{i\alpha}$ (where $\alpha = x, y, z$ and $\mathbf{v}_i = \dot{\mathbf{r}}_i$ is the velocity vector), correspond to the components of the momentum vector

$$\mathbf{p}_i = \frac{\partial L}{\partial \mathbf{v}_i}$$

of the i−th individual particle. Since only the kinetic energy

$$K = \frac{1}{2} \sum_{i=1}^{N} m_i v_i^2$$

in the Lagrangian L depends on the velocities, we obtain after differentiation that $\mathbf{p}_i = m_i \mathbf{v}_i$, which is a familiar expression for the particle momentum.

If the system is subjected to an external field which does not depend on a particular coordinate, e.g. x, then the system will still remain invariant in its translation along this directions. This means that the x component of the momentum, P_x, will be conserved.

An isolated system should also be invariant under an arbitrary rotation around an arbitrary axis, and this must correspond to the conservation of another physical quantity. Consider an axis passing through the centre of the coordinate system which

Fig. 9.8 The vector **r** rotated by angle $\delta\phi$ around the axis drawn along the unit vector **s** goes over into a new vector $\mathbf{r}' = \mathbf{r} + \delta\mathbf{r}$

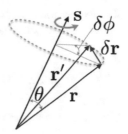

is characterised by the unit vector **s**. A rotation by the angle $\delta\phi$ around this axis will change the position vectors $\mathbf{r}_i$ of the particle i (where $i = 1, \ldots, N$) by

$$\delta\mathbf{r}_i = \delta\phi\,[\mathbf{s} \times \mathbf{r}_i]\ , \tag{9.84}$$

as can be seen from Fig. 9.8. Indeed, the vector **r** in the figure, which makes an angle θ with the axis of rotation, defines a plane perpendicular to the axis when rotated by angle $\delta\phi$ to $\mathbf{r}'$. The projections of both **r** and $\mathbf{r}'$ on that plane have both the length $r \sin\theta$, and the angle between them is exactly $\delta\phi$. For small such angles the length of the difference, $\delta\mathbf{r} = \mathbf{r}' - \mathbf{r}$, is $\delta r = (r \sin\theta)\,\delta\phi = |\mathbf{s} \times \mathbf{r}|\,\delta\phi$ as $|\mathbf{s}| = 1$. Taking into account the directions of the vectors and the definition of the vector product of the two vectors, we arrive at Eq. (9.84) written above.

Correspondingly, as the system rotates, velocities $\mathbf{v}_i$ of its particles change as well:

$$\delta\mathbf{v}_i = \delta\frac{d\mathbf{r}_i}{dt} = \frac{d}{dt}\,(\delta\mathbf{r}_i) = \frac{d}{dt}\delta\phi\,[\mathbf{s} \times \mathbf{r}_i] = \delta\phi\left[\mathbf{s} \times \frac{d\mathbf{r}_i}{dt}\right] = \delta\phi\,[\mathbf{s} \times \mathbf{v}_i]\ .$$

Therefore, the total variation of the Lagrangian due to rotation of the whole system by $\delta\phi$ around the axis given by the unit vector **s** is (the sum over α corresponds to the summation over the Cartesian components):

$$\delta L = \sum_{i=1}^{N}\sum_{\alpha}\left(\frac{\partial L}{\partial r_{i\alpha}}\delta r_{i\alpha} + \frac{\partial L}{\partial v_{i\alpha}}\delta v_{i\alpha}\right) = \sum_{i=1}^{N}\left(\frac{\partial L}{\partial \mathbf{r}_i}\cdot\delta\mathbf{r}_i + \frac{\partial L}{\partial \mathbf{v}_i}\cdot\delta\mathbf{v}_i\right).$$

Using Lagrange equations, this expression can be manipulated further:

$$\delta L = \sum_{i=1}^{N}\left[\frac{d}{dt}\left(\frac{\partial L}{\partial \mathbf{v}_i}\right)\cdot\delta\mathbf{r}_i + \frac{\partial L}{\partial \mathbf{v}_i}\cdot\delta\mathbf{v}_i\right] = \frac{d}{dt}\left(\sum_{i=1}^{N}\frac{\partial L}{\partial \mathbf{v}_i}\cdot\delta\mathbf{r}_i\right)$$

$$= \frac{d}{dt}\sum_{i=1}^{N}\mathbf{p}_i\cdot\delta\mathbf{r}_i = \frac{d}{dt}\sum_{i=1}^{N}\mathbf{p}_i\cdot\delta\phi\,[\mathbf{s} \times \mathbf{r}_i] = \delta\phi\frac{d}{dt}\sum_{i=1}^{N}\mathbf{p}_i\cdot[\mathbf{s} \times \mathbf{r}_i]\ .$$

The expression under the sum contains the triple product $[\mathbf{p}_i, \mathbf{s}, \mathbf{r}_i]$ which is invariant under a cyclic permutation of its components. Therefore, we can write

$$\delta L = \delta\phi \frac{d}{dt} \sum_{i=1}^{N} [\mathbf{p}_i, \mathbf{s}, \mathbf{r}_i] = \delta\phi \frac{d}{dt} \sum_{i=1}^{N} [\mathbf{s}, \mathbf{r}_i, \mathbf{p}_i] = \delta\phi\, \mathbf{s} \cdot \frac{d}{dt} \left(\sum_{i=1}^{N} [\mathbf{r}_i \times \mathbf{p}_i] \right) .$$

Since the Lagrangian must not change, $\delta L = 0$, the expression in the round brackets above,

$$\mathbf{M} = \sum_{i=1}^{N} [\mathbf{r}_i \times \mathbf{p}_i] ,$$

must be conserved. This is the familiar expression for the angular momentum of the system.

So far we have discussed various applications of Euler's equations in mechanics. Let us now consider some applications of Ostrogradsky's equation (9.77) for functions of more than one variable. The simplest example is related to the oscillations of a string considered in Sect. 8.2.1. Consider a string of length l which is set along the x axis with the tension T_0 (per unit length). Each unit element dx of the string oscillates in the perpendicular direction, and its vertical displacement is described by the function $u\,(x, t)$. To construct the equation of motion for the string, we need to write its Lagrangian, i.e. difference of the kinetic and potential energy terms. If ρ is the density (per unit length) of the string, the kinetic energy of its piece of length dx will obviously be

$$dK = \frac{1}{2}\rho \dot{u}^2 dx = \frac{1}{2}\rho \left(u_t' \right)^2 dx .$$

The potential energy of the dx element of the string can be written as a work of the tension force, T_0, to stretch its length from dx (when the string is strictly horizontal, i.e. when not oscillating) to

$$ds = \sqrt{(dx)^2 + (du)^2} = \sqrt{1 + \left(u_x' \right)^2} dx ,$$

i.e.

$$dU = T_0 \left(\sqrt{1 + \left(u_x' \right)^2} - 1 \right) dx = T_0 \left(1 + \frac{1}{2} \left(u_x' \right)^2 + \ldots - 1 \right) dx$$

$$\simeq \frac{1}{2} T_0 \left(u_x' \right)^2 dx ,$$

where we assumed that deformation of the string is small and hence expanded the square root term keeping only the first two terms. Assuming that there is also an external force $\mathcal{F}$ (per unit length) acting on the element dx, there will be an additional term $-(\mathcal{F}dx)\,u$ in the potential energy. Integrating these expressions along the whole string, gives the total kinetic and potential energies, respectively, so that the total Lagrangian of the whole string is

$$L = \int_0^l \left[\frac{1}{2} \rho \left(u_t' \right)^2 - \frac{1}{2} T_0 \left(u_x' \right)^2 + \mathcal{F} u \right] dx . \tag{9.85}$$

The action to optimise is then given by the double integral

$$S = \int_{t_1}^{t_2} L\, dt = \int_{t_1}^{t_2} dt \int_0^l dx \left[\frac{1}{2} \rho \left(u_t' \right)^2 - \frac{1}{2} T_0 \left(u_x' \right)^2 + \mathcal{F} u \right]$$

$$= \int_{t_1}^{t_2} dt \int_0^l dx\, F \left(u, u_t', u_x' \right) ,$$

while the corresponding equation of motion for the transverse displacement is Ostro-gradsky's equation (9.77):

$$\frac{\partial F}{\partial u} - \frac{\partial}{\partial t} \frac{\partial F}{\partial u_t'} - \frac{\partial}{\partial x} \frac{\partial F}{\partial u_x'} = 0 .$$

Calculating the necessary derivatives, we obtain

$$\mathcal{F} - \frac{\partial}{\partial t} \left(\rho u_t' \right) - \frac{\partial}{\partial x} \left(-T_0 u_x' \right) = 0 \quad \Longrightarrow \quad \mathcal{F} - \rho u_{tt}'' + T_0 u_{xx}'' = 0 ,$$

which is the wave equation (8.18).

Problem 9.15 Consider transverse oscillations of a rod of length L and density (per unit length) ρ, which is placed horizontally along the x axis. Let $u(x, t)$ be a displacement of the rod at x at time t. Assuming that the potential energy per unit length of the deformed rod is given by $-\frac{1}{2} \kappa \left(u_{xx}'' \right)^2$, where κ is some material dependent constant, construct the corresponding Lagrangian and the action, and then show that the function $u(x, t)$ satisfies the following PDE:

$$\rho \frac{\partial^2 u}{\partial t^2} = \kappa \frac{\partial^4 u}{\partial x^4} .$$

[Hint: use Eq. (9.78).]

9.3.2 Functional Derivatives

In physics calculus of variations is frequently formulated in a slightly different way via the so-called functional derivatives. Although the mathematics behind those is basically the same as the one described above, it is worth considering it here since the slightly different notations widely used in physics literature may cause confusion.

If we have a functional $L[f(x)]$, then its variation can formally be written as an integral

$$\delta L[f(x)] = \int \frac{\delta L[f(x)]}{\delta f(x')} \delta f(x') \, dx' \,, \tag{9.86}$$

where the quantity $\frac{\delta L[f(x)]}{\delta f(x')}$ is called the functional derivative. The functional L here can be either an algebraic expression involving $f(x)$ or an integral containing $f(x)$; the latter case we assumed everywhere above. If the functional L is just $f(x)$, then obviously, to be consistent, one has to *define* the functional derivative of the function with respect to itself to be the delta function:

$$\frac{\delta f(x)}{\delta f(x')} = \delta(x - x') \,. \tag{9.87}$$

Indeed, in this case

$$\delta f(x) = \int \frac{\delta f(x)}{\delta f(x')} \delta f(x') \, dx' = \int \delta(x - x') \delta f(x') \, dx' = \delta f(x) \,,$$

as required.

For instance, consider $L[f] = f(x)^n$. Using our previous method, we obtain

$$\delta L = L'(0) = \lim_{\alpha \to 0} \frac{(f + \alpha \delta f)^n - f^n}{\alpha} = n f(x)^{n-1} \delta f(x) \,.$$

It is easy to see that the same result is obtained using the functional derivative if we define

$$\frac{\delta F[f(x)]}{\delta f(x')} = \delta(x - x') \left(\frac{dF}{df} \right)(x) \,. \tag{9.88}$$

Here dF/df is a usual derivative with respect to the function f calculated at x. Indeed,

$$\frac{\delta f(x)^n}{\delta f(x')} = \delta(x - x') \left(\frac{df^n}{df} \right)(x) = \delta(x - x') n f(x)^{n-1}$$

$$\implies \delta L = \int \frac{\delta f(x)^n}{\delta f(x')} \delta f(x') \, dx' = n f(x)^{n-1} \delta f(x) \,.$$

Similarly, applying the functional derivative to a functional written as an integral, we have

$$\delta \int_a^b F(f)\,dx = \int_a^b \delta F(f)\,dx = \int_a^b \left[\int \frac{\delta F(f(x))}{\delta f(x')} \delta f(x')\,dx' \right] dx$$

$$= \int_a^b \left[\int_a^b \delta(x - x') \left(\frac{dF}{df} \right) (x) \delta f(x')\,dx' \right] dx$$

$$= \int_a^b \left(\frac{dF}{df} \right) \delta f(x)\,dx\,,$$

as one would expect.

Let us apply now this technique to a more general functional (9.7) which also contains a derivative of the function $f(x)$. In this case the variation of the derivative has also to be included as shown below:

$$\delta L = \delta \int_a^b F\left(x, f, f'\right) dx$$

$$= \int_a^b dx \int_a^b dx' \left\{ \frac{\delta F\left(x, f, f'\right)}{\delta f(x')} \delta f(x') + \frac{\delta F\left(x, f, f'\right)}{\delta f'(x')} \delta f'(x') \right\}$$

$$= \int_a^b dx \int_a^b dx' \left\{ \left[\delta(x - x') \frac{\partial F\left(x, f, f'\right)}{\partial f(x)} \right] \delta f(x') \right.$$

$$\left. + \left[\delta(x - x') \frac{\partial F\left(x, f, f'\right)}{\partial f'(x)} \right] \delta f'(x') \right\}$$

$$= \int_a^b dx \left[\frac{\partial F}{\partial f} \delta f(x) + \frac{\partial F}{\partial f'} \delta f'(x) \right]$$

$$= \int_a^b dx \left[\frac{\partial F}{\partial f} \delta f(x) + \frac{\partial F}{\partial f'} \frac{d\left(\delta f(x)\right)}{dx} \right].$$

Calculating the integral with respect to the second term within the square brackets by parts and using the fact that $\delta f(a) = \delta f(b) = 0$, we obtain

$$\delta L = \int_a^b dx \left[\frac{\partial F}{\partial f} \delta f(x) - \frac{d}{dx}\left(\frac{\partial F}{\partial f'} \right) \delta f(x) \right]$$

$$= \int_a^b dx \left[\frac{\partial F}{\partial f} - \frac{d}{dx}\left(\frac{\partial F}{\partial f'} \right) \right] \delta f(x)\,,$$

which is the same expression as before, see Eq. (9.8). Hence, in the case of a functional L given by an integral the functional derivative $\delta L/\delta f(x)$ is given by an expression in the square brackets above; it does not contain the delta function:

$$\frac{\delta L}{\delta f(x)} = \frac{\delta}{\delta f(x)} \int_a^b F\left(x, f, f'\right) dx = \frac{\partial F}{\partial f} - \frac{d}{dx}\left(\frac{\partial F}{\partial f'} \right). \tag{9.89}$$

Let us work out a useful relationship involving the functional derivatives. Let L be a functional of F (and x), which in turn is a functional of a function $g = g(x)$ (and x), i.e. $L = L[F[g(x)]]$. Then,

$$\delta L = \int \frac{\delta L}{\delta F(x')} \delta F(x')\, dx' \quad \text{and also} \quad \delta F(x') = \int \frac{\delta F(x')}{\delta g(x)} \delta g(x)\, dx ,$$

so that

$$\delta L = \int \frac{\delta L}{\delta F(x')} \frac{\delta F(x')}{\delta g(x)} \delta g(x)\, dx dx' = \int \left[\int \frac{\delta L}{\delta F(x')} \frac{\delta F(x')}{\delta g(x)} dx' \right] \delta g(x)\, dx .$$

On the other hand, the expression in the square brackets above must be the functional derivative $\delta L / \delta g(x)$. Hence, we have

$$\frac{\delta L}{\delta g(x)} = \int \frac{\delta L}{\delta F(x')} \frac{\delta F(x')}{\delta g(x)} dx' . \tag{9.90}$$

This is an analogue of the chain rule we use in usual differentiation.

Finally, we note that the condition that the variation of a functional be zero needed for finding its optimum value can be equivalently formulated as zero of the corresponding functional derivative. For instance, Euler's equation (9.11) follows immediately from setting the derivative (9.89) to zero.

9.3.3 Many-Electron Theory

In quantum mechanics the variational principle plays a central role. Consider a non-relativistic quantum system of N electrons moving in the field of atomic nuclei, e.g. a molecule. In a stationary state the system is characterised by its Hamiltonian operator, $\widehat{H} = \widehat{K} + \widehat{U}$, which is a sum of the kinetic energy,

$$\widehat{K} = \sum_{i=1}^{N} \left(-\frac{1}{2} \Delta_i \right)$$

(we use atomic units for simplicity), and potential energy,

$$\widehat{U} = \sum_{i=1}^{N} V_n(\mathbf{r}_i) + \frac{1}{2} \sum_{i,j=1\,(i \neq j)}^{N} \frac{1}{|\mathbf{r}_i - \mathbf{r}_j|} ,$$

operators. Above, in the kinetic energy operator Δ_i is the Laplacian calculated with respect to the position $\mathbf{r}_i$ of the i-th electron; next, the first term in $\widehat{U}$ describes the interaction of each electron with all atomic nuclei, while the second term stands for the repulsive electron–electron interaction.

The quantum state of the system is described by the Schrödinger equation

$$\widehat{H}\Psi = E\Psi \,, \tag{9.91}$$

or, if written fully,

$$\sum_{i=1}^{N}\left(-\frac{1}{2}\Delta_i\Psi\right) + \sum_{i=1}^{N}V_n\left(\mathbf{r}_i\right)\Psi + \frac{1}{2}\sum_{i,j=1\,(i\neq j)}^{N}\frac{1}{\left|\mathbf{r}_i - \mathbf{r}_j\right|}\Psi = E\Psi \,,$$

where $\Psi = \Psi\left(\mathbf{r}_1, \ldots, \mathbf{r}_N\right)$ is the system wavefunction depending on the coordinates $\mathbf{r}_1, \ldots, \mathbf{r}_N$ of all particles (we omitted spin here for simplicity). The wavefunction is assumed to be normalised to unity,

$$\int \cdots \int \Psi\left(\mathbf{r}_1, \ldots, \mathbf{r}_N\right)^*\Psi\left(\mathbf{r}_1, \ldots, \mathbf{r}_N\right) d\mathbf{r}_1 \cdots d\mathbf{r}_N$$
$$= \int \cdots \int \left|\Psi\left(\mathbf{r}_1, \ldots, \mathbf{r}_N\right)\right|^2 d\mathbf{r}_1 \cdots d\mathbf{r}_N = 1 \,, \tag{9.92}$$

where we used shorthand notations for the volume element of each particle: $d\mathbf{r}_i = dx_i dy_i dz_i$ with $\mathbf{r}_i = (x_i, y_i, z_i)$. The normalisation condition corresponds to the unit probability for the electrons to be found somewhere. Indeed, the probability to find the first electron between $\mathbf{r}_1$ and $\mathbf{r}_1 + d\mathbf{r}_1$, the second between $\mathbf{r}_2$ and $\mathbf{r}_2 + d\mathbf{r}_2$, and so on is given simply by

$$dP = \left|\Psi\left(\mathbf{r}_1, \ldots, \mathbf{r}_N\right)\right|^2 d\mathbf{r}_1 \cdots d\mathbf{r}_N \,,$$

so that the normalisation condition sums all these probabilities over the whole space for each particle resulting in the unity. To simplify our notations, we shall use in the following the vector $\mathbf{R}$ to designate all electronic coordinates $\mathbf{r}_1, \ldots, \mathbf{r}_N$, while the corresponding product of all volume elements $d\mathbf{r}_1 \cdots d\mathbf{r}_N$ will then be written simply as $d\mathbf{R}$. Also a single integral symbol will be used; however, of course, multiple integration is implied as above.

Finally, E in Eq. (9.91) is the total energy of the system. The lowest total energy, E_0, corresponds to the ground state with the wavefunction Ψ_0; assuming that the system is stable in its ground state, we have $E_0 < 0$. There could be more bound (stable) states characterising the system. Their energies E_0, E_1, E_2, etc. are negative and form a sequence $E_0 < E_1 < E_2 < \ldots$; the corresponding wavefunctions are Ψ_0, Ψ_1, Ψ_2, etc. The states with energies higher than the ground state energy E_0 correspond to the excited states of the system. The wavefunctions Ψ_i of different states ($i = 0, 1, 2, \ldots$) are normalised to unity and are orthogonal,

$$\int \Psi_i\left(\mathbf{R}\right)^* \Psi_j\left(\mathbf{R}\right) d\mathbf{R} = \delta_{ij} \,. \tag{9.93}$$

Consider the equation for the ground state, $\widehat{H}\Psi_0 = E_0\Psi_0$. Multiplying both sides by Ψ_0^* from the left (so that the operator $\widehat{H}$ would not act on it) and integrating over all electronic coordinates $\mathbf{R}$, we have

$$\int \Psi_0^*\widehat{H}\Psi_0 d\mathbf{R} = E_0 \int \Psi_0^*\Psi_0 d\mathbf{R} \quad \Longrightarrow \quad E_0 = \int \Psi_0^*\widehat{H}\Psi_0 d\mathbf{R} \ ,$$

where normalisation of the wavefunction, Eq. (9.92), was used. This gives an expression for the ground state energy directly via its ground state wavefunction. This expression has been obtained from the Schrödinger equation. However, an alternative way which is frequently used is to postulate the energy expression,

$$E = \int \Psi^*\widehat{H}\Psi d\mathbf{R} \ ,$$

as a functional of the wavefunction, $E = E\left[\Psi\left(\mathbf{R}\right)\right]$, and then find the best wavefunction, $\Psi\left(\mathbf{R}\right)$, which minimises it subject to the normalisation condition $\int \Psi^*\Psi d\mathbf{R} = 1$. As usual, we define a function $E\left(\alpha\right) = E\left[\Psi + \alpha\delta\Psi\right]$ and the variation of the energy, $\delta E = E'\left(\alpha\right)\big|_{\alpha=0}$. The optimum condition for the wavefunction is established by requiring that $\delta H = 0$, where

$$H = \int \Psi^*\widehat{H}\Psi d\mathbf{R} - \lambda\left(\int \Psi^*\Psi d\mathbf{R} - 1\right)$$

is the auxiliary functional with λ being the appropriate Lagrange multiplier.

Note that both functions, Ψ and Ψ^*, must be varied. However, since the wavefunction is complex, we can vary independently its real and imaginary parts. Alternatively, we can consider Ψ and Ψ^* being independent. Then, varying the Ψ^* function, we get

$$\delta H = \int \left(\delta\Psi^*\right)\widehat{H}\Psi d\mathbf{R} - \lambda \int \left(\delta\Psi^*\right)\Psi d\mathbf{R}$$
$$= \int \left(\delta\Psi^*\right)\left(\widehat{H}\Psi - \lambda\Psi\right)d\mathbf{R} \ .$$

Since $\delta\Psi^*$ is arbitrary, then $\delta H = 0$ leads to the Schrödinger equation $\widehat{H}\Psi - \lambda\Psi = 0$ with λ being equal to the system energy, as above.

The calculation is a bit lengthier if we vary Ψ instead:

$$\delta H = \int \Psi^*\widehat{H}\left(\delta\Psi\right)d\mathbf{R} - \lambda \int \Psi^*\left(\delta\Psi\right)d\mathbf{R} \ . \tag{9.94}$$

The Hamiltonian operator contains the kinetic energy term which involves differentiation, and the potential energy operator which is a multiplication operator. For such an operator one can easily place the function in front of the operator:

$$\int \Psi^* \widehat{U} \, (\delta\Psi) \, d\mathbf{R} = \int (\delta\Psi) \, \widehat{U} \Psi^* d\mathbf{R} \,. \tag{9.95}$$

The situation with the kinetic energy operator $\widehat{K}$ is less trivial. Consider the component $\widehat{K}_i$ of $\widehat{K}$ corresponding to the i-th electron:

$$\int \Psi^* \widehat{K}_i \, (\delta\Psi) \, d\mathbf{R} = -\frac{1}{2} \int \Psi^* \Delta_i \, (\delta\Psi) \, d\mathbf{R}$$

$$= -\frac{1}{2} \underbrace{\int \cdots \int d\mathbf{R}'}_{N-1} \left[\int d\mathbf{r}_i \Psi^* \Delta_i \, (\delta\Psi) \right], \tag{9.96}$$

where in the last passage we separated out the integration over the i-th electron from all the others, which was denoted by $d\mathbf{R}'$. Next, consider the triple integral over $d\mathbf{r}_i$ enclosed in the square brackets above:

$$\int d\mathbf{r}_i \Psi^* \Delta_i \, (\delta\Psi) = \int dz_i \int dy_i \left[\int dx_i \Psi^* \frac{\partial^2}{\partial x_i^2} \, (\delta\Psi) \right]$$

$$+ \int dz_i \int dx_i \left[\int dy_i \Psi^* \frac{\partial^2}{\partial y_i^2} \, (\delta\Psi) \right] \tag{9.97}$$

$$+ \int dx_i \int dy_i \left[\int dz_i \Psi^* \frac{\partial^2}{\partial z_i^2} \, (\delta\Psi) \right] .$$

Each of the integrals in the square brackets can be taken twice by parts, e.g.:

$$\int dx_i \Psi^* \frac{\partial^2}{\partial x_i^2} \, (\delta\Psi) = \underbrace{\Psi^* \frac{\partial}{\partial x_i} \, (\delta\Psi) \Big|_{-\infty}^{\infty}}_{=0} - \int dx_i \frac{\partial \Psi^*}{\partial x_i} \frac{\partial}{\partial x_i} \, (\delta\Psi)$$

$$= - \int dx_i \frac{\partial \Psi^*}{\partial x_i} \frac{\partial}{\partial x_i} \, (\delta\Psi)$$

$$= \underbrace{-\frac{\partial \Psi^*}{\partial x_i} \, (\delta\Psi) \Big|_{-\infty}^{\infty}}_{=0} + \int dx_i \frac{\partial^2 \Psi^*}{\partial x_i^2} \, (\delta\Psi) = \int dx_i \frac{\partial^2 \Psi^*}{\partial x_i^2} \delta\Psi \,.$$

Note that all free terms appearing while integrating by parts disappear as we assume that the wavefunction goes to zero together with its spatial derivatives at infinities. Repeating this integration for the other two integrals in Eq. (9.97), we obtain that[8]

---

[8] When $\int \phi^* \left(\widehat{A}\varphi \right) dx = \int \left(\widehat{A}\phi^* \right) \varphi dx = \int \varphi \left(\widehat{A}\phi^* \right) dx$, i.e. where effectively the functions $\phi(x)$ and $\varphi(x)$ are allowed to change places around the operator $\widehat{A}$, the operator is called self-adjoint. We have just shown that the operator $\frac{d^2}{dx^2}$ is such an operator.

$$\int d\mathbf{r}_i \Psi^* \Delta_i (\delta\Psi) = \int d\mathbf{r}_i (\Delta_i \Psi^*) \delta\Psi .$$

By looking at Eq. (9.96) it becomes clear now that

$$\int \Psi^* \widehat{K}_i (\delta\Psi) d\mathbf{R} = \int (\delta\Psi) \widehat{K}_i \Psi^* d\mathbf{R} ,$$

and hence the same can be done for the whole kinetic energy operator $\widehat{K} = \sum_i \widehat{K}_i$. Combining this result with the similar result for the potential energy operator, Eq. (9.95), we obtain for δH in (9.94):

$$\delta H = \int (\delta\Psi) \widehat{H}\Psi^* d\mathbf{R} - \lambda \int (\delta\Psi) \Psi^* d\mathbf{R}$$
$$= \int (\delta\Psi) (\widehat{H}\Psi^* - \lambda\Psi^*) d\mathbf{R} .$$

Since $\delta H = 0$, we obtain $\widehat{H}\Psi^* - \lambda\Psi^* = 0$, which, after taking its complex conjugate, is nothing but the same Schrödinger equation as before.

To obtain the first excited state of the system the same procedure can be followed. However, in this case one has to make sure that the wavefunction Ψ_1 is orthogonal to Ψ_0 of the ground state, i.e. $\int \Psi_1^* (\mathbf{R}) \Psi_0 (\mathbf{R}) d\mathbf{R} = 0$. This can be achieved by setting up the variational problem with an additional Lagrange multiplier:

$$H = \int \Psi^* \widehat{H}\Psi d\mathbf{R} - \lambda_0 \left(\int \Psi^* \Psi d\mathbf{R} - 1 \right) - \lambda_1 \int \Psi^* \Psi_0 d\mathbf{R} ,$$

so that the condition of orthogonality is imposed. Then, setting $\delta H = 0$ should give quantum states that are orthogonal to the ground one; the lowest in energy state Ψ has to be the first excited state Ψ_1 of the system. It is clear, that this procedure can be generalised to finding any excited state Ψ_n of the system by forcing its orthogonality to all the previous states $\Psi_0, \ldots, \Psi_{n-1}$.

To illustrate the variational principle further, let us consider the so-called Thomas–Fermi (TF) model of an atom. Any atom (apart from the hydrogen atom) is essentially a many-electron system which is very difficult to treat exactly. In this model an attempt was made to propose an expression for the atom energy, E, which is a functional of the atom electron density $\rho(\mathbf{r})$ only. The latter is defined as the probability density to find one (any) electron between points $\mathbf{r}$ and $\mathbf{r} + d\mathbf{r}$. It can be obtained from the wavefunction Ψ by integrating over all electrons but one:

$$\rho(\mathbf{r}) = N \underbrace{\int \cdots \int dwirthm{R}' \left[|\Psi(\mathbf{R})|^2 \right]_{\mathbf{r}_1 = \mathbf{r}}}_{N-1}$$
$$= N \int \cdots \int d\mathbf{r}_2 d\mathbf{r}_3 \cdots d\mathbf{r}_N |\Psi(\mathbf{r}, \mathbf{r}_2, \ldots, \mathbf{r}_n)|^2 .$$

The factor of N appears here because any of the N electrons can contribute to the density. The proposed TF functional has the form:

$$E_{TF}\left[\rho\left(\mathbf{r}\right)\right] = C_F \int d\mathbf{r}\,\rho(\mathbf{r})^{5/3} - Z \int \frac{\rho\left(\mathbf{r}\right)}{r}d\mathbf{r} + \frac{1}{2}\int \int \frac{\rho\left(\mathbf{r}\right)\rho\left(\mathbf{r}'\right)}{|\mathbf{r}-\mathbf{r}'|}d\mathbf{r}d\mathbf{r}'\ .$$
(9.98)

The first term here corresponds to the energy of a free uniform electron gas occupying a volume $d\mathbf{r}$, C_F being a constant pre-factor. The second term corresponds to the attractive interaction of all the electrons of the density ρ to the nucleus of change Z. Finally, the last term describes the electron–electron interaction. It is simply the Coulomb interaction energy of the charge cloud of the density ρ with itself. The energy functional must be supplemented by the normalisation of the density,

$$\int \rho\left(\mathbf{r}\right)d\mathbf{r} = N \int d\mathbf{r} \underbrace{\int \cdots \int}_{N-1} d\mathbf{R}'\,|\Psi\left(\mathbf{R}\right)|^2 = N \int |\Psi\left(\mathbf{R}\right)|^2\,d\mathbf{R} = N\ .$$
(9.99)

Therefore, to find the 'best' electron density that would minimise the atom energy, we need to vary the energy with respect to the density ρ subject to the constraint of the ρ normalisation:

$$\delta H = \delta \left\{ E_{TF}\left[\rho\left(\mathbf{r}\right)\right] - \mu \left(\int \rho\left(\mathbf{r}\right)d\mathbf{r} - N \right) \right\}$$

$$= \delta E_{TF}\left[\rho\left(\mathbf{r}\right)\right] - \mu \int \delta\rho\left(\mathbf{r}\right)d\mathbf{r}\ .$$
(9.100)

Here μ is the Lagrange multiplier.

Problem 9.16 Show using the method based on the replacement $\rho \to \rho + \alpha\delta\rho$ that

$$\delta E_{TF} = \int \left(C_F \frac{5}{3}\rho^{2/3} - \frac{Z}{r} + \int \frac{\rho\left(\mathbf{r}'\right)}{|\mathbf{r}-\mathbf{r}'|}d\mathbf{r}' \right) \delta\rho\left(\mathbf{r}\right)d\mathbf{r}$$

$$= \int \left(C_F \frac{5}{3}\rho^{2/3} - V\left(\mathbf{r}\right) \right) \delta\rho\left(\mathbf{r}\right)d\mathbf{r}\ ,$$

where

$$V\left(\mathbf{r}\right) = \frac{Z}{r} - \int \frac{\rho\left(\mathbf{r}'\right)}{|\mathbf{r}-\mathbf{r}'|}d\mathbf{r}'$$

is the electrostatic potential due to the nucleus and the entire electron cloud.

Problem 9.17 Consider the functional

$$H = E_{TF}\left[\rho\left(\mathbf{r}\right)\right] - \mu \left(\int \rho\left(\mathbf{r}\right) d\mathbf{r} - N\right).$$

Apply the method of functional derivatives to show that

$$\frac{\delta H}{\delta \rho\left(\mathbf{r}\right)} = C_F \frac{5}{3}\rho(\mathbf{r})^{2/3} - V\left(\mathbf{r}\right).$$

Using the above results, the following equation for the electron density is obtained:

$$\delta H = 0 \quad \Longrightarrow \quad C_F \frac{5}{3}\rho(\mathbf{r})^{2/3} - V\left(\mathbf{r}\right) = \mu.$$

This is a rather complex integral equation for the density ρ. It is to be solved together with the normalisation condition (9.99) which is required to determine the Lagrange multiplier μ. The TF model gives plausible results for some of the many-electron atoms, but fails miserably for molecules (which only require some modification of the second term in the energy expression (9.98)), for which no binding is obtained at all.

Still, from historical point of view, the TF model is very important. Essentially, this model avoids dealing with the wavefunction, a very complicated object indeed as depending on many electronic variables; instead, it was proposed to work only with the electron density $\rho\left(\mathbf{r}\right)$ which depends only on three variables. The TF model was the first attempt to implement that idea of replacing Ψ with ρ, and is an early predecessor of the modern density functional theory (DFT) developed by P. Hohenberg, W. Kohn and L. J. Sham (in 1964-5), in which that idea has been successfully developed and implemented into a powerful computational technique which is being widely used nowadays in condensed matter physics, material science and computational chemistry.

In Kohn–Sham DFT the electron density is used as well, however, the variational problem is formulated for some one-particle wavefunctions $\psi_a(\mathbf{r})$ called orbitals which are required to form an orthonormal set:

$$\int \psi_a^*(\mathbf{r})\psi_{a'}\left(\mathbf{r}\right) d\mathbf{r} = \delta_{aa'}.$$

The idea is to map a real interacting many-electron system into an artificial non-interacting system of the same number of electrons N and of the same electron density $\rho\left(\mathbf{r}\right)$; the fictitious system of electrons is subjected to an effective external potential to be determined self-consistently. Because of that mapping, the electron density can be written explicitly as a sum of the densities $\rho_a = |\psi_a\left(\mathbf{r}\right)|^2$ due to each electron:

$$\rho\left(\mathbf{r}\right) = \sum_{a=1}^{N} |\psi_a\left(\mathbf{r}\right)|^2 \ . \tag{9.101}$$

Moreover, the kinetic energy of all electrons can also be calculated as a sum of kinetic energies due to each individual electron:

$$K = \sum_{a=1}^{N} \int \psi_a^*\left(\mathbf{r}\right) \left(-\frac{1}{2}\Delta_a\right) \psi_a\left(\mathbf{r}\right) d\mathbf{r} \ . \tag{9.102}$$

Then, the total energy functional of the electron density is proposed to be of the following form:

$$E_{DFT}\left[\{\psi_a\}\right] = \sum_{a=1}^{N} \int \psi_a^*\left(-\frac{1}{2}\Delta_a\right) \psi_a d\mathbf{r} + \frac{1}{2} \int \int \frac{\rho\left(\mathbf{r}\right)\rho\left(\mathbf{r}'\right)}{|\mathbf{r}-\mathbf{r}'|} d\mathbf{r} d\mathbf{r}'$$
$$+ \int \rho\left(\mathbf{r}\right) V_n(\mathbf{r}) d\mathbf{r} + E_{xc}\left[\rho\right] \ . \tag{9.103}$$

Here the first term is the kinetic energy of the fictitious electron gas, the second term describes the bare Coulomb interaction between electrons, next is the term describing the interaction of the electrons with the potential of the nuclei $V_n(\mathbf{r})$. Finally, the last, the so-called exchange-correlation term, E_{xc}, describes all effects of exchange and correlation in the electron gas. This term also absorbs an error due to replacing the kinetic energy of the real interacting gas with that of the non-interacting gas. The exact expression for E_{xc} is not known, however, good approximations exist. In the so-called local density approximation (LDA)

$$E_{xc}\left[\rho\right] = \int \rho\left(\mathbf{r}\right) \epsilon_{xc}\left[\rho\left(\mathbf{r}\right)\right] d\mathbf{r} \ , \tag{9.104}$$

where ϵ_{xc} is the exchange-correlation energy density which is some known smooth function of the density fitted to calculations of the uniform electron gas of various densities.

The energy expression above is a functional of all orbitals ψ_a, so it has to be varied taking into account the expression (9.101) which relates explicitly the density with the orbitals. All the necessary functional derivatives we need in order to obtain equations for the orbitals are calculated in the following problem:

Problem 9.18 Obtain the following expressions for the functional derivatives of different terms in the energy functional:

$$\frac{\delta\rho\left(\mathbf{r}\right)}{\delta\psi_a^*(\mathbf{r}_1)} = \delta\left(\mathbf{r} - \mathbf{r}_1\right)\psi_a\left(\mathbf{r}_1\right) , \qquad \frac{\delta K}{\delta\psi_a^*(\mathbf{r}_1)} = -\frac{1}{2}\Delta_{\mathbf{r}_1}\psi_a(\mathbf{r}_1) ,$$

$$\frac{\delta}{\delta\psi_a^*(\mathbf{r}_1)} \frac{1}{2}\int\int \frac{\rho\left(\mathbf{r}\right)\rho\left(\mathbf{r}'\right)}{|\mathbf{r}_1 - \mathbf{r}'|}d\mathbf{r}d\mathbf{r}' = \psi_a\left(\mathbf{r}_1\right)\int \frac{\rho\left(\mathbf{r}'\right)}{|\mathbf{r}_1 - \mathbf{r}'|}d\mathbf{r}' ,$$

$$\frac{\delta}{\delta\psi_a^*(\mathbf{r}_1)}\int \rho\left(\mathbf{r}\right)V_n(\mathbf{r})d\mathbf{r} = V_n\left(\mathbf{r}_1\right)\psi_a\left(\mathbf{r}_1\right) ,$$

$$\frac{\delta E_{xc}}{\delta\psi_a^*(\mathbf{r}_1)} = \left[\epsilon_{xc}\left(\rho\right) + \rho\frac{d\epsilon_{xc}}{d\rho}\right]_{\rho=\rho(\mathbf{r}_1)}\psi_a\left(\mathbf{r}_1\right) = V_{xc}\left(\rho\left(\mathbf{r}_1\right)\right)\psi_a\left(\mathbf{r}_1\right) .$$

Here $\Delta_{\mathbf{r}_1}$ is the Laplacian with respect to the point $\mathbf{r}_1$, and V_{xc} is called the exchange-correlation potential.

Once we have all the necessary derivatives, we can consider the optimum of the energy functional (9.103) subject to the condition that the orbitals form an orthonormal set. Consider the appropriate auxiliary functional:

$$H\left[\{\psi_a, \psi_a^*\}\right] = E_{DFT}\left[\{\psi_a, \psi_a^*\}\right] - \sum_{a,a'}\lambda_{aa'}\left(\int \psi_a^*\psi_{a'}d\mathbf{r} - \delta_{aa'}\right) ,$$

where numbers $\lambda_{aa'}$ are the corresponding Lagrange multipliers. They form a symmetric square matrix with the size equal to the number of orbitals we use. This gives

$$\frac{\delta H}{\delta\psi_a^*\left(\mathbf{r}_1\right)} = \left[-\frac{1}{2}\Delta_{\mathbf{r}_1} + \int \frac{\rho\left(\mathbf{r}'\right)}{|\mathbf{r}_1 - \mathbf{r}'|}d\mathbf{r}' + V_n\left(\mathbf{r}_1\right) + V_{xc}\left(\mathbf{r}_1\right)\right]\psi_a(\mathbf{r}_1) - \sum_{a'}\lambda_{aa'}\psi_{a'}\left(\mathbf{r}_1\right) .$$

Setting this functional derivative to zero, results in the equations

$$\left[-\frac{1}{2}\Delta_{\mathbf{r}_1} + \int \frac{\rho\left(\mathbf{r}'\right)}{|\mathbf{r}_1 - \mathbf{r}'|}d\mathbf{r}' + V_n\left(\mathbf{r}_1\right) + V_{xc}\left(\mathbf{r}_1\right)\right]\psi_a(\mathbf{r}_1) = \sum_{a'}\lambda_{aa'}\psi_{a'}\left(\mathbf{r}_1\right) ,$$

$$(9.105)$$

or simply

$$\widehat{F}_{KS}\psi_a = \sum_{a'}\lambda_{aa'}\psi_{a'} , \qquad\qquad (9.106)$$

where the operator $\widehat{F}_{KS}$ is the expression within the square brackets in Eq. (9.105).

Problem 9.19 Repeat the previous steps by calculating the functional derivative of H with respect to $\psi_a(\mathbf{r}_1)$, not $\psi_a^*(\mathbf{r}_1)$, i.e. opposite to what has been done above. Show that in this case we obtain the following equation

$$\widehat{F}_{KS}\psi_a^* = \sum_{a'} \lambda_{a'a}\psi_{a'}^* \tag{9.107}$$

instead of Eq. (9.106).

Taking the complex conjugate of the above equation and comparing it with Eq. (9.106), we see that $\lambda_{aa'} = \lambda_{a'a}^*$, i.e. the matrix $\boldsymbol{\lambda} = (\lambda_{aa'})$ of Lagrange multipliers must be Hermitian.

The obtained equations (9.106) are not yet final. The point is that one can choose another set of orbitals, $\varphi_b(\mathbf{r})$, as a linear combination of the old ones,

$$\varphi_b(\mathbf{r}) = \sum_a u_{ba}\psi_a(\mathbf{r}) \tag{9.108}$$

with the coefficients, u_{ba}, forming a unitary matrix $\mathbf{U} = (u_{ba})$, i.e. $\mathbf{U}^\dagger\mathbf{U} = \mathbf{U}\mathbf{U}^\dagger = \mathbf{E} = (\delta_{aa'})$, or, in components,

$$\sum_b u_{ba}^* u_{ba'} = \sum_b u_{ab} u_{a'b}^* = \delta_{aa'} \ .$$

Then, the old orbitals can easily be expressed via the new ones: multiply both sides of (9.108) by $u_{ba'}^*$ and sum over b:

$$\sum_b u_{ba'}^* \varphi_b(\mathbf{r}) = \sum_a \underbrace{\left(\sum_b u_{ba'}^* u_{ba} \right)}_{\delta_{aa'}} \psi_a(\mathbf{r}) \quad \Longrightarrow \quad \psi_a = \sum_b u_{ba}^* \varphi_b \ .$$

Problem 9.20 Prove that the new orbitals still form an orthonormal set as the old ones.

Then, it is easy to see that the electron density can be expressed via the new orbitals in exactly the same way as when using the old ones:

$$\rho = \sum_a \psi_a^* \psi_a = \sum_a \sum_{bb'} u_{ba} u_{b'a}^* \varphi_b^* \varphi_{b'} = \sum_{bb'} \left(\underbrace{\sum_a u_{ba} u_{b'a}^*}_{\delta_{bb'}} \right) \varphi_b^* \varphi_{b'} \tag{9.109}$$

$$= \sum_b \varphi_b^* \varphi_b = \sum_b |\varphi_b|^2 .$$

We may say that the electron density is invariant with respect to a unitary transformation of the orbitals. Hence, since the density remains in the same form when expressed via either the old or new sets of orbitals, and using the fact that the operator $\widehat{F}_{KS}$ depends entirely on the density, one can rewrite these equations via the new orbitals:

$$\widehat{F}_{KS} \psi_a = \sum_{a'} \lambda_{aa'} \psi_{a'} \quad \Longrightarrow \quad \widehat{F}_{KS} \sum_b u_{ba}^* \varphi_b = \sum_{a'} \lambda_{aa'} \sum_b u_{ba'}^* \varphi_b .$$

Multiply both sides by $u_{b'a}$ and sum over a:

$$\widehat{F}_{KS} \sum_b \underbrace{\left(\sum_a u_{b'a} u_{ba}^* \right)}_{\delta_{bb'}} \varphi_b = \sum_b \left(\sum_{aa'} u_{b'a} \lambda_{aa'} u_{ba'}^* \right) \varphi_b \tag{9.110}$$

$$\Longrightarrow \quad \widehat{F}_{KS} \varphi_{b'} = \sum_b \widetilde{\lambda}_{b'b} \varphi_b ,$$

where

$$\widetilde{\lambda}_{b'b} = \sum_{aa'} u_{b'a} \lambda_{aa'} u_{ba'}^* . \tag{9.111}$$

Using matrix notations, the last equation is simply $\widetilde{\boldsymbol{\lambda}} = \mathbf{U}\boldsymbol{\lambda}\mathbf{U}^\dagger$.

We can now choose the matrix $\mathbf{U}$ in such a way as to diagonalise the matrix of Lagrange multipliers, $\widetilde{\boldsymbol{\lambda}} = \left(\delta_{bb'} \widetilde{\lambda}_b \right)$. In fact, we have seen that the matrix $\boldsymbol{\lambda}$ is Hermitian. Hence, all its eigenvalues $\widetilde{\lambda}_b$ are real numbers. Therefore, one can rewrite Eqs. (9.110) in their final form as

$$\widehat{F}_{KS} \varphi_b = \widetilde{\lambda}_b \varphi_b . \tag{9.112}$$

These are called Kohn–Sham equations. It is said that the orbitals φ_b are eigenfunctions of the Kohn–Sham operator $\widehat{F}_{KS}$ with the eigenvalues $\widetilde{\lambda}_b$ in close analogy to eigenvectors and eigenvalues of a matrix.

These equations are to be solved self-consistently together with Eq. (9.109). First, some orbitals $\{\varphi_b\}$ are assumed. These allow one to calculate the density $\rho(\mathbf{r})$ and hence the Kohn–Sham operator $\widehat{F}_{KS}$. Once this is known, new eigenfunctions $\{\varphi_b\}$ can be obtained by solving the eigenproblem (9.112) which gives an updated density, and so on. The iterative process is stopped when the density does not change any more (within a numerical tolerance). The obtained electron density corresponds to the density of the real electron gas, and the total energy E_{KS} calculated with this density gives the total electron energy of the system. The orbitals φ_a and the corresponding eigenvalues $\widetilde{\lambda}_a$ do not, strictly speaking, have a solid physical meaning; however, in actual calculations they are interpreted as effective one-electron wavefunctions and energies, respectively.

In the so-called Hartree–Fock (HF) method the total energy of a system of electrons is written via a more general object than the density $\rho(\mathbf{r})$ itself, which is called the density matrix (for simplicity, we completely neglect spin here even though it is essential in the formulation of the method):

$$\rho(\mathbf{r}, \mathbf{r}') = \sum_a \psi_a(\mathbf{r}) \psi_a^*(\mathbf{r}') \ .$$

The 'diagonal element' of this object, $\rho(\mathbf{r}, \mathbf{r})$, is the same as the density $\rho(\mathbf{r})$. The corresponding energy functional still is a functional of all the orbitals and reads

$$E_{HF}\left[\{\psi_a, \psi_a^*\}\right] = \int \left[\left(-\frac{1}{2}\Delta_\mathbf{r} + V_n(\mathbf{r})\right)\rho(\mathbf{r}, \mathbf{r}')\right]_{\mathbf{r}' \to \mathbf{r}} d\mathbf{r}$$
$$+ \frac{1}{2}\int\int \frac{d\mathbf{r}d\mathbf{r}'}{|\mathbf{r} - \mathbf{r}'|}\left[\rho(\mathbf{r}, \mathbf{r})\rho(\mathbf{r}', \mathbf{r}') - \rho(\mathbf{r}, \mathbf{r}')\rho(\mathbf{r}', \mathbf{r})\right] \ .$$

The first term describes the kinetic energy of the electrons together with the energy of their interaction with the nuclei (note that the notation $\mathbf{r}' \to \mathbf{r}$ means that after application of the Laplacian to the density matrix one has to set $\mathbf{r}'$ to $\mathbf{r}$); the last term describes both Coulomb and exchange interaction of the electrons with each other.

Problem 9.21 By imposing the condition for the orbitals to form an orthonormal set and setting the variation of the corresponding functional to zero, show that the equations for the orbitals ψ_a are determined from solving the eigenproblem

$$\widehat{F}_{HF}\psi_a = \lambda_a\psi_a \ ,$$

where the so-called Fock operator is

$$\widehat{F}_{HF}(\mathbf{r}) = -\frac{1}{2}\Delta_\mathbf{r} + V_n(\mathbf{r}) + \int \frac{\rho(\mathbf{r}', \mathbf{r}')}{|\mathbf{r} - \mathbf{r}'|}d\mathbf{r}' - \int \frac{\rho(\mathbf{r}, \mathbf{r}')}{|\mathbf{r} - \mathbf{r}'|}P_{\mathbf{r}\mathbf{r}'}d\mathbf{r}' \ .$$

The last term describes the exchange interaction. It contains an exchange operator $P_{\mathbf{r}\mathbf{r}'}$ defined as follows: $P_{\mathbf{r}\mathbf{r}'}\chi(\mathbf{r}) = \chi(\mathbf{r}')$. In other words, the last term in the Fock operator works in such a way that when it acts on a function of $\mathbf{r}$ on the right of it, it changes its arguments to $\mathbf{r}'$ so that it appears inside the integral:

$$\left[\int \frac{\rho(\mathbf{r}, \mathbf{r}')}{|\mathbf{r} - \mathbf{r}'|} P_{\mathbf{r}\mathbf{r}'} d\mathbf{r}' \right] \chi(\mathbf{r}) = \int \frac{\rho(\mathbf{r}, \mathbf{r}')}{|\mathbf{r} - \mathbf{r}'|} \chi(\mathbf{r}') d\mathbf{r}' .$$

Index

A
Abel's formula, 87
Absolute convergence, 318
Advanced solution, 635
Algebraic multiplicity, 108, 191
Analytic continuation, 341, 670
Analytic function, 269, 338
Angular momentum, 898
Anti-Hermitian matrix, 37, 107
Antisymmetric matrix, 25
Aperture, 649
Aperture function, 651
Argument of complex number, 257
Associated Legendre equation, 375, 551
Associated Legendre functions, 550, 560
Atomic orbitals, 231
Atomic vibrations, 213
Augmented matrix, 72
Autocorrelation function, 499, 702

B
Back substitution, 73
Baker–Hausdorff identity, 141
Basis vectors, 6
Bessel differential equation, 564
Bessel function, 594, 597, 653
Bessel function of the first kind, 565
Beta function, 508, 548
Binomial coefficient, 312
Bloch theorem, 591
Bohr magneton, 587
Born, 475
Born series, 136, 160
Bosons, 440

Boundary conditions, 527, 595, 790
Boundary problem, 844
Brachistochrone problem, 863
Branch cut, 278, 288
Branches of multi-valued function, 278
Branch point, 278, 288
Brillouin zone, 476
Brownian particle, 645, 647, 701, 707

C
Canonical form of PDE, 783
Catenary problem, 862
Cauchy–Euler equation, 695
Cauchy principal value, 360
Cauchy-Riemann conditions, 270, 339
Cauchy theorem, 299, 308
Cauchy theorem for multiply connected region, 306
Causality principle, 377
Characterisation of PDE, 783
Characteristic equation, 96
Characteristic polynomial of a matrix, 97
Charge density, 487
Chebyshev polynomials, 535
Classical action, 891
Closed domain in complex plane, 262
Closed region in complex plane, 262
Closed set of functions, 532
Co-factor of matrix, 56
Coloured noise, 705
Commutator of two matrices, 141
Completeness, 240
Completeness condition, 453
Completeness of eigenvectors, 119, 216

© The Editor(s) (if applicable) and The Author(s), under exclusive license to Springer Nature Switzerland AG 2024
L. Kantorovich, *Mathematics for Natural Scientists II*, Undergraduate Lecture Notes in Physics, https://doi.org/10.1007/978-3-031-46320-4

Complete set of functions, 443, 480, 532
Complete set of vectors, 9
Complex conjugate matrix, 26
Complex Fourier series, 435
Complex plane, 258
Complex spherical harmonics, 562
Conditional convergence, 472
Cone, 734
Conservation laws of mechanics, 894
Constraint, 850
Continued fraction, 61, 94, 235
Continuity equation, 500, 793
Continuous function, 265
Convolution, 637, 642
Convolution of functions, 620, 688
Convolution theorem, 621, 688, 691, 847
Coordinate line, 722
Coordinate transformation, 719
Correlation function, 644, 703
Cosine Fourier series, 417
Cotangent power expansion, 434
Coulomb formula, 636
Coulomb potential, 631
Cramer's method, 62
Cramer's rule, 62
Critical temperature, 589
Crystallography, 469
Curl of vector field, 765
Current, 388
Curvilinear coordinate system, 821
Curvilinear coordinates, 719
Cyclic boundary condition, 475
Cyclic permutation, 149
Cycloid, 863
Cylindrical coordinates, 720
Cylindrical symmetry, 719
Cylindrical wave, 799

D
D'Alembert, 632
D'Alembert equation, 632
D'Alembert PDE, 840
D'Alembert's formula, 798, 843
Damped harmonic oscillator, 637
De Moivre's formula, 259
Defective matrix, 103, 104, 191
Degenerate eigenvalue, 102, 123
Delta function, 489, 629, 691, 740
Delta function in 3D, 597
Delta function three-dimensional, 500
Delta sequence, 489, 614
Density functional theory, 908

Density matrix, 913
Density Of States (DOS), 601
Derivative of inverse function, 269
Determinant of matrix, 42
Diagonalisable matrix, 118
Diagonalisation of quadratic form, 146
Diagonal matrix, 20
Dielectric function, 377
Differential arc length, 732
Differential equation of hypergeometric
         type, 540, 568, 578
Differential line element, 730
Differentiation of Fourier series, 429
Diffraction pattern, 649
Diffusion equation, 641, 770, 781, 791
Diffusion of a particle, 590
Dirac, 489
Dirac delta function, 454, 489, 519, 597, 614,
         637
Directed area element, 760
Dirichlet, 419
Dirichlet conditions, 420, 610, 612, 665
Dirichlet theorem, 419, 447
Discontinuity of the first kind, 419, 656, 681
Dispersion relation, 227
Divergence, 760
Divergence theorem, 888
Domain in complex plane, 262
Domain wall, 864
Domains, 864
Dot product of two functions, 11
Dot product of vectors, 2
Double factorial, 503
Double Fourier series, 818
Dynamical matrix, 215, 498
Dyson equation, 135

E
Eigenfunction, 552, 559, 805, 817, 820, 826,
         912
Eigenproblem, 95
Eigenvalue, 95, 552, 559, 804, 820, 826, 912
Eigenvector, 95
Einstein summation convention, 21
Electro-magnetic field, 252
Electron density, 467, 906
Electronic circuits, 697
Elliptic PDE, 785, 787
Energy bands, 590
Energy gaps, 590
Equipartition theorem, 498, 647
Error function, 452, 473, 686, 848

Essential singularity, 337
Euclidean norm, 220
Euler identity, 284, 289
Euler-Mascheroni constant, 508
Euler's equation, 856, 869, 873, 877, 881,
      882, 891, 898, 902
Even function, 417
Even function Fourier transform, 613
Ewald, 472
Ewald formula, 472
Ewald's formula, 625
Exchange-correlation potential, 910
Exchange interaction, 913
Exchange operator, 914
Excited states, 903
Exponential form of complex number, 284
Exponential function, 282
Exponential integral, 474

F

Fermions, 440
Ferromagnetic phase, 589
Feynman, 643
Filtering theorem, 454, 615, 621, 623, 634,
      638
Final width step function, 683
Finite geometric progression, 327
First Kirchhoff's law, 698
Flow of incompressible fluid, 789
Fluctuation-dissipation theorem, 498
Flux of vector field, 760
Fock operator, 913
Force constant matrix, 214
Forced oscillations of string, 809
Fourier coefficients, 415
Fourier cosine integral, 608
Fourier integral, 605, 607
Fourier method for solving PDE, 802, 819,
      825
Fourier series, 415, 591, 807, 810, 818, 838
Fourier series solution of DE, 440
Fourier sine integral, 608
Fourier transform, 612, 836, 840
Fractional coordinates, 468
Fraunhofer, 650
Fraunhofer diffraction, 650
Free energy, 587
Frequency, 418
Frequency spectrum, 613
Fresnel integral, 409, 643
Friction kernel, 702
Frobenius norm, 222

Full Width at Half Maximum (FWHM), 505
Functional, 388, 849
Functional dependence, 849
Functional derivative, 899, 910
Functional of function of many variables,
      884
Functional operator, 612
Functional sequence, 321
Functional series, 321, 411, 413, 450
Functional with constraints, 872
Function of matrix, 131
Fundamental frequency, 418
Fundamental mode, 808
Fundamental set of solutions, 172
Fundamental solution, 637
Fundamental theorem of algebra, 314

G

Galilean transformation, 789
Gamma function, 501, 548, 565, 673
Gaussian delta sequence, 618
Gaussian elimination, 71, 101
Gaussian function, 746
Gaussian integral, 504, 747
General boundary conditions, 813, 833, 843
Generalised binomial coefficients, 536, 548
Generalised coordinate, 890
Generalised eigenproblem, 138
Generalised eigenvectors, 161, 165
Generalised equation of hypergeometric
      type, 537, 551, 580
Generalised Fourier coefficients, 451
Generalised Fourier expansion, 451, 480
Generalised function, 489
Generalised functional Fourier series, 240
Generalised Laguerre polynomials, 535,
      547, 550, 582
Generalised series expansion, 368
Generalised velocities, 890
General power function, 294
General solution of PDE, 796, 820
Generating function, 510, 530, 533, 534,
      572, 598
Geodesic problem, 860
Geometric multiplicity, 108
Geometric progression, 260, 845
Gibbs phenomenon, 423
Gradient, 759
Gram and Schmidt method, 14
Gram–Schmidt procedure, 116
Green's formula, 887
Green's function, 135, 160, 239, 440, 637,
      690, 693, 702, 707, 836

Ground state, 903
Group, 31
Growth kinetics, 212
Growth order parameter, 656, 663

H
Hamiltonian, 392
Hamiltonian operator, 902
Harmonic approximation, 214
Harmonic function, 270, 557
Harmonic oscillator, 440, 577, 893
Harmonic oscillator equation, 804, 805, 826,
    829
Hartree–Fock method, 913
Heat transport PDE, 770, 781, 822, 823, 836,
    843
Heaviside, 496
Heaviside function, 359, 496, 616, 635, 639,
    642
Heisenberg Hamiltonian, 864
Helmholtz, 640
Helmholtz differential equation, 640
Hermite polynomials, 533, 546, 547, 550,
    578
Hermitian matrix, 34, 111
Hessian, 148
Holomorphic function, 269, 338
Holonomic constraints, 872, 891
Homogeneous PDE, 782
Homogeneous system of ODEs, 172
Hydrodynamic equation of motion, 793
Hydrogen atom, 579
Hyperbolic function, 286
Hyperbolic PDE, 785, 787, 796
Hypergeometric differential equation, 537

I
Idempotent matrix, 203
Identity matrix, 20
Image of function in Laplace transform, 656
Indefinite integral, 304
Induction, 742
Infinite numerical series, 417
Infinite product, 507
Infinite string, 795
Inhomogeneous PDE, 782, 800, 810, 813,
    832
Initial conditions, 595, 790
Integral representation of delta function, 615
Integral transform, 605, 655
Integration of Fourier series, 425
Integro-differential equations, 696

Invariance in time, 894
Invariance with rotation, 896
Invariance with translation, 895
Inverse Fourier transform, 613, 835
Inverse functional operator, 613
Inverse matrix, 28
Irregular singular point, 372
Irreversible growth, 212
Ising model, 253
Isolated singularity, 334
Isoparametric problem, 880

J
Jacobian, 470, 738, 741, 758
Jacobi polynomials, 535, 546, 548, 550, 553
Jordan chains, 161, 191
Jordan normal form, 161
Jordan's lemma, 354, 356, 669

K
Kinetic energy, 777
Kirchhoff's laws, 697
Klein–Gordon equation, 250
Kohn–Sham equations, 912
Kramers-Kronig relations, 377
Kronecker symbol, 413, 807
Kronig-Penney model, 590

L
Lagrange equations, 891
Lagrange function, 891
Lagrange multiplier, 876, 878, 880, 891, 904,
    907, 910
Lagrangian, 891
Laguerre polynomials, 534, 546, 549
Lanczos method, 152, 233
Landau, 403
Landau-Zener problem, 403
Laplace equation, 270, 550, 557, 580, 782,
    822, 827, 828, 837
Laplace equation in spherical coordinates,
    557
Laplace transform, 655, 840
Laplacian, 579, 594, 631, 632, 763, 769, 795,
    820, 837, 902
Laurent series, 328, 571
LDE decomposition of a matrix, 158
Left eigenproblem, 108
Left eigenvector, 108
Left triangular matrix, 29, 523
Legendre, 510

Legendre equation, 515, 561
Legendre polynomials, 510, 530, 533, 550, 553, 561, 598
Leibniz, 520
Leibniz formula, 520, 536, 553
Length of curve, 732
Linear antisymmetric function, 53
Linear combination of atomic orbitals, 231
Linear functional, 850
Linear independence of functions, 12, 85, 442
Linear independence of vectors, 4
Linear operator, 657
Linear ordinary differential equation, 87, 90
Linear transformation, 783
Linear vector space, 2
Liouville equation, 134
Lorentz transformation of coordinates, 789

M
Maclaurin, 489
Magnetic field, 208
Magnetic moments, 864
Magnetisation, 588
Main quantum number, 582
Majorana, 403
Matrix, 19
Matrix blocks, 156
Matrix exponential, 206
Matrix exponential function, 132
Matrix multiplication, 20
Matsubara, 440
Matsubara frequencies, 394
Matsubara sums, 394
Maximum of functional, 855
Maximum of function of many variables, 148
Maxwell equations, 380, 632
Maxwell's distribution, 778
Mean square error, 444
Mechanical stability, 217
Memory effects, 702
Meromorphic function, 338
Method of separation of variables, 573, 580, 583, 594, 803, 840
Method of variation of parameters, 180
Metric tensor, 730
Minimum of functional, 855
Minimum of function of many variables, 148
Minkowski space, 732
Minor of matrix, 56
Modal matrix, 117, 784

Modified Bessel function, 715
Modified Bessel function of first kind, 567, 573
Modulation theorem, 617
Molecular dynamics, 223, 777
Molecular self-assembly, 212
Moment of function, 619
Momentum of system, 896
Moore–Penrose conditions, 202
Moore–Penrose pseudo-inverse, 198
Movement under central force, 775
Multi-dimensional space, 6
Multi-valued function, 264
Multi-variable Gaussian functions, 746
Multiple Fourier series, 467
Multiple Fourier transform, 628
Multiply connected region, 264
Multipole expansion of electrostatic potential, 597
Multipole moments of charge distribution, 599

N
N-dimensional sphere, 755
Necessary condition for functional extremum, 855
Neighbourhood of point, 262
Newton-Leibnitz formula, 304
Newton's equation of motion, 793
Newton's equations of motion, 890
Newton's equations of motion in curvilinear coordinates, 771
Noether theorem, 894
Non-Contact Atomic Force Microscopy (NC-AFM), 477
Non-defective matrix, 116, 136
Non-holonomic constraints, 872, 880
Non-homogeneous system of ODEs, 172
Non-isolated singularity, 338
Non-periodic functions, 411, 605
Non-simply connected region, 264
Normal matrices, 108, 114, 118
Normal mode, 216, 808
Norm of function, 11
Numerical series, 318

O
Odd function, 418
Odd function Fourier transform, 613
One-dimensional chain, 224
One-to-one mapping, 264
Open domain in complex plane, 262

Open region in complex plane, 262
Operator, 805
Order of partial differential equation, 782
Order of pole, 336
Order parameter, 587
Ordinary point, 368
Original in Laplace transform, 656
Orthogonal curvilinear system, 725
Orthogonal functions, 11, 413, 435, 450
Orthogonalisation of vectors, 14
Orthogonal matrix, 32
Orthogonal transformation, 33, 726
Orthogonality condition, 807
Orthonormal functions, 12
Orthonormality, 239
Orthonormality of eigenvectors, 216
Orthonormal vectors, 8
Oscillations in circular membrane, 594
Oscillations of square membrane, 815
Ostrogradsky, 887
Ostrogradsky–Gauss theorem, 888
Ostrogradsky's equation, 898
Overlap integral, 11, 522

P
Pair correlation function, 748
Parabolic coordinates, 729, 764
Parabolic cylinder coordinates, 729, 774
Parabolic PDE, 785, 787, 822
Paraboloid of revolution, 734
Parity of permutation, 44, 243
Parseval, 430, 439
Parseval's equality, 447
Parseval's theorem, 429, 439, 444, 451, 622,
    652
Partial Differential Equation (PDE), 558,
    638, 639, 719, 769, 781
Partial Fourier transform, 646
Partial sum, 318, 447, 452
Particular solution of PDE, 790
Path integrals, 643
Pauli exclusion principle, 245
Pauli matrices, 27, 134, 143, 151
Pendulum, 892, 893
Periodic boundary conditions, 225
Periodic chain of atoms, 467
Periodic function, 411, 559
Periodic solid, 467
Period of function, 411
Permanent, 246
Permutation, 43
Phase transition, 586

Phase transition temperature, 589
Phonon, 579
Phonon Green's function, 228
Piecewise continuous function, 419
Pivot, 79
Plancherel, 431, 439, 622
Plancherel's theorem, 431, 439, 452, 622
Plane wave, 379, 476
Plemelj, 495
Point charges, 488
Poisson, 471, 639
Poisson distribution, 713
Poisson equation, 471, 636, 639, 770
Poisson summation formula, 623
Pole, 335
Positive-definite matrix, 139
Power series, 323
Principal value, 495
Principal value integral, 667
Principle of least action, 890
Probability, 777
Probability distribution, 708
Probability of hops, 708
Problems with constraints, 850
Product solution, 558, 802, 825
Propagation matrix, 392
Propagation of wave, 797
Propagator, 237
Pseudo-inverse matrix, 198
Pseudopotentials, 476
Purity of quantum state, 151
Pythagoras, 732
Pythagoras' theorem, 732

Q
Quadratic form, 145, 214
Quantum mechanics, 232, 480, 643

R
Rabi problem, 238
Radius of convergence, 324
Random force, 702
Rank decomposition, 198
Rank of matrix, 81
Ratio test, 319
Real spherical harmonics, 562
Reciprocal lattice vectors, 467, 469, 475
Reduced echelon form, 75
Reduced row echelon form, 79
Reflection matrix, 41
Region in complex plane, 262

Regular singular point, 368, 372
Regularisation, 643
Remainder term, 327
Removable singularity, 335
Repeated eigenvalues, 123
Residence probability, 709
Residue, 342
Resolvent matrix, 160
Resolvent of matrix, 135, 228, 233
Resonance, 442, 477
Response function, 377
Retardation effects, 650
Retarded potential, 632
Retarded solution, 635
Reversible growth, 212
Right eigenproblem, 108
Right-hand screw rule, 765
Rodrigues, 520, 529
Rodrigues formula, 520, 529, 533–535, 543,
    546–548, 552, 553
Roots of Bessel function, 575
Root test, 319
Rope under gravity, 882
Rotation matrix, 39
Row echelon, 171
Row echelon form, 72

S
Saddle point, 602
Scalar field, 721, 761, 763
Scalar product of vectors, 2
Scale factor, 725, 731
Scanning Tunnelling Microscopy (STM),
    391
Schrödinger equation, 384, 392, 404, 480,
    577, 579, 582, 591, 643, 765, 903
Second fluctuation–dissipation theorem, 704
Second Kirchhoff's law, 698
Second mean-value integral formula, 455
Second-order linear PDE, 782
Secular equation, 96
Self-adjoint form of DE, 541, 574
Self-adjoint matrix, 111
Self-adjoint operator, 905
Semi-infinite rod, 847
Separation constant, 558, 803, 816, 820, 825
Separation of variables, 820
Shifted Gaussian function, 753
Shrödinger equation, 246
Signal, 613
Similarity transformation, 116
Similar matrices, 116

Simple cubic lattice, 625
Simple pole, 336
Simple zero of $f(z)$, 337
Simply connected region, 263, 299
Sinc function, 424, 652
Sine Fourier series, 418, 811
Single-valued function, 264
Singularity of function, 334
Singular matrix, 65
Singular point of coordinate transformation,
    720
Skew-Hermitian matrix, 37, 107
Skin effect, 383
Slater determinant, 239
Sokhotski, 495
Sokhotski-Plemelj formula, 495
Solid angle, 562, 598
Solid-state physics, 469
Sound propagation, 794
Special relativity, 732
Spectral analysis, 613
Spectral power density, 645
Spectral theorem, 131, 138, 393
Spherical coordinates for N-dimensional
    space, 757
Spherical functions, 557, 580
Spherical harmonics, 557, 561
Spherically symmetric wave, 798
Spherical symmetry, 719
Spherical wave, 379, 650, 799
Spin, 864
Spinor, 247
Spontaneous magnetisation, 589
Square matrix, 20
Stationary boundary conditions, 814, 827,
    832
Stationary distribution, 822, 837, 839
Stationary PDE, 814
Stationary solution, 827
Statistical ensemble, 644
Stirling's approximation, 585
Stochastic force, 498
String guitar, 808
Stueckelberg, 403
Subgroup, 32
Superposition principle, 805, 818, 820, 826
Surface area of N-dimensional sphere, 757
Symmetric matrix, 25, 111
System of linear differential equations, 178,
    208, 215, 238
System of ordinary DEs, 694

T

Tangent power expansion, 435
Taylor, 510
Taylor's expansion, 326
Taylor's formula, 326, 327
Theta-function transformation, 471, 473, 625
Thomas–Fermi model, 906
Three-dimensional delta function, 597
3D rotation, 205
Time-ordered exponential, 134
Trace of matrix, 149
Transpose matrix, 24
Transverse oscillations of rod, 899
Triangular decomposition of matrix, 157
Triangular matrix, 60
Tridiagonal matrix, 60, 94, 152, 155, 225, 234
Trigonometric form of Fourier integral, 608
Tunnelling, 384

U

Uncertainty principle, 614
Uniform convergence, 310, 321, 610, 714
Unitary matrix, 34, 106
Unitary transformation, 34
Unit base vectors of curvilinear system, 723
Unit impulse function, 614
Unit matrix, 20

V

van Hove singularities, 601

Variation of functional, 851
Vector field, 721, 759, 760, 764, 765
Vector-column, 19
Vector-row, 19
Vector space, 2
Volterra, V., 696
Volume of N-dimensional sphere, 756
Volume of sphere, 755
von Karman, 475

W

Wave, 378, 799
Wave equation, 379, 769, 781, 789–791
Wave frequency, 379
Wavefunction, 231
Wavelength, 379
Wave phase velocity, 379
Wavevector, 227, 379, 629
Weight function, 11, 522
White noise, 705
Wicks theorem, 239
Wiener–Khinchin theorem, 645
Woodbury identity, 159
Wronskian, 85, 373

Z

Zener, 403
Zero boundary conditions, 802, 809, 813–815, 820, 825, 828, 834
Zero of order n of $f(z)$, 337
Zero point energy, 579

Printed in the United States
by Baker & Taylor Publisher Services